# Biophysical Techniques in Photosynthesis
# Volume II

# Advances in Photosynthesis and Respiration

VOLUME 26



The scope of our series, beginning with volume 11, reflects the concept that photosynthesis and respiration are intertwined with respect to both the protein complexes involved and to the entire bioenergetic machinery of all life. *Advances in Photosynthesis and Respiration* is a book series that provides a comprehensive and state-of-the-art account of research in photosynthesis and respiration. Photosynthesis is the process by which higher plants, algae, and certain species of bacteria transform and store solar energy in the form of energy-rich organic molecules. These compounds are in turn used as the energy source for all growth and reproduction in these and almost all other organisms. As such, virtually all life on the planet ultimately depends on photosynthetic energy conversion. Respiration, which occurs in mitochondrial and bacterial membranes, utilizes energy present in organic molecules to fuel a wide range of metabolic reactions critical for cell growth and development. In addition, many photosynthetic organisms engage in energetically wasteful photorespiration that begins in the chloroplast with an oxygenation reaction catalyzed by the same enzyme responsible for capturing carbon dioxide in photosynthesis. This series of books spans topics from physics to agronomy and medicine, from femtosecond processes to season long production, from the photophysics of reaction centers, through the electrochemistry of intermediate electron transfer, to the physiology of whole orgamisms, and from X-ray crystallography of proteins to the morphology or organelles and intact organisms. The goal of the series is to offer beginning researchers, advanced undergraduate students, graduate students, and even research specialists, a comprehensive, up-to-date picture of the remarkable advances across the full scope of research on photosynthesis, respiration and related processes.

*The titles published in the Series are listed at the end of this volume.*

# Biophysical Techniques in Photosynthesis

## Volume II

*Edited by*

**Thijs J. Aartsma**
*Leiden University,*
*The Netherlands*

and

**Jörg Matysik**
*Leiden University,*
*The Netherlands*

Springer

Library of Congress Control Number: 2007943526

ISBN 978-1-4020-8249-8 (HB)
ISBN 978-1-4020-8250-4 (eBook)

---

Published by Springer,
P.O. Box 17, 3300 AA Dordrecht, The Netherlands.

*www.springer.com*

Cover illustration: The figure on the cover is a courtesy of Petra Fromme and Ingo Grotjohann.
It summarizes significant advances in our understanding of the process of photosynthesis
in the recent years.

The camera ready text was prepared by Lawrence A. Orr,
Center for Bioenergy & Photosynthesis, Arizona State University, Tempe, Arizona 85287-1604, U.S.A.

*Printed on acid-free paper*

# In Memoriam

**Jan Amesz**

March 11, 1934 – January 29, 2001

For a tribute, see Photosynth Res (2002) 71: 1–4

**Arnold J. Hoff**

April 30, 1939 – April 22, 2002

For a tribute, see Chem Phys (2003) 294: 223–225

# From the Series Editor

## Advances in Photosynthesis and Respiration

### *Volume 26: Biophysical Techniques in Photosynthesis, Volume II*

I am delighted to announce the publication, in *Advances in Photosynthesis and Respiration* (AIPH) Series, of a book ***Biophysical Techniques in Photosynthesis II***. Two distinguished authorities (Thijs Aartsma and Jörg Matysik, both of Leiden University, The Netherlands) have edited this Volume: Aartsma is an authority on the photophysical properties of biological molecules, using innovative optical spectroscopic techniques, whereas Matysik is an authority on photochemistry and spin chemistry as well as in the development of complementary magnetic resonance methods. This book is produced as a sequel to the outstanding Volume 3 of the Series (*Biophysical Techniques in Photosynthesis*), published in 1996, and edited by Jan Amesz and Arnold Hoff, also of Leiden University.

### Published Volumes (2006–1994)

- *Volume 25* (2006): ***Chlorophylls and Bacteriochlorophylls: Biochemistry, Biophysics, Functions and Applications***, edited by Bernhard Grimm, Robert J. Porra, Wolfhart Rüdiger, and Hugo Scheer, from Germany and Australia. 37 Chapters, 603 pp, Hardcover. ISBN: 978-1-4020-4515-8
- *Volume 24* (2006): ***Photosystem I: The Light-Driven Plastocyanin: Ferredoxin Oxidoreductase***, edited by John H. Golbeck, from USA. 40 Chapters, 716 pp, Hardcover. ISBN: 978-1-4020-4255-3
- *Volume 23* (2006): ***The Structure and Function of Plastids***, edited by Robert R. Wise and J. Kenneth Wise, from USA. 27 Chapters, 575 pp, Softcover: ISBN: 978-1-4020-6570-6, Hardcover, ISBN: 978-1-4020-4060-3
- *Volume 22* (2005): ***Photosystem II: The Light-Driven Water:Plastoquinone Oxidoreductase***, edited by Thomas J. Wydrzynski and Kimiyuki Satoh, from Australia and Japan. 34 Chapters, 786 pp, Hardcover. ISBN: 978-1-4020-4249-2
- *Volume 21* (2005): ***Photoprotection, Photoinhibition, Gene Regulation, and Environment***, edited by Barbara Demmig-Adams, William W. III Adams and Autar K. Mattoo, from USA. 21 Chapters, 380 pp, Hardcover. ISBN: 978-1-4020-3564-7
- *Volume 20* (2006): ***Discoveries in Photosynthesis***, edited by Govindjee, J. Thomas Beatty, Howard Gest and John F. Allen, from USA, Canada and UK. 111 Chapters, 1304 pp, Hardcover. ISBN: 978-1-4020-3323-0
- *Volume 19* (2004): ***Chlorophyll a Fluorescence: A Signature of Photosynthesis***, edited by George C. Papageorgiou and Govindjee, from Greece and USA. 31 Chapters, 820 pp, Hardcover. ISBN: 978-1-4020-3217-2
- *Volume 18* (2005): ***Plant Respiration: From Cell to Ecosystem***, edited by Hans Lambers and Miquel Ribas-Carbo, from Australia and Spain. 13 Chapters, 250 pp, Hardcover. ISBN: 978-1-4020-3588-3
- *Volume 17* (2004): ***Plant Mitochondria: From Genome to Function***, edited by David Day, A. Harvey Millar and James Whelan, from Australia. 14 Chapters, 325 pp, Hardcover. ISBN: 978-1-4020-2399-6
- *Volume 16* (2004): ***Respiration in Archaea and Bacteria:Diversity of Prokaryotic Respiratory Systems***, edited by Davide Zannoni, from Italy. 13 Chapters, 310 pp, Hardcover. ISBN: 978-1-4020-2002-5
- *Volume 15* (2004): ***Respiration in Archaea and Bacteria: Diversity of Prokaryotic Electron Transport Carriers***, edited by Davide Zannoni, from Italy. 13 Chapters, 350 pp, Hardcover. ISBN: 978-1-4020-2001-8
- *Volume 14* (2004): ***Photosynthesis in Algae***, edited by Anthony W. Larkum, Susan Douglas and John A. Raven, from Australia, Canada and UK. 19 Chapters, 500 pp, Hardcover. ISBN:

978-0-7923-6333-0

- *Volume 13* (2003): ***Light-Harvesting Antennas in Photosynthesis***, edited by Beverley R. Green and William W. Parson, from Canada and USA. 17 Chapters, 544 pp, Hardcover. ISBN: 978-0-7923-6335-4
- *Volume 12* (2003): ***Photosynthetic Nitrogen Assimilation and Associated Carbon and Respiratory Metabolism***, edited by Christine H. Foyer and Graham Noctor, from UK and France. 16 Chapters, 304 pp, Hardcover. ISBN: 978-0-7923-6336-1
- *Volume 11* (2001): ***Regulation of Photosynthesis***, edited by Eva-Mari Aro and Bertil Andersson, from Finland and Sweden. 32 Chapters, 640 pp, Hardcover. ISBN: 978-0-7923-6332-3
- *Volume 10* (2001): ***Photosynthesis Photobiochemistry and Photobiophysics***, by Bacon Ke, from USA. 36 Chapters, 792 pp, Softcover: ISBN: 978-0-7923-6791-8. Hardcover: ISBN: 978-0-7923-6334-7
- *Volume 9* (2000): ***Photosynthesis: Physiology and Metabolism***, edited by Richard C. Leegood, Thomas D. Sharkey and Susanne von Caemmerer, from UK, USA and Australia. 24 Chapters, 644 pp, Hardcover. ISBN: 978-0-7923-6143-5
- *Volume 8* (1999): ***The Photochemistry of Carotenoids***, edited by Harry A. Frank, Andrew J. Young, George Britton and Richard J. Cogdell, from UK and USA. 20 Chapters, 420 pp, Hardcover. ISBN: 978-0-7923-5942-5
- *Volume 7* (1998): ***The Molecular Biology of Chloroplasts and Mitochondria in Chlamydomonas***, edited by Jean David Rochaix, Michel Goldschmidt-Clermont and Sabeeha Merchant, from Switzerland and USA. 36 Chapters, 760 pp, Hardcover. ISBN: 978-0-7923-5174-0
- *Volume 6* (1998): ***Lipids in Photosynthesis: Structure, Function and Genetics***, edited by Paul-André Siegenthaler and Norio Murata, from Switzerland and Japan. 15 Chapters, 332 pp, Hardcover. ISBN: 978-0-7923-5173-3
- *Volume 5* (1997): ***Photosynthesis and the Environment***, edited by Neil R. Baker, from UK. 20 Chapters, 508 pp, Hardcover. ISBN: 978-0-7923-4316-5
- *Volume 4* (1996): ***Oxygenic Photosynthesis: The Light Reactions***, edited by Donald R. Ort, and Charles F. Yocum, from USA. 34 Chapters, 696 pp, Softcover: ISBN: 978-0-7923-3684-6. Hardcover: ISBN: 978-0-7923-3683-9
- *Volume 3* (1996): ***Biophysical Techniques in Photosynthesis***, edited by Jan Amesz and Arnold J. Hoff, from The Netherlands. 24 Chapters, 426 pp, Hardcover. ISBN: 978-0-7923-3642-6
- *Volume 2* (1995): ***Anoxygenic Photosynthetic Bacteria***, edited by Robert E. Blankenship, Michael T. Madigan and Carl E. Bauer, from USA. 62 Chapters, 1331 pp, Hardcover. ISBN: 978-0-7923-3682-8
- *Volume 1* (1994): ***The Molecular Biology of Cyanobacteria***, edited by Donald R. Bryant, from USA. 28 Chapters, 916 pp, Hardcover. ISBN: 978-0-7923-3222-0

Further information on these books and ordering instructions can be found at <http://www.springeronline.com> under the Book Series 'Advances in Photosynthesis and Respiration.' Table of Contents of Volumes 1–25 can be found at <http://www.life.uiuc.edu/govindjee/photosynSeries/ttocs.html>. Special discounts are available to members of the International Society of Photosynthesis Research, ISPR (<http://www.photosynthesisresearch.org/>).

## About the Volume 26: Biophysical Techniques in Photosynthesis II

*Biophysical Techniques in Photosynthesis II* has 24 authoritative Chapters, and is authored by 54 international authorities from 10 countries (Australia, Canada, China, France, Germany, Israel, Japan, The Netherlands, United Kingdom, and the United States of America). It is a truly international book and the editors deserve our thanks and our congratulations for giving this gift for our future. Readers of other Volumes in the Series, particularly Volume 22 (Photosystem II), Volume 24 (Photosystem I), Volume 25 (Chlorophylls and Bacteriochlorophylls) and the forthcoming Volume on Purple Phototrophic Bacteria will benefit a great deal by using this Volume as an accompanying Volume.

Since 1996, when the first Volume on *Biophysical Techniques in Photosynthesis*, was published, several new experimental techniques and methods have been devised at a rapid pace. The present book is a sequel to the 1996 book, which was edited by Jan Amesz and Arnold Hoff, to whom the current book is dedicated. It complements that Volume by providing a comprehensive overview of the most important new techniques developed over the past ten years, especially those that are relevant for research on the mechanism and fundamental aspects of photosynthesis. The contributions are written by leading scientists in their field. The book has five sections: Imaging (4 chapters); Structure (5 chapters); Optical and laser spectroscopy (4 chapters); Magnetic resonance (6 chapters); and Theory (5 chapters). Each chapter describes the basic concepts of the technique, practical applications and scientific results. Possibilities and limitations from a technical as well as a scientific point of view are addressed, allowing the reader not only to recognize the potential of a particular method for his/her own quest, but also to assess the resources that are required for implementation. The book is intended for use by both the beginning graduate students and the researchers in photosynthesis as well as in (bio)physics, (bio)chemistry and biology in general.

The readers can easily find the titles and the authors of the individual chapters in the Table of Contents of this book. Instead of repeating this information here, I prefer to thank each and every author by name (listed in alphabetical order) that reads like a "Who's Who in Principles of Biophysical Techniques":

Thijs J. **A**artsma; James P. Allen; Andrei V. Astashkin; Virginijus **B**arzda; Igor V. Borovykh; Claudia Büchel, Francesco Buda; Wenrui **C**hang; Paula C.A. **d**a Fonseca; Eugenio Daviso; Huub J. M. de Groot; Graham R. **F**leming; Petra Fromme; Bas **G**obets; Daniella Goldfarb; Marie Louise Groot; Warwick **H**illier; Martin F. Hohmann-Marriott; Alfred R. Holzwarth; Gunnar **J**eschke; Asako **K**awamori; Jürgen Köhler; Lars Konermann; Ioan Kosztin; Gerd Kothe; Zhenfeng **L**iu; Ying-Zhong **M**a; Nancy Makri; Jörg Matysik; Johannes Messinger; Klaus Möbius; Edward P. Morris; James R. **N**orris; William W. **P**arson; Oleg G. **P**oluektov; Elizabeth L. **R**ead; Thomas Renger; Robert W. Roberson; Simon **S**cheuring; Klaus Schulten; Heinz-Jürgen Steinhoff; Marion C. **T**hurnauer; David M. Tiede; Herbert **v**an Amerongen; Henk Van As; Allison M. L. van de Meene; Rienk van Grondelle; Frank van Mourik; Bart van Oort; Ivo H. M. van Stokkum; Arieh **W**arshel; Carel W. Windt; Donatas **Z**igmantas; and Xiaobing Zuo.

## Complete List of Chapters in Biophysical Techniques in Photosynthesis, AIPH Volume 3, edited by Jan Amesz and Arnold Hoff

As Volume 26 is a sequel to Volume 3, it is beneficial for the readers of the new volume to consult and cite chapters in the earlier volume, I present below complete references to all the chapters in that book. Please note that this volume was published by Kluwer Academic Publishers which was later acquired by Springer, the publishers of the current volumes.

*Chapter 1:* Amesz J (1996) Developments in classical optical spectroscopy. In: Amesz J and Hoff AJ (eds) Biophysical Techniques in Photosynthesis (Advances in Photosynthesis and Respiration, Vol 3), pp 3–10. Kluwer Academic Publishers, Dordrecht

*Chapter 2:* Garab G (1996) Linear and circular dichroism. In: Amesz J and Hoff AJ (eds) Biophysical Techniques in Photosynthesis (Advances in Photosynthesis and Respiration, Vol 3), pp 11–40. Kluwer Academic Publishers, Dordrecht

*Chapter 3:* Sauer K and Debreczeny M (1996) Fluorescence. In: Amesz J and Hoff AJ (eds) Biophysical Techniques in Photosynthesis (Advances in Photosynthesis and Respiration, Vol 3), pp 41–61. Kluwer Academic Publishers, Dordrecht

*Chapter 4:* Jimenez R and Fleming GR (1996) Ultrafast spectroscopy of photosynthetic systems. In: Amesz J and Hoff AJ (eds) Biophysical Techniques in Photosynthesis (Advances in Photosynthesis and Respiration, Vol 3), pp 63–73. Kluwer Academic Publishers, Dordrecht

*Chapter 5:* Holzwarth AR (1996) Data analysis of time-resolved measurements. In: Amesz J and Hoff AJ (eds) Biophysical Techniques in Photosynthesis (Advances in Photosynthesis and Respiration, Vol 3), pp 75–92. Kluwer Academic Publishers, Dordrecht

*Chapter 6:* Inoue Y (1996) Photosynthetic thermoluminescence as a simple probe of Photosystem II election transport. In: Amesz J and Hoff AJ (eds) Biophysical Techniques in Photosynthesis (Advances in Photosynthesis and Respiration, Vol 3), pp 93–107. Kluwer Academic Publishers, Dordrecht

*Chapter 7:* Aartsma TJ, Louwe RJW and Schellenberg P (1996) Accumulated photon echo measurements of excited state dynamics in pigment-protein complexes. In: Amesz J and Hoff AJ (eds) Biophysical Techniques in Photosynthesis (Advances in Photosynthesis and Respiration, Vol 3), pp 109–122. Kluwer Academic Publishers, Dordrecht

*Chapter 8:* Reddy NRS and Small GJ (1996) Spectral hole burning: Methods and applications to photosynthesis. In: Amesz J and Hoff AJ (eds) Biophysical Techniques in Photosynthesis (Advances in Photosynthesis and Respiration, Vol 3), pp 123–136. Kluwer Academic Publishers, Dordrecht

*Chapter 9:* Mäntele W (1996) Infrared and Fourier-transform infrared spectroscopy. In: Amesz J and Hoff AJ (eds) Biophysical Techniques in Photosynthesis (Advances in Photosynthesis and Respiration, Vol 3), pp 137–160. Kluwer Academic Publishers, Dordrecht

*Chapter 10:* Robert B (1996) Resonance Raman studies in photosynthesis — chlorophyll and carotenoid molecules. In: Amesz J and Hoff AJ (eds) Biophysical Techniques in Photosynthesis (Advances in Photosynthesis and Respiration, Vol 3), pp 161–176. Kluwer Academic Publishers, Dordrecht

*Chapter 11:* Boxer SG (1996) Stark spectroscopy of photosynthetic systems. In: Amesz J and Hoff AJ (eds) Biophysical Techniques in Photosynthesis (Advances in Photosynthesis and Respiration, Vol 3), pp 177–189. Kluwer Academic Publishers, Dordrecht

*Chapter 12:* Malkin S (1996) The photoacoustic method in photosynthesis—monitoring and analysis of phenomena which lead to pressure changes following light excitation. In: Amesz J and Hoff AJ (eds) Biophysical Techniques in Photosynthesis (Advances in Photosynthesis and Respiration, Vol 3), pp 191–206. Kluwer Academic Publishers, Dordrecht

*Chapter 13:* Hoff AJ (1996) Magnetic resonance: An introduction. In: Amesz J and Hoff AJ (eds) Biophysical Techniques in Photosynthesis (Advances in Photosynthesis and Respiration, Vol 3), pp 209–210. Kluwer Academic Publishers, Dordrecht

*Chapter 14:* Levanon H (1996) Time-resolved electron paramagnetic resonance spectroscopy—principles and applications. In: Amesz J and Hoff AJ (eds) Biophysical Techniques in Photosynthesis (Advances in Photosynthesis and Respiration, Vol 3), pp 211–233. Kluwer Academic Publishers, Dordrecht

*Chapter 15:* Britt RD (1996) Electron spin echo methods in photosynthesis research. In: Amesz J and Hoff AJ (eds) Biophysical Techniques in Photosynthesis (Advances in Photosynthesis and Respiration, Vol 3), pp 235–253. Kluwer Academic Publishers, Dordrecht

*Chapter 16:* Lubitz W and Lendzian F (1996) ENDOR spectroscopy. In: Amesz J and Hoff AJ (eds) Biophysical Techniques in Photosynthesis (Advances in Photosynthesis and Respiration, Vol 3), pp 255–275. Kluwer Academic Publishers, Dordrecht

*Chapter 17:* Hoff AJ (1996) Optically detected magnetic resonance (ODMR) of triplet states in photosynthesis. In: Amesz J and Hoff AJ (eds) Biophysical Techniques in Photosynthesis (Advances in Photosynthesis and Respiration, Vol 3), pp 277–298. Kluwer Academic Publishers, Dordrecht

*Chapter 18:* De Groot HJM (1996) Magic angle spinning nuclear magnetic resonance of photosynthetic components. In: Amesz J and Hoff AJ (eds) Biophysical Techniques in Photosynthesis (Advances in Photosynthesis and Respiration, Vol 3), pp 299–313. Kluwer Academic Publishers, Dordrecht

*Chapter 19:* Schiffer M (1996) Structure determination of proteins by X-ray diffraction. In: Amesz J and Hoff AJ (eds) Biophysical Techniques in Photosynthesis (Advances in Photosynthesis and Respiration, Vol 3), pp 317–324. Kluwer Academic Publishers, Dordrecht

*Chapter 20*: Boekema EJ and Rögner M (1996) Electron microscopy. In: Amesz J and Hoff AJ (eds) Biophysical Techniques in Photosynthesis (Advances in Photosynthesis and Respiration, Vol 3), pp 325–336. Kluwer Academic Publishers, Dordrecht

*Chapter 21:* Yachandra VK and Klein MP (1996) X-ray absorption spectroscopy: determination of transition metal site structures in photosynthesis. In: Amesz J and Hoff AJ (eds) Biophysical Techniques in Photosynthesis (Advances in Photosynthesis and Respiration, Vol 3), pp 337–354. Kluwer Academic Publishers, Dordrecht

*Chapter 22:* Debrunner PG (1996) Mössbauer spectroscopy. In: Amesz J and Hoff AJ (eds) Biophysical Techniques in Photosynthesis (Advances in Photosynthesis and Respiration, Vol 3), pp 355–373. Kluwer Academic Publishers, Dordrecht

*Chapter 23:* Tiede DM and Thiyagarajan P (1996) Characterization of photosynthetic supramolecular assemblies using small angle neutron scattering. In: Amesz J and Hoff AJ (eds) Biophysical Techniques in Photosynthesis (Advances in Photosynthesis and Respiration, Vol 3), pp 375–390. Kluwer Academic Publishers, Dordrecht

*Chapter 24:* Van Gorkom HJ and Gast P (1996) Measurement of photosynthetic oxygen evolution. In: Amesz J and Hoff AJ (eds) Biophysical Techniques in Photosynthesis (Advances in Photosynthesis and Respiration, Vol 3), pp 391–405. Kluwer Academic Publishers, Dordrecht

## Future AIPH and Other Related Books

The readers of the current series are encouraged to watch for the publication of the forthcoming books (not necessarily arranged in the order of future appearance):

- *Sulfur Metabolism in Phototrophic Organisms* (Editors: Rüdiger Hell, Christiane Dahl, David B. Knaff and Thomas Leustek);

- *The Purple Phototrophic Bacteria* (Editors: C. Neil Hunter, Fevzi Daldal, Marion Thurnauer and J. Thomas Beatty);
- *C-4 Photosynthesis and Related $CO_2$ Concentrating Mechanisms* (Editors: Agepati S. Raghavendra and Rowan Sage);
- *Photosynthesis: Biochemistry, Biophysics, Physiology and Molecular Biology* (Editors: Julian Eaton-Rye and Baishnab Tripathy);
- *Abiotic Stress Adaptation in Plants: Physiological, Molecular and Genomic Foundation* (Editors: Ashwani Pareek, Sudhir K. Sopory, Hans J. Bohnert and Govindjee);
- *The Chloroplast Biochemistry, Molecular Biology and Bioengineering, Part 1. The Chloroplast System: Pigments, Lipids, Pigment-Proteins and Macromolecular Complexes; Part 2. Genes, Genomes, Proteomes, Regulation, Transformation, Bioengineering and Stress* (Editors: Constantin Rebeiz, Hans Bohnert, Christoph Benning, Henry Daniell, Beverley R. Green, J. Kenneth Hoober, Hartmut Lichtenthaler, Archie R. Portis and Baishnab C. Tripathy);
- *Photosynthesis In Silico: Understanding Complexity from Molecules to Ecosystems* (Editors: Agu Laisk, Ladislav Nedbal and Govindjee); and
- *Lipids in Photosynthesis: Essential and Regulatory Functions,* (Editors: Hajime Wada and Norio Murata).

In addition to these contracted books, the following topics, among others, are under consideration:

- Cyanobacteria
- Genomics, Proteomics and Evolution
- Biohydrogen Production
- ATP Synthase and Proton Translocation
- Interactions between Photosynthesis and other Metabolic Processes
- Carotenoids II
- Green Bacteria and Heliobacteria
- Ecophysiology
- Photosynthesis, Biomass and Bioenergy
- Global Aspects of Photosynthesis
- Artificial Photosynthesis

Readers are encouraged to send their suggestions for these and future Volumes (topics, names of future editors, and of future authors) to me by E-mail (gov@uiuc.edu) or fax (1-217-244-7246).

In view of the interdisciplinary character of research in photosynthesis and respiration, it is my earnest hope that this series of books will be used in educating students and researchers not only in Plant Sciences, Molecular and Cell Biology, Integrative Biology, Biotechnology, Agricultural Sciences, Microbiology, Biochemistry, and Biophysics, but also in Bioengineering, Chemistry, and Physics.

I take this opportunity to thank and congratulate both Thijs Aartsma and Jörg Matysik for their outstanding and painstaking editorial work. I thank all the 54 authors (see the list above) of this book in our AIPH Series: without their authoritative chapters, there would be no such Volume. I give special thanks to Larry Orr for typesetting this book: his expertise has been crucial in bringing this book out to completion. We owe Jacco Flipsen and Noeline Gibson (both of Springer) thanks for their friendly working relation with us that led to the production of this book. Thanks are also due to Jeff Haas (Director of Information Technology, Life Sciences, University of Illinois at Urbana-Champaign, UIUC), Evan DeLucia (Head, Department of Plant Biology, UIUC) and my dear wife Rajni Govindjee for constant support.

November 29, 2007

Govindjee
Series Editor, *Advances in Photosynthesis and Respiration*
University of Illinois at Urbana-Champaign,
Department of Plant Biology
Urbana, IL 61801-3707, USA
E-mail: gov@uiuc.edu;
URL: http://www.life.uiuc.edu/govindjee

**Govindjee**

A 2006 photograph of the Series Editor Govindjee in front of the Natural History Building (one of the places where the concept of the two light reaction and two pigment system, in oxygenic photosynthesis, arose during 1956-1960), and the plaque at Urbana, Illinois, honoring two of the early pioneers in Biophysics of Photosynthesis: Robert Emerson (1903–1959) and Eugene I. Rabinowitch [born as Evgenii Isaakovich Rabinovich] (1898–1973). Left to right: Rajni Govindjee (1961 PhD, under Eugene Rabinowitch), Govindjee (1960 PhD, also under Rabinowitch) and Rita Khanna (1980 PhD, under Govindjee)

Govindjee, born in 1932, obtained his B.Sc. (Chemistry, Biology) and M.Sc. (Botany, Plant Physiology) in 1952 and 1954, from the University of Allahabad, India, and his Ph.D. (Biophysics, under Prof. Eugene Rabinowitch), in 1960, from the University of Illinois at Urbana-Champaign (UIUC), IL, U.S.A. He is best known for his research on the excitation energy transfer, light emission, the primary photochemistry and the electron transfer in Photosystem II (PS II). His research, with many collaborators, has included the discovery of a short-wavelength form of chlorophyll (Chl) *a* functioning in the Chl *b*-containing system, now called PS II; of the two-light effects in Chl *a* fluorescence and in NADP (nicotinamide dinucleotide phosphate) reduction in chloroplasts (Emerson Enhancement). Further, he has worked on the existence of different spectral fluorescing forms of Chl *a* and the temperature dependence of excitation energy transfer down to 4 K; basic relationships between Chl *a* fluorescence and photosynthetic reactions; unique role of bicarbonate on the acceptor side of PS II, particularly in protonation events involving the $Q_B$ binding region; the theory of thermoluminescence in plants; picosecond measurement on the primary photochemistry of PS II; and the use of Fluorescence Lifetime Imaging Microscopy (FLIM) of Chl *a* fluorescence in understanding photoprotection against excess light. His current focus is on the 'History of Photosynthesis Research,' in 'Photosynthesis Education,' and in the 'Possible Existence of Extraterrestrial Life.' He has served on the faculty of the UIUC for ~40 years. Since 1999, he has been Professor Emeritus of Biochemistry, Biophysics and Plant Biology at the same institution. His honors include: Fellow of the American Association of Advancement of Science; Distinguished Lecturer of the School of Life Sciences, UIUC; Fellow and Lifetime member of the National Academy of Sciences (India); President of the American Society for Photobiology (1980–1981); Fulbright Scholar and Fulbright Senior Lecturer; Honorary President of the 2004 International Photosynthesis Congress (Montréal, Canada); and the 2006 Recipient of the Lifetime Achievement Award from the Rebeiz Foundation for Basic Biology.

# Contents



## *Imaging*

## *Structure*

## *Optical Spectroscopy*

## *Magnetic Resonance*

## *Theory*

# Preface

Eleven years ago, Jan Amesz and Arnold Hoff edited the book *Biophysical Techniques in Photosynthesis* as volume 3 in the Series *Advances in Photosynthesis and Respiration* (Kluwer Academic Publishers, now Springer). These methods are now well established in the photosynthesis community and beyond, and are used by many laboratories in various scientific fields. At the 13$^{th}$ Photosynthesis Congress in Montréal (2004) it became evident that over the past decade a range of new biophysical methods had been developed at a remarkable pace. Many of these are not covered in the book by Amesz and Hoff. This provided the inspiration and the incentive for the publication of a second volume on biophysical techniques in photosynthesis research, with specific emphasis on the developments over the past ten years.

The present book is intended to continue the successful concept of the previous one. This means that it will also focus on the techniques rather than on the scientific questions involved. The purpose is to make the basic techniques and underlying principles of these newly developed methods accessible to young and experienced researchers alike, and to give information about the practical aspects of the methods. In addition, the discussion of selected results obtained in photosynthesis research provides insight into their potential.

Care has been taken to avoid overlap with and duplication of topics that were already covered in the previous volume. Hence the two books are highly complementary, and together they will provide an excellent entry into biophysical techniques in photosynthesis research. The methods discussed in this book are divided in five categories:

1. Imaging: It is becoming increasingly important in the life sciences, and many of the advancements quickly find their way in photosynthesis research. Methods that are discussed in this section are atomic force microscopy (AFM), nonlinear optical microscopy, three-dimensional electron microscopy, and magnetic resonance imaging.

2. Structure: The structure of proteins is a key to the functional and mechanistic aspects of biological processes at the molecular level. It provides the basis for a quantitative approach in terms of theory and modeling of key parameters of the photosynthetic process. This section reviews major developments in methods for structure determination, from the crystallization of membrane proteins to electron and X-ray crystallography.

3. Optical and Laser Spectroscopy: These have been prime tools for investigating photosynthetic systems, e.g., for the analysis of pigment composition and for unraveling pathways for and dynamics of energy and electron transfer. In the past ten years we have seen the development of femtosecond time-resolved infrared and nonlinear optical spectroscopy, of the simultaneous measurement of the time- and spectral evolution of fluorescence, and of single-molecule techniques, all with important applications in photosynthesis research.

4. Magnetic Resonance: Developments in this technique have revolved on high-field techniques in EPR (Electron Paramagnetic Resonance) and NMR (Nuclear Magnetic Resonance), high time-resolution in EPR, distance measurements by EPR in combination with spin labeling, and magic angle spinning techniques in NMR in conjunction with chemically induced nuclear polarization. The dynamic and structural details revealed by these techniques provide access to the driving forces of photosynthesis at the molecular level.

5. Theory: It plays an increasingly important role in the understanding of photosynthetic mechanisms, a development spurred by the rapidly increasing and detailed structural information that is becoming available. Advancements are found in the calculations of electrostatic energies, the theory for excitation energy transfer, molecular dynamics simulations, and new mathematical and quantum-chemical methods.

Many of the techniques and methods discussed in this book were not in existence ten years ago, providing a rationale for this volume. It is interesting to note the growing importance of the relationship between structure, experiment and theory.

This book has been made possible by the help and effort of many. First of all, we are indebted to the contributing authors for their willingness to spend precious time and energy. Furthermore, we thank Govindjee, the Series Editor, who engendered the idea of this book. He has provided invaluable support and advice along the way. Finally, we acknowledge the assistance of Larry Orr, not only for typesetting, but also for many helpful suggestions, and of Jacco Flipsen and his staff at Springer in producing this book. Finally, we thank our families for their patience when we spent many odd hours on preparing this book.

Thijs J. Aartsma
Department of Biophysics
Huygens Laboratory
Leiden University
P.O. Box 9504
2300 RA Leiden
The Netherlands
E-mail: aartsma@physics.leidenuniv.nl

Jörg Matysik
Leiden Institute of Chemistry
Leiden University
P.O. Box 9502
2300 RA Leiden
The Netherlands
E-mail: j.matysik@chem.leidenuniv.nl

**Thijs J. Aartsma**, born in 1948, has a long term interest in the study of the photophysical properties of biomolecules using optical spectroscopic techniques. After obtaining his M.Sc. degree in 1973 in physical chemistry, he received his PhD degree in 1978 on photon echo relaxation in molecular mixed crystals, at the University of Groningen, The Netherlands, under supervision of Prof. D.A. Wiersma. From 1978–1981 he was a postdoctoral fellow at the University of Washington, Seattle, in the group of Prof. A.L. Kwiram, after which he became an assistant professor at the Florida State University in Tallahassee. He joined the Department of Biophysics at the Leiden University in 1986, first as an associate professor and since 2004 as a full professor. His research in photosynthesis has centered on time-resolved spectroscopy of energy and electron transfer. He has studied the effect of intermolecular interactions on the optical spectra of light-harvesting complexes, and has, among other, established the signature of exciton coherence in the optical properties of the FMO (Fenna-Mathews-Olson) -complex, with successful modeling of the optical spectra. He initiated the application if single-molecule techniques in the investigation of antenna complexes from purple bacteria, providing a unique view on the exciton structure in these systems. His recent work also involves the investigation of the redox turn-over of single metalloproteins and enzymes by fluorescence detection.

**Jörg Matysik** was born in 1964 in Essen (Germany). His research interest is in photochemistry and spin-chemistry as well as in the development of complementary optical magnetic resonance methods. Of especial interest is the spin evolution in the early light-reaction in photosynthesis. Initially, Matysik went for vocational training as Chemielaborant in the Bergbau-Forschung (Institute for Coal Mining Research) in Essen-Kray before he studied Chemistry at the Universität-Gesamthochschule in Essen where he obtained his Chemie-Diplom (1992) in the group of Prof. Bernhard Schrader, working with Fourier-Transform (FT) Raman spectroscopy on tetrapyrroles. For his PhD (1995), he investigated phytochrome with FT-Raman spectroscopy at the Max-Planck-Institut für Stahlenchemie in Mülheim an der Ruhr in the group of Prof. Peter Hildebrandt in the department headed by Prof. Kurt Schaffner. As Japanese Society for Promotion of Science (JSPS) and Humboldt fellow he worked with Raman spectroscopy on heme proteins in Prof. Teizo Kitagawa's group at the Institute for Molecular Sciences in Okazaki. Since 1997 he is at the Leiden Institute of Chemistry, first as Marie-Curie fellow and Casimir-Ziegler awardee in the solid-state nuclear magnetic resonance (NMR) group of Prof. Huub de Groot. He now has his own research group on optical solid-state NMR. He is a recipient of the Jonge Chemici award (2001) and the Vidi award (2003) of the Nederlandse Organisatie voor Wetenschappelijk Onderzoek.

# Author Index

# Color Plates

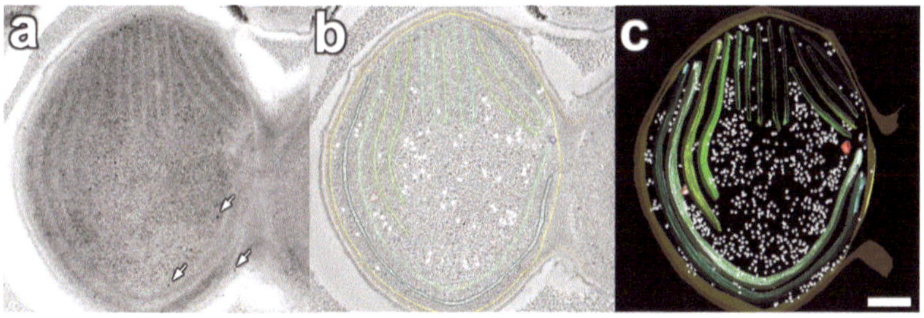

*Fig. 1.* Steps involved in the recording, reconstruction and modeling of a tomogram of a thin section of the cyanobacterium *Synechocystis* PCC 6803. (a) Image that is part of a tilt series. Fiducial markers appear as dark circles and are indicated by arrows. (b) Slice through reconstructed volume with model contours (white: ribosomes; yellow: cytoplasmic membrane; green: thylakoid membranes. (c) Model of section based on the reconstructed volume. Scale ~100 nm. See Chapter 2, p. 26.

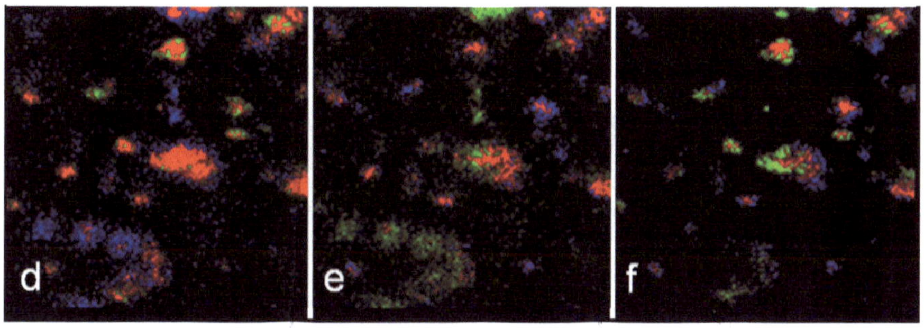

*Fig. 2.* Imaging isolated LHCII with a multimodal microscope. Panels d, e and f show colocalized images obtained by structural image cross-correlation analysis. Panel d compares MPF and SHG images (panels a and b; see Fig. 3, Chapter 3, p. 48), where correlated pixels are presented in red, uncorrelated fluorescence is presented in green and uncorrelated SHG is shown in blue. Panel e compares MPF and THG images (panels a and c; see Fig. 3, Chapter 3, p. 48). The red color represents correlation between MPF and THG while the green and blue shows uncorrelated MPF and THG signals, respectively. The panel f compares SHG and THG (panels b and c). The red color represents correlation between SHG and THG, while the green and blue represents uncorrelated SHG and THG, respectively.

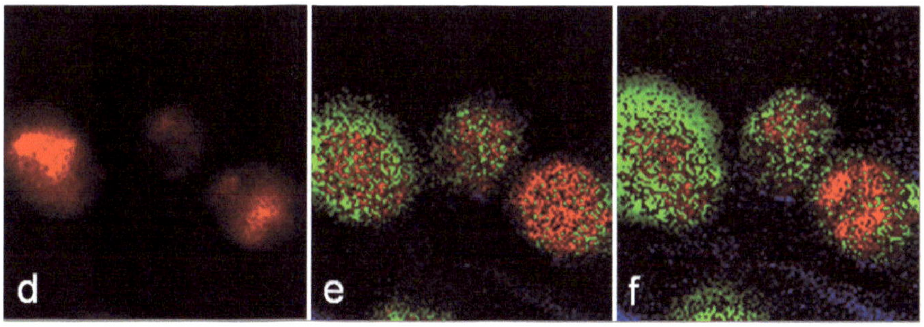

*Fig. 3.* Imaging in situ chloroplasts of *Clivia mineata* with a multimodal microscope. Panels d, e and f show colocalized images obtained by structural image cross-correlation analysis. Panel d compares MPF and SHG images (panels a and b; see Fig. 4, Chapter 3, p. 50), where correlated pixels are presented in red, uncorrelated fluorescence is presented in green and uncorrelated SHG is shown in blue. Panel e compares MPF and THG images (panels a and c; see Fig. 4, Chapter 3, p. 50). The red color represents correlation between MPF and THG while the green and blue shows uncorrelated MPF and THG signals, respectively. The panel f compares SHG and THG (panels b and c). The red color represents correlation between SHG and THG, while the green and blue represents uncorrelated SHG and THG, respectively.

*Thijs J. Aartsma and Jörg Matysik (eds): Biophysical Techniques in Photosynthesis II, pp. CP1–CP7.*

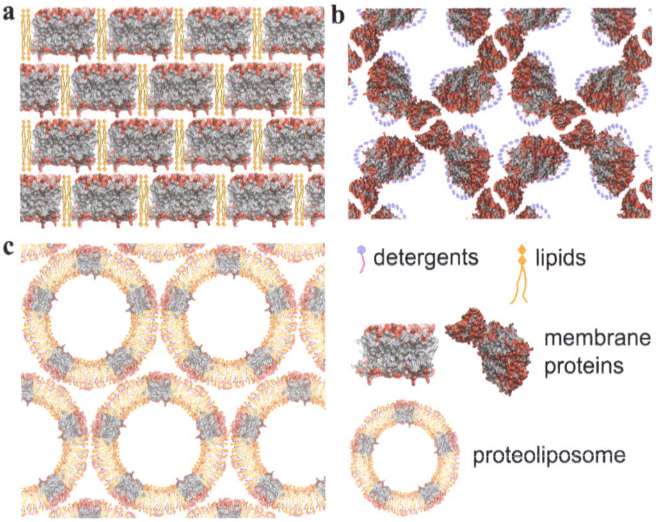


*Fig. 1.* The packing diagrams of three basic types of membrane protein crystals. (a) Type I, stacks of two-dimensional crystals. (b) Type II membrane proteins are crystallized in complex with detergent micelles. (c) Type III, crystals built from proteoliposome vesicles. The nonpolar regions of membrane proteins are in light gray while the polar regions are shown in darker gray. See Chapter 5, p. 80.

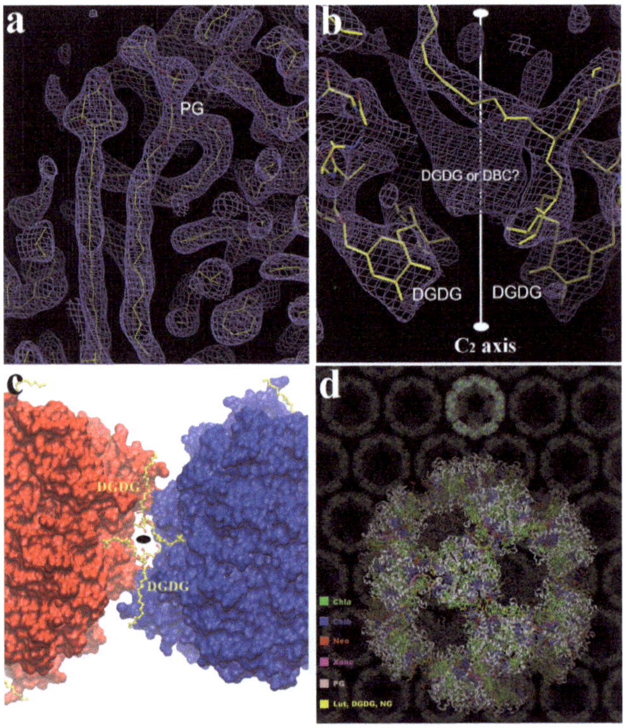


*Fig. 2.* The roles of lipids in the structure of LHC-II-DGDG proteoliposome. (a), (b) The well-defined 2Fo-Fc electron densities ($1.5 \times \sigma$ level) of PG and DGDG, respectively. In between the two DGDG molecules, there is a piece of electron density remaining unassigned. The icosahedral $C_2$ axis runs through it. (c) Top view of the trimer-trimer interface along the icosahedral $C_2$ axis. The two trimers are represented as surface models in the left and right halves of the panel, while the DGDG molecules are shown as ball-and-stick models in light gray. The black ellipse indicates the projection of $C_2$ axis on the paper plane. (d) The detailed structure of LHC-II-DGDG proteoliposome determined at 2.72 Å resolution. The apoproteins are shown as white ribbons. Prosthetic groups are drawn as ball-and-stick models in different gray levels. Chla, chlorophyll *a*; Chlb, chlorophyll *b*; Lut, lutein; Neo, neoxanthin; Xanc, xanthophyll-cycle carotenoid. The background shows a portion of the crystal lattice. See Chapter 5, p. 94.

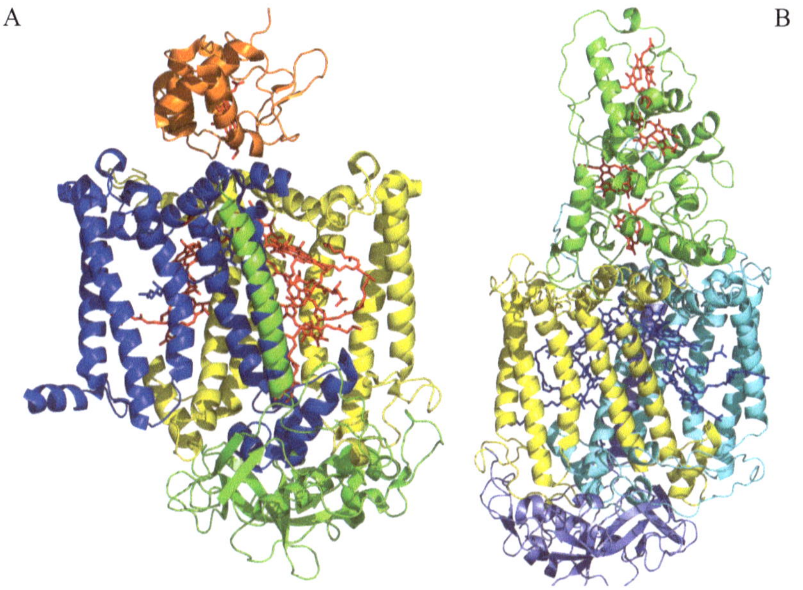


*Fig. 1.* A. Structure of the reaction center from *Rhodobacter sphaeroides* with cytochrome $c_2$ bound (PDB entry 1L9J). The three subunits of the reaction center, L (yellow), M (blue), and H (green) and the bound cytochrome $c_2$ (red) are shown as ribbon diagrams. Cofactors are shown in red. Also shown are the cofactors of the reaction center and the cytochrome heme (red). B. Structure of the reaction center from *Blastochloris viridis* (PDB entry 1PRC). The reaction center consists of four subunits, L (yellow), M (blue-green), H (blue), and a tetraheme cytochrome. Hemes are shown in red, other cofactors in blue. See Chapter 6, p. 115.

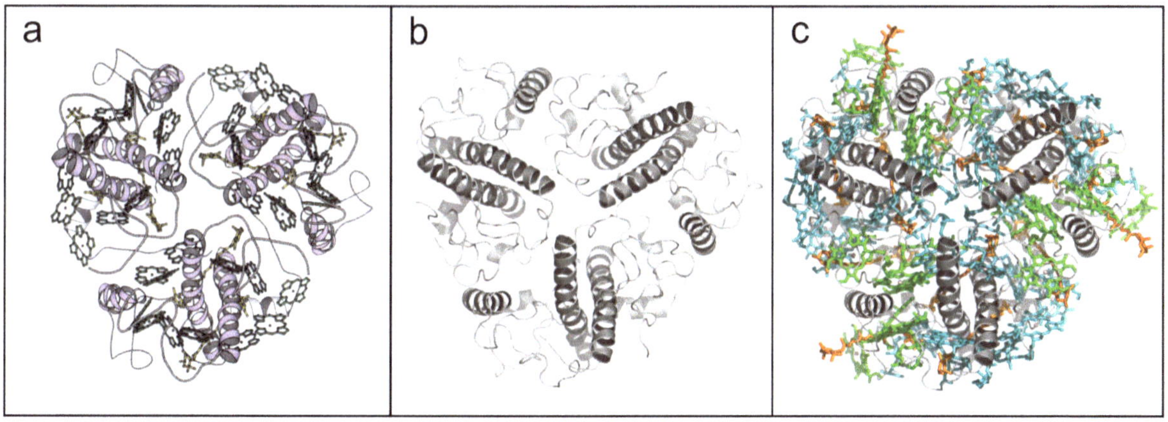


*Fig. 2.* Comparison of the structure of LHC2 as determined by electron crystallography (a) and X-ray crystallography (b and c). For clarity, all pigments were omitted in (b). (Figures courtesy of Professor W. Kühlbrandt). See Chapter 7, p. 143.

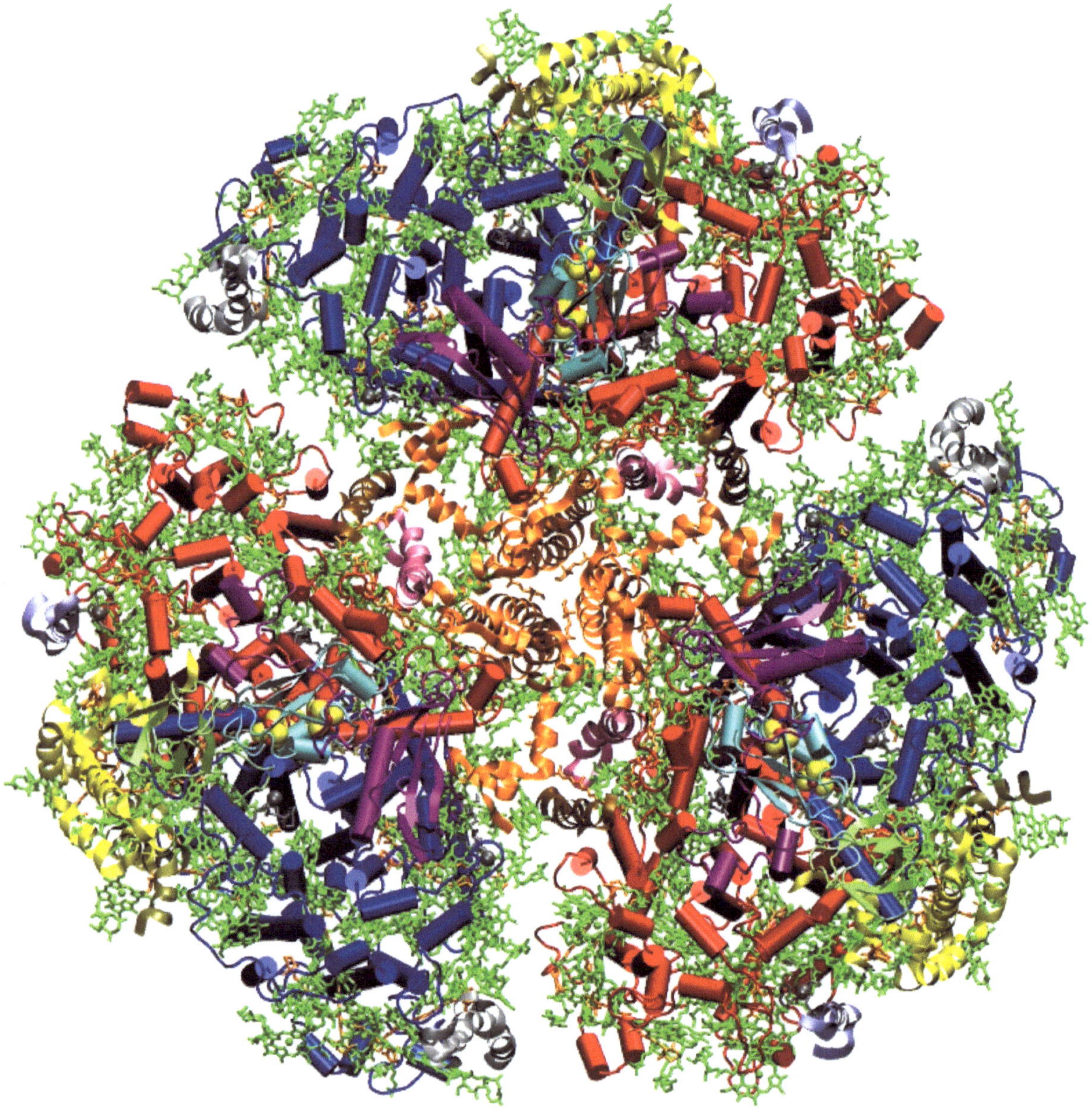

*Fig. 1.* Structure of Photosystem I (PDB entry 1JB0). The molecule is a trimer. View direction is from the stromal side onto the membrane plane. Each monomer consists of 12 proteins subunits, shown in a ribbon presentation. The large subunit PsaA and PsaB are forming the center of each monomer. PsaA and PsaB contain 11 transmembrane helices, each, which are shown as solid columns. These large subunits carry the electron transfer chain and coordinate most of the antenna chlorophylls and carotenoids. 127 cofactors are bound to one monomer of PS I: 96 chlorophylls, 22 carotenoids, 3 4Fe4S clusters, 2 phylloquinones and 1 Ca. PS I contains 7 small membrane intrinsic proteins that are all located peripheral to the core of PsaA and PsaB. They contain 1 to 3 transmembrane helices which are shown in a ribbon representation. The three stromal subunits, PsaC, D and E provide the docking site for ferredoxin. Modified from Jordan et al., 2001. See Chapter 6, p. 119.

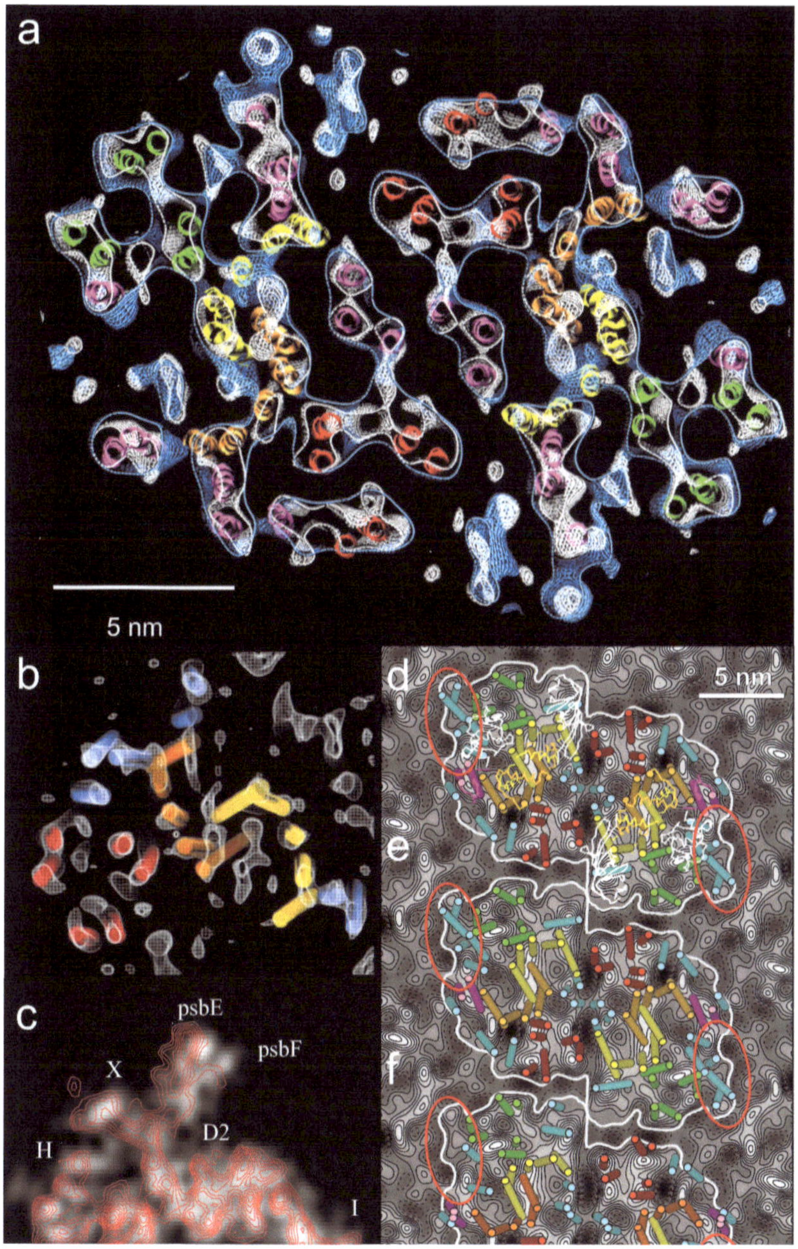


*Fig. 2.* Electron crystallographic studies of PS II. (a) Section of the 10Å 3D electron crystallographic structure of spinach PS II dimeric core complex (Hankamer et al., 2001) viewed from the lumen. Protein density is shown as chicken-wire and fitted transmembrane helices as ribbons which are color coded to identify subunits as follows: D1, yellow; D2, orange; CP47, red; CP43, green; α- and β-subunits of Cyt *b*559 and low-molecular-weight subunits, magenta. (b) Lumenal view of the 8Å 3D electron crystallographic structure of the spinach CP47-RC PS II subcomplex (Rhee et al., 1998). Protein density is shown as chicken wire and fitted transmembrane helices as cylinders, color coded as in (a) except that Cyt *b*559 and the low-molecular-weight subunits are colored blue. (c) 5.5Å electron crystallographic projection structure of the spinach CP47-RC PS II subcomplex (red contours) (Büchel and Kühlbrandt, 2005) compared with a simulated projection map at the same resolution calculated from the relevant subunits in the X-ray crystallographic structure of cyanobacterial PS II (Ferreira et al., 2004). (d-f) Comparison of the structural data on the PS II core dimer from *Synechococcus elongatus* and spinach (da Fonseca et al., 2002). The electron crystallographic 16Å projection structure of PS II from *Synechococcus elongatus*, represented as contoured grayscale, is overlaid with: (d) the full X-ray crystallography structure and (e) the transmembrane helices of *Synechococcus elongatus* (Zouni et al., 2001); (f) the transmembrane helices of the spinach core dimer deduced from electron crystallography (Hankamer et al., 2001). Color coding as in (a) except that small subunits are shown as blue and, Cyt *b*559 as magenta and extrinsic subunits as white ribbon. See Chapter 7, p. 144.

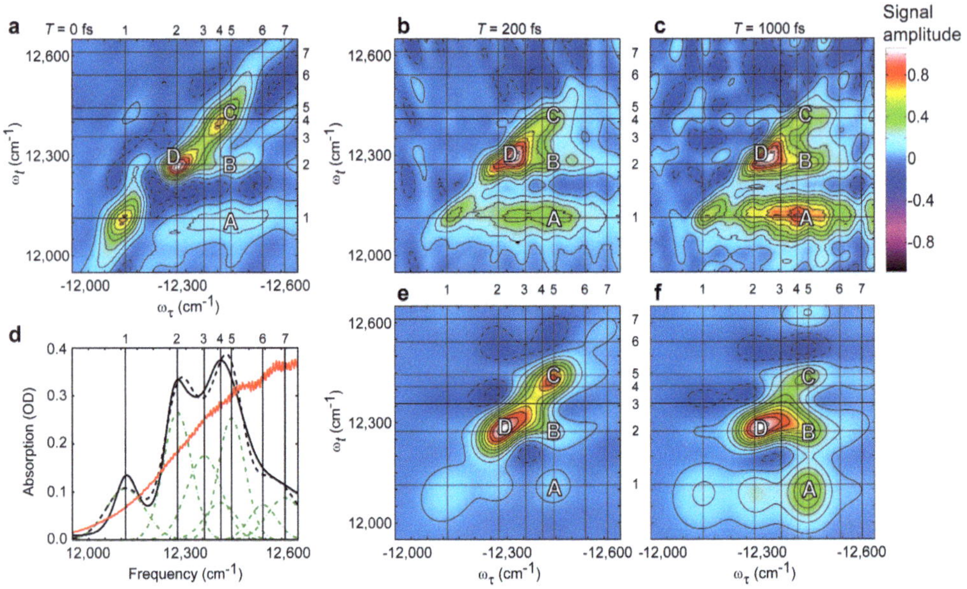

*Fig. 1.* Experimental and simulated real part of the electric field, corrected for radiative line-shape distortions, of the FMO complex at 77K. (a–c), The experimental 2D spectra (upper three panels) are shown for population times $T = 0$ fs (a), $T = 200$ fs (b) and $T = 1$ ps (c). Contour lines are drawn in 10% intervals of the peak amplitude, with solid lines representing positive features and dashed lines negative features. Horizontal and vertical grid lines indicate excitonic levels 1–7 as labeled. (d), The experimental (solid black) and simulated (dashed black) linear absorption spectra with individual exciton contributions as shown (dashed-dotted green). The laser spectrum (red) covers all transition frequencies. (e,f), Simulation of 2D spectra are shown for $T = 200$ fs (e) and $T = 1$ ps (f). Two off-diagonal peaks marked as A and B are indicators of electronic coupling and energy transport, and two diagonal peaks are marked as C and D (Brixner et al., 2005). See Chapter 11, p. 217.

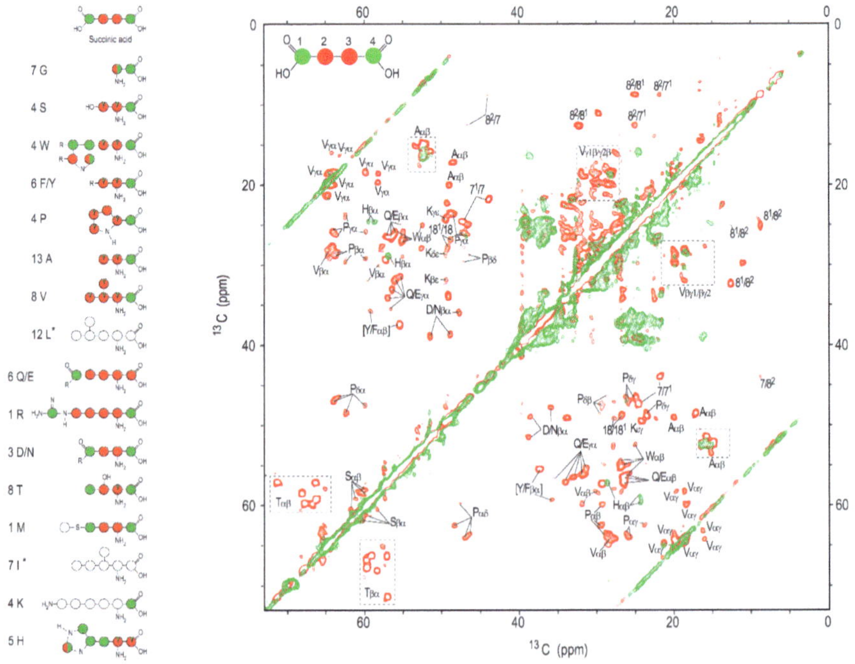

*Fig. 2.* Pattern labeling of LH2 starting from 1,4 (black) or 2,3 (grey) $^{13}C$ labeled succinic acid in *R. acidophila*. On the left the enrichment of the residues is shown, while on the right a superposition of the aliphatic region of two $^{13}C$-$^{13}C$ PDSD datasets is shown, illustrating the spectral simplifications that can be obtained with pattern labeling. See Chapter 18, p. 379.

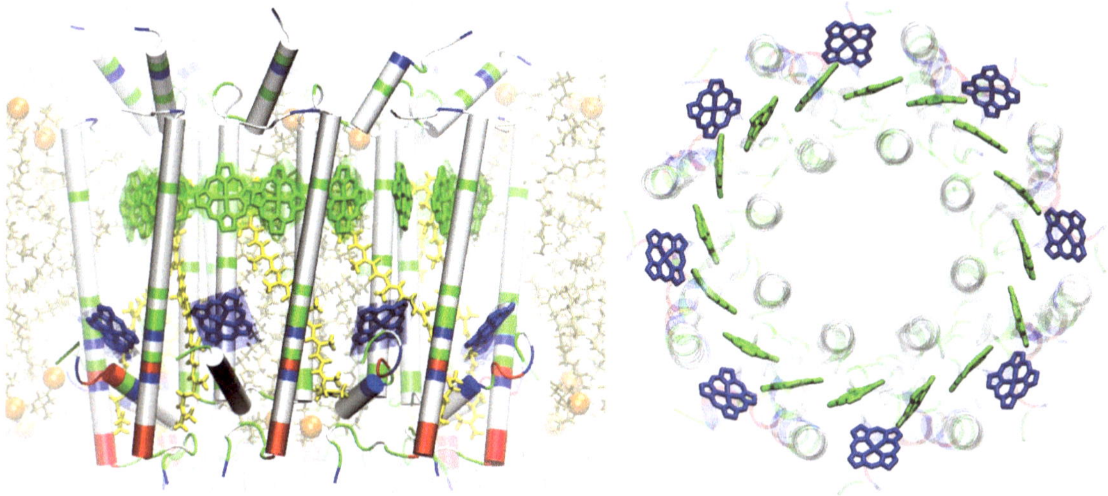

*Fig. 1.* Left: Side view of the LH2 complex from *Rs. molischianum* (PDB entry 1LGH) embedded in a fully solvated POPC lipid bilayer. The transmembrane helices of the apoprotein subunits are shown as cylinders (cartoon representation) and are colored by residue type; dark (light) colors represent hydrophilic (hydrophobic) residues. For clarity only the BChl macrocycles and the back half of the lipids are shown. The clearly visible B800 (B850) ring is surrounded mostly by polar and charged (nonpolar) protein residues. Right: Tilted side view of the quantum system formed by the optically active B800 and B850 rings. Graphics rendered with the program VMD (Humphrey et al., 1996). See Chapter 22, p. 448 and Chapter 23, p. 473.

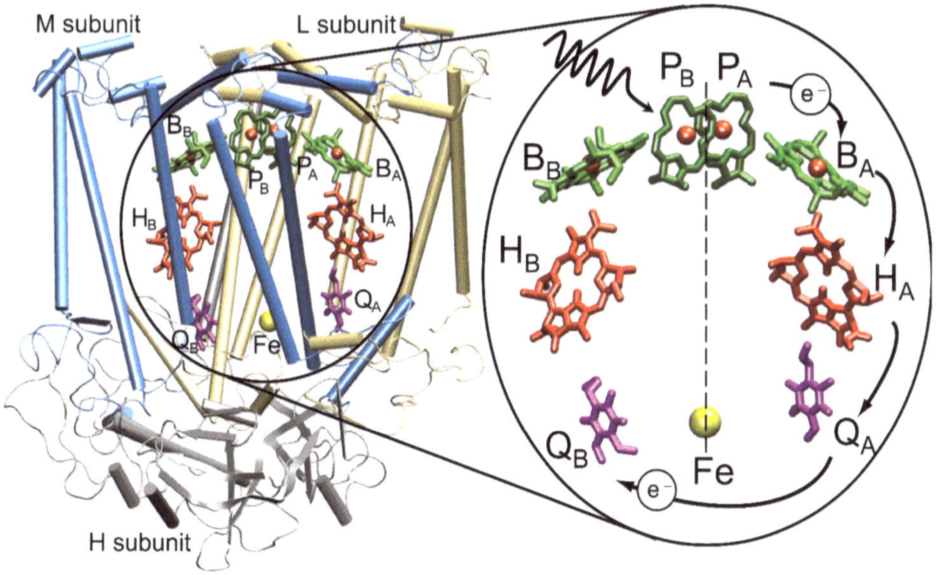


*Fig. 2.* Left: Structure of the PRC of *Rb. sphaeroides* (PDF entry 1PCR). For clarity only the backbone of the protein subunits L, M and H are shown. The protein helices are represented by cylinders. The cofactors are labeled, and for clarity their phytyl tails are not shown. Right: Spatial distribution of the cofactors in the PRC. The path of the electrons through the PRC is indicated by the arrows. Graphics rendered with the program VMD (Humphrey et al., 1996). See Chapter 22, p. 453 and Chapter 23, p. 477.

# Chapter 1

# The Supramolecular Architecture of the Bacterial Photosynthetic Apparatus Studied by Atomic Force Microscopy (AFM)

Simon Scheuring*

*Institut Curie, UMR168-CNRS, 26 Rue d'Ulm, 75248 Paris, France*



## Summary

The atomic force microscope (AFM) has developed into a powerful tool in structural biology allowing topographical information of membrane proteins at submolecular resolution to be acquired. Recently, AFM has been demonstrated to be the unique tool to image the photosynthetic apparatus in native membranes from different photosynthetic bacteria species. This chapter provides rationales how to image at high resolution a native membrane using the AFM, and summarizes the recent results concerning the structure and the supramolecular assembly of the photosynthetic complexes. On the single molecule level, membrane proteins directly studied in the native membrane were never subject to extraction, purification, reconstitution, or crystallization. Hence structural data in a native state and information concerning structural heterogeneity of the individual photosynthetic complexes are contributed. On the level of multi-protein assemblies, experimental images of the supramolecular architecture of the photosynthetic apparatus, its adaptation to environmental factors, and its particularities among species are reported.

*Email: simon.scheuring@curie.fr

*Thijs J. Aartsma and Jörg Matysik (eds): Biophysical Techniques in Photosynthesis II, pp. 1–11.*

## I. Introduction

The atomic force microscope (AFM) (Binnig, 1986) has developed into a powerful tool in membrane protein research and is now a complementary technique to electron microscopy (Chapter 2, Hohmann-Marriott et al.) and X-ray crystallography (Chapter 5, Liu and Chang; Chapter 6, Fromme and Allen). The AFM allows high-resolution topographies of biological samples to be acquired near physiological conditions, i.e,. in buffer solution at room temperature and under normal pressure. The AFM, restricted to surface contouring, will never provide an atomic structure, however it is to date the only technique that allows studying the native supramolecular membrane protein assembly in native membranes at nanometer resolution, and hence provides the pictures into which high-resolution membrane protein structures can be docked.

Anaerobic photosynthetic prokaryotes fuel their metabolism with light energy and have developed for this purpose an astonishingly efficient photosynthetic apparatus consisting of a highly organized nanometric assembly of trans-membrane protein complexes. Absorption of light and its conversion into chemical energy require only four trans-membrane protein-pigment complexes localized in intra-cytoplasmic membranes: a peripheral light harvesting complex II (LH2), a central light harvesting complex I (LH1) that complexes with the photochemical reaction center (RC) and a proton translocating cytochrome (Cyt) $bc_1$ complex. The cascade of molecular events in the photosynthetic chain has been determined in fine detail using time resolved spectroscopy and structural analysis of the protein constituents of the photosynthetic apparatus by X-ray crystallography.

In this chapter, the experimental approaches that have allowed high-resolution AFM imaging of single molecules and their supramolecular architecture in the intact photosynthetic apparatus and corresponding recent results are outlined.

*Abbreviations:* 4Hcyt – tetraheme cytochrome; AFM – atomic force microscopy; ATP – adenosine triphosphate; BChl – bacteriochlorophyll; *Blc.* – *Blastochloris*; Cyt – cytochrome; LH1 – light harvesting complex I; LH2 – light harvesting complex II; *Phsp.* – *Phaeospirillum*; *Rb.* – *Rhodobacter*; RC – reaction center; *Rps.* – *Rhodopseudomonas*; *Rvi.* – *Rubrivivax*

## II. The Proteins of the Photosynthetic Apparatus

The structure determination through X-ray crystallography of the different components of the photosynthetic apparatus, the light harvesting complex 2 (LH2), the light harvesting complex 1 (LH1), the reaction center (RC), the cytochrome $bc_1$ complex, and the ATP-synthase, is almost complete (Chapter 6, Fromme). Briefly, the determination of the three-dimensional structures of the RCs of *Blastochloris* (*Blc.*) *viridis* (Deisenhofer et al., 1984, 1985, 1995; Deisenhofer and Michel, 1989; Lancaster et al., 2000) and of *Rhodobacter* (*Rb.*) *sphaeroides* (Allen et al., 1987) represent the first solved membrane protein structures in general, and have provided deep insight into their functional mechanism. The RC consists of three subunits L, M and H. While subunits L and M are similar and arranged with a quasi two-fold axis, each consisting of five trans-membrane helices, subunit H crosses the membrane only once and has a cytoplasmic domain. In *Blc. viridis*, a non-membranous tetraheme cytochrome (4Hcyt) tops the RC on the periplasmic side. The LH2 from *Rhodopseudomonas* (*Rps.*) *acidophila* (McDermott et al., 1995) and from *Phaeospirillum* (*Phsp.*) *molischianum* (Koepke et al., 1996) revealed nonameric and octameric cylinders of two small polypeptides ($\alpha$ and $\beta$), both span the membrane once. The inner wall of the cylinders is formed by the $\alpha$-polypeptides and the outer wall by the $\beta$-polypeptides, keeping in place three bacteriochlorophyll *a* (BChl *a*) and two carotenoids per subunit. To date, no atomic structure of the LH1-RC core-complex is available. However, medium and low resolution X-ray, EM and AFM data of different core-complexes indicate structural variability of core-complexes among species: The core-complex was described to consist of a RC surrounded by an ellipse of 16 LH1 $\alpha/\beta$ heterodimers (*Rsp. rubrum*, *Blc. viridis*, *Rsp. photometricum*, *Phsp. molischianum*; (Jamieson et al., 2002; Scheuring et al., 2003b; Fotiadis et al., 2004; Gonçalves et al., 2005a; Scheuring and Sturgis, 2005)), or 15 LH1 $\alpha/\beta$ heterodimers plus an yet non-identified W-peptide (*Rps. palustris*; Roszak et al., 2003; Scheuring et al., 2005b), or as a dimer consisting of two RCs surrounded by an S-shaped LH1 assembly constituted of two times 13 LH1 $\alpha/\beta$ heterodimers plus 2 PufXs at the dimer center (*Rb. blasticus, Rb. sphaeroides*; Scheuring, 2004a, 2005a;) or two RCs surrounded by an S-shaped LH1 assembly constituted of two times 14 LH1 α/β heterodimers plus 2 PufXs located at the periphery of the S (*Rb. sphaeroides*; Qian et al., 2005). In general, LH1 α/β heterodimers, considered structurally similar as LH2 α/β heterodimers (McDermott et al., 1995; Koepke et al., 1996; Hu and Schulten, 1998), surround the RC in slightly different manner, i.e. with or without an additional peptide integrated, probably reflecting adaptation to particular environmental conditions of the different species. Despite the diversity of core-complex architectures among species, all structural data agree now in the point that the LH1 assembly forms an ellipsis around the RC. A recent preliminary X-ray structure of the cytochrome $bc_1$ from *Rb. capsulatus* (Berry et al., 2004) indicates structural similarities with the homologous Cyt $bc_1$ complex from mitochondria (Xia et al., 1997), and the cytochrome $b_6f$ complexes from cyanobacteria (Kurisu et al., 2003) and algae (Stroebel et al., 2003). All are dimers with an overall size of ~5 nm × ~10 nm consisting of three subunits with eight, cytochrome *b*, one, the high potential 2Fe-2S Rieske protein, and one, the cytochrome $c_1$, trans-membrane helices. The cytochrome *b* subunit, which contains two *b*-type hemes, is the central core of the complex, to which other subunits are bound, and houses the quinone binding sites. Finally the homologous structures of both the $F_1$ (Abrahams et al., 1994) and $F_O$ (Meier et al., 2005) parts of the ATP-synthase are known.

Despite the wealth of functional and structural information available on the individual membrane proteins, their supramolecular organization in the bacterial photosynthetic membrane is still poorly understood. Numerous models of a bacterial photosynthetic apparatus have been proposed to explain its high efficiency of >95% (Papiz et al., 1996; Sundström et al., 1999; Hu et al., 2002). Finally, the AFM, featuring an exceptional high signal-to-noise ratio at high lateral resolution has proven to be the key technique for the experimental structural investigation of the photosynthetic apparatus in native membranes, and has provided some striking insights into the supramolecular architecture of the bacterial photosynthetic apparatus of different species (Scheuring et al., 2003b, 2004b,c, 2005a,b; Bahatyrova et al., 2004a; Gonçalves et al., 2005a; Scheuring and Sturgis, 2005).

## III. High-Resolution Contact Mode Atomic Force Microscopy (AFM)

### *A. Membrane Preparation*

Photosynthetic bacteria were grown anaerobically and photoheterotrophically on Hutner media under various light conditions. Cells were harvested and washed two times with 1 mM Tris-HCl, pH 7.0 before being broken by a passage through a French pressure cell (30 MPa). Lysates were loaded directly onto 5–60 % sucrose gradients and centrifuged for 1.5 hours. The membranes corresponding to the major pigmented band sedimented to about 40% sucrose and contained the photosynthetic membranes. Sucrose was removed from the membrane samples through dialysis against a sucrose-free buffer. Membranes were kept at 4 °C for AFM analysis.

### *B. Equipment and Experimental Procedure*

For AFM contact mode imaging, it is recommended to prepare Mica as support, freshly cleaved before every experiment (Schabert and Engel, 1994). Immediately after cleavage 40 μL of adsorption buffer is pipetted to the mica surface. Subsequently, a few μl of membrane solution are injected into the adsorption buffer drop. After membrane adsorption (that can take between a few minutes and a few hours depending on membrane concentration and exposed charges on the protein surfaces) the sample is rinsed with 10 volumes of recording buffer. For imaging, buffers that contain 10 mM Tris-HCl, pH 7.5 and variable KCl-content (0 mM–500 mM) should be prepared allowing the optimization of imaging conditions during the experiments. The acquisition of the data presented in this chapter was performed with a commercial Nanoscope-E contact-mode AFM (from Digital Instruments, Santa Barbara, CA, USA) equipped with a low-noise laser, and a 160 μm scanner (J-scanner) using oxide-sharpened $Si_3N_4$ cantilevers with a length of 100 μm (k = 0.09 N/m; Olympus Ltd., Tokyo, Japan). However, it appears that now a variety of commercial AFMs and cantilevers allow the acquisition of high-resolution topographs of biological samples in buffer solution. For imaging minimal loading forces of ~100 pN were applied at scan frequencies of 4–6 Hz using optimized feedback parameters (see Section III.C). The piezo precision can be determined by analyzing bacteriorhodopsin

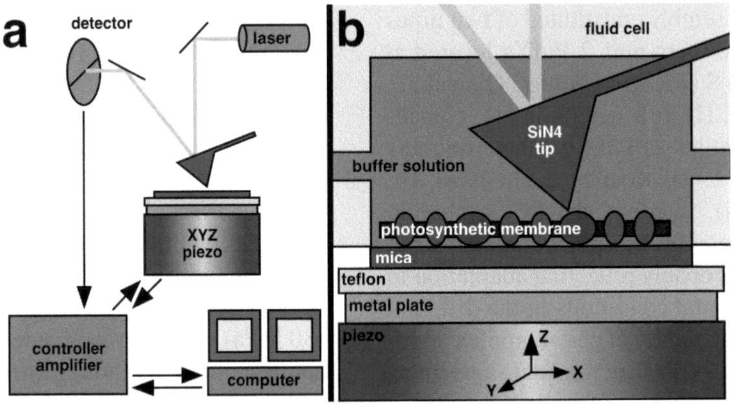


*Fig. 1.* Schematic representation of the AFM set-up. a) A laser beam is focused and deflected on the backside of the AFM cantilever monitoring the relative position of the cantilever on a multi-field detector. This signal is processed by the controller that drives the piezo in X- and Y-dimensions and compensates for the cantilever deflection through movements in Z-dimension. This compensation of the tip deflection by Z-movement of the piezo, which in turn will influence the tip deflection at each position (XY), is termed the feedback loop. The user interacts over a computer with the controller, feeds in the scanning parameters and reads out the piezo movements corresponding to the image, to and from the controller. The faster the feedback loop reacts, the more precise is the surface contouring of the sample. b) Close-up view of the fluid cell. The sample is deposited on mica, which is glued to a Teflon plate (protecting the piezo), which in turn is glued to a magnetic plate. The tip and the sample are permanently immerged in buffer solution during the measurements.

membrane protein 2D-crystals (lattice dimensions: $a = b = 62.5$ Å, $\gamma = 120°$) at scan ranges between 100 nm and 300 nm, and errors in x- and y-dimensions smaller than 2% should be found.

## C. Experimental Rationales

Key elements of an AFM (Binnig et al., 1986) are the cantilever with a pyramidal-shaped tip that touches the sample, an optical system consisting of a laser and a multi-faced photo diode that allows detection of cantilever deflection, a piezo-electric scanner that translates the sample relative to the tip in *x, y, z* directions and a computer that drives the microscope and stores the surface contours (Fig. 1a).

A topograph is recorded by raster scanning the sample below the tip that is attached to the flexible cantilever, while the feed-back-loop drives the piezo in z-dimension corresponding to the measured deflection changes in order to keep the cantilever deflection, i.e., the force, constant. The mica support, the tip and the sample are permanently immerged in buffer solution throughout experiments (Fig. 1b). The optical system resolves cantilever deflections of 0.1 nm, which corresponds to a force difference of typically 10–50 pN. With modern instruments stable contact mode operation is possible at forces of ~100 pN, provided the sample is in aqueous solution with optimized ionic strength (Müller et al., 1999).

Tight sample adsorption is an important parameter for contact mode AFM imaging. Operating the AFM in contact mode induces friction forces, therefore samples need to adhere well to freshly cleaved mica supports. As demonstrated using a variety of biological samples with different surface charge densities (Butt, 1992), sample adsorption is controlled by the nature and concentration of electrolytes in the buffer solution. Immobilization therefore requires buffer conditions in that the Debye layer thickness is minimized to allow adsorption by attractive van der Waals forces (Müller et al., 1997). For the adsorption of native photosynthetic membranes a buffer containing 10 mM Tris-HCl, pH 7.6, 150 mM KCl, 25 mM $MgCl_2$ was often successful. Non-flatly adsorbed membranes, or not completely collapsed vesicles, can present multiple technical problems and forces applied by the AFM tip might be wrongly estimated. Not-tightly-attached membranes can be vertically 'squeezed' by the tip. As a consequence, the 'squeezing' of the object contributes to the apparent spring constant, and the feedback-loop monitoring and compensating for the tip deflection will not drive the piezo appropriately to gain a highly contrasted image. Furthermore, membranes that are not tightly attached can provoke 'smeared' images, probably due to faint displacement of the sample during scanning.

High-resolution imaging involves electrostatic and van der Waals interactions between the sample and the tip (Israelachvili, 1991). For imaging, the electrolyte should be selected to balance electrostatic repulsions and van der Waals attraction. These forces can be quantitatively described and their integration can be evaluated in force-distance curves, allowing optimizing the recording conditions and to adjust the forces applied to the cantilever to ~100 pN (Müller et al., 1999). For high-resolution contact mode AFM imaging of native photosynthetic membranes a buffer containing 10 mM Tris-HCl, pH 7.6, 150 mM KCl is a good starting point for subsequent buffer optimization during the experiment (Scheuring et al., 2005c).

Problems for AFM imaging may be related to the imaging of strongly protruding protein structures. The feedback-loop is more challenged in contouring strongly corrugated surfaces, but also the mobility of those are increased. For both reasons, one generally expects the contouring precision to decrease with increasing height of surface protruding structures. When strongly protruding globular domains are imaged, the tip geometry, in a first approximation a hemi-sphere convolutes the topography significantly; the height measurement of a strongly protruding domain can be precise, the diameter measurement appears enlarged due to tip convolution, hence it seems unreasonable to interpret structural details of large globular topographies. As a consequence of the tip convolution and the feedback loop system, the imaging resolution within one AFM topograph is not identical on all areas of an image. A weakly protruding particle may be highly resolved while a globular structure in proximity is strongly subjected to tip convolution and to non-precision of contouring. The problem becomes more pronounced when the tip geometry deviates from the shape of a perfect hemi-sphere, a phenomenon called 'asymmetric tip' or 'double tip.' This effect makes difficult the analysis of non-oligomeric molecules (in particular in native membranes, where no regular lattice or crystallographic symmetry can be used to estimate tip symmetry) or of molecules of that no alternative structural information is available. The situation gets more complex if one wants considering particular physicochemical surface properties to interfere at each tip-sample contact point (Israelachvili, 1991; Israelachvili and Wennerstrom, 1996).

Another crucial limitation for high-resolution AFM is related to the tip radius. It has to be stressed that only very sharp tips with ~nm-sized asperity can acquire topographs at resolution of ~10Å. Based on image analysis of barrel-shaped molecules with small diameters, like LH2 (Scheuring et al., 2001, 2003a, 2004b; Gonçalves et al., 2005a,b; Scheuring and Sturgis, 2005), a high-resolution tip radius of ~2.5 nm is estimated (Scheuring et al., 2005c). The finding of a tip with such a sharp apex can be considered as another limitation of high-resolution AFM imaging.

In principle, the AFM can provide atomic resolution on solid crystalline surfaces such as mica or graphite (Binnig et al., 1986, 1987). On membranes in buffer solution, the imaging resolution of ~10Å is most probably limited due to protein fluctuation at room temperature. In contact mode high-resolution AFM, images are acquired at scan speeds around 1 ms/nm. It seems reasonable to assume, that some protein fluctuation is averaged during this time scale, limiting the acquisition of more detailed structural information.

## IV. AFM Analysis of Bacterial Photosynthetic Apparatus

The first high-resolution AFM analysis of a constituent of the bacterial photosynthetic apparatus concerned the stoichiometry and sidedness of the LH2 complex of *Rubrivivax* (*Rvi.*) *gelatinosus*. Interesting in this work, in some less extensively purified LH2 samples a minor fraction of LH1 complexes were co-purified and found in membranes assembled with LH2, providing first high-resolution images of different protein complexes in one membrane (Scheuring et al., 2001). A series of following AFM analysis of purified and reconstituted proteins concerned LH2 (Scheuring et al., 2003a; Stamouli et al., 2003; Gonçalves et al., 2005b), LH1 (Bahatyrova et al., 2004b), LH1-RC core-complexes (Fotiadis et al., 2004; Scheuring et al., 2004a; Siebert et al., 2004). Finally native membranes were imaged, first only containing LH1-RC core-complexes (Scheuring et al., 2003b), then containing LH2 and LH1-RC core-complexes (Bahatyrova et al., 2004a; Scheuring et al., 2004b,c, 2005a,b; Gonçalves et al., 2005a; Scheuring and Sturgis, 2005).

### *A. The Photosynthetic Apparatus of Blastochloris viridis*

*Blastochloris* (*Blc.*) *viridis* is LH2 deficient and therefore the core-complexes form hexagonal lattices in chromatophores (Miller, 1982; Jacob and Miller, 1983, 1984). Projection maps (Stark et al., 1984; Engelhardt et al., 1986; Ikeda-Yamasaki et al.,

1998) could not resolve the subunit architecture of the core-complex due to the absence of higher packing order. The AFM features such a high signal-to-noise ratio that the core-complex subunit architecture can readily be analyzed on single molecules in native membranes (Fig. 2) (Scheuring, 2003b).

In *Blc. viridis*, the RC consists of the L, M, H and the tetraheme cytochrome (4Hcyt) subunits (Deisenhofer, 1985), and sixteen LH1 subunits each made of one α and one β-polypeptide. High resolution images of native membranes revealed a lateral resolution of 10Å, as judged from spectral signal-to-noise analysis gained through single particle averaging, and a vertical resolution of ~1Å.

Using the AFM tip as a nano-dissector (Hoh et al., 1991), the strongly protruding (~5 nm) 4Hcyt was removed, and a RC-$LH1_{16}$ complex remains that could be imaged at highest resolution (Fig. 2a). The topography revealed that the LH1 subunit architecture around the RC is elliptical with axes length differences of ~10%. The orientation of the long axis of the $LH1_{16}$ ellipsis coincides with the long axis of the RC topography. $LH1_{16}$ assemblies, where the 4Hcyt and the RC proteins have been removed, relax into closed circles with ~100Å top diameter (Fig. 2a). The ellipticity of the $LH1_{16}$ in complex with the RC, compared to a circular $LH1_{16}$ after removal of the RC reflects flexibility of the LH1 assembly and a strong and specific interaction between the core-complex components. Flexibility and/or fluctuations of ellipticity of the LH1 may favor the quinone/quinol passage through a closed LH1 fence (Scheuring et al., 2003b).

In native *Blc. viridis* membranes, that do not contain LH2, core-complexes are hexagonally packed. In such an arrangement, all core-complexes are equal, i.e., they are structurally identical, and their molecular neighborhood is identical. As a consequence, the chance that a RC is occupied by the primary photoreaction is equal for all RCs, and excitation energy can be transferred between neighboring core-complex without passing over a large energy barrier (Fig. 2b) (Scheuring et al., 2003b).

### *B. The Photosynthetic Apparatus of Rhodospirillum photometricum*

The supramolecular organization of the photosynthetic apparatus of *Rhodospirillum* (*Rsp.*) *photometricum*, was studied in detail and striking novel findings concerning antenna heterogeneity and complex assembly were described (Fig. 3) (Scheuring et al., 2004b,c; Scheuring and Sturgis, 2005).

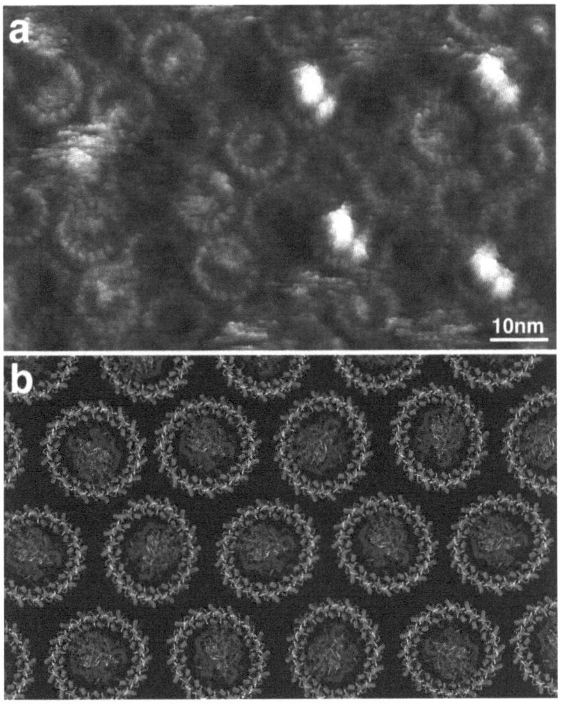


*Fig. 2.* High-resolution AFM analysis of the native photosynthetic membrane of *Blastochloris viridis* (Scheuring et al., 2003b). a) Raw data AFM topograph of hexagonally packed core-complexes in the native photosynthetic membrane of *Blc. viridis*. b) Model of the complex assembly in the native membrane of *Blc. viridis*. LH1 subunits form a closed ellipsis of 16 αβ-heterodimers around the RC subunits topped by the tetraheme cytochrome. For clarity, protein side chains are not shown.

The majority of the LH2 complexes assemble in rings with nine-fold symmetry. However, on closer examination LH2 of various sizes were found present within the native membranes. Image processing analysis showed heterogeneity of the LH2 complex stoichiometry with ~70% of nonameric complexes, and 30% of the complexes that had either significantly smaller or larger diameters, corresponding to eight- or ten-fold symmetry, respectively. This finding was qualitatively corroborated by examination of raw data images. It appeared that size heterogeneity is an inherent feature of the native LH2 assembly (Scheuring, et al. 2004b).

In contrast to the heterogeneity found for LH2 complexes, the RC-LH1 core-complexes appeared uniform in size. The core complex is monomeric with a closed elliptical $LH1_{16}$ assembly, with long and short axes of 95 Å and 85 Å, respectively, around the RC and following the long RC axis.

Analysis of the distribution of the photosynthetic complexes of *Rsp. photometricum* showed significant clustering of both antenna complexes and core-complexes. Membranes, with domains densely packed with photosynthetic proteins and protein-free lipid bilayers were found. Clustering seems reasonable and not very surprising, as each light-harvesting component that segregates away from the system is non-functional since it cannot pass its harvested energy to a neighboring complex and eventually to the RC (Scheuring et al., 2004b).

There is no fixed structural assembly of LH2 and core-complexes. Core-complexes making multiple core-core contacts and core-complexes completely surrounded by LH2 were found (Fig. 3a,b) (Scheuring et al., 2004b; Scheuring and Sturgis, 2005). However, pair correlation function analysis showed that the most frequent assembly was two core-complexes separated by an intercalated LH2. The supramolecular assembly of the photosynthetic complexes in *Rsp. photometricum* membranes from cells grown under different light intensities were studied and compared, since under low-light conditions increased quantities of peripheral LH2 are synthesized. Additional LH2 were not randomly inserted into the membrane but form para-crystalline hexagonally packed antenna domains (Fig. 3c,d). Core-complexes remained in domains in which they are locally much higher concentrated (LH2 rings/core-complex = ~3.5) than the average density under low-light cell growth (LH2 rings/core-complex = ~7). This indicated that complex assembly follows an eutectic phase behavior with an ideal LH2 rings / core-complex ratio ~3.5 independent of the growth conditions, and additional LH2 being synthesized under low-light conditions are integrated in specialized antenna domains. RCs that are grouped together, independent of the growth conditions, and formation of antenna domains when excess LH2 are synthesized, ensures efficient photon capture under low-light conditions and prevents photodamage under high-light conditions (Scheuring and Sturgis, 2005).

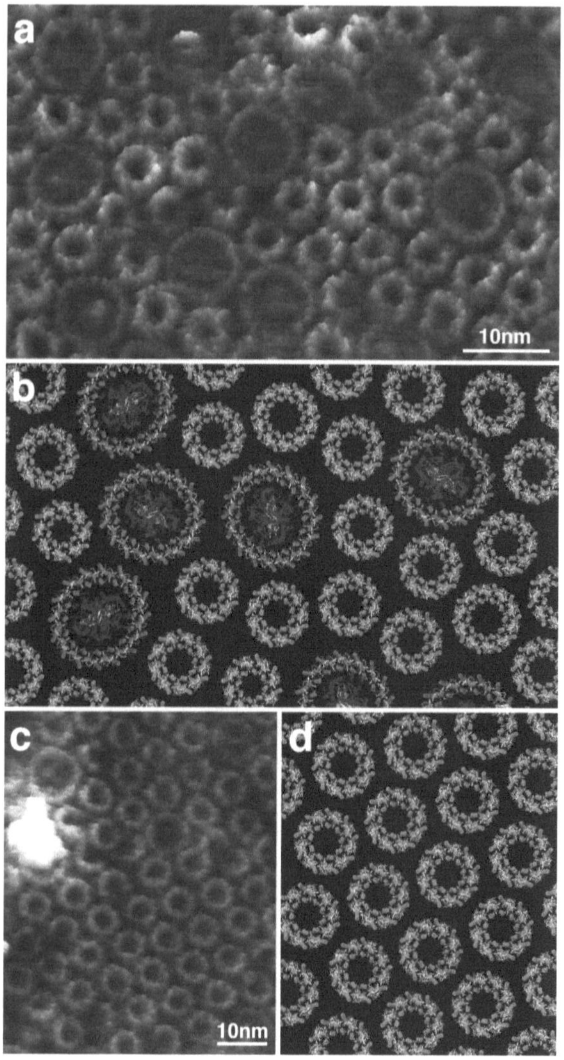


*Fig. 3.* High-resolution AFM analysis of the native photosynthetic membrane of *Rhodospirillum photometricum* (Scheuring et al., 2004b,c; Scheuring and Sturgis, 2005; reprinted with permission from AAAS). a) Raw data AFM topograph revealing the complex mixture of LH2 and core-complexes in high-light adapted native photosynthetic membranes of *Rsp. Photometricum*. b) Model of the complex assembly in the native membrane of *Rsp. photometricum*. LH2 subunits arrange in variable sized rings (mostly 9 mers, but also 8 mers and 10 mers) intercalating the core-complexes consisting of LH1 subunits forming a closed ellipse of 16 $\alpha\beta$-heterodimers around the RC subunits. For clarity, protein side chains are not shown. c) Raw data AFM topograph revealing a LH2 antenna domain in low-light adapted native photosynthetic membranes of *Rsp. Photometricum*. d) Model of the LH2 assembly in antenna domains in native membranes of *Rsp. photometricum*. LH2 forms hexagonal para-crystals with 75Å periodicity. For clarity, protein side chains are not shown.

## C. The Photosynthetic Apparatus of *Phaeospirillum molischianum*

The supramolecular architecture of the photosynthetic complexes in native membranes of *Phaeospirillum (Phsp.) molischianum* (Gonçalves et al., 2005a) is similar to that found in *Rsp. photometricum* (Scheuring et al., 2004b,c; Scheuring and Sturgis, 2005). The proteins of the photosynthetic apparatus show a strong tendency to cluster (Scheuring et al., 2004b).

All photosynthetic proteins are in physical contact assuring efficient energy transfer between the pigment molecules within and between the complexes (Scheuring et al., 2004b). In native membranes of *Phsp. molischianum* core-complexes are monomeric with the RC completely enclosed by an elliptic LH1 assembly. A wide variety of protein arrangements exist. About 30% of core-complexes were completely surrounded by LH2, and about 70% of the core-complexes are connected to other cores (Fig. 4a). This reflects an effort to maximize the efficiency of the photosynthetic machinery. Energy transfer reactions are much faster than the reduction of the RC. As in *Rsp. photometricum* core-core contacts may be important for the function of the apparatus when photons are abundant to prevent photodamage (Fig. 4b).

In agreement with the X-ray structure of *Phsp. molischianum* LH2 (Koepke et al., 1996), mainly octameric LH2 were found in the membranes. Surprisingly, octameric LH2 form hexagonal arrays in native membranes (Fig. 4c), like nonameric LH2 in chromatophores of *Rsp. photometricum* (Scheuring and Sturgis, 2005). Octameric LH2 has a symmetry mismatch with a hexagonal lattice, and a symmetry match with a square lattice. LH2 octamers could be packed in a square lattice with four optimized contacts with adjacent LH2 and a packing density of 0.785. In a hexagonal lattice the density is higher (0.907) but maximally only two optimized contacts with neighboring LH2 can be formed. In contrast, nonameric LH2 are in better symmetry match in a hexagonal lattice than in a square lattice. LH2 nonamers gain higher packing density in a hexagonal lattice and the possibility to form an average of two optimized contacts, versus lower density and maximally only 1 optimized contact in a square lattice. The observation that both eight-fold (*Phsp. molischianum*; Gonçalves et al., 2005a) and nine-fold (*Rsp. photometricum*; Scheuring and Sturgis, 2005) symmetrical LH2 form hexagonal antenna domains, while eight-fold LH2 could double the optimized protein-protein contacts in a square lattice, is strong evidence that high packing density is favored over specific protein-protein contacts (Fig. 4d). This suggests that lipophobic effects, e.g., a solvation mismatch between the membrane protein and the lipid bilayer, drive complex assembly. Such an effect could result from hydrophobic mismatch, or a solvation incompatibility between the branched polyprenoid chains exposed by LH2 and the anisotropic linear aliphatic chains of the phospholipids. Furthermore, as in membranes of

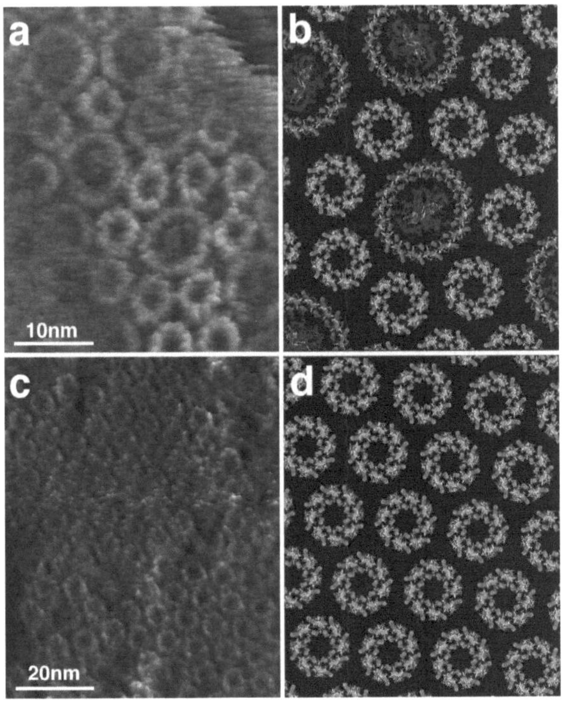


*Fig. 4.* High-resolution AFM analysis of the native photosynthetic membrane of *Phaeospirillum molischianum* (Gonçalves et al., 2005a). a) Raw data AFM topograph revealing the complex mixture of LH2 and core-complexes in the native photosynthetic membrane of *Phsp. molischianum*. b) Model of the complex assembly in the native membrane of *Phsp. molischianum*. LH2 arrange intercalating the core-complexes consisting of LH1 subunits forming a closed ellipse of 16 αβ-heterodimers around the RC subunits. For clarity, protein side chains are not shown. c) Raw data AFM topograph revealing an LH2 antenna domain in native photosynthetic membranes of *Phsp. molischianum*. d) Model of the LH2 assembly in antenna domains in native membranes of *Phsp. molischianum*. LH2 forms hexagonal para-crystals with 74Å periodicity. For clarity, protein side chains are not shown.

*Rsp. photometricum*, evidence for variability of LH2 size was found in native chromatophores of *Phsp. molischianum* (Gonçalves et al., 2005a).

### *D. The Photosynthetic Apparatus of Rhodobacter blasticus*

Membranes of wild type *Rhodobacter* (*Rb.*) *blasticus*, have been imaged by high-resolution AFM. The purified native photosynthetic membrane were uniformly small sized vesicles, with a diameter of ~50 nm. In order to make these membranes amenable to high-resolution AFM analysis, the small vesicles were fused by freeze thawing, leading to the formation of membranes with sizes up to 1μm. Due to the

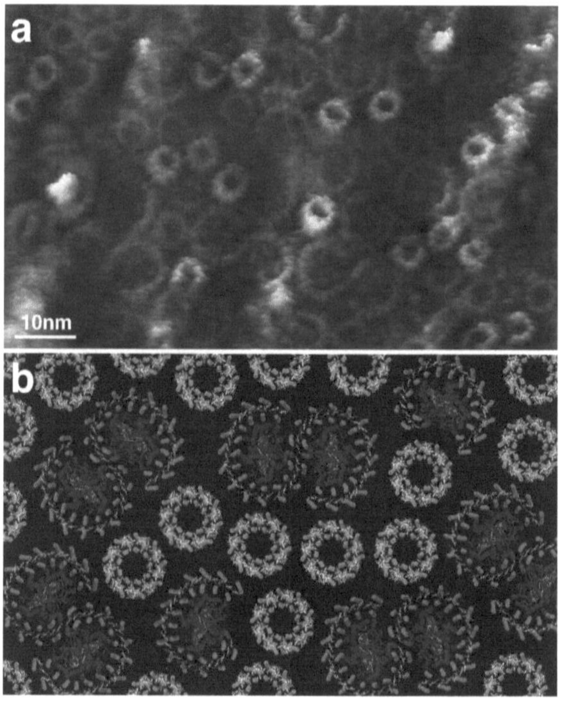


*Fig. 5.* High-resolution AFM analysis of the native photosynthetic membrane of *Rhodobacter blasticus* (from Scheuring et al., 2005a, with permission). a) Raw data AFM topograph revealing the complex mixture of LH2 and core-complexes in the fused native photosynthetic membranes of *Rb. blasticus*. The PufX containing core-complexes form stable dimers (~75%) with a stoichiometry of $(LH1_{13}\text{-RC-PufX})_2$, a smaller fraction (~25%) is monomeric $LH1_{13}$-RC-PufX complexes. b) Model of the complex assembly in the native membrane of *Rb. blasticus*. LH2 nonamers intercalating dimeric (75%) and monomeric (25%) core-complexes consisting of LH1 subunits and RCs. At the core-complex dimer center, two PufX subunits hold the dimeric core-complexes together. For clarity, protein side chains are not shown.

'mechanics' of this fusion process, AFM topographs revealed proteins in 'up and down' orientations (Fig. 5a) (Scheuring et al., 2005a).

The LH2 complexes are nonameric rings, inserted with their nine-fold axis perpendicular to the membrane (Walz et al., 1998; Scheuring et al., 2003a; Gonçalves et al., 2005b ). Importantly, high-resolution topographs of PufX-containing core-complexes have been acquired revealing that core-complexes assemble in S-shaped dimers of ~10 nm × ~20 nm size. Each core complex is composed of a clearly open LH1 assembly, with a protein-free gap of about 25Å width, housing one RC (Fig. 5a). Furthermore, the LH1 assembly of each core is elliptical, with long and short axes of 100 Å and 90 Å, respectively. Compared to other core-complexes, that all feature an elliptical LH1 assembly, the ellipticity is particularly astonishing, since one would expect larger flexibility of an open LH1 assembly. This is strong evidence that the ellipticity results from specific interactions of the LH1 subunits around the RC. At the dimer center an enlarged and depressed topography, in which averaging visualized two tiny protrusions, was found (Fig. 5a) (Scheuring et al., 2005a). Since the LH1 assembly is composed of multiple identical αβ heterodimers and one PufX polypeptide (Francia et al., 1999), the particular topography at the dimer center was attributed to the two PufX peptides from the two cores (Fig. 5b). Based on the high-resolution AFM topographs, and in agreement with previous EM studies of 2D-crystals of dimeric *Rb. sphaeroides* core-complexes (Scheuring et al., 2004a), a structural model was proposed for the subunit organization of the *Rhodobacter* core-complex. The core complex is an S-shaped dimer, composed of two RCs, each encircled by an open ellipsis composed of 13 LH1 α/β -heterodimers, with two PufX subunits, one of each core, holding the dimer together. This model provides explanation for the PufX-induced dimerization of the *Rhodobacter* core-complex that indirectly favors quinol/quinone passage across the ~25 Å wide protein-free gap (Scheuring et al., 2005a).

Although the 'up and down' membrane fusion process impeded a detailed analysis of the position and association of the different photosynthetic complexes in the native membrane, regions corresponding to one orientation showed that all the complexes are in physical contact with neighboring complexes, and no regular assembly of complexes was found in *Rb. blasticus* (Scheuring et al., 2005a). Except for the particular dimeric core-complex architecture, the overall distribution of complexes resembles the complex architectures found in *Rsp. photometricum* (Scheuring et al., 2004b,c; Scheuring and Sturgis, 2005) and *Phsp. molischianum* (Gonçalves et al., 2005a).

## V. Outlook

The most confusing aspect of the studies on the native architectures of the photosynthetic membranes of different photosynthetic bacteria species is the absence of the Cyt $bc_1$ complex in the membrane areas imaged by AFM, while functional measurements demand connectivity, i.e., proximity, of Cyt $bc_1$ complexes and core-complexes. Two explanations may apply:

membrane fragments that do not contain significant amounts of Cyt $bc_1$ complexes adsorb preferentially to the AFM support, or Cyt $bc_1$ complexes are not in proximity of many core-complexes and this concepts must be reviewed. The discrepancy between the photosynthetic complex architecture in native membranes reported by AFM images and functional measurements must be addressed now.

Some technical developments indicate that the AFM will strengthen its abilities to analyze membrane protein supramolecular assemblies in the near future. First, AFM tips with nano-tubes may provide improved reproducibility of high-resolution image acquisition and higher resolution (Cheung et al., 2000). Second, fast scanning AFMs that can acquire images at rates sufficient to track motion of individual molecules may allow the visualization of flexible and diffusing molecules (Viani et al., 2000; Ando et al., 2001).

## Acknowledgments

The author thanks former and present collaborators. This study was supported by the Institut Curie, the INSERM (Institut National de la Santé et la Recherche Médicale), the INSERM Avenir 2005 program, and a 'ACI Nanosciences 2004' grant (NR206).

## References

Abrahams JP, Leslie AGW, Lutter R and Walker JE (1994) Structure at 2.8 Ångstrom resolution of $F_1$-ATPase from bovine heart mitochondria. Nature 370: 621–628

Allen JP, Feher G, Yeates TO, Komiya H and Rees DC (1987) Structure of the reaction center from *Rhodobacter sphaeroides* R-26: The protein subunits. Proc Natl Acad Sci USA 84: 6162–6166

Ando T, Kodera N, Takai E, Maruyama D, Saito K and Toda A (2001) A high-speed atomic force microscope for studying biological macromolecules. Proc Natl Acad Sci USA 98: 12468–12472

Bahatyrova S, Frese RN, Siebert CA, Olsen JD, van der Werf KO, van Grondelle R, Niederman RA, Bullough PA and Hunter CN (2004a) The native architecture of a photosynthetic membrane. Nature 430: 1058–1062

Bahatyrova S, Frese RN, van der Werf KO, Otto C, Hunter CN and Olsen JD (2004b) Flexibility and size heterogeneity of the LH1 light harvesting complex revealed by atomic force microscopy: Functional significance for bacterial photosynthesis. J Biol Chem 279: 21327–21333

Berry EA, Huang L-S, Saechao LK, Pon NG, Valkova-Valchanova M and Daldal F (2004) X-ray structure of *Rhodobacter capsulatus* cytochrome $bc_1$: Comparison with its mitochondrial and chloroplast counterparts. Photosynth Res 81: 251–275

Binnig G, Gerber C, Stoll E, Albrecht TR and Quate CF (1987) Atomic resolution with atomic force microscopy. Europhys Lett 3: 1281–1286

Binnig G, Quate CF and Gerber C (1986) Atomic force microscope. Phys Rev Lett 56: 930–933

Butt H-J (1992) Measuring local surface charge densities in electrolyte solutions with a scanning force microscope. Biophys J 63: 578–582

Cheung CL, Hafner JH and Lieber CM (2000) Carbon nanotube atomic force microscopy tips: Direct growth by chemical vapor deposition and application to high-resolution imaging. Proc Natl Acad Sci USA 97: 3809–3813

Deisenhofer J and Michel H (1989) Nobel Lecture: The photosynthetic reaction centre from the purple bacterium *Rhodopseudomonas viridis*. EMBO J 8: 2149–2170

Deisenhofer J, Epp O, Miki K, Huber R and Michel H (1984) X-ray structure analysis of a membrane protein complex. Electron density map at 3 Å resolution and a model of the chromophores of the photosynthetic reaction center from *Rhodopseudomonas viridis*. J Mol Biol 180: 385–398

Deisenhofer J, Epp O, Miki K, Huber R and Michel H (1985) Structure of the protein subunits in the photosynthetic reaction centre of *Rhodopseudomonas viridis* at 3Å resolution. Nature 318: 618–624

Deisenhofer J, Epp O, Sinning I and Michel H (1995) Crystallographic refinement at 2.3Å resolution and refined model of the photosynthetic reaction centre from *Rhodopseudomonas viridis*. J Mol Biol 246: 429–457

Engelhardt H, Engel A and Baumeister W (1986) Stoichiometric model of the photosynthetic unit of *Ectothiorhodospira halochloris*. Proc Natl Acad Sci USA 83: 8972–8976

Fotiadis D, Qian P, Philippsen A, Bullough PA, Engel A and Hunter CN (2004) Structural analysis of the reaction center light-harvesting complex I photosynthetic core complex of *Rhodospirillum rubrum* using atomic force microscopy. J Biol Chem 279: 2063–2068

Francia F, Wang J, Venturoli G, Melandri B, Barz W and Oesterheld D (1999) The reaction center-LH1 antenna complex of *Rhodobacter sphaeroides* contains one PufX molecule which is involved in dimerization of this complex. Biochemistry 38: 6834–6845

Gonçalves RP, Bernadac A, Sturgis JN and Scheuring S (2005a) Architecture of the native photosynthetic apparatus of *Phaeospirillum molischianum*. J Struct Biol 152: 221–228

Gonçalves RP, Busselez J, Lévy D, Seguin J and Scheuring S (2005b) Membrane insertion of *Rhodopseudomonas acidophila* light harvesting complex 2 (LH2) investigated by high resolution AFM. J Struct Biol 149: 79–86

Hoh JH, Lal R, John SA, Revel J-P and Arnsdorf MF (1991) Atomic force microscopy and dissection of gap junctions. Science 253: 1405–1408

Hu X and Schulten K (1998) Model for the light-harvesting complex I (B875) of *Rhodobacter sphaeroides*. Biophys J 75: 683–694

Hu X, Ritz T, Damjanovic A, Autenrieth F and Schulten K (2002) Photosynthetic apparatus of purple bacteria. Quart Rev Biophys 35: 1–62

Ikeda-Yamasaki I, Odahara T, Mitsuoka K, Fujiyoshi Y and Murata K (1998) Projection map of the reaction center-light harvesting 1 complex from *Rhodopseudomonas viridis* at 10 Å resolution.

FEBS Lett 425: 505–508
Israelachvili J (1991) Intermolecular and Surface Forces. Academic Press, London
Israelachvili J and Wennerstrom H (1996) Role of hydration and water structure in biological and colloidal interactions. Nature 379: 219–225
Jacob JS and Miller KR (1983) Structure of a bacterial photosynthetic membrane. Arch Biochem Biophys 223: 282–290
Jacob JS and Miller KR (1984) Structure of a bacterial photosynthetic membrane: Integrity of reaction centers following proteolysis and detergent solubilization. Biochem Biophys Res Com 120: 164–171
Jamieson SJ, Wang P, Qian P, Kirkland JY, Conroy MJ, Hunter CN and Bullough PA (2002) Projection structure of the photosynthetic reaction centre-antenna complex of *Rhodospirillum rubrum* at 8.5 Å resolution. EMBO J 21: 3927–3935
Koepke J, Hu X, Muenke C, Schulten K and Michel H (1996) The crystal structure of the light-harvesting complex II (B800-850) from *Rhodospirillum molischianum*. Structure 4: 581–597
Kurisu G, Zhang H, Smith JL and Cramer WA (2003) Structure of the cytochrome $b_6f$ complex of oxygenic photosynthesis: Tuning the cavity. Science 302: 1009–1014
Lancaster CR, Bibikova M, Sabatino P, Oesterheld D and Michel H (2000) Structural basis of the drastically increased initial electron transfer rate in the reaction center from a *Rhodopseudomonas viridis* mutant described at 2.00 Å resolution. J Biol Chem 275: 39364–39368
McDermott G, Prince SM, Freer AA, Hawthornthwaite-Lawless AM, Papiz MZ, Cogdell RJ and Isaacs NW (1995) Crystal structure of an integral membrane light-harvesting complex from photosynthetic bacteria. Nature 374: 517–521
Meier T, Polzer P, Diederichs K, Welte W and Dimroth P (2005) Structure of the rotor ring of F-Type $Na^+$-ATPase from *Ilyobacter tartaricus*. Science 308: 659–662
Miller KR (1982) Three-dimensional structure of a photosynthetic membrane. Nature 300: 53–55
Müller DJ, Amrein M and Engel A (1997) Adsorption of biological molecules to a solid support for scanning probe microscopy. J Struct Biol 119: 172–188
Müller DJ, Fotiadis D, Scheuring S, Müller SA and Engel A (1999) Electrostatically balanced subnanometer imaging of biological specimens by atomic force microscopy. Biophys J 76: 1101–1111
Papiz MZ, Prince SM, Hawthornthwaite-Lawless AM, McDermott G, Freer AA, Isaacs NW and Cogdell RJ (1996) A model for the photosynthetic apparatus of purple bacteria. Trends Plant Sci 1: 198–206
Qian P, Hunter CN and Bullough PA (2005) The 8.5 Å projection structure of the core RC-LH1-PufX dimer of *Rhodobacter sphaeroides*. J Mol Biol 349: 948–960
Roszak AW, Howard TD, Southall J, Gardiner AT, Law CJ, Isaacs NW and Cogdell RJ (2003) Crystal Structure of the RC-LH1 Core complex from *Rhodopseudomonas palustris*. Science 302: 1969–1972
Schabert FA and Engel A (1994) Reproducible acquisition of *Escherichia coli* porin surface topographs by atomic force microscopy. Biophys J 67: 2394–2403
Scheuring S and Sturgis JN (2005) Chromatic adaptation of photosynthetic membranes. Science 309: 484–487
Scheuring S, Reiss-Husson F, Engel A, Rigaud J-L and Ranck J-L (2001) High resolution topographs of the *Rubrivivax gelatinosus* light-harvesting complex 2. EMBO J 20: 3029–3035
Scheuring S, Seguin J, Marco S, Lévy D, Breyton C, Robert B and Rigaud J-L (2003a) AFM Characterization of tilt and intrinsic flexibility of *Rhodobacter sphaeroides* light harvesting complex 2 (LH2). J Mol Biol 325: 569–580
Scheuring S, Seguin J, Marco S, Lévy D, Robert B, and Rigaud JL (2003b) Nanodissection and high-resolution imaging of the *Rhodopseudomonas viridis* photosynthetic core-complex in native membranes by AFM. Proc Natl Acad Sci USA 100: 1690–1693
Scheuring S, Francia F, Busselez J, Melandri B, Rigaud J-L and Lévy D (2004a) Structural role of PufX in the dimerization of the photosynthetic core-complex of *Rhodobacter sphaeroides*. J Biol Chem 279: 3620–3626
Scheuring S, Rigaud J-L and Sturgis JN (2004b) Variable LH2 stoichiometry and core clustering in native membranes of *Rhodospirillum photometricum*. EMBO J 23: 4127–4133
Scheuring S, Sturgis JN, Prima V, Bernadac A, Lévy D and Rigaud J-L (2004c) Watching the photosynthetic apparatus in native membranes. Proc Natl Acad Sci USA 101: 11293–11297
Scheuring S, Busselez J and Levy D (2005a) Structure of the dimeric PufX-containing core complex of *Rhodobacter blasticus* by in situ AFM. J Biol Chem 180: 1426–1431
Scheuring S, Gonçalves RP, Prima V and Sturgis JN (2005b) The photosynthetic apparatus of *Rhodopseudomonas palustris*: Structures and organization. J Mol Biol 358: 83–96
Scheuring S, Levy D and Rigaud J-L (2005c) Watching the components of photosynthetic bacterial membranes and their 'in situ' organization by atomic force microscopy. Biochim Biophys Acta 1712: 109–127
Siebert CA, Qian P, Fotiadis D, Engel A, Hunter CN and Bullough PA (2004) Molecular architecture of photosynthetic membranes in *Rhodobacter sphaeroides*: The role of PufX. EMBO J 23: 690–700
Stamouli A, Kafi S, Klein DC, Oosterkamp TH, Frenken JW, Cogdell RJ and Aartsma TJ (2003) The ring structure and organization of light harvesting 2 complexes in a reconstituted lipid bilayer, resolved by atomic force microscopy. Biophys J 84: 2483–2491
Stark W, Kühlbrandt W, Wildhaber I, Wehrli E and Muhlethaler K (1984) The structure of the photoreceptor unit of *Rhodopseudomonas viridis*. EMBO J 3: 777–783
Stroebel D, Choquet Y, Popot JL and Picot D (2003) An atypical haem in the cytochrome $b_6f$ complex. Nature 426: 413–418
Sundström V, Pullerits T and van Grondelle R (1999) Photosynthetic light-harvesting: Reconciling dynamics and structure of purple bacterial LH2 reveals function of photosynthetic unit. J Phys Chem B 103: 2327–2346
Viani MB, Pietrasanta LI, Thompson JB, Chand A, Gebeshuber IC, Kindt JH, Richter M, Hansma HG and Hansma PK (2000) Probing protein-protein interactions in real time. Nature Struct Biol 7: 644–647
Walz T, Jamieson SJ, Bowers CM, Bullough PA and Hunter CN (1998) Projection structures of three photosynthetic complexes from *Rhodobacter sphaeroides*: LH2 at 6 Å, LH1 and RC-LH1 at 25 Å. J Mol Biol 282: 833–845
Xia D, Yu CA, Kim H, Xia JZ, Kachurin AM, Zhang L, Yu L and Deisenhofer J (1997) Crystal structure of the cytochrome $bc_1$ complex from bovine heart mitochondria. Science 277: 60–66

Chapter 2

# Three-Dimensional Electron Microscopy. A Gateway to Photosynthetic Structure

Martin F. Hohmann-Marriott, Allison M. L. van de Meene and Robert W. Roberson*
*Arizona State University, School of Life Sciences, PO Box 874501, Tempe, AZ 85287-4501, U.S.A.*



*Author for correspondence, email: robby2@asu.edu

*Thijs J. Aartsma and Jörg Matysik (eds): Biophysical Techniques in Photosynthesis II, pp. 13–34.*

## Summary

Three-dimensional electron microscopy techniques have emerged as powerful and practical tools in cell and structural biology. These tools carry the potential of unlocking many novel details of cytoplasmic order and macromolecular structure with nanometer resolution. Interfacing this plethora of structural information with genomic and physiological data holds the promise of unprecedented advancement in biology and medicine. Three-dimensional electron microscopy techniques accomplish the generation of a three-dimensional dataset that represents overall shape as well as internal structure of an object. This is achieved by processing transmission electron microscopy (TEM) images at different viewing angles. The resulting dataset can be used to generate random (virtual) sections through the reconstructed volume. One three-dimensional electron microscopy approach, electron tomography, has matured into a technique that is now routinely applied to a variety of biological questions with great success. Using plastic-embedded or ice-embedded samples, biological structures can be investigated at a resolution of 5–10 nm within their cellular context. A second three-dimensional electron microscopy approach, single particle electron microscopy, is pushing the obtainable resolution of isolated macromolecules into the sub-nanometer range, revealing tertiary and secondary structural arrangements. The combination of three-dimensional electron microscopy techniques with cryo-preservation and imaging at cryogenic temperatures is revolutionizing structural biology. This combination permits the investigation of hydrated structures, and thereby allows a direct examination of biological structure, which closely resembles the in vivo state. This chapter gives an introduction into theoretical background of tomography, reconstruction, image formation by transmission electron microscopy and cryo-preservation. The two methodologies, electron tomography and single particle electron microscopy are discussed with detailed technical information. An example for each technique is presented.

## I. Introduction

Tomography (from the Greek words tomos: section, and graphein: to write) accomplishes the generation of a three-dimensional dataset, by mathematical processing of transmission electron micrographs of an object taken at different viewing angles. With the dataset (referred to as the reconstruction or the reconstructed volume), it is possible to obtain virtual slices at any depth or angle within the reconstruction, thus revealing overall shape as well as inner structural details of the object.

The process of tomography is relatively easy to understand as the human brain is an excellent intuitive tomography processor. Our daily experience is that we can determine the three-dimensional shape of an object by examining it visually from different directions. Light does not permeate most objects, so we can usually only deduce the surface outline of an object. Fortunately electrons can permeate many thin biological specimens sufficiently well to yield 'transmission images.' These images can then be used to determine the overall shape as well as the internal features of the investigated object.

The mathematical formalism that describes the tomographic process was established in 1917 (Radon, 1917). However, the union of this mathematical framework with the imaging of specimens of biological and medical interest was only realized some 40 years later (Cormack, 1963). X-ray tomography

*Abbreviations:* ART – algebraic reconstruction techniques; CEMOVIS – cryo-electron microscopy of vitreous sections; CTF – contrast transfer function; FEG – field emission gun; FFT – fast Fourier transform; FRC – Fourier ring correlation; FSC – Fourier shell correlation; FT – Fourier transform; NAD – nonlinear anisotropic diffusion; NMR – nuclear magnetic resonance; PS II – Photosystem II; SIRT – simultaneous iterative reconstruction technique

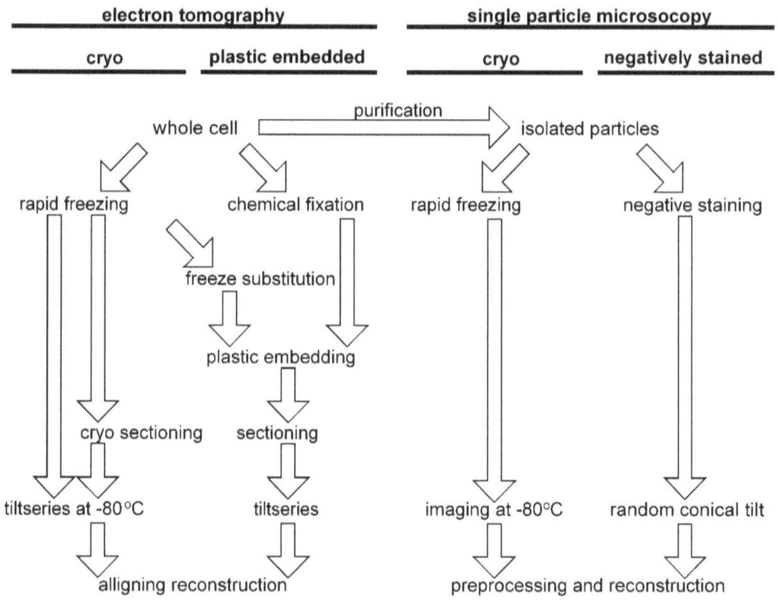


*Fig. 1.* Three-dimensional electron microscopy. Overview of steps involved in electron tomography and single particle electron microscopy.

provided the first demonstration of the power of this approach (Hounsfield, 1973). The application of tomography using electrons as an imaging agent promised increased resolution (Hart, 1968), but also demanded the development of specific instrumentation, computational approaches, and advances in specimen preservation. These initial challenges have been largely overcome, and three-dimensional electron microscopy techniques have evolved into approaches that can be employed routinely with outstanding success.

Within three-dimensional electron microscopy two distinct methodologies have evolved (Fig. 1): electron tomography and single particle microscopy (Frank, 1996). These techniques are based on the same theoretical principles (Section II), but differ in preparation of the specimen (Fig. 1) and processing of acquired data (Fig. 2).

Electron tomography examines structural arrangements within a cell and attempts to describe each element of interest separately as well as in context. The three-dimensional data obtained through electron tomography (5–10 nm resolution) promises to fill the resolution gap that exists between light microscopy (200 nm resolution) and X-ray diffraction and NMR that deliver atomic resolution of biological macromolecules. Electron tomography has the potential of revealing these structures in the context of the cell and their interactions (Medalia et al., 2002). In addition this technique is exceptionally well suited for the investigation of membranous networks (Nicastro et al., 2000; Marsh et al., 2001; van de Meene et al., 2005).

Single particle electron microscopy, on the other hand, determines structural detail of macromolecular complexes, which are not amenable to crystallization, by imaging thousands of purified complexes. This technique can be employed to discern secondary structure of protein conformation states. The ultimate goal of single particle electron microscopy is to achieve atomic resolution (van Heel, 2000).

This goal has already been achieved using electron crystallography (reviewed in Chapter 7, da Fonseca et al.) which takes advantage of the electron diffraction patterns of structures that contain ordered, self-repetitive elements (Henderson et al., 1990; Snyder et al, 2001). Electron crystallography shares methodology with X-ray crystallography (reviewed in Chapter 5, Liu and Chang and Chapter 6, Fromme) and is not further covered in this chapter.

Sample preservation and imaging at cryogenic temperatures is an important extension of electron tomography and single particle electron microscopy. To date the application of cryo-electron microscopy techniques have resulted in great strides towards elucidating the structure of molecules important in

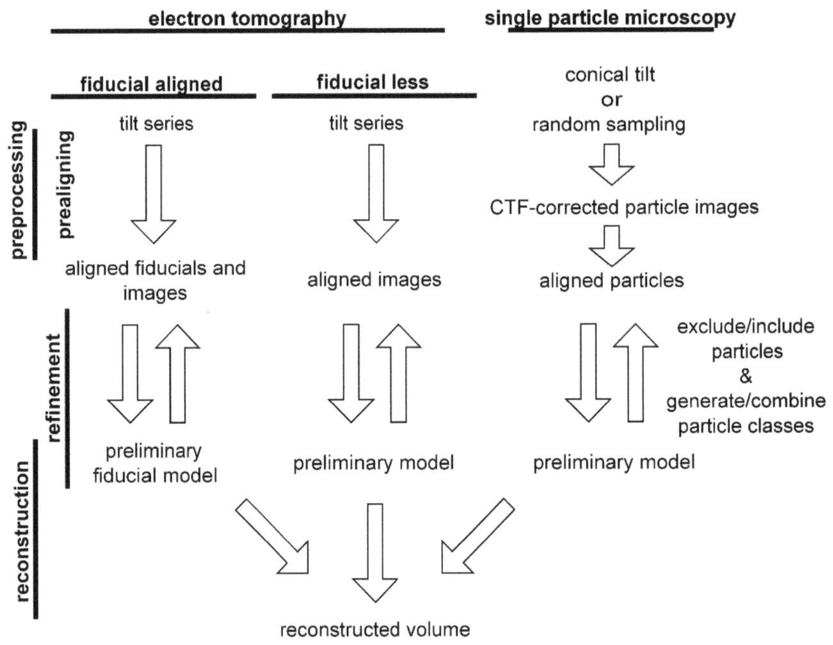


*Fig. 2.* Data processing in three-dimensional electron microscopy. Overview of preprocessing and reconstruction of data for electron tomography and single particle electron microscopy.

gene transcription and protein synthesis (Valle et al., 2003; Allen et al., 2005; Gao et al., 2005), regulation of cytoskeletal and muscle function (Hirose et al., 1999; Arnal et al., 2004; Santarella et al., 2004; Suzuki et al., 2004; Liu et al., 2005, Murakami et al., 2005; Wang and Nogales, 2005), enzyme function (Morais et al., 2003; Tama et al., 2004; Crowther et al., 2005; Falke et al., 2005), and photosynthesis (Marr et al., 1996; Qian et al., 2005). Furthermore, cryo-electron microscopy applications excel in structural investigations of viral particles and the proteins important in their pathology cycle (Jiang et al., 2003a; Morais et al., 2003; Zhang et al., 2003b, 2005; Bubeck et al., 2005; Förster et al., 2005; Pous et al., 2005; Xiao et al., 2005), subcellular organelles (Medalia et al., 2002; Beck et al., 2004; Baumeister, 2005) whole-cell microorganisms (Baumeister, 2004; Lučić et al., 2005), and eukaryotic cells and tissues (Steven and Aebi, 2003; McIntosh et al., 2005; Subramaniam, 2005).

Three-dimensional electron microscopy is an exciting and still evolving field (McEwen and Marko, 2001; McIntosh, 2001; Unger, 2001; Jiang and Ludtke, 2005). Three-dimensional electron microscopy (Frank, 1992; Subramaniam and Milne, 2004), electron tomography (Baumeister, 2002; Lučić et al., 2005; McIntosh et al., 2005), and single particle microscopy (Frank, 1996, 2002, 2006; van Heel, 2000) have been reviewed in detail.

This chapter describes the technical details of sample preparation and data acquisition encountered in three-dimensional electron microscopy, and also has the goal to serve as a short introduction for researchers not familiar with underlying principles and used terminology.

## II. Tomographic Principles

### A. Line Integrals

The logic that goes into deriving a three-dimensional model from a series of projections is straightforward. Assuming, that there is only one point in the space of interest, the two functions of line integrals $i(x)$ and $i(y)$ each with different direction, each summing up the values of all points encountered, are sufficient to determine the position of this point as illustrated in Fig. 3a. Assuming the point has a value of 1, the line integral ($i$) of exactly one line in $x$ and one line in the

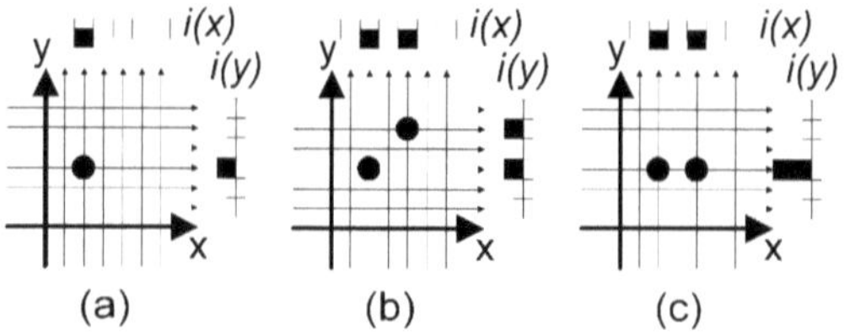


*Fig. 3.* Point positions can be deduced by analyzing line integrals. Line integrals (small arrows along mesh grid) are used to sum up values (dots) along their path. The function of all line integrals in x-direction ($i(x)$) and y-direction ($i(y)$) can be used to deduce the location of one point (a) and two points (b, c).

$y$ direction is 1, all other line integrals are 0 (Fig. 3a). This thought experiment can be expanded to 2 points in the system (Fig. 3b). Even if the two points 'hide' on the same line, looking at both functions of line integrals will reveal the location (Fig. 3c). Having an infinite number of parallel line integrals from an infinite number of directions it will be possible to identify any object with a finite number of points.

However, in real life applications, the mathematical directives (unlimited integral functions and viewing angles) cannot be fully implemented: data acquisition is limited by the tilting angle increment, maximal tilting angle, as well as potential sample damage by extended collection times, sample thickness and low signal-to-noise ratio. With these limitations only an imperfect reconstruction can be performed.

### *B. Radon Transform*

A transform is the conversion from one mathematical reference system to another. The Fourier transform, for instance, takes physical (direct) space data and converts it into Fourier space. The inverse Fourier transform reverses the operation.

The Radon transform, a mathematical representation of the tomography principle was formulated in the beginning of the twentieth century (Radon, 1917). The developed mathematical framework assumes summation of an infinite number of line integrals over infinite viewing angles to accomplish a complete (true to life) reconstruction of an object. To derive the Radon transform the Cartesian coordinate system in which a point is defined by distances along two perpendicular axis ($x$ and $y$), is transformed to a polar coordinate system, in which a point is defined by the viewing angle ($\theta$), and a distance ($r$) (Fig. 4).

When the polar coordinate system is rotated by the angle $\theta$, the new coordinates ($r$, $s$) for a point are related to the coordinates ($x$, $y$) before the rotation

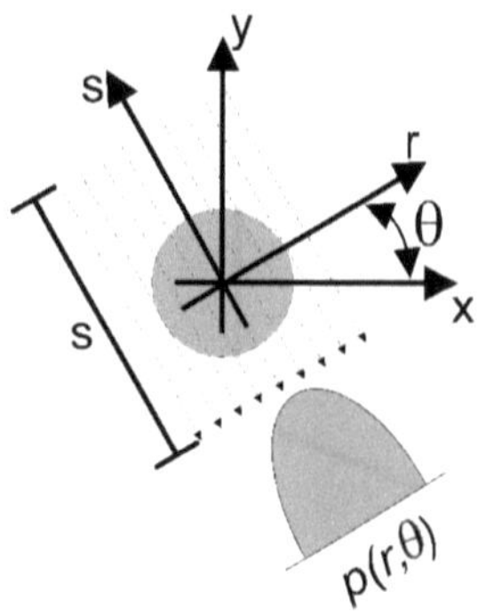


*Fig. 4.* Relationship between Cartesian and polar coordinate systems. Transformation of a Cartesian coordinate system (perpendicular x-axis and y-axis) into a polar coordinate system (r-axis and θ). The s-axis is 90° rotated in respect to the r-axis. For more information see text.

by trigonometric functions as

$$x = r\cos\theta - s\sin\theta$$
$$y = r\sin\theta + s\cos\theta \qquad (1)$$

The transformation of a function $i(x,y)$, representing the line integrals with a length of $\Delta s$, into a function of the form $p(r,\theta)$, representing a projection (Fig. 4), is called the Radon transform [$\Re$]. The mathematical formalism is expressed as

$$p(r,\theta) = \Re\{i(x,y)\} = \sum i(r\cos\theta - s\sin\theta,\ r\sin\theta + s\cos\theta)\,\Delta s \qquad (2)$$

The function $p(r,\theta)$ can be represented as a sinogram (Section II.C.2 and Fig. 6).

### *C. Space and Coordinate Systems*

Processing of tomographic data takes advantage of several different mathematically defined spaces and their specific features. Because the used spaces and nomenclature are an indispensable part of electron tomographic techniques, their knowledge is often assumed in literature. Therefore a brief overview of these spaces the accompanying nomenclature and applications is presented. For further information on nomenclature conventions used in three dimensional electron microscopy the publication by Heymann and coworkers (2005) and a glossary of terms used in single particle electron microscopy is recommended (Wadsworth Center, 2002).

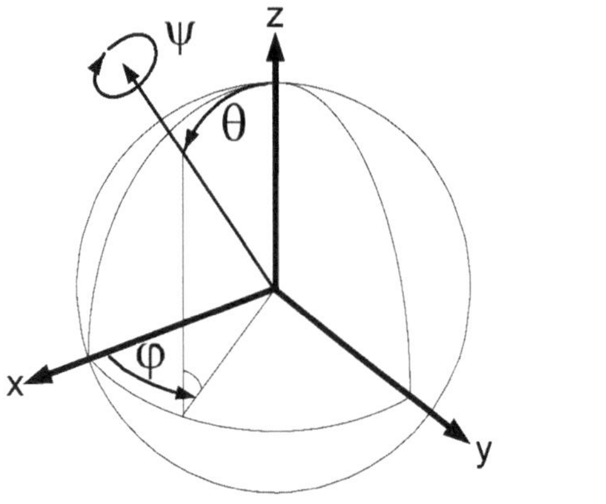


*Fig. 5.* Euler angles. Three Euler angles define the orientation of a particle in space. For more information see text.

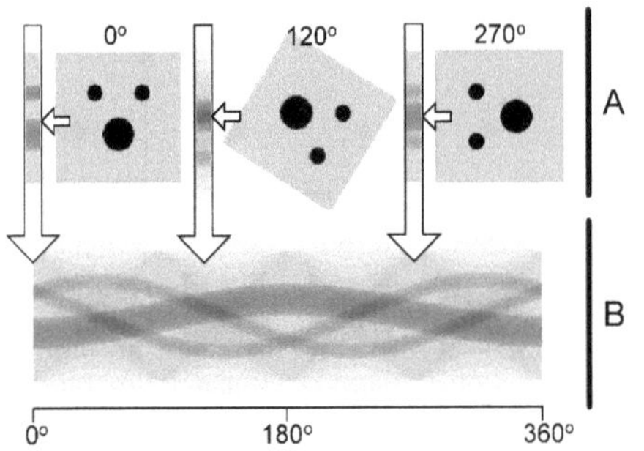


*Fig. 6.* Sinogram. (A) Integration (small left arrows) of image densities at different viewing angles results in projections (down arrows, containing projection density values). (B) Assembly of these projection densities constitutes a sinogram.

#### *1. Direct Space*

A point in three-dimensional direct space is often defined by three distances along three perpendicular axes that meet at a common origin. The axes are usually called $x$, $y$ and $z$ and can be well visualized in a Cartesian coordinate system.

The Euler angles are used to define the orientation and rotation of an object in direct space. A direction (point on a sphere) is defined by two perpendicular angles with a common origin. The first Euler angle is termed $\phi$ and defines the latitude of a sphere by defining elevation from the equator. The second Euler angle $\theta$ defines the longitude of the sphere. A third Euler angle $\psi$ defines the rotation around the axis determined by $\phi$ and $\theta$ (Fig. 5). The Euler angles are used to describe the orientation of single particles in images for single particle microscopy.

#### *2. Radon Space*

The three-dimensional Radon space represents the values of line integrals (Section II.C.2). Usually, line integral values are displayed by two angles and one distance. The angles are termed $\theta$ or $\alpha$ (viewing angle) for the first axis and $\phi$ or $\beta$ for the second axis. The distance is termed 'l' or 'r'. A two-dimensional Radon space can be represented in sinogram form (Edholm et al., 1978) (Fig. 6). Correlating line integrals of object projections images under different viewing angles can be used to determine relative orientation of the projection images by finding a common line (van Heel, 1987). That such a common line (Section II.E) exists is a consequence of the projection theorem (Section II.D).

#### *3. Fourier Space*

Fourier transformation converts a familiar direct space image, which consists of pixels with intensity values, into the magnitude of frequencies encountered in the original image (Fig. 7; 2D Fourier space) and the phase offsets of the waves with respective frequencies. A two-dimensional magnitude image displays the magnitude of each frequency through an intensity value in dependence of the frequency ($x$ and $y$). The frequency decreases with the distance from the origin along the $x$ and $y$ axis. The phase information is usually not displayed or manipulated. Fast Fourier transform (FFT) is a computationally efficient algorithm to compute the discrete Fourier transform (FT).

Data manipulation in Fourier space offers several advantages over direct space data manipulation. Frequency domains that constitute noise can be easily filtered out by eliminating these frequencies from the data in Fourier space (Section V.E). Most three-dimensional electron microscopy software packages give the option of eliminating high and low frequency noise. It is also possible to give frequencies different weight, a strategy used for weighted back projection (Section II.F). Fourier operations are also used to determine the theoretical resolution frequency (Section II.D) and to calculate the resolution limit in single particle reconstruction by the so called Fourier Shell Correlation or Fourier Ring Correlation (Section VI.E).

### *D. Projection Theorem*

The projection theorem states that any two-dimen-

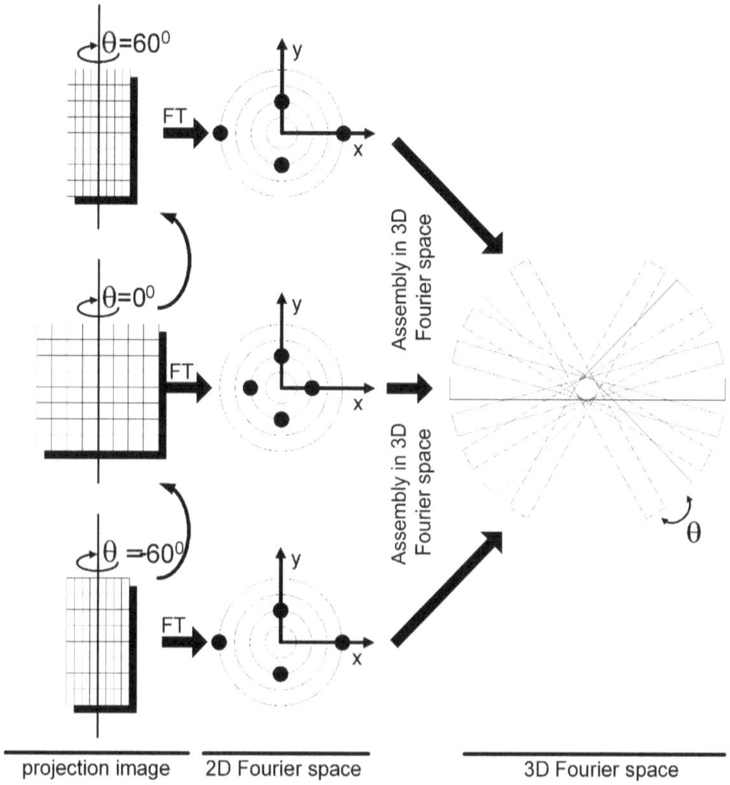


*Fig. 7.* Fourier space reconstruction process. Projection images of an object under different viewing angle increments (θ) are Fourier transformed. The resulting 2D Fourier space slices contain frequency magnitude information (2 D Fourier space). The object can be reconstructed by assembling the 2D Fourier slices in 3D Fourier space (3D Fourier space).

sional Fourier transform of a projection image of one object at a certain tilt angle represents a plane with the respective tilt angle that transects the origin (0 frequency) in a three-dimensional Fourier space reconstruction of this object (DeRosier and Klug, 1968). Calculating the two-dimensional Fourier transforms of images at different tilting angles and assembling the resulting planes in three-dimensional Fourier space is equivalent to generating a three-dimensional reconstruction of the object (Fig. 7 and 8).

A look at the graphic representation of the projection theorem (Fig. 8) reveals that there is missing frequency coverage where sections do not overlap. The maximal theoretical resolution frequency (*f*) is the frequency where slices still overlap. This frequency is indicated by a grey circle in Fig. 8 and is dependent on the sample thickness (1/D) and number (*n*) of projections with identical angle increment (θ). Deriving an approximation for the maximal resolution frequency is straightforward.

The circumference ($2\pi f$) of a circle with radius *f* is equal to number of slices (*n*) times the thickness of the slice ($1/D$) multiplied by two (because each slice covers both sides of a quasi circle). Rearranging of this relationship yields:

$$f = \frac{n}{\pi D} \tag{3}$$

The volume in the three-dimensional Fourier space (Fig. 8), not covered by overlapping sections due to lack of slices at respective angles is called the 'missing wedge' in single axis tomography, and the 'missing pyramid or cone' in dual axis tomography.

A direct space object can be obtained by transforming the assembled slices from three-dimensional Fourier space to direct space. However, this operation poses great computational hurdles because the frequencies of the Fourier slices are arranged in a polar coordinate system and have to be interpolated to intersect with a Cartesian coordinate system to perform an inverse Fourier transformation.

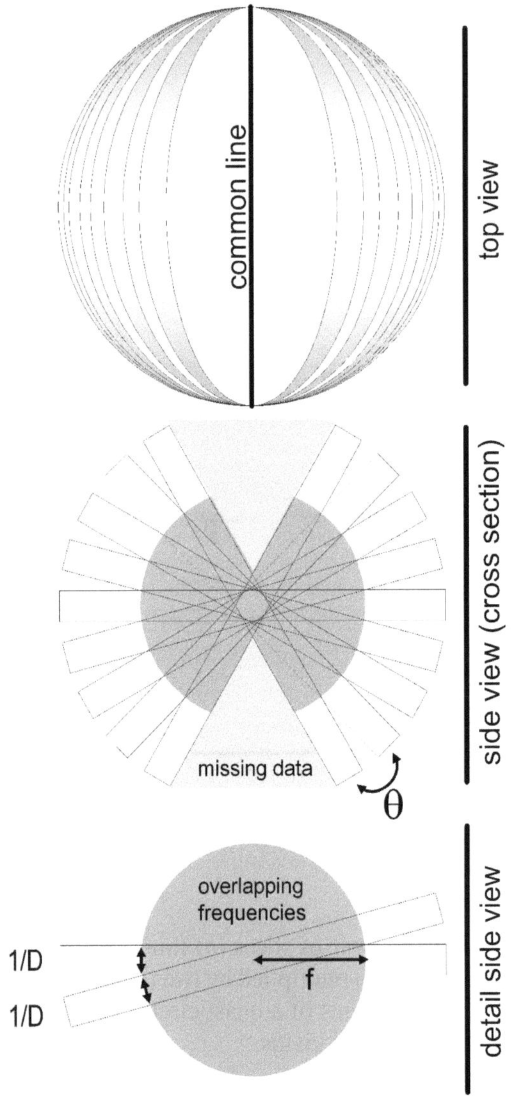


*Fig. 8.* Features of Fourier space reconstructions. The top view of the assembled 3D Fourier space shows that all 2D Fourier slices intersect to form a single common line. The 3D Fourier space does not contain information at viewing angles, where no projection images were recorded (missing data, side view). The frequencies that have overlapping 2D Fourier slices in 3D Fourier space constitute the fully resolved frequencies (*f*). The highest frequency (maximal theoretical resolution frequency) is defined by the object thickness (1/D) and the viewing angle increment (θ) (detail side view).

## E. Common Line Theorem

A consequence of the projection theorem is that two projections of an object have a common line in three-dimensional Fourier space (Fig. 8, top view). The 'common line theorem' or 'central line theorem', as well as its equivalent in Radon space are applied to determine the orientation of particles in single particle reconstruction.

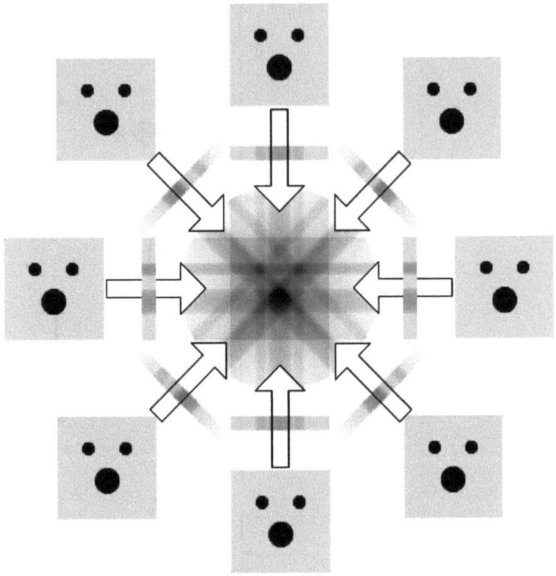

*Fig. 9.* Back projection. Projections through an object at different viewing angles (arrows) result in a line integrals (lines). When these line integrals are projected through a common space at their respective viewing angles (center), object features can be recovered.

## F. Reconstruction

Although the projection theorem is formulated in Fourier space it is rarely used for reconstruction (Section II.D). Radermacher (1994) points out the relatedness between Radon space and Fourier space and demonstrates that reconstruction in Radon space is possible and might even offer advantages. However, due to its practicability and robustness, back projection is the most commonly used reconstruction method in three-dimensional electron microscopy (Fig. 9) (Radermacher, 1992).

### 1. Back Projection

In back projection the recorded object images are projected to a common origin at the respective object viewing angles, and the projected intensities are summed up (Fig. 9). Looking at an example of this procedure, one can see that a reconstruction has star-shaped features.

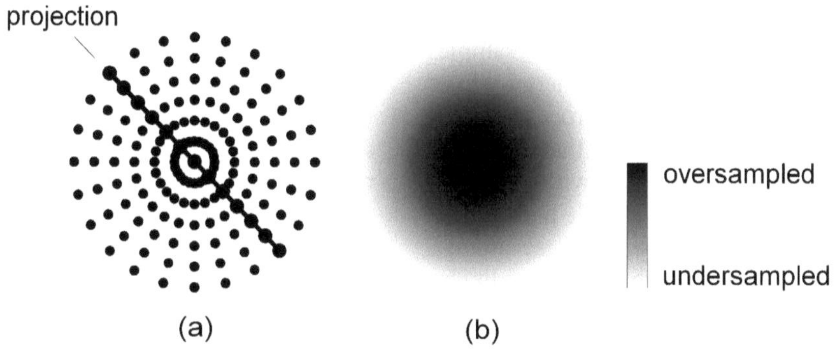


*Fig. 10.* Weighing of Fourier space. (a) Fourier space reconstruction of projections. Uniformly distributed sample points are indicated to show sampling distribution. (b) Overall representation of sample distribution.

The sampling distribution in Fourier space shows that the low frequencies are over-sampled compared to the high frequencies. To compensate for this fact, the tilt series data is transformed into Fourier space and the frequencies that were over-sampled are multiplied with a frequency-dependent factor (filter) to decrease their effect in the image (Fig. 10) (Radermacher, 1992; Harauz and van Heel, 1986). A simple weighing factor can be calculated from the frequencies and tilting angles.

### 2. Iterative Direct Space Techniques

Direct space reconstruction methods, which use a set of linear equations iteratively to optimize the reconstructed volume, were developed (Gordon et al., 1970) and have proven their usefulness. These reconstruction methods are known as ART and its successors SIRT (Gilbert, 1972). A more recent generation of reconstruction methods try to optimize a system of linear equations to find minimal variance (Penczek et al., 1992), maximum entropy (Skoglund et al., 1996) or limited basis functions (Marabini et al., 1997).

### G. Denoising

The signal to noise ratio of images and reconstructions can be improved by denoising procedure at the level of individual images and at the final reconstructed volume. Besides basic filtering methods such as median, mean, frequency and Gaussian filtering more complex and potentially more powerful denoising procedures have been developed. Two such approaches are nonlinear anisotropic diffusion (NAD) (Frangakis and Hegerl, 2001; Fernandez and Li, 2003) and bilateral filtering (Jiang et al., 2003b). An approach based on wavelet transforms has been implemented (Stoscheck and Hegerl, 1997). This method, however, is computationally intensive and currently the direct space approaches are preferred.

## III. Imaging with the Electron Microscope

### A. Electron Source and Lens System

The imaging agents in electron microscopy are electrons that are extracted from an electron source. The voltage that is used accelerate the electrons through the electron microscope column is 100 to 400 kV for most applications.

Ideally, the extracted electrons are redirected via electromagnetic lenses into a coherent beam of uniform intensity and energy, when they encounter the sample. The samples are usually deposited on a carrier layer of carbon or plastic covering a center hole of a thin metal disk (grid) of 3 mm diameter. Electrons that interact with the sample are scattered to a different path and/or lose energy. After the beam interacts with the sample, the beam is refocused and magnified by the objective and projector lens systems respectively, and finally, the magnified image is recorded by photographic film or a digital camera.

### B. Image Formation

Beam electrons can interact with the sample in several ways. Electrons can interact with the sample and change their direction significantly while maintaining all of their energy, a process called elastic scattering. Alternatively, electrons can interact with sample atoms and lose a significant amount of energy, while changing direction only slightly, a process called inelastic scattering. The probability that multiple scattering events will occur increases with the number of atoms encountered in the sample and is also dependent on the atomic number of these atoms. Elements with higher atomic numbers are more likely to interact with beam electrons than elements with lower atomic number. A measure of the probability of electrons interacting with the sample is the 'mean free path for electrons,' which is also dependent on the energy (acceleration voltage) of imaging electrons. The higher the energy of the electron, the less probability there is of the electron to interact with specimen atoms. Therefore thicker specimens can be investigated using higher acceleration voltages. The use of higher acceleration voltages also decreases radiation damage of the sample.

In thick specimens stained with heavy metals, the interaction of the electrons results in a substantial amount of elastically scattered electrons with large scattering angles. Scattering of electrons is the reason for dark areas in the image. This form of contrast is called amplitude contrast.

In unstained thin specimens large scattering that leads to elimination of the electrons from the image is negligible, and phase contrast is the main contrast mechanism. Phase contrast arises due to the wave properties of moving electrons. The waves of a beam electron interact with each other in a constructive and destructive fashion, generating contrast.

The combination of low contrast and a high level of noise in unstained samples usually limit the interpretation of the acquired images. The relative low contrast is a consequence of the sample composition. The bulk elements (C, O, N and H) of most biological samples have relatively low atomic numbers and interact with beam electrons similar to the embedding material (ice, plastic). Contrast can be enhanced by encasing (negative staining in single particle electron microscopy) or staining (transmission electron microscopy of plastic sections) biological structures with heavy metals. Samples imaged at cryogenic temperatures are usually not treated with heavy metals, although such approaches are being pursued and might provide future benefits (Adrian et al., 1998).

The theoretical resolution increases as the acceleration voltage of the electron beam increases. Using an acceleration of 100 kV the theoretical resolution is 0.0037 nm. The attainable resolution in electron microscopy is further limited by the aperture angle that is specific to objective lens system, leading to a minimum resolvable distance of 0.2–0.5 nm. The resolution in three-dimensional electron microscopy is therefore in most cases not limited by the electron microscope.

The resolution in single particle electron microscopy is limited mainly by low specimen contrast, the limited number of particle images, and the resulting low signal to noise. In electron tomography the final resolution is also determined by data acquisition and sample parameters, such as sample thickness and number of tilts (Section II.D.).

The signal-to-noise ratio can be increased using energy filtering methods. When an electron beam passes through electromagnetic lenses, electrons with different energies have different path trajectories. Energy filtering can be used to filter out all electrons that have lost energy through inelastic scattering, thereby selecting only electrons, which have not interacted with the sample, thereby increasing the signal-to-noise ratio (Grimm et al., 1996, 1998). A requirement for this technique is a uniform energy of extracted electrons that is provided most efficiently by a field emission gun (FEG) as an electron source. This electron source is becoming more common in modern electron microscopes and is essential for combating the low signal-to-noise ratios encountered in cryo-electron microscopy.

An additional way to improve the signal-to-noise ratio is averaging. In single particle electron microscopy, thousands of identical structures are imaged and then grouped and computationally realigned before being averaged (van Heel et al., 2000; Frank, 2002). In electron tomography each image in a tilt series can be exposed longer, thereby averaging the signal over time. Long acquisition times, however, expose the samples to a high number of electrons, and damage the delicate sample. The signal improves with the square root of sampling time and number of samples.

### C. Beam Damage

High electron doses alter the specimen by breaking bonds and by generating free radicals, which lead to structural disintegration of the sample. Embedding the sample in plastic and/or staining it with heavy metals improves sample stability, and relatively high electron exposure can be tolerated. However, when cryo-electron microscopy techniques are employed, the resolution expectations are higher because fixation artifacts are largely eliminated. Generally the acceptable amount of electrons is less than 50 e/Å$^2$ over the entire tilt series for cryo-electron tomography (Conway et al., 1993; McEwen et al., 1995), and 10 e/Å$^2$ for the cryo-single particle approach. Automation of data acquisition in conjunction with dose fractionation (McEwen et al., 1995) leads to a substantial reduction in electron exposure.

The embedding medium (plastic or ice) also interacts with the electron beam. The electron beam thins out plastic, where less cellular material is present to support the sample. This thinning can reach 40% in the direction of the beam and 10% in the perpendicular orientation (Luther, 1988, 1992). The thinning of plastic sections poses serious challenges for reconstruction algorithms that can be overcome by refining the model using fiducial markers or sample structure as references (Jing and Sachs, 1991).

In cryo-electron microscopy, shrinking does not occur. However, where the specimen interacts with electrons, the sample can show the formation of 'bubbles' (Lučić et al., 2005). Imaging at liquid nitrogen temperatures (~90 K) has been shown to protect the sample from degrading, allowing a 10 to 30 times higher exposure to electrons (Stark et al., 1996) than at room temperature. Imaging at liquid helium temperatures (~5–12 K) has the promise to increase electron exposure by a fact of 2–3 compared to liquid nitrogen temperatures (Unger, 2001). Recent reports, however, indicate that 'bubble' formation increase and sample stability decreases at liquid helium temperatures, compared to liquid nitrogen temperature (Iancu et al., 2006; Wright et al., 2006).

## IV. Cryo-Preservation

Fixation protocols that start with rapid freezing of the specimen are the best approach to preserve structural features of cells and biological macromolecules (Unger, 2001). The aim is to accomplish the solidification of water into amorphous (vitreous) ice, and thereby eliminate or greatly reduce the formation of ice crystals. Dubochet and McDowall (1981) showed in their pioneering study that water can solidify without crystal formation when a thin layer of water is frozen rapidly at very low temperatures. Different classes of ice can be distinguished that form depending on the time it takes to form the ice, as well as the freezing temperature. Hexagonal ice crystals forms when water is cooled slowly, while rapid freezing to −100 °C results in cubic ice crystals. Only rapid freezing to very low temperatures (below −140°C) accomplishes the formation of crystal-free, amorphous ice. The fast removal of heat from the sample can be accomplished by different approaches.

Plunge or quench freezing, is accomplished by moving the sample that is deposited or grown on a thin carrier membrane (e.g., dialysis membrane), quickly into the cryogen (Costello, 1980; Ryan et al., 1987). Plunge freezing may be accomplished by manually plunging a forceps-held sample into the cryogen, or by using special machines, known as vitrification robots (Cole et al., 1990; Egelhaaf et al., 2000; Frederik et al., 2000). It should be noted that the manipulation of the sample just prior to freezing can stress rapidly growing cells and should be minimized. In some cases, it is desirable to remove excess liquid from the sample by blotting either manually or mechanically shortly before the sample is plunged into the cryogen. A thinner layer of ice freezes more rapidly, thereby decreasing crystal formation. Cooled metal-surfaces (slam freezing) can also act as good heat sinks (Bald, 1983; Allison et al., 1987). A third freezing method takes advantage of small droplets of sample suspensions that are generated by a spray nozzle and subsequently propelled into the liquid cryogen (Plattner and Bachman, 1979). However, all freezing methods presented so far can only achieve rapid freezing without crystal formation to a sample depth of 10–15 μm, a depth suitable for single particles and small single cells, but not suitable for tissues.

The use of 'high-pressure freezing' increases the depth of vitrified ice from 10–15 μm (plunge freezing, slam freezing, and spray freezing) to 100 μm (Moor, 1987; Studer et al., 1989, 2001, Sartori et al., 1993), providing the potential of excellent cryo-fixation of large complex cells and tissues. Several commercial high-pressure freezers are available.

Once the samples are frozen, no further processing is required for cryo-single particles and thin cryo-electron tomography samples. Thicker samples that are plastic embedded or ice embedded require sectioning before they can be investigated in the electron microscope.

## V. Electron Tomography

### A. Sample Preparation

Because one limiting factor for resolution of electron tomography is the thickness of the sample, small specimen such as bacteria and cellular organelles are ideal objects since they do not require sectioning. However, as many researchers are interested in structural aspects of eukaryotes and cellular tissue, the object of interest is often buried within the cell. In order to reveal the internal nature of these complex specimens sectioning of plastic or ice embedded specimens has been pursued.

#### 1. Plastic-Embedding

Dehydration with organic solvents and infiltration of the sample with plastic resin that is then polymerized allows the sample to be sectioned into slices of 30–1000 nm thickness using a diamond knife. Generated sections float on water surface adjacent to the cutting edge. Repeated sectioning can result in a continuous ribbon of sections. Taken together, these sections can contain sample information of several micrometers. The sections are then picked up from the surface of the water onto carrier grids.

At room temperature, the dehydration and infiltration with resin, is usually preceded by a fixation with chemicals such as gluteraldehyde (Sabatini et al., 1963) and osmium tetroxide that cross-connect and thereby stabilize biological structures. It has long been shown that the use of standard aqueous-based fixation protocols can induce structural artifacts (Gilkey and Staehelin, 1986; Kellenberger, 1992).

Many of these artifacts have been overcome by rapid freezing and processing of the sample at ultra-cold temperatures. To obtain a plastic embedded sample, the frozen specimen is dehydrated and cellular components are stabilized by a process called freeze substitution. Here, the frozen sample is incubated in an organic solvent (acetone, methanol, or ethanol) containing fixatives and/or heavy metals, such as gluteraldehyde, uranyl acetate and osmium tetroxide, at –85 °C (Giddings, 2003). During the incubation the water is slowly replaced by the organic solvent. The choice of solvents, chemicals and the duration of freeze substitution used depends on the specimen and the investigated structure (Graham and Beveridge, 1990; McDonald and Morphew, 1993; Walther and Ziegler, 2002; Zechmann et al., 2005).

After freeze-substitution, which can take several days, the sample is infiltrated with resin and polymerized. Resins that polymerize at –50 °C (Lowicryl HM20) to –20 °C (Lowicryl K4M), as well as resins that polymerize at room temperature and/or with heat (Epon Araldite, LR-white, and Spurr's resin) have been used successfully. The choice of the embedding resin can influence preservation and appearance of the sample structures and has therefore to be evaluated for each sample type (Hess, 2004).

Plastic sections are usually post-stained with uranyl acetate (Terzakis, 1968) and lead citrate (Reynolds, 1963), which bind to biological molecules thereby increasing contrast. Colloidal gold particles of 10–20 nm can be used as fiducial markers (Jing and Sachs, 1991), which act as reference points in the reconstruction alignment process. Fiducial markers are applied before the tilt series is recorded.

#### 2. Ice-Embedding

An alternative to embedding specimen in plastic is embedding in ice. Ice-embedded samples have to be imaged at cryogenic temperatures (Dubochet et al., 1988). Because rapid freezing can result in virtually artifact-free specimens, as opposed to chemical fixation, the idea of fixing the sample by rapid freezing and imaging them in their hydrated state at cryogenic temperature is promising.

Researchers recognized that sample preparation and imaging at cryogenic temperatures also opens another possibility: sectioning of frozen samples using the ice as a 'virtual resin' material (Al-Amoudi et al., 2005). Since the original inception, cryo-electron microscopy of vitreous sections (CEMOVIS), i.e., sectioning of samples within vitreous ice and imaging at cryogenic temperatures has progressed tremendously (Al-Amoudi et al., 2004a,b).

Cryo-sectioning remains far from trivial and the progress made is a credit to the persistence and vision of the involved researchers. The most common challenges that have to be overcome are the sticking of the sections to the sectioning knife, as well as compression (Richter, 1994), and chatter of the sections (Al-Amoudi et al., 2004a). Exposing the diamond knife to an ion beam (Michel et al., 1992), and the newly introduced oscillating diamond knife (Studer and Gnägi, 2000), promise to mitigate these problems. Successful sectioning results in slices of 30 nm to 1000 nm thickness. After sectioning, individual section are transferred onto carbon-covered grid (Erk et al., 1996). The carbon-covered grid with the sections is then transferred with a cryo-specimen holder into an electron microscope (Sartori and Salamin Michel, 1994).

### B. Data Acquisition

For electron tomography the most common acquisition geometry is tilting of the sample around an axis. For plastic embedded samples, tilting increments of 1 to 2 degrees are generally used. Ice embedded samples are tilted with larger (2 nm or more) degree increments, in order to reduce the total electron exposure and thus beam damage. The maximal tilt angle is usually limited by the sample holder and typically ranges from ±60 to ±75 degrees. Therefore, a total of 50 to 150 images are usually acquired during each tilt series.

The acquisition of the tilt series has become an automated process, in which the imaged specimen is tracked. The image is automatically refocused and captured after each tilting increment. Automated acquisition is convenient and limits electron exposure of the specimen. Digital cameras have sufficient pixel size and sensitivity to capture images for electron tomography.

Samples that can tolerate a higher electron dose can be imaged by 'dual axis tomography' in which two perpendicular tilt series are acquired (Mastronarde, 1997). The second tilt series can be recorded after removing the grid from the electron microscope and reinserting the grid after is has been manually rotated by 90°. This process has been made less cumbersome by dual-axis holders that enable a dual-axis tilt of the grid without breaking the column vacuum (Iancu et al., 2005). The data from the second tilt series decreases the missing wedge effect (Section II.D.) and improves the overall signal-to-noise ratio (Mastronarde, 1997).

Colloidal gold markers that aide in reconstructing the volume can either be incorporated into the sample or the sample solution.

### C. Alignment and Reconstruction

To obtain the reconstruction of a three dimensional object, recorded tilt series are first aligned. A variety of algorithms can then be employed to generate a tomogram from the aligned images. Since the sample rotates during acquisition with deviations from the predicted tilt axis, images have to be realigned before they are further processed. Image rotations and shifts, changes in the magnification, and lens imperfections have to be computationally determined as well. Fiducial markers can be used to compute these deviations precisely (Mastronarde, 1997). When these reference markers are lacking, cross correlation methods that take advantage of inherent sample features (Frank and McEwen, 1992), or matching of intensity surface features can be employed (Brandt et al., 2001).

The most commonly used method for reconstruction of tomograms in electron tomography is the weighted back projection (Radermacher, 1992), which takes advantage of direct space and Fourier space operations (Section II.F). Publicly available program packages such as IMOD (Kremer et. al., 1996) and TOM (Nickel et al., 2005) as well as commercial products are available that facilitate an integration of data acquisition and reconstruction process.

### D. Segmentation and Model Generation

To extract useful information from the overwhelming amount of data in electron tomograms, it is important to segment out features of interest, such as membranes and filaments, to produce a model for analysis. There are currently two approaches to segmentation: manual outlining of features and computer assisted segmentation.

#### 1. Manual Segmentation

Manually outlining features was originally done with transparencies overlaying 70 nm thin section electron micrographs. Models were then produced by constructing physical structures or by using computer programs to display the integrated data. With advances in computing and the development of electron tomography, manually segmented models are now formed by outlining features of interest on sequential tomographic slices (Fig. 11b). Computer programs are able to converge these individual contours and produce a three-dimensional model of the features that have been segmented: a powerful method for observing true three-dimensional relationships between structures within the cell (Fig. 11c). Whilst this method is extremely powerful, it is based entirely on the subjective viewpoint of the researcher, and is substantially time consuming.

### 2. Computer Assisted Segmentation

Computer assisted segmentation is, in theory, more objective (Frangakis and Hegerl, 2001; Volkmann, 2002) as the researcher does not select and outline structures, but rather the computer program follows the density of the boundary chosen. Contour based approaches require the user to define the boundary at which the computer will segment the object (Volkmann, 2002; Bajaj et al., 2003). This method is sensitive to noise and requires substantial user interaction (Frangakis and Förster, 2004). A region-based approach (Frangakis and Hegerl, 2002) is a larger scale method that is able to segment the main feature from the background. This means a structure of interest can be quickly segmented out from the tomogram and then further segmented using contour based approaches (Frangakis and Förster, 2004).

### E. Electron Tomography Example

In Fig. 11 the major steps in electron tomography are illustrated. A 120 nm (semi-thick) plastic embedded section of the cyanobacterium *Synechocystis* PCC 6803 was imaged at different tilting angles. One such image is shown in Fig. 11a. The dark circles in this image are fiducial markers that are used to align the recorded images. After backprojection of the aligned images a three dimensional reconstruction is obtained (Fig. 11b). The reconstructed three-dimensional volume is used to examine virtual sections and outline structures of interest. In Fig. 11b the cytoplasmic membrane (light grey lines), the membranes comprising the thylakoid system (dark grey lines), lipid bodies (dark gray circles), and the ribosomes (white circles) are outlined. The outlined contours that are generated in different virtual sections can be combined to obtain a tomographic model of the specimen (Fig. 11c).

## VI. Single Particle Electron Microscopy

### A. General Overview

Single particle electron microscopy analyses macromolecular cell components that have been removed from their cellular context. It is required that these cellular components have been isolated and purified, to an extent that it can be assumed that only the particles of interest are contained within the imaged sample. The biochemical procedures to obtain the components have to be chosen carefully, so that structural artifacts are avoided. Different classes of biological molecules have been analyzed by single particle analysis, including DNA, RNA, proteins and protein complexes such as viruses. For photosynthesis research proteins appear to be the most relevant class of macromolecules and we will restrict our discussion to proteins. The classical method for single particle microscopy uses negative staining. However, more recently embedding the protein sample in amorphous ice has become the method of choice.

A variety of three-dimensional structures have been determined at resolutions in the range of 3–1 nm for none symmetrical proteins complexes, and even better resolution for some symmetrical protein structures like viruses. At resolution better

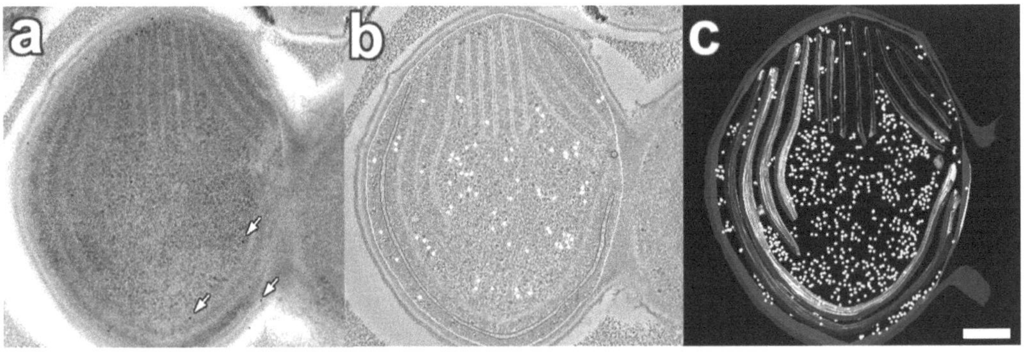


*Fig. 11.* Steps involved in the recording, reconstruction and modeling of a tomogram of a thin section of the cyanobacterium *Synechocystis* PCC 6803. (a) Image that is part of a tilt series. Fiducial marker appear as dark circles and are indicated by arrows. (b) Slice through reconstructed volume with model contours (white: ribosomes; grey: cytoplasmic and thylakoid membranes). (c) Model of section based on the reconstructed volume. Scale ~100 nm. See also Fig. 1, Color Plate 1.

than 0.6 nm, α-helixes and in some cases β-sheets can be distinguished.

One of the factors that limit the resolution of single particle microscopy is heterogeneity in the sample. Heterogeneity can arise due to a lack of protein subunits and different conformational states. Conformational changes can arise from the binding of cofactors to a protein or from modification of the protein by phosphorylation/dephosphorylation that can be controlled by choosing appropriate experimental conditions. If these conditions cannot be defined, multi-variant statistical procedures have been proposed to identify unique subunit composition and conformation (van Heel et al., 2000). However, these statistical methods require a much larger number of samples to reach the desired resolution for the different structures compared to homogeneous uniform samples.

For membrane-embedded structures, like photosynthetic reaction centers and antenna complexes that are difficult to crystallize, single particle electron tomography can be a successful approach to obtain high resolution structural data. Isolated photosynthetic complexes have been imaged by negative staining (Boekema et al., 1995) and embedded in ice (Nield et al., 2000). Jiang and coworkers (2001) demonstrated an approach that takes advantage of the spherical nature of lipid membranes to obtain high resolution structures of membrane proteins.

### *B. Sample Preparation*

#### *1. Negatively-Stained Particles*

For negative staining, macromolecules and small cellular components are encased within a layer of dried staining solution. The staining solution usually consists of 1–2% heavy metal salts of tungsten, molybdenum or uranium. Proteins are deposited on grids by blotting off the excess solution, followed by the application and blotting off of staining solution. An alternative to this two step process is the mixing of particle solution with staining solution, followed by blotting off the excess liquid. Negative staining results in high electron density differences between the sample and stain; internal protein structure, however, is mostly lost. Furthermore, a comparison between structures obtained from negatively stained and ice-embedded proteins shows that the former exhibits significant distortions (Frank, 2002).

#### *2. Ice-Embedded Particles*

The artifacts encountered by negative staining are overcome by rapidly freezing a thin layer of protein solution (Section IV). This procedure maintains proteins in their native, hydrated state. The proteins are usually randomly oriented within the ice allowing for a tilt-free data acquisition approach.

### *C. Data Acquisition*

As with electron tomography, limiting exposure to the electron beam and increased sample throughput are the driving forces in the automation of data acquisition. For single particle reconstruction, photographic film is still the preferred method to record images, although some laboratories have begun using high grade 4000 × 4000 pixel camera systems (Zhang et al., 2003a; Booth et al., 2004) that convert electrons into a sufficiently detectable signal. The 8 Gbyte of usable information obtained through photographic film is indeed astonishing (Cheng and Taylor, 1998). Currently this data depth cannot be matched by digital cameras. Recently, automated image acquisition and processing have been demonstrated (Rouiller et al., 2001; Frederik and Hubert, 2005; Lei and Frank, 2005) that is considerably more time-efficient than the traditional human-centered approach.

Several data collection protocols were developed that are optimized for protein symmetry and protein distribution. When proteins are randomly oriented on the specimen grid, projections of the particle from all orientations will be sampled by acquiring a large number of images. However, if the particles show a preferred orientation, usually due to interaction with the sample grid, the random conical tilt method can be used to obtain a good distribution of orientations. Using this data collection protocol, particles are imaged before and after a large (40–60 degree) tilt, thereby revealing previously unexposed particle views. Usually no tilting is required for single particle approaches at cryogenic temperatures, because the samples are oriented randomly within the vitreous ice.

### *D. Data Preprocessing*

A first step in image preprocessing is the generation of a 'true' object density image by means of the contrast transfer function (CTF). The Fourier transform of the bright field electron microscope image is the product of the Fourier transform of the object and the CTF

(Frank, 2002). The CTF alternates between 1 and –1 over frequencies depending on electromagnetic lens parameters (spherical aberration), electron source size, electron energy distribution, and distance from 'Gaussian focus' (defocus). Dividing the Fourier transform of the electron micrograph by the contrast transfer function restores the Fourier transform of the object image. However, frequencies where the CTF crosses 0 are lost. These lost frequencies can be recovered by recording images under slightly different defocus values in which the CTF crosses 0 at different frequencies (Penczek et al., 1997).

The next step in data processing for single particle electron microscopy is to determine what part of the recorded image constitutes a particle of interest and what features contain noise or contamination artifacts. Particle picking has started to become an automated process, in which the user is only verifying selected areas. Some programs can be 'trained' by a user to improve the recognition of particles.

For reconstruction, it is necessary to know the precise spatial orientation of the particles. Through automatic centering and aligning, identical views are oriented in identical directions. The employed programs usually accomplish centering of the particles by finding the center of gravity, followed by rotational alignment, which is achieved by cross-correlation and/or self correlation (van Heel et al., 2000). The aligned particles are then assigned into classes and all members of a class are then averaged. The absolute orientation of each class is still unknown. Methodology based on the 'common line theorem' (Section II.E) is implemented to determine the absolute orientations, which are usually expressed in Euler angles (van Heel et al., 2000).

## E. Refinement and Reconstruction

Refinement in single particle reconstruction is based on an iterative process in which a preliminary model is computed from the class averages and the determined Euler angles. The model is reprojected into the presumed class orientations and the resulting reprojection is used to refine classes and class members. A new model can then be constructed and reprojected. This process can be repeated until no improvement in structural data can be determined.

While the 'common line' approach can be employed for particles with unknown structure, a different approach can be pursued when a lower resolution structure is known from previous studies. In this case the existing reconstruction can be reprojected and resulting views can be used to define classes and correlate preprocessed particle images. Iterative refinement and reconstruction after this initial assignment can be pursued as outlined above.

If symmetrical particles are investigated the symmetry features can be used to restrain reconstruction parameters. These restraints, when applied correctly, can result in reconstructions with very high resolution using relatively few particle images.

An established method for estimating the resolution of the final reconstruction is the Fourier Ring Correlation (FRC) (Saxton and Baumeister, 1982) or Fourier Shell Correlation (FSC). In both approaches previously determined particle classes are randomly split into two groups. Each of these groups is used to generate a reconstruction, which is transformed into Fourier space. The Fourier transforms are then correlated with each other. A correlation of 1 indicates good correlation of two frequencies, typical for low frequencies. Higher frequencies show lower correlation indicating an increase in noise. At a certain frequency, the contribution of noise outweighs the contribution of the signal. Elimination of higher frequencies will improve signal to noise, but reduce resolution. Different philosophies exist on the assessment of this frequency, called the resolution limit. While some researches use consistently the 0.5 value as the resolution limit, other researcher derive a resolution limit that factors in the nature of the sample.

## F. Single Particle Microscopy Example

Figure 12 shows the successful implementation of single particle cryo-reconstruction and exemplifies the different steps involved in single particle reconstruction. Before beginning with the acquisition of micrographs, sample purity and the absence of contamination must be ensured (Fig. 12a). Cryo-electron micrographs show little contrast close to focus, but more contrast when the defocus value are larger (Figs. 12b and 12c). After selection, alignment and classification of the particle images, class averages are obtained (Fig. 12d). The resulting reconstruction can be reprojected revealing electron density distributions under different viewing angles (Fig. 12e). Surface rendering of the reconstruction are used to construct a model of the reconstructed particle.

The investigated particle is the Photosystem (PS) II supercomplex of spinach (*Spinacia oleraceae*). Each

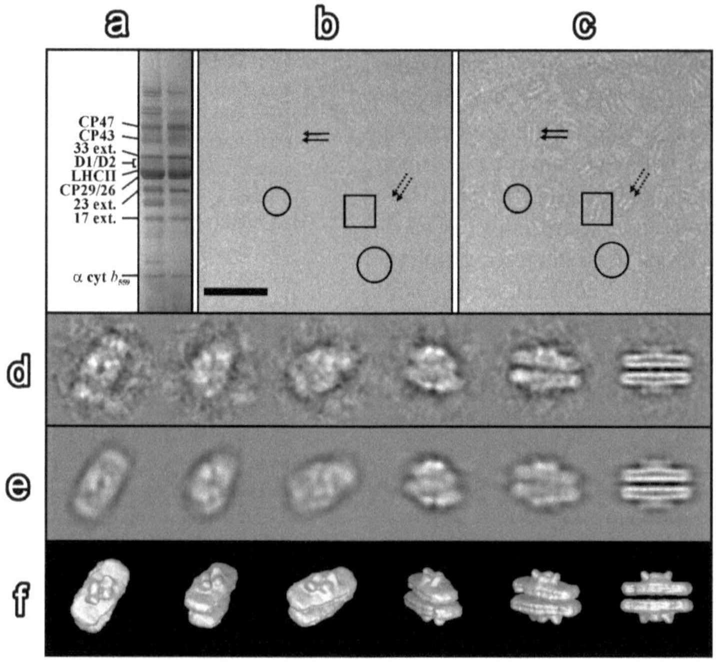


*Fig. 12.* Single particle electron microscopy. (a) Protein gel of the PS II-enriched membranes (lane 1) and purified PS II supercomplexes (lane 2) (b+c) Cryoelectron micrographs of a typical preparation of the PS II supercomplex showing particles randomly orientated in vitreous ice, with under focus of 1.35 mm and 7.2 mm respectively (d) Selection of typical class averages used for the 3D reconstruction. (e) Reprojections of the 3D map in identical orientations with the corresponding class averages. (f) Surface representation of the final 3D map viewed in the same orientation as the class averages. With permission from Dr. James Barber and Science (Nield et al., 2000).

supercomplex consists of two reaction centers and antenna system proteins. The oxygen evolving complex of each supercomplex can be distinguished as two protruding features in the reconstruction model.

## VII. Outlook

Three-dimensional electron microscopy techniques have established themselves as key approaches for high resolution structural biology and are continually maturing on a theoretical and technological level. The three crucial technological advances that propelled these techniques to the current status are the availability of computational resources, the establishment of cryo-methodology, and the automation of image acquisition (Dierksen et al., 1992; Koster et al., 1992).

The thirst for computational resources, which seemed unquenchable a few years ago, can be for the most part satisfied easily (and at low cost) by off the shelf computers or even more powerful computer clusters.

Cryo-specimen preparation and cryo-imaging require special equipment that has now become available and performs reliably. Establishment of cryo-techniques were crucial in overcoming the 'artifact barrier' inherent to traditional techniques, and has become the gold standard for sample preservation and imaging. Advances in cryo-sectioning have the promise to analyze larger specimens including eukaryotic cells and complex tissues.

There are a number of modern electron microscopes that can accommodate cryo-electron microscopy and/or electron tomography needs. Highly coherent electron sources that produce well-defined electron energies, in combination with energy filtering and automation have led to substantial improvements of signal to noise. Imaging at liquid helium temperatures has the potential to decrease radiation damage further, allowing for an increase in signal-to-noise ratio.

With all these technical advances single particle electron microscopy is well on its way to reach a resolution that will allow to match the individual amino acids of a protein sequence to an electron density map (Jiang and Ludtke, 2005).

The electron tomography toolbox is still expanding (Leapman, 2004). Energy-filtered transmission electron microscopy (EFTEM)-tomography of biological samples can generate a three-dimensional elemental map of the investigated sample (Leapman et al., 2004). High angle annular dark field (HAADF) scanning transmission electron microscopy (STEM)-tomography can be used to address interesting biological questions (Buseck et al., 2001). However, both EFTEM and HAADF-STEM do not fulfill low-dose requirements that ensure artifact free specimens.

Cryo-electron tomography contains structural information of macromolecules in their hydrated state within the cellular density map. As the structure of many proteins and macromolecules is already known from x-ray diffraction, nuclear magnetic resonance, or single particle reconstruction, it is possible to dock the geometrical shape of the structure of interest into the electron density of the tomogram (Brooijmans and Kuntz, 2003). Template matching is a larger scale approach where the density information of the particle of interest is matched (Roseman et al., 2001; Frangakis et al., 2002). If this approach can be fully realized, electron tomography data can be used for example, to determine the number and location of all proteins in an organism. However, the cellular matrix is filled with macromolecules and this 'cellular crowding' (Grünewald et al., 2002) makes identification of macromolecules difficult. Computational approaches to determine the interaction of these macromolecular complexes promise to overcome the challenges posed by the densely-packed cytoplasm. The integration of structural data with structures modeled from genomic data (Baumeister and Steven, 2000; Topf and Sali, 2005) will open a new frontier to structural biology.

## Acknowledgments

We would like to thank Dr. Robert E. Blankenship (Department of Chemistry and Biochemistry, Arizona State University) and Wolfgang Stefan (Department of Mathematics and Statistics, Arizona State University) and Maria Aronova (Division of Bioengineering and Physical Sciences, National Institute of Health) for valuable discussion of the manuscript. Martin Hohmann-Marriott acknowledges support through the Research Associate Program of the National Research Council.

## References

Adrian M, Dubochet J, Fuller S and Harris R (1998) Cryo-negative staining. Micron 29: 145–160

Al-Amoudi A, Chang JJ, Leforestier A, McDowall A, Salamin LM, Norlen LPO, Richter K, Blanc NS, Studer D and Dubochet J (2004a) Cryo-electron microscopy of vitreous sections. EMBO J 23: 3583–3588

Al-Amoudi A, Norlen LPO and Dubochet J (2004b) Cryo-electron microscopy of vitreous sections of native biological cells and tissues. J Struct Biol 148: 131–135

Al-Amoudi A, Studer D and Dubochet J (2005) Cutting artifacts and cutting process in vitreous sections for cryo-electron microscopy. J Struct Biol 150: 109–121

Allen GS, Zavialov A, Gursky R, Ehrenberg M and Frank J (2005) The cryo-EM structure of a translation initiation complex from *Escherichia coli*. Cell 121: 703–712

Allison DP, Daw CS and Rorvik MC (1987) The construction and operation of a simple inexpensive slam freezing device for electron microscopy. J Microsc 147: 103–108

Arnal I, Heichette C, Diamantopoulos GS and Chretien D (2004) CLIP-170/tubulin-curved oligomers coassemble at microtubule ends and promote rescues. Curr Biol 14: 2086–2095

Bajaj C, Yu Z and Auer M (2003) Volumetric feature extraction and visualization of tomographic molecular imaging. J Struct Biol 144: 132–143

Bald WB (1983) Optimizing the cooling block for the quick freeze method. J Microsc 131: 11–23

Baumeister W (2002) Electron tomography: Towards visualizing the molecular organization of the cytoplasm. Curr Opin Struct Biol 12: 679–684

Baumeister W (2004) Mapping molecular landscapes inside cells. Biol Chem 385: 865–872

Baumeister W (2005) From proteomic inventory to architecture. FEBS Lett 579: 933–937

Baumeister W and Steven AC (2000) Macromolecular electron microscopy in the era of structural genomics. Trends Biochem Sci 25: 624–631

Beck M, Förster F, Ecke M, Plitzko JM, Melchior F, Gerisch G, Baumeister W and Medalia O (2004) Nuclear pore complex structure and dynamics revealed by cryoelectron tomography. Science 306: 1387–1390

Boekema EJ, Hankamer B, Bald D, Kruip J, Nield J, Boonstra AF, Barber J and Rögner M (1995) Supramolecular structure of the Photosystem II complex from green plants and cyanobacteria. Proc Nat Acad Sci USA 92: 175–179

Booth CR, Jiang W, Baker ML, Zhou ZH, Ludtke, SJ and Chiu W (2004) A 9 Å single particle reconstruction from CCD captured images on a 200 kV electron cryomicroscope. J Struct Biol 147: 116–127

Brandt S, Heikkonen J and Engelhardt P (2001) Automatic alignment of transmission electron microscope tilt series without fiducial markers. J Struct Biol 136: 201–213

Brooijmans N and Kuntz ID (2003) Molecular recognition and docking algorithms. Annu Rev Biophys Biomol Struct 32: 335–373

Bubeck D, Filman DJ and Hogle JM (2005) Cryo-electron microscopy reconstruction of a poliovirus-receptor-membrane complex. Nat Struct Mol Biol 12: 615–618

Buseck PR, Dunin-Borkowski RE, Devouard B, Frankel RB, McCartney MR, Midgley PA, Pósfai, M and Weyland M (2001)

Magnetite morphology on mars. Proc Nat Acad Sci USA, 98: 13490–13495
Cheng Y and Taylor KA (1998) Characterization of the low magnification performance of a Philips CM300-FEG. Ultramicroscopy 74: 209–220
Cole R, Matuszek G, See C and Rieder CL (1990) A simple pneumatic device for plunge-freezing cells grown on electron microscopy grids. J Electron Microsc Tech 16: 167–173
Conway JF, Trus BL, Booy FP, Newcomb WW, Brown JC and Steven AC (1993) The effects of radiation damage on the structure of frozen hydrated HSV-1 capsids. J Struct Biol 111: 222–233
Cormack AM (1963) Representation of a function by its line integrals, with some radiological applications. J Appl Phys 34: 2722–2727
Costello MJ (1980) Ultra-rapid freezing of thin biological samples. Scan Electron Microsc 2: 361–370
Crowther LJ, Yamagata A, Craig L, Tainer JA and Donnenberg MS (2005) The ATPase activity of BfpD is greatly enhanced by zinc and allosteric interactions with other Bfp proteins. J Biol Chem 280: 24839–24848
DeRosier D and Klug A (1968) Reconstitution of 3-dimensional structures from electron micrographs. Nature 217: 130–134
Dierksen K, Typke D, Hegerl R, Koster A and Baumeister W (1992) Towards automatic electron tomography. Ultramicroscopy 40: 71–87
Dubochet J and McDowall AW (1981) Vitrification of pure water for electron microscopy. J Microsc 124: RP3–RP4
Dubochet J, Adrian M, Chang JJ, Homo JC, Lepault J, McDowall AW and Schultz P (1988) Cryo-electron microscopy of vitrified specimens. Q Rev Biophys 21: 129–228
Edholm PR, Hellstrom LG and Jacobsen B (1978) Transverse tomography with incoherent optical reconstruction. Phys Med Biol 23: 90–99
Egelhaaf SU, Schurtenberger P and Muller M (2000) New controlled environment vitrification system for cryo-transmission electron microscopy: Design and application to surfactant solutions. J Microsc 212: 175–185
Erk I, Michel M and Lepault J (1996) Electron cryo-microscopy of vitrified biological specimens — ideal and real structures of water lipid phase. J Microsc 182: 15–23
Falke S, Tama F, Brooks CL 3rd, Gogol EP and Fisher MT (2005) The 13 Å structure of a chaperonin GroEL-protein substrate complex by cryo-electron microscopy. J Mol Biol 348: 219–230
Fernandez JJ and Li S (2003) An improved algorithm for anisotropic nonlinear diffusion for denoising cryotomograms. J Struct Biol 144: 152–161
Förster F, Medalia O, Zauberman N, Baumeister W and Fass D (2005) Retrovirus envelope protein complex structure in situ studied by cryo-electron tomography. Proc Natl Acad Sci USA 102: 4729–4734
Frangakis AS and Förster F (2004) Computational exploration of structural information from cryo-electron tomograms. Curr Opin Struct Biol 14: 325–31
Frangakis AS and Hegerl R (2001) Noise reduction in electron tomographic reconstructions using nonlinear anisotropic diffusion. J Struct Biol 135: 239–250
Frangakis AS and Hegerl R (2002) Segmentation of two- and three-dimensional data from electron microscopy using eigenvector analysis. J Struct Biol 138: 105–113
Frangakis AS, Böhm J, Förster F, Nickell S, Nicastro D Typke D, Hegerl R and Baumeister W (2002) Identification of macromolecular complexes in cryoelectron tomograms of phantom cells. Proc Nat Acad Sci USA 99: 14153–14158
Frank J (ed) (1992) Electron Tomography, Three Dimensional Imaging with the Transmission Electron Microscope. Plenum, New York
Frank J (1996) Three-Dimensional Electron Microscopy of Macromolecular Assemblies. Academic Press, New York
Frank J (2002) Single-particle imaging of macromolecules by cryo-electron microscopy. Annu Rev Biophys Biomol Struct 31: 303–319
Frank J (ed) (2006) Three-Dimensional Electron Microscopy of Macromolecular Assemblies: Visualization of Biological Molecules in Their Native State. Academic Press, New York
Frank J and McEwen BF (1992) Alignment by cross-correlation. In: Frank J (ed) Electron Tomography, Three Dimensional Imaging with the Transmission Electron Microscope, pp 205–213. Plenum, New York
Frederik PM and Hubert DHW (2005) Cryoelectron microscopy of liposomes. Meth Enzymol 391: 431–448
Frederik P, Bomans P, Franssen V and Laeven P (2000) A vitrification robot for time resolved cryo-electron microscopy. In: Cech S and Janish R (eds) Proceedings of the 12th European Congress on Electron Microscopy, Vol. I, pp B383–B384. Reklamní Atelier Kupa, Brno
Gao N, Zavialov AV, Li W, Sengupta J, Valle M, Gursky RP, Ehrenberg M and Frank J (2005) Mechanism for the disassembly of the post-termination complex inferred from cryo-EM studies. Mol Cell 18: 663–674
Giddings TH (2003) Freeze-substitution protocols for improved visualization of membranes in high-pressure frozen samples. J Microsc 212: 53–61
Gilbert PFC (1972) The reconstitution of three-dimensional structure from projections and its application to electron microscopy. II Direct methods. Proc R Soc Lond B 182: 89–102
Gilkey JC and Staehelin LA (1986) Advances in ultrarapid freezing for the preservation of cellular ultrastructure. Electron Microsc Techn 3: 177–210
Gordon R, Bender R and Herman GT (1970) Algebraic reconstruction techniques (ART) for three dimensional electron microscopy and x-ray photography. J Theor Biol 29: 471–481
Graham LL and Beveridge TJ (1990) Evaluation of freeze-substitution and conventional embedding protocols for routine electron microscopic processing of eubacteria. J Bacteriol 172: 2141–2149
Grimm R, Koster AJ, Ziese U, Typke D and Baumeister W (1996) Zero-loss energy filtering under low-dose conditions using a post column energy filter. J Microsc 183: 60–68
Grimm R, Typke D and Baumeister W (1998) Improving image quality by zero-loss energy filtering: Quantitative assessment by means of image cross-correlation. J Microsc 190: 339–349
Grünewald K, Medalia O, Gross A, Steven AC and Baumeister W (2002) Prospects of electron cryotomography to visualize macromolecular complexes inside cellular compartments: Implications of crowding. Biophys Chem 100: 577–591
Harauz G and van Heel M (1986) Exact filters for general geometry three-dimensional reconstruction. Optik 73:146–156
Hart RG (1968) Electron microscopy of unstained biological material: The polytropic montage. Science 159: 1464–1467
Henderson R, Baldwin JM, Ceska TA, Zemlin F, Beckmann E

and Downing KH (1990) Model for the structure of bacteriorhodopsin based on high-resolution electron cryo-microscopy. J Mol Biol 213: 899–929

Hess MW (2003) Of plants and other pets: Practical aspects of freeze-substitution and resin embedding. J Microsc 212: 44–52

Heymann JB, Chagoyen M and Belnap DM (2005) Common conventions for interchange and archiving of three-dimensional electron microscopy information in structural biology. J Struct Biol 151: 196–207

Hirose K, Löwe J, Alonso M, Cross RA and Amos LA (1999) 3D electron microscopy of the interaction of kinesin with tubulin. Cell Struct Func 24: 277–284

Hounsfield GN (1973) Computerized transverse axial scanning tomography: Part I description of the system. Br J Radiol 46: 1016–1022

Iancu CV, Wright ER, Benjamin J, Tivol WF, Dias DP, Murphy GE, Morrison RC, Heymann JB and Jensen GJ (2005) A 'flip-flop' rotation stage for routine dual-axis electron cryotomography. J Struct Biol 151: 288–297

Iancu CV, Wright ER, Heymann JB and Jensen GJ (2006) A comparison of liquid nitrogen and liquid helium as cryogens for electron cryotomography. J Struct Biol 153: 231–240

Jiang W and Ludtke SJ (2005) Electron cryomicroscopy of single particles at subnanometer resolution. Curr Opin Struct Biol 15: 571–577

Jiang QX, Chester DW and Sigworth FJ (2001) Spherical reconstruction: A method for structure determination of membrane proteins from cryo-EM images. J Struct Biol 33: 119–131

Jiang W, Li Z, Zhang Z, Baker ML, Prevelige PE Jr and Chiu W (2003a) Coat protein fold and maturation transition of bacteriophage P22 seen at subnanometer resolutions. Nature Struct Biol 10: 131–135

Jiang W, Baker ML, Wu Q, Bajaj C and Chiu W (2003b) Applications of a bilateral denoising filter in biological electron microscopy. J Struct Biol 144:114–122

Jing Z and Sachs F (1991) Alignment of tomographic projections using an incomplete set of fiducial markers. Ultramicroscopy 35: 37–43

Kellenberger E, Johansen R, Maeder M, Bohrmann B, Stauffer E and Villiger W (1992) Artifacts and morphological changes during chemical fixation. J Microsc 168: 181–201

Koster AJ, Chen H, Sedat JW and Agard DA (1992) Automated microscopy for electron tomography. Ultramicroscopy 46: 207–227

Kremer JR, Mastronarde DN and McIntosh JR (1996) Computer visualization of three-dimensional image data using IMOD. J Struct Biol 116: 71–76

Leapman RD (2004) Novel techniques in electron microscopy. Curr Opin Neurobiol 14: 591–598

Leapman RD, Kocsis E, Zhang G, Talbot TL and Laquerriere P (2004) Three-dimensional distributions of elements in biological samples by energy-filtered electron tomography. Ultramicroscopy 100: 115–125

Lei JL and Frank J (2005) Automated acquisition of cryo-electron micrographs for single particle reconstruction on an FEI Tecnai electron microscope. J Struct Biol 150: 69–80

Liu Z, Wang R, Zhang J, Chen SR and Wagenknecht T (2005) Localization of a disease-associated mutation site in the three-dimensional structure of the cardiac muscle ryanodine receptor. J Biol Chem 280: 37941–37947

Lučić V, Förster F and Baumeister W (2005) Structural studies by electron tomography: From cells to molecules. Annu Rev Biochem 74: 833–865

Luther PK (1992) Sample shrinkage and radiation damage. In: Frank J (ed) Electron Tomography, Three-Dimensional Imaging with the Transmission Electron Microscope, pp 39–60. Plenum Press, New York

Luther PK, Lawrence MC and Crowther RA (1988) A method for monitoring the collapse of plastic sections as a function of electron dose. Ultramicroscopy 24: 7–18

Marabini R, Rietzel E, Schroeder R, Herman G and Carazo J (1997) Three-dimensional reconstruction from reduced sets of very noisy images acquired following a single-axis tilt schema: Application of a new three-dimensional reconstruction algorithm and objective comparison with weighted backprojection. J Struct Biol 120: 363–371

Marr KM, Mastronarde DN and Lyon MK (1996) Two-dimensional crystals of Photosystem II: Biochemical characterization, cryoelectron microscopy and localization of the D1 and cytochrome $b_{559}$ polypeptides. J Cell Biol 132: 823–833

Marsh BJ, Mastronarde DN, Buttle KF, Howell KE and McIntosh JR (2001) Organellar relationships in the Golgi region of the pancreatic beta cell line, HIT-T15, visualized by high resolution electron tomography. Proc Nat Acad Sci USA 98: 2399–2406

Mastronarde DN (1997) Dual axis tomography: An approach with alignment methods that preserves resolution. J Struct Biol 120: 343–352

McDonald K and Morphew MK (1993) Improved preservation of ultrastructure in difficult-to-fix organisms by high pressure freezing and freeze-substitution: I. *Drosophila melanogaster* and *Strongylocentrotus purpuratus* embryos. Microsc Res Tech 24: 465–473

McEwen BF and Marko M (2001) The emergence of electron tomography as an important tool for investigating cellular ultrastructure. J Histochem Cytochem 49: 553–564

McEwen BF, Downing KH and Glaeser RM (1995) The relevance of dose fractionation in tomography of radiation-sensitive specimens. Ultramicroscopy 60: 357–373

McIntosh JR (2001) Electron microscopy of cells: A new beginning for a new century. J Cell Biol 153: F25–F32

McIntosh R, Nicastro D and Mastronarde D (2005) New views of cells in 3D: An introduction to electron tomography. Trends Cell Biol 15: 43–51

Medalia O, Weber I, Frangakis AS, Nicastro D, Gerisch G and Baumeister W (2002) Macromolecular architecture in eukaryotic cells visualized by cryoelectron tomography. Science 298: 1209–1213

Michel M, Gnaegi H and Müller M (1992) Diamonds are a cryosectioners best friend. J Microsc 166: 43–56

Moor H (1987) Theory and practice of high pressure freezing. In: Steinbrecht RA and Zierold K (eds) Cryotechniques in biological electron microscopy, pp 175–191. Springer-Verlag, Berlin

Morais MC, Kanamaru S, Badasso MO, Koti JS, Owen BA, McMurray CT, Anderson DL and Rossmann MG (2003) Bacteriophage phi29 scaffolding protein gp7 before and after prohead assembly. Nature Struct Biol 10: 572–576

Murakami K, Yumoto F, Ohki SY, Yasunaga T, Tanokura M and Wakabayashi T (2005) Structural basis for $Ca^{2+}$-regulated muscle relaxation at interaction sites of troponin with actin and tropomyosin. J Mol Biol 352: 178–201

Nicastro D, Frangakis AS, Typke D and Baumeister W (2000)

Cryo-electron tomography of *Neuraspora* mitochondria. J Struct Biol 129: 48–56

Nickell S, Förster F, Linaroudis A, Del Net W, Beck F, Hegerl R, Baumeister W and Plitzko JM (2005) TOM software toolbox: Acquisition and analysis for electron tomography. J Struct Biol 149: 227–34

Nield J, Orlova EV, Morris EP, Gowen B, van Heel M and Barber J (2000) 3D map of the plant Photosystem II supercomplex obtained by cryoelectron microscopy and single particle analysis. Nature Struct Biol 7: 44–47

Penczek P, Radermacher M and Frank J (1992) Three-dimensional reconstruction of single particles embedded in ice. Ultramicroscopy 40: 33–53

Penczek PA, Zhu J, Schröder R and Frank J (1997) Three-dimensional reconstruction with contrast transfer compensation from defocus series. Scanning Microsc 11: 147–154

Plattner H and Bachmann L (1979) Spray-freezing of cells and cell fractions. Mikroskopie 35: 21–22

Pous J, Chevalier C, Ouldali M, Navaza J, Delmas B and Lepault J (2005) Structure of birnavirus-like particles determined by combined electron cryomicroscopy and X-ray crystallography. J Gen Virol 86: 2339–2346

Qian P, Hunter CN and Bullough PA (2005) The 8.5Å projection structure of the core RC-LH1-PufX dimer of *Rhodobacter sphaeroides*. J Mol Biol 349: 948–960

Radermacher M (1992) Weighted back-projection methods. In: Frank J (ed) Electron Tomography, Three-Dimensional Imaging with the Transmission Electron Microscope, pp 91–115. Plenum Press, New York

Radermacher M (1994) Three-dimensional reconstruction from random projections: Orientational alignment via Radon transforms. Ultramicroscopy 53: 121–36

Radon J (1917) Über die Bestimmung von Funktionen durch ihre Integralwerte längs gewisser Manningfaltigkeiten. Berichte Verhandl Königl Sächs Gesell Wissenschaft Leipzig Math Phys Klasse 69: 262–277

Reynolds VR (1963) The use of lead citrate at high pH as an electron opaque stain in electron microscopy. J Cell Biol 71: 208–212

Richter K (1994) Cutting artifacts on ultrathin cryosections of biological bulk specimens. Micron 25: 297–308

Roseman, AM, Ranson NA, Gowen B, Fuller SD and Saibil HR (2001) Structures of unliganded and ATP-bound states of the *Escherichia coli* chaperonin GroEL by cryoelectron microscopy. J Struct Biol 135: 115–125

Rouiller I, Pulokas J, Butel VM, Milligan RA, Wilson-Kubalek EM, Potter CS and Carragher BO (2001) Automated image acquisition for single-particle reconstruction using p97 as the biological sample. J Struct Biol 133: 102–107

Ryan KP, Purse DH, Robinson SG and Wood JW (1987) The relative efficiency of cryogens used for plunge-cooling biological specimens. J Microsc 145: 89–96

Sabatini DD, K. Bensch and Barrnett R (1963) Cytochemistry and electron microscopy. The preservation of cellular ultrastructure and enzymatic activity by aldehyde fixation. J Cell Biol 17: 19–59

Santarella RA, Skiniotis G, Goldie KN, Tittmann P, Gross H, Mandelkow EM, Mandelkow E and Hönger A (2004) Surface-decoration of microtubules by human tau. J Mol Biol 339: 539–553

Sartori N and Salamin Michel LA (1994) Cryo-transmission electron microscopy of thin vitrified sections. In: Celis JE (ed) Cell Biology: A Laboratory Handbook, pp 177–185. Academic Press, New York

Sartori N, Richter K and Dubochet J (1993) Vitrification depth can be increased more than 10-fold by high-pressure freezing. J Microsc 172: 55–61

Saxton WO and Baumeister W (1982) The correlation averaging of a regularly arranged bacterial cell envelope protein. J Microsc 127, 127–138

Skoglund U, Ofverstedt LG, Burnett RM and Bricogne G (1996) Maximum-entropy three-dimensional reconstruction with deconvolution of the contrast transfer function: A test application with adenovirus. J Struct Biol 117: 173–88

Snyder JP, Nettles JH, Cornett B, Downing KH and Nogales E (2001) The binding conformation of Taxol in beta-tubulin: A model based on electron crystallographic density. Proc Nat Acad Sci USA 98: 5312–5316

Stark H, Zemlin F and Boettcher C (1996) Electron radiation damage to protein crystals of bacteriorhodopsin at different temperatures. Ultramicroscopy 63: 75–79

Steven AC and Aebi U (2003) The next ice age: Cryo-electron tomography of intact cells. Trends Cell Biol 13: 107–110

Stoscheck A and Hegerl R (1997) Denoising of electron tomographic reconstructions using multiscale transformations. J Struct Biol 120: 257–265

Studer D and Gnägi H (2000) Minimal compression of ultrathin sections with use of an oscillating diamond knife. J Microsc 197: 94–100

Studer D, Michel M and Müller M (1989) High-pressure freezing comes of age. Scanning Microsc Suppl 3: 253–269

Studer D, Graber W, Al-Amoudi A and Eggli P (2001) A new approach for cryofixation by high-pressure freezing. J Microsc 203: 285–294

Subramaniam S (2005) Bridging the imaging gap: Visualizing subcellular architecture with electron tomography. Curr Opin Microbiol 8: 316–22

Subramaniam S and Milne JL (2004) Three-dimensional electron microscopy at molecular resolution. Annu Rev Biophys Biomol Struct 33: 141–155.

Suzuki H, Yonekura K and Namba K (2004) Structure of the rotor of the bacterial flagellar motor revealed by electron cryomicroscopy and single-particle image analysis. J Mol Biol 3371: 105–113

Tama F, Miyashita O and Brooks CL (2004) Normal mode based flexible fitting of high-resolution structure into low-resolution experimental data from cryo-EM. J Struct Biol 147: 315–326

Terzakis JA (1968) Uranyl acetate, a stain and a fixative. J Ultrastruct Res 22: 168–169

Topf M and Sali A (2005) Combining electron microscopy and comparative protein structure modeling. Curr Opin Struct Biol 15: 578–585

Unger VM (2001) Electron cryomicroscopy methods. Curr Opin Struct Biol 11: 548–554

Valle M, Zavialov A, Li W, Stagg SM, Sengupta J, Nielsen RC, Nissen P, Harvey SC, Ehrenberg M and Frank J (2003) Incorporation of aminoacyl-tRNA into the ribosome as seen by cryo-electron microscopy. Nature Struct Biol 10: 899–906. Erratum in: Nature Struct Biol 10: 1074

van Heel M (1987) Angular reconstruction: A posteriori assignment of projection directions for 3D reconstruction. Ultramicroscopy 21: 111–124

van Heel M, Gowen B, Matadeen R, Orlova EV, Finn R, Pape

T, Cohen D, Stark H, Schmidt R, Schatz M and Patwardhan A (2000) Single-particle electron cryo-microscopy: towards atomic resolution. Q Rev Biophys 33: 307–369

van de Meene AML, Hohmann-Marriott MF, Vermaas WFJ and Roberson RW (2005) The three-dimensional structure of the cyanobacterium *Synechocystis sp*. PCC 6803. Arch Microbiol 184: 259–270

Volkmann N (2002) A novel three-dimensional variant of the watershed transform for segmentation of electron density maps. J Struct Biol 138: 123–129

Wadsworth Center (2002) Glossary of terms used in image processing and single-particle method. http://www.wadsworth.org/spider_doc/spider/docs/glossary.html (January 29, 2002)

Walther P and Ziegler A (2002) Freeze substitution of high-pressure frozen samples: The visibility of biological membranes is improved when the substitution medium contains water. J Microsc 208: 3–10

Wang HW and Nogales E (2005) Nucleotide-dependent bending flexibility of tubulin regulates microtubule assembly. Nature 435: 911–915

Wright ER, Iancu CV, Tivol WF and Jensen GJ (2006) Observations on the behavior of vitreous ice at ~8 and ~12K. J Struct Biol 153: 241–252

Xiao C, Chipman PR, Battisti AJ, Bowman VD, Renesto P, Raoult D and Rossmann MG (2005) Cryo-electron microscopy of the giant Mimivirus. J Mol Biol 353: 493–496

Zechmann B, Müller M and Zellnig G (2005) Effects of different fixation and freeze substitution methods on the ultrastructural preservation of ZYMV-infected *Cucurbita pepo* (L.) leaves. J Electron Microsc (Tokyo) 54: 393–402

Zhang P, Borgnia MJ, Mooney P, Shi D, Pan M, O'Herron P, Mao A, Brogan D, Milne JL and Subramaniam S (2003a) Automated image acquisition and processing using a new generation of 4K x 4K CCD cameras for cryo electron microscopic studies of macromolecular assemblies. J Struct Biol 143: 135–144

Zhang W, Chipman PR, Corver J, Johnson PR, Zhang Y, Mukhopadhyay S, Baker TS, Strauss JH, Rossmann MG and Kuhn RJ (2003b) Visualization of membrane protein domains by cryo-electron microscopy of dengue virus. Nature Struct Biol 10: 907–912

Zhang W, Heil M, Kuhn RJ and Baker TS (2005) Heparin binding sites on Ross River virus revealed by electron cryo-microscopy. Virology 332: 511–18

# Chapter 3

# Non-Linear Contrast Mechanisms for Optical Microscopy

Virginijus Barzda
*Department of Physics, and Institute for Optical Sciences, University of Toronto, 3359 Mississauga Road, Mississauga, Ontario, L5L 1C6, Canada*



## Summary

Optical microscopy is a primary research tool that is widely used for visualization of cells and subcellular organelles in living photosynthetic organisms. Although being primarily an imaging tool, a microscope can also be used as a spectroscopic instrument that offers means to access spectral properties of individual photosynthetic membranes and pigment-protein complexes with a diffraction limited spatial resolution. When coupled with a femtosecond laser, the optical microscope becomes an imaging tool that differentiates the biological structures based on non-linear light-matter interactions. A large plethora of non-linear optical responses can be employed for imaging, and spatially resolved spectroscopy, rendering structural and functional information about photosynthetic systems.

This chapter will introduce three non-linear contrast mechanisms that have been applied for microscopic investigations of photosynthetic samples: second harmonic generation, third-harmonic generation, and multiphoton excitation fluorescence. Two applications of non-linear microscopy will be presented featuring structural and spectroscopic investigations of individual in situ chloroplasts and isolated aggregates of plant major light-harvesting chlorophyll *a/b* pigment-protein complexes associated with Photosystem II (LHCII). Readers will be exposed to novel tools and possibilities of the rapidly growing field of non-linear microscopy.

*Email: vbarzda@utm.utoronto.ca

*Thijs J. Aartsma and Jörg Matysik (eds): Biophysical Techniques in Photosynthesis II, pp. 35–54.*

## I. Introduction

In photosynthesis, the efficient conversion of solar radiation to chemical energy is largely determined by the architecture of photosynthetic machinery. Highly optimized photosynthetic structures manifest themselves at different levels of organization. Different biophysical techniques are used to study the complexity of these structural levels. Microscopy is usually employed for investigation of organelles inside cells, and localization of different cells in specialized tissues of photosynthetic organisms. For structural investigations below the spatial resolution of optical microscopes, electron or scanning probe microscopy, spectroscopy, and X-ray or electron crystallography is usually employed. The combination of spectroscopy and optical microscopy gives the advantage of both methods by extending the spatial resolution range over which different levels of structural organization can be investigated. Similarly, the technique of non-linear microscopy builds on non-linear spectroscopy and uses non-linear optical responses as imaging contrast mechanisms for obtaining structural information in photosynthetic samples from tissues down to single molecules.

Non-linear microscopy employs femtosecond and picosecond lasers for achieving high photon flux density by focusing the beam onto a sample with a high numerical aperture (NA) microscope objective. By focusing and pulsing, high peak intensities can be achieved while keeping the average excitation power low. This prevents photosynthetic systems from photodestruction. The high light intensities can induce non-linear optical effects. The intensity of non-linear signals scales with the power dependency on the fundamental light intensity. So far, the following non-linear effects have been implemented in an optical microscope: second harmonic generation (SHG) (Hellwart and Christen, 1974; Sheppard et al., 1977; Mizutani et al., 2000), sum-frequency generation (Florsheimer, 1999), third harmonic generation (THG) (Barad et al., 1997; Müller et al., 1998a), multiphoton-excitation fluorescence (MPF) (Denk et al., 1990; Barzda et al., 2001a; Tirlapur and König, 2001), optical Kerr effect (Potma et al., 2001), and coherent anti-Stokes Raman scattering (CARS) (Duncan et al., 1982; Zumbusch et al., 1999; Müller et al., 2000).

In the microscope, non-linear signals originate from a femtoliter focal volume. This reduces the sampled molecular ensemble compared to the macroscopic measurements, where the non-linear responses are observed over the whole beam pass in the cuvette. Microscopic measurements can reveal unique molecular features that are otherwise obscured by the ensemble averaged measurement. For example, SHG is symmetry forbidden in homogeneous suspensions even for non-centrosymmetric molecules (Boyd, 1992). Spatially confined excitation at an interface between two media or materials provides a central symmetry breaking that reveals second order non-linearity of chiral molecules (Shen, 1989; Kriech and Conboy, 2005). Therefore, molecular aggregates and cellular structures that form interfaces can be readily investigated with SHG microscopy.

Non-linear microscopy has several advantages compared to classical microscopy. The main draw back of classical transmission and fluorescence microscopes is their poor axial resolution. This can be improved by introducing confocal detection, or by employing non-linear optical contrast mechanisms (Pawley, 1995; Fung and Theriot, 1998; Hepler and Gunning, 1998; Feijó and Moreno, 2004). Since non-linear optical effects are proportional to the second or third power of the fundamental light intensity, when focusing femtosecond laser beam, only the focal volume reaches enough intensity for non-linear responses to occur. This focal confinement of non-linear signal generation determines the excitation confocality and renders the inherent optical sectioning. Three-dimensional (3D) images can be obtained by rastering the focal volume in the transverse direction and translating the sample or the excitation objective along the optical axis.

A large acceptance angle of high NA microscope objective ensures a high-efficiency collection geometry for optical signals. This is very beneficial for MPF which is emitted in all directions. Although coherent non-linear signals have defined propagation directions due to phase matching conditions, their collection with large acceptance angle objectives provides a very efficient method for detecting weak signals from

*Abbreviations:* 2D – two-dimensional; 3D – three-dimensional; BChl – bacteriochlorophyll; Car – carotenoid; CARS – coherent anti-Stokes Raman scattering; CCD – charge-coupled device; CD – circular dichroism; Chl – chlorophyll; CPL – circularly polarized luminescence; LHCII – light-harvesting chlorophyll *a/b* pigment-protein complex of Photosystem II; MPF – multiphoton excitation fluorescence; NA – numerical aperture; PSF – point spread function; SCIA – structural cross-correlation image analysis; SHG – second harmonic generation; THG – third harmonic generation; TPE – two-photon excitation; TPF – two-photon excitation fluorescence

highly scattering photosynthetic samples.

For non-linear confocality to be achieved, the wavelength of the excitation laser has to be tuned outside the linear absorption bands of the sample. Usually, infrared excitation wavelengths are employed that also provide deeper penetration of light into the tissue, reaching up to a few hundred microns penetration depth in highly scattering plant specimens (Prent et al., 2005a).

Cells and intracellular organelles can be imaged inside the living tissue. This marvelous feature of optical microscopy allows the study of physiological processes without appreciable intervention by the measurement. The non-invasiveness of the MPF microscopy was demonstrated by the real-time imaging of chloroplast division in living bundle sheath cells of *Arabidopsis* (Tirlapur and König, 2001). The advantage of MPF compared to confocal microscopy lies in the focal confinement of the excitation due to the power dependence of the non-linear responses on the fundamental light intensity. The excitation and subsequent photobleaching occurs along the whole beam pass in confocal microscopy, while it is confined only to the focal volume in the non-linear microscope (Denk et al., 1990). The reduced photobleaching in non-linear microscopy appears to be a crucial factor in the investigation of light sensitive photosynthetic specimens.

Although MPF microscopy is the most frequently used non-linear contrast mechanism, the first non-linear microscope was introduced by employing second harmonic generation (Hellwart and Christen, 1974; Sheppard et al., 1977). SHG microscopy has been extensively reviewed in several articles (Yamada and Lee, 1998; Gauderon et al., 2001; Millard et al., 2003; Mertz, 2004; Sun, 2005). The renaissance of non-linear microscopy started with the demonstration of the MPF microscope and its application to biological imaging (Denk et al., 1990). The introduction of stable solid state Ti:Sapphire femtosecond lasers largely facilitated the development of non-linear microscopy. MPF imaging of biological structures have been reviewed in a number of recent papers (König, 2000; So et al., 2000; Williams et al., 2001; Zipfel et al., 2003; Feijó and Moreno, 2004). THG microscopy was introduced by Barad et al. (1997). It is interesting to note that the first demonstration of THG applications for biological imaging was performed on a thin plant leaf section showing strong THG signals originating from chloroplasts in the mesophyll cells (Müller et al., 1998a). THG microscopy has been recently reviewed (Squier and Müller, 2001; Sun, 2005). Other non-linear contrast mechanisms have not been used in photosynthesis research, although CARS microscopy is very promising for studying photosynthetic samples. CARS microscopy was introduced by Duncan et al. (1982). Reviews on CARS microscopy can be found in Cheng and Xie (2004) and Volkmer (2005).

Another trend has developed in the recent years: different non-linear contrast mechanisms have been combined into one multimodal microscope. Since the same excitation pulse may lead to the generation of MPF, SHG, and THG signals, simultaneous detection schemes have been implemented (Campagnola et al., 1999; Chu et al., 2001; Barzda et al., 2005; Sun, 2005). The images can be compared and correlated giving rich information about the structural architecture and molecular distribution in the sample (Barzda et al., 2005; Greenhalgh et al., 2005).

Non-linear microscopy is a rapidly developing field of research. With the exception of MPF microscopy, which is firmly established as a valuable tool for biological imaging, other contrast mechanisms are going through the stage of major technical development. Most of the work on non-linear microscopy has been published in the engineering literature. Biophysical and biological investigations are starting to emerge. The wider spread of harmonic generation microscopy is partly limited by the lack of commercially available instrumentation from major microscope manufacturers. Fortunately, with simple modifications of confocal or MPF microscopes, SHG and THG microscopy can readily be implemented. The applications of non-linear microscopy in photosynthesis research are also starting to emerge. Relatively strong non-linear signals generated in photosynthetic samples will definitely attract more attention in the near future.

This chapter will introduce the physical principles of non-linear optics and describe the contrast mechanisms of most commonly used non-linear imaging modalities: MPF, SHG, and THG microscopies. The instrumentation used in non-linear microscopy will be overviewed and some methods of image analysis will be presented. This chapter will be completed with the examples of non-linear micro-spectroscopy and imaging of photosynthetic samples.

## II Principles of Non-Linear Microscopy

### A. Non-Linear Light-Matter Interaction

The interaction of light with matter can be described via the induced polarization, i.e., the induced dipole moment per unit volume. The ultrafast laser radiation that is used in laser scanning microscopes has a high enough intensity to induce non-linear polarization in the material. For describing non-linear interactions, the polarization vector **P** can be expanded in a power series (Boyd, 1992):

$$\mathbf{P} = \chi^{(1)}\boldsymbol{E} + \chi^{(2)}\boldsymbol{EE} + \chi^{(3)}\boldsymbol{EEE}... \tag{1}$$

where the first term describes the linear response of material to the optical electric field **E**. The induced linear polarization is proportional to the linear susceptibility tensor $\chi^{(1)}$, and represents the effects such as linear absorption, reflection and refraction of light. The second term is proportional to the second order non-linear susceptibility $\chi^{(2)}$ which is a third rank tensor. The second term depends on two optical electric field vectors that can have different orientations and frequencies, or can be the same as in the case of second harmonic generation. The product operator implies the Einstein summation convention over tensor indices. The third term is proportional to the third order non-linear susceptibility $\chi^{(3)}$ and three electric field vectors. The third term represents various four-wave mixing processes such as third harmonic generation, coherent Raman scattering, two-photon absorption, and the optical Kerr effect.

When linear absorption is negligible, waves generated via non-linear interactions can be described by the inhomogeneous wave equation:

$$\nabla^2\mathbf{E} - \frac{n^2}{c^2}\frac{\partial^2\mathbf{E}}{\partial t^2} = \frac{4\pi}{c^2}\frac{\partial^2\mathbf{P}^{NL}}{\partial t^2} \tag{2}$$

where non-linear polarization $\mathbf{P}^{NL}$, (second, third and further terms in Eq. (1)) drives the electric field **E**, leading to the acceleration of charges and the generation of electromagnetic radiation at the driving frequency of the non-linear polarization that is different from the fundamental frequency of the incident radiation; $n$ represents linear refractive index; and $c$ stands for the speed of light in the vacuum.

Non-linear responses, which are most widely used in non-linear microscopy, can be represented with the energy level diagrams in Fig. 1. From the presented optical responses only SHG belongs to the second order, while two-photon excitation fluorescence (TPF) and THG are third order non-linear effects. The solid lines in the figure represent electronic states of molecules while dashed lines show virtual levels that are the superposition of one or more photon radiation fields and an eigenstate of the molecule. Transitions to electronic states are associated with the absorption of one or multiple photons, while virtual states do not result in photon absorption. The transition energy gained by a virtual state is coherently emitted leaving no energy deposition in the system. This implies that higher excitation light intensities can be used for SHG and THG compared to MPF which is more prone to photobleaching and thermal damage effects.

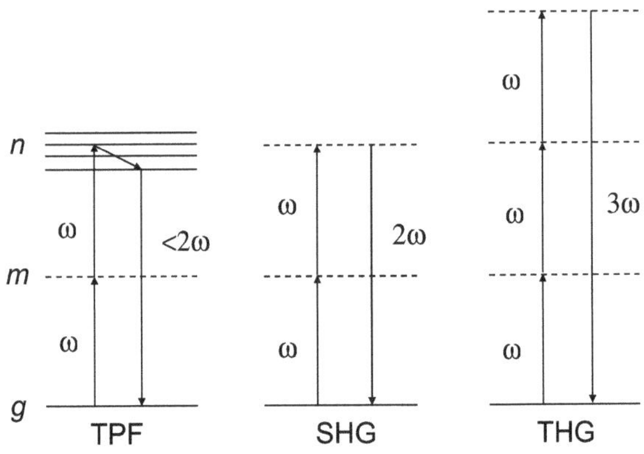


*Fig. 1.* Energy level diagrams for two-photon absorption (TPF), second harmonic generation (SHG), and third harmonic generation (THG) processes. The solid and dashed horizontal lines represent real and virtual energy levels, respectively. Photons with frequency ω are absorbed and photons with < 2ω, 2ω, and 3ω are emitted.

From Fig. 1 it is also apparent that the same laser frequency can induce multiple non-linear effects. While the two-photon absorption frequency is determined by the electronic transitions of molecules, the virtual levels can shift depending on the radiation frequency of the excitation. Therefore, the same laser pulse that is used for fluorescence excitation can induce SHG and THG.

Coherent non-linear effects can readily be enhanced when the excitation frequency appears near the resonance transition of the molecule. Resonance enhancement can be observed in SHG and THG, and is often used in Raman spectroscopy. With resonance enhancement stronger coherent signals can be generated, however, resonance excitation also leads to absorption that brings about photobleaching and thermal damage effects. Therefore, a tradeoff between excitation wavelength and intensity has to be achieved for the optimal imaging conditions in multimodal microscopy. It is very important to know the non-linear absorption spectrum of a sample in order to choose the optimal excitation wavelength.

In the following sections specific imaging conditions will be described for each of the three non-linear contrast mechanisms.

### B. Second Harmonic Generation

SHG is a degenerate case of sum-frequency generation where two interacting electric fields have the same frequency. SHG was first generated in crystalline quartz (Franken et al., 1961). SHG is generated in media with non-central symmetry, i.e., the second order non-linear susceptibility vanishes in the system with inversion symmetry, for example, in random media or cubic lattices (Boyd, 1992). Therefore a non-zero $\chi^{(2)}$ is determined by asymmetric molecules or asymmetric microcrystalline structures. Different strategies have been developed to maximize $\chi^{(2)}$ via optimizing asymmetry of the molecules (Marder et al., 1991) or improving the alignment of non-centrosymmetric molecules in a macroscopic structure (Verbiest et al., 1998). A special case appears when chiral molecules of one enantiomer are suspended in an isotropic solution. The solution exhibits an optical activity and has a non-vanishing $\chi^{(2)}$. The sum and difference frequency generation can be observed in such media, however, second harmonic generation appears to be symmetry forbidden (Giordmaine, 1965). Therefore, SHG is not generated in optically active liquids, for example, sugar solution.

Inversion symmetry can be broken at an interface between two materials (Bloembergen and Pershan, 1962). This provides a powerful tool for studying second order hyperpolarizabilities of molecules adsorbed onto surfaces (Shen, 1989). A sufficient strength of SHG can be generated from a monolayer of molecules at the interface. In macroscopic experiments of harmonic generation at the boundary between two isotropic media, the harmonic signals can be easily interpreted as being generated only from the interface (Shen, 1989).

In addition to the symmetry and structural constraints, phase matching conditions have to be met for efficient SHG to occur in bulk media. The propagation speed of light depends on the frequency dependent refractive index of the material. This creates a phase shift, i.e., a walk-off between the fundamental light and the generated second harmonic (Boyd, 1992). When the out of phase walk-off is reached, SHG diminishes. That occurs periodically at certain propagation distances of the waves into non-linear material (Armstrong et al., 1962). The phase matching conditions are usually achieved by employing uniaxial non-linear crystals that have different refractive indices along the ordinary and extraordinary crystal axes. At a certain orientation of the crystal, both waves propagate at the same speed leading to the efficient conversion of the non-linear signal. For SHG crystals, Type I or Type II phase matching conditions can be observed, where two incident beams have the same or orthogonal polarizations, respectively (Midwinter and Warner, 1965). It is also possible to achieve quasi-phase-matching conditions when alternating layers of material are manufactured with certain periodicities to match the phase walk-off between the fundamental and the generated waves (Byer, 1997). The alternating layers have a periodic inversion of the optical axes each time the phase shift reaches $\pm\pi/2$. The quasi-phase-matching is often used when crystal birefringence is insufficient to compensate for the occurring phase mismatch.

Due to the wide angle cone of light produced with a high NA microscope objective, phase matching conditions are usually satisfied for at least some of the incoming rays. However, the efficiency of SHG generation depends strongly on the structural arrangement of the sample. The presence of multilayer structures, macrochirality and proper orientation of structural axes with respect to the direction of the laser beam propagation largely determines the efficiency of microscopic SHG (Chu et al., 2004; Kolthammer et al., 2005).

SHG was the first non-linear optical response used in microscopy (Hellwart and Christen, 1974). Wide field excitation was employed in the first SHG microscopes where a large area of the specimen was illuminated by a laser, and the generated second harmonic was observed with an optical microscope. The scanning mode for SHG microscopes was implemented by Gannaway and Sheppard (1978). The first biological application of SHG microscopy was demonstrated by imaging a rat tail tendon (Freund and Deutsch, 1986). It was not until the nineties that femtosecond mode-locked lasers were developed providing stable sources of high intensity excitation required for the efficient generation of non-linear responses. SHG microscopy has been applied for a variety of biological investigations. SHG is very efficiently generated in collagen (Freund et al., 1986; Campagnola et al., 2002; Lin et al., 2005), actin-myosin complexes of myocytes (Campagnola et al., 2002; Chu et al., 2004; Barzda et al., 2005), tubulin (Campagnola et al., 2002), plant starch granules and other formations of polysaccharides (Mizutani et al., 2000; Chu et al., 2001; Cox et al., 2005), chloroplasts (Chu et al., 2001; Prent et al., 2005a), and aggregates of light harvesting chlorophyll *a/b* pigment-protein complexes of Photosystem II (LHCII) (Prent et al., 2005b).

The contrast mechanism of SHG microscopy is based on non-centrosymmetric organization in the sample. Microscopic SHG can be generated due to a broken symmetry at an interface or due to a non-centrosymmetric arrangement of bulk structures. Biological membranes are good candidates for SHG generation at an interface. The two leaflets of a lipid bilayer have to be asymmetrically arranged to produce SHG (Moreaux et al., 2000b, 2001). Often this appears to be the case in biological membranes where cellular membranes have different lipid and protein content in the two sides. If membranes are stained with non-linear dye that accumulates only in one leaflet of the membrane, intense SHG can be observed. When two such asymmetric membranes appear adhered to each other to construct a symmetric arrangement, the SHG vanishes (Moreaux et al., 2000b). On the other hand, a symmetric distribution of chiral molecules in the membrane will not give rise to SHG. SHG will be enhanced if an ordered asymmetric distribution of chiral molecules is present in the membranes (Campagnola et al., 1999; Florsheimer, 1999).

Sometimes bulk generation of SHG in biological specimens appears from the microcrystalline structures, for example, in calcite or starch granules (Mizutani et al., 2000; Baconnier and Lang, 2004). The SHG from those structures is generated in a similar way as in non-linear crystals. The main difference between macroscopic and microscopic measurements of SHG is that in microscopic experiments the fundamental light is tightly focused with a high NA objective providing a cone of light with a wide range for incident angles satisfying the phase matching conditions for some of the rays. Correspondingly, phase matching conditions in non-linear crystals are usually obtained by angle or temperature tuning (Boyd, 1992).

Regularly organized biological macrostructures that exhibit optical activity also efficiently generate SHG. The efficient generation of bulk SHG has been observed in myocytes (Campagnola et al., 2002; Chu et al., 2004; Barzda et al., 2005), collagen (Freund and Deutsch, 1986), chloroplasts (Chu et al., 2001; Prent et al., 2005b), and LHCII (Prent et al., 2005b). Highly organized birefringent biological structures can provide phase matching conditions for effective bulk second harmonic generation. By measuring SHG at different sample orientations, the tensor elements of $\chi^{(2)}$ can be deduced describing the structural organization of molecules in a 3D array (Chu et al., 2004). In most cases bulk generated SHG signals appear to be stronger than surface SHG, due to larger amounts of coherently phased molecules present in a focal volume compared to the excited molecules at an interface. The angular distribution of SHG with high NA objectives was modeled by Moreaux and coworkers showing that SHG intensities are distributed in two lobs at a small angle with respect to the optical axis providing evidence that at least the same NA objective is needed for efficient collection of SHG in the forward direction (Moreaux et al., 2000a, 2001).

### C. Third Harmonic Generation

Third harmonic generation (THG) is a third order non-linear process, and therefore it is dipole allowed. The THG can be generated in homogeneous media (Boyd, 1992). THG was initially observed in a calcite crystal (Terhune et al., 1962). During early stages of research on harmonic generation in gases, it was shown that third harmonic generation vanishes under tight focusing conditions although THG has no symmetry restrictions in contrast to SHG (Ward and New, 1969). THG is not observed when the laser beam is focused into a homogeneous material with normal dispersion.

For the focused beams, THG can be described by the paraxial wave equation that is derived from the inhomogeneous wave equation (Eq. 2). In the paraxial wave equation, the assumption is made of a slow spatial variation of electric field amplitudes along the beam propagation direction (z direction). The solution of the paraxial wave equation for the amplitude of third harmonic ($A_{3\omega}$) can be written as follows (Boyd, 1992):

$$A_{3\omega}(z) = \frac{i6\pi\omega}{nc}\chi^{(3)}A_{\omega}^{3}J_{3\omega}$$

$$\text{where } J_{3\omega}(\Delta k, z_0, z) = \int_{z_0}^{z} \frac{e^{i\Delta k z'}}{\left(1 + 2iz'/b\right)^2} dz' \qquad (3)$$

It can be seen that third harmonic amplitude depends on the third power of the amplitude of the fundamental electric field $A_\omega$, and on $\chi^{(3)}$. The phase-matching integral $J_{3\omega}$ can be numerically integrated for any arbitrary structural configuration with $z_0$ being the z value at the entrance of the non-linear medium. The $b$ represents a confocal parameter, and $\Delta k$ is the phase matching criterion expressed as: $\Delta k = 3k_\omega - k_{3\omega}$. For the tightly focused Gaussian beams inside the homogeneous non-linear medium, the limits of integration can be replaced by the $-\infty$ to $\infty$, and the phase-matching integral $J_{3\omega}$ can be analytically integrated leading to the expression:

$$J_{3\omega}(\Delta k, z_0, z) = \begin{cases} 0, & \Delta k \leq 0 \\ \frac{1}{2}\pi b^2 \Delta k e^{-b\Delta k/2}, & \Delta k > 0 \end{cases} \qquad (4)$$

It can be seen that the integral equals zero for the normally dispersive materials, and even for the perfect matching conditions ($\Delta k = 0$). This illustrates that THG is not observed in the far field when the beam is focused into a normal dispersion material. These results can be understood in terms of the phase anomaly or Gouy shift that is experienced by the beam when passing through the focus. The sign reversal of the phase at the focus determines the destructive interference of THG beams originating from the opposite sides of the focal point leading to the annihilation of the signal in the far-field (Ward and New, 1969; Boyd, 1992).

THG can be observed if the focal symmetry is broken by introducing an interface of the two materials with different refractive indexes or third order non-linear susceptibilities. Although THG appears at the interfaces, it is primarily a volume effect since THG is generated in the bulk of the media on both sides of the interface (Saeta and Miller, 2001). Efficient THG generation was initially observed at a dielectric surface (Tsang, 1995). Multilayer structures can enhance the far field THG intensity provided the right periodicity matching for the excitation wavelength is achieved (Tsang, 1995; Kolthammer et al., 2005). The sensitivity of THG to the presence of interfaces has been employed in THG microscopy that was introduced by Silberberg and coworkers (Barad et al., 1997). Soon after the introduction of THG microscopy, it was shown that THG is sensitive to the orientation of the interface with respect to the principal direction of propagation of the laser beam (Müller et al., 1998a). Third harmonic generation from different structural arrangements has been modeled by conducting the numerical integration of the phase-matching integral (Eq. 3) (Naumov et al., 2001; Schins et al., 2002). The THG microscopy was recently reviewed in (Squier and Müller, 2001; Sun, 2005).

For biological applications, THG can be used to visualize biological membranes (Müller et al., 1998a), cell walls (Squier et al., 1998) and multilayer structures such as grana of chloroplasts (Müller et al., 1998a; Millard et al., 1999; Chu et al., 2001; Prent et al., 2005a), aggregates of LHCII (Prent et al., 2005a), and crista in mitochondria (Barzda et al., 2005). THG was also observed in rhizoids from green algae (Squier et al., 1998), erythrocytes (Millard et al., 1999), cultured neurons and yeast cells (Yelin and Silberberg, 1999), glial human cells (Barille et al., 2001), muscle cells (Chu et al., 2004; Barzda et al., 2005), *Drosophila* embryos (Supatto et al., 2005), and sea urchin larval spicules (Oron et al., 2003). Given that THG efficiency depends on the third order non-linear susceptibility of the material, biological samples containing $\chi^{(3)}$ molecules, such as carotenoids, are one of the best candidates for imaging with THG. Strong THG signals observed in LHCII and chloroplasts are largely determined by the $\chi^{(3)}$ of carotenoids and chlorophylls.

### D. Multiphoton Excitation Fluorescence

The two-photon excitation process was first theoretically described by Göppert-Mayer (1931) and experimentally demonstrated in the optical range thirty years later by Kaiser and Garrett (1961). Three-photon absorption was demonstrated a few years later by Singh and Bradley (1964). Two-photon absorption is a third order non-linear effect, whereas three-photon absorption is a fifth order non-linear effect. The energy diagram of two-photon excitation fluorescence is presented in Fig. 1. Whether a sample is excited via two- or three-photon excitation can be determined by a quadratic or cubic dependence of absorption or MPF on the excitation intensity. Non-linear excitation of molecules is followed by the emission of fluorescence and excitation relaxation via non-radiative pathways. The excitation relaxation dynamics proceeds in a similar way as if the excited state was populated via a linear excitation process. If the non-linear excitation process is not clearly determined, then the detected fluorescence signal is commonly denoted as multiphoton-excited fluorescence (MPF).

Multiphoton absorption occurs when the energy of a molecular transition from the ground to an excited electronic state matches the combined energy of two or three photons. Quantum mechanically, in the case of two-photon excitation, the absorption probability is proportional to the two-photon transition moment $M_{ng}$ from the ground state $g$ to the excited state $n$ via intermediate state $m$ (Fig. 1), and can be expressed as follows (Boyd, 1992; Abe, 2001):

$$M_{ng} = \sum_{m(\neq n, \neq g)} \left[ \frac{2(\boldsymbol{\mu}_{nm} \cdot \mathbf{E})(\boldsymbol{\mu}_{mg} \cdot \mathbf{E})}{\hbar(\omega - \omega_{mg})} \right] - \frac{2(\Delta \mathbf{p}_{ng} \cdot \mathbf{E})(\boldsymbol{\mu}_{ng} \cdot \mathbf{E})}{\hbar \omega} \tag{5}$$

where $\boldsymbol{\mu}_{nm}$, $\boldsymbol{\mu}_{mg}$, and $\boldsymbol{\mu}_{ng}$ are the transition dipole moments between the $n$ and $m$, $m$ and $g$, and $n$ and $g$ states, respectively. The term $\Delta \mathbf{p}_{ng} = \mathbf{p}_{nn} - \mathbf{p}_{gg}$ is the change in the static dipole moment between the final $n$ and initial $g$ states. $\omega_{mg}$ is the energy difference between the states $m$ and $g$; and $\omega$ is the frequency of the excitation laser. Equation (5) has two distinct terms: (i) the first term describes the two-photon excitation process via one photon excitation to a virtual state $m$ and subsequent excitation to the final electronic state $n$; (ii) the second term expresses direct two-photon excitation to the final state via the change in the static dipole moment between the $n$ and $g$ states. According to the first process, the ground and the final excited states have the same symmetry, therefore the excitation is one-photon forbidden and the excited state can be populated only via two-photon absorption. This situation usually appears for non-polar molecules. For polar molecules, two-photon absorption can proceed via the second mechanism, if a large change in the dipole moment occurs during excitation. In this case, the two-photon transition moment is proportional to the one-photon transition dipole, thus two-photon absorption bands will be similar to the linear absorption. In conclusion, two-photon absorption spectrum of molecules can have similar shaped bands as one-photon absorption. Moreover, new bands may appear and some bands might be missing in the two-photon absorption compared to the linear absorption spectrum.

A growing number of non-linear spectroscopy studies in photosynthetic samples show that carotenoids (Car) dominate the multiphoton absorption spectrum (for a review, see Polivka and Sundström, 2004) in the range where linear absorption is at a minimum, while chlorophylls (Chl) do not show new two-photon absorption bands compared with linear absorption bands (Walla et al., 2000, and references therein). The measured two-photon absorption bands of Car and Chl in solution and in pigment-protein complexes are summarized in Table 1. Most of the two-photon experiments were targeted to investigate one-photon forbidden and two-photon allowed optical transition of Car from the ground state $1^1A_g^-$ to the first singlet excited state $2^1A_g^-$. (The nomenclature describing the electronic structure of Car is used with analogy to the $C_{h2}$ symmetry structure of all-*trans* polyenes.) The absorption of this transition falls around 1000–1400 nm (double energy at 500–700 nm) for plant/green algae, and around 1100–1500 nm (double energy at 550–750 nm) for bacterial systems (see Table 1). Cars have a very low fluorescence yield; therefore the fluorescence observed in photosynthetic samples originates mostly from carotenoid sensitized chlorophyll or bacteriochlorophyll fluorescence. The excitation energy transfer between Cars and Chls plays a major role in determining the fluorescence yield in photosynthetic samples after non-linear excitation. The energy from Cars to Chls is transferred on a subpicosecond to a few picoseconds time scale (for recent review, see Polivka and Sundström, 2004). The energy transfer and thermal equilibration within antenna Chls is also achieved on a scale of few picoseconds rendering virtually the same Chl emission spectrum as after the one-photon excitation (Barzda et al., 2001a; Walla et al., 2002).

Two-photon excitation fluorescence is the most frequently used non-linear response in today's microscopic investigations. The popularity of TPF is mostly determined by the commercial availability of two-photon excitation microscopes and due to a wide use of its predecessor, the confocal fluorescence microscope. The multiphoton excitation fluorescence contrast mechanism was first implemented by Denk et al. (1990). Inherent sectioning capability with minimal out of focus photobleaching was demonstrated, revealing advantages of MPF compared to the confocal microscopy (Denk et al., 1990). In addition, the

*Table 1.* Measured two-photon absorption ranges for different photosynthetic samples

| Sample | Absorption range nm | Measurement method | References |
|---|---|---|---|
| β-carotene in octane | 1240–1500 (620–750)[1] | TPE[2] pump-probe | (Walla et al., 2002) |
| β-carotene in chloroform | 590–700 (295–350)[1] | z-scan | (Major et al., 2005) |
| Lutein in octanole | 1160–1480 (580–740)[1] | TPE[2] pump-probe | (Walla et al., 2002) |
| Thylakoids, *Phaeodactylum tricornutum* | 980–1360 (490–680)[1] | Sensitized Chl *a* fluorescence | (Shreve et al., 1990) |
| LHCII, *Chlamydomonas reinhardtii*, *Spinacia ol.* | 1000–1320 (500–660)[1] | Sensitized Chl *a* fluorescence | (Walla et al., 2000, 2002) |
| PS I, *Thermosynechococcus elongatus* | 1100–1420 (550–710)[1] | Sensitized Chl *a* fluorescence | (Hilbert et al., 2004) |
| BChl *a* in acetone | 1300–1490 (650–745)[1] | BChl *a* fluorescence | (Krikunova et al., 2002) |
| LH2, *Rhodobacter sphaeroides* | 1100–1400 (550–700)[1] | Sensitized BChl fluorescence | (Krueger et al., 1999) |
| LH2, *Chromatium minutissimum* | 1300–1490 (650–745)[1] | Sensitized BChl fluorescence | (Krikunova et al., 2002) |
| Peridinin in benzene | 920–1200 (460–600)[1] | Peridinin fluorescence | (Zimmermann et al., 2002) |
| PCP, *Amphidinium carterae* | 920–1310 (460–655)[1] | Sensitized BChl fluorescence | (Zimmermann et al., 2002; Shima et al., 2003) |

[1] The wavelengths in parentheses are half the laser wavelength used. [2] TPE: two-photon excitation.

use of infrared lasers provides deeper penetration of the excitation beam into biological specimens due to very low linear absorption and reduced scattering. MPF microscopy has been extensively described in several reviews (for examples, see König, 2000; So et al., 2000; Diaspro and Sheppard, 2002; Zipfel et al., 2003). Imaging with MPF microscopy in plant science was extensively reviewed by Tirlapur and König (2002) and by Feijó and Moreno (2004).

The microscopic contrast mechanism of MPF is based on the non-linear excitation of molecules and the subsequent emission of a photon either by the same or by a sensitized neighboring molecule. In random media, fluorescence is emitted isotropically, therefore forward and backward collection of fluorescence can be implemented. Most often, epi-fluorescence (backward) collection is employed in MPF microscopes, mainly because of an easier optical setup where the same objective can be used for excitation and fluorescence collection.

Fluorescence is not a coherent optical response in contrast to SHG or THG; therefore phase matching and interference effects are not observed in MPF. This simplifies the interpretation of microscopic images where the fluorescence can be directly related to the presence of fluorophores in the sample. However, the direct assignment of fluorescence intensity to the concentration of fluorophores can not be applied. In photosynthetic samples, the Car to Chl energy transfer efficiency varies for different pigment-protein complexes (Polivka and Sundström, 2004). Fluorescence intensity depends on the fluorescence lifetime, which is different for different pigment-protein complexes. Furthermore, the Chl fluorescence lifetime strongly depends on the quenching properties of light-harvesting antenna, which in turn depend on the efficiency of the quenchers, excitation energy migration in the antenna, as well as on the size of connectivity domain (Barzda et al., 2001b). At high excitation intensities that are met in non-linear microscopes, significant singlet-singlet annihilation shortens Chl fluorescence lifetimes. This was observed in LHCII aggregates and chloroplasts (Barzda et al., 2001a; Cisek et al., 2005). In conclusion, some fluorescing domains observed with MPF microscopy might be quenched while others might be highly fluorescent; however, the fluorophore concentration in the focal volume might be very similar. Better understanding of fluorescence properties can be achieved by simultaneous measurement of fluorescence intensities and decays with fluorescence lifetime imaging microscopy (FLIM) (Barzda et al., 2001a). Fluorescence lifetime imaging techniques are extensively reviewed by Becker (2005).

## *E. Multimodal Microscopy*

Ultrafast laser pulses can generate simultaneously several non-linear optical responses. Therefore, different non-linear contrast mechanisms can be used to record parallel images of the same structure. The multimodal microscopes, which utilize simultaneously SHG, THG, and MPF responses, have been recently developed (Chu et al., 2001; Barzda et al., 2005; Prent et al., 2005a). The multimodal microscopy appears to be very beneficial when different optical responses reveal different functional structures of the same biological object. A multimodal SHG and MPF microscope was used to image labeled lipid vesicles (Moreaux et al., 2000a), labeled neuroblastoma cells (Campagnola et al., 1999), muscle and tubulin structures (Campagnola et al., 2002), and labeled neurons (Moreaux et al., 2001). Simultaneous THG and MPF detection was used for imaging of glial human cells (Barille et al., 2001), while THG and SHG microscopy was used to monitor mitosis in a live zebrafish embryo (Chu et al., 2003b). All three contrast mechanisms were implemented to image mitochondria in cardiomyocytes (Barzda et al., 2005), chloroplasts (Chu et al., 2001; Prent et al., 2005a) and LHCII (Prent et al., 2005b).

Parallel images, recorded using different contrast mechanisms, can be directly compared on a pixel by pixel basis. Although, SHG, THG and MPF images originate from the same structure, their image contrast mechanisms are fundamentally different. The comparison of images obtained with coherent and non-coherent contrast mechanisms can be very challenging, because homogeneous structures cannot be visualized in SHG or THG, but might be visible in the fluorescence. Additionally, the differences appear for the signal generation at the structural interfaces. For the SHG and THG signals, the maximum appears at the central position of the interface; whereas for the interface between fluorescing bulk and non-fluorescing structure, only half of the onset intensity will be reached at the interface position. The maximum MPF signal intensity is observed when the full focal volume is immersed in the media. Therefore, image comparisons will always have to be taken with caution.

It is always beneficial to deconvolute the images with the point spread function of particular contrast mechanism. The point spread functions are different for the different non-linear responses. Non-deconvoluted images appear to be blurred due to the finite size of the focal volume that is scanned across the structure. If deconvolution is not performed, comparison may lead to artifacts, where two neighboring structures may be revealed by different contrast mechanisms and may appear to overlap because of the effect of blurring.

For comparing two images obtained simultaneously with different non-linear contrast mechanisms, a method of structural cross-correlation image analysis (SCIA) has been developed (Greenhalgh et al., 2005). To obtain standardized images, segmentation of each optical slice is performed by thresholding and normalizing to the maximum pixel intensity of the image. Thresholding the noise level is somewhat arbitrary; however, for a reasonable estimate an image area unoccupied by the objects can be used. Zero intensity is assigned to the pixels that appear below the threshold values. Care must be taken to analyze the intensity distribution of each image in order to avoid normalization to a high intensity "glitch" that can artificially appear during image recording. Comparison of the standardized images is performed on a pixel by pixel basis. The correlated intensity for each pixel $C(x,y)$ at the image position $x$ and $y$ is obtained from the corresponding pixel intensities $A(x,y)$, $B(x,y)$ of standardized images as follows:

$$C(x,y) = \sqrt{A(x,y) \cdot B(x,y)} \tag{6}$$

where $A(x,y) = \frac{a(x,y)}{a_{\max}}$ and $B(x,y) = \frac{b(x,y)}{b_{\max}}$

Here, $a$ and $b$ denote the pixel intensities of the initial data and $a_{max}$ and $b_{max}$ are the maximum pixel intensities of the corresponding images. During standardization, the thresholded pixels are assigned zero (black) values, therefore, when a pixel with non-zero intensity appears only in one of the images, zero intensity shows up in the correlated image. Two uncorrelated images can be constructed directly from the standardized images where corresponding intensity values of $A(x,y)$ and $B(x,y)$ are assigned to the pixels in the respective images excluding the correlated pixels. The correlated and two uncorrelated images can be combined into one co-localized image by assigning different colors to the pixels from corresponding images. If three dimensional structures are initially recorded with the multimodal microscope, the co-localized 2D images can be reconstructed into 3D images for further rendering and image visualization. Examples of SCIA images of LHCII aggregates and chloroplasts will be discussed in section IV.

## III. Instrumentation of Non-Linear Multimodal Microscope

Non-linear microscopes share many common features with widely used confocal microscopes. Many research groups implemented non-linear excitation fluorescence by coupling femtosecond lasers into a confocal microscope and using the non-descanned port for efficient signal detection (see discussion by Zipfel et al., 2003). The harmonic signals are usually detected in the transmission mode; therefore, larger modifications of a confocal microscope are required for the setup implementation. The easiest way of building a harmonic microscope is by using a high NA condenser and transmission detector existing on some models of laser scanning confocal microscopes (Millard et al., 2003; Moreno et al., 2004). Three-channel detection requires extensive modification of commercial microscopes (Sun, 2005). The instrumentation of non-linear microscopes was extensively reviewed by Squier and Müller (2001) and Sun (2005).

This chapter presents a three-channel multimodal non-linear microscope built in our own laboratory. The outline of the microscope is presented in Fig. 2. The laser is coupled to the microscope via two mirrors schematically shown by M1. Inside the microscope, the beam is expanded with a telescope (lenses L1 and L2) to fill the clearance aperture of the scanning mirrors (SM). The same telescope is also used for shaping the beam with a spatial filter (SF). The expanded beam is coupled to two galvanometric mirrors (SM) that can raster the beam in both the lateral

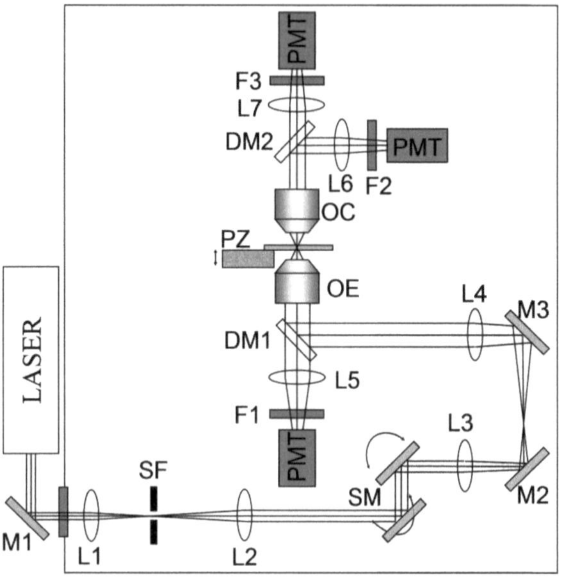


*Fig. 2.* Principal setup of a multimodal non-linear microscope. See the text for a detailed description. The letters indicate: M1-3 mirrors, L1-7 lenses, SF spatial filter, SM scanning mirrors, DM1-2 dichroic mirrors, F1-3 optical filters, OE excitation objective, OC collection objective or condenser, PMT detectors, photomultiplier tubes that are usually used for detection, PZ piezo stage.

$x$ and $y$ directions. Then, the beam is passed through a second telescope consisting of an achromatic lens (L3) and a tube lens (L4). The second telescope is designed to expand the beam to match the entrance aperture of the excitation objective (OE). The tube lens is designed to correct for chromatic aberrations. It is advisable to use the tube lenses that match the objective and are produced by the same manufacturer. After the second telescope, the collimated beam is reflected by the dichroic mirror (DM1) and coupled into the objective. The excitation in non-linear microscopes is confined to the focal volume of the objective; therefore, it is not required to implement the confocal detection. The multiphoton fluorescence is collected in the backward direction with the same microscope objective OE, transmitted through the dichroic mirror (DM1), and focused with the lens (L5) through the filter (F1) onto the detector (PMT). Interference or band pass filters are used in front of the detector for filtering scattered fundamental and harmonic light. The harmonic signals are detected in the forward direction with the high NA objective (OC). The collimated beam is passed through the dichroic mirror (DM2) for separation of the second and third harmonic signals. The second harmonic is focused onto the detector (PMT) by the lens (L7), and filtered by the interference filter (F3). Similarly, the third harmonic is focused onto the detector (PMT) with the lens (L6), and filtered with the interference filter (F2). A photon counting data acquisition card is used to receive the pulses simultaneously from all three detectors. The photon counting is synchronized with the reading of $x$ and $y$ positions from the scanning mirrors. This renders the three simultaneously acquired images with different non-linear contrast mechanisms. For obtaining optical sections at different depth, the sample is translated along the optical axis with the piezo stage (PZ). The setup of non-linear microscope can be divided into three main functional units: the femtosecond laser source, the optical microscope, and the detection system synchronized with the laser beam scanning. In the following, we will give more detailed information about each of these functional units.

### A. Laser Sources

Due to the high intensity excitation requirements for non-linear effects to occur, picosecond and femtosecond lasers have been employed. The ultrafast lasers provide a high flux density during the pulse duration, yet have a low average power that is tolerated by the biological specimen. The non-linear responses have either a quadratic or a cubic dependence on the excitation intensity. Thus, pulses as short as possible are desired for the most efficient non-linear excitation. However, there is a limited pulse peak power that can be applied to a biological specimen. Above this limit dielectric breakdown takes place manifesting itself in the occurrence of cavitations and visible structural changes (Müller et al., 1998a). Therefore, a balance between the highest peak power and the average laser power has to be achieved. In practical terms, pulses of a few hundred femtoseconds work well. Very short pulses (<20 fs) are getting significantly broadened due to a group velocity dispersion when passing through glass material of the optical components. Microscope objectives, in particular, contain a lot of glass leading to a considerable pulse broadening. Dispersion pre-compensation is required for obtaining very short optical pulses at the focal point of the objective (Müller et al., 1998b).

In addition to the pulse duration, the repetition rate of laser pulses provides another parameter for optimization of the non-linear excitation. A repetition rate of around 80 MHz is standard in most commercial Ti: Sapphire lasers. Two trends of decreasing or increasing the repetition rate have been suggested (Barzda et al., 2001a; Chu et al., 2003a). If non-linear excitation appears in the two or three photon absorption range, the duration between the pulses is desired to be kept long enough to permit full relaxation of the excited states. This reduces exciton-exciton annihilation effects and diminishes the generation of triplet states, which lead to the production of singlet oxygen (Barzda et al., 2001a). Singlet oxygen is known to facilitate bleaching of chromophores and oxidation of biomolecules. If the excitation falls outside of the linear and the two- or three-photon absorption bands, excitation to the virtual levels does not depopulate the ground state and does not deposit energy into the sample. Therefore, higher pulse repetition rates can be used. Since the peak intensities of the pulses have to be kept below the critical level, increasing repetition rate linearly increases the signal intensity (Chu et al., 2003a).

The choice of excitation wavelength is governed by several factors. The non-linear excitation should be tuned away from any linear absorption. Near-infrared excitation usually works well for imaging biological samples. The tuning window for excitation lies in the approximate range between 800 nm and 1500 nm. The blue side of the range is limited by the linear absorption of biomolecules and the red side is limited by the water absorption. Another technical limitation arises when THG signals are generated in the ultraviolet wavelength range. This appears when a Ti:Sapphire laser is used. In this case, UV transmitting objectives and signal collection optics must be introduced, and these are not readily available.

The polarization of the laser excitation is also playing a significant role in non-linear excitation. Circularly polarized excitation does not generate third harmonics (Fittinghoff et al., 2005). Therefore, elliptical polarization, that is a combination of linear and circular polarization, produces less THG than linearly polarized excitation.

### B. The Microscope

A non-linear excitation fluorescence microscope has many common features of a laser scanning microscope. Figure 2 depicts the most important parts of the non-linear microscope. For three dimensional imaging, the focal volume of the high NA objective has to be scanned across the 3D region of interest. It is possible to translate the sample with respect to the beam, or raster the beam while keeping the sample fixed. Sample translation is usually accomplished by a piezo stage, with scanning speeds of several frames per second. Faster scanning rates can be reached by rastering the beam in the lateral direction. Galvanometric mirrors are most commonly used for the lateral scanning. If video rate scanning speeds are required, resonance scanners, or rotating polygon mirrors can be used. Axial scanning is most commonly performed by translating the sample, or by translating the excitation objective. For epi-detection, translation of the objective gives satisfactory results. If transmission detection is implemented, the focal points of the excitation and collection objectives have to overlap. Therefore, axial scanning by translating the sample is mandatory in this case.

Almost all non-linear microscopes are constructed using commercially available refractive objectives. Most objectives are designed for the visible spectral range. Only a few objectives work optimally in the infrared region. Users are advised to contact microscope manufacturers for specifications on the use of their objectives with infrared femtosecond lasers.

Another important issue is the uniformity of the excitation beam across the entrance aperture of the objective. Overfilling the entrance aperture often helps to achieve good uniformity and the specified NA of the objective. It is recommended to test the alignment of the microscope on a regular basis by recording the point spread function (PSF). The PSF can be recorded by translating the sample in the axial direction. The easiest way to record PSF is by using a solution of fluorescent dye and measuring how steep is the fluorescence rise at the dye solution and glass interface while translating the sample. For testing the PSF of SHG and THG, air glass interface can be used.

The excited fluorescence is usually collected in the backward direction. Since a confocal pinhole is not used in non-linear microscopes, the fluorescence can be collected without descanning. In addition to fluorescence, scattered second and third harmonic signals can be recorded in the epi-detection mode. However, the strongest harmonic signals are observed in the forward direction. The harmonic signals in forward direction are collected with an objective (Barzda et al., 2005), or a high NA condenser (Millard et al., 2003). Either, one detector with appropriate filters or several detectors detecting different signals separated by a dichroic mirror can be used. Alternatively, the non-linear response signal can be coupled to a monochromator, and a whole spectrum can be recorded at each pixel (Sun, 2005). In this case, the harmonic and fluorescence images are constructed by obtaining the pixel intensities from selected range of the spectrum. Non-linear signals as MPF, SHG and THG are simultaneously generated; therefore, if simultaneously recorded, the images can be directly compared and a statistical analysis of the images can be performed (Barzda et al., 2005).

### *C. Detection Systems*

Detection is a very important part of the microscope. Photomultiplier tubes, avalanche photodiodes, and charge-coupled device (CCD) cameras have been widely used for detection. CCD cameras can be easily interfaced with the microscope. If CCD cameras are not synchronized with the scanners, several scans are required to acquire one image. For imaging scattering samples, point or line rastering synchronized with the detection is used. Photomultipliers are usually employed for the non-descanned mode due to the large area of the photocathode. The descanned detection mode is beneficial when small area avalanche photodiodes are used, or light is coupled to a monochromator for spectral imaging. The descanned mode usually has higher losses compared to the non-descanned configuration.

Signal integration, lock-in, or photon counting methods can be employed for the detection of non-linear responses. Most microscope manufacturers use a signal integration approach. However, the non-linear responses, which originate from biological samples, typically are generated at a very low intensity level. In most cases, less than one photon per excitation pulse is detected. In this situation, photon counting detection becomes the method of choice. However, photon counting can be saturated at high excitation intensities. This has to be taken with caution, but usually does not present a big problem, because the excitation power can be reduced or higher scanning rates can be implemented.

## IV. Application of Non-Linear Microscopy in Photosynthesis

### *A. Non-Linear Microscopy of LHCII*

LHCII is the most abundant pigment-protein complex in photosynthetic membranes of chloroplasts. It is a good model system for revealing the origin of non-linear signals in photosynthetic structures. The aggregation state of LHCII can easily be controlled by a detergent concentration in the suspension (Barzda et al., 1994). This provides an opportunity to investigate the origin of non-linear signals generated from monomers, trimers, and aggregates of LHCII. LHCII can form disordered, or multilamellar chiral aggregates (Simidjiev et al., 1997). Multilamellar chiral aggregates mimic the ultrastructure of the grana. This makes it possible to investigate the influence of aggregate macro-organization on the non-linear signal generation efficiency.

Figure 3 presents images of LHCII recorded simultaneously with MPF, SHG, and THG contrast mechanisms. All three contrast mechanisms visualize an inhomogeneous structure of LHCII aggregates. The focused 800 nm laser pulses predominantly excite Chl pigments via two-photon excitation. Most probably, Cars are not excited with two-photon excitation at 800 nm. The two-photon excitation spectrum of β-carotene has no transition around 800 nm (Major et al., 2005). After excitation, fast interpigment energy transfer determines the fluorescence emission from the spectrally equilibrated states of LHCII (Barzda et al., 2001b). The MPF signals (Fig. 3a) conveniently reveal locations with the presence of LHCII. It is well documented that various degrees of quenching can be observed in LHCII aggregates (Horton et al., 2005). The fluorescence quantum yield of LHCII depends on the size of the exciton migration domain, exciton hopping times between the pigments, the dimensionality of the domain, the concentration of the quenchers in the domain, and exciton-exciton annihilation conditions (Barzda et al., 1999, 2001b). Simultaneous fluorescence intensity and lifetime measurements are necessary for detailed investigation of quenching properties in LHCII aggregates.

Strong SHG signals are obtained in most LHCII aggregates (Fig. 3b). It is interesting to note that disaggregated LHCII shows intense SHG signal as well (R. Cisek and V. Barzda, unpublished). LHCII exhibits strong CD, which is our evidence for a non-centrosymmetric arrangement of pigments in these complexes. If aggregated into chiral structures that are well ordered, LHCII exhibits psi-type CD and strong circularly polarized luminescence (CPL) signals that can be a few times stronger compared to LHCII trimers or monomers (Barzda et al., 1994; Gussakovsky et al., 2000). This non-centrosymmetric macro-organization of LHCII aggregates provides

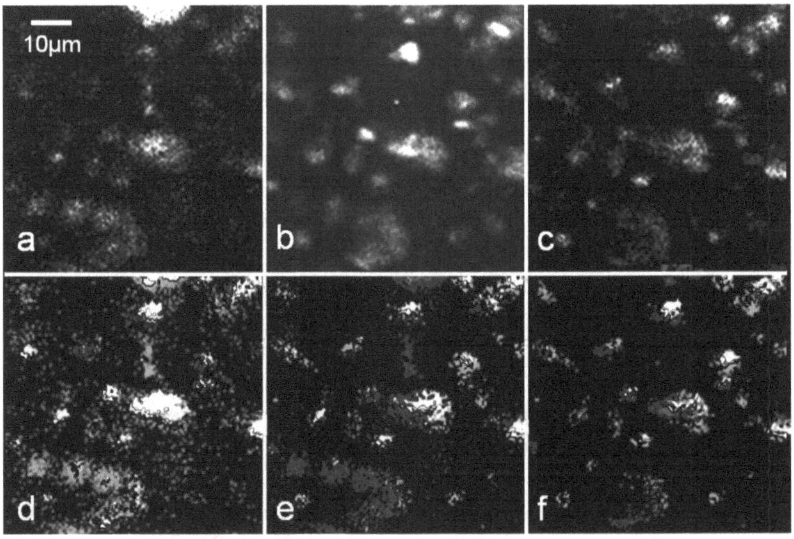


*Fig. 3.* Imaging isolated LHCII with a multimodal microscope. Panels a, b and c represent the MPF, SHG and THG images of LHCII aggregates, respectively. Panels d, e and f show colocalized images obtained by structural image cross-correlation analysis. Panel d compares MPF and SHG images (panels a and b), where correlated pixels are presented in white, uncorrelated fluorescence is presented in dark-gray and uncorrelated SHG is shown in light-gray. Panel e compares MPF and THG images (panels a and c). The white color represents correlation between MPF and THG while the dark-gray and light-gray shows uncorrelated MPF and THG signals, respectively. The panel f compares SHG and THG (panels b and c). The white color represents correlation between SHG and THG, while the dark-gray and light-gray represents uncorrelated SHG and THG, respectively. See also Fig. 2, Color Plate 1.

phasing between the pigment-protein complexes rendering a bulk enhancement of SHG. LHCII aggregates exhibit orientation dependent CD (Garab et al., 1988). Similarly, the SHG efficiency is also strongly dependent on the orientation of macroaggregates with respect to the propagation direction of the excitation beam (Chu et al., 2004). Therefore, SHG is a very sensitive tool for measuring the macrostructural arrangement of aggregates in biological samples.

THG is also efficiently generated in LHCII aggregates (Fig. 3c). Disaggregated LHCII does not show an appreciable third harmonic signal (R. Cisek and V. Barzda, unpublished). This is consistent with the absence of THG in homogeneous media under tight focusing conditions (Boyd, 1992). When aggregated, LHCII forms interfaces that provide the breaking of focal symmetry resulting in observable THG. The presence of Cars, which have very strong third order non-linear susceptibility, enhances the THG signal in LHCII aggregates. Enhancement of THG also appears from multilayer arrangements of the aggregate (Tsang, 1995; Barzda et al., 2005). In addition, the orientation of the multilayer structure perpendicular to the optical axis of the beam renders the strongest THG signal, while orientation of the surface parallel to the optical axis does not generate any THG (Müller et al., 1998a).

A comparison of images obtained with the different non-linear contrast mechanisms can be made with SCIA. Correlation between fluorescence and SHG or THG images provides evidence that harmonics are generated from the pigment-protein complexes (Fig. 3d,e). The comparison of SHG and THG (Fig. 3f) shows a general trend that weaker SHG appears in the areas with stronger THG. The opposite is true for strong SHG signals. This behavior can be explained by the multilamellar structure of LHCII aggregates where THG is most intense when the multilayer structure is oriented perpendicular to the laser beam, while the largest SHG and CD, indicating the largest chirality, is observed in multilayer membranes oriented parallel to the optical axes. A complication in interpreting harmonic signals is added by the heterogeneity of the macrostructures ranging from highly ordered two- or three-dimensional crystals to disordered aggregates. More precise elucidation of the macrostructure is possible by obtaining images of an aggregate at different orientations and deducing the matrix elements of the non-linear susceptibility tensor (Chu et al., 2004).

### *B. Non-Linear Microscopy of in-situ Chloroplasts*

Plant cells and chloroplasts can be imaged in situ and in vivo with non-linear microscopy (Müller et al., 1998a; Chu et al., 2001; Tirlapur and König, 2001; Prent et al., 2005a). Non-linear excitation with femtosecond pulses provides optical sectioning and deep penetration of the infrared beam into the plant tissue. While MPF is emitted isotropically, SHG and THG are preferentially generated in the forward direction of the laser beam propagation. Therefore, simultaneous MPF, SHG, and THG signal detection is mostly performed on thin sections of plant tissue. The MPF is usually recorded in the epi-detection mode, while SHG and THG are detected in the transmission mode. When studying chloroplasts in plant tissue, it is very beneficial to record simultaneously the MPF, SHG and THG signals. The SHG and THG signals can be generated from various structures inside the plant tissue. The colocalization of chlorophyll fluorescence and harmonic signals helps to identify the harmonic signals originating from chloroplasts. Each non-linear signal can provide complementary information about the structure and functional state of the chloroplast. Figure 4 presents in situ chloroplasts of mesophyll cell simultaneously imaged with MPF (a), SHG (b), and THG (c) contrast mechanisms. The lower row of the images in Fig. 4 shows SCIA correlations between each of the two signal channels.

For the MPF image, the signal was collected between 630 and 720 nm wavelength representing Chl *a* and Chl *b* fluorescence. Two-photon excitation was achieved with an 800 nm Ti:Sapphire laser operating at a 27 MHz pulse repetition rate. Three chloroplasts inside a cell can be seen in the MPF image (Fig. 4a). For *Clivia miniata*, chloroplasts are normally found adjacent to the cell walls; therefore, they often assemble into "ring like" formations. The MPF signal has minor variations inside the individual chloroplasts. Fluorescence lifetime imaging renders a larger variation in fluorescence lifetimes and allows discrimination between quenched and unquenched areas of the chloroplast (Cisek et al., 2005). When multimodal detection is used, higher intensities are required for generating harmonic images. The increased excitation intensities shorten fluorescence lifetimes due to exciton-exciton annihilation effects and due to light induced quenching, which to a large extent equalize fluorescence intensities obtained from different grana inside the chloroplast (Barzda et al.,

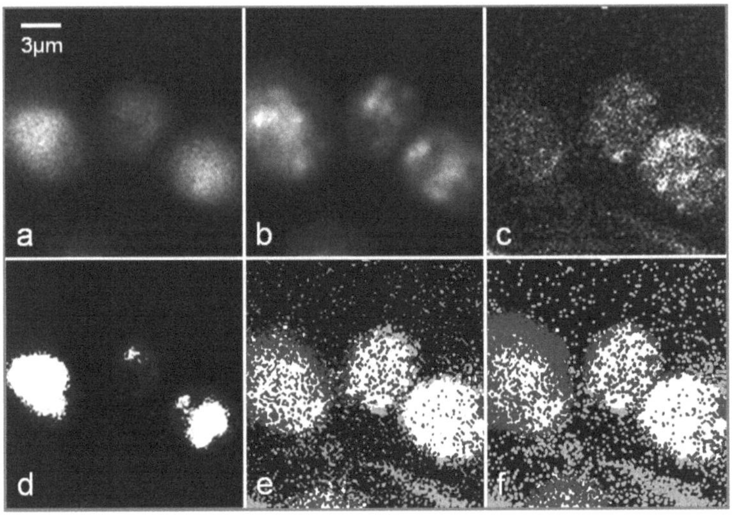


*Fig. 4.* Imaging in situ chloroplasts of *Clivia mineata* with a multimodal microscope. Panels a, b and c represent the MPF, SHG and THG images of chloroplasts, respectively. Panels d, eand f show colocalized images obtained by structural image cross-correlation analysis. Panel d compares MPF and SHG images (panels a and b), where correlated pixels are presented in white, uncorrelated fluorescence is presented in dark-gray and uncorrelated SHG is shown in light-gray. Panel e compares MPF and THG images (panels a and c). The white color represents correlation between MPF and THG while the dark-gray and light-gray shows uncorrelated MPF and THG signals, respectively. The panel f compares SHG and THG (panels b and c). The white color represents correlation between SHG and THG, while the dark-gray and light-gray represents uncorrelated SHG and THG, respectively. See also Fig. 3, Color Plate 1.

2001a; Cisek et al., 2005).

Plant cells generate relatively strong SHG signals (Fig 4b). The SHG originates mostly from cell walls (Cho et al., 2001), starch granules (Mizutani et al., 2000), and chloroplasts (Chu et al., 2001; Prent et al., 2005a). Comparison of MPF and SHG helps to distinguish SHG signal of chloroplasts (Fig. 4d). SHG microscopy presents us with a tool for measuring the macrochirality of biological samples. Similarly, macrochirality in the linear regime can be investigated with CD microscopy. Chloroplasts exhibit very strong psi-type CD signals originating from chiral macro-arrangement in thylakoid membranes (Finzi et al., 1989). The CD and CPL intensities depend on the orientation of chloroplasts in respect to the light propagation direction (Garab et al., 1988; Barzda et al., 2004). Face-aligned chloroplasts (with thylakoid membranes oriented perpendicular to the laser beam propagation) exhibit lower intensity CD signals, while edge aligned chloroplasts (with thylakoid membranes oriented parallel to the laser beam propagation) show a very strong psi-type CD (Finzi et al., 1989). Orientation dependent SHG is also observed in chloroplasts. Starch granules deposited inside chloroplasts can also produce SHG signal (Sun, 2005), however, dark adapted chloroplasts, lacking starch granules, show very strong SHG (Prent et al., 2005a). Therefore the origin of SHG in chloroplasts can be attributed predominantly to the chiral macro-arrangement of the pigment-protein complexes in thylakoid membranes, similarly as in lamellar LHCII aggregates.

THG signals can be generated from membranous structures of plant cells. The most intense THG is generated from cell walls and chloroplasts (Millard et al., 1999). Chloroplast THG is easily distinguishable, especially if it is correlated with the chlorophyll fluorescence (Fig. 4e). It has been shown that multilayer structures enhance THG as compared to corresponding monolayer interfaces (Tsang, 1995). Therefore, the multilamellar structures of the grana in chloroplasts as well as multilamellar aggregates of LHCII enhance the intensity of THG. Large hyperpolarizabilities of Car molecules also contribute to the strong THG signals in chloroplasts (Marder et al., 1997). The orientation of membranes perpendicular to the excitation beam propagation direction renders the largest THG signal, while a parallel orientation of membrane planes with respect to the beam will not generate THG. This is in opposite to the orientation dependent SHG signal, where the strongest SHG is observed in the direction of the largest chirality, which happens to be in the edge aligned chloroplasts, where thylakoid membrane planes are oriented parallel to the excitation beam. Therefore, a trend of strong THG and weak SHG can be observed when comparing both harmonic generation signals of chloroplasts. By recording non-linear signals of chloroplasts at different orientations a three dimensional organization can be obtained. This will allow deducing granum ultrastructure below the diffraction limited resolution of optical microscope.

## V. Perspectives

Non-linear microscopy is a relatively new biophysical technique. While two-photon excitation fluorescence microscopy has been applied for different investigations in plant science and photosynthesis research (Tirlapur and König, 2002; Feijó and Moreno, 2004), harmonic generation microscopy has been featured in only a few published studies (Müller et al., 1998a; Prent et al., 2005a; Sun, 2005), and the CARS contrast mechanism has not yet been used in photosynthesis research. Strong non-linear signals obtained from chloroplasts and aggregates of LHCII open the grounds for further spectroscopic and structural investigations of various photosynthetic structures.

Linear and circular polarization microscopy has been implemented for structural studies of single cellular organelles (Finzi et al., 1989; Steinbach et al., 2005). Harmonic generation microscopy that is sensitive to the structural macro-organization can be applied for similar studies. Due to strong signals, especially for SHG, structural dynamics can be investigated with millisecond time resolution. The possibility to investigate photosynthetic organelles in vivo and in situ constitutes a big attraction for studying physiologically important structural changes in photosynthetic organisms.

Non-linear spectroscopy is very beneficial for studying optically forbidden energy levels of photosynthetic pigments and the role of these levels in exciton relaxation dynamics in photosynthetic systems (Walla et al., 2002). Non-linear microscopes, which inherently utilize non-linear excitation, can be readily used for such non-linear spectroscopic studies.

Harmonic signals are strongly dependent on the excitation wavelength. Distinct multiphoton absorption spectra of different carotenoids offer a good perspective for the use of harmonics as an analytical tool for the determination of carotenoid composition in pigment-protein complexes. Especially promising is excitation at wavelengths above 1 μm, which has very low phototoxicity compared to 800 nm. This permits the conduct of prolonged investigations that render high signal to noise ratios.

It is certain that in the future we will see more investigations of photosynthetic structures with non-linear microscopy. Non-linear microscopy will evolve via the development of new non-linear imaging modalities and the utilization of existing non-linear contrast mechanisms for structural, spectroscopic, and physiological investigations in photosynthesis.

## Acknowledgments

I would like to thank Prof. F. Code, Prof. G. Garab, Prof. J. Squier, and Prof. L. Valkunas for helpful suggestions on the manuscript. I am grateful to my coworkers, Dr. A. Major, R. Cisek, C. Greenhalgh, and N. Prent for experimental support, discussions, and help in preparing the manuscript. The work was supported by grants from the Natural Sciences and Engineering Research Council of Canada, Canada Foundation of Innovation, and Ontario Innovation Trust.

## References

Abe S (2001) Two-photon probe of forbidden exciton states in symmetric aggregates of asymmetric molecules. Chem Phys 264: 355–363

Armstrong JA, Bloembergen N, Ducuing J and Pershan PS (1962) Interactions between light waves in a nonlinear dielectric. Phys Rev 127: 1918–1939

Baconnier S and Lang SB (2004) Calcite microcrystals in the pineal gland of the human brain: Second harmonic generators and possible piezoelectric transducers. IEEE Trans Dielect Electr Insul 11: 203–209

Barad Y, Eisenberg H, Horowitz M and Silberberg Y (1997) Nonlinear scanning laser microscopy by third harmonic generation. Appl Phys Lett 70: 922–924

Barille R, Canioni L, Rivet S, Sarger L, Vacher P and Ducret T (2001) Visualization of intracellular $Ca^{2+}$ dynamics with simultaneous two-photon-excited fluorescence and third-harmonic generation microscopes. Appl Phys Lett 79: 4045–4047

Barzda V, Mustárdy L and Garab G (1994) Size dependency of circular-dichroism in macroaggregates of photosynthetic pigment-protein complexes. Biochemistry 33: 10837–10841

Barzda V, Jennings RC, Zucchelli G and Garab G (1999) Kinetic analysis of the light-induced fluorescence quenching in light-harvesting chlorophyll *a/b* pigment-protein complex of Photosystem II. Photochem Photobiol 70: 751–759

Barzda V, Ionov M, van Amerongen H, Gussakovsky EE and Shahak Y (2004) The effect of pea chloroplast alignment and variation of excitation wavelength on the circularly polarized chlorophyll luminescence. J Fluoresc 14: 207–216

Barzda V, de Grauw CJ, Vroom J, Kleima FJ, van Grondelle R, van Amerongen H and Gerritsen HC (2001a) Fluorescence lifetime heterogeneity in aggregates of LHCII revealed by time-resolved microscopy. Biophys J 81: 538–546

Barzda V, Gulbinas V, Kananavicius R, Cervinskas V, van Amerongen H, van Grondelle R and Valkunas L (2001b) Singlet-singlet annihilation kinetics in aggregates and trimers of LHCII. Biophys J 80: 2409–2421

Barzda V, Greenhalgh C, Aus der Au J, Elmore S, van Beek JHGM and Squier J (2005) Visualization of mitochondria in cardiomyocytes by simultaneous harmonic generation and fluorescence microscopy. Opt Express 13: 8263–8276

Becker W (2005) Advanced Time-Correlated Single Photon Counting Techniques. Springer, Berlin

Bloembergen N and Pershan PS (1962) Light waves at boundary of nonlinear media. Phys Rev 128: 606–622

Boyd RW (1992) Nonlinear Optics. Academic Press, Amsterdam

Byer RL (1997) Quasi-phasematched nonlinear interactions and devices. J Nonlinear Opt Phys 6: 549–592

Campagnola PJ, Wei MD, Lewis A and Loew LM (1999) High-resolution nonlinear optical imaging of live cells by second harmonic generation. Biophys J 77: 3341–3349

Campagnola PJ, Millard AC, Terasaki M, Hoppe PE, Malone CJ and Mohler WA (2002) Three-dimensional high-resolution second-harmonic generation imaging of endogenous structural proteins in biological tissues. Biophys J 82: 493–508

Cheng JX and Xie XS (2004) Coherent anti-Stokes Raman scattering microscopy: Instrumentation, theory, and applications. J Phys Chem B 108: 827–840

Chu SW, Chen IH, Liu TM, Chen PC, Sun CK and Lin BL (2001) Multimodal nonlinear spectral microscopy based on a femtosecond Cr:forsterite laser. Opt Lett 26: 1909–1911

Chu SW, Liu TM and Sun CK (2003a) Real-time second-harmonic-generation microscopy based on a 2-GHz repetition rate Ti:sapphire laser. Opt Express 11: 933–938

Chu SW, Chen SY, Tsai TH, Liu TM, Lin CY, Tsai HJ and Sun CK (2003b) In vivo developmental biology study using noninvasive multi-harmonic generation microscopy. Opt Express 11: 3093–3099

Chu SW, Chen SY, Chern GW, Tsai TH, Chen YC, Lin BL and Sun CK (2004) Studies of $\chi^{(2)}/\chi^{(3)}$ tensors in submicron-scaled bio-tissues by polarization harmonics optical microscopy. Biophys J 86: 3914–3922

Cisek R, Aus der Au J, Squier J and Barzda V (2005) Fluorescence quenching and exciton-exciton annihilation effects in individual chloroplasts investigated by fluorescence lifetime imaging microscopy. In: van der Est A and Bruce D (eds) Photosynthesis: Fundamental Aspects to Global Perspectives, pp 776–778. Allen Press, Lawrence

Cox G, Moreno N and Feijó J (2005) Second-harmonic imaging of plant polysaccharides. J Biomed Opt 10: 0240131–0240136

Denk W, Strickler JH and Webb WW (1990) Two-photon laser scanning fluorescence microscopy. Science 248: 73–76

Diaspro A and Sheppard CJR (2002) Two-photon excitation fluorescence microscopy. In: Diaspro A (ed) Confocal and Two-Photon Microscopy: Foundations, Applications, and Advances, pp 39–43. Wiley, New York

Duncan MD, Reintjes J and Manuccia TJ (1982) Scanning coherent anti-Stokes Raman microscope. Opt Lett 7: 350–352

Feijó JA and Moreno N (2004) Imaging plant cells by two-photon excitation. Protoplasma 223: 1–32

Finzi L, Bustamante C, Garab G and Juang CB (1989) Direct observation of large chiral domains in chloroplast thylakoid membranes by differential polarization microscopy. Proc Natl Acad Sci USA 86: 8748–8752

Fittinghoff DN, Aus der Au J and Squier J (2005) Spatial and temporal characterizations of femtosecond pulses at high-numerical aperture using collinear, background-free, third-harmonic autocorrelation. Opt Commun 247: 405–426

Florsheimer M (1999) Second-harmonic microscopy — A new tool for the remote sensing of interfaces. Phys Status Solidi A 173: 15–27

Franken PA, Hill AE, Peters CW and Weinreich G (1961) Generation of optical harmonics. Phys Rev Lett 7: 118–120

Freund I and Deutsch M (1986) Second-harmonic microscopy of biological tissue. Opt Lett 11: 94–96

Freund I, Deutsch M and Sprecher A (1986) Connective-tissue polarity — optical second-harmonic microscopy, crossed-beam summation, and small-angle scattering in rat-tail tendon. Biophys J 50: 693–712

Fung DC and Theriot JA (1998) Imaging techniques in microbiology. Curr Opin Microbiol 1: 346–351

Gannaway JN and Sheppard CJR (1978) Second-harmonic imaging in scanning optical microscope. Opt Quant Electron 10: 435–439

Garab G, Faludidaniel A, Sutherland JC and Hind G (1988) Macroorganization of chlorophyll *a/b* light-harvesting complex in thylakoids and aggregates — information from circular differential scattering. Biochemistry 27: 2425–2430

Gauderon R, Lukins PB and Sheppard CJR (2001) Optimization of second-harmonic generation microscopy. Micron 32: 691–700

Giordmaine JA (1965) Nonlinear optical properties of liquids. Phys Rev 138: 1599–1606

Göppert-Mayer M (1931) Über Elementarakte mit zwei Quantensprüngen. Ann Phys (Leipzig) 5: 273–294

Greenhalgh C, Cisek R, Prent N, Major A, Aus der Au J, Squier J and Barzda V (2005) Time and structural crosscorrelation image analysis of microscopic volumes, simultaneously recorded with second harmonic generation, third harmonic generation, and multiphoton excitation fluorescence microscopy. Proc SPIE 5969: 59692F-1–59692F-8

Gussakovsky EE, Shahak Y, van Amerongen H and Barzda V (2000) Circularly polarized chlorophyll luminescence reflects the macro-organization of grana in pea chloroplasts. Photosynth Res 65: 83–92

Hellwart R and Christen P (1974) Nonlinear optical microscopic examination of structure in polycrystalline ZnSe. Opt Commun 12: 318–322

Hepler PK and Gunning BES (1998) Confocal fluorescence microscopy of plant cells. Protoplasma 201: 121–157

Hilbert M, Wehling A, Schlodder E and Walla PJ (2004) Two-photon-sensitized fluorescence and excitation spectra of Photosystem I of *Thermosynechococcus elongatus*. J Phys Chem B 108: 13022–13030

Horton P, Wentworth M and Ruban A (2005) Control of the light harvesting function of chloroplast membranes: The LHCII-aggregation model for non-photochemical quenching. FEBS Lett 579: 4201–4206

Kaiser W and Garrett CGB (1961) 2-photon excitation in $CaF_2$-$Eu^{2+}$. Phys Rev Lett 7: 229–231

Kolthammer WS, Barnard D, Carlson N, Edens AD, Miller NA and Saeta PN (2005) Harmonic generation in thin films and multilayers. Phys Rev B 72: 04544601–04544615

König K (2000) Multiphoton microscopy in life sciences. J Microsc 200: 83–104

Kriech MA and Conboy JC (2005) Imaging chirality with surface second harmonic generation microscopy. J Am Chem Soc 127: 2834–2835

Krikunova M, Kummrow A, Voigt B, Rini M, Lokstein H, Moskalenko A, Scheer H, Razjivin A and Leupold D (2002) Fluorescence of native and carotenoid-depleted LH2 from *Chromatium minutissimum*, originating from simultaneous two-photon absorption in the spectral range of the presumed (optically 'dark') S-1 state of carotenoids. FEBS Lett 528: 227–229

Krueger BP, Yom J, Walla PJ and Fleming GR (1999) Observation of the $S_1$ state of spheroidene in LH2 by two-photon fluorescence excitation. Chem Phys Lett 310: 57–64

Lin SJ, Hsiao CY, Sun Y, Lo W, Lin WC, Jan GJ, Jee SH and Dong CY (2005) Monitoring the thermally induced structural transitions of collagen by use of second-harmonic generation microscopy. Opt Lett 30: 622–624

Major A, Yoshino F, Aitchison JS, Smith PWE, Zigmantas D and Barzda V (2005) Picosecond z-scan measurements of the two-photon absorption in beta-carotene solution over 590–790 nm wavelength range. Proc SPIE 5724: 269–276

Marder SR, Beratan DN and Cheng LT (1991) Approaches for optimizing the first electronic hyperpolarizability of conjugated organic-molecules. Science 252: 103–106

Marder SR, Torruellas WE, BlanchardDesce M, Ricci V, Stegeman GI, Gilmour S, Bredas JL, Li J, Bublitz GU and Boxer SG (1997) Large molecular third-order optical nonlinearities in polarized carotenoids. Science 276: 1233–1236

Mertz J (2004) Nonlinear microscopy: New techniques and applications. Curr Opin Neurobiol 14: 610–616

Midwinter JE and Warner J (1965) Effects of phase matching method and of crystal symmetry on polar dependence of third-order non-linear optical polarization. Brit J Appl Phys 16: 1667–1670

Millard AC, Campagnola PJ, Mohler W, Lewis A and Loew LM (2003) Second harmonic imaging microscopy. Methods Enzymol 361: 47–69

Millard AC, Wiseman PW, Fittinghoff DN, Wilson KR, Squier JA and Müller M (1999) Third-harmonic generation microscopy by use of a compact, femtosecond fiber laser source. Appl Optics 38: 7393–7397

Mizutani G, Sonoda Y, Sano H, Sakamoto M, Takahashi T and Ushioda S (2000) Detection of starch granules in a living plant by optical second harmonic microscopy. J Lumin 87–89: 824–826

Moreaux L, Sandre O and Mertz J (2000a) Membrane imaging by second-harmonic generation microscopy. J Opt Soc Am B 17: 1685–1694

Moreaux L, Sandre O, Blanchard-Desce M and Mertz J (2000b) Membrane imaging by simultaneous second-harmonic generation and two-photon microscopy. Opt Lett 25: 320–322

Moreaux L, Sandre O, Charpak S, Blanchard-Desce M and Mertz J (2001) Coherent scattering in multi-harmonic light microscopy. Biophys J 80: 1568–1574

Moreno N, Feijó JA and Cox G (2004) Implementation and evaluation of a detector for forward propagated second harmonic signals. Micron 35: 721–724

Müller M, Squier J, Wilson KR and Brakenhoff GJ (1998a) 3D microscopy of transparent objects using third-harmonic generation. J Microsc 191: 266–274

Müller M, Squier J, Wolleschensky R, Simon U and Brakenhoff GJ (1998b) Dispersion pre-compensation of 15 femtosecond optical pulses for high-numerical-aperture objectives. J Microsc 191: 141–150

Müller M, Squier J, De Lange CA and Brakenhoff GJ (2000) CARS microscopy with folded BoxCARS phasematching. J Microsc 197: 150–158

Naumov AN, Sidorov-Biryukov DA, Fedotov AB and Zheltikov AM (2001) Third-harmonic generation in focused beams as a method of 3D microscopy of a laser-produced plasma. Opt Spectrosc 90: 778–783

Oron D, Tal E and Silberberg Y (2003) Depth-resolved multiphoton polarization microscopy by third-harmonic generation. Opt Lett 28: 2315–2317

Pawley JB (1995) Handbook of Biological Confocal Microscopy. Plenum Publishing, New York

Polivka T and Sundström V (2004) Ultrafast dynamics of carotenoid excited states — from solution to natural and artificial systems. Chem Rev 104: 2021–2071

Potma EO, de Boeij WP and Wiersma DA (2001) Femtosecond dynamics of intracellular water probed with nonlinear optical Kerr effect microspectroscopy. Biophys J 80: 3019–3024

Prent N, Cisek R, Greenhalgh C, Aus der Au J, Squier J and Barzda V (2005a) Imaging individual chloroplasts simultaneously with third- and second-harmonic generation and multiphoton excitation fluorescence microscopy. In: van der Est A and Bruce D (ed) Photosynthesis: Fundamental Aspects to Global Perspectives, pp 1037–1039. Allen Press, Lawrence

Prent N, Cisek R, Greenhalgh C, Sparrow R, Rohitlall N, Milkereit MS, Green C and Barzda V (2005b) Application of nonlinear microscopy for studying the structure and dynamics in biological systems. Proc SPIE 5971: 5971061–5971068

Saeta PN and Miller NA (2001) Distinguishing surface and bulk contributions to third-harmonic generation in silicon. Appl Phys Lett 79: 2704–2706

Schins JM, Schrama T, Squier J, Brakenhoff GJ and Müller M (2002) Determination of material properties by use of third-harmonic generation microscopy. J Opt Soc Am B 19: 1627–1634

Shen YR (1989) Surface-properties probed by second-harmonic and sum-frequency generation. Nature 337: 519–525

Sheppard CJR, Gannaway JN, Kompfner R and Walsh D (1977) Scanning harmonic optical microscope. IEEE J Quantum Elect 13: 912–912

Shima S, Ilagan RP, Gillespie N, Sommer BJ, Hiller RG, Sharples FP, Frank HA and Birge RR (2003) Two-photon and fluorescence spectroscopy and the effect of environment on the photochemical properties of peridinin in solution and in the peridinin-chlorophyll-protein from *Amphidinium carterae*. J Phys Chem A 107: 8052–8066

Shreve AP, Trautman JK, Owens TG and Albrecht AC (1990) Two-photon excitation spectroscopy of thylakoid membranes from *Phaeodactylum tricornutum* — evidence for an in vivo two-photon-allowed carotenoid state. Chem Phys Lett 170: 51–56

Simidjiev I, Barzda V, Mustardy L and Garab G (1997) Isolation of lamellar aggregates of the light-harvesting chlorophyll *a/b*

protein complex of Photosystem II with long-range chiral order and structural flexibility. Anal Biochem 250: 169–175
Singh S and Bradley LT (1964) Three-photon absorption in napthalene crystals by laser excitation. Phys Rev Lett 12: 612–614
So PTC, Dong CY, Masters BR and Berland KM (2000) Two-photon excitation fluorescence microscopy. Annu Rev Biomed Eng 2: 399–429
Squier JA and Müller M (2001) High resolution nonlinear microscopy: A review of sources and methods for achieving optimal imaging. Rev Sci Instrum 72: 2855–2867
Squier JA, Müller M, Brakenhoff GJ and Wilson KR (1998) Third harmonic generation microscopy. Opt Express 3: 315–324
Steinbach G, Bessona F, Pomozi I and Garab G (2005) Differential polarization laser scanning microscopy. Biological applications. Proc SPIE 5969: 59692C59601–59692C59610
Sun CK (2005) Higher harmonic generation microscopy. Adv Biochem Eng Biot 95: 17–56
Supatto W, Debarre D, Moulia B, Brouzes E, Martin JL, Farge E and Beaurepaire E (2005) In vivo modulation of morphogenetic movements in *Drosophila* embryos with femtosecond laser pulses. Proc Natl Acad Sci USA 102: 1047–1052
Terhune RW, Maker PD and Savage CM (1962) Optical harmonic generation in calcite. Phys Rev Lett 8: 404–406
Tirlapur UK and König K (2001) Femtosecond near-infrared lasers as a novel tool for non-invasive real-time high-resolution time-lapse imaging of chloroplast division in living bundle sheath cells of *Arabidopsis*. Planta 214: 1–10
Tirlapur UK and König K (2002) Two-photon near-infrared femtosecond laser scanning microscopy in plant biology. In: Diaspro A (ed) Confocal and Two-Photon Microscopy: Foundations, Applications, and Advances, pp 449–468. Wiley, New York
Tsang TYF (1995) Optical third-harmonic generation at interfaces. Phys Rev A 52: 4116–4125
Verbiest T, van Elshocht S, Kauranen M, Hellemans L, Snauwaert J, Nuckolls C, Katz TJ and Persoons A (1998) Strong enhancement of nonlinear optical properties through supramolecular chirality. Science 282: 913–915
Volkmer A (2005) Vibrational imaging and microspectroscopies based on coherent anti-Stokes Raman scattering microscopy. J Phys D Appl Phys 38: R59–R81
Walla PJ, Linden PA, Ohta K and Fleming GR (2002) Excited-state kinetics of the carotenoid S-1 state in LHC II and two-photon excitation spectra of lutein and beta-carotene in solution: Efficient car $S_1 \rightarrow$ Chl electronic energy transfer via hot $S_1$ states? J Phys Chem A 106: 1909–1916
Walla PJ, Linden PA, Hsu CP, Scholes GD and Fleming GR (2000) Femtosecond dynamics of the forbidden carotenoid $S_1$ state in light-harvesting complexes of purple bacteria observed after two-photon excitation. Proc Natl Acad Sci USA 97: 10808–10813
Ward JF and New GHC (1969) Optical third harmonic generation in gases by a focused laser beam. Phys Rev 185: 57–72
Williams RM, Zipfel WR and Webb WW (2001) Multiphoton microscopy in biological research. Curr Opin Chem Biol 5: 603–608
Yamada S and Lee IYS (1998) Recent progress in analytical SHG spectroscopy. Anal Sci 14: 1045–1051
Yelin D and Silberberg Y (1999) Laser scanning third-harmonic-generation microscopy in biology. Opt Express 5: 169–175
Zimmermann J, Linden PA, Vaswani HM, Hiller RG and Fleming GR (2002) Two-photon excitation study of peridinin in benzene and in the peridinin chlorophyll *a*-protein (PCP). J Phys Chem B 106: 9418–9423
Zipfel WR, Williams RM and Webb WW (2003) Nonlinear magic: Multiphoton microscopy in the biosciences. Nature Biotechnol 21: 1368–1376
Zumbusch A, Holtom GR and Xie XS (1999) Three-dimensional vibrational imaging by coherent anti-Stokes Raman scattering. Phys Rev Lett 82: 4142–4145

# Chapter 4

# Magnetic Resonance Imaging of Plants: Water Balance and Water Transport in Relation to Photosynthetic Activity

Henk Van As[1,2]* and Carel W. Windt[1]
[1]*Laboratory of Biophysics and* [2]*Wageningen NMR Centre, Wageningen University, Dreijenlaan 3, 6703 HA Wageningen, The Netherlands*



## Summary

Nuclear magnetic resonance (NMR) and nuclear magnetic resonance imaging (MRI) techniques are described in relation to applications on intact plants. These techniques are non-destructive and non-invasive and can be used to study the dynamics of plant water relations and water transport, e.g., as a function of environmental (stress) conditions. Non-spatially resolved NMR as well as two-dimensional or even three-dimensional imaging strategies can be applied. The relevant NMR parameters that can be used to discriminate between water in sub-cellular compartments are the relaxation times $T_1$ and $T_2$, and (restricted) diffusion. The information within MRI images can be manipulated in such a way that in addition to anatomical information physico-chemical information like cell compartment size, water membrane permeability, water diffusion and flow, information on chemical composition (water, oils, sugars and other metabolites) are obtained. This information is also available in non-spatially resolved NMR with a much higher signal-to-noise ratio, but the interpretation of such non-spatially resolved measurements can be hampered by heterogeneity of the plant tissue (sub-cellular

*Author for correspondence, email: henk.vanas@wur.nl

*Thijs J. Aartsma and Jörg Matysik (eds): Biophysical Techniques in Photosynthesis II, pp. 55–75.*

compartments, tissue heterogeneity and different types of tissue). Here we present a number of [1]H NMR and MRI techniques that can be used to study plant performance in relation to its photosynthetic activity. In addition, dedicated hardware for intact plant studies is discussed.

## I. Introduction

One of the least well-understood aspects of plant physiology is the integration and coordination of processes at the organ and whole plant level. This deficiency is not surprising; technological developments that would make it possible to routinely and accurately measure performance of the whole plant have so far received little attention. Within the framework of this book we consider the interplay between leaf water content, stomatal opening, water evaporation, xylem water transport, $CO_2$ fixation, chloroplast volume regulation, photosynthetic activity, loading and transport in the phloem. Water fluxes in the phloem are associated with the loading process and transport of assimilates from sources (such as photosynthesizing leaves and storage organs) to sinks (such as meristems, developing tissues and storage organs), where the phloem unloading takes place. The water flux in the xylem is associated with replacement of water lost by transpiration, which is modulated by the leaf conductance for water vapor, which in its turn is intimately connected with $CO_2$ fixation. Understanding mechanism and control of water and solute fluxes in the xylem and the phloem are essential for understanding growth, fruit development and growth inhibition by, for example, drought stress and fungal infection. By being able to measure these fluxes in relation to the plant (cell) water balance, and photosynthetic activity it will be possible to explicitly or implicitly examine many aspects of plant function and plant performance. This is key information to validate biophysical functional-structural plant models based on integrated carbon and water allocation and functions to address water stress-induced effects and growth limitations (Daudet et al., 2002).

In this chapter, we introduce how Nuclear Magnetic Resonance (NMR) and NMR Imaging (NMRI or MRI) can be used to obtain such information. Here we consider MRI to include the NMR Microscopy or MRM, the high spatial resolution analog of MRI. The latter is well known from whole body human and animal applications in medical and biomedical research.

NMR is a non-invasive and non-destructive technique that has a large number of anatomical and physiological applications to plant cells, plant organs and living plants (Walter et al., 1992; Ratcliffe 1994; Shachar-Hill and Pfeffer, 1996; Chudek and Hunter, 1997; Ishida et al., 2000; Köckenberger 2001a; Ratcliffe et al., 2001). [1]H NMR has long been used for the non-invasive measurement of molecular displacement such as flow, diffusion and perfusion and for the characterization of the physical state of water in biological systems. Both [1]H NMR and MRI have been used to study plant anatomy, histochemistry, internal quality of fruits and vegetables, plant development (including seeds and pollens), changes in water content and transport in stems of transpiring plants and in root systems (MacFall and Van As, 1996). Both techniques have also been used to investigate plant cell metabolism (Ratcliffe, 1994; Shachar-Hill and Pfeffer, 1996; Köckenberger, 2001b; Ratcliffe et al., 2001; Mesnard and Ratcliffe, 2005). The possibilities of MRM to study (plant) cells and fungi have recently been reviewed (Ciobanu et al., 2003; Köckenberger et al., 2004).

In terms of absolute spatial resolution MRI generally performs poorly in comparison to other microscopic modalities. However, in addition to anatomical information, MRI techniques provide access to spectroscopic and dynamic (rotational and translational) information of the nuclei under observation. The information within the images can be manipulated in such a way that either anatomical information or, and in addition, physico-chemical information like the microscopic/mesoscopic environment of water (sub-cellular compartments), water membrane permeability, water diffusion and flow, information on chemical composition (water, oils, starch, sugars and other metabolites), is obtained. A number of

*Abbreviations:* 2D-FT – two-dimensional Fourier transform; CVR – chloroplast volume regulation; EPI – echo planar imaging; FID – free induction decay; FLI – fast Laplace inversion; FOV – field of view; FT – Fourier transform; Gd-DTPA – gadolinium diethylene triaminepentaacetic acid; IR – inversion recovery; MRI – magnetic resonance imaging; MRM – magnetic resonance microscopy; MSE – multiple spin-echo; NMR – nuclear magnetic resonance; PAM – photosynthetic activity meter; PFG – pulsed magnetic field gradient; rf – radio frequency; *S/N* – signal to noise ratio; SE – spin-echo; STE – stimulated echo; $T_E$ – echo time; $T_R$ – repetition time

MRI approaches exist to get sub-pixel resolution based on this information. For example, by diffusion tensor imaging the detection of diffusion lengths in the (sub) micrometer range is possible, at any position in a three dimensional space (see special issue *NMR in Biomedicine*, van Zijl and Le Bihan, 2002). Relaxation time imaging and diffusion imaging can be used to measure sub-cellular compartment size and membrane water permeability. Other strategies to further improve resolution have been presented and will be discussed below.

When applying NMR to plants one has to deal with structural and physiological features that are different from those in animals at all levels of biological complexity, from the macromolecule to the organs. Additional compartments such as the chloroplast, the vacuole or the periplasm and unique physiological features, such as photosynthesis or assimilation of various inorganic ions, impose specific constraints on the experimental parameters and the equipment that is used. In general, intact plant experiments require dedicated NMR or MRI equipment and careful control of environmental conditions (Walter et al., 1992).

## II. NMR and MRI Basics Related to Plant (Cell) Structure

Here we introduce some basics of NMR and MRI and the information available by NMR directly related to plant (cell) structure. For more extended descriptions of the NMR technique and MRI strategies see one of the many excellent recent textbooks (Callaghan, 1993; Levitt, 2001) and the reviews cited in the Introduction.

In a homogeneous main magnetic field $B_0$, equal spins (e.g., protons of the water molecules) have identical Larmor precession frequency, and a single resonance line in the frequency spectrum is observed at

$$\nu_0 = \frac{1}{2\pi}\gamma \cdot B_0 \tag{1}$$

where $\gamma$ is the gyromagnetic ratio. When a well-defined constant magnetic field gradient $G$, $G = \partial B_0/\partial r$, is created within the magnet, identical spins at different positions along this gradient have different resonance frequencies, each frequency being characteristic for a particular position $r$:

$$\nu(r) = \frac{1}{2\pi}\gamma \cdot \left(B_0 + \vec{G}.\vec{r}\right) \tag{2}$$

This is the basis for NMR imaging.

### A. Image Contrast and Single Parameter Images

Position labeling by magnetic field gradients can be performed in a variety of ways (Callaghan, 1993). Depending on the actual sequence used, the process of position labeling will take some time. In the frequently used two-dimensional Fourier Transform (2D-FT) spin-echo (SE) sequence, acquisition of the signal occurs at a certain time $T_E$ (echo-time) after the excitation of the spin system (Fig. 1). During that time the signal will decay according to the transverse or $T_2$ relaxation process:

$$S(T_E) = A_0 \cdot e^{-T_E/T_2} \tag{3}$$

$A_0$ is the signal amplitude directly after excitation, and is a direct measure of the amount of spins under observation in the NMR measuring coil. In imaging it is primarily a function of the spin density. The contrast or signal intensity in images therefore strongly depends on spin density, $T_E$ and $T_2$. In order

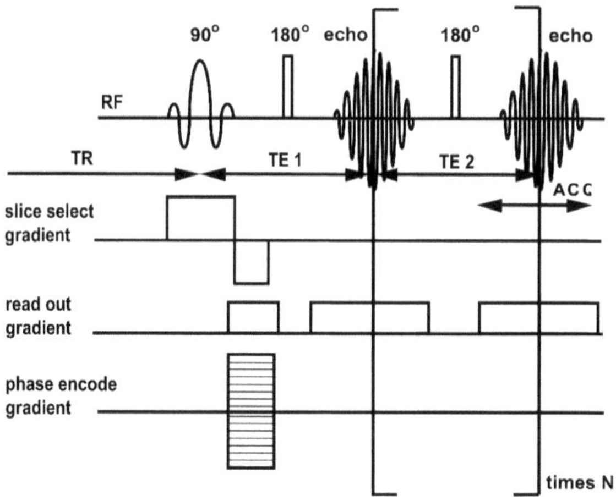


*Fig. 1.* Scheme of a pulse sequence for multiple spin-echo imaging. The echo times $T_{E1}$ and $T_{E2}$ may be different in size. The echoes can be acquired separately to obtain images with different $T_2$ weighting and can be used to calculate local $T_2$ values, or the echoes can be added to obtain a higher signal to noise for the images. To obtain a $N \times N$ image matrix, $N$ points have to be sampled during the acquisition of each echo. The sequence has to be repeated for $N$ different values of the phase encoding gradient, ranging from $-G_{max}$ to $G_{max}$.

to obtain a full two-dimensional image of $N \times N$ pixels, the sequence has to be repeated N times for phase encoding at varying $G$ values, whereas $N$ data points are acquired during each echo acquisition (see Fig. 1). If the repetition time $T_R$ is long enough, the spin system has restored equilibrium along the magnetic field direction: $T_R > 3T_1$. $T_1$ is the decay time in the longitudinal or magnetic field direction. If $T_R < 3T_1$, the effective signal amplitude, $A_{eff}$, does not uniquely represent the spin density in each pixel, but depends on a combination of the spin density and the relaxation time $T_1$: $A_{eff} = A_0 \exp(-T_R/T_1)$. As a result, NMR image intensity usually depends on a combination of these parameters, reflecting spin density, $T_1$, and $T_2$. In addition to these parameters, diffusion behavior of the molecules can also contribute to the contrast (see below).

Many pulse sequences exist by means of which quantitative images are obtained that represent each of these parameters. Multiple Spin-Echo (MSE) MRI (Edzes et al., 1998) and Inversion recovery (IR) MSE MRI (Donker et al., 1996) are examples of such sequences. In MSE a spin-echo series is created by applying a train of 180° rf pulses (Fig. 1). Each echo is acquired in the presence of a read-out or frequency encoding gradient (Eq. 2) and prepared with identical phase encoding gradient. By repeating the experiment as a function of different values of the phase encoding gradient many echo images are obtained. Single parameter images can now be processed from the MSE-experiment by assuming a mono-exponential relaxation decay (Eq. 3) as a function of $n_{echo} \cdot T_E$ in each picture element, pixel. Here $n_{echo}$ is the echo number. The resulting images are: signal amplitude ($A_0$), $T_2$, and $T_1$ (the latter in the case of Inversion Recovery (IR-)MSE, where the MSE sequence is preceeded by a 180° inversion pulse with a varying delay time $t_d$ between the inversion pulse and the start of the MSE part of the sequence). Examples of single parameter images are presented in Fig. 2.

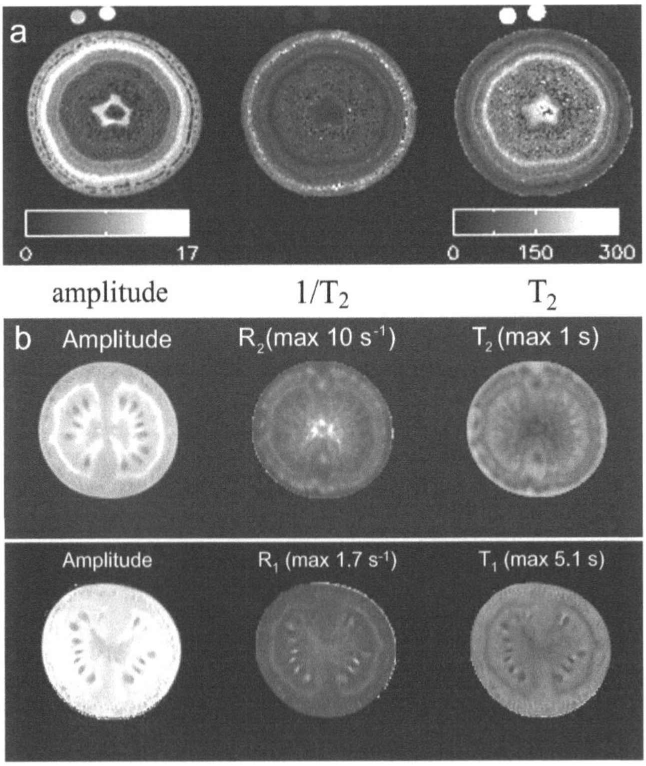


*Fig. 2.* a. Single parameter images of a 3 mm slice through the trunk of a poplar tree (1.8 m total length), obtained at 0.75 Tesla. These images have been calculated based on multi echo images obtained by the imaging sequence presented in Fig. 1, by a mono-exponential fit per pixel of the signal decay as a function of echo time (cf. Eq. 3). b. Single parameter images from a slice through a cherry tomato obtained at 3 Tesla. $R_1 = 1/T_1$, and $R_2 = 1/T_2$. The differences in the amplitude images are due to the different values of $T_E$ and $T_R$ that were used in the $T_2$ and $T_1$ experiments: $T_E = 7.5$ or 15 ms and $T_R = 2$ or 5 s, respectively. At longer $T_E$ (Amplitude image resulting from $T_1$ experiment) some susceptibility artifacts are visible in the outer pericarp.

The observed $T_2$ value in imaging mode with respect to its value in non-imaging mode depends on a number of contributions depending on the details of the image experiment and plant tissue characteristics (Edzes et al., 1998):

$$\frac{1}{T_{2,obs}} = \frac{1}{T_2} + \frac{1}{T_2'} + \frac{1}{T_2''} \tag{4}$$

Here $T_2$ represents the original $T_2$ of the liquid in the tissue, which is also observed in non-imaging mode experiments. $T_2'$ and $T_2''$ originate from diffusion effects.

In plants, especially in leaves and woody tissue, air spaces in the order of a few µm up to 100 µm are present. Due to the difference in the magnetic permeability of air and tissue/water local magnetic field gradients, $g_z$, originate. These local field gradients lead to a rapid dephasing of the magnetization in the $x,y$-plane and will affect the decay rate of the FID directly after the excitation pulse: $1/T_2^* = 1/T_2 + f(g_z)$. Spin echo experiments correct for these effects as long as spins remain stationary. However, displacement through these local gradients due to diffusion results in dephasing effects and will affect the echo amplitude and translate to a reduction of the observed $T_2$ value:

$$\frac{1}{T_2'} \propto \gamma^2 \cdot \langle g_z^2 \rangle \cdot T_E^2 \cdot D \tag{5}$$

Here $D$ is the (self-)diffusion coefficient of water in the tissue. The strength of these field inhomogeneities is linearly proportional to $B_0$ (Lüdecke et al., 1985; for an overview of the more general problem of susceptibility effects, see Durrant et al., 2003). Minimizing this effect can be achieved by imaging at relatively low $B_0$ and at short $T_E$ values. In non-spatially resolved $T_2$ measurements this $T_E$ value can be chosen to be very short (in the order of 200–400 µs), whereas in imaging mode $T_E$ is in the order of a few ms or longer. In quantitative $T_2$ and proton density studies at $B_0 = 0.47$ Tesla, $T_2$-values in big vacuolated cells were found to approach the value of pure water (>1.5 s) (Edzes et al., 1998). With such long $T_2$-values many spin-echoes can be recorded (up to 1000 in a cherry tomato (Edzes et al., 1998)) increasing the total signal-to-noise ratio, *S/N*. At higher field strength the observed $T_2$ normally decreases. The optimal field strength, obtaining the highest *S/N* with minor susceptibility artifacts, depends on the size, the shape, the concentration and distribution of air spaces in the sample.

In leaves, where most of the photosynthetic activity is located, we expect strong effects of air spaces. These can strongly affect the results. An illustration of the effects can be found in mushrooms where the cap contains 50% by volume of air spaces (Donker et al., 1996, 1997). At a magnetic field strength of 0.47 Tesla the transverse relaxation of water in the mushrooms could be characterized based on a large number of echoes and the correct water distribution was obtained by extrapolation of the decay curve to zero time. A slight decrease of $T_2$ values due to susceptibility effects was observed at longer echo times $T_E$. However, at higher fields (4.7 and 9.4 Tesla) the $T_2$ values that were found were very short and decreased with increasing $T_E$ values. The actual water distribution could not be reproduced at all. The local field gradient could be removed by filling the air spaces with a contrast solution, for example a Gd-DTPA solution, removing the local field gradients (Donker et al., 1999). Gd-DTPA cannot pass intact membranes. At high concentration of Gd-DTPA the $T_2$ of this solution in the extra-cellular space is too short to be observed in the images and images of the mushroom itself were observed, with identical water distribution but longer $T_2$ values as observed before vacuum infiltration. If membranes start to become leaky this is immediately observed in a decrease in the observed $T_2$ of the water in the cells, which now can exchange with protons of the infiltrate, with a very short $T_2$.

In addition to the effect of local field gradients, diffusion through the repeatedly applied read out gradients in a multiple spin-echo experiment contributes to the observed relaxation time:

$$\frac{1}{T_2''} = \gamma^2 G^2 \delta^2 \left(1 - \frac{4\delta}{3T_E}\right) D \tag{6}$$

Here $2\delta$ is the duration of the applied read gradient. In practice, $G$ and $\delta$ are dictated by the choice of pixel resolution, image size and actual $T_E$. To image relaxation times in the order of 1 s, as a rule of thumb a lower limit of 100 µm for the pixel size can be used (Edzes et al., 1998). At smaller pixel sizes the diffusive attenuation in the read gradients becomes the leading contribution to the observed transverse relaxation times, and the information available from the actual $T_2$ is lost and mixed up with (restricted) diffusion information.

Thus the information content and contrast of the MR images can be manipulated to represent different, single NMR parameters of different physical or metabolic processes. In this way a combination of anatomical and physiological (functional) information is obtained, referred to as functional imaging. Examples are the signal amplitude — representing proton density per volume = tissue density times water content — the relaxation times $T_1$ and $T_2$, the diffusion coefficient, $D$, flow (see below), magnetization transfer effects (Bendel et al., 2001), and metabolic information based on $^1$H, $^{13}$C or $^{31}$P NMR.

Because of the relatively poor spatial resolution, most pixels within an image will contain information that originates from different sub-cellular compartments or even cells. This is called the partial volume problem. Obviously, this situation is also encountered in non-spatially resolved NMR measurements. In general the low $S/N$ per pixel in imaging (typically in the order of 10–50) does not allow multi-exponential fitting, which might provide a way to resolve sub-cellular information. Such information is available from non-spatially resolved measurements, with a much higher $S/N$ (1000 or higher). Therefore the information presented in images mostly is called 'apparent': e.g., apparent $T_2$ , $T_{2,\mathrm{app}}$, or apparent $D$, $D_{\mathrm{app}}$.

### *B. Information Available from Low Field $^1$H NMR*

#### *1. Signal Amplitude and Relaxation Times*

A variety of interactions between the magnetic moments of the observed spins and the surrounding nuclei and electrons contribute to the relaxation times $T_1$ and $T_2$. These interactions make it possible to probe the physicochemical properties of the spin environment using NMR relaxation measurements. The protons in water molecules experience an intramolecular dipolar interaction between the two proton spins within one and the same molecule, as well as an intermolecular interaction with protons of neighboring water molecules. Both interactions fluctuate when the molecules rotate or translate. When the rotation correlation time $\tau_c$ of the molecules is short, as is the case for free water molecules ($\tau_c \sim 10^{-12}$ s), both $T_1$ and $T_2$ are equal and relatively long (~2 sec.). Water close to macromolecules or to solid surfaces generally have slower tumbling rates ($\tau_c \sim 10^{-12}$–$10^{-10}$ s), which leads to a reduction in both relaxation times. In addition, exchange of protons between water and other molecules, such as sugars and proteins, also influences (shortens) the relaxation times (Hills and Duce, 1990).

Some general practical observations of water relaxation in plant tissue, containing (vacuolated) plant cells, are that the observed $T_2$ is much shorter than $T_1$. $T_2$ in plants is shorter or even much shorter than in pure water, and in general multi-exponential decay curves are observed. In biological systems multi-exponential relaxation has been used to obtain information on the relative proportions of water in different environments, e.g., (sub)cellular compartments (Snaar and Van As, 1992; van Dusschoten et al., 1995).

There is a good, though not always quantitative, correlation between the NMR data and the classical measurements of water content, relative water content, tissue hydration and water potential. For the $T_1$ data the correlation with water content and water potential has been suggested to be high, whereas for $T_2$ data this correlation is more complex. Depending on experimental parameters (e.g., magnetic field strength, inter echo time ($T_E$), number of data points, $S/N$) a number of exponentials are observed, which can be assigned more or less uniquely to either water in the vacuole (longest $T_1$ and $T_2$), cytoplasm ($T_1 >$ $T_2$, both shorter than vacuolar $T_1$ and $T_2$), and cell wall/extra cellular space ($T_2$ depends strongly on the water content in this compartment, and ranges from about 5 ms up to hundreds of ms). In leaves chloroplasts can also be discriminated (see Section III.A). Within these compartments diffusive exchange results in single exponential behavior. Due to exchange of water protons with protons of the biomass (which have a small chemical shift difference with respect to water), the intrinsic $T_2$ values in the compartments depend on the magnetic field strength. This is manifest especially for the vacuolar water.

Proton exchange over the plasmalemma and the tonoplast membrane affects the observed relaxation times. Due to the effect of exchange, which depends on the difference in the relaxation times in the exchanging compartments, $T_1$ and $T_2$ results are in general different, even for the number of observed exponentials. Differences in $T_1$ for the different compartments are relatively small, and exchange between the compartments results in averaging over the compartments. Therefore $T_1$ values relate with water content more directly. Differences in $T_2$ are more pronounced, and exchange over membranes only results in partial averaging (depending on size of compartments and membrane water permeability). Compartments therefore can still be discriminated based on $T_2$ values (Snaar and Van As, 1992; van Dusschoten et al., 1995) (for membrane permeability, see also Section B3).

### 2. Diffusion and Restricted Diffusion

All molecules in a fluid are subject to Brownian motion. The extent of this motion depends on the temperature and the viscosity of the fluid, which are incorporated in the bulk diffusion coefficient of the fluid. When an ensemble of molecules is followed in time, the root mean square distance traveled increases with time as long as no boundaries are encountered, according to the Einstein relation:

$$\sqrt{\langle r^2 \rangle} = \sqrt{2Dt} \tag{7}$$

If water experiences a barrier to diffusion, for example a cell membrane, the cell dimensions determine the maximum displacement (in case of an impermeable barrier) or to what extent it is hindered (permeable membrane). As a result the NMR diffusion experiments result in apparent values of diffusion coefficients, $D_{app}$, which are smaller than the intrinsic $D$.

$D_{app}$ can be measured using a so-called pulsed field gradient (PFG) experiment. In this experiment a sequence of two magnetic field gradient pulses of duration δ and equal magnitude $G$ but opposite sign (or equal sign but separated by an 180° rf pulse) label the protons as a function of their position. If the spins remain at exactly the same position the effect of the gradient pulses compensate each other. However, as soon as translational (displacement) motion occurs, the gradients do not exactly compensate each other anymore, resulting in attenuation of the signal amplitude. The amount of this attenuation is determined by the length and amplitude of the gradient pulses, and by the mean translational distance traveled during the interval Δ between the two pulses. The NMR-signal amplitude $S(G)$ vs. $G$ normalized to the signal amplitude $S(0)$ at $G = 0$ for free diffusion at a given Δ is given by (Stejskal and Tanner, 1965; Norris, 2001):

$$\frac{S(G,\Delta)}{S(0,\Delta)} = e^{-\gamma^2\delta^2G^2D(\Delta-\delta/3)} \tag{8}$$

The distance traveled depends on the bulk diffusion coefficient of the fluid in the compartment and Δ (Eq. 7 with t = Δ). The compartment dimensions define the (maximum) free diffusion distance. If $\sqrt{\langle r^2 \rangle}$ is in the order of the compartment size, diffusion becomes restricted. For restricted diffusion Eq. (8) does no longer hold and becomes dependent on the geometry and size of the compartment (Callaghan et al., 1999; Norris, 2001; van der Weerd et al., 2002b).

### 3. Cell Size, Membrane and Tissue Water Permeability

Membrane permeability is frequently determined using the NMR relaxation times of intracellular water protons (Conlon and Outhred, 1972; Brownstein and Tarr, 1979; Snaar and Van As, 1992; Zhang and Jones, 1996; van der Weerd et al., 2001). The observed transverse relaxation time $T_2$ of water in a confined compartment such as a vacuole can be described as a function of the bulk $T_2$, $T_{2,bulk}$, and the probability to reach the membrane and to loose magnetization at the membrane, either by a direct interaction with the membrane (acting as a surface sink for relaxation) or by passing the membrane and entering a compartment with a (much) shorter relaxation time. The probability to reach the membrane is defined by the diffusion time and thus directly related to the compartment radii. No evidence has been found that membranes itself act as a relaxation sink (McCain, 1995b; van der Weerd et al., 2002a). The net loss of magnetization in a vacuole therefore depends on the membrane water permeability of the tonoplast and the effective relaxation in the cytoplasm (van der Weerd et al., 2002b; L. van der Weerd, J.E.M Snaar, F.J. Vergeldt and H. Van As, unpublished). As a result, although relaxation mainly originates from dipolar interactions, the observed relaxation time depends in addition to $T_{2,bulk}$, to the radii of the compartment along the $x$, $y$, and $z$ directions ($R_{x,y,z}$) and the net loss of magnetization at the compartment boundary, the so called magnetization sink strength $H$ (van der Weerd et al., 2001):

$$\frac{1}{T_{2,obs}} = H\left(\frac{1}{R_x}+\frac{1}{R_y}+\frac{1}{R_z}\right)+\frac{1}{T_{2,bulk}} \quad (9)$$

$H$ is linearly related to the actual membrane permeability (van der Weerd et al., 2002b, 2006). In addition to $T_2$ measurements, diffusion measurements by a PFG experiment can be used to access membrane permeability of water. In such PFG experiments diffusion coefficients can be measured as a function of the diffusion labeling time, or observation time, $\Delta$, the time between the two gradient pulses. In this way the distance over which the spins can diffuse freely and to what extend they can pass over the membrane, which restricts the diffusion process, are observed. The latter is a direct measure of the permeability of the surrounding wall (membranes). In this way the size and the membrane permeability of, e.g., the vacuoles in vacuolated plant tissue have been measured, even within single pixels of an MRI image (T. Sibgatullin, F.J. Vergeldt, E. Gerkema and H. Van As, unpublished). At long observation times cell-to-cell transport becomes visible.

## 4. Flow

Not only diffusion, but all types of molecular displacements, including flow and tissue perfusion, can be traced by means of NMR experiments. Both non-imaging and imaging methods have been developed and applied to plants.

The first (non-imaging) method to measure xylem water transport in plants was presented some twenty years ago (Van As and Schaafsma, 1984; Reinders et al., 1988; Schaafsma et al., 1992). That method is based on a series of equidistant identical rf pulses (in the range of 30°–180°) applied in the presence of a static magnetic field gradient. Due to displacement of the flowing spins in the direction of the gradient the transversal magnetization develops a phase shift during the rf pulse train as a function of time. Stationary water does not develop such a phase shift. By detecting the out-of-phase signal component, flowing water is uniquely observed even in the presence of an excess (40–50 fold) amount of stationary water, as is the situation in plant stems. However, this repetitive pulse method needs calibration: the results depend on the actual flow profile. Flow in a single xylem vessel in general is assumed to be laminar, but velocities within the different xylem vessels in a stem will be different, depending on vessel diameter. So the flow profile within the total cross section of a stem is therefore not known a priori. In addition, xylem and phloem flow can not be discriminated, only the sum of both is observed.

To accurately quantify (long distance) xylem and phloem transport one needs to determine both the direction of the flow, and the actual flow profile, from which the flow velocity, the flux and the flow conducting area can be obtained. At the same time diffusion and flow have to be discriminated. These goals can best be obtained by use of pulsed magnetic field gradient (PFG) techniques. In a PFG experiment a net unidirectional displacement $R$ of an ensemble of identical spins in the direction of that gradient results in a net phase shift $\phi$ of the NMR signal:

$$\phi = \gamma\delta GR \quad (10)$$

Here $G$ is the amplitude of the gradient. If a number of ensembles of spins contribute to the NMR signal the observed signal $S$ is given by

$$S = \iint p(r_0)P(r_0|r_0+R,\Delta)e^{i\gamma\delta G.R}\, d(r_0+R)dr_0 \quad (11)$$

where $p(r_0)$ is the probability to find a spin at $r_0$ at time zero, and $P(r_0|r_0{+}R,\Delta)$ is the conditional probability that a spin originally at $r_0$ at time zero (the time of the first magnetic field gradient pulse) will be at $r_0{+}R$ at $t = \Delta$, which is the time between the first and second field gradient pulse. The integration is over all starting and end positions. Doing so for the conditional probability results in an averaged probability or propagator $P(R, \Delta)$, the probability that a spin at any initial position is displaced by $R$ over time $\Delta$.

To correctly quantify the unknown displacement-behavior of an observed ensemble of spins, one has to measure the NMR-signal $S(G)$ as a function of $G$. By defining a reciprocal space $q = \gamma\delta G/2\pi$, Eq. (11) can be rewritten as:

$$S(q) = \int P(R,\Delta)e^{i2\pi q.R}\, dR \quad (12a)$$

$$S(R,\Delta) = \int S(q)e^{-i2\pi q.R}\, dR \quad (12b)$$

This demonstrates the Fourier relationship between $S(q)$ and $P(R,\Delta)$. Thus, by monitoring $S(q)$, the NMR-signal as a function of $q$, a Fourier transform of $S(q)$ results in the position averaged probability distribution of displacements of all spins. For flow, dividing the displacement axis $R$ by $\Delta$ results in the flow profile

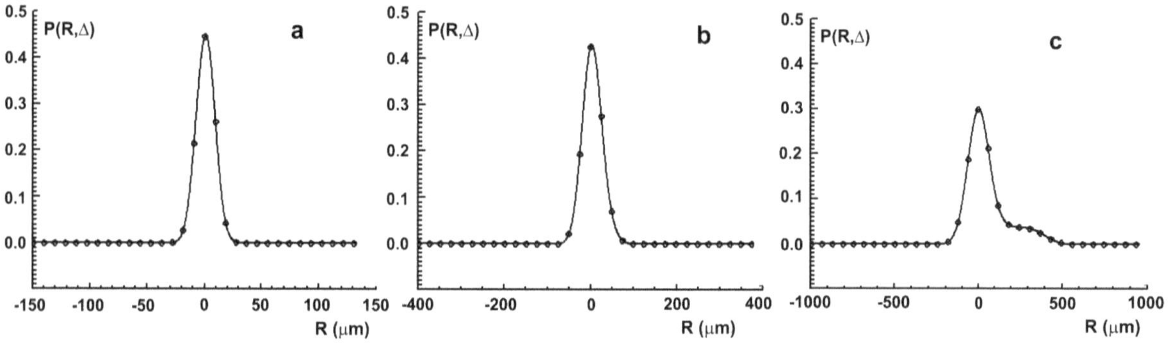


*Fig. 3.* Simulated propagators for a system consisting of 75% stationary water and 25 % flowing water, with a laminar flow profile; $D = 2.2 \times 10^{-9}$ m²/s. The mean linear average flow velocity is 0.2 mm/s (typical for phloem). A: $\Delta$ = 15 ms ($\sigma$ = 8.1 μm, $r$ = 3.0 μm); B: $\Delta$ = 100 ms ($\sigma$ = 21 μm, $r$ = 20 μm); $\Delta$ = 1000 ms ($\sigma$ = 66 μm, $r$ = 200 μm). After Scheenen et al. (2001), with permission from Elsevier.

$P(v)$. This type of measurements is referred to as displacement or $q$-space imaging, and was introduced by Kärger and Heink (1983) and combined with spatial imaging by Callaghan, Eccles and Xia (1988). Displacement imaging has been applied to measure flow in many porous systems (for some introductions and reviews see Fukushima, 1999; Callaghan et al., 1999; Mantle et al., 2003; Stapf and Han, 2005), including plants (see Section III.B).

The propagator for free, unhindered, diffusing water has a Gaussian shape, centered around $R = 0$ (Fig. 3). The root mean square displacement, $\sigma$, of diffusing protons, observed by NMR, is proportional to the square root of the corrected displacement labeling time, $(\Delta - \delta/3)$, and the diffusion coefficient, $D$, of the observed protons:

$$\sigma = \sqrt{2D(\Delta - \delta/3)} \qquad (13)$$

$\sigma$ is directly related to the width of the Gaussian distribution. In contrast, the mean displacement r of flowing protons is linearly proportional to $\Delta$ itself:

$$r = v_{av}\,\Delta \qquad (14)$$

where $v_{av}$ is the average velocity of the flowing protons. The labeling time between the two PFGs has to be long (in the order of 150 ms or longer) to discriminate between slow flow (50 μm/s) and diffusion ($D$ in the order of $2.10^{-9}$ m²/s) (Scheenen et al., 2001). At lower velocities $\Delta$ has to be increased further, see Fig. 3.

The choice between using a SE or a stimulated echo (STE) version of the PFG pulse sequence (Fig. 4)

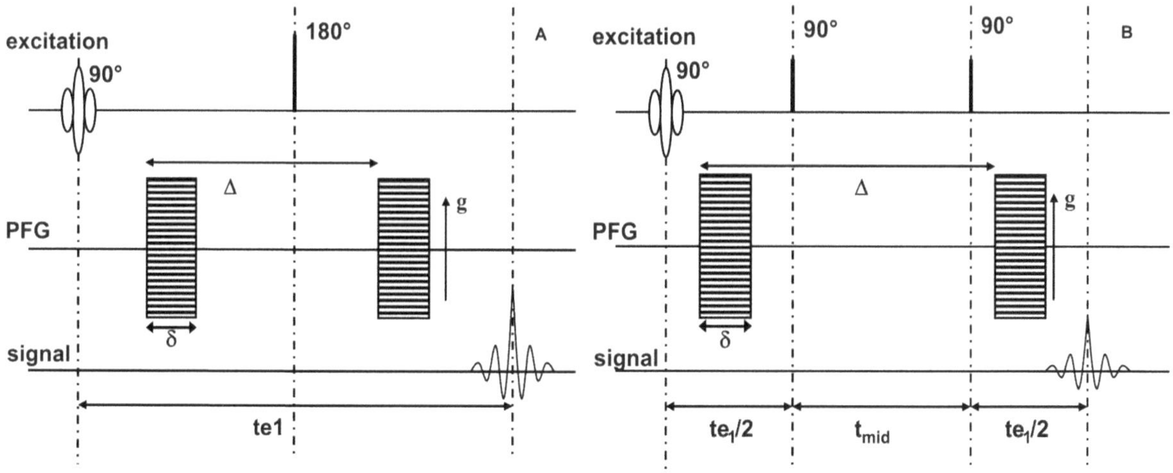


*Fig. 4.* Schematic representation of a PFG-SE (A) and PFG-STE (B) sequence.

depends on the time from excitation to the first echo (containing Δ), the spin-lattice relaxation time $T_1$ and spin-spin relaxation time $T_2$ of the observed protons. In a SE experiment (Fig. 4a) the NMR signal $S(t)$ is in the *xy*-plane throughout the time from excitation to detection ($T_E$), where it decays according to the $T_2$ (Eq. 3). In a STE experiment, the signal is stored along the *z*-axis during $t_{mid}$ (Fig. 4b), where it decays according to the spin-lattice relaxation time $T_1$, which is longer (often substantially) than $T_2$:

$$S\left(T_E\right)_{STE} = \frac{1}{2} A_0 \cdot e^{-(t_{el}/T_2)-(t_{mid}/T_1)} \tag{15}$$

in which $t_{el}$ is the time that magnetization is in the *xy*-plane (Fig. 4b). The maximum amplitude of the stimulated echo is intrinsically half the initial signal amplitude, but can still be larger than the spin echo amplitude at the same time after excitation. Propagators of slowly flowing water (averaged linear velocity in the order of 0.2–0.7 mm/s) in a cut-off pedicel of a tomato and phloem flow in intact plants have been measured with the stimulated echo version of the PFG TSE pulse sequence (Scheenen et al., 2001; Windt et al., 2006). If linear flow velocities are high, short labeling times can be used for which the spin echo version of the sequence is more suitable.

To discriminate between diffusing and flowing water, even within a single pixel, we make use of the fact that stationary (only diffusing) water results in a propagator that is symmetrical around zero displacement. The signal in the non-flow direction can be mirrored around the displacement axis and subtracted from the signal in the flow direction to produce the flow profile of the flowing as well as the stationary water (Fig. 5). The resulting flow profiles can then be used to calculate the flow conducting area, the average velocity of the flowing water, and by taking the integral of the propagator of the flowing water, the volume flow per pixel or in any selected area in an image (Fig. 6) (Scheenen et al., 2000b).

For a reliable determination of the displacement distribution the NMR signal (echo envelope) has to be measured as a function of at least 32 different gradient values. It is clear that this takes time. A two dimensional SE image consisting of $N \times N$ pixels requires $N$ acquisitions to be repeated. Combined with displacement measurements using 32 gradient steps, this results in $32 \times N$ acquisitions. If the time between repeated measurements is 2 s and $N = 128$, a scan time of at least 136 minutes is needed. To reduce

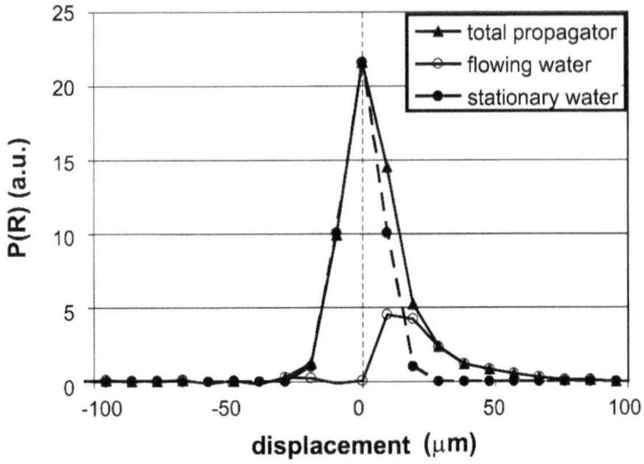


*Fig. 5.* Total propagator (▲) of flowing and stationary (only diffusing) water measured in the xylem region of the stem of a tomato plant. By taking the part of the total propagator that does not contain a flow contribution (here the left hand part), and noting that the propagator of stationary water is symmetrical around zero displacement, the propagator in the non-flow direction can be mirrored with respect to $R$=0 to obtain the propagator of stationary water (●). The propagator of flowing water (○) is the difference between the total propagator and propagator of stationary water.

the acquisition time displacement imaging has been combined with fast imaging techniques. For turbo-SE this results in a N/$m$ times reduction in scan time as compared to a standard $N \times N$ SE image sequence (Scheenen et al., 2000a). Here $m$ is the turbo factor, equal to the number of echoes that can be acquired in one scan (Fig. 7). It is clear that the number of pixels, $N$, directly determines both spatial and temporal resolution, but acquisition times are in the order of 5–10 minutes. Alternatively, $q$-space measurements may be combined with other (very) fast imaging sequences like spin echo based Echo Planar Imaging (MBEST-EPI) (Howseman et al., 1988) or FLASH (Rokitta el al., 1999) to further improve temporal resolution. Pure EPI methods probably cannot be used, because they are based on the detection of the FID signal and thus limited by $T_2$* effects due to local magnetic field gradients (cf. Section II.A), which may be limiting in plant tissues.

### *C. Spatial and Temporal Resolution*

Clearly, the highest temporal resolution is obtained in the non-imaging (non-spatially resolved) measurement mode. For relaxation time measurements this mode results in the best discrimination of different water fractions (high $S/N$ ratio), but interpretation of results can be hard if samples or tissues are heteroge-

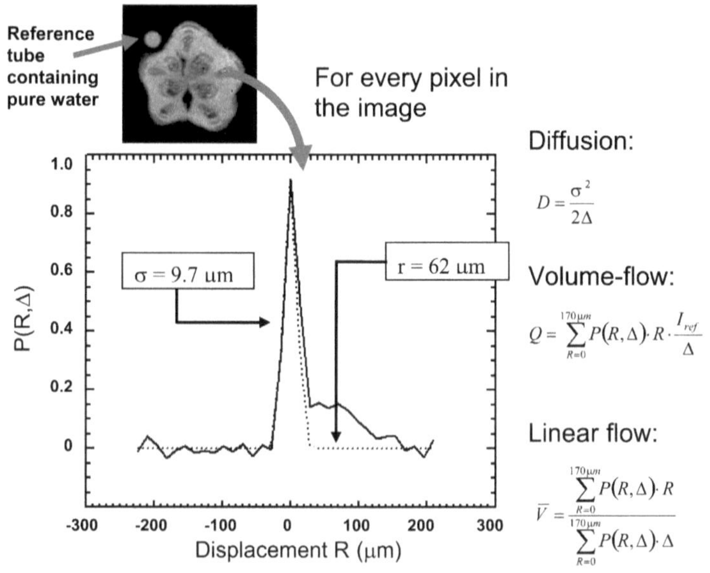


*Fig. 6.* Example of a single pixel propagator in the xylem region of a slice trough the stem of a cucumber plant (see image on top). From such a propagator the diffusion coefficient of stationary water, $D$, the volume flow rate or flux, $Q$, and the average linear velocity can be calculated is indicated. For $Q$ one needs a reference signal to calibrate the intensity. Here the intensity of a pixel in a tube containing pure water is used. The summation in the calculations is up to the maximum displacement, here 170 μm. For details, see Scheenen et al., 2000b.

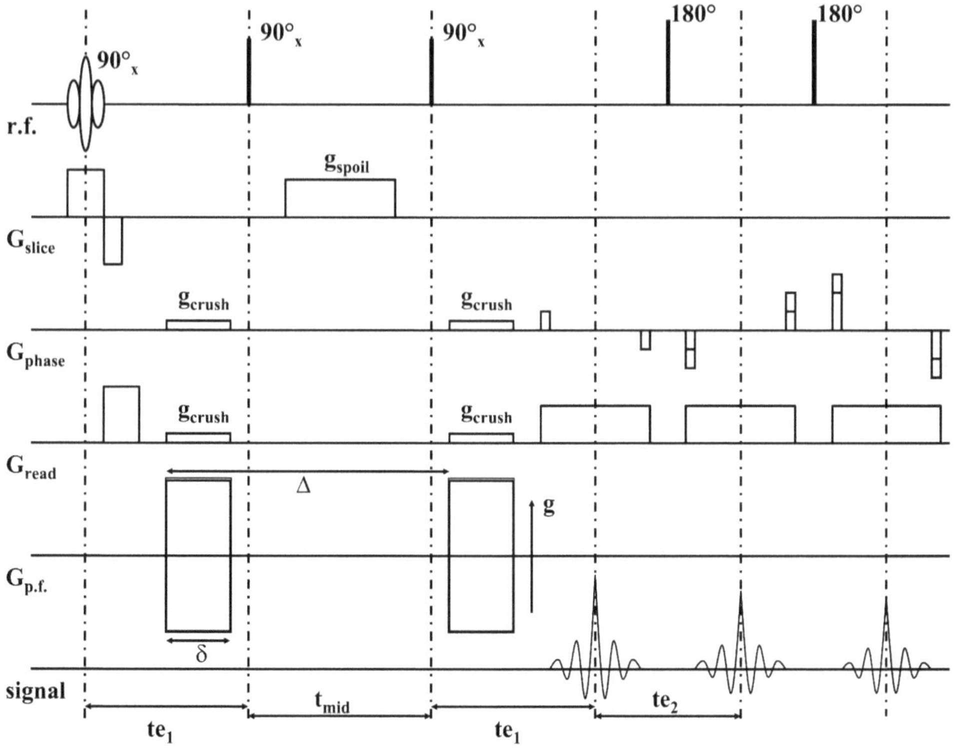


*Fig. 7.* PFG-STE-turbo Spin Echo pulse sequence for fast imaging. Now a number of $m$ echoes are used from one echo train for phase encoding to reduce the image acquisition time by a factor m. For details, see Scheenen, et al., 2000a (reproduced with permission from Elsevier).

neous. For moving water, stationary and flowing water (including direction) can be discriminated by $q$-space measurements. In plant stems the ratio between stationary water and flowing water is very unfavorable, in the order of 40–50. Under these circumstances, the best way to discriminate both is by considering the out of phase or imaginary part of the signal: stationary water does not result in a phase shift, whereas flowing water does. However, discrimination of phloem and xylem water can become problematic when the two are flowing in different directions.

In imaging mode, the spatial resolution (pixel dimension) that can be obtained depends on the magnetic field strength, $B_0$, the radius of the rf measuring coil (detector), $r$, and details of the experiment:

$$S/N \propto \frac{V}{r} \cdot B_0^{7/4} \left(\Delta f \cdot N_{av} \cdot N_{echo}\right)^{1/2} \tag{16}$$

Here $V$ is the pixel volume, and is defined by the number of pixels $N$ within the Field-of-View (FOV) of the image. $N_{av}$ is the number of averages, $N_{echo}$ the number of echoes used to construct or calculate the image. $\Delta f$ is the spectral width, representing the frequency range over the given FOV. It is inversely related to the dwell time, the time between successive sampled data points. The dwell time times $N$ is the time needed to detect the signal and determines the minimal echo time $T_E$. $\Delta f$ divided by the FOV defines the magnetic field gradient strength.

A number of different approaches can be followed to increase the spatial resolution (minimal $V$) at a certain $S/N$, at the same time trying to avoid increasing the measurement time (Scheenen, 2001). The $S/N$ of a pixel in an NMR image depends on the amount of water in that pixel. This is the product of tissue water content, tissue density and pixel volume: the larger a pixel, the lower the spatial resolution of the image, and the higher the $S/N$ of the pixel. In plant stems the high degree of tissue symmetry along the longitudinal axis provides a means to increase $S/N$. By increasing the slice thickness pixel volume is increased without affecting in plane spatial resolution.

Gain can also easily be obtained by optimizing $r$ with respect to (part of) the object to be measured. The smaller $r$, the higher $S/N$ (which may allow a smaller pixel volume $V$ at a fixed $S/N$). The best approach therefore is to construct rf detector coils that closely fit the object (Scheenen et al., 2002). Real microscopy therefore is limited to small objects. But small parts on even tall plants can be selected for MRI. In this way, e.g., anthers and seed pods, stems, (leaf) petioles and (fruit) stalks, still attached on intact plants, can be imaged with high spatial resolution. At increasing object size $r$ will increase and at the same time $N$ has to be increased if one would like to fix $V$. This will result in an increase of measurement time.

Another way to increase $S/N$ is to use higher field strength $B_0$. However, for plant tissues with extra-cellular air spaces this results in increased susceptibility artifacts, as discussed earlier. These artifacts can be overcome by increasing $\Delta f$ (and thus maximum $G$), but this would result in a decrease in $S/N$ and loss of $T_2$ information (see Section II.A). At higher $B_0$ the effective $T_2$ is (much) shorter than at lower field strength, limiting the number of measurable echoes ($N_{echo}$), again resulting in lower $S/N$.

Because both a good spatial resolution and a high $S/N$ per pixel are desirable, as well as an acceptable measurement time, every imaging experiment is a compromise between spatial resolution, $S/N$ and measurement time. The main consideration in this compromise should be the question what information needs to be extracted from the experiment. Quantitative $T_2$ imaging can even be severely hampered by a high spatial resolution (see Section II.A).

At lower resolution, the $S/N$ of one pixel can be sufficiently high for a meaningful multi-exponential fit (i.e., with acceptable standard deviations of the fitted parameters). This procedure results in two or more water fractions and corresponding relaxation times, which can be assigned to water in sub-cellular compartments (e.g., vacuolar and cytoplasmic water) within one pixel, thus giving access to a kind of sub-pixel resolution. Transient changes in $T_2$-values of the fractions in the tissues can be related to changes in cell size and/or exchange of water over the membranes separating the fractions (the water permeability of the vacuolar and plasmalemma membrane) (van der Weerd et al., 2001, 2002a).

Not only in quantitative $T_2$ imaging, but also in a MR flow imaging experiment, a high spatial resolution is not always necessary. The acquisition of propagators enables discrimination between stationary and flowing water at pixel level (see II.B.4). Then, another compromise has to be made between spatial resolution and the number of gradient steps encoding for flow. The choice depends on the question: what information is more important: an exact localization of flow, an accurate flow profile or a high temporal resolution?

## III. Applications Related to Photosynthesis Research

### A. Chloroplast Volume Regulation (CVR)

Photosynthetic oxygen production is sensitive to changes in chloroplast volume. Chloroplasts below or above an optimum volume produce less $O_2$ (Robinson, 1985; Sen Gupta and Berkowitz, 1988). CVR is the process by which chloroplasts import or export osmolytes to maintain a constant volume in a changing environment. Both light and water stress can result in such volume changes. It has been found that the ability to keep chloroplast volume constant extended to lower water potentials in chloroplasts isolated from plants that had experienced water stress than that that had not (Santakumari and Berkowitz, 1991). Below a critical value of the water potential CVR failed. Photosynthetic rates dropped suddenly at the same water potential as did chloroplast volumes. This critical water potential value is lower in the dark in sun leaves than in shade leaves (McCain, 1995a), and also depends on physiological levels of osmotic solutes (McCain, 1992).

NMR can be used to measure such changes in chloroplast water content, in combination with measurements of photosynthetic activity (McCain, 1992, 1995b). Such studies are especially of interest for plant performance studies under stress conditions. To study CVR in vivo two different NMR approaches are available: by using chemical shift differences of chloroplast and non-chloroplast water as observed in leaves of some deciduous trees, or by making use of combined spin relaxation and diffusion behavior to discriminate between chloroplast and non-chloroplast water.

#### 1. Using Chemical Shift Information

The proton spectrum of a plant leaf predominantly consists of the spectrum of water. Therefore we may expect the NMR spectrum to be a single peak. However, the leaves of some species display complex spectral patterns, which are dependent on the orientation of the leaf with respect to the main magnetic field (McCain et al., 1984). The origin of these spectra has been suggested to be heterogeneous magnetic susceptibilities and local dipolar magnetic fields in chloroplasts, in combination with the distribution and orientation of the chloroplasts in the leaf structure (McCain and Markley, 1985). In favorable cases a shift of chloroplast water resonance results. In such leaves chloroplast water and non-chloroplast water can be discriminated and quantified by using their respective resonance peaks. However, this method is certainly not applicable to most plant leaves.

#### 2. More Generalized Approach: Combined Diffusion—$T_2$ or $T_1$–$T_2$ Measurements

Water in different cell compartments can be discriminated on basis of the differences in relaxation behavior ($T_1$, $T_2$) and (restricted) diffusion behavior. An early demonstration has been presented by McCain, who studied spin relaxation and restricted diffusion behavior of chloroplast water, in combination with the spectral information observed in maple leaves (McCain 1995b). By combined relaxation and diffusion measurements, together with a recently developed efficient and stable two-dimensional fitting procedure based on a Fast Laplace Inversion algorithm (Hürlimann et al., 2002; Venkataramanan et al., 2002), two-dimensional correlation plots between $T_1$ and $T_2$ or $D$ and $T_2$ can now be generated, which greatly enhances the discrimination of different water pools in sub-cellular compartments. In this way an unambiguous correlation between compartment size and relaxation time can be obtained, resulting in a general approach to quantify water in the different cell compartments. This approach is very promising in non-spatially resolved measurements (Qiao et al, 2005), but can be applied to obtain sub-pixel information in images as well (H. Van As, F.J. Vergeldt, C.W. Windt, unpublished).

### B. Phloem and Xylem Transport

By being able to measure phloem flow in relation to xylem flow and plant water balance, phloem loading and unloading mechanisms under normal and stress conditions can be studied in relation to photosynthesis activity, e.g., monitored by PAM techniques. NMR imaging techniques have been presented which allow the measurement of water transport processes non-invasively in stems of plants (for overviews see MacFall and Van As, 1996; Köckenberger, 2001a). Several groups used different MRI methods. Most of them are based on (modified) PFG methods, either by a limited number of G steps or by (difference) propagator approaches (Callaghan et al., 1994). Also flow measurements, based on uptake and transport by (paramagnetic) tracers, have been used (Link and

Seelig, 1990; Clearwater and Clark, 2003).

The first NMR non-imaging method was presented by Van As and Schaafsma (Van As and Schaafsma, 1984; Reinders et al., 1988; Schaafsma et al., 1992). Based on that method a (trans)portable NMR bio-flowmeter has been developed with a u-shaped permanent magnet (open access from one side, Fig. 10a) and an openable, hinged, rf coil. It has been used in greenhouse situations on intact plants (Van As et al., 1994).

With flow-sensitive PFG NMR imaging, it was shown to be possible to measure xylem water flow in single vascular bundles of maize plants (Kuchenbrod et al., 1998). Using a different version of the PFG method, Köckenberger et al. (1997) were the first to measure xylem and phloem transport in an intact six day old castor bean seedling, growing in the dark. However, the approach used by Köckenberger et al. was very time consuming with a measurement time of 4.5 hours. A more rapid flow imaging method called PFG-FLASH was developed by Rokitta et al. (1999), which was shown to be capable of measuring phloem and xylem flow in 40 day old castor bean plants, placed horizontally in a high field super-conducting magnet (Peuke et al., 2001). However, the short measurement times of PFG-FLASH (between 4 and 7 minutes) did come at a price. The method yielded a quantitative average velocity per pixel, but did not give information about the velocity profile, nor did it allow the volume flow to be quantified in absolute units. Also, spatial resolution was reduced to speed up the imaging part of the NMR sequence.

At the same time, a *q*-space flow imaging method was developed that allowed the flow profile of every pixel in an image to be recorded quantitatively, with a relatively high spatial resolution (around 100 × 100 μm in plane resolution) and while keeping measurement times down to 15 to 30 minutes. This was done by combining flow encoding with a rapid (turbo-Spin Echo) NMR imaging scheme (Scheenen et al., 2000a, 2001) (Fig. 7). The quantitative flow profiles allowed the following parameters to be calculated on a per pixel basis: the amount of stationary water, the flow conducting area, the average velocity of the flowing water, and the volume flow (flux). The *q*-space flow imaging approach was used in combination with a low field MRI setup (see Section III.C) to visualize and quantify xylem flow in tomato (Scheenen et al., 2000a), in stem pieces of chrysanthemum (Scheenen et al., 2000b) and large cucumber plants (Scheenen et al., 2002). While in the last study the authors were able to visualize phloem sap movement, they were not yet able to quantify phloem flow in the same manner as was demonstrated for xylem flow. Xylem flow quantification was shown to be satisfactorily accurate. A comparison between the fluxes measured by different methods demonstrated that the MRI flow results are in close agreement with the water uptake by the roots. In addition, in a pixel containing a single vessel the vessel diameter could be calculated from the flow conducting area as measured by flow MRI. Assuming a cylindrical vessel, the calculated vessel diameter was within 10% agreement with the diameter of the same vessel as determined by microscopy after the MRI inspection (Scheenen, 2001).

Windt et al. (2006) further optimized this method as well as the hardware (see Section III.C). Doing so, they were the first to routinely and quantitatively measure detailed flow profiles (propagators) of xylem and phloem flow, on large and fully developed plants: a poplar tree, tomato, tobacco, and castor bean plants. An example of phloem and xylem flow propagator and images obtained in castor bean is presented in Fig. 8. Poplar showed clear differences in day and night phloem flow profiles, but the total volume flow remained unchanged. In tomato only small differences between day and night phloem flow profiles were observed. The velocity and the flow-conducting area appeared to decrease slightly at night, accompanied by slight differences between day and night time fluxes. The phloem flow characteristics of castor bean and tobacco did not exhibit any day-night differences. The phloem fluxes and flow conducting areas showed large differences that roughly corresponded with plant size. The differences in phloem flow velocities between the four species were remarkably small (0.25–0.40 mm/s). Diurnal xylem transport exhibited large day-night differences, in which changes in flux were accompanied by a decrease in velocity, but also by a decrease in flow-conducting area. The phloem to xylem flux ratio reflects the fraction of xylem water that is used for phloem transport, also known as Münch's counterflow. This ratio was surprisingly large at night for poplar (0.19), castor bean (0.37) and tobacco (0.55), but low in tomato (0.04) (Windt et al., 2006).

The plants could be placed upright in the dedicated NMR imager where they continued to grow and were measured for extended periods of time. The NMR-measurement can be combined with gas exchange measurements (Rokitta et al., 1999).

The combination of dedicated NMR/MRI equip-

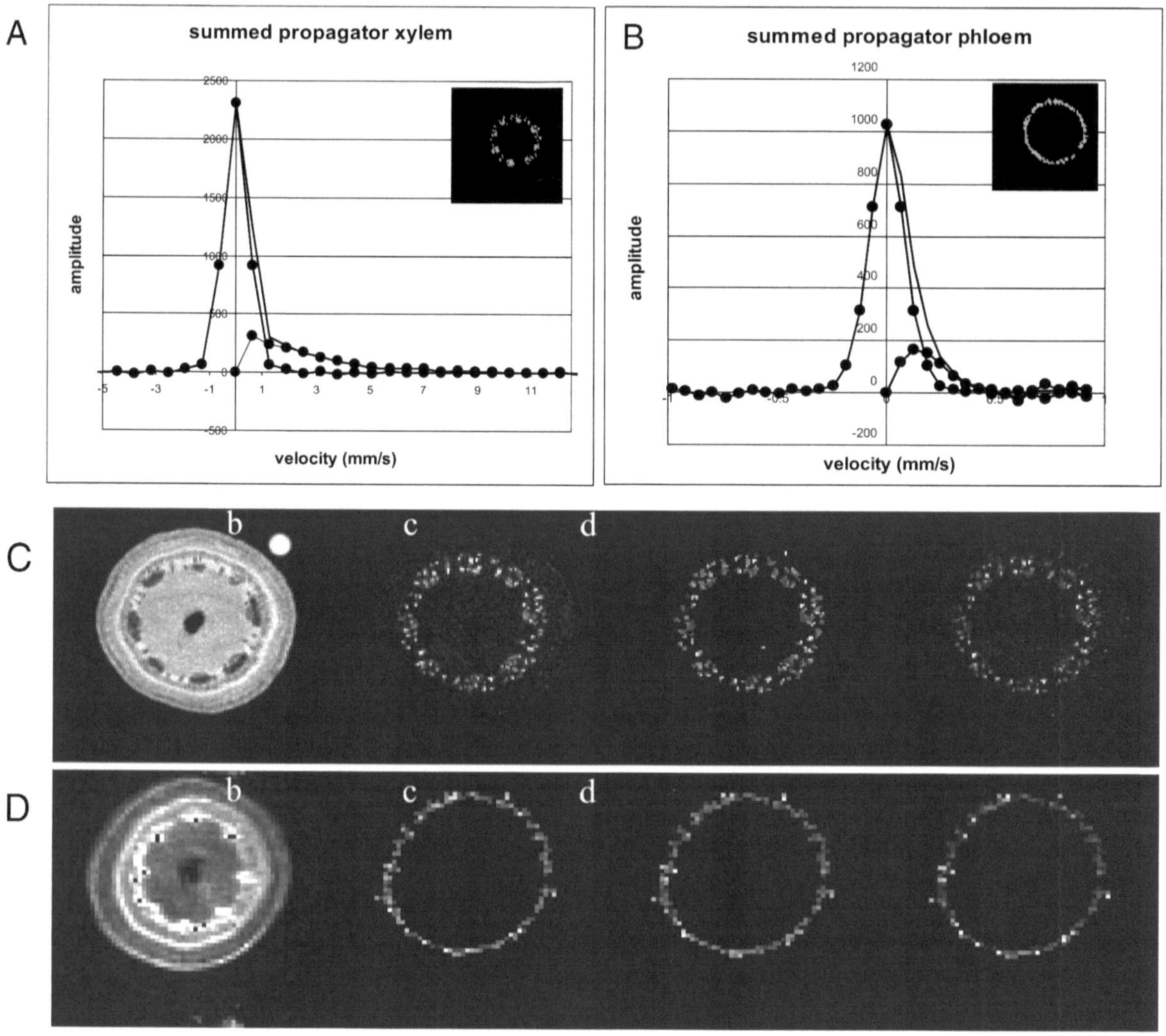


*Fig. 8.* A. Total propagators, presented as flow profiles, of the xylem and phloem region in the stem of a *Ricinus* plant. The images indicate the regions that have been selected for the summation. Note the difference in flow direction. B and C. Calculated images of total water (amplitude, a), amount of flowing water (flow conducting area per pixel, b), average linear velocity (c), and volume flow rate (d), for xylem (B) and phloem (C) flow. The propagators for xylem and phloem flow have been obtained at different values of Δ, which explains the difference in propagator axes in A and the amplitude images in B and C. (Reproduced from Peuke et al. (2006) with permission of Blackwell Publishing.)

ment (see next section) and NMR flow imaging methods is now available to routinely measure and quantify phloem as well as xylem transport in plants up to a size of several meters and over periods of weeks. Plant responses as a function of changes in environmental conditions can now be studied. The method was used by Peuke et al. (2006) to study the effects of cold treatment on mass flow in the phloem. A first example of the effect of an extended dark period (trying to stop photosynthesis and phloem loading) on phloem flow in castor bean (*Ricinus*) is shown in Fig. 9. The method has been applied to study day-night rhythm in stem, dynamics in phloem in response to stem cold girdling, root anoxia, and long dark periods, the flow (and changes therein) in the stalk of a tomato truss during a 5-weeks period of fruit development, and xylem air embolism induction and refilling.

### *C. Hardware Considerations*

In many cases dedicated hardware is required in order to image intact plants. NMR imagers that are used for plant work often have been developed for other

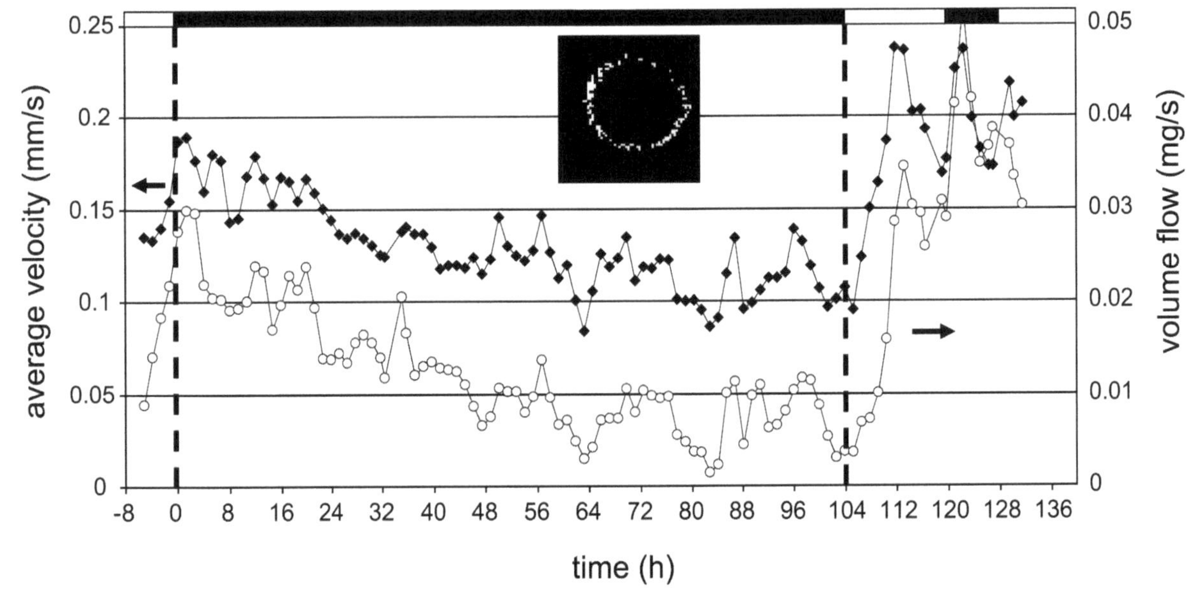


*Fig. 9.* Response observed in phloem flow (both volume flow and average velocity) in a *Ricinus* plant to an extended dark period of over four days. The position of the phloem is indicated by the image of the phloem flow mask.

purposes, such as medical imaging. As a result, many machines consist of horizontal bore superconducting magnets with cylindrical geometry of the magnetic field gradient coils. For plant work this is not an ideal setup. In such a system plants have to be placed horizontally instead of vertically, and fitting the shoot or roots of a plant through the narrow cylindrical bore of a gradient set can be stressful and damaging for the plant. Better solutions for minimal invasive studies on intact plants, with a (potted) root system and extended shoot (leaves), requires dedicated hardware: open access magnets, open access gradient coil systems, e.g., plan parallel gradient plates or openable gradient coils (see below), and climate control.

Normally, only small parts of the plant (e.g., stem, a leaf, petioles, seed pots, fruit stalk, etc) will be chosen for study. If so, an optimal *S/N* is obtained by optimizing the radius of the rf coil, *r*, with respect to that part of the plant to be measured (Eq. 16). The best approach is to construct rf detector coils that closely fit that part of the plant or tree to be imaged. Also, when the distance between the magnetic field gradient coils is made the smaller, higher gradient strengths can be obtained. This is important to obtain a high spatial resolution, fast switching of gradients and measure low flow velocities: small displacements require a high *q* or *G* value to be detected (Eqs. 11 and 12).

Some hardware solutions for intact plant NMR and MRI are presented in Figs. 10 and 11. In Fig. 10 two examples of open access low field magnets are presented. In Fig. 10a the permanent magnet that has been used in the NMR bioflowmeter (0.24 T) (Van As et al., 1994) is shown. It combines a transportable (30 kg) open access u-shaped permanent magnet with an openable (hinged Helmholtz) rf coil. Plants can easily be inserted into the magnet from the front. In Fig. 10b and c a low field (0.73 T) imaging system based on an open access electromagnet is shown. On top of the magnet a climate chamber for environmental control is placed. Maximal access into the center of the magnet is obtained by use of plan parallel gradient plates (maximum gradient strength 1 T/m) (Fig. 10c). To maximize the filling factor a (solenoid) rf coil is wrapped directly around the stem of the plants. A solenoid coil results in about a factor 3 more sensitivity than Helmholtz or birdcage coils of the same diameter. This configuration allowed us to easily place large plants, up to a size of two meters, upright in the NMR magnet. Placement of plants can be undertaken without causing much stress to the subject, other than the stress that is caused by moving and handling it. Potentially stressful actions like mounting an RF coil or, when necessary, the removal of a branch or leaf can be undertaken well in advance.

In Fig. 11 a dedicated intact plant 3 T MRI system is shown. The cryo-magnet has a 50 cm vertical free bore (Fig. 11a). One of the gradient coils is openable.

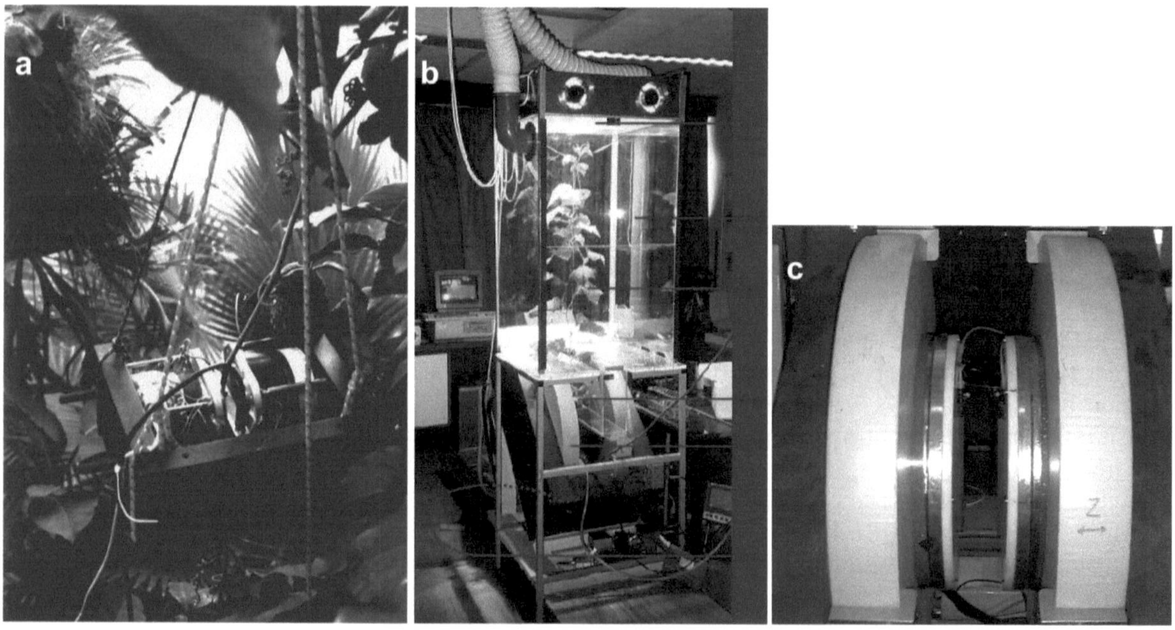

*Fig. 10.* a. U-shaped permanent magnet in action to measure flow in different heights in a liana in a tropical greenhouse. b and c. Magnet and climate control unit of a 0.7 Tesla imager based on an electromagnet. The plan parallel gradient coils plates results in maximum access to the center of the magnet (c).

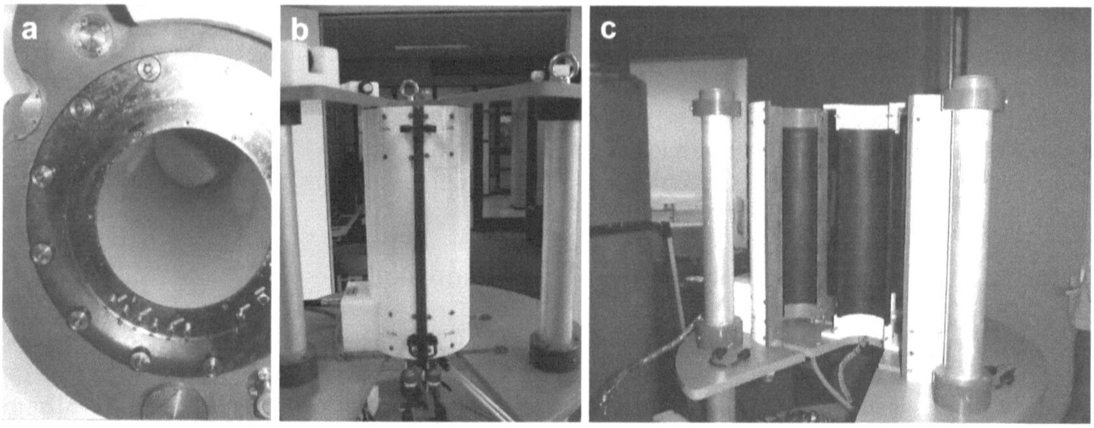

*Fig. 11.* a. 50 cm diameter vertical free bore of the cryo-magnet of an intact plant 3 Tesla MRI system. Within this bore air temperature, humidity and gas composition and light intensity can be varied and controlled. b and c. Gradient coil insert that can be opened (c) and closed around, e.g., a plant stem. An rf coil that consists of two parts, completes the fully openable construction.

It consists of four parts and can easily be mounted around the stem of a plant or trunk of a tree (Figs. 11b and c). The maximum gradient strength that can be generated is 1 T/m. An rf coil (4 cm inner diameter) that consists of two parts completes the fully openable construction. Inside the bore of the magnet the climate is controlled by use of a remote climate control unit. This 3-T MRI system provides an excellent and unique infrastructure for MRI on intact plants and trees.

## IV. Future of Functional Plant MRI

NMR and MRI techniques certainly deliver an abundant amount of information in addition to a reasonable spatial and temporal resolution. Part of this information is very difficult to measure, or cannot be measured at all using other techniques. Functional imaging by flow MRI offers exciting new possibilities for the physiological mapping of intact plants. For the first time, MRI allows the physiological state of individual plant organs to be described, as well as their inter-relationship at the whole plant level. By use of dedicated hardware xylem and phloem flow can be studied, and its mutual interaction, in relation to the plant water balance. This is key information to test old theories about transport processes (Cohesion Tension (CT) theory for xylem transport; Münch pressure flow hypothesis for phloem) within plants and trees. The CT theory is heavily under debate (Steudle, 2001; Zimmermann et al., 2004). For the first time by using flow MRI, Münch counter flow was quantified and the flow profiles and dynamics of xylem and phloem in intact plants observed (Windt et al., 2006).

In addition to water balance and transport, metabolic information like distribution and flow of nutrients (e.g., sugars) is key information to study plant performance in relation to, e.g., photosynthesis. At higher magnetic fields (3 Tesla and higher) it is possible to measure spatially resolved metabolite concentration by chemical shift (spectroscopic) information (Tse et al., 1996; Köckenberger, 2004). Using current 'state of the art' methods it is possible to measure soluble carbohydrates/sugars and amino acids *in planta*. Direct $^{1}$H-NMR spectroscopic imaging (Metzler et al., 1995; Vertsch et al., 1998; Wolf et al., 2000), $^{13}$C spectroscopic imaging (Kalusche et al., 1999) or $^{1}$H detected $^{13}$C (fast) imaging (Hudson et al., 2002) have been demonstrated in plants. Non-spatially resolved flow encoded (PFG) NMR spectroscopy for metabolite transport has also been reported (Szimtenings et al., 2003). These methods are far from optimized, and should be further employed. The next step to be realized is to combine water balance and transport studies with metabolite balance and transport measurements non-invasively in the intact plant situation. Understanding plant physiology at this level will allow many as yet unanswered questions about plant performance and productivity, development and stress responses to be answered, and open up exciting new ways to understand the relationships between growth, productivity, stress tolerance and competitiveness of plants. In addition, functional MRI is expected to become a powerful tool to characterize functionality of gene-expression, by allowing the action of certain proteins to be studied. In this way it will contribute to resolve to role of e.g., aquaporins in plant water balance and stress tolerance (Chrispeels and Maurel, 1994; Tournaire-Roux et al., 2003; Lee et al., 2005).

Currently MRI in plant science is still far from being a routine tool. Some reasons for this could be the high costs of MRI equipment, the difficult theoretical fundamentals on which the technique is based, the horizontal orientation of most of the standard imaging setups, the limited accessibility (for large plants) of the magnetic field's iso-center of most setups, and the specific (climatic) requirements which have to be met when measuring plants. It is to be expected, however, that relatively cheap imaging setups based on permanent magnet systems are now becoming available (Rokitta et al., 2000; Haishi et al., 2001).

For NMR flow measurements to be applicable in situ (greenhouses, field situations), quantitative non-spatially resolved (non-imaging) methods with specifically designed magnets have to be developed. The basic principles and components are now readily available to develop (reasonably priced) NMR plant flow meters, as are specially designed magnets (Raich and Blümler, 2004). We can also think about small NMR or MRI detectors by making use of rather small magnets, as presented by Blümich and others (Blümich et al., 2002). Such systems can readily be combined with PAM equipment for combined NMR and photosynthesis (activity) monitoring.

## Acknowledgments

The research described in part herein was supported by the Dutch Technology Foundation (STW), Applied Science Division of the Netherlands Organization for Scientific Research (NWO) (projects WAW07.0068, WBI 3493 and WBI 4803). The funding of the dedicated intact plant 3 Tesla MRI facility of Wageningen NMR Centre by NWO, the Dutch Ministery of Agriculture, Nature and Food quality (LNV) and the Wageningen University and Research Centre is greatly acknowledged.

HVA is indebted to his coworkers Frank Vergeldt and Edo Gerkema, PhD students Natalia Homan and Timur Sibgatullin, MSc students Paul ter Horst and Kim de Lange, and former PhD students Drs Dagmar van Dusschoten, Tom Scheenen, Louise van der Weerd, Angelien Snaar, Johan Reinders, and last, but not least, Prof. Tjeerd Schaafsma, who all contributed to the development of the research reported in this Chapter. Our late colleague Adrie de Jager strongly contributed to the hardware developments and the definition of the 3 Tesla MRI system.

Trans-European cooperation on this research has been stimulated by financial support of the EU Transnational Access to Research Infrastructure programs in the period 1994–2004.

## References

Bendel P, Zemah H, Kamenetsky R, Vergeldt F and Van As H (2001) Magnetization transfer and double-quantum filtered imaging as probes for motional restricted water in tulip bulbs. Magn Reson Imaging 19: 857–865

Blümich B, Anferov V, Anferova S, Klein M, Fechete R, Adams M, Casanova F (2002) Simple NMRMOUSE with a bar magnet. Concepts Magn Reson 15B:255–261

Brownstein KR and Tarr CE (1979) Importance of classical diffusion in NMR studies of water in biological cells. Phys Rev A 19: 2446–2453

Callaghan PT (1993) Principles of Nuclear Magnetic Resonance Microscopy. Clarendon Press, Oxford

Callaghan PT, Eccles CD and Xia Y (1988) NMR microscopy of dynamic displacements: k-space and q-space imaging. J Phys E: Sci Instrum 21: 820–822

Callaghan PT, Köckenberger W and Pope JM (1994) Use of difference propagators of imaging of capillary flow in the presence of stationary fluid. J Magn Reson B 104: 183–188

Callaghan PT, Codd SL and Seymour JD (1999) Spatial coherence phenomena arising from translational spin motion in gradient spin echo experiments. Concepts Magn Reson 11: 181–202

Chrispeels MJ and Maurel C (1994) Aquaporins: The molecular basis of facilitated water movement through living plant cells? Plant Physiol 105: 9–13

Chudek JA and Hunter G. (1997) Magnetic resonance imaging of plants. Progr Nucl Magn Reson Spectrosc 31: 43–62

Ciobanu L, Webb AG and Pennington CH (2003) Magnetic resonance imaging of biological cells. Progr Nucl Magn Reson Spectrosc 42: 69–93

Clearwater MJ and Clark CJ (2003) In vivo magnetic resonance imaging of xylem vessel contents in woody lianas. Plant Cell Environm 26: 1205–1214

Conlon T and Outhred R (1972) Water diffusion permeability of erythrocytes using an NMR technique. Biochim Biophys Acta 288:354–361

Daudet FA, Lacointe A, Gaudillère JP, and Cruiziat P (2002) Generalized Münch coupling between sugar and water fluxes for modeling carbon allocation as affected by water status. J Theor Biol 214: 48–498

Donker HCW and Van As H (1999) Cell water balance of while button mushrooms (*Agaricus bisporus*) during its post-harvest lifetime studied by quantitative magnetic resonance imaging. Biophys Biochim Acta 1427: 287–297

Donker HCW, Van As H, Edzes HT and Jans AWH (1996) NMR imaging of white button mushroom (*Agaricus bisporus*) at various magnetic fields. Magn Reson Imaging 14: 1205–1215

Donker HCW, Van As H, Snijder HJ and Edzes HT (1997) Quantitative 1H-NMR imaging of water in white button mushrooms (*Agaricus bisporus*). Magn Reson Imaging 15: 113–121

Durrant CJ, Hertzberg MP and Kuchel PW (2003) Magnetic susceptibility: Further insights into macroscopic and microscopic fields and the sphere of Lorentz. Concepts Magn Reson A 18: 72–95

Edzes HT, van Dusschoten D and Van As H (1998) Quantitative $T_2$ imaging of plant tissues by means of multi-echo MRI microscopy. Magn Reson Imaging 16: 185–196

Fukushima E (1999) Nuclear magnetic resonance as a tool to study flow. Ann Rev Fluid Mechan 31: 95–123

Haishi T, Uematsu T, Matsuda Y and Kose K (2001) Development of a 1.0 T MR microscope using a Nd-Fe-B permanent magnet. Magn Reson Imaging 19: 875–880

Hills BP and Duce SL (1990) The influence of chemical and diffusive exchange on water proton transverse relaxation in plant tissues. Magn Reson Imaging 8: 321–332

Howseman AM, et al. (1988) Improvements in snapshot nuclear magnetic imaging. Br J Radiol 61: 822–828

Hudson A, Köckenberger W, Heidenreich M, Chandrakumar N, Kimmich R and Bowtell R (2002) $^1$H detected $^{13}$C Echo Planar Imaging. J Magn Reson 155: 64–71

Hürlimann MD, Venkataramanan L and Flaum C (2002) The diffusion-spin relaxation time distribution as an experimental probe to characterize fluid mixtures in porous media. J Chem Phys 117: 10223–10232

Ishida N, Koizumi M and Kano H (2000) The NMR microscope, a unique and promising tool for plant science. Ann Bot 86: 259–278

Kalusche B, Verscht J, Gebauer G, Komor E and Haase A (1999) Sucrose unloading in the hypocotyl of the *Ricinus communis* L-seedling measured by C-13-nuclear magnetic resonance spectroscopy in vivo. Planta 208: 358–364

Kärger J and Heink W (1983) The propagator representation of molecular transport in microporous crystallites. J Magn Reson 51: 1–7

Köckenberger W (2001a) Functional imaging of plants by magnetic resonance experiments. Trends Plant Sci 6: 286–292

Köckenberger W (2001b) Nuclear magnetic resonance microimaging in the investigation of plant cell metabolism. J Exp Bot 52: 641–652

Köckenberger W, Pope JM, Xia Y, Jeffrey KR, Komor E and Callaghan PT (1997) A non-invasive measurement of phloem and xylem water flow in castor bean seedlings by nuclear magnetic resonance microimaging. Planta 201: 53–63

Köckenberger W, De Panfilis C, Santoro D, Dahiya P and Rawsthoine S (2004) High resolution NMR microscopy of plants and fungi. J Microsc 214: 182–189

Kuchenbrod E, Kahler E, Thurmer F, Deichmann R, Zimmermann U and Haase A (1998) Functional magnetic resonance imaging in intact plants — Quantitative observation of flow in plant vessels. Magn Reson Imaging 16: 331–338

Lee SH, Chung GC and Steudle E (2005) Low temperature and mechanical stresses differently gate aquaporins of root cortical cells of chilling-sensitive cucumber and -resistant figleaf gourd. Plant Cell Environ 28: 1191–1202

Levitt M (2001) Spin Dynamics: Basics of Nuclear Magnetic Resonance. Wiley, Chichester

Link J and Seelig J (1990) Comparison of deuterium NMR

imaging methods and applications to plants. J Magn Reson 89: 310—330
Lüdecke KM, Röschmann P and Tischler R (1985) Susceptibility artifacts in NMR imaging. Magn Reson Imaging 3: 329–343.
MacFall JJ and Van As H (1996) Magnetic resonance imaging of plants. In: Shachar-Hill Y and Pfeffer PE (eds) Nuclear Magnetic Resonance in Plant Biology, pp 33–76. The American Society of Plant Physiologists, Rockville
Mantle MD and Sederman AJ (2003) Dynamic MRI in chemical process and reaction engineering. Prog Nucl Magn Reson Spectrosc 43: 3–60
McCain D (1992) In vivo study of chloroplast volume regulation. Biophys J 61: 1207–1212
McCain D (1995a) Combined effects of light and water stress on chloroplast volume regulation. Biophys J 69: 1105–1110
McCain D (1995b) Nuclear magnetic resonance study of spin relaxation and magnetic field gradients in maple leaves. Biophys J 69: 1111–1116
McCain D and Markley JL (1985) A theory and model for interpreting the proton NMR spectra of water in plant leaves. Biophys J 48: 687–694
McCain D, Selig TC Govindjee and Markley JL (1984) Some plant leaves have orientation-dependent EPR and NMR spectra. Proc Natl Acad Sci USA 81: 748–752
Mesnard F and Ratcliffe RG (2005) NMR analysis of plant nitrogen metabolism. Photosynth Res 83: 163–180
Metzler A, Izquierdo M, Ziegler A, Köckenberger W, Komor E, Von Kienlin M, Haase A and Decorps M (1995) Plant histochemistry by correlation peak imaging Proc Natl Acad Sci USA 92: 11912–11915
Norris DG (2001) The effects of microscopic tissue parameters on the diffusion weighted magnetic resonance imaging experiment. NMR in Biomed 14: 77–93
Peuke AD, Rokitta M, Zimmermann U, Schreiber L and Haase A (2001) Simultaneous measurement of water flow velocity and solute transport in xylem and phloem of adult plants of *Ricinus communis* over a daily time course by nuclear magnetic resonance spectrometry. Plant Cell Environm 24: 491–503
Peuke AD, Windt C and Van As H (2006) Effects of cold-girdling on flows in the transport phloem in *Ricinus communis*: Is mass flow inhibited? Plant Cell Environm 29: 15–25
Qiao Y, Galvosas P and Callaghan PT (2005) Diffusion correlation NMR spectroscopic study of anisotropic diffusion of water in plant tissue. Biophys J 89: 2899–2905
Raich H and Blümler P (2004) Design and construction of a dipolar Halbach array with a homogeneous field from Identical bar Magnets: NMR Mandhalas. Concepts Magn Reson 23B: 16–25
Ratcliffe RG (1994) In vivo NMR studies of higher plants and algae. Adv Bot Res 20: 42–123
Ratcliffe RG, Roscher A and Shachar-Hill Y (2001) Plant NMR spectroscopy. Progr Nucl Magn Reson Spectrosc 39: 267–300
Reinders JEA, Van As H, Schaafsma TJ, Sheriff DW (1988) Water balance in *Cucumis* plants, measured by nuclear magnetic resonance, I & II. J Exp Bot 39: 1199–1220
Robinson SP (1985) Osmotic adjustment by intact chloroplasts in response to osmotic stress and its effect on photosynthesis and chloroplast volume. Plant Physiol 79: 996–1002
Rokitta M, Peuke AD, Zimmermann U and Haase A (1999) Dynamic studies of phloem and xylem flow in fully differentiated plants by fast nuclear-magnetic-resonance microimaging. Protoplasma 209: 126–131
Rokitta M, Rommel E, Zimmermann U and Haase A (2000) Portable nuclear magnetic resonance imaging system. Rev Sci Instrum 71: 4257–4262
Santakumari M and Berkowitz GA (1991) Chloroplast volume: Cell water potential relationships and acclimation of photosynthesis to leaf water deficits. Photosynth Res 28: 9–20
Schaafsma TJ, Van As H, Palstra WD, Snaar JE, de Jager PA (1992) Quantitative measurement and imaging of transport processes in plants and porous media by $^1$H NMR. Magn Reson Imaging 10: 827–36
Scheenen TWJ (2001) Nuclear magnetic resonance imaging of water motion in plants. PhD Thesis (dissertation nr. 3032), Wageningen University
Scheenen TWJ, van Dusschoten D, de Jager PA and Van As H (2000a) Microscopic displacement imaging with pulsed field gradient turbo spin-echo NMR. J Magn Reson 142: 207–215
Scheenen TWJ, van Dusschoten D, de Jager PA and Van As H (2000b) Quantification of water transport in plants with NMR imaging. J Exp Bot 51: 1751–1759
Scheenen TWJ, Vergeldt FJ, Windt CW, de Jager PA and Van As H (2001) Microscopic imaging of slow flow and diffusion: A pulsed field gradient stimulated echo sequence combined with turbo spin echo imaging. J Magn Reson 151: 94–100
Scheenen TWJ, Heemskerk AM, de Jager PA, Vergeldt FJ and Van As H (2002) Functional imaging of plants: A Nuclear Magnetic Resonance study of a cucumber plant. Biophys J 82: 481–492
Sen Gupta A and Berkowitz GA (1988) Chloroplast osmotic adjustment and water stress effects on photosynthesis. Plant Physiol. 88: 200–206
Shachar-Hill Y and Pfeffer PE (eds) (1996) Nuclear Magnetic Resonance in Plant Biology. American Society of Plant Physiologists, Rockville
Snaar JEM and Van As H (1992) Probing water compartment and membrane permeability in plant cells by proton NMR relaxation measurements. Biophys J 63: 1654–1658
Stapf S and Han S-I (eds) (2005) NMR Imaging in Chemical Engineering. Wiley, Weinheim
Stejskal EO and Tanner JE (1965) Spin diffusion measurements: Spin echoes in the presence of a time-dependent field gradient. J Chem Phys 42: 288–292
Steudle E (2001) The cohesion-tension mechanism and the acquisition of water by plant roots. Ann Rev Plant Physiol Plant Mol Biol 52: 847–875
Szimtenings M, Olt S and Haase A (2003) Flow encoded NMR spectroscopy for quantification of metabolite flow in intact plants. J Magn Reson 161: 70–76
Tournaire-Roux C, Sutka M, Javot H, Gout E, Gerbeau P, Luu DT, Bligny R and Maurel C (2003) Cytosolic pH regulates root water transport during anoxic stress through gating of aquaporins. Nature 425: 393–397
Tse TY, Spanswick RM and Jelinski LW (1996) Quantitative evaluation of NMR and MRI methods to measure sucrose concentrations in plants. Protoplasma 194: 54–62
Van As H and Schaafsma TJ (1984) Noninvasive measurement of plant water-flow by nuclear magnetic resonance. Biophys J 45: 496–472
Van As H, Reinders JEA, de Jager PA, van der Sanden PACM and

Schaafsma TJ (1994) In situ plant water-balance studies using a portable NMR spectrometer. J Exp Bot 45: 61–67

van der Weerd L, Claessens MMAE, Ruttink T, Vergeldt FJ, Schaafsma TJ and Van As H (2001) Quantitative NMR microscopy of osmotic stress responses in maize and pearl millet. J Exp Bot 52: 2333–2343

van der Weerd L, Claessens MMAE, Efdé C and Van As H (2002a) Nuclear Magnetic Resonance imaging of membrane permeability changes in plants during osmotic stress. Plant Cell Environ 25: 1538–1549

van der Weerd L, Melnikov SM, Vergeldt FJ, Novikov EG and Van As H (2002b) Modelling of self-diffusion and relaxation time NMR in multi-compartment systems with cylindrical geometry. J Magn Reson 156: 213–221

van Dusschoten D, de Jager PA, Van As H (1995) Extracting diffusion constants from echo-time-dependent PFG NMR data using relaxation-time information. J Magn Reson A 116: 22–28

van Zijl, PC and Le Bihan D (eds) 2002 Diffusion Tensor Imaging and Axonal Mapping–State of the Art. Special Issue. NMR Biomed 15 (7-8): 431–601

Venkataramanan L, Song YQ, Hürlimann MD (2002) Solving Fredholm integrals of the first kind with tensor product structure in 2 and 2.5 dimensions. IEEE Transaction on Signal Procession 50: 1017–1026

Verscht J, Kalusche B, Kohler J, Köckenberger W, Metzler A, Haase A, Komor E (1998) The kinetics of sucrose concentration in the phloem of individual vascular bundles of the *Ricinus communis* seedling measured by nuclear magnetic resonance microimaging. Planta 205: 132–139

Walter L, Callies R and Altenburger R (1992) Studies of plant systems by in vivo NMR spectroscopy. In: de Certaines JD, Bovée WMMJ and Podo F (eds) Magnetic Resonance Spectroscopy in Biology and Medicine, pp 573–610. Pergamon Press, Oxford

Windt CW, Vergeldt FJ, de Jager PA and Van As H (2006) MRI of long distance water transport: A comparison of the phloem and xylem flow characteristics and dynamics in poplar, castor bean, tomato and tobacco. Plant Cell Environ 29: 1715–1729

Wolf K, van der Toorn A, Hartmann K, Schreiber L, Schwab W, Haase A and Bringmann G (2000) Metabolite monitoring in plants with double-quantum filtered chemical shift imaging. J Exp Bot 51: 2109–2117

Zhang WH and Jones G.P (1996) Water permeability in wheat root protoplasts determined from nuclear magnetic resonance relaxation times. Plant Science 118: 97–106

Zimmermann U, Schneider H, Wegner LH and Haase A (2004) Water ascent in tall trees: Does evolution of land plants rely on a highly metastable state? New Phytol 162: 575–615

# Chapter 5

# Crystallization Methods of Membrane Proteins: Practical Aspects of Crystallizing Plant Light-Harvesting Complexes

Zhenfeng Liu[†] and Wenrui Chang*
*National Laboratory of Biomacromolecules, Institute of Biophysics, Chinese Academy of Sciences, 15 Datun Road, Chaoyang District, Beijing 100101, People's Republic of China*



## Summary

The light reactions in photosynthesis involve a series of membrane protein complexes residing in photosynthetic membranes. The three-dimensional structures of these complexes are indispensable for a better understanding of the underlying molecular mechanism of light reactions. The key step in the structure determination of these complexes by X-ray crystallography is crystallization. Here we review the major aspects concerning membrane protein crystallization. Membrane protein crystals can be classified into three basic types according to their different packing patterns. Membrane proteins are purified and crystallized in detergent solutions. The basic properties of detergents and some considerations about detergents are discussed. A homogeneous membrane protein sample with detergent concentration properly controlled is a good start for crystallization.

*Author for correspondence, email: wrchang@sun5.ibp.ac.cn
[†]Present address: Howard Hughes Medical Institute, California Institute of Technology, Mail Code 114-96, Pasadena, CA 91125, U.S.A.

*Thijs J. Aartsma and Jörg Matysik (eds): Biophysical Techniques in Photosynthesis II, pp. 77–96.*

It is not difficult to find initial crystallization conditions for membrane proteins, but further optimization for high-quality crystals suitable for structure determination is a challenging task. The optimization strategies include additive screen, detergent exchange and protein modification, in addition to adjustments on regular factors like temperature, pH and precipitants. In the second part of this chapter, we present a practical case of photosynthetic membrane protein crystallization and optimization. The plant major light-harvesting complex of Photosystem II was crystallized in four different forms. One of them was highly ordered in all three dimensions and could diffract X-rays to 2.5 Å resolution. It was obtained after careful optimization of detergent, lipid and additives. This crystal form belongs to a novel type of membrane protein crystal, type III, which is built from icosahedral proteoliposome vesicles. The formation of rigid and homogenous proteoliposomes is a prerequisite for the growth of well-ordered type III crystals.

## I. Review of Membrane Protein Crystallization

### A. Overview

Light reactions in photosynthesis are catalyzed by a series of membrane protein complexes residing in the photosynthetic membranes. These involve the absorption and transmission of light energy by light-harvesting complexes, electron transfer within and between photosynthetic reaction centers, pumping of protons across the membrane by cytochrome $b_6f$ in plants or $bc_1$ complex in photosynthetic bacteria and the production of ATP by ATP synthase powered by the transmembrane proton gradient. A full understanding of the mechanisms underlying these processes requires the determination of high-resolution structures of these membrane protein complexes. Presently, three major methods have been successfully applied in solving the three-dimensional structures of biological macromolecules, namely X-ray crystallography, electron crystallography and NMR spectroscopy. For structure determination with X-ray crystallography it is indispensable to obtain well-ordered three-dimensional crystals. This is one of the rate-limiting steps of X-ray crystallography, especially of membrane proteins. There are actually two major hurdles in getting crystals suitable for structure determination, one in the preparation of enough pure material for crystallization experiments and the other in finding optimal conditions for crystallization.

As we know, photosynthetic and respiratory membrane complexes are highly abundant in natural sources. Therefore, it is relatively easy to purify them in large amount. However, this is not the case for most other membrane proteins, which are only expressed at low level in nature. It is almost impossible to purify them in milligram quantity from natural sources. Thanks to the advance of molecular biology techniques, researchers are now able to over-express some membrane proteins in heterologous cellular systems, like *Escherichia coli*. With these techniques, we now have much more freedom than before to select membrane proteins of special interest to work on. On the other side, the current concern about the issue of membrane protein expression is that most of the successful cases are still limited to prokaryotic membrane proteins expressed in prokaryotic systems. Heterologous expression of eukaryotic membrane proteins with proper folding and function remains to be an especially challenging task. Intense functional studies have been performed on eukaryotic membrane proteins for many years and are becoming more and more hampered by the lack of structural information nowadays. This is mostly due to the unavailability of enough pure sample for structural study.

Membrane proteins are endowed with a unique amphipathic feature on their surfaces. This characteristic enables them to adapt to biological membranous

---

*Abbreviations:* $C_8E_4$ – n-octyl-tetraoxyethylene; $C_8E_5$ – n-octyl-pentaoxyethylene; $C_{10}E_5$ – n-decyl-pentaoxyethylene; $C_{12}E_8$ – n-dodecyl-octaoxyethylene; $C_{12}E_9$ – n-dodecyl-nonaoxyethylene; Chl – chlorophyll; CMC – critical micelle concentration; CTAB – cetyl trimethyl-ammonium bromide; DBC – N,N-bis-(3-D-Gluconamidopropyl) deoxycholamide (deoxy-bigchap); DDAO – n-decyl-N,N-dimethylamine-N-oxide; DDM – n-dodecyl-β-D-maltopyranoside; DG – n-decyl-β-D-glucopyranoside; DGDG – digalactosyl diacylglycerol; DM – n-decyl-β-D-maltopyranoside; FC-14 – n-tetradecyl-phosphocholine (Fos-choline 14); LDAO – n-dodecyl-N,N-dimethylamine-N-oxide; LHC II – plant major light-harvesting complex of Photosystem II; LLG – 1-linolenoyl-rac-glycerol; MEGA-10 – decanoyl-N-methylglucamide; MGDG – monogalactosyl diacylglycerol; MscL – mechanosensitive channel of large conductance; NG – n-nonyl-β-D-glucopyranoside; NMR – nuclear magnetic resonance; NTM – n-nonyl-β-D-thiomaltopyranoside; OG – n-octyl-β-D-glucopyranoside; OM – n-octyl-β-D-maltopyranoside; OTG – n-octyl-β-D-thioglucopyranoside; PA – phosphatidic acid; PAGE – polyacrylamide gel electrophoresis; PC – phosphatidyl choline; PG – phosphatidyl glycerol; SDS – sodium dodecyl sulfate; UDAO – n-undecyl-N,N-dimethylamine-N-oxide; UDM – n-undecyl-β-D-maltopyranoside

environment and exert their function properly within the membrane. The extensive hydrophobic surface of membrane proteins is embedded in the lipid bilayer of the membrane, and is covered by the hydrophobic tails of lipid molecules. The purification procedures involve the dissociation of membrane proteins from the membrane, followed by several additional steps of biochemical work in aqueous solution. The problem is that membrane proteins would remain insoluble if they were left in aqueous solution without proper treatments. To address this problem, an amphiphilic reagent called detergent can be introduced in the solution to assist in the solubilization of membrane proteins. An ideal detergent should be able to fully solubilize membrane proteins without disturbing their structures and preserve the original activity of membrane proteins. The detergent factors, including species and concentration, must be taken into consideration when one is dealing with membrane proteins. Thus they add to the complexity and difficulty of membrane protein purification and crystallization.

### B. Three Different Types of Membrane Protein Crystals

Membrane protein crystals obtained so far can be classified into three basic types according to their different packing patterns. Type I and II crystals are originally described by Michel (1983). In a type I crystal (Fig. 1a), membrane proteins form a two-dimensional lattice and then the 2-D crystals stack in the third dimension to form a 3-D crystal. Lipid molecules may be involved in the formation of the 2-D lattice by mediating the interactions between membrane proteins. They are supposed to form a bilayer structure just as in the membrane. Two forces, hydrophobic and polar interactions, respectively, stabilize the crystal packing within the 2-D plane. In the third dimension, only polar interactions between hydrophilic surfaces of membrane proteins and/or lipids are involved in the crystal packing. In practice, it is very difficult to grow large and ordered type I crystals since there seems to be no easy way to increase both hydrophobic and polar interactions in a crystallization trial (Michel, 1983). Therefore, type I crystals are frequently discovered to bear an anisotropic nature. They are usually well-ordered in the 2-D plane, but disordered or only partially ordered in the third dimension due to the weak packing force in this dimension. Morphologically, they often occur as thin plates. In spite of this problem, there are still chances of getting high quality type I crystals suitable for structure determination. For example, type I crystals of bacteriorhodopsin (Luecke et al., 1999, 2001) and calcium ATPase (Toyoshima et al., 2000) diffracted to 1.5 Å and 2.6 Å resolution, respectively. The former one was crystallized in lipidic cubic phase (Landau and Rosenbusch, 1996), while crystallization of the latter was achieved by dialyzing the mixture of protein and phospholipid against crystallization solution. The application of a lipidic cubic phase in membrane protein crystallization has been shown to be fairly successful in producing type I crystals of sensory rhodopsin (Luecke et al., 2001; Royant et al., 2001), halorhodopsin (Kolbe et al., 2000) and bacterial photosynthetic reaction center (Katona et al., 2003), in addition to bacteriorhodopsin (Landau and Rosenbusch, 1996). These crystals have all yielded X-ray structure at high resolution.

A high percentage of membrane protein crystals adopt type II packing. As shown in Fig. 1b, the building blocks of type II crystal are the complexes of membrane protein and detergent micelle. The crystal packing is mainly reinforced by the polar interactions between the hydrophilic surfaces of membrane proteins. In an ideal type II crystal, the detergent micelles which embrace membrane proteins should have a proper size, and should fit into a crystal lattice without interfering with interactions between proteins. Therefore, screening for the most suitable detergent for the crystallization of a specific membrane protein is a critical step in obtaining well-ordered type II crystal. There are several strategies one can employ to enlarge the hydrophilic regions of membrane proteins and improve type II crystal quality consequently. The first idea is to add some small amphiphilic molecules into the detergent micelles to form mixed micelles (Michel, 1983). This may reduce the size of the detergent belt around membrane protein and lead to the exposure of more hydrophilic regions available for crystal contacts (Timmins et al., 1991). The second idea is to find some soluble proteins, which can bind to membrane proteins specifically, such as the Fv or Fab fragment of monoclonal antibody (Iwata et al., 1995; Zhou et al., 2001). This will enlarge the hydrophilic parts of membrane proteins greatly and provide a high possibility of getting new crystal contact sites.

A novel type of crystal packing was described for the first time in the report of the crystal structure of spinach LHC II (Liu et al., 2004). It was categorized as type III since it is completely different from type I

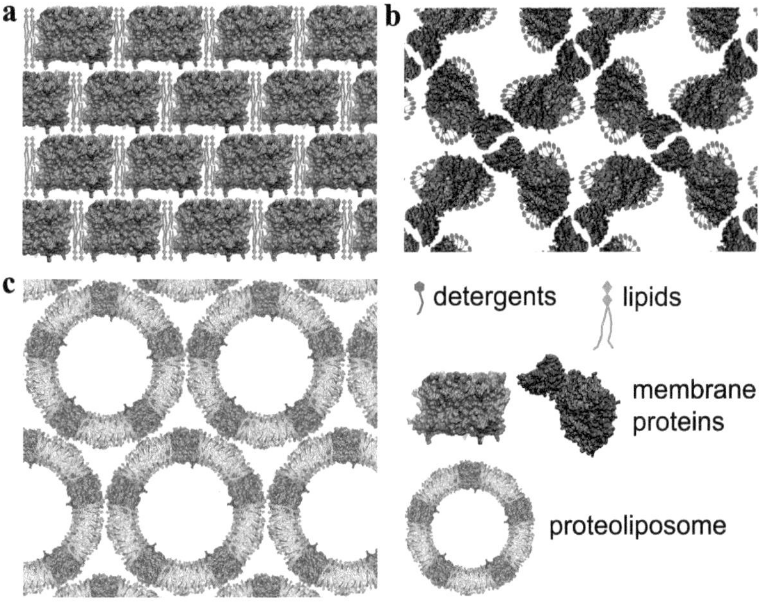


*Fig. 1.* The packing diagrams of three basic types of membrane protein crystals. (a) Type I, stacks of two-dimensional crystals. (b) Type II membrane proteins are crystallized in complex with detergent micelles. (c) Type III, crystals built from proteoliposome vesicles. The non-polar regions of membrane proteins are in light gray while the polar regions are shown in darker gray. See also Fig. 1, Color Plate 2.

and II. Sixty LHC II monomers together with numerous lipid and detergent molecules are assembled into a vesicular proteoliposome with an outer diameter of about 261 Å and inner diameter of around 160 Å (Fig. 1c). The LHC II trimers are held together by the lipid molecules located at their interfaces. The hydrophobic surfaces of membrane proteins are supposed to be covered by the presumed lipid bilayer within the proteoliposome. The two hydrophilic ends of embedded membrane proteins are exposed at the outer and inner surfaces of the proteoliposome respectively. Such a large proteoliposome with extensive hydrophilic surfaces serves as the basic structural element of the crystal. The crystal contacts are made through the polar interactions between the stromal ends of membrane proteins exposed on the outer surfaces of the proteoliposome. Under the conditions reported by Kouyama et al. (1994), bacteriorhodopsin could also be organized into a proteoliposome structure and then crystallized. However, the crystal was poorly ordered and only diffracted X-ray to around 30 Å.

### C. Detergents

Membrane proteins are purified and crystallized in detergent solutions. One thing special about the detergents is their amphipathic property. They have both a hydrophilic head and a hydrophobic alkyl tail in their structure. Above the critical micelle concentration, detergent molecules in solution will aggregate into micelles with their hydrophilic heads facing outwards and hydrophobic tails buried inside the micelle. When a detergent micelle binds to a membrane protein, it forms a belt surrounding the hydrophobic surface of the membrane protein. This results in a soluble complex of membrane protein and detergent, which is just what we want to crystallize. Obviously, the size and shape of the detergent micelle will influence the crystallization behavior of the protein-detergent complex.

There are lots of different species of detergents available commercially. For a new membrane protein, it will be almost impossible to try out all of them. Therefore, one is always encouraged to try out those detergents that have been frequently used before in membrane protein crystallization (Table 1). Detergents can be classified as ionic, non-ionic and zwitterionic types according to the properties of their hydrophilic head groups. Those with net charges are called ionic (either anionic or cationic) detergents. For instance, SDS and bile acid salts are anionic detergents, while CTAB is a cationic detergent. Ionic detergents are usually harsh and effective in solubilizing membrane proteins, but they will often

*Table 1.* The most commonly used detergents in membrane protein crystallization

| Detergent | Molecular weight | CMC (mM) | Aggregation number δ | Type* |
|---|---|---|---|---|
| $C_8E_4$ | 306.44 | 8.5 | 82 | N |
| $C_8E_5$ | 350.50 | 7.1 | – | N |
| $C_{10}E_5$ | 378.55 | 0.81 | 73 | N |
| $C_{12}E_8$ | 538.77 | 0.09 | 90–120 | N |
| $C_{12}E_9$ | 582.81 | 0.08 | – | N |
| DDAO | 201.36 | 10.48 | – | Z |
| DDM | 510.6 | 0.17 | 78–149 | N |
| DM | 482.6 | 1.8 | 69 | N |
| FC-14 | 379.5 | 0.12 | – | Z |
| LDAO | 229.41 | 1.0 | 76 | Z |
| MEGA-10 | 349.5 | 6 | – | N |
| NG | 306.4 | 6.5 | – | N |
| OG | 292.4 | 18 | 78 | N |
| OM | 454.4 | 19.5 | 47 | N |
| UDAO | 215.38 | 3.21 | – | Z |
| UDM | 496.6 | 0.59 | 74 | N |

Aggregation number δ: the number of detergent monomers contained in a single micelle. * N – non-ionic; Z – zwitterionic.

cause structural disturbance to membrane proteins and even result in denaturation. They are seldom used in membrane protein crystallization. Non-ionic detergents possess uncharged polar head groups, like polyoxyethylene moieties in Triton or glycosidic groups in OG and DDM. Zwitterionic detergents have the combined properties of ionic and non-ionic detergents. Their head groups have both positive and negative charges at the same time, but like non-ionic detergents, they do not show any net charges. Both, non-ionic and zwitterionic detergents are relatively mild for membrane proteins. They are suitable for the extraction of membrane proteins and can be used in the subsequent purification and crystallization steps. Among the detergents listed in Table 1, OG, LDAO, DDM and $C_8E_4$ are the four most successful detergents in membrane protein crystallization (Michel, 2005). These four are strongly recommended for first trials.

For detergents with the same alkyl chain length, those with bigger polar head groups tend to be milder. For example, OM is milder than OG. On the other hand, if the detergent head groups are too big, they will possibly interfere with crystal packing by covering the hydrophilic parts of membrane proteins. Therefore, we want to choose detergents with relatively smaller head group but not to destabilize membrane proteins at the same time. Detergents with longer hydrophobic tails will be milder than the ones with shorter tails provided that they have the same polar head. For instance, NG is milder than OG but harsher than DG. Meanwhile, the solubility of detergents decreases as the alkyl tail gets longer. Glucoside detergents will become hardly soluble in water when the alkyl tail reaches 12 hydrocarbons (dodecyl).

It will be helpful to learn about detergent phase behavior at different temperatures. Alkyl polyoxyethylene solution tends to be partitioned into two immiscible phases at higher temperature, with one phase enriched with detergent and the other phase depleted with detergent. This kind of phenomenon is called phase separation. In contrast, phase separation of alkyl glucoside occurs at lower temperature. Frequently, crystallization of membrane protein takes place near the consolution boundary between the two immiscible phases and the micelle phase (Zulauf, 1991). During the crystallization process, the solution mixture with membrane protein, detergent, precipitant, etc., starts at the micelle state and then is driven gradually towards consolution boundary as the concentrations of protein and/or precipitant are increased through vapor diffusion or dialysis methods. If the consolution boundary is crossed, phase separation will occur. Severe phase separation should be avoided since it will sometimes induce a shower of small crystals instead of large single crystals. Moreover, the phase enriched with detergent will absorb most of the membrane protein. This phase might denature the membrane protein or destroy the order of crystal packing if the crystals grow at the border or inside the detergent-rich phase. In case that the detergent is responsible for the phase separation problem, it may be solved by changing the temperature or by substituting the detergent with those bearing shorter alkyl chain or bigger polar head. Phase separation may also be related to lipid contents in the sample, precipitant, salt, additives or even pH.

Detergent concentration should be carefully controlled. If it is too low, protein will not be well solubilized and some aggregates might form. If too high, protein might become denatured and phase separation tends to occur instead of crystallization. Some membrane proteins, like the LHC II, crystallize in a very narrow range of detergent concentration. Some are selective for a specific species of detergent. LHC II can only be crystallized with NG but not with OG whose alky chain is smaller by only one carbon length (Kühlbrandt, 1988), while bacteriorhodopsin selects OG instead of NG for its crystallization (Schertler et al., 1993). Some other proteins, such as cytochrome $bc_1$ complex (Huang et al., 2003), are less selective for detergents. They will crystallize with several different detergents, but the crystal form may turn out to be different.

### D. Sample Preparation

High homogeneity is a common requirement for the crystallization of both soluble and membrane proteins. For membrane proteins, besides the polypeptide homogeneity that can be determined by SDS polyacrylamide gel electrophoresis, one may also want to analyze the lipid contents in the sample. Excess lipids, loosely bound to the surfaces of membrane proteins, may introduce heterogeneity into the sample and make it reluctant to crystallize. Thin-layer chromatography can be used to analyze lipid content in the membrane protein sample. Usually, the column chromatography techniques applied in membrane protein purification, like affinity, ion exchange and gel filtration chromatography, will result in partial or full delipidation of the sample in addition to the improvement of protein purity. There are several cases that lipid molecules bind to membrane protein specifically and they can actually be an integral part of membrane protein complexes. In both, cyanobacterial Photosystem I (Jordan et al., 2001) and LHC II (Liu et al., 2004), there exists a phosphatidylglycerol molecule that coordinates a chlorophyll *a* molecule. It is hard to remove this lipid without loss of the chlorophyll. We do not recommend removing lipids like this, since doing so will disturb the protein structure. Monodispersity of the sample in solution will be a good sign for crystallization. This can be verified by doing gel filtration or dynamic light scattering experiment. It is noteworthy that membrane proteins will behave differently in different detergent solutions. They may appear monodisperse in one detergent solution but polydisperse in another. Therefore, it is necessary to screen for a detergent that will maintain the monodisperse state of a specific membrane protein.

The last sample-processing step before setting up crystallization trials is concentrating the solution. The most popular way is by using a concentrator with ultrafiltration membrane. This can be done by centrifugation or under pressure exerted by a compressed nitrogen gas tank. It should be noticed that if the molecular weight cut-off of the membrane is smaller than the size of detergent micelle, detergent itself will get concentrated along with the protein. The final concentration of detergent in the sample will be variable between batches if the folds of concentration vary between batches. For those membrane proteins that will only crystallize over a narrow range of detergent concentration, it may become difficult to get crystals out of this sample. Even if it did crystallize by chance, it would be hard to reproduce the crystals. As shown by Strop and Brunger (2005), DDM, FC-14 and $C_{12}E_9$ solutions will be concentrated by 3.0–4.7 fold in a 30-kDa cut-off concentrator, while no concentration effect was observed with 100-kDa cut-off concentrator. Their micelle molecular weights are 70, 46 and 82 kDa, respectively. LDAO has a smaller micelle size (17 kDa) and will pass through a 30-kDa cutoff membrane. For other detergents, one should be able to calculate their micelle sizes by multiplying the molecular weight of monomeric detergent by aggregation numbers and then choose concentrators with molecular weight cut-off larger than the empty detergent micelles but smaller than the protein-detergent complexes. This should avoid the build-up of detergent in the concentrator while concentrating the protein.

### E. Crystallization and Optimization

#### 1. Initial Screening

Like for soluble proteins, the initial screening for crystallization conditions of membrane proteins can be done with most commercially available kits based on sparse matrix screens, such as Crystal Screen, Crystal Screen II, MembFac Screen, Index Screen, PEG/Ion Screen from Hampton Research, MBClass and MBClass II Suites from Nextal, as well as some other screening kits designed by experienced researchers (Iwata, 2003). Protein concentration usually starts at 10–20 mg/mL. Vapor diffusion methods, either with sitting drops or hanging drops, is the most popular for setting up a crystallization screen. Protein solution is mixed with precipitant solution and then the drops are equilibrated with reservoir solution in a sealed chamber. Microbatch under-oil method is especially useful for setting up high-throughput crystallization by using a robotic system. It allows setting up crystallization under thousands of conditions within minutes while costing relatively small amounts of protein sample. In addition, protein solutions can be dialyzed against precipitant solution in order to trigger crystallization. The dialysis method based on a reverse salting-in protocol has been successfully applied in the crystallization of Photosystem I (Fromme and Witt, 1998). Whatever method is used, it is advisable to set up initial screens at several different temperatures, typically at 4, 10, 15 and 20 °C, as proteins and detergents usually have different behavior at different temperatures.

### *2. Optimization*

It is not difficult to get hits on initial screens. However, it is quite common that the initial membrane protein crystals only diffract poorly at low resolution or do not diffract X-rays at all. Further optimization for crystals suitable for X-ray structure determination may require more time, efforts and utmost patience. There are a lot of parameters one can adjust for optimization. Temperature, pH and precipitant are the major factors that will influence protein crystallization. Once having an initial hit, one would start to optimize the condition by varying temperature and making a pH/precipitant grid screen first. If the precipitant is polyethylene glycerol (PEG), different molecular weight species of PEG should be tried. At this stage, the purpose is to obtain conditions under which single crystals would grow well reproducibly. During optimization, several different crystal forms might appear at different conditions. It is necessary to do X-ray diffraction experiments on crystals grown under every different condition to identify the best one for structure determination. If no well-ordered crystal can be obtained at this stage, one may want to move on to screen for additives, exchange current detergent for other species or make some changes on the protein itself.

#### *a. Additives*

There are a group of small molecule additives called amphiphiles that are fairly useful in the optimization of membrane protein crystallization. Organic small molecules with one end polar and the other end hydrophobic are theoretically able to serve as amphiphiles, like alkyl-polyol and alkyl-n-oxide (Michel, 1983). Two amphiphiles, heptane-1,2,3-triol and benzamidine hydrochloride have received the most frequent success so far. The first successful application of heptane-1,2,3-triol in combination with detergent LDAO led to high-quality crystals of the purple bacterial photosynthetic reaction center, which diffracted X-rays to 2.3 Å resolution (Deisenhofer and Michel, 1989). Since then, there have been 13 successful cases with heptane-1,2,3-triol (Michel, 2005). Benzamidine was successfully used in four cases, in particular the crystallization of bacterial reaction centers (Allen, 1994), of bacterial light-harvesting complexes (McDermott et al., 1995), of bacteriorhodopsin (Schertler et al., 1993) and of fumarate reductase (Lancaster et al., 1999). Others like cyclooctane-1,2,3,4-tetraol, octane-1,2,3-triol, hexane-1,2,3-triol, triethylammonium and D-phenylalanine were also suggested to be tried as amphiphiles in optimization (Michel, 1983). How do amphiphiles work to improve the quality of membrane protein crystals? Amphiphiles do not form micelles by themselves but they can incorporate into the detergent micelles and lead to the formation of mixed micelles with smaller size than the pure detergent micelles (Timmins et al., 1991). Membrane proteins, with such mixed micelles bound on their surface, may have larger hydrophilic regions exposed and available for crystal contacts. This would consequently promote the order of crystal packing. In some cases, amphiphiles may replace detergent molecules at specific locations that would be unfavorable for crystal packing. They can bind to membrane proteins at these sites like detergent molecules, while they have smaller polar head groups so that they can fit better into the crystal lattice (Michel, 1983). Amphiphiles also work to inhibit phase separation to some extent. We have tried heptane-1,2,3-triol and D-phenylalanine in the optimization of LHC II crystallization. Eventually we found out that both work to inhibit phase separation and help to grow large single type III crystals out of clear drops.

Extensive purification of membrane proteins will result in over-delipidation of the sample. Even if such a preparation is pure in polypeptide composition, sometimes it may still be reluctant to crystallize or crystallize with poor quality. Addition of some native or synthetic lipids to the purified membrane protein sample may restore its crystallization capability. Over-delipidated LHC II samples from both pea (Nußberger et al., 1993) and spinach (see part II) would not crystallize until a native lipid, DGDG, is supplemented back into the sample. Crystallization of $Ca^{2+}$-ATPase with phosphatidylcholine yielded well-ordered type I crystals (Toyoshima et al., 2000). Dramatic increase of crystallization efficiency and great improvement in diffraction quality were observed on cyanobacterial cytochrome $b_6f$ after the purified complex was augmented with synthetic dioleoyl-phosphatidylcholine (Zhang et al., 2003). Lipids might take effect in assisting membrane protein crystallization in two ways. Firstly, the binding of lipids at the periphery of membrane proteins would restore the lateral pressure that is present in the lipid bilayers of native membranes. This might help to improve the conformational homogeneity of membrane protein preparations (Zhang et al., 2003). Secondly, lipid molecules may bind to membrane protein at a specific site and trigger the formation of crystal contacts.

Heavy atom reagents can also be used as additives for improving the diffraction power of protein crystals. The native crystals of a mechanosensitive ion channel of large conductance (MscL) from *Mycobacterium tuberculosis* only diffract to about 7 Å. After the crystals were soaked in $Na_3Au(S_2O_3)_2$ solutions, the diffraction resolution was greatly improved to 3.5 Å (Chang et al., 1998). Soaking the crystals of MsbA (an ATP binding cassette transporter) in $OsCl_3$ solution also led to improvement of resolution from about 6.2 Å to 4.5 Å (Chang and Roth, 2001). In both MscL and MsbA structures, heavy atoms were found to bind at crystal packing sites and thus improve the crystal order.

Some membrane proteins have been crystallized in a mixture of different detergents and the crystals turn out to be much better than those obtained with a single species of detergent. For instance, a mixture of DDM/$C_{12}E_9$ in combination with divalent salt $SrCl_2$ or $MgCl_2$ yielded well-ordered crystals of the Glycerol-3-phosphate transporter (Glpt) from *Escherichia coli* that diffracted X-rays isotropically to a resolution of around 3.2 Å (Lemieux et al., 2003). A mixture of different detergents will lead to a mixed micelle with adjusted size and shape and will perhaps fit better into the crystal lattice. We would suggest a strategy of holding one detergent as a major detergent and then screen for a second detergent as additives for crystallization. The concentration of the second detergent would start around CMC and then subject to optimization later.

Some other additives, like multivalent cations, organic small molecules, reducing agents, etc, which are commonly used to optimize soluble protein crystals, are also applicable to membrane protein crystallization.

### *b. Detergent Exchange*

Frequently the detergent used for purification is not necessarily the most suitable for crystallization. Instead of replacing detergent *ab initio*, one can exchange detergents at the final step of protein purification. Dialysis methods are suitable for exchange of detergents with high CMC value and small micelle size. For detergents with low CMC value and large micelles, it will take a long time (even weeks) to achieve complete exchange and it is better to do it in other ways. Detergent exchange on gel filtration column will result in a mixture of both detergents instead of complete exchange. Anyway, such a mixture is helpful for the crystallization of some membrane proteins, like the glycerol-3-phosphate transporter (Lemieux et al., 2003). To attain complete exchange, one would bind protein to an ion exchange or affinity column, wash it extensively with the new detergent solution and then elute it. Precipitation methods are useful for LHC II, as it will remain intact after it precipitates out from solution. The idea is to dilute the protein-detergent solution to a point below the CMC and then the membrane protein will precipitate out. The pellets can be collected by centrifugation. After being washed for several times with detergent free buffer, the pellets can be resolubilized in a new detergent solution. This method is not applicable to those membrane proteins that would become denatured if they were stripped of detergents.

Once getting new protein samples in new detergents, one might want to try previous conditions on the new samples and see if they will crystallize under similar conditions. Make sure that a gradient of different detergent concentrations has been tried. This can be done by introducing different amounts of detergent stock solution into the initial protein sample, or by using SM2 Bio-Beads (Biorad) to remove excess detergents (Dahout-Gonzalez et al., 2003). If the new samples do not crystallize under previous conditions, one may want to start over with a series of sparse matrix screens to find new conditions.

### *c. Protein Modification*

The optimization strategy of modifying target proteins is based on the idea that modified proteins will sometimes crystallize easier or better than the wild-type proteins. There are basically two ways to make change on the target proteins. One is to search for natural homologues of the target proteins. Homologous proteins from different species have more or less different surface properties that will make them behave differently during crystallization. A successful example of membrane protein homologues screening is shown in the case of mechanosensitive channel of large conductance (MscL) (Chang et al., 1998). The authors cloned, expressed and purified MscL homologues from nine prokaryotic species. They tried to crystallize all nine homologues and found one that would yield crystals of the best diffraction quality. A second way is to make some artificial changes on the target proteins, either by proteolytic cleavage or recombinant techniques. The flexible termini or loops of proteins will sometimes interfere with crystallization. Controlled proteolytic cleavage on the target proteins will yield a rigid part of the protein that may have better crystallization behavior. The KcsA potassium channel was crystallized after thirty-five carboxyl amino acid residues were cleaved by chymotrypsin proteolysis (Doyle et al., 1998). Recombinant methods can be employed to make all kinds of changes on the target protein, such as truncation, mutation and fusion of affinity tag proteins with the target proteins. Rational mutations on the protein surfaces have been shown to be effective in improving the quality of some soluble protein crystals (Derewenda, 2004). A semi-directed mutagenesis at the polar surfaces of membrane protein OmpA and OmpX led to mutant crystals that could diffract to better than 2 Å resolution, while the wild type crystals only diffract to 3.2 and 3.3 Å, respectively (Pautsch et al., 1999).

Fusion of large affinity tags, such as the glutathione-S-transferase, thioredoxin or maltose-binding protein to the protein of interest facilitates the expression and purification of target protein. Usually, these tags will be removed by proteolytic cleavage before crystallization, as the flexible linkage between the tag and target protein will introduce conformational heterogeneity that will hamper crystallization. If a rigid fusion, via a short three to five amino acid spacer, were made between the tag and target, it would be possible to crystallize the fusion protein. Thus far, it has worked out for several soluble proteins as reviewed by Smyth et al. (2003). This strategy has an obvious advantage for membrane proteins because it will increase the size of polar surface areas of membrane proteins. This may be helpful in getting well-ordered type II crystal on condition that a rigid fusion is achieved. Byrne et al. (2000) reported successful crystallization of a fusion protein between protein Z and membrane protein cytochrome $bo_3$, but the crystal did not diffract to high resolution. Another way to enlarge the hydrophilic region of membrane protein is through specific binding of antibody fragments or substrate proteins to the hydrophilic surface of target membrane protein. This idea has received success in the crystallization of three species of membrane proteins, namely cytochrome *c* oxidase (Iwata et al., 1995; Ostermeier et al., 1997), cytochrome $bc_1$ complex (Hunte et al., 2000; Lange and Hunte, 2002) and potassium channel KcsA (Zhou et al., 2001) and KvAP (Jiang et al., 2003). Two kinds of antibody fragments were used. Recombinant Fv fragments were used for the first two cases, while Fab fragments generated by proteolysis of monoclonal antibody were used for the potassium channels. Antibody fragments have important roles in forming stable crystal contacts within these complex crystals. The Fab-KcsA crystal is a good example, where the Fab fragments form all the crystal contacts. The membrane protein KcsA was totally suspended without taking part in the crystal contacts. The crystals are of very high quality and diffract to 2.0 Å (Zhou et al., 2001). It is worth mentioning two advantages of antibody strategy. By engineering an affinity tag to the antibody fragment, it will simplify the purification of membrane proteins to a single step of affinity chromatography (Ostermeier et al., 1995). If the antibody fragment is larger than the membrane protein it binds, there is a good chance that the structure can be solved by molecular replacement method, which is a much easier and faster way than other phasing methods, as indicated in the structure determination of Fab-KcsA crystal structure (Zhou et al., 2001).

### F. Conclusions

Back in 1980, two membrane proteins were crystallized for the first time. They are bacteriorhodopsin (Michel and Oesterhelt, 1980) and porin (Garavito and Rosenbusch, 1980). Before that, it was once thought impossible to crystallize membrane proteins. After five years, the first X-ray structure of a membrane protein, the photosynthetic reaction center from purple bacteria, was reported (Deisenhofer et al., 1985). Since then, great advances have been made in this field. There have already been 96 unique membrane protein structures solved so far (White, 2005). Such an achievement would not be possible without the availability of all kinds of detergents, various crystal optimization techniques, overexpression methods of functional membrane proteins and the high-intensity synchrotron radiation sources. The third generation synchrotron radiation beam lines installed with microfocus setups has allowed data collection with microcrystals of a few microns (Cusack et al., 1998; Riekel et al., 2005). This is remarkably useful for membrane protein crystallography since it is sometimes extremely difficult to grow large membrane protein single crystals. In spite of the advancement, the growth rate of membrane protein structures lags behind that of soluble proteins during the equivalent time period (White, 2004). The bottleneck has gradually moved upstream from crystallization to protein expression and purification. Preparation of enough functional membrane protein material, especially from eukaryotic species, is the rate-limiting step in the structure determination of membrane proteins. A breakthrough in the technique of heterologous over-expression of eukaryotic membrane proteins is direly needed. Just recently, two mammalian membrane proteins, both over-expressed in yeasts, have been successfully crystallized. One is the voltage-dependent potassium channel Kv1.2 from rat (Long et al., 2005) and the other is rabbit $Ca^{2+}$-ATPase (Jidenko et al., 2005). They were expressed in two different species of yeasts, namely *Pichia pastoris* and *Saccharomyces cerevisiae,* respectively. Their crystal structures have been solved at 2.9 Å and 3.3 Å resolution. These examples represent ways to address the issue of over-expressing eukaryotic membrane proteins for crystallization purpose.

## II. Crystallization of the Major Light-Harvesting Complex of Plants

### A. Introduction

The major light-harvesting complex of Photosystem II in plants, termed LHC II, is the most abundant integral membrane protein in chloroplasts. It binds half of the thylakoid chlorophylls and exists as a trimer in the native state. Every monomeric LHC II is composed of a polypeptide of about 232 amino acid residues, 13–15 chlorophylls (Peter and Thornber, 1991), 3–4 carotenoids (Ruban et al., 1999), and one tightly bound phospholipid (Nußberger et al., 1993). LHC II serves as the principal solar energy collector in the photosynthesis of green plants. Moreover, it is involved in the non-radiative dissipation (NRD) of excess excitation energy formed under high-light conditions, a photoprotective mechanism of plants (Horton et al., 1996; Elrad et al., 2002). The structure of LHC II from pea was determined by electron crystallography at 3.4 Å resolution (Kühlbrandt et al., 1994). The basic structural features of LHC II were revealed in this report for the first time. However, the structural information about the pigments and pigment binding environments in this model were not detailed enough to provide accurate data for the follow-up functional studies. A higher-resolution structure has been eagerly expected since then. Apparently, X-ray crystallography is the most feasible way to determine the high-resolution structure of LHC II.

At the moment when we started to crystallize LHC II in 1998, there had already been several reports on the crystallization of LHC II before. Early in 1987, Kühlbrandt (1987) reported two different crystal forms of LHC II, one as small hexagonal plate and the other as octahedron. The hexagonal plates were stacks of highly ordered two-dimensional crystals. Hence, they belong to type I membrane protein crystals. The octahedral crystals had an exceptionally large cubic unit cells that were comparable to those of virus crystals. They only diffract to low resolution despite their large sizes. Neither the hexagonal plates nor the octahedra were suitable for X-ray structure determination at that time. Nußberger et al. (1993) found that the presence of lipid DGDG is essential for the crystallization of LHC II. Furuichi et al. (1997) crystallized a monomeric spinach LHC II preparation in Triton X-100 or n-nonyl-β-D-glucoside. The crystals appeared as thin plates, but no diffraction data were available in that report. Since the polar areas of the LHC II at both ends of the complex are extremely small and flat, it is very difficult for the LHC II to form stable type II crystal contacts. Here we describe the details about the crystallization of spinach LHC II in several different crystal forms including the type III, a novel type of membrane protein crystal.

### B. Protein Purification

The recipes for the solutions used in LHC II purification are listed below. Solution A: 2 mM $MgCl_2$, 10 mM NaCl, 0.4 M sucrose, 50 mM Tricine (pH7.8). Solution B: 5 mM $MgCl_2$, 0.2 M sucrose, 20 mM Tricine (pH7.8). Solution C: 5 mM $MgCl_2$, 15 mM NaCl, 20 mM MES (pH6.0). Solution D: 35 mM OG, 10 mM $CaCl_2$, 1 M NaCl, 0.4 M sucrose, 50 mM MES (pH6.0). Solution E: 15 mM NaCl, 5 mM EDTA, 20 mM MES (pH 6.0). Solution F: 0.8% OG, 0.02% $NaN_3$, 20 mM Tris-HCl (pH 7.5).

The following purification protocol is based on early publications (Xu et al., 1994; Lou et al., 1995) with some modifications to meet the requirement of crystallization. Spinach leaves were dark adapted for more than 20 h at 277 K, and then homogenized in solution A. The suspension was filtered through 8-layer gauze and the filtrate was centrifuged at 6,000 g for 10 min to collect chloroplasts. The pellets were suspended in solution B and then centrifuged to spin down pellets. The chloroplast preparation was resuspended in solution C to a final concentration of 2 mg Chl $mL^{-1}$. Triton X-100 was slowly added to give a molar ratio of Triton to Chl of 25:1. The solution was stirred for 30 min on ice and then two volumes of solution C were added. The diluted solution was centrifuged at 10,000 g for 5 min to remove debris. The supernatant is subject to centrifugation at 40,000 g for 30 min. The pellets were resuspended in a small volume of solution C and then solubilized for 15 min in solution D to a Chl concentration of 1.5 mg $mL^{-1}$ with constant stirring on ice. After solubilization, the solution was mixed with 2 volume of detergent-free solution D and then centrifuged at 40,000 g for 90 min. The pellets were resuspended in solution E and centrifuged again (40,000 g for 40 min). At this stage, the pellets were suspended in a small volume of water. This led to a crude preparation of LHC II.

The following procedure is to refine the LHC II preparation. (i) The crude preparation was solubilized in 1.5 % Triton X-100 for 30 min on ice to give 0.8 mg $mL^{-1}$ Chl. The solution is then centrifuged at 30,000 g for 40 min. (ii) The pellets were collected and resuspended in a small volume of water. Step 1 was repeated once and then the supernatant was pooled this time after centrifugation. (iii) Solid KCl and $MgCl_2$ were added to the supernatant to give final concentrations of 100 mM and 20 mM, respectively. The pellets were collected by centrifugation at 40,000 g for 40 min. (iv) The purified LHC II pellets were washed once with 100 mM KCl and once with water by repeating resuspension and centrifugation. The washed pellets were dissolved in 1.0% OG, 20 mM Tris-HCl (pH 7.5) to 4 mg $mL^{-1}$ Chl.

The sample was applied to Hiload 16/60 Superdex 200 pg column (Pharmacia Biotech) and eluted with solution F. For each run, 0.5 mL sample was injected to the column and the flow rate was set at 0.2 mL $min^{-1}$. The trimer peak was pooled and diluted ten-fold by adding 100 mM KCl to precipitate LHC II. The pellets were collected by centrifugation at 16,000 g for 15 min and washed three times in 100 mM KCl and twice in water. Finally, the pellets were solubilized in 1.0% NG to 5–6 mg $mL^{-1}$ Chl. After centrifugation at 13,000 g for 15 min, the supernatant was stored in liquid nitrogen for later crystallization trials.

### C. Crystallization

All the crystallization experiments were set up with hanging drop or sitting drop vapor diffusion method. The former was used in screening for initial crystallization conditions, while the latter was for producing large crystals for data collection. In hanging drops, 1 μL of protein sample was mixed with 1 μL of crystallization solution. Sitting drops were made by mixing 3.0 μL of protein sample with 1.8 μL of crystallization solution on the polypropylene microbridges (Hampton Research). Both hanging drops and sitting drops were equilibrated with one milliliter of reservoir solution.

At first, we tried to crystallize the sample without doing gel filtration. Instead of solubilizing the washed pellets in OG buffer, we dissolved it in 1.0% NG solution and at 4 mg $mL^{-1}$ Chl concentration. The choice of NG as the detergent for LHC II crystallization was made according to previous reports (Kühlbrandt, 1987; Nußberger et al., 1993). Such a preparation was used to set up a sparse matrix screen with the MembFac kit from Hampton Research. Fortunately, we got a hit on the first screening trial. The condition

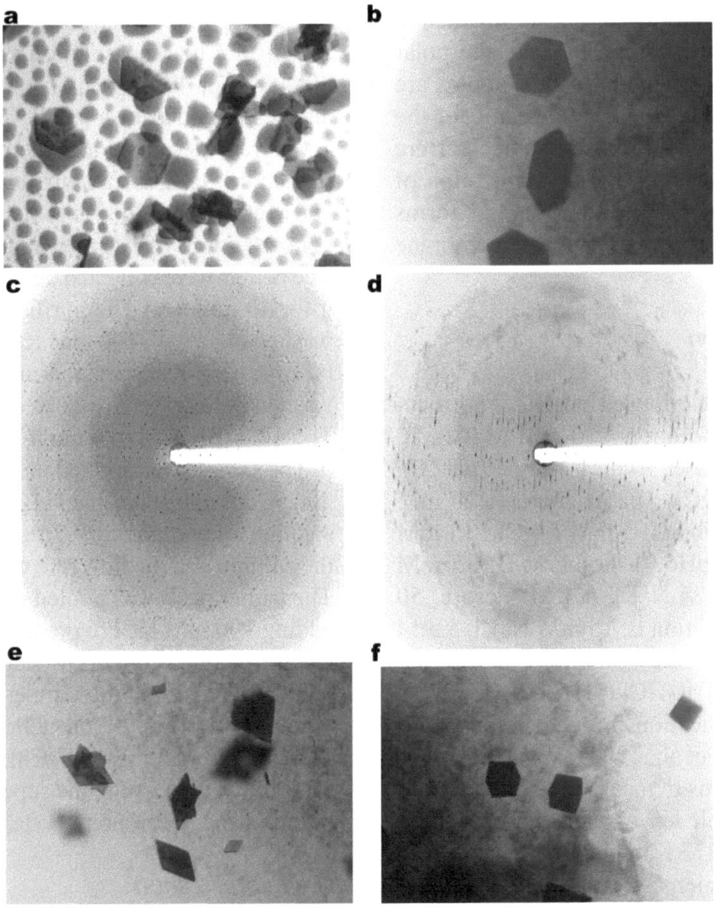


*Fig. 2.* Microscopic photographs and X-ray diffraction patterns of the LHC-II crystals. (a) Thin hexagonal plate crystals of spinach LHC-II grown after phase separation. (b) After optimization, the hexagonal plates grew before phase separation. (c, d) Typical X-ray diffraction patterns of a hexagonal plate crystal of spinach LHC-II showing the anisotropic packing order of this crystal form. The incident X-ray beam was perpendicular and parallel to the hexagonal plane respectively. (e) Rhombic plate crystals of spinach LHC-II grown with low DGDG concentration. (f) Dark cubic crystals of cucumber LHC-II.

(Table 2) was very much different from those for pea LHC II (Kühlbrandt, 1987; Nußberger et al., 1993). The crystals appeared as hexagonal plates (Fig. 2a). They would reach the largest dimensions of $0.3 \times 0.3 \times \sim 0.01$ mm$^3$ after optimization against precipitant concentration. X-ray diffraction experiments on these crystals showed an anisotropic packing order. They are indeed well-ordered in two dimensions but disordered in the third dimension. When the X-rays were incident perpendicular to the hexagonal plane, the diffraction spots were very sharp and could extend to around 3 Å resolution (Fig. 2c). However, if the crystal was rotated by 90° to make the incident beam parallel to the hexagonal plane, the diffraction became streaky and only extended to low resolution (Fig. 2d). It was impossible for us to index and scale the data correctly due to poor data quality. This crystal form may be similar to the hexagonal plate crystals of pea LHC II reported by Kühlbrandt (1987). Both are type I membrane protein crystals.

Lipid analysis by thin-layer chromatography indicated that there were at least four species of lipids contained in the LHC II preparation before gel filtration. Besides the tightly bound PG, three other species of lipid were detected in the sample, including DGDG, MGDG and another unidentified lipid. The presence of excess lipids led to the phase separation in the drop (Fig. 2a) and the crystals usually appeared after phase separation. We suspected that it might be the phase separation that interfered with the crystal packing at the third dimension. Then, we tried to remove excess lipids by doing gel filtration. Almost all galactolipids were removed after gel filtration, while the PG and the unknown lipid remained tightly bound

*Table 2.* Crystallization conditions for spinach LHC-II

| Conditions | Crystals in Fig. 2b | Crystals in Fig. 2e | Crystals in Fig. 3a |
|---|---|---|---|
| Solution a | 1.0–1.2% NG, 1.5–2.0 mg $mL^{-1}$ DGDG, 4.0 mg $mL^{-1}$ Chl | 1.0% NG, 0.3 mg $mL^{-1}$ DGDG, 4.0 mg $mL^{-1}$ Chl | 0.8–0.9% NG, 2.0 mg $mL^{-1}$ DGDG, 4.0 mg $mL^{-1}$ Chl |
| Solution b | Well solution | Well solution | Well solution with 0.2% DBC |
| Drops (hanging or sitting) | 1 μL a + 0.6 μL b | 1 μL a + 0.6 μL b | 3 μL a + 1.8 μL b |
| Well solutions | 0.96–1.09 M trisodium citrate, 0.02% $NaN_3$, 100 mM HEPES-NaOH, pH 7.5 | 1.02–1.09 M trisodium citrate, 0.02% $NaN_3$, 100 mM HEPES-NaOH, pH 7.5 | 0.9–1.15 M trisodium citrate, 0.02% $NaN_3$, 66.5 mM HEPES-NaOH, pH 7.5 |
| Temperature | 289 K | 289 K | 291 K |

to the LHC II. The same crystallization condition was tried on the new preparation after gel filtration, but the sample turned out to be uncrystallizable. According to Nußberger et al. (1993), addition of exogenous DGDG to the partially delipidated LHC II sample fully restored its crystallization capability. This finding inspired us to add DGDG back to the gel-filtrated sample. At DGDG concentration of 0.3 mg $mL^{-1}$, rhombic plate crystals were grown (Fig. 2e). Under this condition, one monomeric LHC II binds about one DGDG molecule. These crystals frequently appeared as stacks of many crystals. It was very hard to obtain single rhombic plate crystal. They only diffracted to low resolution (worse than 10 Å). We did not pursue them further. When the DGDG concentration reached 1.5–2.0 mg $mL^{-1}$, large hexagonal plates with dimensions of $0.3 \times 0.3 \times \sim 0.01$ $mm^3$ (Fig. 2b) grew readily and reproducibly before phase separation. Diffraction pattern showed that these crystals were still disordered in the third dimension even though they looked much better than those grown under phase separation.

A series of strategies have been tried in order to improve the crystal packing order in the third dimension. To limit the number of variables for optimization, we fixed the protein concentration at 4 mg/mL Chl as well as the ingredients of well solution (Table 2) and then set out to optimize temperature, detergents, lipids and search for additives. The temperature for growing type I crystals was found to be optimal at 289 K. If the temperature is higher than 293 K, lots of brownish precipitates formed in the drop and crystals grew too fast. At 277 K, it would become easy to get phase separation before crystallization. As far as detergent was concerned, we tried to exchange NG with other detergents by the method of precipitation, wash and resolubilization. Three other detergents were found to be able to support LHC II crystallization. They were OTG, DG and MEGA-10. The crystals all appeared to be hexagonal plates. As OTG and DG are more hydrophobic than NG, crystals would grow after phase separation. MEGA-10 is quite similar to NG in the way of hydrophobicity and hydrophilicity balance. However, crystals with MEGA-10 did not have sharp edges and they would usually grow into multicrystals in a few days after they appeared. OG has a shorter alkyl chain than NG. Thus, it is more hydrophilic than NG. Nevertheless, the LHC II samples solubilized in OG with concentrations varying from 0.8% to 1.4% have been tried but none of them yielded crystals. This observation was consistent with previous reports (Kühlbrandt, 1988). The LHC II preparations solubilized in DDM, UDM, NTM and $C_{12}E_8$ were not crystallizable either. After all these trials, we assumed that the optimal detergent for LHC II crystallization is nothing but NG. As for concentration, 1.0–1.2% NG for solubilizing LHC II at 4mg/mL Chl would be suitable for type I crystal growth.

After fixing the detergent factor, we went on trying some other lipids as substitutes for DGDG. The lipids have various polar heads and fatty acid chains. These tries included PG (dipalmitoyl and distearoyl), PC (dipalmitoyl), PA (dipalmitoyl and dioleoyl), MGDG and LLG. Aliquots of lipid solutions in chloroform were measured out and dried under a nitrogen gas stream. Then the partially delipidated (by gel filtration) protein sample was added to dry lipids to dissolve them. After being incubated at 16 °C for about 4 hours, these lipid-augmented LHC II samples were subject to crystallization trials. We found that phospholipids were not good at inducing LHC II crystallization. With MGDG, the LHC II could be crystallized as rhombic and hexagonal plates. However, when the concentration of MGDG was higher than 1 mg/mL, phase separation would occur. It was reported that the MGDG and DGDG from plant thylakoid membranes contained a high percentage of α-linolenic acid (all-*cis*-18:3$\Delta^{9,12,15}$) (Dormann et al., 1999). LLG was brought to our attention because it has a linolenoyl chain in its structure but lacks a galactosyl head. Severe phase separation was observed in the drop with LLG due to its high hydrophobicity but very thin hexagonal plate crystals still appeared in the oily phase or at the phase boundary. We infer that the hydrophobic interactions, between lipid fatty acid chains and membrane protein hydrophobic surfaces, are the major stabilizing forces for the packing of LHC II molecules within the hexagonal plane. The galactosyl head group does not seem to be necessary for the type I crystal growth. Anyway, these crystals with MGDG or LLG were not better than the crystals grown with DGDG. In the following optimization experiments, we consistently added 2 mg $mL^{-1}$ DGDG in the NG-LHC II sample. The lipid factor was fixed in the following experiments.

Considering the homologue screening strategy, we tried to crystallize cucumber LHC II purified in a similar way as the spinach LHC II. The crystallization conditions were almost the same as those of spinach LHC II. However, the crystallization behavior of cucumber LHC II is more dependent on seasonal factors. Preparations made on summer time were predominantly crystallized as dark cubes or octahedra (Fig. 2f), while those made in late autumn tended to be crystallized as hexagonal plates. The cubic or octahedral crystals only diffracted to around 10 Å and the hexagonal plates exhibited the same anisotropic diffraction as spinach LHC II crystals.

A large-scale additive screen was eventually found to be effective in greatly improving the quality of LHC II crystal. Small amphiphiles, like heptane-1,2,3-triol, benzamidine hydrochloride, octane-1,2,3-triol, hexane-1,6-diol, D-phenylalanine, L-phenylalanine, etc, have been tried at various concentrations. The detergent screen kits I, II, III from Hampton Research were also screened as additives for optimizing the LHC II crystals. Out of more than 80 additives we have tried, three of them, heptane-1,2,3-triol, D-phenylalanine and DBC were found to be most effective. Cucumber LHC II was best crystallized with D-phenylalanine, while heptane-1,2,3-triol and DBC were most effective for spinach LHC II. These additives were used at low concentration, 0.75% for heptane-1,2,3-triol and 0.2 % for DBC in precipitant solutions, 0.5% for D-phenyalanine in protein sample. They will denature LHC II if used at high concentrations. We finally settled for DBC because crystallization of LHC II with this additive is the most reproducible in yielding crystals diffracting better than 3 Å. A little adjustment on the temperature and NG concentration was done in order to grow large single crystals. The optimal temperature and NG concentration were 291 K and 0.8–0.9%, respectively. Other factors were set at the same values as those for type I crystals.

### D. Preliminary Crystallographic Analysis

Crystals were soaked in a cryoprotectant solution (0.4% NG, 0.15% DBC, 1.0 mg $mL^{-1}$ DGDG, 1.15 M sodium citrate, 11% saturated sucrose, 50 mM HEPES-NaOH pH 7.5) for 5 minutes and then dipped into liquid nitrogen quickly. A proper cryoprotectant solution is extremely important for preserving the crystal quality. Crystals soaked in glycerol solution never diffracted better than 10 Å. Introducing 11% saturated sucrose into the crystals will preserve their original diffraction quality very well upon freezing. Data were collected at beam line BL6B of the Photon Factory (PF) (Tsukuba, Japan) and beam line 3W1A of the Beijing Synchrotron Radiation Facility (BSRF) (Beijing, China). Rigaku R-AXIS IV 300 mm imaging plate detector (at PF) and Mar Research 345 mm imaging plate detector (at BSRF) were used during data collection. All data were collected at 100 K. A large crystal-to-detector distance of 300–350 mm and small oscillation at 0.5° were necessary for collecting data better than 3 Å due to the small spacing between spots. Data were processed with DENZO/SCALEPACK (Otwinowski and Minor, 1996).

To our surprise, crystals grown with heptane-1,2,3-triol, D-phenylalanine and DBC all belong to a new crystal form although their morphology frequently appeared as thick hexagonal plates (Fig. 3a) that look quite like the type I crystal (Fig. 2b). The space groups of these new crystals were all determined to be *R*32 and the unit cells were exceptionally large with similar cell parameters. Crystals with DBC were of the highest quality. The best one would diffract X-rays to 2.5 Å (Fig. 3b). A native data set was collected and

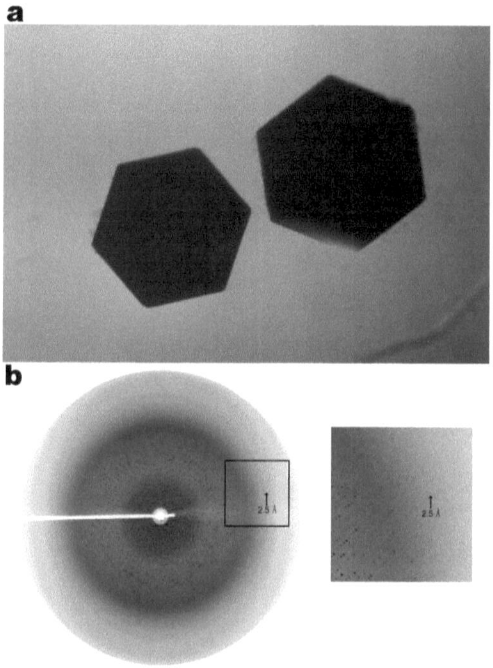


*Fig. 3.* The photograph (a) and diffraction (b) of high-quality spinach LHC-II crystals used in structure determination. These crystals look similar to the hexagonal plates in Fig. 2 but belong to a different crystal form. One of the best crystals could isotropically diffract to a highest resolution of 2.5 Å.

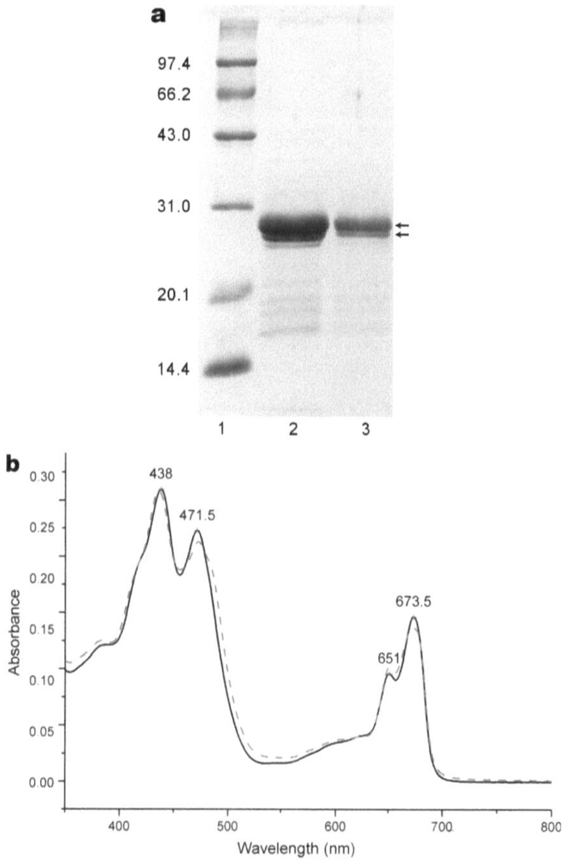


*Fig. 4.* Biochemical analysis of spinach LHC-II crystals shown in Fig. 3a. (a) SDS-PAGE image of the LHC-II sample used for crystallization (lane 2) and the crystal sample (lane 3). Molecular weight markers (unit, kDa) are shown in lane 1. The crystal sample contains two major polypeptides of LHC-II, namely Lhcb1 and Lhcb2 with molecular weights of 28 kDa and 27 kDa at the positions of arrows. (b) Room temperature absorption spectra of the LHC-II sample before and after crystallization, shown as solid line and dash line respectively.

*Table 3.* Data statistics for the spinach LHCII crystal shown in Fig. 3. The values in parentheses are for the outmost shell.

| | |
|---|---|
| Unit cell parameters | $a = b = 261.8$ Å, $c = 660.5$ Å |
| Space group | R32 |
| Resolution | 25–2.7Å |
| Observations | 1,104,148 |
| Unique reflections | 211,079 |
| $R_{merge}$ [1] | 8.2% (36.8%) |
| Completeness | 90.8% (79.6%) |
| $I/\sigma(I)$ | 14.4 (2.5) |
| Redundancy | 5.2 |

[1] $R_{merge} = \Sigma | I_j - \langle I_j \rangle | / \Sigma \langle I_j \rangle$, where $I_j$ is the intensity of reflection $j$ and $\langle I_j \rangle$ is the average intensity of reflection $j$.

processed to 2.7 Å. The data have a good statistics, as shown in Table 3. The mosaicity was 0.47°, which is a reasonable value considering data collection at cryotemperature (100 K).

About twenty big crystals were picked out and washed in an artificial mother liquor (50 mM HEPES-NaOH pH 7.5, 0.6% NG, 0.1% DBC, 1.5 mg mL$^{-1}$ DGDG, 1.09 M sodium citrate). They were finally dissolved in 0.4% NG and then used for SDS-PAGE and room temperature absorption spectral analysis. As shown by the SDS-PAGE result (Fig. 4a), the LHC II crystals contain two major polypeptides, Lhcb1 and Lhcb2 with molecular weight of 28 kDa and 27 kDa, and also trace amount of 26 kDa Lhcb3 polypeptide. The room temperature absorption spectrum of the crystal component showed up as a typical LHC II absorption spectrum (Fig. 4b).

Self rotation function was calculated by using the GLRF program (Tong and Rossmann, 1990). The result indicated the presence of a closed '532' point group symmetry in the crystal (Fig. 5a). It means

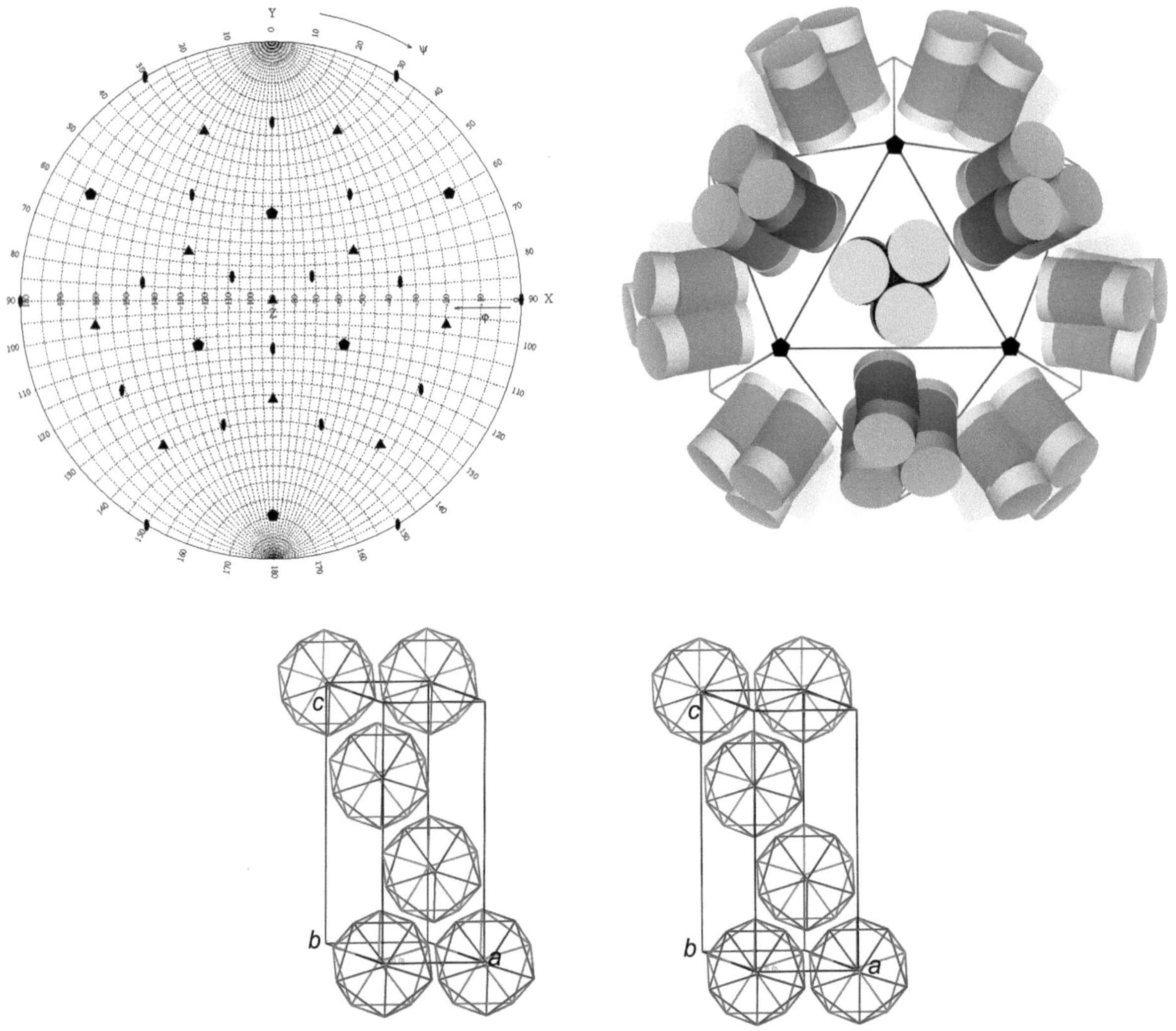


*Fig. 5.* Organization and packing of the icosahedral units in the R32 LHC-II crystal. (a) Stereographic projection showing the presence of a closed "532" point group in the LHC-II crystal and the orientation of icosahedral units in the crystal. (b) A schematic drawing of a T=1 icosahedral assembly constituted by twenty LHC-II trimers. View along a $C_3$ axis of the icosahedron. Each cylinder represents a LHC-II monomer. The spaces between cylinders might be filled by lipid, detergent and additive molecules. (c) A stereo view showing the packing of icosahedra in a hexagonal unit cell. Note that the diameter of icosahedron should be about the size of crystallographic axis a or b.

that the crystal was built from icosahedral structure units (Fig. 5b). The orientation of icosahedra in the crystal was determined by calculating locked self-rotation function (Tong and Rossmann, 1990) and analyzing the crystal packing. Figure 5c shows the packing diagram of icosahedral units in a hexagonal unit cell. One of the icosahedral '32' subgroups overlaps with the crystallographic '32' point group. The diameters of the icosahedra would be no larger than the dimensions of crystallographic a/b axes that are 261.8 Å. We were curious about how the LHC II complexes were organized into such an icosahedral unit. As the LHC IIs exist as trimers in solution, we would expect the local $C_3$ axes of LHC II trimers to be superimposed with the icosahedral $C_3$ axes. Assuming that each LHC II trimer occupies a *T*=1 icosahedral triangular face, there will be twenty trimers within one icosahedron (Fig. 5b). The gaps between LHC II trimers may be filled by lipid and detergent molecules, which probably form a curved lipid bilayer. Such an icosahedron made of LHC II, lipid and detergents is supposed to be a vesicular proteoliposome. We calculated that one LHC II monomer binds about 7 DGDG molecules. The molecular weight for a monomer, with all the bound pigments and lipids taken into account, would be 49 kDa. The molecular weight for the icosahedral vesicle containing 60 LHC II monomers is 2.94 MDa. We expected three vesicles in a hexagonal unit cell, so there will be 1/6 vesicle in an asymmetric unit. The resulting Matthews coefficient $V_M$ (Matthews, 1968) is 4.4 Å$^3$/Da and the solvent content (including detergent) is around 72%. These values are reasonable since relatively high $V_M$ and solvent content are quite common for membrane protein crystals (Soulimane et al., 1995; Jormakka et al., 2002).

## E. Discussion

Recently, the X-ray structure of pea LHC II was determined by Kühlbrandt's group (Standfuss et al., 2005). They used typical type I crystals which turned out to be stacks of 2D crystals. Three basic residues on the lumenal surfaces of one trimer form salt bridges with three acidic residues on the stromal surfaces of the other trimer below. These provide the stacking force between layers and are important for the formation of well-ordered type I LHC II crystals. The crystallization condition is quite different from ours. We noticed that their conditions include low ionic strength precipitant solution (10–15% polyethylene glycol 350 monomethyl ether, 50 mM morpholinoethanesulphonic acid buffer pH 5.2–5.6, 10–15% glycerol, 20 mM NaCl), which may allow the formation of strong ionic interactions between the charged residues of LHC II. Since we used high concentration of trisodium citrate as the major precipitant for crystallization, such kind of salt bridges might be destabilized under our conditions. For this reason, the stacking forces between the layers in our type I crystals are much weaker. This might explain why they are always disordered in the third dimensions.

Fortunately, we found our way to crystallize LHC II in a completely different manner. Structure determination with the new crystal form revealed the structural details about the organization of icosahedral proteoliposome within this novel LHC II crystal (Liu et al., 2004). The crystal packing (Fig. 1c) shows that this is indeed a new type of membrane protein crystal, type III. Most of the lipids and detergents in the proteoliposome are disordered except that one DGDG (Fig. 6a) and one phospholipid (Fig. 6b) per monomer are clearly distinguishable in the electron density map. The phospholipid molecules are bound at the monomer-monomer interfaces within the LHC II trimer. They are critical for the trimerization of LHC II (Nußberger et al., 1993). The DGDG molecules are located at the trimer-trimer interface (Fig. 6c). Their digalactosyl head groups interact with two adjacent trimers on both sides through hydrogen bonds. The two alkyl chains extend to the hydrophobic region and form hydrophobic interactions with the hydrophobic surfaces of two adjacent LHC II trimers respectively (Fig. 6c). Thus, twenty LHC II trimers were chained together by the DGDG molecules and then assembled into icosahedral vesicle (Fig. 6d). This kind of specific and network-like lipid-protein interactions prevent the rotation, inversion or translation of LHC IIs on the proteoliposome. The LHC IIs are actually firmly fixed on the vesicle.

In between the two DGDG molecules around icosahedral $C_2$ axis, there is still some electron density that we could not assign properly (Fig. 6a). They might belong to another DGDG molecule or most likely a deoxy-bigchap molecule. As the $C_2$ axis runs through this molecule, the crystallographic and non-crystallographic molecular averaging led to the fuzzy appearance of its density. As for the roles of deoxy-bigchap, we noticed that the deoxy-bigchap molecule has a rigid steroid part just like the cholesterol. It was found that incorporation of cholesterol into phosphatidylcholine vesicles would increase their rigidity (Liang et al., 2004; Liu et al., 2000) and the cholesterol-modified vesicle has much higher bending modulus compared with pure phosphatidylcholine vesicles (Liang et al., 2004). Therefore, we would expect a similar role for deoxy-bigchap in changing the property of lipid bilayer. Without deoxy-bigchap, the LHC II in complex with lipids would tend to form type I crystal with lamella structure. The insertion of deoxy-bigchap molecule into the lipid bilayer may cause the curvature of bilayer and trigger the formation of liposome. Upon the incorporation of deoxy-bigchap, probably at the junction points between LHC II trimers, the lipid bilayer structure become mismatched and the planar two-dimensional structures would curve into a more stable spherical structures, proteoliposomes. Like cholesterol, deoxy-bigchap molecules would decrease the deformability of proteoliposomes and hence increases their rigidity. This effect is reinforced by the specific lipid-protein interactions between DGDG molecules and the LHC II trimers. All these lead to highly stable and rigid proteoliposomes that would satisfy the requirements for building well-ordered crystals.

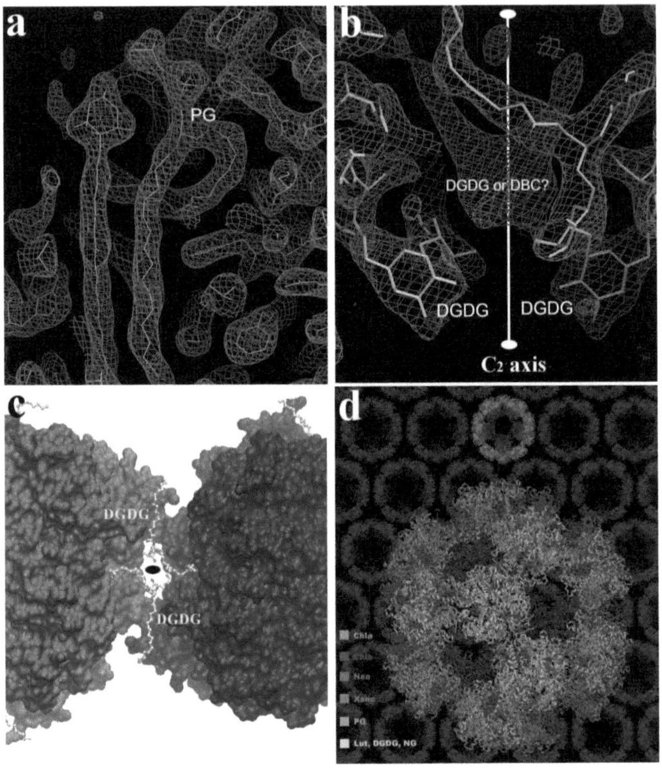


*Fig. 6.* The roles of lipids in the structure of LHC-II-DGDG proteoliposome. (a), (b) The well-defined 2Fo-Fc electron densities ($1.5 \times \sigma$ level) of PG and DGDG, respectively. In between the two DGDG molecules, there is a piece of electron density remaining unassigned. The icosahedral $C_2$ axis runs through it. (c) Top view of the trimer-trimer interface along the icosahedral $C_2$ axis. The two trimers are represented as surface models in the left and right halves of the panel, while the DGDG molecules are shown as ball-and-stick models in light gray. The black ellipse indicates the projection of $C_2$ axis on the paper plane. (d) The detailed structure of LHC-II-DGDG proteoliposome determined at 2.72 Å resolution. The apoproteins are shown as white ribbons. Prosthetic groups are drawn as ball-and-stick models in different gray levels. Chla, chlorophyll *a*; Chlb, chlorophyll *b*; Lut, lutein; Neo, neoxanthin; Xanc, xanthophyll-cycle carotenoid. The background shows a portion of the crystal lattice. See also Fig. 2, Color Plate 2.

Our results demonstrate that it is feasible to obtain well-ordered type III membrane protein crystals. For those membrane proteins that are reluctant to form ordered type I or type II crystals, there is a third way for them to crystallize. We think the key for growing highly ordered type III membrane protein crystal is to find a condition under which a homogeneous proteoliposome sample could form before crystallization occurs. The orientation of membrane proteins must be firmly fixed on the proteoliposomes through the direct and/or indirect (mediated by lipid molecules) interactions between membrane proteins. The proteoliposome should have a rigid instead of deformable structure. Practically, it requires a fine control of mixing ratios among proteins, detergents, lipids and additives. We would suggest multifactor screening in order to find the optimal type III crystal growth condition. The presence of high order of non-crystallographic symmetry is one advantage of type III crystals. It will aid in structure determination, especially in solving a set of highly accurate phases through iterative phase refinement and extension steps, leading to a high quality electron density map that is extremely helpful in structural model building process.

## Acknowledgments

This work was financially supported by grants from the National Key Research Development Project of China (G1998010110), the National Natural Science Foundation of China (30170202 and 30530210), the Knowledge Innovation Project of Chinese Academy of Sciences (KJCX2-SW-N06) and the President Special Foundation of Chinese Academy of Sciences.

## References


Allen JP (1994) Crystallization of the reaction center from *Rhodobacter sphaeroides* in a new tetragonal form. Proteins 20: 283–286

Byrne B, Abramson J, Jansson M, Holmgren E and Iwata S (2000) Fusion protein approach to improve the crystal quality of cytochrome $bo_3$ ubiquinol oxidase from *Escherichia coli*. Biochim Biophys Acta 1459: 449–455

Chang G and Roth CB (2001) Structure of MsbA from *E. coli*: A homolog of the multidrug resistance ATP binding cassette (ABC) transporters. Science 293: 1793–1800

Chang G, Spencer RH, Lee AT, Barclay MT and Rees DC (1998) Structure of the MscL homolog from *Mycobacterium tuberculosis*: A gated mechanosensitive ion channel. Science 282: 2220–2226

Cusack S, Belrhali H, Bram A, Burghammer M, Perrakis A and Riekel C (1998) Small is beautiful: Protein micro-crystallography. Nat Struct Biol 5: 634–637

Dahout-Gonzalez C, Brandolin G and Pebay-Peyroula E (2003) Crystallization of the bovine ADP/ATP carrier is critically dependent upon the detergent-to-protein ratio. Acta Crystallogr D 59: 2353–2355

Deisenhofer J and Michel H (1989) Nobel lecture. The photosynthetic reaction centre from the purple bacterium *Rhodopseudomonas viridis*. EMBO J 8: 2149–2170

Deisenhofer J, Epp O, Miki K, Huber R and Michel H (1985) Structure of the protein subunits in the photosynthetic reaction centre of *Rhodopseudomonas viridis* at 3 Å resolution. Nature 318: 618–624

Derewenda ZS (2004) Rational protein crystallization by mutational surface engineering. Structure 12: 529–535

Dormann P, Balbo I and Benning C (1999) *Arabidopsis* galactolipid biosynthesis and lipid trafficking mediated by DGD1. Science 284: 2181–2184

Doyle DA, Morais CJ, Pfuetzner RA, Kuo A, Gulbis JM, Cohen SL, Chait BT and MacKinnon R (1998) The structure of the potassium channel: Molecular basis of $K^+$ conduction and selectivity. Science 280: 69–77

Elrad D, Niyogi KK and Grossman AR (2002) A major light-harvesting polypeptide of Photosystem II functions in thermal dissipation. Plant Cell 14: 1801–1816

Fromme P and Witt HT (1998) Improved isolation and crystallization of Photosystem I for structural analysis. Biochim Biophys Acta 1365: 175–184

Furuichi M, Nishimoto E, Koga T, Takase A and Yamashita S (1997) Detergent effects on the light-harvesting chlorophyll *a/b*-protein complex crystallization revealed by fluorescence depolarization. Biochim Biophys Res Comm 233: 555–558

Garavito RM and Rosenbusch JP (1980) Three-dimensional crystals of an integral membrane protein: An initial X-ray analysis. J Cell Biol 86: 327–329

Horton P, Ruban AV and Walters RG (1996) Regulation of light harvesting in green plants. Annu Rev Plant Phys 47: 655–684

Huang LS, Cobessi D and Berry EA (2003) Crystallization of cytochrome $bc_1$ complex. In: S Iwata (ed) Methods and Results in Crystallization of Membrane Proteins, pp 203–226. International University Line, La Jolla

Hunte C, Koepke J, Lange C, Rossmanith T and Michel H (2000) Structure at 2.3 Å resolution of the cytochrome $bc_1$ complex from the yeast *Saccharomyces cerevisiae* co-crystallized with an antibody Fv fragment. Struct Fold Des 8: 669–684

Iwata S (2003) Crystallization informatics of membrane proteins. In: S Iwata (ed) Methods and Results in Crystallization of Membrane Proteins, pp 280–297. International University Line, La Jolla

Iwata S, Ostermeier C, Ludwig B and Michel H (1995) Structure at 2.8 Å resolution of cytochrome *c* oxidase from *Paracoccus denitrificans*. Nature 376: 660–669

Jiang Y, Lee A, Chen J, Ruta V, Cadene M, Chait BT and MacKinnon R (2003) X-ray structure of a voltage-dependent $K^+$ channel. Nature 423: 33–41

Jidenko M, Nielsen RC, Sørensen TL, Møller JV, le Maire M, Nissen P and Jaxel C (2005) Crystallization of a mammalian membrane protein overexpressed in *Saccharomyces cerevisiae*. Proc Natl Acad Sci USA 102: 11687–11691

Jordan P, Fromme P, Witt HT, Klukas O, Saenger W and Krauss N (2001) Three-dimensional structure of cyanobacterial Photosystem I at 2.5 Å resolution. Nature 411: 909–917

Jormakka M, Tornroth S, Abramson J, Byrne B and Iwata S (2002) Purification and crystallization of the respiratory complex formate dehydrogenase-N from *Escherichia coli*. Acta Crystallogr D 58: 160–162

Katona G, Andreasson U, Landau EM, Andreasson LE and Neutze R (2003) Lipidic cubic phase crystal structure of the photosynthetic reaction centre from *Rhodobacter sphaeroides* at 2.35 Å resolution. J Mol Biol 331: 681–692

Kolbe M, Besir H, Essen LO and Oesterhelt D (2000) Structure of the light-driven chloride pump halorhodopsin at 1.8 Å resolution. Science 288: 1390–1396

Kouyama T, Yamamoto M, Kamiya N, Iwasaki H, Ueki T and Sakurai I (1994) Polyhedral assembly of a membrane protein in its three-dimensional crystal. J Mol Biol 236: 990–994

Kühlbrandt W (1987) Three-dimensional crystals of the light-harvesting chlorophyll *a/b* protein complex from pea chloroplasts. J Mol Biol 194: 757–762

Kühlbrandt W (1988) Three-dimensional crystallization of membrane proteins. Quart Rev Biophys 21: 429–77

Kühlbrandt W, Wang DN and Fujiyoshi Y (1994) Atomic model of plant light-harvesting complex by electron crystallography. Nature 367: 614–621

Lancaster CRD, Kröger A, Auer M and Michel H (1999) Structure of fumarate reductase from *Wolinella succinogenes* at 2.2 Å resolution. Nature 402: 377–385

Landau EM and Rosenbusch JP (1996) Lipidic cubic phases: A novel concept for the crystallization of membrane proteins. Proc Natl Acad Sci USA 93: 14532–14535

Lange C and Hunte C (2002) Crystal structure of the yeast cytochrome $bc_1$ complex with its bound substrate cytochrome *c*. Proc Natl Acad Sci USA 99: 2800–2805

Lemieux MJ, Song J, Kim MJ, Huang Y, Villa A, Auer M, Li XD and Wang DN (2003) Three-dimensional crystallization of the *Escherichia coli* glycerol-3-phosphate transporter: A member of the major facilitator superfamily. Protein Sci 12: 2748–2756

Liang X, Mao G and Ng KY (2004) Mechanical properties and stability measurement of cholesterol-containing liposome on mica by atomic force microscopy. J Colloid Interf Sci 278: 53–62

Liu DZ, Chen WY, Tasi LM and Yang SP (2000) Microcalorimetric and shear studies on the effects of cholesterol on the physical stability of lipid vesicles. Colloid Surface A 172: 57–67

Liu ZF, Yan HC, Wang KB, Kuang TY, Zhang JP, Gui LL, An XM and Chang WR (2004) Crystal structure of spinach major light-harvesting complex at 2.72 Å resolution. Nature 428: 287–292

Long SB, Campbell EB and Mackinnon R (2005) Crystal structure of a mammalian voltage-dependent Shaker family $K^+$ channel. Science 309: 897–903

Lou SQ, Wang KB, Zhao FH, Xu CH and Kuang TY (1995) A comparative study on PS II light harvesting chlorophyll *a*/*b* protein complexes between spinach and cucumber. Acta Bot Sin 37: 192–197

Luecke H, Schobert B, Richter HT, Cartailler JP and Lanyi JK (1999) Structure of bacteriorhodopsin at 1.55 Å resolution. J Mol Biol 291: 899–911

Luecke H, Schobert B, Lanyi JK, Spudich EN and Spudich JL (2001) Crystal structure of sensory rhodopsin II at 2.4 Å: Insights into color tuning and transducer interaction. Science 293: 1499–1503

Matthews BW (1968) Solvent content of protein crystals. J Mol Biol 33: 491–497

McDermott G, Prince SM, Freer AA, Hawthornthwaite-Lawless AM, Papiz MZ, Cogdell RJ and Isaacs NW (1995) Crystal structure of an integral membrane light-harvesting complex from photosynthetic bacteria. Nature 374: 517–521

Michel H (1983) Crystallization of membrane proteins. Trends Biochem Sci 8: 56–59

Michel H (2006) Membrane proteins of known structure. http://www.mpibp-frankfurt.mpg.de/michel/public/memprotstruct.html (March 30, 2006)

Michel H and Oesterhelt D (1980) Three-dimensional crystals of membrane proteins: Bacteriorhodopsin. Proc Natl Acad Sci USA 77: 1283–1285

Nußberger S, Dorr K, Wang DN and Kühlbrandt W (1993) Lipid-protein interactions in crystals of plant light-harvesting complex. J Mol Biol 234: 347–356

Ostermeier C, Iwata S, Ludwig B and Michel H (1995) Fv fragment-mediated crystallization of the membrane protein bacterial cytochrome *c* oxidase. Nat Struct Biol 2: 842–846

Ostermeier C, Harrenga A, Ermler U and Michel H (1997) Structure at 2.7 Å resolution of the *Paracoccus denitrificans* two-subunit cytochrome *c* oxidase complexed with an antibody Fv fragment. Proc Natl Acad Sci USA 94: 10547–10553

Pautsch A, Vogt J, Model K, Siebold C and Schulz GE (1999) Strategy for membrane protein crystallization exemplified with OmpA and OmpX. Proteins 34: 167–172

Peter GF and Thornber JP (1991) Biochemical composition and organization of higher plant Photosystem II light-harvesting pigment-proteins. J Biol Chem 266: 16745–16754

Riekel C, Burghammer M and Schertler G (2005) Protein crystallography microdiffraction. Curr Opin Struc Biol 15: 556–562

Royant A, Nollert P, Edman K, Neutze R, Landau EM, Pebay-Peyroula E and Navarro J (2001) X-ray structure of sensory rhodopsin II at 2.1 Å resolution. Proc Natl Acad Sci USA 98: 10131–10136

Ruban AV, Lee PJ, Wentworth M, Young AJ and Horton P (1999) Determination of the stoichiometry and strength of binding of xanthophylls to the Photosystem II light harvesting complexes. J Biol Chem 274: 10458–10465

Schertler GF, Bartunik HD, Michel H and Oesterhelt D (1993) Orthorhombic crystal form of bacteriorhodopsin nucleated on benzamidine diffracting to 3.6 Å resolution. J Mol Biol 234: 156–164

Smyth DR, Mrozkiewicz MK, McGrath WJ, Listwan P and Kobe B (2003) Crystal structures of fusion proteins with large-affinity tags. Protein Sci 12: 1313–1322

Soulimane T, Gohlke U, Huber R and Buse G (1995) Three-dimensional crystals of cytochrome-*c* oxidase from *Thermus thermophilus* diffracting to 3.8 Å resolution. FEBS Lett 368: 132–134

Standfuss J, Terwisscha van Scheltinga AC, Lamborghini M and Kühlbrandt W (2005) Mechanisms of photoprotection and nonphotochemical quenching in pea light-harvesting complex at 2.5 Å resolution. EMBO J 24: 919–928

Strop P and Brunger AT (2005) Refractive index-based determination of detergent concentration and its application to the study of membrane proteins. Protein Sci 14: 2207–2211

Timmins PA, Hauk J, Wacker T and Welte W (1991) The influence of heptane-1,2,3-triol on the size and shape of LDAO micelles. Implications for the crystallisation of membrane proteins. FEBS Lett 280: 115–120

Tong L and Rossmann MG (1990) The locked rotation function. Acta Crystallogr A 46: 783–792

Toyoshima C, Nakasako M, Nomura H and Ogawa H (2000) Crystal structure of the calcium pump of sarcoplasmic reticulum at 2.6 Å resolution. Nature 405: 647–655

White SH (2004) The progress of membrane protein structure determination. Protein Sci 13: 1948–1949

White SH (2006) Membrane Proteins of Known 3D structure. http://blanco.biomol.uci.edu/Membrane_Proteins_xtal.html (June 8, 2006)

Xu CH, Zhao FH, Wang KB, Yang DH and Dai YL (1994) Isolation and spectroscopic analysis on the major light-harvesting chlorophyll *a*/*b* binding protein of Photosystem II. Bot Res 7: 202–207

Zhang H, Kurisu G, Smith JL and Cramer WA (2003) A defined protein-detergent-lipid complex for crystallization of integral membrane proteins: The cytochrome $b_6f$ complex of oxygenic photosynthesis. Proc Natl Acad Sci USA 100: 5160–5163

Zhou Y, Morais-Cabral JH, Kaufman A and MacKinnon R (2001) Chemistry of ion coordination and hydration revealed by a $K^+$ channel-Fab complex at 2.0 Å resolution. Nature 414: 43–48

Zulauf M (1991) Detergent phenomena in membrane protein crystallization. In: H Michel (ed) Crystallization of Membrane Proteins, pp 53–72. CRC Press, Boca Raton

# Chapter 6

# X-ray Crystallography of Photosynthetic Proteins

Petra Fromme and James P. Allen*
*Department of Chemistry and Biochemistry and Center for Bioenergy and Photosynthesis, Arizona State University, P.O. Box 871604, Tempe, AZ 85287-1604, U.S.A.*



*Author for correspondence, email: jallen@asu.edu

*Thijs J. Aartsma and Jörg Matysik (eds): Biophysical Techniques in Photosynthesis II, pp. 97–124.*

## Summary

This chapter provides an overview on the present status of the X-ray structure analysis of photosynthetic proteins. The methods and problems of crystallization and X-ray structure analysis of the large photosynthetic protein complexes are discussed. In the second part of the chapter, selected structures of the protein complexes are presented and the function of the proteins is briefly discussed based on the structural information.

## I. Introduction

Photosynthesis is the main process on earth that converts solar energy into chemical energy. The reactions of photosynthesis can be divided into the light reactions, which take place in the photosynthetic membrane -the thylakoids- and the dark reactions that mainly take place in the cytosol. In the light reactions, solar energy is captured by antenna systems and the excitation energy is transferred to the reaction centers, which catalyze the light induced electron transfer across the membrane. Oxygenic photosynthetic organisms such as plants, green algae and cyanobacteria use water as an electron source and contain two photosystems, Photosystem I and II, which function in series, whereas non-oxygenic photosynthetic organisms contain only either a type I or type II reaction center (RC). In oxygenic photosynthesis, the two RCs are functionally coupled by a mobile quinone pool, the cytochrome $b_6f$ complex and the soluble electron carriers plastoquinone or cytochrome $c_6$. The electron is transferred from Photosystem I to the soluble electron carriers ferredoxin or flavodoxin and finally transferred to the ferredoxin: $NADP^+$:reductase (FNR) which reduces $NADP^+$ to NADPH. The electron transfer reactions build up an electrochemical proton gradient that is used by ATP-synthase for the synthesis of ATP from ADP and $P_i$. In the dark reactions, the high energy products ATP and reduced hydrogen in form of NADPH are used to fix $CO_2$ and build up glucose in the Calvin cycle. Some organisms have also established active $CO_2$ enrichment reactions.

The light reactions are catalyzed by membrane proteins, for which structural information is rare. Whereas more than 35,000 structures of soluble proteins have been determined by X-ray structure analysis, only 100 different membrane protein structures have been unraveled so far.

The present status of the X-rays structure analyses of photosynthetic proteins is very fortunate as structures of the majority of large membrane proteins that are involved in these processes are known. The primary reactions of photosynthesis are thereby the only membrane-bound process, where at least partial structures of all major proteins and protein complexes have been determined by X-ray structure analysis. The field of membrane protein structure determination started with the successful crystallization of the purple bacterial reaction center by H. Michel and J. Deisenhofer in 1985 for which they received the Nobel Award together with R. Huber in 1988.

Figure 1 shows an overview of the structures of all major proteins involved in the electron transfer chain of oxygenic photosynthesis: the large membrane protein complexes Photosystem II (Ferreira et al., 2004), the cytochrome $b_6f$ complex (Kurisu et al., 2003; Stroebel et al., 2003; Zhang et al., 2003), Photosystem I (Jordan et al., 2001) and the ATP-synthase (Abrahams et al., 1994; Stock et al., 1999) as well as the soluble electron carriers plastocyanine (Schirmer et al., 1985; Redinbo et al., 1993), cytochrome $c_6$ (Frazao et al., 1995; Kerfeld et al., 1995), ferredoxin (Adman et al., 1976), flavodoxin (Watenpaugh et al., 1972) and the FNR (Bruns and Karplus, 1995).

Detailed structural information on the more ancient RCs of anaerobic bacteria is only available for type II RCs from purple bacteria (Deisenhofer et al., 1984; Deisenhofer and Michel, 1991), but none of the ancient type I RCs from neither heliobacteria or green-sulfur bacteria has been crystallized so far. The primary reactions of photosynthesis also involve antenna complexes, which differ strongly between different photosynthetic species. Higher plants contain membrane integral antenna proteins, the LHC II and LHC I. The structure of the LHC II has been determined at atomic resolution; (Chapter 5, Liu and Chang), whereas the LHC I structure has

*Abbreviations:* ATP – Adenosine triphosphate; CMC – critical micellar concentration; FNR – Ferredoxin:$NADP^+$:reductase; IP – isoelectric point; LH1 – light-harvesting complex 1; LH2 – light-harvesting complex 2; MAD – multiple anomalous diffraction; MIR – multiple isomorphous replacement; NADP – nicotinamide-adenine dinucleotide phosphate; PEG – polyethylene glycol; Pi – phosphate; RC – reaction center; SAD – single-wavelength anomalous dispersion

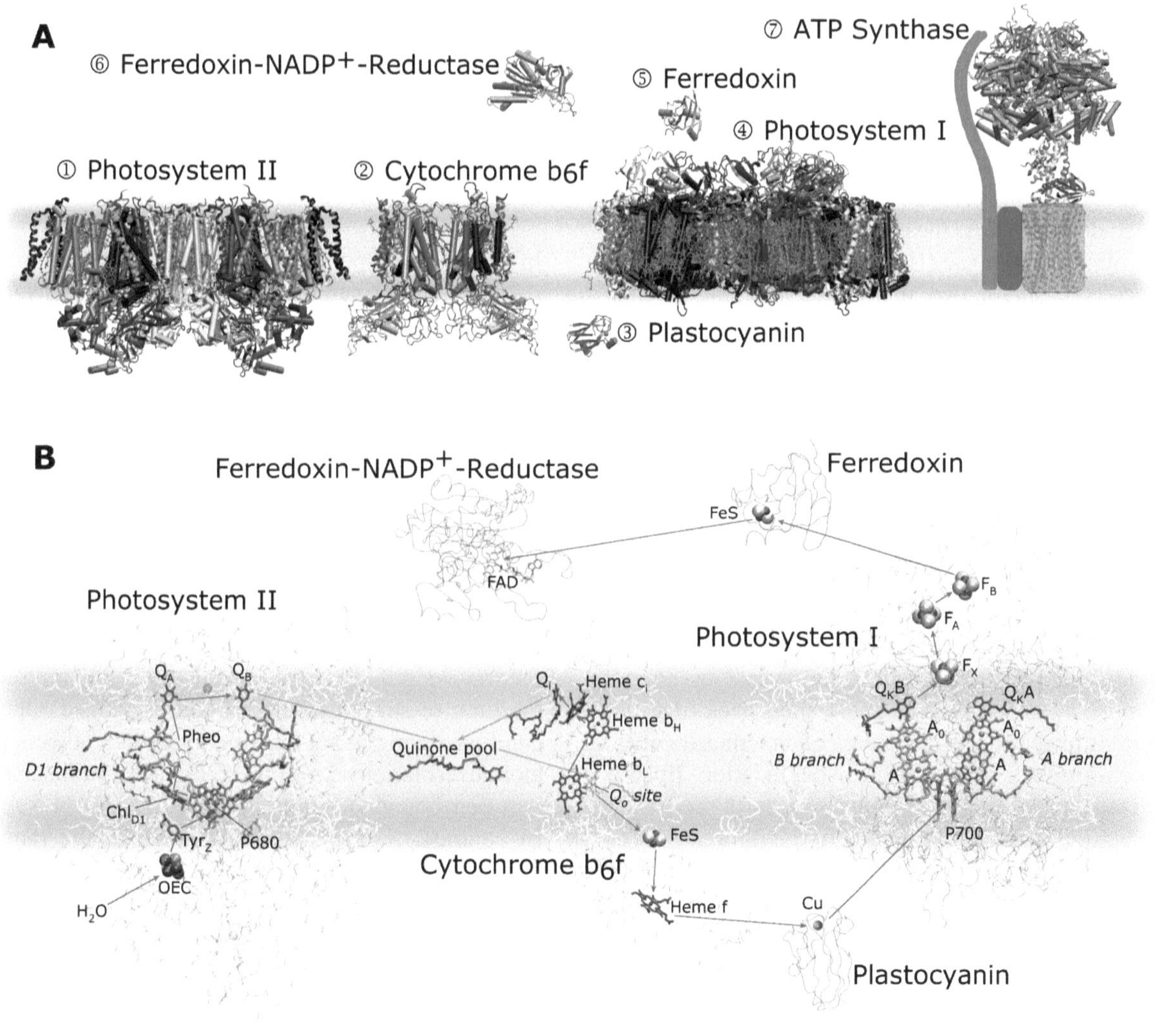


*Fig. 1.* A. Overview of the structures of photosynthetic proteins of the electron transfer chain: Photosystem II (Ferreira et al., 2004), cytochrome $b_6f$ complex (Stroebel et al., 2003), plastocyanin (Schirmer et al., 1985), Photosystem I (Jordan et al., 2001), ferredoxin (Adman et al., 1976), FNR (Bruns and Karplus, 1995) and the ATP-synthase (Abrahams et al., 1994; Stock et al., 1999) are shown. B. Structural depiction of the cofactors of the electron transfer chain. Electrons are extracted from water in Photosystem II and transferred to the plastoquinone pool that transfers the electron to the cytochrome $b_6f$ complex. From there, the electron is transferred to the soluble electron transfer protein plastocyanine, which reduces P700$^+$ in Photosystem I. Photosystem I catalyzes the light induced transmembrane electron transfer from plastocyanine to ferredoxin, which transfers the electrons to the FNR, where they are finally used for the reduction of NADP$^+$ to NADPH. The electrical and proton gradient that are formed drive the synthesis of ATP by the ATP-synthase.

been determined at medium resolution in a complex with monomeric plant Photosystem I (Ben-Shem et al., 2003). Cyanobacteria contain large membrane-attached but extrinsic antenna complexes, the phycobilisomes. The structures of all major pigment-containing proteins of the phycobilisomes have been determined: phycocyanin, phycoerythrin and allophycocyanin (Schirmer et al., 1985; Ficner and Huber, 1993; Brejc et al., 1995). However, no reports have been published on the crystallization and structure determination of the intact phycobilisome complex including the linkers, nor has the site been identified, where the phycobilisomes are attached to the photosystems.

The anoxygenic photosynthetic organisms show a large variation of peripheral antenna. Green sulfur bacteria and non-green sulfur bacteria contain chlorosomes. They may be regarded as a prokaryotic light harvesting organelle and consist of huge assemblies of chlorophylls covered by a membrane.

These chlorosomes might be functionally coupled to the reaction center by the FMO protein, which is soluble and contains seven chlorophylls. The FMO protein was the first (bacterio)chlorophyll-binding photosynthetic protein, for which the structure had been determined (Matthews et al., 1979). Another soluble antenna protein, for which the structure has been determined, is the peridinin-Chl protein, which is found in red algae (Hofmann et al., 1996). Purple bacteria contain two membrane integral antenna proteins — the LH2 and LH1 — which form ring structures. The structure of LH2 has been determined at atomic resolution (McDermott et al., 1995; Prince et al., 1997), and the LH1 have been crystallized in form of an open ring, surrounding the reaction center (Law et al., 1998; Roszak et al., 2003).

One of the two enzymes from the Calvin cycle that have been determined from plant origin are rubisco (for review, see Spreitzer and Salvucci, 2002) and the fructose-1,6-bisphosphate phosphatase (Chiadmi et al., 1999). Most of the other enzymes have been determined from their bacterial, yeast or animal counterparts. There is an astonishing lack of structural information on structures of enzymes involved in chlorophyll biosynthesis or degradation. Even if this is the largest global process on earth there is no structural information on any of the photosynthesis specific enzymes available, except for one protein subunit of the Mg-chelatase (Fodje et al., 2001).

In the following parts of the chapter, we will give a brief introduction into the biophysical method of X-ray crystallography, discuss the problems and advancements of crystallization of the photosynthetic proteins, summarize techniques and progress and pitfalls in data collection, phase determination and discuss model building. The chapter will also give a brief description of the structures of the photosynthetic proteins that have been determined.

## II. Principles of X-ray Diffraction

In the following paragraphs, we will briefly summarize the main principles of X-ray structure analysis. The reader should be aware that this summary can only serve as a short introduction into the general principles, as they are useful for the understanding of the X-ray structures of photosynthetic proteins and their limitations. The general principle of X-ray crystallography is summarized in Fig. 2.

For a detailed description of the technique, the reader is referred to textbooks on X-ray crystallography (Stout and Jensen, 1989; Giacovazzo et al., 1992; Abrahams and Ban, 2003). Large objects can be visualized by optical microscopy, however, the wavelength of visible light at 400 to 800 nm sets the limit of structures that can be visualized. Optical microscopy is not suitable to resolve objects to atomic resolution taking into account that the distance between bounded atoms is only 0.15 nm (1.5 Å). In contrast, X-rays with wavelength in the range between 0.8 and 2 Å can be diffracted even by the smallest molecules. Since X-ray diffraction is based on the scattering of the X-rays by the electron cloud of each atom, the strength of the diffraction depends on the electron density of the individual atoms in the unit cell, which is the smallest building block of the crystal. However, the interaction of X-rays with

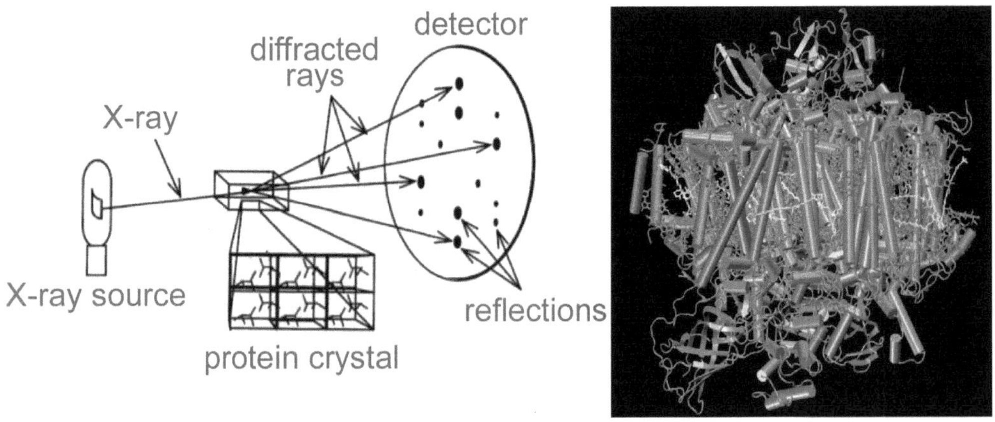


*Fig. 2.* General principle of X-ray diffraction.

atoms in a protein is very week so that X-ray scattering by a single molecule cannot be detected. This is the reason why the protein must be crystallized. A protein crystal contains about $10^{12}$ molecules, aligned in a three-dimensional lattice. The scattered waves interfere and produce a diffraction pattern, where each diffracted wave contains contributions from each atom. The electron density in the unit cell can be calculated from the fully description of the scattered waves by Fourier transformation. Each wave is characterized by three parameters: (i) wavelength, (ii) amplitude and (iii) phase. As the scattering of X-rays is elastic, the wavelengths of the scattered waves are identical to the wavelength of the X-rays source. The amplitude can be determined from the intensity of the reflections. However, the phase of the scattered waves is lost and has to be determined indirectly. For large molecules (like proteins) direct methods can still not be used. There are three methods that can be used for phase determination: experimental phases have to be determined for new structures for which no homology models exist. Experimental phases can be determined by multiple isomorphous replacement (MIR), multiple anomalous diffraction (MAD) or single anomalous diffraction (SAD). As the large membrane protein complexes from photosynthesis have all been isolated from the natural organism, all structures of these large macromolecular assemblies have initially been determined using the time consuming method of finding heavy metal atom derivatives and solving the phases with MIR. After the initial structures of the membrane proteins have been solved all structures of mutant complexes have been determined by molecular replacement. The same is valid for most of the soluble photosynthetic proteins. Only recently, the structures of several proteins, some of them being cytochromes, have been solved by SAD (Wang et al., 2004). After the initial electron density map has been determined, further methods as solvent flattening and the use of non-crystallographic symmetries have been very useful for the improvement of the electron density map (Krauss et al., 1993). This was especially important for the membrane protein structures as these generally show a high solvent content. Non-crystallographic symmetry was very important for the structure determination of oligomeric proteins such as the LH2 of purple bacteria (Prince et al., 1997).

Recently automated model building programs are increasingly more powerful. They can also be used for small photosynthetic proteins with a low cofactor content. However, the models of the large protein complexes with many cofactors, as Photosystems I and II, were all built by hand-fitting of the molecules in the electron density as the automated programs can still not handle a high cofactor content in combination with a medium resolution electron density map.

Now we want to briefly summarize the main essentials and terms of X-rays crystallography:

### *A. The Unit Cell*

The *unit cell* is the smallest building block of a crystal. The dimensions of the unit cell are designated by the length of the three *unique edges a, b, and c* and the three *unique angles* α, β and γ. The crystals can have the following space groups indicated in Table 1.

The largest assembly of molecules that does not show crystallographic symmetry but can be transferred into other identical assemblies of molecules in the unit cell by crystallographic symmetry operations is called the *asymmetric unit*. As proteins are chiral molecules the only allowed symmetry operations are: *rotation* around a rotation axis, *translation*, i.e., movement by a specific distance, and *screw axes* which are a combination of translation and rotation. The rotation is symbolized by the symbol n and by a number 360/n (n = 2 for rotation by 180°, 3 for rotation by 120°, 4 for rotation by 90°, 6 for rotation by 60°). Translation is defined by a shift of the asymmetric unit by a specific distance. A distance equal to length of one axis is called unit translation. The unit cells often show symmetry elements that entail translation by a fraction of the axial length such as c/6. The *screw axes* result from a combination of rotation and translation and are symbolized by $n_m$, which stands for a n fold screw axis with a translation of m/n of the unit cell. As an example, the unit cell of Photosystem I, $P6_3$, is shown in Fig. 3. P stands for a primitive lattice with a $6_3$ screw axis parallel to c.

The *International Tables for X-ray Crystallography* contain the symmetry parameters for all possible 230

*Table 1.* Crystal forms and unit cells

| | | |
|---|---|---|
| triclinic | $a \neq b \neq c$ | $\alpha \neq \beta \neq \gamma$ |
| monoclinic | $a \neq b \neq c$ | $\alpha = \gamma = 90°$ $\beta > 90°$ |
| hexagonal | $a = b$ | $\alpha = \beta = 90°$ $\gamma = 120°$ |
| cubic | $a = b = c$ | $\alpha = \beta = \gamma = 90°$ |
| tetragonal | $a = b \neq c$ | $\alpha = \beta = \gamma = 90°$ |
| orthorhombic | $a \neq b \neq c$ | $\alpha = \beta = \gamma = 90°$ |
| trigonal | $a = b \neq c$ | $\alpha = \beta = \gamma$ |

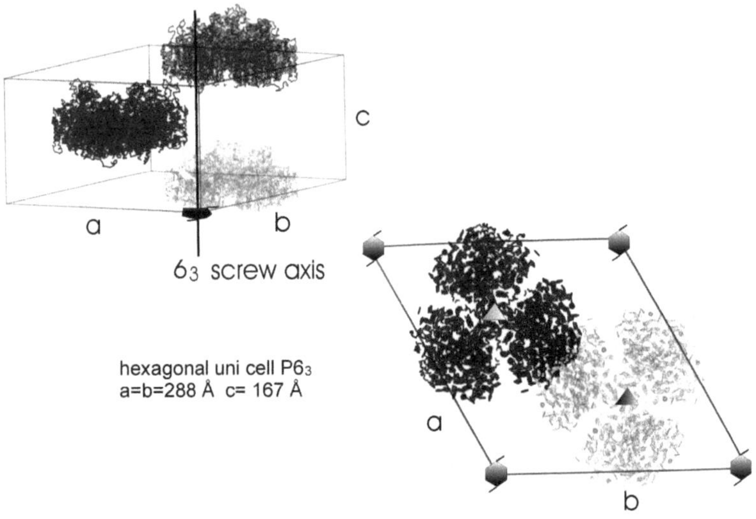


*Fig. 3.* Unit cell of Photosystem I crystals.

unit cells. However, no mirror planes exist in proteins crystals. This limits the possible number of unit cells of protein crystals to 64.

## B. Indices and Atomic Planes in a Crystal

The unit cell can be divided into sets of planes in the crystalline lattice. The most apparent sets of planes are the ones that are determined by the faces of the unit cell. These planes are the sources of diffraction and are called *lattice indices* (or Miller-Indices) and are indicated with the symbols *hkl*. The index *h* gives the number of parts into which the set of planes cut the a edge of the unit cell, the indices *k* and *l* give the number of parts into which the set of planes cut the b and c edges of the unit cell, respectively. The set of planes that is parallel to the bc face of the unit cell is designated to *hkl* (100), the set of planes that is parallel to the ac face is designated to *hkl* (010) and the set of planes that is parallel to the ab face of the unit cell is designated to *hkl* (001). These are the easiest set of planes. As an example the planes that cut the a-axis in two parts and the b-axis in three parts would be designed with the *hkl* indices (230). In the language of crystallography, reflections that are cause by a set of planes that cut the edges of the unit cell more than once or called higher order reflections. The higher the order is, the smaller is the *interplanar spacing* $d_{hkl}$ between in individual planes in the set of planes.

## C. Bragg's Law

In the *Bragg model of diffraction*, any set of planes is the source of one diffracted X-ray beam. A set of planes with indices *hkl* and the interplanar spacing $d_{hkl}$ produces one diffracted beam when X-rays of the wavelength λ impinge upon the set of planes at an angle θ and are reflected at the same angle if θ meets the condition of Bragg's law:

$$2\, d_{hkl} \sin\theta = n\,\lambda \text{ (with } n = \text{integer)} \quad (1)$$

The geometric construction Bragg's law is shown in Fig. 4.

If the difference in path length for X-rays from successive planes is equal to an integer number of wavelength of the impinging X-rays (i.e., when they follow Bragg's law) the rays reflected from successive planes are in phase with each other. Thereby, only X-rays that impinge the crystal at specific angles that that fulfill the Bragg law interfere constructively and produce a diffracted beam, which is detected as a reflection; all other interfere destructively and do not led to a reflection. The angle of diffraction is inversely related to the interplanar spacing $d_{hkl}$. This leads to the assumption that large unit cells with large spacing give small angles of diffraction, i.e., the spacing between the reflections is reciprocal proportional to the size of the unit cell. This often leads to serious problems in the separation between reflections on diffraction images of crystals with large unit cells, as it is the case for all four large membrane protein

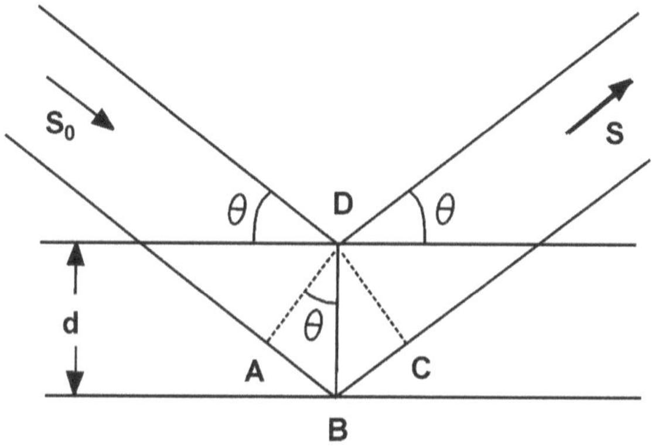


*Fig. 4.* Reflection conditions that satisfy Bragg's law.

complexes of photosynthesis. The problem is even more pronounced when the crystals show a mosaic spread, i.e., high degree of internal disorder, which leads to broadening of the reflections.

## *D. The Real and Reciprocal Lattice*

For the description of the diffracted beams we will now define the reciprocal lattice and the reciprocal unit cell. For the easiest case, the cubic unit cell, the reciprocal unit axes a*, b* and c* are lying along the real unit cell axes a, b and c. However, the length of the reciprocal axes a*, b* and c* are reciprocals of the length of the real axes, so that a* = 1/a, b*= 1/b and c* = 1/c. This leads to the fact that a small real unit cell with small interplanar spacing corresponds to a large reciprocal unit cell with large spacing between the reflections, whereas large unit cells with large interplanar spacing $d_{hkl}$ lead to a small reciprocal unit cell with small spacing between reflections. The relationship between real and reciprocal unit cells is shown in Fig. 5.

## *E. Bragg's Law in Reciprocal Space: The Ewald Sphere*

When a crystal is rotated in the X-ray beam various reciprocal lattice points come into contact with the Sphere of reflection (or Ewald Sphere) as depicted in Fig. 6. According to Bragg's law, the *hkl* reflection is caused by diffraction from a set of equivalent, parallel real space planes *hkl*. In the reciprocal space, this set of planes corresponds to the lattice point $P_{hkl}$ and a reflection is produced when the reciprocal lattice point $P_{hkl}$ contacts the sphere of reflections. That means that the direction and the number of reflections only depends on the unit cell and the cell dimensions, i.e., the diffraction pattern is independent of the content of the unit cell and can be predicted just from the space group and unit cell dimension. This is the reason why all data evaluation programs can easily predict the location of all reflections and indicate them as 'peak boxes.' However, the intensity of the reflection *hkl* depends on the values of the electron density p(x,y,z) on the planes *hkl* and thereby contains structural information about the content of the unit cell.

## *F. The Total Number of Reflections and Resolution of the X-ray Diffraction Data Sets*

The total number of reflections can be predicted from the space group and the cell dimensions. Given that the sphere of reflections has a radius of 1/λ, any lattice point within a distance of 2/λ of the origin can be rotated into contact with the Ewald sphere

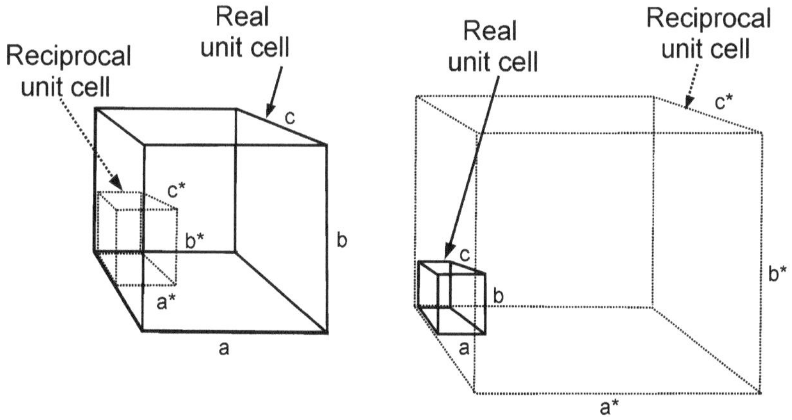


*Fig. 5.* Dimensions of real and reciprocal unit cells.

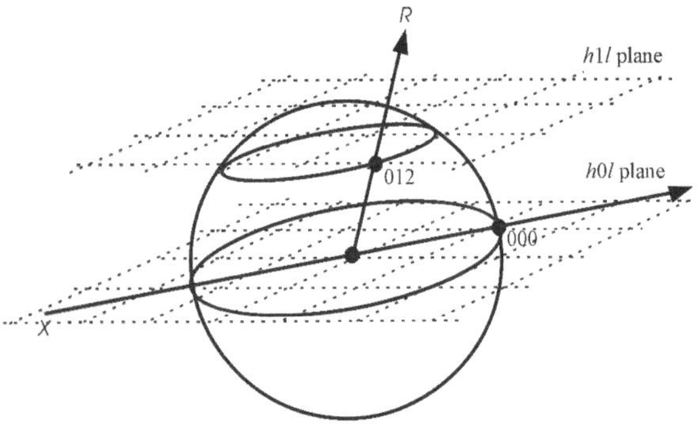


*Fig. 6.* The Ewald sphere.

of reflections, thereby defining the *limiting sphere.* The number of reflections that can be obtained by rotating a crystal through all possible orientations in the X-ray beam is equal to the number of *reciprocal lattice points* $P_{hkl}$ in the limiting sphere. Thereby, the unit cell and the wavelength of the X-ray beam determine the maximal number of reflections that can be observed. Shorter wavelengths will result in a larger sphere of reflections, i.e., more measurable reflections. Larger unit cells with smaller reciprocal unit cells will contain more reflections in the limiting space, leading to more measurable reflections. The number of the possible reflections $N$ is approximately equal to the volume of the limiting space divided by the volume of one reciprocal cell:

$$N = \frac{(4\pi/3)\cdot(2/\lambda)^3}{V_{\text{recip}}} \qquad (2)$$

with $V = 1/V_{\text{recip}}$ ($V$ = volume of real unit cell). The number of maximal measurable reflections is

$$N = 33.5 \cdot V/\lambda^3 \qquad (3)$$

An example is given for a protein with a large unit cell (Photosystem I) with a = b = 288 Å, c = 167 Å and data collection at $\lambda = 1$ Å. The theoretical description leads to a theoretical number of $4.6 \times 10^8$ reflections. However, not all reflections will be unique due to the symmetry of the space group. Furthermore, the maximal resolution of a data set has to be taken into account and limits the number of observed reflections. As an example, the native Photosystem I data set at 6 Å resolution contained 49,832 reflections with 18,703 unique reflections (at a completeness of 94%) (Krauss et al., 1993). The amount of data increases dramatically with an increase in resolution. The native data set at 2.5 Å resolution (with the same unit cell) contained 233,377 unique reflections (the completeness was 97.1%) (Jordan et al., 2001).

A frequently asked question deals with the resolution of a structure. The resolution of an X-ray crystallography model at 2.5 Å resolution means that the X-ray structure analysis included reflections out to a distance of 1/2.5 Å in the reciprocal lattice (counted from the center of the diffraction pattern), which are caused by diffraction from sets of equivalent parallel planes with a spacing of 2.5 Å. The maximal resolution depends mainly on the quality of the crystals.

In a crystal with no disorder, the reflected beam would lead to a very intense and small point in the diffraction image. However, real crystals contain defects, which lead to the broadening of the reflections. This is indicated as the mosaic spread. For example, a mosaic spread of 1° leads to a broadening of the reflections over a range of 1°. A high mosaic spread is a serious problem of many crystals of large photosynthetic membrane proteins. It leads to severe problems during data collection and evaluation. If we assume that data are collected in increments of 0.5° image then one unique reflection would be distributed over 2 to 3 images. Thereby, the number of reflections on one image is increased dramatically. Together with the fact that the spacing between reflections is already very short, due to the small reciprocal and large real unit cell. The mosaicity causes an overlap of the reflections. Even if the data evaluation software for Laue-diffraction data can de-convolute partially overlapping reflections, the software packages for single wavelength X-ray data collection and evaluation do not yet include de-convolution software.

## G. Fourier Series and Fourier Transformation

The electron density $\rho(x,y,z)$ is described by the Fourier series

$$\rho(x,y,z) = \frac{1}{V}\cdot\sum_h\sum_k\sum_l F_{hkl}\cdot e^{-2\pi i(hx+ky+lz)} \qquad (4)$$

The electron density map contains the triple sum of all structure factors $F_{hkl}$. Thereby an electron density map can be only calculated if the structure factors $F_{hkl}$ of most of the reflections *hkl* of a data set are known. $F_{hkl}$ is a diffracted X-ray and therefore a periodic function that can be described by three parameters: the frequency, the amplitude and the phase. The frequency is that of the X-ray source, and the amplitude of is the square root of the intensity of the reflection, i.e., the amplitudes can be calculated from the intensities of the measured reflections. However, the phase can not be detected from the diffraction pattern and must be determined in an indirect way, either by multiple isomorphous replacement (MIR), multiple anomalous dispersion (MAD), single-wavelength anomalous dispersion (SAD) or, if a homologous structure is available, by the method of molecular replacement. We will describe the use of these methods for the structure determination of photosynthetic proteins after we have discussed the crystallization and data collection processes.

## III. Crystallization of Photosynthetic Proteins

The crystallization of proteins depends on a large number of parameters, which can be divided into two groups of biological/biochemical and physical/chemical parameters. We want to briefly discuss the biological parameters first. A large number of biological and biochemical parameters play an important role for the crystallization of photosynthetic proteins; as examples we want to mention the organism, the physiological status of the organism, the tertiary structure, the subunit composition and the binding of ligands.

### *A. Influence of Biological and Biochemical Parameters*

#### *1. The Organism*

At present, most structures of soluble proteins are determined using crystals that have been grown from protein that is over-expressed in *Escherichia coli.* Other organisms that play an important role in the structure determination are yeast and, more recently, many structures have been determined from thermophilic organisms as *Thermotoga maritima*. From this latter organism, 463 structures have been determined so far and are deposited at the PDB database (http://www.rcsb.org). For most of the enzymes involved in the Calvin cycle, structures have been determined from a variety of organism but only few structures, mainly the plant specific enzymes as rubisco, have been determined from a plant origin (Schneider et al., 1986; Chapman et al., 1988; Spreitzer and Salvucci, 2002). The vast majority of photosynthetic proteins and all photosynthetic membrane proteins have been isolated and crystallized from the natural organism. The first membrane protein that have been crystallized was the reaction center (RC) of the purple bacteria in 1984 (Deisenhofer et al., 1984). Different species of purple bacteria have been the basis for the structure determination of the photosynthetic RC and the bacterial antenna proteins LH1 and LH2 (Allen et al. 1987; Deisenhofer et al., 1984; Prince et al., 1997; Roszak et al., 2003). Thermophilic cyanobacteria play a dominant role in the structure determination of membrane proteins of oxygenic photosynthesis. The structures of Photosystems I and II have been determined from *Thermosynechococcus elongatus* (Jordan et al., 2001; Zouni et al., 2001; Ferreira et al., 2004; Loll et al., 2005). One of the Photosystem II structures has been determined using Photosystem II isolated from the closely related thermophilic cyanobacterium *Thermosynechococcus vulcanus* (Kamiya and Shen, 2003). The structure of the cytochrome $b_6f$ complex has been determined from another thermophilic cyanobacterium *Mastigocladus laminosus* (Kurisu et al., 2003) and from the green algae *Chlamydomonas reinhardtii* (Stroebel et al., 2003). The 4.4 Å structure from the supercomplex of Photosystem I with the LHCI from pea by Ben Shem et al. (2003) was the first structure of a plant membrane protein, determined by X-ray crystallography. Shortly after, the structure of the plant LHC II was determined from spinach (Liu et al., 2004) (see detailed description in Chapter 5, Liu and Chang). The proteins from thermophilic organism are generally more stable and are therefore better suitable for crystallization, but these examples show that it is also possible to isolate and crystallize membrane proteins from green algae and plants, when the organism can be grown under reproducible conditions in the laboratory. The change of the source of the organism does not only influence the stability of the proteins but also has a great influence on the crystallization conditions and the solubility of the protein. As an example, we can look at the solubility of Photosystem I from different cyanobacteria. Whereas Photosystem I from the thermophilic cyanobacterium *Synechococcus elongatus* has been crystallized by dialysis against low ionic strength finally leading to crystals diffracting to 2.5 Å resolution (Fromme, 1998; Jordan et al., 2001). Photosystem I from the mesophilic cyanobacterium *Synechocystis PCC 6803* has a solubility of >150mg/mL at low ionic strength, so that it can not be crystallized by dialysis against low ionic strength

(Jekow et al., 1995). Furthermore, the stability of the trimeric Photosystem I and dimeric Photosystem II in the thermophilic organism is much higher so that these are the preferred photosynthetic organisms for crystallization attempts on Photosystems I and II.

## 2. The Physiological Status of the Organism

All photosynthetic organisms are able to adapt to different environmental conditions (light, salt, nutrients) by global changes of the membrane- and protein composition of the whole cell (Allen et al., 1995; Fey et al., 2005; Scheuring and Sturgis, 2005; Walters, 2005). Furthermore, the protein composition also depends on the growth phase of the cells (Neisser et al., 1994). The protease level in the cells is low in the logarithmic phase and increases dramatically in the stationary growth phase. A synchronized growth of the cells under standardized laboratory conditions has been proven to be a critical prerequisite for the successful crystallization of photosynthetic proteins as Photosystem I (Fromme, 1998).

## 3. The Quaternary Structure and the Subunit Composition

Many soluble photosynthetic proteins and many photosynthetic membrane proteins, for which the structures have been determined exist in the photosynthetic membranes in form of oligomers (Table 2).

Some proteins exist in the membrane in different oligomeric forms like Photosystem I, which changes the ratio of monomers to trimers as a function of the light intensity. The oligomeric state has a very profound influence on the solubility of proteins. For membrane proteins, the solubility of the monomeric forms is generally higher than the solubility of the multimer, because of the increased ratio of detergent to protein in the monomer. A very drastic example is the solubility difference of trimeric and monomeric Photosystem I from *Thermosynechococcus elongatus* at low ionic strength. The trimer is crystallized by dialysis against low ionic strength, with a solubility of less than 1 μg/mL, whereas monomeric Photosystem I is soluble up to a protein concentration of 150 mg/mL under identical conditions, i.e., has a solubility that is increased by more than five orders of

*Table 2.* Oligomeric states of photosynthetic proteins

| Protein | Oligomeric state in vivo | Oligomeric state in the crystals |
|---|---|---|
| bacterial reaction center | monomer, surrounded by LH 1 | monomer (Deisenhofer et al., 1984) monomer, surrounded by LH1 (Roszak et al., 2003) |
| LH 1 | open ring of 15 subunits, surrounding the RC | open ring of 15 subunits surrounding the RC (Roszak et al., 2003) |
| LH 2 | ring of 9 subunits | ring of 9 subunits (McDermott et al., 1995) |
| Photosystem I (*Thermosynechococcus elongatus*) | monomer or trimer (depending on light conditions) | trimer (Jordan et al., 2001) |
| Photosystem I and LHC I (*pea*) | monomer with 4 to 10 LHCI proteins and LHC II attached (number and attachment of LHC II depends on light conditions) | monomer with 4 LHC I proteins (Lhca1, Lhca2, Lhca3, Lhca4) attached to one site of the monomer (Ben-Shem et al., 2003) |
| Photosystem II (*TS elongatus and TS vulcanus*) | monomer (under photodamage and repair) or dimer representing the intact functional Photosystem II | dimer (Zouni et al., 2001) |
| LHC II | trimer | trimer (Liu et al., 2004) |
| $b_6f$ complex (*Mastiglocadus and Chlamydomonas rh.*) | dimer | dimer (Kurisu et al., 2003; Stroebel et al., 2003) |
| Rubisco | octamer | octamer 8 small and 8 large subunits (Schneider et al., 1986; Chapman et al., 1988) |
| FMR | monomer | monomer (Karplus et al., 1991; Serre et al., 1996) |

magnitude. A mixture of different oligomeric forms of the protein often negatively influences the crystallization attempts. It has been for example shown that the formation of 1 monomer per 10,000 trimeric molecules has already a strong negative effect for the growth of large well ordered single crystals of trimeric Photosystem I (Fromme, 1998).

### 4. Binding of 'Ligands' and Redox State

Many photosynthetic proteins are involved in electron transfer and /or interact with other proteins. The redox state and the binding of other proteins can strongly influence the crystallization behavior of both soluble and membrane proteins (for review, see Ostermeier and Michel, 1997). Binding of hydrophilic ligands increases the solubility in the case of membrane proteins as has been shown for the cocrystals of Photosystem I with ferredoxin (Fromme et al., 2002), the cocrystals of the bacterial reaction center (RC) with cytochrome $c_6$ (Axelrod et al., 2002) and the cocrystals of the $bc_1$ complex with cytochrome (Lange and Hunte, 2002).

Furthermore, the ligand can induce conformational changes of the protein, which then in a secondary effect changes the stability and solubility of the protein. Besides direct conformational changes, the stability of the protein can be modified. In addition, the ligand can change the crystal packing by forming new interactions and/or shielding parts of the protein which have been involved in crystal contact before ligand binding.

## B. Influence of Physical and Chemical Parameters

Several physical and chemical parameters influence the solubility of all proteins, including photosynthetic proteins. The most important parameters are ionic strength, pH, temperature, crystallization agents and (for membrane proteins) detergent and lipids. For the optimization of crystals further parameter as diffusion, convection and the use of seeding techniques may also play an important role.

### 1. Ionic Strength

The solubility of proteins is highest at moderate ionic strength (in the range between 50 to 500 mM). It is well known that the solubility of proteins decreases with increasing ionic strength and many photosynthetic proteins have been crystallized by increasing the ionic strength, among them are the crystals of the purple bacterial reaction center (Deisenhofer et al., 1984) (see Table 3). At high ionic strength, there is a competition between the protein and the ions for the water molecules, which are necessary for solvation. This leads to a decrease of protein-solubility. Furthermore, all charged surface exposed groups are saturated with counter-ions, so that the molecule behaves as a neutral dipole in solution. Under these conditions, van der Waals contacts between hydrophobic surface exposed groups between the individual proteins can be formed, which could lead to formation of crystals.

At the other side of the solubility curve, at low ionic strength the solubility of the protein is reduced by depletion of the counter-ions from the surface of the protein. In this case negatively and positively charged groups of neighboring proteins can get in direct contact forming protein-protein interacting sites which could lead to crystal formation as shown for Photosystem I, which was crystallized by decreasing ionic strength (Fromme, 1998). Regarding the protein as a polyvalent ion, the solubility of proteins as a function of the ionic strength can be described by the Debye-Hückel theory as:

$$\log S - \log S_0 = \frac{Az_+z_-\sqrt{\mu}}{1 + aB \cdot \sqrt{\mu}} \qquad (5)$$

In this formula is: $S$ = solubility of the protein, $S_0$ = solubility of the protein at zero ionic strength in water, $A$, $B$ = constants, depending on the temperature and the dielectric constant, $z_+$ and $z_-$ = valences of the ions, $\mu$ = the ionic strength ($\mu = \frac{1}{2} \sum c_j z^2_j$), and $a$ = the diameter of the ion. The term $aB\sqrt{\mu}$ takes into account that the ions have a finite diameter, so that they polarize the media to a lower extend than a point-charge.

The *ab initio* calculation of the solubility of proteins based on equation 1 is difficult, because the simplification of a protein as a polyvalent ion could hold for small, nearly globular proteins, but is not sufficient to describe the complicated phase behavior of a large anisotropic proteins or membrane-protein-detergent micelles. The determination of the phase diagram can be very helpful for the crystallization of proteins but phase diagrams for the crystallization have been reported only for very few photosynthetic proteins (Photosystem I and the PbRC) have been reported (Odahara et al.; 1994; Fromme, 2003).

*Table 3.* Photosynthetic membrane protein structures and crystallization conditions

| reference | resolu-tion | PDB entry | organism | growth T | detergent | crystallization agent | additives |
|---|---|---|---|---|---|---|---|
| **Photosystem II** | | | | | | | |
| (Zouni et al., 2001) | *3.8* | *1FE1* | *Thermo-synechococcus elongatus* | 56 | β-DDM | PEG 1500 | glycerole |
| (Ferreira et al., 2004) | *3.5* | *1S5L* | | | β-DDM | PEG 1450 | glycerole |
| (Loll et al., 2005) | *3.0* | *2AXT* | | | β-DDM | PEG 2000 | glycerole |
| (Kamiya and Shen, 2003) | *3.7* | *1IZL* | *Thermo-synechococcus vulcanus* | 55 | β-DDM | PEG 4000 | none |
| **Photosystem I** | | | | | | | |
| (Jordan et al., 2001) | *2.5* | *1JB0* | *Thermo-synechococcus elongatus* | 56 | β-DDM | none, dialysis against low ionic strength | none |
| **Photosystem I in a complex with 4 LHC I proteins** | | | | | | | |
| (Ben-Shem et al., 2003) | *4.4* | *1QZV* | *Pisum sativum alaska* | 22–24 | β-DDTM | PEG 6000 & ammonium citrate | none |
| Amunts et al., 2007 | *3.4* | *2O01* | | 24 | | | |
| **Reaction center of purple bacteria (only some of the many variants selected)** | | | | | | | |
| (Deisenhofer et al., 1985) | *3.0* | *none* | *Rhodopseudo-monas viridis* | 30 | LDAO | ammonium sulfate | HT |
| (Deisenhofer et al., 1995) | *2.3* | *1PRC* | | | LDAO | ammonium sulfate | |
| (Lancaster and Michel, 1997) | *2.4* | *3PRC* | | | LDAO | ammonium sulfate | |
| (Lancaster et al., 2000) | *2.0* | *1DXR* | | | LDAO | ammonium sulfate | |
| (Allen et al., 1987) | *2.8* | 4*RCR* | *Rhodobacter sphaeroides* | | LDAO | PEG 4000 | HT |
| (Chang et al., 1991) | *3.1* | *2RCR* | | | β-OG | PEG 4000 | HT |
| (Ermler et al., 1994) | *2.65* | *1PCR* | | | LDAO | potassium phos-phate | HT & dioxane |
| (Stowell et al., 1997a) | *2.2* | *1AIG* | | | β-OG | PEG 4000 | BA & HT |
| (McAuley et al., 1999) | *2.1* | *1QOV* | | | LDAO | sodium citrate | HT |
| (Katona et al., 2003) | *2.35* | *1OGV* | | | crystallized in lipid cubic phases of monoolein | ammonium sulfate | none |
| (Nogi et al., 2000) | *2.2* | *1EYS* | *Thermo-chromatium tepidum* | 48–50 | β-OG | PEG 4000 & NaCl | none |
| **Reaction center of purple bacteria in the complex with cytochrome $c_2$** | | | | | | | |
| (Axelrod et al., 2002) | *3.25* | *1L9J* | *Rhodobacter sphaeroides* | 25–30 | LDAO | PEG 4000 | HT |
| **Light harvesting complex II from plants (LHC II)** | | | | | | | |
| (Liu et al., 2004) | *2.72* | *1RWT* | *spinach* | 20–25 | β-NG & N,N-bis-(3-D-gluconamido propyl)deoxy-cholamide | sodium citrate | none |
| **Light harvesting complex II from purple bacteria (LH 2)** | | | | | | | |
| (McDermott et al., 1995) | *2.5* | *1KZU* | *Rhodopseudo-monas acidophila* | 25–30 | β-OG | phosphate | BA |
| (Papiz et al., 2003) | *2.0* | *1NKZ* | | | β-OG | phosphate | BA |
| (Koepke et al., 1996) | *2.4* | *1LGH* | *Rhodospiril-lum molischia-num* | 25 | LUAO | ammonium sulfate | HT |

*Table 3.* Continued

| reference | resolution | PDB entry | organism | growth T | detergent | crystallization agent | additives |
|---|---|---|---|---|---|---|---|
| **Light harvesting complex III from purple bacteria (LH 3)** | | | | | | | |
| (McLuskey et al., 2001) | *3.0* | *1IJD* | *Rhodopseudomonas acidophila* | 25–30 | β-OG | phosphate | BA |
| **Cytochrome $b_6f$ complex** | | | | | | | |
| (Kurisu et al., 2003) | | *1UM3* *1VF5* | *Mastigocladus laminosus* | 45 | β-DM | PEG 400 | DPC |
| (Stroebel et al., 2003) | *3.1* | *1Q90* | *Chlamydomonas reinhardtii* | 30 | β-DDM | PEG- monomethylether 350 | glycerole |
| **Cytochrome $bc_1$ complex** | | | | | | | |
| (Berry et al., 2004) | *3.5* | *1ZRT* | *Rhodobacter capsulatus* | 32 | β-UDM | PEG 400 | HT |

*Abbreviations of chemicals:* BA, benzamidine; DPC, dioleylphosphadidyl - choline; β-OG, octyl-β-D-glucopyranoside; β-NG, nonyl-β-D-glucopyranoside; β-DM, decyl-β-D-maltoside, β-UDM, undecyl-β-D-maltoside,; β-DDM, dodecyl-β-D-maltoside; β-DDTM, dodecyl-β-D-thiomaltoside; HT, heptane-1,2,3-triol; LDAO, N,N-dimethyldodecylamine-N-oxide; LUAO, N,N-dimethyl-undecylamine-N-oxide

## 2. Anions and Cations

The solubility of proteins depends not only on the ionic strength, but is also a function of the nature of the ions. Furthermore, many proteins (like Photosystem II) need essential cations (like $Ca^{2+}$ or $Cl^{-}$) for the stability and functional integrity of the protein. Cations and anions can also be directly involved in crystal contact, change the binding affinity of ligands or stabilize flexible loops. On the 'dark' site, some cations or anions may have also a very negative influence on the protein and the crystallization process. They can cause denaturation of the protein, unfolding of domains, inactivate the protein or even oxidize or reduce the protein or its ligands. The oxidation/reduction effect has to be carefully taken into account when the crystallization of photosynthetic proteins is considered.

## 3. The pH Value

The solubility of proteins depend on the pH; the solubility reaches a minimum at the isoelectric point (IP). For soluble stable proteins crystallization is often favorable at the isoelectric point, whereas photosynthetic membrane proteins are in most cases crystallized at a pH away form the IP at a pH where a reasonable solubility can be achieved and where the proteins shows the highest stability in the detergent micelles.

## 4. Temperature

The solubility of all proteins depends on the temperature. The majority of proteins shows an increase of solubility with increasing temperature, but examples of the opposite behavior of proteins with a well defined temperature minimum of the solubility are known. The crystallization by temperature gradient is commonly and successfully used for crystallization of inorganic molecules, whereas proteins crystallized by temperature shift are rare. It has not been used for the crystallization of photosynthetic proteins, with the exception of one report of crystallization of Photosystem I (Witt et al., 1988). The most frequently used temperatures for the crystallization of photosynthetic proteins are 20 °C and 4 °C. There is no special reason for the choice of these two temperatures except for he convenience that rooms at 4 °C and 20 °C are normally available in most research institutions. As many thermophilic proteins have been used for crystallization, one would expect that the crystallization would also have been tested at elevated temperature, however, all reported crystals have been grown at temperatures that are significantly lower than the growth temperature of the organism. Examples are Photosystems I and II crystals from *Thermosynechococcus elongatus* that have been grown at 4 °C and 18 °C respectively, even if the optimal growth of the organism is at 56 °C. There are several advantages to lower the temperature during isolation and crystallization as they are: low rate of proteolytic activity, increased protein stability, and decrease of mobility of flexible regions. Another positive effect of the lower temperature is the observation that the nucleation rate is decreased by decreasing temperature, leading to smaller amounts and increased sizes of crystals.

The optimal temperature for the crystallization experiments is not only dependent on the protein, but the method of crystallization and the solvent composition also greatly influence the choice of the best temperature range. Polyethyleneglycol (PEG), for example, can undergo phase transition between 18 °C and 4 °C, which can lead to problems with crystallization in PEG at 4 °C a complex behavior that also depends on the size of the PEG. This is the main reason why most proteins crystals that were grown with PEG are crystallized at 20–25 °C.

### 5. Detergents

Detergents are essential for the crystallization of photosynthetic membrane proteins. They are used for the solubilization of the proteins from the membranes, as well as the protein isolation and crystallization. In most cases, the membrane proteins are crystallized as a protein-detergent micelle, but there is also one example of successful crystallization of the purple bacterial reaction center in lipid cubic phase (Yano et al., 2005). The detergent, that is optimal for crystallization must not be the same detergent that is optimal for solubilization and isolation. For the solubilization and isolation, the following criteria must be met as a prerequisite for successful crystallization: (i) sufficient amount of extracted protein, (ii) functional integrity, (iii) homogenous protein and cofactor composition, (iv) stability of the oligomeric state of the protein.

For the crystallization, the exchange of the detergent or the use of detergent mixtures can be useful to reduce the size of the detergent micelle. Non-ionic-detergents with large alkyl chains and low critical micellar concentration (CMC) like β-dodecylmaltoside are very mild detergents and superior for the maintenance of the integrity and function of the photosynthetic membrane proteins. This is the reason why nearly half of all alpha helical membrane proteins have been isolated and crystallized with β-dodecylmaltoside, among the photosynthetic membrane proteins photosystems I and II. Other popular detergents that have been used for crystallization of Photosynthetic membrane proteins are LDAO and β-octylgycoside. Table 3 summarizes the crystallization conditions and detergents used for the crystallization of Photosynthetic membrane proteins.

### 6. Crystallization Agents

Crystallization agents like polyethylene glycol are commonly used for the crystallization of proteins, and have also very frequently been used for the crystallization of photosynthetic proteins like Photosystem II, the Photosystem I-LHC I complex and the cytochrome $b_6f$ complex (see Table 3). They generally decrease the solubility of proteins by competing with the protein for water molecules. The hydrophobic protein-protein interactions are strongly increased in the presence of PEG and the effect increases with the size/molecular weight of the PEG molecules. The use of PEG in the presence of detergent leads to very complex phase diagrams as the solubility of the detergent is decreased in the presence of PEG (Hitscherich et al., 2001). In general, one can conclude that much lower PEG concentrations are suitable for crystallization of photosynthetic membrane proteins than for soluble photosynthetic proteins. Further improvement of the X-ray diffraction quality of photosynthetic membrane protein crystals was in many cases achieved by addition of small amounts of additives like benzamidine or hexanetriol (see Table 3).

Additives and crystallization agents that are alcohols as MPD or hexanetriole must be used with care as they can also cause partial denaturation of the protein, the de-assembly of oligomers and removal of pigments as chlorophylls and carotenoids.

### 7. Nucleation and Seeding Techniques

In order to further increase the quality of the crystals for X-ray structure analysis, homogeneous seeding techniques (micro- and macroseeding) were successfully used for the crystallization of Photosystem I (Fromme and Witt, 1998; Jordan et al., 2001). In the first step, small Photosystem I crystals were grown by spontaneous nucleation. In the second step, small Photosystem I crystals were used for microseeding. The protein solution was brought into the meta-stable zone by decreasing the salt concentration by microdialysis. Seeding crystals were added and medium size crystals (diameter ≈ 0.1mm, length ≈ 0.1–0.3 mm) grew within 2 days. The third crystallization step uses the medium size crystals for macroseeding. The salt concentration of the protein solution was slightly reduced to a concentration, at which the solution is still unsaturated but close to border of the saturation curve. When the solution had been equilibrated, one single crystal of medium size was added. Dialysis against lower salt concentration is started immediately after addition. First, the crystal began to dissolve, but due to the parallel decrease of the salt concentration the solution becomes supersaturated and the growth of the crystal was observed. Using this procedure one large single crystal of up to 3 mm grows within 2–3 days. Photosystem I crystals diffracting X-rays to 2.5 Å resolution were grown, using this seeding procedure which were the basis for the high resolution structure of Photosystem I (Jordan et al., 2001).

## IV. Freezing of Crystals

The successful freezing of crystals was a major breakthrough in X-ray crystallography, because it reduces radiation damage due to strong X-ray exposure at synchrotron sources and is one of the major breakthroughs responsible for in the exponential increase of X-ray structures (Garman, 1999). The most popular cryo-protectants are glycerol, polyethyleneglycol 400, ethyleneglycol and various sugars such as sucrose, glucose or threhalose. The crystals are dipped shortly in the cryo-protectant, scooped into small loops and then either frozen directly in the cool-stream or on liquid nitrogen. However, significant damage of crystals can still occur in the X-ray beam even in the frozen state (Burmeister, 2000). One of the most prominent examples of severe damage of the protein structure by X-ray exposure is Mn cluster in Photosystem II, which is reduced and partially disassembled by radiation damage (Yano et al., 2005).

Freezing is well established for crystals of small soluble proteins, but is more difficult for crystals with large unit cells, high solvent content and for membrane protein crystals. There are two parameters, which are important for freezing of large single crystals: (i) The stabilization of the crystals in the cryo-solution and (ii) the speed of freezing.

(i) In large single crystals, the diffusion of the cryo-protectant into the protein crystal is an important factor, which has to be taken into account. Freezing experiments with large photosystem-I crystals (> 1 mm), frozen for EPR and X-ray experiments (Brettel et al., 1992; Kamlowski et al., 1997; Jordan et al., 2001) have shown that large single crystals can be successfully frozen. However, it takes at least 1 hour until the solvent cavities inside the crystal are equilibrated with the cryo-protectant. Photosystem I has a high solvent content of 78%, so diffusion occurs via large pores. That means that crystals with lower solvent content may show even a slower diffusion rate and may therefore need even a longer time for fully equilibration. Another important point is that the mechanical stress to the crystals has to be limited by slow change of the concentration of the cryo-protectant inside the crystal, which can be ideally achieved by growth of the crystals in the presence of cryo-protectant. Alternatively, crystals can also been brought into the cryo-protectant solution by stepwise increase of the cryo-protectant concentration or dialysis against the cryo-protectant solution.

(ii) The velocity of the freezing process is also very critical for the successful freezing of single crystals. Crystals are in most cases frozen directly in the cold nitrogen stream or frozen in liquid nitrogen. In critical cases, the freezing of the single crystals in liquid propane can increase the velocity of freezing. Propane has a much higher thermal conductivity than air or liquid nitrogen, because propane does not show the 'Leidenfrost phenomena' as found in liquid nitrogen (Chandra and Aziz, 1994). The heat-transfer coefficient of liquid nitrogen and liquid propane, differ by a factor of 30 (Serp et al., 2002a,b).

## V. X-ray Data Collection

The technical developments of X-ray diffraction have been powerful and have led to the determination of over 35,000 protein structures with a rate that continues to grow exponentially. As discussed above, the lack of structures of membrane proteins is largely due to our limited understanding of the processes involved in crystallization that leads to a lack of strategies for obtaining well-diffracting crystals. Once crystals of a protein have been obtained, current high-throughput structural genomics approaches that have been developed can be utilized to determine the three-dimensional structure of a protein. An outline of the techniques is presented here with an emphasis on how recent technical developments have helped in overcoming the difficulties in determining the structures of large protein complexes and membrane proteins from photosynthesis. For more details concerning the technical aspects of X-ray diffraction, the reader is invited to examine one of the many books or recent reviews (Carter and Sweet, 1997; Dauter, 2006; Girard et al., 2006).

Once well-ordered single crystals have been obtained then the goal is to obtain high-quality diffraction data by exposing the crystals to monochromatic X-rays (for diffraction patterns, see Chapter 5, Liu and Chang). In a typical system, the detector is fixed and the crystal is rotated through a set of angles in order to collect a complete data set. The range of angles depends upon the space group (Table 1) as higher symmetry crystals require a smaller range of angles. For example, after a 90° rotation the diffraction repeats due to symmetry in orthorhombic crystals while a complete 180° rotation is needed for P1 crystals that have no intrinsic symmetry. Since the crystal will decay during exposure, there is a balance between measuring all possible reflections with long exposures and using only the data before a significant decay has occurred. As a measure of the quality of the measurement, data is collected through angles greater than the minimal range and the intensities of equivalent diffraction points obtained at different times are compared (called the Rsym). The data is collected as a series of exposures with the crystal being rotated through a small oscillation angle for each exposure. For small crystals that are well ordered, the crystals may be rotated through a few degrees for each measurement. The data collection of large complexes and membrane proteins can take much more time than the data collection for soluble protein crystals due to a high mosaic spread (Section II.F). To reduce the overlap of reflections, small oscillation angles of 0.3° to 0.5° per image can be used for data collection. Despite being frozen (Section IV), the crystals will still decay during data collection requiring the use of multiple crystals or the repositioning of larger crystals during data collection. In order to optimize data collection, multiple data sets are usually measured with some being set to measure a very complete data set at a limited resolution limit. These are less complete but contain accurate measurements of the weaker higher resolution reflections.

For most laboratory work, the X-ray beam is generated using a rotating copper anode with the beam being focused using X-ray mirrors into a 0.1–0.3 mm diameter. These generators are limited in the intensity of the X-rays due to heating of the anode. Synchrotrons provide X-ray radiation that has a significantly greater intensity than laboratory sources. The more intense beam allows data to be collected over a much shorter period of time. Perhaps more importantly, the diffraction quality from the synchrotron source are usually improved as the beam is focused into a much smaller area than a typical laboratory beam providing more radiation onto small crystals and correspondingly a sharper diffraction spot. For crystals of less than 0.1 mm, special microfocus beamlines are now available at synchrotrons. A strong and well-focused beam may be the essential for data collection on small, weakly diffracting crystals of large complexes and membrane proteins. There are a number of synchrotrons providing beam lines dedicated for protein crystallography, including the Advanced Light Source, Argonne National Laboratory, Stanford Synchrotron Radiation Laboratory, Brookhaven National Laboratory, the ESRF, the DESY and the Swiss Light Source.

For both laboratory and synchrotron sources, image-plates or charge-coupled devices are used as detectors to measure the diffraction pattern. These detectors provide a real-time assessment of the quality of the data. A variety of programs are available for the integration of the data, such as DENZO and Scalepack from the HKL package (HKL research, Charlottesville, NC, USA) and Mosflm from the CCP4 package (Collaborative Computational Project, Number 4, 1994; Potterton et al., 2002). While highly diffraction crystals can be rapidly analyzed, weakly diffracting crystals require special care in the analysis as do crystals that have certain disorders, such as twinning in which two orientations of a crystal contribute to the diffraction.

## VI. Phase Determination

The measurement of the intensities of each diffraction point is not sufficient for the determination of the three dimensional structure due to the lack of any experimental information concerning the phase at each point. There are several techniques that are utilized to overcome the phase problem. For small molecules, the phases can be estimated using procedures known as direct methods. Despite the great improvements in these techniques and in the increase of the speed and ability of computer-technology, proteins are beyond the approach for these techniques due to the large number of atoms in a protein. Instead, the phases are determined experimentally using the following approaches.

### A. Multiple Isomorphous Replacement

Multiple isomorphous replacement (MIR) is the original technique that was used to solve the structures of hemoglobin and myoglobin (Kendrew et al., 1960; Perutz et al., 1960) and still represents the approach used for the large majority of proteins, including most structures of photosynthetic proteins, such as the RC from *Blastochlorosis viridis*, Photosystem I and Photosystem II (Deisenhofer et al., 1985; Jordan et al., 2001; Zouni et al., 2001). In MIR, heavy atom derivatives are prepared that have a few metals attached to a protein at specific locations. These derivatives are usually prepared by soaking each crystal in a solution containing a certain metal compound and then screening the crystals for systematic changes in the diffraction. Because the testing is empirical, the number of possible derivatives that must be screened can vary tremendously. In order to accurately determine the phases, multiple derivatives must be identified and their phases combined. The commonly used derivatives were designed to bind to water-soluble proteins. Their effectiveness for membrane proteins, with their large layers of bound detergents, appears to be somewhat ineffective many structural investigations have reported needing very large surveys of possible derivatives with the best being marginal in quality. When the protein has a metal cluster, the electron density is much stronger than for other regions allowing identification of the cluster location even when the individual metal atoms are not resolved (see Figs. 12B and C).

### B. Molecular Replacement

Molecular replacement (MR) can be used if there is a homologous structure to the one being solved. In technique, existing model is rotated and translated in the cell of the unknown structure until the calculated diffraction (actually the structure factors) match those measured. Once the model is rotated and positioned, the differences between the model and true structure must be determined by examining differences between the model and calculated electron density. In the absence of any other phase information, identifying regions of the model that must be altered significantly can be challenging.

Because the approach requires the availability of an existing model, its use has been limited for membrane proteins. One example is that the reaction center from *Rhodobacter sphaeroides* was solved using the structure of the reaction center from *Blastochloris viridis* (Allen et al., 1986). The level of homology required for the approach to be successful is difficult to predict as systematic shifts of the backbone of one protein relative to the model may prevent the successful application. For example, the light harvesting II complex from *Rhodopseudomonas acidophila* (McDermott et al., 1995) could not be successfully used to determine the structure of the light harvesting II complex from *Rhodospirillum molischianum* (Koepke et al., 1996). This approach is most useful when a protein of known structure has been modified in a specific manner as the differences the structures can be readily identified as changes in the electron density map. Thus, in many cases the structures of mutants are readily solved immediately after the diffraction data has been obtained.

### C. Multi-Wavelength Anomalous Dispersion

Multi-wavelength anomalous dispersion (MAD) is now a routinely used technique for phase determination (Walsh et al., 1999). This technique is based upon the change in the phase known of the scattered X-rays, known as anomalous scattering, when the wavelength is near the absorption edge of a metal. The anomalous scatterers can be metal cofactors of the protein or heavy metals attached to the protein as derivatives. Alternatively, while the sulfur atom present in the amino acid methionine has a relatively weak signal, this residue can be replaced by selenomethionine providing a much more favorable selenium atom when the protein is obtained from an expression system.

By measuring the diffraction from a protein crystal at the absorption edge as well and away from the edge, it is possible to attribute the differences in the reflection intensities measured at different wavelengths to the anomalous scattering and consequently phases can be assigned. Such measurements require the use of special beamlines at synchrotrons that allow the X-rays to be systematically adjusted through a broad range of wavelengths. An increasing number of structures have been solved based on single-wavelength anomalous dispersion (SAD) due to the advantage that there are more single wavelength data-collection facilities. SAD can be performed at any wavelength where anomalous scattering can be observed and in some cases obtained using laboratory data.

A powerful outcome of the anomalous dispersion data is its use in the determination of metal clusters.

The electron density calculated, after phases have been determined, does not distinguish atom types. However, since each metal has a unique absorption, use of the anomalous data results in electron density that can be unambiguously assigned as arising from specific metal ions. For example, anomalous data was utilized to position a calcium atom in the Mn cluster of Photosystem II (Fig. 12C) (Ferreira et al., 2004).

## VII. Refinement and Model Building

Once the phases have been determined, the structural model must be built. Usually clear secondary structure features are first identified, such as alpha helices, and then less defined features such as connecting loops are modeled. For small proteins with high resolution data, the electron density maps are clear and the model building is straightforward but for large complexes with poor electron density considerable judgment is required on the part of the crystallographer. After an initial model has been built then the position and geometry is refined by use of computer algorithms, such as CNS and the CCP4 package (Collaborative Computational Project, Number 4, 1994; Brunger and Rice, 1997). New electron density maps can then be calculated and the process repeats until the model agrees with the electron density while maintaining proper geometry.

The structures of most proteins to date have been built manually but automated model building is increasingly becoming more powerful and is used not only for small proteins, but also to complete loops in large protein complexes. For the membrane protein structures solved so far, model building has been done manually due to the complexity of the proteins, including the large number of cofactors, as well as the limited resolution limits for most membrane proteins. As computers get more powerful this new tool will also be very helpful in the determination of proteins as new programs are developed, such as PHENIX (Python-based Hierarchical ENvironment for Integrated Xtallography) a new package that is designed to proceed from reduced intensity data to a refined molecular model (Adams et al., 2002).

## VIII. Structures of Photosynthetic Complexes

A tremendous number of structures of proteins from photosynthetic organisms have been determined since the late 1970s and structural investigations remain active areas of research as summarized in two issues of Photosynthesis Research (Allen and Knaff, 2004, 2005). Among the earliest of protein structures to be solved was cytochrome $c_2$ from *Rhodospirillum rubrum* (Salemme et al., 1973). This protein has approximately 110 amino acid residues, depending upon the species, that surrounds most of the heme with a small portion exposed to the solvent. Cytochrome $c_2$ serves as mobile electron carrier in purple bacteria that transfer electrons between the reduced hemes of the cytochrome $bc_1$ complex and the reaction center as part of the Q cycle (Tiede et al., 1978; Axelrod and Okamura, 2005). To perform the electron transfer reaction, the cytochrome first binds to the membrane protein, either the reaction center or the cytochrome $bc_1$. The binding is primarily facilitated by electrostatic interactions involving lysine residues on the cytochrome and carboxylates on the reaction center, although hydrophobic interactions also contribute to the binding. The structure of the cytochrome $c_2$ bound to the reaction center from *Rhodobacter sphaeroides* shows heme bound directly above the bacteriochlorophyll dimer of the reaction center (Fig. 7A) (Axelrod and Okamura, 2005). The close distance of 8.4 Å between the closest atoms of the two cofactors allows the rapid 1 μs transfer of an electron from the reduced heme to the light-induced oxidized bacteriochlorophyll dimer.

Ferredoxins are also water-soluble proteins that serve as electron carriers with the redox active cofactor in plants being a center consisting of two irons and two sulfurs. These enzymes are present in the stroma of chloroplast of higher plants as well as the cytoplasm of cyanobacteria where their major function is to transfer electrons between Photosystem I and the FNR where NADPH is produced for $CO_2$ assimilation. The structure of a plant-type ferredoxin was first determined from *Spirulina platensis* (Fukuyama et al., 1980). The protein has a core β-sheet with a flanking α-helix (Fig. 8A). The metal cluster is located near the surface of the protein and bridged by four conserved cysteine residues. Subsequent determinations of ferredoxins from other organisms showed that the structures were highly conserved (Fukuyama, 2004). In addition, ferredoxin carries electron to several ferredoxin-dependent enzymes for nitrogen and sulfur assimilation (Knaff and Hirasawa, 1991). For example, ferredoxin carries electron to nitrite reductase that carries out the multi-electron and

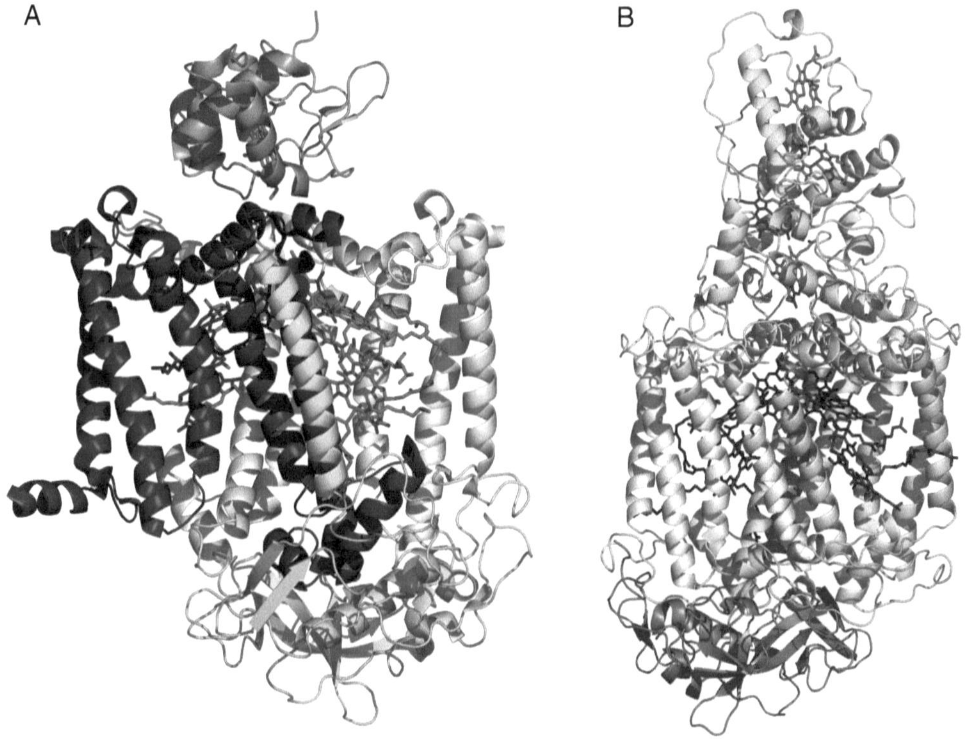


*Fig. 7.* A. Structure of the reaction center from *Rhodobacter sphaeroides* with cytochrome $c_2$ bound. The three subunits of the reaction center, L, M, and H and the bound cytochrome $c_2$, are shown as ribbon diagrams with different shades. Also shown are the cofactors of the reaction center and the cytochrome heme. B. Structure of the reaction center from *Blastochloris viridis*. The reaction center consists of four subunits, L, M, H, and a tetraheme cytochrome that have different shades. Also shown as a dark shade are the six tetrapyrroles, the primary quinone, non-heme iron, and four hemes. See also Fig. 1, Color Plate 3.

proton conversion of nitrite to ammonia. This enzyme has a single subunit that surrounds a four iron-four sulfur center coupled to a siroheme where the conversion occurs (Fig. 8B) (Swamy et al., 2005).

Phycobilisomes are large peripheral antenna complexes attached to the surface of the membranes in cyanobacteria and red algae that contain pigment-protein complexes that harvest the light energy. The most studied phycobilisomes consist of three types of complexes known as biliproteins along with a number of additional proteins known as linkers. Some of the biliproteins are arranged into rods that attach in a fanlike arrangement to cores attached to the stromal side of the thylakoid membrane. The structures of several biliproteins have been determined, including the structure of C-phycocyanin (Schirmer et al., 1985, 1986). The protein consists of a dimer of $\alpha$ and $\beta$ peptides which form a trimer of dimers. An analysis of the spectral and energy transfer properties of these proteins have shown that energy transfer proceeds through these proteins in order of their relative spectral energies (MacColl, 1998).

Green anoxygenic bacteria are the only photosynthetic organisms that contain the unique peripheral membrane complex known as chlorosomes. These complexes are approximately 200 nm in length and each chlorosome contains thousands of bacteriochlorophylls that are organized to capture light and direct the energy to the photosynthetic complexes in the membrane (Olson, 1998). Associated with both chlorosomes and the photosystem is the FMO complex, named after Fenna-Matthews-Olson, that was first discovered in 1962 and was the first chlorophyll containing protein structure determined (Li et al., 1997; Matthews et al., 1979). The FMO complex is a trimer with each individual subunit having identical structures consisting of a series of $\beta$-strands that form two sheets enveloping seven bacteriochlorophyll molecules in a 'taco shell' arrangement with six short helices completing the open end of the taco (Fig. 8C). The central magnesiums of the bacteriochlorophylls are coordinated by five histidines, a carbonyl

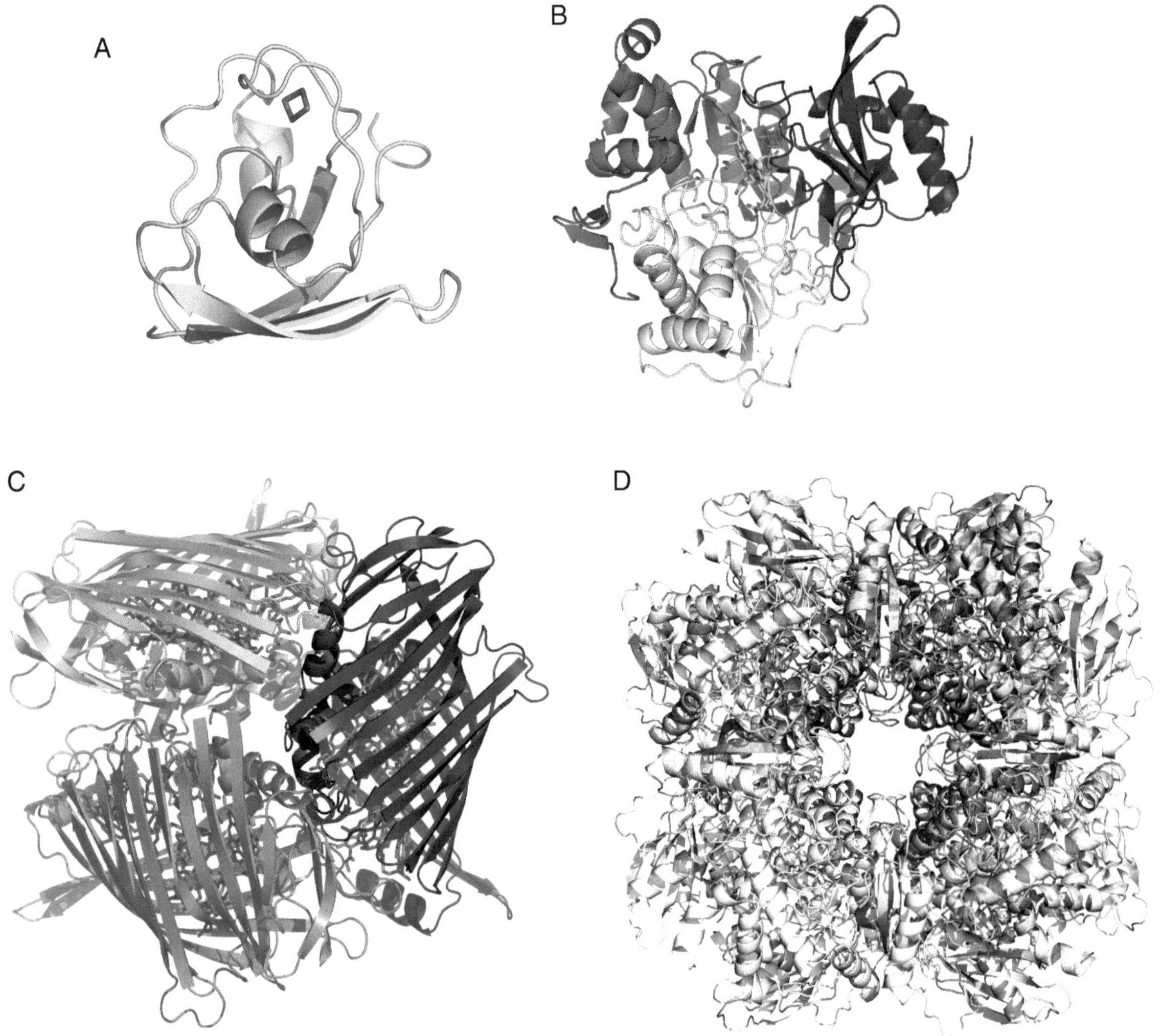


*Fig. 8.* A. Structure of ferredoxin from *Spirulina platensis*. Shown is the backbone with two small alpha helices and 4 beta sheets with the $Fe_2S_2$ cofactor. B. Structure of nitrite reductase from spinach. Nitrate reductase has a single polypeptide chain that can be divided into three domains that are each shaded differently. Also shown in the center of the structure are the cofactors, a siroheme that is coupled to a $Fe_4S_4$ cofactor. C. Structure of the FMO complex from *Chlorobium tepidum*. The protein consists of two beta sheets that fold around the seven bacteriochlorophylls. Shown for clarity is one subunit; the biological unit is a trimer of three identical subunits. D. Structure of rubisco from plants. The eight large and small subunits are shown in different shades.

backbone, and a water molecule. The bacteriochlorophylls are highly interacting and the energy transfer among the cofactors are highly directed to one of the bacteriochlorophylls as recently shown by two-dimensional transient optical spectroscopy (Brixner et al., 2005).

Among the enzymes participating in the Calvin cycle, ribulose 1,5 bisphosphate carboxylase, Rubisco, has been well-characterized and shown to carry out carboxylation during carbon fixation. Rubisco is a complex enzyme that is highly regulated and can exhibit different functional activities. The enzyme from higher plants and most photosynthetic bacteria consists of eight copies of a dimer formed by a L (large) and S (small) subunit forming a large quaternary structure (Fig. 8D) (Andersson et al., 1989). The L subunits contain the catalytic site and the function of the S subunit is not well understood although it probably stabilizes the complex. Before rubisco becomes active, a lysine residue in the active site must be carbamylated by another enzyme, rubisco activase. Rubisco in photosynthetic bacteria has a quite different structure with the complex having only two subunits which resemble the L subunit found in plants.

The first integral membrane protein whose structure was determined using protein crystallography was the reaction center from *Blastochloris viridis* (Fig. 7B) (Deisenhofer, 1985). Soon afterwards the structure of the better characterized reaction center from *Rhodobacter sphaeroides* was determined (Allen et al., 1987). The number of protein subunits for the reaction center ranges from 2 to 4 in bacteria, with all reaction centers having the L and M subunits serving as the core of the protein. These two subunits each have five transmembrane helices that surround the cofactors and have a pseudo-two fold symmetry about the center of the protein. The cofactors were arranged symmetrically into two branches although only one of the branches is photochemically active. In purple bacteria, reaction centers also have a H subunit that has a large extra-membraneous domain and one transmembrane helix and in some cases, as found in *Blastochloris viridis*, the reaction centers have a fourth tetraheme containing subunit. Since the original structure determinations, there have been a tremendous number of mutant structures determined for the reaction center from *Rhodobacter sphaeroides*.

The cytochrome $bc_1$ complex is a large multisubunit membrane protein in the intracytoplasmic membrane coupled to the reaction center by cytochrome $c_2$ and quinones in the Q cycle (Berry et al., 2000). The complex consists of a minimum of three subunits, known as cytochrome *b*, cytochrome $c_1$, and the Rieske iron-sulfur subunit. The structures of the complex from a number of organisms have been determined, including *Rhodobacter capsulatus* (Berry et al., 2004). The cytochrome *b* has eight transmembrane helices and two noncovalently associated heme *b* cofactors, the cytochrome *c* has a covalently attached heme, and the Rieske subunit has a high potential two iron–two sulfur cluster (Fig. 9A). A comparison of the structures from different organisms showed that the Rieske subunit to be in different positions relative to the two cytochrome subunits, suggesting that the subunit, and hence the position of the two iron two sulfur center, changes during catalysis in order to drive the reaction forward. The $bc_1$ complex shows similarities to the core of the cytochrome $b_6f$ complex that couples electron transfer from Photosystem II to Photosystem I in oxygenic photosynthesis. The struc-

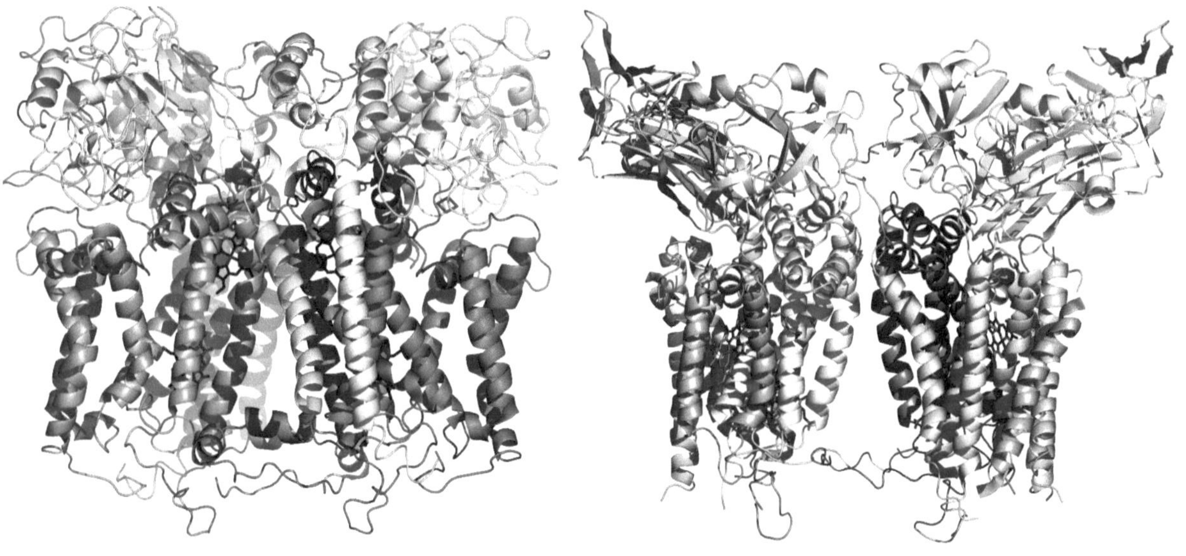

*Fig. 9.* A. Structure of the cytochrome $bc_1$ from *Rhodobacter capsulatus*. The structure is a homo-dimer with each unique element consisting of three protein subunits, cytochrome *b*, cytochrome *ca*, and the Rieske iron subunit. Also shown as a dark shade are the cofactors, the two hemes and the $Fe_2S_2$ cofactor. B. Structure of the cytochrome $b_6f$ complex from *Mastigocladus laminosus.* The structure is a homodimer with the Rieske FeS protein crossing over between the two monomers. The cytochrome $b_6f$ complex shows homologies to the $bc_1$ complex, with high homology in the cytochrome *b* subunits, but low/ no homology concerning the minor membrane intrinsic subunits and the unique cytochrome *f*. The cytochrome $b_6f$ complex contains an additional heme as well as a chlorophyll and a carotenoid as cofactors.

ture of this complex was determined in 2003 from green algae and thermophilic cyanobacteria (Kurisu et al., 2003; Stroebel et al., 2003) (Fig. 9B).

In purple bacteria, two integral membrane complexes serve to collect light and transfer the energy to the RC (Cogdell et al., 2004). Both of these complexes have been solved, although the low resolution limit of 4.8 Å for the LH1 complex prevented a detailed analysis. The LH2 complexes are circular structures composed of 8–9 pairs of $\alpha/\beta$ subunits enclosing the bacteriochlorophylls and carotenoids with the large center of the ring being empty of protein or cofactors and containing only lipids (Fig. 10). The LH1 complex forms an elliptical structure composed of 16 pairs of $\alpha/\beta$ subunits that has a central region large enough to fit the reaction center. Spectroscopic studies have established the rates of energy transfer among the complexes on the membrane that range from 50 fs to 50 ps. The structure of LH1 was broken at one point by the presence of a third, unidentified protein subunit that is thought to correspond to the PufX protein in *Rhodobacter capsulatus* and *Rhodobacter sphaeroides*.

Photosystem I is a trimeric complex, each monomer consists of 12 proteins to which more than 100 cofactors (96 chlorophylls, 22 carotenoids, 4 lipids, 3 4Fe4S, 2 phylloquinones) are non-covalently attached (Figs. 1 and 11). The structure has now been determined to a resolution limit of 2.5 Å (Jordan et al., 2001). Excitation begins in the antenna pigments and is rapidly transferred to the primary donor after which electron transfer proceeds to the terminal FeS cluster FB from where the electron is transferred to ferredoxin and finally used for the reduction of $NADP^+$ to NADPH (see Fig. 1). Photosystem I is a joint reaction center and core antenna system. The main subunits of Photosystem I, PsaA and PsaB carry the electron transfer chain and most of the core antenna system. All small membrane intrinsic subunits are located peripheral to PsaA and PsaB and stabilize the antenna system. Three subunits PsaC, D and E are located at the stromal hump of Photosystem I and provide the docking site for ferredoxin or flavodoxin. The electron transfer chain of Photosystem I consists of 6 chlorophylls, 2 phylloquinones and 3 FeS clusters (see Fig. 1). As is true for the bacterial reaction center,

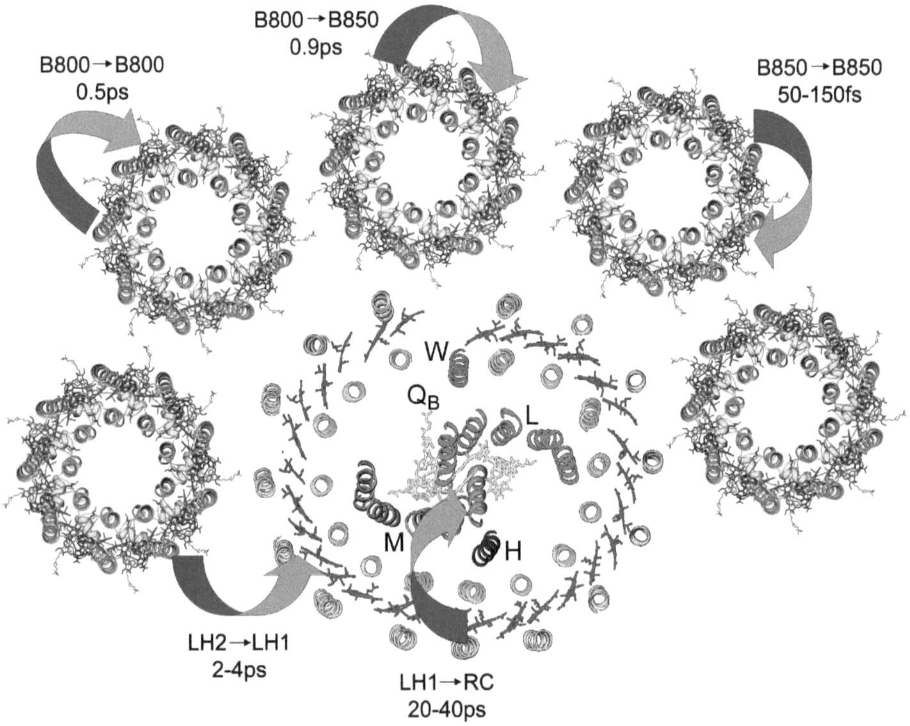


*Fig. 10.* Model of the cellular arrangement of the light harvesting-2 (LH2), light harvesting-1 (LH1), and reaction center (RC) complex based upon their three dimensional structures. The times for energy transfer between the different components are also shown. (Courtesy of R. J. Cogdell, A. W. Roszak and A. T. Gardiner.)

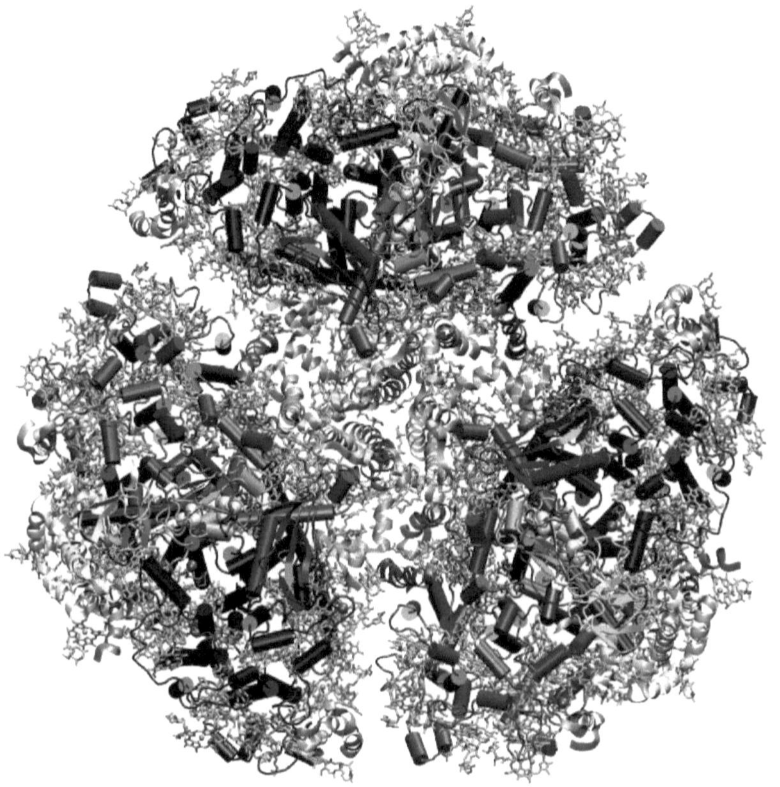

*Fig. 11.* Structure of Photosystem I. The molecule is a trimer. View direction is from the stromal side onto the membrane plane. Each monomer consists of 12 proteins subunits, shown in a ribbon presentation. The large subunit PsaA and PsaB are forming the center of each monomer. PsaA and PsaB contain 11 transmembrane helices, each, which are shown as solid columns. These large subunits carry the electron transfer chain and coordinate most of the antenna chlorophylls and carotenoids. 127 cofactors are bound to one monomer of PS I: 96 chlorophylls, 22 carotenoids, 3 4Fe4S clusters, 2 phylloquinones and 1 Ca. PS I contains 7 small membrane intrinsic proteins that are all located peripheral to the core of PsaA and PsaB. They contain 1 to 3 transmembrane helices which are shown in a ribbon representation. The three stromal subunits, PsaC, D and E provide the docking site for ferredoxin. Modified from Jordan et al., 2001. See also Color Plate 4.

the involvement of the two branches of cofactors is still debated, although the electron always ends up on the first iron sulfur cluster, $F_X$, and then the other two iron-sulfur clusters of the complex. In addition to the cyanobacterial structures, the structure of the plant Photosystem I has been determined at 4.4 and 3.4 Å resolution. It is a monomer with 4 LHCI proteins attached to one side of the monomer (Ben-Shem et al., 2003; Amunts et al., 2007).

One of the key questions concerning oxygenic photosynthesis center is the molecular mechanism of water oxidation that occurs at the manganese cluster in Photosystem II. During the S cycle, Photosystem II is able to convert the energy from four sequential light absorptions to produce molecular oxygen in this formidable process. Several structures have now been published at 3.8 to 3.0 Å resolution (Zouni et al., 2001; Kamiya and Shen, 2003; Ferreira et al., 2004; Loll et al., 2005). These structures have established the overall organization of the 17–19 protein subunits and the arrangements of the chlorophylls and other cofactors (Fig. 12A).

Notable is the arrangement of the core protein subunits, D1 and D2, as well as the core tetrapyrroles in a twofold symmetry also evident in the reaction center and Photosystem I (see Fig. 1). While the position of the manganese cluster is known, the molecular structure of the cluster remains elusive due to the limited resolution and the reduction of the cluster during data collection (Yano et al., 2005). Thus, efforts to obtain more accurate structures remain ongoing goals of several research groups.

Once a structure of a protein has been determined, the determination of the protein in different functions states is desirable. For example, knowing both the reduced and oxidized states provides the basis

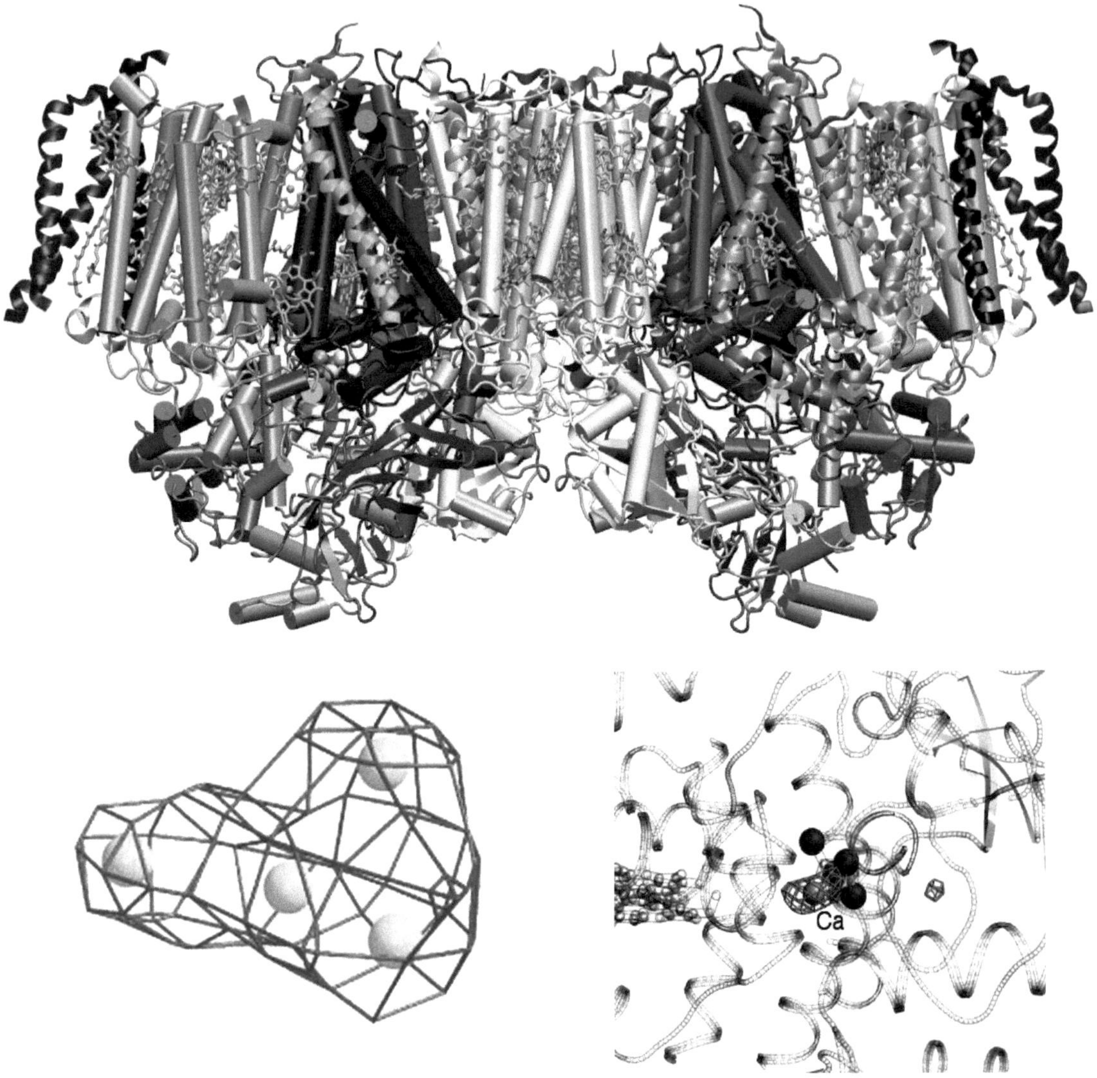


*Fig. 12.* A. Structure of Photosystem II. Photosystem II is a dimer. The view direction is along the membrane plane. Helices are shown as columns. The core of Photosystem II consists of the subunits D1 and D2, which carry the cofactors of the electron transfer chain. The Mn cluster, that is responsible for oxygen evolution is located at the lumenal side of the D1 protein. The large antenna proteins CP43 and CP47 flank the core at both sites. Photosystem II contains 10–12 small membrane intrinsic proteins that are located at the periphery. Three extrinsic subunits are located at the lumenal side of Photosystem II. Modified from Ferreira et al., 2005. The picture was generated using the VMD program and the PDB data file 1S5L. B. Electron density of the Mn cluster from the first structure of Photosystem II at 3.8 Å resolution (Zouni et al., 2001). Four Mn atoms were modeled into the 'papaya' shaped electron density. The model revealed for the first time the '3+1' organization of the Mn ions, but due to the limited resolution no individual atoms or bridging oxygen ligands can be identified. C. Anomalous density of the $Ca^{2+}$ ion that is located close to the Mn cluster in Photosystem II. (Modified from Ferreira et al., 2004; reprinted with permission from AAAS).

for understanding the electron transfer function of electron carriers such as cytochromes. For electron transfer proteins, determination of the structures of light-induced states are revealing. Two approaches can be utilized to generate the light-induced state that is stable during the period of data collection. Traditionally, the different functional states are trapped by quick freezing at low temperatures, for instance in liquid nitrogen. For the RC from *Rhodobacter sphaeroides*, the structure of the trapped light-induced state showed that the predominant effect was a movement of the secondary quinone (Stowell et al., 1997). This shift in the position of the secondary quinone suggests that electron transfer from the primary quinone is regulated by a physical movement of the quinone from the distal to proximal position. An alternate approach is to irradiate the crystal with a broad range of wavelengths rather than using monochromatic X-rays. The simultaneous radiation over a range alleviates the need to rotate the crystal and allows the entire diffraction pattern to be determined with only a few exposures. The tremendous number of reflections on each exposure makes this approach, known as Laue diffraction, a potentially powerful but challenging technique. Application of Laue diffraction to the reaction center of *Rhodobacter sphaeroides* confirmed the positioning of the secondary at two locations although the extent of the light-induced shift was not as evident (Baxter et al., 2004). Both approaches are expected to play critical roles in the elucidation of the mechanism of water oxidation in Photosystem II in the future.

## Acknowledgments

This material is based upon work supported by the National Science Foundation NSF under grant no (MCB 0640002) and United States Department of Agriculture (2003-02149) to Jim Allen and from grants from the National Science foundation (MCB-0417142), the United States Department of Agriculture USDA (grant no. 2003-35318-13573) and the National Institute of Health (grant no. R01GM71619-01) to Petra Fromme.

## References

Abrahams JP and Ban N (2003) X-ray crystallographic structure determination of large asymmetric macromolecular assembies. Meth Enzymol 374: 163–188

Abrahams JP, Leslie AG, Lutter R and Walker JE (1994) Structure at 2.8 Å resolution of F1-ATPase from bovine heart mitochondria. Nature 370: 621–628

Adams PD, Grosse-Kunstleve RW, Hung LW, Ioerger TR, McCoy AJ, Moriarty NW, Read RJ, Sacchettini JC, Sauter NK and Terwilliger TC (2002) PHENIX: Building new software for automated crystallographic structure determination. Acta Crystal D 58: 1948–1954

Adman ET, Siefker LC and Jensen LH (1976) Structure of *Peptococcus aerogenes* ferredoxin. Refinement at 2 Å resolution. J Biol Chem 251: 3801–3806

Allen JF, Alexciev K and Hakansson G (1995) Photosynthesis. Regulation by redox signalling. Curr Biol 5: 869–872

Allen JP, Feher G, Yeates TO, Rees DC, Deisenhofer J, Michel H and Huber R (1986) Structural homology of reaction centers from *Rhodopseudomonas sphaeroides* and *Rhodopseudomonas viridis* as determined by X-ray-diffraction. Proc Natl Acad Sci USA 83: 8589–8593

Allen JP, Feher G, Yeates TO, Komiya H and Rees DC (1987) Structure of the reaction center from *Rhodobacter sphaeroides* R-26: The protein subunits. Proc Natl Acad Sci USA 84: 6162–6166

Amunts A, Drory O and Neslon N (2007) The struture of a plant Photosystem I supercomplex at 3.4 Å resolution. Nature 447: 58–63

Andersson I, Knight S, Schneider G, Lindqvist Y, Lundqvist T, Branden CI and Lorimer GH (1989) Crystal structure of the active-site of ribulose-bisphosphate carboxylase. Nature 337: 229–234

Axelrod HL and Okamura MY (2005) The structure and function of the cytochrome $c_2$:reaction center electron transfer complex from *Rhodobacter sphaeroides*. Photosynth Res 85:101–114

Axelrod HL, Abresch EC, Okamura MY, Yeh AP, Rees DC and Feher G (2002) X-ray structure determination of the cytochrome $c_2$:reaction center electron transfer complex from *Rhodobacter sphaeroides*. J Mol Biol 319: 501–515

Baxter RHG, Ponomarenko N, Srajer V, Pahl R, Moffat K and Norris JR (2004) Time-resolved crystallographic studies of light-induced structural changes in the photosynthetic reaction center. Proc Nat Acad Sci USA 101: 5982–5987

Ben-Shem A, Frolow F and Nelson N (2003) Crystal structure of plant Photosystem I. Nature 426: 630–635

Berry EA, Guergova-Kuras M, Huang LS and Crofts AR (2000) Structure and function of cytochrome *bc* complexes. Ann Rev Biochem 69: 1005–1075

Berry EA, Huang LS, Saechao LK, Pon NG, Valkova-Valchanova M and Daldal F (2004) X-ray structure of *Rhodobacter capsulatus* cytochrome $bc_1$: Comparison with its mitochondrial and chloroplast counterparts. Photosynth Res 81: 251–275

Brejc K, Ficner R, Huber R and Steinbacher S (1995) Isolation, crystallization, crystal structure analysis and refinement of allophycocyanin from the cyanobacterium *Spirulina platensis* at 2.3 Å resolution. J Mol Biol 249: 424–40

Brettel K, Siekmann I, Fromme P, van der Est A and Stehlik D (1992) Low-temperature EPR on single crystals of Photosystem I: Study of the iron-sulfur center $F_A$. Biochim Biophys Acta 1098: 266–270

Brixner T, Stenger J, Vaswani HM, Cho M, Blankenship RE and Fleming GR (2005) Two-dimensional spectroscopy of electronic couplings in photosynthesis. Nature 434: 625–628

Brunger AT and Rice LM (1997) Crystallographic refinement by simulated annealing: Methods and applications. Macromol Crystallogr B 277: 243–269

Bruns CM and Karplus PA (1995) Refined crystal structure of spinach ferredoxin reductase at 1.7 Å resolution: Oxidized, reduced and 2′-phospho-5′-AMP bound states. J Mol Biol 247: 125–145

Burmeister WP (2000) Structural changes in a cryo-cooled protein crystal owing to radiation damage. Acta Crystallogr D 56: 328–341

Carter CJ and Sweet R (1997) Macromolecular Crystallography. Academic Press, New York

Chandra S and Aziz SD (1994) Leidenfrost evaporation of liquid-nitrogen droplets. J Heat Tran 116: 999–1006

Chang CH, el-Kabbani O, Tiede D, Norris J and Schiffer M (1991) Structure of the membrane-bound protein photosynthetic reaction center from *Rhodobacter sphaeroides*. Biochemistry 30: 5352–5360

Chapman MS, Suh SW, Curmi PM, Cascio D, Smith WW and

Eisenberg DS (1988) Tertiary structure of plant RuBisCO: Domains and their contacts. Science 241: 71–74

Chiadmi M, Navaza A, Miginiac-Maslow M, Jacquot JP and Cherfils J (1999) Redox signalling in the chloroplast: Structure of oxidized pea fructose-1,6-bisphosphate phosphatase. EMBO J 18: 6809–6815

Cogdell RJ, Gardiner AT, Roszak AW, Law CJ, Southall J and Isaacs NW (2004) Rings, ellipses and horseshoes: How purple bacteria harvest solar energy. Photosynth Res 81: 207–214

Collaborative Computational Project, Number 4 (1994) The CCP4 suite: Programs for protein crystallography. Acta Crystallogr D 50: 760–763

Dauter Z (2006) Current state and prospects of macromolecular crystallography. Acta Crystallogr D 62: 1–11

Deisenhofer J and Michel H (1991) High-resolution structures of photosynthetic reaction centers. Annu Rev Biophys Biophys Chem 20: 247–266

Deisenhofer J, Epp O, Miki K, Huber R and Michel H (1984) X-ray structure analysis of a membrane protein complex. Electron density map at 3 Å resolution and a model of the chromophores of the photosynthetic reaction center from *Rhodopseudomonas viridis*. J Mol Biol 180: 385–398

Deisenhofer J, Epp O, Miki K, Huber R and Michel H (1985) Structure of the Protein subunits in the photosynthetic reaction center of *Rhodopseudomonas viridis* at 3 Å resolution. Nature 318: 618–624

Deisenhofer J, Epp O, Sinning I and Michel H (1995) Crystallographic refinement at 2.3 Å resolution and refined model of the photosynthetic reaction centre from *Rhodopseudomonas viridis*. J Mol Biol 246: 429–457

Ermler U, Michel H and Schiffer M. 1994. Structure and function of the photosynthetic reaction center from *Rhodobacter sphaeroides*. J Bioenerg Biomembr 26: 5–15

Ferreira KN, Iverson TM, Maghlaoui K, Barber J and Iwata S (2004) Architecture of the photosynthetic oxygen-evolving center. Science 303: 1831–1838

Fey V, Wagner R, Brautigam K and Pfannschmidt T (2005) Photosynthetic redox control of nuclear gene expression. J Exp Bot 56: 1491–1498

Ficner R and Huber R. 1993. Refined crystal structure of phycoerythrin from *Porphyridium cruentum* at 0.23-nm resolution and localization of the gamma subunit. Eur J Biochem 218: 103–106

Fodje MN, Hansson A, Hansson M, Olsen JG, Gough S, Willows RD and Al-Karadaghi S (2001) Interplay between an AAA module and an integrin I domain may regulate the function of magnesium chelatase. J Mol Biol 311: 111–122

Frazao C, Soares CM, Carrondo MA, Pohl E, Dauter Z, Wilson KS, Hervas M, Navarro JA, De la Rosa MA and Sheldrick GM (1995) *Ab initio* determination of the crystal structure of cytochrome $c_6$ and comparison with plastocyanin. Structure 3: 1159–1169

Fromme P (1998) Crystallization of Photosystem I for structural analysis [Habilitation]. Technical University of Berlin, Berlin, Germany

Fromme P (2003) Crystallization of Photosystem I. In: Iwata S (ed) Methods and Results in Crystallization of Membrane Proteins. pp 134–173. International University Line, La Jolla

Fromme P and Witt HT (1998) Improved isolation and crystallization of Photosystem I for structural analysis. Biochim Biophys Acta 1365: 175–184

Fromme P, Bottin H, Krauss N and Sétif P (2002) Crystallization and EPR characterization of a functional complex of Photosystem I with its natural electron acceptor ferredoxin. Biophys J 83: 1760–1773

Fukuyama K (2004) Structure and function of plant-type ferredoxins. Photosynth Res 81: 289–301

Fukuyama K, Hase T, Matsumoto S, Tsukihara T, Katsube Y, Tanaka N, Kakudo M, Wada K and Matsubara H (1980) Structure of S-platensis [2fe-2s] ferredoxin and evolution of chloroplast-type ferredoxins. Nature 286: 522–524

Garman E (1999) Cool data: Quantity and quality. Acta Crystallogr D 55: 1641–1653

Giacovazzo C, Monaco H, Viterbo D, Scordari F, Gilli G, Zanotte G and Cati M (1992) In: Giacovazzo C (ed) Fundamentals of Crystallography, p 654. University Press, New York, Oxford

Girard E, Legrand P, Roudenko O, Roussier L, Gourhant P, Gibelin J, Dalle D, Ounsy M, Thompson AW, Svensson O, Cordier MO, Robin S, Quiniou R and Steyer JP (2006) Instrumentation for synchrotron-radiation macromolecular crystallography. Acta Crystallogr D 62: 12–18

Hitscherich C, Jr., Aseyev V, Wiencek J and Loll PJ (2001) Effects of PEG on detergent micelles: Implications for the crystallization of integral membrane proteins. Acta Crystallogr D 57: 1020–1029

Hofmann E, Wrench PM, Sharples FP, Hiller RG, Welte W and Diederichs K (1996) Structural basis of light harvesting by carotenoids: Peridinin-chlorophyll-protein from *Amphidinium carterae*. Science 272: 1788–1791

Jekow P, Fromme P, Witt H and Saenger W (1995) Photosystem I from *Synechococcus elongatus*: Preparation and crystallization of monomers with varying subunit compositions. Biochim Biophys Acta 1229: 115–120

Jordan P, Fromme P, Witt HT, Klukas O, Saenger W and Krauss N (2001) Three-dimensional structure of cyanobacterial Photosystem I at 2.5 Å resolution. Nature 411: 909–917

Kamiya N, Shen JR (2003) Crystal structure of oxygen-evolving Photosystem II from *Thermosynechococcus vulcanus* at 3.7 Å resolution. Proc Natl Acad Sci USA 100: 98–103

Kamlowski A, van der Est A, Fromme P and Stehlik D (1997) Low temperature EPR on Photosystem I single crystals: Orientation of the iron sulfur centers $F_A$ and $F_B$. Biochim Biophys Acta 1319: 185–198

Karplus PA, Daniels MJ and Herriott JR (1991). Atomic structure of ferredoxin-NADP$^+$ reductase: Prototype for a structurally novel flavoenzyme family. Science 251: 60–66

Katona G, Andreasson U, Landau EM, Andreasson LE and Neutze R (2003) Lipidic cubic phase crystal structure of the photosynthetic reaction centre from *Rhodobacter sphaeroides* at 2.35 Å resolution. J Mol Biol 331: 681–692

Kendrew JC, Dickerson RE, Strandberg BE, Hart RG, Davies DR, Phillips DC and Shore VC (1960) Structure of myoglobin: A 3-dimensional Fourier synthesis at 2 Å resolution. Nature 185: 422–427

Kerfeld CA, Anwar HP, Interrante R, Merchant S and Yeates TO (1995) The structure of chloroplast cytochrome $c_6$ at 1.9 Å resolution: Evidence for functional oligomerization. J Mol Biol 250: 627–647

Knaff DB and Hirasawa M (1991) Ferredoxin-dependent chloroplast enzymes. Biochim Biophys Acta 1056: 93–125

Koepke J, Hu XC, Muenke C, Schulten K and Michel H (1996) The crystal structure of the light-harvesting complex II (B800-850)

from *Rhodospirillum molischianum*. Structure 4: 581–597

Krauss N, Hinrichs W, Witt I, Fromme P, Pritzkow W, Dauter Z, Betzel C, Wilson KS, Witt HT and Saenger W (1993) Three-dimensional structure of system I of photosynthesis at 6 Å resolution. Nature 361: 326–361

Kurisu G, Zhang H, Smith JL and Cramer WA (2003) Structure of the cytochrome $b_6f$ complex of oxygenic photosynthesis: Tuning the cavity. Science 302: 1009–1014

Lancaster CR, Bibikova MV, Sabatino P, Oesterhelt D and Michel H (2000) Structural basis of the drastically increased initial electron transfer rate in the reaction center from a *Rhodopseudomonas viridis* mutant described at 2.00-Å resolution. J Biol Chem 275: 39364–39368

Lancaster CR and Michel H (1997) The coupling of light-induced electron transfer and proton uptake as derived from crystal structures of reaction centres from *Rhodopseudomonas viridis* modified at the binding site of the secondary quinone, $Q_B$. Structure 5: 1339–1359

Lange C and Hunte C (2002) Crystal structure of the yeast cytochrome $bc_1$ complex with its bound substrate cytochrome *c*. Proc Natl Acad Sci USA 99: 2800–2805

Law CJ, Prince SM and Cogdell RJ (1998) Crystallising the LH1-RC 'core' complex of purple bacteria. Biochem Soc Trans 26: 160S.

Li YF, Zhou WL, Blankenship RE and Allen JP (1997) Crystal structure of the bacteriochlorophyll *a* protein from *Chlorobium tepidum*. J Mol Biol 271: 456–471

Liu Z, Yan H, Wang K, Kuang T, Zhang J, Gui L, An X and Chang W (2004) Crystal structure of spinach major light-harvesting complex at 2.72 Å resolution. Nature 428: 287–292

Loll B, Kern J, Saenger W, Zouni A and Biesiadka J (2005) Towards complete cofactor arrangement in the 3.0 Å resolution structure of Photosystem II. Nature 438: 1040–1044

MacColl R (1998) Cyanobacterial phycobilisomes. J Struc Biol 124: 311–334

Matthews BW, Fenna RE, Bolognesi MC, Schmid MF and Olson JM (1979) Structure of a bacteriochlorophyll *a*-protein from the green photosynthetic bacterium *Prosthecochloris aestuarii*. J Mol Biol 131: 259–285

McAuley KE, Fyfe PK, Ridge JP, Isaacs NW, Cogdell RJ and Jones MR (1999) Structural details of an interaction between cardiolipin and an integral membrane protein. Proc Natl Acad Sci USA 96: 14706–14711

McDermott G, Prince SM, Freer AA, Hawthornthwaite-Lawless AM, Papiz MZ, Cogdell RJ and Isaacs NW (1995) Crystal-structure of an integral membrane light-harvesting complex from photosynthetic bacteria. Nature 374: 517–521

McLuskey K, Prince SM, Cogdell RJ and Isaacs NW (2001) The crystallographic structure of the B800-820 LH3 light-harvesting complex from the purple bacteria *Rhodopseudomonas acidophila* strain 7050. Biochemistry 40: 8783–8789

Neisser A, Fromwald S, Schmatzberger A and Peschek GA (1994) Immunological and functional localization of both F-type and P-type ATPases in cyanobacterial plasma membranes. Biochem Biophys Res Commun 200: 884–892

Nogi T, Fathir I, Kobayashi M, Nozawa T and Miki K (2000) Crystal structures of photosynthetic reaction center and high-potential iron-sulfur protein from *Thermochromatium tepidum*: thermostability and electron transfer. Proc Natl Acad Sci USA 97:13561–13566

Odahara T, Ataka M and Katsura T (1994) Phase diagram determination to elucidate the crystal growth of the photoreaction center from *Rhodobacter sphaeroides*. Acta Crystallogr D Biol Crystallogr 50: 639–642

Olson J (1998) Chlorosome origin and function in green photosynthetic bacteria. Photochem Photobiol 67: 61–75

Ostermeier C and Michel H (1997) Crystallization of membrane proteins. Curr Opin Struct Biol 7: 697–701

Papiz MZ, Prince SM, Howard T, Cogdell RJ and Isaacs NW (2003). The structure and thermal motion of the B800-850 LH2 complex from *Rps. acidophila* at 2.0 Å resolution and 100 K: New structural features and functionally relevant motions. J Mol Biol 326: 1523–1538

Perutz MF, Rossmann MG, Cullis AF, Muirhead H, Will G and North ACT (1960) Structure of haemoglobin — 3-dimensional Fourier synthesis at 5.5-Å resolution, obtained by X-ray analysis. Nature 185: 416–422

Potterton E, McNicholas S, Krissinel E, Cowtan K and Noble M (2002) The CCP4 molecular-graphics project. Acta Crystallogr D 58: 1955–1957

Prince SM, Papiz MZ, Freer AA, McDermott G, Hawthornthwaite-Lawless AM, Cogdell RJ and Isaacs NW (1997) Apoprotein structure in the LH2 complex from *Rhodopseudomonas acidophila* strain 10050: Modular assembly and protein pigment interactions. J Mol Biol 268: 412–423

Redinbo MR, Cascio D, Choukair MK, Rice D, Merchant S and Yeates TO (1993) The 1.5-Ångström crystal-structure of plastocyanin from the green-alga *Chlamydomonas reinhardtii*. Biochemistry 32: 10560–10567

Roszak AW, Howard TD, Southall J, Gardiner AT, Law CJ, Isaacs NW and Cogdell RJ (2003) Crystal structure of the RC-LH1 core complex from *Rhodopseudomonas palustris*. Science 302: 1969–1672

Salemme FR, Freer ST, Xuong NH, Alden RA and Kraut J (1973) Structure of oxidized cytochrome $c_2$ of *Rhodospirillum rubrum*. J Biol Chem 248: 3910–3921

Scheuring S, Sturgis JN (2005) Chromatic adaptation of photosynthetic membranes. Science 309: 484–487

Schirmer T, Bode W, Huber R, Sidler W and Zuber H (1985) X-ray crystallographic structure of the light-harvesting biliprotein C-phycocyanin from the thermophilic cyanobacterium *Mastigocladus laminosus* and its resemblance to globin structures. J Mol Biol 184: 257–277

Schirmer T, Huber R, Schneider M, Bode W, Miller M and Hackert ML (1986) Crystal-structure analysis and refinement at 2.5-Å of hexameric *c*-phycocyanin from the cyanobacterium *Agmenellum quadruplicatum* — the molecular-model and its implications for light-harvesting. J Mol Biol 188: 651–676

Schneider G, Lindqvist Y, Branden CI and Lorimer G (1986) Three-dimensional structure of ribulose-1,5-bisphosphate carboxylase/oxygenase from *Rhodospirillum rubrum* at 2.9 Å resolution. EMBO J 5: 3409–3415

Serp D, Mueller M, von Stockar U and Marison IW (2002a) Low-temperature electron microscopy for the study of polysaccharide ultrastructures in hydrogels. I. Theoretical and technical considerations. Biotechnol Bioeng 79: 243–252

Serp D, Mueller M, von Stockar U and Marison IW (2002b) Low-temperature electron microscopy for the study of polysaccharide ultrastructures in hydrogels. I. Theoretical and technical considerations. Biotechnol Bioeng 79: 243–252

Serre L, Vellieux FM, Medina M, Gomez-Moreno C, Fontecilla-Camps JC and Frey M (1996) X-ray structure of the ferredoxin:

NADP+ reductase from the cyanobacterium *Anabaena* PCC 7119 at 1.8 Å resolution, and crystallographic studies of NADP+ binding at 2.25 Å resolution. J Mol Biol 263: 20–39

Spreitzer RJ and Salvucci ME (2002) Rubisco: Structure, regulatory interactions, and possibilities for a better enzyme. Annu Rev Plant Biol 53: 449–475

Stock D, Leslie AG and Walker JE (1999) Molecular architecture of the rotary motor in ATP synthase. Science 286: 1700–1705

Stout G and Jensen L (1989) X-Ray Structure Determination — A Practical Guide. Wiley-Interscience, New York

Stowell MHB, McPhillips TM, Rees DC, Soltis SM, Abresch E and Feher G (1997) Light-induced structural changes in photosynthetic reaction center: Implications for mechanism of electron-proton transfer. Science 276: 812–816

Stroebel D, Choquet Y, Popot JL and Picot D (2003) An atypical haem in the cytochrome $b_6f$ complex. Nature 426: 413–418

Swamy U, Wang MT, Tripathy JN, Kim SK, Hirasawa M, Knaff DB and Allen JP (2005) Structure of spinach nitrite reductase: Implications for multi-electron reactions by the iron-sulfur: siroheme cofactor. Biochemistry 44: 16054–16063

Tiede DM, Leigh JS and Dutton PL (1978) Structural Organization of reaction center and associated *c*-cytochromes in *Chromatium vinosum*. Biophys J 21: A196–A196

Walsh MA, Evans G, Sanishvili R, Dementieva I and Joachimiak A (1999) MAD data collection — current trends. Acta Crystallogr D 55: 1726–1732

Walters RG (2005) Towards an understanding of photosynthetic acclimation. J Exp Bot 56, 435–447

Wang JW, Chen JR, Gu YX, Zheng CD, Jiang F, Fan HF, Terwilliger TC and Hao Q (2004) SAD phasing by combination of direct methods with the SOLVE/RESOLVE procedure. Acta Crystallogr D 60: 1244–1253

Watenpaugh KD, Sieker LC, Jensen LH, Legall J and Dubourdieu M (1972) Structure of the oxidized form of a flavodoxin at 2.5-Ångström resolution: Resolution of the phase ambiguity by anomalous scattering. Proc Natl Acad Sci USA 69: 3185–3188

Witt I, Witt H, DiFiore D, Rögner M, Hinrichs W, Saenger W, Granzin J, Betzel C and Dauter Z (1988) X-ray characterization of single crystals of the reaction center I of water splitting photosynthesis. Ber Bunsenges Phys Chem 92: 1503–1506

Yano J, Kern J, Irrgang KD, Latimer MJ, Bergmann U, Glatzel P, Pushkar Y, Biesiadka J, Loll B, Sauer K, Messinger J, Zouni A and Yachandra VK (2005) X-ray damage to the $Mn_4Ca$ complex in single crystals of Photosystem II: A case study for metalloprotein crystallography. Proc Natl Acad Sci USA 102: 12047–12052

Zhang H, Kurisu G, Smith JL and Cramer WA (2003) A defined protein-detergent-lipid complex for crystallization of integral membrane proteins: The cytochrome $b_6f$ complex of oxygenic photosynthesis. Proc Natl Acad Sci USA 100: 5160–5163

Zouni A, Witt HT, Kern J, Fromme P, Krauss N, Saenger W and Orth P (2001) Crystal structure of Photosystem II from *Synechococcus elongatus* at 3.8 Å resolution. Nature 409: 739–743

# Chapter 7

## Electron Crystallography in Photosynthesis Research

Paula C.A. da Fonseca and Edward P. Morris
*Section of Structural Biology, The Institute of Cancer Research, Chester Beatty Laboratories, 237 Fulham Road, London SW3 6JB, U.K.*

Claudia Büchel*
*Institute of Molecular Biosciences, University of Frankfurt, Siesmayerstr. 70, D-60323 Frankfurt, Germany*



*Author for correspondence, email: c.buechel@bio.uni-frankfurt.de

*Thijs J. Aartsma and Jörg Matysik (eds): Biophysical Techniques in Photosynthesis II, pp. 125–150.*

## Summary

Electron crystallography covers the analysis of crystals by electron microscopy based methods in order to infer into the structure of the crystallized molecules. Electron crystallography of two-dimensional crystals has become a powerful alternative to X-ray crystallography and NMR spectroscopy for the determination of the structure of proteins or protein complexes. The electron crystallography approach is particularly suitable for the study of membrane proteins, which due to their amphiphilic properties are stable in lipid bilayers, their natural environment, where they can form crystalline arrays. This method of structural analysis has also proved suitable for the study of soluble proteins, which can be induced to form two-dimensional lattices, as well as the study of very thin three-dimensional crystals. The preparation of two-dimensional crystals normally requires less protein and at a lower concentration than required for three-dimensional crystallization. Overall, the structural analysis based on electron microscopy is more amenable to the study of conformational variations associated with different functional states of membrane proteins than other techniques used in structural biology. Furthermore, the analysis of two-dimensional crystals enables the calculation of preliminary structural maps at relatively low resolutions, which at early stages in the optimization crystallization and image analysis may render valuable information on the overall protein architecture. Here we present a summary of the methods involved in protein electron crystallography, regarding both the crystallization process itself and the fundaments of electron microscopy and image analysis required. We also show how electron crystallography has greatly contributed for the development of the present knowledge on the structural organization of photosynthetic protein complexes, focusing on the studies involving transmembrane pigmented proteins and protein complexes.

## I. Introduction

Electron crystallography exploits the diffraction of electrons by regular arrays of molecules. The electron diffraction is the result of the interaction of electrons with matter and can be used for structure determination in material (Dorset, 1996; Leslie et al., 1999; Caldes et al., 2001) as well as life sciences.

This review focuses on the use of electron beams in structural biology, especially electron crystallography. The concept of using electron microscopes and Fourier mathematics in the structural analyzes of biological specimens can be traced to the late 1960s, when DeRosier and Klug (1968) published their work on a helical structure, the tail of the bacteriophage T4. Since then, technological developments concerning electron microscopes, especially the introduction of cryo-electron microscopy, as well as in computational equipment and software have led to an immense progress in the methodology used and now there is a range of approaches available to determine structures of biological samples using transmission electron microscopy. Overall three methods have emerged: electron tomography (see Chapter 2, Hohmann-Marriott et al.), single particle analysis, and electron crystallography.

For single particle analysis, isolated proteins or protein complexes in solution are examined. Ideally these particles adopt random orientations in the support film (normally a layer of carbon or vitreous ice) on the electron microscope grid thus yielding different projection views of equivalent particles. Alignment procedures and statistical analysis of these projection images can be used in order to group images with

*Abbreviations:* 2D – two-dimensional; 3D – three-dimensional; CCD – charge coupled device; Chl – chlorophyll; cmc – critical micellar concentration; Cyt – cytochrome; DGDG – digalactosyl diglyceride; IQ – integer quality; LH – bacterial light harvesting complex; LHC – light harvesting complex; MGDG – monogalactosyl diglyceride; NMR – nuclear magnetic resonance; pCTF – phase contrast transfer function; PS I – Photosystem I; PS II – Photosystem II; RC – reaction center; *R.* – *Rhodospirillum*; *Rb.* – *Rhodobacter*; *Rh.* – *Rhodopseudomonas*

equivalent orientations which are then averaged to produce images with improved signal-to-noise ratio. The relative orientation of these noise reduced images can then be determined and this information used to calculate a three-dimensional (3D) reconstruction of the particle. Ideally, proteins should have a certain size (>200 kDa, Henderson, 1995) to allow for this method, since they have to be directly visualized by electron microscopy. High symmetry and increasing size greatly facilitates the analysis. This method is now widely used especially for the study of large proteins or protein complexes to obtain information about their structure, the arrangement and interaction of subunits, and conformational changes associated with different functional states. The methods of single particle analysis have been extensively reviewed (van Heel et al., 2000; Frank, 2002, 2005; Orlova and Saibil, 2004; Jiang and Ludtke, 2005).

Membrane proteins account for a large proportion of all proteins in a cell. For instance in yeast about one third of the sequenced genes are predicted to code for membrane proteins, and an even larger fraction for membrane associated proteins (Stahlberg et al., 2001). Electron crystallography uses so-called two-dimensional (2D) crystals, i.e., crystals consisting of a single layer of protein arranged in a regular lattice. 2D crystals of membrane proteins usually contain lipids as in naturally occurring bilayers. In these bilayers, membrane proteins are more likely to be kept in their natural configuration, compared for example to the situation when they are solubilized in detergent. Thus, electron crystallography has its distinct advantages for the structural studies of membrane proteins, and studies of these proteins account for the majority of applications of this methodology (for electron crystallography on soluble proteins, see Ellis and Herbert, 2001).

In analyzing protein structures within 2D crystals, electron crystallography exploits diffraction patterns as well as real space images of the protein (see Section III.A). Because images can be recorded, phase information is available: this represents an important distinction from other diffraction methods such as X-ray crystallography in which only amplitudes are directly available and phases have to be obtained by indirect and complex methods. A number of structures of membrane proteins have been obtained by electron crystallography with sufficient resolution for atomic resolution models to be built. Bacteriorhodopsin has been studied intensively by electron crystallography and the projection map at 2.6 Å (Grigorieff et al., 1995) and the 3D structure at 3 Å (Kimura et al., 1997) represent examples of the highest resolution structures achieved by this method. Recently, the structure of aquaporin 0 was determined at 1.9 Å by electron diffraction analysis and molecular replacement (Gonen et al., 2005). This represents the highest resolution structure achieved to date by electron crystallography of a biological sample and reveals not only detailed information on the protein organization but also the structure of the lipid molecules present in the crystal.

Many excellent reviews about electron crystallography have been written over the last decade (Fujiyoshi, 1998; Walz and Griegorieff, 1998; Glaeser, 1999; Kühlbrandt and Williams, 1999; Yeager et al., 1999; Auer, 2000; Stahlberg et al., 2001; Unger, 2001; Werten et al., 2002). In this chapter we introduce the underlying concepts, possibilities and limitations of electron crystallography and discuss current methodology and its application in the structural analysis of photosynthetic pigmented membrane proteins and protein complexes involved in photosynthesis.

## II. Two-Dimensional Crystals

### A. Types of Two-Dimensional Crystals

Structure determination by electron crystallography requires 2D crystals which ideally should be highly ordered, flat, large (>1 μm) and single layered, although 2D crystals which do not meet all of these requirements can be successfully analyzed by electron crystallography as discussed below. 2D crystals of membrane proteins generally consist of the protein arranged in a crystal lattice embedded in a lipid bilayer. Some occur naturally, but such crystals are often not sufficiently well ordered and/or not large enough to be useful for structure determination at high resolution. Although in a number of cases it is possible to improve the quality of native crystals in vitro, it has become more common to use 2D crystals prepared using purified protein. Different types of 2D crystals can be distinguished: (single layered) sheets, tubes, vesicles and multilayered crystals (Fig. 1). From these, crystalline sheets are the most difficult to obtain but are the most useful since they are normally bigger and flatter. More commonly 2D crystals occur as vesicular structures which may adopt tubular conformations. Narrow tubes, such as crystals of the acetyl choline receptor (Toyoshima and Unwin, 1990; Unwin, 2005) and crystals of Photosystem II (PS II, Tsiotis et al., 1996), have the advantage of presenting the protein in different orientations along the tube and may thus allow for 3D reconstruction of the structure without tilting of the sample (see Section III.E.2). Vesicles or wide tubes (Fig. 1g), such as those obtained with the sub-complex formed from PS II reaction centers (RC) together with CP47 (CP47-RC), analyzed by Rhee et al. (1997, 1998), are more easily obtained. Such vesicles or tubes when visualized by electron microscopy using a carbon film as support resemble a collapsed balloon or cylinder and are generally characterized by two overlapping lattices formed from crystalline arrays in the upper and lower sides of the vesicle or tube. It is usually possible to analyze such lattices independently. Multilayered crystals are stacks of 2D crystals, whereby the lattices can be either perfectly in register or only close to register. In the former case crystals are true small 3D crystals (Fig. 1f), whereas the latter might be just adhesion of different 2D crystals. These are more difficult to analyze but nevertheless successful structure determination using multilayered crystals was reported (Cyrklaff et al., 1995; Auer et al., 1999).

During 2D crystallization by in vitro reconstitution methods, proteins can get inserted into the membrane either all with the same polarity (Fig. 1a, e.g., Photosystem II cores in Hankamer et al., 1999) or in an up-and-down configuration (Fig. 1b, e.g., CP47-RC in Rhee et al., 1997, 1998). Whereas the former more commonly tends to form tubes because of the curvature imposed (Fig. 1c), the latter is more likely to yield flat sheets, although the CP47-RC 2D crystals were generally tubular.

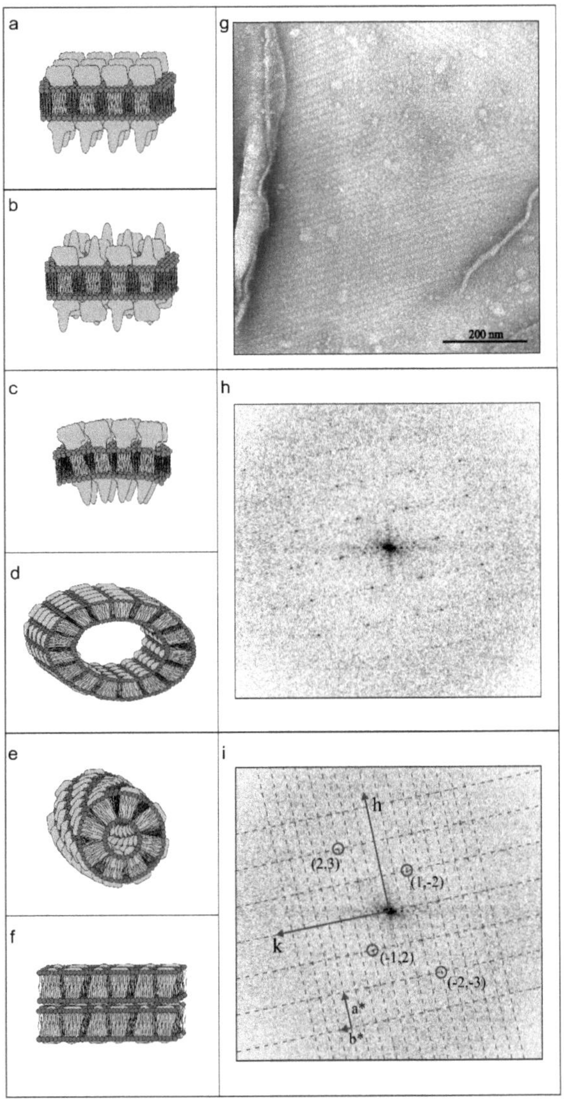


*Fig. 1.* Different types of 2D crystals can be formed depending on the proteins involved. (a) Single layer crystals with proteins arranged all with the same polarity. (b) Single layer crystal with proteins inserted with opposite polarities. (c) Same as (a) but demonstrating the possible curvature induced by packing of the protein in the crystal. (d) Wide tubular vesicle. (e) Narrow tube with helical symmetry, allowing different views of the protein in a single projection image. (f) Multilayer crystal. (g) Electron microscopy image of a 2D crystal of PS II core complexes from *Thermosynechococcus elongatus* (da Fonseca et al., 2002, with permission). (h) Diffraction pattern of the crystal shown in (g). (i) Indexing the diffraction pattern: the *h,k* indices of the circled diffraction spots are shown together with the reciprocal unit cell vectors a* and b*.

## B. Requirements for 2D Crystallization and Protein Purification Methods

One decisive factor in achieving crystals suitable for high resolution analysis is the quality of the protein preparation, which has to meet the following conditions: purity, high quantity (although usually much less is required than for the preparation of 3D crystals for X-ray diffraction analysis), functionality and homogeneity concerning its oligomeric state.

In general, several methods can be used for purification of membrane proteins for electron crystallography. Generally, the same principles apply as for the preparation of 3D crystals. Proteins which occur only in very small amounts in vivo can be over-expressed in the various expression systems available today such as bacteria (the most common being *E. coli*), yeast, insect or mammalian cells. A system which resembles the natural system most closely should be preferable, which would make bacteria the obvious choice for proteins coded for in organelles and eukaryotic systems for nuclear encoded genes. This is supposed to yield the best results in terms of structure and functionality of the protein, i.e., the correct fold and post-translational modifications. However, such procedures are generally less suited for photosynthetic protein complexes which frequently consist of different protein subunits and usually bind substantial amounts of cofactors (such as chlorophylls (Chl), FeS centers, cytochromes (Cyt) to name only a few). Thus, the correct assembly is not easy to achieve using heterologous expression systems since in vitro refolding is required. Only some photosynthetic proteins have been refolded successfully in vitro so far, all belonging to the family of light harvesting complexes (LHC) (Paulsen et al., 1993; Hobe et al., 1994; Sandona et al., 1998). Another approach is the homologous over-expression, i.e., using the original organism for expression of the proteins by putting the gene of choice under control of a strong promoter and also adding a tag (like His- or Strep-tags) facilitating purification later on. Nuclear encoded photosynthetic proteins can be homologously over-expressed by transformation of higher plants using Ti-plasmids, which lead to insertion of additional genes (or modified gene copies) in the nuclear genome. Such attempts included over-expression of LHCII in tobacco (Flachmann and Kühlbrandt, 1995), where unfortunately expression levels never met expectations. In case of genes of the plastom, transformation using ballistic methods can substitute the original genes by the modified version. For example, Photosystem II preparations with His-tags added to psbH in *Chlamydomonas* (Büchel et al., 2001) and to Cyt *b*559 in tobacco (Büchel, unpublished) were successfully produced. However, most modifications such as adding tags to various proteins of the thylakoid membrane were done in the cyanobacteria *Synechocystis* and *Thermosynechococcus*. In principle, adding a tag makes it possible to selectively bind the protein of choice to tag-specific affinity chromatography media, allowing a quick removal of a huge background of unwanted proteins which greatly decreases the length of time otherwise required for the purification procedures. If genetic modifications are not practical the protein has to be purified from the native material, which, in the case of photosynthetic proteins, is often facilitated by the high abundance of these proteins and the biochemical experience accumulated over the years.

### *C. Principles of 2D Crystallization of Membrane Proteins*

The thermodynamic driving force for the insertion of membrane proteins into lipid bilayers is the segregation of hydrophobic regions of the protein into the lipid phase. In the case of detergent solubilized membrane proteins this can be achieved by detergent removal in the presence of suitable lipids. In the lipid bilayer, the formation of a crystalline lattice, opposed to a random distribution, is favored by the same principles as in 3D crystallization, i.e., a net gain in entropy, but with the energy involved likely to be smaller than in 3D crystallization (Kühlbrandt, 1992).

Different methods of 2D crystallization have been reviewed (Hasler et al., 1998; Rigaud et al., 2000; Mosser, 2001; Rosenbusch et al., 2001). The lipids involved in crystallization may co-purify with the protein in sufficient amounts to support crystallization, as in the case of LHCII (Kühlbrandt et al., 1994) and the CP47-RC subcomplex of PS II (Rhee et al., 1997). More commonly lipids have to be added to the detergent solubilized protein to form lipid bilayers (e.g., in the case of PS II cores, Hankamer et al., 1997). Here, micelles of protein and detergent are mixed with micelles of lipid plus detergent, with the formation of a crystalline lattice driven by the protein reconstitution into lipid bilayers by decreasing the amount of detergent under appropriate conditions. There is still an ongoing debate as to whether crystallization and reconstitution into the lipid bilayer occur simultaneously or whether proteins arrange in a crystalline lattice after the bilayer is formed (Rigaud et al., 2000).

### *D. Strategies for Detergent Removal*

In the 2D crystallization by reconstitution into lipid bilayers the detergent can be removed either by simple dilution, by the use of adsorbent beads or by dialysis. Dilution lowers the protein, detergent and lipid concentrations simultaneously. Lower protein and lipid concentrations may not favor crystallization and this method is rarely used. Polystyrene beads (Biobeads, Biorad, Rigaud et al., 1997) have been successfully used to produce 2D crystals of cyanobacterial Photosystem I (PS I, Böttcher et al., 1992), and PS II from plants (Hankamer et al., 1999, 2001) and cyanobacteria (da Fonseca et al., 2002). These beads are substantially selective for detergent adsorption, but protein and lipid adsorption may also occur leading to interference with 2D crystal formation. In these experiments, the rate of detergent removal is controlled by adjusting the quantity of beads: this can be achieved by counting or weighing, although the differing size of the beads tends to interfere with the reproducibility in each case. Dialysis is the most commonly used method for detergent removal since it has the advantage that it is rather reproducible and it is easy to perform. Due to the necessity of testing a wide range of crystallization conditions, a single trial will usually have a volume of no more than 50–100 μL. For this purpose either commercial micro-dialysis devices (Yeager et al., 1999) or custom devices such as hockey-sticks (Kühlbrandt, 1992) or the top of Eppendorf tubes (Yeager et al., 1999) have been used.

### *E. Factors Affecting Crystallization: Lipids*

In order to achieve optimal conditions to favor the protein contacts required for the formation of 2D crystals, protein-protein and protein-lipid interactions should be considered. In this regard the nature of the lipids, the lipid-to-protein ratio and the choice of detergent are the most important parameters in 2D crystallization. Usually lipid to protein ratios varying from 10–100 (mol/mol) are chosen as a starting point for screening. The lipids should be selected to recreate the native environment of the protein taking into account the nature of the lipid head group and fatty acid chain. The length of the fatty acid chain influences both the thickness of the lipid bilayer (the most common chain length in biological membranes is 16 or 18 carbon atoms) and its fluidity. Bilayer-forming lipids show a phase transition from a rather rigid phase of lamellar appearance to a liquid-crystalline phase that is more fluid and allows for protein diffusion. Thus, crystallization can only occur at temperatures above the transition temperature of the lipid or lipid mixture used. This temperature not only depends on the hydrocarbon chain length, but also on its degree of saturation: the more it is saturated, the higher is its transition temperature (for details, see Table 2 in Kühlbrandt, 1992). Thylakoid membranes are quite unusual in containing mainly glycolysated lipids, digalactosyl diglyceride (DGDG) and monogalactosyl diglyceride (MGDG), the latter tending to produce hexagonal rather than bilayer structures (Garab et al., 2000). These lipids might be important for keeping proteins in their right quaternary structure and indeed highly diffracting 3D crystals of LHCII do contain DGDG molecules in stoichiometric amounts, which remained bound to LHCII during preparation (Standfuss et al., 2005). This and other examples, such as Cyt $bc_1$ complex (Lange et al., 2001) demonstrate that in membrane protein purification care should be taken to avoid removal of essential lipids.

### *F. Factors Affecting Crystallization: Detergent*

The choice of the detergent used both in the preparation of the protein sample and in the formation of lipid-detergent micelles is critical in two ways: first, the functionality of a protein can be greatly impaired if the wrong detergent is chosen; second, the lower the critical micellar concentration (cmc) of the detergent the slower is its removal by bead adsorption or dialysis. Usually non-ionic detergents are used and generally the longer the hydrocarbon chain is, the lower the cmc (more detailed information in Rigaud et al., 2000; le Maire et al, 2001; Rosenbusch et al., 2001).

### *G. Other Factors Affecting Crystallization*

Besides the choice of lipid, the lipid-to-protein ratio and the detergent used, other factors influence 2D crystallization and crystal growth. Buffers, and their pH, and salts not only influence the functional state of the protein but also crystallization. Furthermore, temperature is a decisive factor because of the lipid fluidity as discussed above. In some cases temperature ramping has been successful in the production of 2D crystals (e.g., CP47-RC, Büchel and Kühlbrandt, 2005), but care has to be taken because not all proteins resist temperatures above 4 °C. However, when already incorporated into the lipid bilayer, proteins are usually more stable than in a detergent micelle and thus after membrane insertion higher temperatures can be used.

The protein concentration used for successful 2D crystallization is usually lower than in 3D trials: concentrations of around 1 mg protein/mL are typical for 2D crystallization although there are also examples of higher protein concentrations (~10 mg/mL) being required (Rhee et al., 1997; Büchel and Kühlbrandt, 2005). Very often protein and membrane stabilizing agents such as glycerol have successfully been used.

Although different parameters affecting 2D crystallization can be listed and used as guidelines, the conditions required for the formation of crystals completely depend on the protein sample used. Therefore no standard advice can be given on which conditions to use and trials under a range of different conditions have to be prepared and evaluated.

### H. Other Methods of 2D Crystallization

As ideal 2D crystals should be flat and large, some attempts were made to use different surfaces as a template for 2D crystallization. One of these attempts was to directly crystallize the protein on the carbon film of the electron microscope grid. Unfortunately, this usually leads to thin 3D crystals (Cyrklaaf et al., 1995; Auer et al., 1999; Büchel et al., 2000), which are more difficult to analyze by electron crystallography.

Another method uses functionalized lipids, e.g., lipids with $Ni^{2+}$-groups to bind to His-tags, which are spread on the water-air interface of the crystallization trial and induce a localized increase in the concentration of protein, oriented according to the affinity label used (reviewed by Lévy et al., 1999). This method has been used quite successfully for the 2D crystallization of soluble proteins.

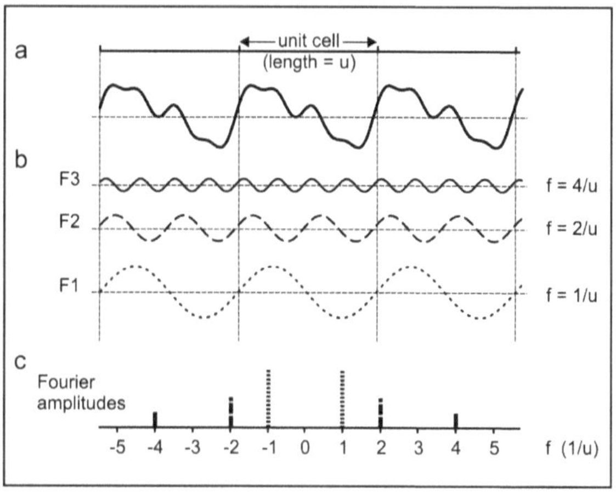


Fig. 2. Fourier transform: the deconvolution of a function into its sine components. Any function can be described by a sum of sine waves. (a) One-dimensional function which repeats at $u$ (unit cell length) intervals. (b) Component sine waves (F1, F3 and F3), differing in their amplitudes and phases, the sum of which results in the function represented in (a). (c) Representation of the Fourier amplitudes of F1, F2 and F3 as a function of frequency ($f$). Note that the Fourier transform is centrosymmetric and therefore results in positive and negative frequency components, each with an amplitude A/2, where A is the amplitude of the correspondent sine wave.

## III. Electron Microscopy

Many excellent books and reviews exist (Buseck et al., 1988; Amelincks et al., 1997; Reimer, 1997; Zemlin, 1998) about the fundamentals of electron microscopy. Here are presented only the essentials for the understanding of electron crystallography of biological samples.

### A. Images and Diffraction—Real Space and Reciprocal Space

Diffraction is the deflection and spreading of waves that occur when they interact with an object: light and electron beams are examples of wave systems of relevance here. Under appropriate conditions the diffraction of a wave leads to the formation of a diffraction pattern. Much of the image processing in electron crystallography involves the analysis of diffraction patterns: these may be obtained physically or alternatively with the aid of a computer by calculating Fourier transforms of images.

The projection images of 2D crystals recorded in the electron microscope are characterized by a 2D lattice arising from the size, shape and symmetry of the unit cell of the crystal. Such an image is a distribution of intensities in 2D space which as any other function can be considered the sum of a series of sine waves, as illustrated in Fig. 2 for a simplified one-dimensional repetitive function. The Fourier transform of a function corresponds to the mathematical decomposition into its sinusoidal components. The representation of the square of the Fourier amplitudes against frequency is known as power spectrum. In the Fourier transform of a repetitive function, such as a projection image of a 2D crystal, the amplitude of the sine waves with the same frequency as that of the repeating units (and their multiples) of the crystalline lattice will tend to be strong, while other amplitudes will cancel themselves out. As a result, the power spectrum of a 2D crystal is characterized by a lattice of spots, in which higher order spots correspond to smaller spacings in the original real space image.

Diffraction patterns of 2D crystals can be obtained directly by optical methods, either by using the electron microscope in electron diffraction mode or by recording images on photographic film which are then exposed, in an optical diffractometer, to a very coherent beam of visible light. In optical and electron diffraction the regular array of unit cells within the

2D crystal acts as a diffraction grating, such that the interference between the resulting waves gives rise to a 2D lattice of diffraction spots. This 2D lattice is equivalent to the power spectrum of the 2D crystal calculated by Fourier transformation (Fig. 2). However, while the calculation of Fourier transforms preserves the information on the amplitude and phases of the diffraction spots, in optical and electron diffraction patterns the phase information is lost.

### B. Underlying Principles of Electron Microscopy

Transmission electron microscopes are very often compared to light microscopes. This comparison principally holds as the major optical principles apply to light as well as electron beams. However there are two important differences: image formation (see Section III.C) and depth of focus. The focal depth is much larger in an electron microscope than in a light microscope. Accordingly it is not possible to use focal settings of the lens system to selectively image individual planes of the object normal to the optical axis, as is the practice in confocal light microscopy, for example. In reality, to a close approximation electron microscope images correspond to the projection of the object densities along the optical axis, thus each image contains contributions from all the details of the molecule under investigation (see Section IV.B).

In structural analysis, the maximal theoretical resolution that can be obtained either by conventional imaging or diffraction methodology is limited by the wavelength of the radiation used, whether light, electrons, or X-rays are used for investigation. In electron microscopy the wavelength of the incident beam is inversely proportional to the acceleration voltage and is in the range of 0.001–0.005 nm (compared to X-ray beams of around 0.1 nm and visible light of 400–700 nm). Thus, both electron microscopy and X-ray crystallography can potentially provide structural information to atomic resolution. Compared to X-rays, electron beams have the advantage that they can be easily focused using magnetic lenses, and therefore in electron microscopy real space images of an object can be recorded in addition to the recording of its electron diffraction pattern. Unlike the light microscope or X-ray diffraction analysis, where it is possible to achieve resolutions close to the wavelength of the beam used, the highest resolution from electron microscopy based structural analysis, obtained from inorganic materials such as crystalline gold, extend to 0.05–0.1 nm, a value which is about 50 times larger than the wavelength of the electrons. This arises mostly from limitations in the properties of magnetic lenses. Moreover, to date the resolution obtained by electron crystallography of biological samples falls short of that obtained with inorganic material. The major reason for this is the sensitivity of unstained biological samples to radiation damage (see Section III.C). So far the highest resolution protein structure calculated by electron crystallography is that of aquaporin 0 (Gonen et al., 2005), which at 1.9 Å is comparable to those normally obtained by X-ray crystallography.

### C. Image Formation

Two electron microscopy modes used in electron crystallography have to be distinguished: bright-field imaging and diffraction modes. In the former a real space image of the object is recorded, while in the latter optics are set to directly record electron diffraction patterns. The diffraction mode is usually used only on the analysis of regular objects such as crystals.

Image formation in bright-field mode results mainly from the scattering of the incident electron beam rather than its absorption by the object. The scattering of electrons can be elastic or inelastic. Whereas in elastic scattering the kinetic energy of the electrons does not change, inelastic scattering involves loss of energy due to interactions with the sample. While a beam of elastically scattered electrons contains information about the scattering object, for inelastically scattered electrons this information is degraded. This is due to their random energy transfer to the sample and therefore their unknown energy and angle of incidence in the imaging plane, which is translated into noise in the image. In addition, the energy transfer to protein samples associated with inelastic interactions leads to radiation damage, due to the formation of free radicals and ionization of groups within the molecules, which eventually result in the random loss of their structure integrity (Glaeser, 1971). For biological samples the ratio of inelastic to elastic scattering is about three (Henderson, 1995) and it cannot be avoided. However, radiation damage can be reduced by working at lower temperatures (liquid nitrogen or helium) in order to reduce the diffusion of free radicals, and by exposing the samples to low electron doses (Section V.C.2). A reduction in beam damage by a factor of about 2 associated with data collection using a liquid helium cooled microscope, compared to operation at liquid nitrogen temperatures, has been reported (Fujiyoshi et al., 1991).

In image formation the amplitude contrast, or scattering contrast, results mostly from the electrons deflected to higher angles not reaching the imaging plane. This can be accentuated by the use of apertures and consequently the objective aperture is also known as contrast aperture. High resolution transmission electron microscopy relies on the analysis of unstained protein samples (see Section III.D). Since these samples generate mainly low angle scattering of the electron beam, they are intrinsically associated with low amplitude contrast and therefore low contrast images. However, at the imaging plane the scattered electron beam is characterized not only by its amplitude contrast, but also by its phase contrast, or interference contrast, caused by the interferences arising from differences in phase between the scattered and unscattered electron beams in the image plane. If the electron microscope magnetic lens systems were perfect and the images were recorded in true focus the effect of phase contrast on the amplitudes of the electron wave front at the imaging plane, and therefore image intensities, would be negligible. This arises both because only a very small fraction of electrons is scattered by the sample and because at the imaging plane the phase shifts between scattered and unscattered electron beams would be too small to give rise to effective interference. However, in the electron microscope additional frequency dependent phase shifts are generated by parameters such as the spherical aberration of the lens system and by the deliberate recording of out of focus images (usually underfocus). These phase shifts result in an effective interference between elastically scattered and unscattered electron waves at the imaging plane and hence in a significant amplitude modulation. The effect of the enhanced interferences generated by the electron microscope optics on the amplitudes of the electron wave front (and therefore contrast in the image) is described by the phase contrast transfer function (pCTF).

The pCTF oscillates between positive and negative contrast as a function of frequency with points of zero contrast where the function crosses the x-axis (zero crossings, Fig. 3). The effect of the pCTF is commonly observed in computed or optical diffraction patterns of electron micrographs as rings of intensity known as Thon-rings (inset Fig. 3b). In practice, in order to use the information at frequencies higher than that of the first zero crossing, the pCTF for each image has to be carefully characterized in order to allow for an accurate correction of the modulations associated with it. The exact shape of the pCTF depends on various parameters. For example, with increasing underfocus each of the zero crossings is shifted to lower spatial frequencies (Fig. 3a), while higher accelerating voltage shifts the zero crossings to higher spatial frequencies. The pCTF envelope function describes the damping of the amplitudes towards higher spatial frequencies which constrains the resolution limit achievable in the analysis of a particular image (Fig. 3b). The coherence of the electron source, the stability of the high tension, the stability

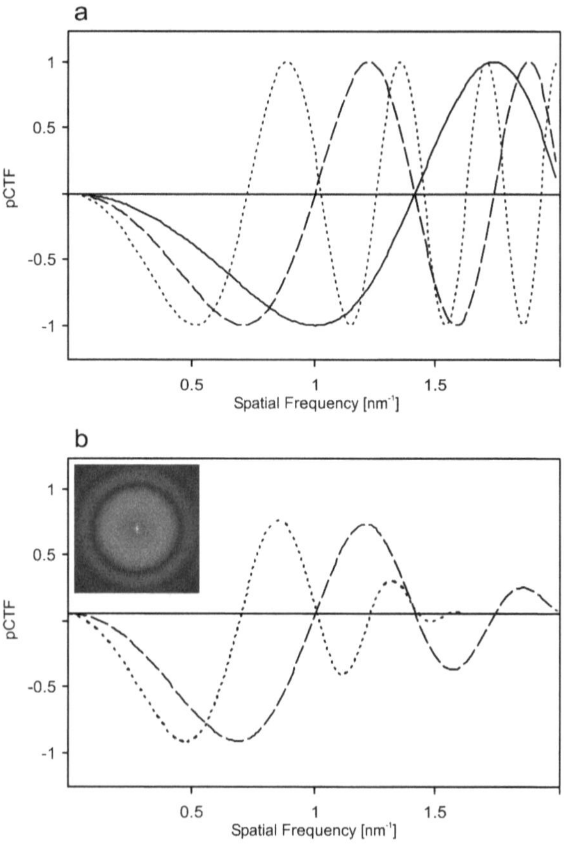


*Fig. 3.* The phase contrast transfer function. pCTFs calculated according to Zemlin (1999) for a spherical aberration of 1.2 mm. (a) Theoretical pCTFs calculated for images recorded at 100 kV acceleration voltage and 500 nm underfocus (dotted line), 300 kV acceleration voltage and 500 nm underfocus (dashed line), and 300 kV acceleration voltage and 250 nm underfocus (solid line). (b) As in (a), but additionally showing the effect of pCTF envelope functions. Inset shows a power spectrum calculated from an image of a carbon film showing Thon rings characteristic of an underfocus image. The zero-crossings of the pCTF correspond to the darker rings.

of the objective lens current and the spherical and chromatic aberrations of the lenses all contribute to the envelope function of the pCTF. Furthermore, the Thon-rings in diffraction patterns of recorded images can be distorted by astigmatism (the differential focus in orthogonal directions which gives rise to elliptic Thon-rings) or specimen drift (which leads to the fade out of the amplitudes with increasing frequency in the direction of the drift). In electron crystallography the effects of the pCTF and its envelope function, associated with the use of the electron microscope in imaging mode, can be avoided by the use of diffraction mode, which thus allows the highly accurate recording of the intensities of the diffraction spots arising from the crystal lattice.

### *D. Sample Preparation*

The prerequisite for the preparation of protein samples for electron microscopy is the preservation of the protein native structure under the high vacuum in the microscope. Two methods of sample preparation are used: negative staining and flash freezing into vitreous ice (Dubochet et al, 1988). For in-depth discussion of the different preparation techniques for cryo-electron microscopy see Walz and Grigorieff (1998).

Negative staining is mainly used for screening purposes as it is easy to perform and negatively stained samples have enhanced amplitude contrast due to the electron scattering by the heavy metals used as stain. Moreover negatively stained samples are less sensitive to radiation damage. The most commonly used stain is uranyl acetate. For the staining a droplet of sample is applied to a carbon coated copper grid, which can be previously glow discharged in order to render the carbon film more hydrophilic thus enhancing its interaction with the sample and facilitating the even distribution of stain (Aebi et al., 1987). After a brief incubation of the sample on the carbon the staining solution is applied, surplus liquid is removed by gentle blotting with filter paper and the grid is allowed to dry in the air before being mounted in the electron microscope. As a result of this procedure stain-accessible features on the protein structure are filled, producing a negative imprint of the protein which is imaged in the electron microscope. Because not all details of the protein structure are accessible by the stain the resolution limit for the structural analysis of images from negatively stained samples is about 15 Å. Furthermore, although this method has the advantage of high contrast, artifacts can arise. For example, with 2D crystals it is quite commonly observed that the crystal surface in contact with the carbon film will stain differently from the exposed surface. This phenomenon is most readily detected in 2D crystals where molecules are packed in the unit cell with alternating orientations with respect to the plane of the membrane (Fig 1b): under these circumstances molecular images with opposite orientations will show systematic departure from the mirror image relationship which should arise if they are perfectly contrasted. A second type of distortion is the flattening effect commonly observed in negative stained samples (Berriman et al., 1986) which impairs the use of tilted data for 3D analysis (see Section III.E.2).

Electron microscopy under cryogenic conditions is used for high resolution studies of biological samples, since these are maintained in an aqueous environment and the data are derived from the protein itself rather than a heavy metal stain shell. Specimens prepared in vitreous ice are more difficult to handle and show little contrast when imaged in the electron microscope, however, the protein structure is optimally preserved and the potential for high resolution analysis is greatly enhanced. Vitreous ice, as opposed to crystalline ice, yields an amorphous background for imaging. It can be formed by rapid cooling (at about $10^5$ °C $s^{-1}$) of aqueous solutions to temperatures of −140° C or lower, without allowing time for the rearrangement of water molecules into crystalline arrays, a method originally introduced for virus particles (Adrian et al., 1984). In practice, this rapid cooling is achieved by plunging the electron microscope grid with the sample applied into a cold liquid (below −160° C). For this purpose liquid ethane or propane are used, rather than liquid nitrogen, since they do not form an insulating layer of vapor around the sample during freezing. The fact that the vapor pressure of water becomes negligible at these temperatures avoids drying of the samples and thus the associated artifacts. One essential point when preparing samples for cryo-electron microscopy is the thickness of the layer of ice created, which should be thin enough to allow for little background but thick enough to cover the proteins completely. All handling and storage of the grids prepared has to be carried under liquid nitrogen or in a dry low temperature environment in order to avoid devitrification, sublimation and condensation of atmospheric water on the sample.

Attention should be given to the fact that the cryogenic temperatures also affect the grid itself: copper grids tend to decrease the flatness of the sample due to differences in the thermal expansion coefficient between the carbon support film and copper. To reduce this 'cryo-crinkling' effect (Booy and Pawley, 1993), molybdenum grids have been used. Furthermore, in order to reduce beam induced movement and charging of the sample a sandwich technique has successfully been applied, i.e., a carbon film on top as well as underneath the sample (Gyobu et al., 2004).

For the analysis of some 2D crystals the addition to the sample, prior to freezing, of small molecules acting as cryo-protectants, such as tannin (e.g., for LHCII crystals, Wang and Kühlbrandt, 1991; Kühlbrandt et al., 1994), trehalose or glucose (e.g., for bacteriorhodopsin, Unwin and Henderson, 1975), has improved results in preserving high resolution structural information.

### *E. Data Acquisition*

#### *1. Screening of Crystallization Trials*

Electron crystallography requires the preparation of a large number of crystallization trials which have to be screened by electron microscopy for the formation of areas of highly ordered protein crystals eventually suitable for high resolution studies. Usually screening is carried out on negatively stained samples. The search for the formation of appropriate lipid layers and crystal lattices can be performed directly in image mode. Here CCD (charge coupled device) cameras can be very useful because they allow for the immediate calculation of Fourier transforms of images of any crystalline area, allowing an immediate evaluation of the quality of the diffraction spots (sharpness and brightness) and the maximal order of diffraction. When samples with crystals suitably large and ordered for high resolution analysis are found data acquisition requires the use of cryogenic conditions. Ideally these samples should contain a high proportion of good crystals in order to increase the chances of successful identification in the low magnification screening mode of low dose microscopy (see Section III.E.2), where crystallinity can only be evaluated after image recording.

#### *2. High Resolution Data Acquisition*

While the screening of negatively stained crystallization trials can be performed in relatively simple entry-level electron microscopes, more sophisticated instruments are usually used for cryo-electron microscopy and high resolution data collection. Ideally, the electron source should be a field-emission gun since it provides a beam with higher spatial and temporal coherence, leading to pCTF envelope functions with better preservation of high frequency information, than conventional sources. The stability of the grid holder is of particular importance as minimal specimen drift should occur whilst recording images. The sample is maintained in the microscope at low temperature, either close to liquid nitrogen or close to liquid helium temperatures. Moreover low dose procedures (Williams and Fisher, 1970) are required in order to minimize the effects of radiation damage (Section III.C). Typically for low dose microscopy the sample is screened at low magnification (4000–8000×) in order to identify regions likely to contain crystalline lattices. The low magnification means that the intensity of the illuminating beam can be set at a value ~100× lower than when recording images. The low beam intensity combined with minimizing the size of the illuminated area should ensure that during screening the electron dose is no more than a few percent of that when images of the crystal are recorded. Focusing, which is inevitably associated with much higher electron doses, is carried out on an adjacent region of the grid, using a deflector system to introduce appropriate displacements of the region illuminated by the beam and the region from which an image is to be recorded. Finally an image is recorded from the area of interest restricting the electron dose typically to 10–20 electrons per Å$^2$.

The restricted exposure to the electron beam has the effect of making the resultant images intrinsically noisy. In order to extract high or intermediate resolution structural information from such images it is necessary to boost the signal-to-noise ratio. This can be achieved by combining information from a large number of molecular images. The minimum number of molecular images required for high resolution analysis (~0.3 nm) has been considered (Henderson, 1995; Glaeser, 1999), leading to estimates of between 1,400 (for ideal images) and 1,000,000 (corresponding to actual experimental practice). While the 3D crystals analyzed by X-ray crystallography have an optimal size of approximately 0.2 to 0.4 mm in any of their three dimensions, 2D crystals analyzed by electron crystallography are formed by a single layer of protein normally extending to no more than 3 μm in their two dimensions. Consequently, inherent to the analysis of 3D crystals by X-ray crystallography is the averaging of information provided by a much larger number of unit cells than those available in a 2D crystal. Therefore electron crystallography requires the merging of information obtained from different crystals sharing the same lattice parameters.

During image recording beam induced specimen movements can be reduced by spotscan procedures (Henderson and Glaeser, 1985; Bullough et al., 1991), in which small regions of the sample from which an image is to be recorded are sequentially exposed, instead of the conventional method in which the entire desired field is exposed with a single flat beam. A high sensitivity camera system (video or CCD) is almost invariably used for searching, focusing and alignment procedures. Unfortunately, the number of pixels in a CCD image remains significantly lower than can be achieved by recording on photographic film and digitization with a high quality densitometer. However, although most high resolution data acquisition is still carried out on photographic films, recently introduced large format CCD cameras have proved very effective for collection of electron diffraction data (see, for example, Gonen et al., 2004).

The choice of magnification for image recording on photographic films depends on the target resolution and the step size with which the film is digitized: for example, if the film is digitized with an effective step size of 7 μm, the minimum magnification required for a target resolution of 0.3 nm is 46,667×. The necessity to limit the electron dose to 10 to 20 electron/$Å^2$ to avoid radiation damage (see Section III.C) gives rise to a competing constraint: at higher magnifications the film is subjected to lower exposure and therefore, due to the limited film sensitivity, images have increased noise. In practice, magnifications of ~ 50–70,000× are commonly adopted as a useful compromise when recording images on film. The optimal focal setting is also very important for high resolution analysis: some defocus is necessary to generate contrast, but working closer to focus leads to a more favorable pCTF envelope function. In practice, given current instrumentation, a setting which places the first zero crossing of the pCTF between 7–10 Å is the most effective for high resolution electron crystallography studies.

The images recorded by electron microscopy approximate to projections of the object densities in the direction of the electron beam, as described above (Section III.B). Consequently, 3D reconstructions require the recording of data of the object from as many directions in the 3D space as possible. In the case of the analyzes of 2D crystals, which lie on the plane of the carbon film, to record data from directions other than normal to the support film it is necessary to tilt the specimen in the microscope (one exception to this is the analyzes of narrow tubular 2D crystals (Section II.A)). There are several considerations to account for when recording data from tilted samples, such as the flatness of the crystals (if the crystals are not perfectly flat resolution is lost at high tilt angles in the direction of the tilt (Glaeser et al., 1991)), and at high tilt angles the size and abundance of the crystals on the grid becomes more critical, since the proportion of grid that can be viewed is reduced with increased tilt angles due to the thickness of the grid bars holding the support film.

When crystals are large (at least 1 μm in each direction of the crystal plane) it becomes possible to record electron diffraction patterns from which accurate amplitudes for the diffraction spots can be measured (Section III.C). The amplitudes obtained directly from electron diffraction patterns are computationally merged with the phases of the corresponding diffraction spots determined by Fourier analysis of recorded images and the combined data are used to calculate an improved 3D structure.

## IV. Data Processing

### *A. Data Evaluation and Digitalization*

The first step in data analysis is the evaluation of the recorded images, normally by analysis of photographic films in an optical diffractometer. To this end, laser illumination and lenses are used to generate a diffraction pattern from regions of interest in the film. The 2D crystal images on the films selected for further processing should show sharp and intense spots to high resolution in all directions. Additionally, the Thon rings are normally assessed to check for appropriate focusing and to identify images that should be rejected on the basis of directional attenuations due to specimen drift during image recording.

Photographic films selected for further analysis are digitized using high resolution scanners with a step size of 10 μm or less. For example, 7 μm step size would result in a pixel size of 1 Å on the specimen level if a magnification of 70,000× was used. However, since the minimum resolvable distance is twice the pixel resolution, the films should be sampled at a spacing corresponding to at least half the resolution expected to be achieved. In practice the films are often scanned to give a pixel size corresponding to ¼ of the desired final resolution in order to allow for loss of resolution during interpolation steps in image processing.

## *B. Zero Tilt Data and the First Projection Map*

The program suite developed at the MRC in Cambridge (Crowther et al., 1996), including updates and extensions, is the most widely used software package for the data analysis in electron crystallography of protein samples. The procedures are described in detail by Baldwin and Henderson (1984) as well as in Ceska and Henderson (1990).

The first step of the analysis of digitized images is the determination of the crystal lattice parameters. For this a Fourier transform is calculated and the resulting diffraction spots in the power spectrum are indexed (Fig. 1i). The reciprocal lattice is defined in terms of two vectors a* and b* and the angle between them, $\gamma$. Each diffraction spot can be defined in terms of Miller indices, which are the number of translations of a* and b*, from the origin, required to reach that spot. The correct indexing yields the unit cell parameters, a*, b* and $\gamma$, which gives first hints about the crystal packing.

Using the information from the crystal parameters a first preliminary noise-reduced image can be calculated. This is achieved by defining a small radius around each diffraction spot and by setting all other areas, which contain only noise, to zero, followed by the calculation of a filtered real space image using an inverse Fourier transform. Further improvements on the first Fourier filtered images can be gained by unbending methods which are used to eliminate crystal distortions (for details, see Henderson et al., 1986). Distortions in the crystal lattice can be determined from a cross-correlation map between a filtered image and a selected reference area. This reference area should contain at least one unit cell, normally two, and be selected from the area of the crystal with the more ordered lattice, although it can also be generated from existing structure maps during refinement procedures (Kunji et al., 2000). The cross-correlation map can be used to interpolate points in the original image in order to produce a new corrected or unbent image. It is important to note that this procedure corrects only for translation distortions in the crystal lattice and cannot account for any rotational disorders or unit cell imperfections within the crystal. The procedures of crystal unbending are iterated until no further improvement in the quality of the calculated diffraction spots can be gained. After the unbending procedures the images are carefully corrected for the pCTF (see Section III.C).

The progress of the analysis is monitored on the basis of the quality of the diffraction spots. The amplitudes of the diffraction spots are compared to their surrounding noise and expressed in terms of integer quality (IQ) (Henderson et al., 1986). IQ values are defined on a scale 1 to 8, such that an IQ value of 1 corresponds to a signal-to-noise of 7:1, IQ 2 represents 3.5:1, and an IQ of 7 corresponds to a signal-to-noise ratio of 1, i.e., the noise level is as strong as the signal. These values can be plotted analogously to a power spectrum, with the IQ values replacing the diffraction spots. Figure 4a-c gives examples of the IQ-plots obtained during the different steps of analysis of a CP47-RC crystal and illustrates the improvement obtained by these procedures. Such plots are assessed both in terms of the IQ values and the completeness of high quality data.

The end result of the procedures of unbending and pCTF correction is a list of numbers (the index of diffraction spot and its amplitude and phase) which can be used to calculate a projection map (Fig. 4d) by Fourier synthesis (the generation of an image by inverse Fourier transform using the amplitudes and phases of selected diffraction spots). At this stage usually only spots with IQ values $\leq 4$ are used. It is important to note that an enhancement of the signal-to-noise of the diffraction data can be achieved by the identification of the plane group symmetry of the crystal (i.e., the symmetry operations imposed on the object in order to recreate the crystal unit cell) and by imposing the relevant symmetry operations.

Although these procedures lead to substantial improvement in the resolution and signal-to-noise ratio, the projection maps calculated from a single 2D crystal remain somewhat noisy and some of the structural information is missing due to the pCTF zeros. Further improvements in signal-to-noise ratio can be obtained by merging data from different crystals. This procedure has the additional advantage that if data are merged from images recorded at slightly different focus levels information lost due to the pCTF in one image will be recovered from other images. In order to merge information from different 2D crystals their structure factors have to be placed on a common phase origin (this corresponds to the alignment of the lattice of the image to be merged to

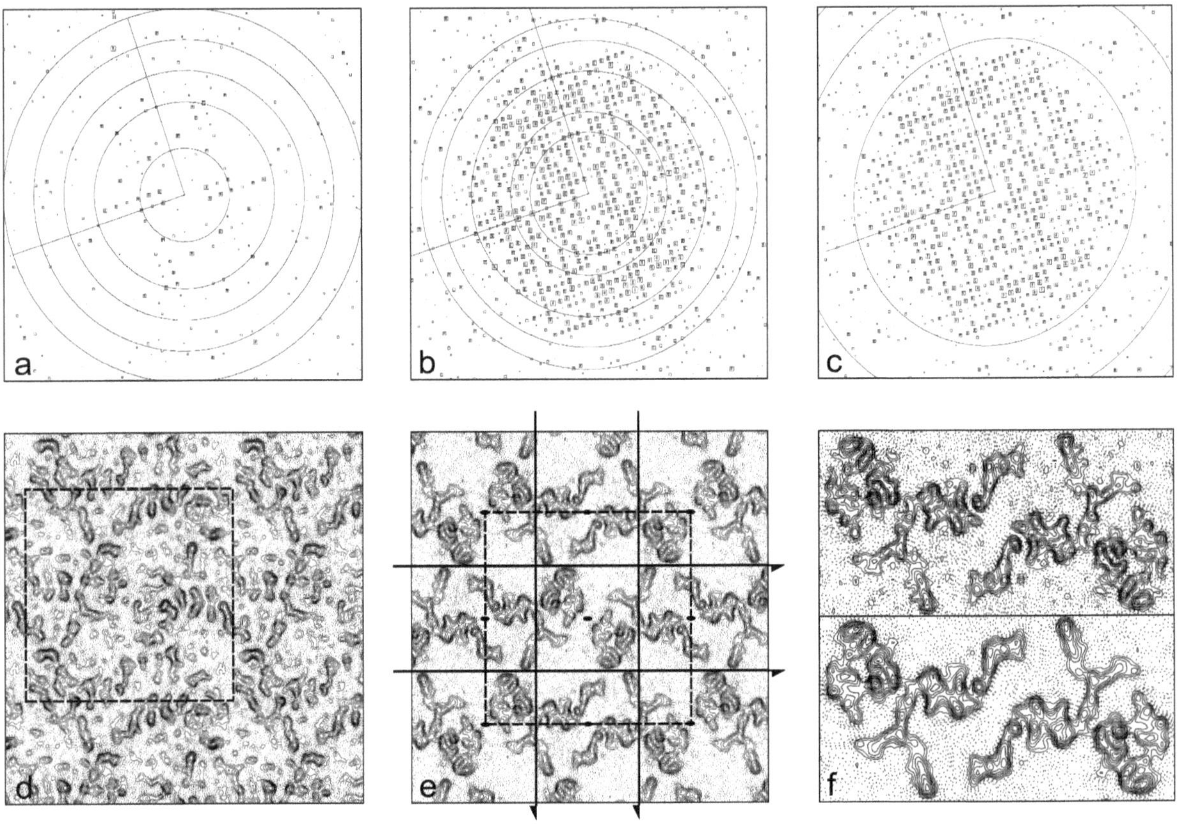


*Fig. 4.* Data analysis of a CP47RC-crystal. (a) IQ-plot of the Fourier components calculated from the original recorded image of a 2D crystal. Circles correspond to 40, 20, 15, 12 and 10Å, respectively. (b) IQ-plot calculated from the analysis of the same image after several rounds of unbending. Circles correspond to 20, 10, 8, 6 and 5Å, respectively. This plot shows that after processing of the crystal image the diffraction spots with good IQ values are isotropic and complete to a resolution of approximately 8Å. (c) Same as in (b) but, instead of circles of resolution, the pCTF zeros are marked. Note that these are slightly elliptical indicating astigmatism in the recorded image. (d) Projection map of the crystal calculated by Fourier synthesis of the data in (b), displaying $p22_12_1$ symmetry. A unit cell is marked by a dashed line. (e) Merged data from 12 projection maps displayed at a resolution of 5.5Å. Rotational and screw axes defined by the crystal symmetry are shown. One unit cell is marked by a dashed line. (f) Magnification of part of a unit cell as shown in (e). Upper panel displays the map drawn using a temperature factor of –150, the map of the lower panel was calculated without imposing any amplitude enhancement and is identical to that shown in (e).

that of the original image). The merging procedure provides an opportunity to evaluate the structural information. Phase statistics provide the best estimate of the quality of the merged data, with phase errors directly reflecting how well details of the structure agree in the independent data sets. Note, however, that there are several formulations for the presentation of phase errors, as discussed by Unger (2000).

If image derived amplitudes are used, i.e., if electron diffraction amplitudes are not available, they typically suffer from the attenuation of the pCTF at high spatial frequencies (Section III.C), leading to the higher resolution details of the structure being obscured. To compensate for this, an inverse temperature factor (B factor) can be used. Since this will boost the high resolution terms and associated noise (Fig. 4f), these methods should only be used in combination with a careful evaluation of the high resolution spots by their phase error.

### C. Data from Tilted Specimens and 3D Maps

Projection maps at zero tilt can already be useful for the interpretation of the overall organization of the molecules in the 2D crystal. However, a 3D structure is frequently the final aim and to this end data has to be recorded at different tilt angles in order to obtain data along the z axis (Section III.E.2). However, the maximum tilt angle which is practical to record data is limited to a value of between 60° and 70°.

Moreover, due to departures from flatness of the 2D crystals, the diffraction of tilted samples commonly results in a pattern with uneven distribution of spots, i.e., with more high resolution spots along the tilt axis than perpendicular to it (Glaeser et al., 1991). This implies that usually, for a certain tilt plane, the higher the tilt angle the larger the number of images that have to be recorded in order to obtain statistically reliable isotropic data.

The fact that data are not usually collected beyond a tilt angle of 60° to 70° results in the so-called 'missing cone,' i.e. a conical volume in reciprocal space which is not sampled. As a consequence, although the missing cone accounts for only 13% of the structural information if 60° tilt data are available (Glaeser et al., 1989), the resolution in electron crystallography is anisotropic, with the resolution in plane (along the crystal x and y axes) being higher than perpendicular to it (along z).

Images recorded from a tilted specimen are in principle analyzed as described above for zero tilt images. However, the apparent unit cell parameters change under the different tilt angles, which for higher values of tilt can be used to calculate the precise tilt angle and the location of the tilt axis (Shaw and Hills, 1981). After the procedures of unbending and pCTF correction (the latter requiring a careful determination of the tilt angles and axes) the phase origins are determined and the structure factors from the tilted data merged with those from zero tilt data, or that from an already existing preliminary 3D map. Here it should be noted that, while the diffraction pattern of a projection image (all intensities are projected in a 2D plane) of a 2D crystal is a lattice of spots (Section III.A), in reality since a 2D crystal of a protein is not a flat 2D surface, but has a non-regular component along the *z*-axis (the 'height' of the protein, or protein layers in a multi-layer crystal), its 3D diffraction pattern is a lattice of parallel lines, perpendicular to the plane of the crystal. Along these reciprocal lattice lines the amplitudes and phases fluctuate reflecting the real 3D density distribution. The lattice of spots obtained on the diffraction pattern of projection images of a 2D crystal, recorded at different tilt angles, are no more than slices (always crossing the coordinate origin) through these lines at the relevant angles. Consequently, when structural factors from tilted data are merged, the distributions of the recorded amplitudes and phases along each lattice line are fitted by a least-squares method and the resulting amplitude and phase functions are taken as the reciprocal lattice lines of the 3D volume. Here, for each line the statistics of the phase fitting are the best measure for data quality (Unger, 2000). Fourier synthesis is used to generate a 3D map by sampling the calculated lattice lines with a spacing of at least 1/T (where T is the estimated specimen thickness).

### *D. Data Evaluation and Limitations of the Method*

#### *1. Resolution of the 3D Map*

A final evaluation of the effective resolution limit of a 3D map has to be carried out in order to establish to what frequencies data can safely be interpreted. Two factors have to be considered when determining resolution: the reliability of the (merged) data and its completeness. Usually shells around the coordinate origin of the reciprocal lattice, corresponding to different bands of resolution, are considered. The highest frequency shell where significant data are available corresponds to the resolution limit of the 3D map.

As noted in Section IV.C, data from tilted images is frequently anisotropic with better resolution along the tilt axis, which results in lack of completeness in the data. For this reason Cheng and Yeager (2004) proposed a method for tracking the progress of 3D data accumulation in which tilted images are individually assessed for their contribution to the analysis. The diffraction spots at a ±45° wedge from the tilting direction, where the data is weaker, are scored for quality and completeness. By taking the contributions from all merged data, this method allows for the determination of the tilt angles and axes for which information is still required in order to improve the quality of the 3D analysis.

However, there is still no universally accepted resolution criterion for the 3D structural analysis by electron crystallography. This applies in particular to the estimation of the resolution perpendicular to the crystal plane, where data are intrinsically missing due to the 'missing cone.' No fixed relation exists between the resolution in plane and perpendicular to it, as the latter only depends on the quality of the tilted data, i.e., the quality of the information along a single tilt angle and the maximum effective tilt angle obtained. Thus, the vertical resolution of a 3D map at 6 Å resolution in the *x-y* plane can easily vary from ~7 Å up to 25 Å (Unger, 2000). Methods to evaluate vertical resolution are discussed in detail by Unger (2000).

### 2. Interpretation of Structure Maps (2D and 3D)

Although the structural information in 2D projection maps of proteins is incomplete, their interpretation can render valuable information. In 2D projections, even at intermediate resolution, the crystal symmetry can be determined and within the unit cell the symmetry of the protein or protein complex, its dimensions along the 2D crystal plane and its oligomeric state can be estimated with great accuracy. It is also often possible to infer the location of α-helices running perpendicular, or near perpendicular, to the crystal plane, especially if these correspond to transmembrane helices in complexes which are predicted to have low extrinsic densities. Examples of the direct identification of α-helices in intermediate resolution projection maps can be found in the analysis of the bacterial reaction centers with its antenna complex (Karrasch et al., 1995; Qian et al., 2005) and are described in Section V.A.

Another fruitful approach is the comparison of projection maps with data available from other sources. These may be structures at atomic resolution of homologous proteins or subunits or fragments of the system being studied. For this purpose projection maps can be simulated to match the available resolution of the current electron crystallographic analysis. Alternatively the high resolution coordinates can be simply superimposed into the projection map. This approach was fundamental in the development of the understanding of the structure of PS II (Hankamer et al, 1999; da Fonseca et al., 2002), as described in Section V.D.

Insights into the biological functioning of a protein can be attained by the comparison of projection maps obtained from 2D crystals of a protein in different conformational states induced by different physiologic conditions, such as the maps calculated from 2D crystals obtained in the presence and absence of ligands or comparative studies of wild-type and mutant proteins. Sometimes the functional rearrangements of the protein may be too subtle to be directly visualized. In such cases the calculation of difference maps can be used to reveal these effects. However, for difference analysis to be meaningful it is important that the unit cell parameters of both crystals are maintained, or that internal references are available (like helix location) for an accurate alignment.

Ultimately the objective of electron crystallography is to achieve 3D structural data at the highest possible resolutions, aiming at the fitting of atomic models into the calculated densities and the interpretation of the resulting atomic structures in terms of the intricate mechanisms of protein function. Secondary structure elements like α-helices can be fitted rather accurately at intermediate resolutions of ≤8 Å. At this level of resolution bulky cofactors, like chlorophylls, also become visible, although the precise orientation from which detailed information such as the transition dipole moment can only be established at higher resolution.

When resolutions of about 3–4 Å or higher are obtained, the tracing of the amino acid chain can be made using the same procedures as in the modeling of X-ray data, i.e., densities of individual side chains can be discriminated and the protein sequence can be fitted to the densities obtained. This procedure has been followed in the analysis of the structure of LHCII (Kühlbrandt et al., 1994), as presented in Section V.B.

## V. Examples in Photosynthesis

Electron microscopy, particularly electron crystallography, has been crucial for the understanding of the structural organization of the photosynthetic apparatus. Described here is how electron crystallography has been used for the study of photosynthetic pigmented membrane proteins and complexes, namely reaction center cores and intrinsic antenna proteins.

### A. Photosynthetic Bacterial Proteins (Reaction centers, LH1 and LH2)

Electron microscopy studies have shown that photosynthetic membranes of a number of purple bacteria like *Rhodopseudomonas* (*Rh.*) *viridis* contain crystalline arrays made up of the photosynthetic reaction center (RC) and the surrounding antenna (LH1) (Miller, 1982; Stark et al., 1984). The structure of the reaction center originally determined by X-ray crystallography in the pioneering work of Deisenhofer et al. (1985) is now known to better than 2 Å resolution (Fritzsch et al., 2002), but the structure of the LH1 antenna and their complexes with the RC has remained a fruitful area of electron crystallography. In electron crystallography studies, 2D crystals of purified and reconstituted LH1 were used to obtain an 8.5 Å projection map showing a 16-fold ring of subunits. The densities which correspond to the inner ($\alpha$) and outer ($\beta$) helices were clearly resolved along with additional densities assigned to bacteriochlorophylls (Karrasch et al., 1995).

The arrangement in which the reaction center is surrounded by a ring of LH1 antenna protein subunits was originally proposed on the basis of the appearance of native crystalline arrays of *Rh. viridis* (Miller, 1982; Stark et al., 1984). These RC-LH1 complexes have been the subject of a number of electron crystallographic studies. 2D crystals of reconstituted RC-LH1 complexes from *Rh. viridis*, *Rhodospirillum* (*R.*) *rubrum* and *Rhodobacter* (*Rb.*) *sphaeroides* gave rise to projection maps with a central density consistent with a single RC complex surrounded by a ring of densities assigned to the LH1 antenna (Walz and Gosh, 1997; Ikeda-Yamsaki et al., 1998; Stahlberg et al., 1998; Walz et al., 1998). More recently, Jamieson et al. (2002) succeeded in resolving 16 individual LH1 subunits and their component helices in the antenna ring as well as a number of densities consistent with eight of the transmembrane helices of the L and M subunits of the RC in crystals from *R. rubrum*.

However, native arrays, obtained from *Rb. sphaeroides*, have given a somewhat different picture of the distribution of LH1 subunits around the RC. Electron crystallography revealed a dimeric organization of the RC-LH1 complex in an S-shaped elongated supercomplex (Jungas et al., 1999; Siebert et al., 2004). The dimerization was attributed to the presence of the PufX subunit in the complexes and discussed in terms of electron transport (Siebert et al., 2004).

Recently, a 4.8 Å resolution map has been determined by X-ray crystallography of the monomeric RC-LH1 complex of *Rh. palustris* (Roszak et al., 2003), showing one reaction center and an open ring of 15 $\alpha/\beta$ subunits together with one additional single helix. The additional helix has not been identified but may correspond to a PufX-orthologue (Fig. 5a). The overall arrangement in the X-ray structure is on the one hand consistent with the earlier work by Jamieson et al. (2002) by electron crystallography, but on the other hand reveals an open ring first seen in the dimeric complexes from native membranes. However, in the early maps of the RC-LH1 dimers (Jungas et al., 1999; Siebert et al., 2004) resolution

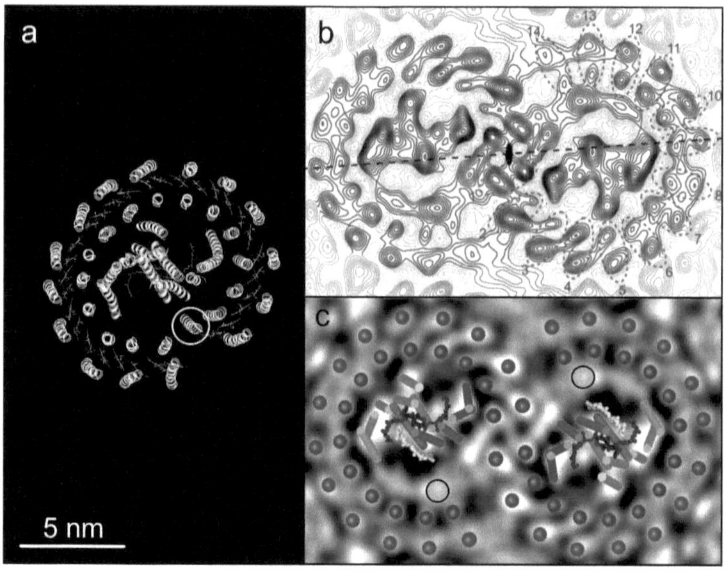


Fig. 5. Structural studies of the bacterial reaction center and its LH1 antenna proteins. (a) X-ray crystallographic structure of the monomeric RC-LH1 complex from *Rps. palustris* (Roszak et al., 2003) viewed from the cytoplasmic side of the membrane. The image was created using the pdb-file from the Protein Data Bank with the accession number 1PHY using the program Molscript (Kraulis, 1991) and shows the ring of inner $\alpha$- and outer ß-subunits of LH1 as well as the transmembrane helices of the RC proteins, represented as ribbons, and cofactors, drawn as sticks. A circle indicates the helix attributed to a possible PufX orthologue. (b) A contour map of the averaged projected densities of a 2D crystal of RC-LH1-PufX of *Rb. sphaeroides* (Qian et al., 2005). (c) A schematic representations of the components of the complex shown in (b) on the same scale as (a) viewed from the cytoplasmic side of the membrane. The $\alpha$ and $\beta$ subunits of LH1 are marked by spheres, the helices of the reaction center are drawn as cylinders. Two chlorophylls are depicted as well. The favored locations of the PufX protein are indicated by circles (b & c: courtesy of Professor Per Bullough).

was too low to identify individual LH1 subunits in the S-shaped outer complex or to allow unambiguous identification of the RC. An electron crystallographic 8.5 Å projection structure of a dimeric RC-LH1-PufX complex from *Rb. sphaeroides* recently published by Qian et al. (2005) for the first time revealed the 28 LH1 subunits in the outer complex and the orientation of the two RC complexes within the dimer (Fig. 5b and c). The crystals used for this analysis were formed by reconstitution from purified RC-LH1-PufX complexes. The observed arrangement is quite similar to the crystal structure of the monomeric RC-LH1 complex of *Rh. palustris* (Fig. 5a), although Qian et al. (2005) draw attention to the displacement of LH1 subunits 6 to 14 to higher radius in their structure giving each half of the dimer a more open outer ring than in the monomeric complex.

The outer antenna of the bacterial photosynthetic membranes, LH2, was intensively studied by electron crystallography (Montoya et al., 1995; Oling et al., 1996; Savage et al., 1996; Walz et al., 1998; Ranck et al., 2001). This might be surprising as the first high-resolution structure of LH2 determined by X-ray crystallography was already published in 1995 for *Rh. acidophila* (McDermott et al., 1995) followed by that for *R. molischianum* (Koepke et al., 1996). Nevertheless, electron crystallography turned out to be especially useful in determining the oligomeric state of the LH2 rings, which differ according to the species analyzed.

### *B. Antenna Proteins in Higher Plants and Algae (LHC)*

The two photosystems in eukaryotes are composed of many more subunits than bacterial photosystems. The core complexes consist of the RCs, the inner antenna proteins (CP47 and CP43 in PS II, psaA and psaB that also function as RC proteins in PS I) and a range of small subunits. These cores are surrounded by a variety of outer antenna proteins, the LHC proteins.

One of the higher resolution 3D structures determined by electron crystallography is that of the main antenna complex of PS II in higher plants, LHCII. This 3D structure, at 3.4 Å, was published in 1994 (Kühlbrandt et al., 1994) and revealed the organization of the LHCII trimer together with the detection, in each monomer, of three transmembrane and a short amphiphilic α-helix, 12 out of 14 Chl and 2 out of 4 carotenoids (Fig. 6a). In this study the distinction between Chl *a* and Chl *b* was done on the basis of energetic considerations since a methyl and a formyl group could not be distinguished at this resolution. Recent X-ray crystallographic structures confirmed these findings (Fig. 6b and c), albeit at higher resolution and thus providing more details including the location of all pigments (Liu et al., 2004; Standfuss et al., 2005). It is interesting to note that one of these structures was solved using phases derived from the electron crystallographic data (Standfuss et al., 2005). Up to now LHCII is the only intrinsic antenna complex which has been solved to atomic resolution.

### *C. Photosystem I in Higher Plants and Cyanobacteria*

Analysis of the structure of Photosystem I (PS I) has for the most part focused on the study of complexes from cyanobacteria, namely *Synechococcus* (today: *Thermosynechococcus*). Structural analysis by electron crystallography and X-ray diffraction proceeded in parallel in the early stages. Böttcher et al. (1992) published the first 3D structure of *Synechococcus* PS I calculated from negatively stained 2D crystals at relatively low resolution (15–18 Å). Karrasch et al. (1996) published a second, broadly similar 3D analysis alongside a somewhat higher resolution cryo-electron crystallography projection map. In each case the crystals were made up of PS I monomers, rather than the trimeric form characteristic of native PS I in cyanobacteria. Electron crystallography has often been successfully used to determine the correct quaternary structure of protein complexes, but there are a number of examples where monomers were crystallized in 2D, especially when using purified protein samples and reconstitution methods, where monomerization may be the consequence of the purification procedures used (see Section V.A). These electron crystallographic studies of cyanobacterial PS I although of limited resolution yielded structures which were broadly consistent with those calculated by X-ray crystallography determined initially at 6 Å by Krauss et al. (1993) and subsequently at 2.5 Å by Jordan et al. (2001).

The structure of higher plant PS I was rather rarely studied, maybe because it is less stable than that from thermophilic cyanobacteria and its purification is more complex due to the less favorable PS II/PS I ratio in eukaryotes. A structural analysis of a higher plant PS I complex by electron crystallography has been reported (Kitmitto et al., 1998). In 2003, an

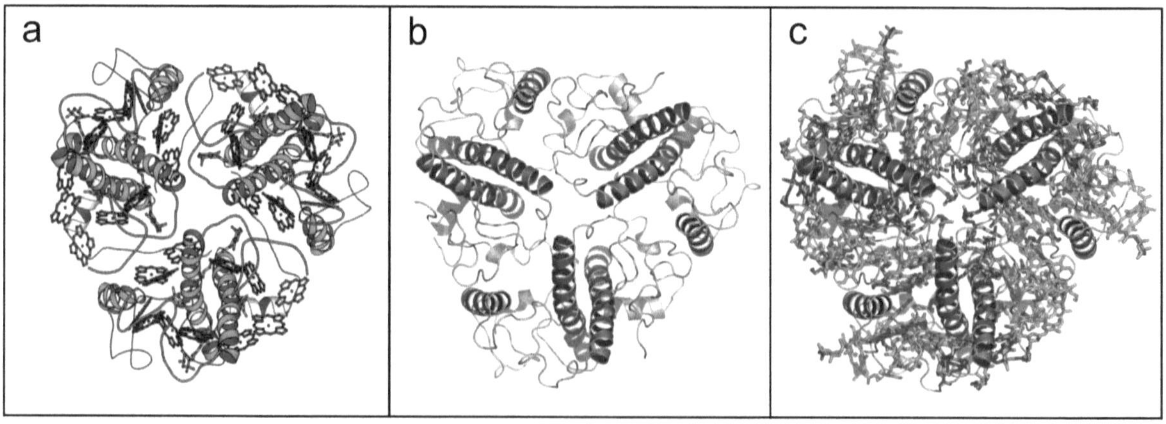


*Fig. 6.* Comparison of the structure of LHC2 as determined by electron crystallography (a) and X-ray crystallography (b and c). For clarity, all pigments were omitted in (b). (Figures courtesy of Professor W. Kühlbrandt). See also Fig. 2, Color Plate 3.

X-ray crystallographic analysis at 4.4 Å of the plant PS I-LHCI complex revealed the arrangement of the LHCI on one side of the PS I monomers (Ben-Shem et al., 2003).

### *D. Photosystem II in Higher Plants and Cyanobacteria*

Electron crystallography, single particle approaches, X-ray and spectrometric methods have all been used to elucidate the structures of PS II cores and super-complexes. A recent review (Bumba and Vácha, 2003) describes the efforts using electron microscopy based methods, which will be summarized here. It is interesting to note that attempts to produce PS II 3D crystals for X-ray studies almost exclusively used cyanobacterial complexes, whereas studies using electron crystallography were mainly done on preparations from higher plants. The latter approaches used either native crystalline arrays of PS II or 2D crystals from in situ procedures.

The first 2D crystals from whole PS II complexes were reported by Bassi et al. (1989), using delipidated granal membranes, and were shortly followed by Dekker et al. (1990), reporting crystals of purified CP47-RC complexes reconstituted into lipid membranes. These crystals of CP47-RC complexes were used to calculate, from negatively stained specimens, a projection map at 25 Å resolution revealing a monomeric complex. This work was followed by a wealth of analyzes of different 2D crystals obtained of the whole PS II core complexes, mostly done using data collected from negatively stained samples (Lyon et al., 1993; Santini et al., 1994; Marr et al., 1996a,b; Tsiotis et al., 1996; Morris et al., 1997). In most studies, a dimeric nature of the PS II core was elucidated, but the debate about whether PS II is a monomer or a dimer in vivo was fuelled by a map derived from crystals of whole PS II complexes by Holzenburg et al. (1993) and later publications from the same group. Features were interpreted as PS II monomers despite the symmetry properties and sizes being identical to dimers as nicely demonstrated by Bumba and Vácha (2003).

The main achievement from the electron crystallography studies on higher plant PS II has been the elucidation of the overall organization of subunits within PS II complexes. 2D crystals of CP47-RC subcomplexes from spinach have been analyzed by cryo-electron microscopy first by Nakazato et al. (1996) and later on by Rhee and co-workers. Their analyzes, first in projection (Rhee et al., 1997) and then in 3D at 8 Å (Rhee et al., 1998) is still the highest resolution 3D structure of a plant PS II complex available (Fig. 7b). A total of 23 transmembrane helices were identified in each monomeric CP47-RC complex (Rhee et al, 1998). Among those it was realized that two groups of five helices form, in projection, an S-shaped feature with local near two-fold symmetry. These were assigned to the RC proteins D1 and D2 based on their amino acid sequence and predicted structural homology to the M and L subunits of the bacterial RC. An adjacent group of three pairs of helices was assigned to CP47, which was previously predicted to have six transmembrane segments and to be located close to the D2 protein. This arrangement of the D1, D2 and CP47 was supported by the structural similarity to the eleven helices of PsaA and

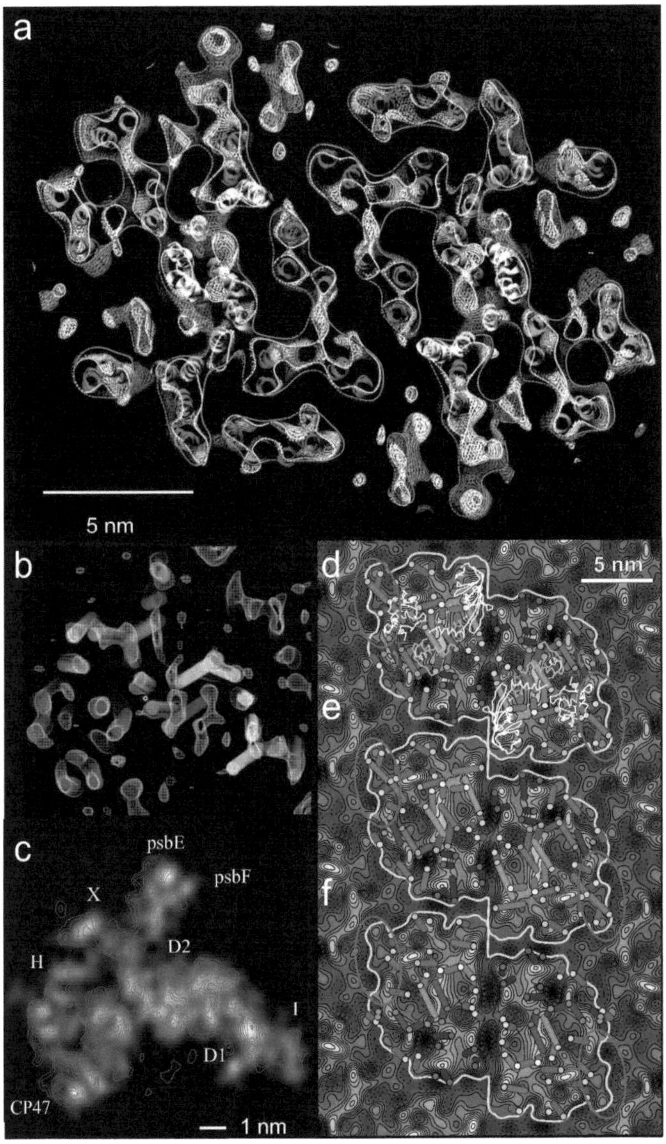


Fig. 7. Electron crystallographic studies of PS II. (a) Section of the 10Å 3D electron crystallographic structure of spinach PS II dimeric core complex (Hankamer et al., 2001) viewed from the lumen. Protein density is shown as chicken-wire and fitted transmembrane helices as ribbons. (b) Lumenal view of the 8Å 3D electron crystallographic structure of the spinach CP47-RC PS II subcomplex (Rhee et al., 1998). Protein density is shown as chicken wire and fitted transmembrane helices as cylinders. (c) 5.5Å electron crystallographic projection structure of the spinach CP47-RC PS II subcomplex contours (Büchel and Kühlbrandt, 2005) compared with a simulated projection map at the same resolution calculated from the relevant subunits in the X-ray crystallographic structure of cyanobacterial PS II (Ferreira et al., 2004). (d-f) Comparison of the structural data on the PS II core dimer from *Synechococcus elongatus* and spinach (da Fonseca et al., 2002, with permission). The electron crystallographic 16Å projection structure of PS II from *Synechococcus elongatus*, represented as contoured grayscale, is overlaid with: (d) the full X-ray crystallography structure and (e) the transmembrane helices of *Synechococcus elongatus* (Zouni et al., 2001); (f) the transmembrane helices of the spinach core dimer deduced from electron crystallography (Hankamer et al., 2001). See also Color Plate 5.

PsaB proteins of PS I resolved by X-ray crystallography. In the CP47-RC complex it was also possible to assign six tetrapyrrole head groups of chlorophylls or pheophytins to six densities located within the ten helices of the D1/D2 heterodimer, which revealed that these are located too far apart to form a 'special pair' of chlorophylls similar to that found in bacterial RC. Another 14 densities between the six helices of CP47 were also assigned to chlorophylls. More recently, Hankamer et al. (1999) reported a projection map with a resolution of about 9 Å for dimeric PS II cores from spinach and a 3D map of the same complex, with an in-plane resolution of 10 Å (Fig. 7a) (Hankamer et al., 2001). These analyzes resulted in the assignment of the helices of the CP47-RC together with the six helices of CP43 and the identification of 11 further unassigned transmembrane helices (Hankamer et al., 1999; 2001). A proposal of the location of the Cyt *b*559 was put forward and the densities corresponding to the large extrinsic lumenal loops of CP43 and CP47 were visualized.

At about the same time, structures of the core of cyanobacterial PS II became available by X-ray crystallography at high resolution, first at 3.8 Å (Zouni et al., 2001) and later on at 3.7 Å (Kamiya and Shen, 2003). Lately, structures at 3.5 Å (Ferreira et al., 2004) and 3.2 Å (Biesiadka et al., 2004; Kern et al., 2005, Zouni et al.; 2005) revealed details on the cofactors, like $Q_B$, carotenoids in the reaction center and the Mn-cluster. The higher resolution also allowed the assignment of small subunits of the PS II core. These X-ray crystallographic structures confirmed the overall features previously identified in the 3D maps from electron crystallography.

The only electron crystallography study on cyanobacterial core complexes resulted in a 2D map at a resolution of 16 Å which allowed a comparison to the maps derived from spinach and highlighted differences in the outer part of the core due to different subunit composition between higher plants and cyanobacteria (da Fonseca et al., 2002) (Fig. 7 d–f). Lately, a projection map of the CP47-RC complex at 5.5 Å was published and at this resolution first differences in the inner part of PS II between higher plants and cyanobacteria could be detected (Büchel and Kühlbrandt, 2005) (Fig. 7c).

The interaction of the PS II core and the surrounding LHC proteins have been mainly analyzed using single particle approaches and is thus outside the scope of this review. However, the reader is referred to two recent reviews, one about PS II by Bumba and Vácha (2003) as mentioned above and the other about the supramolecular organization of all proteins of the thylakoid membrane by Dekker and Boekema (2005).

## VI. Outlook

### A. Advantages of Electron Crystallography

Electron crystallography possesses a number of advantages over other methods of structure determination for membrane proteins. Membrane proteins in 2D crystals are retained in the lipid bilayer close to their native environment thus reducing the likelihood for non-native conformations to be adopted. Additionally, the retention of the lipid bilayer provides the potential to observe the native interaction between lipid and protein. In comparison, both X-ray crystallography and single particle analysis of cryo-electron microscope images involve the analysis of detergent solubilized protein. It should be noted, however, that electron crystallography is frequently time consuming and may involve years of work in order to achieve the highest resolution. In this respect it compares to X-ray crystallography of membrane proteins and suffers from the same problem, i.e. the difficulty of obtaining good crystals. When single particles are analyzed by cryo-electron microscopy the procedure is usually faster, providing the sample is suitable. However, to date the ultimate resolution obtained using single particle approaches is still insufficient to solve structures at atomic resolution, whereas this has been achieved by electron crystallography.

### B. Future Technical Developments

As noted above, a major bottleneck in electron crystallography is obtaining suitable 2D crystals for analysis. Finding the appropriate crystallization conditions involves the evaluation of a large number of crystallization trials by electron microscopy. The time required to evaluate each trial is much greater than, for example, a 3D protein crystallography trial. A number of laboratories, therefore, have started to develop procedures for automating the steps involved in the assessment of crystallization trials in the electron microscope. Potter et al. (2004), for example, have created a robotic grid loading system that allows for the unattended loading of trays of 96 grids in order to enable automated screening of 2D crystal and helical crystal formation.

In addition to developments in automation, technical improvements in electron microscopes and ancillary equipment continue to play an important role in the development of electron crystallography. The highest resolution electron crystallographic studies of recent years have nearly all involved data collection with 300 kV electron microscopes with field emission guns and liquid helium cooled stages. The higher beam coherence and specimen stability of these instruments contribute to images with better signal-to-noise ratio at high resolution. It is likely that refinement of these microscopes will lead to improved performance. An area with potential for considerable improvement is the generation of contrast. As illustrated in Fig. 3b, the use of underfocus to generate contrast comes at the expense of an envelope function leading to the fade out of the amplitudes at increasing resolution. This can be minimized by working as close to focus as possible, but substantial fade is still generated at all practical levels of underfocus. The use of a phase plate has the potential to generate contrast with little or no underfocus, thereby reducing amplitude fade out. This approach is currently under development using carbon films to introduce phase shifts (Hosokawa et al., 2005) or by tuning spherical aberration correctors (Lentzen, 2004).

The use of electronic detector systems to record images for electron crystallography would allow faster recording of data contributing to a streamlining of the data collection procedure and/or the selection of higher quality images for analysis from a larger pool. As noted in Section III.E.2 current CCD cameras compare unfavorably to photographic film in terms of the number of pixels available, which is particularly relevant to electron crystallography where large crystalline arrays are required for high resolution analysis. In future it is highly likely that CCD cameras will be available with more pixels. Further technical development is also likely to focus on electronic detector systems with improved point spread function.

These and other developments in cryo-electron microscope technology are substantially driven by the explosion in interest in single particle analysis over recent years. It is likely, however, that electron crystallography will both benefit from these developments and retain its advantages in the structural analysis of membrane proteins.

## Acknowledgments

The authors would like to acknowledge Professor J. Barber and Professor W. Kühlbrandt for their support and encouragement in the structural analysis of photosynthetic complexes over a number of years and Professor Agostinho R. da Fonseca for comments on the manuscript. CB gratefully acknowledges funding by an EU MCTRN grant (Interdisciplinary Training and Research Network on Photosystem II, Intro2) and in addition this work was supported by the DFG (SFB 472, 'Molecular Bioenergetics'). EPM gratefully acknowledges support from the Wellcome Trust and from Cancer Research UK.

## References

Adrian M, Dubochet J, Lepault J and McDowall AW (1984) Cryo-electron microscopy of viruses. Nature 308: 32–36

Aebi U and Pollard TD (1987) A glow discharge unit to render electron microscope grids and other surfaces hydrophilic. J Electron Microsc Tech 7: 29–33

Amelincks S, van Dyck D, van Landuyt J and van Tendeloo G (1997) Electron Microscopy. VCH Wiley, Weinheim

Auer M (2000) Three-dimensional electron cryo-microscopy as a powerful structural tool in molecular medicine. J Mol Med 78: 191–202

Auer M, Scarborough GA and Kühlbrandt W (1999) Surface crystallization of the plasma membrane $H^+$-ATPase on a carbon support film for electron crystallography. J Mol Biol 287: 961–968

Baldwin J and Henderson R (1984) Measurement and evaluation of electron diffraction patterns from two-dimensional crystals. Ultramicroscopy 14: 319–336

Baldwin JM, Henderson R, Beckmann E and Zemlin F (1988) Images of purple membrane at 2.8Å resolution obtained by cryo-electron microscopy. J Mol Biol 202: 585–591

Bassi R, Magaldi AG, Tognon G, Giacometti GM and Miller KR (1989) Two-dimensional crystals of the Photosystem II reaction center complex from higher plants. Eur J Cell Biol 50: 84–93

Ben-Shem A, Frolow F and Nelson N (2003) Crystal structure of plant Photosystem I. Nature 426: 630–635

Berriman J and Leonard KR (1986) Methods for specimen thickness determination in electron microscopy. II. Changes in thickness with dose. Ultramicroscopy 19: 349–366

Biesiadka J, Loll B, Kern J, Irrgang KD and Zouni A (2004) Crystal structure of cyanobacterial Photosystem II at 3.2Å resolution: A closer look at the Mn cluster. Phys Chem Chem Phys 6: 4733–4736

Booy FP and Pawley JB (1993) Cryo-crinkling: What happens to carbon films on copper grids at low temperature. Ultramicroscopy 56: 241–252

Böttcher B, Gräber P and Boekema EJ (1992) The structure of Photosystem I from the thermophilic cyanobacterium *Synechococcus* sp. determined by electron microscopy of two-dimensional crystals. Biochim Biophys Acta 1100: 125–136

Büchel C and Kühlbrandt W (2005) Structural differences in the inner part of Photosystem II between higher plants and cyanobacteria. Photosynth Res 85: 3–13

Büchel C, Morris E and Barber J (2000) Crystallization of CP43, a chlorophyll binding protein of Photosystem II: An electron microscopy analysis of molecular packing. J Struct Biol 131: 181–186

Büchel C, Morris E, Orlova E and Barber J (2001) Localization of the psbH-subunit in Photosystem II — a new approach using labelling of His-tags with a $Ni^{2+}$-NTA-goldcluster and single particle analysis. J Mol Biol 312: 371–379

Bullough PA and Tulloch PA (1991) Spot-scan imaging of microcrystals of an influenza neuraminidase-antibody fragment complex. Ultramicroscopy 35: 131–143

Bumba L and Vácha F (2003) Electron microscopy in structural studies of Photosystem II. Photosynth Res 77: 1–19

Buseck P, Cowley J and Eyring L (1988) High-Resolution Transmission Electron Microscopy and Associated Techniques. Oxford University Press, Oxford

Caldes MT, Deniard P, Zou XD, Marchand R, Diot N and Brec R (2001) Solving modulated structures by X-ray and electron crystallography. Micron 32: 497–507

Ceska TA and Henderson R (1990) Analysis of high-resolution patterns from purple membrane labelled with heavy atoms. J Mol Biol 213: 539–560

Cheng A and Yeager M (2004) A graphical representation of image quality for three-dimensional structure analysis of two-dimensional crystals. Acta Cryst A 60: 351–354

Crowther RA, Henderson R and Smith JM (1996) MRC image processing programs. J Struct Biol 116: 9–16

Cyrklaaf M, Auer M, Kühlbrandt W and Scarborough GA (1995) 2D structure of the *Neurospora crassa* plasma membrane ATPase as determined by electron cryomicroscopy. EMBO J 14: 1854–1857

Da Fonseca P, Morris EP, Hankamer B and Barber J (2002) Electron crystallographic study of Photosystem II of the cyanobacterium *Synechococcus elongatus*. Biochemistry 41: 5163–5167

Deisenhofer J, Epp O, Miki K, Huber R and Michel H (1985) Structure of the protein subunits in the photosynthetic reaction center of *Rhodopseudomonas viridis* at 3 Å resolution. Nature 318: 618–624

Dekker JP and Boekema EJ (2005) Supramolecular organization of thylakoid membrane proteins in green plants. Biochim Biophys Acta 1706: 12–39

Dekker JP, Betts SD, Yocum CF and Boekema EJ (1990) Characterization by electron microscopy of isolated particles and two-dimensional crystals of the CP47-D1-D2-cytochrome *b*-559 complex of Photosystem II. Biochemistry 29: 3220–3225

De Rosier D and Klug A (1968) Reconstruction of 3-dimensional structures from electron micrographs. Nature 217: 130–134

Dorset DL (1996) Electron crystallography. Acta Cryst B 52: 753–769

Dubochet J, Adrian M, Chang JJ, Homo JC, Lepault J, McDowall AW and Schultz P (1988) Cryo-electron microscopy of vitrified specimens. Q Rev Biophys 21: 129–228

Ellis MJ and Hebert H (2001) Structure analysis of soluble proteins using electron crystallography. Micron 32: 541–550

Ferreira KN, Iverson TM, Maghlaoui K, Barber J and Iwata S (2004) Architecture of the photosynthetic oxygen-evolving center. Science 303: 1831–1838

Flachmann R and Kühlbrandt W (1995) Accumulation of plant antenna complexes is regulated by post-transcriptional mechanisms in tobacco. Plant Cell 7: 149–160

Frank J (2002) Single-particle imaging of macromolecules by cryo-electron microscopy. Annu Rev Biophys Biomol Struc 31: 303–319

Frank J (2005) Three-Dimensional Electron Microscopy of Macromolecular Assemblies. Oxford University Press, Oxford

Fritzsch G, Koepke J, Diem R, Kuglstatter A and Baciou L (2002) Charge separation induces conformational changes in the photosynthetic reaction center of purple bacteria. Acta Crystallogr D 58: 1660–1663

Fujiyoshi Y, Mizusaki T, Morikawa K, Yamagishi H, Aoki Y, Kihara H and Harada Y (1991) Development of a superfluid helium stage for high-resolution cryo electron microscopy. Ultramicroscopy 38: 241–251

Fujiyoshi Y (1998) The structural study of membrane proteins by electron crystallography. Advan Biophys 35: 25–80

Garab G, Löhner K, Laggner P and Farkas T (2000) Self-regulation of the lipid content of membranes by non-bilayer lipids: A hypothesis. Trends Plant Sci 5: 489–494

Glaeser RM (1971) Limitations to significant information in biological electron microscopy as a result of radiation damage. J Ultrastruct Res 36: 466–482

Glaeser RM (1999) Electron crystallography: Present excitement, a nod to the past, anticipating the future. J Struct Biol 128: 3–14

Glaeser RM, Tong L and Kim SH (1989) 3-dimensional reconstructions from incomplete data — interpretability of density maps at atomic resolution. Ultramicroscopy 27: 307–318

Glaeser RM, Zilker A, Radermacher M, Gaub HE, Hartmann T and Baumeister W (1991) Interfacial energies and surface-tension forces involved in the preparation of thin, flat crystals of biological macromolecules for high-resolution electron microscopy. J Microsc 161: 21–45

Gonen T, Sliz P, Kistler J, Cheng Y and Walz T (2004) Aquaporin-0 membrane junctions reveal the structure of a closed water pore. Nature 429: 193–197

Gonen T, Cheng Y, Sliz P, Hiroaki Y, Fujiyoshi Y, Harrison SC and Walz T (2005) Lipid-protein interactions in double-layered two-dimensional AQP0 crystals. Nature 438: 633–638

Grigorieff N, Beckmann E and Zemlin F (1995) Lipid location in deoxycholate-treated purple membrane at 2.6 Ångström. J Mol Biol 254: 404–415

Gyobu N, Tani K, Hiroaki Y, Kamegawa A, Mitsuoka K and Fujiyoshi Y (2004) Improved specimen preparation for cryo-electron microscopy using a symmetric carbon sandwich technique. J Struct Biol 146: 325–333

Hankamer B, Boekema E and Barber J (1997) Structure and membrane organization of Photosystem II of green plants. Annu Rev Plant Physiol Mol Biol 48: 641–671

Hankamer B, Morris EP and Barber J (1999) Revealing the structure of the oxygen-evolving core dimer of Photosystem II by cryoelectron crystallography. Nature Struct Biol 6: 561–564

Hankamer B, Morris E, Nield J, Gerle C and Barber J (2001) Three-dimensional structure of the Photosystem II core dimer of higher plants determined by electron microscopy. J Struct Biol 135: 262–269

Hasler L, Heymann JB, Engel A and Walz T (1998) 2D crystallization of membrane proteins: Rationales and examples. J Struct Biol 121: 162–171

Henderson R (1995) The potential and limitations of neutrons, electrons and X-rays for atomic resolution microscopy of unstained biological molecules. Q Rev Biophys 28: 171–193

Henderson R and Glaeser RM (1985) Quantitative analysis of image contrast in electron-micrographs of beam-sensitive crystals. Ultramicroscopy 16: 139–150

Henderson R, Baldwin JM, Downing KH, Lepault J and Zemlin F (1986) Structure of the purple membrane from *Halobacte-*

*rium halobium:* Recording, measurement and evaluation of electron-micrographs at 3.5Å resolution. Ultramicroscopy 19: 147–178

Hobe S, Prytulla S, Kühlbrandt W and Paulsen H (1994) Trimerization and crystallization of reconstituted light-harvesting chlorophyll *a/b* complex. EMBO J 13: 3423–3429

Holzenburg A, Bowley MC, Wilson FH, Nicholson WV and Ford RC (1993) Three-dimensional structure of Photosystem II. Nature 363: 470–472

Hosokawa F, Danev R, Arai Y and Nagayama K (2005) Transfer doublet and an elaborated phase plate holder for 120 kV electron-phase microscope. J Electron Microsc 54: 317–324

Ikeda-Yamasaki I, Odahara T, Mitsuoka K, Fujiyoshi Y and Murata K (1998) Projection map of the reaction center light-harvesting 1 complex from *Rhodopseudomonas viridis* at 10Å resolution. FEBS Lett 425: 505–508

Jamieson SJ, Wang P, Quian P, Kirkland JY, Conroy MJ, Hunter CN and Bullough PA (2002) Projection structure of the photosynthetic reaction center-antenna complex of *Rhodospirillum rubrum* at 8.5Å resolution. EMBO J 21: 3927–3935

Jiang W and Ludtke SJ (2005) Electron cryomicroscopy of single particles at subnanometer resolution. Curr Opin Struct Biol 15: 571–577

Jordan P, Fromme P, Witt HT, Klukas O, Saenger W and Krauss N (2001) Three-dimensional structure of cyanobacterial Photosystem I at 2.5 Ångstrom resolution. Nature 411: 909–917

Jungas C, Ranck JL, Rigaud JL, Joliot P and Vermeglio A (1999) Supramolecular organization of the photosynthetic apparatus of *Rhodobacter sphaeroides*. EMBO J 18: 534–542

Kamiya N and Shen JR (2003) Crystal structure of oxygen-evolving Photosystem II from *Thermosynechococcus vulcanus* at 3.7 Å resolution. Proc Natl Acad Sci USA 100: 98–103

Karrasch S, Bullough PA and Ghosh R (1995) A 8.5 Å projection map of the light-harvesting complex I from *Rhodospirillum rubrum* reveals a ring composed of 16 subunits. EMBO J 14: 631–638

Karrasch S, Typke D, Walz T, Miller M, Tsiotis G and Engel A (1996) Highly ordered two-dimensional crystals of Photosystem I reaction center from *Synechococcus* sp.: Functional and structural analyzes. J Mol Biol 262: 336–348

Kern J, Loll B, Zouni A, Saenger W, Irrgang KD and Biesiadka J (2005) Cyanobacterial Photosystem II at 3.2Å resolution — the plastoquinone binding pockets. Photosynth Res 84: 153–159

Kimura Y, Vassylyev DG, Miyazawa A, Kidera A, Matsushima M, Mitsuoka K, Murata K, Hirai T and Fujiyoshi Y (1997) Surface structure of bacteriorhodopsin revealed by high resolution electron crystallography. Nature 389: 206–211

Kitmitto A, Mustafa AO, Holzenburg A and Ford RC (1998) Three-dimensional structure of higher plant Photosystem I determined by electron crystallography. J Biol Chem 273: 29592–29599

Koepke J, Hu X, Muenke C, Schulten K and Michel H (1996) The crystal structure of the light-harvesting complex II (B800-B850) from *Rhodospirillum molischianum*. Structure 4: 581–597

Kraulis PJ (1991) Molscript — A program to produce both detailed and schematic plots of protein structures. J Appl Crystallogr 24: 946–950

Krauss N, Hinrichs W, Witt I, Fromme P, Pritzkow W, Dauter Z, Betzel C, Wilson KS, Witt HT and Saenger W (1993) Three-dimensional structure of system I of photosynthesis at 6 Å resolution. Nature 361: 326–331

Kunji ERS, von Gronau S, Oesterhelt D and Henderson R (2000) The three-dimensional structure of halorhodopsin to 5 Ångstrom by electron crystallography: A new unbending procedure for two-dimensional crystals by using a global reference structure. Proc Natl Acad Sci USA 97: 4637–4642

Kühlbrandt W (1992) Two-dimensional crystallization of membrane proteins. Q Rev Biophys 25: 1–49

Kühlbrandt W and Williams KA (1999) Analysis of macromolecular structure and dynamics by electron cryo-microscopy. Curr Opin Chem Biol 3: 537–543

Kühlbrandt W, Wang DN and Fujiyoshi Y (1994) Atomic model of plant light-harvesting complex by electron crystallography. Nature 367: 614–621

Lange C, Nett JH, Trumpower BL and Hunte C (2001) Specific roles of protein-phospholipid interactions in the yeast cytochrome $bc_1$ complex structure. EMBO J 20: 6591–6600

Lentzen M (2004) The tuning of a Zernike phase plate with defocus and variable spherical aberration and its use in HRTEM imaging. Ultramicroscopy 99: 211–220

le Maire M, Champeil P and Møller JV (2001) Interaction of membrane proteins and lipids with solubilizing detergents. Biochim Biophys Acta 1508: 86–111

Leslie C, Landree E, Collazo-Davila C, Bengu E, Grozea D and Marks LD (1999) Electron crystallography in surface structure analysis. Microsc Res Tech 46: 160–177

Lévy D, Mosser G, Lambert O, Moeck GS, Bald D and Rigaud JL (1999) Two-dimensional crystallization on lipid layer: A successful approach for membrane proteins. J Struct Biol 127: 44–52

Liu Z, Yan H, Wang K, Kuang T, Zhang J, Gui L, An X and Chang W (2004) Crystal structure of spinach major light-harvesting complex at 2.72Å resolution. Nature 428: 287–292

Loll B, Kern J, Zouni A, Saenger W, Biesiadka J and Irrgang KD (2005) The antenna system of Photosystem II from *Thermosynechococcus elongatus* at 3.2Å resolution. Photosynth Res 86: 175–184

Lyon MK, Marr KM and Furcinitti PS (1993) Formation and characterization of two-dimensional crystals of Photosystem II. J Struct Biol 110: 133–140

Marr KM, Mastronarde D and Lyon MK (1996a) Two-dimensional crystals of Photosystem II: Biochemical characterization, cryo-electron microscopy and localization of the D1 and cytochrome *b*559 polypeptides. J Cell Biol 132: 823–833

Marr KM, McFeeters RL and Lyon MK (1996b) Isolation and structural analysis of two-dimensional crystals of Photosystem II from *Hordeum vulgare viridis* $zb^{63}$. J Struct Biol 17: 86–98

McDermott G, Prince SM, Freer AA, Hawthornthwaite-Lawless AM, Papiz MZ, Cogdell RJ and Isaacs NW (1995) Crystal structure of an integral membrane light-harvesting complex from photosynthetic bacteria. Nature 374: 517–521

Miller KR (1982) Three-dimensional structure of a photosynthetic membrane. Nature 300: 53–55

Montoya G, Cyrklaff M and Sinning I (1995) Two-dimensional crystallization and preliminary structure analysis of light harvesting II (B800-B850) complex from the purple bacterium *Rhodovulum sulfidophilum*. J Mol Biol 250: 1–10

Morris EP, Hankamer B, Zheleva D, Friso G and Barber J (1997) The three-dimensional structure of a Photosystem II core complex determined by electron crystallography. Structure 5: 837–849

Mosser G (2001) Two-dimensional crystallogenesis of transmembrane proteins. Micron 32: 517–540

Nakazato K, Toyoshima C, Enai I and Inoue Y (1996) Two-dimensional crystallization and cryoelectron microscopy of Photosystem II. J Mol Biol 257: 225–232

Oling F, Boekema EJ, de Zarate IO, Visschers R, van Grondelle R, Keegstra W, Brisson A and Picorel A (1996) Two-dimensional crystals of LH2 light-harvesting complexes from *Ectothiorhodospira* sp. and *Rhodobacter capsulatus* investigated by electron microscopy. Biochim Biophys Acta 1273: 44–50

Orlova EV and Saibil HR (2004) Structure determination of macromolecular assemblies by single-particle analysis of cryo-electron micrographs. Curr Opin Struct Biol 14: 584–590

Paulsen H, Finkenzeller B and Kühlen N (1993) Pigments induced folding of light-harvesting chlorophyll *a/b*-binding protein. Eur J Biochem 215: 809–816

Potter CS, Pulokas J, Smith P, Suloway C and Carragher B (2004) Robotic grid loading system for a transmission electron microscope. J Struct Biol 146: 431–40

Qian P, Hunter CN and Bullough PA (2005) The 8.5 Å projection structure of the core RC-LH1-PufX dimer of *Rhodobacter sphaeroides*. J Mol Biol 349: 948–960

Ranck JL, Ruiz T, Pehau-Arnaudet G, Arboux B and Reiss-Husson F (2001) Two-dimensional structure of the native light-harvesting complex LH2 from *Rubrivivax gelatinosus* and of a truncated form. Biochim Biophys Acta 1506: 67–78

Reimer R (1997) Transmission Electron Microscopy. Springer Publisher, Wien, New York

Rhee KH, Morris EP, Zheleva D, Hankamer B, Kühlbrandt W and Barber J (1997) Two-dimensional structure of plant Photosystem II at 8-Å resolution. Nature 389: 522–526

Rhee KH, Morris E, Barber J and Kühlbrandt W (1998) Three-dimensional structure of the Photosystem II reaction center at 8 Å resolution. Nature 396: 283–286

Rigaud JL, Mosser G, Lacapere JJ, Olofsson A, Levy D and Ranck JL (1997) Bio-beads: An efficient strategy for two-dimensional crystallization of membrane proteins. J Struct Biol 118: 226–235

Rigaud JL, Chami M, Lambert O, Levy D and Ranck JL (2000) Use of detergents in two-dimensional crystallization of membrane proteins. Biochim Biophys Acta 1508: 112–128

Rosenbusch JP, Lustig A, Grabo M, Zulauf M and Regenass M (2001) Approaches to determining membrane protein structures to high resolution: Do selections of subpopulations occur? Micron 32: 75–90

Roszak AW, Howard TD, Southall J, Gardiner AT, Law CL, Isaacs NW and Cogdell RJ (2003) Crystal structure of the RC-LH1 core complex from *Rhodopseudomonas palustris*. Science 30: 1969–1972

Sandona D, Croce R, Pagano A, Crimi M and Bassi R (1998) Higher plants light harvesting proteins. Structure and function as revealed by mutation analysis of either protein or chromophore moieties. Biochim Biophys Acta 1365: 207–214

Santini C, Tidu V, Tognon G, Magladi A and Bassi R (1994) Three-dimensional structure of the higher plant Photosystem II reaction center and evidence for its dimeric organization in vivo. J Biochem 221: 307–315

Savage H, Cyrklaff M, Montoya G, Kühlbrandt W and Sinning I (1996) Two-dimensional structure of light harvesting complex II (LHII) from the purple bacterium *Rhodovulum sulfidophilum* and comparison with LHII from *Rhodopseudomonas acidophila*. Structure 4: 243–252

Shaw PJ and Hills GJ (1981) Tilted specimen in the electron microscope: A simple specimen holder and the calculation of tilt angles for crystalline specimens. Micron 12: 279–282

Siebert CA, Qian P, Fotiadis D, Engel A, Hunter CN and Bullough PA (2004) Molecular architecture of photosynthetic membranes in *Rhodobacter sphaeroides*: The role of PufX. EMBO J 23: 690–700

Stahlberg H, Dubochet J, Vogel H and Gosh R (1998) Are light-harvesting I complexes from *Rhodospirillum rubrum* arranged around the reaction center in a square geometry? J Mol Biol 282: 819–831

Stahlberg H, Fotiadis D, Scheuring S, Rémigy H, Braun T, Mitsuoka K, Fujiyoshi Y and Engel A (2001) Two-dimensional crystals: A powerful approach to assess structure, function and dynamics of membrane proteins. FEBS Lett 504: 166–172

Standfuss J, van Scheltinga ACT, Lamborghini M and Kühlbrandt W (2005) Mechanisms of photoprotection and nonphotochemical quenching in pea light-harvesting complex at 2.5 Å resolution. EMBO J 24: 919–928

Stark W, Kühlbrandt W, Wildhaber H, Wehrli E and Mühlethaler K (1984) The structure of the photoreceptor unit of *Rhodopseudomonas viridis*. EMBO J 3: 777–783

Toyoshima C and Unwin N (1990) Three-dimensional structure of the acetylcholine receptor by cryoelectron microscopy and helical image reconstruction. J Cell Biol 111: 2623–2635

Tsiotis G, McDermott G and Ghanotakis D (1996) Progress towards structural elucidation of Photosystem II. Photosynth Res 50: 93–101

Unger V (2000) Assessment of electron crystallographic data obtained from two-dimensional crystals of biological specimen. Acta Cryst D 56: 1259–1269

Unger VM (2001) Electron cryomicroscopy methods. Curr Opin Struct Biol 11: 548–554

Unwin N (2005) Refined structure of the nicotinic acetylcholine receptor at 4 Ångström resolution. J Mol Biol 346: 967–989

Unwin PNT and Henderson R (1975) Molecular structure determination by electron microscopy of unstained crystalline specimens. J Mol Biol 94: 425–440

Van Heel M, Gowen B, Matadeen R, Orlova EV, Finn R, Pape T, Cohen D, Stark H, Schmidt R, Schatz M and Patwardhan A (2000) Single-particle electron cryo-microscopy: Towards atomic resolution. Quart Rev Biophys 33: 307–369

Walz T and Gosh R (1997) Two-dimensional crystallization of the light-harvesting I-reaction center photounit from *Rhodospirillum rubrum*. J Mol Biol 265: 107–111

Walz T and Gregorieff N (1998) Electron crystallography of two-dimensional crystals of membrane proteins. J Struct Biol 121: 142–161

Walz T, Jamieson SJ, Bowers CM, Bullough PA and Hunter CN (1998) Projection structures of three photosynthetic complexes from *Rhodobacter sphaeroides*: LH2 at 6 Å, LH1 and RC-LH1 at 25 Å. J Mol Biol 282: 833–845

Wang DN and Kühlbrandt W (1991) High-resolution electron crystallography of light-harvesting chlorophyll *a/b*-protein complex in three different media. J Mol Biol 217: 691–699

Werten PJL, Rémigy HW, de Groot BL, Fotiadis D, Philippsen A, Stahlberg H, Grubmüller H and Engel A (2002) Progress in the analysis of membrane protein structure and function. FEBS Lett 529: 65–72

Williams RC and Fisher HW (1970) Electron microscopy of

tobacco mosaic virus under conditions of minimal beam exposure. J Mol Biol 52: 121–123
Yeager M, Unger VM and Mitra AK (1999) Three-dimensional structure of membrane proteins determined by two-dimensional crystallization. Electron cryomicroscopy and image analysis. Meth Enzymolog 294: 135–179
Zemlin F (1998) Image formation in high-resolution electron microscopy. Cryst Res Technol 33: 1097–1111
Zouni A, Witt HT, Kern J, Fromme P, Krauß N, Saenger W and Orth P (2001) Crystal structure of Photosystem II from *Synechococcus elongatus* at 3.8 Å resolution. Nature 409: 739–744

Chapter 8

# X-ray Scattering for Bio-Molecule Structure Characterization

*David M. Tiede* and Xiaobing Zuo*
*Chemistry Division, Argonne National Laboratory, 9700 South Cass Avenue, Argonne, Illinois 60439-4833, U.S.A.*



## Summary

One of the challenges in photosynthesis research lies in resolving the molecular basis for function with atomic-scale precision across the full time-scale of photosynthetic events. While crystallography provides the most detailed and accurate measurement of molecular structure, structure determination is necessarily confined to specific crystalline states. Critical issues that can be left unresolved by crystallography include the range, amplitudes, and time-scales of molecular motions, and a determination of how these dynamic events are linked to biological function. Advances in synchrotron X-ray scattering techniques offer new opportunities for characterization of photosynthetic structure and dynamics in non-crystalline media that build upon crystallographic, NMR, and molecular dynamics databases, but are applied to conditions most closely relevant to in-situ function. Recent advances in the application of synchrotron X-ray scattering techniques include the extension to the high-angle, high-resolution domain, where measurements can be routinely made to a spatial resolution of 1 Å or better, and the development of coordinate-based analyses of X-ray scattering data that allows scattering data to be analyzed in terms of detailed coordinate models for structure and dynamics. Promising new directions include application of element specific anomalous X-ray scattering and ultrafast time-resolved scattering techniques. This chapter will review recent progress in the application of synchrotron scattering techniques for the in-situ characterization of molecular structure. This work has focused primarily on proof-of-principle and technique development using model compounds and molecular systems. These studies

*Author for correspondence, email: tiede@anl.gov

*Thijs J. Aartsma and Jörg Matysik (eds): Biophysical Techniques in Photosynthesis II, pp. 151–165.*

demonstrate opportunities for using in-situ solution diffraction for the resolution of conformational landscapes for photosynthetic complexes in non-crystalline states and for extending these studies to the picosecond time domain using pulsed synchrotron techniques.

## I. Introduction

A central challenge in photosynthesis research lies in correlating structure and dynamics of photosynthetic complexes to function. Experimental aspects of this challenge are being met by success in the determination of three-dimensional protein structure using X-ray crystallography and NMR techniques. While crystallography provides the most detailed and accurate measurement of molecular structure, structural information is necessarily confined to the crystalline state (see Chapters 5, Liu and Chang; 6, Fromme and Allen; 7, da Fonseca et al.). Conformational landscapes of proteins in crystals can differ from those in non-crystalline states and in physiologically relevant media. Questions concerning the role of protein structural landscapes are particularly critical for problems in protein folding, protein-protein interactions, biological signaling, and energy transduction. Critical issues that can be left unresolved by crystallography include measurements of the range, amplitudes, and timescales of molecular motions, and a determination of how these conformational landscapes are coupled to biological function. Nuclear magnetic resonance (NMR) approaches overcome limitations in crystallography by resolving molecular structure both in liquids and in solids (Chapters 18, de Groot; 19, Daviso et al.), but NMR measurements are focused on a set of short-range distance constraints that do not directly determine complete molecular structure. Molecular dynamics (MD) simulations (Chapter 22, Schulten and Kosztin) and Monte Carlo (Chapter 23, Makri) configurational searches are frequently used in conjunction with crystallographic and NMR data to provide a detailed picture of protein conformational landscapes. However, restrictions in the range of feasible simulation timescales and uncertainties in the applicability of empirical force fields that are largely developed using crystallographic data leave open the question whether MD simulations are sufficiently accurate to model the full range biological macromolecule motion in solution.

Advances in synchrotron X-ray scattering techniques offer opportunities for direct characterization of protein conformational landscapes in non-crystalline media. X-ray scattering measures rotationally averaged molecular diffraction. Wide-angle X-ray scattering (WAXS) patterns are one-dimensional 'fingerprints' of three-dimensional structure that are directly related to atomic configuration by Fourier transform (Svergun et al., 1995; Zhang et al., 2000; Tiede et al., 2004; Zuo and Tiede, 2005). It is this direct connection between X-ray scattering and coordinate models that allows X-ray scattering patterns to serve as benchmarks for quantitative testing of crystallographic, NMR, and MD models for solution state protein structure. X-ray scattering provides a global measure of atomic configuration in disordered media that complements and extends both crystallographic and NMR analyses.

The concepts of X-ray scattering have been known and practiced since the, 1950s (Vaughan et al., 1950; Guinier and Fournet, 1955; Svergun et al., 1987; Glatter, 1991). However, the development of third generation synchrotrons offer new picosecond pulsed X-ray light sources of exceptional brilliance, energy tunability, and stability (Riekel et al., 1996; Hellemans, 1997). Furthermore, X-ray source developments have been coupled with parallel achievements in the development of large area CCD and image plate X-ray detectors. The advances in X-ray source and detector capabilities enable scattering patterns for relatively dilute molecular solutions in the millimolar concentration range (0.1 mM to 1 mM) to be measured to 1 Å spatial resolution and beyond, with high X-ray energy resolution, and ultrafast time resolution. These capabilities significantly enhance the ability of scattering measurements to characterize structure and dynamics in solution and to discriminate between competing coordinate models for solution-state structure. Small angle X-ray scattering

---

*Abbreviations:* 1D – one dimensional; 2D – two dimensional; 3D – three dimensional; *B.* – *Blastochloris*; CCD – charge-coupled device; DNA – deoxyribonucleic acid; *E.* – *Escherichia*; HEPES – N-(2-hydroxyethyl)-piperazine-N′-2-ethanesulfonic acid; keV – kiloelectron volt; MD – molecular dynamics; mm – millimeter; mM – millimolar; MOPS – 3-(N-morpholino) propanesulfonic acid; MSP – magnesium stabilizing protein; NMR – nuclear magnetic resonance; PDDF – pair distance distribution function; PYP – photoactive yellow protein; *Rb.* – *Rhodobacter*; SANS – small angle neutron scattering; SAXS – small angle X-ray scattering; Tris – Tris hydroxymethylaminoethane; WAXS – wide angle X-ray scattering; μL – microliter; μm – micrometer

for biomolecular structure characterization has been reviewed extensively (Guinier and Fournet, 1955; Cantor and Schimmel, 1980; Chen and Bendedouch, 1986; Svergun et al., 1987; Moore, 1988; Glatter, 1991; Chacon et al., 2000; Svergun and Koch, 2003). Important new concepts include the development of low-resolution *ab initio* structure reconstruction from small angle scattering data (Chacon et al., 2000; Svergun and Koch, 2003; Zhou et al., 2005). This chapter will review advances in the application of wide-angle solution X-ray scattering techniques and coordinate-based structural analyses, and discuss opportunities for the application of these techniques for the resolution of structure and dynamics in photosynthesis.

## II. X-ray Scattering Theory

### A. Coordinate-Based Analyses

Coordinate-based computational approaches for calculating scattering patterns from molecules in solution have been described previously (Svergun et al., 1995; Zhang et al., 2000; Tiede et al., 2002; Zuo and Tiede, 2005) and are briefly outlined here. Solution X-ray scattering patterns are produced by the instantaneous X-ray scattering from all atoms in the ensemble. Figure 1 illustrates the simple case of X-ray plane wave scattering for a pair of atoms, A and B, separated by a vector, **r**. For coherently scattered X-rays neither the energy nor the phase is changed by the scattering event, only the direction is changed, typically described by the momentum transfer vector, **q** having a length $(4\pi/\lambda)\sin\theta$, where $\lambda$ is the X-ray wavelength and $2\theta$ is the scattering angle. However, because the path lengths traced by the rays passing through atoms A and B differ, the phases are shifted by $2\pi(\mathbf{q}\bullet\mathbf{r})/\lambda$ with respect to each other and a detector placed at a remote distance would measure a sinusoidally varying interference pattern of summed intensity as function of the vector **q** (Warren, 1990). For molecules, the scattering pattern can be calculated from the pair-wise sum of the position of $N$ atoms in the molecular assembly:

$$I(\mathbf{q}) = \sum_{j}^{N}\sum_{k}^{N} A_j(q)A_k(q)e^{i\mathbf{q}\cdot\mathbf{r}_{j,k}} \quad (1)$$

where $A_j(q)$ are the atomic form factors (Ibers and Hamilton, 1973) for the $j$th and $k$th atoms, or they are composite form factors for atomic groups that combine the form factors for hydrogen atoms with those of the associated heavy atoms in chemical groups like $CH_x$, $NH_x$, OH, etc. (Svergun et al., 1995; Zhang et al., 2000). $r_{j,k}$ are the distances between the $j$th and $k$th atoms. The format of expression 1 shows that the molecular scattering patterns are the Fourier transforms of the atomic pairs in the molecule. In amorphous media molecular scattering is orientationally averaged, yielding the following form:

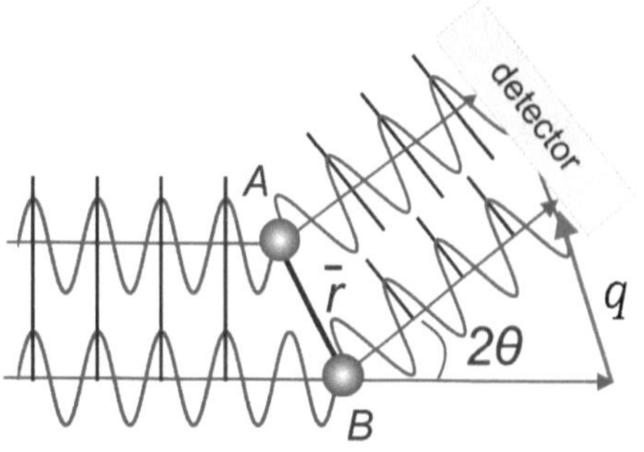


*Fig. 1.* Illustration of position-dependent interference produced by X-ray scattering from a pair of atoms, A and B, separated by a position vector, **r**, and measured at a scattering angle 2θ that is characterized by the momentum transfer vector, **q**.

$$I(q) = \langle I(\mathbf{q})\rangle_{\Omega} = \sum_{j}^{N}\sum_{k}^{N} A_j(q)A_k(q)\frac{\sin(qr_{j,k})}{qr_{j,k}} \quad (2)$$

In solution scattering spatially resolved electron density variation arises not just from the atomic structure associated with the solute, but also from the displacement of solvent by the solute and the imperfect packing of solvent molecules around the solute. One computationally tractable method for simulating the electron density variation across the solvent-solute interface has been to introduce 'dummy' atoms that are centered on the solute atoms, but have a form factor of a Gaussian sphere with zero electron density (Fraser et al., 1978; Svergun et al., 1995). The resulting atomic form factors are then expanded to have the form:

$$A_j(q) = f_j(q) - \rho_0 g_j(q) \quad (3)$$

where $f_j(q)$ are the atomic X-ray scattering form factors of the atoms or atom groups, $\rho_0$ is solvent electron density, and $g_j(q)$ are the form factors for the dummy atoms or atom groups with volume $V_j$:

$$g_j(q) = G(q)V_j e^{-q^2 V_j^{2/3}/4\pi} \quad (4)$$

and $G(q)$ is a volume expansion factor written as

$$G(q) = \frac{V_o}{V_m} e^{-q^2(V_o^{2/3} - V_m^{2/3})/4\pi} \quad (5)$$

where $V_o$ is the adjustable atomic volume of the dummy atom, and $V_m$ is the average atomic volume for the atomic group. The volumes are defined by the factors:

$$V_x = \pi^{3/2} r_x^3, \text{ where } x = j, m, o \quad (6)$$

$$r_m = \frac{1}{N}\sum_j^N r_j \quad (7)$$

where $r_o$ and $r_j$ are the adjustable Gaussian sphere radii of the dummy and $j$th atoms, and $r_m$ is the average Gaussian sphere radius of all atoms in the molecule.

The dummy atom form factor defines a solvent-excluded volume occupied by the solute atoms and uses a Gaussian form factor to describe the transition in electron density from the solute atoms to the solvent mean value. Average atomic radii or volumes have been developed for proteins (Nina et al., 1997; Tsai et al., 1999) and nucleic acids (Nadassy et al., 2001) in water. However, the atomic solvent excluded volumes can be expected to vary with the physical properties of the solvent molecule, specific solvent-solute interactions, and steric constraints imposed by the positions of neighboring atoms in the supramolecular architecture. In order to account for variations in the structure of the molecular interface for different solvents and solute molecules, the volume ratio can be treated as an adjustable parameter using a volume expansion factor, $R_{om} = r_o/r_m$, empirically parameterized by comparing calculated and experimental scattering patterns (Svergun et al., 1995; Zhang et al., 2000). For example, Fig. 2 shows the variation in the calculated scattering pattern calculated from the coordinates for cytochrome *c* (Bushnell et al., 1990) using Eq. (2), but with expansion factors that vary in the range 0.95 to 1.05. Even the illustrated 5% adjustments in the expansion factor are seen to have significant effects on the scattering intensities at low angles, but have only minor effects at high angle.

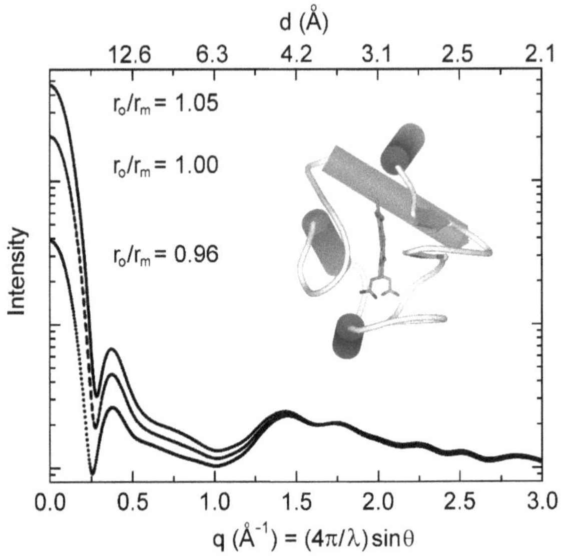


*Fig. 2.* Effect of variation in the atomic solvent excluded volume ($r_o/r_m$) on X-ray scattering patterns calculated for cytochrome *c*, Protein Data Bank entry 1HRC (Bushnell et al., 1990).

The spatial resolution achieved in small angle scattering measurements is low, $d = 2\pi/q$, and scattering is measured as an average of the solute and solvent excluded volumes. Adjustments in $R_{om}$ have the effect of altering the scattering contrast. At high angle the spatial resolution is sufficient to directly resolve the atomic pair correlations 'inside' the solvent excluded volume. Interestingly, values of approximately 1.02 have been found to generally work well for proteins (Svergun et al., 1995; Zhang et al., 2000; Tiede et al., 2002) while a value near 0.96 has been found to work better for DNA (Zuo and Tiede, 2005; Zuo et al., 2006). These expansion factor differences may reflect a closer average packing of solvent molecules around DNA compared to proteins, or possibly differences in the disposition of counter ions around the highly anionic DNA polymer (Zuo et al., 2006). A counter ion 'cloud' will similarly contribute significantly to small angle scattering (Das et al., 2003; Morfin et al., 2004), but because of the absence of short-range order, as will be discussed in the following section, will have progressively diminished effects at high angle.

An example of the X-ray scattering data progression is illustrated in Fig. 3. X-ray scattering from an ensemble of randomly oriented 3D molecular structures produces a 2D scattering pattern that is isotropic about the beam position and imaged by an area detector. The 2D scattering pattern is azimuth-

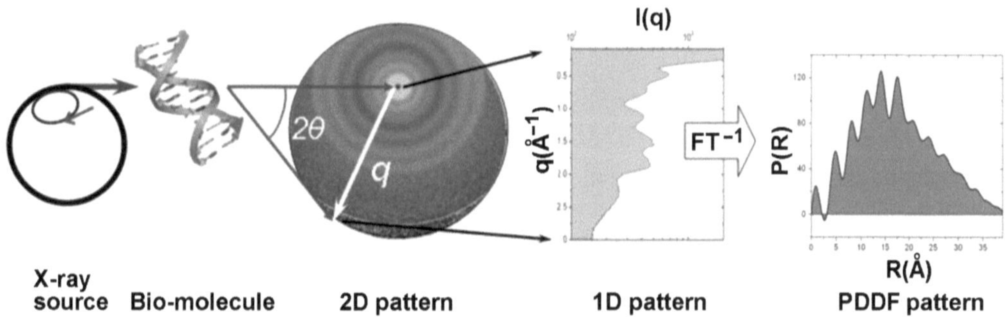


*Fig. 3.* Wide-angle X-ray scattering experiment.

ally averaged to yield the final 1D scattering pattern where scattered intensity is plotted as a function of the magnitude of reciprocal space vector $q$. Inverse Fourier transform of the reciprocal space scattering pattern produces the real space pair distance distribution function, PDDF, which is the sum of the atomic pair distances in the molecular structure (Svergun et al., 1988; Glatter, 1991; Svergun, 1992). For small molecules, composed of about ten non-hydrogen atoms or less, measurement of the atom pair distances is sufficient for 3D structure determination (Vaughan et al., 1950; Williamson et al., 1997; Ihee et al., 2001). For more complex biomolecules, as will be shown in the remainder of this review, scattering and corresponding PDDF patterns provide 'fingerprints' of 3D structure that are a direct measurement of the structure and the structural dispersion in the solution state, summarized in the form of the spatially resolved atomic pair correlations. Comparisons between experiment and model can either be made at the level of the reciprocal space scattering patterns or the corresponding real space PDDF. Typically we have chosen to use the reciprocal space scattering patterns as fingerprints for comparison of models with experiment. The scattering patterns can be calculated rigorously by Fourier transform of coordinate data as described above, and comparison of experimental and model scattering patterns avoids the possible introduction of artifacts by inverse transformation of imperfect experimental data into PDDF.

### B. Structure Fluctuations, Dynamics and Disorder

Structure fluctuations, dynamics, and disorder are critical parameters that distinguish molecules in the liquid from crystalline states. One of the most powerful opportunities provided by the wide-angle scattering technique is the ability to directly detect structural dispersion and dynamics in solution. Wide-angle scattering provides new opportunities to investigate dynamic landscapes of bio-molecules and to examine linkages between dynamic events and biological function. The effects of structural disorder and dynamics on wide-angle X-ray scattering patterns have been examined for detergent micelles (Zhang et al., 1999), proteins (Tiede et al., 2002; Hirai et al., 2004), DNA (Zuo and Tiede, 2005; Zuo et al., 2006), and covalently-linked supramolecular assemblies (Tiede et al., 2004). The effects of static and dynamic disorders on wide-angle scattering patterns can be understood in the same way that static mosaic disorder and dynamic thermal disorder broaden Bragg diffraction peaks in crystallography (Warren, 1990). Since atomic scattering is an instantaneous event, both static and dynamic sources for disorder produce equivalent effects on scattering patterns. The effects of disorder-induced structural dispersion have been analyzed using both analytical rigid-body and numerical methods.

An analytical approach was developed to account for rigid-body motions in supramolecular architectures by dividing the molecular structure into composite groups, and treating the motion of the group as a perturbation that affects the amplitude of the scattering cross-product between atom pairs in different groups (Zhang et al., 1999; Tiede et al., 2004). The groups can be subsets within a single molecule, or the atoms of two different molecules in a solution. The motion of each atom in the ensemble is characterized by $\mathbf{u}_j$, the displacement vector of the atom from its average positions. Each $\mathbf{u}_j$ is assumed to have a harmonic (Gaussian) distribution. Then, the configurationally averaged molecular scattering can be written as follows:

$$I(\mathbf{q}) = \sum_j^N \sum_k^N A_j(q) A_k(q) e^{i\mathbf{q}\cdot\mathbf{r}_{j,k}} e^{-\frac{1}{2}\left\langle\left(\mathbf{q}\cdot\mathbf{u}_{j,k}\right)^2\right\rangle} \quad (8)$$

where the magnitude of the vector $\mathbf{u}_{j,k} = (\mathbf{u}_j - \mathbf{u}_k)$ reflects the spatial coherence of the atom pair during the rigid-body movement. Equation 8 can be extended to include the orientational average to yield an expression for isotropic solution scattering:

$$I(q) = \langle I(\mathbf{q})\rangle_\Omega = \sum_j^N \sum_k^N A_j(q) A_k(q) \left\langle e^{i\mathbf{q}\cdot\mathbf{r}_{j,k}} e^{-\frac{1}{2}\langle(\mathbf{q}\cdot\mathbf{u}_{j,k})^2\rangle} \right\rangle_\Omega \quad (9)$$

This expression allows the effects of rigid body motion on scattering to be calculated by mapping group motion onto configurationally averaged interatomic displacements between groups, $\mathbf{u}_{j,k}$. Note that for atom pairs in the same rigid-body group, $\mathbf{u}_{j,k} = 0$. In the limit of a strictly rigid molecular structure with no dispersion in atomic group positions, i.e., for all atom pairs $\mathbf{u}_{j,k} = 0$, equation 9 reduces to equation 2 that describes the scattering for a molecular structure with fixed coordinates. In the limit of complete disorder between atomic groups, $\mathbf{u}_{j,k} \rightarrow \infty$ for all atomic pairs between groups but remains $\mathbf{u}_{j,k} = 0$ within the group, and the scattering calculated from Eq. (9) reduces to that of the sum of the individual atomic groups. In the region of group disorder between these two limits, equation 9 treats rigid-body group motion in a manner analogous to that of the atomic thermal factors. In a dilute solution, if interactions between solute and solvent molecules are so weak that $\mathbf{u}_{j,k}$ of solute and solvent atoms are large enough to make the cross scatterings negligible, solute and solvent molecules will scatter independently.

In the case of fully isotropic rigid body motion the inter-atomic displacements that make a constant projection on $\mathbf{q}$, Eq. (9) reduces to:

$$I(q) = \sum_j^N \sum_k^N A_j(q) A_k(q) \frac{\sin qr_{j,k}}{qr_{j,k}} e^{-\frac{1}{2}q^2\left\langle u_{j,k}^2\right\rangle} \quad (10)$$

Equation 10 illustrates the analogy between isotropic rigid body motions and the Debye-Waller treatment of atomic thermal motions (Warren, 1990). This analogy can be extended by defining a term describing the atom-pair spatial coherence:

$$G_{j,k} = 8\pi^2\langle \boldsymbol{u}^2_{j,k}\rangle \quad (11)$$

which allows Eq. (10) to be written:

$$I(q) = \sum_j^N \sum_k^N A_j(q) A_k(q) \frac{\sin qr_{j,k}}{qr_{j,k}} e^{-G_{j,k}q^2/16\pi^2} \quad (12)$$

The analytical approach for rigid-body broadening of wide-angle scattering patterns is the most accurate for the cases when $u_{j,k} \ll r_{j,k}$. However, the utility of this approach is that it provides a method for parameterizing experimental scattering patterns in terms of an atom-pair coherence factor, $G_{j,k}$. For example, porphyrin group motions within synthetic light-harvesting arrays have been characterized by rigid-body simulations of configurationally broadened experimental wide-angle scattering patterns (Tiede et al., 2004). This application illustrates the ability to test concepts of atomic group and domain motions within a supramolecular assembly by comparing analytical models for rigid-body broadening in calculated scattering patterns with configurationally broadened experimental data.

An alternate method for treating configurational disorder in experimental wide-angle scattering data is by direct simulation of the conformational ensembles. Molecular dynamics (MD) simulations are widely used as the primary tool for characterizing the conformational landscapes of biological macromolecules in solution and underlie structural determination using solution nuclear magnetic resonance (NMR) techniques. The ability of MD simulations to accurately represent a solution-state landscape for bio-molecules can be directly tested with wide-angle X-ray scattering measurements. Fourier transforms of coordinate ensembles of MD simulations are the exact numerical analogues of experimental solution wide-angle X-ray scattering data. Scattering patterns calculated from the MD coordinate ensembles are the predicted experimental observables, and the calculation and experimental measurement can be quantitatively compared.

For example, wide-angle solution X-ray scattering has been used to carry out a detailed comparison of MD simulations with experimental solution-state structures for adenine-rich duplex DNA sequences (Zuo et al., 2006). Figure 4 illustrates an example of the results comparing a scattering pattern measured

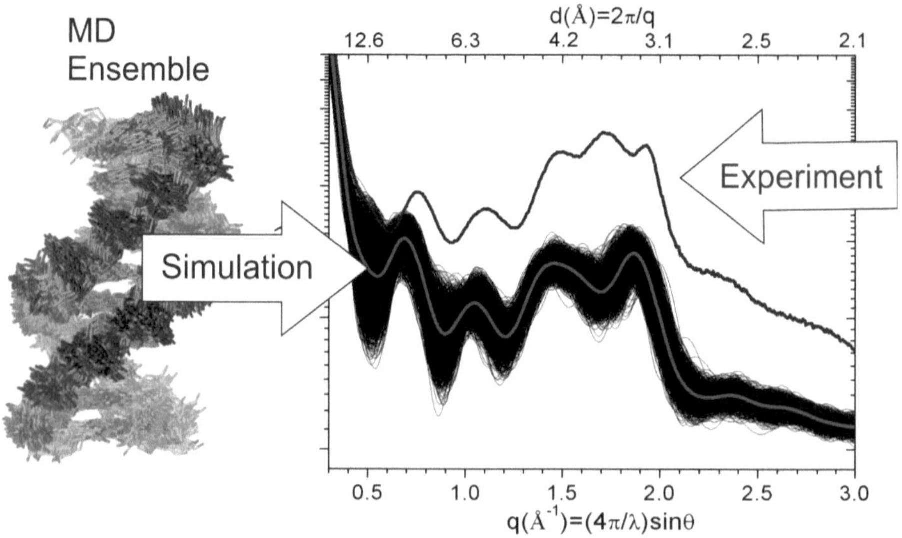


*Fig. 4.* Comparison of the experimental wide-angle X-ray scattering pattern for the DNA duplex, poly(A)-poly(T), with those calculated from MD simulation.

for poly$(A)_{20}$-poly$(T)_{20}$ duplex DNA (i.e., a 20-mer of poly(deoxyadenosine)-poly(deoxythymidine)) in solution with those calculated from an MD ensemble produced by 6 ns AMBER simulation. WAXS patterns for duplex DNA consist of a series of configurationally broadened diffraction peaks that are the rotationally averaged analogues of fiber diffraction data. Although not discussed here, the diffraction peaks can be shown to be sensitively derived from conformer-specific structural features in the atomic structure of the double helix DNA (Zuo and Tiede, 2005; Zuo et al., 2006). Scattering patterns are plotted for each conformer in the ensemble together with the ensemble-averaged scattering pattern. The scattering patterns calculated for individual conformers show considerable variation that reflects the amplitude of the structural variation. The scattering patterns for the individual conformers differ in both the number of resolved diffraction peaks and their positions. The distribution of the WAXS patterns is a direct expression of the conformational dispersion in the ensemble. Since the experimental X-ray scattering pattern arises from the sum of X-ray scattering from all DNA conformers sampled along the beam path, it represents the average of an experimental ensemble that can be directly compared to the ensemble-averaged scattering pattern calculated from MD simulation.

Clear differences are seen in Fig. 4 between the experimentally measured scattering pattern and the ensemble averaged scattering pattern calculated from the MD simulation. Compared to experiment, MD simulation shortcomings for poly$(A)_{20}$-poly$(T)_{20}$ duplex DNA were found to include both mis-matches in simulated conformer structures and number population within the ensembles (Zuo et al., 2006). Closer accord between experiment and simulation were found for duplex DNA composed of self-complementary poly$(AT)_{10}$ sequences (Zuo et al., 2006). These findings demonstrated that the MD force fields for DNA are skewed toward B-form conformations adopted by poly(AT) sequences. More generally, this work demonstrated the wide-angle scattering approach to be a widely-applicable, robust method for quantitatively evaluating the accuracy of MD simulations and other coordinate models to simulate biopolymer structure in solution, and suggests opportunities for the use of solution wide-angle scattering data as experimental benchmarks for developing supramolecular force fields optimized for a range of in-situ applications.

## II. X-ray Scattering Measurements

### *A. Sample Conditions*

Synchrotron beam lines equipped with undulator insertion devices provide X-ray light sources of exceptional brilliance and energy tunability that are well-suited for X-ray scattering measurements (Seifert et al., 2000; Svergun and Koch, 2003). X-ray energies in the range of 12 keV ($\lambda$ = 1.0 Å) to 20 keV ($\lambda$ = 0.62 Å) are convenient for wide-angle X-ray scattering measurements. Energies in this range allow the sample-to-detector distances to be adjusted to readily achieve scattering measured across the range of momentum transfer 0.01 $Å^{-1} < q < 10$ $Å^{-1}$ that includes spatial resolution in the range of 500 Å to 0.6 Å. This energy range is also far from the absorption edges of most elements in bio-molecular samples and largely avoids problems associated with X-ray fluorescence overlapping the scattering patterns. Depending upon the design of the X-ray focusing mirrors, X-ray spot sizes on the sample are routinely in the range of 20 μm to 200 μm (Seifert et al., 2000). In addition, hard X-ray focusing optics are being developed to permit diffraction and scattering measurements to be made on restricted domains in the 30 nm to 150 nm size range (Xiao et al., 2005).

X-ray scattering pattern amplitudes are proportional to the number of scatterers in the X-ray beam path. Bio-molecule concentrations needed for measurement of wide-angle scattering patterns are in the 0.1 mM to 5 mM millimolar range using X-ray sample path lengths in the range of 1 mm to 2 mm. The small X-ray beam sizes and scattering path lengths result in sample volumes needed for scattering measurements to be quite small, typically a microliter or less with 2D scattering image acquisition times that are typically in the range of 2 s to 200 s. However, radiation induced damage to bio-molecular structure is a significant factor with high intensity X-ray beams. In the distilled water or solutions with inorganic phosphate buffers, the effects of radiation damage on protein samples can be seen after a few tenths of seconds of high intensity X-ray exposure (Fischetti et al., 2003). Radiation damage is detected by the exposure time dependent loss of wide-angle scattering features in the high angle region, and by shifts in the small angle scattering patterns toward unfolded random coil conformations. Radiation damage can be prevented by flowing samples at a sufficient rate to exchange the X-ray irradiated volume with a fresh sample in less than a second. Convenient sample flow cells can be built from commercially available thin-walled (10 μm) quartz or glass capillaries. In addition we have found that the addition of common organic pH buffers, such as Tris, HEPES, MOPS, extends the lifetime of proteins and DNA in X-ray irradiated aqueous solutions several fold, presumably by scavenging reactive oxygen species produced by X-ray radiation of aqueous samples. For example, we have found that with 20 mM buffer a relatively modest flow range of 20 μL/min to 50 μL /min is sufficient to prevent radiation damage in bio-molecular samples (Tiede et al., 2002; Zuo and Tiede, 2005; Zuo et al., 2006). Even further radiation protection is conferred when supramolecular assemblies are dissolved in organic solvents (Tiede et al., 2004). With capillary flow cells sandwiched between Pelletier thermoelectric coolers, wide-angle scattering measurements can be made throughout the temperature range of liquid samples to investigate thermally accessible conformational landscapes (Tiede et al., 2002; Zuo and Tiede, 2005; Zuo et al., 2006), and with capillary flow cells housed in cryostats and cryogenically cooled gas jets, wide-angle scattering measurements can be made from cryogenically cooled liquid glass samples.

### *B. Solvent Background Subtraction and Data Corrections*

If the interactions between molecular solute and solvent molecules and the perturbation on solvent structures are negligible, which is often the case in dilute biomolecular samples, solute and solvent molecules scatter X-rays independently and solute scattering patterns can be obtained by solvent background subtraction. The measurement of bio-molecular scattering from dilute solutions is ultimately limited by the ability to detect macromolecular scattering on top of the scattering background produced by the solvent. Liquid solvents have well-defined short range order determined by hydrogen bonding and van der Waals contacts that produce broad peaks in wide-angle scattering patterns (Badyal et al., 2000). Scattering from molecular solutes in the millimolar concentration range typically produce patterns that have intensities 1% or less of the solvent scattering intensity in the high-angle $q$-range. Detection of wide angle scatting patterns for bio-molecules in solution requires techniques that allow for accurate solvent background subtraction.

An example of DNA solution scattering, water background scattering, and background solvent subtraction for 1D scattering patterns is shown in Fig. 5. The top traces show the overlapping scattering pattern for the DNA solution and its solvent background. At this scale the two traces are nearly indistinguishable. The scattering patterns are dominated by water scattering that has a broad, asymmetric peak near $q = 2.0$ $Å^{-1}$ that arises from the O-O atomic pair correlation in hydrogen bonded water molecule networks (Ba-

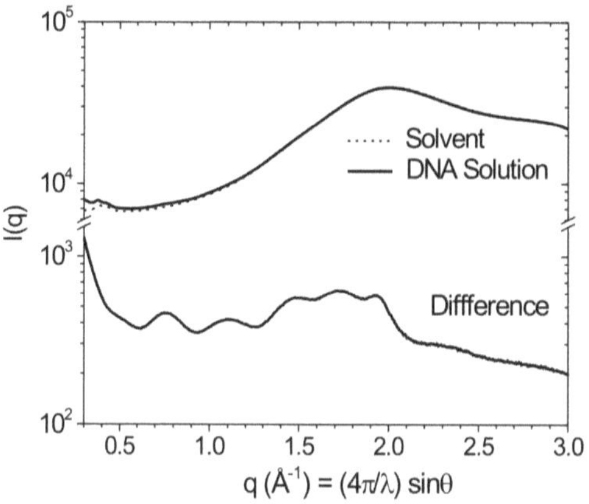


*Fig. 5.* Comparison of X-ray scattering patterns for a DNA solution sample and the corresponding aqueous solvent sample (top two curves), with the difference pattern reflecting the solute DNA scattering pattern (lower curve) using the normalization procedure described in the text.

dyal et al., 2000). The lower trace in Fig. 5 shows the residue excess scattering intensity resolved following subtraction of the solvent background from the DNA solution scattering pattern (Zuo and Tiede, 2005; Zuo et al., 2006). In this subtraction a scaling factor was applied to the solvent background to account for the volume fraction occupied by the DNA molecules in the solution. In principle the scaling factor can be determined from the DNA solution concentration and measured X-ray transmissions for the solution and solvent samples. In practice, it has been found necessary to make slight adjustments (±0.5%) to the scaling factor in order to prevent over-subtraction, as indicated by anomalously steep slopes or negative scattering intensities at the high $q$ limits, or under-subtraction, as indicated by the presence of a discernable solvent peak in the difference scattering patterns. DNA model-base scattering calculations were used to provide a target for diffuse scattering slopes in the region $q > 2.2$ Å$^{-1}$. The normalization constant for background subtraction was finely adjusted to achieve a slope in the high $q$ region that approximated slopes in model calculations. In general, the broadness of the solvent diffraction peak compared to those of protein and DNA are found to make measurements of bio-molecule solution diffraction peak positions and linewidths relatively insensitive to uncertainties in the accuracy of solvent background subtraction.

An example of the solvent normalization and background subtraction procedure is shown applied to the 2D scattering images recorded for the DNA solution and the aqueous solvent background is

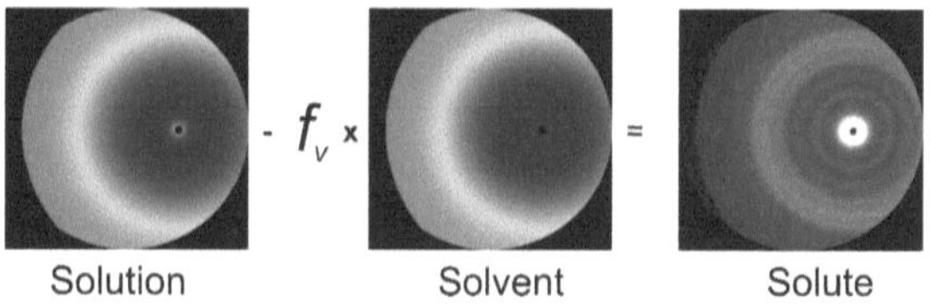


*Fig. 6.* Comparison of 2D scattering images for a DNA solution, the solvent background and the background subtracted solution image.

shown in Fig. 6. The water diffraction peak appears as a ring of prominent scattering intensity in both the solution and solvent 2D patterns. However the additional small angle scattering due to the presence of DNA in the solution sample is discernable in the 2D solution scattering image by a ring of extra scattered intensity around the beam stop that is present in the solution sample but is absent in solvent background. However, weaker DNA high angle scattering features can not be discerned by inspection of the 2D solution scattering image due to the high intensity scattering of the water background. Following subtraction of the normalized solvent background pattern, the concentric rings of the DNA high angle scattering pattern can be resolved.

Measured scattering patterns were adjusted by a factor of $\cos(2\theta)^{-3}$ to account for the angle-dependent variation in the scattering solid angle subtended by each pixel in the linear detector. Calculations were made for corrections that account for cylindrical sample cell X-ray absorption, Compton scattering, fluorescence, X-ray beam polarization, but each of these was found to be negligible in the experimental $q$-range, in part due to the high energy of the X-ray (20 keV) beam and the absence of high Z elements in the aqueous DNA samples.

## III. Protein Structure Fingerprinting

A number of reports have demonstrated the ability to acquire high-angle X-ray scattering patterns as 1D 'fingerprints' of 3D protein structure in solution, and to use these scattering patterns for distinguishing between different protein structures and conformational states (Fedorov and Denesyuk, 1978; Hirai et al., 2002; Tiede et al., 2002; Fischetti et al., 2003, 2004; Hirai et al., 2004; Kim et al., 2004; Svensson et al., 2004). Figure 7 shows an example of the high-angle scattering pattern measured for the photoactive yellow protein, PYP, compared to the scattering pattern

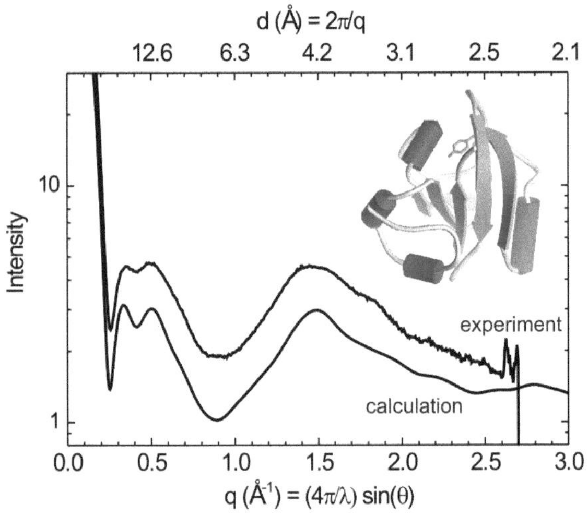


*Fig. 7.* Comparison of an experimental wide-angle solution scattering pattern for the photoactive yellow protein (Kim et al., 2004) with a scattering pattern calculated from the crystal coordinates, Protein Data Bank entry 2PHY (Borgstahl et al., 1995). The inset shows the protein structure and 4-hydroxycinnamyl chromophore.

calculated from crystallographic coordinates (Kim et al., 2004). The scattering pattern in the region 0.3 $Å^{-1} < q < 1.0$ $Å^{-1}$ is a fingerprint of the protein tertiary structure that is sensitive to the details of the protein folding and packing of the peptide domains. In this scattering range the spatial resolution is not sufficient to resolve electron density variation within protein segments. The scattering pattern in the region 1.0 $Å^{-1} < q < 3.0$ $Å^{-1}$ contains a fingerprint of the protein secondary structure. Secondary protein components such as alpha helix, beta sheet and loop segments each have characteristic atomic pair correlations that can be resolved in this scattering range. Scattering patterns measured in this $q$-region consist of a complex set of contributions from each of the protein segments and the scattering cross products between them. The experimental scattering pattern measured for PYP in solution is seen to be in good accord with the scattering features calculated from the crystal coordinates. The most noticeable difference is that the experimental scattering pattern tends to show a slight broadening of scattering features throughout the wide-angle scattering range. This would suggest a slight configuration disorder, or 'breathing' motions across multiple length scales. The wide-angle scattering data can be used to provide a quantitative experimental marker for testing the accuracy of MD trajectories to simulate the experiments conformation ensembles (Zuo et al., 2006). In addition to the general agreement for the experimental and calculated scattering patterns for PYP, minor variations between the two patterns can be discerned in the symmetry and presence of shoulders on the broad scattering peaks. These variations suggest that solution and crystal state structures for PYP differ in the details of secondary structure packing.

The important concept of discerning differences between solution and crystal conformational states has been demonstrated previously by comparisons of experimental solution state scattering patterns with scattering patterns calculated from crystal coordinate models. In the case of the N-terminal domains of calmodulin, differences between the crystallographic coordinates and experimental small-angle solution scattering data were traced to the effects of crystal packing forces distorting protein domain positions (Vigil et al., 2001). For ribonuclease $T_1$ small-angle X-ray scattering data were used to demonstrate that the protein exists in a 'swollen' conformation in solution compared to the crystal state, while retaining most of the secondary protein features (Kojima et al., 2004). Variations between experimental and calculated wide-angle scattering patterns for myoglobin suggest that the tertiary protein structure is different in solution and crystal states (Fedorov and Denesyuk, 1978; Fischetti et al., 2003).

In an investigation of oxidation-state induced conformational changes in cytochrome *c*, a comparison of experimental wide-angle X-ray scattering to scattering patterns calculated from crystallographic and NMR coordinate models allowed the two contradictory models for oxidation-induced conformation change in cytochrome *c* to be resolved, and showed that the crystallographic model provided the best approximate match to data (Tiede et al., 2002). In addition, this analysis showed that the error in the NMR model originated from an incorrect determination of the oxidized cytochrome *c* NMR structure (Tiede et al., 2002). One of the useful applications of the wide-angle X-ray scattering approach will be to resolve discrepancies between crystallographic and NMR models for protein structures.

Wide-angle X-ray scattering has also been used to characterize the solution state structure of the manganese stabilizing protein, MSP, of the oxygen-evolving complex from Photosystem II, and to compare structures for a range of physiologic states, including states isolated from native photosynthetic membranes, recovered for reconstituted oxygen

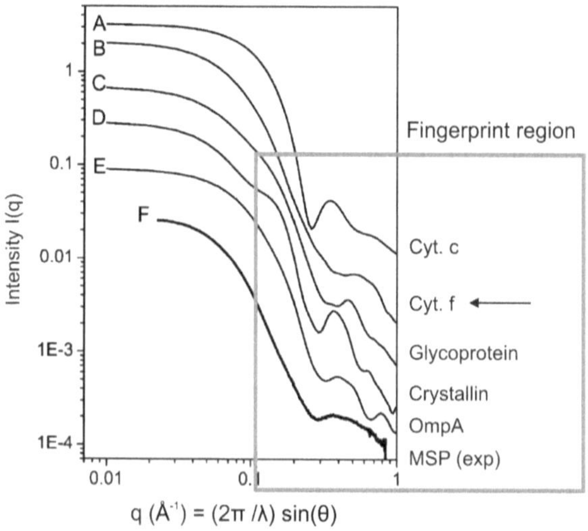


*Fig. 8.* Comparison of the experimental wide-angle scattering pattern measured for the manganese stabilizing protein, MSP, and the scattering patterns calculated for a series of protein models with beta-sheet protein domain structures, redrawn from (Svensson et al., 2004). The beta-sheet containing proteins are cytochrome *f* (1EWH), an adhesion glycoprotein (1HNF), crystallin (2BB2), and outer membrane protein A (1BXW). Also for reference is the scattering pattern for cytochrome *c* (1HRC).

evolving complexes, and expressed from *E. coli* (Svensson et al., 2002, 2004). Figure 8 shows an example of a comparison of an experimental scattering pattern for the MSP with X-ray scattering patterns calculated from a series of proteins containing beta-sheet protein domain motifs. Although the experimental data did not extend into the secondary protein fingerprint region, the data demonstrate that the beta-barrel domain of cytochrome *f* provides the closest starting model for the ternary structure of the MSP (Svensson et al., 2004). Furthermore, the scattering data demonstrated that the conformation of MSP differs when expressed in *E. coli* or isolated from native photosynthetic membranes, and that the conformation change induced upon reconstitution of active oxygen-evolving complexes are retained when the MSP is recovered from the reconstituted complexes (Svensson et al., 2004). The ability to use wide-angle scattering data to identify native and mis-folded conformations of recombinant proteins has also been demonstrated for a multiheme *c*-cytochrome from *Geobacter* expressed in *E. coli* (Londer et al., 2002).

The success in the measurement of wide-angle scattering for proteins and the quantitative comparison of these patterns to coordinate models demonstrate the significant opportunity of using wide-angle scattering data to 'refine' crystallographic, NMR, and MD coordinate models to obtain more accurate models for solution state protein structures. For example, small-angle X-ray scattering data have been used successfully as constraints in restricted MD simulations and achieved significantly improved models for solution state protein conformation (Vigil et al., 2001; Kojima et al., 2004). Similarly, a combination of NMR and small angle X-ray scattering data has been successfully used as combined constraints in refinement of the structural model for γS crystalline (Grishaev et al., 2005). The inclusion of the small angle X-ray scattering data significantly altered the refined model compared to the use of NMR distance constraints alone, and resulted in an improvement in the fit with a crystallographic model structure (Grishaev et al., 2005). It is anticipated that the inclusion of high-angle scattering data will add further precision to the model refinement (Zuo and Tiede, 2005).

## IV. Protein Complexes

X-ray scattering techniques also offer opportunities for the solution-state characterization of function protein complexes and protein-protein interactions. One example is the dynamic bi-molecular electron transfer complex formed between cytochrome $c_2$ and the bacterial reaction center (Tiede and Dutton, 1993; Axelrod and Okamura, 2005). Structures for two reaction center-cytochrome $c_2$ co-crystals of the complex have been determined and show different structures and positioning of the cytochrome $c_2$ on the periplasmic surface of the reaction center (Adir et al., 1996; Axelrod et al., 2002). Issues raised by the crystallographic data include roles of steric constraints and neighboring reaction center contacts imposed by the crystalline unit cell on cytochrome positioning. Monte Carlo and electron transfer pathway modeling suggests a dynamic range of structures with an exchange between a broad distribution of encounter complex conformations with smaller subset of transition state and electron transfer conformations (Miyashita et al., 2004). Solution X-ray scattering measurements provide a method for experimentally characterizing the solution state structure and for making quantitative comparison to simulated conformational landscapes. Low resolution small angle neutron scattering (SANS) measurements of the radius of gyration for the reaction center-cytochrome $c_2$

solution state complex have been made, and the results favored the model of an encounter complex with the cytochrome shifted towards the M-subunit side of the reaction center (Tiede et al., 2000). The relatively low signal-to-noise in the SANS experiments prevented alternative models from being completely ruled out. However scattering calculations (Tiede et al., 2000) show that the improved signal-to-noise and higher scattering angles achievable in X-ray scattering measurements are capable of more accurately characterizing the solution state complex and distinguishing between the various coordinate models.

These studies demonstrate the widely-applicable opportunity to use in-situ X-ray scattering measurements for the solution-state characterization of functional protein complexes and protein-protein interactions. Recent advances include the development of coordinate-based global rigid-body modeling of macromolecular complexes using small angle scatter data (Petoukhov and Svergun, 2005). Extension of these methods to include rigid body motions will permit direct, quantitative evaluation of Monte Carlo and MD simulations.

## VI. Time-resolved Reaction Coordinate Studies

The pulsed nature of synchrotron radiation has been exploited for picosecond time-resolved X-ray structural studies of photochemical reaction coordinates using crystallographic (Moffat, 2002; Hummer et al., 2004; Ihee et al., 2005b), spectroscopic (Chen et al., 2001), and wide-angle solution scattering approaches (Plech et al., 2004; Ihee et al., 2005a). Resolution of structural rearrangements along photosynthetic reaction coordinates is critical for achieving a fundamental understanding of the molecular basis for photosynthesis. One intensively investigated, but currently unresolved area has been the characterization of the molecular motions responsible for gating electron transfer between quinone acceptors in bacterial reaction centers (Breton, 2004).

Frozen, charge-separated trapped and time-resolved X-ray crystallographic approaches have been used to investigate structural reorganization accompanying electron transfer in bacterial reaction centers with differing results. Low temperature (90 K–100 K) static structures have been determined for the *Rhodobacter* (*Rb.*) *sphaeroides* reaction center trapped in states that stabilize the light-induced, charge-separated state between the bacteriochlorophyll special pair, P, and the first quinone acceptor, $Q_A$, (Katona et al., 2005), and in the charge-separated state involving the light-oxidized P and reduced second quinone acceptor, $Q_B$ (Stowell et al., 1997; Fritzsch et al., 2002). Compared to the dark-adapted reaction center structure, reaction centers trapped in the $P^+Q_B^-$ charge-separated state showed a prominent shift and rotation of the $Q_B$ cofactor that was proposed to be a central part of the gating motion that regulates electron transfer between the $Q_A$ and $Q_B$ acceptors (Stowell et al., 1997). For reaction centers trapped in a light-adapted conformational state that stabilizes the $P^+Q_A^-$ charge separation, small conformational changes were seen clustered within the H-subunit (Katona et al., 2005). In contrast, millisecond time-resolved and low temperature trapped crystallographic studies of the state in *Blastochloris* (*B.*) *viridis* showed no discernable structural change associated with formation of the $P^+Q_B^-$ state (Baxter et al., 2004, 2005). The crystallographic studies in the *B. viridis* reaction center are in accord with spectroscopic evidence for *Rb. sphaeroides* reaction centers in solution states (Breton, 2004). These discrepancies raise questions about the applicability of crystallographic models for light-induced structural change in the *Rb. sphaeroides* reaction center to non-crystalline states.

Time-resolved wide-angle scattering can provide a direct method for measuring reaction-linked, light-induced structural reorganization in photosynthesis, and for testing the applicability of crystallographic models for light-induced structural change to non-crystalline states. For example, Fig. 9 shows wide-angle X-ray scattering patterns calculated from the coordinates for the *Rb. sphaeroides* reaction center in low-temperature dark-adapted and trapped $P^+Q_B^-$ states (Stowell et al., 1997), together with the calculated difference scattering pattern. The X-ray scattering patterns for the reaction center in the protein fingerprint region show a broad peak with numerous overlapping minor peaks that vary in amplitude between the two conformational states of the reaction center. The difference pattern is a 1D fingerprint of the conformational change based on these coordinate models. The difference pattern calculated from dark-adapted and $P^+Q_A^-$ adapted reaction center coordinates (Katona et al., 2005) are of smaller amplitudes and clearly distinguished from the pattern generated by the coordinates for the trapped $P^+Q_B^-$ state. These calculations suggest that wide-angle scattering patterns can be used as markers for testing coordinate models for

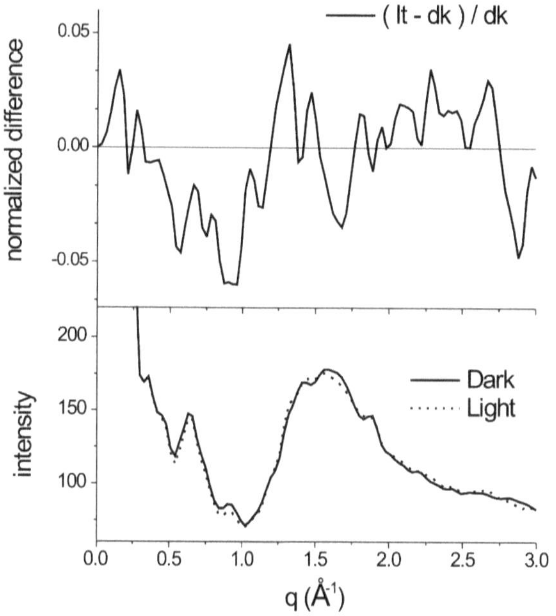


*Fig. 9.* Wide-angle X-ray scattering patterns calculated from the dark adapted (1AIJ) and the trapped $P^+Q_B^-$ (1AIG) reaction center structures (Stowell et al., 1997) plotted in the lower panel, and the normalized difference scattering pattern plotted in the upper panel.

reaction-linked structure change. Wide-angle scattering measurements on a variety of model compounds have demonstrated that when macromolecules have sufficient rigidity in solution, the X-ray scattering fine features of the magnitude displayed in Fig. 9 can readily be resolved (Dimitrijevic et al., 2004; Tiede et al., 2004; Zuo and Tiede, 2005; Saponjic et al., 2005). Protein conformational dispersion can be a major factor that precludes measurement of predicted X-ray scattering fine features based on static crystallographic coordinates. Detailed comparison of experimental wide-angle scattering patterns with those calculated from crystallographic and model structures will be significant for determining the magnitude of protein conformational dispersion in solution and for interpreting how protein dynamics factor into proposed mechanisms for protein function.

## VII. Prospects for the Future

Exciting prospects for application of wide-angle X-ray scattering techniques include ultrafast fingerprinting of protein conformational intermediates along reaction coordinates and the experimental characterization protein conformational landscapes and dynamics in solution. Wide angle X-ray scattering methods have been demonstrated to accurately resolve intermediate structures along ultrafast excited state reaction coordinates for small molecules (Anfinrud and Schotte, 2005). In this review we have described the ability of wide-angle X-ray scattering patterns to provide fingerprints of protein conformation states in solution that are sensitive to the detail of protein structure and that can be quantitatively compared to coordinate models. Prospects for applying ultrafast X-ray fingerprinting of protein conformational landscapes include laser-triggered electron transfer, pH-jump, and temperature jump reactions. Next-generation X-ray light sources are targeted at providing femtosecond time-resolution (Cornacchia et al., 2004; Miao et al., 2004), and designs are being conceptualized for 'desktop' synchrotrons that would offer opportunities for routinely accessible home laboratory tunable X-ray light sources (Giles, 2004).

## Acknowledgments

Supported by the U.S. Department of Energy, Office of Basic Energy Sciences, Division of Chemical Sciences, Geosciences, and Biosciences, under contract DE-AC02-06CH11357 to Argonne National Laboratory. We thank Drs. A. Goshe and R. Zhang for insightful discussions on wide-angle X-ray scattering measurements and modeling.

## References

Adir N, Axelrod HL, Beroza P, Isaacson RA, Rongey S, Okamura MY and Feher G (1996) Co-crystallization and characterization of the photosynthetic reaction center-cytochrome $c_2$ complex from *Rhodobacter sphaeroides*. Biochemistry 35: 2535–2547

Anfinrud PA and Schotte F (2005) X-ray fingerprinting of chemical intermediates in solution. Science 309: 1192–1193

Axelrod HL and Okamura MY (2005) The structure and function of the cytochrome $c_2$: Reaction center electron transfer complex for *Rhodobacter sphaeroides*. Photosynth Res 85: 101–114

Axelrod HL, Abresch EC, Okamura MY, Yeh AP, Rees DC and Feher G (2002) X-ray structure determination of the cytochrome $c_2$: Reaction center electron transfer complex from *Rhodobacter sphaeroides*. J Mol Biol 319: 501–515

Badyal YS, Saboungi M-L, Price DL, Shastri SD, Haeffner DR and Soper AK (2000) Electron distribution in water. J Chem Phys 112: 9206–9208

Baxter RHG, Ponomarenko N, Srajer V, Pahl R and Moffat K (2004) Time-resolved crystallographic studies of light-induced structural changes in the photosynthetic reaction center. Proc Natl Acad Sci USA 101: 5982–5987

Baxter RHG, Seagle B-L, Ponomarenko N and Norris JR (2005) Cryogenic structure of the photosynthetic reaction center of *Blastochloris viridis* in the light and dark. Acta Crystallogr D61: 605–612

Borgstahl GEO, Williams DR and Getzoff ED (1995) 1.4 Ångström structure of photoactive yellow protein, a cytosolic photoreceptor: Unusual fold, active site, and chromophore. Biochemistry 34: 6278–6287

Breton J (2004) Absence of large-scale displacement of quinone $Q_B$ in bacterial photosynthetic reaction centers. Biochemistry 43: 3318–3326

Bushnell GW, Louie GV and Brayer GD (1990) High-resolution three-dimensional structure of horse heart cytochrome *c*. J Mol Biol 214: 585–595

Cantor CR and Schimmel PR (1980) Biophysical Chemistry, Part II: Techniques for the Study of Biological Structure and Function. W. H. Freeman and Co., San Francisco

Chacon P, Diaz JF, Moran F and Andreu JM (2000) Reconstruction of protein form with X-ray solution scattering and a genetic algorithm. J Mol Biol 299: 1289–1302

Chen LX, Jager WJH, Jennings G, Gosztola DJ, Munkholm A and Hessler JP (2001) Capturing a photoexcited molecular structure through time-domain X-ray absorption fine structure. Science 292: 262–264

Chen SH and Bendedouch D (1986) Structure and interactions of proteins in solution studied by small-angle neutron scattering. Meth Enzymol 130: 79–116

Cornacchia M, Arthur J, Bane K, Bolton P, Carr R, Decker FJ, Emma P, Galayda J, Hastings J, Hodgson K, Huang Z, Lindau I, Nuhn H-D, Paterson JM, Pellegrini C, Reiche S, Schlarp H, Stohr J, Stupakov G, Walz D and Winick H (2004) Future possibilities of the Linac coherent light source. J Synchrotron Rad 11: 227–238

Das R, Mills TT, Kwok LW, Maskel GS, Millett IS, Doniach S, Finkelstein KD, Herschlag D and Pollack L (2003) Counterion distribution around DNA probed by solution X-ray scattering. Phys Rev Let 90: 188103-1–188103-4

Dimitrijevic NM, Rajh T, Saponjic Z, de la Garza L and Tiede DM (2004) Light-induced charge separation and redox chemistry at the surface of $TiO_2$/host-guest hybrid nanoparticles. J Phys Chem B 108: 9105–9110

Fedorov BA and Denesyuk AI (1978) Large-angle X-ray diffuse scattering, a new method for investigating changes in the conformation of globular proteins in solution. J Appl Cryst 11: 473–477

Fischetti RF, Rodi DJ, Mirza A, Irving TC, Kondrashkina E and Makowski L (2003) High-resolution wide-angle X-ray scattering of protein solutions: Effect of beam dose on protein integrity. J Synchrotron Rad 10: 398–404

Fischetti RF, Rodi DJ, Gore DB and Makowski L (2004) Wide-angle X-ray solution scattering as a probe of ligand-induced conformational changes in proteins. Chem Biol 11: 1431–1443

Fraser RDB, MacRae TP and Suzuki E (1978) An improved method for calculating the contribution of solvent to the X-ray diffraction pattern of biological molecules. J Appl Cryst 11: 693–694

Fritzsch G, Koepke J, Diem R, Kuglstatter A and Baciou L (2002) Charge separation induces conformational changes in the photosynthetic reaction centre of purple bacteria. Acta Crystallogr D58: 1660–1663

Giles J (2004) Lasers bend beams for desktop X-ray source. Nature 428: 789

Glatter O (1991) Small-angle scattering and light scattering. In: Lindner P and Zemb T (eds) Neutron, X-ray and Light Scattering, pp 33–82. Elsevier Science Publishers B.V., Amsterdam

Grishaev A, Wu J, Trewhella J and Bax A (2005) Refinement of multidomain protein structures by combination of solution small-angle X-ray scattering and NMR data. J Am Chem Soc 127: 16621–16628

Guinier A and Fournet G (1955) Small Angle Scattering of X-rays. Wiley, New York

Hellemans A (1997) X-rays find new ways to shine. Science 277: 1214–1215

Hirai M, Iwase H, Hayakawa T, Miura K and Inoue K (2002) Structural hierarchy of several proteins observed by wide-angle solution scattering. J Synchrotron Rad 9: 202–205

Hirai M, Koizumi M, Hayakawa T, Takahashi H, Abe S, Hirai H, Miura K and Inoue K (2004) Hierarchical map of protein unfolding and refolding at thermal equilibrium revealed by wide-angle X-ray scattering. Biochemistry 43: 9036–9049

Hummer G, Schotte F and Anfinrud PA (2004) Unveiling functional protein motions with picosecond X-ray crystallography and molecular dynamics simulations. Proc Natl Acad Sci USA 101: 15330–15334

Ibers JA and Hamilton WC (eds) (1973) International Tables for X-ray Crystallography, Vol IV. Kynoch Press, Birmingham

Ihee H, Lobastov VA, Gomez UM, Goodson BM, Srinivasan R, Ruan C-Y and Zewail AH (2001) Direct imaging of transient molecular structures with ultrafast diffraction. Science 291: 458–462

Ihee H, Lorenc M, Kim TK, Kong QY, Cammarata M, Lee JH, Bratos S and Wulff M (2005a) Ultrafast X-ray diffraction of transient molecular structures in solution. Science 309: 1223–1227

Ihee H, Rajagopal S, Srajer V, Pahl R, Anderson S, Schmidt M, Schotte F, Anfinrud PA, Wulff M and Moffat K (2005b) Visualizing reaction pathways in photoactive yellow protein from nanoseconds to seconds. Proc Natl Acad Sci USA 102: 7145–7150

Katona G, Snijder A, Gourdon P, Andreasson J, Hansson O, Andreasson L-E and Neutze R (2005) Conformational regulation of charge recombination reactions in a photosynthetic bacterial reaction center. Nat Struct Mol Biol 12: 630–631

Kim TK, Zuo X, Tiede DM and Ihee H (2004) Exploring fine structures of photoactive yellow protein in solution using wide-angle X-ray scattering. Bull Korean Chem Soc 25: 1676–1680

Kojima M, Timchenko AA, Higo J, Ito K, Kihara H and Takahashi K (2004) Structural refinement by restrained molecular-dynamics algorithm with small-angle X-ray scattering constraints for a biomolecule. J Appl Cryst 37: 103–109

Londer YY, Pokkuluri P, Tiede DM and Schiffer M (2002) Production and preliminary characterization of a recombinant triheme cytochrome $c_7$ from *Geobacter sulfurreducens* in *Escherichia coli*. Biochim Biophys Acta 1554: 202–211

Miao J, Chapman HN, Kirz J, Sayre D and Hodgson KO (2004) Taking X-ray diffraction to the limit: Macromolecular structures from femtosecond X-ray pulses and diffraction microscopy of cells with synchrotron radiation. Annu Rev Biophys Biomol Struct 33: 157–176

Miyashita O, Onuchic JN and Okamura MY (2004) Transition

state and encounter complex for fast association of cytochrome $c_2$ with bacterial reaction center. Proc Natl Acad Sci USA 101 (46): 16174–16179

Moffat K (2002) The frontiers of time-resolved macromolecular crystallography: Movies and chirped X-ray pulses. Faraday Disc 122: 65–77

Moore PB (1988) On the role of small-angle scattering in biological research. J Appl Cryst 21: 675–680

Morfin I, Horkay F, Basser PJ, Bley F, Hecht A-M, Rochas C and Geissler E (2004) Adsorption of divalent cations on DNA. Biophys J 87: 2897–2904

Nadassy K, Tomas-Oliveira I, Alberts I, Janin J and Wodak SJ (2001) Standard atomic volumes in double-stranded DNA and packing in protein-DNA interfaces. Nucl Acids Res 29: 3362–3376

Nina M, Beglov D and Roux B (1997) Atomic radii for continuum electrostatic calculations based on molecular dynamics free energy simulations. J Phys Chem B 101: 5239–5248

Petoukhov MV and Svergun DI (2005) Global rigid body modeling of macromolecular complexes against small-angle scattering data. Biophys J 89: 1237–1250

Plech A, Wulff M, Bratos S, Mirloup F, Vuilleumier R, Schotte F and Anfinrud PA (2004) Visualizing chemical reactions in solution by picosecond X-ray diffraction. Phys Rev Let 92: 125505-1–5

Riekel C, Bosecke P, Diat O and Engström P (1996) New opportunities in small-angle X-ray scattering and wide-angle X-ray scattering at a third generation synchrotron radiation source. J Molec Struc 3883: 291–302

Saponjic ZV, Dimitrijevic NM, Tiede DM, Goshe A, Zuo X, Chen LX, Barnard AS, Zapol P, Curtiss L and Rajh T (2005) Shaping nanoscale architectures through surface chemistry. Adv Mat 17: 965–972

Seifert S, Winans RE, Tiede DM and Thiyagarajan P (2000) Design and performance of an ASAXS instrument at the Advanced Photon Source. J Appl Cryst 33: 782–784

Stowell MHB, McPhillips TM, Rees DC, Soltis SM, Abresch E and Feher G (1997) Light-induced structural changes in photosynthetic reaction center: Implications for mechanism of electron-proton transfer. Science 276: 812–816

Svensson B, Tiede DM and Barry BA (2002) Small-angle X-ray scattering studies of the manganese stabilizing subunit in Photosystem II. J Phys Chem B 106: 8485–8488

Svensson B, Tiede DM, Nelson DR and Barry BA (2004) Structural studies of the manganese-stabilizing subunit in Photosystem II. Biophys J 86: 1807–1812

Svergun DI (1992) Determination of the regularization parameter in indirect-transform methods using perceptual criteria. J Appl Cryst 25: 495–503

Svergun DI and Koch MHJ (2003) Small-angle scattering studies of biological macromolecules in solution. Rep Prog Phys 66: 1735–1782

Svergun DI, Feigin LA and Taylor GW (1987) Structure Analysis by Small Angle X-ray and Neutron Scattering, pp 1–335. Plenum Press, New York

Svergun DI, Semenyuk AV and Feigin LA (1988) Small-angle scattering data treatment by the regularization method. Acta Crystallogr A 44: 244–250

Svergun D, Barberato C and Koch MHJ (1995) CRYSOL — a program to evaluate X-ray solution scattering of biological macromolecules from atomic coordinates. J Appl Cryst 28: 768–773

Tiede DM and Dutton PL (1993) Electron transfer between bacterial reaction centers and mobile *c*-type cytochromes. In: Deisenhofer H and Norris JR (eds) The Photosynthetic Reaction Center, pp 257–288. Academic Press, New York

Tiede DM, Littrell K, Marone PA, Zhang R and Thiyagarajan P (2000) Solution structure of a biological bimolecular electron transfer complex: Characterization of the photosynthetic reaction center-cytochrome $c_2$ protein complex by small angle neutron scattering. J Appl Cryst 33: 560–564

Tiede DM, Zhang R and Seifert S (2002) Protein conformations explored by difference high-angle solution X-ray scattering: Oxidation state and temperature dependent changes in cytochrome *c*. Biochemistry 41: 6605–6614

Tiede DM, Zhang R, Chen LX, Yu L and Lindsey JS (2004) Structural characterization of modular supramolecular architectures in solution. J Am Chem Soc 126: 14054–14062

Tsai J, Taylor R, Chothia C and Gerstein M (1999) The packing density in proteins: Standard radii and volumes. J Mol Biol 290: 253–266

Vaughan PA, Sturdivant JH and Pauling L (1950) The determination of the structures of complex molecules and ions from X-ray diffraction by their solutions: The structures of the groups $PtBr_6^{--}$, $PtCl_6^{--}$, $Nb_6Cl_{12}^{++}$, $TaBr_{12}^{++}$, and $Ta_6Cl_{12}^{++}$. J Am Chem Soc 72: 5477–5486

Vigil D, Gallagher SC, Trewhella J and Garcia AE (2001) Functional dynamics of the hydrophobic cleft in the N-domain of calmodulin. Biophys J 80: 2082–2092

Warren BE (1990) X-Ray Diffraction. Dover Publications, New York

Williamson JC, Cao J, Ihee H, Frey H and Zewail AH (1997) Clocking transient chemical changes by ultrafast electron diffraction. Nature 386: 159–162

Xiao Y, Cai Z, Wang ZL, Lai B and Chu YS (2005) An X-ray nanodiffraction technique for structural characterization of individual nanomaterials. J Synchrotron Rad 12: 124–128

Zhang R, Marone PA, Thiyagarajan P and Tiede DM (1999) Structure and molecular fluctuations of n-alkyl-β-D-glucopyranoside micelles determined by X-ray and neutron scattering. Langmuir 22: 7510–7519

Zhang R, Thiyagarajan P and Tiede DM (2000) Probing protein fine structures by wide angle solution X-ray scattering. J Appl Cryst 33: 565–568

Zhou J, Deyhim A, Krueger S and Gregurick SK (2005) LORES: Low resolution shape program for the calculation of small angle scattering for biological macromolecules in solution. Computer Phys Comm 170: 186–204

Zuo X and Tiede DM (2005) Resolving conflicting crystallographic and NMR models for solution-state DNA with solution X-ray diffraction. J Am Chem Soc 127: 16–17

Zuo X, Cui G, Mertz KM, Zhang L, Lewis FD and Tiede DM (2006) X-ray diffraction 'fingerprinting' of DNA structure in solution for quantitative evaluation of molecular dynamics simulation. Proc Natl Acad Sci USA 102: 3534–3539

# Chapter 9

# Mass Spectrometry-Based Methods for Studying Kinetics and Dynamics in Biological Systems

Lars Konermann
*Department of Chemistry, The University of Western Ontario, London, Ontario, N6A 5B7, Canada*

Johannes Messinger*
*Max-Planck-Institut für Bioanorganische Chemie, Stiftstraße 34 – 36, D-45470 Mülheim an der Ruhr, Germany*

Warwick Hillier
*Research School of Biological Sciences, The Australian National University, Canberra ACT 0200, Australia*



*Author for correspondence, email: messinger@mpi-muelheim.mpg.de

*Thijs J. Aartsma and Jörg Matysik (eds): Biophysical Techniques in Photosynthesis II, pp. 167–190.*

## Summary

In recent years, mass spectrometry (MS) has become one of the most widely used analytical techniques. MS allows studies on compounds ranging in size from single atoms to mega-Dalton biomolecular assemblies. This chapter provides an overview of recent MS applications in biophysical chemistry. The focus of our discussion is on 'time-resolved' techniques for tracking changes in complex biological reaction mixtures on time scales of milliseconds to days, thereby providing important structural and mechanistic insights. After a general introduction to biological MS, we discuss practical aspects of time-resolved membrane inlet mass spectrometry (MIMS), such as membrane properties and the use of different sample chambers. The MIMS technique allows online detection of dissolved gases and volatile compounds. It is particularly useful for resolving competing biochemical reactions involving common reactants, because isotopic labeling of substrates can be performed. As examples we present mechanistic studies on Photosystem II, carbonic anhydrase and hydrogenase. In the third part of this chapter we discuss the kinetics and mechanisms of protein folding and unfolding in solution, which can be explored via electrospray ionization mass spectrometry (ESI-MS). On-line coupling of ESI-MS with continuous-flow rapid mixing devices allows monitoring conformational changes of polypeptide chains with millisecond time resolution, as well as the detection and characterization of (un)folding intermediates. Due to its 'softness' the ESI process retains even weakly bound noncovalent complexes during the transition into the gas phase, such that protein-protein and protein-ligand interactions can be monitored directly. Additional insights into the conformational dynamics of proteins can be obtained by using time-resolved ESI-MS in conjunction with hydrogen/deuterium exchange methods. It is hoped that this chapter will stimulate the application of time-resolved MS techniques to a wide range of hitherto unexplored research areas.

## I. Introduction

### *A. Mass Spectrometry Fundamentals*

Mass spectrometry (MS) has evolved into an essential research tools for a wide range of biophysical applications. MS is capable of providing information on biological systems that is complementary to other commonly used methods such as X-ray crystallography, NMR, X-ray absorption, EPR/ENDOR, and optical or vibrational spectroscopy (Kaltashov and Eyles, 2005). Whereas spectroscopic methods generally involve the detection of electromagnetic radiation, MS measures signals induced by gas phase ions interacting with a suitable detector. To account for this fundamental difference, the term 'mass spectroscopy' should be avoided, and the correct term 'mass spectrometry' should be used instead. Nonetheless, there are analogies between the two areas. Optical spectra are plots of signal intensity vs. wavelength, whereas mass spectra display signal intensities as a function of mass-to-charge ratio ($m/z$). The focusing and reflection of a light beam is achieved by lenses and mirrors; an ion beam inside the vacuum chamber of a mass spectrometer can be manipulated through interactions with 'ion optics', i.e., devices employing magnetic, electric, or radio frequency fields.

Mass spectrometers encompass a few key elements, namely (*i*) ion source, (*ii*) vacuum chamber

*Abbreviations:* CEM – Channel electron multiplier; Da – Dalton (1 Da = 1 g/mol); DEMS – Differential electrochemical mass spectrometry; EI – electron impact (or electron ionization); ENDOR – electron nuclear double resonance; EPR – electron paramagnetic resonance; ESI – electrospray ionization; FTIR – Fourier transform infrared; HDPE – High density poly ethylene; HDX – hydrogen-deuterium exchange; MALDI – matrix-assisted laser desorption/ionization; MCP – multi channel plate; MIMS – membrane inlet mass spectrometry; MS – mass spectrometry; $m/z$ – mass-to-charge ratio; NMR – nuclear magnetic resonance; SS – stainless steel; S/N – signal-to-noise; TOF – time-of-flight.

with pumping system, (*iii*) mass analyzer, and (*iv*) ion detector. The ion source is required to generate charged species in the gas phase from the analytes of interest. Neutral species are undetectable in MS. Ionization can be achieved by several methods, some of which will be discussed in Section I. B. Following the ionization event, the analytes are separated either in space or time within the mass analyzer. This term is somewhat misleading, because mass spectrometers separate ions according to *m/z*, not according to mass. The various types of analyzers currently being used include sector instruments (Roboz, 1968), quadrupoles and quadrupole ion traps (Douglas et al., 2005), time-of-flight (TOF) instruments (Chernushevich et al., 2001) and Fourier-transform ion cyclotron resonance mass spectrometers (Marshall et al., 1998). A discussion of the principles of these different analyzers is beyond the scope of this chapter. The requirement for vacuum is due to the fact that excessive collisions with atmospheric gas molecules would interfere with ion trajectories from the source to the detector. Elevated gas pressures would also compromise the long term stability of filaments and ion detectors.

After separation in the analyzer, the ions are detected by a suitable device. Channel electron multipliers (CEM) are the most common type of ion detector for quadrupoles and ion traps, whereas TOF instruments employ multi-channel plates (MCPs). Modern sector instruments are equipped with array detectors that allow the simultaneous monitoring of multiple analytes. The individual detection devices within these arrays can be Faraday cups (robust but with limited sensitivity) or electron multipliers (high sensitivity but less stable). In all cases, the detection principle involves the impact of ions on a metal surface which leads to the release of secondary electrons. In the case of a Faraday cup, these electrons are collected directly. CEMs and MCP detectors amplify the initial signal by several orders of magnitude before it is read out into a computer.

MS is an extraordinarily powerful tool for both quantitative and qualitative studies. The signal response from the detector (peak height or peak area) reflects the amount of material available for ionization. The linear range of detection can cover several orders of magnitude. One major advantage of MS, when compared to optical methods, is the extremely high selectivity of the technique. Optical spectra of chromophoric compounds in solution, for example, may exhibit broad absorption bands leading to spectral overlap. In contrast, the very sharp peaks observed in MS allow the detection and quantitation of multiple coexisting species in a single spectrum. The high spectral resolution of most MS techniques also permits the discrimination between various isotopically labeled forms of a single compound.

### B. Ionization Techniques

Virtually any analyte, from single atoms all the way to cell organelles and intact virus particles, is amenable to mass spectrometric analysis (Heck and Van den Heuvel, 2004). The main key to a successful MS experiment is the choice of ionization method for the analyte of interest. Over the years, numerous ionization schemes have been devised, all of which ultimately result in a charged gaseous species.

One of the 'classical' approaches is referred to as electron impact (or electron ionization, EI) (Siuzdak, 1996). This method is most suitable for analyzing gases and organic compounds that readily evaporate upon heating. Exposure of gaseous analytes to an electron beam results in the formation of radical cations, schematically $M + e^- \rightarrow M^{\bullet+} + 2e^-$. In addition to intact $M^{\bullet+}$ molecular ions, EI mass spectra typically show a host of other peaks at lower *m/z* that correspond to charged fragments. The fragmentation patterns in EI-MS (also referred to as 'cracking patterns') are highly reproducible and relatively independent of the instrument used. Fingerprinting methods can therefore be employed to identify unknowns through comparison with reference spectra that are stored in computerized databases.

For several decades, MS studies were limited to relatively small compounds, ranging in size up to several hundred Daltons. The extensive fragmentation occurring with EI and other traditional ionization methods precludes studies on biological macromolecules, such as proteins and nucleic acids. A ground-breaking development was the invention of matrix-assisted laser desorption/ionization (MALDI) (Karas and Hillenkamp, 1988) and electrospray ionization (ESI) (Bruins et al., 1987; Fenn et al., 1989) in the late 1980s. The salient feature of both ionization techniques is their 'softness,' a term used to describe the fact that little or no fragmentation occurs during ion formation. Analyte charging occurs by protonation or de-protonation, thus leading to ions of the composition $[M + nH]^{n+}$ or $[M - mH]^{m-}$. Both MALDI-MS and ESI-MS are highly versatile, providing efficient means for the analysis of species ranging from low molecular weight compounds all the way to mega-Dalton species. As a result, MS has become one of the most important analytical tools in numerous areas of chemistry, biochemistry, pharmacology, as well as for clinical research and diagnostic applications. One half of the 2002 Nobel Prize in Chemistry was awarded '*for the development of soft desorption ionisation methods for mass spectrometric analyses of biological macromolecules*' (Fenn, 2003; Tanaka, 2003).

For MALDI-MS, the analyte is embedded in a crystalline matrix that typically consists of a UV-absorbing low molecular weight compound. Exposure to a nanosecond laser pulse leads to the desorption of matrix and analyte; subsequent analyte charging occurs by proton-transfer reactions in the MALDI plume (Tanaka, 2003). For most MALDI instruments, desorption and ionization occur within the vacuum chamber of the mass spectrometer, however, the use of atmospheric pressure MALDI methods is becoming increasingly popular (Laiko et al., 2000).

ESI allows ionization to occur directly from the liquid phase. Analyte solution is infused into a metal capillary that is held at a potential of several kV. This arrangement leads to the formation of highly charged solvent droplets at the capillary tip. In case of a capillary sprayer with a positive potential, the excess charge on these droplets is primarily due to protons. Rapid solvent evaporation leads to droplet shrinkage and subsequent droplet fission, ultimately resulting in multiply protonated analyte ions in the gas phase (Kebarle and Ho, 1997). ESI occurs at atmospheric pressure. The ions generated by this process, therefore, have to be sampled and transferred into the vacuum chamber of the mass spectrometer by means of a differentially pumped interface.

### *C. Kinetic Studies in Solution: General Considerations*

Common to kinetic studies carried out in the time domain is the use of a trigger that initiates the process of interest. The trigger event results in non-equilibrium conditions within the sample, thus giving rise to relaxation phenomena that allow the measurement of rate constants and the observation of intermediates. Optical triggers are, obviously, the method of choice for monitoring photochemical processes. Many other types of solution-phase processes can be triggered by mixing of two solutions containing the initially separated reaction partners which may be enriched isotopically above natural abundance. The use of turbulent flow or diffusion-based devices can result in mixing times as short as a few microseconds (Knight et al., 1998; Shastry et al., 1998). Manual mixing often provides sufficient temporal resolution for processes that occur on slower time scales. Kinetic experiments with MS detection may be carried out in different ways. *On-line* studies require the direct coupling of a reaction vessel to the ion source of the mass spectrometer. The composition of the reaction mixture is monitored directly, as the process of interest proceeds in solution. *Off-line* experiments usually require the availability of a quenching mechanism that stops the process of interest at well defined time points. Quenching can be achieved in various ways, e.g., by a pH-jump or by a rapid temperature change (Gross and Frey, 2002). Off-line experiments are generally more time-consuming and labor-intensive, but they allow the incorporation of sample clean-up or derivatization steps that may be required for the analysis. Due to its high sensitivity and selectivity, MS is a very attractive detection method for kinetic studies employing either on-line or off-line approaches (Houston et al., 2000; Liesener and Karst, 2005). The following sections provide examples that illustrate the wide range of possible MS applications in biophysical chemistry.

## II. Time-Resolved Membrane Inlet Mass Spectrometry (TR-MIMS)

Membrane inlet mass spectrometry (MIMS) allows continuous on-line sampling of gaseous analytes (either dissolved in solution or directly from the gas phase) with a temporal resolution of a few seconds. The center piece of a MIMS experiment is a semipermeable membrane, which separates the sample matrix from the vacuum and allows gases, but not liquids to enter the mass spectrometer. MIMS obviates the need for time-consuming off-line sampling and/or gas reprocessing (e.g., the conversion of $O_2$ to $CO_2$). It is therefore ideally suited for on-line studies of photosynthesis, respiration and many other biological and technical reactions that involve gaseous reactants such as $H_2$, $CH_4$, CO, $CO_2$, HCN, $N_2$, $NH_3$, $N_2O$, NO, $NO_2$, $O_2$, $H_2S$, or $SO_2$. In addition, volatile organic molecules such as $CH_3OH$, $C_2H_5OH$, $(CH_3)_2S$ (DMS), $(CH_3)_2SO$ (DMSO) can be detected by MIMS. One particular advantage of MIMS over other techniques such as voltammetry and amperometry is that isotopically labeled compounds can be employed to distinguish between fluxes of competing reactions; for example oxygen production ($^{16}O_2$ from $H_2{}^{16}O$) and oxygen consumption (from $^{18}O_2$) in photosynthetic algae (Radmer and Ollinger, 1980a).

### A. History of TR-MIMS Instrumentation

In the early 1960s George Hoch and Bessel Kok were based in Maryland with the Martin Marietta Corporation and were studying the action spectrum of photosynthesis, when they struck upon the notion of using a mass spectrometer with a membrane to provide a liquid/vacuum interface (Hoch and Kok, 1963). This system adopted elements from the membrane approach used in 'Clark' $O_2$ electrodes and was developed to avoid the relatively slow equilibration processes of gas molecules at gas/liquid interfaces that are rate limiting in time resolved manometry experiments. Kok and colleagues applied the new MIMS technique to several interesting areas of research and produced many important papers in photosynthesis (Govindjee et al., 1963; Hoch et al., 1963; Radmer and Ollinger, 1980b,c, 1981, 1982, 1983, 1986). As a spin-off they also developed isotopic assays for respiration and photosynthesis, thereby providing a potential basis for discovering extraterrestrial life (Kok and Varner, 1967; Radmer and Kok, 1971; Martin et al., 1975; Radmer et al., 1976). This no doubt pleased their corporate hosts who were awarded with a prime contract from NASA to build the Viking Lander and examine life on Mars. In 1976 two Viking missions did land on Mars with mass spectrometers on board, and they did perform experiments to test for life. Ironically, however, these instruments were not MIMS systems.

### B. MIMS Design Considerations

The general setup of a MIMS system is shown in Fig. 1. It consists of (*i*) a membrane-covered gas inlet system that for many applications is integrated within a sample chamber, (*ii*) a fore-line vacuum system containing two isolation valves ($V_1$, $V_2$) and a cryogenic vapor trap, and (*iii*) the mass spectrometer. In the following we discuss the function of these components and various design options.

#### 1. Mass Spectrometer and Vacuum System

The mass analyzers used for MIMS usually employ either quadrupoles (e.g., Agilent, Waters, Thermo Electron, Shimadzu) or magnetic sector ion optics with array detection, as realized in isotope ratio mass spectrometers (e.g., GV Instruments, Thermo Electron). Quadrupole mass analyzers have several advantages; they are compact, relatively cheap, robust, easy to clean, fast scanning, require only a moderate vacuum, and are easy to adapt to varying applications. Magnetic sector field instruments with

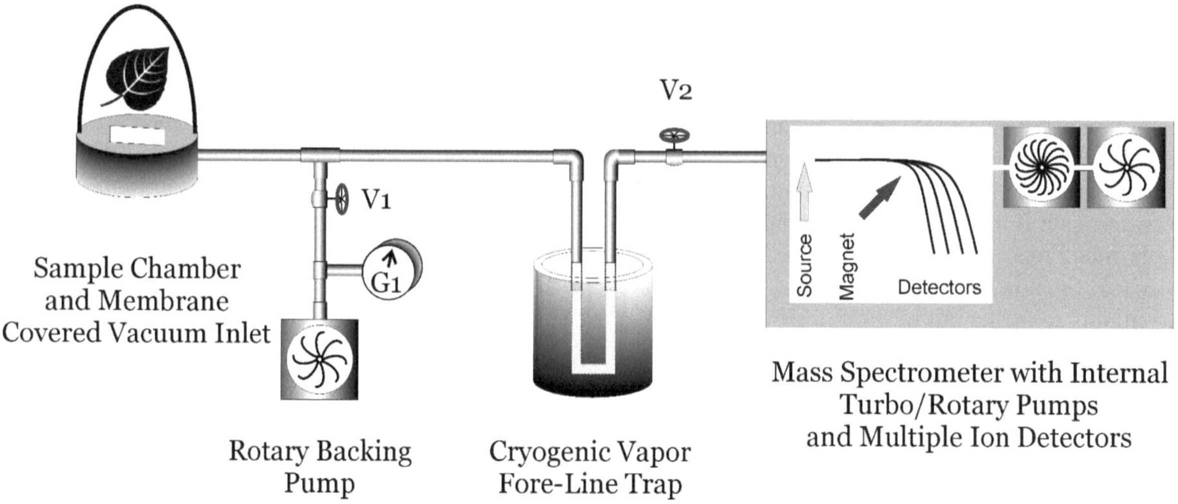


*Fig. 1.* An on-line MIMS system with sample chamber and membrane inlet. This system may be used with any of the cells or inlets shown in Fig. 3. A cryogenic fore-line trap removes water vapor that inadvertently penetrates through the semi-permeable membrane into the vacuum line ('fore-line') before the gases enter the mass spectrometer. The function of the isolation valves ($V_1$, $V_2$) is outlined in the text. $G_1$ is a vacuum gauge.

array detectors have the advantages of truly simultaneous detection of several masses and of greater sensitivity and signal stability. These points become important for isotope ratio studies. For the gaseous analytes monitored in MIMS experiments, EI is the ionization method of choice.

For the operation of MIMS systems at least two valves are required (Fig. 1). Initially valve $V_2$ closes the connection to the mass spectrometer. A pre-vacuum is established in the fore-line via opening valve $V_1$ that controls the connection to a rotary backing pump. This first step also removes water or other solvents that may still be in the cooling loop from previous measurements. Then the cryogenic vapor fore-line trap ('cryogenic trap') is engaged, by closing $V_1$ and slowly opening $V_2$. It is advisable to perform this opening with the ionization voltage switched off in order to avoid damage to the filament in case of a sudden increase in source pressure. Subsequently, data acquisition can commence.

The cryogenic trap removes trace amounts of water that inadvertently penetrate the semi-permeable membrane (see Section II.B.2). This helps to maintain an adequate vacuum within the analyzer region. The trap usually consists of a small loop in the vacuum line that runs through a Dewar containing a suitable coolant such as dry ice/ethanol (~200 K) or liquid nitrogen (77 K). One further feature of the cryogenic trap is that it safeguards against the potentially disastrous consequences of a possible membrane failure by freezing out the aqueous contents of the sample chamber before they can enter the mass spectrometer. Some care has to be taken in selecting the proper temperature and cooling loop length as to not trap the analyte(s). Condensation and boiling temperatures of some typical analytes are listed in Table 1.

For time resolved MIMS experiments the response time of the whole setup is important. Under many conditions it is the permeability of the semi-permeable membrane that is rate limiting for the signal rise (see Section II.B.2). Nevertheless, it is advisable to minimize the length of the diffusion path for the analyte by using narrow gauge stainless steel (SS) tubing (1/16–1/8′) with a total length of ≤ 60 cm to directly couple the sample chamber to the ionization chamber (Bader et al., 1983; Baltruschat, 2004). High flow turbo-molecular pumps (~200 L/s) in the mass spectrometer are also useful as high pumping rates shorten the settling time between measurements.

### 2. Membrane Properties

The semi-permeable membrane is the 'heart' of every MIMS set up. It is this membrane that allows the fast detection of the dissolved gases in solution by circumventing the slow transition across the liquid-gas interface. Nonetheless, its permeability often determines the response time of the whole set up, which usually is in the range of 3–10 s. The transmission of gas occurs via pervaporation. This process involves (*i*) adsorption of the analyte to the membrane surface, (*ii*) permeation through the membrane and (*iii*) desorption into the vacuum (Silva et al., 1999; Johnson et al., 2000). The gas transmission rate constant ($k_{trans}$) through the membrane is given by Fick's law (Hoch and Kok, 1963; Johnson et al., 2000):

$$k_{trans} = (P\,A\,\Delta p)/l \qquad (1)$$

where $P$ is the gas permeability constant (a product of the diffusion constant $D$ and solubility coefficient of the gas in the membrane), $A$ is the area of the membrane inlet, $\Delta p$ is the partial pressure difference across the membrane, and $l$ is the membrane thickness. As the partial pressure of the analyte on the low pressure side of the membrane is very small, $k_{trans}$ is proportional to the analyte concentration in the liquid phase.

The gas transmission properties of two MIMS membranes are shown in Fig. 2 as a response of the $m/z = 32$ signal to the injection of 6 nmol dissolved $O_2$ into the 160 µL sample chamber containing degassed water. Curve A was measured with a 25 µm thick silicone Membrane (MEM-213, Mem Pro), while trace B was obtained with a 12.5 µm thick Teflon membrane (S4, Hansatech). A range of other membranes may also be considered that might include Teflon films such as FET or AF (DuPont), silicone rubber, oxygen electrode membranes[1], or HDPE plastic films (various sources). Other possible membrane materials are listed in a recent review (Johnson et al., 2000). However, in our experience, and as shown in Fig. 2, the silicone MEM-213 membrane is one of the most permeable non-porous membrane films, resulting in fast signal rise and large signal amplitudes (see also Radmer, 1979).

The response of a MIMS system can be modeled by a series of first order rate equations. In the case of

[1] YSI provides a 12.5 µm high sensitivity and a 25.5 µm standard sensitivity Teflon membrane, Hansatech a 25 µm Teflon membrane.

Table 1: Typical mass spectral 'cracking patterns' for a number of gasses and small molecules under natural isotope enrichment (90 eV ionization energy). The peak heights are given in percent of the largest signal amplitude. (Mao and Leck, 1987).

| | Hydrogen ($H_2$) | Helium (He) | Methane ($CH_4$) | Ammonia ($NH_3$) | Water ($H_2O$) | Carbon Monoxide (CO) | Nitrogen ($N_2$) | Nitric oxide (NO) | Methanol ($CH_3OH$) | Oxygen ($O_2$) | Hydrogen Sulfide ($H_2S$) | Argon (Ar) | Carbon Dioxide ($CO_2$) | Nitrogen Dioxide ($NO_2$) | Ethanol ($C_2H_5OH$) |
|---|---|---|---|---|---|---|---|---|---|---|---|---|---|---|---|
| Nominal mass | 2 | 4 | 16 | 17 | 18 | 28 | 28 | 30 | 32 | 32 | 34 | 40 | 44 | 46 | 46 |
| 1 | 3 | | 17 | | 2 | | | | | | | | | | |
| 2 | 100 | | | | | | | | | | 0.2 | | | | |
| 4 | | 100 | | | | | | | | | | | | | |
| 12 | | | 3 | | | | | | | | | | 10 | | |
| 13 | | | 8 | | | | | | | | | | 0.1 | | |
| 14 | | | 19 | 2 | | 0.8 | 14 | 8 | | | | | | 10 | |
| 15 | | | 36 | 8 | | | | 2 | | | | | | | |
| 16 | | | 100 | 80 | 2 | 3 | | 2 | | 18 | | | 16 | 22 | |
| 17 | | | 1 | 100 | 26 | | | | | | | | | | |
| 18 | | | | 0.4 | 100 | | | | 2 | | | | | | 6 |
| 19 | | | | | | | | | | | | | | | 2 |
| 20 | | | | | | | | | | | | 23 | | | |
| 22 | | | | | | | | | | | | | 2 | | |
| 26 | | | | | | | | | | | | | | | 8 |
| 27 | | | | | | | | | | | | | | | 24 |
| 28 | | | | | | 100 | 100 | | 6 | | | | 13 | | 7 |
| 29 | | | | | | 0.7 | 1 | | 65 | | | | 0.1 | | 23 |
| 30 | | | | | | | | 100 | 0.8 | | | | | 100 | 6 |
| 31 | | | | | | | | 0.4 | 100 | | | | | | 100 |
| 32 | | | | | | | | 0.2 | 67 | 100 | 44 | | | | |
| 33 | | | | | | | | | 1 | 0.1 | 42 | | | | |
| 34 | | | | | | | | | | 0.4 | 100 | | | | |
| 35 | | | | | | | | | | | 3 | | | | |
| 36 | | | | | | | | | | | 4 | 0.3 | | | |
| 37 | | | | | | | | | | | | | | | |
| 38 | | | | | | | | | | | | 0.1 | | | |
| 40 | | | | | | | | | | | | 100 | | | |
| 42 | | | | | | | | | | | | | | | 3 |
| 43 | | | | | | | | | | | | | | | 8 |
| 44 | | | | | | | | | | | | | 100 | | |
| 45 | | | | | | | | | | | | | 1 | | 34 |
| 46 | | | | | | | | | | | | | 0.2 | 37 | 16 |
| Boiling temperature (K) | 20 | 4 | 112 | 241 | 373 | 81 | 77 | 121 | 338 | 90 | 212 | 87 | 195 | 294 | 352 |
| Melting temperature (K) | 14 | – | 91 | 195 | 273 | 68 | 63 | 109 | 176 | 54 | 152 | 84 | 216 | 262 | 159 |

rapid mixing (< 10 ms; see vertical sample chambers Section II.B.3.c) the injection profile is given by a step function representing the quasi-instantaneous rise in analyte concentration to $c_i$ in the MIMS sample chamber, the overall response rate constant ($k_{rise}$) of the MIMS system to the change in analyte concentration, and the rate constant for the gas transfer ($k_{leak}$) from the sample chamber out across the membrane. The overall change in signal amplitude $I(t)$ in response to an injection at $t_0$ is therefore given at $t > t_0$ by:

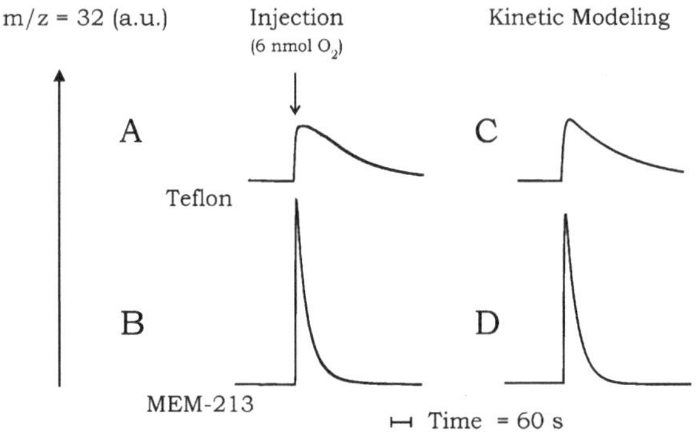


*Fig. 2.* Response of the $m/z = 32$ signal to the injection and rapid mixing of 6 nmol dissolved $O_2$ into 160 mL degassed water at 10 °C. Two different membranes were used: A, Teflon membrane (Hansatech, thickness 12.5 μm); B, MEM-213 silicone membrane (MemPro, thickness 25μm). The data can be fitted using Eq. (2), thereby providing the system response rate, $k_{rise}$, and the $O_2$ leak rate into the MS vacuum, $k_{leak}$. The kinetic modeling resulted in the following rate constants: C, Teflon membrane, $k_{rise} = 0.12\ s^{-1}$, $k_{leak} = 0.006\ s^{-1}$; D, Mem-213, $k_{rise} = 0.37\ s^{-1}$, $k_{leak} = 0.03\ s^{-1}$.

$$I(t) = I_0 + c_i e^{-k_{leak}(t-t_0)} K(1 - e^{-k_{rise}(t-t_0)}) \quad (2)$$

where $I_0$ is the constant background signal level just before the injection, and $K$ is a constant that accounts for the overall sensitivity of the spectrometer for a certain analyte (see also (Calvo et al., 1981). For the MIMS system used in Fig. 2, the value of $k_{rise}$ is limited by the membrane permeability and therefore equals $k_{trans}$ (Eq. 1) of a membrane (neglecting the effect of the boundary layer). The other kinetic term in Eq. (2) is the membrane consumption or leak rate constant ($k_{leak}$); this term will also depend on $k_{trans}$ of the membrane, but is smaller because it is also inversely proportional to the sample volume, i.e. a smaller sample volume degasses faster than a larger one, if the same membrane area is used.

### a. Membrane-Induced Artifacts

All factors that affect the permeability of the membrane will influence the signal intensity and stability. Outside of events leading to membrane fouling or membrane damage, the two main factors to be considered are temperature and stirring. The temperature is important because the permeability constant $P$ in Eq. (1) is highly temperature dependent and heating will generally lead to a signal increase. Therefore, the sample chamber and membrane require efficient temperature regulation. In photochemical experiments, this may be exacerbated by the intensity and spectral properties of the actinic light, so care is needed in order to avoid heat-induced signal amplitude changes. One option to reduce light-induced artifacts is a 90° orientation between illumination and membrane.

Stirring speed is an important factor for measurements of liquid samples, because stirring critically effects the thickness of the boundary layer and thereby $k_{trans}$ (Section II.B.3.b,c). For achieving good $S/N$ ratios a constant stirring speed is essential. To this end, a good magnetic coupling between stir bar and the drive is required.

Artifacts induced by heat, light or stirring can be discovered by simultaneously monitoring the concentration of an inert internal standard such as argon, since its concentration should, in the absence of artifacts, simply decline with $k_{leak}$.

### b. Isotope Discrimination and Calibration

When performing highly sensitive isotope-ratio measurements with a MIMS setup it is important to consider possible isotope discrimination events that might be caused by differential analyte diffusion across the membrane. Discrimination against mass (Grahams law) will occur because the thermal velocity of a gas and therefore its rate of diffusion is proportional (via $D$ in Eq. 1) to the inverse square root of its molecular weight (see also Hoch and Kok, 1963). For the diffusion of $^{32}O_2$ and $^{34}O_2$ we obtain, for example :

$$\frac{k_{trans}(^{34}O_2)}{k_{trans}(^{32}O_2)} = \frac{\sqrt{32}}{\sqrt{34}} = 0.97 \quad (3)$$

It should be noted that $K$ (Eq. 2) can also be isotope dependent. Therefore, careful calibrations are required for all quantitative studies. For most gases the saturation levels for water solubility are tabulated for various temperatures and pressures (Chemical Rubber Company, 2005). Calibrations can be made by injections of known volumes of such solutions into degassed buffer in the sample chamber. For $CO_2$, calibration buffers can also be prepared with known concentrations of bicarbonate or carbonate at well defined pH values (enough buffer capacity is important).

### c. Background Signal Determination

Background signals (especially for $CO_2$) can be caused for instances by sorbed gasses that leach from the SS tubing, or by volatile molecules such as $CH_3OH$, $C_2H_5OH$, DMSO that readily penetrate the membrane and may be detected as molecular ions, or as fragmented species such as $CO_2$ or $SO_2$. Table 1 lists some possible cross contributions for some *m/z* ratios that are important for biological studies. The degree to which this is happening will depend on the abundance of the parent molecules in the sample (concentration and isotope enrichment) and the degree to which double ionization, fragmentation or gas phase reactions occur (the latter factors can be influenced by the ionization voltage). The background levels can be determined either after extensive degassing of the sample, or by selectively removing a specific gas species. $O_2$ can be quickly removed by a glucose, glucose oxidase, catalase oxygen scrubbing system (dithionate is not recommended because it can create a strong signal at $m/z = 32$). Solutions free of $CO_2$ can be obtained by preparing a dilute solution of KOH. Bubbling with argon or nitrogen within the cell are other options (Radmer and Ollinger, 1980a).

### d. Membrane Support

Typical MIMS membranes are rather fragile and cannot withstand a ~1 atm pressure difference across a large inlet area without mechanical support. Ideally, this support should not produce an additional diffusion barrier. Different types of materials can be used, such as sintered glass or steel frits, porous ceramic, or porous plastic such as Teflon. In many cell designs the magnetic stirrer is in direct contact with the membrane (see Section II.B.3). This has the advantage of reducing boundary layer effects, but imposes mechanical stress.

## 3. Principles of Sample Chamber and Gas Inlet Design

The above criteria provide guidelines for designing MIMS systems and enable the user to tailor sample volume, membrane and inlet area for specific applications. For example, in cases where a dissolved gas is the substrate rather than the product it is best to use a sample chamber with a large volume to inlet area ratio or a membrane with low gas-permeability (Section II.B.2). Below we discuss several types of sample chamber and gas inlet designs that have been employed in laboratory and field studies. We place particular emphasis on stirring, mixing (time resolution), illumination and temperature control of these systems, but also report on capillary inlets (Section II.B.3.e) that directly probe bulk samples.

### a. Unstirred Cell

The unstirred, horizontal cell is the simplest design and operates via sampling a quasi-equilibrium of the liquid phase. To improve sensitivity and increase response rate and stability experiments are usually conducted with material sedimented onto the membrane (Radmer, 1979; Radmer and Ollinger, 1982; Bader et al., 1987). The construction of such cells is typically similar to horizontally stirred cells described in detail below (Section II.B.3.b), with the exception that usually a larger inlet area is used. A variation of this design type consists of capillary inlets described in Section II.B.3.d.

### b. Horizontally Stirred Cells

These cells consist of an upper sample chamber part and a lower inlet section. Both contain provisions for effective temperature control. The membrane is held and sealed at the smooth metal/metal (or plastic/metal) interface between these two components. The inner sleeve of the sample chamber is made of SS or glass and is incorporated into a brass or plastic water jacket for temperature control. Typical volumes are 1–3 mL with a ~8 mm stirrer flea (Fig. 3A). A plastic plug with a Perspex center serves as a window to the sample and closes the sample chamber to prevent atmospheric equilibration. Injections can be made via a small hole in the plug and the centre of the plug is made to accommodate an optical fiber that enables simultaneous illumination and fluorescence measurements to be performed while recording $O_2$ and $CO_2$ signals (Badger and Andrews, 1982; Hanson et al., 2003). In case of a glass inner sleeve and a transparent water jacket (commercially available for example for Clark-type electrodes from Hansatech) illumination can be made from the side. These systems can also be operated non-stirred as a gas phase chamber (Canvin et al., 1980; Maxwell et al., 1998; Ruuska et al., 2000) such as schematically depicted in Fig. 1, analogous to the Hansatech LD1/2 systems.

The inlet section (lower part) of the horizontally stirred cell is typically made out of brass or stainless steel. An insert is cut in the top for a gas-permeable membrane support and the surrounding metal surface is highly polished for a gas tight seal with the membrane (a small amount of high-vacuum grease can

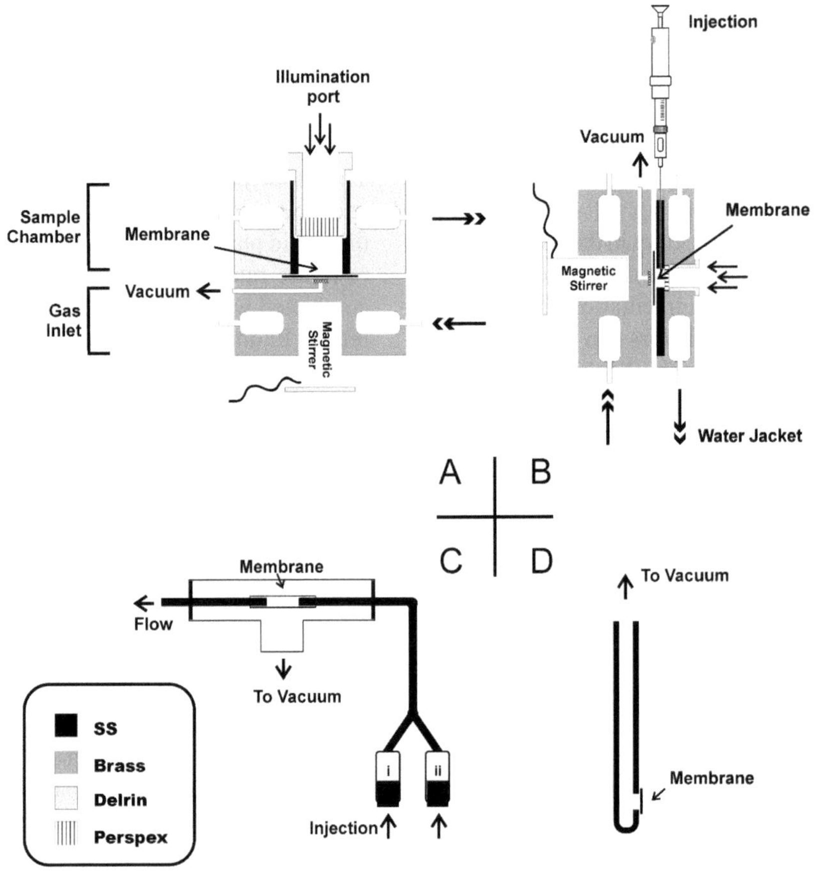


*Fig. 3.* Sample chambers and inlets for MIMS experiments as described in II.B.3: (A) vertically stirred cell with a removable port on the top able to accept illumination (arrows) and water jacket for temperature control (double arrows); (B) the horizontal small-volume stirred cell for the fast mixing and flash experiments; (C) a flow tube; and (D) a microprobe inlet.

be used to facilitate membrane adhesion). A vacuum line connects the cavity underneath the membrane support with the MS. This central arrangement may also contain provisions for attaching the stirrer in a defined position.

### c. Vertically Stirred Cells

The vertical cell is similar to the horizontal cell but turned by 90°. This design (Messinger et al., 1995) allows experiments on photosynthetic samples with significantly reduced sample volumes (~150 μL), and it enables rapid injection/mixing experiments with isotopically labeled substrates at affordable costs. Mixing within the sample chamber is achieved via a thinned cross type stirrer bar and the rapid injection (mixing halftime ~4 ms) is made by using a modified spring loaded Hamilton (CR series) syringe that is computer triggered (Messinger et al., 1995; Hillier et al., 1998). The overall experimental time resolution of this system is in the millisecond range and is, therefore, considerably faster than that determined by $k_{rise}$ (several seconds). This resolution is achieved via the use of a pump probe technique in which the injection of labeled substrate is followed with a variable time prior to a flash that generates the analyte (see Section II.C.1 for further details). The schematic of this sample chamber is shown in Fig. 3B. It appears feasible to achieve a similar time-resolution for non photochemical reactions by combining such a setup with a second injection that either rapidly quenches the observed reaction or releases the product.

### d. Flow Tubes

Continuous flow MIMS experiments can be performed by using a set up in which the solution traverses through a SS tubing into the MS, where the metal tubing has a gap that is bridged by a gas permeable tube or membrane (see Fig. 3C). With liquid samples, a time resolution around 2–15 s has been reported (Silverman, 1982; Johnson et al., 2000). The time resolution depends on the length of the line to the MS after the mixing point of the reactants (or of enzyme with substrate) and the flow rate (up to a few mL/min). This approach is related to the time-resolved ESI-MS experiments described in Section III.

#### e. Capillary Inlets

Another variation are low leak rate ('microinvasive') probes (Lloyd et al., 2002). These sampling devices operate using a semi-permeable membrane sealed to the end of a SS or quartz capillary (ID 0.1–2 mm, length up to 10 m) as inlet system. With such a probe it is possible to obtain MIMS data with spatial resolution as well as temporal resolution (Fig. 3D; Lloyd et al., 2002). By connecting such a probe to a simple quadropole, ion trap or TOF MS it is possible to assemble portable systems for the three-dimensional mapping of gases in complex ecosystems such as lakes, sediments or soils. In other applications the distribution and activity of colonies of microorganisms in food or on biofilms have been studied. A spatial resolution of less than 1 mm at a time resolution of 10–100 s (depending on membrane type and tubing length) have been reported (Lloyd et al., 2002).

### 4. Differential Electrochemical Mass Spectrometry (DEMS)

A very attractive experimental approach is the combination of a MIMS instrument with an electrochemical cell. This was first attempted by Bruckenstein and Gadde (Bruckenstein and Gadde, 1971), and than further developed by Wolter and Heitbaum (Wolter and Heitbaum, 1984) who also coined the name differential electrochemical mass spectrometry (DEMS). A recent overview of new technical developments in this area has been given by (Baltruschat, 2004).

A time resolution around 5 ms can be achieved by omitting cryogenic traps and by using a porous Teflon membrane. These membranes became commercially available in 1977 as Gore-Tex™ and allow the permeation of $H_2O$ vapor while retaining liquid water. The pore size used is approximately 0.02 μm and the thickness of the membrane is 75 μm. Under these conditions the time resolution is determined by the pumping speed (differential pumping is required) and the volume of the ionization chamber (Baltruschat, 2004).

Several different cell designs have been implemented. In the simplest case the working electrode is created by painting or depositing electrode material onto the porous Teflon membrane. However, also porous rotating disc electrodes have been successfully employed as inlets (Tegtmeyer et al., 1989). Alternatively, a thin layer design with large area electrodes can be used (Hartung and Baltruschat, 1990; Baltruschat and Schmiemann, 1993). These cells are usually coupled to a quadrupole mass spectrometer.

So far the main applications of DEMS have been related to fuel cell technologies. However, in principle this technology or MIMS cells coupled with electrochemistry should also be very interesting for functional studies on redox-active proteins that produce or consume gaseous or volatile molecules.

## C. Examples of MIMS Applications

Since its introduction in 1963, MIMS has revolutionized many fields with its selectivity, accuracy and its ability of continuous on-line sampling. Applications include studies on whole ecosystems and organisms, as well as mechanistic investigations on isolated enzymatic or chemical reactions. In the biophysical realm we cover three MIMS applications: photosynthetic oxygen evolution, carbonic anhydrase and hydrogenase. Other notable biophysical and biological applications include (chloro)respiration (Cournac et al., 2000), alternative oxidase (Ribas-Carbo et al., 2005), nitrogen fixation (Bader and Roben, 1995; Prior et al., 1995), denitrification (Cartaxana and Lloyd, 1999; An et al., 2001), nitric oxide (Bethke et al., 2004; Conrath et al., 2004), depth profiles of dissolved gases in oceanic waters (Tortell, 2005), and determinations of gas exchange rates of peat cores (Beckmann and Lloyd, 2001; Beckmann et al., 2004). MIMS will also be an essential tool for testing catalysts designed for artificial solar water splitting into $H_2$ and $O_2$ (Poulsen et al., 2005). Outside of the biological realm there are other fields where online MIMS sampling has made major contributions including soil, water and air analysis, fermentation and chemical reactors. Many of these more technical and environmental applications have recently been reviewed (Johnson et al., 2000) and will not be further discussed here.

### *1. Photosystem II Water Oxidation Chemistry*

Photosystem II is a multisubunit protein complex that catalyses the light-driven oxidation of water to molecular oxygen. The reaction proceeds after four charge separation events have accumulated four oxidizing equivalents in the catalytic oxygen evolving complex (OEC) (Renger, 2001; Britt et al., 2004; McEvoy and Brudvig, 2004; Messinger, 2004; Hillier and Messinger, 2005; Yano et al., 2006). The most direct and unambiguous evidence for substrate water binding to the OEC is obtained using MIMS experiments of $^{18}O$-water exchange. The measurements involve the preflashing of the sample into the desired redox state (S state) with zero to three single turn-over flashes in normal buffer media, then a small amount of $H_2{}^{18}O$ is injected and rapidly ($t_{1/2}$ = 4 ms) mixed. This is followed by another group of one to four closely spaced (10 ms) detecting flashes which induce oxygen evolution. The kinetics of substrate water exchange and thereby properties of the substrate water binding sites can be probed by recording the amount of labeled oxygen (*m/z* = 34 and 36) as a function of the delay between injection and the detecting flash sequence. The millisecond time resolution in these experiments therefore results from the fast mixing and signal induction (by flashes) and is not compromised by the relatively slow (seconds) detection, because the rapid photochemical production of oxygen is followed by a long dark time (10–20 s) in which no further product can be formed. The first experiments of this type were preformed using an open, unstirred sample cell (Section II.B.3.a). This resulted in a relatively long (>30 s) stabilization time between the $^{18}O$-water injection and the actual photochemically measured $O_2$ (Radmer and Ollinger, 1980c, 1986; Bader et al., 1993). Later, a closed chamber system was developed that enabled the actual $^{18}O$ ligand exchange reaction of the water to be resolved (Messinger et al., 1995) (Fig. 4). In this system a stirred chamber (Section II.B.3.b; Fig. 3B) was used in conjunction with an injection system capable of equilibrating the $^{18}O$-water throughout the sample chamber with $t_{1/2}$ ~4 ms. The use of an $O_2$ scavenging system (glucose/glucose oxidase/catalase) ensured that the 10–25 μL injection of $^{18}O$-water was made under anaerobic conditions. This work led to the conclusion that the two substrate water molecules are bound in the OEC at two different sites (Messinger et al., 1995; Hillier et al., 1998). This discovery has

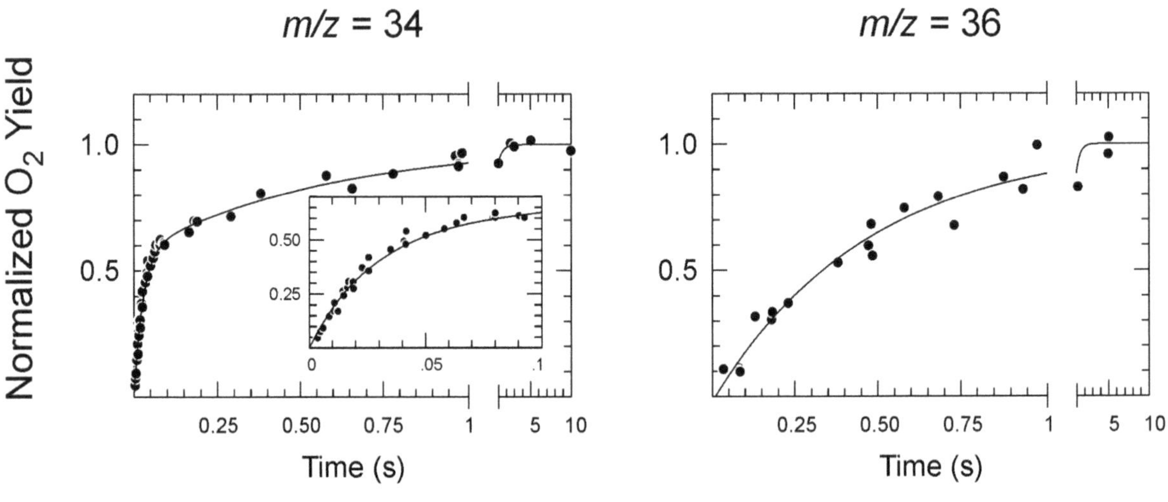


*Fig. 4.* MIMS-based determination of substrate water exchange rates in the $S_3$ state of Photosystem II in spinach thylakoids. The $S_3$ state is produced by two preflashes. This is followed by the injection and rapid mixing of a small amount of $H_2{}^{18}O$ (final enrichment 12%). $O_2$ is then produced by giving a third, detection flash at varying delay times (0–10 sec) after the $H_2{}^{18}O$ injection. Measurements were made at *m/z* = 34 (left) for the mixed labeled $^{16,18}O_2$ and at *m/z* = 36 (right) for the double labeled $^{18,18}O_2$ at 10°C. Solid lines show first-order kinetic fits yielding rate constants of 38 ± 2 $s^{-1}$ for the fast phase (left panel) and 1.8 ± 0.2 $s^{-1}$ for the slow phase (left and right panel). At *m/z* = 36 only the slow phase is observed, because both substrate molecules need to exchange for this signal to appear and therefore its rise is kinetically limited by the exchange at the slow exchanging substrate water binding site. This also proves that the two kinetic phases at *m/z* = 34 are not due to sample heterogeneity. All $O_2$ signals are normalized to the respective values obtained after complete exchange. (Hillier et al. ,1998; Hillier and Messinger, 2005, with permission).

led to further experiments that have been summarized elsewhere (Hillier and Wydrzynski, 2004; Hillier and Messinger, 2005).

MIMS has also been used to investigate other aspects of the water oxidation chemistry in Photosystem II. This includes the reactivity of the catalytic site to small reductants ($NH_2OH$, $NH_2NH_2$) that are chemically oxidized to liberate $N_2$ (Radmer, 1979; Radmer and Ollinger, 1982; Kretschmann and Witt, 1993) and were extended to generate a steric profile of the substrate entry site with the use of organic derivatives of hydrazine or hydroxylamine of different size (Radmer and Ollinger, 1983). One MIMS experiment has also probed the chemical oxidation of $H_2O_2$ to $O_2$ (Mano et al., 1987) using $^{18}O$-labeled substrate, and others have been performed with $^{18}O$-bicarbonate to examine the intrinsic rates of carbonic anhydrase activity in Photosystem II and to show that bicarbonate is not a significant substrate of the OEC (Radmer and Ollinger, 1980c; Clausen et al., 2005; Hillier et al., 2006).

### 2. Carbonic Anhydrase

Carbonic anhydrase (CA) catalyzes the equilibration of dissolved $CO_2$ with bicarbonate via the carbonic acid intermediate, according to Eq. (4) (Lindskog and Coleman, 1973).

$$HCO_3^- + H^+ \leftrightarrow H_2CO_3 \underset{k_2}{\overset{k_1}{\rightleftarrows}} CO_2 + H_2O \quad (4)$$

This reaction plays an extremely important role for the metabolism of both plant and animal organisms (Badger and Price, 1994; Christianson and Fierke, 1996), because it facilitates the rapid chemical equilibration of dissolved $CO_2$ and bicarbonate in living cells, a process that is the cornerstone of much of the biology associated with respiration and photosynthetic carbon fixation.

The application of MIMS to study the CA enzymology was introduced by Silverman and colleagues in the early 1970s (Silverman and Tu, 1976; Silverman, 1982) using a flow tube approach (Section III.B.3.d; Fig. 3C). This technique analyses the activity of CA by monitoring the relatively slow isotopic equilibration between bicarbonate and $CO_2$ in solution that follows the much more rapid chemical equilibration. This is done by adding a known amount of $HC^{18}O_3^-$ and following the time dependent speciation of the $CO_2$ signals. Directly after the addition of labeled bicarbonate the $m/z = 48$ signal represents the amount of $C^{18}O^{18}O$ formed by the rapid chemical equilibration between $CO_2$ and bicarbonate. Then as the CA catalyzed isotopic equilibration proceeds, the $m/z = 48$ signal decays forming first the $m/z = 46$ ($C^{18}O^{16}O$) species and ultimately all $^{18}O$ label is diluted into the large excess of water and the $m/z = 44$ ($C^{16}O^{16}O$) species dominates. With this technique Silverman and colleagues continue to unravel the enzymatic properties of CA enzymes (Duda et al., 2005; Fisher et al., 2005), while others use this approach for elucidating mechanisms of $CO_2$ concentration in cyanobacteria and algae (Badger and Price, 1989; Badger et al., 1994; So et al., 1998). MIMS is one of the few techniques capable of measuring CA activity under conditions of chemical equilibrium.

### 3. Hydrogenase

Hydrogenases are found in many organisms, including photosynthetic ones like *Chlamydomonas reinhardtii* and *Synechocystis* PCC 6803. In presence of suitable electron donors or acceptors, hydrogenases catalyze the interconversion of $H_2$ into 2 $H^+$ and 2 $e^-$. Three different types exist: FeFe, NiFe and 'Fe-free' hydrogenases (the latter were recently shown to also contain one Fe) (Shima et al., 2005). In light of a possible future $H_2$ economy such organisms are prime targets for studying photochemical $H_2$-production. In absence of exogenous electron carriers H/D exchange between $D_2$ and $H_2O$ can be observed that is characteristic for the activity of the catalytic centre alone, while $H_2$ production/consumption rates may also depend on the properties of the product and substrate channels and the electron transfer chain, which consists of FeS clusters. Both types of reactions can be conveniently followed by MIMS by recording the $m/z = 2$, 3 and 4 signals. Since most hydrogenases are $O_2$ sensitive it is useful to also monitor the concentration of this gas (for details see Krasna, 1978; Vignais, 2005). These types of assays allow a comparison of the activity of hydrogenases from various organisms under different conditions, and to study the mechanism of activation/deactivation in response to inhibitors such as $O_2$, acetylene or CO. The assay can either be performed with whole microorganisms or with highly purified protein complexes. Together with mutagenesis and crystallography, MIMS forms an essential tool for structure-function analyses in hydrogenases. In principle, one should also be able to use MIMS for studying the unusual inorganic ligands of the catalytic sites of hydrogenases, which depending on sample condition and species may include CO, $CN^-$, SO, $S^{2-}$, $O^{2-}$, $OH^-$ or $O_2$. Using gas chromatography, the release of a small amount of $H_2S$ during the activation with $H_2$ was reported for hydrogenase isolated from *Desulfovibrio vulgaris Miyazaki F* and by pyrolysis-MS and TOF-secondary ion MS small amounts of SO and $SO_2$ were only released from active hydrogenases, leading to the proposal that SO is a ligand to the active site (Higuchi and Yagi, 1999; Higuchi et al., 2000).

## III. Time-Resolved Electrospray Mass Spectrometry

The ESI process provides a direct bridge between solution-phase chemistry and analyte detection in the gas-phase by MS. Thus, ESI-MS has enormous potential for on-line kinetic studies (Konermann and Douglas, 2002). Following the initiation of a process by mixing of two or more reactants, solution-phase kinetics can be monitored by direct injection of the reaction mixture into the ESI source, such that the relative concentrations of multiple reactive species can be recorded as a function of time. Application of this type of experiment include mechanistic studies on bio-organic processes (Meyer et al., 2003), and enzymatic reactions (Zechel et al., 1998; Norris et al., 2001; Li et al., 2003; Wilson and Konermann, 2004).

One possible method for carrying out time-resolved ESI-MS experiments is to interface a stopped-flow device to the ESI source of the mass spectrometer (Northrop and Simpson, 1997; Kolakowski and Konermann, 2001). However, a more common approach is to carry out time-resolved ESI-MS studies in continuous-flow mode. Fig. 5 shows one possible implementation of an ESI-MS-coupled continuous-flow setup (Wilson and Konermann, 2003). This system represents a concentric capillary mixer with adjustable reaction chamber volume. In contrast to previously described continuous-flow devices (Shastry et al., 1998), the setup operates under laminar flow conditions, a fact that has to be taken into account for data analysis (Konermann, 1999; Wilson and Konermann, 2003). Reactant solutions are continuously expelled from two syringes. Syringe A is connected to the inner capillary, whereas the solution delivered by syringe B flows through the outer capillary. Mixing occurs at the end of the inner capillary. The reaction proceeds while the mixture flows towards the outlet of the apparatus, where ESI takes place. The reaction time is determined by the solution flow rate, the diameter of the outer tube, and by the distance between the mixing point and the outlet. The latter can be adjusted continuously. Kinetic experiments are performed by initially positioning the mixer directly at the ESI source. The mass spectrometer monitors the ions emitted from the sprayer, while the assembly — consisting of syringe A, inner capillary, and mixer — is slowly pulled back. These experiments provide data in three dimensions; the time axis is determined by the mixer position, the

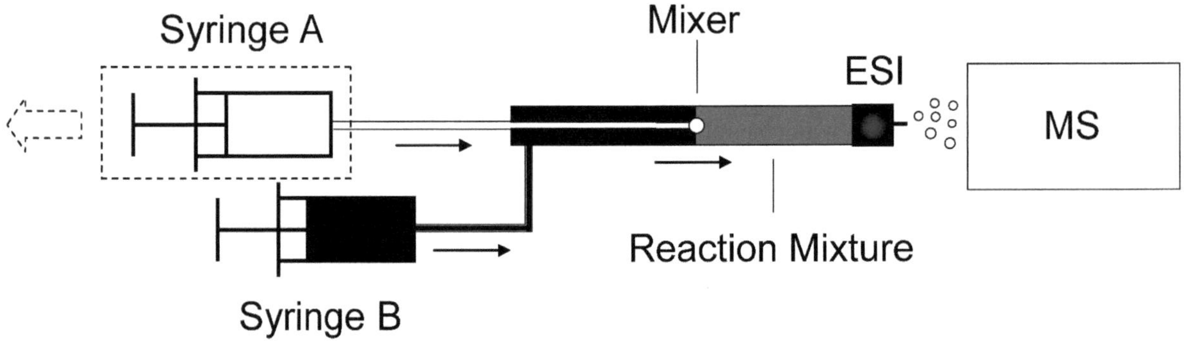


*Fig. 5.* Schematic depiction of a concentric capillary mixer with adjustable reaction chamber volume for time-resolved ESI-MS. Reactants are continuously expelled from syringes A and B. Mixing of the solutions from the two syringes at the outlet of the inner capillary initiates the process of interest. The reaction time can be controlled by adjusting the distance between mixer and outlet of the ESI source. This distance can be altered during the experiment by pulling back syringe A along with the inner capillary and the mixer, as indicated by the dashed block arrow. As a result, ESI-MS data can be recorded for a wide range of reaction times. Thin arrows indicate the direction of liquid flow.

$m/z$ axis provides information on the identity of the species in the reaction mixture, and the intensity axis is related to the concentration of each of the various solution-phase species. The kinetic data obtained in this way can be visualized by either displaying mass spectra for specific reaction times, or by plotting intensity-time profiles for selected ionic species. In principle, it would also be possible to control the time axis of the experiment by altering the solution flow rate. However, this approach is not advisable because it may result in artifactual changes of analyte ion abundances (Konermann et al., 2001). One limitation of on-line ESI-MS studies is that solvent additives such as salts, buffers, and detergents which may be present in the reaction mixture can interfere with the ionization process. The use of electrosonic sprayers, fused droplet ESI, or ultra-rapid sample clean-up procedures provides possible ways to overcome this problem (Wilson and Konermann, 2005).

Global analysis methods are a powerful tool for dissecting kinetic data obtained by time-resolved ESI-MS. For chemical processes obeying first-order kinetics a master equation of the general form (Berberan-Santos and Martinho, 1990)

$$\frac{d\vec{x}}{dt} = \langle A \rangle \vec{x} \tag{5}$$

can be established, where $\vec{x} = x_1(t), \ldots, x_n(t)$ is a vector containing all time-dependent concentrations of the $n$ species that are involved in the reaction. $\langle A \rangle$ is the $n \times n$ rate constant matrix of the system with eigenvalues $\lambda_1, \ldots \lambda_n$. Except for $\lambda_n$ (which is zero), the apparent rate constants $\lambda_j$ are functions of all the microscopic rate constants in the matrix. Equation (5) can be solved for any set of initial conditions, resulting in multi-exponential expressions for the concentration profiles $x_j(t)$. Accordingly, any intensity profiles, $I_i(m/z, t)$, accompanying the kinetics can be expressed as

$$I_i(m/z, t) = \sum_{j=1}^{n-1} C_{ij}(m/z) \exp(-t/\tau_j) + C_{in}(m/z) \tag{6}$$

The ($n$-1) relaxation times, $\tau_j$, in Eq. (6) are given by $\tau_j = (\lambda_j)^{-1}$. These $\tau_j$ values are common to all the intensity profiles $I_i(m/z, t)$, and the kinetics observed across the $m/z$ range differ only in the amplitudes $C_{ij}(m/z), \ldots, C_{in}(m/z)$ (Beechem et al., 1985; Holzwarth, 1995).

The time resolution of the system depicted in Fig. 5 allows the measurement of relaxation times in the range of ca. 1 min down to less than 10 ms (Wilson and Konermann, 2003).

The following sections illustrate some applications of time-resolved ESI-MS for studying the folding and conformational dynamics of proteins. Space limitations preclude a discussion of kinetic studies on enzyme mechanisms, which represent another fascinating area of research (Zechel et al., 1998; Norris et al., 2001; Li et al., 2003; Wilson and Konermann, 2004). The specific examples chosen here are for water-soluble model proteins from animal organisms. It is an interesting question in how far the methodologies described here will be applicable to membrane proteins, such as those involved in photosynthetic light reactions. Initial attempts in this direction appear to be very encouraging (Whitelegge et al., 1999; Demmers et al., 2000).

### A. Protein Folding and Unfolding

The three-dimensional structure adopted by the polypeptide chain of a protein depends on its solvent environment. Under physiological conditions of pH and temperature, and in the absence of chemical denaturants, most proteins fold into a unique, highly ordered, and compact structure. This ‘native’ conformation represents the biologically active state of a protein. Denaturing agents such as acid or heat can induce a transition to a largely disordered conformation. Many unfolded proteins spontaneously refold to their native structures once the denaturant is removed. This remarkable observation implies that all of the information required to attain the native structure is contained within the amino acid sequence of the polypeptide chain. The native state of a protein corresponds to the conformation with the lowest overall free energy, taking into account the contributions from the polypeptide chain and the surrounding solvent (Anfinsen, 1973). Although a lot of progress has been made in protein folding research over the last forty years, the question of how and why proteins fold remains one of the major unsolved problems in biophysical chemistry (Pain, 2000). The ‘protein folding problem’ is not only a fascinating challenge from an intellectual point of view, it also has important biomedical implications. For example, misfolding and aggregation are related to diseases like Alzheimer’s, Parkinson’s, Creutzfeldt-Jakob syndrome, type II diabetes, BSE (‘mad cow disease'), and many others (Dobson, 2003).

Of particular importance for the mechanistic understanding of folding and unfolding is the detection and structural characterization of transient intermediates. This task is not straightforward, mostly because the lifetimes of these species are often very short, spanning a range from microseconds to seconds, which prevents the use of X-ray or NMR methods for direct structural studies (Bachmann and Kiefhaber, 2001). Another important aspect that has received surprisingly little attention is the fact that many proteins adopt their biologically active conformation only after binding metal ions, prosthetic groups, or other proteins. Very little is known about the interplay of folding and binding, and the interactions of noncovalent binding partners with short-lived folding intermediates. Only very recently have these important questions come to the forefront of research (Shoemaker et al., 2000; Wittung-Stafshede, 2002). Time-resolved ESI-MS has the potential to address these and other questions, thus contributing to a better understanding of protein folding and unfolding in general.

As discussed in Section I.B, positive ion ESI produces intact, multiply protonated ions directly from proteins in solution. The charge states (protonation states) observed in ESI-MS are strongly dependent on the protein structure in solution. Unfolded proteins exhibit wide distributions of highly protonated ions. In contrast, tightly folded, compact structures give rise to much lower charge states and narrow distributions. While the physical reasons underlying this empirical relationship are still a matter of debate, ESI charge state distributions have become a widely used probe for the overall compactness of protein solution-phase conformations (Chowdhury et al., 1990; Konermann et al., 1997; Kaltashov and Eyles, 2002a).

Figure 6 illustrates how time-resolved ESI-MS can provide 'snapshots' of a protein folding reaction for selected time points. Ubiquitin, a small (8.6 kDa) protein, is a common model system for folding studies. Its compact native structure breaks down in acidic solutions containing organic cosolvents such as methanol, to form an extended A state that possesses an extensive non-native α-helicity (Brutscher et al., 1997). For the data depicted in Fig. 6, the transition from the denatured state back to the native protein was triggered by mixing protein solution containing a high concentration of methanol and acetic acid with excess water. The initial spectrum, recorded for $t = 20$ ms (Fig. 6A), shows a relatively broad charge state distribution, with the 12+ peak having the highest intensity. As refolding proceeds, the relative abundance of highly charged protein ions decreases, and that of the 8+ and 7+ ions increases (Figs. 6B, C). Refolding of ubiquitin under these conditions appears to be a simple two-state process that does not involve any kinetic intermediates.

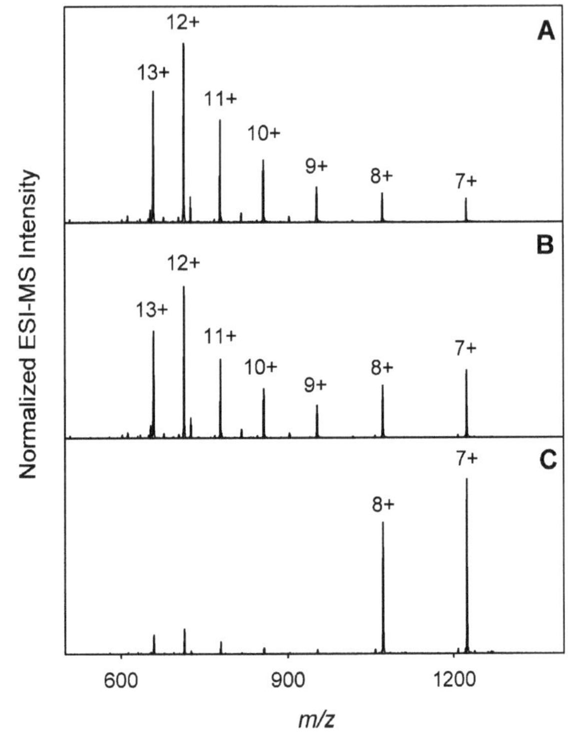


*Fig. 6.* Time-resolved ESI mass spectra recorded during refolding of ubiquitin. The protein was initially denatured by exposing it to 50% methanol and 4% acetic acid. Refolding was triggered by a mixing with water, resulting in a drop of the methanol and acid content to of 16% and 1.6%, respectively. Spectra were recorded 20 ms (A), 160 ms (B), and 2.1 s (C) after mixing. Notation: '$n^+$' represents intact gas-phase protein ions of the composition [ubiquitin + $n$H]$n^+$. (Wilson and Konermann, 2003, with permission).

### *B. Assembly and Disassembly of Protein Complexes*

Due to its very gentle nature, ESI allows not only the ionization of intact biopolymers, but also of multi-component complexes that are held together by weak noncovalent interactions. Thus, assemblies involving proteins and metal ions, prosthetic groups, inhibitors, substrates, nucleic acids, and other types of ligands are amenable to ESI-MS (Heck and Van den Heuvel, 2004; Schermann et al., 2005). The ubiquitin example discussed in the previous paragraph illustrates a folding transition that does not involve any intermolecular interactions. However, structural changes of many other proteins are closely intertwined with the binding or loss of ligands. Time-resolved ESI-MS can simultaneously provide data on protein conformation and ligand binding state as a function of time.

Mammalian hemoglobins possess a heterotetrameric structure, comprising two pairs of heme-containing α and β subunits in a tetrahedral arrangement. Comparatively little effort has been directed towards the processes by which hemoglobin is formed from (or broken down into) its monomeric constituents. This is partially due to the structural heterogeneity of the protein, which makes studies on association/dissociation processes by conventional spectroscopic methods challenging. Hemoglobin exists in equilibrium between several quaternary structures, notably as monomeric (α, β), dimeric (αβ), and tetrameric ($\alpha_2\beta_2$) species. Adding to this complexity is the fact that, in principle, each of the subunits can exist in the heme-bound holo-globin form or as apo-globin (Griffith and Kaltashov, 2003).

The native state binding equilibria of hemoglobin are reflected by the presence of tetramers, dimers, and

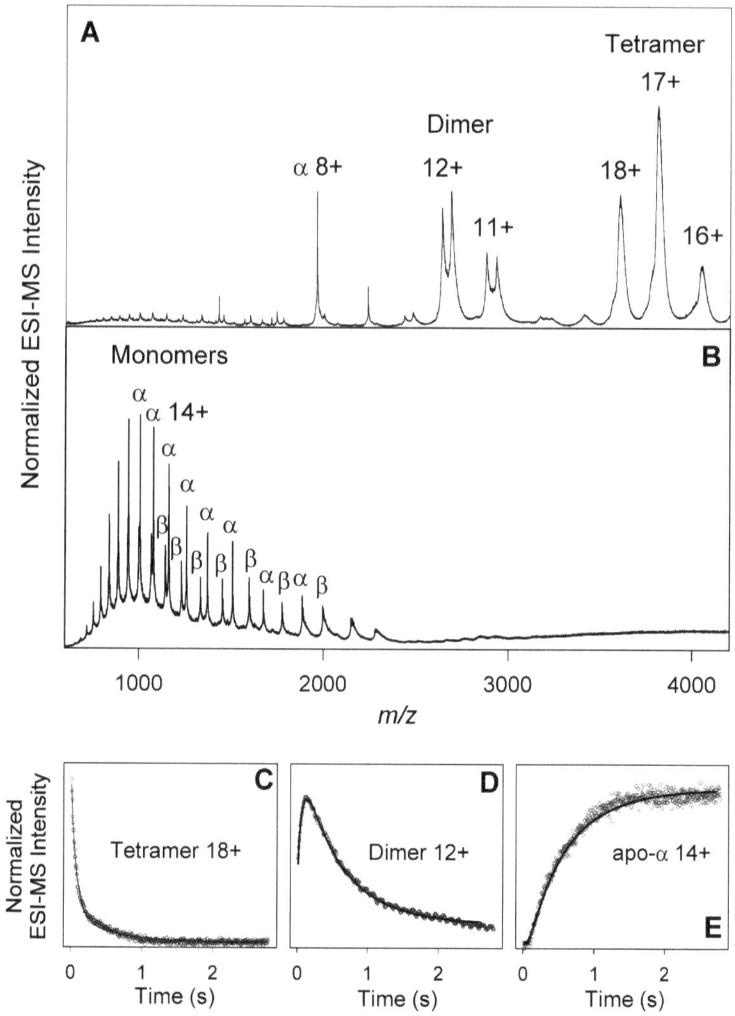


*Fig. 7.* (A) ESI mass spectrum of hemoglobin recorded under native solvent conditions (pH 6.8). The native tetramer appears in charge states 16+–18+. Heterodimeric species in charge states 11+ and 12+ appear as doublets due to partial loss of heme from the β subunit. (B) ESI mass spectrum of acid-denatured hemoglobin at pH 2.8. Both α and β subunits appear exclusively in their apo-forms. Panels C-E display intensity-time profiles for selected ionic species. Solid lines are the results of a global analysis procedure based on Eq. (6). (Simmons et al., 2004, with permission).

monomers in the ESI mass spectrum of Fig. 7A. Markedly different data are obtained after acid exposure of the protein, resulting in a spectrum dominated by highly charged ions that correspond to monomeric apo-α and apo-β subunits in highly unfolded conformations (Fig. 7B). These data reveal that denaturation of hemoglobin under acidic conditions results in the breakdown of the protein's quaternary structure, along with unfolding of the individual subunits and loss of heme. A detailed view of this process is obtained by using time-resolved ESI-MS for recording intensity-time profiles of all the ionic species in the spectrum (Simmons et al., 2004). The resulting data are consistent with an overall reaction mechanism proceeding from folded tetramers to dimeric species to unfolded apo-α and apo-β monomers. Accordingly, tetrameric assemblies exhibit a very rapid decay (Fig. 7C), that is concomitant with an intensity rise of dimeric species (Fig. 7D). These dimers undergo a decay process on a somewhat slower time scale, matching the rise of unfolded monomers (Fig. 7E). When considered in the context of other recent studies on the unfolding kinetics of large protein complexes (Chen and Smith, 2000; Wilson et al., 2005), the hemoglobin data discussed here suggest that the occurrence of complex reaction mechanisms involving several short-lived intermediates is a common feature for the denaturation of large multiprotein complexes.

### C. Studying Protein Conformational Dynamics by Hydrogen Exchange Methods

Many proteins are known to adopt partially folded conformations under mildly denaturing solvent conditions. Studies on these semi-denatured states are of great importance for understanding a wide range of biological processes such as folding, amyloid formation, signal transduction, ligand-binding, and protein transport across membranes. Recent work has shown that some proteins exhibit a significant degree of disorder, even under physiological conditions where they are biologically active (Dyson and Wright, 2002).

Partially folded proteins are structurally heterogeneous ensembles, undergoing conformational fluctuations on time scales ranging from sub-nanoseconds to seconds. Experimental techniques capable of probing the dynamic nature of these species are crucial for obtaining a better understanding of their biophysical properties. One important approach in this area is based the use of hydrogen-deuterium exchange (HDX) (Krishna et al., 2004; Busenlehner and Armstrong, 2005). HDX at sites that are sterically shielded from the solvent, and/or involved in stable hydrogen bonds, is mediated by structural fluctuations of the protein. The exchange kinetics observed upon exposure of a protein to a $D_2O$-containing solvent system are, therefore, related to the conformational dynamics of the polypeptide chain. Brief opening events mediating exchange may correspond to global or subglobal unfolding/refolding transitions, or to local fluctuations (Krishna et al., 2004). The exchange mechanism in a strongly $D_2O$-enriched solvent system is usually described by

$$X-H_{closed} \underset{k_{close}}{\overset{k_{open}}{\rightleftharpoons}} X-H_{open} \xrightarrow[D_2O]{k_{ch}} X-D \qquad (7)$$

where $k_{open}$ and $k_{close}$ are the rate constants for the opening and closing, respectively, of a particular exchangeable site. The chemical exchange rate constant, $k_{ch}$, represents the kinetics that would be expected for a fully unprotected hydrogen. For amide hydrogens, the $k_{ch}$ value of any individual site is strongly affected by pD, and by the nature of neighboring amino acid side chains. Reference data obtained from dipeptide model compounds are available that allow the estimation of amide $k_{ch}$ for any set of solvent conditions (Bai et al., 1993). In the so-called EX2 limit, characterized by $k_{close} >> k_{ch}$, the overall exchange rate constant, $k_{ex}$, is given by $k_{ex} = K_{open}k_{ch}$, where $K_{open} = (k_{open}/k_{close})$ is the equilibrium constant of the opening reaction. Under EX2 conditions, most sites have to visit the open conformation many times before exchange occurs. Conversely, if $k_{ch} >> k_{close}$ (the so-called EX1 regime), complete labeling will occur with the first opening event, such that $k_{ex} = k_{open}$.

NMR spectroscopy has traditionally been the method of choice for analyzing proteins in HDX studies (Krishna et al., 2004). However, in recent years, the use of ESI-MS for this purpose has become increasingly popular (Engen and Smith, 2001; Kaltashov and Eyles, 2002a). Every individual exchange event increases the mass of the protein by one Da. In contrast to NMR, ESI-MS data are not averaged over all the protein molecules in the sample. Instead, co-existing species can be detected, and their HDX properties can be monitored individually. One particularly attractive feature of ESI-MS is the possibility to clearly distinguish EX1 from EX2 exchange events. In the former case, spectra recorded at different times exhibit peaks corresponding to the fully labeled and the unlabeled protein, respectively. As time proceeds, the intensity ratio of these peaks changes, but not their individual mass values. In contrast, EX2 kinetics give rise to a single peak that gradually moves to higher mass as HDX proceeds (Miranker et al., 1996; Konermann and Simmons, 2003; Ferraro et al., 2004).

Many previous MS-based studies have analyzed the HDX pattern of partially labeled proteins using proteolytic digestion approaches (Smith et al., 1997; Engen and Smith, 2001). These experiments provide spatially resolved information, i.e., the degree of structural flexibility can be directly mapped to specific regions along the polypeptide backbone. However, these studies rely on the use of acid quenching, which normally causes the complete breakdown of all protein-protein and protein-ligand interactions. As a consequence, it is not possible to correlate the observed HDX characteristics with individual binding states of a protein in solution. This problem can be circumvented by employing on-line ESI-MS approaches, analogous to the time-resolved experiments described in the preceding sections. Spatial resolution in these on-line studies may be obtained by using top-down gas-phase fragmentation techniques (Kaltashov and Eyles, 2002b; Xiao and Kaltashov, 2005).

Native holo-myoglobin (hMb, 17.5 kDa) has eight α-helices that form a hydrophobic pocket into which a heme group is bound. Unfolding causes a disruption of the heme-protein interactions, thereby generating apo-myoglobin (aMb). In a recent HDX study, the conformational dynamics of myoglobin have been explored in the presence of a moderate concentration of acetonitrile (27% v/v) at pD 9.3 (Simmons et al., 2003). The value of $k_{ch}$ at this pD is around $10^3$ $s^{-1}$. The ESI mass spectrum of the protein recorded under these semi-denaturing conditions reveals the presence of hMb and aMb in various conformations and heme binding states (Fig. 8A). It is noted that this spectrum represents an equilibrium situation; it does not change over time. Nonetheless, it is clear that any equilibrium is dynamic, such that the protein species represented by the different types of ions will be continuously involved in interconversion processes.

This interconversion, along with the structural dynamics of the various conformational species, can be visualized by tracking the mass distributions of the corresponding ionic signals. As an example, $aMb^{14+}$ represents a relatively expanded apo-protein conformation in solution (Fig. 8B-D). With increasing labeling time these ions shift to higher mass,

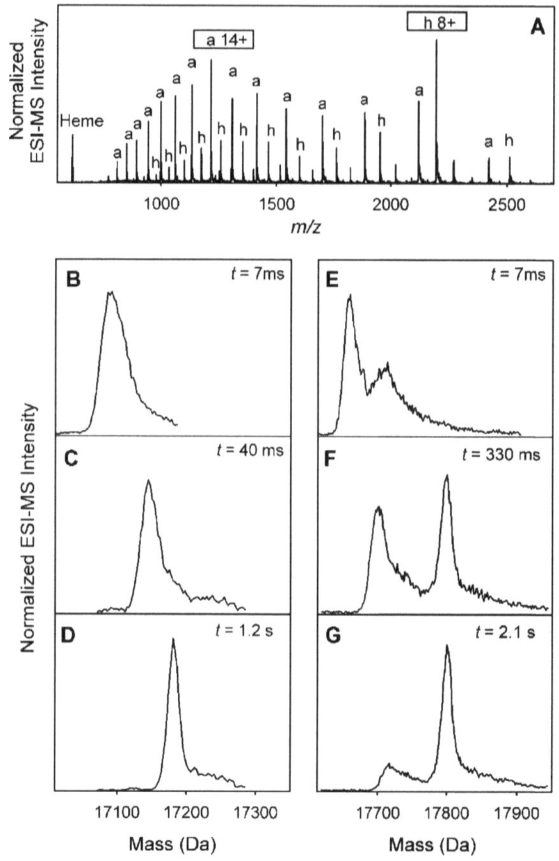


*Fig. 8.* Structural dynamics of myoglobin studied by time-resolved hydrogen-deuterium exchange (HDX) ESI-MS. (A) myoglobin spectrum recorded under semi-denaturing conditions (27% acetonitrile, pH 9.3). Notation: 'h' represents heme-bound holo-myoglobin (hMb), 'a' denotes heme-free apo-myoglobin (aMb). (B-D) mass distributions of the $aMb^{14+}$ ionic species for selected time points. (E-G) mass distributions of the $hMb^{8+}$ ionic species for selected time points. (Simmons et al., 2004, with permission).

indicating EX2 exchange, i.e., the occurrence of rapid unfolding/refolding events with closing rate constants $k_{cl} \gg 10^3$ $s^{-1}$. A more complex exchange behavior is exhibited by $hMb^{8+}$ (Fig. 8E-G), which represents a native-like conformation of the holo-protein. For labeling times up to roughly 2 s, these protein ions show bimodal mass distributions. The relative intensity of the low mass peak decreases, and that of the high mass peak increases (EX1 behavior). This phenomenon indicates the occurrence of a slow aMb/hMb interconversion process in solution. In addition, a gradual shift to higher mass is observed for both peaks, which shows that the proteins also undergo more rapid structural fluctuations (EX2 exchange) (Simmons et al., 2003).

Another interesting approach is the use of double-mixing sequences, e.g., for combining time-resolved ESI-MS studies of the type discussed in III.A with on-line pulsed HDX. Experiments of this kind can provide even more information on protein conformational changes during folding. This is due to the fact that the HDX pattern and the ESI charge state distribution represent non-redundant probes of protein structure in solution. The charge state distribution reflects the overall compactness of the protein, whereas pulsed HDX reports on the intactness of the overall hydrogen bonding network, and on the solvent accessibility of exchangeable sites. The complementary nature of the two probes allows the detection of short-lived intermediates in cases where a cursory analysis indicates two-state behavior (Pan et al., 2005). A detailed discussion of these pulse-labeling approaches is beyond the scope of this chapter. Interested readers should consult the recent review of Konermann and Simmons (2003).

## IV. Conclusions

In this chapter we described how time resolved mass spectrometry can be applied to biophysical questions like enzyme mechanisms and the dynamics and intermediates of protein (un)folding. Some specific examples are given to illustrate the range of different strategies that have been employed in this rapidly growing field. The key advantages of MS over many spectroscopic techniques are certainly the isotope specificity, the large time span that can be analyzed in on-line experiments and the enormous variety of different analytes that can be studied at physiological temperatures in solution. In this way, unique information can be gathered that, together with complementary spectroscopic data, is crucial for developing a comprehensive understanding of biophysical and biochemical reaction mechanisms.

## Acknowledgments

LK thanks his students and post-docs that were involved in the work summarized in this chapter, in particular Douglas A. Simmons, Derek J. Wilson, and Jingxi Pan. Financial support for LK's laboratory was provided by the Natural Sciences and Engineering Research Council of Canada (NSERC), the Canada Foundation for Innovation (CFI), the Provincial Government of Ontario, The University of Western Ontario, and by the Canada Research Chairs Program. JM thanks his PhD student K. Beckmann for critically reading the manuscript and her contributions to the described research. He acknowledges support from the Deutsche Forschungsgemeinschaft (DFG; Me 1629/2-3) and the Max-Planck-Gesellschaft (MPG). WH acknowledges informative discussions with George Hoch, Murray Badger, Richard Radmer and support from the Human Frontiers Science Program Organization (RGP0029/2002).

## References

An SM, Gardner WS and Kana T (2001) Simultaneous measurement of denitrification and nitrogen fixation using isotope pairing with membrane inlet mass spectrometry analysis. Appl Environ Microb 67: 1171–1178

Anfinsen CB (1973) Principles that govern the folding of protein chains. Science 181: 223–230

Bachmann A and Kiefhaber T (2001) Apparent two-state tendamistat folding is a sequential process along a defined route. J Mol Biol 306: 375–386

Bader KP and Roben A (1995) Mass spectrometric detection and analysis of nitrogen fixation in *Oscillatoria chalybea*. Z Naturforsch C 50: 199–204

Bader KP, Thibault P and Schmid GH (1983) A study on oxygen evolution and on the S-state distribution in thylakoid preparations of the filamentous blue-green alga *Oscillatoria chalybea*. Z Naturforsch C 38: 778–792

Bader KP, Thibault P and Schmid GH (1987) Study on the properties of the S3 state by mass spectrometry in the filamentous cyanobacterium *Oscillatoria chalybea*. Biochim Biophys Acta 893: 564–571

Bader KP, Renger G and Schmid GH (1993) A mass spectrometric analysis of the water splitting reaction. Photosynth Res 38: 355–361

Badger MR and Andrews TJ (1982) Photosynthesis and inorganic carbon usage by the marine cyanobacterium *Synechococcus* Sp. Plant Physiol 70: 517–523

Badger MR and Price GD (1989) Carbonic anhydrase activity associated with the cyanobacterium *Synechococcus* PCC7942. Plant Physiol 89: 51–60

Badger MR and Price GD (1994) The role of carbonic anhydrase in photosynthesis. Annu Rev Plant Phys 45: 369–392

Badger MR, Palmqvist K and Yu JW (1994) Measurement of $CO_2$ and $HCO_3^-$ fluxes in cyanobacteria and microalgae during steady state photosynthesis. Physiol Plantarum 90: 529–536

Bai Y, Milne JS, Mayne L and Englander SW (1993) Primary structure effects on peptide group hydrogen exchange. Proteins: Struct Funct Genet 17: 75–86

Baltruschat H (2004) Differential electrochemical mass spectrometry. J Am Soc Mass Spectr 15: 1693–1706

Baltruschat H and Schmiemann U (1993) The adsorption of unsaturated organic species at single crystal electrodes studied by differential electrochemical mass spectrometry. Ber Bunsen

Phys Chem 97: 452–460
Beckmann M and Lloyd D (2001) Mass spectrometric monitoring of gases ($CO_2$, $CH_4$, $O_2$) in a mesotrophic peat core from Kopparas Mire, Sweden. Global Change Biol 7: 171–180
Beckmann M, Sheppard SK and Lloyd D (2004) Mass spectrometric monitoring of gas dynamics in peat monoliths: Effects of temperature and diurnal cycles on emissions. Atmos Environ 38: 6907–6913
Beechem JM, Ameloot M and Brand L (1985) Global and target analysis of complex decay phenomena. Anal Instrum 14: 379–402
Berberan-Santos MN and Martinho JMG (1990) The integration of kinetic rate equations by matrix methods. J Chem Ed 67: 375–379
Bethke PC, Badger MR and Jones RL (2004) Apoplastic synthesis of nitric oxide by plant tissues. Plant Cell 16: 332–341
Britt RD, Campbell KA, Peloquin JM, Gilchrist ML, Aznar CP, Dicus MM, Robblee J and Messinger J (2004) Recent pulsed EPR studies of the Photosystem II oxygen evolving complex: Implications as to water oxidation mechanisms. Biochim Biophys Acta 1655: 158–171
Bruckenstein S and Gadde RR (1971) Use of a porous electrode for in situ mass spectrometric determination of volatile electrode reaction products. J Am Chem Soc 93: 793–794
Bruins AP, Covey TR and Henion JD (1987) Ion spray interface for combined liquid chromatography/atmospheric pressure ionization mass spectrometry. Anal Chem 59: 2642–2646
Brutscher B, Brüschweiler R and Ernst RR (1997) Backbone dynamics and structural characterization of the partially folded A state of ubiquitin by $^{1}H$, $^{13}C$ and $^{15}N$ nuclear magnetic resonance spectroscopy. Biochemistry 36: 13043–13053
Busenlehner LS and Armstrong RN (2005) Insights into enzyme structure and dynamics elucidated by amide H/D exchange mass spectrometry. Arch Biochem Biophys 433: 34–46
Calvo KC, Weisenberger CR, Anderson LB and Klapper MH (1981) Permeable membrane mass spectrometric measurement of reaction kinetics. Anal Chem 53: 981–985
Canvin DT, Berry JA, Badger MR, Fock H and Osmond CB (1980) Oxygen exchange in leaves in the light. Plant Physiol 66: 302–307
Cartaxana P and Lloyd D (1999) $N_2$, $N_2O$ and $O_2$ profiles in a Tagus estuary salt marsh. Estuar Coast Shelf S 48: 751–756
Chemical Rubber Company (2005) CRC Handbook of Chemistry and Physics. Chemical Rubber, Cleveland
Chen J and Smith DL (2000) Unfolding and disassembly of the chaperonin GroEL occurs via a tetradecameric intermediate with a folded equatorial domain. Biochemistry 39: 4250–4258
Chernushevich IV, Loboda AV and Thomson BA (2001) An introduction to quadrupole time-of-flight mass spectrometry. J Mass Spectrom 36: 849–865
Chowdhury SK, Katta V and Chait BT (1990) Probing conformational changes in proteins by mass spectrometry. J Am Chem Soc 112: 9012–9013
Christianson DW and Fierke CA (1996) Carbonic anhydrase: Evolution of the zinc binding site by nature and by design. Acc Chem Res 29: 331–339
Clausen J, Beckmann K, Junge W and Messinger J (2005) Evidence that bicarbonate is not the substrate in photosynthetic oxygen evolution. Plant Physiol 139: 1444–1450
Conrath U, Amoroso G, Kohle H and Sultemeyer DF (2004) Non-invasive online detection of nitric oxide from plants and some other organisms by mass spectrometry. Plant J 38: 1015–1022
Cournac L, Redding K, Ravenel J, Rumeau D, Josse EM, Kuntz M and Peltier G (2000) Electron flow between Photosystem II and oxygen in chloroplasts of Photosystem I deficient algae is mediated by a quinol oxidase involved in chlororespiration. J Biol Chem 275: 17256–17262
Demmers JAA, Haverkamp J, Heck AJR, Koeppe RE and Killian A (2000) Electrospray ionization mass spectrometry as a tool to analyze hydrogen/deuterium exchange kinetics of transmembrane peptides in lipid bilayers. Proc Natl Acad Sci USA 97: 3189–3194
Dobson CM (2003) Protein folding and misfolding. Nature 426: 884–890
Douglas DJ, Frank AJ and Mao D (2005) Linear ion traps in mass spectrometry. Mass Spectrom Rev 24: 1–29
Duda DM, Tu CK, Fisher SZ, An HQ, Yoshioka C, Govindasamy L, Laipis PJ, Agbandje-McKenna M, Silverman DN and McKenna R (2005) Human carbonic anhydrase III: Structural and kinetic study of catalysis and proton transfer. Biochemistry 44: 10046–10053
Dyson HJ and Wright PE (2002) Coupling of folding and binding for unstructured proteins. Curr Op Struct Biol 12: 54–60
Engen JR and Smith DL (2001) Investigating protein structure and dynamics by hydrogen exchange mass spectrometry. Anal Chem 73: 256A–265A
Fenn JB (2003) Electrospray wings for molecular elephants (Nobel Lecture). Angew Chem Int Ed 42: 3871–3894
Fenn JB, Mann M, Meng CK, Wong SF and Whitehouse CM (1989) Electrospray ionization for mass spectrometry of large biomolecules. Science 246: 64–71
Ferraro DM, Lazo ND and Robertson AD (2004) EX1 hydrogen exchange and protein folding. Biochemistry 43: 587–594
Fisher Z, Prada JAH, Tu C, Duda D, Yoshioka C, An HQ, Govindasamy L, Silverman DN and McKenna R (2005) Structural and kinetic characterization of active site histidine as a proton shuttle in catalysis by human carbonic anhydrase II. Biochemistry 44: 1097–1105
Govindjee, Owens OV and Hock G (1963) A mass-spectroscopic study of the Emerson enhancement effect. Biochim Biophys Acta 75: 281–284
Griffith WP and Kaltashov IA (2003) Highly asymmetric interactions between globin chains during hemoglobin assembly revealed by electrospray ionization mass spectrometry. Biochemistry 42: 10024–10033
Gross JW and Frey PA (2002) Rapid mix-quench MALDI-TOF mass spectrometry for analysis of enzymatic systems. Meth Enzymol 354: 27–49
Hanson DT, Franklin LA, Samuelsson G and Badger MR (2003) The *Chlamydomonas reinhardtii* cia3 mutant lacking a thylakoid lumen-localized carbonic anhydrase is limited by $CO_2$ supply to rubisco and not Photosystem II function in vivo. Plant Physiol 132: 2267–2275
Hartung T and Baltruschat H (1990) Differential electrochemical mass spectrometry using smooth electrodes: Adsorption and H/D-exchange reactions of benzene on Pt. Langmuir 6: 953–957
Heck AJR and Van den Heuvel RHH (2004) Investigation of intact protein complexes by mass spectrometry. Mass Spectrom Rev 23: 368–389

Higuchi Y and Yagi T (1999) Liberation of hydrogen sulfide during the catalytic action of *Desulfovibrio* hydrogenase under the atmosphere of hydrogen. Biophys Res Comm 255: 295–299

Higuchi Y, Toujou F, Tsukamoto K and Yagi T (2000) The presence of a SO molecule in [NiFe] hydrogenase from *Desulfovibrio vulgaris Miyazaki* as detected by mass spectrometry. J Inorg Biochem 80: 205–211

Hillier W and Messinger J (2005) Mechanism of photosynthetic oxygen production. In: Wydrzynski T and Satoh K (eds) Photosystem II: The Water/Plastoquinone Oxidoreductase in Photosynthesis (Advances in Photosynthesis and Respiration, Vol 22) pp 567–608. Springer, Dordrecht

Hillier W and Wydrzynski T (2004) Substrate water interactions within the Photosystem II oxygen evolving complex. Phys Chem Chem Phys 6: 4882–4889

Hillier W, Messinger J and Wydrzynski T (1998) Kinetic determination of the fast exchanging substrate water molecule in the $S_3$ state of Photosystem II. Biochemistry 37: 16908–16914

Hillier W, McConnell I, Badger MR, Boussac A, Klimov VV, Dismukes GC and Wydrzynski T (2006) Quantitative assessment of intrinsic carbonic anhydrase activity and the capacity for bicarbonate oxidation in Photosystem II. Biochemistry 5: 2094–102

Hoch G and Kok B (1963) A mass spectrometer inlet system for sampling gases dissolved in liquid phases. Arch Biochem Biophys 101: 160–170

Hoch G, Owens OHV and Kok B (1963) Photosynthesis and Respiration. Arch Biochem Biophys 101: 171–180

Holzwarth AR (1995) Time-resolved fluorescence spectroscopy. Meth Enzymol 246: 334–362

Houston CT, Taylor WP, Widlanski TS and Reilly JP (2000) Investigation of enzyme kinetics using quench-flow techniques with MALDI TOF mass spectrometry. Anal Chem 72: 3311–3319

Johnson RC, Cooks RG, Allen TM, Cisper ME and Hemberger PH (2000) Membrane introduction mass spectrometry: Trends and applications. Mass Spectrom Rev 19: 1–37

Kaltashov IA and Eyles SJ (2002a) Crossing the phase boundary to study protein dynamics and function: Combination of amide hydrogen exchange in solution and ion fragmentation in the gas phase. J Mass Spectrom 37: 557–565

Kaltashov IA and Eyles SJ (2002b) Studies of biomolecular conformations and conformational dynamics by mass spectrometry. Mass Spectrom Rev 21: 37–71

Kaltashov IA and Eyles SJ (2005) Mass spectrometry in biophysics. John Wiley and Sons, Hoboken

Karas M and Hillenkamp F (1988) Laser desorption ionization of proteins with molecular masses exceeding 10,000 Daltons. Anal Chem 60: 2299–2301

Kebarle P and Ho Y (1997) On the mechanism of electrospray mass spectrometry. In: Cole RB (ed) Electrospray Ionization Mass Spectrometry, pp 3–63. John Wiley and Sons, New York

Knight JB, Vishwanath A, Brody JP and Austin RH (1998) Hydrodynamic focusing on a silicon chip: Mixing nanoliters in microseconds. Phys Rev Lett 80: 3863–3866

Kok B and Varner JE (1967) Extraterrestial life detection based on oxygen isotope exchange reactions. Science 155: 1110–1112

Kolakowski BM and Konermann L (2001) From small-molecule reactions to protein folding: Studying biochemical kinetics by stopped-flow electrospray mass spectrometry. Anal Biochem 292: 107–114

Konermann L (1999) Monitoring reaction kinetics by continuous-flow methods: The effects of convection and molecular diffusion under laminar flow conditions. J Phys Chem A 103: 7210–7216

Konermann L and Douglas DJ (2002) Pre-steady-state kinetics of enzymatic reactions studied by electrospray mass spectrometry with on-line rapid-mixing techniques. Meth Enzymol 354: 50–64

Konermann L and Simmons DA (2003) Protein-folding kinetics and mechanisms studied by pulse-labeling and mass spectrometry. Mass Spectrom Rev 22: 1–26

Konermann L, Collings BA and Douglas DJ (1997) Cytochrome *c* folding kinetics studied by time-resolved electrospray ionization mass spectrometry. Biochemistry 36: 5554–5559

Konermann L, Silva EA and Sogbein OF (2001) Electrochemically induced pH changes resulting in protein unfolding in the ion source of an electrospray mass spectrometer. Anal Chem 73: 4836–4844

Krasna AI (1978) Oxygen stable hydrogenase and assay. Meth Enzymol 53: 296–314

Kretschmann H and Witt HT (1993) Chemical reduction of the water splitting enzyme system of photosynthesis and its light induced reoxidation characterized by optical and mass spectrometric measurements: A basis for the estimation of the states of the redox active manganese and of water in the quaternary oxygen evolving S-state cycle. Biochim Biophys Acta 1144: 331–345

Krishna MMG, Hoang L, Lin Y and Englander SW (2004) Hydrogen exchange methods to study protein folding. Methods 34: 51–64

Laiko VV, Baldwin MA and Burlingame AL (2000) Atmospheric pressure matrix-assisted laser desorption/ionization mass spectrometry. Anal Chem 72: 652–657

Li Z, Sau AK, Shen S, Whitehouse C, Baasov T and Anderson KS (2003) A snapshot of enzyme catalysis using electrospray mass spectrometry. J Am Chem Soc 125: 9938–9939

Liesener A and Karst U (2005) Monitoring enzymatic conversions by mass spectrometry: A critical review. Anal Bioanal Chem 382: 1451–1464

Lindskog S and Coleman JE (1973) Catalytic mechanism of carbonic anhydrase. Proc Natl Acad Sci USA 70: 2505–2508

Lloyd D, Thomas KL, Cowie G, Tammam JD and Williams AG (2002) Direct interface of chemistry to microbiological systems: Membrane inlet mass spectrometry. J Microbiol Meth 48: 289–302

Mano J, Takahashi M-A and Asada K (1987) Oxygen evolution from hydrogen peroxide in Photosystem II: Flash induced catalytic activity of water oxidizing Photosystem II membranes. Biochemistry 26: 2495–2501

Mao FM and Leck JH (1987) The quadrupole mass spectrometer in practical operation. Vacuum 37: 669–675

Marshall AG, Hendrickson CL and Jackson GS (1998) Fourier transform ion cyclotron resonance mass spectrometry: A primer. Mass Spectrom Rev 17: 1–35

Martin JP, Johnson RD, Kok B and Radmer R (1975) Unified mars life detection system. J Astronaut Sci 23: 99–119

Maxwell K, Badger MR and Osmond CB (1998) A comparison of $CO_2$ and $O_2$ exchange patterns and the relationship with chlorophyll fluorescence during photosynthesis in $C_3$ and CAM plants. Aust J Plant Physiol 25: 45–52

McEvoy JP and Brudvig GW (2004) Structure based mechanism of photosynthetic water oxidation. Phys Chem Chem Phys 6: 4754–4763

Messinger J (2004) Evaluation of different mechanistic proposals for water oxidation in photosynthesis on the basis of $Mn_4O_xCa$ structures for the catalytic site and spectroscopic data. Phys Chem Chem Phys 6: 4764–4771

Messinger J, Badger M and Wydrzynski T (1995) Detection of one slowly exchanging substrate water molecule in the $S_3$ state of Photosystem II. Proc Natl Acad Sci USA 92: 3209–3213

Meyer S, Koch R and Metzger JO (2003) Investigation of reactive intermediates of chemical reactions in solution by electrospray ionization mass spectrometry: Radical cation chain reactions. Angew Chem Int Edit 42: 4700–4703

Miranker A, Robinson CV, Radford SE and Dobson CM (1996) Investigation of protein folding by mass spectrometry. FASEB J 10: 93–101

Norris AJ, Whitelegge JP, Faull KF and Toyokuni T (2001) Analysis of enzyme kinetics using electrospray ionization mass spectrometry and multiple reaction monitoring: Fucosyltransferase V. Biochemistry 40: 3774–3779

Northrop DB and Simpson FB (1997) Beyond enzyme kinetics: Direct determination of mechanisms by stopped-flow mass spectrometry. Bioorg Med Chem 5: 641–644

Pain RH (2000) Mechanisms of Protein Folding. Oxford University Press, New York

Pan JX, Wilson DJ and Konermann L (2005) Pulsed hydrogen exchange and electrospray charge-state distribution as complementary probes of protein structure in kinetic experiments: Implications for ubiquitin folding. Biochemistry 44: 8627–8633

Poulsen AK, Rompel A and McKenzie CJ (2005) Water oxidation catalyzed by a dinuclear Mn complex: A functional model for the oxygen-evolving center of Photosystem II. Angew Chem Int Edit 44: 6916–6920

Prior JJ, Christie PD, Murray RJ, Ormejohnson WH and Cooney CL (1995) Continuous monitoring of nitrogenase activity in *Azotobacter vinelandii* fermentation using off-gas mass spectrometry. Biotechnol Bioeng 47: 373–383

Radmer R (1979) Mass spectrometric determination of hydroxylamine photo-oxidation by illuminated chloroplasts. Biochim Biophys Acta 546: 418–425

Radmer R and Kok B (1971) Unified procedure for detection of life on mars. Science 174: 233–239

Radmer R and Ollinger O (1980a) Measurement of the oxygen cycle: The mass spectrometric analysis of gases dissolved in a liquid phase. Meth Enzymol 69: 547–560

Radmer R and Ollinger O (1980b) Light driven uptake of oxygen, carbon dioxide, and bicarbonate by the green algae *Scenedesmus*. Plant Physiol 65: 723–729

Radmer R and Ollinger O (1980c) Isotopic composition of photosynthetic $O_2$ flash yields in the presence of $H_2{}^{18}O$ and $HC^{18}O_3^-$. FEBS Lett 110: 57–61

Radmer R and Ollinger O (1981) Mass spectrometric studies of hydrazine photooxidation by illuminated chloroplasts. Biochim Biophys Acta 637: 80–87

Radmer R and Ollinger O (1982) Nitrogen and oxygen evolution by hydroxylamine treated chloroplasts. FEBS Lett 144: 162–166

Radmer R and Ollinger O (1983) Topography of the $O_2$ evolving site determined with water analogs. FEBS Lett 152: 39–43

Radmer R and Ollinger O (1986) Do the higher oxidation states of the photosynthetic $O_2$ evolving system contain bound $H_2O$. FEBS Lett 195: 285–289

Radmer RJ, Kok B and Martin JP (1976) System for biological and soil chemical tests on a planetary lander. J Spacecraft Rockets 13: 719–726

Renger G (2001) Photosynthetic water oxidation to molecular oxygen: Apparatus and mechanism. Biochim Biophys Acta 1503: 210–228

Ribas-Carbo M, Robinson SA and Giles L (2005) The application of oxygen isotope technique to respiratory pathway partitioning. In: Lambers H and Ribas-Carbo M (eds) Plant Respiration: From Cell to Ecosystem (Advances in Photosynthesis and Respiration, Vol 18), pp 31–42. Springer, Dordrecht

Roboz J (1968) Mass spectrometry: Instrumentation and techniques. Wiley, New York

Ruuska SA, Badger MR, Andrews TJ and von Caemmerer S (2000) Photosynthetic electron sinks in transgenic tobacco with reduced amounts of Rubisco: Little evidence for significant Mehler reaction. J Exp Bot 51: 357–368

Schermann SM, Simmons DA and Konermann L (2005) Mass spectrometry-based approaches to protein-ligand interactions. Exp Rev Proteomics 2: 475–485

Shastry MCR, Luck SD and Roder H (1998) A continuous-flow mixing method to monitor reactions on the microsecond time scale. Biophys J 74: 2714–2721

Shima S, Lyon EJ, Thauer RK, Mienert B and Bill E (2005) Mössbauer studies of the iron-sulfur cluster-free hydrogenase: The electronic state of the mononuclear Fe active site. J Am Chem Soc 127: 10430–10435

Shoemaker BA, Portman JJ and Wolynes PG (2000) Speeding molecular recognition by using the folding funnel: The fly-casting mechanism. Proc Natl Acad Sci USA 97: 8868–8873

Silva ACB, Augusti R, Dalmazio I, Windmoller D and Lago RM (1999) MIMS evaluation of pervaporation processes. Phys Chem Chem Phys 1: 2501–2504

Silverman DN (1982) Carbonic anhydrase $^{18}O$ exchange catalyzed by an enzyme with rate contributing proton transfer steps. Methods in Enzymology 87: 732–752

Silverman DN and Tu CK (1976) Carbonic anhydrase catalyzed hydration studied by $^{13}C$ and $^{18}O$ labeling of carbon dioxide. J Am Chem Soc 98: 978–984

Simmons DA, Dunn SD and Konermann L (2003) Conformational dynamics of partially denatured myoglobin studied by time-resolved electrospray mass spectrometry with online hydrogen-deuterium exchange. Biochemistry 42: 9248–9248

Simmons DA, Wilson DJ, Lajoie GA, Doherty-Kirby A and Konermann L (2004) Subunit disassembly and unfolding kinetics of hemoglobin studied by time-resolved electrospray mass spectrometry. Biochemistry 43: 14792–14801

Siuzdak G (1996) Mass spectrometry for biotechnology. Academic Press, New York

Smith DL, Deng Y and Zhang Z (1997) Probing the noncovalent structure of proteins by amide hydrogen exchange mass spectrometry. J Mass Spectrom 32: 135–146

So AKC, Van Spall HGC, Coleman JR and Espie GS (1998) Catalytic exchange of $^{18}O$ from $^{13}C^{18}O$-labelled $CO_2$ by wild-type cells and *ecaA, ecaB,* and ccaA mutants of the cyanobacteria *Synechococcus* PCC7942 and *Synechocystis* PCC6803. Can J Bot 76: 1153–1160

Tanaka K (2003) The origin of macromolecule ionization by laser irradiation (Nobel Lecture). Angew Chem Int Edit 42: 3861–3870

Tegtmeyer D, Heindrichs A and Heitbaum J (1989) Electrochemical on line mass spectrometry on a rotating electrode

inlet system. Ber Bunsen Phys Chem 93: 201–206
Tortell PD (2005) Dissolved gas measurements in oceanic waters made by membrane inlet mass spectrometry. Limnol Oceanogr Meth 3: 24–37
Vignais PM (2005) H/D exchange reactions and mechanistic aspects of the hydrogenases. Coordin Chem Rev 249: 1677–1690
Whitelegge JP, Le Coutre J, Lee JC, Engel CK, Prive GG, Faull KF and Kaback HR (1999) Toward the bilayer proteome, electrospray ionization-mass spectrometry of large, intact transmembrane proteins. Proc Natl Acad Sci USA 96: 10695–10698
Wilson DJ and Konermann L (2003) A capillary mixer with adjustable reaction chamber volume for millisecond time-resolved studies by electrospray mass spectrometry. Anal Chem 75: 6408–6414
Wilson DJ and Konermann L (2004) Mechanistic studies on enzymatic reactions by electrospray ionization MS using a capillary mixer with adjustable reaction chamber volume for time resolved measurements. Anal Chem 76: 2537–2543
Wilson DJ and Konermann L (2005) Ultrarapid desalting of protein solutions for electrospray mass spectrometry in a microchannel laminar flow device. Anal Chem 77: 6887–6894
Wilson DJ, Rafferty SP and Konermann L (2005) Kinetic unfolding mechanism of the inducible nitric oxide synthase oxygenase domain determined by time-resolved electrospray mass spectrometry. Biochemistry 44: 2276–2283
Wittung-Stafshede P (2002) Role of cofactors in protein folding. Acc Chem Res 35: 201–208
Wolter O and Heitbaum J (1984) Differential electrochemical mass spectroscopy (DEMS) — a new method for the study of electrode processes. Ber Bunsen Phys Chem 88: 2–6
Xiao H and Kaltashov IA (2005) Transient structural disorder as a facilitator of protein ligand binding: Native H/D exchange-mass spectrometry study of cellular retinoic acid binding protein I. J Am Soc Mass Spectrom 16: 869–879
Yano J, Kern J, Sauer K, Latimer MJ, Pushkar Y, Biesiadka J, Loll B, Saenger W, Messinger J, Zouni A and Yachandra VK (2006) Where water is oxidized to dioxygene: Structure of the photosynthetic $Mn_4Ca$ cluster. Science 314: 821–825
Zechel DL, Konermann L, Withers SG and Douglas DJ (1998) Pre-steady state kinetic analysis of an enzymatic reaction monitored by time-resolved electrospray ionization mass spectrometry. Biochemistry 37: 7664–7669

Chapter 10

# Femtosecond Time-Resolved Infrared Spectroscopy

Marie Louise Groot* and Rienk van Grondelle
*Department of Physics and Astronomy, Faculty of Sciences, Vrije Universiteit, De Boelelaan 1081, 1081 HV Amsterdam, The Netherlands*



## Summary

In this chapter we describe how femtosecond time-resolved infrared spectroscopy is useful for the study of the dynamics of pigment-protein complexes, and what the technical requirements are to perform such experiments. We further discuss a few examples of experiments performed on photosynthetic complexes in more detail.

## I. Introduction

The vibrational spectrum of a protein or a protein-bound chromophore contains a wealth of information about its structure, the interaction with the environment and the electronic properties of the molecule. The chlorophylls are spectroscopically 'visible' in the IR through their absorption due to the C=O stretching modes, the skeletal C-C and C-N stretch modes and the $CH_3$C-H bending modes (see Fig. 1). For instance, the precise frequency of the keto C=O mode of a chlorophyll molecule depends on the presence (and strength) of a hydrogen bond, the polarity of the environment, whether the chlorophyll is in the ground or in the excited state, and whether it is in the neutral, anion or cation state. Therefore IR spectroscopy is a powerful tool that can reveal many of the dynamic structural details of chlorophylls involved in the primary events in photosynthesis, such as excitation energy transfer and electron transfer. Of course, the same is true for other (photoactive) proteins. In addition, IR spectroscopy can reveal the response of those parts of the protein that are affected by the reactions that are taking place. Following reac-

*Author for correspondence, email: ml.groot@few.vu.nl

*Thijs J. Aartsma and Jörg Matysik (eds): Biophysical Techniques in Photosynthesis II, pp. 191–200.*

tion-induced infrared absorption changes *in real time* offers the exciting possibility to observe in real time the dynamic changes of the chromophore and protein residues and relate those, with the help of (static) X-ray diffraction structures, to physical reaction mechanisms and protein structural changes.

The spectral region that is most informative for chlorophyll molecules and protein residues is between 2500 and 800 $cm^{-1}$. In this region also the backbone of the protein absorbs: The amide I (mainly C=O stretching), II (C-N stretching coupled with N-H bending) and III (C-N stretching, N-H bending and C-C stretching) are at 1620–1690 $cm^{-1}$, ~1550 $cm^{-1}$ and ~1300 $cm^{-1}$, respectively, with the precise frequency being dependent on the secondary structure of the protein. For example the amount of α-helix or β-sheet in a protein complex can be deduced from the precise position and shape of the amide I and II bands (for a review, see Barth and Zscherp, 2002). In a typical photosynthetic pigment-protein the absorption in the mid-infrared will be dominated by absorption of the protein, while the absorption of the chlorophyll molecules will usually make only a very minor contribution of ~$10^{-3}$ – $10^{-4}$, depending on the size of the protein. The extinction coefficient of a peptide C=O in an α-helix is between 600–800 $M^{-1}cm^{-1}$ (Venyaminov and Kalnin, 1990). The vibrational extinction coefficients are much lower than those of its electronic transitions. An estimate from our own data from CP47 (a light-harvesting complex associated with Photosystem II (PS II)) yields an extinction coefficient of about 500 $M^{-1}cm^{-1}$ for the Chl keto C=O group (Groot et al., 2004), similar to that estimated for the Chls of P700 (the primary donor) in Photosystem I (PS I) (Hamacher et al., 1996), although that of BChl seems to be larger (M. L. Groot, unpublished). As a consequence we are faced with the situation that even the signals from the strongest

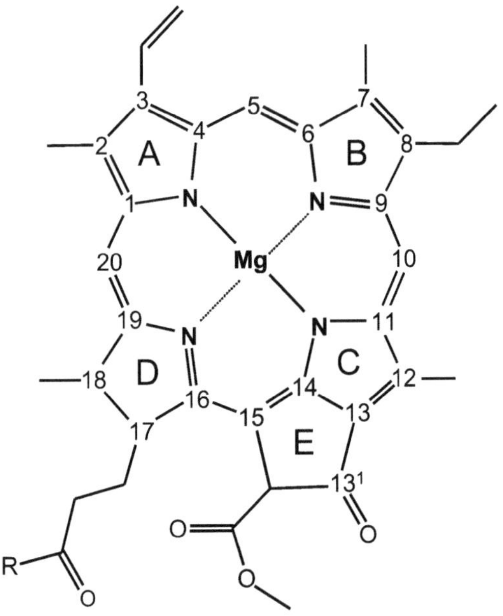


*Fig. 1.* Molecular structure of chlorophyll *a*. The keto and ester groups, located on pyrrole ring E, are important for IR spectroscopy.

chlorophyll modes, those due to the keto group (see Fig. 1), are about 200 times smaller than those of the electronic transitions, whereas the total absorption of the protein sample in the IR may be similar to that of the electronic transition. Transient ΔOD signals are therefore typically 1 mOD or smaller. The classical method of time-resolved IR spectroscopy is step-scan Fourier Transform Spectroscopy where, by the multiplex advantage of interferometry, high signal to noise ratios can be obtained. This technique was reviewed by Mäntele (1996) in Volume 3 of this series (*Advances in Photosynthesis and Respiration*), and one of the most recent reviews is that by (Kotting and Gerwert, 2005), where more references to this subject can be found. However, to gain access to the femto-, pico- and nanosecond time scale, the so-called pump-probe absorption difference technique must be used, where the first ultra-short laser pulse starts the photoreaction, while a second fs-IR pulse, delayed by a certain time interval precisely determined by a scanning delay line, sets the moment of probing. The consequences of shot-to-shot instabilities of the laser to the data are reduced through detection of a full difference spectrum for each laser shot, achieved by dispersing the probe after the sample and measuring the intensity at each color using an array of infrared detectors. The spectral window in this type of experiment is determined by the bandwidth of the IR

*Abbreviations:* BBO – β-$BaB_2O_4$; BChl – bacteriochlorophyll; BR – bacteriorhodopsin; Chl – Chlorophyll; CP47 – light-harvesting complex associated with PS II; CW – continuous wave; DFG – difference frequency generation; fs – femtosecond; FTIR – Fourier-transform infrared; H – bacteriopheophytin; IR – Infrared; MCT – mercury cadmium telluride (HgCdTe); NMR – nuclear magnetic resonance; NOPA – non-collinear optical parametric amplifier; OD – optical density; OPA – optical parametric amplifier; OPG – optical parametric generator; P – primary donor; ps – picosecond; P700 – primary donor of PS I; PS I – Photosystem I; PS II – Photosystem II; PYP – Photoactive Yellow Protein; Q – quinone; RC – reaction center; RP1, RP2 – radical pair 1 (2); S/N – signal to noise; THF – tetrahydrofuran; Ti:Sapph – titan-sapphire

pulse, which in turn is related to the pulse duration (for a Gaussian pulse, the time bandwidth product is $\Delta\omega\Delta t \geq 0.44$).

## A. History

In the following we will give a brief (and certainly not complete) overview of the development of femtosecond vibrational techniques with the emphasis on its application to photosynthetic proteins.

For photosynthetic complexes the visible pump/midIR probe method was pioneered by Hochstrasser and coworkers who in 1993 published a study on bacterial reaction centers in the region from 1800 to 1550 $cm^{-1}$ with a time resolution of 60 ps (Maiti et al., 1993). They measured single wavelength traces using a tunable CW CO laser as probe, which was upconverted by a short 800-nm laser pulse in a nonlinear crystal, and detected the intensity changes with a photomultiplier. Subsequently they improved the time resolution to 0.3–0.4 ps (Maiti et al., 1994, Walker et al., 1994), which allowed observation of the P* state and the formation of the charge separated state $P^+H^-$ in 3 ps, as well as the characteristics of electronic states corresponding to a P* interexciton transition and that of the $P^+$ electron-hole transition between the two halves of the dimer at around 2400 $cm^{-1}$. It was proposed that in response to the formation of this radical pair state, small modifications of (many) protein peptide frequencies gave rise to a 1665 $cm^{-1}$ band shift, but no major protein response was observed. In 1995, the group of Zinth employed an array of 10 infrared detectors after dispersion of the IR-probe pulse in a grating spectrometer, monitoring the full bandwidth of the IR probe pulses (65 $cm^{-1}$) directly after each laser shot. Also in this work, the difference spectra of P*, $P^+H_A^-$ and $P^+Q_A^-$ were measured, of which the latter showed good agreement with those obtained by steady state FTIR techniques (Hamm et al., 1995).

At this point it is worthwhile to mention that at the same time Anfinrud and coworkers used visible pump/IR probes spectroscopy to monitor the photodissociation of CO in myoglobin. They used a 250 $cm^{-1}$ broad IR probe pulse that was sent through a monochromator and recorded with a single IR detector (Lim et al., 1995a,b). With the Stanford free electron laser as a light source, IR photon echo spectroscopy was used to study the vibrational dynamics of CO-myoglobin (Hill et al., 1994, 1996). More recently two new IR spectroscopic techniques were developed to study the CO photodissociation dynamics in myoglobin. The first technique, coherent infrared emission spectroscopy, uses interference of infrared emission with a single cycle IR pulse, by which the IR emitted field of the vibrationally excited dipoles of the heme and protein residues can be measured, and information on the phase and amplitude of the oscillating IR dipoles is retrieved (Groot et al., 2002). In the second method, photodissociation of CO is induced by vibrational-ladder climbing in the CO stretch of carboxyhemoglobin using a chirped infrared excitation pulse as pump (Ventalon et al., 2004). Also bacteriorhodopsin was one of the earliest proteins to be investigated with picosecond IR spectroscopy (Diller et al., 1992). In both bacteriorhodopsin (BR) and the bacterial signal transductor, Photoactive Yellow Protein (PYP), femtosecond-IR spectroscopy has recently yielded direct information on the isomerization time of the chromophores: 0.5 ps for BR (Herbst et al., 2002) and 2 ps for PYP (Groot et al., 2003), respectively. In addition, detailed information was obtained on the changed interaction of the PYP chromophore with its hydrogen bond network (Groot et al., 2003) and the ultrafast change in dipole moment of BR upon excitation (Groma et al., 2004).

In the above described techniques, the photo-induced spectral changes in the mid-IR are measured to profit from the high information density concerning the electronic nature (such as for the BChl molecules in the bacterial reaction center) or the structure of the molecules under study (such as the retinal and p-coumaric acid molecules in BR and PYP, respectively), and to measure the response of the protein environment to the reaction. Another line of research is to obtain information on the structure of molecules, such as polypeptides, by measuring the electronic coupling between vibrational dipoles, using an approach analogous to coherent pulse-techniques in NMR spectroscopy. By exciting one vibrational mode, the response of another mode to this excitation reveals that it is sufficiently close by to sense the change in energy. By scanning the excitation laser over the entire spectrum of interest and monitoring the induced absorption changes at each excitation frequency, a two-dimensional image is obtained that reveals the couplings between the different oscillators. This method was pioneered by the groups of Hochstrasser and Hamm (Hamm et al., 1998), and is mainly used to study small polypeptides (Demirdoven et al., 2004; Bredenbeck et al., 2005; Kim and Hochstrasser, 2005, Rubtsov et al., 2005), but can also be used to study water dynamics and that of organic molecules in solvents (Eaves et al., 2005). The extraction of the relevant parameters from the experiments is made possible by the development of the corresponding theory (Scheurer and Mukamel, 2002; Dreyer et al., 2003; Hahn et al., 2005). The appealing aspect of the technique is that a *dynamic* structure is obtained on a time scale much faster than NMR.

With the further development of titanium sapphire lasers into all-solid state, compact laser systems with high short- and long term stability, femtosecond visible pump/mid-IR probe spectroscopy has become possible on protein complexes with inherently smaller signals (due to lower extinction coefficients and larger protein sizes) such as the chlorophyll *a* containing PS II reaction center from plants and its core antenna proteins (Groot et al., 2004, 2005), and even on the PS I and PS II core complexes.

In the following we will first discuss the nonlinear methods most commonly used to generate the mid-infrared probe pulses, describe the time-resolved vibrational spectroscopy setup that we have developed in the Laser Centre of the Vrije Universiteit in Amsterdam especially for the study of proteins and enzymes, and then briefly discuss some results obtained on photosynthetic proteins in more detail.

## II. Technique

### A. Lasers, Nonlinear Methods and Detection

Pulses tunable in the 3500–800 $cm^{-1}$ range can be obtained through difference frequency mixing (DFG) in an appropriate nonlinear crystal (e.g., $AgGaS_2$) of two pulses of appropriate frequency. These two pulses are obtained through optical parametric generation and amplification using the 800 nm fundamental output of a regeneratively amplified titanium-sapphire (Ti:Sapph) laser system. Alternatively a white light-seeded optical parametric amplifier (OPA) configuration can be used. Generally the nonlinear crystal(s) used in the OPA is BBO ($\beta$-$BaB_2O_4$). The output of such a OPA/DFG combination is typically a few microjoules when pumping with several hundred μJ of the 800 nm fundamental of the amplified Ti: Sapph laser. The spectral width of the mid-IR pulse is determined by the bandwidth of the 800 nm pulse and the phase matching conditions of the crystal. In the laser system we employ in our experiments, we typically obtain a spectral width in the mid-IR of more than 200 $cm^{-1}$ for an 800 nm pulse of 85 fs.

The currently operative setup at our institute consists of an integrated Ti:Sapph oscillator-regenerative amplifier laser system (Hurricane, Spectra Physics Inc.) working at 1 kHz, which produces 85 fs pulses of 0.8 mJ at 800 nm. A portion of this 800-nm light is used to pump a non-collinear optical parametric amplifier (Cerullo et al., 1997, 1998; Wilhelm et al., 1997; Shirakawa et al., 1998) to produce excitation pulses with a center wavelength tunable between 470 and 900 nm. The excitation pulses are not compressed, since the mid-infrared pulse length and inherent vibrational line widths of the samples are limiting for the instrument response function of the experiment, and have a duration of about 60 fs. They are focused with a 20 cm lens into the sample. The polarization of the excitation pulse with respect to the IR probe pulses is set with a Berek polarizer. The intensity of the visible pulses is attenuated to a value which depends on the experiment performed, typically 100 nJ or less. Experiments as a function of excitation density indicated that in the CP47 antenna complex, which contains 14 pigments connected by fast energy transfer, an excitation energy of 40 nJ/(150 μm)$^2$ is very close to annihilation free conditions.

The mid-IR probe pulses are generated with a second portion of the 800-nm light. About 450 μJ is used to pump the optical parametric generator and amplifier with difference frequency generator (TOPAS, Light Conversion), yielding pulses tunable between 3 and 12 μm with energy of ~1–4 μJ. The pulses are attenuated to an intensity of about 1 nJ, focused with a 5 cm lens and spatially overlapped with the excitation beam in the sample. The focusing lens of the mid-IR pulse is chosen 4× times stronger than that of the visible pulse to better match the size of the foci, which vary approximately with $f \times \lambda$. Usually the lens for the visible pulse is aligned such that the focus is slightly in front of or beyond the sample to get the size of the pump spot as large as or, preferably, larger than that of the probe. After overlap in the sample, the mid-IR probe pulses are dispersed in a spectrograph (Chromex) and imaged onto a 32-element MCT detector (Infrared Associates). The spectrograph is equipped with three gratings with 150, 75 and 50 grooves per mm, to obtain a dispersion of ~6 $cm^{-1}$/channel when used in the 3000–2000, 2000–1400 and 1400–900 $cm^{-1}$ range, respectively, or appropriately higher or lower when one of the other gratings is used in a particular frequency range. The slits of the spectrograph are reduced in opening to achieve a spectral resolution that is higher than the sampling resolution. The signals of the detector array are individually amplified (Infrared Associates) and fed into 32 home-built integrate-and-hold devices that are read out with a National Instruments acquisition card (PCI6031E) after every shot. A phase-locked chopper operating at 500 Hz is used to ensure that every other shot the sample is excited and the change in transmission and hence optical density between two consecutive shots can be measured. In case a second detector array is employed the spectrum of a reference beam can be measured simultaneously and a ΔOD spectrum can be determined in a single shot. Which of the two methods is preferred depends on whether the noise is determined by the laser, or by fluctuations in the transmission of the sample.

With the noise of the NOPA, a non-collinear optical parametric amplifier, being 5% (at the edge of the spectrum) or less, the stability of the experiment is determined by the noise in the probe pulse. The stability in the IR depends critically on the dispersion of the 800 nm pulse, of course on the alignment of the OPG/OPA, and on the time delay between the signal and idler pulse in the difference frequency generator (Hamm et al., 2000). After careful alignment, the stability of the mid-IR pulses is such that when a spectral selection on sets of 500 spectra is applied and only those spectra that deviate less than 2% from the average are selected, typical acceptance percentages of 80–100% are obtained. Thus in 1 minute of data collection the typical noise level of one single spectrum is $10^{-5}$ OD (see Fig. 2A), implying a noise level of 30 μOD for a full data set consisting of 80 time points in a few hours (see Fig. 2B). However, in practice, the movement of the sample (see below), when it contains bubbles or small heterogeneities or when it is strongly absorbing, is the limiting factor in the *S/N* ratio. The setup is partially contained in a box which is flushed with nitrogen or dry air to reduce the absorption of the laser beam by water vapor, which leads to distortion of the time profile of the pulse and to pulse-to-pulse instabilities.

The overlap of visible and mid-IR pulses in time and space is obtained in a slab of GaAs of 50 μm thickness. The visible excitation pulse creates free electron carriers in this sample and thus a change in the index of refraction that results in huge changes in the transmission of the IR pulse, corresponding to several hundreds of mOD of absorption change. The cross correlation of the visible and IR pulses as measured in GaAs is typically about 180 fs. In a single experiment a spectral probe window of about 180 $cm^{-1}$ (when a spectral resolution of 6 $cm^{-1}$ is sufficient) is covered, so partly overlapping spectral regions must be measured to obtain a wider window.

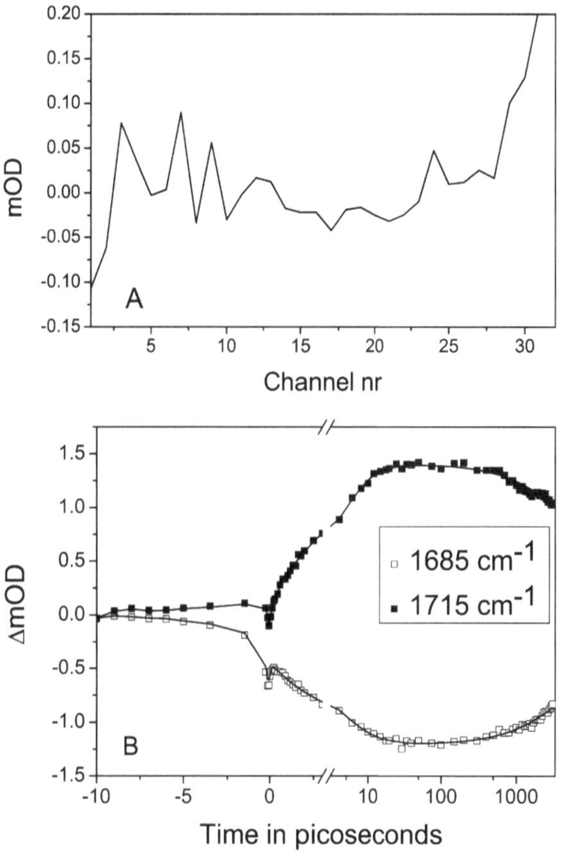


*Fig. 2.* (A). Noise level expressed in milli optical density (mOD), obtained after 1 minute of data acquisition. (B). Time dependent absorption change of the keto-group of the primary donor of the bacterial reaction center, at 1685 $cm^{-1}$ and 1715 $cm^{-1}$ upon excitation at 600 nm, noise level 30 μOD, the sample was measured in the Lissajous scanner. The solid line through the data points is a fit with $t_1 = 3.8$ ps, $t_2 = 16$ ps, $t_3 = 4$ ns and $t_5 = \infty$. The time scale is linear up to 3 ps and logarithmic thereafter.

The excitation laser gives rise to a thermal lens effect in the sample, leading (due to its relatively long life time) to an offset in the channels with respect to the IR spectra without excitation. This effect is small and varies, depending on the alignment, between 0.1 and 1 mOD. However, compared to the small transient pump-probe signals this is still relatively large. Experiments with different alignments and thus varying thermal baseline effects showed us that this effect could be corrected for by subtracting the IR signal at −10 ps, before the pump pulse. Perturbation by the pump pulse of the free induction decay signals of vibrational oscillators resulting from interaction with the probe pulse, leads to observation of signals in the time-window between about −2 ps and 0 ps, as described by (Joffre et al., 1988; Hamm, 1995).

All data sets are analyzed using global analysis (Van Stokkum et al., 2004). Since part of the noise is correlated, i.e., baseline noise or amplitude noise of the whole spectrum, this kind of analysis is particularly suited to extract more reliable information from the data than a single-trace analysis. If the data contains sufficient information, or when extra information is available, a target analysis is applied (i.e., a physical model is fitted to the data) from which the spectra of physical states and the intrinsic rates connecting them can be obtained.

### *B. Samples*

Typical time scales of the photo-cycle of many of the proteins and enzymes of interest are longer than 1 ms. Therefore, to ensure a fresh sample spot for each laser shot, we designed a Lissajous sample scanner. In short, the sample is moved perpendicular to the beam on three eccentrics which leads to a scan pattern filling a 20 × 20 mm surface. The velocity in the corners and at the sides is too low to refresh the sample after each laser shot, therefore these shots are discarded (~16–18%). The laser comes back at the same sample position after about a minute. The samples are contained in a demountable cell consisting of 2 $CaF_2$ plates separated by a Teflon spacer of 10 or 20 μm. The sample cell is filled through two metal tubes drilled through and glued to the first $CaF_2$ plate. The sample cell is held on to the Lissajous scanner by three magnetic supports.

The samples are prepared at very high concentrations to reduce the amount of water in the sample, since otherwise water would dominate the absorption spectrum. Water has absorption throughout the entire 3500–800 $cm^{-1}$ region with two main bands at 1650 and 3500 $cm^{-1}$. The H-O-H bending mode at 1650 $cm^{-1}$ carries only a small extinction coefficient of ~181 $M^{-1}cm^{-1}$. However, since the concentration of water is 55 $M^{-1}$, even a 10 μm layer will lead to an absorbance of 1 (Mäntele, 1996). Therefore the protein samples are concentrated in a centrifuge, and when measurements are done around 1650 or 3500 $cm^{-1}$, the $H_2O$ buffer is exchanged for $D_2O$ that has its main absorption bands at 1250 and 2500 $cm^{-1}$, respectively. Note that in FTIR experiments often dry (humidified) films of protein are used; however, in our experiments where the sample is moved after each laser shot, this would lead to large variations in the transmitted intensity (and thus to noise). For this reason our preparations always consist of liquid in order to obtain the most homogeneous sample possible. Typically, for photosynthetic proteins the samples have an $OD_{max}$ in the visible ($Q_Y$ region) between 0.2 and 0.5 per 20 μm path length, and a similar absorption around 1650 $cm^{-1}$.

## III. Visible Pump/MidIR Probe Spectroscopy on Photosynthetic Systems

### *A. Chlorophyll and Pheophytin*

Figure 3 shows the result of an absorption difference measurement between 1775 and 1585 $cm^{-1}$ on chlorophyll *a* and pheophytin *a* in tetrahydrofuran (THF), upon excitation at 550 nm. In this solvent the keto- and the ester group are free from hydrogen bond interactions and absorb at 1694 and 1743 $cm^{-1}$ for Chl in the ground state, respectively. In the excited state these frequencies downshift to 1660 and 1731 $cm^{-1}$. The signal from the keto group is larger than that of the ester since it is more strongly coupled to the π electron system of the chromophore. The precise frequency of the keto mode is dependent on the presence of hydrogen bonds and the polarity of the environment; the stronger the hydrogen bond or the polarity of the environment, the lower the frequency, which can be as low as 1650 $cm^{-1}$.

### *B. The Bacterial Photosynthetic Reaction Center*

Upon excitation with light the bacterial reaction center drives an ultrafast electron transfer between the ‘special pair’, a bacteriochlorophyll dimer (P), and the bacteriopheophytin in the active branch of cofactors ($H_A$) in about 3–4 ps. On a time scale of a few hundreds of ps this initial charge separation is stabilized by electron transfer from $H_A$ to the bound quinone ($Q_A$). The time evolution of the IR absorption changes in the 1600–1800 $cm^{-1}$ region for a bacterial reaction center protein clearly reflects these early electron transfer events. Figure 4 shows the decay of the down-shifted keto mode at ~1660 $cm^{-1}$ and the appearance of the up-shifted keto mode in the Chl- cation state reflecting the electron transfer from P to $H_A$ to take place with a time constant of 4.4 ps. The

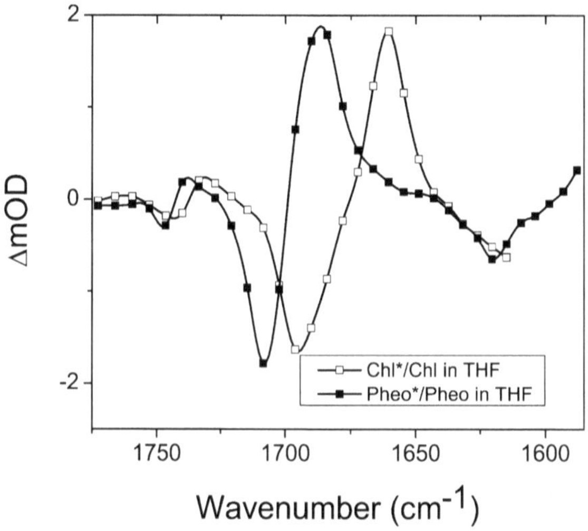


*Fig. 3.* Absorption difference spectra of chlorophyll *a* and pheophytin *a* in tetrahydrofuran in the region of the keto and ester modes, measured 10 ps after excitation.

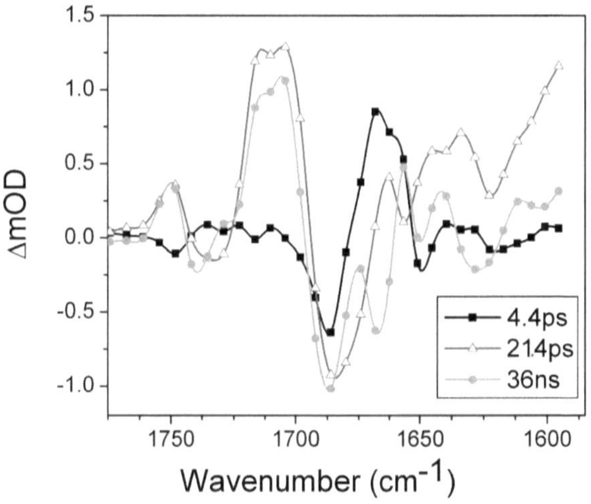


*Fig. 4.* Evolution associated difference spectra of reaction centers from *Rhodobacter sphaeroides*, excited at 600 nm. Global analysis of the data set revealed three exponential time constants, corresponding to $P^* \rightarrow P^+H_A^- \rightarrow P^+Q_A^-$, in 4.4 and 214 ps, respectively. The lifetime of the last state is too long to be determined in this experiment.

double peaked structure at 1714 and 1704 $cm^{-1}$ is due to the keto of $P_L^+$ and $P_M^+$, respectively (Nabedryk et al., 1993, 1998, 2000), which are at ~1680 $cm^{-1}$ in the P ground state. After 214 ps the spectrum evolves into that typical for the $P^+Q_A^-$ state, which contains, besides the $P^+/P$ and $Q_A^-/Q_A$ spectral features, an amide C=O band-shift signal at 1650 $cm^{-1}$.

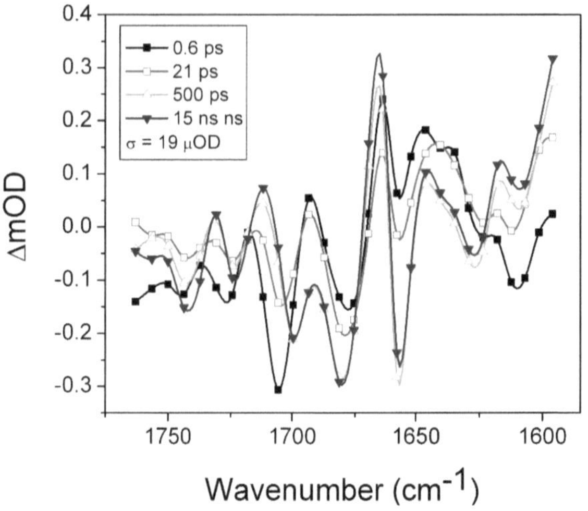


*Fig. 5.* Evolution associated difference spectra of reaction centers from Photosystem II, excited at 680 nm. Global analysis of the data set revealed four exponential time constants.

## C. The D1D2 Reaction Center of Plant Photosystem II

The D1D2 reaction center is the minimal unit of plant Photosystem 2 that still performs a light-driven charge separation. D1D2 contains a bacterial RC-like core with 4 chlorophylls and 2 pheophytins, where, like in the bacterial RC, light-induced electron transfer occurs along the A- or D1-branch of pigments. D1D2 contains two additional chlorophylls bound to the periphery of the complex, which transfer excitation energy slowly into the core.

The time evolution of the mid-infrared spectrum of isolated D1D2 reaction centers of Photosystem II, following excitation at 680 nm, reveals the presence of several different keto groups of which the relative population changes as energy and electron transfer proceed (Fig. 5). As in Fig. 3, in the excited state the keto modes shift to the wide 1660 $cm^{-1}$ region, but as time progresses and charge separation occurs, the specific signature of the Chl cation and pheophytin anion states, as identified in steady state FTIR spectroscopy (Nabedryk et al., 1990; Breton et al., 1997; Noguchi et al., 2001), can be recognized. In particular we can, as in the bacterial RCs, distinguish the keto $P^+$ band at 1711 $cm^{-1}$ and the pheophytin ester bands at 1739 (–), 1731 (+) and 1722 (–) $cm^{-1}$, which are very strong in the anion state. Since the energy and electron transfer processes are reversible, a target analysis with a model that takes into account the equilibria between the different states is necessary

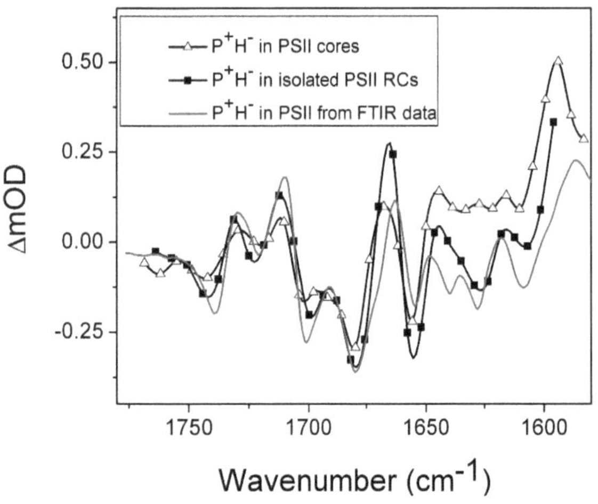


*Fig. 6.* Comparison of $P^+H_A^-/PH_A$ radical pair spectra. Spectrum 1 (open triangles) results from a target analysis of PS II core data, spectrum 2 (closed squares) from a target analysis of PS II RC data, and spectrum 3 is the sum of steady state FTIR $P^+/P$ and $H_A^-/H_A$ spectra (Nabedryk et al., 1990; Breton et al., 1997).

to extract the difference spectra of the excited and radical pair states. A comparison of the target RP 1 and RP 2 spectra (see Fig. 6) with the steady state FTIR difference spectra of $P^+/P$ and $H^-/H$ revealed that prior to $P^+$ formation, $H^-/H$ was present already within 1 ps. We could therefore conclude that in D1D2 not P is the initial electron donor, but the chlorophyll on the $D_1$ branch located between P and $H_{D1}$. In contrast, $P^+$ is only formed on a timescale of a few ps, probably by electron transfer to $Chl_{D1}^+$ (Groot et al., 2005).

### *D. Cores of Photosystem II*

Experiments are even possible on the larger core particles of PS II, containing the CP43 and CP47 antenna proteins and the D1D2 RC with in total 36 chlorophyll and 2 pheophytin pigments. Of course, upon excitation of this complex most of the excitations will be created on the two core antenna complexes, CP43 and CP47, implying that the observed kinetics of charge separation will be limited by the amount of time it takes to transfer excitations from CP43/CP47 to the RC. As can be seen in Fig. 6, the $P^+H^-/PH$ difference spectra in isolated PS II RCs and in the core particles (as obtained from a target analysis on the data) are within the noise identical (note though that there is probably some contribution from Chl* in the core spectrum, since the signal in the 1660–1620 $cm^{-1}$ region is overall larger). Both spectra agree well with the sum of $P^+/P$ and $H^-/H$ spectra measured with steady state FTIR spectroscopy (Nabedryk et al., 1990; Breton et al., 1997). Since the precise frequency of the different modes is very sensitively dependent on the hydrogen bond strength that the group is engaged in, and on the polarity of the environment, this is a very strong indication that the chromophores in the isolated RC protein are not disturbed by the isolation procedure.

### *E. The Core Antenna Complex CP47*

CP47 is one of the two core antenna proteins of Photosystem II that feed excitations into the PS II RC. Recently a detailed structural model was published for the core of Photosystem II (Zouni et al., 2001; Ferreira et al., 2004; Loll et al., 2005) in principle allowing a detailed structure-function analysis. However, in chlorophyll-proteins such as CP47, the fact that all the pigments absorb in a single band centered at 675 nm makes a detailed connection between this structure and the role of the different pigments in the flow of energy through the complex, very difficult. A simultaneous measurement of the visible absorption changes with those in the mid-IR, as a function of excitation density, can then provide valuable clues to unraveling which chlorophyll absorb at which wavelength and the rate at which each chlorophyll transfers its excitation energy. The absorption changes in the mid-IR upon excitation of purified CP47 at 590 nm revealed a set of different keto frequencies in the spectrum, of which the relative population changed as energy transfer and annihilation processes took place (Groot et al., 2004). A future comparison of the mid-IR data with a high resolution structure of the CP47 complex, which would reveal the hydrogen bond strength of the keto group of each of the Chls with the protein, will connect this structure with the observed dynamics and energy transfer function of CP47, and provide an important verification of models for the function of the antenna complex (de Weerd et al., 2002; Novoderezhkin et al., 2004).

## IV. Conclusions

Since the demonstration of the essential contributions that steady-state and time-resolved FTIR spectroscopy have made to the understanding of the function of photosynthetic proteins, it makes only sense to extend the time scale of vibrational spectroscopy to that needed to capture the processes of energy transfer and excitation transfer. Initial studies were perhaps performed with the expectation to find protein signals which could explain the relaxation that takes place in bacterial RCs when the charge separated state $P^+H_A^-$ has been formed. During the lifetime of $P^+H_A^-$ a large decay in the remaining excited state population is observed (by time resolved fluorescence and absorption spectroscopy) which was ascribed to a protein-induced relaxation and a concomitant drop in free energy of $P^+H_A^-$. However, then such protein signals were not found (Maiti et al., 1994; Hamm et al., 1995). Nevertheless, the relative ease with which these experiments are performed nowadays allows them to be used as an information-rich technique that can reveal minor responses of the protein; characterize energy- and electron-transfer pathways in plant and bacterial pigment proteins with the aim to understand the driving forces in energy and electron transfer; help to resolve the electron transfers steps in Photosystem I and II, and so on and also to revisit the earlier-mentioned relaxation problem in the state $P^+H_A^-$ mentioned above. With the increasing resolution of the structures of Photosystem I and II (Loll et al., 2005), we see a very large role for femto-IR as a unique technique that can provide the link between the ultrafast energy- and electron transfer dynamics and the structure.

## Acknowledgment

This research was supported by the Netherlands Organization for Scientific Research (NWO) via the Dutch Foundation of Earth and Life Sciences (ALW).

## References

Barth A and Zscherp C (2002) What vibrations tell us about proteins. Q Rev Biophys 35: 369–430

Bredenbeck J, Helbing J, Kumita JR, Woolley GA and Hamm P (2005) α-Helix formation in a photoswitchable peptide tracked from picoseconds to microseconds by time-resolved IR spectroscopy. Proc Natl Acad Sci USA 102: 2379–2384

Breton J, Hienerwadel R and Nabedryk E (1997) Hydrogen bonding to the primary electron donor in reaction centers from *Rb. sphaeroides*: FTIR characterization of a series of mutants at residue M160. In: Carmona P (ed) Spectroscopy of Biological Molecules: Modern Trends, pp 101–106. Kluwer Academic Publishers, Dordrecht

Cerullo G, Nisoli M and De Silvestri S (1997) Generation of 11 fs pulses tunable across the visible by optical parametric amplification. Appl Phys Lett 71: 3616–3618

Cerullo G, Nisoli M, Stagira S and De Silvestri S (1998) Sub-8-fs pulses from an ultrabroadband optical parametric amplifier in the visible. Opt Lett 23: 1283–1285

de Weerd FL, van Stokkum IHM, van Amerongen H, Dekker JP and van Grondelle R (2002) Pathways for energy transfer in the core light-harvesting complexes CP43 and CP47 of Photosystem II. Biophys J 82: 1586–1597

Demirdoven N, Cheatum CM, Chung HS, Khalil M, Knoester J and Tokmakoff A (2004) Two-dimensional infrared spectroscopy of antiparallel beta-sheet secondary structure. J Am Chem Soc 126: 7981–7990

Diller R, Iannone M, Cowen BR, Maiti S, Bogomolni RA and Hochstrasser RM (1992) Picosecond dynamics of bacteriorhodopsin, probed by time-resolved infrared-spectroscopy. Biochemistry 31: 5567–5572

Dreyer J, Moran AM and Mukamel S (2003) Coherent three-pulse spectroscopy of coupled vibrations in a rigid dipeptide: Density functional theory simulations. J Phys Chem B 107: 5967–5985

Eaves JD, Loparo JJ, Fecko CJ, Roberts ST, Tokmakoff A and Geissler PL (2005) Hydrogen bonds in liquid water are broken only fleetingly. Proc Natl Acad Sci USA 102: 13019–13022

Ferreira KN, Iverson TM, Maghlaoui K, Barber J and Iwata S (2004) Architecture of the photosynthetic oxygen-evolving center. Science 303: 1831–1838

Groma GI, Colonna A, Lambry JC, Petrich JW, Varo G, Joffre M, Vos MH and Martin JL (2004) Resonant optical rectification in bacteriorhodopsin. Proc Natl Acad Sci USA 101: 7971–7975

Groot ML, Vos MH, Schlichting I, van Mourik F, Joffre M, Lambry JC and Martin JL (2002) Coherent infrared emission from myoglobin crystals: An electric field measurement. Proc Natl Acad Sci USA 99: 1323–1328

Groot ML, van Wilderen L, Larsen DS, van der Horst MA, van Stokkum IHM, Hellingwerf KJ and van Grondelle R (2003) Initial steps of signal generation in photoactive yellow protein revealed with femtosecond mid-infrared spectroscopy. Biochemistry 42: 10054–10059

Groot ML, Breton J, van Wilderen L, Dekker JP and van Grondelle R (2004) Femtosecond visible/visible and visible/mid-IR pump-probe study of the Photosystem II core antenna complex CP47. J Phys Chem B 108: 8001–8006

Groot ML, Pawlowicz NP, van Wilderen L, Breton J, van Stokkum IHM and van Grondelle R (2005) Initial electron donor and acceptor in isolated Photosystem II reaction centers identified with femtosecond mid-IR spectroscopy. Proc Natl Acad Sci USA 102: 13087–13092

Hahn S, Kim SS, Lee C and Cho M (2005) Characteristic two-dimensional IR spectroscopic features of antiparallel and parallel beta-sheet polypeptides: Simulation studies. J Chem Phys 123: 084905 (10 pages).

Hamacher E, Kruip J, Rögner M and Mäntele W (1996) Characterization of the primary electron donor of Photosystem I, P700, by electrochemistry and Fourier-transform infrared (FTIR) difference spectroscopy. Spectrochim Acta A 52: 107–121

Hamm P (1995) Coherent effects in femtosecond infrared-spectroscopy. Chem Phys 200: 415–429

Hamm P, Zurek M, Mäntele W, Meyer M, Scheer H and Zinth W (1995) Femtosecond infrared-spectroscopy of reaction centers from *Rhodobacter sphaeroides* between 1000 and 1800 $cm^{-1}$. Proc Natl Acad Sci USA 92: 1826–1830

Hamm P, Lim MH and Hochstrasser RM (1998) Structure of the amide I band of peptides measured by femtosecond nonlinear-infrared spectroscopy. J Phys Chem B 102: 6123–6138

Hamm P, Kaindl RA and Stenger J (2000) Noise suppression in femtosecond mid-infrared light sources. Opt Lett 25: 1798–1800

Herbst J, Heyne K and Diller R (2002) Femtosecond infrared spectroscopy of bacteriorhodopsin chromophore isomerization. Science 297: 822–825

Hill JR, Tokmakoff A, Peterson KA, Sauter B, Zimdars D, Dlott DD and Fayer MD (1994) Vibrational dynamics of carbon-monoxide at the active-site of myoglobin — Picosecond infrared free-electron laser pump-probe experiments. J Phys Chem 98: 11213–11219

Hill JR, Dlott DD, Rella CW, Smith TI, Schwettman HA, Peterson KA, Kwok A, Rector KD and Fayer MD (1996) Ultrafast infrared spectroscopy in biomolecules: Active site dynamics of heme proteins. Biospectroscopy 2: 277–299

Joffre M, Hulin D, Migus A, Antonetti A, Laguillaume CBA, Peyghambarian N, Lindberg M and Koch SW (1988) Coherent effects in pump probe spectroscopy of excitons. Opt Lett 13: 276–278

Kim YS and Hochstrasser RM (2005) Chemical exchange 2D IR of hydrogen-bond making and breaking. Proc Natl Acad Sci USA 102: 11185–11190

Kotting C and Gerwert K (2005) Proteins in action monitored by time-resolved FTIR spectroscopy. Chem Phys Chem 6: 881–888

Lim M, Jackson TA and Anfinrud PA (1995a) Binding of CO to myoglobin from a heme pocket docking site to form nearly linear Fe-C-O. Science 269: 962–966

Lim MH, Jackson TA and Anfinrud PA (1995b) Midinfrared vibrational spectrum of CO after photodissociation from heme evidence for a ligand docking site in the heme pocket of hemoglobin and myoglobin. J Chem Phys 102: 4355–4366

Loll B, Kern J, Saenger W, Zouni A and Biesiadka J (2005) Towards complete cofactor arrangement in the 3.0 Ångström resolution structure of Photosystem II. Nature 438: 1040–1044

Maiti S, Cowen BR, Diller R, Iannone M, Moser CC, Dutton PL and Hochstrasser RM (1993) Picosecond infrared studies of the dynamics of the photosynthetic reaction center. Proc Natl Acad Sci USA 90: 5247–5251

Maiti S, Walker GC, Cowen BR, Pippenger R, Moser CC, Dutton PL and Hochstrasser RM (1994) Femtosecond coherent transient infrared-spectroscopy of reaction centers from *Rhodobacter sphaeroides*. Proc Natl Acad Sci USA 91: 10360–10364

Mäntele W (1996) Infrared and Fourier-Transform Infrared Spectroscopy. In: Amesz J and Hoff AJ (eds) Biophysical Techniques in Photosynthesis (Advances in Photosynthesis, Vol 3), pp 137–157. Kluwer Academic Publishers, Dordrecht

Nabedryk E, Andrianambinintsoa S, Berger G, Leonhard M, Mäntele W and Breton J (1990) Characterization of bonding interactions of the intermediary electron-acceptor in the reaction center of Photosystem-II by FTIR spectroscopy. Biochim Biophys Acta 1016: 49–54

Nabedryk E, Allen JP, Taguchi AKW, Williams JC, Woodbury NW and Breton J (1993) Fourier-transform infrared study of the primary electron-donor in chromatophores of *Rhodobacter sphaeroides* with reaction centers genetically-modified at residue M160 and residue L131. Biochemistry 32: 13879–13885

Nabedryk E, Breton J, Williams JC, Allen JP, Kuhn M and Lubitz W (1998) FTIR characterization of the primary electron donor in double mutants combining the heterodimer HL(M202) with the LH(L131), HF(L168), FH(M197), or LH(M160) mutations. Spectrochim Acta A 54: 1219–1230

Nabedryk E, Schulz C, Müh F, Lubitz W and Breton J (2000) Heterodimeric versus homodimeric structure of the primary electron donor in *Rhodobacter sphaeroides* reaction centers genetically modified at position M202. Photochem Photobiol 71: 582–588

Noguchi T, Tomo T and Kato C (2001) Triplet formation on a monomeric chlorophyll in the Photosystem II reaction center as studied by time-resolved infrared spectroscopy. Biochemistry 40: 2176–2185

Novoderezhkin VI, Palacios MA, van Amerongen H and van Grondelle R (2004) Energy-transfer dynamics in the LHCII complex of higher plants: Modified redfield approach. J Phys Chem B 108: 10363–10375

Rubtsov IV, Kumar K and Hochstrasser RM (2005) Dual-frequency 2D IR photon echo of a hydrogen bond. Chem Phys Lett 402: 439–443

Scheurer C and Mukamel S (2002) Infrared analogs of heteronuclear nuclear magnetic resonance coherence transfer experiments in peptides. J Chem Phys 116: 6803–6816

Shirakawa A, Sakane I and Kobayashi T (1998) Pulse-front-matched optical parametric amplification for sub-10-fs pulse generation tunable in the visible and near infrared. Opt Lett 23: 1292–1294

Van Stokkum IHM, Larsen DS and Van Grondelle R (2004) Global and target analysis of time-resolved spectra. Biochim Biophys Acta 1657: 82–104

Ventalon C, Fraser JM, Vos MH, Alexandrou A, Martin JL and Joffre M (2004) Coherent vibrational climbing in carboxyhemoglobin. Proc Natl Acad Sci USA 101: 13216–13220

Venyaminov SY and Kalnin NN (1990) Quantitative IR spectrophotometry of peptide compounds in water ($H_2O$) solutions 2. Amide absorption-bands of polypeptides and fibrous proteins in alpha-coil, beta-coil, and random coil conformations. Biopolymers 30: 1259–1271

Walker GC, Maiti S, Cowen BR, Moser CC, Dutton PL and Hochstrasser RM (1994) Time Resolution of electronic-transitions of photosynthetic reaction centers in the infrared. J Phys Chem 98: 5778–5783

Wilhelm T, Piel J and Riedle E (1997) Sub–20-fs pulses tunable across the visible from a blue-pumped single-pass noncollinear parametric converter. Opt Lett 22: 1494–1496

Zouni A, Witt HT, Kern J, Fromme P, Krauss N, Saenger W and Orth P (2001) Crystal structure of Photosystem II from *Synechococcus elongatus* at 3.8 Ångström resolution. Nature 409: 739–743

# Chapter 11

## Nonlinear Femtosecond Optical Spectroscopy Techniques in Photosynthesis

Donatas Zigmantas, Ying-Zhong Ma, Elizabeth L. Read and Graham R. Fleming*
*Department of Chemistry, University of California, Berkeley, Physical Biosciences Division, Lawrence Berkeley National Laboratory, B77 Hildebrand Hall, Berkeley, CA 94720, U.S.A*



### Summary

Over the past ten years, the techniques and sophistication of analysis of ultrafast spectroscopy have advanced to a remarkable extent. Following the demonstration and validation of new methods using simple dilute dye solutions, many applications of these new techniques have been to photosynthetic systems. The reasons for this are not hard to find: photosynthetic pigment protein complexes function through a delicate interplay of interpigment and pigment-environment interactions, and standard methods provide an unsatisfactory level of microscopic insight because signals are dominated by inhomogeneous effects, or the methods themselves are insensitive to the interactions between states of interest. The developments described in this chapter are designed to address all these issues. Multiphoton transient absorption spectroscopy helps to unravel complex energy transfer and relaxation pathways. Two-photon excitation spectroscopy prepares states that are not ac-

*Author for correspondence, email: GRFleming@lbl.gov

*Thijs J. Aartsma and Jörg Matysik (eds): Biophysical Techniques in Photosynthesis II, pp. 201–222.*

cessible directly from the ground state. Photon echo methods defeat inhomogeneous broadening, measure it, and exploit it to observe energy transfer between chemically identical donors and acceptors. Two-dimensional techniques explicitly reveal electronic couplings and the pathways of energy flow. New methods based on Raman spectroscopy reveal vibrational frequencies in excited states, and indeed a wide range of transient species.

## I. Introduction

The technology of generating and characterizing tunable, intense, ultrashort light pulses has matured, and experimentalists have now turned to using these pulses to provide more incisive and detailed probes of photosynthetic systems. For example, the system can be subjected to more than one excitation pulse prior to the probing event. The transfer of population between potential energy surfaces at a series of different times can be very useful in disentangling complex relaxation pathways. The electric field strength of pulses tens of femtoseconds in duration is necessarily high, permitting observation of nonlinear processes such as two-photon absorption. This allows direct population of states whose optical transition from the ground state is forbidden in one-photon absorption, such as the lowest excited singlet states of polyenes and in particular carotenoids in photosynthetic light-harvesting. Photon echo based methods, in particular the peak shift and two-dimensional Fourier transform electronic spectroscopy, provide new insights into the design principles of light-harvesting pigment-protein complexes. The peak shift method enables defeat of, quantification of, and use of the inhomogeneous energy level distribution of the ensemble of chromophores to reveal energy transfer over a very broad range of timescales. The newly developed two-color peak shift technique has potential to reveal electronic couplings, wavefunction overlaps, and pathways of energy flow. Perhaps the technique with the greatest potential to reveal the microscopic design principles of natural light-harvesting is the method of two-dimensional Fourier transform photon echo spectroscopy. This technique, analogous to two-dimensional NMR methods, reveals electronic couplings between individual molecules directly. A series of such spectra, recorded at different time intervals after the initial pulse, map out the relaxation pathways in time and (implicitly) in space. This has already led to new insights into the mechanism of energy flow within the Fenna-Matthews-Olson (FMO) complex of green sulfur bacteria, and seems likely to produce real advances in our understanding of light-harvesting in plant and bacterial systems.

Electronic spectroscopy of large molecules such as chlorophyll in condensed phase environments do not provide any information on the vibrational spectra of transient species, which can be crucial to determining the electronic structure and molecular geometry of the molecule. Newly developed methods in visible pump-infrared probe (covered in Chapter 10, Groot and van Grondelle) and two-dimensional infrared spectroscopy and stimulated Raman spectroscopy combine both high time resolution with high spectral resolution to provide detailed insight into the nuclear motions involved in the relaxation processes, thereby offering an important complement to the spectroscopy of electronic states.

In this chapter, we describe all of these new techniques and illustrate them with examples of application to photosynthetic systems.

*Abbreviations*: (B)Chl – (bacterio)chlorophyll; 2C3PEPS – two-color three pulse photon echo peak shift; 2D – two-dimensional; 3PEPS – three pulse photon echo peak shift; Car – carotenoid; ESA – excited state absorption; FMO – Fenna-Matthews-Olson; FROG – frequency-resolved optical gating; fs – femtosecond; FSRS – femtosecond stimulated Raman spectroscopy; FWHM – full-width half-maximum; ICT – intramolecular charge transfer; IVR – intramolecular vibrational relaxation; LH1 – primary light-harvesting complex; LH2 – peripheral light-harvesting complex; LHC II – light-harvesting complex II; LO – local oscillator; MP-TA – multipulse transient absorption; NOPA – noncollinear optical parametric amplifier; OPA – optical parametric amplifier; PAP – pump-action-probe; PDP – pump-dump-probe; PEPS – photon echo peak shift; PP – pump-probe; PrPP – pump-repump-probe; SPIDER – spectral phase interferometry for direct electric-field reconstitution; TP – two-photon; TPE – two-photon fluorescence excitation

## II. Developments in Laser Technology and Pulse Measurement Techniques

Recent years have seen remarkable progress in solid state ultrafast laser technology. Since the invention of the self-mode-locked Ti:Sapphire laser in 1990 (Spence et al., 1991), the field has evolved very rapidly: now 10–20 fs pulses are produced routinely

(Asaki et al., 1993) and 5-fs pulses can be generated directly from Ti:Sapphire oscillators using specially designed chirped mirrors to compensate for dispersion introduced by the Ti:Sapphire crystal in the cavity (Morgner et al., 1999; Kartner et al., 2004). Pulses with a few to under two optical cycles can also be generated from amplified or cavity-dumped low-repetition Ti:Sapphire lasers systems employing different pulse compression techniques (Baltuska et al., 1997; Nisoli et al., 1997; De Silvestri et al., 2004). There is also a range of optical parametric amplifiers able to generate sub-10 fs pulses at all possible wavelengths in the visible and near-IR regions. For example, the noncollinear optical parametric amplifier (NOPA) (Wilhelm et al., 1997; Butkus et al., 2004) is widely used in a variety of ultrafast spectroscopy experiments. Significant advances have been made not only in laser technology, but also in the field of ultrashort pulse characterization. Interferometric techniques like frequency-resolved optical gating (FROG) (Trebino et al., 1997) and spectral phase interferometry for direct electric-field reconstitution (SPIDER) (Iaconis and Walmsley, 1999) deliver a full characterization of pulses in electric field and phase. Fully computerized tunable laser systems with full pulse characterization are commercially available. Fiber lasers (Fermann et al., 1997) and supercontinuum generation in microstructure optical fibers (Ranka et al., 2000) are among the most promising developments towards low power and substantially less expensive lasers systems. This array of high performance laser systems and pulse characterization techniques has enabled a variety of new time-resolved spectroscopic methods with unprecedented time resolution, sensitivity and flexibility.

## III. Multipulse Transient Absorption Spectroscopy

### A. Technique

Multipulse transient absorption (MP-TA) assumes various forms such as pump-dump-probe (PDP) (Gai et al., 1997; Larsen et al., 2003) stimulated emission pumping (Kovalenko et al., 1998; Ruhman et al., 2002), two-pump and probe (Logunov et al., 2001), and pump-repump-probe (PrPP) (Larsen et al., 2003). This class of techniques involves using an additional pulse in the standard pump-probe (PP) scheme. In MP-TA experiment, an intense pump pulse produces a nonequilibrium population of excited states whose evolution is perturbed with a variably delayed, intense action pulse, resulting in partial depletion of the population in a particular state or states. The pump-induced differential absorption ($\Delta OD$) of a sample and its change arising from the interaction of the action pulse is monitored by a weak probe pulse at either a selected wavelength or over a broad spectral range. Data acquisition usually involves measuring the $\Delta OD$ signals in the presence (pump-action-probe (PAP)) and in the absence (PP) of the action pulse by varying the time delay between the pump and probe pulses ($t$) while keeping the delay between the pump and action pulses ($\tau$) fixed. Occasionally, measurements of $\Delta OD$ signals as a function of $\tau$ for a fixed $t$ may be also performed, enabling direct access to the temporal dependence of the $\Delta OD$ changes caused by the action pulse (Larsen et al., 2003). A successive detection of the $\Delta ODPAP$ and $\Delta ODPP$ signals at each time delay, $t$ or $\tau$, is preferable to ensure identical experimental conditions. This detection can be readily realized by introduction of a mechanical shutter into the action beam, or by simultaneous modulation of the pump and the action beams with, for instance, two optical choppers operating at different frequencies (Larsen et al., 2004). In the latter case, the signal induced by the action pulse, $\Delta ODAP$, should also be measured, especially when this pulse is spectrally resonant with the ground state absorption of the sample.

The most convenient way to differentiate the $\Delta OD$ change caused by the action pulse is to subtract the PP signal, $\Delta ODPP$, from the $\Delta ODPAP$ signal, and the result is usually defined by a double difference absorption signal (Larsen et al., 2004; Papagiannakis et al., 2004):

$$\Delta\Delta OD(\lambda,t,\tau) = \Delta OD_{PAP}(\lambda,t,\tau) - \Delta OD_{PP}(\lambda,t) \quad (1)$$

where $\lambda$ denotes the probe wavelength. In addition, the $\Delta ODAP$ signal should also be subtracted when the action beam is modulated. From Eq. (1), it is clear that the $\Delta\Delta OD$ signal is opposite in sign to the PP signal, i.e., it has a negative sign for ESA but a positive sign for bleaching and stimulated emission (Kovalenko et al., 1998; Larsen et al., 2004).

A successful application of the MP-TA technique relies largely on proper selection of the frequency of the action pulse. An exclusive or preferential selection of a well-characterized excited-state by tuning the frequency of the action pulse can usually greatly simplify data analysis. For a chosen excited state, different frequencies of the action pulse can be used to match a transition to a final state with either higher or lower energy. However, a criterion should be fulfilled in selecting the final state: the relaxation to the perturbed state must proceed slowly enough to allow time resolution of the $\Delta OD$ change induced by the action pulse. Furthermore, successive perturbation of different excited states by varying the frequency of the action pulse is often needed for a complex system (Larsen et al., 2003).

### *B. Applications*

Applications of the MP-TA technique to condensed phase systems have demonstrated its capability for unraveling overlapping spectral features (Kovalenko et al., 1998; Larsen et al., 2003) and for disentangling complex pathways of excited state relaxation (Gai et al., 1997; Logunov et al., 2001; Ruhman et al., 2002; Papagiannakis et al., 2004). Until now, the application of this technique to photosynthesis has been focused on identifying new electronic states of carotenoids (Cars) in solution. Larsen and coworkers performed the MP-TA experiment on a hexane solution of β-carotene (Larsen et al., 2003). After excitation of its S2 state at 400 nm, an 800 nm pulse was applied to preferentially repump the population from the S1 state to a higher excited-state. The resulting −ΔΔOD spectrum characterizing the ESA spectrum of the Car $S_1$ state is markedly different from the PP data obtained at the same time delay *t*, indicating that the latter contains contributions from additional state(s). In another experiment, after excitation at 400 nm a 530 nm pulse, spectrally resonant with both the $S_2$ emission and the $S_1$ ESA bands, was applied to simultaneously dump the population of the $S_2$ state to the ground state and to repump the $S_1$ state. Target analysis of the data enabled the authors to ascribe the observed feature to a new electronic state, $S^{\ddagger}$ (Fig. 1).

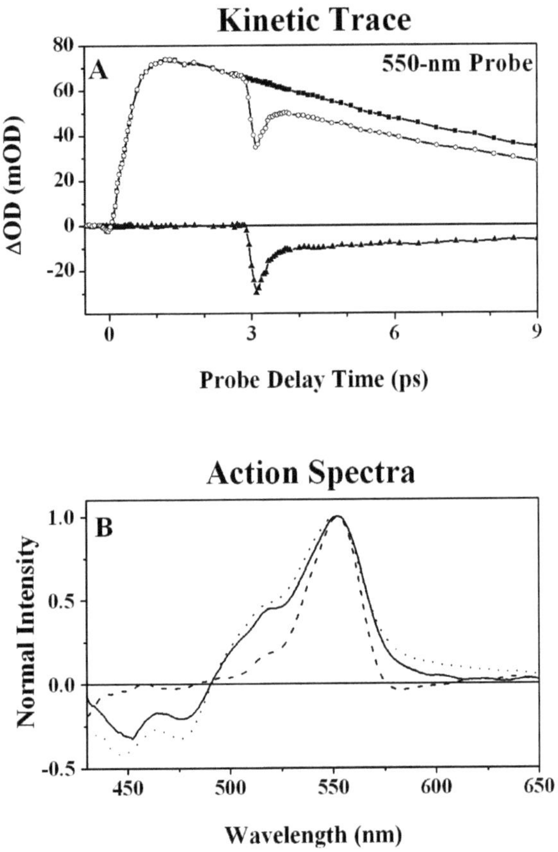


*Fig. 1.* (a) TA kinetics of β-carotene in hexane measured upon pumping at 400 and probing at 550 nm in the absence (filled squares) and in the presence (open circles) of a 530 nm pulse applied at 3 ps with respect to the pump pulse. The resulting ΔΔOD profile (filled triangles) is shown at the bottom of the panel. (b) Comparison of the 3 ps pump-probe spectrum (dotted line) with the spectra obtained from a target analysis of the ΔΔOD data collected at different probe wavelength as a function of τ, the delay between the first two pulses. The probe delay was set to $t = 3$ ps. The data analysis resulted in two spectra, inverted by multiplying by (−1), with lifetimes of ~300 fs and ~10 ps, characterizing the $S_2$ dumping (solid line) and the $S_1$ repumping (dashed line), respectively. All three spectra were normalized at the maximum (Larsen et al., 2003). (Fig. courtesy of D.S. Larsen, Univ. of California, Davis)

The second application concerns peridinin, a carbonyl Car with spectral properties depending strongly on the polarity of the environment. A change of solvent from nonpolar to polar gives rise to markedly different spectra and lifetimes, which was attributed to stabilization of an intramolecular charge transfer (ICT) state in the polar medium. To gain a better understanding about the relation between the $S_1$ and the ICT state, Papagiannakis and coworkers performed PDP measurements on a methanol solution of peridinin (Papagiannakis et al., 2004). In this experiment, peridinin was excited at 530 nm, an 800 nm dump pulse was applied to preferentially transfer population from the ICT state to ground state, and the induced change was monitored by a broadband probe pulse. The kinetics measured at different probe wavelengths clearly show that the dump pulse selectively interacts with the ICT state but not the $S_1$ state, and this observed difference demonstrates that the $S_1$ and the ICT states are distinct (Fig. 2). Target analysis of the experimental data further enabled the authors to propose a network of complex relaxation pathways involving $S_1$, $S_2$, ICT, and a ground state intermediate.

Future applications of this technique to photosynthetic systems, in which the pigments are electronically coupled, are expected to further advance our understanding of the energy transfer pathways, especially those difficult to assess by conventional PP and fluorescence upconversion techniques. However, special care should be taken in these applications. First, the overall intensity of the pump and the action pulses should be kept at a level that is well below the initiation threshold of non-linear dynamic processes, e.g., singlet-singlet and singlet-triplet excitation annihilation (van Amerongen et al., 2000). Occurrence of these processes and their concomitant changes of spectral shapes caused, for instance, by a local heating (Valkunas and Gulbinas, 1997) will complicate data analysis, and in certain cases may even prohibit extracting useful information from the experimental data. Second, prolonged illumination of a sample by two intense laser pulses may cause permanent degradation; therefore, independent experiments must be conducted in order to assess the integrity of the sample.

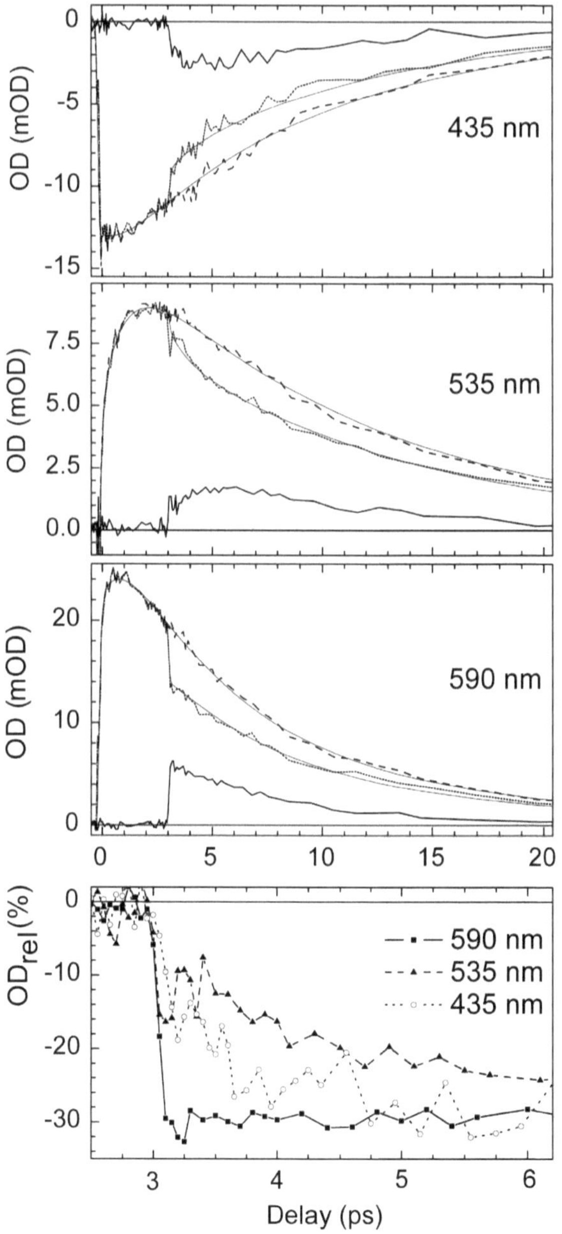


*Fig. 2.* Pump-probe (dashed lines) and pump-dump-probe (dotted lines) kinetic traces of peridinin in methanol measured at 435, 535 and 590 nm, monitoring the population evolution in the ground state, $S_1$ and ICT state, respectively. The thin solid lines through the data denote the target analysis fit, while the thick solid lines correspond to the depletion ($-\Delta\Delta OD$) signal. The lowest panel shows the relative loss signals ($\Delta\Delta OD_{rel}$) at 435, 535 and 590 nm, which is defined by $\Delta\Delta OD_{rel} = \Delta\Delta OD(\lambda,t,\tau)/\Delta OD(\lambda,t)$. From the $-\Delta\Delta OD$ and $\Delta\Delta OD_{rel}$ signals, it is clear that the interaction of a 800 nm dump pulse at 3 ps results in an instantaneous population depletion from the ICT state (590 nm curve), whereas the responses of the $S_1$ (535 nm curve) and $S_0$ (435 nm curve) states to this perturbation are significantly slower (Papagiannakis et al., 2004). (Fig. courtesy of R. van Grondelle, Vrije Universiteit, Amsterdam)

## IV. Two-Photon Fluorescence Excitation and Pump-Probe Spectroscopy

Two-photon (TP) absorption involves an electronic transition from an initial to a final state by simultaneous absorption of two quanta of electromagnetic radiation. When the final state fluoresces or undergoes relaxation by transferring energy to a fluorescent acceptor, highly sensitive detection of TP absorption can be achieved by the two-photon fluorescence excitation (TPE) technique. Since a TP transition follows selection rules that are different from those of a one-photon process, the TPE serves as a sensitive tool for probing excited states that cannot be directly reached by one-photon excitation.

Over the past four decades, TP absorption has become an important tool in the field of molecular spectroscopy. For an in-depth description of the theory, experimental techniques, and various early applications, we refer to the monographs and reviews by McClain (1974), McClain and Harris (1977), Friedrich and McClain (1980), Birge (1983), Shen (1984), and the references therein. In this section, we will focus on the techniques of TPE and TP-excited pump-probe (PP) that are based on femtosecond (fs) laser excitation, and their latest applications to photosynthetic systems.

### A. Femtosecond Two-Photon Fluorescence Excitation Spectrophotometer

The basic setup of a single beam fs TPE spectrophotometer consists of a tunable fs laser source for exciting a sample, a reference detector and a signal detector for monitoring the intensities of the two-photon excitation light and its induced fluorescence emission, respectively, and an auto-correlator for measuring continuously the temporal length of the laser pulses. Because TPE is a second-order process in interaction with the electric field, the experimentally measured fluorescence intensity at a chosen frequency $\omega_1$, $F(\omega_1)$, is proportional to the square of the incident power of the excitation pulse tuned to $\omega_2$, $P(\omega_2)$. When the variation of pulsewidth at different wavelengths is negligible, a TPE spectrum can be readily obtained by measuring $F(\omega_1)/p_2(\omega_2)$ at different values of $\omega_2$. Complications arise when the pulsewidth is not constant in the spectral range of interest, and because of its dramatic effect on the emission intensity, a proper correction is necessary in order to obtain a correct TPE spectrum. For pulses with simple analytical temporal profiles, such as Gaussian or hyperbolic secant squared (sech$^2$), the pulse width effect can be corrected by simply multiplying $F(\omega_1)/p_2(\omega_2)$ with the corresponding pulse full-width half-maximum (FWHM), $\tau(\omega_2)$. When the pulse shape is not of a simple analytical form, correction of this pulse width effect can be done numerically.

A schematic diagram of an experimental set up for TPE experiments in our laboratory is shown in Fig. 3 (Zimmermann et al., 2002). Because the cross section of TP absorption is many orders of magnitude smaller than that of the typical one-photon absorption (Fischer et al., 1995; Xu and Webb, 1996), and the quantum yield of fluorescence emission is often small, detection of fluorescence emission upon TP excitation usually requires a tight focus of the excitation beam in combination with efficient emission collection.

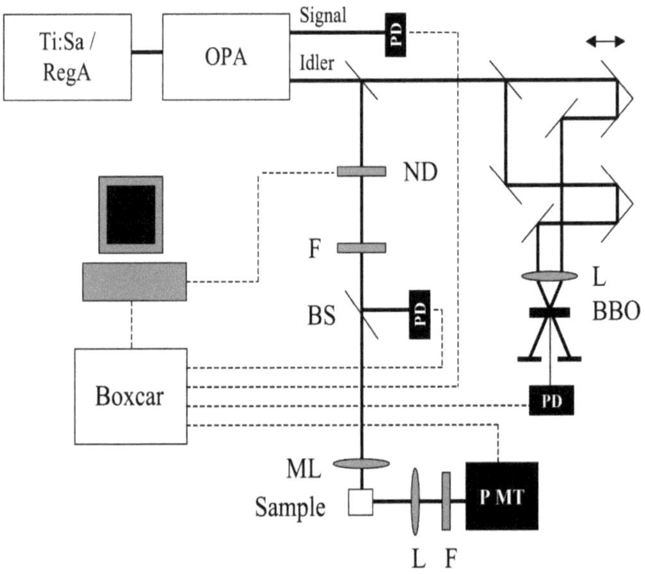


*Fig. 3.* Schematic diagram of a single beam TPE spectrophotometer. The excitation source is a 2 kHz Ti:Sapphire regenerative amplifier pumped OPA, and its idler output provides near-infrared pulses tunable from 900 to 1600 nm, with a typical pulse energy of 20 nJ and a temporal width of 95 fs. A beam splitter (BS) is used to separate a small portion of the excitation light for monitoring the excitation intensity with a calibrated photodiode (PD). The majority of the excitation light is focused to the sample with a microscope objective (Nikon E10, N.A. 0.25, focal length: 6 mm), denoted by ML. The emitted fluorescence light is collected at 90 degrees by a 5 cm lens (L) and focused onto a photomultiplier (PMT). The outputs of the detectors were preamplified and measured by two boxcar integrators (Stanford Research SR250), synchronized using the signal output of the OPA. A neutral density filter wheel (ND) is used to continuously vary the intensity of excitation light, and emission wavelength was selected using a bandpass filter (F). Before each experiment the excitation beam is sent into an autocorrelator, which was used to measure the pulsewidth and to determine the spectral shape of the pulse by doubling it in the autocorrelator BBO crystal (Zimmermann et al., 2002).

Gated or lock-in detection must be utilized in order to effectively suppress any stray light. Furthermore, when high energy pulses are used, care should be exercised to ensure that the tightly focused beam does not generate a white-light continuum from the walls of the cuvette and/or the sample solution, which can cause direct one-photon excitation of the sample and produce much more intense emission.

Measurement of a TPE spectrum involves the following steps: (1) tuning the wavelength of the OPA; (2) determining the pulse width; (3) measuring the power dependence of the TPE fluorescence, fitting the data with a second-order polynomial, and determining the amplitude of the second-order term (which provides a quantitative measure of the relative two-photon absorption intensity); (4) correcting the effect of pulse width, if necessary, and finally (5) constructing the TPE spectrum. In addition to the TPE spectrum measurements, time-resolved TPE experiments can be performed by employing the fluorescence upconversion technique.

### *B. Two-Photon Excitation Pump-Probe*

While conceptually identical to the conventional PP technique based on one-photon excitation (Jimenez and Fleming, 1996), several essential prerequisites must be met in setting up a successful TP-excitation-based PP experiment. The first prerequisite is to generate a detectable concentration of excited molecules, especially when the pulse energy for TP excitation is low, e.g., on the order of 10 nJ or less. Possible ways to do this include (1) tightly focusing the pump beam with a lens or objective with a short focal length, thereby increasing the photon density in the excitation volume; (2) using a concentrated sample for efficiently capturing available photons. The collinear geometry for the pump and probe beams enables detection of all molecules in the excitation volume, and an enhancement of the detected signal with a signal-to-noise ratio comparable to that observed in the PP data upon one-photon excitation has been demonstrated (Linden et al., 2004).

The second prerequisite is to choose a suitable test sample for finding the PP signal and optimizing the sensitivity of the set-up. Inexpensive choices include commonly used organic solvents, e.g., acetone, ethanol, methanol, etc., and a criterion for selecting a particular solvent is that its transmission cut-off wavelength (Lide, 1996) must be equal to or longer than the wavelength corresponding to the energy of three-photon absorption, i.e., two photons from the pump pulse and one photon from the probe pulse. This solvent response is instantaneous, with a temporal profile similar to the cross-correlation function of the pump and probe pulses, and therefore a time step of a fraction of the pulse width should be used to scan the delay between the pump and probe pulses. A properly selected solvent is capable of producing a very large signal, which, for sub-100 fs pump pulses, is approximately two orders of magnitude larger than the signal from Cars dissolved in organic solvents or bound in pigment-protein complexes. The extremely low signal from molecules under study, on the other hand, requires high-sensitivity detection such as lock-in amplification and effective noise suppression, in order to obtain high-quality data. A simple noise suppression method involves simultaneous monitoring of the intensity fluctuations of the pump and probe pulses with separate detectors, and normalizing the PP data.

Verification that the detected signal truly arises from TP excitation becomes crucial when additional, unwanted electronic states, such as the triplet state of (bacterio)chlorophyll ((B)Chl) or the $S_2$ state of certain Cars, can contribute. In addition to quadratic dependence on the excitation power (McClain and Harris, 1977), the ratio of the signals obtained by circularly versus linearly polarized excitation serves as an additional mark of a TP process (Birge and Zhang, 1990). Since it depends on the symmetries of the molecule's excited- and ground-state wave functions (McClain, 1971), knowledge about this ratio must be available from either theoretical calculations or independent measurements on the given molecule or one that is structurally similar to it.

### *C. Applications*

A significant application of the TPE technique to photosynthesis involves the verification of the involvement of the Car $S_1$ state in energy transfer processes to Chl or BChl molecules. Indirect evidence for such an involvement has been found by comparing the $S_1$ lifetimes of Cars in solution and in light-harvesting (LH) complexes, and by resolving the rise times of the relevant energy acceptor Chl or BChl molecules (Polivka and Sundström, 2004). Because these experiments are performed by exciting directly the Car $S_2$ state, interpretation of the collected data is often complicated by multiple potential energy transfer pathways such as Car

$S_2$ to (B)Chl and between (B)Chls, as well as possible contributions from vibrational relaxation within the acceptor molecules. A further complication, albeit less significant than the complication arising from the $S_2$ state, is due to the dependence of the $S_1$ lifetime on the solvent (Macpherson et al., 2001), which results in considerable difficulty of defining a 'standard' lifetime for comparison with the in situ lifetime of the same Car. Selective excitation of a Car $S_1$ state with TP absorption in combination with detection of its acceptor fluorescence greatly simplifies this verification. Krueger and coworkers have measured the TPE spectrum of spheroidene (Fig. 4), the principal Car in the LH2 of purple bacterium *Rhodobacter* (*Rb.*) *sphaeroides* (Krueger et al., 1999). By directly exciting its $S_1$ transition with TP absorption, observation of the fluorescence emission from the BChls enabled the authors to verify unambiguously for the first time the involvement of the Car $S_1$ state in the Cars → BChls energy transfer process. Furthermore, a detailed analysis of the TPE spectral band shape allowed an estimate of 13 900 ±150 cm$^{-1}$ to be made for the $S_0 \rightarrow S_1$ transition energy.

Direct TP excitation of the Car $S_1$ state with fs pulses allows observation of subsequent dynamics with a probe pulse tuned to its $S_1 \rightarrow S_n$ transition. Thus, interference arising from other excited states of the Car can be eliminated. Walla and coworkers have performed PP measurements on three LH complexes of purple bacteria using a probe pulse centered at 550 nm (Walla et al., 2000). The experimental data can be satisfactorily described by a mono-exponential decay plus a constant offset. An example of the kinetic decays measured upon TP excitation at 1310 nm is shown in Fig. 5a. The TP nature of the experimentally measured signals was verified by a quadratic intensity dependence of the amplitude associated with the exponential decay (Fig. 5b), and a change of the BChl fluorescence emission intensity with the polarization of the excitation beam (linear vs. circular) (Fig. 5d). In addition, the linear power dependence of the offset signal (Fig. 5c) indicated a distinct one-photon origin of this signal. In another experiment the TPE spectrum of LHC II, the major light-harvesting complex of Photosystem II in green plants, was determined by Walla, et al. (Walla et al., 2000). Also, the time dependence of Chl fluorescence after TPE measured by fluorescence upconversion showed a very fast (~250 fs ) Chl fluorescence rise, corresponding to the energy transfer from the Car $S_1$ state to the Chls.

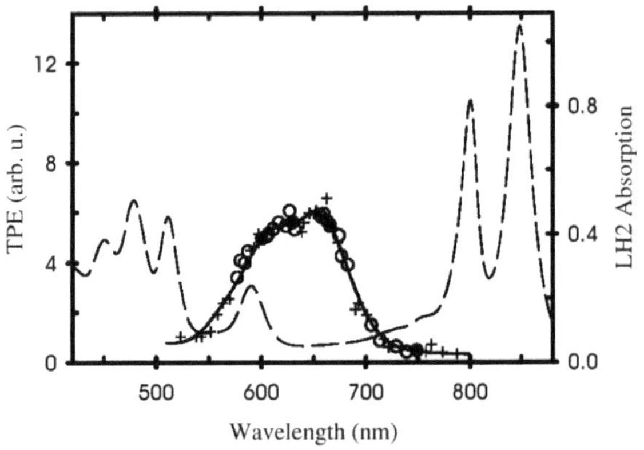


*Fig. 4.* The TPE results obtained for the LH2 of *Rb. sphaeroides*. The experiments were performed using a 250 kHz Ti:Sapphire regenerative amplifier pumped OPA in combination with lock-in detection. The TPE data (symbols) and fit (solid line) are shown on a wavelength scale together with the one-photon absorption spectrum (dashed line). The TPE data were obtained from three independent measurements, and the points from individual datasets are given by pluses. In combining three separate datasets in a single TPE spectrum, points averaged from all three datasets are shown with filled circles, while points averaged from two datasets are shown with open circles (Krueger et al., 1999, with permission).

Currently, measurement of the TPE spectrum of Car $S_1$ state is restricted to those Cars with $S_2$ states that are well separated from $S_1$ or strictly TP forbidden. When an $S_2$ state is accessible by TP absorption and the energies of the $S_1$ and $S_2$ states are similar as observed for peridinin, the resulting TPE spectrum contains limited information about the $S_1$ state (Zimmermann et al., 2002; Shima et al., 2003). However, most Cars are nonpolar molecules and do not carry a static dipole moment in the $S_2$ state and therefore their $S_2$ state is TP forbidden by the symmetry. On the other hand, TP-excited PP has so far been limited to detection of kinetics at the selected probe wavelength. Frequency-resolved measurements, particularly in the spectral region of energy acceptor absorption, await future explorations.

## V. Three Pulse Photon Echo Peak Shift Spectroscopy

### A. Principles of Photon Echo Spectroscopy

If the dynamical processes contributing to the broadening of a spectral line can be separated into very rapid (homogeneous) and ultraslow (inhomogeneous) contributions on the experimental timescale and if

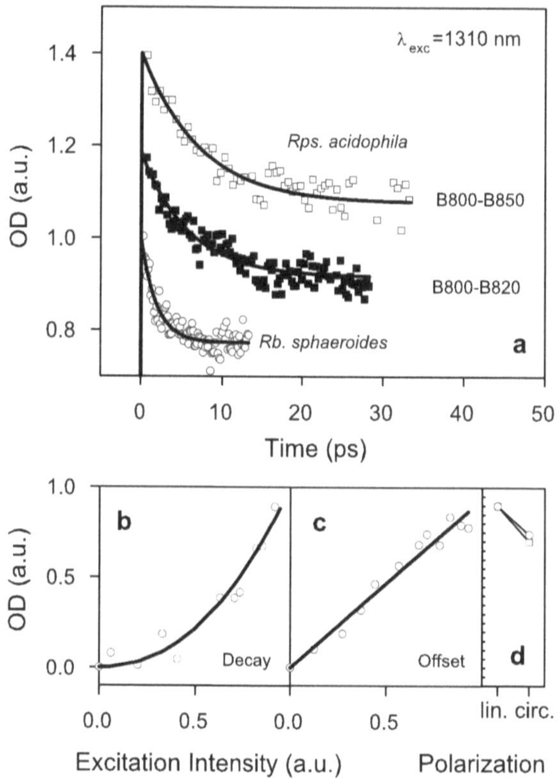


*Fig. 5.* Time dependence of ESA signals detected at 550 nm for different LH complexes of purple bacteria: (□) B800-B850 and (■) B800-B820 from *Rps. acidophila* and (○) LH2 from *Rb. sphaeroides*, and the solid lines show the corresponding mono-exponential fits with lifetimes of 7 ± 0.5, 6 ± 0.5 and 1.9 ± 0.5 ps, respectively. The traces were vertically offset for clarity. The pump and probe beams were generated using a 250 kHz Ti:Sapphire regenerative amplifier pumped OPA, arranged in a non-collinear geometry. (b) Power dependence of the amplitude associated with the mono-exponential fit of the data of *Rb. sphaer*oides, and the solid line shows the fit by an exponent of 2.2 ± 0.3. (c) Linear power dependence of the constant offset obtained from the fit to the data of *Rb. sphaeroides*. (d) Relative intensity of the fluorescence measured after TPE at 1310 nm with linear and circular polarized light of LH2 from *Rps. acidophila* (□) and *Rb. sphaeroides* (○), with a corresponding ratio of 0.80 ± 0.04 and 0.84 ± 0.04, respectively (Walla et al., 2000).

the inhomogeneous broadening is very large, (as is the case in nuclear magnetic resonance experiments), then simply measuring the decay of the intensity of the photon echo signal can provide valuable information on the homogeneous ($T_2$) timescale of the system (Aartsma et al., 1996). Early attempts to use ultrashort pulses in a photon echo experiment on a condensed phase system were made on this notion. However, condensed phase systems in general, and photosynthetic systems in particular, have a continuous range of timescales associated with them and do not have effectively infinite inhomogeneous broadening, so meaningful information cannot be obtained in this way. A decade ago Cho and Fleming and coworkers found that a different quantity, the three photon echo peak shift (PEPS) is directly related to the solvation dynamics of dye molecules in solution (Cho et al., 1996). Related work was also carried out by Wiersma and co-workers (deBoeij et al., 1996). The one-color PEPS (using three pulses (3PEPS)) was extended to the study of population transfer and incoherent energy transfer (Yang and Fleming, 1999a; Yang et al., 1999), and to more strongly coupled systems where coherent effects play a role (Yang and Fleming, 2000; Ohta et al., 2001). Yang and Fleming proposed a two-color version of the peak shift for the study of electronic coupling in dimers (Yang and Fleming, 1999b) and the method was experimentally demonstrated (Agarwal et al., 2002a; Prall et al., 2004). Cho and Fleming explored the application of one- and two-color peak shifts to coupled chromophore systems and provided a comparison with the two-dimensional photon echo spectroscopy (Cho and Fleming, 2005) described in Section VI of this chapter.

### *B. One-Color Three Pulse Photon Echo Peak Shift Spectroscopy*

Three pulse stimulated photon echoes are generated by focusing a series of three identical, variably delayed, ultrashort laser pulses onto a sample (for a detailed description of the experiment, see Joo et al., 1996). In the perturbative (weak-field) limit, the effect of the laser field on the sample can be thought of as follows. The first pulse prepares the system in a coherent superposition state, and after a time delay $\tau$ (the coherence time), the second pulse transfers the system to a population state. Then, after waiting a time $T$ (the population time), the third pulse returns the system to a coherent state. The reader new to this field may wonder how the effects of the 3 pulses can be differentiated in this manner — this is done by the experimental geometry which utilizes interference (or equivalently momentum conservation) to enforce the sequence of events just described for a signal detected in a specific direction, $\mathbf{k}_S = -\mathbf{k}_1 + \mathbf{k}_2 + \mathbf{k}_3$, where the various **k**'s are the wavevectors of the signal (S) and pulses 1–3, respectively. This ability to separate signals spatially is a unique advantage of optical spectroscopy over NMR. In a disordered sample, rephasing can occur during the second coherence period to reverse the effect of dephasing during the first coherence period, and in this case a macroscopic polarization, marked by emission of a photon echo, is generated in the sample. Dynamics in the system ranging from a few femtoseconds to hundreds of picoseconds are tracked by varying the population time. In a 3PEPS measurement, the time-integrated photon echo signal is collected for a fixed population time as the coherence delay is scanned. The peak shift is considered to be the coherence delay at which the maximum echo signal is recorded. The fact that integration is performed over the final time period, $t$ (Fig. 6), (as well as over the frequency range of the signal in general) makes the 3PEPS technique particularly simple to apply, as it does not require interferometric stability in the optical delays or complex signal processing to extract the required data.

The peak shift is a useful parameter because it is essentially a measure of the time correlation function of the electronic transition frequencies in an ensemble. This memory function, $M$(t) (Yan and Mukamel, 1990), is given by:

$$M(t) = \frac{\langle \delta\omega(t)\delta\omega(0) \rangle}{\langle \delta\omega^2(0) \rangle} \quad (2)$$

where $\delta\omega(t)$ describes the fluctuations of the electronic transition frequency of each chromophore and the angular brackets represent ensemble averages. The transition frequency as a function of time of the $i$th chromophore is given by:

$$\omega_i(t) = \delta\omega(t) + \langle\omega\rangle + \varepsilon_i \quad (3)$$

where $\varepsilon_i$ is the energetic offset in the inhomogeneous distribution and $\langle\omega\rangle$ is the average electronic transition frequency. The transition frequency is affected by interactions with the solvent, intramolecular vibrations, and energy transfer processes. Simultaneous fitting of the linear absorption spectrum and peak shift decay with a model $M$(t) therefore yields a wealth of information about system dynamics. In particular, the initial peak shift value is a measure of

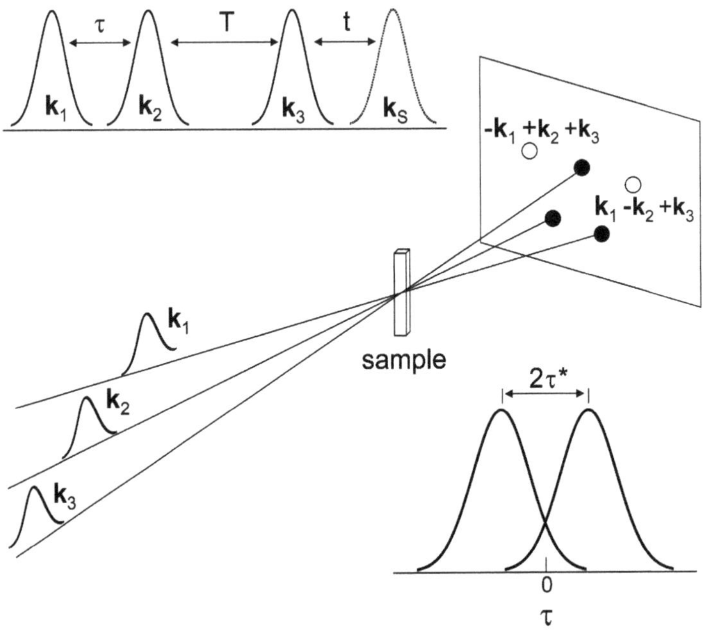


*Fig. 6.* Pulse sequence, phase matching, and peak shift for the three-pulse photon echo. The three pulses, incident on the sample with wave vectors $\mathbf{k}_1$, $\mathbf{k}_2$, and $\mathbf{k}_3$, generate two mirror-image echo signals in the phase matched directions $\mathbf{k}_1-\mathbf{k}_2+\mathbf{k}_3$ and $-\mathbf{k}_1+\mathbf{k}_2+\mathbf{k}_3$, as indicated in the diagram by the open circles. The experimentally controllable time delays between the first and second pulses and between the second and third pulses are the coherence time ($\tau$) and population time ($T$), respectively. After waiting time t following interaction with the third pulse, the photon echo signal is emitted. The lower diagram is a schematic of the integrated photon echo signal intensity versus coherence delay. The peak shift ($\tau$*) is defined as the coherence delay at which the echo signal is maximum, or, more precisely, half the difference between the corresponding signals collected in both phase matched directions.

the system's coupling to the environment, long-time values are indicative of disorder in the system that is static on the timescale of the experiment, and the decay profile contains information about dynamical processes. It has been demonstrated that the peak shift is particularly sensitive to energy transfer processes inside the laser bandwidth, and relatively insensitive to inter-band transfer processes that remove population from the pulse window (Yang and Fleming, 1999a). The reason for this insensitivity is that in a dilute system with inhomogeneous broadening the peak shift does not decay to zero at very large values of population time, *T*, because the excitation is unable to 'average' over the entire distribution of energy gaps. With energy transfer the situation changes: now the excitation can visit a representative collection of sites and the peak shift will decay to zero with a timescale set by the energy transfer timescale.

A key experimental condition of 3PEPS is that all transitions of interest must lie within the laser pulse bandwidth. Two-color 3PEPS, in which the third pulse is a different color from the first two, has been developed in recent years to circumvent this requirement. The one-color experiment utilizes a three beam interferometer, where the equal-intensity beams are split from the same fs laser source, and two beams are intercepted by computer-controlled delay stages. Lock-in processing is done to minimize scatter, and phase-matching conditions can be exploited to separate the photon echo signal from other signals and background. When the three input beams are arranged noncollinearly in an equilateral triangle geometry (from a head-on view), the photon echoes are emitted from the sample in phase-matched directions $\mathbf{k}_1 - \mathbf{k}_2 + \mathbf{k}_3$ and $-\mathbf{k}_1 + \mathbf{k}_2 + \mathbf{k}_3$ (see Fig. 6). These signals contain identical information and are mirror images of each other. Both are collected, and half the distance between them is taken as a precise measure of the peak shift.

3PEPS experiments on photosynthetic light-harvesting complexes reveal interactions among pigments and between pigments and the membrane proteins that surround them. Jimenez, et al., showed that the initial peak shift values (~25 fs) for primary light-harvesting complex (LH1) and LH2 of purple photosynthetic bacteria are higher than observed for typical laser dyes in polar solvents, indicating weak coupling of the complexes to their protein-solvent surroundings (Jimenez et al., 1997). This weak coupling may be a key aspect of efficient light-harvesting, because dissipation of energy to the surroundings

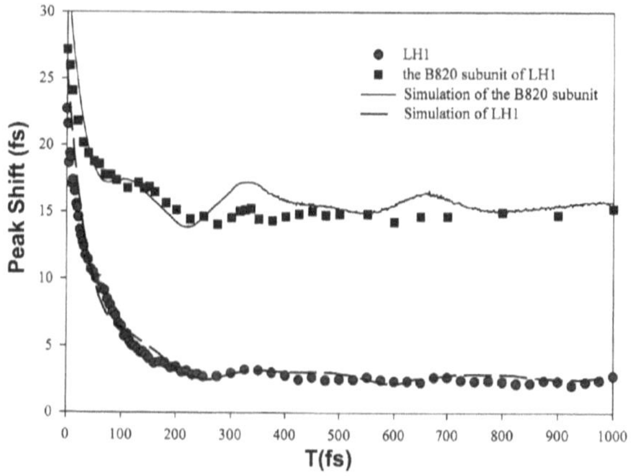


*Fig. 7.* 3PEPS measurements of LH1 of *Rb. sphaeroides* (circles) and the B820 subunit from LH1 of *Rhodospirillum (Rs.) rubrum* (squares). The solid lines represent two simulations with identical input parameters except that the energy transfer rate is set to zero for the B820 sample (Yu et al., 1997, with permission).

during the 50 ps transfer time from LH2 to LH1 to the reaction center is minimized. That the rapid (~150 fs) decay in the peak shift for the detergent isolated LH1 complex is due to energy transfer around the ring is nicely demonstrated in Fig. 7 in which the peak shift decay of an intact LH1 ring is compared with the peak shift decay for the B820 subunit of LH1 which contains just two BChl molecules (Yu et al., 1997). The solid lines represent two calculations with identical input parameters except that the energy transfer rate is set to zero for the B820 sample. The large long-time peak shift value for the B820 subunit, both observed and predicted, results from the inhomogeneous broadening, which is averaged over by energy transfer in the 28 BChl LH1 complex. Direct observation of specific energy transfer processes through peak shift measurements was also demonstrated clearly by Agarwal and coworkers in a 3PEPS study comparing isolated LH2 complexes with complexes in native membranes (Agarwal et al., 2002b). The isolated sample was detergent-solubilized so that the inter-complex distance was large, while the membrane sample, prepared from a mutant strain that contained only LH2, mimicked the wild-type membrane where LH2 rings come into close contact and energy can transfer between them. The results of the peak shift measurement using 850 nm excitation are shown in Fig. 8. The samples show similar initial decay timescales of 100–200 fs, but the membrane sample shows an additional decay component of ~5 ps. The ultrafast initial decay is ascribed to intra-complex exciton relaxation, while the 5 ps component is attributed to inter-complex (LH2-LH2) energy transfer in the membrane. Additionally, the difference between these two peak shift traces demonstrates that the static disorder of LH2 apparent in the linear spectrum is a combination of intra- and inter-complex disorder (Yang et al., 2001). That is, there are two distributions of energies corresponding to the energies of BChls within individual rings and the energies of the LH2 complexes in the membrane. Energy transfer within the ring destroys the memory of intra-complex disorder, just as energy transfer through the membrane destroys the memory of inter-complex disorder. The presence of the long 5 ps decay in the membrane sample indicates the timescale of LH2-LH2 energy transfer, as well as confirming that two levels of energetic disorder are present in the photosynthetic complexes.

### *C. Two-Color Three Pulse Photon Echo Peak Shift Spectroscopy*

In Two-color three-pulse photon echo peak shift (2C3PEPS) spectroscopy, the first two pulses have the same spectrum ('color') and the third pulse has a different, non-overlapping spectrum ('color'). Clearly, two versions of the method are possible — 'downhill' in which the second color is of longer wavelength then the first and the 'uphill' in which the reverse applies. At first sight the method seems an almost trivial extension of one-color 3PEPS, but the technique turns out to be a complementary method to 2D photon echo spectroscopy, providing information on the spatial overlap of electronic states, mixing coefficients between coupled chromophores, and remarkably, pathway-dependent information on the spectral evolution of a dynamical system.

A particularly interesting use of one- and two-color 3PEPS lies in the ratio of the one-and two-color peak shifts $\tau^*(T)$ as a function of population time, $T$ (Yang and Fleming, 1999b)

$$\frac{\tau^*_{ONE}(T)}{\tau^*_{ONE}(T)+\tau^*_{TWO}(T)} = 2\left\{1-\frac{2A_C(T)}{A_1(T)}\right\}\cos^2\theta\sin^2\theta \quad (4)$$

where $\theta$ is the mixing angle,

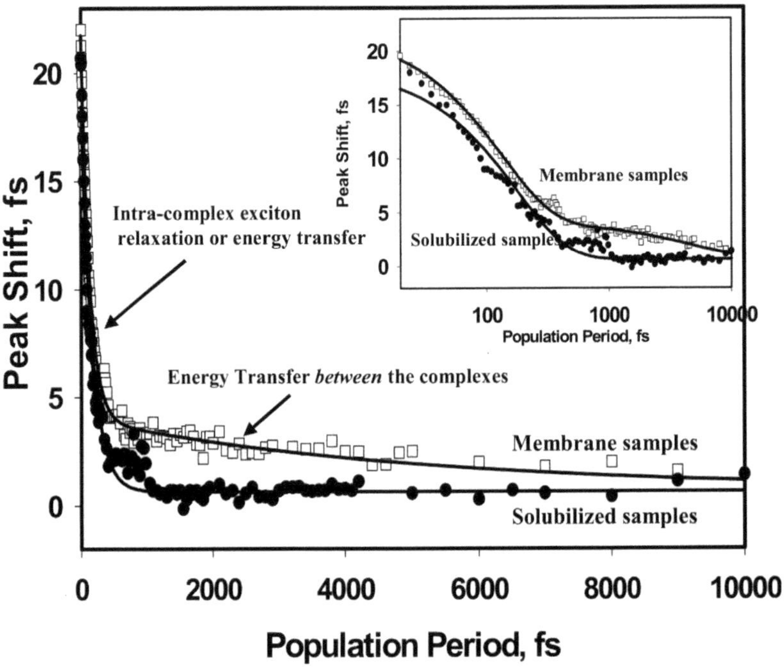


*Fig. 8.* 3PEPS data for the B850 band of LH2 from *Rb. sphaeroides* at room temperature. The circles and squares correspond to the detergent solubilized samples and native membrane samples, respectively. The membrane samples are obtained from a mutant strain containing LH2 as the sole BChl-protein complex. The inset shows the same data on a logarithmic scale. The solid lines represent exponential fits to the data. The data for the solubilized samples are best described by a single-exponential decay component (95%, 160 fs), whereas the membrane samples are best fit by two exponential components (80%, 180 fs, 16%, 5 ps) (Agarwal et al., 2002b, with permission).

$$\theta = \frac{1}{2}\arctan\left(\frac{2J}{\varepsilon_1 - \varepsilon_2}\right) \quad (5)$$

with $J$ the coupling and $\varepsilon_1$ and $\varepsilon_2$ the monomer site energies. $A_1(t)$ is the real part of site energy fluctuation correlation function, the un-normalized form of $M(t)$ and $A_c(t)$ is the identical quantity describing the fluctuations in $J$. Thus for a homodimer, by measuring both one-color 3PEPS and 2C3PEPS and calculating the ratio in Eq. (4) as a function of $T$, one can estimate (1) the mixing angle, (2) the coupling constant, $J$ and (3) the ratio of $A_c(T)/A_1(t)$. This approach has been used by (Prall et al., 2004) to successfully extract the coupling constant for a phthalocyanine dimer system.

The two-color peak shift method is still under active development and the reader is referred to couple of references (Cho and Fleming, 2005; Prall et al., 2005) for the most recent ideas.

## VI. Femtosecond Two-Dimensional Fourier Transform Electronic Spectroscopy

### *A. Principles of Two-Dimensional Spectroscopy*

At the end of the nineties, multidimensional femtosecond spectroscopy emerged as a powerful new tool to study ultrafast dynamics, interactions and transient structures of complex molecular systems. This new technique was inspired by the tremendous success of multidimensional Fourier transform magnetic resonance spectroscopy, e.g., in structure determination of large molecules and magnetic resonance imaging. Advances in ultrafast lasers (see Section II) have allowed the adoption of many two-dimensional (2D) techniques from magnetic resonance for applications in infrared and optical spectroscopy. The basic concept of two-dimensional spectroscopy is as follows. The behavior of the spectrum is probed across a range of single excitation frequencies, yielding two dimensions of frequency information and thus revealing connections between molecular transitions corresponding to specific frequencies. More complex 2D techniques use spectral interferometry and Fourier transform (FT) tools to obtain full amplitude and phase information of the system polarization. The measured electric field is separated into its real and imaginary parts, conveying information about the absorption (real part) and refractive properties (imaginary part) of the system under investigation. The real part can be interpreted as the transient field amplitude at probe frequency $\omega_t$ (the detection frequency) due to excitation at frequency $\omega_\tau$ (the coherence frequency) after waiting time $T$ (the population time). The imaginary part describes transient changes in the refractive index in an analogous way. The time resolution is defined by the length of the excitation pulses, and frequency resolution is limited only by the detection system and uncertainty principle set by the resonances in the sample itself, rather than the laser pulses. Pioneering 2D experiments for vibrational transitions were performed using dynamic hole burning (Hamm et al., 1998) and heterodyne-detected photon echoes (Asplund et al., 2000; Hochstrasser et al., 2000; Golonzka et al., 2001). 2D Fourier transform spectroscopy in the visible-domain is closely related to its infrared-domain counterpart, although the technique is somewhat more difficult. 2D Fourier transform electronic spectroscopy (we will use '2D electronic spectroscopy' from now on) is complicated by difficult-to-achieve phase stability at shorter, visible wavelengths and generally broad electronic transitions resulting in the overlap of various peaks in the 2D spectra. 2D spectroscopy for electronic transitions was introduced and discussed in detail in a series of excellent publications by Jonas and coworkers (Hybl et al., 1998; Faeder and Jonas, 1999; Jonas, 2003). Foundations for the theoretical explorations of vibrational and electronic 2D spectroscopy were laid by Mukamel (1995, 2000).

2D electronic spectroscopy is especially well suited for studies of multi-chromophoric protein complexes in photosynthesis because it probes electronic couplings between chromophore molecules and explores the distribution of instantaneous molecular environments and geometries.

### *B. Experiment and Analysis*

The technique of 2D electronic spectroscopy discussed here is based on measuring three pulse photon echoes, which are described in detail in section V of this chapter. The goal of 2D electronic spectroscopy is to obtain the complete third-order optical response function of investigated systems, which can include molecular aggregates or any excitonically coupled system. In contrast to the 3PEPS experiment, heterodyne detection of the photon echo are required

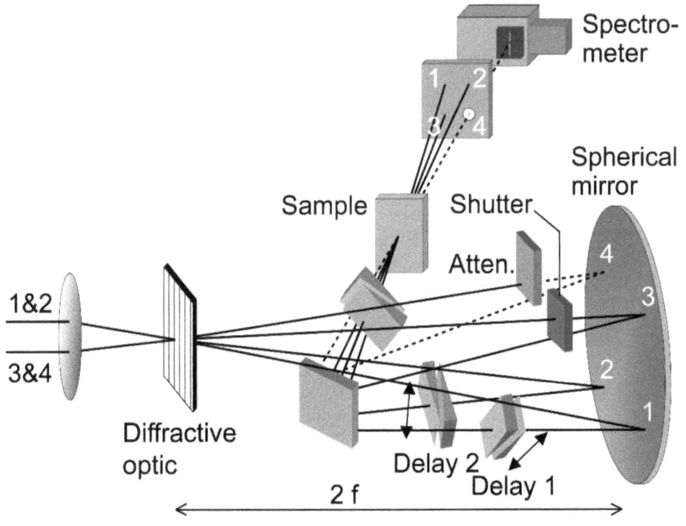


*Fig. 9.* Inherently phase stabilized 2D spectroscopy setup. The two parallel beams are focused onto the diffractive optic (DO). The first diffraction orders emerge with high efficiency providing all four beams used in the experiments. Beam 4 is the local oscillator. The spherical mirror (2f = 50 cm) creates an image of the pulse overlap in the sample cell via a plane folding mirror. The delays in beam 1 and 2 are provided by computer-controlled movable fused silica wedges. Full characterization of signal field is carried out by spectral interferometry with the attenuated LO. An automated beam shutter is used for subtraction of scattering contributions (Brixner et al., 2004a).

for 2D electronic spectroscopy. The four pulses used in the experiment are arranged in a box geometry (Fig. 9), so that the signal, obeying the phase matching relationship $\mathbf{k}_S = -\mathbf{k}_1 + \mathbf{k}_2 + \mathbf{k}_3$, is emitted in the direction of the fourth local oscillator (LO) pulse $\mathbf{k}_{LO}$. As mentioned above, one major difficulty in electronic 2D spectroscopy is to accomplish phase stability on the timescale of the experiment. To achieve this goal, a diffractive optic (DO) setup (shown in Fig. 9) with passive phase stabilization was built in our laboratory (Brixner et al., 2004a, 2004b). It can be shown that the overall signal phase is stabilized if $-\Delta t_1+\Delta t_2+\Delta t_3-\Delta t_4 = 0$, where $\Delta t_i$ is a small time shift of the i$^{th}$ beam. This condition is fulfilled in our setup, because the combination of a DO and special geometry compensates for the small timing/phase differences in the four beams introduced by vibrations of the big spherical mirror or folding mirror (for details, see Brixner et al., 2004b). Also, the setup is very compact to guarantee uniform environmental conditions and minimize the effect of any vibrations. The (interferometric) accuracy in the coherence time ($\tau$) delay, which is necessary for Fourier transform spectroscopy, is accomplished by moving fused silica glass wedges with nanometer accuracy in the paths of beams 1 and 2 (Fig. 9). In this way, timing precision of 2.7 as is achieved over the total coherence time scanning range of 400 fs. Complete characterization of the third order signal electric field is obtained by spectral interferometry (Dorrer et al., 2000) of the signal heterodyned (mixed) with the LO. For every population time $T$, the coherence time $\tau$ is typically scanned from −400 to 400 fs by moving pulse 1 from $-(\tau+T)$ to $-T$, and then moving pulse 2 from $-T$ to $-(\tau+T)$ (see Fig. 10). For each population time T and coherence time $\tau$, the total interferometric signal is recorded, given by:

$$\mathrm{I}_{SI}(\omega_t) = |\mathrm{E}_S(\omega_t)|^2 + |\mathrm{E}_{LO}(\omega_t)|^2 + 2\mathrm{Re}E^*_{LO}(\omega_t)\mathrm{E}_S(\omega_t)\exp(i\omega_t t_4) \qquad (6)$$

where $E_S(\omega_t)$ is the signal field and $E_{LO}(\omega_t)$ is the local oscillator field for time delay $t_4$.

The measurement is very sensitive to any scattering signals arriving at the same time as the true photon echo signal. Therefore additional measurements are performed while blocking some of the beams, and background, linear terms and unwanted scattering signals are subtracted. Thus only $S(\omega_t)$, the interference signal between the photon echo and local oscillator corresponding to the third term in Eq. (6), is extracted (Brixner et al., 2004b). The inverse Fourier transform of the interferogram $S(\omega_t)$ is computed in a rotating frame with the central frequency of the spectrum, and a window is applied which keeps only the desired

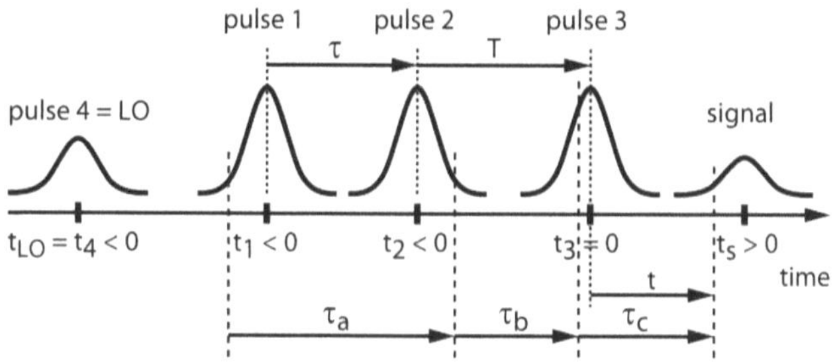


*Fig. 10.* Pulses and time variables in the heterodyne-detected three photon echo experiment. Time zero is set at the center of the third excitation pulse. The first two excitation pulses are separated by the coherence time τ. The population time T is the separation between the second and third excitation pulses. Nonlinear third order polarization at time t is induced by field interaction at times $\tau_a+\tau_b+\tau_c$, $\tau_b+\tau_c$ and $\tau_c$ earlier, which occur under the excitation pulse envelopes. The photon echo signal is observed with an average arrival time $t_S$ that is similar to coherence time. The local oscillator (LO) pulse arrives first at time $t_4$ (Brixner et al., 2004b).

heterodyne contribution around time $t_S-t_4$, which is typically 600–800 fs. Here $t_S$ and $t_4$ are the signal and LO arrival times, correspondingly. After applying the window, a Fourier transform is performed back to the frequency domain and the result is divided by the LO field, obtained in a separate measurement. Thus the complex electric field of the signal for each population and coherence time is equal to

$$E_S(\omega_t) = \frac{F(W(t)F^{-1}S(\omega_t))\exp(-i\omega_t t_4)}{E_{LO}(\omega_t)} \qquad (7)$$

where $W(t)$ is a window function. Using Maxwell's equations, it can be seen that the signal electric field in the frequency domain is related to the sample polarization by

$$iP^{(3)}(\tau,T,\omega_t) \sim \frac{n(\omega_t)}{\omega_t} E_S(\tau,T,\omega_t) \qquad (8)$$

Here $n(\omega_t)$ is the linear refractive index. Radiative line-shape distortions in the recovered electric field signal are corrected for by multiplication with a factor $n(\omega_t)/\omega_t$. Finally, Fourier transformation along the coherence time τ leads to the 2D spectrum

$$S_{2D}(\omega_\tau,T,\omega_t) = \int_{-\infty}^{\infty} iP^{(3)}(\tau,T,\omega_t)\exp(i\omega_\tau \tau)\,d\tau \qquad (9)$$

with two frequency axes corresponding to coherence frequency $\omega_\tau$ and detection frequency $\omega_t$. However, the 2D spectrum is determined in the relative phase, and therefore the final step of data analysis is obtaining the absolute phase. This is accomplished by determining a constant phase factor following the projection-slice theorem, which states that the real part of the 2D spectrum projected onto the $\omega_t$ axis is equal to the spectrally resolved transient electric field spectrum (sometimes called pump-probe, even though it is calculated in a different way than a conventional pump-probe spectrum), which is measured separately (Jonas, 2003).

In order to extract relevant parameters (e.g., electronic couplings between the chromophores) from the 2D spectra, rigorous theoretical simulation of the 2D spectra is necessary. Here we describe just very basic points of the theoretical simulations. For a detailed description see (Brixner et al., 2004b; Cho et al., 2005). Using perturbation theory developed for nonlinear spectroscopy (Mukamel, 1995), the third order polarization in the sample giving rise to the photon echo at time $t$ can be calculated as follows:

$$\begin{aligned} P^{(3)}(t) = &\int_0^\infty \int_0^\infty \int_0^\infty S^{(3)}(\tau_a,\tau_b,\tau_c) \\ &\times E(t-\tau_a-\tau_b-\tau_c)E(t-\tau_b-\tau_c) \\ &\times E(t-\tau_c)\,d\tau_a d\tau_b d\tau_c \end{aligned} \qquad (10)$$

The polarization is induced by the third-order time domain sample response function $S^{(3)}(\tau_a,\tau_b,\tau_c)$ after three electric field interactions with pulses at times $t-\tau_a-\tau_b-\tau_c$, $t-\tau_b-\tau_c$ and $t-\tau_c$ (see Fig. 10 for the ordering of the pulses and time variables). The temporal electric field of laser pulses interacting with the sample is given by

$$E(t) = \tilde{A}(t-t_1)e^{-\omega_0(t-t_1)+i\vec{k}_1\vec{r}} + \tilde{A}(t-t_2)e^{-\omega_0(t-t_2)+i\vec{k}_2\vec{r}}\,\tilde{A}(t-t_3) \times e^{-\omega_0(t-t_3)+i\vec{k}_3\vec{r}} + c.c. \quad (11)$$

with laser carrier frequency $\omega_0$ and complex field amplitude envelopes $\tilde{A}(t)$. Inserting the laser fields of Eq. (11) into Eq. (10) results in 216 terms, however only six terms generate signal in the $-\mathbf{k}_1 + \mathbf{k}_2 + \mathbf{k}_3$ direction along which measurements are performed. The third order response function $S^{(3)}_{rw}(\tau_a,\tau_b,\tau_c)$ is calculated under the rotating-wave approximation (RWA), i.e. the fast oscillating terms, which contribute much less than slowly varying terms after integration, are ignored. The two-dimensional Fourier transformation of $iP^{(3)}_{rw}(\tau,T,t)$ with respect to variables $\tau$ and $t$ produces a two-dimensional spectrum for each population time $T$, which is directly comparable to the 2D spectrum obtained from the experiment.

### C. Two-Dimensional Spectroscopy of Light-Harvesting Complexes

Light-harvesting complexes in photosynthesis are multi-chromophoric systems with highly-optimized structures. Information on pigment couplings is crucial for understanding energy transfer pathways resulting in extremely efficient light-harvesting, as well as important quenching mechanisms essential to the survival of photosynthetic organisms under stressful, high-light conditions. 2D electronic spectroscopy was applied to the (FMO) protein complex (Fenna and Matthews, 1975) of green sulfur bacteria *Chlorobium tepidum*, which serves as an antenna and mediator for directing excitations from chlorosome antennae to the reaction center (Brixner et al., 2005; Cho et al., 2005) and to light-harvesting complex III from purple bacteria *Rhodopseudomonas acidophila* (Zigmantas et al., 2006). Herein we discuss the FMO study to illustrate 2D electronic spectroscopy explorations of photosynthetic systems. The FMO complex is a trimer, with each monomer containing seven BChl *a* molecules. Because the inter-complex BChl interactions in the trimer are very weak, only interactions within the monomer were considered. The FMO complex was extensively studied and used as a model system for excitonic interactions because of its relatively simple structure. Approximate coupling constants between BChls, energy levels, and location of excitons in the monomer were estimated in previous experimental and theoretical studies.

Experimental and simulated linear and 2D spectra of FMO are shown in Fig. 11. Note that in order to explore connections between different transitions, the laser spectrum has to cover all transitions of interest in the FMO linear absorption spectrum (Fig. 11d). In 2D spectra, positive features correspond to 'more light', given by stimulated emission and bleaching, and negative features correspond to 'less light', as a result of excited state absorption. Excitonic energy levels are indicated by horizontal and vertical lines. The 2D spectrum of the FMO complex at population time $T = 0$ (Fig. 11a) features diagonal peaks corresponding to linear absorption bands. Elongation of these features indicates inhomogeneous spectral broadening, since it corresponds to the correlation between excitation and emission frequencies within the same pigment. Moreover, analysis of these features can reveal homogeneous linewidths of the corresponding transitions. Most interesting features appear off the diagonal as cross-peaks, e.g., A and B in Fig. 11a. Because of coherent electronic couplings between pigments, nonlinear optical transitions involving two different exciton states are allowed and thus produce cross-peaks already at $T = 0$. Using the assignment of excitonic states, coherent correlations between the excitons can be identified. Cross-peak A indicates that the BChls making up excitons 1 and 5 are coherently coupled, and cross-peak B shows the same for the excitons 2 and 5. Detailed state-to-state energy transfer pathways in the FMO complex can be followed by analyzing the 2D spectra for population times of 200 fs and 1 ps as shown in Fig. 11b, c. It is easily seen that the amplitude of diagonal peak C decreases, and the main diagonal peak D shifts to progressively lower energies at 200 fs and 1 ps as population is transferred downwards. Similarly, downhill transfer is indicated by a concentration of features below the diagonal at detection frequencies $\omega_t$ corresponding to the lowest exciton band. Qualitative details of energy transfer pathways can be recognized directly from the spectra. For example, focusing on the two dominant cross-peaks A and B, it is clear that exciton states 4 and 5 relax to exciton state 2 (B) and state 1 (A), skipping state 3. For quantitative information, simulations of the experimental results were performed. A Frenkel exciton Hamiltonian was used with electronic coupling constants and site energies obtained by simultaneously fitting linear absorption and 2D spectra. A single ohmic spectral density and

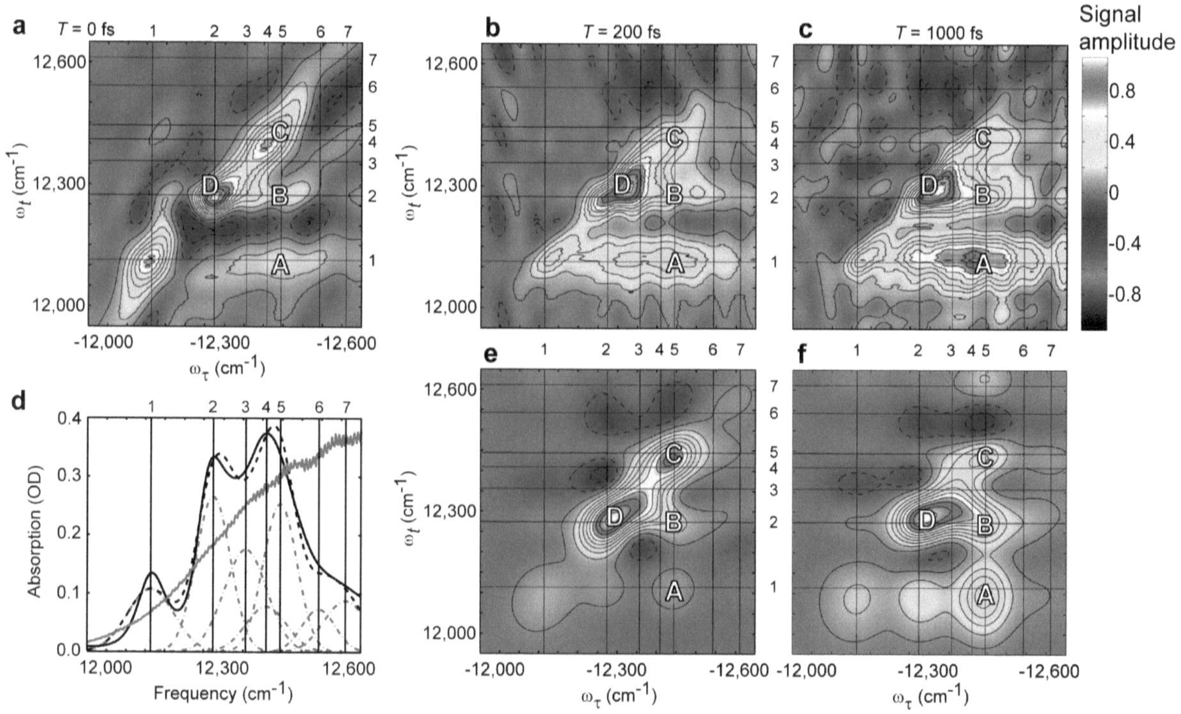


*Fig. 11.* Experimental and simulated real part of the electric field, corrected for radiative line-shape distortions, of the FMO complex at 77 K. (a-c), The experimental 2D spectra (upper three panels) are shown for population times $T = 0$ fs (a), $T = 200$ fs (b) and $T = 1$ ps (c). Contour lines are drawn in 10% intervals of the peak amplitude, with solid lines representing positive features and dashed lines negative features. Horizontal and vertical grid lines indicate excitonic levels 1–7 as labeled. (d), The experimental (solid black) and simulated (dashed black) linear absorption spectra with individual exciton contributions as shown (dashed-dotted grey). The laser spectrum (solid grey) covers all transition frequencies. (e, f), Simulation of 2D spectra are shown for $T = 200$ fs (e) and $T = 1$ ps (f). Two off-diagonal peaks marked as A and B are indicators of electronic coupling and energy transport, and two diagonal peaks are marked as C and D. (Brixner et al., 2005. Reprinted by permission from Macmillan Publishers Ltd.). See also Fig. 1, Color Plate 6.

modified Förster/Redfield theory were employed for self-consistent calculations of the exciton transfer rates, the linear spectrum, and time-dependent 2D spectra (Fig. 11e, f). Considering the complexity of the system, agreement with the experimental 2D spectra is good. Simulations taking into account the location of different exciton states on different pigments showed that there are two distinct pathways of energy flow after excitation to higher excitonic states. In one pathway, energy from exciton 7 goes directly to exciton 3, then to 2 and finally to the lowest-energy exciton 1. In the second pathway, energy from exciton 6 goes to 5, and then either via 4 or directly to 2, and eventually to 1. Thus the energy is not transferred stepwise down the energy ladder as suggested earlier, but instead follows specific pathways.

2D electronic spectroscopy enables direct determination of the mechanism underlying the evolution of population in complex photosynthetic systems. Furthermore, since 2D electronic spectroscopy is a coherent technique, coherence dynamics on sub-100 fs time can be explored. In summary, the combination of 2D spectroscopy experiments and self-consistent calculations enables the determination of electronic couplings, energy transfer pathways and rates, which depend on spatial distribution of the chromophores in the complex. This leads to the combination of temporal and spatial resolution of complex physical system such as light-harvesting complexes.

## VII. Femtosecond Stimulated Raman Spectroscopy

### *A. Method and Apparatus*

Time-resolved Raman spectroscopy has proved to be a very useful tool in tackling dynamic structural

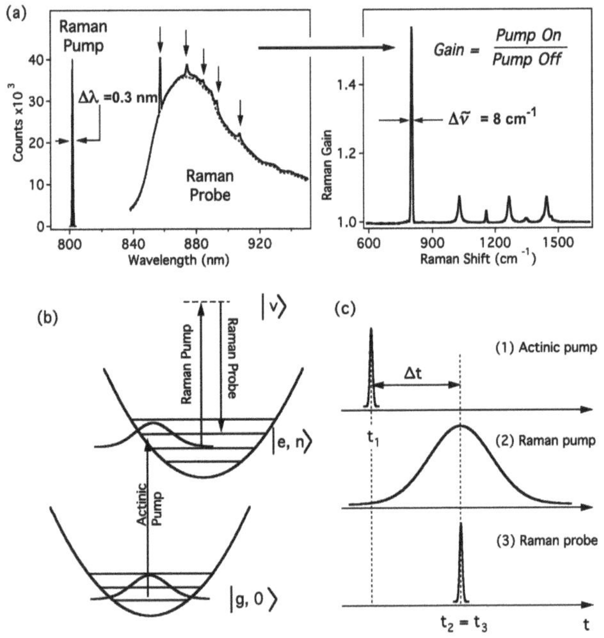


*Fig. 12.* Illustrative diagrams of the FSRS technique. (a) The spectra of the Raman pump and probe pulses for the experiment on cyclohexane. In the presence of the Raman pump pulse, the stimulated Raman effect amplifies the probe continuum at wavelengths (arrows) shifted by Raman frequencies from the Raman pump wavelength. (b) Energy level diagram showing electronic and vibrational states and vibrational resonances of the three laser pulses. (c) Timing diagram of the FSRS pulses. The Raman pump (2) and Raman probe (3) are optically delayed from the actinic pump (1) by $\Delta t$ (McCamant et al., 2004). (Fig. courtesy of R.A. Mathies, Univ. of California, Berkeley).

changes in molecules on an ultrafast time scale, as different vibrational modes are very sensitive to molecular conformations and other bond structure changes. Time-resolved Raman techniques were established in the nanosecond regime and subsequently extended to the picosecond regime, but these experiments lacked the time resolution necessary to resolve initial structural changes after excitation, which frequently happen on a fs time scale. Early fs Raman experiments, however, had poor spectral resolution because of the uncertainty principle, which implies that ultrashort duration and narrow spectral bandwidth cannot be combined in a single pulse. The elegant technique of femtosecond stimulated Raman spectroscopy (FSRS) was developed to overcome the problems of time and spectral resolution by introduction of an additional pulse (Yoshizawa and Kurosawa, 2000; Lee et al., 2004; McCamant et al., 2004). In FSRS <100 fs temporal resolution and <10 cm$^{-1}$ spectral resolution can be achieved (McCamant et al., 2004), resulting in an instrument response product of ~1 cm$^{-1}$ps, which is more than an order of magnitude better than the transform limit of spontaneous Raman (15 cm$^{-1}$ ps).

The scheme of the FSRS technique is shown in Fig. 12. First the molecules in the sample are promoted to an excited electronic state by a resonant actinic pulse of ≤50 fs duration. This pulse can be tuned to match any molecular electronic transition. Then, after a time delay Δt, in order to separate time and spectral resolutions, Raman transitions are probed by two optical fields: the Raman pump at frequency $\omega_p$ and the Raman probe at frequency $\omega_S$. The spectral resolution of the experiment is defined by the narrow-bandwidth (3–17 cm$^{-1}$), picosecond Raman pump pulse at ~800 nm and the inherent spectral resolution of the detection system. The broadband Raman probe pulse is a continuum generated in a sapphire plate or corresponding optical element and compressed to ~30 fs. The spectral region to the red of the Raman pump pulse from 820 nm to 1060 nm can be probed, allowing collection of the full Raman spectrum from 300 to 3050 cm$^{-1}$ (typically a window of 1500 cm$^{-1}$ is used). All three beams (Fig. 12(c)), actinic pump, Raman pump and probe are sent to the sample collinearly, although a crossed geometry may be implemented. The reference beam for shot-to-shot normalization of measured spectra to laser fluctuations is obtained by splitting off part of the Raman probe beam as a reference and measuring it in parallel with the Raman probe beam in the spectrograph. The time resolution is defined by the length of the actinic and Raman probe pulses and the ability to resolve the time delay between the excitation pulse and initiation of the Raman transition by the probe. Stokes frequency lines are probed at frequencies $\omega_{vib}=\omega_p-\omega_S$, and appear in the measurement as sharp spikes in the Raman probe spectrum (Fig. 12(a)). The Raman gain spectrum is calculated by dividing Raman pump-on and pump-off spectra after correcting for the background and fluctuations in the continuum using the reference:

$$Raman\ gain = \frac{[(probe-bkgnd)/(ref-bkgnd)]_{RamanPumpOn}}{[(probe-bkgnd)/(ref-bkgnd)]_{RamanPumpOff}} \quad (12)$$

To obtain excited-state vibrational Raman spectra, contributions from the solvent and electronic ground state vibrational spectra as well as transient absorption of electronic transitions have to be subtracted. Since FSRS measures relatively small changes in the intense Raman probe beam, the technique is fluorescence background free.

## B. Time-Resolved Raman Studies of β-Carotene

Understanding the photochemical and photophysical properties of Cars is essential for grasping the crucial role they play in photosynthetic complexes. The femtosecond stimulated Raman technique has been applied to the intensely-studied carotenoid β-carotene (Yoshizawa et al., 2001; McCamant et al., 2003; Kukura et al., 2004). In these experiments β-carotene was excited to the second singlet excited state, $S_2$ ($1B_u^+$ in $C_{2h}$ symmetry point group notation). The unique combination of spectral and temporal

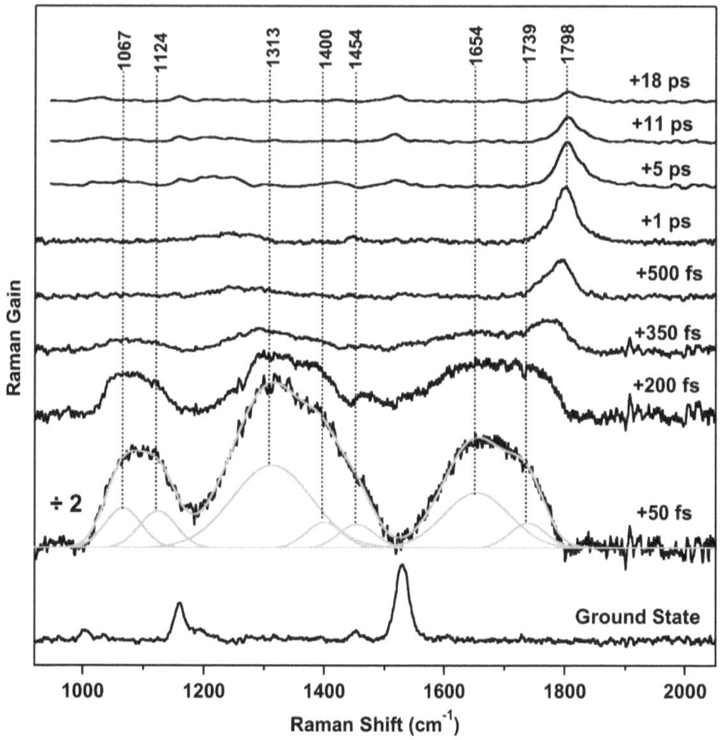


*Fig. 13.* FSRS spectra of β-carotene in cyclohexane after excitation at 492–497 nm presented as Raman gain versus Raman frequency. Pictured at the bottom is the β-carotene Raman ground state spectrum. The $S_2$ state Raman spectrum peaks at 50 fs and completely disappears within 500 fs. At the same time, a peak corresponding to the central C=C stretch in the $S_1$ state appears at 1798 cm$^{-1}$. A possible decomposition of the signal at 50 fs into Gaussian bands is indicated as grey shapes with corresponding central frequencies (McCamant et al., 2003; Kukura et al., 2004). (Fig. courtesy of R.A. Mathies, Univ. of California, Berkeley).

resolution of FSRS makes it possible to follow relaxation of β-carotene through the manifold of electronic and vibrational states.

FSRS spectra of β-carotene in cyclohexane at selected delay times between the actinic and Raman probe pulses are shown in Fig. 13 (McCamant et al., 2003; Kukura et al., 2004). The ground state β-carotene Raman spectrum is dominated by the methyl rock ($C-CH_3$) at ~1005 cm$^{-1}$, the carbon single-bond stretch (C–C) at ~1161 cm$^{-1}$ and the carbon double-bond stretch (C=C) at ~1528 cm$^{-1}$. The broad features observed for time delays <400 fs are attributed to the second singlet excited state $S_2$ and decay with time constants of 134 – 161 fs (Kukura et al., 2004). These lifetimes match the β-carotene $S_2$ state lifetime as measured by transient absorption spectroscopy. At 350 fs, another feature appears on the high-frequency side of the C=C stretch at ~1760 cm$^{-1}$, which subsequently blue-shifts, narrows, and increases in intensity. It is well known that the C=C stretching frequency in the Car $S_1$ state is blue-shifted compared to the ground state and appears at 1770 – 1800 cm$^{-1}$. This shift has been explained in terms of the vibronic coupling between the $S_1$ and $S_2$ states through the totally symmetric C=C stretch mode. Thus the peak appearing at ~1760 cm$^{-1}$ was assigned to the C=C vibration in the $S_1$ state. The evolution of this peak shows rich dynamics as it rises with a ~160 fs lifetime and decays biexponentially with time constants of ~200 fs and 9 ps (McCamant et al., 2003). It also blue-shifts and narrows with a time constant of ~450 fs. The rise of ~160 fs and decay of ~9 ps correspond to $S_2$–$S_1$ internal conversion and the lifetime of the $S_1$ state. The other two sub-fs time constants were attributed to two-step intramolecular vibrational relaxation (IVR), whereby excitation in the C=C mode is first distributed to only the strongly coupled vibrational modes in ~200 fs and then equilibrated throughout the complete set of normal modes in ~450 fs. Note that unlike time resolved absorption, where only overall vibrational relaxation in 400–700 fs was identified, employing FSRS enabled the assignment of two IVR processes to specific vibrational cooling mechanisms.

## VIII. Outlook

Over the next ten years, new optical spectroscopy methods will undoubtedly be developed and applied to photosynthetic systems. At the most straightforward, the development of polarization-based multidimensional spectroscopies should be very powerful tools to constrain the products of dipole directions and coupling strengths that determine the strength of cross peaks in two-dimensional spectra of multicomponent complexes (Cho et al., 2005; Cho and Fleming, 2006). Taking a bigger step, perhaps methods will be developed to switch on or off particular interactions in multichromophore complexes, and to measure interactions underlying electron transfer to complement the current advances in understanding of energy transfer (Scholes and Fleming, 2006).

## Acknowledgments

This work was supported by the director, Office of Science, Office of Basic Energy Sciences, Chemical Sciences Division of the U.S. Department of Energy under Contract DE-AC03-76SF00098.

## References

Aartsma TJ, Louwe RJW and Schellenberg P (1996) Accumulated photon echo measurements of excited state dynamics in pigment-protein complexes. In: Amesz J and Hoff AJ (eds) Biophysical techniques in photosynthesis (Advances in Photosynthesis, Vol 3), pp 109–122. Kluwer Academic Publishers, Dordrecht

Agarwal R, Prall BS, Rizvi AH, Yang M and Fleming GR (2002a) Two-color three pulse photon echo peak shift spectroscopy. J Chem Phys 116: 6243–6252

Agarwal R, Rizvi AH, Prall BS, Olsen JD, Hunter CN and Fleming GR (2002b) Nature of disorder and inter-complex energy transfer in LH2 at room temperature: A three pulse photon echo peak shift study. J Phys Chem A 106: 7573–7578

Asaki MT, Huang CP, Garvey D, Zhou JP, Kapteyn HC and Murnane MM (1993) Generation of 11-fs pulses from a self-mode-locked Ti:sapphire laser. Opt Lett 18: 977–979

Asplund MC, Zanni MT and Hochstrasser RM (2000) Two-dimensional infrared spectroscopy of peptides by phase-controlled femtosecond vibrational photon echoes. Proc Natl Acad Sci USA 97: 8219–8224

Baltuška A, Wei ZY, Pshenichnikov MS and Wiersma DA (1997) Optical pulse compression to 5 fs at a 1-MHz repetition rate. Opt Lett 22: 102–104

Birge RR (1983) One-photon and two-photon excitation spectroscopy. In: Kliger DS (ed) Ultrasensitive Laser Spectroscopy, pp 109–171. Academic Press, New York

Birge RR and Zhang C-F (1990) Two-photon double-resonance spectroscopy of bacteriorhodopsin. Assignment of the electronic and dipolar properties of the low-lying $^1A_g$-like and $^1B_u^{*(+)}$-like π,π* states. J Chem Phys 92: 7178–7195

Brixner T, Stiopkin IV and Fleming GR (2004a) Tunable two-dimensional femtosecond spectroscopy. Opt Lett 29: 884–886

Brixner T, Mancal T, Stiopkin IV and Fleming GR (2004b) Phase-stabilized two-dimensional electronic spectroscopy. J Chem Phys 121: 4221–4236

Brixner T, Stenger J, Vaswani HM, Cho M, Blankenship RE and Fleming GR (2005) Two-dimensional spectroscopy of electronic couplings in photosynthesis. Nature 434: 625–628

Butkus R, Danielius R, Dubietis A, Piskarskas A and Stabinis A (2004) Progress in chirped pulse optical parametric amplifiers. Appl Phys B 79: 693–700

Cho MH and Fleming GR (2005) The integrated photon echo and solvation dynamics. II. Peak shifts and two-dimensional photon echo of a coupled chromophore system. J Chem Phys 123: 1–18

Cho MH and Fleming GR (2006) Two-dimensional electronic spectroscopy of molecular complexes. J Chin Chem Soc 53: 15–24

Cho MH, Yu JY, Joo TH, Nagasawa Y, Passino SA and Fleming GR (1996) The integrated photon echo and solvation dynamics. J Phys Chem 100: 11944–11953

Cho MH, Vaswani HM, Brixner T, Stenger J and Fleming GR (2005) Exciton analysis in 2D electronic spectroscopy. J Phys Chem B 109: 10542–10556

De Silvestri S, Nisoli M, Sansone G, Stagira S and Svelto O (2004) Few-cycle pulses by external compression. In: Kärtner FX (ed) Few-Cycle Laser Pulse Generation and its Applications (Topics in Applied Physics, Vol 95), pp 137–177. Springer, Berlin/Heidelberg

deBoeij WP, Pshenichnikov MS and Wiersma DA (1996) System-bath correlation function probed by conventional and time-gated stimulated photon echo. J Phys Chem 100: 11806–11823

Dorrer C, Belabas N, Likforman JP and Joffre M (2000) Spectral resolution and sampling issues in Fourier-transform spectral interferometry. J Opt Soc Am B 17: 1795–1802

Faeder SMG and Jonas DM (1999) Two-dimensional electronic correlation and relaxation spectra: Theory and model calculations. J Phys Chem A 103: 10489–10505

Fenna RE and Matthews BW (1975) Chlorophyll arrangement in a bacteriochlorophyll protein from *Chlorobium-limicola*. Nature 258: 573–577

Fermann ME, Galvanauskas A, Sucha G and Harter D (1997) Fiber-lasers for ultrafast optics. Appl Phys B 65: 259–275

Fischer A, Cremer C and Stelzer EHK (1995) Fluorescence of coumarins and xanthenes after two-photon absorption with a pulsed titanium-sapphire laser. Appl Opt 34: 1989–2003

Friedrich DM and McClain WM (1980) Two-photon molecular electronic spectroscopy. Annu Rev Phys Chem 31: 559–577

Gai F, McDonald JC and Anfinrud PA (1997) Pump-dump-probe spectroscopy of bacteriorhodopsin: Evidence for a near-IR excited state absorbance. J Am Chem Soc 119: 6201–6202

Golonzka O, Khalil M, Demirdoven N and Tokmakoff A (2001) Vibrational anharmonicities revealed by coherent two-dimensional infrared spectroscopy. Phys Rev Lett 86: 2154–2157

Hamm P, Lim MH and Hochstrasser RM (1998) Structure of the amide I band of peptides measured by femtosecond nonlinear-infrared spectroscopy. J Phys Chem B 102: 6123–6138

Hochstrasser RM, Asplund MC, Hamm P and Ge NH (2000) Femtosecond two-dimensional infrared spectroscopy. J Chinese Chem Soc 47: 843–853

Hybl JD, Albrecht AW, Faeder SMG and Jonas DM (1998) Two-dimensional electronic spectroscopy. Chem Phys Lett 297: 307–313

Iaconis C and Walmsley IA (1999) Self-referencing spectral interferometry for measuring ultrashort optical pulses. IEEE J Quantum Elect 35: 501–509

Jimenez R and Fleming GR (1996) Ultrafast spectroscopy of photosynthetic systems. In: Amesz J and Hoff AJ (eds) Biophysical Techniques in Photosynthesis (Advances in Photosynthesis, Vol 3). Kluwer Academic Publishers, Dordrecht

Jimenez R, vanMourik F, Yu JY and Fleming GR (1997) Three-pulse photon echo measurements on LH1 and LH2 complexes of *Rhodobacter sphaeroides*: A nonlinear spectroscopic probe of energy transfer. J Phys Chem B 101: 7350–7359

Jonas DM (2003) Two-dimensional femtosecond spectroscopy. Annu Rev Phys Chem 54: 425–463

Joo TH, Jia YW, Yu JY, Lang MJ and Fleming GR (1996) Third-order nonlinear time domain probes of solvation dynamics. J Chem Phys 104: 6089–6108

Kartner FX, Morgner U, Schibli T, Ell R, Haus HA, Fujimoto JG and Ippen EP (2004) Few-cycle pulses directly from a laser. In: Kärtner FX (ed) Few-Cycle Laser Pulse Generation and its Applications (Topics in Applied Physics, Vol 95), pp 137–177. Springer, Berlin/Heidelberg

Kovalenko SA, Ruthmann J and Ernsting NP (1998) Femtosecond hole-burning spectroscopy with stimulated emission pumping and supercontinuum probing. J Chem Phys 109: 1894–1900

Krueger BP, Yom J, Walla PJ and Fleming GR (1999) Observation of the $S_1$ state of spheroidene in LH2 by two-photon fluorescence excitation. Chem Phys Lett 310: 57–64

Kukura P, McCamant DW and Mathies RA (2004) Femtosecond time-resolved stimulated Raman spectroscopy of the $S_2$ ($^1B_u^+$) excited state of beta-carotene. J Phys Chem A 108: 5921–5925

Larsen DS, Papagiannakis E, van Stokkum IHM, Vengris M, Kennis JTM and van Grondelle R (2003) Excited state dynamics of beta-carotene explored with dispersed multi-pulse transient absorption. Chem Phys Lett 381: 733–742

Larsen DS, Vengris M, van Stokkum IHM, van der Horst MA, de Weerd FL, Hellingwerf KJ and van Grondelle R (2004) Photoisomerization and photoionization of the photoactive yellow protein chromophore in solution. Biophys J 86: 2538–2550

Lee SY, Zhang DH, McCamant DW, Kukura P and Mathies RA (2004) Theory of femtosecond stimulated Raman spectroscopy. J Chem Phys 121: 3632–3642

Lide DR (1996) CRC Handbook of Chemistry and Physics. CRC Press, Boca Raton

Linden PA, Zimmermann J, Brixner T, Holt NE, Vaswani HM, Hiller RG and Fleming GR (2004) Transient absorption study of peridinin and peridinin-chlorophyll *a*-protein after two-photon excitation. J Phys Chem B 108: 10340–10345

Logunov SL, Volkov VV, Braun M and El-Sayed MA (2001) The relaxation dynamics of the excited electronic states of retinal in bacteriorhodopsin by two-pump-probe femtosecond studies. Proc Natl Acad Sci USA 98: 8475–8479

Macpherson AN, Arellano JB, Fraser NJ, Cogdell RJ and Gillbro T (2001) Efficient energy transfer from the carotenoid $S_2$ state in a photosynthetic light-harvesting complex. Biophys J 80: 923–930

McCamant DW, Kukura P and Mathies RA (2003) Femtosecond time-resolved stimulated Raman spectroscopy: Application to the ultrafast internal conversion in β-carotene. J Phys Chem A 107: 8208–8214

McCamant DW, Kukura P, Yoon S and Mathies RA (2004) Femtosecond broadband stimulated Raman spectroscopy: Apparatus and methods. Rev Sci Instrum 75: 4971–4980

McClain WM (1971) Excited state symmetry assignment through polarized two-photon absorption studies of fluids. J Chem Phys 55: 2789–2796

McClain WM (1974) Two-photon molecular spectroscopy. Acc Chem Res 7: 129–135

McClain WM and Harris RA (1977) Two-photon molecular spectroscopy in liquids and gases. In: Lim EC (ed) Excited states, Vol 3, pp 1–56. Academic Press, New York

Morgner U, Kartner FX, Cho SH, Chen Y, Haus HA, Fujimoto JG, Ippen EP, Scheuer V, Angelow G and Tschudi T (1999) Sub-two-cycle pulses from a Kerr-lens mode-locked Ti:sapphire laser. Opt Lett 24: 411–413

Mukamel S (1995) Principles of Nonlinear Optical Spectroscopy. Oxford University Press, New York

Mukamel S (2000) Multidimensional femtosecond correlation spectroscopies of electronic and vibrational excitations. Annu Rev Phys Chem 51: 691–729

Nisoli M, DeSilvestri S, Svelto O, Szipocs R, Ferencz K, Spielmann C, Sartania S and Krausz F (1997) Compression of high-energy laser pulses below 5 fs. Opt Lett 22: 522–524

Ohta K, Yang M and Fleming GR (2001) Ultrafast exciton dynamics of J-aggregates in room temperature solution studied by third-order nonlinear optical spectroscopy and numerical simulation based on exciton theory. J Chem Phys 115: 7609–7621

Papagiannakis E, Larsen DS, van Stokkum IHM, Vengris M, Hiller RG and van Grondelle R (2004) Resolving the excited state equilibrium of peridinin in solution. Biochemistry 43: 15303–15309

Polivka T and Sundström V (2004) Ultrafast dynamics of carotenoid excited states — from solution to natural and artificial systems. Chem Rev 104: 2021–2071

Prall BS, Parkinson DY, Fleming GR, Yang M and Ishikawa N (2004) Two-dimensional optical spectroscopy: Two-color photon echoes of electronically coupled phthalocyanine dimers. J Chem Phys 120: 2537–2540

Prall BS, Parkinson DY and Fleming GR (2005) Probing correlated spectral motion: Two-color photon echo study of Nile blue. J Chem Phys 123: 054515 (1–13)

Ranka JK, Windeler RS and Stentz AJ (2000) Visible continuum generation in air-silica microstructure optical fibers with anomalous dispersion at 800 nm. Opt Lett 25: 25–27

Ruhman S, Hou BX, Friedman N, Ottolenghi M and Sheves M (2002) Following evolution of bacteriorhodopsin in its reactive excited state via stimulated emission pumping. J Am Chem Soc 124: 8854–8858

Scholes GD and Fleming GR (2006) Energy transfer and photosynthetic light-harvesting. In: Rice SA (ed) Advances in Chemical Physics, Vol 132, pp 57–129. Wiley, New York

Shen YR (1984) The Principles of Nonlinear Optics. John Wiley & Sons, New York

Shima S, Ilagan RP, Gillespie N, Sommer BJ, Hiller RG, Sharples FP, Frank HA and Birge RR (2003) Two-photon and fluorescence spectroscopy and the effect of environment on the photochemical properties of peridinin in solution and in the peridinin-chlorophyll-protein from *Amphidinium carterae*. J Phys Chem A 107: 8052–8066

Spence DE, Kean PN and Sibbett W (1991) 60-fsec pulse generation from a self-mode-locked Ti:sapphire laser. Opt Lett 16: 42–44

Trebino R, DeLong KW, Fittinghoff DN, Sweetser JN, Krumbugel MA, Richman BA and Kane DJ (1997) Measuring ultrashort laser pulses in the time-frequency domain using frequency-resolved optical gating. Rev Sci Instrum 68: 3277–3295

Valkunas L and Gulbinas V (1997) Nonlinear exciton annihilation and local heating effects in photosynthetic antenna systems. Photochem Photobiol 66: 628–634

van Amerongen H, Valkunas L and van Grondelle R (2000) Photosynthetic Excitons. World Scientific, Singapore

Walla PJ, Linden PA, Hsu CP, Scholes GD and Fleming GR (2000) Femtosecond dynamics of the forbidden carotenoid S-1 state in light-harvesting complexes of purple bacteria observed after two-photon excitation. Proc Natl Acad Sci USA 97: 10808–10813

Wilhelm T, Piel J and Riedle E (1997) Sub-20-fs pulses tunable across the visible from a blue-pumped single-pass noncollinear parametric converter. Opt Lett 22: 1494–1496

Xu C and Webb WW (1996) Measurement of two-photon excitation cross sections of molecular fluorophores with data from 690 to 1050 nm. J Opt Soc Am B 13: 481–491

Yan YJ and Mukamel S (1990) Femtosecond pump-probe spectroscopy of polyatomic-molecules in condensed phases. Phys Rev A 41: 6485–6504

Yang M and Fleming GR (1999a) Third-order nonlinear optical response of energy transfer systems. J Chem Phys 111: 27–39

Yang M and Fleming GR (1999b) Two-color three-pulse photon echoes as a probe of electronic coupling in molecular complexes. J Chem Phys 110: 2983–2990

Yang M and Fleming GR (2000) Third-order nonlinear optical response and energy transfer in static disordered systems. J Chem Phys 113: 2823–2840

Yang M, Ohta K and Fleming GR (1999) Three-pulse photon echoes for model reactive systems. J Chem Phys 110: 10243–10252

Yang M, Agarwal R and Fleming GR (2001) The mechanism of energy transfer in the antenna of photosynthetic purple bacteria. J Photochem Photobiol A 142: 107–119

Yoshizawa M and Kurosawa M (2000) Femtosecond time-resolved Raman spectroscopy using stimulated Raman scattering. Phys Rev A 61: 013808 (1–6)

Yoshizawa M, Aoki H and Hashimoto H (2001) Vibrational relaxation of the $^2A_g^-$ excited state in all-trans-β-carotene obtained by femtosecond time-resolved Raman spectroscopy. Phys Rev B 63: 180301 (1–4)

Yu JY, Nagasawa Y, van Grondelle R and Fleming GR (1997) Three pulse echo peak shift measurements on the B820 subunit of LH1 of *Rhodospirillum rubrum*. Chem Phys Lett 280: 404–410

Zigmantas D, Read EL, Mančal T, Brixner T, Gardiner AT, Cogdell RJ and Fleming GR (2006) Two-dimensional electronic spectroscopy of the B800-B820 light-harvesting complex. Proc Natl Acad Sci USA 103: 12672–12677

Zimmermann J, Linden PA, Vaswani HM, Hiller RG and Fleming GR (2002) Two-photon excitation study of peridinin in benzene and in the peridinin chlorophyll a-protein (PCP). J Phys Chem B 106: 9418–9423

# Chapter 12

# (Sub)-Picosecond Spectral Evolution of Fluorescence Studied with a Synchroscan Streak-Camera System and Target Analysis

Ivo H. M. van Stokkum[1]*, Bart van Oort[2], Frank van Mourik[3], Bas Gobets[4] and Herbert van Amerongen[2]

*[1]Department of Physics and Astronomy, Faculty of Sciences, Vrije Universiteit, De Boelelaan, 1081 HV Amsterdam, The Netherlands; [2]Laboratory of Biophysics, Wageningen University, PO Box 8128, 6700 ET Wageningen, The Netherlands; [3]École Polytechnique Fédérale de Lausanne, Laboratory of Ultrafast Spectroscopy, Institut de Sciences et Ingéniérie Chimiques, Lausanne-Dorigny, Switzerland; [4]Image Science Institute, University Medical Center Utrecht, Utrecht, The Netherlands*



## Summary

A synchroscan streak camera in combination with a spectrograph can simultaneously record temporal dynamics and wavelength of fluorescence representable as an image with time and wavelength along the axes. The instrument response width is about 1% of the time range (of typically 200 ps to 2 ns). The spectral window of 250 nm may lie between 250 and 850 nm. Such spectrotemporal measurements using low excitation intensities

*Author for correspondence, email: ivo@nat.vu.nl

*Thijs J. Aartsma and Jörg Matysik (eds): Biophysical Techniques in Photosynthesis II, pp. 223–240.*

have become routine. Sophisticated data analysis methods are mandatory to extract meaningful physicochemical parameters from the wealth of information contained in the streak image. In target analysis a kinetic scheme is used in combination with assumptions on the spectra of the species to describe the system. In this chapter the principals of operation of a streak-camera setup are described, along with the fundamental and technical limitations that one encounters. The correction and calibration steps that are needed as well as data processing and analysis are discussed. Several case studies of bioluminescence are presented, with a particularly in-depth analysis of trimeric Photosystem I core particles of the cyanobacterium *Spirulina platensis*.

## I. Introduction

Time-resolved fluorescence spectroscopy has proven to be extremely useful in photosynthesis research in the past decades (Sauer and Debreczeny, 1996). Time-correlated single photon timing (TCSPT) has often been the method of choice, since it is relatively cheap, provides excellent signal-to-noise ratios and is rather standardized. It is particularly useful for determining the overall charge-separation time of a variety of photosynthetic systems, and can even be applied to entire cells and chloroplasts. Like all methods, TCSPT has its limitations: the instrument response time is several tens of picoseconds, which is a serious draw-back, for instance when studying individual pigment-protein complexes where relevant processes occur on sub-ps and ps time scales. Extremely careful measurements and deconvolution of the time traces are needed to resolve a time constant of at best ~5 ps. A second limitation of TCSPT is that, commonly, one selects one detection wavelength at a time, and recording the spectral evolution of the fluorescence requires subsequent measurements at different wavelengths. This restriction determines to a large extent the minimum time for data recording. The temporal instrument response of a synchroscan streak-camera system has a full width at half maximum (FWHM) of a few picoseconds. With deconvolution it is even possible to measure at sub-ps time resolution, which is approaching the resolution of fluorescence up-conversion (Jimenez and Fleming, 1996). Although the streak camera is generally used to record time-resolved fluorescence, it has also been applied to measure time-resolved absorption spectra in the range from ps to ns (Ito et al., 1991). In this chapter, we discuss fluorescence detection with a streak camera in combination with a spectrograph. This allows for simultaneous registration of both the time of emission of a fluorescence photon and the emission wavelength, reducing the measuring time substantially. The fluorescence photons eventually lead to a two-dimensional image on a CCD camera, of which the vertical position indicates the emission time, whereas the horizontal position corresponds to the emission wavelength. An example of such an image (Gobets et al., 2001b) is given in Fig. 1, in which the grey levels reflect the fluorescence intensity as a function of time and wavelength. Such an image contains a wealth of information and it will be discussed in detail how this information can be extracted. Throughout this chapter we will refer to these data as 'the PS I trimer data'.

Two decades ago Campillo and Shapiro (1983) wrote an excellent review on the history and possibilities of the streak camera, including its application to photosynthesis. Measurements were performed without wavelength dispersion and only in a few cases several wavelengths were probed. In the same year Freiberg and Saari (1983) published a detailed article on the possibilities and limitations of obtaining simultaneously time and wavelength information. Ohtani et al. (1990) performed one of the first fluorescence experiments in photobiology in which excellent time resolution (3 ps) was combined with measuring complete spectra, studying bacteriorhodopsin from purple membranes of *Halobacterium halobium*. There are several later reports on similar preparations (Ohtani et al., 1994, 1999; Kamiya et al., 1997; Haacke et al., 2001; van Stokkum et al., 2006). Such single-chromophore systems are generally easier to study than chlorophyll-containing photosynthetic complexes: in photosynthetic systems excitation-energy transfer between chromophores takes place, which in case of too high excitation energies can result in singlet-singlet annihilation,

*Abbreviations:* Chl – chlorophyll; cw – continuous wave; DAS – decay associated spectrum; EAS – evolution associated spectrum; EET – excitation energy transfer; f – focal length; FWHM – full width at half maximum; IRF – instrument response function; LHC – light harvesting complex; MA – magic angle; MCP – micro-channel plate; OPA – optical parametric amplifier; PCP – peridinin-chlorophyll protein; PS – Photosystem; RegA – regenerative amplifier; SAS – species associated spectrum; SVD – singular value decomposition; TCSPT – time-correlated single photon timing

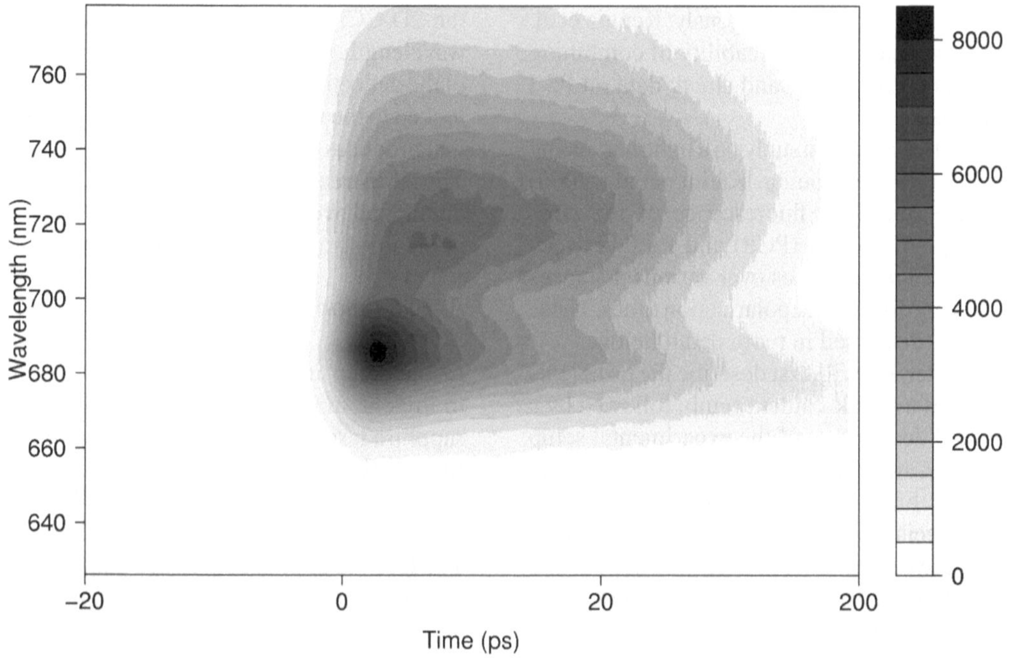


*Fig. 1.* Filled contour plot of emission data from trimeric core particles of PS I of *Spirulina platensis* (from Gobets et al., 2001b) after excitation at 400 nm. Note that the time axis is linear from –20 to +20 ps relative to the maximum of the IRF, and logarithmic thereafter.

a process that can distort the fluorescence kinetics (Sauer and Debreczeny, 1996).

Gilmore et al. (2000, 2003a,b) nicely demonstrated the application of the streak camera to obtain time-resolved fluorescence spectra of leaves. Spectral and kinetic differences between Photosystems I and II could be discerned but relevant spectral evolution was only observed for times longer than 100 ps. Donovan et al. (1997) used a streak camera with 4–9 ps time resolution to study isolated PS II reaction centers. Measuring at multiple wavelengths they concluded that the charge separation time should be either faster than 1.25 ps or slower than 20 ps. Later studies by van Mourik et al. (2004) and Andrizhiyevskaya et al. (2004b) on isolated PS II reaction centers revealed at least four different lifetimes. For excitation at 681 nm lifetimes of 6 ps, 34 ps, 160 ps and 7 ns were observed. The corresponding decay-associated spectra (DAS) were all different except for the 160 ps and 7 ns DAS, indicating that relatively slow excitation energy transfer (EET) takes place. The 34 ps component was assigned to partly represent EET. Further evidence for slow EET was obtained by the fact that excitation at 690 nm resulted in different DAS. In addition, the data indicated that charge separation is ultrafast (<1 ps) and that relatively slow radical pair relaxation takes place.

The streak camera has been particularly useful for the study of fast kinetics in PS I (Gobets et al., 2001a,b; Kennis et al., 2001; Ihalainen et al., 2002, 2005c,d; Andrizhiyevskaya et al., 2004a). Much spectral evolution occurs on a time scale of several ps and higher, which makes PS I an ideal candidate for streak-camera measurements. Below we will make use of some of these results to demonstrate the experimental possibilities of the setup and the power of advanced data analysis.

The streak camera was also used for the study of light-harvesting complexes. It was for instance used to measure lifetimes on the order of many hundreds of ps to several ns and fluorescence quantum yields (Monshouwer et al., 1997; Palacios et al., 2002; Ihalainen et al., 2005a,b). Gobets et al. (2001a) studied LHC-I by fluorescence up-conversion at five different wavelengths (IRF 150 fs, time range 5 ps) and with the streak camera (IRF 3 ps/20 ps, time range 200 ps/2.2 ns) and at common wavelengths the kinetic traces of both techniques joined smoothly in the overlapping time interval. A multitude of decay times was observed ranging from 150 fs to 2 ns (four orders of magnitude) and the corresponding spectra revealed many pathways of EET between carotenoids, Chls *b*, Chls *a* and 'red' Chls *a*, the fluorescence of which is shifted to the red by tens of nm as compared to 'normal' Chls *a*. Analogously, Kennis et al. (2001) demonstrated the applicability of combining fluorescence up-conversion and streak data on PS I core complexes.

One more streak-camera study on a light-harvesting complex is worth mentioning. Kleima et al. (2000) measured the polarized fluorescence of the peridinin-chlorophyll protein (PCP) and EET between isoenergetic Chl *a* molecules over various distances was reflected by different depolarization times. These results will be discussed in more detail below.

In this chapter, we will first describe the principals of operation of a streak-camera setup, followed by a more detailed description of the experimental setup in Wageningen and a discussion of the fundamental and technical limitations that one encounters. In particular, special precautions have to be taken to prevent sample degradation and one has to be aware of the possible occurrence of unwanted nonlinear effects such as singlet-singlet annihilation. In order to exploit the full potential of the setup and the recorded data, several correction and calibration steps are needed as well as advanced data processing and fitting, which will be discussed subsequently.

## II. Principle of Operation of the Streak-Camera Setup

The basic goal of the streak-camera setup (Fig. 2) is to determine the wavelength and time of emission of each fluorescence photon detected. A pulsed light source induces fluorescence photons from the sample, which are diffracted by a grating in a horizontal plane after which they hit a horizontal photocathode, producing photo-electrons. These photo-electrons from the photocathode are accelerated and imaged by electrostatic or magnetic lenses onto a 2D detector consisting of a micro-channel plate (MCP) electron multiplier, a phosphor screen, and a cooled CCD camera. On their way from the cathode to the MCP the electrons produced at different times experience a time-dependent vertical electric field (the deflection field or sweep field). Thus photo-electrons generated at different times experience a changed electric field, and therefore hit the MCP at different vertical positions. In the MCP each accelerated photo-electron causes a cascade of electrons (electron multiplication) which in turn hit the phosphor screen, causing a number of photons then detected by the CCD camera. Thus, the vertical and horizontal axes of the 2D CCD-image code respectively for time and wavelength. The time-dependence of the magnitude of the deflection field is sinusoidal and its frequency is locked to the frequency of the same optical oscillator that produces the exciting laser pulses (synchroscan). Thus the streak image on the CCD camera can be accumulated over many successive laser pulses, whilst maintaining a good temporal resolution.

### A. Excitation

In the setup in Wageningen, which is comparable to the one in Amsterdam, a mode-locked titanium-sapphire laser, pumped by a 5-W cw diode pumped frequency doubled Nd:$YVO_4$ laser, provides light pulses at a repetition rate of 75.9 MHz, wavelength 800 nm, 1 W average power and 0.2 ps pulse width. The laser beam is split into two paths: Path 1 is used for synchronization of the deflection field. Path 2 enters a regenerative amplifier (RegA), pumped by a 10 W cw diode-pumped frequency-doubled Nd: $YVO_4$ laser. The amplifier increases the pulse energy to ~4 μJ at a repetition rate of 250 kHz (0.2 ps, 800 nm). These pulses are fed into an Optical Parametric Amplifier (OPA). In the OPA the beam is split: it is partially frequency-doubled and partially used to generate white light. Mixing of these beams leads to selective and tunable amplification of light at any selected wavelength in the range of 470 to 700 nm. This light can be used directly for excitation or after frequency doubling to 235–350 nm. Alternatively, the Ti:sapphire laser can be tuned in the range from 700 to 1000 nm and applying frequency doubling, this allows excitation at 'all' wavelengths longer than 235 nm. The excitation light is directed through a Berek variable waveplate to control its polarization direction and is focused into the sample by a lens of 15 cm focal length, leading to a focal spot of ~100 μm diameter.

### B. Polarization

Anisotropic measurements can be performed in two ways: by adjusting the polarization of either the detected light, or the exciting light. In the first case one excites with vertically polarized light and turns a polarizer in the detection branch either horizontally of vertically, to obtain the perpendicular and parallel components of the emission. However, in this case one needs to correct for the difference in sensitivity of the detection system for horizontally and verti-

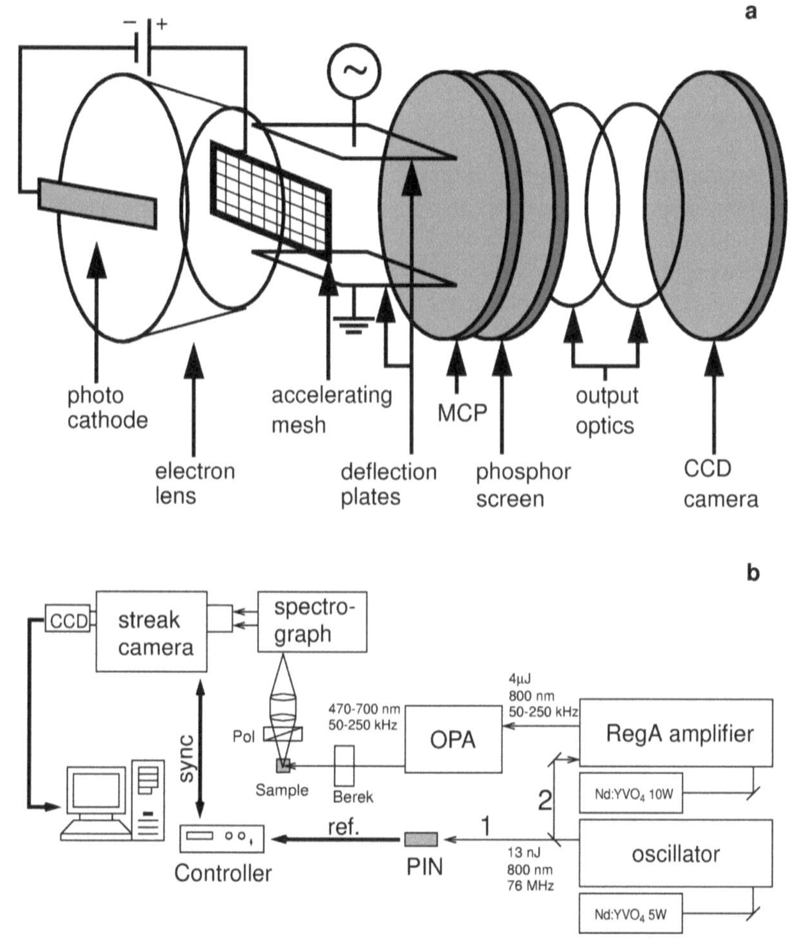


*Fig. 2.* Schematic representation of a streak camera (a) and of the synchroscan streak-camera setup (b). PIN: photodiode. Further explanation in text.

cally polarized light. In particular the gratings of the spectrograph may introduce such a polarization-dependence of the sensitivity of the detection. The second way to record anisotropic measurements is by detecting only the vertical component of the emission, and using the Berek variable waveplate to turn the polarization of the excitation light to either horizontal or vertical, to obtain the perpendicular and parallel components of the emission. The advantage of this method is that one does not have to correct for the polarization-dependence of the detection, however, great care has to be taken not to move the excitation beam by adjusting the variable waveplate, since a change of the position of the focus in the sample will lead to unwelcome intensity changes. For isotropic measurements, one uses vertically polarized excitation light, and a detection polarizer set to the magic angle (54.74°). Finally, if the sample is contained in a rotating cell (see below) that is placed at an angle with the exciting light, one has to be aware that due to the refraction in the sample the direction of both the exciting light and the fluorescence is changed, which will affect both anisotropic and isotropic measurements.

### C. Detection

Light following path 1 hits a reference diode, a tunnel diode, which as a consequence oscillates with a frequency forced to the repetition rate of the laser oscillator. The output of the tunnel diode is used to phase-lock the sweep frequency of the streak camera to the pulses of the laser oscillator. For timing stability on a timescale of minutes, cancellation of drift of the timing is necessary (Uhring et al., 2003). The deflection field is a sine function of time, with period 1/75.9 MHz = 13.2 ns. The controller can phase-shift the deflection field to move the relevant part of the fluorescence decay into the time window recorded on the CCD camera. The controller can also change the amplitude of the signal to set the time range. In our setup four time-windows can be selected ranging from 180 ps to 2 ns. Instead of phase-locking the deflection field frequency to the frequency of the laser oscillator, like in this setup, the opposite is also feasible: the laser oscillator could be phase locked to the frequency of the streak camera, in a way similar to the way lasers are being synchronized to synchrotrons or free-electron lasers (Knippels et al., 1998).

Light from the sample is collected at right angle to the excitation beam through an achromatic lens and the detection polarizer, and focused by a second achromatic lens onto the input slit of a modified Czerny-Turner polychromator. This is equipped with a turret of three gratings with different blazing (spectral window 250 nm) which together span the wavelength range of 250–850 nm. Using concave mirrors after the slit the light is collimated towards the grating and after that the diffracted light is focused onto the photocathode, where the photons induce photo-electrons. These electrons are accelerated by an accelerating mesh and then deflected by the sweeping field. Since the amplitude and sign of the deflection field are functions of time (varying between +V and –V), the extent of deflection depends on the time of arrival of the photon at the photocathode. Only electrons traveling through a field between $+V_c$ and $-V_c$ ($V_c$ = critical deflection field strength) reach the MCP, all other electrons are deflected too much. The electric field is within the detection range every half period of the oscillation frequency, with alternating field sweep direction, so the overall MCP signal is the sum of multiple forward and backward decay trace fragments. This is the so-called backsweep effect. The photons arriving during the backsweep contain information on longer-lived species.

### D. Sample Cell

Stable fluorescent chromophores can be measured in a normal cuvette but photosynthetic samples usually require special measuring cells to prevent photo damage and/or a build-up of long-lived triplet or charge-separated states. We mention two types of cells that can be used to measure photosynthetic preparations: a flow-through cell, and a spinning cell. In the case a flow-cell is used, the solute is pumped through a 1 × 1 mm cuvette with a typical speed of 100 mL/min. Using a repetition rate of 250 kHz, the sample is hit by 15 pulses while passing the excitation spot.

In the case a spinning cell (diameter ~0.1 m, 20–50 Hz rotation) is used, also under the repetition rate of 250 kHz, the sample is hit by 1.5 pulses while passing the excitation spot. This allows for higher intensities and triplet (typical lifetimes μs-ms) build-up is easily avoided. However, the sample returns to the same position with a frequency of 50 Hz, so the build-up of longer-lived (>10 ms) species may still occur. Also the cell is not suitable for larger particles like thylakoid membranes, since the centrifugal forces will spin the particles to the rim of the cell.

## E. Fundamental and Technical Limitations

First we will estimate the number of photons detected per laser shot, which is the motivation for synchroscan averaging. Then we will investigate the different sources of time broadening of the instrument response function (IRF), which ultimately result in an IRF width of about 1% of the selected time range.

### 1. Light Limitations

The detection with a streak camera in combination with a spectrograph (polychromator) puts an important restriction on the size of the illuminated spot of the sample. First of all the horizontal slit of the streak camera typically needs to be closed down to less than 100 μm in order to obtain an instrument response width of a few ps. Alternatively, a narrow width photocathode (70 μm) can be used. This restricts the spot from which fluorescence is collected vertically. The vertical slit of the spectrograph restricts the spot horizontally. In order to maintain good temporal resolution low dispersion gratings, typically 50 grooves/mm, must be used. In order to obtain the desired spectral resolution, also the entrance slit of the spectrograph must be closed down (for a 1/4-m spectrograph, with 50 grooves/mm the dispersion is ~60 nm/mm) (note that some imaging spectrographs enlarge the image of the entrance slit onto the output focal plane by 20%). Therefore, the spot in the sample that is monitored by the detection system typically has a diameter of 100 μm.

The spectrograph also dictates the light collection optics. Typically a numerical aperture of *f*/4 is used (*f* is the focal length of the spectrograph).

To get an idea of the best case performance we presume front face detection, of a concentrated sample (in practice detection under an angle of 90 degrees is used, which reduces the detection efficiency significantly). How much light can we get in and out of the small spot monitored in the sample?

For isotropic emission, *f*/4 optics collect < 0.5% of the emitted light. Given the restrictions imposed by the spot-sizes and slit-widths, deviating from 1:1 imaging of the fluorescence would not help, because a larger collection angle that can be attained would be spoiled by the magnification of the spot onto the entrance slit. Larger collection angles would require a spectrograph with a larger numerical aperture. This can be reached by using larger mirrors and gratings, but, as will become clear in Section II.E.2, the broadening in the spectrometer is proportional to the size of the beam inside the spectrometer. The other way to get a larger collection angle would be to use a shorter focal length spectrograph, but this would further reduce the spectral resolution. The saturation fluence for a laser dye is typically 1 mJ/cm$^2$, which corresponds to about 75 nJ for a 100 μm spot size. Of course this excitation density cannot be used in a proper fluorescence experiment (except when studying lasing phenomena) and one typically needs to stay at least one order of magnitude below this value. Things are even worse for most photosynthetic systems where annihilation and other non-linear effects can occur. To avoid these effects, we will make some estimates for 1 nJ excitation pulses. Around 500 nm this corresponds to $2.5 \times 10^9$ photons. If these are all absorbed, the initial fluorescence intensity will be $\sim 2.5 \times 10^5$ photons/ps (assuming a strongly emitting molecule with a radiative lifetime of 10 ns). Less than 0.5% of these photons is collected using with *f*/4 optics, so we are left with $\sim 10^3$ photons/ps entering the slit of the spectrograph. The efficiency of the spectrograph is typically 10–50%, and the quantum yield of the photocathode of the streak camera is 1–20%, so this leaves us with about 1–100 photoelectrons per ps, spread out along the spectral axis.

This demonstrates that substantial averaging is required in order to get good spectro-temporal data, i.e., a large number of shots are required. This is where the main difference between single shot and synchroscan streak cameras comes to light. Single shot devices are optically triggered by the excitation laser and the deflection field is directly generated by a fast photoconductive switch. The maximal switching frequency of such a device is in the kHz range. In a synchroscan camera the deflection field is an oscillatory function synchronized to the repetition rate of the laser oscillator. Therefore, repetition rates up to 76 MHz (and of course sub harmonics of this frequency) can be employed. With a high repetition rate system like the RegA (250 kHz), fluorescence signals from laser dyes can be obtained within seconds, and emission data from less luminant samples in tens of minutes.

### 2. Time Resolution

Together with the electronic contribution of the setup, the major limitation to the time resolution (on the fastest time base) comes from the dispersion in the spectrograph. This phenomenon is related to pulse broadening in pulse stretchers/compressors (Martinez, 1987), and stems from the fact that after angular dispersion the wave front of a light pulse exhibits a tilt (with respect to the phase front) given by (Hebling, 1996):

$$\tan(\theta) = -\lambda \frac{\partial \alpha}{\partial \lambda} \tag{1}$$

where $\lambda$ is the average wavelength of the light, and $\alpha$ the wavelength-dependent dispersion angle. This can be included in the grating equation

$$\alpha = \arcsin\left(\frac{m\lambda}{d} - \sin(\beta)\right) \tag{2}$$

in which $m$ represents the order of diffraction (for all practical purposes here $\pm 1$), $\beta$ the angle of incidence, and $d$ the grating constant. This gives

$$\tan(\theta) = -\lambda \frac{m}{d}\left(\frac{1}{\sqrt{1-\left(m\lambda/d - \sin(\beta)\right)^2}}\right) \tag{3}$$

In other work (Schiller and Alfano, 1980; Wiessner and Staerk, 1993), the term between brackets is ignored, which corresponds with taking the phase velocity of the light instead of the group velocity. For the example given here the difference is insignificant, but this would not be the case for more dispersive gratings.

The total spatial stretch that occurs is $W\tan(\theta)$, where $W$ represents the width of the beam after the grating. This is where the numerical aperture of the detection and spectrograph enter. For an 1/4 m spectrograph, with *f*/4 optics, $W = \sim 5$ cm, and for $\lambda = 600$ nm, $m = 1$, $\beta = 0$, and d = 20 μm (50 grooves/mm), this amounts to a spread $\Delta_{dispersion}$ of 1.5 mm, which corresponds to a temporal spread of 5 ps. Therefore, even when using a 50 grooves/mm grating one needs to reduce the *f*-number of the spectrograph (or the light collection) to get a time response that is close to the limits of the electronic part as described below.

The time resolution limits of the streak camera itself are given by the spread in transit time of the photoelectrons in the streak tube. The transit time spread is mainly generated in the region near the photocathode where the electrons still have a relatively low speed (Zavoiski and Fanchenko, 1965; Bradley and New, 1974; Campillo and Shapiro, 1983). The resulting distribution of transit times has a half width of

$$\Delta\tau_c = m\frac{\Delta v}{eE} \tag{4}$$

where $m$ and $e$ are the mass and charge of the electron, $\Delta v$ is the halfwidth of the initial photoelectron velocity distribution, and $E$ is the field strength in the vicinity of the photo-cathode. Clearly it is important to have a high acceleration voltage near the photocathode, typically fields of ~10 kV/cm are used. For this extraction field a kinetic energy spread of 1 eV (a blue photon on a red-sensitive photocathode) would lead to a time spread of ~4 ps, which is significant when operating the streak camera on the fastest time base. Near the cut-off wavelength of the photocathode the energy spread becomes much smaller (and fortunately most fluorescence experiments are performed there), but in general there is a noticeable increase of the width of the instrument response when detecting blue photons.

For a higher time resolution higher extraction fields are required, but this comes at the cost of field emission (field induced dark current from the photocathode) and reduced reliability. Significantly higher pulsed extraction voltages can be used for single shot devices but this is not possible at the sweep rate of synchroscan streak cameras. Moreover, once below the 1 ps resolution other factors start to become limiting, like the quality of the imaging of the photocathode onto the MCP. Any aberrations of the electrostatic or electromagnetic electron lens (like the chromatic aberration $\Delta\tau_c$ caused by differences in electron speeds) will have adverse effects on the width of the instrument response.

The timing errors described here are independent. Therefore the error calculus for the total temporal instrument response width $\Delta$ becomes

$$\Delta^2 = (\Delta\tau_c)^2 + (\Delta_{imaging})^2 + (\Delta_{dispersion})^2 + (\Delta_{width})^2 \tag{5}$$

Where $\Delta_{imaging}$ is the imaging error due to the electrostatic or magnetic lenses, $\Delta_{dispersion}$ is the dispersion error, and $\Delta_{width}$ is the error due to the streak-slitwidth or the cathode width. At time bases larger than 400 ps $\Delta_{imaging}$ and $\Delta_{width}$ dominate. Imaging a 70 μm photocathode on a 7 mm wide CCD yields an IRF width of ≈1% of the time base used. At shorter time bases the contribution of the other two terms becomes appreciable, resulting in an IRF width of 3 ps (at 700 nm) for the 200 ps time base (which is 1.5%). At short wavelengths the IRF has broadened to 4 ps because of the larger $\Delta\tau_c$, see Fig. 3a. The broadening of the IRF (to 24 ps) with the 2.2 ns time base is clearly visible in Fig. 6.

## F. Averaging and Correction of Images

Typically, the sample is excited with pulses of 0.2 ps FWHM at a repetition rate of 50–250 kHz, which is much lower than the laser oscillator frequency (typically 76 MHz). Synchroscan streak cameras are generally chosen for their ability to do signal integration over extended periods, in which case the time resolution will generally be limited by the drift between the laser and the camera clock. Often individual datasets contain internal tell-tales for absolute timing and of drift, e.g., Rayleigh or Raman scattering signals of the excitation pulse can be used to pin down the exact timing of the dataset. A fool-proof method for eliminating all sources of electronic drifts and jitter consists of directly illuminating a spot of the photocathode with the excitation pulse so as to obtain a fiducial, which is an absolute timing reference (Jaanimagi et al., 1986; Uhring et al., 2003). Using drift compensation electronics, up to 1000 seconds of accumulation on the CCD chip can be performed without significant deterioration of the temporal resolution. Typically, the full time and wavelength ranges are 200 ps and 250 nm, respectively. In the dark the CCD chip accumulates a dark current, which can be minimized by using a Peltier cooling element. Data must be corrected by subtracting the measured dark current contribution. The sensitivity of the entire detection system is quite strongly position dependent. In particular at the edges of the streak-image the signal shows a pronounced drop. To account for this spatial variation of the sensitivity, streak images are divided by a shading image. This shading image consists of a streak image of the light emitted by a halogen lamp, which is directed into the spectrograph. This shading correction directly accounts for the sensitivity variation along the time-axis, since the intensity of the lamp is constant in time. For the sensitivity variation along the wavelength-axis the emission spectrum of the lamp has to be taken into account, which is done prior to the analysis of the data. Thus, the data is also corrected for the spectral sensitivity of the system.

Because of the limited wavelength resolution of the spectrograph (7 nm FWHM), the curvature-corrected and averaged images can be reduced to a matrix of $\approx$1000 points in time and 30–60 points in wavelength. In this same averaging step outliers (e.g., resulting from cosmic rays) can be removed. To deal with the small remaining drift after compensation multiple data sets can be collected. Instead of averaging e.g. 30 minutes and suffering from drift induced time broadening, it is better to collect six averages of 5 minutes and correct them for slow drift of time zero. Then, after scrutinous inspection, to check for trends like sample degradation, the series of images can be averaged. Figure 1 depicts a filled contour plot of the PS I trimer data derived from 48 traces between 625 and 785 nm, resulting from an average of 20 images, which will be globally analyzed below. Other visualizations of these data can be found in Gobets (2002).

### G. Calibrations

Because of the sinusoidal nature of the deflection field, the 'time per pixel' is not a constant, but varies over time. For the shortest (200 ps) time range the time per pixel is practically constant (because the sinusoid is practically linear near the zero-crossing), but for longer time ranges the time per pixel varies significantly. Calibration of the time base can be done by fitting the train of imaged pulses from an etalon, to estimate a polynomial function that describes the time per pixel over the whole time base. Calibration of the wavelength axis can be done with the help of the lines of a calibration lamp, to estimate a linear function. Images of continuous narrow-band sources are also instrumental for checking that the sweep axis is parallel to the vertical axis of the CCD. A crucial procedure for the analysis of the two-dimensional data sets is the characterization of the curvature of the image, i.e., the spatial dependence of 'time zero' on the CCD image, caused by the different path lengths of the photo-electrons in the streak camera. Additionally, the light-collecting optics and the spectrograph cause wavelength-dependent temporal shifts. To assess this curvature, scattering of the white light from the OPA is recorded with the streak camera, resulting in an IRF limited curved line on the streak image. In its turn, the intrinsic dispersion of the white light itself is measured using the optical Kerr signal in carbon disulphide (Greene and Farrow, 1983). The combination of both these measurements yields the spatial dependence of time zero during a measurement.

### H. Further Exploitation of the Horizontal Dimension

Streak tubes generally contain a second set of deflection plates, to facilitate the horizontal deflection of photoelectrons. In commercial instruments these plates are used, e.g., for blanking (blocking) the detection in between sweeps, or during the back sweep of the camera. These horizontal plates can also be exploited to perform 2D experiments other than the ones we focused on above, and consequently, spectral information will be lost. In Bühler et al. (1998) and Ohtani et al. (1999) the horizontal sweep plates were used to provide a secondary, slow, time axis. In combination with a stopped-flow apparatus, the time-evolution of the ps fluorescence lifetime of a sample could thus be measured on a ms time scale. In van Mourik et al. (2003) the horizontal sweep direction was synchronized to the electric field applied in a Stark fluorescence experiment, and thus the effect of the Stark field on the fluorescence intensity and lifetime of the sample could be measured.

## III. Data Analysis

When the streak image has been corrected for the instrumental curvature it is ready for data analysis. The aim is to obtain a model-based description of the full data set in terms of a model containing a small number of precisely estimated parameters, of which the rate constants and spectra are the most relevant. With polarized-light experiments also anisotropy parameters come into play. Description of the basic ingredient of kinetic models, the exponential decay, will be given first, followed by a description of how to use these ingredients for global and target analysis (for reviews, see Holzwarth, 1996 and van Stokkum et al., 2004) of the full data. Our main assumption here is that the time and wavelength properties of the system of interest are separable, which means that spectra of species or states are constant. For details on parameter estimation techniques the reader is also referred to the above cited reviews and references cited therein, and to van Stokkum (2005). Software issues are discussed in van Stokkum and Bal (2006). We will describe in depth the analysis of typical streak data, with the analysis of the PS I trimer data serving as the main example.

### A. Modeling an Exponential Decay

Here an expression is derived for describing the contribution of an exponentially decaying component to the streak image. The instrument response function (IRF) $i(t)$ can usually adequately be modeled with a Gaussian with parameters $\mu$ and $\Delta$ for, respectively, location and full width at half maximum (FWHM):

$$i(t) = \frac{1}{\tilde{\Delta}\sqrt{2\pi}} \exp(-\ln(2)(2(t-\mu)/\Delta)^2) \tag{6}$$

where $\tilde{\Delta} = \Delta / (2\sqrt{2\ln(2)})$ . The adequacy of the Gaussian approximation of the IRF shape is depicted in Fig. 3a. The convolution (indicated by an *) of this IRF with an exponential decay (with rate $k$) yields an analytical expression which facilitates the estimation of the IRF parameters $\mu$ and $\Delta$:

$$\begin{aligned} c(t,k,\mu,\Delta) &= \exp(-kt) * i(t) \\ &= \frac{1}{2}\exp(-kt)\exp\left(k\left(\mu + \frac{k\tilde{\Delta}^2}{2}\right)\right) \\ &\quad \times \left\{1 + \operatorname{erf}\left(\frac{t-(\mu + k\tilde{\Delta}^2)}{\sqrt{2}\tilde{\Delta}}\right)\right\} \end{aligned} \tag{7}$$

The periodicity of the synchroscan results in detection of the fluorescence that remains after multiples of half the synchroscan period $T$ (typically $T \approx 13$ ns). Therefore, if lifetimes longer than ~1 ns occur in a sample, the above expression should be extended with a summation over the signal contributions that result from forward and backward sweeps:

$$\begin{aligned} c(t,k,T) &= \sum_{n=0}^{\infty} e^{-kTn}\left(e^{-k(t-\mu+T)} + e^{-k(T/2-t-\mu)}\right) \\ &= \left(e^{-k(t-\mu+T)} + e^{-k(T/2-t-\mu)}\right) \Big/ \left(1 - e^{-kT}\right) \end{aligned} \tag{8}$$

Note that it is assumed here that time zero of the time base corresponds to the zero crossing of the sweep, and that the convolution with the IRF is no longer necessary at times longer than $T/2$. Adding the previous expressions provides the full model function for an exponential decay recorded with a synchroscan streak camera and will henceforth be denoted by $c^I(k)$:

$$c^I(k) \equiv c(t,k,\mu,\Delta,T) = c(t,k,\mu,\Delta) + c(t,k,T) \tag{9}$$

Examples of $c^I(k)$ are depicted in Fig. 4c, and fits of traces with linear combinations of decays are shown in Fig. 3b, where an ultrafast lifetime of 1.2 ps is detected, and Fig. 3c, which is dominated by a 17 ns lifetime (note the huge backsweep signal apparent from the signal 'before time zero'). Figure 4a and b depict the fits of two traces from the data shown in Fig. 1, using 5 lifetimes. The simultaneous estimation of up to 5 lifetimes in the range of (sub)ps to ns is more or less routine.

Because fluorescence samples are relatively dilute, elastic scattering or Raman scattering of the excitation light by water (or of other solvents) can complicate the measurement, if they occur within the analyzed wavelength interval. Such contributions can be modeled with an extra component with a time course identical to the IRF $i(t)$. Usually it is possible to restrict the contribution of scattering to a limited wavelength region.

If the streak image has not been corrected for the instrumental curvature the wavelength dependence of the IRF location $\mu$ can be modeled with a polynomial (usually a parabola is adequate). Sometimes the IRF shape is better described by a superposition of two Gaussians, leading to a superposition description of the exponential decay (van Stokkum, 2005).

### B. Global and Target Analysis

The basis of global analysis is the superposition principle, which states that the measured data $\psi(t,\lambda)$ result from a superposition of the spectral properties $\varepsilon_l(\lambda)$ of the components present in the system of interest

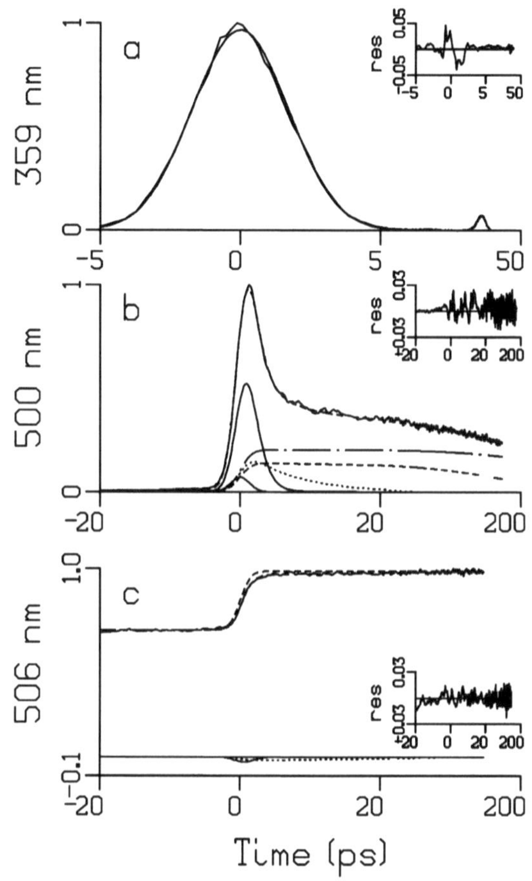


*Fig. 3.* (a) IRF of streak scope measured from scattered white light fitted with a Gaussian. Estimated FWHM Δ = 4 ps, note a small (7%) reflection after 26 ps. Detection wavelength (in nm) indicated along the ordinate. Dashed lines indicate fit. Insets show residuals. Note that the time axis is linear from –5 to +5 ps relative to the maximum of the IRF, and logarithmic thereafter. In (b) and (c) it is linear from –20 to +20 ps. (b) Emission from thioredoxin reductase mutant C138S (from van den Berg et al., 2001) showing a dominant 1.2 ps decay (depicted by solid line). Other contributions to the fit have lifetimes of 7.3 ps (dotted), 0.18 ns (dashed), 0.74 ns (dot dashed), and pulse follower (chain dashed). The sum of these contributions which is the fit of the trace is shown as a dashed line. (c) Emission from lumazine protein (from Petushkov et al., 2003) showing a dominant 17 ns decay (depicted by dashed line). Other contributions to the fit have lifetimes of 0.7 ps (solid) and 24 ps (dotted).

weighted by their concentration $c_l(t)$,

$$\psi(t,\lambda) = \sum_{l=1}^{n_{comp}} c_l(t)\varepsilon_l(\lambda) \tag{10}$$

The $c_l(t)$ of all $n_{comp}$ components are described by a compartmental model, that consists of first-order differential equations, with as solution sums of exponential decays. We will consider three types of compartmental models: (1) a model with components decaying mono-exponentially in parallel, which yields Decay Associated Spectra (DAS), (2) a sequential model with increasing lifetimes, also called an unbranched unidirectional model, giving Evolution Associated Spectra (EAS), and (3) a full compartmental scheme which may include possible branchings and equilibria, yielding Species Associated Spectra (SAS). The latter is most often referred to as target analysis, where the target is the proposed kinetic scheme, including possible spectral assumptions.

(1) With parallel decaying components the model reads:

$$\psi(t,\lambda) = \sum_{l=1}^{n_{comp}} c^I(k_l) DAS_l(\lambda) \tag{11}$$

The DAS thus represent the estimated amplitudes of the above defined exponential decays $c^l(k_l)$. The DAS estimated from the PS I trimer data are shown in Fig. 4d. Several observations can be made: the 0.4 ps DAS (solid) represents the rise due to the relaxation from the initially excited Soret state (a higher excited state, of which the emission is outside the detection range) to the $Q_y$ emission (lowest excited state). The next DAS of 3.9 ps (dotted) is conservative, i.e., the positive and negative areas are more or less equal. It represents decay of more blue and rise of more red emission, and can be interpreted as energy transfer from bulk to red chlorophyll *a* (Chl *a*), i.e. Chl *a* that absorb at wavelengths longer than the primary electron donor P700. The 15 ps DAS (dashed) is not conservative, although it does show some rise above 730 nm. Apparently some trapping of excitations takes place on this time scale, concurrently with energy transfer. The 50 ps DAS (dot dashed) represents the trapping spectrum. The long lived (4.9 ns) DAS (chain dashed) is attributed to a small fraction of free Chl *a* in the preparation. Clearly, the first three DAS do not represent pure species, and they are interpreted as linear combinations (with positive and negative contributions) of true species spectra.

(2) A sequential model reads:

$$\psi(t,\lambda) = \sum_{l=1}^{n_{comp}} c_l^{II} EAS_l(\lambda) \tag{12}$$

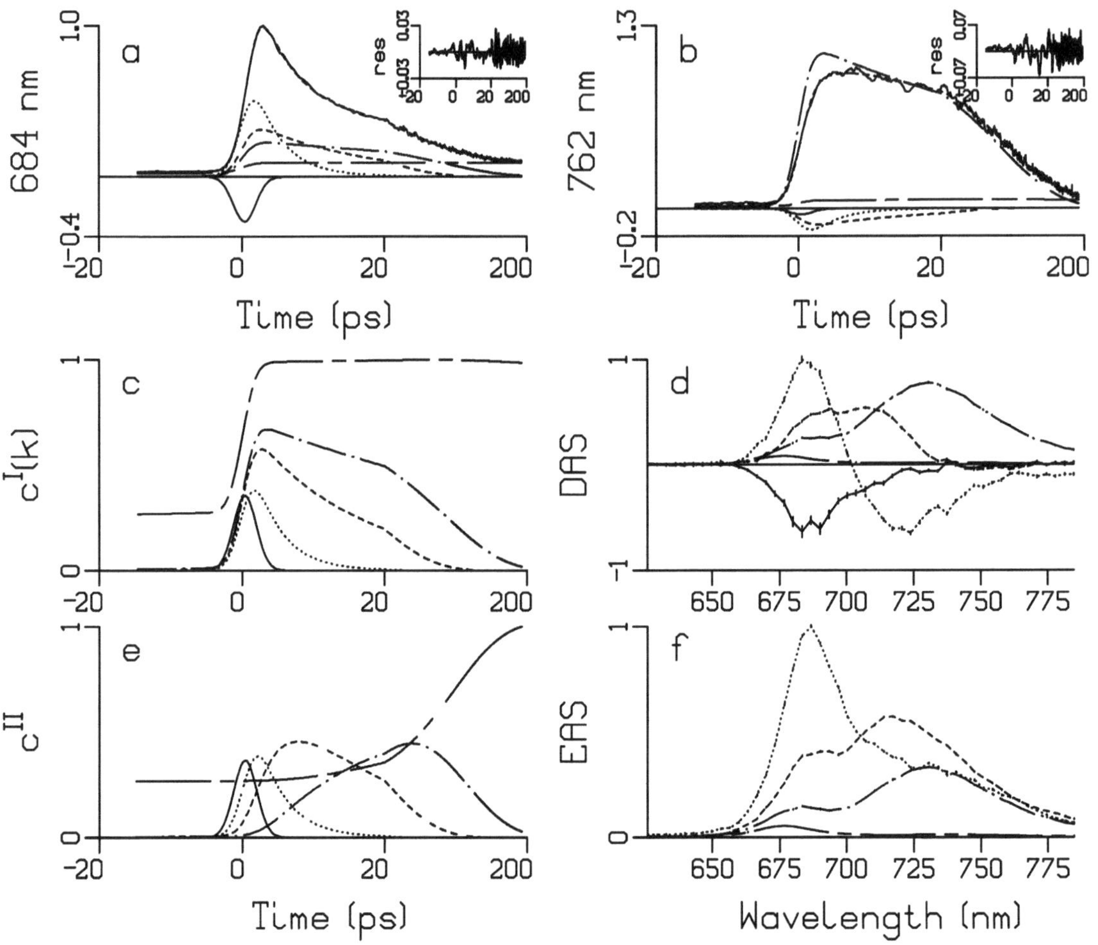


*Fig. 4.* Results from global analysis of PS I data depicted in Fig. 1. Note that in a–c and e the time axis is linear from –20 to +20 ps relative to the maximum of the IRF, and logarithmic thereafter. Insets in a, b show residuals. (a) Fit of bulk Chl *a* emission trace showing multiexponential decay. Contributions of the five exponential decays with different lifetimes (shown in c) are indicated by line type. (b) Fit of red Chl *a* emission trace showing multiexponential rise and decay. (c) Exponential decays $c^I(k_l)$. Estimated lifetimes: 0.4 ps (solid), 3.9 ps (dotted), 15 ps (dashed), 50 ps (dot dashed), and 4.9 ns (chain dashed). (d) Decay Associated Spectra (DAS), note that the first DAS which represents overall rise has been multiplied by 0.2. Vertical bars indicate estimated standard errors. (e) Evolutionary concentration profiles $c_l^{II}$ (assuming a sequential kinetic scheme with increasing lifetimes). (f) Evolution Associated Spectra (EAS). Note that the first EAS is zero, since excitation was in the Soret band.

where each concentration is a linear combination of the exponential decays,

$$c_l^{II} = \sum_{j=1}^{l} b_{jl} c^I(k_l) \tag{13}$$

and the amplitudes $b_{jl}$ are given by $b_{11} = 1$ and for $j \leq l$:

$$b_{jl} = \frac{\prod_{m=1}^{l-1} k_m}{\prod_{n=1, n\neq j}^{l} \left(k_n - k_j\right)} \tag{14}$$

Examples of $c_l^{II}$ are depicted in Fig. 4e, whereas the EAS estimated from the PS I trimer data are shown in Fig. 4f. With increasing lifetimes, and thus decreasing rates $k_l$, the first EAS (equal to the sum of DAS) corresponds to the spectrum at time zero with an ideal infinitely small IRF, $i(t) = \Delta(t)$. In Fig. 4f, this first EAS is zero in the $Q_y$ region. The second EAS (dotted), which is formed in 0.4 ps and decays in 3.9 ps, represents the sum of the spectra of all excitations that have arrived from the Soret region, and is dominated by bulk Chl *a*. The third EAS, which is formed in 3.9 ps and decays in 15 ps, is already dominated by red Chl *a* emission, which is even more the case with the fourth EAS (dot dashed, formed in 15 ps, decays in 50 ps). The final EAS (chain dashed, formed in 50 ps) is proportional to the final DAS, and represents the spectrum of the longest living component (4.9 ns). Clearly, these EAS do not represent pure species, except for the final EAS, and they are interpreted as a weighted sum (with only positive contributions) of true species spectra.

(3) When neither of these two simple models is applicable, a full kinetic scheme may be appropriate. The problem with such a scheme is that, while the kinetics are described by microscopic rate constants, the data only allows for the estimation of decay rates (or lifetimes). Thus additional information is required to estimate the microscopic rates, which can be spectral constraints (zero contribution of SAS at certain wavelengths) or spectral relations. This is explained in detail in van Stokkum et al. (2004).

Now the model reads:

$$\psi(t,\lambda) = \sum_{l=1}^{n_{comp}} c_l^{III}\, SAS_l(\lambda) \tag{15}$$

where the concentrations $c_l^{III}$ are again linear combinations of the exponential decays, with coefficients that depend upon the microscopic rate constants that describe the transitions between all the compartments. Figure 5a depicts the kinetic scheme that was applied to the trimeric PS I data of Fig. 1. The concentrations of all compartments are collated in a vector $c(t) = [c_1(t)\ c_2(t)\ \ldots\ c_{n_{comp}}(t)]^T = [S(t)\ B(t)\ R_1(t)\ R_2(t)\ F(t)]^T$, which obeys the differential equation:

$$\frac{d}{dt}c(t) = Kc(t) + j(t) \tag{16}$$

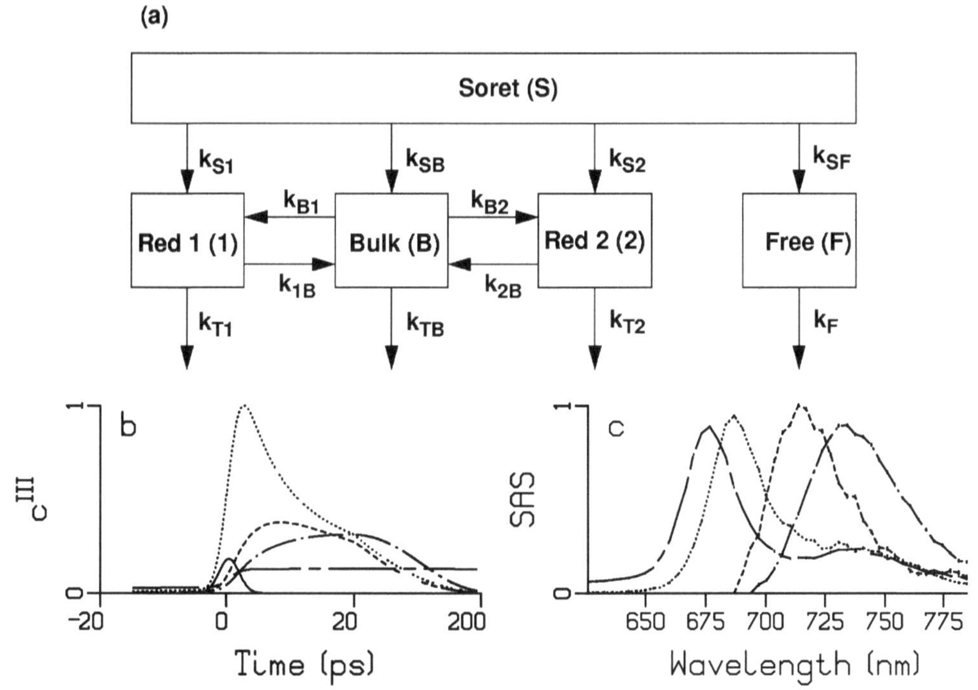


*Fig. 5.* (a) Kinetic scheme used for the target analysis of PS I data depicted in Fig. 1. After excitation in the Soret band four compartments are populated: bulk Chl *a* (B), two pools of red Chl *a* (1 and 2) and a small fraction of free Chl *a* (F). The first three compartments equilibrate, and excitations are trapped with different rates. (b) Concentration profiles $c_l^{III}$, note that the time axis is linear from –20 to +20 ps relative to the maximum of the IRF, and logarithmic thereafter. (c) Species Associated Spectra (SAS). Key in (b) and (c): bulk Chl *a* (dotted), red Chl *a* 1 (dashed), red Chl *a* 2 (dot dashed), free Chl *a* (chain dashed).

where the transfer matrix $K$ contains off-diagonal elements $k_{pq}$, representing the microscopic rate constant from compartment $p$ to compartment $q$. The diagonal elements contain the total decay rates of each compartment. The input to the compartments is $j(t) = i(t)[1\ 0\ 0\ 0\ 0]^T$. The $K$ matrix from Fig. 5a reads:

$$K = \begin{bmatrix} -(k_{SB}+k_{S1}+k_{S2}+k_{SF}) & & & & \\ k_{SB} & -(k_{TB}+k_{B1}+k_{B2}) & k_{1B} & k_{2B} & \\ k_{S1} & k_{B1} & -(k_{T1}+k_{1B}) & & \\ k_{S2} & k_{B2} & & -(k_{T2}+k_{2B}) & \\ k_{SF} & & & & -k_F \end{bmatrix} \quad (17)$$

In Fig. 5b, the $c_l^{III}$ have been drawn, calculated from the estimated parameters, whereas the estimated SAS are shown in Fig. 5c. Note that it has been assumed that the two red Chl *a* compartments only contribute above 690 and 697 nm, respectively. Therefore, the forward and backward rate constants between the bulk Chl *a* compartment and both compartments of red Chl *a* can be estimated from the multi-exponential decay of the bulk Chl *a*. The SAS in Fig. 5c are considered satisfactory, because the shapes of the bulk and red SAS resemble the free Chl *a* SAS, and the areas, and thus the oscillator strengths, of the different Chls *a* are equal within 10%. This area constraint was instrumental in determining the branching ratios from Soret to the four different Chl *a* pools, and the trapping ratios.

### 1. Target Analysis of Anisotropic Data

When in addition to magic angle (MA) data also parallel (VV) and perpendicular (VH) data are collected, more information is available to disentangle the complex kinetics, and estimate the SAS. In such an extended target analysis the magic angle concentrations $c^{III}_l$ are multiplied by the anisotropic properties of the components.

$$\begin{bmatrix} MA(t,\lambda) \\ VV(t,\lambda) \\ VH(t,\lambda) \end{bmatrix} = \sum_{l=1}^{n_{comp}} c_l^{III} SAS_l(\lambda) \begin{bmatrix} 1 \\ 1+2r_l \\ 1-r_l \end{bmatrix} \quad (18)$$

Note that here the anisotropy $r_l$ is assumed to be constant. When an anisotropy decay rate is present, each isotropic exponential decay has to be multiplied by the associated anisotropy decay rate before the convolution with the IRF (Beechem, 1989; Yatskou et al., 2001). Figure 6 shows a representative trace

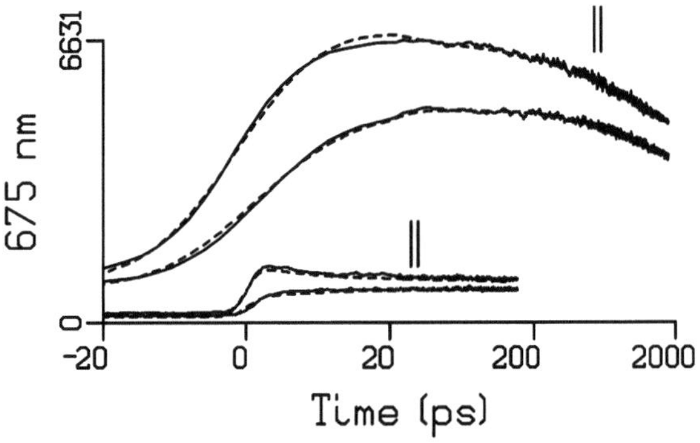


*Fig. 6.* Parallel (||, upper curves) and perpendicular (lower curves) time traces measured at 675 nm after exciting PCP at 660 nm, from Kleima et al. (2000). The smaller curves were measured on the shortest time base. The dashed lines indicate the fit. Note that the time axis is linear from –20 to +20 ps relative to the maximum of the IRF, and logarithmic thereafter.

from Kleima et al. (2000) who used a bi-exponential anisotropy decay

$$r(t) = A_1 \exp(-t/\tau_1) + A_2 \exp(-t/\tau_2) + r_\infty \quad (19)$$

An isotropic lifetime of ≈4.2 ns was estimated from this target analysis, in combination with depolarization times of about 7 and 350 ps, which are clearly visible in the data measured on the different time scales.

## C. Spectral Modeling

SAS can sometimes be fitted with a spectral model consisting of a skewed Gaussian in the energy domain ($\bar{\nu} = 1/\lambda$):

$$SAS(\bar{\nu}) = \bar{\nu}^5 S_{\max} e^{(-\ln(2)\{\ln(1+2b(\bar{\nu}-\bar{\nu}_{\max})/\Delta\bar{\nu})/b\}^2)} \quad (20)$$

where the parameter $\bar{\nu}_{max}$ is the Franck-Condon wavenumber of maximum emission. The FWHM is given by $\Delta\bar{\nu}_{1/2} = \Delta\bar{\nu}\ \sinh(b)/b$. Note that with skewness parameter $b$ equal to zero the expression simplifies to a Gaussian. The average wavenumber of this function is given by

$$\bar{\nu}_{av} = \bar{\nu}_{\max} + \frac{\Delta\bar{\nu}}{2b}\left(\exp\left(-3b^2 \big/ 4\ln(2)\right) - 1\right) \quad (21)$$

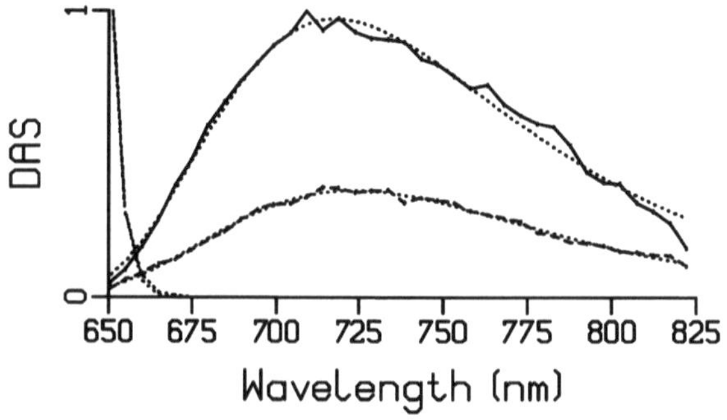


*Fig. 7.* Decay Associated Spectra from global analysis of bacteriorhodopsin mutant D85S excited at 635 nm, from van Stokkum et al. (2006). Key: 5.2 ps (solid), 19.1 ps (dashed), scatter (dotted). Fits of the DAS using a skewed Gaussian shape are indicated by dots. The estimated $\bar{\nu}_{av}$ were both 13000 cm$^{-1}$ and the FWHM was 2540 cm$^{-1}$.

The spectral evolution description of solvation approximates a gradual change with an average spectral change associated with a time constant. Alternatively, solvation occurring on sub-ps timescales can be described using a time-dependent shift of $\bar{\nu}_{max}$ (Horng et al., 1995; Vilchiz et al., 2001). This requires data with a higher time and wavelength resolution, e.g. from fluorescence up-conversion (Horng et al., 1995; Pal et al., 2002; Vengris et al., 2004), for which excitation intensities are required that are too high for the study of photosynthetic systems.

Figure 7 shows DAS estimated from the multi-exponential decay of the excited state of the D85S mutant of bacteriorhodopsin. Both DAS possessed almost identical shapes, and thus show no evidence for solvation on the picosecond timescale. The DAS were well described by a skewed Gaussian, and the multi-exponentiality is ascribed to heterogeneity of the protein.

### D. Usage of the Singular Value Decomposition

The matrix structure of the streak data enables the usage of matrix decomposition techniques, in particular the singular value decomposition (SVD). Formally the data matrix can be decomposed as

$$\psi(t,\lambda) = \sum_{l=1}^{m} u_l(t) s_l w_l(\lambda) \tag{22}$$

Where $u_l$ and $w_l$ are the left and right singular vectors, $s_l$ the sorted singular values, and $m$ is the minimum of the number of rows and columns of the data matrix. The singular vectors are orthogonal, and provide an optimal least squares approximation of the matrix.

From the SVD the rank of the data matrix can be estimated, as judged from the singular values and singular vector pairs significantly different from noise. This rank corresponds to the number of spectrally and temporally independent components. When

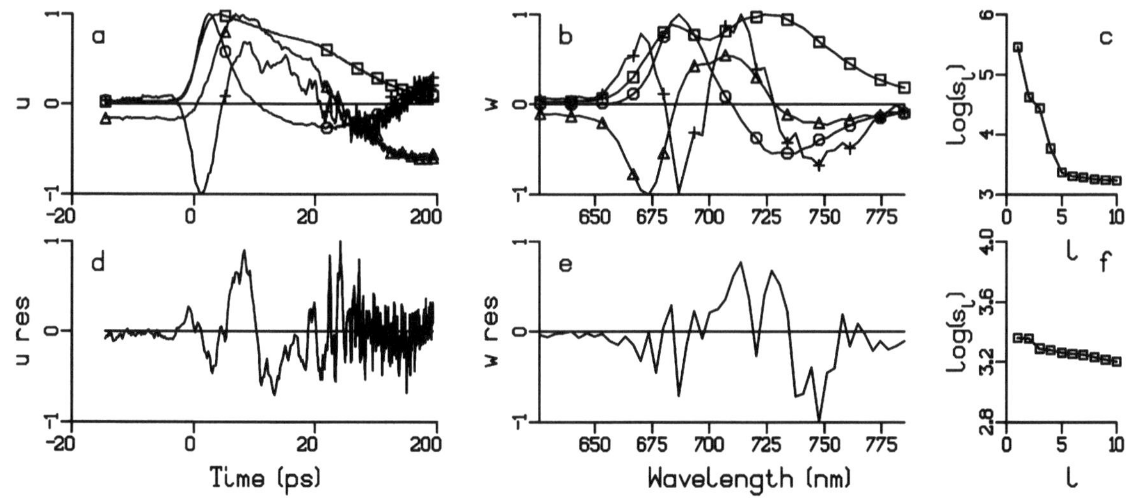


*Fig. 8.* SVD of the PS I trimer data matrix (top) and matrix of residuals (bottom). (a) First four (order squares, circles, triangles, plus symbols) left singular vectors $u_l$, (b) first four right singular vectors $w_l$, (c) first ten singular values $s_l$ on a logarithmic scale, (d) first left singular vector $u_{res,1}$, (e) first right singular vector $w_{res,1}$, (f) first ten singular values $s_{res,1}$ on a logarithmic scale.

the data matrix has not been corrected for dispersion, this is no longer true. Furthermore SVD of the residual matrix is useful to diagnose shortcomings of the model used, or systematic errors in the data. Figure 8a-c depicts the SVD of the trimeric PS I data, where four singular values and singular vector pairs are significantly different from noise. These first four singular values account for 99.923% of the variance of the data matrix. The left and right singular vectors are both linear combinations of the true concentration profiles and SAS, and are hard to interpret. The first pair (squares) represents a kind of average. The SVD of the residual matrix (shown in Fig. 8d-f) shows that its singular values are comparable to the noise singular values in Fig. 8c, and that there is no clear structure in the first singular vector pair. The sum of squares of the residuals is 0.088% of the variance of the data matrix, indicating a small lack of fit. The root mean square error of the fit was 41, which is 0.5% of the peak in Fig. 1.

## IV. Conclusions

When comparing the present state of the art with the excellent review of Campillo and Shapiro (1983) the most striking developments are the utilization of the horizontal dimension, in particular using a spectrograph, and the improvement of the data analysis methods. The collection and analysis of true spectrotemporal measurements with (sub)ps time resolution using low excitation intensities have become routine, and the promises of the technique have largely been fulfilled. It has now become possible to functionally describe the complicated energy transfer and trapping processes in photosynthetic complexes with the help of a compartmental model, characterized by SAS and microscopic rate constants. The streak measurements of spectral evolution of fluorescence can be combined with fluorescence up-conversion measurements (Gobets et al., 2001a; Kennis et al., 2001) to extend the number of time scales covered, or with femtosecond difference absorption measurements (Groot et al., 2005) to uncover also non-emitting states. The complementary information contained in data obtained with different techniques is optimally extracted in a simultaneous target analysis.

## Acknowledgments

Marloes Groot, Kate Mullen, Arie van Hoek, Janne Ihalainen and Rienk van Grondelle are thanked for critically reading the text. This research was supported by the Netherlands Organization for Scientific Research (NWO) via the Dutch Foundation of Earth and Life Sciences (ALW) and via the Netherlands Organization of Fundamental Research of Matter (FOM).

## References

Andrizhiyevskaya EG, Frolov D, van Grondelle R, and Dekker JP (2004a) Energy transfer and trapping in the Photosystem I complex of *Synechococcus* PCC 7942 and in its supercomplex with IsiA. Biochim Biophys Acta 1656: 104–113

Andrizhiyevskaya EG, Frolov D, van Grondelle R, and Dekker JP (2004b) On the role of the CP47 core antenna in the energy transfer and trapping dynamics of Photosystem II. Phys Chem Chem Phys 6: 4810–4819

Beechem JM (1989) A second generation global analysis program for the recovery of complex inhomogeneous fluorescence decay kinetics. Chem Phys Lipids 50: 237–251

Bradley DJ and New GHC (1974) Ultrashort pulse measurements. Proc IEEE 62: 313–345

Bühler CA, Graf U, Hochstrasser RA and Anliker M (1998) Multidimensional fluorescence spectroscopy using a streak camera based pulse fluorimeter. Rev Sci Instrum 69: 1512–1518

Campillo AJ and Shapiro SL (1983) Picosecond streak camera fluorometry — a review. IEEE J Quantum Electron QE-19: 585–603

Donovan B, Walker LA, Kaplan D, Bouvier M, Yocum CF and Sension RJ (1997) Structure and function in the isolated reaction center complex of Photosystem II. I. Ultrafast fluorescence measurements of PS II. J Phys Chem B 101: 5232–5238

Fleming GR, Morris JM and Robinson GW (1977) Picosecond fluorescence spectroscopy with a streak camera. Austr J Chem 30: 2338–2352

Freiberg A and Saari P (1983) Picosecond spectrochronography. IEEE J Quantum Electron QE-19: 622–630

Gilmore AM, Itoh S and Govindjee (2000) Global spectral-kinetic analysis of room temperature chlorophyll *a* fluorescence from light-harvesting antenna mutants of barley. Phil Trans R Soc Lond B 355: 1371–1384

Gilmore AM, Matsubara S, Ball MC, Barker DH and Itoh S (2003a) Excitation energy flow at 77 K in the photosynthetic apparatus of overwintering evergreens. Plant Cell Environ 26: 1021–1034

Gilmore AM, Larkum AWD, Sallh A, Itoh S, Shibata Y, Bena C, Yamasaki H, Papina M and Woesik R (2003b) Simultaneous time resolution of the emission spectra of fluorescent proteins and zooxanthellar chlorophyll in reef-building coral. Photochem Photobiol 77: 515–523

Greene BJ and Farrow RC (1983) The subpicosecond Kerr effect in $CS_2$. Chem Phys Lett 98: 273–276

Gobets B (2002) The life and times of Photosystem I. Excitation energy transfer and trapping unravelled. Thesis, Vrije Universiteit Amsterdam

Gobets B, Kennis JTM, Ihalainen JA, Brazzoli M, Croce R, van Stokkum IHM, Bassi R, Dekker JP, van Amerongen H, Fleming GR and van Grondelle R (2001a) Excitation energy transfer in dimeric Light Harvesting Complex I: A combined streak-camera/fluorescence upconversion study. J Phys Chem B 105: 10132–10139

Gobets B, van Stokkum IHM, Rögner M, Kruip J, Schlodder E, Karapetyan N, Dekker JP and van Grondelle R (2001b) Time-resolved fluorescence emission measurements of Photosystem I particles of various cyanobacteria: A unified compartmental model. Biophys J 81: 407–424

Gobets B, van Stokkum IHM, van Mourik F, Dekker JP and van Grondelle R (2003) Excitation wavelength dependence of the fluorescence kinetics in Photosystem I particles from *Synechocystis* PCC 6803 and *Synechococcus elongatus*. Biophys J 85: 3883–3898

Groot ML, Pawlowicz NP, van Wilderen LJGW, Breton J, van Stokkum IHM, and van Grondelle R (2005) Initial electron donor and acceptor in isolated Photosystem II reaction centers identified with femtosecond mid-IR spectroscopy. Proc Natl Acad Sci USA 102: 13087–13092

Haacke S, Vinzani S, Schenkl S and Chergui M (2001) Spectral and kinetic fluorescence properties of native and nonisomerizing retinal in bacteriorhodospin. Chem Phys Chem 2: 310–315

Hebling J (1996) Derivation of the pulse front tilt caused by angular dispersion, Opt Quantum Electron 28: 1759–1763

Holzwarth AR (1996) Data analysis of time-resolved measurements. In: Amesz J and Hoff AJ (eds) Biophysical Techniques in Photosynthesis (Advances in Photosynthesis and Respiration, Vol. 3), pp 75–92. Kluwer Academic Press, Dordrecht.

Horng ML, Gardecki JA, Papazyan A and Maroncelli M (1995) Subpicosecond measurements of polar solvation dynamics: Coumarin 153 revisited. J Phys Chem 99: 17311–17337

Ihalainen JA, Jensen PE, Haldrup A, van Stokkum IHM, van Grondelle R, Schneller HV and Dekker JP (2002) Pigment organization and energy transfer dynamics in isolated Photosystem I (PS I) complexes from *Arabidopsis thaliana* depleted of the PS I-G, PS I-K, PS I-L or PS I-N subunit. Biophys J 83: 2190–2201

Ihalainen JA, Croce R, Morosinotto T, van Stokkum IHM, Bassi R, Dekker JP, van Grondelle R (2005a) Excitation decay pathways of Lhca proteins – A time-resolved fluorescence study. J Phys Chem B 109: 21150–21158

Ihalainen JA, D'Haene S, Yeremenko N, van Roon H, Arteni AA, Boekema EJ, van Grondelle R, Matthijs HCP, and Dekker JP (2005b) Aggregates of the chlorophyll-binding protein IsiA (CP43′) dissipate energy in cyanobacteria. Biochemistry 44: 10846–10853

Ihalainen JA, Klimmek F, Ganeteg U, van Stokkum IHM, van Grondelle R, Jansson S, and Dekker JP (2005c) Excitation energy trapping in Photosystem I complexes depleted in Lhca1 and Lhca4. FEBS Lett 579: 4787–4791

Ihalainen JA, van Stokkum IHM, Gibasiewicz K, Germano M, van Grondelle R and Dekker JP (2005d) Kinetics of excitation trapping in intact Photosystem I of *Chlamydomonas reinhardtii* and *Arabidopsis thaliana*. Biochim Biophys Acta 1706: 267–275

Ito T, Hiramatsu M, Hosoda M and Tsuchiya Y (1991) Picosecond time-resolved absorption spectrometer using a streak camera. Rev Sci Instrum 62: 1415–1419

Jaanimagi PA, DaSilva L, Gregory GG, Hestdalen C, Kiikka CD, Kotmel R, and Richardson MC (1986) Optical fiducials for X-ray streak cameras at LLE. Rev Sci Instrum 57: 2189–2191

Jimenez R and Fleming GR (1996) Ultrafast spectroscopy of photosynthetic systems. In: Amesz J and Hoff AJ (eds) Biophysical Techniques in Photosynthesis (Advances in Photosynthesis and Respiration, Vol 3), pp 63–73. Kluwer Academic Press, Dordrecht

Kamiya N, Ishikawa M, Kasahara K, Kaneko M, Yanamoto N and Ohtani H (1997) Picosecond fluorescence spectroscopy of the purple membrane of *halobacterium halobium* in alkaline suspension. Chem Phys Lett 265: 595–599

Kennis JTM, Gobets B, van Stokkum IHM, Dekker JP, van Grondelle R and Fleming GR (2001) Light harvesting by chlorophylls and carotenoids in the Photosystem I core complex of *Synechococcus elongates*: A fluorescence upconversion study. J Phys Chem B 105: 4485–4494

Kleima FJ, Hofmann E, Gobets B, van Stokkum IHM, van Grondelle R, Diederichs K and van Amerongen H (2000) Förster excitation energy transfer in peridinin-chlorophyll-*a*-protein. Biophys J 78: 344–353.

Knippels GMH, van de Pol MJ, Pellemans HPM, Planken PCM, and van der Meer AFG (1998) Two-color facility based on a broadly tunable infrared free-electron laser and a subpicosecond-synchronized 10-fs-Ti:Sapphire laser. Opt Lett 23: 1754–1756

Krishnan RV, Saitoh H, Terada H, Centonze VE and Herman B (2003) Development of a multiphoton fluorescence lifetime imaging microscopy system using a streak camera. Rev Sci Instrum 74: 2714–2721.

Lin S and Knox RS (1988) Time resolution of a short-wavelength chloroplast fluorescence component at low temperature. J Lumin 40/41: 209–210.

Lin S and Knox RS (1991) Studies of excitation-energy transfer within the green-alga *Chlamydomonas-reinhardtii* and its mutants at 77-K. Photosynth Res 27: 157–168.

Martinez OE (1987) 3000 Times grating compressor with positive group-velocity dispersion: Application to fiber compensation in 1.3–1.6 μm region. IEEE J Quantum Electron QE-23: 59–64

Monshouwer R, Abrahamsson M, van Mourik F and van Grondelle R (1997) Superradiance and exciton delocalization in bacterial photosynthetic light-harvesting systems. J Phys Chem B 101: 7241–7248.

Ohtani H, Ishikawa M, Itoh H, Takiguchi Y, Urakami T and Tsuchiya Y (1990) Picosecond fluorescence spectroscopy of purple membrane in *halobacterium halobium* with a photon-counting streak camera. Chem Phys Lett 168: 493–498

Ohtani H, Kaneko M, Ishikawa M, Kamiya N and Yamamoto N (1999) Picosecond-millisecond dual-time base spectroscopy of fluorescent photointermediates formed in the purple membrane of *halobacterium halobium*. Chem Phys Lett 299: 571–575

Pal, SK, Peon J and Zewail AH (2002) Biological water at the protein surface: Dynamical solvation probed directly with femtosecond resolution. Proc Natl Acad Sci USA 99: 1763–1768

Palacios MA, de Weerd FL, Ihalainen JA, van Grondelle R, and van Amerongen H (2002) Superradiance and exciton (de)localization in light-harvesting complex II from green plants? J Phys Chem B 106: 5782–5787

Pellegrino F, Dagen A, Sekuler P and Alfano RR (1983) Temperature-dependence of the 735-nm fluorescence kinetics from spinach measured by picosecond laser-streak camera system. Photobiochem Photobiophys 6: 15–23

Petushkov VN, van Stokkum IHM, Gobets B, van Mourik F, Lee J, van Grondelle R, and Visser AJWG (2003) Ultrafast fluorescence relaxation spectroscopy of 6,7-dimethyl-(8-ribityl)-lumazine and riboflavin, free and bound to antenna proteins from bioluminescent bacteria. J Phys Chem B 107: 10934–10939

Sauer K and Debreczeny M (1996) Fluorescence. In: Amesz J and Hoff AJ (eds) Biophysical Techniques in Photosynthesis (Advances in Photosynthesis and Respiration, Vol 3), pp 41–61. Kluwer Academic Press, Dordrecht

Schiller NH and Alfano RR (1980) Picosecond characteristics of a spectrograph measured by a streak camera/video readout system. Opt Commun 35: 451–454

Sowinska M, Heisel F, Miehe JA, Lahg M, Lichtenthaler HK and Tomasini F (1996) Remote sensing of plants by streak camera lifetime measurements of the chlorophyll *a* emission. J Plant Physiol 148: 638–644

Tars M, Ellervee A, Wasielewski MR and Freiberg A (1998) Biomolecular electron transfer under high hydrostatic pressure. Spectrochim Acta A 54: 1177–1189

Uhring W, Zint CV, Summ P, and Cunin B (2003) Very high long-term stability synchroscan streak camera. Rev Sci Instrum 74: 2646–2653

van den Berg PAW, Mulrooney SB, Gobets B, van Stokkum IHM, van Hoek A, Williams CH Jr, and Visser AJWG (2001) Exploring the conformational equilibrium of thioredoxin reductase: Characterization of two catalytically important states by ultra-fast flavin fluorescence spectroscopy. Protein Science 10: 2037–2049

van Mourik F, Frese RN, van der Zwan G, Cogdell RJ, and van Grondelle R (2003) Direct observation of solvation dynamics and dielectric relaxation in the photosynthetic light-harvesting-2 complex of *Rhodopseudomonas acidophila*. J Phys Chem B 107: 2156–2161

van Mourik F, Groot M-L, van Grondelle R, Dekker JP and van Stokkum IHM (2004) Global and target analysis of fluorescence measurements on Photosystem II reaction centers upon red excitation. Phys Chem Chem Phys 6: 4820–4824

van Stokkum IHM (2005) Global and target analysis of time resolved spectra. Lecture notes Troisième Cycle de la Physique en Suisse Romande, March 14–24, 88 pp.

van Stokkum IHM and Bal HE (2006) A problem solving environment for interactive modelling of multiway data. Concurrency Computat: Pract Exper 18: 263–269

van Stokkum IHM, Larsen DS, and van Grondelle R (2004) Global and target analysis of time resolved spectra. Biochim Biophys Acta 1657, 82–104, and 1658, 262 (Erratum)

van Stokkum IHM, Gobets B, Gensch T, van Mourik F, Hellingwerf KJ, van Grondelle R and Kennis JTM (2006) (Sub)-picosecond spectral evolution of fluorescence in photoactive proteins studied with a synchroscan streak camera system. Photochem Photobiol 82: 380–388

Vengris M, van der Horst MA, Zgrablić G, van Stokkum IHM, Haacke S, Chergui M, Hellingwerf KJ, van Grondelle R and Larsen DS (2004) Contrasting the excited-state dynamics of the photoactive yellow protein chromophore: Protein vs. solvent effects. Biophys J 87: 1848–1857

Vilchiz VH, Kloepfer JA, Germaine AC, Lenchenkov VA and Bradforth SE (2001) Map for the relaxation dynamics of hot photoelectrons injected into liquid water via anion threshold photodetachment and above threshold solvent ionization. J Phys Chem A 105: 1711–1723

Watanabe M, Koishi M, Fujiwara M, Takeshita T and Cieslik W (1994) Development of a new fluorescence decay measurement system using two-dimensional single-photon counting. J Photochem Photobiol A: Chem 80: 429–432

Wiessner A and Staerk H (1993) Optical design considerations and performance of a spectro-streak apparatus for time-resolved fluorescence spectroscopy. Rev Sci Instrum 64: 3430–3439

Yatskou MM, Koehorst RBM, van Hoek A, Donker H, Schaafsma TJ, Gobets B, van Stokkum IHM and van Grondelle R (2001) Spectroscopic properties of a self-assembled zinc porphyrin tetramer. II. Time-resolved fluorescence spectroscopy. J Phys Chem A 105: 11432–11440

Zavoiski EK and Fanchenko SD (1965) Image converter high-speed photography with $10^{-9}$–$10^{-14}$ sec time resolution. Appl Optics 4: 1155–1167

# Chapter 13

## Optical Spectroscopy of Individual Light-Harvesting Complexes

Thijs J. Aartsma*
*Department of Biophysics, Huygens Laboratory, Leiden University, P.O. Box 9504, 2300 RA Leiden, The Netherlands*

Jürgen Köhler
*Universität Bayreuth, Lehrstuhl für Experimentalphysik IV, Universitätsstr. 30, D-95447 Bayreuth, Germany*



### Summary

In this contribution we present an overview of the results of detailed spectroscopic investigations of individual light-harvesting complexes of photosynthetic purple bacteria at 1.2 K. By applying single-molecule detection techniques the properties of the electronically excited states of the assemblies of bacteriochlorophyll *a* (BChl *a*) pigment molecules in the individual complexes are revealed, without ensemble averaging. The results show that the excited states of the B800 ring of pigments in LH2 are largely localized on individual BChl *a* molecules, although accidental degeneracy gives rise to an excitonic character of the B800 excited states. In contrast, the absorption of a photon by the B850 ring can be consistently described in terms of an excitation that is more or less delocalized over the ring. This property is believed to contribute to the high efficiency of energy transfer

*Author for correspondence, email: aartsma@physics.leidenuniv.nl

*Thijs J. Aartsma and Jörg Matysik (eds): Biophysical Techniques in Photosynthesis II, pp. 241–266.*

in these photosynthetic complexes. The analysis of the spectra in terms of disorder is reviewed in some detail. The results of LH2 are compared with those of similar complexes, LH3 and low-light LH2. In addition, we consider the fluorescence spectra of individual LH2 and LH3 complexes.

## I. Introduction

Photosynthetic organisms have developed highly efficient systems to collect sunlight and to use this light energy as the driving force for their metabolic processes (Blankenship, 2002). This energy conversion process, known as photosynthesis, requires synergism in the interaction between tens of different proteins. The structures of many of these proteins have been resolved (see Chapters 5, Liu and Chang, and 6, Fromme and Allen, and references there in), providing a basis for detailed studies of structure-function relationships of these proteins. The intrinsic properties of photosynthetic systems are intimately related to their natural interaction with light. This interaction proceeds via highly specialized pigment-protein complexes which play a key role in the processes of light absorption, energy transfer and charge separation. Optical spectroscopy has been, and still is, essential to study these processes in ever greater detail. It is the primary experimental technique to analyze and identify important constituents of the photosynthetic apparatus, including their role and place in the sequence of events following absorption of light. Advances in technology continue to sustain the development of new and exciting spectroscopic methods, allowing researchers to probe deeper into the mechanistic aspects of the photosynthetic apparatus, guided by structural information that becomes available at an increasing rate.

In this chapter we present new experimental methods which are aimed at the optical spectroscopy and the characterization of individual pigment-protein complexes. Conventional spectroscopy is typically performed on bulk samples, and it allows, therefore, only access to the ensemble-averaged properties of molecular systems. At the single-molecule level, however, ensemble-averaging is completely eliminated. Thus it is possible to explore and study the extent to which variations in molecular conformation affect the photophysics of individual pigment-protein complex. The intrinsic heterogeneity of photosynthetic systems can be mapped in terms of the distribution of experimentally measured parameters. Moreover, enhanced spectral resolution can be achieved by eliminating static disorder effects. In the absence of spectral diffusion one obtains the homogeneous line shape from which lifetime values and vibronic coupling parameters can be derived. Thus, single-molecule techniques can help to elucidate the electronic structure of the light-harvesting (LH) complexes, and provide unique access to disorder parameters that are important for understanding the intermolecular interactions in these systems.

Based on the structural information of the LH complexes it is possible to calculate the intermolecular interaction between the pigments, and to develop extensive quantitative quantum-mechanical models. The high symmetry of the LH complexes from purple bacteria imposes stringent rules on the collective molecular properties, such as selection rules and distribution of oscillator strength over electronic states. This helps tremendously in the assessment and analysis of spectroscopic details in terms of the electronic structure and other system parameters. Thus the combination of. structure, experiments and theory provide unique insights into the photophysics of light-harvesting systems.

We will focus on LH complexes from photosynthetic purple bacteria which form one of the four classes of anoxygenic photosynthetic bacteria. They have a photosynthetic unit (PSU) embedded in a cytoplasmic membrane which contains besides the reaction center (RC) only two types of antenna complexes, i.e., the light-harvesting complex 1 (LH1) and light-harvesting complex 2 (LH2). LH1 is believed to form a more or less ring-like structure which encloses the RC (Gall et al., 1994; Karrasch et al., 1995; Roszak et al., 2003), whereas LH2 is not in direct contact with the RC, but transfers the energy to the RC via the LH1 complex (van Grondelle et al., 1994; Hoff and Deisenhofer, 1997). Some purple bacteria can grow a third type of light-harvesting system, which is called the B800-820 or LH3 complex (McLuskey et al., 2001). Depending on growth conditions this pigment-protein complex is expressed at the expense

*Abbreviations:* BChl – bacteriochlorophyll; FMO – Fenna-Matthews-Olson complex; FWHM – full width at half maximum; LH – light harvesting; LL – low light; *NA* – numerical aperture; PSB – phonon sideband; PSU – photosynthetic unit; *Rb.* – *Rhodobacter*; RC – reaction center; *Rps.* – *Rhodopseudomonas*; ZPL – zero-phonon line

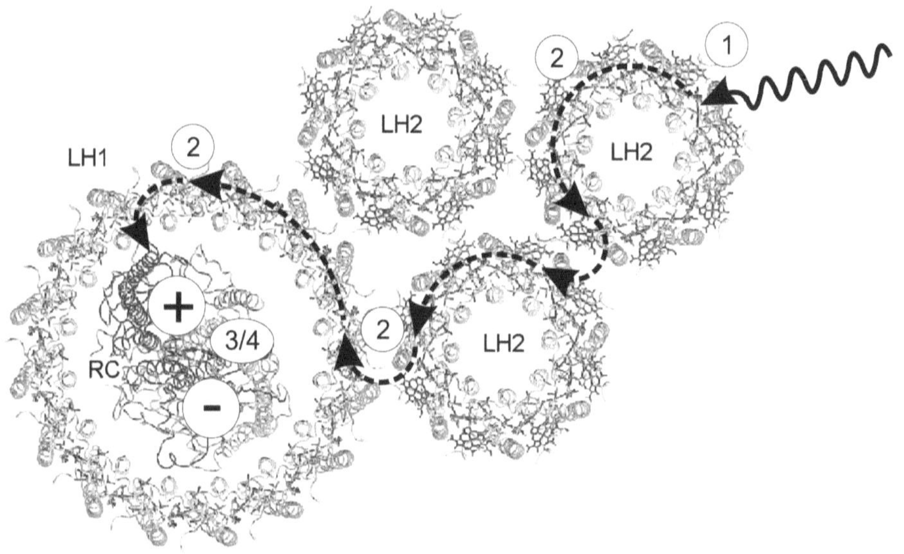


*Fig. 1.* Model for the photosynthetic unit (PSU) of the purple bacterium *Rhodobacter* (*Rb.*) *sphaeroides*, showing the two types of protein-pigment complexes: the photosynthetic reaction center (RC) and the light-harvesting (LH) complexes. The main function of the LH complexes is to gather light energy (process 1) and to transfer this energy to the RC (process 2) for the light-induced redox processes (processes 3 and 4). Figure adapted from Hu et al. (1998).

of LH2. It is believed to have a similar function as LH2. Fig. 1 shows a model of the bacterial PSU of the purple bacteria *Rhodobacter* (*Rb.*) *sphaeroides* where the organization of the different pigment-protein complexes is depicted. The numbers refer to the four different steps of the photosynthetic light reactions as previously described.

Figure 2 shows the low-temperature absorption spectrum of membrane fragments of the purple bacterium *Rhodopseudomonas* (*Rps.*) *acidophila*. In the near infrared it shows broad bands at 800 and 870 nm and a smaller band at 910 nm. The first two bands are assigned to the LH2 antenna, whereas the third band is dominated by LH1. The absorption bands of the RC are largely hidden by the antenna absorption. A cascaded energy transfer occurs from the high-energy pigments to the lower-energy pigments, which are physically closer to the RC. The small amount of energy lost in each step provides a degree of irreversibility to the process. The transfer of energy from LH2 to LH1 and subsequently to the RC occurs in vivo on an over-all time scale of 40 to 50 ps (van Grondelle et al., 1994).

In the following section an elementary description is given of the geometrical and electronic structure of the characteristic ring-shaped LH complexes of *Rps. acidophila* with some emphasis on symmetry aspects. In subsequent sections the requirements

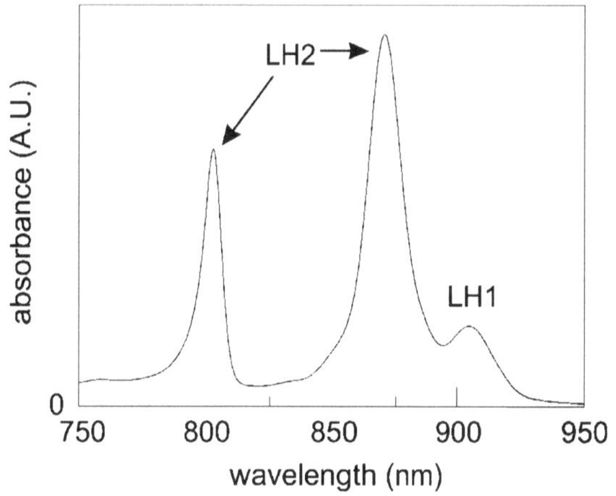


*Fig. 2.* Low-temperature (10 K) absorption spectrum of membrane fragments of *Rps. acidophila*.

for successful single-molecule experiments are presented and the experimental set up is described. Finally, an overview is given of experimental results, and conclusions that can be derived from them. The excitation spectra of single LH2 complexes in the region of the B800 and B850 band are discussed. The analysis shows that the excitations of the B850 systems are largely delocalized over the whole ring of 18 BChl *a* molecules, i.e., the excitations are properly described as Frenkel excitons. The observed splitting and polarization of the main bands in the B850 region

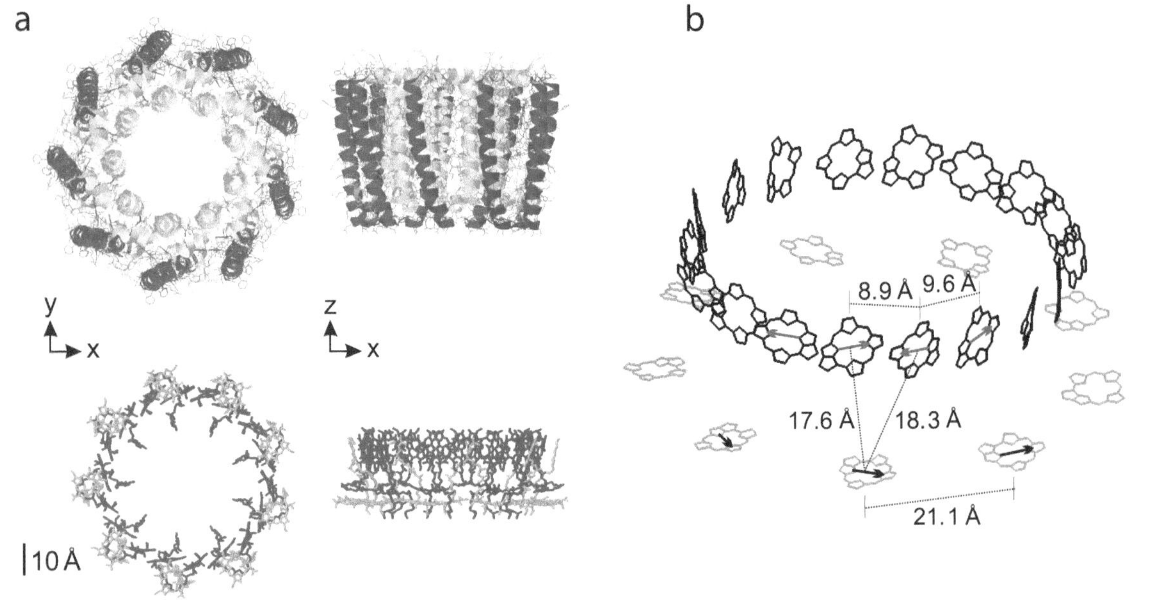


*Fig. 3.* (a) Structure of the LH2 complex from *Rps. acidophila* as determined with X-ray diffraction (McDermott et al., 1995). The left part of the panel displays a top view of the assembly and the right part a side view. The upper part shows the whole protein-pigment complex, the lower part only the BChl *a* pigments. The pigments are arranged in two concentric rings commonly termed B800 (light-gray) and B850 (dark-gray). (b) The spatial arrangement of the BChl *a* pigments in LH2 of *Rps. acidophila*, but now in a tilted side-view. The numbers indicate the center to center distances of the pigments in Å. The arrows indicate the direction of the $Q_y$ transition moments. The phytol chains of the BChl *a* pigments are omitted for clarity. Compare also Fig. 1, Color Plate CP7.

support the contention that the B850 ring is subject to a perturbation that reduces the symmetry of the complex. Finally we will summarize results on a few other systems, in comparison with LH2. Fluorescence spectra of individual LH complexes are reviewed in the last section.

## II. Circular Excitons

The high-resolution X-ray structures of the LH2 (McDermott et al., 1995) and LH3 (McLuskey et al., 2001) complexes of *Rps. acidophila* show a remarkable symmetry in the arrangement of the light-absorbing pigments in their protein matrix. The basic building block of both complexes is a protein heterodimer (αβ), which binds noncovalently three BChl *a* pigments and one carotenoid molecule. The complexes consist of nine of such αβ-polypeptide heterodimers. The 3-dimensional structure of LH2 is depicted in Fig. 3a, where the upper part of the figure shows the pigment-protein complex as a whole while the lower part shows only the BChl *a* pigments. The structure of LH3 shows a similar pigment arrangement which is virtually indistinguishable from that of LH2. A pronounced feature of these complexes is the fact that, at least in crystalline form, they have a nine-fold rotational symmetry (point group $C_9$). Fig. 3b depicts the spatial arrangement of the BChl *a* pigments in LH2 in a tilted side-view. The numbers indicate the center-to-center distances of the pigments. The arrows indicate the direction of the $Q_y$ transition-dipole moment.

Two rings of BChl *a* pigments can be distinguished. One ring consists of nine pigments, which have their molecular plane perpendicular to the symmetry axis. These pigments contribute to the absorption around 800 nm (Fig. 4a) and are therefore labeled B800. The second ring consists of nine repeating pairs of one α- and one β-bound pigment, which are tightly organized as the blades of a turbine with their molecular planes parallel to the symmetry axis. Since these pigments mainly contribute to the absorption around 850 nm at room temperature these are labeled B850 (Fig. 4a). Upon excitation, energy is transferred from the B800 to B850 pigments in 1 to 2 ps (Reddy et al., 1991; De Caro et al., 1994; Monshouwer et al., 1995; Kennis et al., 1996; Wu et al., 1996), while energy transfer among the B850 molecules is an order of magnitude faster (Jimenez et al., 1996; Chachisvilis et al., 1997;

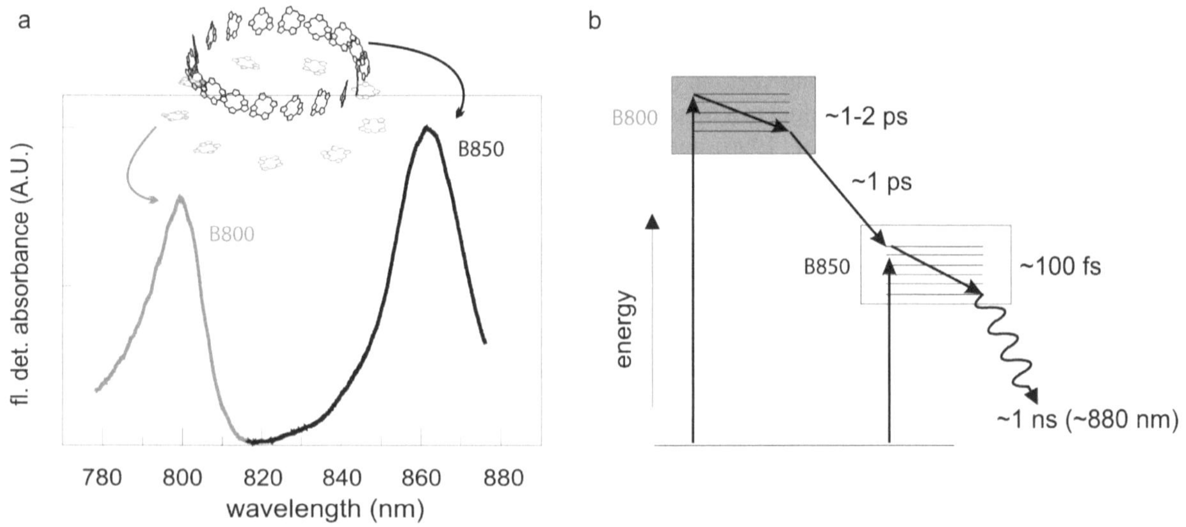


*Fig. 4.* (a) Fluorescence-detected absorption spectrum of an ensemble of LH2 complexes at 1.2 K. The spectrum was measured at 20 W/cm² with LH2 dissolved in a PVA-buffer solution and spincoated on a LiF substrate. (b) Schematic representation of the energy flow in the LH2 complex.

Vulto et al., 1999) (Fig. 4b). The excited-state decay has a relatively long lifetime of approximately 1 ns (Monshouwer et al., 1997). For a recent review on LH2, see Sundström et al. (1999).

The electronically excited states of the collection of BChl *a* molecules in the B800 and B850 band are described by the following Hamiltonian:

$$H = \sum_{n=1}^{N} E_n |n\rangle\langle n| + \sum_{n=1}^{N}\sum_{n\neq m} V_{nm}\{|n\rangle\langle m| + H.c.\} \quad (1)$$

Here $E_n$ denotes the excitation energy of an individual pigment (the site energy) and $V_{nm}$ the interaction between molecules *n* and *m*. *H.c.* stands for the Hermitian conjugate of the preceding term. In the conjugated π-electron system of the chlorin macrocycle of BChl *a* four major electronic transitions are present. In our description we only consider the lower energetic transition, which has its transition-dipole moment polarized along the y-axis of the BChl *a* pigment running along the $N_{III}$-$N_I$ nitrogen atoms and is labeled $Q_y$ (Blankenship, 2002).

The electronic interactions between the BChl *a* pigments are mainly determined by the Coulombic interactions of electrons and nuclei of one pigment with those of neighboring molecules and by exchange interactions caused by overlap of molecular orbitals of adjacent pigments. In both cases the interaction energy ($V_{ij}$) between two molecules, *i* and *j* can be described by:

$$V_{ij} = \frac{1}{4\pi\varepsilon}\sum\frac{q_i q_j}{r_{ij}} \quad (2)$$

where ε is the dielectric permitivity and $r_{ij}$ the distance between the electronic and nuclear charges $q_i$ and $q_j$ on both molecules. When the Coulombic interactions are much larger than the exchange interactions, i.e., when the intermolecular separation is large compared to the molecular size, it is useful to expand the interaction energy into a power series. In this multipole expansion, terms for monopole-monopole, monopole-dipole, dipole-dipole and dipole-quadrupole interaction can be distinguished. For uncharged molecules only the terms for dipole-dipole and dipole-quadrupole interactions are important. If we further assume the dipole-dipole interaction to be dominant and that it can be approximated by point-dipoles, the interaction between the molecules *i* and *j* can be described by:

$$V_{ij} = \frac{1}{4\pi\varepsilon}\frac{\mu^2}{r_{ij}^3}\kappa_{ij},$$

$$\kappa_{ij} = \frac{1}{\mu^2}\left[\left(\vec{\mu}_i\cdot\vec{\mu}_j\right) - 3\frac{\left(\vec{\mu}_i\cdot\vec{r}_{ij}\right)\left(\vec{\mu}_j\cdot\vec{r}_{ij}\right)}{r_{ij}^2}\right] \quad (3)$$

where $\vec{\mu}$ is the transition-dipole moment (see also Chapter 21, Renger and Holzwarth). The distance between the transition dipoles $\vec{\mu}_i$ and $\vec{\mu}_j$ is given by

$r_{ij}$. The mutual orientation and distance of the transition-dipole moments determine the orientation factor $\kappa_{ij}$, which is defined in terms of unit vectors. Using $|\vec{\mu}| = 2.046 \times 10^{-29}$ Cm (corresponding to a transition moment of 6.13 Debye) (Sauer et al., 1996), and $r_{i,i+1}$ = 2.13 nm, one finds an interaction strength of −24 $cm^{-1}$ between nearest neighbors in the B800 ring. The interaction between B850 pigments is estimated to be ~300 $cm^{-1}$.

It may be questioned whether a point-dipole approximation is appropriate for calculating the interaction energies given the small intermolecular separation, especially in the case of the B850 ring. This has prompted more elaborate calculations, such as a point-monopole approach (Sauer et al., 1996), the transition density cube method (Krüger et al., 1998), and quantum chemical calculations (Alden et al., 1997; Scholes et al., 1999). All these methods give similar values for the B850 intermolecular interactions, ranging from 250–400 $cm^{-1}$.

In Eq. (1) the site energy $E_n$ of equivalent sites would be the same in the absence of disorder. In actual practice this energy varies among the pigments as a result of variations in the local electrostatic interactions within the protein. In LH2 of *Rps. acidophila* the site energy of the BChl *a* pigment is largely determined by the interaction of a histidine residue of the protein with the magnesium core of the pigments as well as by hydrogen bonds between the protein and the pigment (Cogdell et al., 1997). Similar variations in the site energy occur in the B820 ring of the related LH3 complex, where a change in the hydrogen bonding pattern together with a conformational change of the C-3-acetyl group of the pigments induces a large shift in the site energy (McLuskey et al., 2001). This implies that the zero-order energy $E_n$ is different for each pigment in the ring. This is taken into account in the Hamiltonian by introducing a random shift $\Delta E_n$ around a mean value of the site energy $E_0$:

$$H = \sum_{n=1}^{N} \left(E_0 + \Delta E_n\right)|n\rangle\langle n| + \sum_{n=1}^{N}\sum_{n\neq m} \left(V_0 + \Delta V_{nm}\right)\left\{|n\rangle\langle m| + H.c.\right\} \quad (4)$$

This disorder is referred to as diagonal disorder or site heterogeneity and is usually modeled by a Gaussian distribution of site energies (Sundström et al., 1999). It gives rise to inhomogeneous broadening of optical transitions. Any structural disorder or deformation also results in a variation of the interactions around a mean value $V_0$, represented by $\Delta V_{nm}$.

Two limits of the dipolar interaction can be distinguished, that of the weak and the strong coupling. As an example we consider in Fig. 5 two interacting dipoles with wave functions, $|\alpha\rangle$ and $|\beta\rangle$, a situation which corresponds to $N = 2$ in Eq. 4. In the weak-coupling case (Fig. 5a) the interaction between the transition-dipole moments is much smaller than the difference in site energies of the two pigments ($V_i << \Delta E_{\alpha\beta} = |E_\alpha - E_\beta|$). As a result the excited state wave functions are mainly localized on individual pigments. The spread of the absorption bands reflects directly the spread in site energies.

In the strong-coupling limit (Fig. 5b) the interaction is much larger than the difference in site energies of the two pigments. As a result the eigenfunctions are given as combinations of the excited state wavefunctions of the individual pigments, and describe collective properties of the molecules involved. These Frenkel-excitons correspond to two energy levels as indicated in Fig. 5b. If the site energies of the two pigments are the same, the wave functions, $|\psi_{\alpha\beta}{}^{s}\rangle$ and $|\psi_{\alpha\beta}{}^{as}\rangle$, have equal amplitudes on both pigments and are given by,

$$\left|\psi^{s}_{\alpha\beta}\right\rangle = \frac{1}{\sqrt{2}}\left\{|\alpha\rangle + |\beta\rangle\right\}$$
$$\left|\psi^{as}_{\alpha\beta}\right\rangle = \frac{1}{\sqrt{2}}\left\{|\alpha\rangle - |\beta\rangle\right\} \quad (5)$$

The splitting between the exciton states is twice the electronic interaction between the monomer pigments. The distribution of oscillator strength over the two exciton levels depends on the mutual orientation of the transition-dipole moments. As a rule of thumb, if the interaction strength is much smaller than the inhomogeneous broadening, the weak coupling limit applies. The other extreme corresponds to the strong coupling limit. The B850 ring is somewhere in between these two limits, the interaction strength being about the same as the inhomogeneous broadening of ~200 $cm^{-1}$.

The LH2 complex of *Rps. acidophila* (Fig. 3a) belongs to the pure-rotational point group $C_9$. This point group contains one symmetry element, the ninefold axis of rotation. The nine symmetry operations generated by this element are: E ($\equiv C_9^9$), $C_9$, $C_9^2$, $C_3$, $C_9^4$, $C_9^5$, $C_3^2$, $C_9^7$ and $C_9^8$, which are all rotations within the plane of the complex over angles of $m \times (2\pi/9)$

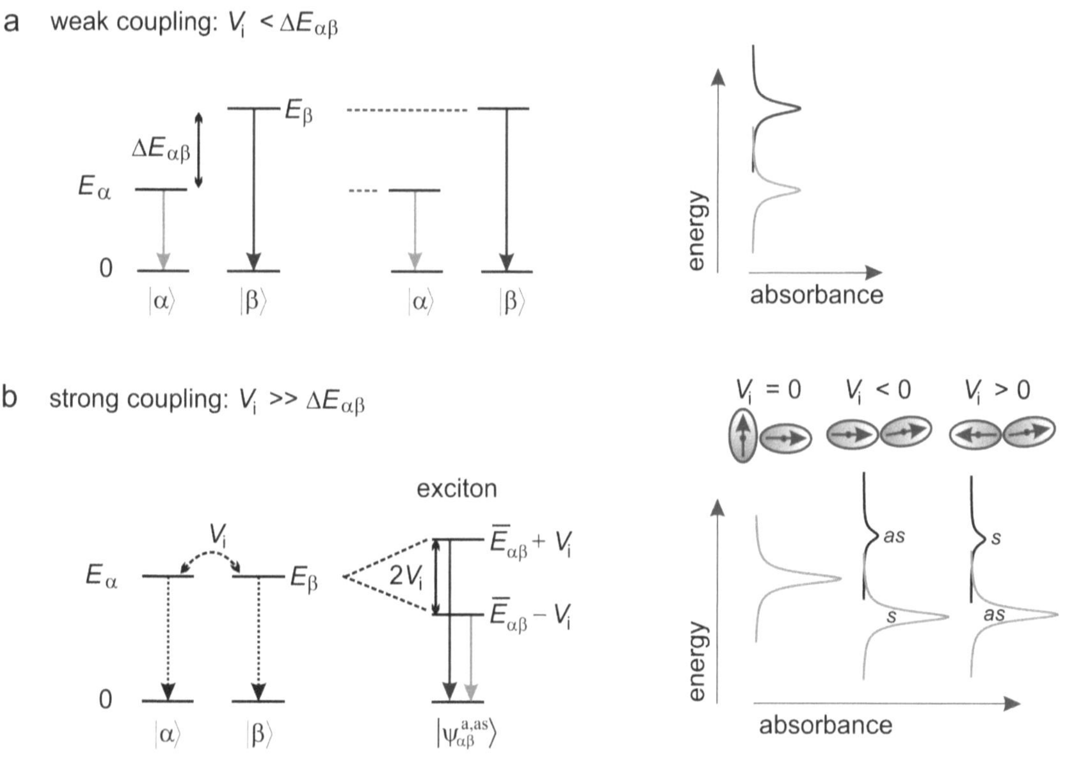


*Fig. 5.* Effects of the intra-dimer interaction ($V_i$) and site-energy difference ($\Delta E_{\alpha\beta} = |E_\alpha - E_\beta|$, $E_{\alpha\beta} = \frac{1}{2}(\bar{E}_\alpha + E_\beta)$) on the energy levels and spectra of two monomer pigments $|\alpha\rangle$ and $|\beta\rangle$. (a) In the weak-coupling limit the interaction between the pigments is weak compared to their site-energy difference and the resulting absorption spectrum is independent of the mutual orientation of the pigments. (b) In the strong-coupling limit the difference in site-energies in negligible compared to the interaction strength and as a result two new, delocalized exciton states are created belonging to the dimer $1/\sqrt{2}\ \{|\alpha\rangle + |\beta\rangle\}$. Spectra are shown for three different orientations of the transition-dipole moments. The small arrow indicates the direction of the dipole moment.

radians with $m = 1,2,\ldots 9$. The character table for the $C_9$-point group is depicted in Table 1.

Group theory is particularly helpful for evaluating whether matrix elements are zero or correspond to a finite value. For instance, consider the matrix elements

$$\langle i|\Omega|j\rangle = \int \Psi_i \Omega \Psi_j d\tau \qquad (6)$$

is zero unless the integrand ($\Psi_i\Omega\Psi_j$) is invariant under all operations of the symmetry group to which the molecule belongs. In other words, the integrand should form a basis for the totally symmetric representation ($A$) of the group. This is equivalent to saying that the direct product of the irreducible representations to which the functions $\Psi_i$, $\Psi_j$ and the operator $\Omega$ belong contains the totally symmetric representation. In Table 2 the direct products of all the irreducible representations of the $C_9$ point group are depicted. It is then straightforward to see that the integral in Eq. (6) will only be nonzero if the representation of the direct product of $\Psi_i$ and $\Psi_j$ contains the irreducible representation to which $\Omega$ belongs.

Linear combinations of the localized symmetric or antisymmetric wave functions of the αβ-dimers, $|\psi_{\alpha\beta}{}^{s}\rangle$ and $|\psi_{\alpha\beta}{}^{as}\rangle$, can be constructed in such a way that they form bases for the irreducible representations of $C_9$. These linear combinations can be found using the projection-operator technique. Fortunately this technique is very simple for pure rotational groups. It can be shown that the characters of the group are the coefficients of the linear combination of all the localized wave functions. After normalization one obtains the following delocalized wave functions, $|\Psi_k^i\rangle$,

$$|\Psi_k^i\rangle = \frac{1}{\sqrt{9}} \sum_{n=1}^{9} e^{in\left(\frac{\pi k}{9}\right)} |\psi_{\alpha\beta}^i\rangle \qquad (7)$$

defined by a quantum number $k$ equal to $k = 0, \pm 1,$

*Table 1.* Character table of the $C_9$ point group. See Tinkham (2003) for more information on the representation of cyclic groups.

| $C_9$ | E | $C_9$ | $C_9^2$ | $C_3$ | $C_9^4$ | $C_9^5$ | $C_3^2$ | $C_9^7$ | $C_9^8$ | $\varepsilon = e^{2\pi i/9}$ |
|---|---|---|---|---|---|---|---|---|---|---|
| $A$ | 1 | 1 | 1 | 1 | 1 | 1 | 1 | 1 | 1 | $z, R_z$ |
| $E_1$ { | 1 | $\varepsilon$ | $\varepsilon^2$ | $\varepsilon^3$ | $\varepsilon^4$ | $\varepsilon^{4*}$ | $\varepsilon^{3*}$ | $\varepsilon^{2*}$ | $\varepsilon^*$ | $(x,y), (R_x,R_y)$ |
| | 1 | $\varepsilon^*$ | $\varepsilon^{2*}$ | $\varepsilon^{3*}$ | $\varepsilon^{4*}$ | $\varepsilon^4$ | $\varepsilon^3$ | $\varepsilon^2$ | $\varepsilon^1$ | |
| $E_2$ { | 1 | $\varepsilon^2$ | $\varepsilon^4$ | $\varepsilon^{3*}$ | $\varepsilon^*$ | $\varepsilon$ | $\varepsilon^3$ | $\varepsilon^{4*}$ | $\varepsilon^{2*}$ | |
| | 1 | $\varepsilon^{2*}$ | $\varepsilon^{4*}$ | $\varepsilon^3$ | $\varepsilon$ | $\varepsilon^*$ | $\varepsilon^{3*}$ | $\varepsilon^4$ | $\varepsilon^2$ | |
| $E_3$ { | 1 | $\varepsilon^3$ | $\varepsilon^{3*}$ | 1 | $\varepsilon^3$ | $\varepsilon^{3*}$ | 1 | $\varepsilon^3$ | $\varepsilon^{3*}$ | |
| | 1 | $\varepsilon^{3*}$ | $\varepsilon^3$ | 1 | $\varepsilon^{3*}$ | $\varepsilon^3$ | 1 | $\varepsilon^{3*}$ | $\varepsilon^3$ | |
| $E_4$ { | 1 | $\varepsilon^4$ | $\varepsilon^*$ | $\varepsilon^3$ | $\varepsilon^{2*}$ | $\varepsilon^2$ | $\varepsilon^{3*}$ | $\varepsilon$ | $\varepsilon^{4*}$ | |
| | 1 | $\varepsilon^{4*}$ | $\varepsilon$ | $\varepsilon^{3*}$ | $\varepsilon^2$ | $\varepsilon^{2*}$ | $\varepsilon^3$ | $\varepsilon^*$ | $\varepsilon^4$ | |
| $\Sigma$ | 9 | 0 | 0 | 0 | 0 | 0 | 0 | 0 | 0 | |

*Table 2.* Direct-product table for the $C_9$ point group.

| $C_9$ | A | $E_1$ | $E_2$ | $E_3$ | $E_4$ |
|---|---|---|---|---|---|
| A | A | $E_1$ | $E_2$ | $E_3$ | $E_4$ |
| $E_1$ | $E_1$ | $2A+E_2$ | $E_1+E_3$ | $E_2+E_4$ | $E_3+E_4$ |
| $E_2$ | $E_2$ | $E_1+E_3$ | $2A+E_4$ | $E_1+E_4$ | $E_2+E_3$ |
| $E_3$ | $E_3$ | $E_2+E_4$ | $E_1+E_4$ | $2A+E_3$ | $E_1+E_2$ |
| $E_4$ | $E_4$ | $E_3+E_4$ | $E_2+E_3$ | $E_1+E_2$ | $2A+E_1$ |

±2, ±3, ±4, for 9 dimers in the ring. They correspond to the irreducible representations A, $E_1$, $E_2$, $E_3$, and $E_4$, respectively The superscript $i$ designates the symmetric (s) or antisymmetric (as) dimer wave function. Additional mixing occurs between $|\psi_k^s\rangle$ and $|\psi_k^{as}\rangle$ states that belong to the same irreducible representation. A detailed analysis has been presented by Matsushita et al. (2001), who derived the following expression for the energy of the excitons states in the absence of disorder:

$$E_k^{\pm} = \bar{E} + 2\bar{W}\cos(k\varphi) \pm \sqrt{\left[\frac{1}{2}E_{\alpha\beta} + W_{\alpha\beta}\cos(k\varphi)\right]^2 + V_i^2 + V_e^2 + 2V_iV_e\cos(k\varphi)} \quad (8)$$

with

$$\bar{E} = (E_\alpha + E_\beta)/2, \qquad E_{\alpha\beta} = (E_\alpha - E_\beta),$$
$$\bar{W} = (W_\alpha + W_\beta)/2, \qquad W_{\alpha\beta} = (W_\alpha - W_\beta) \quad (9)$$

where $\varphi = 2\pi/9$. $E_\alpha$ and $E_\beta$ are the zero-order excitation energies of the α- and β-bound pigments, $V_i$ and $V_e$ the nearest neighbor intra- and inter-dimer interactions, respectively (see Fig. 6). Due to the second-neighbor (interdimer) interactions, $W_\alpha$ and $W_\beta$, the overall manifold is asymmetric around $\bar{E}$.

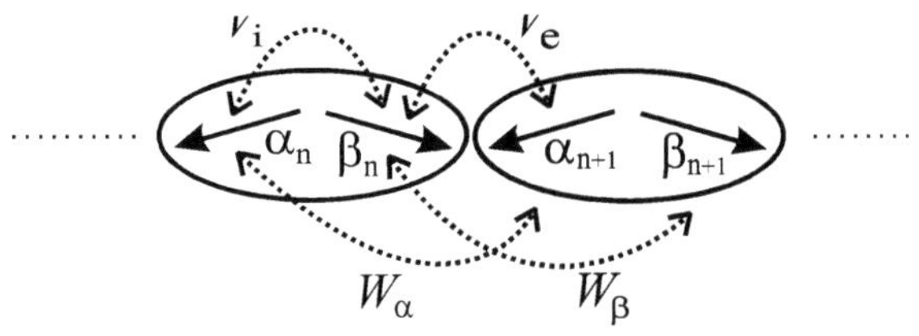


*Fig. 6.* Schematic representation of the orientation of the transition-dipole moments of the individual BChl *a* molecules in two adjacent dimers and their mutual interactions. $V_i$ and $V_e$ are the nearest neighbor intra- and inter-dimer interactions, respectively; $W$ designates the second neighbor interaction.

The energies of the exciton states are depicted in Fig. 7 and show the two manifolds originating from the symmetric and antisymmetric state of the dimer (Eq. 5). Each consists of one nondegenerate and four degenerate states. The levels are denoted by their quantum number $k$ and the symmetry species, A or $E_1$-$E_4$ of their irreducible representation.

It is instructive to examine how the signs of the wave functions of the dimers vary around the ring for the different exciton states. This is depicted in Fig. 8a for the A, $E_1$ and $E_2$ irreducible representations. The A representation does not have any nodal plane in the ring, the $E_1$ representation one and the $E_2$ representation two. By looking at the signs of the

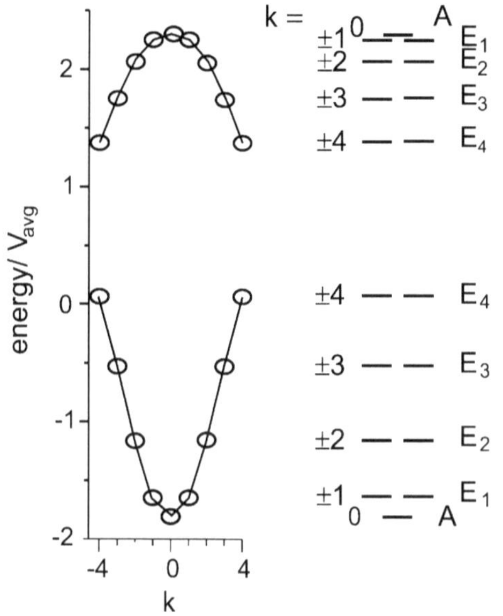


*Fig. 7.* The energy-level scheme of the excited-state manifold of the B850 ring of LH2 (*Rps. acidophila*). For a detailed description of the manifold see (Matsushita et al., 2001). $E_i$ are the irreducible representations of the exciton states with quantum number *k*.

wave function it is clear that multiplying the two $E_1$ states will give a wave function which spans the $E_2$ irreducible representation (Fig. 8b), cf. the multiplication Table 2.

Of the two non degenerate (denoted as $k = 0$) and eight pair-wise degenerate ($k^s$ and $k^{as} = \pm1, \pm2, \pm3, \pm4$) exciton states in the B850 manifold, only the low-energy degenerate pair $k^{as} = \pm1$ will carry appreciable oscillator strength. The upper exciton components $k^s = \pm1$ carry less than 3% of the oscillator strength and give rise to a weak absorption around 790 nm (Sauer et al., 1996). Note that the total oscillator strength is conserved, so that each of the $k^{as} = \pm1$ states has an oscillator strength of about 9 times that of one BChl *a* molecule. Upon introducing diagonal disorder in the ring, the pair-wise degeneracies will be lifted, and the oscillator strength is redistributed over adjacent exciton states (Alden et al., 1997). The transition dipole moments associated with the $k^{as} = \pm1$ transitions will have orthogonal polarizations. This orthogonality is maintained when disorder is introduced, assuming that the diagonal disorder is dominated by variations in electrostatic interactions and possibly intermolecular distances, rather than by changes in the orientations of the BChl *a* molecules.

Small perturbations of the $C_9$-symmetric structure of the B850 ring will have large implications for the energy-level scheme of its excited-state manifold. By classifying the symmetry of these perturbations according to the irreducible representations of the $C_9$-point group it is straightforward to deduce which states are affected by a specific perturbation, using the direct-product Table 2. This has been discussed in detail by Wu and Small (1997).

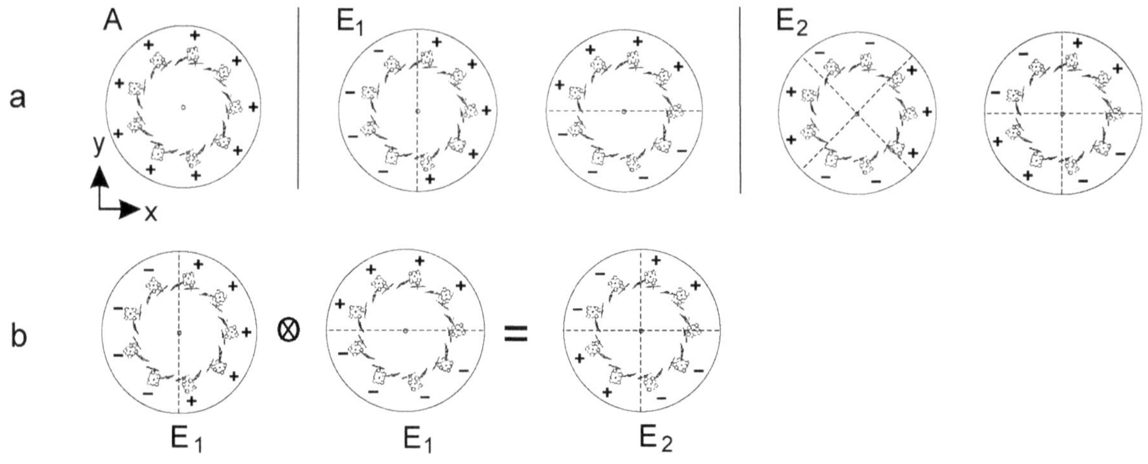


*Fig. 8.* (a) Variation of the signs of the wave functions around the B850 ring of LH2. Depicted are only the three irreducible representations A, $E_1$ and $E_2$. (b) Direct product of the two $E_1$ irreducible representations, which gives the $E_2$ representation. The dashed lines represent the nodal planes in the ring.

## III Experimental Aspects of Single-Molecule Spectroscopy

An important aspect of the optical properties of an ensemble of photosynthetic pigment-protein complexes is the intrinsic heterogeneity in these types of systems. One way to examine this heterogeneity in more detail is by studying the optical spectrum of individual complexes. The great advantage of this method is that the statistical distribution of a particular parameter of the complex can be analyzed rather than just measuring the average value of the distribution (Ambrose et al., 1991; Basché et al., 1997; Köhler, 1999; van Oijen et al., 1999a).

### A. Prerequisites

In principle, the spectrum of a single molecule corresponds to the homogeneous line shape if spectral diffusion can be neglected. Figure 9 shows a schematic representation of the absorption spectrum of the lowest, vibration-free electronic transition of a single chromophore at low temperature. The spectral shape is composed of an extremely narrow line convoluted with a blue-shifted broad, vibronic band. The narrow feature corresponds to the purely electronic transition of the chromophore and is called the zero-phonon line (ZPL). The broad band originates from concomitant excitation of low-frequency vibrations (phonons) in the surrounding matrix and is called the phonon sideband (PSB). The homogenous linewidth of the ZPL, denoted by the full width at half maximum $\Gamma_{hom}$, is related to the total dephasing time $T_2$:

$$\Gamma_{hom}(T) = \frac{1}{\pi T_2(T)} = \frac{1}{2\pi T_1} + \frac{1}{\pi T_2^*(T)} \tag{10}$$

$T_2$ consists of two distinct contributions. $T_1$ denotes the lifetime of the excited state of the chromophore (fluorescence lifetime), which is composed of both radiative and non-radiative decay channels. $T_2^*$ is called the pure dephasing time and represents the time interval in which the coherence of the electronic transition is lost as a result of low-frequency modes (phonons) that couple to the electronic transition of the molecule. In other words, $T_2^*$ reflects the interaction of the chromophore with its environment, which is strongly temperature ($T$) dependent. At sufficiently low temperatures (< 2 K) most of the phonon modes are not populated and $T_2^* >> T_1$. The ZPL then becomes very narrow, its width being determined by the lifetime of the excited states ($T_1$).

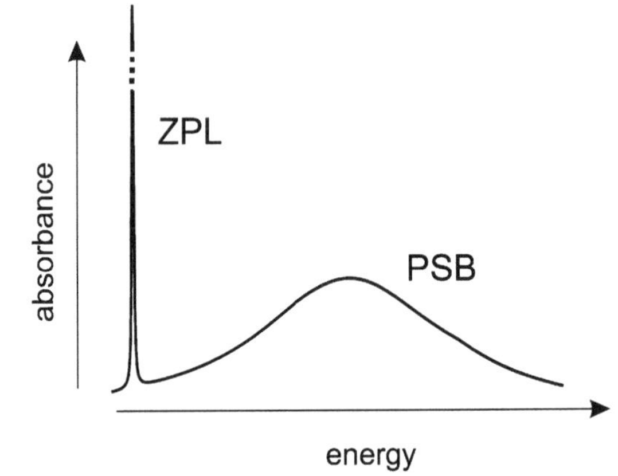


*Fig. 9.* Schematic representation of the absorption spectrum of a single molecule at low temperature. ZPL = zero-phonon line and PSB = phonon sideband. See text for more details.

The probability that a molecule will absorb an incident photon is proportional to $\sigma/A$, where $\sigma$ is the peak absorption cross-section of the molecule and $A$ the cross-sectional area of the focused laser beam. $\sigma$ can be described by the following relation (Rebane, 1994):

$$\sigma(T) = C_{FC}C_{DW}(T) \cdot \frac{T_2(T)}{2\tau_{rad}} \cdot \frac{\lambda_0^2}{2\pi} \cdot \beta(\theta) \tag{11}$$

where $\lambda_0$ is the wavelength at the maximum of the ZPL and $\beta(\theta)$ a geometrical factor accounting for the mutual orientation of the transition-dipole moment and the exciting field. $C_{FC}$ is the Franck-Condon factor of the 0-0 transition, which describes the distribution of the oscillator strength over the vibrational sidebands. $C_{DW}$ is the Debye-Waller factor, which describes the relative intensity of the ZPL compared to the phonon wing. $\tau_{rad}$ is the radiative lifetime, which is equal to $T_1/\phi_f$ with $\phi_f$ the fluorescence quantum yield.

To measure an optical spectrum of a single molecule a high signal-to-noise ratio of the emitted fluorescence is crucial, hence the peak absorption cross-section of the molecule should be high. To achieve this situation one should choose a molecule with a high radiative decay rate ($\tau_{rad}$ small). In addition $T_2$ should be long (Eq. 10). Therefore it is advantageous to work at low temperature (< 2 K) where $T_2^*$ becomes extremely long and consequently $T_2$ equals $2T_1$ (Eq. 10). Furthermore, the Franck-Condon factor

($C_{FC}$) should be large. This condition applies when the geometry of the chromophore is the same in both ground and excited state. Finally, the Debye-Waller factor ($C_{DW}$) should be as large as possible, best achieved at low temperature, which means a weak electron-phonon coupling.

Optical excitation typically involves transitions between singlet-states. Relaxation of the exited state proceeds via radiationless processes (internal conversion and intersystem crossing) and fluorescence or phosphorescence. Thus, there is a finite probability that the absorbed photon will not be emitted as fluorescence, but instead by a transfer (intersystem crossing) from the excited singlet state $S_1$ to the lower triplet state $T_1$. The lifetime of this state is usually much longer than the lifetime of $S_1$ and during this time the molecule will remain dark. Thus, the triplet state may act as a bottleneck in the excitation-relaxation cycle. The expression for the saturated fluorescence-emission rate, $R_\infty$, of a single molecule at low temperature (Ambrose et al., 1991; Plakhotnik et al., 1997) is given by:

$$R_\infty = \frac{k_f}{2+\left(k_{ISC}/k_T\right)}, \qquad \text{with } k_f = \frac{\phi_f}{T_1} \qquad (12)$$

To achieve an optimum emission of the single molecule the following remarks can now be made. First the radiative lifetime, $\tau_{rad}$ ($= 1/k_f$), should be short to ensure a high excitation turnover. Further the intersystem-crossing rate, $k_{ISC}$, should be slow and the triplet lifetime, $\tau_T$ ($= 1/k_T$), short to minimize the triplet bottleneck, i.e., the molecule should have a high fluorescence quantum yield, $\phi_f$. Finally, the photobleaching quantum yield should be small. This effect is usually much reduced at low temperature.

To estimate the maximum emission rate of a single LH2 complex the photophysical parameters of both, the BChl *a* molecules and the carotenoid molecules, should be considered. The latter molecules act as photoprotective agents, quenching the triplet excited states of the BChl *a* molecules and preventing photodestructive reactions in the presence of oxygen. The mechanism for this is the rapid transfer of the BChl *a* triplet excitation to the carotenoid molecule, reducing the lifetime of the BChl *a* triplet state by about two orders of magnitude (Monger et al., 1976; Cogdell et al., 1981) . For a recent review see Cogdell et al. (2000). The efficiency of this quenching is close to 100% (Bittl et al., 2001) and thus the triplet state of the carotenoid acts as the bottleneck in the excitation-relaxation cycle of this complex. The carotenoid triplet state is also known to be a very efficient trap for singlet excitations in the B850 ring (Monger and Parson, 1977; Jimenez et al., 1996), so during the lifetime of the carotenoid triplet no photons are emitted by the complex. We can now estimate the maximum emission rate $R_\infty$ for LH2 at low temperature by using the following parameters: fluorescence yield $\phi_f = 14$ %, fluorescence lifetime $T_1 = 1.25$ ns (Monshouwer et al., 1997), triplet yield $\phi_T = 5$ % (Cogdell et al., 1981) and carotenoid triplet lifetime $\tau_T = 5$ µs (Bittl et al., 2001). With Eq. 12 this gives a maximum emission rate of roughly 500,000 photons per second.

## B. Experimental

To observe a single pigment-protein complex two important requirements should be fulfilled. First of all, only one complex should exist within the excitation volume. Such a requirement can be met by creating a very small excitation volume and a low concentration of complexes. Secondly the fluorescence signal arising from the complex should be larger than the background signal. Since in general the fluorescence signals of a single complex are weak this puts severe demands on the experimental setup and the fluorescence yield of the complex.

Single-molecule spectroscopy is usually carried out using confocal-optical microscopy, a technique for increasing the contrast of microscope images (Pawley, 1995; Webb, 1996). The principle is schematically depicted in Fig. 10. Light from an exciting laser emanating from the excitation pinhole is focused on the sample by the objective. This objective forms an image of the excited volume on the pinhole in front of the detector. In the figure three point sources A, B and C are shown in the excitation volume. From the figure it will be clear that only source A will be properly focused on the detection pinhole, since only this point source is confocal with both the excitation and detection pinhole. The other sources B and C, will not be properly imaged (dashed and dotted lines) and will be attenuated compared to the confocal point A.

The ultimate spatial resolution of the microscope is limited by diffraction. From diffraction theory (Hecht, 1998) it is known that a point source produces a characteristic diffraction pattern in the image plane called, the Airy pattern. This pattern is also known as

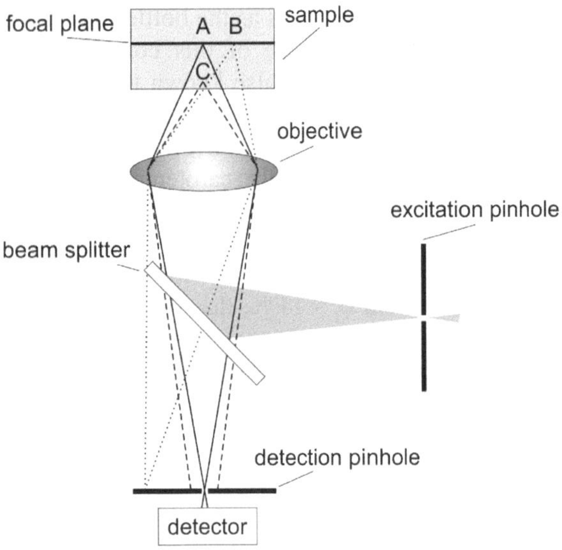


*Fig. 10.* Confocal principle. Excitation and detection path in an epitaxial alignment. of a confocal microscope with three point sources, A, B and C. See text for more details.

the point-spread function of the objective. The size of the Airy pattern depends on the magnification of the objective. Using the Rayleigh criteria, the maximal lateral resolution $r_{resol}$ is given by,

$$r_{resol} = 0.61\frac{\lambda}{NA} \tag{13}$$

here $\lambda$ denotes the wavelength of the light and $NA$ the numerical aperture of the objective.

For the setup as described in the next section the theoretical lateral resolution can be estimated using $\lambda$ = 800 nm and $NA$ = 0.9, which gives for the diameter of the diffraction limited spot of $r_{resel} \cong 540$ nm. If we compare this spatial resolution of 540 nm to the size of e.g., the individual LH2 complex of ~7 nm diameter, it is clear that in order to excite only one complex in our excitation volume we need a sample in which the different LH2 complexes are spatially well separated. This is achieved by using very low concentrations of protein solution to which a detergent is added to prevent aggregation of the complexes.

In Fig. 11, a schematic representation is given of the experimental arrangement used for the single-molecule experiments. With this arrangement both fluorescence microscopy and fluorescence-excitation spectroscopy can be performed. The sample, consisting of a polymer film with a low concentration of pigment-protein complexes on a LiF substrate, is mounted in the cryostat, rapidly cooled down to 1.2 K and illuminated with a continuous-wave, tunable Ti-Sapphire laser (Spectra Physics 3900S). A fluorescence-excitation spectrum of an individual light-harvesting complex is now obtained in two steps.

First a wide-field image is taken of the sample by slightly defocusing the output of the Ti-Sapphire laser onto the sample by inserting a long-focal length lens in the light path, creating a 100 μm large spot on the sample (Fig. 11), and simultaneously directing the fluorescence towards the CCD camera by inserting a mirror in the detection path. The complexes are excited around 800 nm for LH2 and LH3 and 870 nm for LH1-RC and emitted light is collected by an objective lens present in the cryostat close to the sample, and focused onto a CCD-camera (Princeton Instruments, Trenton, USA). The emission light is selected by using appropriate band-pass filters (Dr. Hugo Anders GmbH, Nabburg, Germany), which only transmit a ~20 nm (FWHM) spectral window. This filtering strongly reduces the probability of observing emission caused by impurities. On the lower left-hand side of Fig. 11 a wide-field image is depicted. The peaks represent diffraction-limited images of individual LH complexes, which are spatially well isolated from each other.

From the wide-field image one complex is selected. Next, a fluorescence-excitation spectrum of this complex is obtained by first removing the long-focal length lens and mirror out of the excitation and detection paths, respectively. The spectrum is acquired by scanning the excitation wavelength, while detecting fluorescence with a single-photon counting avalanche photodiode (EG&G, Quebec, Canada) (Fig. 11b). In this mode the excitation light is passed through the objective lens in the cryostat, illuminating a diffraction-limited excitation volume (less than 1 $\mu m^3$) of the sample. As an objective lens a commercial microscope objective ($NA$ = 0.9; $f$ = 2.45 mm (Microthek GmbH, Hamburg, Germany)) is used.

To select a complex from the wide-field image the scan mirror in the microscope is tilted such that the excitation volume exactly coincides with the position of a selected complex observed in the wide-field image. The scan mirror in combination with the pair of telecentric lenses provides a way to displace the focus laterally in a precise and well-controlled way while maintaining alignment with the confocal aperture. The fluorescence is collected by the same objective lens, and is focused on the confocally aligned

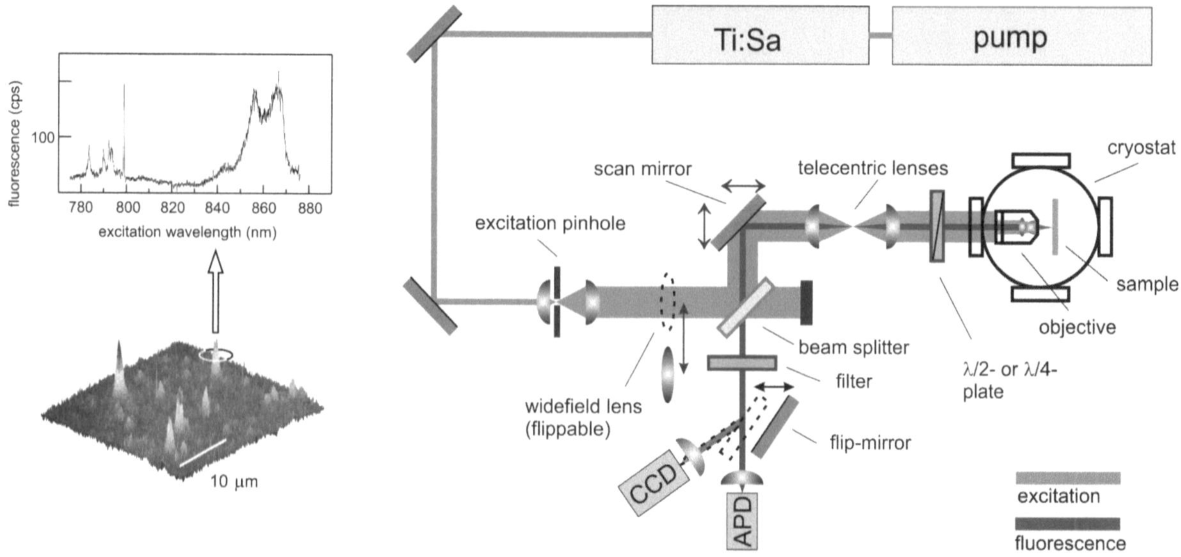


*Fig. 11.* Schematic representation of the experimental set-up used for the single-molecule experiments. The set up can be switched from confocal to wide-field mode by inserting a lens into the excitation beam and a mirror in the detection path redirecting the fluorescence towards the CCD camera. APD, avalanche photodiode. See text for more details.

avalanche photodiode. In the confocal-detection mode the superior background suppression allowed recording of fluorescence-excitation spectra with high signal-to-background ratios (Fig. 11, top-left) (van Oijen et al., 1999b).

To minimize light-induced fluctuations of the fluorescence intensity on a time scale of a single scan, the spectra are obtained by rapidly scanning the whole spectral range and storing the different traces separately. With a scan speed of the laser of 3 nm per second and an acquisition time of 10 ms per data point, this yields a nominal resolution of 0.5 $cm^{-1}$ ensuring that the spectral resolution is limited by the spectral bandwidth of the laser (1 $cm^{-1}$).

To examine the polarization dependence of the spectra, a ½λ plate can be mounted in the confocal excitation path in a computer controlled rotation stage.

## IV. Fluorescence-Detected Excitation Spectra

### A. The B850 Ring of LH2

The BChl *a* molecules in LH2 are arranged in two concentric rings labeled B800 and B850 according to their room temperature absorption maxima in the near infrared. For *Rps. acidophila* the B800 ring consists of nine well-separated (21 Å) BChl *a* molecules arranged in $C_9$ symmetry that have their bacteriochlorin plane perpendicular to the symmetry axis. The B850 ring consists of nine repeating dimer pairs of BChl *a* molecules with their molecular planes parallel to the symmetry axis resembling in their arrangement the blades of a turbine.

The different arrangement of the BChl *a* molecules in the two rings and the concomitant differences in the strength of the mutual interactions drastically influences the character of the electronically excited states. While the excitations in the B800 ring are mainly localized on individual chromophores those of the B850 ring are more appropriately described in terms of Frenkel excitons which are delocalized over a substantial part of the ring. For zero disorder the exciton manifold features two non-degenerate and eight pairwise degenerate states. The out-of-plane components of the monomer transition moments give rise to a small oscillator strength in the non-degenerate $k = 0$ exciton state whereas the in-plane transition moment is concentrated predominantly in the low-energy degenerate pair $k = \pm 1$. The two transition-dipole moments within the pair are of equal strength and oriented perpendicular with respect to each other. In the presence of energetic disorder, represented by $\Delta E_n$ and/or $\Delta V_{nm}$ in Eq. 4, the degeneracy of the exciton states is lifted and oscillator strength is redistributed among the exciton states. Typical spectra

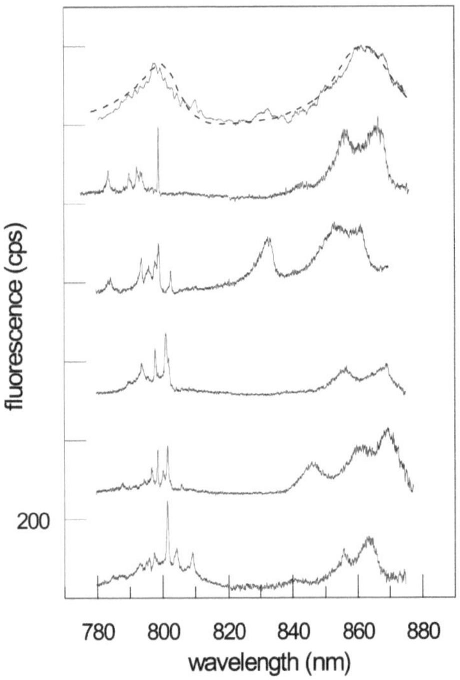


*Fig. 12.* Comparison of fluorescence-excitation spectra for an ensemble of LH2 complexes (top trace) and several individual LH2 complexes at 1.2 K. The vertical scale applies to the bottom spectrum; all other spectra are offset for clarity. cps stands for counts per second. The spectra were recorded with an excitation intensity of 20 W/cm².

are show in Fig. 12, which all show two prominent, orthogonally polarized absorption bands in the B850 region, which are assigned to the $k = \pm 1$ transitions. Occasionally a relatively narrow band — the $k = 0$ state — is observed at lower energy, but this state is in general difficult to detect because of the relatively low oscillator strength, and because it is affected by spectral diffusion (Ketelaars et al., 2001).

The influence of the various types of disorder on the B850 exciton states has been analyzed extensively. As the effect of random off-diagonal disorder is difficult to distinguish from that of random diagonal disorder (Fidder et al., 1991) many studies have focused on the latter (Sauer et al., 1996; Alden et al., 1997; Dempster et al., 2001). But also combinations of random diagonal and correlated off-diagonal disorder (Mostovoy and Knoester, 2000) as well as combinations of on- and off-diagonal disorder were investigated (Jang et al., 2001). Predictions based on these models differ with respect to subtle details in the spectra such as the extent and distribution of the energy separation between the $k = \pm 1$ states and the intensity ratio of the respective transitions. For a review of the theory of energy transfer and exciton interactions, see also Chapter 21, Renger and Holzwarth.

Employing single-molecule spectroscopy we have been able to verify the gross features of the random disorder model (van Oijen et al., 1999a; Ketelaars et al., 2001; Matsushita et al., 2001). The main effects that have been observed are the lifting of the degeneracy as well as an unequal intensity of the $k = \pm 1$ transitions. However, introducing only random diagonal disorder could not explain the data even if higher exciton states were included. Instead, correlated off-diagonal disorder of $C_2$ symmetry had to be introduced, which was taken into account as a regular modulation of the interaction strength in the Hamiltonian to explain the observed large splitting of the $k = \pm 1$ states (van Oijen et al., 1999a; Ketelaars et al., 2001; Matsushita et al., 2001). The dominant effect of such a modulation is a coupling between exciton states that differ in their quantum number by $\Delta k = \pm 2$. In combination with random diagonal disorder the model explained all experimental observations with reasonable accuracy: (i) the splitting of the $k = \pm 1$ states and the distribution of this parameter, (ii) the intensity ratio of the two $k = \pm 1$ transitions and (iii) their mutual orthogonal transition-dipole

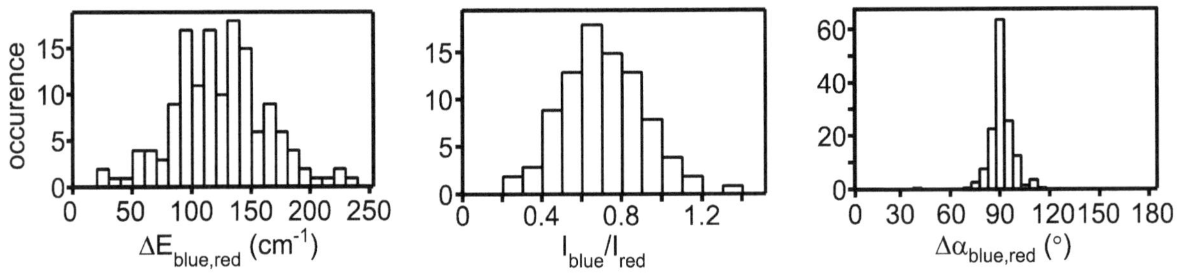


*Fig. 13.* Histograms of parameters of the $k = \pm 1$ states derived from spectra of single LH2 complexes. From left to right: energetic separation $\Delta E_{\text{blue,red}}$ for 144 complexes, intensity ratio $I_{\text{blue}}/I_{\text{red}}$ for 88 complexes, and relative orientation of the transition-dipole moments $\Delta\alpha_{\text{blue,red}}$ for 144 complexes.

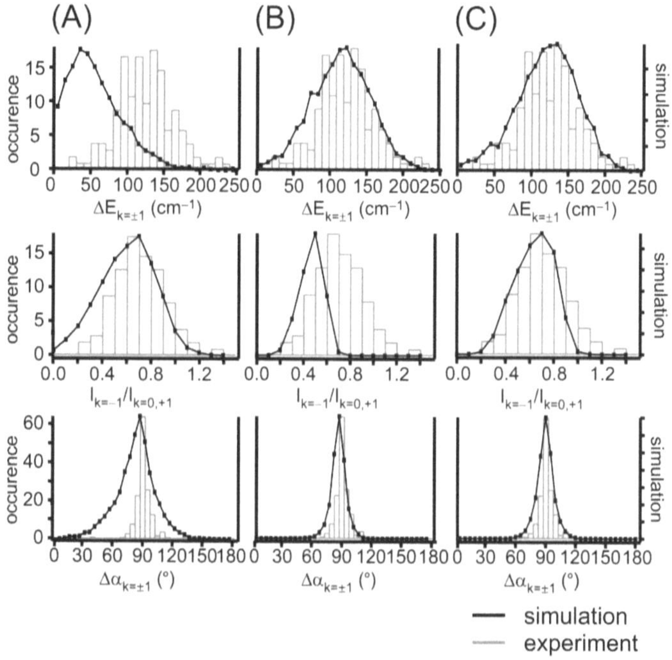


*Fig. 14.* Comparison of the experimental distributions (histograms) for the energetic separation $\Delta E_{blue,red}$ (top), the intensity ratio $I_{blue}/I_{red}$ (center), and the relative orientation of the transition dipole moments $\Delta\alpha_{blue,red}$ (bottom) of the two broad transitions in the B850 spectra of single LH2 complexes from *Rps. acidophila* for (A) random diagonal disorder, (B) random diagonal and correlated off-diagonal disorder, and (C) random and correlated diagonal disorder. The experimental data refer to the left vertical scale and the simulations refer to the right vertical scale. For more information see text and Table 1.

moments, and (iv) the occasional observation of a higher exciton state, presumably with $k = \pm 3$ (with much lower splitting).

A more elaborate study of the statistical distribution of spectra features allowed a more detailed analysis of the properties of the LH2 electronic transitions in terms of disorder (Hofmann et al., 2004), and to discriminate between the models of random and correlated disorder. The results are summarized in Fig. 13. From the experimental data we obtained the energetic separation of the $k = \pm 1$ states, $\Delta E_{blue,red}$, the ratio of the integrated intensities $I_{blue}/I_{red}$, and the relative orientation of the transition-dipole moments $\Delta\alpha_{blue,red}$, where blue (red) refers to the energetically higher (lower) absorption band of the $k = \pm 1$ states. The distributions of these parameters are shown in the histograms of Fig. 13 from left to right. The distribution of the energetic separation $\Delta E_{blue,red}$ of the $k = \pm 1$ states is centered at 126 cm$^{-1}$ and features a width of 101 cm$^{-1}$ (FWHM). Similarly, we find for the centre value (FWHM) of the integrated intensity ratio $I_{blue}/I_{red}$ 0.73 (0.54), and for the mutual orientation of the transition-dipole moments $\Delta\alpha_{blue,red}$ is 91° (19°), respectively.

In numerical simulations of the experimental data in Fig. 13 we compared three different types of disorder: (A) random diagonal disorder, (B) random diagonal disorder together with correlated off-diagonal disorder, and (C) random and correlated diagonal disorder. The simulations are based on Monte Carlo calculations averaged over 5000 iterations. In Fig. 14 the results for the three approaches are shown together with the experimental data.

In Fig. 14a we show the histograms that result for $\Delta = 420$ cm$^{-1}$, a value for the Gaussian distribution of site energies (FWHM) from the literature (Wu and Small, 1997) that was based on simulation of the B850 ensemble absorption spectrum. The magnitude of the $k = \pm 1$ splitting in this model is about half of the experimentally observed value. To reproduce the observed splitting of 126 cm$^{-1}$ by only random diagonal disorder requires a value of $\Delta = 950$ cm$^{-1}$ which is difficult to reconcile with all the spectral features of the LH2 complex.

The large splitting of $k = \pm 1$ can be well explained by a reduced symmetry of the B850 ring. In the case of $C_2$ symmetry of the ring, it can be shown, on the basis of symmetry arguments, that only the $k = \pm 1$

exciton states will be split. This is consistent with the absence in the spectra of a splitting and a polarization effect of the higher exciton state that is occasionally observed. The eigenfunctions of the $k = \pm 1$ states reflect the two symmetry axes of the $C_2$ point group, and hence exhibit orthogonal polarization of their transition moments, as observed experimentally. A possible explanation for such a reduced symmetry is an elliptical deformation of the ring. The effect of reduced symmetry can be taken into account by a modulation of either the interaction energies or the site energies along the ring.

The combination of random diagonal disorder with correlated off-diagonal disorder of $C_2$ symmetry. is taken into account by introducing an additional shift $\delta V_{nm}$ which varies with the angular position according to

$$\delta V_{nm} = V_{\mathrm{mod}} \cos\left(2\varphi\left(n + \tfrac{1}{2}\right)\right)\delta(n-m) + W_{\mathrm{mod}} \cos\left(2\varphi\left(n+1\right)\right)\delta(n-m-1) \qquad (14)$$

The modulation of the nearest- and second-nearest-neighbor interactions is represented by $V_{\mathrm{mod}}$ and $W_{\mathrm{mod}}$ respectively, where the δ-symbols on the right-hand side represent the Kronecker-function, $\varphi = 2\pi/18$, and $n$ varies from 1 to 18, the latter being the total number of pigments in the ring. The result of this approach is compared in Fig. 14b with the experimental histograms. The simulations are in reasonable agreement with the experiments and the model explains the average splitting of the $k = \pm 1$ states, the width of the distribution of this parameter, and the distribution for the orientation of the $k = \pm 1$ transition-dipole moments. However, for the intensity ratio of the $k = \pm 1$ transitions this model does neither explain the centre value nor the width of its distribution.

The alternative is to include, both, random and diagonal disorder in the distribution of site energies. This is achieved by defining $\delta E_n = \Delta E_n + E_{\mathrm{mod}}\cos(2\varphi(n+1/2))$ and setting $\Delta V_{nm} = 0$. As before $\Delta E_n$ refers to the random diagonal disorder chosen from a Gaussian distribution whereas the second term introduces the correlated disorder of $C_2$ symmetry into the site energies of the pigments. For this model the energetic splittings of the $k = \pm 1$ states and the width of its distribution could best be reproduced by assuming a width of $\Delta = 250\ \mathrm{cm}^{-1}$ for the Gaussian distribution of diagonal random disorder and a modulation amplitude $E_{\mathrm{mod}} = 180\ \mathrm{cm}^{-1}$. As can be seen in Fig. 14c this model is in agreement with all experimental observations. In table 3 we have summarized the results of the experiments and of all simulations.

Several factors may contribute to the distribution and modulation of site energies. Any deviation from planarity and/or reorientation of side groups results in spectral shifts of up to 500 cm$^{-1}$ (Gudovska-Novak et al., 1990). Moreover, the energy of the electronically excited states is sensitive to electrostatic interactions within the binding pocket of the chromophore. Modifying hydrogen bonding of the BChl *a* results in a shift of the electronic excitation energy by 150 cm$^{-1}$ (340 cm$^{-1}$) for a single (double) mutation (Fowler et al., 1992). These numbers are commensurate with

*Table 3.* Summary of the experimentally (experiment) and numerically (simulation) obtained values for the energy separation of the two $k = \pm 1$ transitions $\Delta E_{k=\pm 1}$, the intensity ratio $I_{k=-1}/I_{k=0,+1}$ and the mutual angle between the $k = \pm 1$ transition-dipole moments $\Delta\alpha_{k=\pm 1}$. For all values the average of the distribution is given together with the width (FWHM) in parentheses. The simulations were performed by taking into account three different types of disorder: (A) random diagonal disorder, (B) random diagonal and correlated off-diagonal disorder, (C) random and correlated diagonal disorder. The values for the parameters $\Delta$, $V_{\mathrm{mod}}$, $E_{\mathrm{mod}}$ and $\delta r/r_0$ are given below.

| | | simulation | | |
|---|---|---|---|---|
| parameter | experiment | A | B | C |
| $\Delta E_{k=\pm 1}$ (cm$^{-1}$) | 126 (101) | 122 (210) | 122 (97) | 126 (99) |
| $I_{k=-1}/I_{k=0,+1}$ | 0.73 (0.54) | 0.53 (0.52) | 0.52 (0.24) | 0.69 (0.38) |
| $\Delta\alpha_{k=\pm 1}$ (°) | 91 (19) | 73 (83) | 89 (19) | 90 (21) |
| $\Delta$ (cm$^{-1}$) | | 950 | 250 | 250 |
| $E_{\mathrm{mod}}$ (cm$^{-1}$) | | 0 | 0 | 180 |
| $V_{\mathrm{mod}}$ (cm$^{-1}$) | | 0 | 75 | 0 |
| $\delta r/r_0$ (%) | | 0 | 10 | 0 |

the range of parameters that were deduced from the single-LH2 data.

### B. The B800 Ring of LH2

As can be seen in Fig. 12, the B800 band of an individual LH2 complex consists of several relatively narrow spectral lines. As mentioned above, the dipolar coupling between the BChl *a* molecules in the B800 ring is predicted to be small with respect to the variation in site energy and it is expected that the excitation energy is mainly localized on individual B800 BChl *a*. This conjecture is corroborated by the strong dependence of the relative intensities of these lines on the polarization of the incident radiation (van Oijen et al., 2000). This is what one would expect if the excitations are more or less localized on individual BChl *a* molecules because their $Q_y$ transition dipole moments are arranged in a circular manner and, as a result of this, all have different orientations. From a more detailed examination of the B800 spectra it was found that typically only 6–7 absorption lines could be discerned, supporting the idea that occasionally delocalization of the excited state occurs between neighboring BChl *a* molecules in the B800 manifold and a concomitant redistribution of oscillator strength. This would be particularly significant in the case of (near) degeneracy of adjacent pigments.

#### 1. B800 Disorder

In general we have to consider two independent contributions to the spectral distribution of B800 transitions. The first one is the variation in site energies of BChl *a* molecules within the same LH2 complex which is referred to as intracomplex heterogeneity or diagonal disorder. For different complexes, the changes in the spectral position of the center of mass of the whole spectrum, i.e., intercomplex heterogeneity or sample inhomogeneity, has to be taken into account. Obviously, the study of individual LH2 complexes allows to discriminate between these two contributions and to study them separately.

In order to find a measure for the intercomplex heterogeneity we have defined the spectral mean value, $\bar{\nu}$, of the fluorescence-excitation spectrum of a single LH2 complex by

$$\bar{\nu} = \frac{\sum_i I(i) \cdot \nu(i)}{\sum_i I(i)} \tag{15}$$

where $I(i)$ denotes the fluorescence intensity at data point $i$, $\nu(i)$ the spectral position corresponding to datapoint $i$, and the sum runs over all data points of the spectrum. The respective histogram of $\bar{\nu}$, obtained from the spectra of 46 complexes is depicted in Fig.15a and has a width of about 120 cm$^{-1}$.

The intracomplex heterogeneity or diagonal disorder is extracted from the data by calculating the standard deviations $\sigma_\nu$ of the intensity distributions in the individual spectra

$$\sigma_\nu = [\overline{\nu^2} - \bar{\nu}^2]^{1/2} \tag{16}$$

where $\bar{\nu}^2$, is given by

$$\overline{\nu^2} = \frac{\sum_i I(i) \cdot [\nu(i)]^2}{\sum_i I(i)} \tag{17}$$

The result is shown in Fig. 15b. The distribution for $\sigma_\nu$ is centered at a value of about 55 cm$^{-1}$. To obtain the full width at half maximum of the distribution of site energies this value has to be multiplied by a factor of 2.36 to obtain a value of 130 cm$^{-1}$ for the

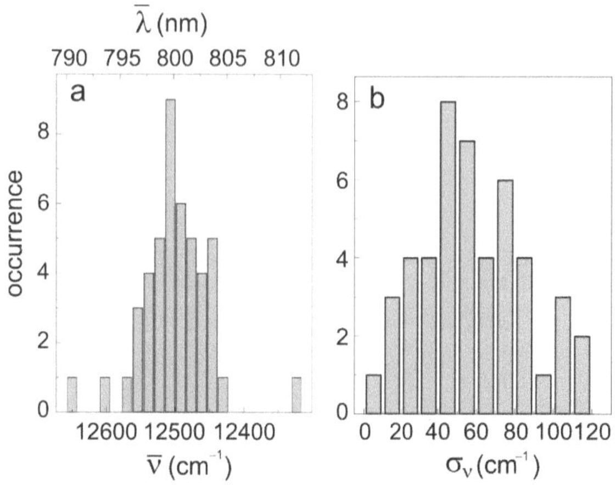


*Fig. 15.* (a) Distribution of the spectral mean of the B800 fluorescence-excitation spectrum for 46 LH2 complexes from *Rps. acidophila* featuring the amount of intercomplex heterogeneity. (b) Distribution of the standard deviations for the spread of absorption lines in the individual fluorescence-excitation spectra for the same 46 LH2 complexes.

diagonal disorder. Clearly, an ensemble spectrum reflects the convolution of both contributions to the heterogeneity. From our data we expect for the B800 band a total inhomogeneous linewidth of about 180 $cm^{-1}$, in excellent agreement with the results from bulk spectra of LH2 of *Rps. acidophila* taken at 1.2 K. It is worth pointing out that only the intracomplex disorder is of relevance for the delocalization of excitation energy.

## 2. B800 Homogeneous Line Shape

In principle, single-molecule spectra will record the homogeneous line shape, usually characterized by a narrow zero-phonon line (ZPL) and a phonon sideband (PSB). However, it was found that the B800 spectra are subjected to spectral diffusion, i.e., fluctuations of the peak positions on a time scale of seconds. The extent of spectral diffusion depended on the laser intensity, indicating that the effect is light-induced. It results in temporal averaging of the spectrum over the time scale of the measurement, by which the relevant details are lost. To circumvent this problem, fluorescence-excitation spectra were recorded by scanning the laser wavelength at a speed of 3 nm per second while maintaining a spectral resolution of 1 $cm^{-1}$. The upper panel on the left-hand side of Fig. 16 shows the result of a sequence of 200 of such fast scans with an excitation intensity of 20 $W/cm^2$ (left) and 80 $W/cm^2$ (right), with the time-averages in the panels below. Each horizontal row of pixels represents a single spectrum in which the fluorescence intensity is grey coded. Successive spectra are displaced vertically along the time axis. As is evident from the collection of fast scans, sudden spectral jumps of the absorptions occur and the averaged fluorescence-excitation spectrum shows significant broadenings of the spectral lines due to the spectral diffusion effects. We think that the spectral diffusion is caused by internal conversion, B800-B850 energy transfer and intersystem crossing, and subsequent dissipation of vibrational energy in the complex. This

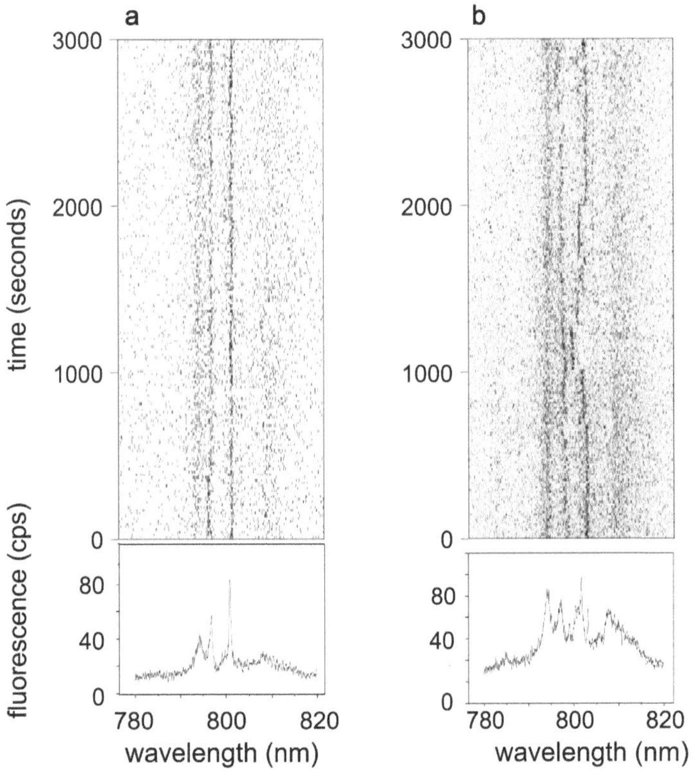


*Fig. 16.* (a) Stack of 200 fluorescence-excitation scans through the B800 band recorded at a scan speed of 3 nm/s and an excitation intensity of 20 $W/cm^2$ (upper panel). (b) Stack of 200 fluorescence-excitation scans recorded at the same scan speed with 80 $W/cm^2$ (upper panel). In both cases the fluorescence intensity is indicated by the gray code and the average of the 200 spectra is displayed in the lower panel.

energy is dumped as heat in the protein surrounding of the pigments inducing conformational changes which in turn give rise to changes of the absorption frequencies of the chromophores.

By fitting all B800-transition in every single sweep with a Lorentzian, the spectral position of each transition was determined as a function of time. Analysis of the statistics of the fluctuations shows that the spectral motion of a particular absorption appears to be correlated with its spectral position within the B800 band (van Oijen et al., 2000). The spectral diffusion increases towards the wings of the spectral distribution of absorptions. Apparently, the probability for conformational changes in the immediate environment is larger for pigments which show absorption frequencies with large deviations from the mean.

The homogeneous line shape of the ZPL could be recovered (i) by Lorentzian fitting of all transitions in each scan, (ii) followed by alignment of the fitted peak positions in subsequent scans by shifting their spectral positions, and finally (iii) averaging the shifted raw spectra. Practically, this procedure corrects for the effects of spectral diffusion at time scales longer than 50 ms as determined by the scan speed of the laser. The dependence of the ZPL linewidth and the emission rate for a particular B800 transition on the excitation power reflected the saturation behavior of two-level systems (Ambrose et al., 1991). The homogeneous linewidth of a particular absorption shows a strong dependence on its spectral position within the B800 band, as shown in Fig. 17. To exclude intercomplex heterogeneity the inhomogeneous linewidth is plotted as a function of the spectral separation from the spectral mean rather than as a function of the absolute spectral position. The homogeneous linewidth decreases from about 10 $cm^{-1}$ on the blue side of the B800 band to less than 2 $cm^{-1}$ in the center and to the red side of this band. This observation is in agreement with results from hole-burning and transient absorption studies (De Caro et al., 1994; Joo et al., 1996; Kennis et al. 1997), and is indicative of energy transfer from energetically higher B800 states to lower-lying states within the B800 band.

A more elaborate method, multivariate statistical analysis with pattern recognition (van Heel et al., 2000), was applied by Hofmann et al. (2005) in the analysis of the B800 spectra of individual LH2 complexes from *Rsp. molischianum*. The statistical analysis of thousands of spectra from a single LH2

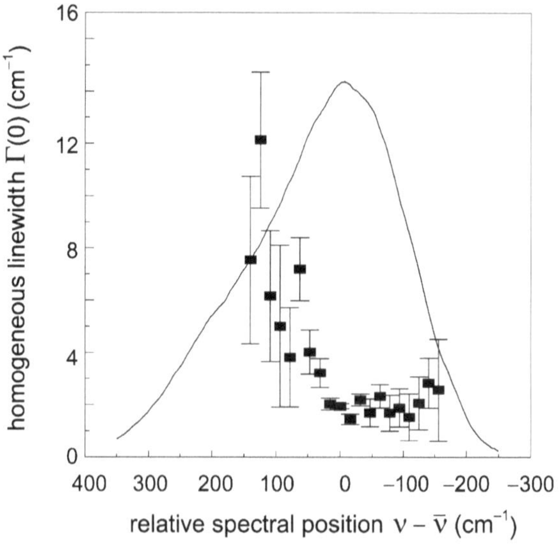


*Fig. 17.* The dependence of the homogeneous linewidth of the observed absorption lines on the spectral position in the B800 band with respect to the spectral mean. The spectra were recorded with an excitation intensity of 20 W/$cm^2$. The solid line is the bulk absorption spectrum of the B800 ring in LH2 from *Rps. acidophila*.

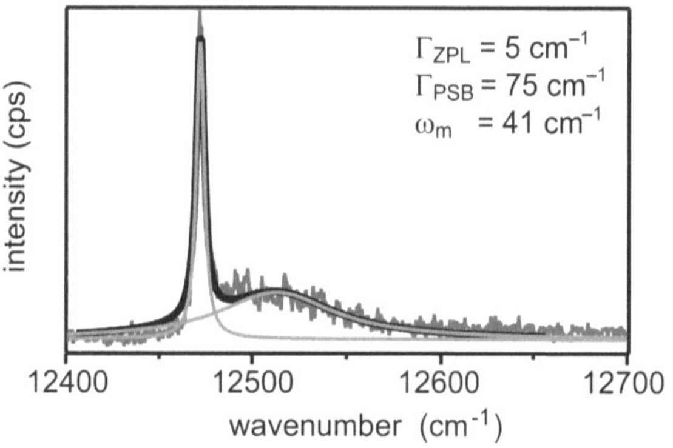


*Fig. 18.* The dominant homogeneous line shape by multivariate statistical analysis with pattern recognition (van Heel et al., 2000) of a large number of single-complex spectra of LH2 from *Rps. acidophila* consists of a two-fold Lorentzian. The experimental data is shown by the noisy grey line, whereas the fitted data is shown in black. The black line is the sum of the two Lorentzians curves that are drawn as smooth gray lines.

complex uncovered the full homogeneous line shape, including the phonon sideband. The result is shown in Fig. 18. The sharp feature is associated with the ZPL of an individual BChl *a* molecule, the broad wing towards higher energy with the PSB. The electron phonon coupling is usually described by the Huang-Rhys factors, $S$, defined as $e^{-S} = I_{ZPL}/(I_{ZPL} + I_{PSB})$, where $I_{ZPL}$ and $I_{PSB}$ refer to the integrated intensity of the

ZPL and PSB, respectively. For the B800 molecules in *Rsp. molischianum* a value of $S = 0.44$ was determined for the Huang-Rhys factor, which reflects a weak electron-phonon coupling strength.

### 3. B800 Intra-Band Energy Transfer

The intermolecular interaction between nearest neighbors in the B800 ring, about $-24$ cm$^{-1}$, is an order of magnitude smaller than in the B850 ring. Therefore, since $V_{n,n+1} < \Delta$, the excitations of the B800 system are often assumed to be largely localized. To examine this in more detail, a detailed analysis of the polarization dependence of the bands in the B800 spectrum of *Rps. molischianum* was performed (Hofmann et al., 2003b). The LH2 from *Rps. molischianum* differs from that of *Rps. acidophila* in that it consists of 8 αβ-subunits in stead of 9 (Koepke et al., 1996). This implies that, if the excitations are completely localized in the 8-membered ring, the relative polarization of the optical transitions would be at multiples of 45°. Fig. 19 shows a histogram of relative polarization angles of the B800 transitions determined experimentally from about 500 individual LH2 complexes of *Rps. molischianum*. What is actually observed are maxima at intervals of 22.5°. It can be shown (Hofmann et al., 2003b) that the intermittent angles of 22.5° and 67.5° are obtained if two adjacent B800 molecules are accidentally degenerate. In other words, these features in the polarization distribution are evidence that exciton coherence in the B800 ring cannot be neglected. The probability of (near) degeneracy of more than two B800 pigments is relatively small. Hence, the B800 systems should be considered as a collection of monomers and (nearly) degenerate dimers.

Earlier it was suggested that intra-band energy transfer from the energetically higher B800 states to lower-lying B800 molecules is driven by a Förster-like process (Joo et al., 1996; Kennis et al., 1997) because the dipole-dipole interaction between neighboring pigments is small compared to their difference in transition energy. The energy-transfer rates in the Förster picture are then determined by the spectral overlap between donor and acceptor molecules. However, the electron-phonon coupling is rather weak and the zero-phonon lines are spectrally distributed over a region about 25 times their homogeneous linewidth, both observations implying small spectral overlap between donor and acceptor. Wu et al. (1996) pointed out that under such conditions Förster transfer would occur with a transfer time of tens of picoseconds, i.e., much slower than that derived from homogeneous linewidth measurements. This leads to the conclusion that Förster processes cannot explain the observed B800 intraband energy transfer. Moreover, a comparison of the average difference in nearest neighbor site energies of 30–60 cm$^{-1}$ (van Oijen et al., 2000) with the interaction strength of about $-24$ cm$^{-1}$ (point-dipole approximation), shows that the dipole-dipole coupling between neighboring BChl *a* molecules in the B800 ring is not necessarily weak.

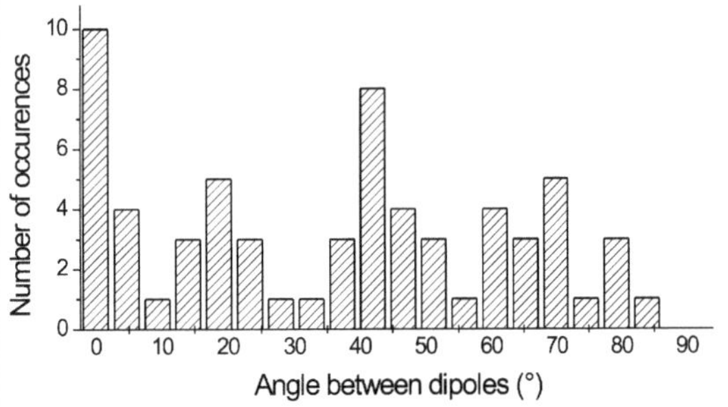


*Fig. 19.* Histogram of the relative orientations of the transition dipole moments from 88 absorption lines in 24 individual complexes in *Rps. molischianum*.

From experiments and modeling on the Fenna-Matthews-Olson (FMO) antenna complex from green sulfur bacteria it is known that similar ratios of the diagonal disorder and the interaction strength lead to delocalization of the excitation over 2–3 pigments (Vulto et al., 1999b; Brixner et al., 2005). In this situation it has been shown that the dynamical properties of the excited states are governed by exciton-phonon interactions and that the lifetimes of individual levels are determined by vibronic relaxation to the lower states in the exciton manifold. We believe that a similar description applies to the dynamic properties of the levels in the B800 band of the LH2 complex, i.e., that the observed increase in the homogeneous linewidth towards the high-energy side of the B800 absorption band is caused by vibronic relaxation between (mostly dimer) exciton states rather than a Förster-type energy transfer. Of course, this process competes with energy transfer from B800 to B850, both of which occur on the same time scale of about 1 ps.

This conjecture was substantiated in a recent study of quantum coherence in the B800-ring by Cheng and Silbey (2006) taking into account molecule-bath interactions. Their results confirm that B800 intraband energy transfer is due to phonon-induced relaxation in the exciton manifold, and that phonon induced fluctuations in the interaction strength play a major role in the process. The experimentally observed energy transfer rate from B800 to B850 is reproduced in a simplified model (see also Jang et al., 2007). A slight delocalization of the excitation energy in the B800 ring appears to improve the robustness of the B800 $\rightarrow$ B850 energy transfer.

### C. Other Light-Harvesting (LH) Complexes

#### 1. LH3

Some strains of *Rps. acidophila* are capable of synthesizing multiple types of light-harvesting complexes depending on growth conditions (Angerhofer et al., 1986; Gardiner et al., 1993). In strain 7750 almost all LH2 complexes are replaced by LH3 when grown at low-temperature and low-light conditions. Both, LH2 and LH3, have a major absorption band at 800 nm but differ in the position of their long-wavelength band, which is at around 850 nm for LH2 and 820 nm for LH3. Despite the spectral differences, LH2 and LH3 have essentially identical structures (McDermott et al., 1995; McLuskey et al., 2001), especially with regard to the arrangement of the BChl *a* molecules.

The fluorescence-excitation spectra of individual LH3 complexes largely resemble those of the individual LH2 complexes (de Ruijer et al., 2006). The differences in spectral features between the 800 and 820 nm regions of the spectra, the number of bands, the bandwidths and the typical polarization behavior confirm the similarity of the two types of antenna complexes. A detailed analysis of the spectra (de Ruijer et al., 2006) however reveals various interesting aspects, which were not or to a lesser extent observed in the LH2 experiments. First of all, 40% of the spectra show a very heterogeneous spectral behavior in the B820 region, which prohibits an unambiguous assignment of the observed absorption bands. The proximity of the B800 and B820 bands leads to spectral congestion which further complicates the assignment. Secondly, a significant structure in both the 800 and 820 nm region of the sum spectrum is observed. Finally, large spectral diffusion effects in both the B800 band and the B820 band are observed.

For only 60% of the complexes we can identify the $k = \pm 1$ transitions in the B820 region, i.e., two broad, orthogonally polarized bands. The splitting of these bands is centered at 160 $cm^{-1}$ and has a distribution of which is significantly broader than that of the LH2 (Hofmann et al., 2004).

The B820 fluorescence-excitation spectra of individual LH3 complexes thus exhibit much more spectral heterogeneity than those of LH2. In additions, spectral diffusion appears to be more prominent in the case of LH3. This indicates a different sensitivity and/or stability of the LH3 complexes compared to LH2, at least under single-molecule experimental conditions.

The observed heterogeneous spectral behavior, especially in the B820 band, indicates that the B820 pigments of LH3 are sensitive to pronounced, light-induced local conformational changes. The main difference between LH2 and LH3 is that the C-3-acetyl group of the BChl *a* pigment bound to the β-subunits of LH3 is not hydrogen-bonded, unlike LH2. This presumably results in an enhanced freedom of rotation of the C-3-acetyl group of these BChl *a* molecules in LH3. Gudowska-Nowak et al. (1990) have shown that rotation of the C-3-acetyl group of BChl *a* can induce changes in the site energy of the pigment of up to 25 nm (~360 $cm^{-1}$). Therefore, it is suggested that a (light-induced) rotation of the C-3-acetyl chain of a BChl *a* pigment bound to the β-subunit is the origin of the conformational flexibility in LH3, affecting the optical properties of the whole pigment-protein complex.

#### 2. Low-Light LH2 (LH4)

The low-light adapted B800 light harvesting complex 4 (LH4, also known as LL-LH2) from *Rhodopseudomonas palustris* is a complex in which the arrangement of the BChl *a* pigments seems to be very different from the well-known LH2 complex of other purple bacteria. The structural model of LH4 is based on an electron density map at 7.5 Å resolution (Hartigan et al., 2002), and accounts for the near infrared absorption and CD spectra and which was also consistent with the electron density map. Main differences with LH2 are: (i) LH4 is octameric, with 4 pigments per subunit, (ii) it has a more radial orientation of the pigments in the densely packed B-α/B-β ring (the equivalent of B850 in LH2), (iii) the model contains two B800-type rings, resulting in a tighter packing of BChl *a*, (iv) intra-subunit interactions are larger than the inter-subunit interactions. Because of the relatively low resolution of the electron density map there is still room for discussion about the validity of

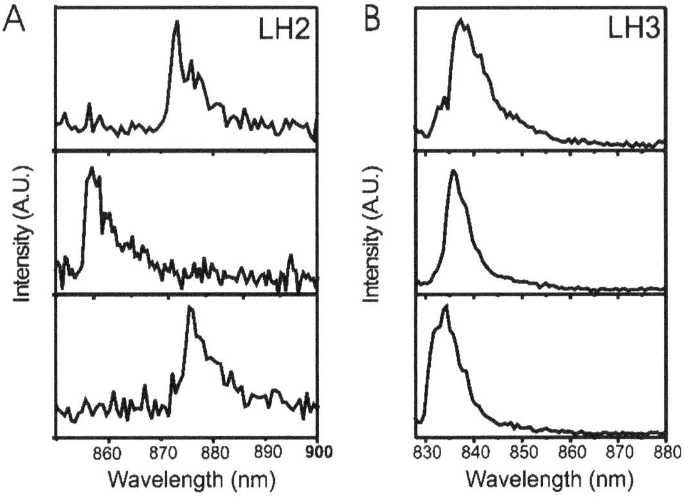


*Fig. 20.* Low-temperature (1.2 K) fluorescence-emission spectra of (A) individual LH2 complexes and (B) individual LH3 complexes. The spectra have been normalized at the maximum of the fluorescence (de Ruijter, 2005).

the proposed structural model of this complex.

Optical properties predicted by this model were tested by single-molecule spectroscopy, especially in comparison with LH2 (de Ruijter et al., 2004). There is reasonable agreement with the model concerning the number of bands observed in the spectrum, the widths of the bands (which are narrower than those of the LH2 complex as a result of the smaller inter-subunit interaction strengths), and the polarization characteristics of the observed transitions. It confirms that the excited state properties of LH4 are unlike that of other bacterial LH complex studied so far. For more details we refer to de Ruijter et al. (2004).

## V. Fluorescence Spectra

In this section we address the fluorescence-emission spectra of single LH2 and LH3 complexes. Downward relaxation in the exciton manifold of these complexes typically occurs on a picosecond time scale (Bergström et al., 1988; Trautman et al., 1990; Ma et al., 1998) . Hence, in these systems the fluorescent state, $k = 0$, differs from the states that determine the absorption spectra. Therefore, the emission spectra provide information that is not readily accessible in fluorescence detected absorption spectra.

Fluorescence-emission spectra of individual LH2 and LH3 complexes were measured at low temperature (1.2 K) and typical spectra are displayed in Fig. 20. The majority of the LH2 (Fig. 20A) and LH3 spectra (Fig. 20B) exhibit a single band with a

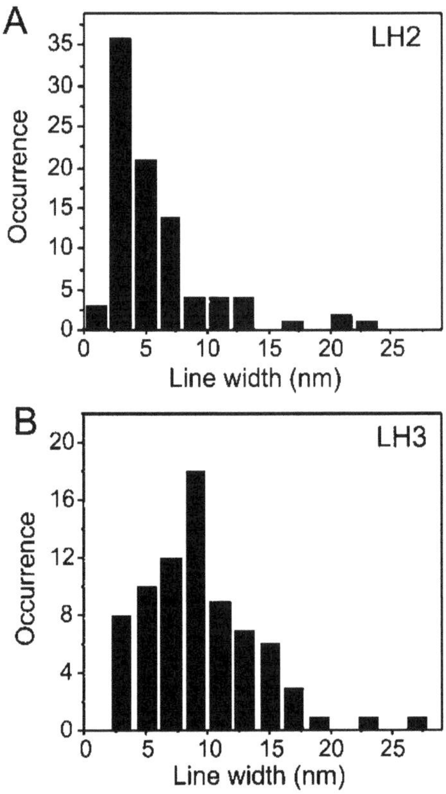


*Fig. 21.* Histogram of the FWHM of the emission bands of (A) 90 individual LH2 complexes and (B) 76 individual LH3 complexes. In total 124 LH2 and 100 LH3 complexes were recorded, but the analysis was only performed on spectra with a sufficiently high signal-to-noise ratio (de Ruijter, 2005).

relatively steep rise on the blue edge and a shoulder on the red side to the fluorescence band. For LH3, we could very occasionally observe two spectrally separated bands (spectra not shown). The full width at half maximum (FWHM) of the analyzed spectra is shown in Fig. 21.

The most striking feature of the current results is the large FWHM of the bands in the LH2 and LH3 spectra (Fig. 21). The shape of the fluorescence bands is presumably governed by a superposition of a very narrow zero-phonon line (ZPL) and a broad phonon sideband (PSB). The average width of the experimental spectra, however, is much broader than that of a lifetime-limited spectrum, taking into account the fluorescence lifetime of about 1 ns for LH2 and LH3 (Monshouwer et al, 1997). We conclude that the linewidth of the spectra is determined by light-induced spectral diffusion of the $k = 0$ exciton state. This conclusion is in line with the earlier assumption by Ketelaars et al. (2001) that spectral diffusion hinders the observation of the $k = 0$ state in fluorescence-detected absorption spectra of individual LH2 complexes, corroborated by the observation of light-induced spectral jumps in these spectra (Hofmann et al., 2003a; Ketelaars et al., 2006).

It is known that excited state relaxation in LH2, and probably also in LH3, is dominated by radiationless relaxation, since the fluorescence yield of LH2 is only 14% (Monshouwer et al., 1997). Therefore, most absorbed photons are dissipated in vibrational modes of the system, which is essentially a local heating effect of the B850 and B820 rings. Note that energy dissipation in the B800 ring is negligible because of the short lifetime of its excited state. The local heating effect has been described earlier by Neu and Silbey (1997) in an analysis of spectral diffusion measured in stimulated photon echo experiments by Wiersma and co-workers (Thorn Leeson et al., 1994; Thorn Leeson and Wiersma, 1995). Spectral diffusion then occurs when the vibrationally excited state relaxes to another nuclear configuration, presumably caused by instabilities in the conformation of the pigment binding pocket or its direct environment. This effect is exacerbated by delocalization of the excitations in the B850 and B820 rings, respectively, which introduces a sensitivity of the exciton transitions to site energy changes of each individual pigment in the ring. We note that large spectral jumps in fluorescence spectra in single LH2's at room temperature have been interpreted as coupling between electronic transitions and slow nuclear motions (Rutkauskas et al., 2004, 2005). However, this mechanism does not apply at the low temperature in the present experiments on LH2 and LH3, where the spectral dynamics are strictly light-induced.

The LH3 spectra are clearly broader than the LH2 spectra, which suggests an additional mechanism, distinct from LH2. This is likely related to the lager heterogeneity in the absorption spectra of LH3 (see section IV.C.1), and is likely related to the fact that the C-3-acetyl group of the βB820 pigment in LH3 has increased rotational freedom due to the lack of an H-bond (McLuskey et al., 2001). We suggest that the C-3-acetyl group can assume different out-of-plane angular orientations due to the dissipation of excitation energy, and that this phenomenon is the determining factor for the large FWHM of the LH3 fluorescence spectra as compared to LH2 spectra.

It was also found that the Stokes shifts of the ensemble spectra in glycerol/$H_2O$ is substantially larger than the sum-spectra of the single-molecules embedded in PVA, and is more pronounced for LH3 than for LH2 (de Ruijter, 2005). The ensemble fluorescence-emission spectrum of LH2 has a maximum at 890 nm. The sum spectrum, however, that was obtained from averaging all single-molecule emission spectra that were measured in a PVA film at 1.2 K, shows a maximum at 874 nm. For LH3 the maximum of the ensemble spectrum is at 878 nm, while that of the sum spectrum is at about 838 nm.

Simulations of the spectral distribution of the position of the $k = 0$ state could reproduce the difference in the Stokes shift by only varying the degree of static disorder (de Ruijter et al., 2007). For LH2 and LH3 in a PVA matrix the diagonal disorder parameter (in terms of the width of a Gaussian distribution of site energies) was in good agreement with the values derived from the absorption characteristics of individual complexes. For LH2 (LH3) a FWHM of the diagonal disorder of 425 (610) $cm^{-1}$ was required to reproduce the maximum of the ensemble spectra in the glycerol/$H_2O$ environment. It is well established that LH2 in solution is "dressed" by a layer of detergent molecules, especially the hydrophobic domains (Prince et al., 2003). Possibly this layer and/or the effect that it has on the conformational sub-states of LH2 is affected by the PVA matrix used in the single-complex measurements. It appears from the simulations that the spectral red-shift of the fluorescence maximum is a sensitive indicator for the degree of disorder.

These results explain some differences in the literature concerning the values of the disorder parameters of LH2. In particular, various (bulk) experiments have reported distributions of LH2 site energies with a width of 350–450 $cm^{-1}$, based on fluorescence quantum yield measurements (Monshouwer et al., 1997), 3-pulse echo peak shift measurements (Jimenez et al., 1997), and the analysis of absorbance difference spectra (Chachisvilis et al., 1997; Novoderezhkin et al., 1999). This value was at variance with the results from single-complex measurements on LH2 (van Oijen et al., 1999a). Apparently the disorder is reduced when LH2 is embedded in a PVA matrix, under conditions of the single-complex measurements. It shows that the solvent matrix may have a significant effect on the spectroscopic and dynamic properties of chromophores in pigment-protein complexes.

## VI. Outlook

This work demonstrates that single-molecule spectroscopy is a powerful tool to reveal in detail the factors determining the electronic structure of pigment-protein complexes and, more generally, of molecular aggregates. Various manifestations of disorder can be probed directly, providing valuable information for the theoretical modeling of energy-transfer processes in these systems, a better understanding of the structure of these biologically important systems, and an understanding of how these systems function.

The light harvesting systems are perfect examples of the clever exploitation of quantum mechanics to produce an efficient energy conversion system. They are excellent test beds for exploring and expanding theories of intermolecular interactions and dynamics associated with the absorption and dissipation of photon energy. The detailed structural information makes it possible to establish a quantitative relationship between theory and experiments.

## Acknowledgments

This work is supported by the 'Stichting voor Fundamenteel Onderzoek der Materie' (FOM) with financial aid from the 'Nederlandse Organisatie voor Wetenschappelijk Onderzoek' (NWO) and is further supported by the 'Section Earth and Life Sciences' (ALW) of NWO and by the Volkswagen-Stiftung (Hannover, Germany).

## References

Alden RG, Johnson E, Nagarajan V, Parson WW, Law CJ and Cogdell RJ (1997) Calculations of spectroscopic properties of the LH2 bacteriochlorophyll-protein antenna complex from *Rhodopseudomonas acidophila*. J Phys Chem B 101: 4667–4680

Ambrose WP, Basché Th and Moerner WE (1991) Detection and spectrocopy of single pentacene molecules in a p-terphenyl crystal by means of fluorescence excitation. J Chem Phys 95: 7150–7162

Angerhofer A, Cogdell RJ and Hipkins MF (1986) A spectral characterization of the light-harvesting pigment-protein complexes from *Rhodopseudomonas acidophila*. Biochim Biophys Acta 848: 333–341

Basché Th, Moerner WE, Orrit M and Wild UP (eds) (1997) Single-molecule Optical Detection, Imaging and Spectroscopy. VCH Verlagsgesellschaft, Weinheim

Bergström H, Sundström V, van Grondelle R, Gillbro T, Cogdell RJ (1988) Energy transfer dynamics of isolated B800-B820 pigment-protein complexes of *Rhodobacter sphaeroides* and *Rhodopseudomonas acidophila*. Biochim Biophys Acta 936: 90–98

Bergström H, Westerhuis WHJ, Sundström V, van Grondelle R, Niederman RA and Gillbro T (1988) Energy transfer within the isolated B875 light-harvesting pigment-protein complex of *Rhodobacter sphaeroides* at 77 K studied by picosecond absorption spectroscopy. FEBS Lett 233: 12–16

Bittl R, Schlodder E, Geisenheimer I, Lubitz W and Cogdell RJ (2001) Transient EPR and absorption studies of carotenoid triplet formation in purple bacterial antenna complexes. J Phys Chem B 105: 5525–5535

Blankenship RE (2002) Molecular Mechanisms of Photosynthesis. Blackwell Science, Oxford

Brixner T, Stenger J, Vaswani HM, Cho M, Blankenship RE and Fleming GR (2005) Two-dimensional spectroscopy of electronic couplings in photosynthesis. Nature 434: 625–628

Chachisvilis M, Kühn O, Pullerits T and Sundström V (1997) Excitons in photosynthetic purple bacteria: Wavelike motion or incoherent hopping? J Phys Chem 101: 7275–7283

Cogdell RJ, Hipkins MF, MacDonald W and Truscott TG (1981) Energy transfer between the carotenoid and the bacteriochlorophyll within the B800-850 light-harvesting pigment-protein complex of *Rhodopseudomonas sphaeroides*. Biochim Biophys Acta 634: 191–202

Cogdell RJ, Isaacs NW, Freer AA, Arrelano J, Howard TD, Papiz MZ, Hawthornthwaite-Lawless AM and Prince S (1997) The structure and function of the LH2 (B800-850) complex from the purple photosynthetic bacterium *Rhodopseudomonas acidophila* strain 10050. Prog Biophys Molec Biol 68: 1–27

Cogdell RJ, Howard TD, Bittl R, Schlodder E, Geisenheimer I and Lubitz W (2000) How carotenoids protect bacterial photosynthesis. Phil Trans R Soc Lond B 355: 1345–1349

De Caro C, Visscher R, van Grondelle R and Völker S (1994) Inter- and intraband energy transfer in LH2-antenna complexes of purple bacteria. A fluorescence line-narrowing and hole-burning study. J Phys Chem 98: 10584–10590

de Ruijter WPF (2005) Photodynamics of light-harvesting systems. PhD Thesis, Leiden University

de Ruijter WPF, Oellerich S, Segura JM, Lawless AM, Papiz M

and Aartsma TJ (2004) Observation of the energy level structure of the low light adapted B800 LH4 complex by single molecule spectroscopy. J Phys Chem 87: 3413–3420

Dempster SE, Jang S, Silbey RJ (2001) Single molecule spectroscopy of disordered circular aggregates: A perturbation analysis. J Phys Chem 114: 10015–10023

Fidder H, Knoester J and Wiersma DA (1991) Optical properties of disordered molecular aggregates: A numerical study. J Chem Phys 95: 7880–7890

Fowler GJS, Visschers RW, Grief GG, van Grondelle R and Hunter CN (1992) Genetically modified photosynthetic antenna complexes with blueshifted absorbance bands. Nature 355: 848–850

Gall A (1994) Purification, characterisation and crystallisation from a range of rhodospirillineae pigment-protein complexes. PhD Thesis, University of Glasgow

Gardiner AT, Cogdell RJ and Takaichi S (1993) The effect of growth conditions on the light-harvesting apparatus in *Rhodopseudomonas acidophila*. Photosynth Res 38: 159–168

Gudowska-Nowak E, Newton MD and Fajer J (1990) Conformational and environmental-effects on bacteriochlorophyll optical spectra — correlations of calculated spectra with structural results. J Phys Chem 94: 5795–5801

Hartigan N, Tharia HA, Sweeney F, Lawless AM and Papiz MZ (2002) The 7.5 Å electron density and spectroscopic properties of a novel low-light B800 LH2 from *Rhodopseudomonas palustris*. Biophys J 82: 963–977

Hecht E (1998) Optics. Addison-Wesley Longman Inc, Reading

Hoff AJ and Deisenhofer J (1997) Photophysics of photosynthesis Structure and spectroscopy of reaction centers of purple bacteria. Phys Reports 287: 2–24

Hofmann C, Aartsma TJ, Michel H and Köhler J (2003a) Direct observation of tiers in the energy landscape of a chromoprotein: A single-molecule study. Proc Nat Acad Sci USA 100: 15534–15538

Hofmann C, Ketelaars M, Matsushita M, Michel H, Aartsma TJ and Köhler J (2003b) Single-molecule study of the electronic couplings in a circular array of molecules: Light-harvesting-2 complex from *Rhodospirillum molischianum*. Phys Rev Lett 90: 013004–013001

Hofmann C, Aartsma TJ and Köhler J (2004) Energetic disorder and the B850-exciton states of individual light-harvesting 2 complexes from *Rhodopseudomonas acidophila* Chem Phys Lett 395: 373–378

Hofmann C, Michel H, van Heel M and Köhler J (2005) Multivariate analysis of single-molecule spectra: Surpassing spectral diffusion. Phys Rev Lett 94: 95501–95504

Hu X, Damjanovic A, Ritz R and Schulten K (1998) Architecture and mechanism of the light-harvesting apparatus of purple bacteria. Proc Natl Acad Sci USA 95: 5935–5941

Jang S, Dempster SE and Silbey RJ (2001) Characterization of the static disorder in the B850 band of LH2. J Phys Chem B 105: 6655–6665

Jang J, Newton MD and Silbey RJ (2007) Multichromophoric Förster resonance energy transfer from B800 to B850 in the light harvesting complex 2: Evidence for subtle energetic optimization by purple bacteria. J Chem Phys 10.1021/jp0701111

Jimenez, R, Dikshit SN, Bradforth SE and Fleming GR (1996) Electronic excitation transfer in the LH2 complex of *Rhodobacter sphaeroides* J Phys Chem 100: 6825–6834

Jimenez R, van Mourik F, Yu JY and Fleming GR (1997) Three pulse photon echo measurements on LH1 and LH2 complexes of *Rhodobacter sphaeroides*. J Phys Chem B 101: 7350–7359

Joo T, Jia Y, Yu JY, Jonas DM and Fleming GR (1996) Dynamics in isolated bacterial light harvesting antenna (LH2) of *Rhodobacter sphaeroides* at room temperature. J Phys Chem 100: 2399–2409

Karrasch S, Bullough PA and Ghosh R (1995) The 85 Ångström projection map of the light-harvesting complex I from *Rhodospirillum rubrum* reveals a ring composed of 16 subunits EMBO J 14: 631–638

Kennis JTM, Streltsov AM, Aartsma TJ, Nozawa T and Amesz J (1996) Energy transfer and exciton coupling in isolated B800-850 complexes of the photosynthetic purple sulfur bacterium *Chromatium tepidum* The effect of structural symmetry on bacteriochlorophyll excited states J Phys Chem B 100: 2438–2442

Kennis JTM, Streltsov AM, Vulto SIE, Aartsma TJ, Nozawa T and Amesz J (1997) Femtosecond dynamics in isolated LH2 complexes of various species of purple bacteria. J Phys Chem B 101: 7827–7834

Ketelaars M, van Oijen AM, Matsushita M, Köhler J, Schmidt J and Aartsma TJ (2001) Spectroscopy on the B850 band of individual light-harvesting 2 complexes of *Rhodopseudomonas acidophila*; I. Experiments and Monte Carlo simulations. Biophys J 80: 1591–1603

Ketelaars M, Segura JM, Oellerich S, de Ruijter WP, Magis G, Aartsma TJ, Matsushita M, Schmidt J, Cogdell RJ and Köhler J (2006) Probing the electronic structure and conformational flexibility of individual light-harvesting 3 complexes by optical single-molecule spectroscopy. J Phys Chem B 110: 18710–18717

Koepke J, Hu XC, Muenke C, Schulten K and Michel H (1996) The crystal structure of the light-harvesting complex II (B800-850) from *Rhodospirillum molischianum*. Structure 4: 581–597

Köhler J (1999) Magnetic resonance of a single molecular spin. Phys Reports 310: 261–339

Krüger BP, Scholes GD and Fleming GR (1998) Calculation of couplings and energy-transfer pathways between the pigments of LH2 by the *ab initio* transition density cube method. J Phys Chem B 102: 5378–5387

Ma Y-Z, Cogdell RJ and Gillbro T (1997) Energy transfer and exciton annihilation in the B800-850 antenna complex of the photosynthetic purple bacterium *Rhodopseudomonas acidophila* (strain 10050). A femtosecond transient absorption study. J Phys Chem B 101:1087–1095

Ma Y-Z, Cogdell RJ and Gillbro T ( 1998) Femtosecond energy-transfer dynamics between bacteriochlorophylls in the B800-820 antenna complex of the photosynthetic purple bacterium *Rhodopseudomonas acidophila* (Strain 7750). J Phys Chem B 102: 881–887

Matsushita M, Ketelaars M, van Oijen AM, Köhler J, Aartsma TJ and Schmidt J (2001) Spectroscopy on the B500 band of individual light-harvesting 2 complexes of *Rhodopseudomonas acidophila*. II. Exciton states of an elliptically deformed ring aggregate. Biophys J 80, 1604–1614

McDermott G, Prince SM, Freer AA, Hawthornthwaite-Lawless AM, Papiz MZ, Cogdell RJ and Isaacs NW (1995) Crystal structure of an integral membrane light-harvesting complex from photosynthetic bacteria. Nature 374: 517–521

McLuskey K, Prince SM, Cogdell RJ and Isaacs NW (2001) The crystallographic structure of the B800-820 LH3 light-harvesting complex from the purple bacteria *Rhodopseudomonas acidophila* Strain 7050. Biochemistry 40: 8783–8789

Monger TG, Cogdell RJ and Parson WW (1976) Triplet states of bacteriochlorophyll and carotenoids in chromatophores of photosynthetic bacteria. Biochim Biophys Acta 449: 136–153

Monger TG and Parson WW (1977) Singlet-triplet fusion in *Rhodopseudomonas sphaeroides* chromatophores: A probe of the organization of the photosynthetic apparatus. Biochim Biophys Acta 460: 393–407

Monshouwer R, I Ortiz de Zarate I, F van Mourik F and R van Grondelle R (1995) Low-intensity pump-probe spectroscopy on the B800 to B850 transfer in the light harvesting 2 complex of *Rhodobacter sphaeroides.* Chem Phys Lett 246: 341–346

Monshouwer R, Abrahamsson M, van Mourik F and van Grondelle R (1997) Superradiance and exciton delocalisation in bacterial photosynthetic light-harvesting systems. J Phys Chem B 101: 7241–7248

Mostovoy MV and Knoester J (2000) Statistics of optical spectra from single ring aggregates and its application to LH2. J Phys Chem B 104: 12355–12364

Novoderezhkin VI, Monshouwer R and van Grondelle R (1999) Exciton (de)localization in the LH2 antenna of *Rhodobacter sphaeroides* as revealed by relative difference absorption measurements of the LH2 antenna and the B820 subunit. J Phys Chem B 103: 10540–10548

Neu P and Silbey RJ (1997) Nonequilibrium spectral diffusion due to laser heating in stimulated-photon-echo spectroscopy of low-temperature glasses. Phys Rev B 56: 11571–11578

Pawley JB, ed (1995) Handbook of Biological Confocal Microscopy. Plenum Press, New York, London

Plakhotnik T, Donley EA and Wild UP (1997) Single-molecule spectroscopy. Annu Rev Phys Chem 48: 181–212

Prince SM, Howard TD, Myles DAA, Wilkinson C, Papiz MZ, Freer AA, Cogdell RJ and Isaacs NW (2003) Detergent structure in crystals of the integral membrane light-harvesting complex LH2 from *Rhodopseudomonas acidophila* strain 10050. J Mol Biol 326: 307–315

Rebane KK (1994) Zero-phonon line as the foundation stone of high-resolution matrix spectroscopy, persistent spectral hole burning, single impurity molecule spectroscopy. Chem Phys 189: 139–148

Reddy NRS, Small GJ, Seibert M and Picorel R (1991) Energy transfer dynamics of the B800-B850 antenna complex of *Rhodobacter sphaeroides*: A hole burning study. Chem Phys Lett 181: 391–399

Roszak AW, Howard TD, Southall J, Gardiner AT, Law CJ, Isaacs NW and Cogdell RJ (2003) Crystal Structure of the RC-LH1 core complex from *Rhodopseudomonas palustris.* Science 302: 1969–1972

Rutkauskas D, Novoderezkhin R, Cogdell RJ and van Grondelle R (2004) Fluorescence spectral fluctuations of single LH2 complexes from *Rhodopseudomonas acidophila* strain 10050. Biochemistry 43: 4431–4438

Rutkauskas D, Novoderezhkin V, Cogdell RJ and van Grondelle R (2005) Fluorescence spectroscopy of conformational changes of single LH2 complexes. Biophys J 88: 422–435

Sauer K, Cogdell RJ, Prince SM, Freer A, Isaacs NW and Scheer H (1996) Structure-based calculations of the optical spectra of the LH2 bacteriochlorophyll-protein complex from *Rhodopseudomonas acidophila*. Photochem Photobiol 64: 564–576

Scholes GD, Gould IR, Cogdell RJ and Fleming GR (1999) *Ab initio* molecular orbital calculations of electronic couplings in the LH2 bacterial light-harvesting complex of *Rhodopseudomonas acidophila.* J Phys Chem B 103: 2543–2553

Sundström V, Pullerits T and van Grondelle R (1999) Photosynthetic light-harvesting: Reconciling dynamics and structure of purple bacterial LH2 reveals function of photosynthetic unit. J Phys Chem B 103: 2327–2346

Thorn Leeson D and Wiersma DA (1995) Real time observation of low-temperature protein motions. Phys Rev Lett 74: 2138–2141

Thorn Leeson D, Berg O and Wiersma DA (1994) Low-temperature protein dynamics studied by the long-lived stimulated photon echo. J Phys Chem 98: 3913–3916

Tinkham M (2003) Group Theory and Quantum Mechanics. Courier Dover Publications, New York

Trautman JK, Shreve AP, Violette CA, Frank HA, Owens TG and Albrecht AC (1990) Femtosecond dynamics of energy transfer in B800-850 light-harvesting complexes of *Rhodobacter sphaeroides*. Proc Nat Acad Sci USA 87: 215–219

van Grondelle R, Dekker JP, Gillbro T and Sundström V (1994) Energy transfer and trapping in photosynthesis Biochim Biophys Acta 1187: 1–65

van Heel M, Gowen B, Matadeen R, Orlova EV, Finn R, Pape T, Cohen D, Stark H, Schmidt R, Schatz M and Patwardhan A (2000) Single-particle electron cryo-microscopy: Towards atomic resolution. Q Rev Biophys 33: 307–369

van Oijen AM, Ketelaars M, Köhler J, Aartsma TJ and Schmidt J (1998) Spectroscopy of single light-harvesting complexes from purple photosynthetic bacteria at 1.2 K. J Phys Chem B 102: 9363–9366

van Oijen AM, Ketelaars M, Köhler J, Aartsma TJ and Schmidt J (1999a) Unraveling the electronic structure of individual photosynthetic pigment-protein complexes. Science 285: 400–402

van Oijen AM, Ketelaars M, Köhler J, Aartsma TJ and Schmidt J (1999b) Spectroscopy of individual LH2 complexes of *Rhodopseudomonas acidophila*: localized excitations in the B800 band. Chem Phys 247: 53–60

van Oijen AM, Ketelaars M, Köhler J, Aartsma TJ and Schmidt J (2000) Spectroscopy of Individual LH2 complexes of *Rhodopseudomonas acidophila*: Diagonal disorder, sample heterogeneity, spectral diffusion and energy transfer in the B800 band. Biophys J 78: 1570–1577

Vulto SIE, Kennis JTM, Streltsov AM, Amesz J and Aartsma TJ (1999a) Energy relaxation within the B850 absorption band of the isolated light-harvesting complex LH2 from *Rhodopseudomonas acidophila* at low temperature. J Phys Chem B 103: 878–883

Vulto SIE, de Baat MA, Neerken S, Nowak FR, van Amerongen H, Amesz J and Aartsma TJ (1999b) Excited state dynamics in FMO antenna complexes from photosynthetic green sulfur bacteria: A kinetic model. J Phys Chem B 103: 8153–8161

Webb RH (1996) Confocal optical microscopy. Rep Progr Phys 59: 427–471

Wu HM, Savikhin S, Reddy NRS, Jankowiak R, Cogdell RJ, Struve WS and Small GJ (1996) Femtosecond and hole-burning studies of B800's excitation energy relaxation dynamics in the LH2 antenna complex of *Rhodopseudomonas acidophila* (Strain 10050). J Phys Chem 100: 12022–12033

Wu HM and Small GJ (1997) Symmetry adapted basis defect patterns for analysis of the effects of energy disorder on cyclic arrays of coupled chromophores. Chem Phys 218: 225–234

# Chapter 14

# High-Field/High-Frequency Electron Paramagnetic Resonance Involving Single- and Multiple-Transition Schemes

Klaus Möbius*
*Department of Physics, Free University Berlin, Arnimallee 14, 14195 Berlin, Germany*

Daniella Goldfarb
*Department of Chemical Physics, Weizmann Institute of Science, 76100 Rehovot, Israel*



## Summary

During the last decade a tremendous growth in the combined efforts of biologists, chemists and physicists to understand the dominant factors determining the specificity and directionality of transmembrane transfer processes in photosynthetic proteins has been observed. Among the large variety of experimental techniques used in these efforts, electron paramagnetic resonance (EPR) spectroscopy at high magnetic fields turned out to

*Author for correspondence, email: klaus.moebius@physik.fu-berlin.de

*Thijs J. Aartsma and Jörg Matysik (eds): Biophysical Techniques in Photosynthesis II, pp. 267–304.*

be particularly powerful. In conjunction with site-specific mutation strategies and advanced quantum-chemical computation methods for data interpretation in terms of protein structure and dynamics, new insights in biological processes have been have been obtained at the atomic and molecular levels. In this chapter, a few large paradigmatic biosystems from photosynthesis are surveyed as examples which have been explored lately by high-frequency/high-field EPR techniques. Taking advantage of the improved spectral and temporal resolution of EPR at 95 GHz/3.4 T and 360 GHz/12.9 T, as compared to conventional X-band EPR (9.5 GHz/0.34 T), the proteins are characterized with respect to electronic properties, structure and dynamics. In this chapter, the following protein systems are primarily discussed: (1) In non-oxygenic photosynthesis, light-induced electron-transfer intermediates in wild-type and mutant reaction centers from photosynthetic bacteria as well as light-driven proton-transfer intermediates of site-specifically nitroxide spin-labeled mutants of bacteriorhodopsin proteins, (2) in oxygenic photosynthesis, specific cofactors of the electron-transfer chains in Photosystem I and Photosystem II. The information obtained is complementary to that of protein crystallography, solid-state NMR, infrared and optical spectroscopy techniques. In addition, a few multifrequency EPR experiments on biomimetic donor-acceptor model systems are reviewed. As a unique strength, high-field EPR can probe structure and dynamics of transient intermediates of proteins while staying in their working states on biologically relevant time scales. The chapter describes the underlying strategies for extending conventional EPR to high-field/high-frequency techniques, and highlights those aspects of structure-dynamics-function relations that are revealed from combining high-field EPR with genetic engineering techniques to study site-specific mutants. The importance of DFT-based quantum-chemical interpretation of the experimental data (g, zero-field-splitting and hyperfine tensors) is emphasized. The chapter concludes with a summary of specific advantages of high-field EPR and an outlook to major challenges this type of magnetic resonance spectroscopy faces in protein research.

## I. Introduction

During the last decade, the chemistry, biology and physics communities have witnessed a boost of new electron paramagnetic resonance (EPR) applications because of technological breakthroughs in pulsed microwave, sweepable cryomagnets and fast data acquisition instrumentation. These allowed to introduce single- and multiple-pulse EPR techniques, as well as multifrequency high-field EPR, as powerful spectroscopic tools with unique potential in elucidating structure and dynamics of complex systems. Both EPR and NMR (nuclear magnetic resonance) were originally invented by continuous wave (cw) experiments, EPR in 1944 by Zavoisky at Kazan University (Zavoisky, 1944, 1945a,b), NMR in 1946 by Purcell, Torrey and Pound at Harvard (Purcell et al., 1946) and by Bloch, Hansen and Packard at Stanford (Bloch et al., 1946). Up to the 1960s, NMR and EPR remained as cw methods, i.e., the samples placed in a static magnetic field were irradiated by continuous radiofrequency (rf) and microwave (mw) electromagnetic fields to drive NMR and EPR transitions, respectively. But it was as early as 1949 when Hahn at Urbana (Hahn, 1949, 1950) applied rf pulses and invented the nuclear spin-echo detection. This, with the introduction of Fourier-transform (FT) NMR, opened the arena for pulse NMR with all its potential for multi-dimensional spectroscopy of complex biosystems in the liquid and solid state. Nowadays, pulse NMR has completely replaced cw NMR, and high-field cryomagnets and resonance frequencies close to 1 GHz for protons have dramatically improved the detection sensitivity and chemical-shift separations. It took almost a decade before Blume at Columbia (Blume, 1958) observed for the first time electron spin echoes analogous to Hahn's nuclear spin echoes, and electron spin echo-detected EPR (ESE) techniques turned out to revolutionize EPR spectroscopy, but decades later.

*Abbreviations:* A – absorptive; BR – bacteriorhodopsin; cw – continuous wave; DF – dual frequency; DFT – density functional theory; E – emissive; *E. – Escherichia*; ELDOR – electron electron double resonance; ENDOR – electron nuclear double resonance; EPR – electron paramagnetic resonance; ESE – electron spin echo; ET – electron transfer ; FJ – field jump; FT – Fourier transform; hfc – hyperfine coupling constant; IF – intermediate frequency; J – exchange coupling; $k_B$ – Boltzmann constant; MTS – (1-oxil-2,2,5,5-tetramethyl-pyrroline-3-methyl)methane-thiosulfonate; mw – microwave; PS I – Photosystem I; PS II – Photosystem II; $Q_L$ – loaded quality factor; rf – radiofrequency; SCRP – spin-correlated radical pair; SDSL – site-directed spin-labeling; tr EPR – transient EPR; TRIPLE – electron-nuclear-nuclear triple resonance; $T_Z$ – Zeeman temperature; VMT – variable mixing time; ZFS – zero-field splitting

Despite the breakthroughs in mm and sub-mm microwave technologies in the last decade, for EPR, unlike NMR, the prognosis is that a peaceful coexistence between cw and pulse EPR will continue to persist. Which option to choose will be determined entirely by the characteristics of the sample under study, i.e., the sample relaxation times ultimately dictate the preference for either a cw or pulse experiment. The electronic transverse relaxation times ($T_2$) are typically in the μs range or shorter and, consequently, in EPR the mw pulses have to be as short as a few ns. To generate them poses great technical problems even today in terms of mw sources and fast electronic components and computers, likewise to detect and handle the transient signals in the ns time scale. Pulse EPR spectroscopy has many founders, and a rich variety of pulse sequences and sophisticated experiments is found in recent reviews and text books, for example in Schweiger and Jeschke (2001). Pioneering work was done among others by Mims at Bell Telephone Laboratories, Tsvetkov in Novosibirsk, Freed at Cornell and Schweiger at ETH Zürich.

In 1956, another benchmark progress was achieved by Feher at Bell Labs (Feher, 1956), who combined the advantages of EPR (high detection sensitivity) with those of NMR (high resolution capability) by inventing ENDOR (electron-nuclear double resonance). He had applied cw irradiation fields at mw and rf frequencies, and this experiment was technically feasible only because the sample, phosphorus doped silicon, was studied at low temperature, where all the relaxation times are sufficiently long to easily obtain the necessary saturation of the EPR and NMR transitions. For radicals in liquid solution, however, these relaxation times are much shorter, in the order of $10^{-5}-10^{-7}$ s, and, consequently, ENDOR-in-solution experiments are technically more demanding as much larger saturating mw and rf fields have to be applied. Hyde and Maki (1964) were the first to succeed with cw ENDOR experiments on an organic radical solution. This technique was shown later to be very powerful in elucidating electronic and spatial structures of the donor and acceptor radical ions in the primary electron-transfer processes in photosynthetic reaction centers (Lubitz and Lendzian, 1996; Levanon and Möbius, 1997). The cw ENDOR technique was later extended to ESE-detected pulse versions by Mims (1965) and Davies (1974).

In many applications of cw ENDOR the assignment of hyperfine lines to molecular positions can be greatly facilitated by applying a second rf field thereby creating electron-nuclear-nuclear triple resonance (TRIPLE). The first experimental realization of cw TRIPLE at 77 K was achieved by Cook and Whiffen (1964). They called it 'double ENDOR,' and applied it to X-ray irradiated organic crystals to determine relative signs of hyperfine couplings. For organic radicals in liquids, cw triple resonance was first realized by Dinse, Biehl and Möbius (Dinse et al., 1974) applying the two rf fields at frequencies symmetrically placed around the nuclear Larmor frequency ('Special TRIPLE') to enhance the signal intensity, and by Biehl, Plato and Möbius (Biehl et al., 1975) with independently variable frequencies of the two rf fields, one pumping a selected ENDOR transition while sweeping the other through the ENDOR spectrum ('General TRIPLE') to determine relative signs of the hyperfine couplings. The pulsed version of triple resonance was introduced by Mehring and collaborators (Mehring et al., 1987).

Distance measurements in biological systems for the scale of a few nm are an important area of application for pulse EPR spectroscopy (Berliner et al., 2000; Chapters 16, Astashkin and Kawamori, and 17, Borovykh and Steinhoff). They are based on selective measurements of the electron-electron dipolar coupling between two spin-carrying domains which is a function of their interspin distance and relative orientation. The main advantage of pulse versus cw EPR in this endeavor is the ability to separate the electron-electron coupling from other interactions, such as electron-nuclear hyperfine interactions, and to reduce inhomogeneous line broadening, thereby extending the distance range that can be probed to 7–8 nm. Milov and collaborators at Novosibirsk (Milov et al., 1981, 1984) established the 3-pulse electron-electron double resonance technique (pulsed ELDOR or DEER). Later it was extended to a 4-pulse sequence for dead-time free detection (Pannier et al., 2000; Schweiger and Jeschke, 2001). Other powerful pulse sequences for measuring electron-electron frequencies and, thereby, distances include the '2n+1' pulse train (Kurshev et al., 1989), 'double-quantum-coherence' DEER (Saxena and Freed, 1997) or RIDME (relaxation-induced dipolar modulation enhancement) (Kulik et al., 2002).

Concerning transient states, besides pulse EPR techniques also specific cw EPR techniques can provide time-resolved signals by invoking suitable field-modulation or photoexcitation strategies. With field modulation at frequencies as high as 1 MHz the time resolution could be extended to the μs range. A decisive step forward to higher time resolution was achieved for photoreactions by abandoning field modulation at all and generating the time-dependent EPR signal via wavelength-selective pulsed laser excitation and subsequent direct detection of the transient EPR signal at fixed Zeeman-field value, employing fast data-acquisition systems. This established the 'tr EPR' technique. The accompanying loss of detection sensitivity can be compensated to a large extent by signal averaging and, fortunately, by the signal enhancement via electron-spin polarization effects inherent in many photoreactions. These effects occur in reactions with instantaneously generated excited states and subsequent fast detection of the transient reaction intermediates, e.g., triplets, radicals, radical pairs, before spin-lattice relaxation can thermalize them. For overviews of electron-spin polarization effects, see Hoff (1989), McLauchlan and Yeung, (1994) and Stehlik and Möbius (1997).

After the pioneering work by Weissman and co-workers (Kim and Weissman, 1979), the time resolution of tr EPR has been pushed to the 10 ns range. By now, tr EPR has proved to be extremely powerful in a broad range of mw frequencies from S-band (4 GHz), X-band (9 GHz), K-band (24 GHz), Q-band (35 GHz) up to the high-field EPR frequencies 95 GHz, 120 GHz and 240 GHz (Prisner et al., 1994; Stehlik and Möbius, 1997; Savitsky et al., 2004; Tol et al., 2005). For many applications in photochemistry a multifrequency approach of tr EPR experiments turned out to be essential for the detailed analysis of spin-polarized spectra.

In the high-field domain of cw EPR spectroscopy, the pioneering work was done by Lebedev in Moscow (Lebedev, 1994, 1998). Together with his co-workers he paved the road for this novel type of EPR spectroscopy with important applications in biology, chemistry and physics. The first pulsed high-field EPR experiment was performed in Leiden (Disselhorst et al., 1995) and the first high-field tr EPR experiment in Berlin (Prisner et al., 1994). The extension to high-field cw ENDOR was achieved first in Berlin (Burghaus et al., 1988), to high-field pulse ENDOR in Leiden (Disselhorst et al., 1995). These early high-field experiments represent milestone progress in modern EPR spectroscopy in the endeavor to establish powerful and versatile tools for structure and dynamics elucidation. Figure 1 shows the mw and rf irradiation schemes of a variety of cw and pulse high-field EPR techniques that are dealt with in this chapter.

To conclude the Introduction, we refer to recent

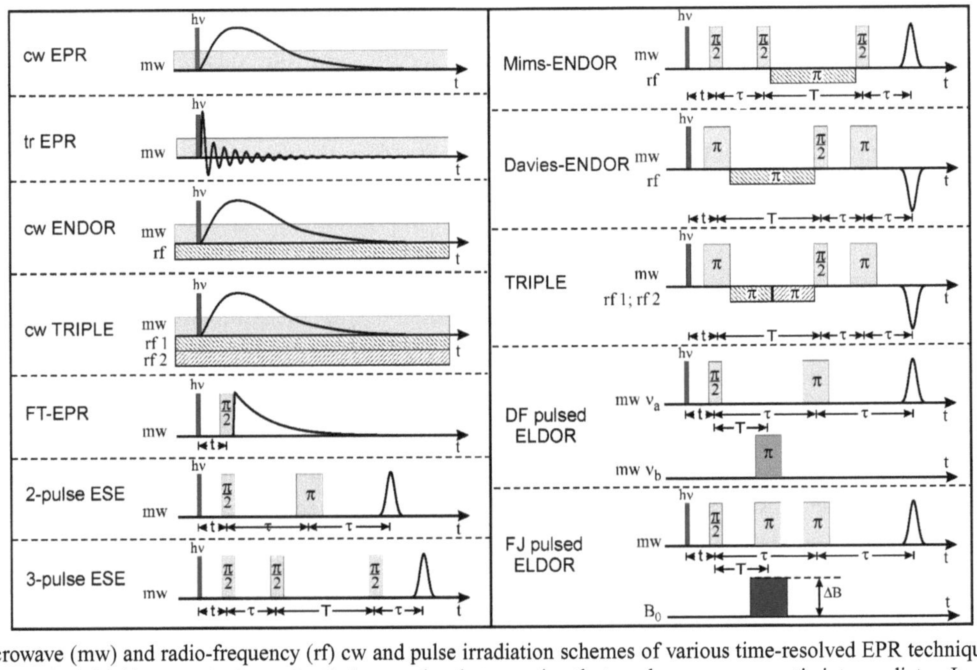


*Fig. 1.* Microwave (mw) and radio-frequency (rf) cw and pulse irradiation schemes of various time-resolved EPR techniques covered in this chapter. The initial laser excitation pulse (hν) starts the photoreaction that produces paramagnetic intermediates. In stable paramagnetic systems the laser pulse is omitted. Abbreviations: cw (continuous wave), tr (transient), FT (Fourier transform), ESE (electron spin echo), DF (dual frequency), FJ (field jump). For details and references, see text.

overview articles on multifrequency EPR spectroscopy containing examples of high-field EPR and ENDOR instrumentation and/or application in relation to biology in general and photosynthesis in particular (Möbius, 1995, 2000, 2001; Weber, 2000; Lubitz et al., 2002, 2005; Grinberg and Berliner, 2004; Lubitz, 2004; Möbius et al., 2005a). Our chapter in this volume is understood as a continuation of the magnetic-resonance chapters in *Biophysical Techniques in Photosynthesis*, previously published in this series (Amesz and Hoff, 1996), in particular of the chapters on electron-spin-echo methods (Britt, 1996), time-resolved EPR (Levanon, 1996), and ENDOR spectroscopy (Lubitz and Lendzian, 1996).

## II. Principles of High-Field/High-Frequency EPR Techniques

In the following we will restrict ourselves to a phenomenological description of EPR techniques that are representative for high-field applications in biophysics and biochemistry. For a more theoretical treatment we refer to textbooks, for example (Hoff, 1989; Atherton, 1993; Schweiger and Jeschke, 2001).

### A. Single-Resonance EPR

We first describe the basic spin Hamiltonians with interaction terms that determine the frequencies of the EPR transitions and thereby the characteristics of the EPR spectrum.

#### 1. Spin Hamiltonian of Organic Radicals or Low-Spin Transition-Metal Ions (S=1/2)

If the EPR sample contains molecular radicals or transition-metal ions with unpaired electron spins S = 1/2, the electron and nuclear spins will align with respect to the total magnetic field. This is composed of the external Zeeman field, $\boldsymbol{B_0}$, the local field originating from spin-orbit coupling in the paramagnetic center (leading to effective g-tensor components shifted from the free-electron value) and local hyperfine fields from nearby magnetic nuclei, for example protons with nuclear spin I = 1/2 or $^{14}$N nuclei with I = 1.

For the simple case of an electron spin S = 1/2 coupled to $n$ nuclei with I = 1/2, the static spin Hamiltonian, $\hat{H}_0$, that describes the time-independent spin-interaction energies, consists of three terms

$$\hat{H}_0/h = \frac{\mu_B}{h}\boldsymbol{B}_0 \cdot \tilde{g} \cdot \hat{\boldsymbol{S}} - \sum_i \frac{g_{ni}\mu_K}{h}\boldsymbol{B}_0 \cdot \hat{\boldsymbol{I}}_i + \sum_i \hat{\boldsymbol{S}} \cdot \tilde{A}_i \cdot \hat{\boldsymbol{I}}_i \tag{1}$$

i.e., the field-dependent electron and nuclear Zeeman interactions and the field-independent electron-nuclear hyperfine interactions ($h$: Planck constant; $\mu_B$, $\mu_K$: Bohr and nuclear magnetons; $g_n$: nuclear g-factors; $\hat{\boldsymbol{S}}$, $\hat{\boldsymbol{I}}$: electron and nuclear spin vector operators; the summation is over all nuclei).

The interaction tensors $\tilde{g}$ and $\tilde{A}_i$ probe the electronic structure (wave function) of the molecule either globally (g-tensor) or locally (hyperfine tensors). The tensors contain isotropic and anisotropic contributions. In isotropic fluid solution, only the isotropic values, $1/3 \cdot \mathrm{Tr}(\tilde{g}) = g_{iso}$ and $1/3 \cdot \mathrm{Tr}(\tilde{A}_i) = A_{iso}$, are observed. In frozen solutions, powders or single crystals, on the other hand, also anisotropic component contributions become observable, provided appropriate spectral resolution conditions prevails. For this situation, the information content of the EPR spectrum is considerably enhanced, for example in terms of molecular orientation with respect to $\boldsymbol{B_0}$ and electron-nuclear distances.

In the strong-field approximation, the energy eigenvalues of Eq. (1) (in frequency units) are classified by the magnetic spin quantum numbers, $m_S$ and $m_I$, and are given, to first order, by

$$E_{m_S m_i}/h = \frac{g'\mu_B}{h} B_0 m_S - \sum_i \frac{g_{ni}\mu_K}{h} B_0 m_{I_i} + \sum_i A_i' m_S m_{I_i} \tag{2}$$

where the scalar quantities $g'$ and $A'$ contain the desired information about magnitude and orientation of the interaction tensors (Atherton, 1993). When irradiating the sample with microwaves at a fixed frequency and varying the external field, EPR transitions occur following the selection rules $\Delta m_S = \pm 1$, $\Delta m_{I_i} = 0$. Second-order contributions from the nuclear Zeeman energy term can normally be neglected because it is considerably smaller than the other two terms. The intensity distribution of the hyperfine lines is determined by the number of symmetry-equivalent nuclei.

For single-crystal samples, the complete tensor information can be extracted from the angular dependence of the resonance lines when the crystal is rotated in its three symmetry planes ('rotation patterns'). If the tensors $\tilde{g}$ and $\tilde{A}_i$ are collinear, i.e., have the same principal axes system ($\alpha,\beta,\gamma$), their rotation patterns will have the same angular dependence. When the crystal is mounted with its rotation axis parallel to the $x$-axis of the molecular axes system ($x,y,z$) and perpendicular to the $B_0$ direction, the rotation occurs in the $yz$-plane defining an angle $\theta$ between the $z$-axis and the field direction. Then, $g'$ takes the form:

$$g' = \sqrt{g_{yy}^2 \sin^2\theta + g_{zz}^2 \cos^2\theta + g_{yz}^2 \sin^2\theta} \qquad (3)$$

(notice that $g_{ij}^2$ denotes the components of the squared $\tilde{g}$ tensor). An analogue expression holds for $A_i'$ in the limit of small g-anisotropy, as is typical for bioorganic systems as long as they do not contain paramagnetic transition-metal ions. For the other two symmetry planes, $zx$ and $xy$, the corresponding-tensor components are found by cyclic permutation $y \rightarrow z \rightarrow x \rightarrow y$.

For many bioorganic samples, in particular in frozen solution, conventional X-band (9.5 GHz) EPR spectra suffer from limited spectral resolution. This is because several radical species or different magnetic sites of rather similar g-values may be present, or because a small g-tensor anisotropy does not allow canonical orientations of the powder spectrum to be resolved. Analogous to modern NMR spectroscopy, the spectral resolution can be improved by applying higher magnetic fields and correspondingly higher microwave frequencies. Moreover, the spectral analysis is generally simpler in high-field EPR because the first-order approximation for the energies often applies. The question arises, what defines 'high' in high-field EPR, i.e., how large the Zeeman field ought to be for a particular sample?

For all cases of delocalized spin systems, in which unresolved hyperfine interactions dominate the inhomogeneous EPR linewidth, a 'true' high-field experiment must fulfill the condition

$$\frac{\Delta g}{g_{\text{iso}}} B_0 > \Delta B_{1/2}^{hf} \qquad (4)$$

which relates properties of the spectrometer with properties of the sample, i.e., the anisotropic electron Zeeman interaction must exceed the inhomogeneous line broadening, $\Delta B_{1/2}^{hf}$. For example, for deuterated samples, Q-band EPR may already fulfill this condition in the case of semiquinone radicals with rather large g-anisotropy (Brink et al., 1994; Lubitz and Feher, 1999), whereas for protonated samples with inherently larger linewidths, it does not. On the other hand, in the case of chlorophyll ion radicals, due to their small g anisotropy, even W-band EPR may not meet the high-field condition for protonated samples. Then, deuteration of the sample will be necessary or, as an alternative, a further increase of the mw frequency and $B_0$ field, for instance by resorting to 360-GHz EPR (see below).

Except for transition-metal complexes, most bioorganic systems have g-values close to 2, and relative g variations and anisotropies, $\Delta g/g$, rarely exceed $10^{-4}$–$10^{-3}$. At X-band frequencies the corresponding $\Delta B_0$ values are only 0.03–0.3 mT, which can easily be masked in disordered samples with typical EPR linewidths around 1 mT. Fortunately, for many non-metallic protein systems the increase of $\Delta B_0$ with increasing Zeeman field directly translates into an increase of spectral resolution, because often no noticeable line broadening due to 'g-strain' effects occurs with increasing $B_0$. For the primary donor cation radical in RCs from *Rb. sphaeroides*, for example, fields up to 24 T were applied (Bratt et al., 2003), and g-strain broadening was found to be negligible.

In addition to the improved g resolution mentioned above, high-field EPR can also improve the sensitivity. In this respect one has to distinguish between the absolute and relative sensitivities because the amount of sample that can be introduced into the cavity of high-field EPR spectrometers is usually significantly smaller than in standard X-band spectrometers. Consequently, the amount of sample available has to be taken into account and, when the amount of sample is limited like in single crystals of proteins, high-field EPR is advantageous.

### *2. Spin Hamiltonian of High-Spin Transition-Metal Ions or Triplet States (S > 1/2)*

Also for high-spin systems (S > 1/2), such as enzymes with one or several transition metal cofactors, high-field EPR might be of advantage. For such systems a 'fine-structure' term $\hat{H}_{SS}$ has to be added to the spin Hamiltonian, i.e., Eq. (1) has to be extended by

$$\hat{H}_{SS}/h = \hat{\mathbf{S}}\cdot\tilde{D}\cdot\hat{\mathbf{S}} - J\hat{\mathbf{S}}_1\cdot\hat{\mathbf{S}}_2 \qquad (5)$$

with the total spin $\mathbf{S}_1 + \mathbf{S}_2 = \mathbf{S}$, to consider the interactions between a pair of paramagnetic ions with spins $\mathbf{S}_1$ and $\mathbf{S}_2$ coupling to the total spin S. Here, $\tilde{D}$ is the traceless zero-field splitting (ZFS) tensor, and $J$ is the isotropic exchange interaction parameter (in standard convention $J < 0$ corresponds to antiferromagnetic coupling, $J > 0$ to ferromagnetic coupling). In the principal axes system of $\tilde{D}$, the anisotropic part in Eq. (5) is normally rewritten in terms of the zero-field parameters $D$, $E$:

$$\hat{\boldsymbol{S}} \cdot \tilde{D} \cdot \hat{\boldsymbol{S}} = D\left(\hat{\boldsymbol{S}}_z^2 - \frac{1}{3}S(S+1)\right) + E\left(\hat{\boldsymbol{S}}_x^2 - \hat{\boldsymbol{S}}_y^2\right) \quad (6)$$

with $D = (3/2)D_{zz}$, $E = (D_{xx} - D_{yy})/2$, where $D_{xx}$, $D_{yy}$ and $D_{zz}$ are the principal values of the ZFS tensor. For a triplet state they are related to the zero-field energy levels as follows:

$$E_x/h = (1/3)D - E, \quad E_y/h = (1/3)D + E,$$
$$E_z/h = -(2/3)D$$

Although in the spin Hamiltonian the fine-structure term is field independent it leads, in combination with the electron Zeeman term, to field-dependent mixing of the electron spin eigenfunctions. This becomes very severe when the electron Zeeman splittings becomes comparable in magnitude to the fine-structure splittings. This intermediate region requires elaborate calculations to analyze the EPR spectrum. Hence, one more reason to perform high-field EPR is to simplify the spectral analysis for high-spin systems. Moreover, it might happen that EPR transitions of high-spin systems with large zero-field splittings cannot be observed at all at standard X-band frequencies because the energy of the mw quantum is too small. For such cases the higher quantum energy of high-frequency microwaves can drive the transitions (Krzystek et al., 2001; Mantel et al., 2003).

High-field EPR is particularly useful for half-integer high-spin systems ($S = (2n+1)/2$, $n = 1, 2, \ldots$), such as Mn(II) and Fe(III) with $S = 5/2$, and Gd(III) with $S = 7/2$. For such systems the inhomogeneous linewidth of orientationally disordered samples, $\Delta B_{1/2}$, of the central $|-½\rangle \rightarrow |½\rangle$ EPR transition is determined by second-order contributions from the ZFS term according to $\Delta B_{1/2} \propto D^2/B_0$. Hence, when $g\mu_B B_0/h \gg D$, the broadening becomes negligible and narrow signals are obtained also in orientationally disordered samples, thus leading to increased sensitivity and resolution. For example, 'needle sharp' manganese hyperfine lines are often observed in 95-GHz high-field EPR spectra of disordered $Mn^{2+}$ protein complexes (Zoleo et al., 2004). In this case the first-order approximation is valid and the resonance fields are given by:

$$B(m_S \rightarrow m_S + 1) = B_0 - \frac{h}{g\mu_B} A_{iso} m_I + \frac{h}{g\mu_B}(2m_S + 1)\nu_D$$

$$\nu_D = \frac{D}{2}\left[(3\cos^2\vartheta - 1) + \eta \sin^2\vartheta \cos 2\phi\right] \quad (7)$$

where $A_{iso}$ is the isotropic hyperfine coupling of $^{55}$Mn (in frequency units), $m_I$ is the corresponding nuclear spin projection and $\eta = 3E/D$. The angles $\vartheta$ and $\phi$ describe the orientation of the magnetic field relative to the principal axes system of the ZFS tensor.

Another important aspect of high-field EPR on high-spin systems, for example organic triplet states and radical pairs with $S = 1$, is the possibility to determine the absolute sign of the zero-field parameter $D$ by sufficient thermal polarization of the triplet levels already at moderately low temperatures (Hornig and Hyde, 1963; Chemerisov et al., 1994). If the temperature is low enough to fulfill the condition $k_B T > g\mu_B B_0$, the lowest spin level, corresponding to $m_S = -1/2$, is predominantly populated, and the EPR spectra of disordered samples become asymmetric. The 'Zeeman temperature,' $T_Z = g\mu_B B_0/k_B$, at which this asymmetry becomes noticeable, i.e., higher intensity either on the high-field side ($D > 0$) or on the low-field side ($D < 0$), is defined accordingly. Approximate values at different EPR frequencies are: $T_Z = 0.4$ K (at 9.5 GHz), 4 K (at 95 GHz), 6.5 K (at 140 GHz) and 15.5 K (at 360 GHz). The sign of $D$ is indicative of the shape of the dipolar tensor. Organic triplet states generally have disk-shaped dipolar tensors ($D > 0$), whereas weakly coupled radical pairs have cigar-shaped ones ($D < 0$).

### *B. Multiple-resonance EPR techniques*

We now turn to high-field/high-frequency multiple-resonance experiments, such as ENDOR, TRIPLE, and pulsed ELDOR, in more detail to show what can be additionally learned about biological systems when going beyond conventional X-band frequencies.

### 1. High-field ENDOR and TRIPLE Resonance

Thorough accounts of high-field/high-frequency ENDOR spectroscopy have recently been given by Goldfarb and Krymov (2004) and Maniero in (Brunel et al., 2004).

For large low-symmetry radicals, such as those occurring in photosynthesis, with each set of inequivalent nuclei the number of EPR lines increases in a multiplicative way, according to the selection rules $\Delta m_S = \pm 1$, $\Delta m_{I_i} = 0$. This results in strongly inhomogeneously broadened EPR spectra because individual hyperfine lines can no longer be resolved in the available spectral range (which is fixed due to the normalization condition for the unpaired electron spin density). For such cases, by resorting to ENDOR techniques the spectral resolution can be greatly improved.

In cw ENDOR, the sample is irradiated simultaneously by two electromagnetic fields, a mw field (to drive EPR transitions $\Delta m_S = \pm 1$) and an rf field (to drive NMR transitions $\Delta m_{I_i} = \pm 1$). Under appropriate experimental conditions, which are more stringent for cw than for pulse irradiation schemes (Dorio and Freed, 1979; Möbius et al., 1982; Schweiger and Jeschke, 2001), ENDOR signals are observed by monitoring the changes of EPR line amplitudes when sweeping the rf field through the nuclear resonance frequencies. Thus, every group of equivalent nuclei — no matter how many nuclei are involved and what their spin is — contributes only two ENDOR lines because, within an $m_S$ manifold, the hyperfine levels are equidistant to first order. Hence, in essence, ENDOR is a variant of NMR on paramagnetic systems, the unpaired electron serving as highly sensitive detector for the NMR transitions. Double resonance excitation thus offers the advantage of detecting low-intensity rf transitions via high-intensity mw transitions, i.e., by means of quantum transformation.

The gain in resolution of ENDOR versus EPR, therefore, becomes very drastic for low-symmetry molecules because, with increasing number of groups of symmetry-related nuclei, the number of ENDOR lines increases only in an additive way. The resolution enhancement is particularly pronounced when nuclei with different magnetic moments are involved. Their ENDOR lines appear in different frequency ranges, and from their Larmor frequencies these nuclei can be immediately identified. In the case of accidental overlap of ENDOR lines from different nuclei at X-band (9.5 GHz, 0.34 T) they can be separated when working at higher mw frequencies and Zeeman fields, for instance at 95 GHz, 3.4 T (W-band) or even at 360 GHz, 12.9 T (see Fig. 2).

For a doublet radical, the ENDOR lines of a particular group of equivalent nuclei with $I = 1/2$, appear, to first order, at

$$\nu^{\pm}_{i,\mathrm{ENDOR}} = \left|\nu_n \pm A'_i/2\right| \tag{8}$$

with the nuclear Larmor frequency $\nu_n = (g_n\mu_K/h)B_0$, and $A'$ being defined analogous to Eq. (3). For the simple case of an axially symmetric hyperfine interaction $A' = A_{iso} + T_\perp(3\cos^2\theta - 1)$, where the principal components of the anisotropic part are given by $(-T_\perp, -T_\perp, 2T_\perp)$ and $\theta$ is the angle between the external magnetic field and the principal $Z$ axis of the hyperfine tensor. When the point-dipole approximation applies $T_\perp = (\mu_0/4\pi h)(\mu_B\mu_K g g_n/r^3)$, where $r$ is the electron nuclei distance and $\mu_0$ is the vacuum permeability. If the molecule under study contains nuclei with $I > 1/2$ with a quadrupole moment, Eq. (8) is extended by a contribution $3/2(2m_I - 1)P$ (with $m_I = I, I-1, \geq 0$) from the traceless nuclear quadrupole tensor which splits

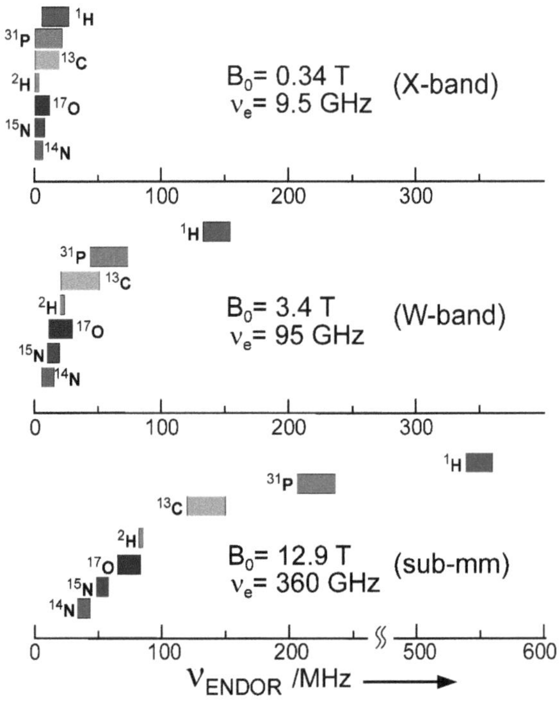


*Fig. 2.* Gain in ENDOR resolution for doublet-state systems ($S = 1/2$, $g = 2$) with increasing Zeeman field $B_0$ and microwave frequency $\nu_e$. Spectral lines of typical nuclei in organic biomolecules, largely overlapping at traditional X-band ENDOR ($B_0 = 0.34$ T, $\nu_e = 9.5$ GHz), become completely separated at 360 GHz/12.9 T.

or shifts the ENDOR lines provided the molecular system is in an anisotropic environment (Biehl et al., 1977; Mayas et al., 1978; Lubitz and Feher, 1999; Flores et al., 2004). *P* is a function of the quadrupole coupling constant $e^2qQ/h$, the asymmetry parameter $\eta$ and the relative orientation of the quadrupole tensor with respect to the Zeeman field.

Deuterium quadrupole couplings in paramagnetic molecular states are sensitive probes for local-structure information on H-bond patterns. Empirical relations between $e^2qQ/h$ values and O-H distances (a $1/R^3_{\mathrm{O...H}}$ correlation) have been established long ago (Soda and Chiba, 1969), but have been verified also by semiempirical MO (Mayas et al., 1978) and DFT calculations (Sinnecker et al., 2004). It is noteworthy that the DFT calculations also reproduce a $1/R^2_{\mathrm{O...H}}$ correlation for the $g_{xx}$ tensor component.

ENDOR measurements can be carried out also in pulsed mode. Although the Mims-type and Davies-type ENDOR pulse sequences (shown in Fig. 1) were introduced already several decades ago, they are still the 'working horses' of pulse ENDOR spectroscopy. In both experiments, the ENDOR spectrum is obtained by recording the echo intensity as a function of the frequency of the rf pulse. A change in the echo amplitude occurs when the rf is on-resonance with an NMR (ENDOR) transition, thereby generating an ENDOR signal. The Davies-ENDOR experiment is based on the selective mw excitation of only one of the EPR multiplet transitions and, therefore, the first mw $\pi$-pulse has to be selective with respect to the hyperfine splitting. Consequently, this sequence is most suited for systems with medium to large hyperfine couplings ($A' > 2$ MHz). Mims-ENDOR, on the other hand, does not require selective pulses but suffers from 'blind spots.' This is because the signal efficiency is scaled by $1/2\sin^2(\pi A'\tau)$. Hence, maximum ENDOR efficiency is obtained for $\tau = (2n+1)/2A'$, whereas blind spots (zero ENDOR efficiency) occur at $\tau = n/A'$ with $n = 0, 1, 2, \ldots$ (Schweiger and Jeschke, 2001). As a consequence, a maximum hyperfine coupling, $A'_{\mathrm{max}}$, exists for which the Mims-ENDOR spectrum remains undistorted by blind spots. It is given by $A'_{\mathrm{max}} = 1/2\tau$. Accordingly, Mims-ENDOR is usually applied when the hyperfine splitting is small, $A' < 2$ MHz, where typical $\tau$ values of 0.15–0.25 µs place the blind spots well outside the ENDOR spectral range of interest. If lower values of $\tau$ are required and cannot be used due to spectrometer 'dead time,' it is possible to apply the remote-detection Mims (ReMims)-ENDOR sequence (Doan and Hoffman, 1997). In general, $^1$H spectra are usually recorded by the Davies-ENDOR sequence, whereas Mims-ENDOR is preferred for $^2$H measurements.

The intensity of the ENDOR signal, i.e., the ENDOR effect, is defined as:

$$F_{\mathrm{ENDOR}} = \frac{1}{2}\frac{I_{\mathrm{rf}}(\mathrm{off}) - I_{\mathrm{rf}}(\mathrm{on})}{I_{\mathrm{rf}}(\mathrm{off})} \tag{9}$$

where, in the case of pulse ENDOR, $I_{\mathrm{rf}}(\mathrm{on})$ and $I_{\mathrm{rf}}(\mathrm{off})$ correspond to the spin-echo intensity with the rf pulse switched on and off, respectively. Usually the ENDOR effect is the same for the $\alpha$ ($m_S = +1/2$) and $\beta$ ($m_S = -1/2$) electron spin manifolds. The maximum ENDOR effect is obtained with rf pulses for 180° rotation.

When the hyperfine coupling becomes large, the rf nutation frequency, $\omega_2$, depends also on the magnitude of the hyperfine coupling. For the simple case of an isotropic interaction, it is given by (Schweiger and Jeschke, 2001)

$$\omega_2 = \gamma E B_2, \qquad E = \left(1 + \frac{m_S A_{\mathrm{iso}}}{\omega_I}\right) \tag{10}$$

where $E$ is the so-called hyperfine-enhancement factor and $B_2$ is the rf-field amplitude. This shows that when the hyperfine coupling is large compared to the nuclear Larmor frequency, $\omega_I$, $E$ becomes large, and a lower rf power is required for generating an rf $\pi$ pulse. For low-$\gamma$ nuclei, that exhibit large enough hyperfine couplings, like $^{14}$N and $^{55}$Mn, the hyperfine-enhancement factor becomes very useful as it shortens the required rf pulse time for a 180° rotation. Also in cw ENDOR the hyperfine-enhancement factor plays an important role for the ENDOR intensity since it amplifies the rf amplitude seen by the nucleus (Kurreck et al., 1988).

The majority of pulse ENDOR experiments are carried out at low temperatures because long enough electron spin-lattice relaxation times, $T_1$, are required to allow insertion of rf pulses, typically 10–40 µs, during the period $T$, without a considerable loss of the electron polarization generated by the first pulses. For this reason, pulse ENDOR experiments are usually not applicable to liquid samples because of their short $T_1$'s. For nitroxide and other organic radicals, liquid-nitrogen cooling is sufficient, whereas for paramagnetic transition-metal ions, measurements in the range of 1.2–15 K are usually necessary and require liquid helium cooling. At high mw frequencies, as 95 and 140 GHz, the cavity tuning and the phase of the signal are highly sensitive to subtle changes in the temperature and He gas flow. Consequently, in addition to the desired resonance effect, the rf pulse can also cause undesired local heating. This results in cavity detuning and leads to an additional decrease in the echo intensity which interferes with the observation of the ENDOR signal. This is particularly problematic for broad signals. The problem is overcome by acquiring the ENDOR spectrum while varying the rf frequency randomly rather than sequentially, thus allowing for an averaging of the baseline distortions (Epel et al., 2003).

In solids the advantages of pulse over cw ENDOR are often dramatic in terms of the distortion-free lineshapes the pulse techniques produce. Moreover, they have the prospect of introducing a variety of one- and two-dimensional experiments for resolution, assignment and correlation purposes, quite in the spirit of established NMR techniques.

Certain drawbacks of ENDOR concerning sensitivity and relative line intensities can be overcome by extending it to electron-nuclear-nuclear triple resonance techniques, such as Special and General TRIPLE (Möbius and Biehl, 1979). The advantages of TRIPLE over ENDOR — enhanced sensitivity and resolution, information about multiplicity and relative signs of hyperfine couplings from line intensity variations — often justify the extra experimental efforts inherent in the triple resonance spectroscopy.

Both the Special and General TRIPLE experiments have their analogue pulse sequences (Mehring et al., 1987; Epel and Goldfarb, 2000; Epel et al., 2003), see Fig. 1. The Special TRIPLE technique requires two rf pulses, one exciting $\nu_\alpha$ and the other the corresponding $\nu_\beta$. Hence, the relation between the two ENDOR frequencies should be known a priori. This condition is often satisfied when the two ENDOR frequencies are symmetrical with respect to the nuclear Larmor frequency, as occurring in liquid samples. For solids, the frequency doublet is symmetrical only when the first-order approximation applies, as is often encountered at a high Zeeman field. Using Special TRIPLE, a maximum of two-fold increase in the ENDOR effect can be achieved (Epel et al., 2003). In the General TRIPLE version, one rf frequency is set to a particular ENDOR transition while the other is scanned (Mehring et al., 1987). As in the cw experiment, the resulting spectrum shows which ENDOR signals belong to the same $m_S$ manifold as that excited by the first rf pulse, thus yielding the relative sign of the hyperfine coupling. When a number of paramagnetic centers contribute to the ENDOR spectrum, the General ENDOR sequence correlates signals belonging to the same center and the same $m_S$ manifold. Scanning the frequency of both rf pulses produces a two-dimensional spectrum with cross peaks between ENDOR lines belonging to the same $m_S$ manifold (Epel and Goldfarb, 2000). In orientationally disordered samples the shape of the cross peaks can provide the relative orientation between the principal directions of the hyperfine interactions involved (Goldfarb et al., 2004).

The sign of the hyperfine coupling is an important parameter. For example, in the case of the isotropic hyperfine constant, $A_{iso}$, the sign provides additional insight into the electronic structure producing unpaired electron density at the nucleus, either by a spin-polarization mechanism ($A_{iso} < 0$, for example $\alpha$ protons) or by conjugation/hyperconjugation mechanisms ($A_{iso} > 0$, for example $\beta$ protons). Moreover, from practical aspects, the sign can be used to obtain unique sets of simulation parameters leading to a more accurate determination of the hyperfine parameters. Finally, it provides a good reference when DFT calculations are carried out for interpreting the hyperfine interaction in terms of structure. It was shown that DFT predicts very well the sign of the interaction (Baute and Goldfarb, 2005; Baute et al., 2004). The experimental determination of the absolute sign of the hyperfine coupling requires large thermal polarization. This is usually not achieved at X-band frequencies but is rather easily obtained at high Zeeman fields and moderately low temperatures. For high-spin system ($S > 1/2$) the sign can be determined by proper selection of the EPR transition (Epel et al., 2001a). This approach, however, does not apply to $S = 1/2$ systems.

For this situation, a variant of the Davies (or Mims)-ENDOR sequence, called variable-mixing-time (VMT)-ENDOR, shown in Fig. 3, can be used for determining absolute signs of hyperfine couplings (Bennebroek and Schmidt, 1997; Epel et al., 2001a,b). In the VMT-ENDOR experiment an additional time interval, $t_{mix}$, which allows for relaxation to take place, is introduced after the rf pulse and before the echo-detection sequence, and the ENDOR spectrum is measured as a function of $t_{mix}$. For a short $t_{mix}$ and a long enough repetition time, which allows the system to return to equilibrium between consecutive sequences

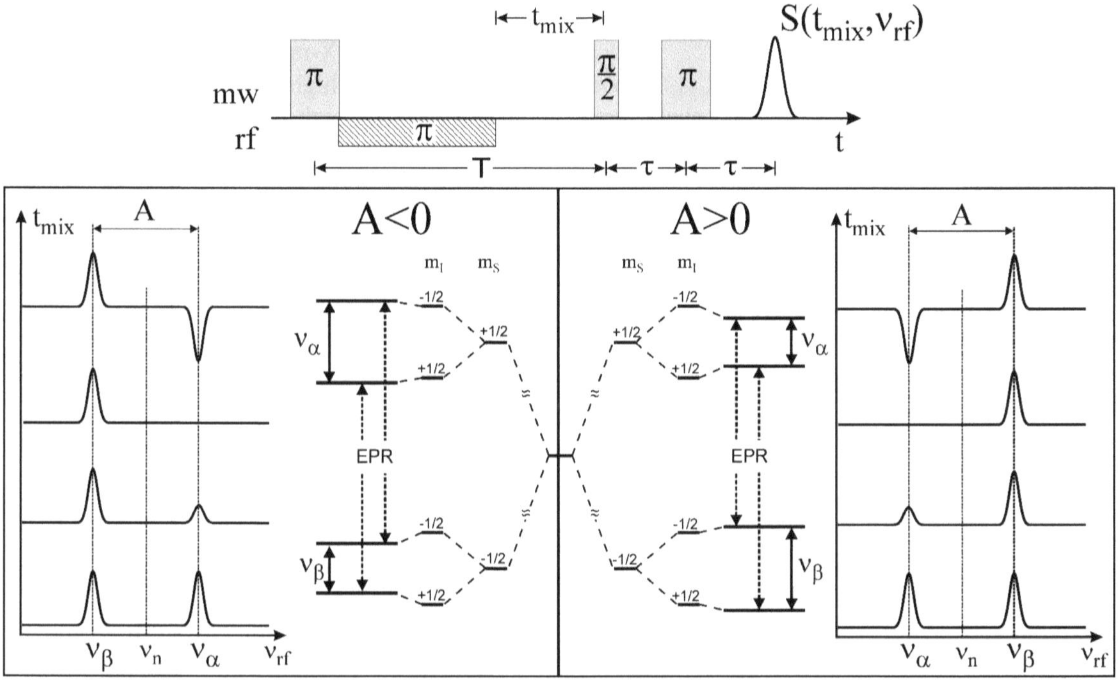


*Fig. 3.* Pulse irradiation scheme of variable-mixing-time (VMT)-ENDOR along with the energy level diagram for S=1/2, I=1/2 and the resulting spectra. In the VMT-ENDOR experiment an additional time interval, $t_{mix}$, which allows for relaxation to take place, is introduced after the rf pulse and before the echo-detection sequence, and the ENDOR effect is measured as a function of $t_{mix}$. For details and references, see text.

(the conditions of standard pulse ENDOR experiments), the $\nu_\alpha$ and $\nu_\beta$ signals have the same intensity ($\nu_\alpha$ and $\nu_\beta$ are the NMR transition frequencies within the $m_S = +1/2$ and $m_S = -1/2$ electron spin manifolds, respectively). As $t_{mix}$ increases and becomes comparable with the electron spin-lattice relaxation time, $T_1$, the signal intensity at $\nu_\alpha$ decreases and can even become negative, while that of $\nu_\beta$ remains positive. This intensity asymmetry is observed, however, only under sufficient thermal polarization and when the electron-nuclear cross-relaxation time, $T_x$, and the nuclear relaxation time, $T_n$, are long relative to $T_1$. Unequal doublet intensities can be observed even at short $t_{mix}$, provided $T_x$ and $T_n$ are longer than the repetition time, $t_R$ ($t_{mix} \ll T_1 \ll t_R$ and $T_n, T_x > t_R$), i.e., saturation of the nuclear transitions prevails (Epel et al., 2001b). An example of the application of VMT-ENDOR can be found in Baute et al. (2004). In principle this experiment should apply to any spin system, and negative ENDOR effects have indeed been observed for high-spin $^{57}Fe(III)$ (Goldfarb et al., 1996).

Extension to high-field ENDOR or TRIPLE has the additional advantage of providing single-crystal like hyperfine information in the reference frame of the g-tensor, even from disordered samples with very small g anisotropy. In section V of this chapter (see below), as an example of Zeeman magnetoselection by W-band pulse EPR and ENDOR, the canonic orientations of radical anions of ubiquinone-10 in frozen solution will be discussed (Rohrer et al., 1998).

In the case of transition-metal complexes, the large hyperfine anisotropy may provide this orientation selectivity for the random distribution of the molecules in the sample. In the case of organic radicals with small hyperfine interactions one has to resort to the anisotropy of the Zeeman interaction. At high $B_0$ fields this can become large enough to provide the desired orientation selectivity for ENDOR experiments on disordered systems. The best approach for elucidating molecular structure and orientation in detail is, of course, to study single-crystal samples. Unfortunately, to prepare them is often difficult or even impossible for large biological complexes. Nonetheless, a few such EPR and ENDOR studies have been carried out at high fields to take advantage of the high sensitivity for size-limited samples, the high resolution and the simplification of the spectra

(Klette et al., 1993; Coremans et al., 1996; Carmieli et al., 2001).

Another virtue of high-field ENDOR is the possibility to detect weakly coupled, low-γ nuclei such as $^2H$. Their ENDOR signals appear around the corresponding Larmor frequency, which at X-band is very low (~2 MHz) and usually hard to detect. Therefore, X-band $^2H$ ENDOR is not often reported. At high fields, like in W-band ENDOR, these signals appear around 20 MHz (see Fig. 2) and are easily detected (Baute et al., 2004). It should be noted that $^2H$ is most useful as a probe for exchangeable protons. It can be used to derive additional structural information if the $^2H$ quadrupolar splittings can be resolved. Such resolution can be obtained if the ENDOR signals are recorded at the canonical position $g_{\|}$ of the EPR spectrum (Kababya et al., 2006).

Finally, at high fields the ENDOR frequencies can often be described using first-order approximations, thereby simplifying the spectral analysis. This is important in the case of quadrupolar nuclei with half-integer spins, for instance $^{17}O$, which have many potential applications for local-structure determination of biological systems. In this case the ENDOR frequencies are given by Eq. (8), including the quadrupolar splitting term. For nuclei with half-integer nuclear spin, such as $I = 3/2$ and 5/2, this term becomes zero for the ENDOR transitions corresponding to $m_I = 1/2 \rightarrow -1/2$. Thus, they are independent of the nuclear quadrupole interaction to first order. To second order, they exhibit an anisotropy which is proportional to $(1/\nu_n)(3e^2qQ/4hI(2I-1))^2$. Consequently, the higher the magnetic field, the simpler is the ENDOR spectrum, and a better resolution is expected for orientationally disordered systems. Similarly, simpler $^{55}Mn$ ENDOR spectra are expected at high fields (Goldfarb et al., 2005), which may be useful for applications to the oxygen evolving complex in Photosystem II.

Below, we give some representative examples of applications of high-field EPR and ENDOR on proteins in general, which demonstrate the power of the methods for analyzing structure and dynamics beyond X-ray crystallography.

The degree of dynamic and structural information on $Mn^{2+}$ sites in proteins that can be derived from single-crystal high-field EPR and ENDOR was demonstrated on the protein concanavalin A (Carmieli et al., 2001, 2003; Goldfarb et al., 2005). This protein is a member of the plant lectin family, a ubiquitous group of proteins that bind saccharides. Concanavalin A has been extensively studied in the past, and its 3D X-ray structure was determined at an exceptionally high resolution, 0.94 Å (Deacon et al., 1997). The high-field EPR/ENDOR investigation emphasizes its complementarities with respect to X-ray crystallography data, namely focusing on local motions and the location of protons. The protein contains two metal binding sites, one occupied by $Mn^{2+}$ and the other by $Ca^{2+}$. The $Mn^{2+}$ site has a slightly distorted octahedral geometry in which the $Mn^{2+}$ is coordinated to three carboxyl groups, one imidazole and two water molecules. The role of the transition-metal ion in this protein is structural, serving as stabilizer for the loops that constitute the saccharide binding site (Kalb et al., 2000).

The ZFS tensor of the $Mn^{2+}$ was determined from W-band EPR single-crystal rotation patterns acquired at room temperature and at 4.5 K (Carmieli et al., 2003). The highly resolved spectra allowed to distinguish two types of $Mn^{2+}$ sites, $Mn_A^{2+}$ and $Mn_B^{2+}$, with different $D$ and $E$ values. This is in contrast to room-temperature measurements, which revealed only one type of $Mn^{2+}$ site, as expected from the X-ray structure. Temperature dependent EPR measurements showed that as the temperature increases, the two well-resolved sextets of $Mn_A^{2+}$ and $Mn_B^{2+}$ shift and gradually coalesce into a single sextet at room temperature. These changes were analyzed in terms of a two-site exchange, with rates in the range of $10^7$–$10^8$ s$^{-1}$ for the temperature range of 200–266 K and an activation energy of 23.8 kJ/mol. This dynamic process was attributed to a conformational equilibrium within the $Mn^{2+}$ binding site which freezes into two conformations at low temperatures (Carmieli et al., 2003).

W-band single-crystal ENDOR measurements, which determined the position of the water protons in the $Mn^{2+}$ sites, were carried out as well (Carmieli et al., 2001). The 3D structure shows that the histidine and water protons are the closest to the $Mn^{2+}$ ions, and therefore they dominate the $^1H$ ENDOR spectrum. $^1H$ ENDOR rotation patterns were collected in two crystallographic planes. The signals of the water and the imidazole ring protons were assigned by comparing the $^1H$ ENDOR spectra of crystals grown in $H_2O$ and $D_2O$, where the $^1H$ water signals are absent from the spectra of $D_2O$ crystals. The signs of the hyperfine splittings were determined through the assignment of the ENDOR signals to their respective $m_S = \pm 1/2$ manifolds. This was achieved by recording the ENDOR spectra at magnetic field positions where the contributions of the $|-1/2\rangle \rightarrow |1/2\rangle$ transitions are negligible and those of the $|-3/2\rangle \rightarrow |-1/2\rangle$ and $|-5/2\rangle \rightarrow |-3/2\rangle$ are substantial (Manikandan et al., 2000). Under such conditions the contribution of the $m_S = 1/2$ ENDOR line is considerably diminished.

Simulations of the rotation patterns (Carmieli et al., 2001) provided the principal components of the hyperfine tensors and their direction relative to the crystallographic axes. The ENDOR spectra recorded at magnetic fields corresponding to $Mn_A^{2+}$ and $Mn_B^{2+}$ were the same. This shows that the Mn-H distances and orientations are the same for the two sites and, therefore, the water and the ligands are not involved in the local motion described above, but rather one (or more) of the carboxylate ligands. Using the point-dipole approximation the coordinates of the water and imidazole protons as well as the O-H distances were calculated and added to the X-ray determined 3D structure. The resulting coordinates of the protons reveal that they participate in hydrogen bonds, showing that besides acting as ligands, the water molecules also play a role in stabilizing the structure of the protein.

Another indication for the source of difference between $Mn_A^{2+}$ and $Mn_B^{2+}$ found at low temperature can be obtained from $^{55}$Mn ENDOR, which provides the $^{55}$Mn hyperfine and nuclear quadrupole interactions. While the former has already been determined through EPR measurements (Carmieli et al., 2003) and was found to be insensitive to the difference between the two sites, the latter turned out to be a better probe. ENDOR measurements were carried out at a crystal orientation where the signals of $Mn_A^{2+}$ and $Mn_B^{2+}$ are well resolved (Goldfarb et al., 2005).The spectra reveal significant shifts, indicating that at this orientation the quadrupole interactions of $Mn_A^{2+}$ and $Mn_B^{2+}$ are different, $P = 0.57$ and 0.26 MHz, respectively. This shows that if the $^{55}$Mn quadrupole interaction can be resolved and fully determined, it can provide further important information regarding the spatial and electronic structure of the protein site.

Other instructive examples of high-field EPR and ENDOR applications were published recently on flavin radicals in DNA photolyase from *Escherichia* (*E.*) *coli* in frozen-solution samples. This protein, though not photosynthetic, uses light-induced electron-transfer reactions to repair UV damages in certain organisms, involving flavin radical intermediates. The neutral flavin-adenin dinucleotid radical cofactor FADH$^\bullet$ has an extremely small g-anisotropy and, hence, 360 GHz/12.8 T EPR was necessary to resolve the canonical g-tensor components in the powder-type EPR spectra (Fuchs et al., 2002). However, in these experiments the orientation of the *x*- and *y*-axes could not be unambiguously determined. The orientation of the g-tensor of FADH$^\bullet$ could, however, be established by pulse ENDOR at 95 GHz and 3.5 T (Kay et al., 2005), taking advantage of the sufficiently high Zeeman magnetoselection to obtain single-crystal like ENDOR data from the disordered sample. This work was extended by an X- and W-band pulsed Davies-ENDOR study at 80 K to probe the N(5)–H bond of the isoalloxazine moiety of the flavin cofactor FADH$^\bullet$ in DNA photolyase (Weber et al., 2005), examining both the protonated and deuterated forms of this bond. By comparing the anisotropic hyperfine couplings measured for the protonated and deuterated samples, after scaling them by the magnetogyric ratio of the deuteron and the proton, subtle differences were revealed concerning the respective deuteron couplings as obtained by 95-GHz ENDOR on H $\rightarrow$ D buffer-exchanged samples. These differences are attributed to the different lengths of N(5)-H and N(5)-D bonds owing to the different masses of H and D. From the distance dependence of the dipolar hyperfine coupling it was estimated that the N(5)-D bond is about 2.5% shorter than the N(5)-H bond.

## 2. High-field pulsed ELDOR

The determination of distance and orientation of protein domains and their changes in the course of biological action are a primary concern in proteomics in the attempt to elucidate the relation between structure, dynamics and function. Hence, a variety of biophysical techniques is used to measure distances (and orientations) in large biosystems. Often, they are available only as disordered samples such as frozen solutions so that X-ray crystallography is not applicable. For such disordered systems, EPR spectroscopy offers a powerful approach to structural information over wide distance ranges depending which anisotropic spin interaction is measured, thereby complementing established techniques like fluorescence energy transfer (FRET) or solid-state NMR. The accessible distance range determined by anisotropic electron-nuclear hyperfine interactions measured by ENDOR stays well below 1 nm. Measurements of the dipolar electron-electron interaction by specialized pulse sequences extends this range dramatically; in ideal cases to about 8 nm (Schweiger and Jeschke, 2001), in proteins realistically to about 5 nm, i.e., still across the photosynthetic membrane. For large interspin distances of well localized electron spins A, B, for which the point-dipole approximation holds and the exchange coupling, $J$, can be assumed to be isotropic, the electron-electron coupling frequency, $\nu_{AB}$, is given by

$$\nu_{AB} = \nu_d(3\cos^2\theta - 1) + J \qquad (11a)$$

with the dipole-dipole coupling constant

$$\nu_d = \frac{\mu_0}{4\pi h} \cdot \frac{\mu_B^2 g_A g_B}{r_{AB}^3} \qquad (11b)$$

Here, $g_A$ and $g_B$ represent the g-values of the radical partners A and B of the pair (principal components of their g-tensors weighted according to their orientation in the pair), $r_{AB}$ their distance and $\theta$ the angle between the Zeeman field $B_0$ and the interspin distance vector $r_{AB}$. For an isotropic distribution of angles $\theta$, a characteristic powder-type spectrum is obtained from which both $r_{AB}$ and $J$ can be deduced, provided $g_A$ and $g_B$ are known from independent EPR experiments. If $J$ can be neglected, as is often the case for radical pairs in photosynthesis, the powder spectrum turns into a Pake pattern, and $r_{AB}$ can be directly obtained from the singularities of the Pake pattern using Eq. (11b). The methodogical challenge is to devise an EPR strategy to separate the electron-electron coupling from other interactions, such as electron-nuclear hyperfine interactions and inhomogeneous line broadening. Among the various methods established so far in this endeavor the 3-pulse electron-electron double resonance technique (pulsed ELDOR or DEER) (Milov et al., 1981, 1984) has proven to be very powerful for large protein systems (Berliner et al., 2000; Schweiger and Jeschke, 2001). Since in this volume an entire chapter by Kawamori and Astashkin (Chapter 16) is devoted to pulsed ELDOR, here we will summarize only those aspects of pulsed ELDOR at high Zeeman fields that are pertinent for measuring both distance and relative orientation of radical-pair partners in photosynthesis.

Orientational information in disordered solids is, of course, available only when a sufficient degree of orientation selectivity is afforded by the EPR spectrum. Therefore, performing pulsed ELDOR experiments in high-field EPR spectroscopy with sufficient Zeeman magnetoselectivity appears to be a promising extension of the pulsed ELDOR technique (Dubinskii et al., 2002). Moreover, interferences from nuclear modulations are usually eliminated at high field. Nonetheless, the separation of the electron-electron dipolar couplings from other modulation or line-broadening effects becomes more demanding at high-field/high-frequency conditions. An obvious problem is the excitation bandwidth which must exceed the dipolar coupling, since both lines of the dipolar doublet of a given spin have to be excited. In pulsed ELDOR, the pump pulse at the second frequency $\nu_B$ must excite a significant fraction of B spins that are dipolarly coupled to the observer spins A, whose resonance frequency is $\nu_A$. In high-field EPR the spectral width of the two radicals in the weakly coupled pair increases in proportion to their difference in g-values and applied Zeeman field. Fortunately, the bandwidth of the EPR resonator also increases with high mw frequency: even for a single-mode cavity with high Q a bandwidth of 100 MHz was achieved at 95 GHz (W-band) and Q = 1000 (Dubinskii et al., 2002). This is sufficient for accommodating both $\nu_A$ and $\nu_B$ for donor-acceptor radical pairs in photosynthesis, but not for nitroxide spin labeled radical pairs. W-band EPR spectra of nitroxide radicals are typically spread over a range of 400 MHz as compared to 200 MHz at X-band. Hence, dual-frequency (DF) pulsed ELDOR (see Fig. 1) at W-band is applicable for $P_{865}^{\bullet+}Q_A^{\bullet-}$ radical pairs in bacterial photosynthesis, and detailed information of interspin distance and orientation of the charge-separated cofactors in the photocycle could be obtained (A.A. Dubinskii, M. Flores, W. Lubitz and K. Möbius, unpublished). For nitroxide biradicals, on the other hand, W-band pulsed ELDOR at a single mw frequency, but with additional fast field jumps (FJ pulsed ELDOR, see Fig. 1) turned out to be an adequate solution of the bandwidth problem (Dubinskii et al., 2002; A.A. Dubinskii, M. Plato, H. Zimmermann and K. Möbius, unpublished). The pros and cons of two-frequency versus field-step pulsed ELDOR techniques have been discussed in the literature (see examples in Schweiger and Jeschke, 2001) and will not be repeated here. At the FU Berlin, a sub-μs field-jump device for W-band high-field pulsed ELDOR has been developed, it is described elsewhere (Dubinskii et al., 2002).

## III. Instrumentation

With the specific advantages of high-field/high-frequency EPR in mind, over the last 20 years a small number of dedicated laboratories met the technological challenge to construct mm and sub-mm high-field EPR and ENDOR spectrometers, thereby opening a promising new research area. The physical principles and technical aspects have been published by the laboratories involved. Appropriate references to them are included in recent overview articles, for instance (Stehlik and Möbius, 1997; Lebedev, 1998; Freed, 2000, 2004; Möbius, 2000; Smith and Riedi, 2000; Prisner et al., 2001; Riedi and Smith, 2002; Grinberg and Berliner, 2004). Since 1995, a W-band high-field EPR spectrometer is commercially available from Bruker GmbH (Karlsruhe), other companies have followed with spectrometer components in the meantime. The main features of the laboratory-built 95-GHz and 360-GHz EPR/ENDOR spectrometers at FU Berlin and at the Weizmann Institute (Rehovot) are summarized below:

The computer-controlled 95-GHz spectrometer at FU Berlin (Burghaus et al., 1992; Prisner et al., 1994; Savitsky et al., 2004; Möbius et al., 2005a) is equipped with a superconducting magnet that provides a $B_0$ field of up to 6 T in a warm-bore with a diameter of 114 mm. The field homogeneity of the solenoid magnet is 1 ppm over a spherical volume of 10 mm diameter. The magnetic field can be linearly swept up to ±0.1 T by controlling the current in additional superconducting sweep coils inside the magnet Dewar. The microwave bridge with a 95-GHz solid-state oscillator source (maximum output power 230 mW) and heterodyne detection operates at an intermediate frequency (IF) of 4 GHz. The quadrature signal is guided either to a lock-in amplifier or to a fast digitizing scope. Several probehead configurations have been constructed including multi-mode Fabry-Perot resonators and single-mode $TE_{011}$ cylindrical cavities. They have been designed and optimized for EPR or ENDOR on a variety of different samples, e.g., lossy aqueous or non-lossy frozen-solution or single-crystal samples. For cw and pulsed ENDOR as well as for field-jump pulsed ELDOR experiments, the $TE_{011}$ cavity is slotted to reduce possible eddy currents. For light-excitation experiments either a laser beam or the light of a halogen lamp is guided to the cavity by means of a quartz fiber. Temperature control of the sample between 90 and 350 K is achieved by a stabilized nitrogen gas-flow system in the probehead. For single-crystal samples, a W-band Fabry-Perot resonator has been constructed containing a goniometer with three rotation axes (Klette et al., 1993) to measure the complete rotation patterns. When using small crystal samples in the $TE_{011}$ cavity, the cw EPR sensitivity is about $10^8$ spins/mT at 1 Hz detection bandwidth, i.e., about two orders of magnitude higher than at standard X-band frequencies. For direct-detection tr EPR measurements, the time resolution of the heterodyne detection channel is about 2 ns. Together with the short ringing time of the EPR cavity (loaded quality factor $Q_L = 2000$) of about 6 ns an excellent time resolution is thus provided that is one order of magnitude higher than in X-band tr EPR. The shortest measured $\pi/2$-pulse length of 22 ns is achieved with a 95-GHz power of only 100 mW at the $TE_{011}$ cavity. In the quadrature detection configuration, the dead time is measured to be about $Q_L/100$ ns when using the full excitation mw power (Möbius et al., 2005a,b).

The computer-controlled W-band EPR/ENDOR spectrometer at the Weizmann Institute (Gromov et al., 1999; Epel et al., 2005) is rather similar to that of the Berlin lab, using the same type of superconducting magnet with a solenoid-coil configuration that can reach a maximum field of 6 T. The major difference between the two spectrometers stems from the fact that the Weizmann spectrometer is used more for measurements of metal-ion complexes and as such requires a wider field sweep and lower temperatures. Hence, the warm bore of 90 mm diameter is equipped with a superconducting sweep coil with permanent current leads to produce a sweep range of ±0.3 T. The magnet is fitted with a variable temperature insert with a temperature range from 2.5 to 300 K using liquid or gaseous helium for cooling. The microwave bridge features two pulse channels and a coherent heterodyne receiver with an IF of 1.82 GHz. The two channels have independent power/phase controls and their outputs can be combined to increase the incident power to the EPR cavity. The second channel can operate in cw mode as well. The maximum power of the 94.9-GHz pulses at the bridge output with the power of the two channels combined is about 200 mW. The probehead set-up is based on a cylindrical $TE_{011}$ cavity, and the rf coil arrangement is similar to that described by the Berlin group. In addition, there is a mechanism that allows for the rotation of the sample holder plunger around the cavity axis for single-crystal measurements. A relatively facile sample exchange for a solenoid configuration, also at low temperature, is obtained by splitting the probehead into two parts (Krymov et al., 1980). Typical sample sizes are 0.8 mm o.d. tubes with a length of about 10 mm, out of which ~3 mm are in the active area of the cavity. With full output power of the bridge the shortest $\pi/2$ pulse obtained is 40 ns, corresponding to a $B_1$ of 0.22 mT. This pulse length is sufficient for all ENDOR experiments, but for ELDOR-type experiments one would probably like to have more power to allow for shorter pulses, especially for the pump $\pi$ pulse. Similarly, for ESEEM (electron spin echo envelope modulation) applications, that for specific magnitudes of $^{14}N$ hyperfine couplings are possible at W-band (Bloeß et al., 1998), shorter pulses may be needed. Typical rf $\pi$ pulses for protons (~144 MHz) are in the range of 12 – 16 μs, whereas for deuterons it is ~40 μs.

The computer-controlled 360-GHz EPR spectrometer built at FU Berlin (Fuchs et al., 1999) uses quasi-optical mw components of the heterodyne mw bridge and a corrugated waveguide in the transmission line to the probehead. This is indispensable, since the losses in conventional fundamental-mode waveguides become prohibitive at mw wavelengths in the sub-mm range, while in a quasi-optical set-up free-space Gaussian-beam propagation leads to very low losses. In an induction-mode detection scheme, the EPR signal is discriminated from the excitation microwave by a wire-grid polarizer. The excitation mw at 360 GHz is generated by a phase-locked tripled 120-GHz Gunn source. In the heterodyne detection channel, an IF signal of 1.2 GHz is generated and handled by a subsequent power-detection or quadrature-detection stage. The probehead contains a semi-confocal Fabry-Perot resonator consisting of a free-standing copper mesh that constitutes the flat mirror, and a spherical gold-coated quartz mirror on which the sample normally is placed. The resonator operates in a cylindrically symmetric $TEM_{006}$ mode. The superconducting magnet provides a $B_0$-field strength of up to 14 T (12.9 T for $g = 2$ systems) with a homogeneity of 3 ppm in a 10 mm sphere and a warm-bore diameter of 88 mm. The integrated superconducting sweep coil provides a sweep-range of ±0.1 T. A gas-flow cryostat enables experiments between 4 K and room temperature, using liquid helium or nitrogen as coolants. The overall detection sensitivity of the spectrometer has been measured to be $1.5 \times 10^{10}$ spins/mT at 1 Hz detection bandwidth.

The basic cw version of the 360-GHz spectrometer has been extended recently for operation both in pulsed EPR and ENDOR modes. For cw ENDOR measurements, the probehead can be fitted with an additional rf circuit, consisting of a single loop coil around the FP resonator and an rf-oscillator circuit with tuneable capacitors to allow for impedance matching of the ENDOR coil. For pulsed EPR operation, the multiplied cw Gunn-diode source can be exchanged for a novel pulsed vacuum-tube source, a 360-GHz Orotron. A detailed description of the operating principle and technical realization of the pulsed Orotron has been presented in a recent publication (Grishin et al., 2004).

## IV. Data Interpretation

The development of high-resolution multifrequency EPR spectroscopy as a powerful tool to explore molecular structure and dynamics has prompted significant improvements of DFT-based quantum chemistry theory for calculating magnetic interactions in large (bio)systems. Reliable computational methods have been developed recently to calculate the interaction parameters of the spin Hamiltonian, in particular zero-field splittings, electron-nuclear hyperfine-, quadrupole- and g-tensor components (Kaupp et al., 2004). It is this combination of new technologies for measuring different molecular parameters and increasing the computing power that now allows biophysicists to describe quantitatively proteins in actions, and to visualize complex biosystems in their three-dimensional structure and functional dynamics. It is noted that, in comparison to hyperfine tensors, the theory for g-tensor calculations of large molecular systems was rather neglected for a long time because experimental g-tensor data of sufficient precision were missing. This has changed owing to the improved spectral resolution and information content of high-field EPR and ENDOR.

## V. Selected Applications of High-Field EPR in Photosynthesis

In this chapter, we focus on a selection of protein systems for which high-resolution X-ray structures are now available. We have chosen (i) bacterial and plant photosynthetic reaction centers for light-induced electron transfer across the membrane and (ii) bacteriorhodopsin, the light-driven transmembrane proton pump in certain halophilic archaebacteria. These proteins have been characterized in detail over the last years by many spectroscopic techniques including ultra-fast laser spectroscopy, FT-IR, solid-state NMR (de Groot, 2000; Mason et al., 2004) and multifrequency EPR (Lebedev, 1994; Möbius, 2000; Riedi and Smith, 2002; Steinhoff, 2002; Lubitz, 2004; Möbius et al., 2005a,b). In addition, sophisticated theoretical studies had been performed to elucidate their light-induced electron and proton transfer characteristics (Jortner and Bixon, 1999). Hence, these proteins represent paradigmatic systems of general interest. They are well suited for new multifrequency high-field EPR experiments to study functionally important transient states during biological action. In photosynthetic electron transfer, these states are directly detectable by EPR via their transient paramagnetism as, for instance, applies for the radical-ion and radical-pair states of cofactors involved. The proton-pump photocycle of bacteriorhodopsin, however, does not involve paramagnetic intermediates. Nevertheless, transient intermediates become EPR-detectable, admittedly somewhat indirectly, in site-specifically nitroxide spin-labeled mutants of bacteriorhodopsin. Generally, this approach of site-directed spin labeling has turned out to be very powerful in protein research (see below).

Most of the recent high-field EPR work on the non-oxygenic photosynthetic reaction centers of purple bacteria and of spin-labeled bacteriorhodopsin presented in this overview was performed at FU Berlin in collaboration with the research groups of W. Lubitz (Mülheim) and H.-J. Steinhoff (Osnabrück), respectively. The examples for high-field EPR on Photosystems I and II in oxygenic photosynthesis mostly refer to recent work of other groups as described in the literature.

### A. Non-Oxygenic Photosynthesis

#### 1. Electron Transfer in Bacterial Reaction Centers

Three billion years before green plants evolved, photosynthetic energy conversion could already be achieved by certain bacteria, for instance the purple bacterium *Rb. sphaeroides*. These early one-cellular photosynthetic organisms contain protein-bound donor-acceptor cofactor complexes, representing one type of reaction center (RC) for light-induced charge-separation of the cofactors by means of electron-transfer (ET) processes. These bacteria cannot split water, but rather use hydrogen sulfide or organic compounds as electron donors to reduce $CO_2$ to carbohydrates with the help of sunlight and bacteriochlorophyll as biocatalyst.

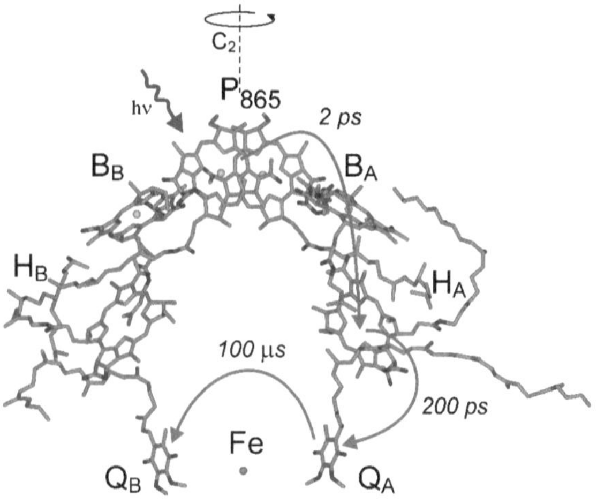


*Fig. 4.* X-ray structural model of the reaction center (RC) from *Rb. sphaeroides* (Chirino et al., 1994) with the cofactors $P_{865}$, (bacteriochlorophyll dimer), B (bacteriochlorophyll monomer), H (bacteriopheophytin), Q (ubiquinone) and $Fe^{2+}$. Light-induced electron transfer proceeds predominantly along the A branch of the cofactors ('unidirectionality' enigma) despite the approximate $C_2$ symmetry of the cofactor arrangement. The ET time constants range from 2 ps to 100 μs in the cascade of transmembrane charge-separation steps. For details, see text.

According to the <3 Å resolution X-ray structures (Allen et al., 1987a,b; Chirino et al., 1994), the cofactors are embedded in the L, M, H protein domains forming two ET branches, A and B. The RC of the carotenoid-less mutant R26 of *Rb. sphaeroides* contains nine cofactors (see Fig. 4): the primary donor $P_{865}$ 'special pair' (a bacteriochlorophyll *a* (BChl *a*) dimer), two accessory BChls ($B_A$, $B_B$), two bacteriopheophytins *a* (BPhe: $H_A$, $H_B$), two ubiquinones ($Q_A$, $Q_B$), one non-heme iron ($Fe^{2+}$). In the context of this chapter it is appropriate to remember that it was EPR and ENDOR by which the primary donor was originally identified as a dimer of BChl molecules (Norris et al., 1971; Feher et al., 1975), long before the first X-ray structure of an RC was published (Deisenhofer et al., 1985). As a dominant motif in the evolution of photosynthetic bacteria, an approximate $C_2$ symmetry of the cofactor arrangement in the RC prevails. It is intriguing that, despite this apparent two-fold local symmetry, the primary ET pathway is one-sided along the A branch, as indicated by the arrows in Fig. 4. The origin of this 'unidirectionality' enigma of bacterial ET is not fully understood yet, although numerous elaborate studies, both experimentally and theoretically, have been performed over the last 25 years. As a matter of fact, from many genetic engineering experiments it turned out that this unidirectionality in the RC is very robust against point mutations of the amino-acid environment of the cofactors (Lancaster and Michel, 2001). It was demonstrated only recently that specific double-site mutations in the vicinity of the primary donor and an accessory BChl can significantly change the partition of ET between the A and B branches (Haffa et al., 2003, 2004). This is a strong indication that in the wild-type system the breakage of symmetry in the ET pathways is largely due to the finely tuned energetics and electronic couplings of the primary donor and the intermediary acceptors (for reviews, see for example, Bixon et al., 1992; Hoff and Deisenhofer, 1997; Lubitz, 2002).

### *a. 360-GHz EPR on the Primary Donor Cation $P^{\bullet+}$ in Mutant RCs*

As indicated above, the unidirectional nature of the primary ET route is probably not determined by a single structural feature, but rather by the concerted effects of small contributions of several factors optimizing, for example, the energetics of the various intermediate states and the coupling scheme of the cofactor wavefunctions (current theoretical concepts invoke 'overlap' and 'superexchange' coupling mechanisms (Jortner and Bixon, 1999)). Both the wavefunctions and the energetics of the cofactors involved in the ET process can be systematically varied by selectively exchanging neighboring amino-acid residues of the protein skeleton by means of site-specific mutation. This can be accomplished, for instance, by introducing or disrupting H-bonds between cofactors and proteins or by changing the ligation of the Mg in the chlorophyll macrocycles of the primary donor P. The effect on the electronic structure by such mutations can be measured, for example, via characteristic shifts of g-tensor and hyperfine-tensor components resolved by high-field EPR (Klette et al., 1993; Huber and Törring, 1995; Fuchs et al., 2003) and ENDOR (Rautter et al., 1995; Huber et al., 1996; Artz et al., 1997; Müh et al., 1998, 2002), respectively. In the course of X-band ENDOR and TRIPLE room-temperature experiments on the R26 mutant and wild-type RCs (Feher, 1992; Lendzian et al., 1993; Hoff and Deisenhofer, 1997; Huber, 1997), a pronounced 2:1 asymmetry of the spin density distribution over the two dimer halves in $P_{865}^{\bullet+}$ was deduced from the hyperfine coupling constants (hfc's), favoring the BChl molecule that is axially ligated by the L-subunit ($BChl_L$). The electronic structure of $P_{865}^{\bullet+}$ in the R26 mutant RC at room temperature turned out to be very similar to that of the carotenoid containing wild-type RC (Geßner et al., 1992).

ENDOR/TRIPLE experiments provide hfc's of the various magnetic nuclei in the $P^{\bullet+}$ molecule which represent local probes of the electronic wavefunction. The assignment of the individual hfc's to molecular positions is generally a difficult task. It involved selective isotope labeling and even additional measurements on RC single crystals (Lendzian et al., 1993). A less time-consuming approach would be desirable to obtain information on the symmetry properties of the electronic structure. This is offered by measurements of the g-tensor. In contrast to the hfc probes, the g-tensor represents a global probe of the electronic wavefunction of the unpaired electron. However, for many organic biosystems the g-tensor anisotropies $\Delta g$ are very small and cannot be resolved by standard EPR at 9.5 GHz (X-band) or 35 GHz (Q-band).

$P^{\bullet+}$ in photosynthetic RCs exhibits extremely small g-tensor anisotropies of $\sim 10^{-3}$. For frozen-solution RCs, therefore, this is a prime example of a situation where even high-field EPR at 3.4 T/95 GHz can only partially resolve the powder pattern of the isotropically disordered samples (Burghaus et al., 1993; Klette et al., 1993; Wang et al., 1994; Huber and Törring, 1995; Huber et al., 1995). To move these samples into the high-field regime already at W-band frequencies (see Eq. (4)), one has to reduce the linewidth by deuteration or by using single-crystal samples. Hence, in an attempt to determine the characteristic symmetry properties of the primary donor electronic structure via the global g-tensor probe, 95-GHz high-field EPR on illuminated single-crystal RCs of *Rb. sphaeroides* was performed at 12 °C, i.e., at physiological temperatures (Klette et al., 1993). The single-crystal EPR linewidth is so small that at a field of 3.3 T even the magnetically inequivalent sites in the unit cell of the RC crystal could be resolved. The rotation patterns of the RC single crystals, i.e., the angular dependence of their g factors in the three symmetry planes of the crystal, have been measured and analyzed by applying Eq. (3). The remarkable result is that the principal directions of the g-tensor are significantly tilted in the molecular axes system. This reveals a breaking of the local $C_2$ symmetry of the electronic structure of $P_{865}^{\bullet+}$, and is consistent with the ENDOR/TRIPLE results for the hyperfine couplings of $P_{865}^{\bullet+}$.

When extending these studies to mutant RCs with

altered amino acids around the primary donor P, the expected shifts of the g-tensor components upon mutation will be even smaller (in the order of $10^{-4}$) as has been shown by previous W-band measurements on similar mutants (Huber and Törring, 1995; Huber et al., 1995). The further increase of EPR frequency and field by a factor of 4 to 360 GHz and 12.9 T, however, provides the spectral resolution necessary to fully resolve all three principal g-tensor components of $P^{\bullet+}$ randomly oriented in frozen solution (Bratt et al., 1999) and to measure their mutation-induced shifts (Fuchs et al., 2003) with high precision. To gain further insight into the origins and consequences of mutation on the electronic structure, various site-directed mutants of the RC have been investigated by 360-GHz high-field EPR. In some of these mutants the ligands to the magnesium of the bacteriochlorophylls were altered, for example by exchange of the histidine His(M202) by leucine (L) or glutamic acid (E) to generate the mutants HL(M202) and HE(M202), respectively (see Fig. 5). Previous experimental results on the M202 mutants from X-ray structural analyses (Yeates et al., 1988; Chirino et al., 1994), ENDOR/TRIPLE (Schulz et al., 1998) and optical and FT-IR measurements (Nabedryk et al., 2000) strongly indicate that the primary donors of the HL(M202) and the HE(M202) mutants are heterodimers of bacteriochlorophyll *a* and bacteriopheophytin *a*, BChl:BPhe, in contrast to the homodimeric species, $(BChl)_2$, of wild-type and R26 RCs. Both EPR studies and energetic considerations then lead to the conclusion that this structural modification of the primary donor in its oxidized paramagnetic state is characterized by an almost complete localization of the spin density on the $BChl_L$ half bound to the L protein subunit (Huber et al., 1996; Schulz et al., 1998).

Figure 6 shows the frozen-solution 360-GHz EPR spectrum of $P^{\bullet+}$ from R26 in comparison with the spectra of the HL(M202) and HE(M202) mutants. The mutation induced g-shifts are clearly resolved. The direction of these shifts is surprising: While for $P^{\bullet+}$ from HL(M202) the overall g-tensor anisotropy $\Delta g = (g_{xx} - g_{zz})$ becomes smaller (more like that of the monomer), for HE(M202) $\Delta g$ increases to a value considerably larger than for R26. This behavior of $\Delta g$ suggests a local structural change of $BChl_L$ as a consequence of the M202 ligand mutations.

To deduce absolute values of the g-tensor components from the experimental spectra, minimum least-squares fits to theoretical spectra were performed invoking a model spin Hamiltonian that includes only the Zeeman interaction of radicals with an isotropic orientation distribution (dotted lines). The $\Delta g$ values obtained from these fits (Fuchs et al., 2003) have an the error of only $2{\cdot}10^{-5}$, and the small shifts in the order of $10^{-4}$, that are induced by the mutations of the protein environment of $P^{\bullet+}$, are therefore detected with high significance.

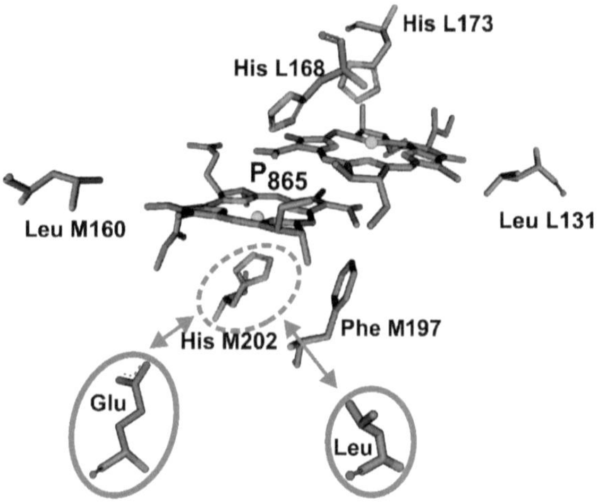


*Fig. 5.* X-ray structure of the primary-donor special pair $P_{865}$ and its immediate amino-acid environment in RCs from *Rb. sphaeroides* (Chirino et al., 1994). In the site-directed mutants HL (M202) and HE (M202) the histidine His M202 is replaced by a leucine Leu and by a glutamic acid Glu, respectively. For details, see text.

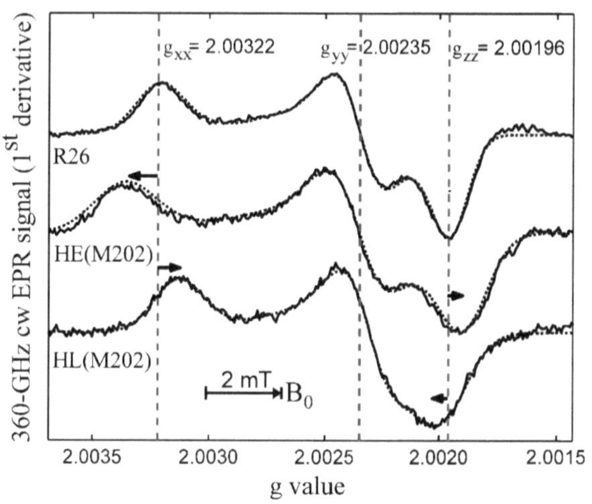


*Fig. 6.* 360-GHz cw EPR signal of $P^{\bullet+}$ in RCs from *Rb. sphaeroides* mutants at 160 K. The most prominent shifts of the g-tensor components $g_{xx}$ and $g_{zz}$ due to the mutation at M202 are indicated with solid arrows. Minimum least squared fits for each spectrum are overlaid with dotted lines, see Fuchs et al. (2003).

To rationalize these experimental g-shifts, quantum-chemical calculations were performed both at semiempirical MO and DFT levels of approximation (Fuchs et al., 2003; Möbius et al., 2005a). The essential conclusion from the current computations is that the observed shifts can be qualitatively understood in terms of mutation induced changes of the rotation angle of the BChl acetyl group, thereby perturbing the planarity of the electronic π-system. This primarily leads to changes in $g_{zz}$. To model the observed strong shift of the $g_{xx}$ component, one needs to take H-bonding interactions with the environment into account. An H-bond of the acetyl-group oxygen atom to the neighboring histidine shifts $g_{xx}$ to larger values, and we conclude that the strength of this H-bond varies in the three mutant species.

At the end of this section on the primary donor, we refer to 130-GHz high-field EPR investigations on the triplet state of P in RCs from *Rb. sphaeroides* (and of the primary donor in Photosystem II) that were started still under the supervision of A. J. Hoff at Leiden University (Paschenko et al., 2001, 2003; Pachtchenko, 2002). The photoexcited triplet state $^3$P was generated in quinone-depleted RCs, and the ESE-detected EPR spectrum revealed a significantly larger g anisotropy than that of the cation radical state $P_{865}^{\bullet+}$. Assuming coaxiality of the g and ZFS tensors, the principal values of the g-tensor could be determined by simulation of the spin-polarized spectra with a precision of ±0.0001. The $^3$P spectral lineshape reveals an orientational anisotropy of the triplet quantum yield. This anisotropy is explained by invoking the participation of the intermediary electron acceptor $I_A$ (a bacteriopheophytin molecule) to form a spin-correlated radical pair $P^{\bullet+} I_A^{\bullet-}$ of ns lifetime as precursor state, thus increasing the probability of formation of the triplet state $^3$P (Pashenko et al., 2001). An extension of this 130-GHz ESE work to the triplet state of Photosystem II reaction centers between 50 and 90 K (Pashenko et al., 2003) showed that at low temperature the ZFS parameters are similar to those of the chlorophyll triplet molecule, in agreement with earlier results. At higher temperatures, the shape of the spin-polarized spectra changes significantly. This is explained by triplet excitation hopping involving the donor chlorophylls, $P_A$ and $P_B$, the accessory chlorophyll $B_A$ and the pheophytin acceptor, $I_A$. In this context, a re-examination of spin-orbit coupling in the triplet state $^3$P in bacterial RCs should be mentioned (Zeng and Budil, 2003) which suggests that singlet-triplet intersystem crossing by spin-orbit coupling plays a significant role in determining the electronic structure of $^3$P. Furthermore, transient (tr) EPR at 240 GHz was used to study the temperature dependence of the g-tensor of the primary donor triplet state $^3$P in photosynthetic RCs of *Rb. sphaeroides* in the temperature range 10 – 230 K (Zeng et al., 2003). In contrast to $P^{\bullet+}$, $^3$P exhibits significant temperature dependence of its g-tensor. Models are discussed to explain the experimental results including the observed $^3$P yield anisotropy.

### *b. 95-GHz EPR on Structure and Dynamics of the Acceptors $Q_A^{\bullet-}$, $Q_B^{\bullet-}$*

Quinones play an important role in many biological systems; prominent examples are the light-driven ET processes of photosynthesis. In purple photosynthetic bacteria, the primary and secondary quinones, $Q_A$ and $Q_B$, act as one- and two-electron gates, respectively. In *Rb. sphaeroides*, $Q_A$ and $Q_B$ are the same ubiquinones-10. Obviously, their different functions in the ET processes are induced by different interactions with the protein environment of their binding sites.

High-field EPR and ENDOR experiments on quinone radical anions in frozen RC and isopropanol solutions were performed at W-band frequencies to measure anisotropic g- and hyperfine-tensor components, respectively (Rohrer et al., 1995, 1998). The aim was to learn about the anisotropic interactions with the protein environment, for example hydrogen bonding to specific amino-acid residues, and about the motional dynamics of the quinones in the H-bond network of their binding sites. Owing to the high Zeeman magnetoselection capability of W-band EPR, a high degree of orientational selectivity is achieved that is inaccessible by X-band EPR (compare Figs. 7A and 7B). The measured g-tensor components follow the sequence $g_{xx} > g_{yy} > g_{zz}$, where $x$ is along the C=O bond direction and $z$ is perpendicular to the quinone plane. This is further exploited by performing W-band pulsed Davies-type ENDOR at the well-separated $B_0$ values corresponding to the canonical g-tensor peaks to obtain dominant proton hyperfine couplings. At least for the $g_{xx}$ and $g_{zz}$ canonical field positions, the ENDOR spectra are single-crystal like with narrow lines. Accordingly, the representations of the orientational selections of molecules show narrow distributions (see right side of Fig. 7C). These representations follow from the simulations of the spectra on the basis of the spin Hamiltonian of Eq. (1). The orientational distribution of molecules

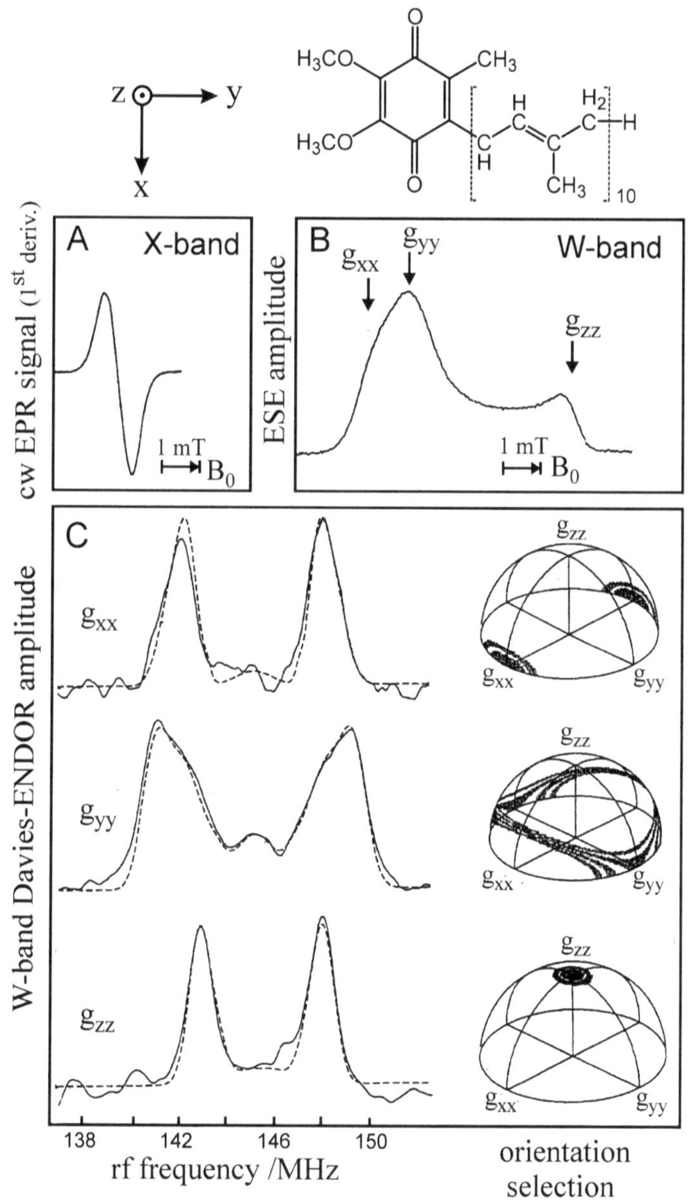


*Fig. 7.* (A) X-band cw-detected and (B) W-band ESE-detected EPR spectra of ubiquinone-10 anion radicals in frozen perdeuterated propan-2-ol solution (T = 115 K). (C) Davies-ENDOR spectra taken at the three $B_0$ positions marked in (B) representing the canonical g-tensor orientations with $g_{xx}$, $g_{yy}$, $g_{zz}$. Dashed lines show the simulated ENDOR spectra from which the degree of orientation selection of contributing molecules (right) was derived (Rohrer et al., 1998). On top the chemical structure of ubiquinone-10 is given together with the coordinate system fixed to the quinone-ring plane.

is considerably broader for the $g_{yy}$ value, which reflects its still insufficient resolution by W-band EPR. When varying the solvent (protic and aprotic, with and without perdeuteration), characteristic changes of hyperfine- (predominantly along the $y$-direction) and g-tensor components (predominantly along the $x$-direction) could be discerned. They are attributed to hydrogen-bond formation at the lone-pair orbitals of the oxygens (Rohrer et al., 1998). Dipolar hyperfine interactions with the solvent protons will result in line broadening along the oxygen lone-pair direction, i.e., of the $g_{yy}$ part of the EPR spectrum. On the other hand, H-bonding induced changes in the lone-pair excitation energy $\Delta E_{n\pi^*}$ and/or spin density $\rho_\pi^O$ at the oxygen will predominantly shift the $g_{xx}$ component of the g-tensor (Rohrer et al., 1995).

Magnetic resonance cannot only probe the structural details of a molecule but also details of its dynamic properties (Thomann et al., 1990; Saxena and Freed, 1997). If the motion is on the time scale of the EPR experiment, additional spin relaxation and concomitant line broadening can be observed in the EPR spectrum. In many cases the analysis of this effect on cw EPR spectra is obscured by static ('inhomogeneous') broadening from unresolved hyperfine interactions or g-strain effects. Hence, pulsed spin-echo (ESE) techniques, which can separate dynamic from static contributions to the spectrum, are often the methods of choice to study molecular motion (Millhauser and Freed, 1984; Dzuba et al., 1992).

As is a common feature in functional proteins, also in photosynthetic RCs molecular dynamics of the protein matrix and of the cofactors plays an important role in the kinetics of the electron transfer (Parak et al., 1980; Kleinfeld et al., 1984). To investigate the molecular motion of the ubiquinone-10 ($UQ_{10}$) acceptor anions, $Q_A^{\bullet-}$ and $Q_B^{\bullet-}$, in RCs of *Rb. sphaeroides* R26, high-field/high-frequency (3.4 T/95 GHz) two-pulse echo-detected EPR experiments have been performed (Rohrer et al., 1996; Schnegg et al., 2002). In the RCs investigated the paramagnetic $Fe^{2+}$ was replaced by diamagnetic $Zn^{2+}$ to reduce the linewidth of the EPR spectra (Debus et al., 1986). Then the g-tensor anisotropy is left as the linewidth-determining relaxation contribution to the W-band spectrum. Because the g-tensor anisotropy of the quinone anion radicals amounts to only $4 \times 10^{-3}$, and the inhomogeneous linewidth in frozen-solution is about 0.5 mT, W-band EPR is required (Burghaus et al., 1993) to spread the spectrum sufficiently (over almost 10 mT) and to separate the $x$, $y$ and $z$ components of the g-tensor (see Fig. 7). Hence, those molecules in the sample can be spectroscopically selected that are oriented with one of their principal g-tensor axes along $B_0$, and the orientation dependent phase-memory time $T_{mem}$ can be measured by the W-band ESE technique (Rohrer et al., 1996), see Fig. 8. $T_{mem}$ is the time constant of the echo decay and, since this was found to be monoexponential, $T_{mem}$ can be identified as the transverse relaxation time $T_2$. At high magnetic fields, the dominant anisotropic $T_2$ relaxation is induced by molecular motion modulating the effective g-value of the tumbling radical. The anisotropic $T_2$ contributions have minima (longest $T_2$) along the canonical orientations of the g-tensor. Thus, by determining $T_2$ as a function of the resonance-field position, high-field ESE decay curves provide information about the directions and amplitudes of molecular motions and their correlation times. Analysis of the data obtained at different temperatures (Rohrer et al., 1996; Schnegg et al., 2002) shows that the relaxation times of $Q_A^{\bullet-}$ and $Q_B^{\bullet-}$ at the resonance positions for $g_{xx}$, $g_{yy}$ and $g_{zz}$ are clearly different, but their orientation dependence is strikingly similar despite their different binding pockets: At 120 K, $T_2$ reaches its maximum at the $g_{xx}$ position of the spectrum while the shortest relaxation times are found at the intermediate position between $g_{yy}$ and $g_{zz}$. From the temperature dependence of the $T_2$ anisotropy of the ESE decays of both $Q_A^{\bullet-}$ and $Q_B^{\bullet-}$ it is concluded that at 120 K, on the EPR time scale, the quinone cofactors librate stochastically around an H-bond axis pointing in the molecular x direction of the g-tensor to an amino acid of their protein pockets.

The $T_2$ values in the various molecular directions mirror the local restrictions to the motion of the quinone in its binding environment. For $Q_A^{\bullet-}$ and $Q_B^{\bullet-}$ the H-bond directions are approximately parallel to $g_{xx}$, as is reflected by their similar relaxation pattern. For $Q_A^{\bullet-}$, this result can be immediately understood from the X-ray structure of the RC in its ground state $Q_A$ (Ermler et al., 1994). The x-axis of the quinone,

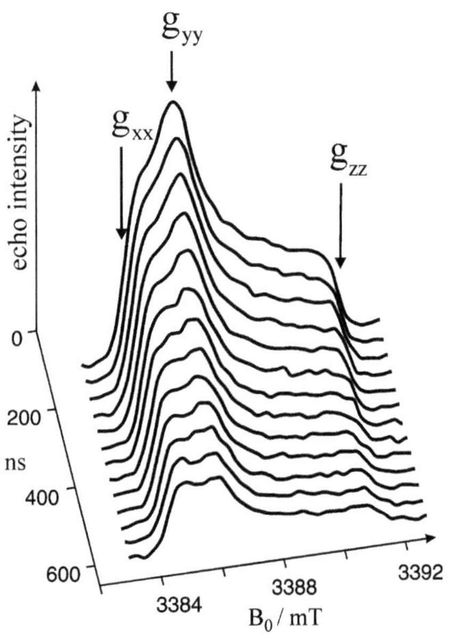


*Fig. 8.* W-band ESE-detected EPR spectra of $Q_B^{\bullet-}$ in RCs ($Fe^{2+}$ substituted by $Zn^{2+}$) from *Rb. sphaeroides* at 120 K. The spectra are measured with the two-pulse echo sequence shown in Fig. 1 for different values of $\tau$. For longer $\tau$, the echo amplitudes decrease and the spectral shape changes because of the orientation dependence of the phase-memory time, see Schnegg et al. (2002).

which is along the C=O bonds, points to the nearby histidine residue, His M219, and to the more distant alanine, Ala M260, allowing a strong and a weak H-bond to be formed between the imidazole and peptide nitrogens and the two respective carbonyl oxygens. At 120 K this asymmetric H-bond pattern is obviously preserved in the anion excited state $Q_A^{\bullet-}$, that is probed by EPR, and is indistinguishable within the accuracy limits from that of the ground-state $Q_A$. This conclusion is in accordance with the H-bond situation observed for $^{13}$C-labeled $Q_A$ by NMR (Breton et al., 1994; van Liemt et al., 1995) and FT-IR (Brudler et al., 1994) and for $^{13}$C-labeled $Q_A^{\bullet-}$ by X- and Q-band EPR (Brink et al., 1994; Bosch et al., 1995; Isaacson et al., 1995; Lubitz and Feher, 1999). In other words: The $Q_A$ binding site does not noticeably change its structure upon light-induced charge separation.

According to recent X-ray diffraction and FT-IR studies (Stowell et al., 1997; Nabedryk et al., 2003; Kötting and Gerwert, 2005) the situation is different for the $Q_B$ binding site. However, the extent of structural changes of the $Q_B$ site to ensure conformational gating for primary electron transfer is still controversially discussed in the literature quoted. Hence, an important question remains to be answered: What are the structural details of the $Q_B^{\bullet-}$ site after charge separation? In the ground-state X-ray structure (Ermler et al., 1994) the C=O bonds of $Q_B$ do not point to any nearby H-bonding amino-acid candidate. Thus, our high-field ESE result, that also $Q_B^{\bullet-}$ undergoes uniaxial librational motion at 120 K around the quinone x-axis, suggests that structural changes of the $Q_B$ binding site upon charge separation occur in such a way that an H-bond candidate like a histidine is shifted towards the C=O bond of the quinone and gets close enough to form an H-bond. This would be in line with structural changes of the $Q_B$ site that have been deduced from recent high-resolution (1.9 Å) X-ray crystallography experiments (Stowell et al., 1997). They have detected pronounced displacements of $Q_B^{\bullet-}$ in the light-adapted X-ray structure in comparison to the dark-adapted one, i.e., between the charge-separated and the ground-state structures.

To summarize: The present results on anisotropic relaxation due to librational motions of the quinones in their specific binding pockets show that low-temperature high-field ESE measurements of the relaxation times at different spectral positions can be used to reveal important structural information on distance and orientation of cofactors in their H-bond network of proteins at work. At temperatures higher than 150 K, the motional modes lose their directional preference.

### *c. 95-GHz EPR on the Spin-Correlated Radical Pair $P_{865}^{\bullet+}Q_A^{\bullet-}$*

Pulsed W-band high-field ESE experiments on the laser-pulse generated short-lived $P_{865}^{\bullet+}Q_A^{\bullet-}$ radical pair in frozen RC solution of *Rb. sphaeroides* were performed with the aim to determine, via spin-polarization effects, the three-dimensional structure of the charge-separated donor-acceptor system (Prisner et al., 1995). This excited-state structure might differ from the ground-state structure and, as was pointed out above, upon illumination drastic changes have been observed in the X-ray structure of the $Q_B$ binding site of *Rb. sphaeroides* in comparison with the dark-adapted X-ray structure (Stowell et al., 1997). The high-field EPR spectra were recorded using the field-swept two-pulse ESE technique. To avoid fast spin relaxation of the $Q_A^{\bullet-}$, the non-heme $Fe^{2+}$ ion was replaced by $Zn^{2+}$. The charge-separated radical pairs $P_{865}^{\bullet+}Q_A^{\bullet-}$ were generated by 10-ns laser flashes. Their time-resolved EPR spectrum is strongly electron-spin polarized because the transient radical pair is suddenly born in a spin-correlated non-eigenstate of the spin Hamiltonian with pure singlet character. Such spin-polarized spectra with lines in enhanced absorption and emission (see Fig. 9) originate from the 'Spin-Correlated-Radical-Pair' (SCRP) mechanism (for key references, see Prisner et al. (1995) and Chapter 19, Daviso et al.). They contain important structural information of magnitude and orientation of the g-tensors of the two radical partners, $P_{865}^{\bullet+}$ and $Q_A^{\bullet-}$, with respect to each other and to the dipolar axis $z_d$ connecting the two radicals (see Fig. 9). Several parameters critically determine the lineshape of the SCRP-polarization pattern, such as the principal values and orientations of the g- and electron dipolar-coupling-tensors, the exchange coupling $J$, and the inhomogeneous linewidths of both radicals. These parameters were determined by independent experiments, i.e., they were not treated as adjustable parameters, in order to obtain meaningful simulations of the spin-polarized lineshapes.

From earlier time-resolved EPR measurements on the radical pair $P_{865}^{\bullet+}Q_A^{\bullet-}$ at X-band (9.5 GHz), K-band (24 GHz) and Q-band (35 GHz), g-tensor orientations could not be extracted unambiguously from spectra simulations (for references, see (Stehlik and Möbius, 1997)). This was mainly because of strongly overlap-

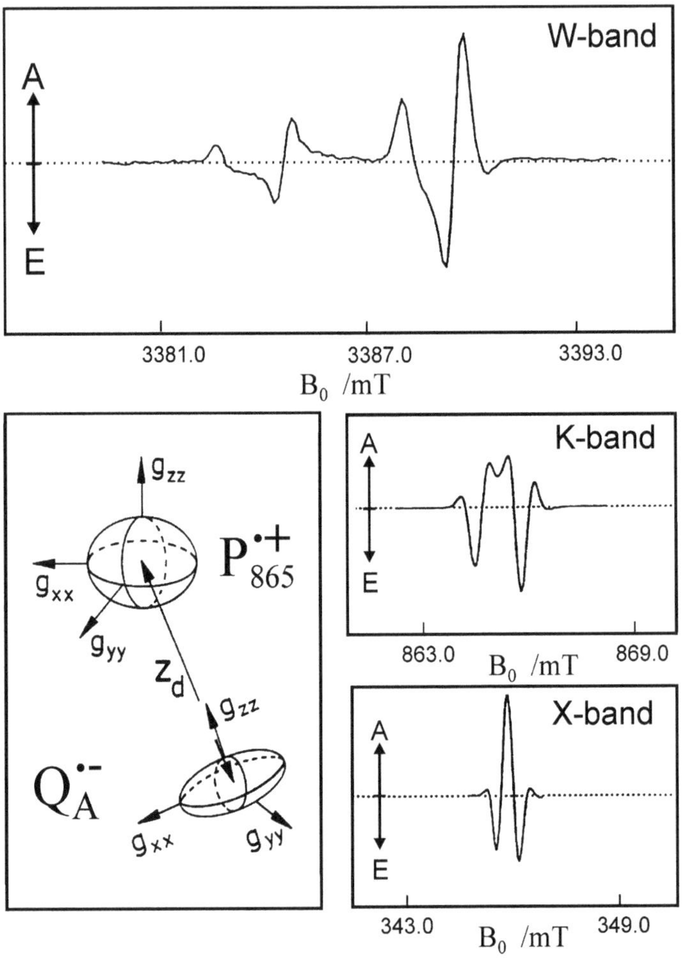


*Fig. 9.* Schematic representation of the relative orientation of the g-tensors and dipolar axis $z_d$ of the transient radical pair $P_{865}^{\bullet+}Q_A^{\bullet-}$ in deuterated frozen RC solution from *Rb. sphaeroides.* The spin-polarized EPR spectra are recorded at various settings of mw frequency and Zeeman field. A and E stand for absorption and emission, respectively. For details, see (Prisner et al., 1995).

ping lines, even when deuterated samples were used to reduce hyperfine contributions. In the pulsed W-band ESE experiments, however, the Zeeman field is strong enough to largely separate the spectral contributions from $P_{865}^{\bullet+}$ and $Q_A^{\bullet-}$. Thus, the overall spectrum is dominated by the characteristics of the two g-tensors, and its interpretation is simplified. It allows an unambiguous analysis of the tensor orientations. The most important result of this high-field ESE study is that, within an error margin of 0.3 Å, no light-induced structural changes of the $Q_A^{\bullet-}$ site with respect to $P_{865}^{\bullet+}$ occur, as compared to the ground-state configuration $P_{865}Q_A$. This finding is in accordance with recent results from various other studies, including X-ray crystallography (Stowell et al., 1997), and contrasts with the $Q_B^{\bullet-}$ situation (see above).

It is noted that also the spin-correlated radical pair $P_{700}^{\bullet+}A_1^{\bullet-}$ in Photosystem I (PS I) from thermophilic *Synechococcus (S.) elongatus* has been studied using similar pulsed W-band ESE techniques (Est et al., 1997). A comparison of the spin-polarized spectra shows that the g-tensors of the quinone acceptors are oriented differently with respect to those of $P^{\bullet+}$ in the *Rb. sphaeroides* RC and in PS I. Taking the spin-correlated radical pair $P_{700}^{\bullet+}A_1^{\bullet-}$ as an example, the extension to pulsed Davies-ENDOR at X-band has been demonstrated (Fursman et al., 2002; Teutloff et al., 2004). The transient ENDOR spectra exhibit both absorptive (A) and emissive (E) lines whose E/A character is strongly field dependent. For the interpretation of the spin-polarized ENDOR spectra, the SCRP model was adapted for systems containing a pair of spin-spin coupled electrons A and B (representing the radical-pair partners), one of them interacting with an I = 1/2 nucleus. This 3-spin radical-pair model gives rise to ENDOR transitions in addition to those usually found in ENDOR of a single radical coupled to a nucleus with I = 1/2. Computer simulations of the experimental time-resolved ENDOR spectra on the basis of the extended SCRC model show that the spin-polarized spectra are sensitive to the relative signs of the hyperfine couplings. The analysis of pulsed X-band ENDOR spectra of spin-correlated radical pairs that are recorded at specific field positions, is further complicated by the strongly overlapping contributions of the individual radicals of the pair, $P_{700}^{\bullet+}$ and $A_1^{\bullet-}$, at X-band. Very recently, in a related high-field ENDOR experiment on the weakly coupled transient spin-correlated radical pair $P_{865}^{\bullet+}Q_A^{\bullet-}$ in RCs from *Rb. sphaeroides*, the contributions of the individual radicals of the pair could be disentangled (Poluektov et al., 2004, 2005). This was achieved by Mims-ENDOR at 130 GHz and 4.6 T. At this high field there is no appreciable mixing of singlet and triplet states of the radical pair, thereby simplifying the spectra analysis. It is noted that in their theoretical treatment the authors included hyperfine interaction with both correlated electrons of the pair, thus extending the model of Fursman and co-workers (Fursman et al., 2002).

We conclude that by pulsed high-field EPR and ENDOR experiments on spin-correlated coupled radical pairs a detailed picture of the electronic structure and spin dynamics of the ET partners can be obtained. Moreover, ET-induced structural changes in the relative orientation of donor and acceptor can be detected with high precision, even for disordered samples. Such information is important for a deeper understanding of the ET characteristics of charge-recombination processes on the molecular level. The charge-separated radical-pair state represents the initial state for ET recombination, i.e., it is one of the working states of the photocycle in the RC.

## 2. *Proton Transfer in Bacteriorhodopsin*

Apparently, nature has invented photosynthesis twice: The strategy to use sunlight as energy source for synthesizing ATP culminated in the photosynthetic reaction-center protein complexes of bacteria, algae and plants performing light-induced electron transfer between primary donor and acceptor cofactors, and in the bacteriorhodopsin (BR) protein complex of certain archaebacteria. They perform light-induced proton transfer between amino-acid residues, controlled by conformational changes of the only cofactor, the retinal. During the photocycle of BR no paramagnetic intermediates occur, i.e., neither radicals and radical pairs nor triplet states. To enable the application of EPR, doublet-state spin labels ($S$ = 1/2) can be introduced to the protein. The site-directed spin-labeling (SDSL) technique in combination with X-band EPR spectroscopy was pioneered by Hubbell (Hubbell and Altenbach, 1994; Altenbach and Hubbell, 1998). The SDSL technique requires selective cysteine-substitution mutagenesis of the protein with subsequent modification of the unique sulfhydryl group of cysteine with a nitroxide reagent, for example (1-oxil-2,2,5,5-tetramethyl-pyrroline-3-methyl)methane-thiosulfonate, commonly abbreviated as MTS spin label. The recent extension to SDSL/high-field EPR has opened new perspectives for studying structure and dynamics of large nitroxide-labeled proteins during biological action (Earle et al., 1994; Lebedev, 1994; Steinhoff et al., 2000; Wegener et al., 2001; Plato et al., 2003; Savitsky et al., 2004; Marsh et al., 2002; Möbius et al., 2005b). Figure 10 demonstrates the remarkable gain in resolution of nitroxide-radical spectra, i.e., the separation of the $g_{xx}$, $g_{yy}$, $g_{zz}$ components in relation to the $A_{zz}$ nitrogen hyperfine-tensor component, when increasing the Zeeman field from X-band EPR to 95-GHz and 360-GHz EPR. A detailed coverage of the spin-label EPR technique is presented in the chapter 17 by Borovykh and Steinhoff in the current volume.

Bacteriorhodopsin is a 26 kDa protein complex located in the cell membrane of halophilic archaes such as *Halobacterium salinarium*. High-resolution (1.6 Å) X-ray crystallography provided atomic coordinates for the ground-state structure (Luecke et al., 1999), see Fig. 11. Seven transmembrane helices (A–G) enclose the chromophore retinal which is covalently attached to the amino acid lysine, K216, on helix G via a protonated Schiff base. Absorption of 570-nm photons initiates the all-trans to 13-*cis* photoisomerization of the retinal. The Schiff base then releases a proton to the extracellular medium and is subsequently reprotonated from the cytoplasm. Transient intermediates of this catalytic photocycle can be distinguished by the different absorption properties of the retinal, and a sequence of intermediates (J, K, L, M, N and O) has been characterized by time-resolved absorption spectroscopy (Váró and Lanyi, 1991; Druckmann et al., 1992; Hessling et al., 1997). During this photocycle conformational

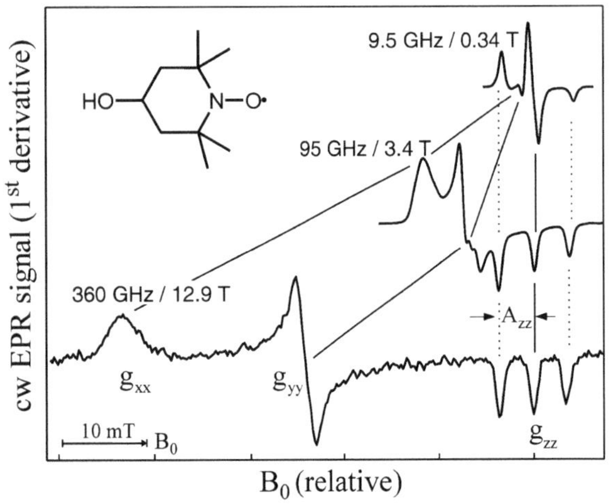


*Fig. 10.* cw EPR spectra of anitroxide radical (OH-TEMPO) in frozen water solution at different mw frequency/$B_0$ settings (Fuchs et al., 1999; Möbius et al., 2005a; with permission). The spectra are plotted relative to the fixed $g_{zz}$ value.

changes of the protein (and the retinal) occur, as has been detected by a variety of experimental techniques (for a review, see (Haupts et al., 1999)). Such changes ensure that release and uptake of protons do not occur from the same side of the membrane, but rather enable BR to work as a vectorial transmembrane proton pump. Detailed analyses of the nature of the conformational changes and their functional role have been performed, for example by neutron diffraction, electron microscopy, X-ray diffraction, solid-state NMR (Hu et al., 1998; Moltke et al., 1999) and EPR spectroscopy (Thorgeirsson et al., 1997; Rink et al., 2000; Steinhoff et al., 2000; Xiao et al., 2000; Radzwill et al., 2001; Wegener et al., 2001). Here, we restrict our discussion to recent 95-GHz high-field EPR experiments on site-specifically spin-labeled BR mutants to measure the hydrophobic barrier for transmembrane vectorial proton transfer (Steinhoff et al., 2000).

For systematic studies of the BR proton pump, a set of SDSL mutants was constructed, each containing a single nitroxide-containing amino-acid side chain, differing by position in the protein sequence. The photocycle of all spin-labeled mutants was checked to ensure that the overall function of the BR protein is retained (Steinhoff et al., 2000; Wegener et al., 2001; Savitsky et al., 2004). To determine the hydrophobic barrier of the putative proton-transfer channel, the polarity profile was probed by means of the g- and

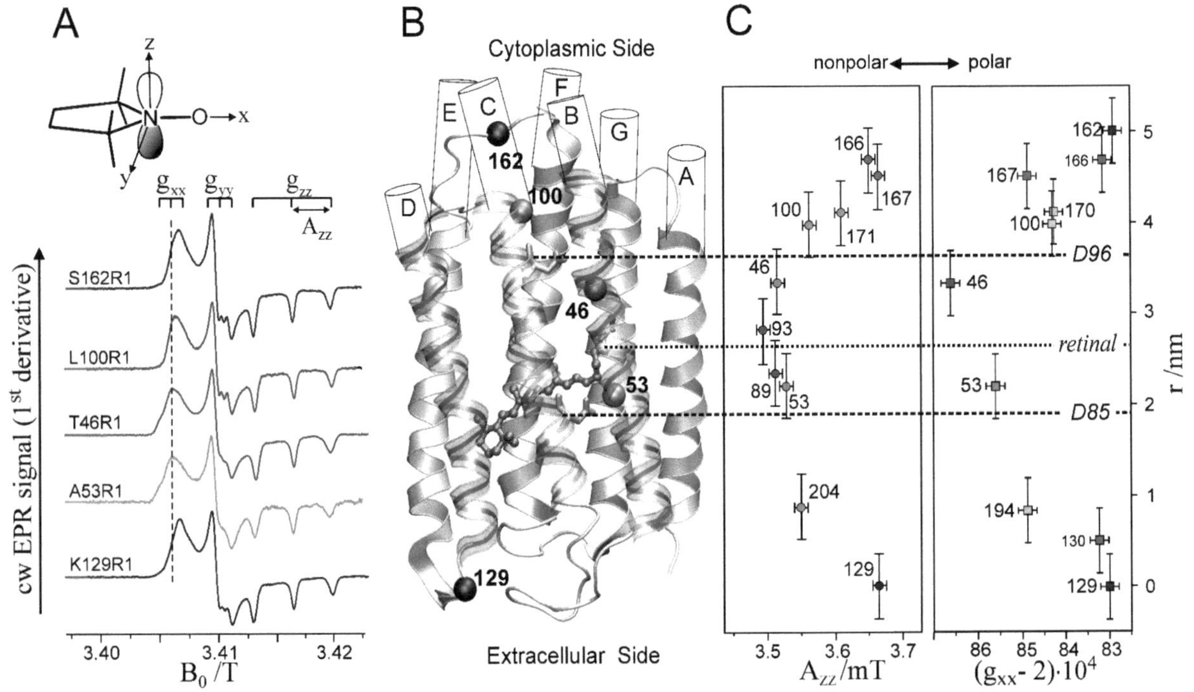


*Fig. 11.* A: Experimental W-band cw EPR spectra for a set of bacteriorhodopsin (BR) mutants spin labeled with the nitroxide side-chain (R1). B: Structural model of BR with the seven helices A–G. The $C_\alpha$ atom of the spin-labeled residues, the chromophore retinal, and the aspartic acids D96 and D85 participating in the $H^+$ transfer are indicated. C: The magnitude of tensor elements $g_{xx}$ and $A_{zz}$ of the spin labels as function of the R1 location r in the protein with respect to position 129 (r = 0).

hyperfine-tensor components of a series of 10 site-specifically nitroxide spin-labeled BR mutants, using MTS spin label as the reporter side chain $R_1$. Previous studies of a large number of spin-labeled proteins have shown that the $A_{zz}$ component and the $g_{xx}$ component are particularly sensitive probes of the microenvironment of the nitroxide side chain $R_1$. They allow to measure changes in polarity and proticity, i.e., $g_{xx}$ and $A_{zz}$ probe the local electric fields and the availability of H-bond forming partners of nearby amino-acid residues or water molecules (Earle et al., 1994; Steinhoff et al., 2000; Wegener et al., 2001; Marsh et al., 2002; Plato et al., 2002; Smirnov, 2002, Möbius et al., 2005b). Moreover, the dynamic properties of the nitroxide side chain and, thus, the EPR spectral lineshapes contain direct information about constraints to molecular motion that are introduced by the secondary and tertiary structures of the protein in the vicinity of the nitroxide binding site. The W-band EPR spectra were recorded at temperatures below 200 K to avoid motional averaging of the anisotropic magnetic tensors. At these temperatures, R1 can be considered as immobilized on the EPR time scale. The spectra of selected mutants are included in Fig. 11. They exhibit the typical nitroxide powder-pattern lineshape expected for an isotropic distribution of diluted radicals. The spectra are clearly resolved into three separate regions corresponding to the components $g_{xx}$, $g_{yy}$ and $g_{zz}$, the latter with resolved $A_{zz}$ splitting. The variations of $g_{xx}$ and $A_{zz}$ with the nitroxide binding site can be measured with high precision. The plots of $g_{xx}$ and $A_{zz}$ versus $R_1$-position $r$ along the proton channel (Fig. 11) demonstrate distinct variations in the polarity and proticity of the nitroxide microenvironment. The plots directly reflect the hydrophobic barrier the protons have to overcome on their way though the protein channel.

The analysis of both tensor components, $g_{xx}$ and $A_{zz}$, allows to characterize the R1 environment in terms of protic and aprotic surroundings. Theoretically, both $g_{xx}$ and $A_{zz}$ are expected to be linearly dependent on the π-spin density $\rho_\pi^O$ at the oxygen atom of the nitroxide group. For $g_{xx}$, however, apart from a direct proportionality to $\rho_\pi^O$, there is an additional dependence on specific properties of the oxygen lone-pair orbitals. The lone-pair orbital energy $E_n$ affects $g_{xx}$ via the excitation energy $\Delta E_{n\pi^*} = E_{\pi^*} - E_n$ and is known to be sensitive to the polarity of the environment. It is particularly sensitive to H-bonding of the lone pairs to water or to polar amino-acid residues. Thus, the plot of $g_{xx}$ versus $A_{zz}$ should indicate the presence or

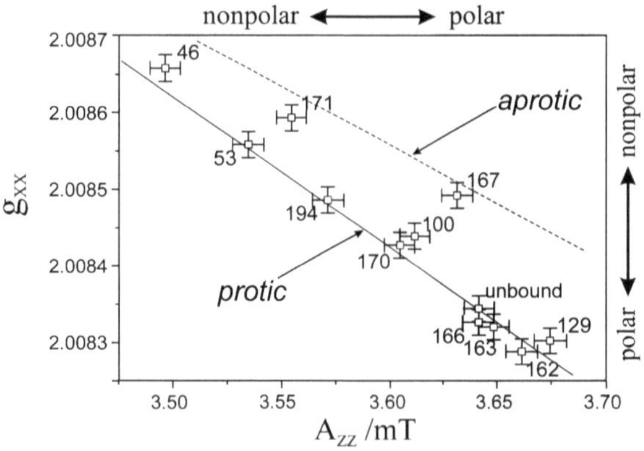


*Fig. 12.* Plot of $g_{xx}$ versus $A_{zz}$ of the nitroxide side-chains for various spin-label positions in BR (see text). The 'protic' and 'aprotic' limiting cases are placed with reference to theory, see (Plato et al., 2002).

absence of H-bonds in the spin-label environment, i.e., its proticity. This dependence is plotted in Fig. 12 for various spin-label positions in BR (Steinhoff et al., 2000).

Obviously, two straight-line correlations can be deduced. The points corresponding to positions 46, 171 and 167 belong to a line whose slope is different from that for the remaining points. These three positions can be classified to be exposed to an aprotic environment (Kawamura et al., 1967; Griffith et al., 1974), the other ones to a protic environment. This allows to characterize the hydrophobic barrier of the BR proton channel in terms of different accessibilities of the respective protein regions to water molecules. The remarkable sensitivity of resolved tensor components $g_{xx}$ and $A_{zz}$ of nitroxide labels to the polarity and proticity of their microenvironment has been recognized also by other high-field EPR research groups studying nitroxides in frozen solution (Ondar et al., 1985) or phospholipid membranes (Earle et al., 1994; Lebedev, 1994; Marsh et al., 2002).

A theoretical interpretation of the observed $g_{xx}$ versus $A_{zz}$ dependence in nitroxide spin labeled BR was performed on the basis of semiempirical MO theory and by using simple models for varying solute-solvent interactions (Plato et al., 2002). The goal was to extract a measure for the different magnitudes of polar and protic contributions from the spin-label environment to $g_{xx}$ and $A_{zz}$ and to find a way to discriminate between these two effects. The variations of $g_{xx}$ due to solute-solvent interactions can be understood qualitatively by analyzing the dominant constituents of this component according to the theory of Stone (Stone, 1963; Törring, 1995,

1997). These are approximately given by

$$g_{xx} = g_e + \frac{2\zeta(O)\rho_\pi^O c_{ny}^2}{\Delta E_{n\pi^*}} \qquad (12)$$

where $g_e = 2.0023$ is the free-electron g-value; $\zeta(O)$ is the oxygen spin-orbit coupling parameter; $\rho_\pi^O$ is the $\pi$ spin density on the oxygen $2p_z$ atomic orbital; $c_{ny}$ is the LCAO coefficient of the $2p_y$ atomic orbital contributing to the oxygen lone-pair MO n; and $\Delta E_{n\pi^*}$ is the $n \rightarrow \pi^*$ excitation energy. Solute-solvent interactions can affect all three quantities $\rho_\pi^O$, $c_{ny}^2$ and $\Delta E_{n\pi^*}$, the first two by electron density rearrangements, and the latter by energy-level shifts either by superimposed electric fields or by H-bond formation. Thus, any change in $g_{xx}$ can be broken down into three contributions:

$$\frac{\delta \Delta g_{xx}}{\Delta g_{xx}} = -\frac{\rho_\pi^N}{\rho_\pi^O} \cdot \frac{\delta A_{zz}}{A_{zz}} - \frac{\delta \Delta E_{n\pi^*}}{\Delta E_{n\pi^*}} + \frac{\delta c_{ny}^2}{c_{ny}^2} \qquad (13)$$

where $\Delta g_{xx} = g_{xx} - g_e$ and where $\delta\rho_\pi^O$ has been substituted by the corresponding change in $A_{zz}$, making use of the sum condition $\rho_\pi^O + \rho_\pi^N \cong 1$ and of the proportionality between $A_{zz}$ and $\rho_\pi^N$, the $\pi$ spin density at the nitrogen atom. Eq. (13) correctly predicts the observed negative slope of the plot $g_{xx}$ versus $A_{zz}$. The last two terms are responsible for additional vertical deviations. A detailed discussion of the theoretical results is given by Plato et al. (2002) and Möbius et al. (2005a).

## B. Oxygenic Photosynthesis

Our present understanding of the structure-function relationship for the primary processes in plant photosynthesis is still considerably lower than in bacterial photosynthesis owing to the much higher complexity of the charge-transfer photomachine of interconnected Photosystems I and II (PS I, PS II). Moreover, until very recently the three-dimensional structures of PS I and PS II were not known with sufficient accuracy. This situation was in contrast to that of RCs of purple photosynthetic bacteria, but has dramatically improved now thanks to the recent progress in high-resolution X-ray crystallography of PS I and PS II (Jordan et al., 2001; Zouni et al., 2001).

### 1. Photosystem I

Taking PS I as an example, the 2001 X-ray structure has reached 2.5 Å resolution (Jordan et al., 2001), while until then only 4 Å resolution could be achieved (Klukas et al., 1999), and it was impossible to even locate the quinone acceptor $A_1$, a phylloquinone. It was, therefore, a challenge to apply the whole arsenal of modern multifrequency time-resolved EPR and ENDOR techniques to PS I with the goal to determine the $A_1^{\bullet-}$ location and orientation (for recent overviews, see Bittl and Zech (2001); Teutloff et al., (2001); Lubitz et al. (2002)).

Additional variations on the PS I theme have been published recently with special emphasis on the advantages of high-field EPR. A detailed g-tensor analysis has been made for the cation radicals and triplet states of the primary donor $P_{700}$ in PS I and chlorophyll *a* in vitro (Poluektov et al., 2002). Time-resolved W-band EPR with direct detection (tr EPR) has been used to study the transient radical pair $P_{700}^{\bullet+}A_1^{\bullet-}$ in oriented multilayer samples (Schnegg et al., 2002). For these experiments special variants of Fabry-Perot resonators have been developed to measure the orientation dependence of spin-polarized spectra from frozen-solution PS I preparations.

To summarize: combining all pieces of information from the various state-of-the-art EPR and ENDOR experiments, finally the goal of localizing $A_1$ in PS I was achieved. Hence, multifrequency time-resolved EPR proved again to be particularly suited for the characterization of the radical and radical-pair intermediates in the RCs of different photosynthetic organisms. It is, of course, satisfying that the results obtained by EPR are in agreement with the most recent structures from high-resolution X-ray crystallography (Jordan et al., 2001).

Aspects of mutation-induced changes of the PS I complex and their characterization motivated a number of pulsed high-field EPR and ENDOR studies (Zybailov et al., 2000; Xu et al., 2003). For example, specific biosynthetic pathways were developed to recruit foreign quinones into the $A_1$ site of PS I. From multifrequency pulsed EPR and ENDOR experiments the structural and dynamic characteristics of the reconstituted $A_1$ site were determined.

### 2. Photosystem II

Taking PS II as an example, one notices a growing interest in advanced EPR studies of the tyrosyl radicals in the $D_1$ and $D_2$ polypeptides. In particular, the determination of individual g-tensor components in frozen PS II preparations by high-field/high-frequency EPR proved to be very informative in probing the different hydrogen-bonding interactions of the $Y_Z$ and $Y_D$ tyrosyl and the $Q_A$ and $Q_B$ quinone radicals (see publications cited in Stehlik and Möbius (1997); Un et al., (2001)). Their frozen-solution (Stehlik and Möbius, 1997; Un et al., 2001) and single-crystal (Hofbauer et al., 2001) spectra exhibit resolved Zeeman and, in some cases, hyperfine structure. Similar cw high-field EPR studies have been performed to characterize the pheophytin anion radical in the wild type and $D_1$-E130 mutants of PS II from *Chlamydomonas* (*C.*) *reinhardtii* (Dorlet et al., 2001). The mutants were constructed site-specifically to weaken or remove the hydrogen bond of the carbonyl group at ring V of the pheophytin molecule to the $D_1$ polypeptide. A strong mutation-induced shift of the $g_{xx}$ component of the g-tensor was observed. DFT calculations were used to rationalize the observed trend in the variations of the $g_{xx}$ component. The results again demonstrate the sensitivity of g-values in probing the protein microenvironment of cofactor radicals.

We conclude this section on PS II by referring to recent $^{55}$Mn pulse Davies-ENDOR experiments at Q-band (34 GHz) of the $S_0$ and $S_2$ states of the oxygen-evolving complex (OEC) (Kulik et al., 2005a,b). Normally, Q-band EPR is not considered to be a 'true' high-field experiment (see Eq. (4)). However, despite the small orientational magnetoselection of Q-band EPR on PS II, the $^{55}$Mn-ENDOR spectrum allowed the determination of the principal values of the hyperfine tensors for all four Mn ions. The hyperfine components were found to be slightly different in the $S_0$ and $S_2$ states. The ultimate goal of such experiments is to derive the electronic and spatial structure of the OEC. Although this goal was not yet achieved by these experiments, their results already safely exclude all models in the literature proposing two magnetically uncoupled dimers as structural elements of the OEC in the $S_2$ state.

## VI. Conclusions and Perspectives

In this chapter it is shown that multifrequency EPR spectroscopy at high magnetic fields provides detailed information about structure and dynamics of transient radicals and radical pairs occurring in light-induced biological electron and proton transfer processes. Thereby our understanding of the relation between structure, dynamics and function in photosynthesis is considerably improved. This holds with respect to the fine-tuning of electronic properties of donor and acceptor cofactors by weak interactions with their protein and lipid environment, such as H-bonding to specific amino-acid residues in photosynthetic reaction centers. This also holds with respect to the fine tuning of polarity and proticity regions along the proton channel in the bacteriorhodopsin protein complex. A survey of biological magnetic-resonance literature shows a strong increase of high-field EPR applications. Important information is revealed by high-field EPR spectroscopy on nm and sub-nm structural details of proteins and on the dynamics of their transient states while in biological action. New insights are obtained that are complementary to what can be learned from other spectroscopic techniques. Sometimes, high-field/high-frequency EPR provides even unique insights that remain inaccessible by other spectroscopy or diffraction techniques. This applies, for example, to large biosystems that lack long-range order, such as glasses and frozen solutions, for which electronic structures, distances and orientations of cofactors or protein fragments can be measured over a broad distance range, say $0.2\,\text{nm} < r < 5\,\text{nm}$, even in the time-resolved mode for short-lived intermediates in enzymatic reaction cycles. Below, some summarizing conclusions and perspectives are given, relevant to biological systems in general:

1. Many organic cofactors in electron-transfer proteins have only small g-anisotropies and, therefore, disordered samples require much higher magnetic fields than available in X-band EPR to resolve the canonical g-tensor orientations in their powder spectra. Thereby, orientation-selective hydrogen bonding and polar interactions in the protein binding sites can be traced which provide information complementary to what is available from high-resolution X-ray diffraction of protein single crystals.

2. Several organic radical species are often generated as transient intermediates during electron-transfer processes in proteins. To distinguish them by the small differences in their g-factor and hyperfine interactions, high-field EPR becomes the method of choice.

3. The common problem of small concentra-

tion of paramagnetic states in protein samples is particularly dramatic for small protein single crystals or for dilute solutions of site-directed spin labeled mutants or isotopically labeled cofactors. Accordingly, to study them by EPR or ENDOR, very high detection sensitivity is needed, still posing a technical challenge when constructing high-field/high-frequency spectrometers. They require incorporation of optimized mw resonators in the probehead, for instance single-mode cavities or Fabry-Perot resonators. Best concentration sensitivity appears to be attainable in the W-band region. At much higher mw frequencies, sometimes a price has to be paid for the gain in spectral resolution in terms of loss of sensitivity by lower resonator quality factor or even by working without resonator at all.

4. High-field/high-frequency cw EPR generally provides, by lineshape analysis, shorter time windows for molecular motion than X-band EPR does, down to the ps range. This allows correlation times and fluctuating local fields to be studied over a wide temperature range. They are associated with characteristic dynamic processes, such as protein, cofactor or lipid motion and protein folding. Optimized time windows can be chosen by a multifrequency EPR approach to disentangle different modes of motion.

5. Time-resolved high-field/high-frequency EPR techniques such as two-dimensional field-swept ESE or direct-detection tr EPR, provide real-time access to specific cofactor and/or protein motions in the ns time scale. Motional relaxation anisotropy can be resolved in the field-dependent echo decays. It is governed by anisotropic interactions, such as hydrogen bonding along a specific molecular axis within the binding site.

6. ENDOR at high Zeeman fields takes additional advantage of the orientation selection of molecular sub-ensembles in powder or frozen-solution samples. Thereby, even in the case of small g anisotropies, ENDOR can provide single-crystal like information about hyperfine interactions, including the directions of hydrogen bonds in the protein.

7. In metallo-protein high-spin systems, such as the $Mn^{2+}$ proteins, the EPR spectrum analysis can be drastically simplified at high Zeeman fields due to the suppression of second-order effects. This is also accompanied by a considerable increase in sensitivity. In case of large zero-field splittings, EPR transitions might not be observable at all at X-band, but become accessible at higher quantum energies of mm or sub-mm microwaves.

8. Another important advantage of high Zeeman fields for high-spin systems, such as triplet states or radical pairs, is the possibility to determine the absolute sign of the zero-field parameter $D$ (and thereby the shape of the zero-field tensor) by thermal polarization already at moderately low temperatures.

9. In principle the advantages of high-field EPR in terms of resolution and sensitivity automatically translate into advantages in high-field ENDOR. In addition, the high field allows a better resolution of different elements and isotopes. Low-$\gamma$ nuclei with small couplings, such as $^2H$ and $^{17}O$, can easily be detected. The ENDOR frequencies can often be described using first-order approximations, thereby simplifying the spectral analysis. Finally the large thermal polarization achieved at low temperatures allows the determination of the absolute sign of the hyperfine interaction. This sign is yet another parameter which provides insight into the electronic structure of the paramagnetic center.

10. High-field EPR noticeably extends the applicability of the site-directed spin-labeling technique in bioengineering. Owing to the resolution of both the hyperfine- and g-tensor components of nitroxide spin probes, polarity and proticity profiles of the protein microenvironment can be identified, even the modulation of their profiles by conformational changes during biological action. They provide key information on molecular switches that control vectorial transmembrane transport processes in functional proteins. In this respect, the recent progress made in the DFT-based quantum theoretical interpretation of the EPR observables, in particular of the g and hyperfine tensors, is well timed.

11. Further developments in the computation capabilities for magnetic resonance parameters of large biosystems are envisaged for the near future. The progress in measuring molecular parameters with high precision will continue to cross-fertilize the efforts of theoreticians to interpret the experimental

data in terms of molecular structure and dynamics, and vice versa.

12. Cross-fertilization of ideas is also apparent with respect to molecular biochemists in their efforts to construct point mutants with site-specifically altered properties. They are often strategically designed for elucidating structure and dynamics of biosystems by magnetic-resonance techniques. This also holds for the synthetic chemists to construct novel donor-acceptor complexes to mimic photosynthetic electron-transfer processes. Such biomimetic model complexes represent a most promising area of applications for multifrequency high-field EPR (see, for example, Levanon and Möbius, 1997; Berg et al., 1999; Fuhs et al., 2000; Valentin et al., 2004).

13. High-field EPR adds substantially to the capabilities of 'classical' spectroscopic and diffraction techniques for determining structure-dynamics-function relations of biosystems because transient intermediates can be observed in their working states at biologically relevant time scales.

14. High-field EPR instrumentation development will remain a challenging task. This refers, for example, to fast-flow probeheads for chemically instable transient (Savitsky et al., 2001) or to sufficient output power of mm and sub-mm microwave sources for fast manipulation of the electron spins by ns mw pulses (Freed, 2000; Grishin et al., 2004).

15. This also refers to high-field pulsed ELDOR techniques to determine distance and orientation of spin-carrying domains in proteins, e.g. electron-transfer radical pairs, transition-metal ions or nitroxide spin labels attached at specific sites for triangulation. Even from disordered samples, relevant new structural information can be obtained covering the distance range 0.2 nm < nm < 5 nm. This is complementary to what is obtained by solid-state NMR and FRET techniques probing very different distance ranges.

To conclude this chapter on high-field/high-frequency EPR spectroscopy: Fascinating biological applications and concomitant instrumental developments are envisaged for the near future. Bright prospects, indeed, 60 years after Zavoisky's discovery of the electron-paramagnetic-resonance phenomenon at only 5 mT and 133 MHz!

## Acknowledgments

With this chapter we pay tribute to the memory of late Arnold J. Hoff (Leiden) for his eminent contributions to magnetic resonance in photosynthesis. K. M. is particularly obliged to Arnold Hoff who has introduced him to the beauties of photosynthesis (Hoff and Möbius, 1978). Over the years, numerous co-workers – students, postdocs, colleagues – from different parts of the world have contributed to our interdisciplinary high-field EPR work at FU Berlin. They are cited in the references. We have greatly benefited from these inspiring cooperations which have initiated so many stimulating discussions. To all of them we want to express our gratitude, in particular to Harry Kurreck (Berlin), Haim Levanon (Jerusalem), Kev Salikhov (Kazan) and Dietmar Stehlik (Berlin). With respect to the more recent work presented in this chapter, which focuses on high-field EPR on bacterial photosynthetic reaction centers and spin-labeled bacteriorhodopsin, K. M. particularly enjoyed collaborating with Wolfgang Lubitz (Mülheim/Ruhr) and Heinz-Jürgen Steinhoff (Osnabrück). Likewise, it is a pleasure to acknowledge the essential contributions to our recent high-field work by Martin Fuchs, Michael Fuhs, Martin Plato, Anton Savitsky and Alexander Schnegg. The contributions by Yuri Grishin (Novosibirsk) and Alexander Dubinskii (Moscow) are most important for our methodological and instrumental developments of high-field EPR spectroscopy. In earlier stages of our high-field EPR work at FU Berlin essential contributions in terms of instrumental development and/or application to photosynthetic systems were made by Olaf Burghaus, Martina Huber, Chris Kay, Robert Klette, Thomas Prisner, Martin Rohrer, Jens Toerring and Stefan Weber. We are grateful for the assistance and suggestions Anton Savitsky provided while preparing the manuscript. D. G. acknowledges the essential contributions of the co-workers involved in the high-field ENDOR applications and instrumental developments: Igor Gromov, Boris Epel, Raanan Carmieli, P. Manikandan, Patrick Carl, Dafna Arieli, Debbie Baute, Shifi Kababya, K. Narasimhulu. The technical support of Yehoshua Gorodeski and Koby Zibziner is highly appreciated and so is the help of Vladimir Krymov. Our laboratories have received sustaining financial support by the Volkswagen-Foundation (K. M.), the Deutsche Forschungsgemeinschaft (DFG), in recent years in particular by the DFG priority program SPP 1051 (K. M. and D. G.), the DFG collaborative research center SFB 498 and the DFG grant MO132/19-2 (K. M.), the Israel Science Foundation, administered by the Israeli Academy of Sciences and Humanities (D. G.), the German-Israeli Science Foundation (D. G.), the MINERVA foundation (D. G.) which are all gratefully acknowledged.

## References


Allen JP, Feher G, Yeates TO, Komiya H and Rees DC (1987a) Structure of the reaction center from *Rhodobacter sphaeroides* R-26: I. The cofactors. Proc Natl Acad Sci USA 84: 5730–5734

Allen JP, Feher G, Yeates TO, Komiya H and Rees DC (1987b) Structure of the Reaction center from *Rhodobacter sphaeroides* R-26: The protein subunits. Proc Natl Acad Sci USA 84: 6162–6166

Altenbach C and Hubbell WL (1998) Site-directed spin labeling: A strategy for determination of structure and dynamics of proteins. In: Eaton GR, Eaton SS and Salikhov KM (eds) Foundations of Modern EPR, pp 423–435. World Scientific, Singapore

Amesz J and Hoff AJ (eds) (1996) Biophysical Techniques in Photosynthesis (Advances in Photosynthesis, Vol 3). Kluwer Academic Publishers, Dordrecht

Artz K, Williams JC, Allen JP, Lendzian F, Rautter J and Lubitz W (1997) Relationship between the oxidation potential and electron spin density of the primary electron donor in reaction centers from *sphaeroides*. Proc Natl Acad Sci USA 94: 13582–13587

Atherton NM (1993) Principles of Electron Spin Resonance. Ellis Horwood, New York

Baute D and Goldfarb D (2005) The $^{17}O$ hyperfine interaction in $V^{17}O(H_2{}^{17}O)_5^{2+}$ and $Mn(H_2{}^{17}O)_6^{2+}$ determined by high field ENDOR aided by DFT calculations. J Phys Chem A 109: 7865–7871

Baute D, Arieli D, Neese F, Zimmermann H, Weckhuysen BM and Goldfarb D (2004) Carboxylate binding in copper histidine complexes in solution and in zeolite Y: X- and W-band pulsed EPR/ENDOR combined with DFT calculations. J Am Chem Soc 126:11733–11745

Bennebroek MT and Schmidt J (1997) Pulsed ENDOR spectroscopy at large thermal spin polarizations and the absolute sign of the hyperfine interaction. J Magn Reson 128:199–206

Berg A, Shuali Z, Asano-Someda M, Levanon H, Fuhs M, Möbius K, Wang R, Brown C and Sessler JL (1999) A first high-field EPR study of photoinduced electron transfer in a base-paired porphyrin-dinitrobenzene supramolecular complex. J Am Chem Soc 121: 7433–7434

Berliner LJ, Eaton SS and Eaton GR (eds) (2000) Distance Measurements in Biological Systems by EPR (Biological Magnetic Resonance, Vol 19). Kluwer Academic Publishers/Plenum, New York

Biehl R, Plato M and Möbius K (1975) General TRIPLE resonance on free radicals in solution. Determination of relative signs of isotropic hyperfine coupling constants. J Chem Phys 63: 3515–3522

Biehl R, Lubitz W, Möbius K and Plato M (1977) Observation of deuterium quadrupole splittings of aromatic free radicals in liquid crystals by ENDOR and TRIPLE resonance. J Chem Phys 66: 2074–2078

Bittl R and Zech SG (2001) Pulsed EPR spectroscopy on short-lived intermediates in Photosystem I. Biochim Biophys Acta 1507: 194–211

Bixon M, Fajer J, Feher G, Freed JH, Gamliel D, Hoff AJ, Levanon H, Möbius K, Nechushtai R, Norris JR, Scherz A, Sessler JL and Stehlik D (1992) Primary Events in photosynthesis: Problems, speculations, controversies, and future trends. Israel J Chem 32: 369–518

Bloch F, Hansen W and Packard ME (1946) Nuclear induction. Phys Rev 69: 127

Bloeß A, Möbius K and Prisner TF (1998) High-frequency/high-field electron spin echo envelope modulation study of nitrogen hyperfine and quadrupole interactions on a disordered powder sample. J Magn Reson 134: 30–35

Blume RJ (1958) Electron spin relaxation times in sodium-ammonia solutions. Phys Rev 109: 1867–1873

Bosch M, Gast P, Hoff AJ, Spoyalov AP and Tsvetkov YD (1995) The primary acceptor quinone $Q_A$ in reaction centers of *Rhodobacter sphaeroides* R26 is hydrogen bonded to the N(1)-H of His M219. An electron spin echo study of $Q_A^{\bullet-}$. Chem Phys Lett 239: 306–312

Bratt PJ, Ringus E, Hassan A, van Tol H, Maniero A-L, Brunel L-C, Rohrer M, Bubenzer-Hange C, Scheer H and Angerhofer A (1999) EPR on biological samples beyond the limits of superconducting magnets—The primary donor cation of purple bacterial photosynthesis. J Phys Chem B 103: 10973–10982

Bratt PJ, Heathcote P, Hassan A, van Tol J, Brunel L-C, Schrier J and Angerhofer A (2003) EPR at 24 T of the primary donor radical cation from *Blastochloris viridis*. Chem Phys 294: 277–284

Breton J, Boullais C, Burie J-R, Nabedryk E and Mioskowski C (1994) Binding sites of quinones in photosynthetic bacterial reaction centers investigated by light-induced FTIR difference spectroscopy: Assignment of the interactions of each carbonyl of $Q_A$ in *Rhodobacter sphaeroides* using site-specific $^{13}C$-labeled ubiquinone. Biochemistry 33: 14378–14386

Britt RD (1996) Electron spin echo methods in photosynthesis research. In: Amesz J and Hoff AJ (eds) Biophysical Techniques in Photosynthesis (Advances in Photosynthesis, Vol 3). Kluwer Academic Publishers, Dordrecht

Brudler R, de Groot HJM, van Liemt WBS, Steggerda WF, Esmeijer R, Gast P, Hoff AJ, Lugtenburg J and Gerwert K (1994) Asymmetric binding of the 1- and 4-C=O groups of $Q_A$ in *Rhodobacter sphaeroides* R26 reaction centres monitored by Fourier transform infra-red spectroscopy using site-specific isotopically labeled ubiquinone-10. EMBO J 13: 5523–5530

Brunel L-C, van Tol J, Angerhofer A, Hill S, Krzystek J and Maniero AL (2004) Modern developments and prospects in multi-frequency high field EMR. In: Grinberg O and Berliner LJ (eds) Very High Frequency (VHF) ESR/EPR (Biological Magnetic Resonance, Vol 22), pp 465–505. Kluwer Academic Publishers/Plenum Publishers, New York

Burghaus O, Toth-Kischkat A, Klette R and Möbius K (1988)

Proton ENDOR at a microwave frequency of 97GHz. J Magn Reson 80: 383–388
Burghaus O, Rohrer M, Götzinger T, Plato M and Möbius K (1992) A novel high-field/high-frequency EPR and ENDOR spectrometer operating at 3 mm wavelength. Meas Sci Technol 3: 765–774
Burghaus O, Plato M, Rohrer M, Möbius K, MacMillan F and Lubitz W (1993) 3-mm High-field EPR on semiquinone radical anions $Q^{\bullet-}$ related to photosynthesis and on the primary donor $P^{\bullet+}$ and acceptor $Q^{\bullet-}$ in reaction centers of *Rhodobacter sphaeroides* R-26. J Phys Chem 97: 7639–7647
Carmieli R, Manikandan P, Kalb AJ and Goldfarb D (2001) Proton positions in the $Mn^{2+}$ binding site of concanavalin A as determined by single-crystal high-field ENDOR spectroscopy. J Am Chem Soc 123: 8378–8386
Carmieli R, Manikandan P, Epel B, Kalb (Gilboa) J, Schnegg A, Savitsky A, Möbius K and Goldfarb D (2003) Dynamics in the $Mn^{2+}$ Binding site in single crystals of concanavalin as revealed by high field EPR spectroscopy. Biochemistry 42:7863–7870
Chemerisov SD, Grinberg OY, Tipikin DS, Lebedev YS, Kurreck H and Möbius K (1994) Mechanochemically-induced radical pair formation in porphyrin-quinone and related donor-acceptor mixtures. Unusual stability and zero-field splittings. Chem Phys Lett 218: 353–361
Chirino AJ, Lous EJ, Huber M, Allen JP, Schenck CC, Paddock ML, Feher G and Rees DC (1994) Crystallographic analyses of site-directed mutants of the photosynthetic reaction center from *Rhodobacter sphaeroides*. Biochemistry 33: 4584–4593
Cook RJ and Whiffen DH (1964) Relative signs of hyperfine coupling constants by a double ENDOR experiment. Proc Phys Soc 84: 845–848
Coremans JWA, Poluektov OG, Groenen EJJ, Canters GW, Nar H and Messerschmidt A (1996) A W-band electron nuclear double resonance study of single crystals of $^{14}N$ and $^{15}N$ azurin. J Am Chem Soc 118: 12141–12153
Davies ER (1974) A new pulse ENDOR technique. Phys Lett A 47: 1–2
Deacon AT, Gleichmann T, Kalb (Gilboa) AJ, Price HJ, Raftery J, Brabbrook G, Yariv J and Helliwell JR (1997) The structure of concanavalin A and its bound solvent determined with small-molecule accuracy at 0.94 Å resolution. J Chem Soc Faraday Trans 93: 4305–4312
Debus RJ, Feher G and Okamura MY (1986) Iron-depleted reaction centers from *Rhodopseudomonas sphaeroides* R-26. 1: Characterization and reconstitution with $Fe^{2+}$, $Mn^{2+}$, $Co^{2+}$, $Ni^{2+}$, $Cu^{2+}$ and $Zn^{2+}$. Biochemistry 25: 2276–2287
de Groot HJM (2000) Solid-state NMR spectroscopy applied to membrane proteins. Curr Opin Struct Biol 10: 593–600
Deisenhofer J, Epp O, Miki K, Huber R and Michel H (1985) Structure of the protein subunits in the photosynthetic reaction centre of *Rhodopseudomonas viridis* at 3 Å resolution. Nature 318: 618–624
Dinse KP, Biehl R and Möbius K (1974) Electron nuclear triple resonance of free radicals in solution. J Chem Phys 61: 4335–4341
Disselhorst JAJM, van der Meer H, Poluektov OG and Schmidt J (1995) A pulsed EPR and ENDOR spectrometer operating at 95 GHz. J Magn Reson Ser A 115: 183–188
Doan PE and Hoffman BM (1997) Making hyperfine selection in Mims ENDOR independent of deadtime. Chem Phys Lett 269: 208–214
Dorio MM and Freed JH (eds) (1979) Multiple Electron Resonance Spectroscopy. Plenum Press, New York
Dorlet P, Xiong L, Sayre RT and Un S (2001) High field EPR study of the pheophytin anion radical in wild type and D1-E130 mutants of Photosystem II in *Chlamydomonas reinhardtii*. J Biol Chem 276: 22313–22316
Druckmann S, Friedmann N, Lanyi JK, Needleman R, Ottolenghi M and Shewes M (1992) The back photoreaction of the M intermediate in the photocycle of bacteriorhodopsin: Mechanism and evidence for two M species. Photochem Photobiol 56: 1041–1047
Dubinskii AA, Grishin YA, Savitsky AN and Möbius K (2002) Submicrosecond field-jump device for pulsed high-field ELDOR. Appl Magn Reson 22: 369–386
Dzuba SA, Tsvetkov YD and Maryasov AG (1992) Echo-induced EPR spectra of nitroxides in organic glasses: Model of orientational molecular motions near equilibrium position. Chem Phys Lett 188: 217–222
Earle KA, Moscicki JK, Ge M, Budil DE and Freed JH (1994) 250-GHz electron spin resonance studies of polarity gradients along the aliphatic chains in phospholipid membranes. Biophys J 66: 1213–1221
Epel B and Goldfarb D (2000) Two-dimensional pulsed TRIPLE at 95 GHz. J Magn Reson 146:196–203
Epel B, Manikandan P, Kroneck PMH, Goldfarb D (2001a) High field ENDOR and the sign of the hyperfine coupling. Appl Magn Reson 21: 287–297
Epel B, Pöppl A, Manikandan P, Vega S and Goldfarb D (2001b) The effect of spin relaxation on ENDOR spectra recorded at high magnetic fields and low temperatures. J Magn Reson 148: 388–397
Epel B, Arieli D, Baute D and Goldfarb D (2003) Improving W-band pulsed ENDOR sensitivity — random acquisition and pulsed special TRIPLE. J Magn Reson 164:78–83
Epel B, Gromov I, Stoll S, Schweiger A and Goldfarb D (2005), Spectrometer Manager: A versatile control software for pulse EPR spectrometers. Concepts Magn Reson B 26: 36–45
Ermler U, Fritzsch G, Buchanan SK and Michel H (1994) Structure of the photosynthetic reaction center from *Rhodobacter sphaeroides* at 2.65 Å resolution. Structure 2: 925–936
Feher G (1956) Observation of nuclear magnetic resonances via the electron spin resonance line. Phys Rev 103: 834–835
Feher G (1992) Identification and characterization of the primary donor in bacterial photosynthesis: A chronological account of an EPR / ENDOR investigation. J Chem Soc Perkin Trans 2: 1861–1874
Feher G, Hoff AJ, Isaacson RA and Ackerson LC (1975) ENDOR experiments on chlorophyll and bacteriochlorophyll in vitro and in the photosynthetic unit. Ann NY Acad Sci 244: 239–259
Flores M, Isaacson RA, Calvo R, Feher G and Lubitz W (2004) Probing hydrogen bonding to quinone anion radicals by $^{1}H$ and $^{2}H$ ENDOR spectroscopy at 35 GHz. Chem Phys 294: 401–413
Freed JH (2000) New technologies in electron spin resonance. Annu Rev Phys Chem 51: 655–689
Freed JH (2004) The development of high-field/high-frequency ESR. In: Grinberg O and Berliner LJ (eds) Very High Frequency (VHF) ESR/EPR (Biological Magnetic Resonance, Vol 22), pp 19–43. Kluwer Academic Publishers/Plenum Publishers, New York

Fuchs MR, Prisner TF and Möbius K (1999) A high-field/high-frequency heterodyne induction mode electron paramagnetic resonance spectrometer operating at 360 GHz. Rev Sci Instrum 70: 3681–3683

Fuchs MR, Schleicher E, Schnegg A, Kay CWM, Törring J, Bittl R, Bacher A, Richter G, Möbius K and Weber S (2002) g-tensor of the neutral flavin radical cofactor of DNA photolyase revealed by 360-GHz electron paramagnetic resonance spectroscopy. J Phys Chem B 106: 8885–8890

Fuchs MR, Schnegg A, Plato M, Schulz C, Müh F, Lubitz W and Möbius K (2003) The primary donor cation $P^{+\bullet}$ in photosynthetic reaction centers of site-directed mutants of *Rhodobacter sphaeroides*: g-tensor shifts revealed by high-field EPR at 360 GHz/12.8 T. Chem Phys 294: 371–384

Fuhs M, Elger G, Möbius K, Osintsev A, Popov A and Kurreck H (2000) Multifrequency time-resolved EPR (9.5 GHz and 95 GHz) on covalently linked porphyrin-quinone model systems for photosynthetic electron transfer: Effect of molecular dynamics on electron spin polarization. Mol Phys 98: 1025–1040

Fursman CE, Teutloff C and Bittl R (2002) Pulsed ENDOR studies of short-lived spin-correlated radical pairs in photosynthetic reaction centers. J Phys Chem B 106: 9679–9686

Geßner C, Lendzian F, Bönigk B, Plato M, Möbius K and Lubitz W (1992) Proton ENDOR and TRIPLE resonance investigation of $P^{+\bullet}_{865}$ in photosynthetic reaction center single crystals of *Rb. sphaeroides* wild type 2.4.1. Appl Magn Reson 3: 763–777

Goldfarb D and Krymov V (2004) W-band pulse ENDOR of transition metal centers in orientationally disordered systems and single crystals. In: Grinberg O and Berliner LJ (eds) Very High Frequency (VHF) ESR/EPR (Biological Magnetic Resonance, Vol 22), pp 306–346. Kluwer Academic Publishers/Plenum Publishers, New York

Goldfarb D, Strohmaier KG, Vaughan DEW, Thomann H, Poluektov O and Schmidt J (1996) Studies of framework iron in zeolites by pulsed ENDOR at 95 GHz. J Am Chem Soc 118: 4665–4671

Goldfarb D, Epel B, Zimmermann H, Jeschke G (2004) 2D TRIPLE in orientationally disordered samples — a means to resolve and determine relative orientation of hyperfine tensors. J Magn Reson 168: 75–87

Goldfarb D, Kuppala VN and Carmieli R (2005) Dynamics and Structure in the $Mn^{2+}$ site of concanavalin A as determined by high field EPR and ENDOR spectroscopy. Magn Reson Chem 43: S40–S50

Griffith OH, Dehlinger PJ and Van SP (1974) Shape of the hydrophobic barrier of phospholipid bilayers. Evidence for water penetration in biological membranes. J Membr Biol 15: 159–192

Grinberg O and Berliner LJ (eds) (2004) Very High Frequency (VHF) ESR/EPR (Biological Magnetic Resonance, Vol 22). Kluwer Academic Publishers/Plenum Publishers, New York

Grishin YA, Fuchs MR, Schnegg A, Dubinskii AA, Dumesh BS, Rusin FS, Bratman VL and Möbius K (2004) Pulsed Orotron — A new microwave source for submillimeter pulse high-field electron paramagnetic resonance spectroscopy. Rev Sci Instrum 75: 2926–2936

Gromov I, Krymov V, Manikandan P, Arieli D, Goldfarb D (1999) A W-band pulsed ENDOR spectrometer: Setup and application to transition metal centers. J Magn Reson 139: 8–17

Haffa ALM, Lin S, Williams JC, Taguchi AKW, Allen JP and Woodbury NW (2003) High yield of long-lived B-side charge separation at room temperature in mutant bacterial reaction centers. J Phys Chem B 107: 12503–12510

Haffa ALM, Lin S, Williams JC, Bowen BP, Taguchi AKW, Allen JP and Woodbury NW (2004) Controlling the pathway of photosynthetic charge separation in bacterial reaction centers. J Phys Chem B 108: 4–7

Hahn EL (1949) An accurate nuclear magnetic resonance method for measuring spin-lattice relaxation times. Phys Rev 76: 145–146

Hahn EL (1950) Nuclear induction due to free Larmor precession spin echoes. Phys Rev 80: 580–594

Haupts U, Tittor J and Oesterhelt D (1999) Closing in on bacteriorhodopsin: Progress in understanding the molecule. Annu Rev Biophys Biomol Struct 28: 367–399

Hessling B, Herbst J, Rammelsberg R and Gerwert K (1997) Fourier transform infrared double-flash experiments resolve bacteriorhodopsin's M1 to M2 transition. Biophys J 73: 2071–2080

Hofbauer W, Zouni A, Bittl R, Kern J, Orth P, Lendzian F, Fromme P, Witt HT and Lubitz W (2001) Photosystem II single crystals studied by high frequency EPR spectroscopy at 94 GHz: the tyrosine radical $Y_D$. Proc Nat Acad Sci USA 98: 6623–6628

Hoff AJ (ed) (1989) Advanced EPR. Applications in Biology and Biochemistry. Elsevier, Amsterdam

Hoff AJ and Deisenhofer J (1997) Photophysics of photosynthesis. Structure and spectroscopy of reaction centers of purple bacteria. Phys Rep 287: 1–247

Hoff AJ and Möbius K (1978) Nitrogen electron nuclear double resonance and proton triple resonance experiments on the bacteriochlorophyll cation in solution. Proc Natl Acad Sci USA 75: 2296–2300

Hornig AW and Hyde JS (1963) Paramagnetic resonance in triplet naphthalene at liquid helium temperatures. Mol Phys 6: 33–41

Hu JG, Sun BQ, Bizounok M, Hatcher ME, Lansing JC, Raap J, Verdegem PJ, Lugtenburg J, Griffin RG and Herzfeld J (1998) Early and late M intermediates in the bacteriorhodopsin photocycle: A solid state NMR study. Biochemistry 37: 8088–8096

Hubbell WL and Altenbach C (1994) Investigation of structure and dynamics in membrane proteins using site-directed spin labeling. Curr Opin Struct Biol 4: 556–573

Huber M (1997) On the electronic structure of the primary electron donor in bacterial photosynthesis — The bacteriochlorophyll dimer as viewed by EPR/ENDOR methods. Photosynth Res 52: 1–26

Huber M and Törring JT (1995) High-field EPR on the primary electron donor cation radical in single crystals of heterodimer mutant reaction centers of photosynthetic bacteria — first characterization of the g-tensor. Chem Phys 194: 379–385

Huber M, Törring JT, Plato M, Finck U, Lubitz W, Feick R, Schenck CC and Möbius K (1995) Investigation of the electronic structure of the primary donor in bacterial photosynthesis — Measurements of the anisotropy of the electronic g-tensor using high-field/high-frequency EPR. Sol Energy Mater Sol Cells 38: 119–126

Huber M, Isaacson RA, Abresch EC, Gaul D, Schenck CC and Feher G (1996) Electronic structure of the oxidized primary electron donor of the HL (M202) and HL (L173) heterodimer mutants of the photosynthetic bacterium *Rhodobacter sphaer-*

*oides*: ENDOR on single crystals of reaction centers. Biochim Biophys Acta 1273: 108–128

Hyde JS and Maki AH (1964) ENDOR of a free radical in solution. J Chem Phys 40: 3117–3118

Isaacson RA, Abresch EC, Lendzian F, Boullais C, Paddock ML, Mioskowski C, Lubitz W and Feher G (1995) Asymmetry of the binding sites of $Q_A^-$ and $Q_B^-$ in reaction centers of *Rb. sphaeroides* probed by Q-Band EPR with $^{13}$C-labeled quinones. In: Beyerle ME (ed) The Reaction Center of Photosynthetic Bacteria, pp 353–368. Springer, Berlin

Jordan P, Fromme P, Klukas O, Witt HT, Saenger W and Krauß N (2001) Three-Dimensional Structure of Photosystem I at 2.5 Å Resolution. Nature 411: 909–917

Jortner J and Bixon M (eds) (1999) Electron Transfer — From isolated molecules to biomolecules, Part 1 and Part 2 (Advances in Chemical Physics, Vol 106 and 107). Wiley, Hoboken

Kalb (Gilboa) AJ, Habash J, Hunter NS, Price HJ, Raftery J and Helliwell JR (2000) Manganese (II) in concanavalin A and other lectin proteins. In: Sigel H and Sigel A (eds) Manganese and Its Role in Biological Processes (Metal Ions in Biological Systems, Vol. 37), pp. 279–304. Dekker, New York

Kababya S, Nelson J, Calle C, Neese F, Goldfarb D (2006) Electronic structure of binuclear mixed valence azacryptates derived from integrated advanced EPR and DFT calculations. J Am Chem Soc 128: 2017–2029

Kaupp M, Bühl M and Malkin VG (eds) (2004) Calculation of NMR and EPR Parameters. Wiley-VCH, Weinheim

Kawamura T, Matsunami S and Yonezawa T (1967) Solvent effects on the g-value of di-t-butyl nitric oxide. Bull Chem Soc Japan 40: 1111–1115

Kay CWM, Bittl R, Bacher A, Richter G and Weber S (2005) Unambiguous determination of the g-matrix orientation in a neutral flavin radical by pulsed electron-nuclear double resonance at 94 GHz. J Am Chem Soc 127: 10780–10781

Kim SS and Weissman SI (1979) Transient magnetization following photoexcitation. Rev Chem Intermed 3: 107–120

Kleinfeld D, Okamura MY and Feher G (1984) Electron-transfer kinetics in photosynthetic reaction centers cooled to cryogenic temperatures in the charge-separated state: Evidence for light-induced structural changes. Biochemistry 23: 5780–5786

Klette R, Törring JT, Plato M, Möbius K, Bönigk B and Lubitz W (1993) Determination of the g-tensor of the primary donor cation radical in single crystals of *Rhodobacter sphaeroides* R-26 reaction centers by 3-mm high-field EPR. J Phys Chem 97: 2015–2020

Klukas O, Schubert W-D, Jordan P, Krauss N, Fromme P, Witt HT and Saenger W (1999) Localization of two phylloquinones, $Q_K$ and $Q_K'$, in an improved electron density map of Photosystem I at 4-Å Resolution. J Biol Chem 274: 7361–7367

Kötting C and Gerwert K (2005) Proteins in action monitored by time-resolved FTIR spectroscopy. Chem Phys Chem 6: 881–888

Krymov VN, Kurochkin VI, Oranskii LG, Lebedev YaS and Grinberg OYa (1982) Bulletin of Inventions #13 (in Russian), USSR Patent #918832 (Priority of 1980)

Krzystek J, Telser J, Hoffman BM, Brunel LC and Licoccia S (2001) High-frequency and field EPR investigation of (8,12-diethyl-2,3,7,13,17,18-hexamethylcorrolato)manganese(III). J Am Chem Soc 123 : 7890–7897

Kulik LV, Paschenko SV and Dzuba SA (2002) 130 GHz ESEEM induced by electron–electron interaction in biradical. J Magn Reson 159: 237–241

Kulik LV, Epel B, Lubitz W and Messinger J (2005a) $^{55}$Mn pulse ENDOR at 34 GHz of the $S_0$ and $S_2$ states of the oxygen-evolving complex in Photosystem II. J Am Chem Soc 127: 2392–2393

Kulik LV, Epel B, Messinger J and Lubitz W (2005b) Pulse EPR, $^{55}$Mn-ENDOR and ELDOR-detected NMR of the $S_2$-state of the oxygen evolving complex in Photosystem II. Photosynth Res 84: 347–353

Kurreck H, Kirste B and Lubitz W (eds) (1988) Electron Nuclear Double Resonance Spectroscopy of Radicals in Solution. Application to Biological and Inorganic Chemistry. VCH Publishers, Weinheim

Kurshev VV, Raitsimring AM and Tsvetkov YD (1989) Selection of dipolar interaction by the 2+1 pulse train ESE. J Magn Reson 81: 441–454

Lancaster CRD and Michel H (2001) Photosynthetic reaction centers of purple bacteria. In: Messerschmidt A, Huber R, Poulos T and Wiegard K (eds) Handbook of Metalloproteins, Vol 1, pp 119–135. John Wiley, Chichester

Lebedev YS (1994) High-field ESR. In: Atherton NM, Davies MJ and Gilbert BC (eds) Electron Paramagnetic Resonance, A Specialist Periodical Report, Vol 14, pp 63–87. The Royal Society of Chemistry, Cambridge

Lebedev YS (1998) Very-high-field EPR. In: Eaton SS and Salikhov KM (eds) Foundations of Modern EPR, pp 731–740. World Scientific, Singapore

Lendzian F, Huber M, Isaacson RA, Endeward B, Plato M, Bönigk B, Möbius K, Lubitz W and Feher G (1993) The electronic structure of the primary donor cation radical in *Rhodobacter sphaeroides* R-26: ENDOR and TRIPLE resonance studies in single crystals of reaction centers. Biochim Biophys Acta 1183: 139–160

Levanon H (1996) Time-resolved electron paramagnetic resonance spectroscopy — principles and applications. In: Amesz J and Hoff AJ (eds) Biophysical Techniques in Photosynthesis (Advances in Photosynthesis, Vol 3). Kluwer Academic Publishers, Dordrecht

Levanon H and Möbius K (1997) Advanced EPR spectroscopy on electron transfer processes in photosynthesis and biomimetic model systems. Annu Rev Biophys Biomol Struct 26: 495–540

Lubitz W (2002) Pulse EPR and ENDOR studies of light-induced radicals and triplet states in Photosystem II of oxygenic photosynthesis. Phys Chem Chem Phys 4: 5539–5545

Lubitz W (2004) EPR in Photosynthesis. In: Gilbert BC, Davies MJ and Murphy DM (eds) Electron Paramagnetic Resonance, A Specialist Periodical Report, Vol 19, pp 174–242. The Royal Society of Chemistry, Cambridge

Lubitz W and Feher G (1999) The primary and secondary acceptors in bacterial photosynthesis III. Characterization of the quinone radicals $Q_A^-$ and $Q_B^-$ by EPR and ENDOR. Appl Magn Reson 17: 1–48

Lubitz W and Lendzian F (1996) ENDOR Spectroscopy. In: Amesz J and Hoff AJ (eds) Biophysical Techniques in Photosynthesis (Advances in Photosynthesis, Vol 3). Kluwer Academic Publishers, Dordrecht

Lubitz W, Lendzian F and Bittl R (2002) Radicals, radical pairs and triplet states in photosynthesis. Acc Chem Res 35: 313–320

Lubitz W, Möbius K and Dinse K-P (eds) (2005) High-field EPR in Biology, Chemistry and Physics (Magnetic Resonance in

Chemistry, Vol 43). Wiley, London
Luecke H, Schobert B, Richter H-T, Cartailler J-P and Lanyi JK (1999) Structure of bacteriorhodopsin at 1.55 Å resolution. J Mol Biol 291: 899–911
Mantel C, Hassan AK, Pecaut J, Deronzier A, and Duboc-Toia C (2003) A high-frequency and high-field EPR study of new azide and fluoride mononuclear Mn(III) complexes, J Am Chem Soc 125: 12337–12344
Marsh D, Kurad D and Livshits VA (2002) High-field electron spin resonance of spin labels in membranes. Chem Phys Lipids 116: 93–114
Mason AJ, Grage SL, Straus SK, Glaubitz C and Watts A (2004) Identifying anisotropic constraints in multiply labeled Bacteriorhodopsin by $^{15}$N MAOSS NMR: A general approach to structural studies of membrane proteins. Biophys J 86: 1610–1617
Mayas L, Plato M, Winscom CJ and Möbius K (1978) Deuterium quadrupole coupling constants in hydrogen bonded dicarboxylic acids. A distant ENDOR study. Mol Phys 36: 753–764
McLauchlan KA and Yeung MT (1994) Time resolved ESR studies of free radicals. In: Atherton NM, Davies MJ and Gilbert BC (eds) Electron Spin Resonance, A Specialist Periodical Report, Vol 14, pp 32–62. The Royal Society of Chemistry, Cambridge
Mehring M, Höfer P and Grupp A (1987) Pulsed electron double and triple resonance schemes. Ber Bunsenges Phys Chem 91: 1132–1137
Millhauser GL and Freed JH (1984) Two-dimensional electron spin echo spectroscopy and slow motions. J Chem Phys 81: 37–48
Milov AD, Salikhov KM and Shirov MD (1981) Application of ELDOR in electron-spin echo for paramagnetic center space distributions in solids. Fiz Tverd Tela 23: 975–982
Milov AD, Ponomarev AB and Tsvetkov YD (1984) Electron-electron double resonance in electron spin echo: Model biradical systems and the sensitized photolysis of decalin. Chem Phys Lett 110: 67–72
Mims WB (1965) Pulsed endor experiments. Proc Roy Soc Lond 283: 452–457
Möbius K (1995) High-field/high-frequency EPR/ENDOR—a powerfull new tool in photosynthesis research. Appl Magn Reson 9: 389–407
Möbius K (2000) Primary processes in photosynthesis: What do we learn from high-field EPR spectroscopy? Chem Soc Rev 29: 129–139
Möbius K (ed) (2001) High-Field and High-Frequency Electron Paramagnetic Resonance (Applied Magnetic Resonance, Vol 21). Springer, Wien
Möbius K and Biehl R (1979) Electron-nuclear-nuclear TRIPLE resonance of radicals in solution. In: Dorio MM and Freed JH (eds) Multiple Electron Resonance Spectroscopy, pp 475–508. Plenum Press, New York
Möbius K, Plato M and Lubitz W (1982) Radicals in solution studied by ENDOR and TRIPLE resonance spectroscopy. Phys Rep 87: 171–208
Möbius K, Savitsky A, Schnegg A, Plato M and Fuhs M (2005a) High-field EPR spectroscopy applied to biological systems: Characterization of molecular switches for electron and ion transfer. Phys Chem Chem Phys 7: 19–42
Möbius K, Savitsky A, Wegener C, Plato M, Fuchs M, Schnegg A, Dubinskii AA, Grishin YA, Grigor'ev IA, Kühn M, Duché D, Zimmermann H and Steinhoff H-J (2005b) Combining high-field EPR with site-directed spin labeling reveals unique information on proteins in action. Magn Reson Chem 43: S4–S19
Moltke S, Wallat I, Sakai N, Nakanishi K, Brown MF and Heyn MP (1999) The angles between the C(1)-, C(5)-, and C(9)-methyl bonds of the retinylidene chromophore and the membrane normal increase in the M intermediate of bacteriorhodopsin: Direct determination with solid-state (2)H NMR. Biochemistry 38: 11762–11772
Müh F, Bibikova M, Lendzian F, Oesterhelt D and Lubitz W (1998) Pigment-protein interactions in reaction centers of *Rhodopseudomonas viridis*: ENDOR study of the oxidized primary donor in site-directed mutants. In: Garab G (ed) Photosynthesis: Mechanisms and Effects, Vol 2, pp 763–766. Kluwer Academic Publishers, Dordrecht
Müh F, Lendzian F, Roy M, Williams JC, Allen JP and Lubitz W (2002) Pigment-protein interactions in bacterial reaction centers and their influence on oxidation potential and spin density distribution of the primary donor. J Phys Chem B 106: 3226–3236
Nabedryk E, Schulz C, Müh F, Lubitz W and Breton J (2000) Heterodimeric versus homodimeric structure of the primary electron donor in *Rhodobacter sphaeroides* reaction centers genetically modified at position M202. Photochem Photobiol 71: 582–588
Nabedryk E, Breton J, Sebban P and Baciou L (2003) Quinone ($Q_B$) binding site and protein structural changes in photosynthetic reaction center mutants at Pro-L209 revealed by vibrational spectroscopy. Biochemistry 42: 5819–5827
Norris JR, Uphaus RA, Crespi HL and Katz JJ (1971) Electron spin resonance of chlorophyll and the origin of signal I in photosynthesis. Proc Natl Acad Sci USA 68: 625–628
Ondar MA, Grinberg OY and Lebedev YS (1985) Study of the effect of the medium on the magnetic resonance parameters of nitroxide radicals by high resolution EPR spectroscopy. Sov J Chem Phys. 3: 781–792
Pachtchenko S (2002) Primary Electron Donor Triplet States in Photosynthetic Reaction Centers as Studied by High-Field EPR. Thesis, Leiden University, Leiden
Pannier M, Veit S, Godt A, Jeschke G and Spiess HW (2000) Dead-time free measurement of dipole-dipole interactions between electron spins. J Magn Reson 142: 331–340
Parak F, Frolov EN, Kononenko AA, Mössbauer RL, Goldanskii VI and Rubin AB (1980) Evidence for a correlation between the photoinduced electron transfer and dynamic properties of the chromatophore membranes from *Rhodospirillum rubrum*. FEBS Lett 117: 368–372
Paschenko SV, Gast P and Hoff AJ (2001) A D-Band (130 GHz) EPR study of the primary electron donor triplet state in photosynthetic reaction centers of *Rhodobacter sphaeroides* R26. Appl Magn Reson 21: 325–334
Paschenko SV, Proskuryakov II, Germano M, van Gorkom HJ and Gast P (2003) Triplet state in Photosystem II reaction centers as studied by 130 GHz EPR. Chem Phys 294: 439–449
Plato M, Steinhoff H-J, Wegener C, Törring JT, Savitsky A and Möbius K (2002) Molecular orbital study of polarity and hydrogen bonding effects on the g and hyperfine tensors of site directed NO spin labeled bacteriorhodopsin. Mol Phys 100: 3711–3721
Plato M, Krauß N, Fromme P and Lubitz W (2003) Molecular

orbital study of the primary electron donor P700 of Photosystem I based on a recent X-ray single crystal structure analysis. Chem Phys 294: 483–499

Poluektov OG, Utschig LM, Dubinskij AA and Thurnauer MC (2004) ENDOR of spin-correlated radical pairs in photosynthesis at high magnetic field: A tool for mapping electron transfer pathways. J Am Chem Soc 126: 1644–1645

Poluektov OG, Utschig LM, Schlesselman SL, Lakshmi KV, Brudvig GW, Kothe G and Thurnauer MC (2002) Electronic structure of the $P_{700}$ special pair from high-frequency EPR spectroscopy. J Phys Chem B 106: 8911–8916

Poluektov OG, Utschig LM, Dubinskij AA and Thurnauer MC (2005) Electron transfer pathways and protein response to charge separation in photosynthetic reaction centers: Time-resolved high-field ENDOR of the spin-correlated radical pair $P^{+}_{865}Q^{-}_{A}$. J Am Chem Soc 127: 4049–4059

Prisner TF, Rohrer M and Möbius K (1994) Pulsed 95 GHz high-field EPR heterodyne spectrometer with high spectral and time resolution. Appl Magn Reson 7: 167–183

Prisner TF, van der Est A, Bittl R, Lubitz W, Stehlik D and Möbius K (1995) Time-resolved W-band (95 GHz) EPR spectroscopy of Zn-substituted reaction centers of *Rhodobacter sphaeroides* R-26. Chem Phys 194: 361–370

Prisner T, Rohrer M and MacMillan F (2001) Pulsed EPR spectroscopy: Biological applications. Annu Rev Phys Chem 52: 279–313

Purcell EM, Torrey HG and Pound RV (1946) Resonance absorption by nuclear magnetic moments in a solid. Phys Rev 69: 37–38

Radzwill N, Gerwert K and Steinhoff H-J (2001) Time-resolved detection of transient movement of helices F and G in doubly spin-labeled bacteriorhodopsin. Biophys J 80: 2856–2866

Rautter J, Lendzian F, Schulz C, Fetsch A, Kuhn M, Lin X, Williams JC, Allen JP and Lubitz W (1995) ENDOR studies of the primary donor cation radical in mutant reaction centers of *Rhodobacter sphaeroides* with altered hydrogen-bond interactions. Biochemistry 34: 8130–8143

Riedi PC and Smith GM (2002) Progress in high-field EPR. In: Atherton NM, Davies MJ, Gilbert BC and McLauchlan KA (eds) Electron Paramagnetic Resonance, A Specialist Periodical Report, Vol 18, pp 254–303. The Royal Society of Chemistry, London

Rink T, Pfeiffer M, Oesterhelt D, Gerwert K and Steinhoff HJ (2000) Unraveling photoexcited conformational changes of bacteriorhodopsin by time resolved electron paramagnetic resonance spectroscopy. Biophys J 78: 1519–1530

Rohrer M, Plato M, MacMillan F, Grishin Y, Lubitz W and Möbius K (1995) Orientation-selected 95 GHz high-field ENDOR spectroscopy of randomly oriented plastoquinone anion radicals. J Magn Reson Ser A 116: 59–66

Rohrer M, Gast P, Möbius K and Prisner TF (1996) Anisotropic motion of semiquinones in photosynthetic reaction centers of *Rhodobacter sphaeroides* R26 and in frozen isopropanol solution as measured by pulsed high-field EPR at 95 GHz. Chem Phys Lett 259: 523–530

Rohrer M, MacMillan F, Prisner TF, Gardiner AT, Möbius K and Lubitz W (1998) Pulsed ENDOR at 95 GHz on the primary acceptor ubisemiquinone in photosynthetic bacterial reaction centers and related model systems. J Phys Chem B 102: 4648–4657

Savitsky AN, Galander M and Möbius K (2001) W-band time resolved electron paramagnetic resonance spectroscopy on transient organic radicals in solution. Chem Phys Lett 340: 458–466

Savitsky A, Kühn M, Duché D, Möbius K and Steinhoff HJ (2004) Spontaneous refolding of the pore-forming colicin a toxin upon membrane association as studied by X-Band and W-Band high-field electron paramagnetic resonance spectroscopy. J Phys Chem B 108: 9541–9548

Saxena S and Freed JH (1997) Two dimensional electron spin resonance and slow motions. J Phys Chem A 101: 7998–8008

Schnegg A, Fuhs M, Rohrer M, Lubitz W, Prisner TF and Möbius K (2002) Molecular dynamics of $Q_A^{\bullet-}$ and $Q_B^{\bullet-}$ in photosynthetic bacterial reaction centers studied by pulsed high-field EPR at 95 GHz. J Phys Chem B 106: 9454–9462

Schulz C, Müh F, Beyer A, Jordan R, Schlodder E and Lubitz W (1998) Investigation of *Rhodobacter sphaeroides* reaction center mutants with changed ligands to the primary donor. In: Garab G (ed) Photosynthesis: Mechanisms and Effects, Vol 2, pp 767–770. Kluwer Academic Publishers, Dordrecht

Schweiger A and Jeschke G (2001) Principles of Pulse Electron Paramagnetic Resonance. University Press, Oxford

Sinnecker S, Reijerse E, Neese F and Lubitz W (2004) Hydrogen bond geometries from electron paramagnetic resonance and electron-nuclear double resonance parameters: Density functional study of quinone radical anion-solvent interactions. J Am Chem Soc 126: 3280–3290

Smirnov AI (2002) Spin labeling in high field ESR. In: Gilbert BC, Davies MJ, McLauchlan KA (eds) Electron Paramagnetic Resonance, A Specialist Periodical Report, Vol 18, pp 109–136. The Royal Society of Chemistry, London

Smith GM and Riedi PC (2000) Electron Paramagnetic Resonance. In: Atherton NM, Davies MJ, Gilbert BC, McLauchlan KA (eds) Electron Paramagnetic Resonance, A Specialist Periodical Report, Vol 17, pp 164–204. The Royal Society of Chemistry, London

Soda G and Chiba T (1969) Deuteron magnetic resonance study of cupric formate tetrahydrate $Cu(DCOO)_2{\cdot}4D_2O$. J Phys Soc Jpn 26: 249–261

Stehlik D and Möbius K (1997) New EPR methods for investigating photoprocesses with paramagnetic intermediates. Annu Rev Phys Chem 48: 745–784

Steinhoff H-J (2002) Methods for study of protein dynamics and protein-protein interaction in protein-ubiquitination by electron paramagnetic resonance spectroscopy. Front Biosci 7: 97–110

Steinhoff H-J, Savitsky A, Wegener C, Pfeiffer M, Plato M and Möbius K (2000) High-field EPR studies of the structure and conformational changes of site directed spin labeled bacteriorhodopsin. Biochim Biophys Acta 1457: 253–262

Stone AJ (1963) Gauge invariance of the g-tensor. Proc Roy Soc Lond A 271: 424–434

Stowell MHB, McPhillips TM, Rees DC, Soltis SM, Abresch E and Feher G (1997) Light-induced structural changes in photosynthetic reaction center: Implications for mechanism of electron-proton transfer. Science 276: 812–816

Teutloff C, Hofbauer W, Zech SG, Stein M, Bittl R and Lubitz W (2001) High-frequency EPR studies on cofactor radicals in Photosystem I. Appl Magn Reson 21: 363–379

Teutloff C, Bittl R and Lubitz W (2004) Pulse ENDOR studies on the radical pair $P^{\bullet+}_{700}A_1^{\bullet-}$ and the photoaccumulated quinone acceptor $A_1^{\bullet-}$ of Photosystem I. Appl Magn Reson 26: 5–21

Thomann H, Dalton LR and Dalton LA (1990) Biological applications of time domain ESR. In: Berliner LJ (ed) Biological Magnetic Resonance, Vol 6, pp 143. Springer, New York

Thorgeirsson TE, Xiao WZ, Brown LS, Needleman R, Lanyi JK and Shin YK (1997) Transient channel-opening in bacteriorhodopsin — an EPR study. J Mol Biol 273: 951–957

Törring JT (1995) Untersuchungen zum g-Tensor des primären Donors in bakteriellen Reaktionszentren. PhD thesis, Free University Berlin, Berlin

Törring JT, Un S, Knüplig M, Plato M and Möbius K (1997) On calculation of g-tensors of organic radicals. J Chem Phys 107: 3905–3913

Un S, Dorlet P and Rutherford AW (2001) A high-field EPR tour of radicals in photosystems I and II. Appl Magn Reson 21: 341–361

Valentin MD, Bisol A, Agostini G, Fuhs M, Liddell PA, Moore AL, Moore TA, Gust D and Carbonera D (2004) Photochemistry of artificial photosynthetic reaction centers in liquid crystals probed by multifrequency EPR (9.5 and 95 GHz). J Am Chem Soc 126: 17074–17086

van der Brink JS, Spoyalov AP, Gast P, van Liemt WBS, Raap J, Lugtenburg J and Hoff AJ (1994) Asymmetric binding of the primary acceptor quinone in reaction centers of the photosynthetic bacterium *Rhodobacter sphaeroides* R26, probed with Q-band (35 GHz) EPR spectroscopy. FEBS Lett 353: 273–276

van der Est A, Prisner TF, Bittl R, Fromme P, Lubitz W, Möbius K and Stehlik D (1997) Time-Resolved X-, K-, W-band EPR of the radical pair state $P_{700}^{\bullet+}A_1^{\bullet-}$ of Photosystem I in comparison with $P_{865}^{\bullet+}Q_A^{\bullet-}$ in bacterial reaction centers. J Phys Chem B 101: 1437–1443

van Liemt WBS, Boender GJ, Gast P, Hoff AJ, Lugtenburg J and de Groot HJM (1995) $^{13}$C magic angle spinning NMR characterization of the functionally asymmetric $Q_A$ binding in *Rhodobacter sphaeroides* R26 photosynthetic reaction centers using site-specific $^{13}$C-Labeled ubiquinone-10. Biochemistry 34: 10229–10236

van Tol J, Brunel L-C and Wylde RJ (2005) A quasioptical transient electron spin resonance spectrometer operating at 120 and 240 GHz. Rev Sci Instrum 76: 074101–074108

Váró G and Lanyi JK (1991) Kinetic and spectroscopic evidence for an irreversible step between deprotonation and reprotonation of the Schiff base in the bacteriorhodopsin photocycle. Biochemistry 30: 5008–5015

Wang W, Belford RL, Clarkson RB, Davis PH, Forrer J, Nilges MJ, Timken MD, Walczak T, Thurnauer MC, Norris JR, Morris AL and Zhang Y (1994) Very high frequency EPR — 94 GHz instrument and applications to primary reaction centers from photosynthetic red bacteria and to other disordered systems. Appl Magn Reson 6: 195–215

Weber S (2000) Recent EPR studies on the bacterial photosynthetic reaction centre. In: Gilbert BC, Davies MJ and McLauchlan KA (eds) Electron Paramagnetic Resonance, A Specialist Periodical Report, Vol 17, pp 43–77. The Royal Society of Chemistry, London

Weber S, Kay CWM, Bacher A, Richter G and Bittl R (2005) Probing the N(5)H Bond of the Isoalloxazine Moiety of Flavin Radicals by X- and W-Band Pulsed Electron-Nuclear Double Resonance. Chem Phys Chem 6: 292–299

Wegener C, Savitsky A, Pfeiffer M, Möbius K and Steinhoff H-J (2001) High-Field EPR-detected shifts of magnetic tensor components of spin label side chains reveal protein conformational changes: The proton entrance channel of bacteriorhodopsin. Appl Magn Reson 21: 441–452

Xiao W, Brown LS, Needleman R, Lanyi JK and Shin Y-K (2000) Light-induced rotation of a transmembrane α-helix in bacteriorhodopsin. J Mol Biol 304: 715–721

Xu W, Chitnis P, Valieva A, Est Avd, Pushkar YN, Krzystyniak M, Teutloff C, Zech SG, Bittl R, Stehlik D, Zybailov B, Shen G and Golbeck JH (2003) Electron transfer in cyanobacterial Photosystem I: I. Physiological and spectroscopic characterization of forward electron transfer in a putative electron transfer pathway from $A_0$ through $A_1$ to $F_X$. J Biol Chem 278: 27864–27875

Yeates TO, Komiya H, Chirino A, Rees DC, Allen JP and Feher G (1988) Structure of the reaction center from *Rhodobacter sphaeroides* R-26 and 2.4.1: Protein-cofactor (bacteriochlorophyll, bacteriopheophytin, and carotenoid) interactions. Proc Natl Acad Sci USA 85: 7993–7998

Zavoisky EK (1944) Парамагнитная абсорбция в перпендикулярных и параллельных полях для солей, растворов и металлов (Paramagnetic absorption in perpendicular and parallel fields for salts, solutions and metals, in Russian). Thesis, Kazan State University, Kazan

Zavoisky EK (1945a) Парамагнитная абсорбция в растворах при параллельных полях (Paramagnetic absorption in solutions at parallel fields, in Russian). Zh Eksp Teor Fiz 15: 253–257

Zavoisky EK (1945b) Парамагнитная релаксация в жидких растворах при параллельных полях (Paramagnetic relaxation in liquid solutions at perpendicular fields, in Russian). Zh Eksp Teor Fiz 15: 344–350

Zeng R and Budil DE (2003) A re-examination of spin-orbit coupling in the triplet state of the primary donor in photosynthetic reaction centers. Chem Phys 294: 347–358

Zeng R, van Tol J, Deal A, Frank HA and Budil DE (2003) Temperature dependence of the primary donor triplet state g-tensor in photosynthetic reaction centers of *Rhodobacter sphaeroides* R-26 observed by transient 240 GHz electron paramagnetic resonance. J Phys Chem B 107: 4624–4631

Zoleo A, Contessi S, Lippe G, Pinato L, Brustolon M, Brunel LC and Dabbeni-Sala F, Maniero AL (2004) High-affinity metal-binding site in beef heart mitochondrial F(1)ATPase: An EPR spectroscopy study, Biochemistry 43: 13214–13224

Zouni A, Witt HT, Kern J, Fromme P, Krauß N, Saenger W and Orth P (2001) Crystal structure of Photosystem II from *Synechococcus elongatus* at 3.8 Å resolution. Nature 409: 739–743

Zybailov B, Est Avd, Zech SG, Teutloff C, Johnson TW, Shen G, Bittl R, Stehlik D, Chitnis PR and Golbeck JH (2000) Recruitment of a foreign quinone into the $A_1$ site of Photosystem I. J Biol Chem 275: 8531–8539

# Chapter 15

# High-Time Resolution Electron Paramagnetic Resonance Study of Quantum Beat Oscillations Observed in Photosynthetic Reaction Center Proteins

Gerd Kothe*
*Department of Physical Chemistry, University of Freiburg, Albertstr. 21, D-79104 Freiburg, Germany*

James R. Norris
*Department of Chemistry, The University of Chicago, 5735 South Ellis Avenue, Chicago, Il 60637, U.S.A.*

Oleg G. Poluektov and Marion C. Thurnauer
*Chemistry Division, Argonne National Laboratory, 9700 S. Cass Ave., Argonne, Il 60439, U.S.A.*



## Summary

The primary energy conversion steps of bacterial and plant photosynthesis proceed via light-induced radical pairs as short-lived intermediates. Time-resolved EPR experiments of the photosynthetic reaction centers monitor the key charge separated state between the oxidized electron donor and reduced acceptor $P_{700}^{\bullet+}A_1^{\bullet-}$ of Photosystem I.

*Author for correspondence, email: gerd.kothe@physchem.uni-freiburg.de

*Thijs J. Aartsma and Jörg Matysik (eds): Biophysical Techniques in Photosynthesis II, pp. 305–323.*

The observed EPR spectra of $P_{700}^{\bullet+}A_1^{\bullet-}$ are indicative of a spin-correlated radical pair that is populated from the excited singlet state of the primary electron donor. Importantly, the spin-correlated radical pair nature of the charge-separated state is a common feature of all photosynthetic reaction centers, which gives rise to several interesting spin-phenomena such as quantum beats, observed at short delay times after optical excitation. In this chapter we present details of the quantum beat phenomena studied by high-time resolution transient and pulsed EPR of Photosystem I reaction centers. The understanding and analysis of the quantum beat oscillations in combination with high-field EPR allows us to obtain the three-dimensional structure of $P_{700}^{\bullet+}A_1^{\bullet-}$ as well as its arrangement in the membrane. We expect that this is of general interest, because the detailed structure of radical pair intermediates can be determined on a nanosecond time scale. In addition, by monitoring the quantum beat oscillations with a pulsed EPR approach, we shed new light on the electronic structure of the oxidized primary electron donor, $P_{700}^{\bullet+}$.

## I. Introduction

The ability to monitor chemical and physical processes on a multitude of time and length scales continues to be a major challenge for addressing critical problems of the 21st century. Optical spectroscopy remains the most direct and widely applied technique to study transient species. Although optical techniques provide superior time resolution, recently approaching atto seconds, they do not readily provide the spectral resolution required to extract dynamic structural information. Thus, there is strong interest in time-resolved methods that can also provide details of dynamic structure.

Time-resolved electron paramagnetic resonance (EPR) techniques were first explored in the 1960s and 1970s to study transient intermediates created by pulsed radiolysis and photolysis in liquid solutions (for reviews see: Trifunac and Thurnauer, 1979; Norris et al., 1980; McLauchlan, 1989). Although, the time resolution of EPR is limited to the nanosecond time scale, the technique yields considerable details of structure as related to function in a chemical system. The challenge in pursuing these methods is to have sufficient sensitivity to attain the highest possible time resolution, wherein both of these factors involve the interplay between instrumental and sample characteristics. Thus, time-resolved EPR techniques are not straightforward, but very powerful because of the possibilities to apply the optimal method (e.g., continuous wave vs. pulsed EPR) for approaching a specific problem. These points are well illustrated by the application for over thirty years of time-resolved EPR to study the primary photochemistry of the integral membrane reaction center proteins of anoxygenic and oxygenic photosynthesis (for reviews see: Snyder and Thurnauer, 1993; Angerhofer and Bittl, 1996; Stehlik and Möbius, 1997; Chapter 14, Möbius). The photosynthetic reaction center proteins continue to provide a consistent testbed for time-resolved EPR methods, both in its evolving developments and in understanding the spin phenomena associated with these experiments. Simultaneously, time-resolved EPR studies of reaction centers have delivered significant new insights into the mechanism of photosynthetic energy conversion, i.e., the separation of a positively charged electron donor and negatively charged electron acceptor across a biological membrane (see Fig. 1 for a reaction scheme of this primary photochemistry in Photosystem I of oxygenic photosynthesis).

In time-resolved EPR experiments, the sample is excited by a laser pulse (typically ~5 ns) and the paramagnetic transients thus formed are monitored by either continuous wave (transient) or pulsed EPR. Much research has gone into developing and understanding the advantages of these different technical approaches to time-resolved EPR, and these studies have been reviewed (Norris et al., 1980; Stehlik et al., 1989a).

In general, pulsed EPR was considered to be most advantageous for time-resolved EPR because it is inherently well-suited for measuring dynamic properties of a spin system. The signal rise time is determined by the width of the initial microwave pulse and the signal intensity is directly related to the transient magnetization created by the laser pulse (Trifunac and Norris, 1978). On the other hand, pulsed methods suffer from the so-called instrumental dead time and necessity for broad band detection. Both of these latter factors lower the sensitivity of pulsed techniques (Norris et al., 1980). In the case of tran-

*Abbreviations:* $A_1$ – phylloquinone acceptor in Photosystem I; Chl – chlorophyll; ENDOR – electron nuclear double resonance; EPR – electron paramagnetic resonance; ESEEM – electron spin echo envelope modulation; $P_{700}$ – primary donor in Photosystem I; *S. lividus* – *Synechococcus lividus*

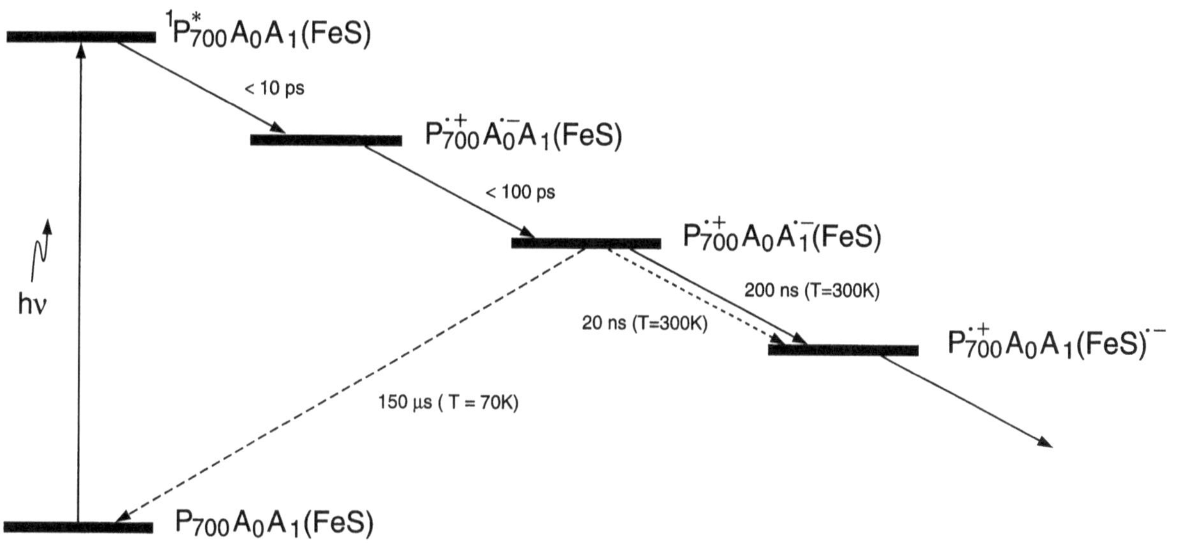


*Fig. 1.* The initial photosynthetic energy conversion reactions involve a photoinitiated sequence of efficient electron-transfer steps resulting in charge separation across a biological membrane, thus converting light into an electrochemical potential (Hoff and Deisenhofer, 1997). These reactions occur between cofactors held within integral membrane reaction center proteins. In Photosystem I of oxygenic photosynthesis, photoexcitation of $P_{700}$, the primary electron donor (two chlorophyll molecules), initiates sequential electron-transfer from the excited singlet state of $P_{700}$ through $A_0$, a chlorophyll acceptor, to $A_1$, a phylloquinone acceptor, resulting in the membrane-spanning charge-separated state, $P^{\bullet+}_{700}A_1^{\bullet-}$. From $A_1^{\bullet-}$ the electron proceeds to three iron-sulfur clusters represented by FeS. At low temperatures a fraction of reaction centers exhibit reversible electron-transfer from the charge-separated state, $P^{\bullet+}_{700}A_1^{\bullet-}$ (Brettel, 1997). The time-resolved EPR experiments monitor this spin-correlated radical pair. Although differing in details, time-resolved EPR experiments on the reaction centers of Photosystem II of oxygenic photosynthesis and purple photosynthetic bacteria of anoxygenic photosynthesis exhibit similar characteristic spin-correlated radical pair electron spin polarization from an oxidized chlorophyll electron donor and reduced quinone electron acceptor. Note that in case of correlated radical pair electron spin polarization, the difference in intensity between deuterated and protonated samples is larger than for non-spin polarized signals (Morris et al., 1995).

sient EPR, the situation is more complex because of the continuous presence of the microwave field. The signal intensity is not simply related to the photo-induced magnetization unless the spin system is in steady state with the applied magnetic field, where approaching steady state depends on the relaxation properties of the sample exposed to the microwave field. However, it was demonstrated that for EPR signals having inhomogeneously broadened lines, as in photosynthetic reaction centers, the signal rise time is independent of the microwave field and determined by the inhomogeneous linewidth (Furrer et al., 1980; Furrer and Thurnauer, 1981; Stehlik et al., 1989a). Thus, with the provision of broadband detection schemes and use of special microwave resonators (both of which lower the sensitivity) the time resolution of transient and pulsed EPR techniques are comparable and typically on the order of tens of nanoseconds. Therefore, the two approaches, transient and pulsed EPR techniques, can be applied to obtain complementary information. As described in this chapter, this point is particularly true in time-resolved EPR studies of photosynthetic reaction centers where the signals monitored exhibit electron spin polarization of correlated radical pairs (see below).

An important aspect of time-resolved EPR experiments is the observed electron spin polarization characterized by a non-Boltzmann population of the electron spin levels (for reviews see: Atkins and McLauchlan, 1973; Trifunac and Thurnauer, 1979; Hoff, 1984; McLauchlan, 1989; Turro and Khudyakov, 1999). The electron spin polarization not only compensates for lowered sensitivity of time-resolved EPR, but also reflects the mechanism of formation of the transient paramagnetic species. For example, time-resolved EPR experiments of photosynthetic reaction centers monitor the key charge separated state between oxidized electron donor and reduced acceptor $P^{\bullet+}_{700}A_1^{\bullet-}$ in Photosystem I (see Fig. 1). The electron spin polarization reported by $P^{\bullet+}_{700}A_1^{\bullet-}$ is indicative of a spin-correlated radical pair that is populated from the excited singlet state of the primary electron donor, ${}^1P^*_{700}$ (Thurnauer and Norris, 1980; Buckley et al., 1987; Closs et al., 1987; Stehlik et al., 1989b;

Norris et al., 1990; Snyder and Thurnauer, 1993). Importantly, the spin-correlated radical pair nature of the charge-separated state is a common feature of all photosynthetic reaction centers. Based on the correlated radical pair model several interesting spin-phenomena have been explained and/or predicted, i.e., quantum beats, observed at short delay times after optical excitation (Salikhov et al., 1990; Bittl and Kothe, 1991; Kothe et al., 1991); out-of-phase modulation of the electron spin echo signal, which is due to dipole-dipole and exchange interactions in the correlated radical pairs and allows for distance measurements (Salikhov et al., 1992; Tang et al., 1994; Dzuba et al., 1995); multiple quantum coherence in photo-induced radical pairs, which allows for direct measurements of coherence decays (Tang and Norris, 1995; Borovykh et al., 2001); sequential electron transfer polarization mechanism, which allows for the study of the electron transfer dynamics (Norris et al., 1990; Morris et al., 1995; Kandrashkin et al., 1998); unusual line shapes in time-resolved high-frequency electron nuclear double resonance (ENDOR), which allows for mapping of electron transfer pathways (Poluektov et al., 2004, 2005).

In this chapter we present details of the quantum beat phenomena observed from the spin-correlated radical pairs formed in Photosystem I reaction centers of oxygenic photosynthesis. The understanding and analysis of the quantum beat oscillations in combination with high-field EPR (Möbius, 2000; Thurnauer et al., 2004) allows us to obtain the three-dimensional structure of $P_{700}^{\bullet+}A_1^{\bullet-}$, as well as its arrangement in the membrane. In addition, by monitoring the quantum beat oscillations with a pulsed EPR approach we shed new light on the electronic structure of the oxidized primary electron donor, $P_{700}^{\bullet+}$.

## II. Formation of Quantum Beats

### A. Basic Theory

The formation of quantum beats in the spin-correlated radical pairs of photosynthesis is a purely quantum mechanical phenomenon. It can be rationalized using an analytical EPR model developed for these short-lived intermediates (Kothe et al., 1998). The model is based on the density matrix approach,

$$\rho(\Omega,t)=\exp[-(i/\hbar)H(\Omega)t]\cdot\rho(\Omega,0)\cdot\exp[(i/\hbar)H(\Omega)t] \tag{1}$$

where $\rho(\Omega,0)$ specifies the initial configuration of the radical pair immediately after the laser pulse. The spin Hamiltonian employed,

$$H(\Omega) = H_Z(\Omega) + H_{EX} + H_D(\Omega) + H_{HF}(\Omega) + H_{NZ} \tag{2}$$

considers Zeeman, exchange, dipolar, hyperfine and nuclear Zeeman interactions of the radical pair. For computational simplicity, hyperfine interactions are restricted to one I = ½ nucleus in the donor. In the presence of a microwave field, a microwave term, $H_{MW}$, must be added to the spin Hamiltonian (2).

The crucial point is the specification of the initial condition of the secondary radical pair at the instant of the laser pulse. In native photosynthetic reaction centers the lifetime of the primary radical pair is short (Fig. 1). Thus, even the secondary radical pair is generated in a virtually pure singlet state,

$$\rho(\Omega,0) = \frac{1}{2}\left\{\left|S+\frac{1}{2}\right\rangle\left\langle S+\frac{1}{2}\right| + \left|S-\frac{1}{2}\right\rangle\left\langle S-\frac{1}{2}\right|\right\} \tag{3}$$

determined by the spin multiplicity of the excited primary donor. The calculated energy level diagram for such a radical pair is shown in Fig 2. One sees that a singlet radical pair with hyperfine coupling to one I = ½ nucleus is formed with spin-correlated population of only four of the eight eigenstates (Buckley et al., 1987; Closs et al., 1987). This gives rise to high electron spin polarization. In addition, there are coherences between the eigenstates of the radical pair. Analysis reveals that zero quantum electron (Salikhov et al., 1990; Bittl and Kothe, 1991; Kothe et al., 1991, 1994a) and single quantum nuclear coherences (Jeschke, 1997; Weber et al., 1997; Kothe et al., 1998) are involved (dotted lines in Fig. 2).

Interestingly, the mechanism for the formation of these coherences is different. The analytical model shows that the nuclear coherences are based on electron-electron-nuclear three-spin mixing (Jeschke, 1997; Weber et al., 1997; Kothe et al., 1998). This particular mechanism implies that the nuclear modulation amplitude sensitively depends on the static magnetic field of the EPR experiment. Generally, large modulation amplitudes are predicted for X-band studies (9.5 GHz/350 mT). At higher microwave frequencies, however, the modulation amplitude rapidly decreases with increasing magnetic field strength. Thus, at Q- (34 GHz) and D-band (130 GHz), only zero quantum electron precessions are expected.

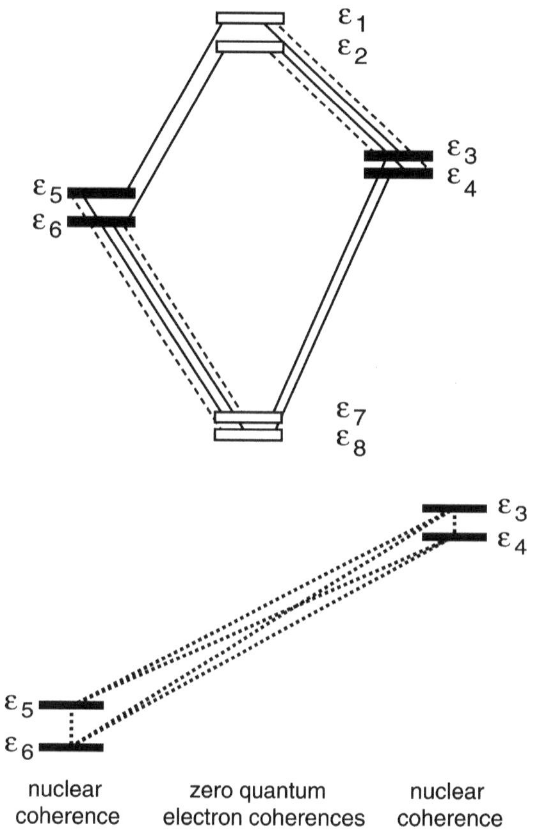


*Fig. 2.* Energy level diagram for a spin-correlated radical pair with hyperfine coupling to one $I = ½$ nucleus in the donor. Note that only four of the eight eigenstates are populated immediately after the laser pulse. Full lines: Allowed EPR transitions. Dashed lines: Forbidden EPR transitions. Dotted lines: Zero quantum electron and singe quantum nuclear coherences.

The light-induced formation of these coherences can be rationalized in terms of the non-adiabatic change of the spin Hamiltonian at the instant of the laser pulse. At time zero, the radical pair is created in a singlet state as a consequence of spin conservation in an ultra-fast photochemical reaction. Since the singlet is not an eigenstate of the corresponding spin Hamiltonian,

$$[\rho(\Omega,0), H(\Omega)] \neq 0 \qquad (4)$$

the radical pair starts out in a coherent superposition of eigenstates, which can manifest itself as quantum beats in an EPR experiment with adequate time resolution.

## B. Zero Quantum Electron Coherences

Under slow-motional conditions, the frequency of the zero quantum electron precessions, $\omega_{ZQ}$, critically depends on the orientation, Ω, of the radical pair in the laboratory frame. As shown previously (Kothe et al., 1994a,b; Kiefer et al., 1999; Link et al., 2001; Heinen et al., 2002), $\omega_{ZQ}$ is given by

$$\omega_{ZQ} = (1/\hbar)\{[2/3D^{zz}(\Omega) - 2J_{ex}]^2 + [(g_1^{zz}(\Omega) - g_2^{zz}(\Omega))\mu_B B_0 + \sum_k A_{1k}^{zz}(\Omega) M_{1k}^{i} - \sum_l A_{2l}^{zz}(\Omega) M_{2l}^{j}]^2\}^{1/2} \qquad (5)$$

where $D^{zz}(\Omega)$, $J_{ex}$, $g_i^{zz}(\Omega)$, $\mu_B$, $B_0$ and $A_{ij}^{zz}(\Omega)$ are the *zz* component of the dipolar coupling tensor, the strength of the isotropic exchange interaction, the *zz* component of the g-tensor of radical *i*, the Bohr magneton, the static magnetic field and the secular part of the hyperfine interaction between nucleus *j* and radical *i*, respectively. Inspection of Eq. (5) reveals that the frequency of the zero quantum electron coherences is determined by the spin-spin interactions of the radical pair and the difference in the Zeeman and hyperfine interactions of the constituent radicals. For the secondary radical pairs of photosynthesis, the Zeeman term provides the largest contribution even in X-band EPR experiments.

The weak $B_1$ field, commonly employed in transient EPR, allows for only a small range of orientations to meet the resonance condition. Consequently, the frequency of the zero quantum electron precessions varies significantly with $B_0$ across the powder EPR spectrum. The pronounced variation can be used to extract the orientation of the magnetic tensors in the radical pair (Kiefer et al., 1999; Link et al., 2001, 2005; Heinen et al., 2002, Thurnauer et al., 2004).

In deriving Eq. (5), all pseudo-secular terms of the hyperfine interactions have been neglected. As a result, light-induced nuclear coherences (Jeschke, 1997; Weber et al., 1997; Kothe et al., 1998) are neglected. The latter is certainly a good approximation for Q- and D-band studies, since the modulation amplitude of these coherences decreases strongly with the applied magnetic field. For the analysis of X-band experiments, however, consideration of pseudo-secular hyperfine interactions is essential.

## C. Nuclear Coherences

In the weak coupling limit, relevant to the secondary radical pairs of photosynthesis, the non-secular elements of the electron spin-spin interactions are usually small compared to energy differences between different electron spin states. Therefore, these elements can be considered by perturbation theory (Weber et al., 1997; Kothe et al., 1998), which predicts two nuclear coherences for each hyperfine coupled nucleus of the radical pair (Fig. 2),

$$\omega_{\mp} = (1/\hbar)[(\frac{1}{2}B_{ij}(\Omega))^2 + (\hbar\omega_N \mp \frac{1}{2}A_{ij}^{zz}(\Omega))^2]^{1/2} \quad (6)$$

Here, $B_{ij}(\Omega)$ is the pseudo-secular part of the hyperfine interaction between nucleus $j$ and radical $i$ and $\omega_N$ denotes the nuclear Larmor frequency. The modulation amplitude of these nuclear coherences sensitively depends on the static magnetic field employed. According to the three-spin mixing mechanism, the maximum modulation amplitude is obtained if the nuclear Zeeman energy matches the anisotropic hyperfine interaction (Kothe et al., 1998). This is satisfied in X-band (9.5 GHz/350 mT) EPR studies of the secondary radical pairs of photosynthesis. Generally, the formation of light-induced nuclear coherences is associated with the generation of high nuclear spin polarization (Jeschke, 1997; Kothe et al., 1998; Jeschke and Matysik, 2003). This greatly increases the detection sensitivity. Moreover, it enables extraction of reliable values for the hyperfine parameters of the radical pair.

## III. Detection of Quantum Beats by Transient EPR

### A. Basic Experiment

The first successful detection of quantum beats in photosynthetic reaction centers was achieved using transient EPR in combination with pulsed laser excitation (Kothe et al., 1991). In this experiment, the sample is irradiated with a short laser pulse and the time evolution of the transverse magnetization is monitored in the presence of a weak microwave magnetic field. Basically, the light pulse generates the spin-correlated radical pair and initiates the formation of quantum coherences. The continuous microwave field, applied in transient EPR, has two effects. First, it converts the longitudinal magnetization associated with the population differences between neighboring eigenstates into transverse magnetization (Zwanenburg and Hore, 1993). This gives the Torrey oscillations (Torrey, 1949). Secondly, it converts the zero quantum electron and nuclear coherences into single quantum electron precessions or quantum beats, observable in EPR experiments with adequate time resolution.

All our experiments were performed using fully deuterated samples. Deuteration provides the greatly enhanced signal-to-noise ratio, necessary for a successful detection of quantum beats. Furthermore, it slows down the rapid decay of the zero quantum electron precessions due to the difference in hyperfine interactions of the radical ions (Eq. (5), last term). To resolve the frequency of the quantum beats, a fast detection system and a low Q microwave resonator are required (Link et al., 2001). Since $\omega_{ZQ}$ increases linearly with $B_0$ (Eq. (5), second term), the bandwidth of the resonator should be increased accordingly. Generally, for X-band studies a resonator bandwidth of 25 MHz is sufficient, whereas for Q-band studies a bandwidth of 70 MHz is necessary. So far, successful quantum beat experiments at higher microwave frequencies have not been reported.

Generally, a complete data set consists of transient signals taken at equidistant field points covering the total spectral width. This yields a two-dimensional variation of the signal intensity with respect to both the magnetic field and the time axis. Such a complete data set for the secondary radical pair of Photosystem I, $P_{700}^{\bullet+}A_1^{\bullet-}$, measured at Q-band microwave frequency ($\omega/2\pi$ = 33.9893 GHz) is shown in Fig. 3. The data set refers to the deuterated and $^{15}$N-substituted cyanobacteria *Synechococcus lividus (S. lividus)*, a microwave field of $B_1$ = 0.029 mT and $T$ = 70 K. Note that a positive signal indicates absorptive (a) and a negative emissive (e) spin polarization. Transient spectra can be extracted from this plot at any fixed time after the laser pulse as slices parallel to the magnetic field axis. Likewise, the time evolution of the transverse magnetization may be obtained for any given field as a slice along the time axis.

Note the pronounced modulations in the transverse magnetization. These consist of fast initial oscillations which disappear 100 ns after the laser flash, and slow persisting oscillations with frequencies of a few MHz. Basically, the fast initial oscillations represent zero quantum electron precessions associated with the spin-correlated generation of the radical pair (Salikhov et al., 1990; Bittl and Kothe, 1991; Kothe et al., 1991, 1994b; Bittl et al., 1994). Notably, the frequency of the quantum beats varies significantly with $B_0$ across the powder spectrum. This pronounced

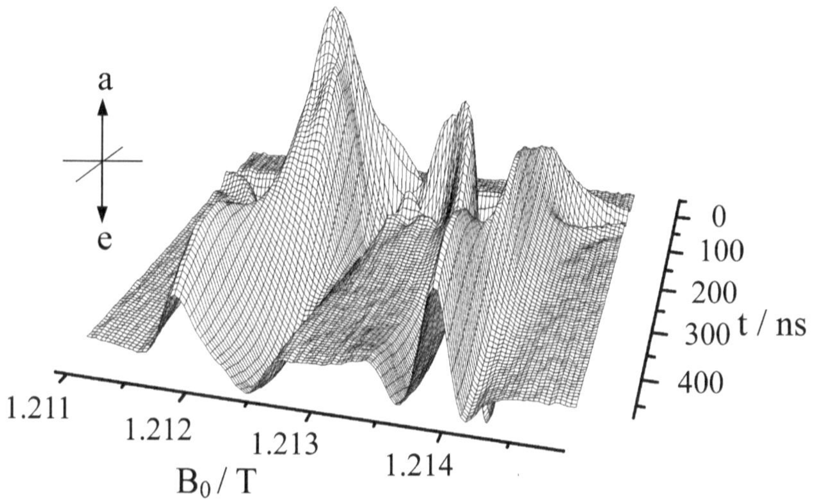


*Fig. 3.* Transient Q-band EPR data set of the light-induced radical pair $P_{700}^{\bullet+}A_1^{\bullet-}$ in plant Photosystem I of deuterated and $^{15}$N-substituted cyanobacteria *S. lividus*. Positive and negative signals indicate absorptive and emissive polarizations, respectively. The fast initial oscillations represent quantum beat oscillations associated with the spin-correlated generation of the radical pair. Microwave frequency: $\omega/2\pi = 33.9893$ GHz. Microwave field: $B_1 = 0.029$ mT. $T = 70$ K.

variation can be used to evaluate the orientation of the magnetic tensors (g-tensors, dipolar tensor) of the radical pair with respect to a magnetic reference system (Kiefer et al., 1999; Link et al., 2001, 2005; Heinen et al., 2002; Thurnauer at al., 2004).

### B. Selected Applications

Multifrequency quantum beat studies and high-field EPR of the spin-correlated radical pair $P_{700}^{\bullet+}A_1^{\bullet-}$ in Photosystem I of oxygenic photosynthesis can provide the three-dimensional structure of the short-lived intermediate in the photosynthetic membrane (Link et al., 2001). This is possible because during the course of high-field EPR experiments on $P_{700}^{\bullet+}A_1^{\bullet-}$ we discovered that the electron spin polarization pattern of the EPR spectra is sensitive to the sample orientation in the magnetic field of the spectrometer (magneto-orientation) (Berthold et al., 1999). Our strategy for obtaining structure of the short-lived radical pair intermediates involves a multifrequency EPR approach with data collection at three microwave frequencies. The orientation of the magnetic tensors of $P_{700}^{\bullet+}A_1^{\bullet-}$ is obtained from analysis of the pronounced anisotropy of the zero quantum electron coherences in a transient Q-band EPR experiment. The orientation of the cofactors of the primary donor is then determined by analyzing transient X-band EPR spectra, extracted from a two-dimensional data set. Finally, time-resolved D-band (130 GHz) spectra of a magnetically aligned sample provide detailed information on the arrangement of the cofactors $P_{700}^{\bullet+}$ and $A_1^{\bullet-}$ in the photosynthetic membrane. These experiments were performed using deuterated and $^{15}$N-substituted cyanobacteria *S. lividus*. Deuteration provides the enhanced signal-to-noise ratio, necessary for quantum beat studies and $^{15}$N-substitution facilitates analysis of the time-dependent EPR spectra at X-band frequency.

#### 1. Orientation of the Cofactors in $P_{700}^{\bullet+}A_1^{\bullet-}$

The orientation of the magnetic tensors of $P_{700}^{\bullet+}A_1^{\bullet-}$ (g-tensors, dipolar tensor) with respect to a magnetic reference system (g-tensor of $A_1^{\bullet-}$) can be described by five Euler angles. Values for these angles were obtained from a computer fit of the two-dimensional Q-band EPR experiment shown in Fig. 3 (Link et al., 2001). Figure 4 depicts the short time behavior of the transverse magnetization of $P_{700}^{\bullet+}A_1^{\bullet-}$ measured at four selected field positions ((a) – (d)) along the powder spectrum (solid lines). Evidently, there are fast initial oscillations which disappear 100 ns after the laser pulse. Note that the phase and frequency of these oscillations varies significantly across the powder spectrum. Generally, oscillation frequencies from 10 to 50 MHz can be extracted from the corresponding power spectra. Basically, these oscillations represent Q-band quantum beats associated with the spin-correlated generation of the radical pair (see Eq. (5)).

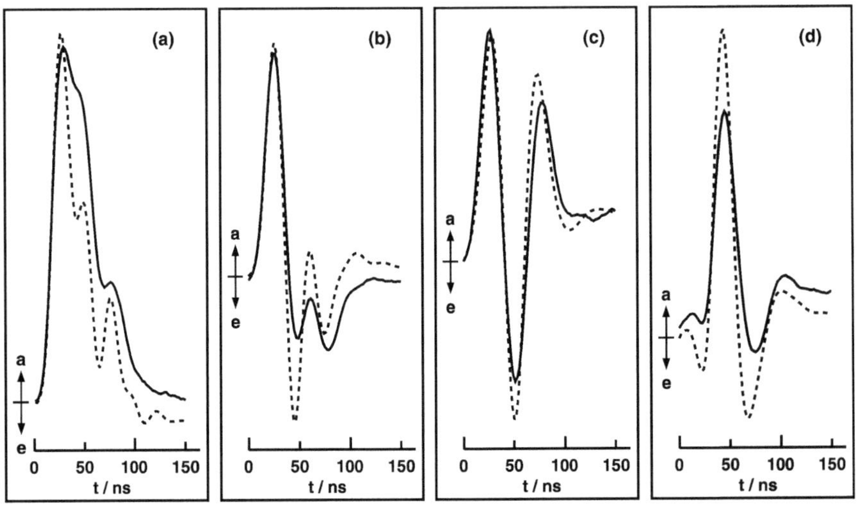


*Fig. 4.* Quantum beat oscillations of the light-induced radical pair $P^{\bullet +}_{700}A^{\bullet -}_1$ in plant Photosystem I measured with transient Q-band EPR. The time profiles refer to four selected field positions, (a), (b), (c) and (d) along the powder spectrum. Positive and negative signals indicate absorptive and emissive polarizations, respectively. Microwave frequency: $\omega/2\pi = 33.9893$ GHz. Microwave field: $B_1 = 0.029$ mT. Full lines: Experimental time profiles from the deuterated and $^{15}$N-substituted cyanobacteria *S. lividus* measured at $T = 70$ K. Dashed lines: Best fit simulation. (Link et al., 2004; © by Oldenbourg Wissenschaftsverlag.)

The pronounced anisotropy of the Q-band quantum beats was used to evaluate the orientation of the magnetic tensors in $P^{\bullet +}_{700}A^{\bullet -}_1$ (Link et al., 2001). Typically, a set of 22 calculated time profiles, covering the total spectral width, was simultaneously fitted to the experimental profiles by varying the parameters of the tensor orientations. This provided values for the five Euler angles, characterizing the magnetic tensor orientations in $P^{\bullet +}_{700}A^{\bullet -}_1$ (Link et al., 2001). The dashed lines in the Fig. 4 represent best fit simulations. Generally, the agreement achieved is very good. The uniqueness of the fit was tested by running the fit procedure with different experimental data sets and different starting values (Link et al., 2001). Within the error limits, the same or magnetically equivalent tensor orientations were obtained.

Evaluation of the three-dimensional structure of $P^{\bullet +}_{700}A^{\bullet -}_1$ requires knowledge of the cofactor orientation relative to the g-tensor of $P^{\bullet +}_{700}$ or $A^{\bullet -}_1$. This information exists for the quinone acceptor, $A^{\bullet -}_1$, in which the molecular axes are collinear with the g-tensor axes (Isaacson et al., 1995). The cofactor orientation of the primary donor in the g-tensor of $P^{\bullet +}_{700}$ can be determined by analyzing transient X-band EPR spectra, extracted from a two-dimensional data set. The analysis is based on anisotropic $^{15}$N hyperfine interactions, giving rise to nuclear modulations in the transverse magnetization (Kothe et al., 1998). ENDOR experiments indicate a highly asymmetric spin density distribution for $P^{\bullet +}_{700}$ over the two chlorophylls that constitute the primary donor (Käss et al., 1995; Mac et al., 1998). It is therefore reasonable to consider only four axially symmetric $^{15}$N hyperfine tensors, whose symmetry axes are collinear. The orientation of this axis in the g-tensor system of $P^{\bullet +}_{700}$ is characterized by the two Euler angles $\Theta_{1j}$ and $\Psi_{1j}$.

Typical X-band line shapes of $P^{\bullet +}_{700}A^{\bullet -}_1$ observed 50, 70 and 200 ns after the laser pulse, are shown in Fig. 5 (solid lines) (Link et al., 2001). The spectra refer to deuterated and $^{15}$N-substituted cyanobacteria *S. lividus*, a microwave frequency of $\omega/2\pi = 9.5738$ GHz, $B_1 = 0.020$ mT and $T = 70$ K. Evidently, the spectral shape changes significantly with time. The variation can be used to extract reliable values for $\Theta_{1j}$ and $\Psi_{1j}$. Typically, a set of 27 calculated spectra, selected at various times after the laser pulse, was simultaneously fitted to the experimental spectra by varying $\Theta_{1j}$ and $\Psi_{1j}$ (Link et al., 2001). The dashed lines in Fig. 5 represent best fit simulations. Evidently, the agreement achieved is very good. The uniqueness of the fit was tested by running the fit procedure with different starting values. Within the error limits, the same angular values were obtained (Link et al., 2001).

The evaluated Euler angles, $\Theta_{1j}$ and $\Psi_{1j}$, characterize the orientation of the symmetry axis of the $^{15}$N hyperfine tensors in the g-tensor system of $P^{\bullet +}_{700}$. Because of the local character of the hyperfine interactions, it is reasonable to identify this symmetry axis with the chlorophyll normal. Thus, the cofactor orientation

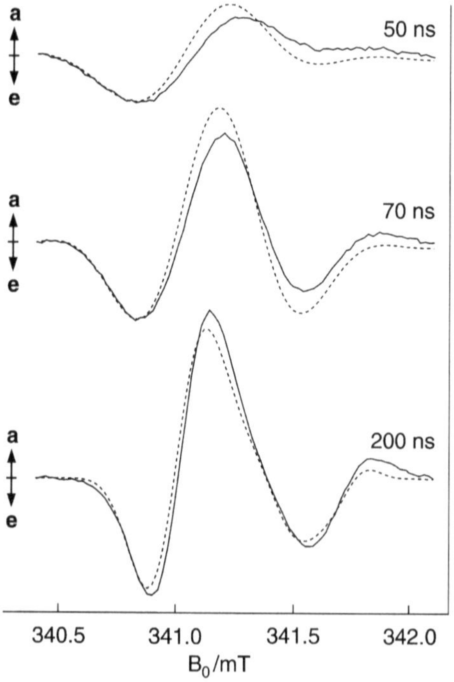


*Fig. 5.* Transient X-band EPR spectra of the light-induced radical pair $P_{700}^{\bullet+}A_1^{\bullet-}$ in plant Photosystem I at various times after the laser pulse. Positive and negative signals indicate absorptive and emissive polarizations, respectively. Microwave frequency: $\omega/2\pi = 9.5738$ GHz. Microwave field: $B_1 = 0.020$ mT. Full lines: Experimental spectra from the deuterated and $^{15}N$-substituted cyanobacteria *S. lividus* measured at $T$ = 70 K. Dashed lines: Best fit simulation. (Link et al., 2004; © by Oldenbourg Wissenschaftsverlag.)

of the primary donor in the g-tensor system of $P_{700}^{\bullet+}$ is obtained. Combining this result with the angular information from the Q-band study gives the mutual orientation of the cofactors in $P_{700}^{\bullet+}A_1^{\bullet-}$.

## *2. Three-Dimensional Structure of $P_{700}^{\bullet+}A_1^{\bullet-}$*

The cofactor arrangement of $P_{700}^{\bullet+}A_1^{\bullet-}$ in the photosynthetic membrane can be determined by analyzing time-resolved high-field EPR spectra observed for whole cells of the deuterated cyanobacteria (Berthold et al., 1999). In a typical experiment, the sample is placed in the static magnetic field of the EPR spectrometer and then cooled to low temperatures ($T$ = 90 K).

Representative D-band (130 GHz) results are shown in Fig. 6 (full lines). The spin-polarized spectra of $P_{700}^{\bullet+}A_1^{\bullet-}$ were recorded by monitoring the electron spin echo from a two microwave pulse sequence as a function of the magnetic field (Heinen et al., 2004a). The EPR spectrum in Fig. 6a refers to the original sample orientation immediately after cooling ($\rho = 0°$). The spectrum in Fig. 6b was taken after the frozen sample was rotated by 90° about an axis perpendicular to the magnetic field ($\rho = 90°$). Pronounced spectral differences are observed particularly in the high-field region.

Evidently, the reaction centers of Photosystem I are aligned in the static magnetic field of the D-band

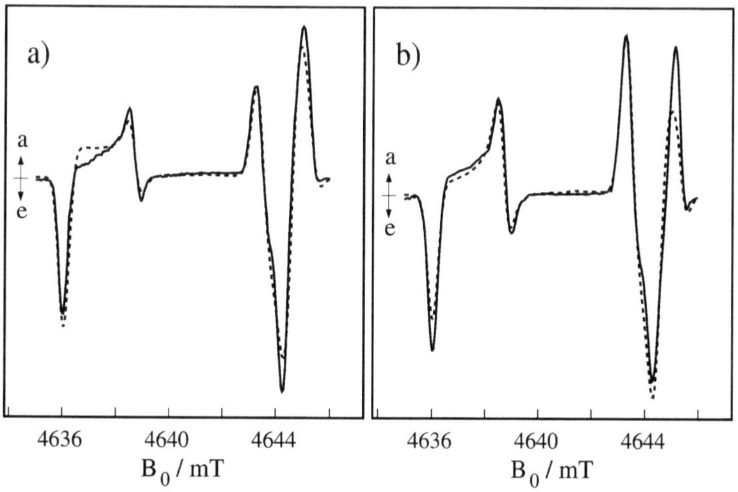


*Fig. 6.* Spin-polarized D-band (130 GHz) EPR spectra of the light-induced radical pair $P_{700}^{\bullet+}A_1^{\bullet-}$ in plant Photosystem I at two different orientational distributions with respect to the laboratory frame. (a) Sample axis parallel to the magnetic field, $\rho = 0°$. (b) Sample axis perpendicular to the magnetic field, $\rho = 90°$. Positive and negative signals indicate absorptive and emissive polarizations, respectively. Full lines: Experimental spectra from the deuterated and $^{15}N$-substituted cyanobacteria *S. lividus* measured at $T$ = 75 K. Dashed lines: Best fit simulation. Magneto-orientation of the sample was achieved at room temperature with an alignment field of $B_0$ = 7 T. The EPR spectra were recorded using the time-resolved electron spin echo technique. (From Link et al., 2005. Copyright John Wiley & Sons Limited. Reproduced with permission.)

EPR spectrometer. What is the physical origin of this magneto-orientation? Theoretical studies indicate that peptide bonds in proteins are a source of anisotropic diamagnetic susceptibility, responsible for the alignment of the polypeptide helix axes parallel to the magnetic field (Worcester, 1978; Pauling, 1979). In cyanobacterial Photosystem I, as well as in most other membrane proteins, a total of about 30 transmembrane α-helices are aligned to within a small angle of the membrane normal (Jordan et al., 2001), and this presents a natural choice for the principal axis of the susceptibility tensor.

If one assumes that the susceptibility tensor is axially symmetric in its principal axis system, the orientational distribution of the spin-correlated radical pair can be described by an order parameter, $S_{Z'Z'}$, and two Euler angles, $\vartheta$ and $\psi$, characterizing the orientation of the membrane normal in a magnetic reference system (Berthold et al., 1999). Values for these parameters can be obtained by fitting the two angular-dependent D-band EPR spectra, measured with the sample axis either parallel ($\rho = 0°$) or perpendicular ($\rho = 90°$) to the magnetic field. The result is shown in Fig. 6, where the dashed lines represent best fit simulations. The fit provides a small positive value for the order parameter, i.e., $S_{Z'Z'} = 0.12$. The positive sign of $S_{Z'Z'}$ implies that the susceptibility tensor axes orient preferentially parallel to the magnetic field. This is clear evidence for membrane proteins being the major source of the anisotropy in the diamagnetic susceptibility (Worcester, 1978; Pauling 1979). If lipid membranes were the dominant source for the susceptibility anisotropy, one would expect that the susceptibility tensor axes orient mainly perpendicular to the magnetic field, i.e., the order parameter $S_{Z'Z'}$ would be negative.

To obtain further information on the mechanism of the magneto-orientation process, values for $S_{Z'Z'}$ were evaluated as a function of the applied magnetic field (Heinen et al., 2004a). For this purpose, time-resolved D-band EPR spectra of $P_{700}^{\bullet+}A_1^{\bullet-}$ were measured for four different alignment fields varying between 0 and 7 T. A computer analysis of the observed angular-dependent line shapes then provided values for the order parameter $S_{Z'Z'}$. In Fig. 7 this order parameter is plotted as a function of the alignment field. The experimental values (solid squares) refer to the cyanobacterium *S. lividus*. One sees that $S_{Z'Z'}$ first increases gradually, then rises steeply and finally levels off to a constant limiting value. Thus, the magnetic field dependence of $S_{Z'Z'}$ can be described by a saturation

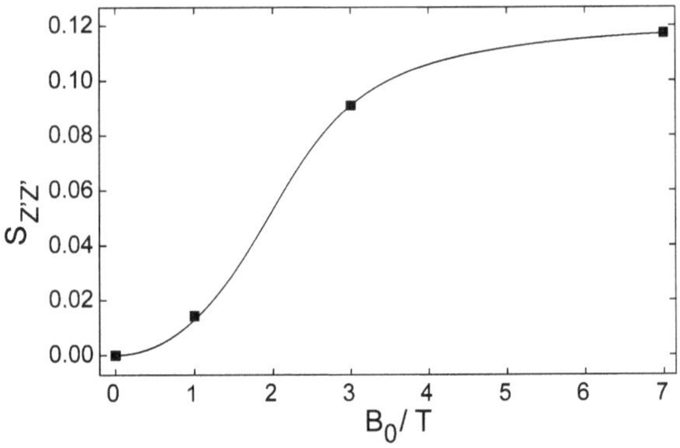


*Fig. 7.* Magnetic field dependence of the order parameter, $S_{Z'Z'}$, that characterizes the alignment of the symmetry axis of the susceptibility tensor along the magnetic field. The experimental values (solid squares) were extracted from spin-polarized D-band EPR spectra of the light-induced radical pair, $P_{700}^{\bullet+}A_1^{\bullet-}$, in plant Photosystem I. The solid line represents a best fit simulation using a simple model for magneto-orientation. The extracted value for the anisotropy of the diamagnetic susceptibility is $\Delta\chi V = 5.7 \times 10^{-27}$ m$^3$. (From Link et al., 2005. Copyright John Wiley & Sons Limited. Reproduced with permission.)

curve, as derived previously (Geacintov et al., 1972). The major adjustable parameter of this curve is the anisotropy of the diamagnetic susceptibility $\Delta\chi V$. It was determined by fitting calculated order parameters $S_{Z'Z'}$ to the experimental ones. The solid line in Fig. 7 represents the best fit simulation based on a parameter value of $\Delta\chi V = 5.7 \times 10^{-27}$ m$^3$. The large value for the anisotropy of the diamagnetic susceptibility suggests that whole bacterial cells are oriented by the magnetic field. Model calculations for $\Delta\chi V$, using information from an electron microscopy study, support this notion (Heinen et al., 2004b).

Evidently, whole cells are aligned in the magneto-orientation process. A prerequisite for this alignment is an anisotropic shape of the bacterial cell and the thylakoid system. Figure 8 depicts an electron micrograph of a *S. lividus* cell at high magnification. One sees that the thylakoid membranes of cyanobacteria form cylinders with hemispherical caps on both ends. The embedded membrane proteins define the symmetry axis of the susceptibility tensor collinear with the local membrane normal. If the long axes of the cells were oriented parallel to the magnetic field, all bilayer normals in the cylinder part would be orthogonal to the field implying a negative order parameter $S_{Z'Z'}$. This is not observed in the D-band EPR experiments. We therefore expect that the cyanobacterial cells align perpendicular to the magnetic field. Recently, this alignment was confirmed by

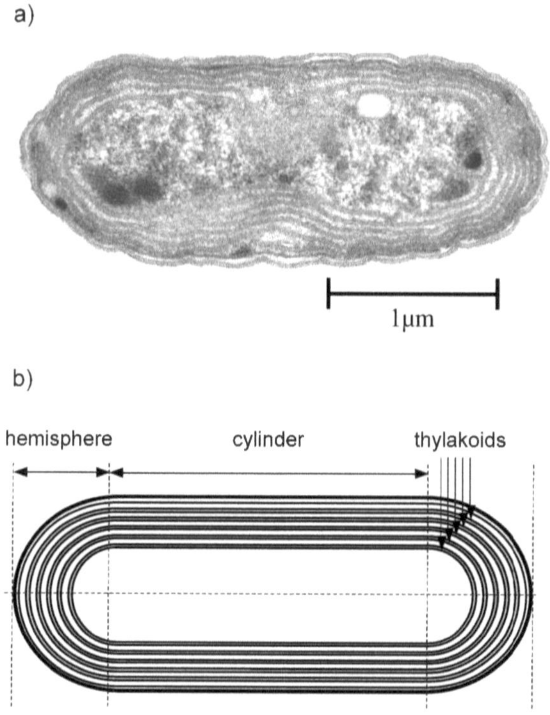


*Fig. 8.* (a) Electron micrograph of an ultrathin section of a *S. lividus* cell at high magnification. By courtesy of J.R. Golecki. (b) Schematic of a cyanobacterial cell with five thylakoid layers.

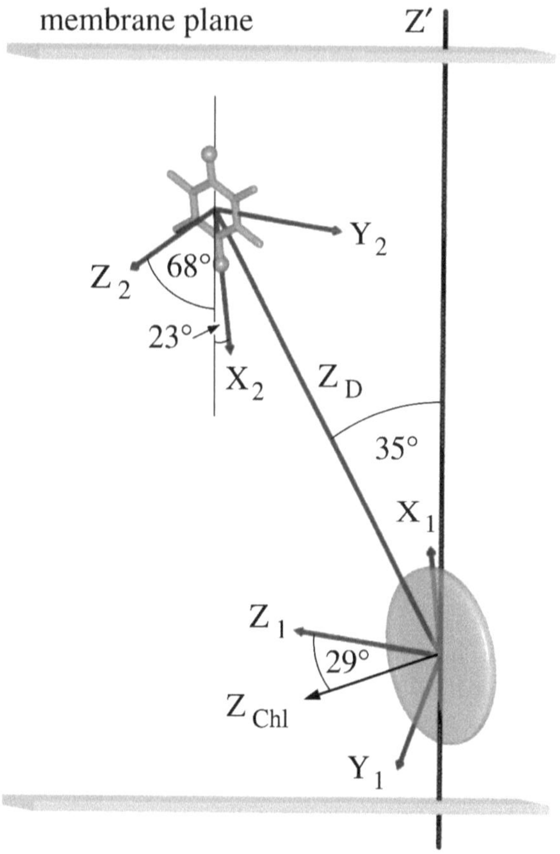


*Fig. 9.* Structure of the secondary pair $P^{•+}_{700}A^{•-}_1$ in plant Photosystem I as determined by high-time resolution multifrequency EPR employing quantum beat oscillations and a magnetically aligned sample. The view direction is parallel to the membrane plane. The shaded disk represents one of the two chlorophyll molecules of the primary donor, which carries the major part of the unpaired spin. $\mathbf{X}_1$, $\mathbf{Y}_1$, $\mathbf{Z}_1$ = principal axis system of the g-tensor of $P^{•+}_{700}$. $\mathbf{X}_2$, $\mathbf{Y}_2$, $\mathbf{Z}_2$ = principal axis system of the g-tensor of $A^{•-}_1$. $\mathbf{Z}_D$ = symmetry axis of the dipolar tensor. $\mathbf{Z}'$ = membrane normal.

electron microscopy (Heinen et al., 2004b).

On the basis of this model, one can rationalize the extracted value for the order parameter $S_{Z'Z'}$. If we assume axially symmetric distribution functions, the total order parameter $S$ can be written as a product of two order parameters $S_1$ and $S_2$ (Heinen et al., 2004b). Here $S_1$ characterizes the ordering of the bilayer normals with respect to the long axis of the cell while $S_2$ describes the alignment of the cells relative to the magnetic field. From electron microscopy data, we estimate $S_1 \approx -0.34$ (Heinen et al., 2004b). At $B_0$ = 7 T, the cells are completely aligned, corresponding to $S_2 \approx -0.5$. This gives a total order parameter of $S = S_1S_2 \approx 0.17$, which compares favorably with the limiting order parameter of $S_{Z'Z'} = 0.12$ observed by D-band EPR. Evidently, the proposed mechanism for magneto-orientation of photosynthetic reaction centers is correct.

The analysis of the time-resolved D-band EPR spectra of $P^{•+}_{700}A^{•-}_1$ also provides the orientation, $\vartheta$ and $\psi$, of the membrane normal in a magnetic reference system. This structural information is unique, because it is not easily available by other EPR techniques. Knowledge of this orientation makes it possible to determine the three-dimensional structure of $P^{•+}_{700}A^{•-}_1$ in its native membrane, as shown in Fig. 9 (Link et al., 2001). The structure describes the orientation of the g-tensor of the primary donor $P^{•+}_{700}$ as well as the position and orientation of the reduced acceptor, $A^{•-}_1$.

In the structural model for $A^{•-}_1$, the quinone plane is found to be inclined by 68° relative to the membrane plane, while the O-O axis makes an angle of 23° with the membrane normal. These values are in substantial agreement with those reported in a high resolution X-ray structure (Jordan et al., 2001). Nevertheless, some characteristic deviations exist. They might reflect a reorientation of the reduced quinone in its binding pocket. It should be noted that the depicted structure refers to the charge separated, $P_{700}^{\bullet+}A_1^{\bullet-}$, observed at low temperature, where forward electron transfer is largely blocked (Sétif et al., 1984; Schlodder et al., 1998).

Figure 10 depicts a three-dimensional representation of the g-tensor orientation of $P_{700}^{\bullet+}$. For simplicity, we show only one of the two chlorophyll molecules of the primary donor, which is the chlorophyll *a* species coordinated by the Psa B protein subunit (Krabben et al., 2000; Link et al., 2001). To describe this orientation, we define the chlorophyll-based reference system $\mathbf{X}_{Chl}$, $\mathbf{Y}_{Chl}$, $\mathbf{Z}_{Chl}$. The $\mathbf{Z}_{Chl}$-axis is the chlorophyll normal, the $\mathbf{Y}_{Chl}$-axis is the projection of the membrane normal onto the chlorophyll plane. The $\mathbf{X}_{Chl}$-axis then lies in the chlorophyll plane perpendicular to $\mathbf{Y}_{Chl}$. For simple π radicals, the principal direction corresponding to the smallest g-tensor component, $g_Z$, is expected to lie near the normal of the molecular plane (Stone, 1963). Inspection of Fig. 10 reveals a significant deviation of 29° between the $g_Z$-direction and the chlorophyll normal (Link et al., 2001), in reasonable agreement with EPR results based on Photosystem I single crystals (Zech et al., 2000).

The new structural information presented above is based on the analysis of quantum beat oscillations in combination with high-field EPR of a magnetically oriented sample (Link et al., 2001; Thurnauer et al., 2004; Link et al., 2005). With these techniques, it is possible to evaluate the three-dimensional structure of the short-lived radical pair intermediates following light excitation of photosynthetic proteins in their native membranes. Thus, one can also obtain the cofactor arrangement of the radical pair with respect to the membrane. The complementary information obtained from quantum beats and a magnetically aligned sample is a powerful structural tool. We expect that this is of general interest, particularly to researchers in the broad fields related to solar energy conversion and storage, because the detailed structure of radical pair intermediates can be determined on a nanosecond time scale (Link et al., 2001).

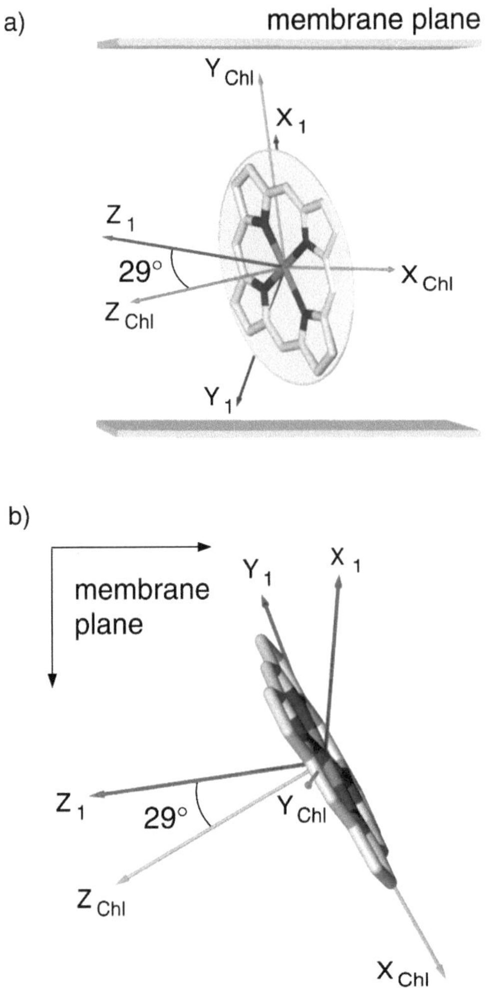


*Fig. 10.* Orientation of the g-tensor of the primary donor $P_{700}^{\bullet+}$ in plant Photosystem I as determined by high-time resolution multifrequency EPR employing quantum beat oscillations and a magnetically aligned sample. For simplicity, we show only one of the two chlorophyll molecules, which carries the major part of the unpaired spin. (a) View direction parallel to the membrane plane. (b) View direction onto the membrane plane. $\mathbf{X}_1$, $\mathbf{Y}_1$, $\mathbf{Z}_1$ = principal axis system of the g-tensor of $P_{700}^{\bullet+}$. $\mathbf{X}_{Chl}$, $\mathbf{Y}_{Chl}$, $\mathbf{Z}_{Chl}$ = chlorophyll based reference system. The orientation of the chlorophyll molecule in the ($\mathbf{X}_{Chl}$, $\mathbf{Y}_{Chl}$) plane was chosen according to a recent X-ray structure of Photosystem I at 2.5 Å resolution (Jordan et al., 2001).

## IV. Detection of Quantum Beats by Pulsed EPR

### A. Basic Experiment

The first pulsed EPR detection of quantum beats in photosynthetic reaction centers was achieved using an electron spin echo sequence (Dzuba et al., 1996). In the following we describe a single microwave pulse EPR experiment, well-suited to probe light-induced quantum coherences in spin-correlated radical pairs (Salikhov et al., 1992; Kothe et al., 1998). The pulse sequence employed,

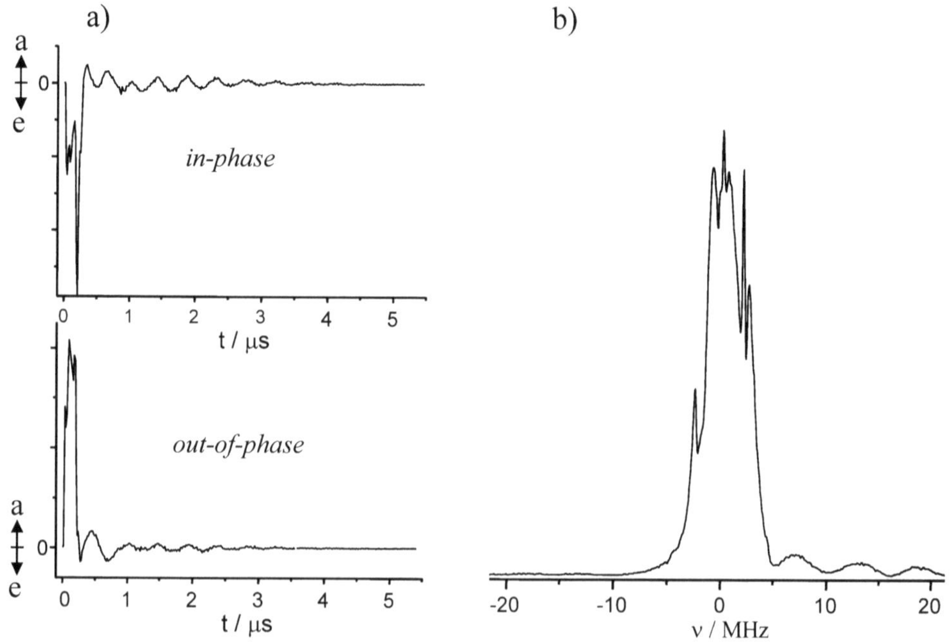


*Fig. 11.* (a) Photochemically induced quantum coherences of the secondary radical pair, $P_{700}^{\cdot+}A_1^{\cdot-}$, in plant Photosystem I detected by a pulsed X-band EPR experiment. The EPR time profiles refer to the deuterated and $^{15}N$-substituted cyanobacteria *S. lividus* and $T = 70$ K. The quantum coherences were recorded by monitoring the FID as a function of the delay, $t$, between the laser flash and a $\pi/2$ microwave pulse using the sequence, laser – $t$ – $(\pi/2)_x$ – $\tau$. Microwave frequency: $\omega/2\pi = 8.4693$ GHz. Coherence transfer was optimized by using a matched $\pi/2$ microwave pulse, applied at $B_0 = 338.0$ mT. (b) Power spectrum, obtained by a complex Fourier transformation of the EPR time profiles. The broad peaks represent zero quantum electron precessions. The sharp lines can be assigned to $^2H$ and $^{15}N$ nuclear coherences.

$$\text{laser} - t - (\pi/2)_x - \tau$$

consists of a short laser pulse at time zero, followed by a variable evolution period, $t$. At the end of this period, a $\pi/2$ microwave pulse is applied. The integrated free-induction decay (FID) is then monitored as a function of successively incremented values of $t$. In this sequence, the laser pulse generates the spin-correlated radical pair and initiates the formation of quantum coherences, which are then transferred to transverse electron magnetization by the $\pi/2$ microwave pulse (Kothe et al., 1998). Thus, we expect that the integrated FID has an oscillatory dependence on the delay, $t$, between the laser and the microwave pulse.

The above pulse sequence enables the study of light-induced quantum coherences in correlated radical pairs independent of any instrumental dead time. However, in contrast to transient EPR, the quantum beat oscillations are measured point by point in the time domain. This markedly lowers the detection sensitivity. Thus, in general, only deuterated samples ensure the improved signal-to-noise ratio, necessary for a successful experiment. It should be noted that signal enhancement by time averaging requires synchronization of laser and microwaves on a nanosecond time scale. In our experiments, the Q-switch of the Nd:YAG laser is triggered by the EPR spectrometer, resulting in a maximum jitter of laser flash and microwave pulse of 1 ns (Kothe et al., 1998). Generally, the modulation depth of the nuclear coherences can be enhanced by applying matched microwave pulses (Jeschke and Schweiger, 1996).

Typical results for $P_{700}^{\cdot+}A_1^{\cdot-}$ in the deuterated and $^{15}N$-substituted cyanobacteria *S. lividus* are shown in Fig. 11 (Kothe et al., 1998). They refer to a microwave frequency of $\omega/2 = 9.4693$ GHz and $T = 70$ K. Fig. 11a depicts the dependence of the in-phase and out-of-phase EPR signals on the delay, $t$, between the laser flash and the $\pi/2$ microwave pulse. Evidently, there are fast initial oscillations which disappear 250 ns after the laser flash. In addition, slow persisting oscillations with frequencies of a few MHz can be observed. Fig. 11b shows the corresponding power spectrum, obtained by a complex Fourier transformation of the time profiles. The broad peaks represent zero quantum electron precessions associated with the spin-correlated generation of the radical pair. The remaining sharp peaks can be assigned to photochemically induced $^2H$ and $^{15}N$ nuclear coherences (Kothe et al., 1998).

In the present experiment, the modulation depth of the nuclear coherences was optimized by using a matched microwave pulse (Jeschke and Schweiger, 1996), applied at $B_0$ = 338.0 mT in the high-field emissive part of the EPR spectrum (see Fig. 5). The matching condition, $\omega_1 \cong \omega_N$, was satisfied by using a selective $\pi/2$ microwave pulse of 128 ns length. Compared to detection with a hard pulse (i.e., 8 ns $\pi/2$ pulse length), the intensity gain is a factor of 4 to 5.

### *B. Selected Applications*

The electronic structure of the primary donor $P_{700}^{\bullet+}$ in Photosystem I is still the subject of numerous discussions. Previous studies, based on $^{15}N$ ENDOR and electron spin echo envelope modulation (ESEEM) experiments, favor a dimeric electronic structure (Käss et al., 1995). More recent ESEEM and ENDOR results, obtained in combination with specific isotope labeling, suggest a monomeric character for $P_{700}^{\bullet+}$ (Mac et al., 1998). To shed some light on this controversial matter, the pulsed quantum beat experiment is analyzed. At X-band (9.5 GHz/350 mT) formation of the $^{15}N$ nuclear coherences is associated with high nuclear spin polarization corresponding to an average enhancement factor of ~12000. This greatly increases the detection sensitivity. Moreover, it enables extraction of reliable values for the $^{15}N$ hyperfine coupling parameters in $P_{700}^{\bullet+}A_1^{\bullet-}$.

#### *1. $^{15}N$ Hyperfine Interactions in $P_{700}^{\bullet+}A_1^{\bullet-}$*

To illustrate the analysis of the pulsed EPR experiment, numerical calculations are presented for a model radical pair with *anisotropic* hyperfine coupling to a single $^{15}N$ nucleus. The employed values for the g-tensors, the electron dipolar coupling and the orientation of the magnetic tensors are typical for a secondary radical pair of photosynthesis. Hyperfine interactions are simplified by considering four equivalent $^2H$ nuclei in the acceptor (*isotropic* coupling constant $a/h$ = 1.54 MHz) and a single $^{15}N$ nucleus in the donor. For convenience, an axially symmetric $^{15}N$ hyperfine tensor is assumed ($A_\perp/h$ = 1.60 MHz, $A_\parallel/h$ = 3.47 MHz). The calculations refer to an X-band microwave frequency of $\omega/2\pi$ = 9.8 GHz. Coherence transfer is modeled by a matched $\pi/2$ microwave pulse (Jeschke and Schweiger, 1996). Inhomogeneous broadening is considered by convolution with a Gaussian.

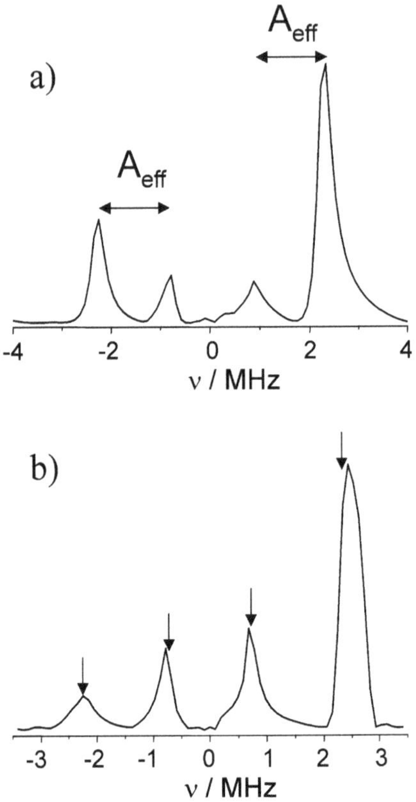


*Fig. 12.* Calculated power spectra for a model radical pair with *anisotropic* hyperfine coupling to a single $^{15}N$ nucleus. The numerical calculations simulate a pulsed X-band EPR experiment designed to study photochemically induced quantum coherences in spin-correlated radical pairs (Kothe et al., 1998). Hyperfine interactions are simplified by considering four equivalent $^2H$ nuclei in the acceptor (*isotropic* $a/h$ = 1.54 MHz) and a single $^{15}N$ nucleus in the donor ($A_\perp/h$ = 1.60 MHz, $A_\parallel/h$ = 3.47 MHz). The calculations refer to an X-band microwave frequency of $\omega/2\pi$ = 9.8 GHz. (a) Power spectrum for a single orientation. (b) Power spectrum for a random distribution of the model radical pair. The arrows indicate $^{15}N$ nuclear transitions, whose hyperfine tensor is oriented perpendicular to the magnetic field.

In Fig. 12a we show the power spectrum of the in-phase and out-of-phase EPR signals, calculated for a given orientation of the radical pair in the laboratory frame. The two lines at 0.9 MHz and 2.3 MHz indicate the $^{15}N$ nuclear coherences expected for the model radical pair. *Isotropic* hyperfine couplings do not show up in nuclear quantum beat experiments (Kothe et al., 1998). The frequency of the lines favor-

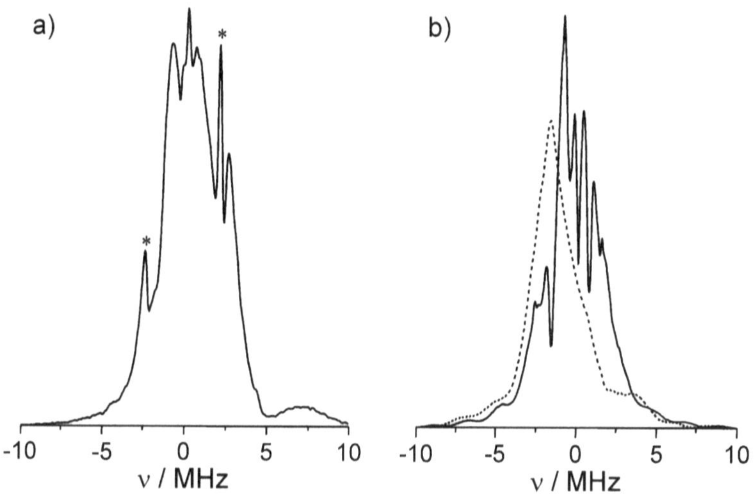


*Fig. 13.* (a) Experimental power spectrum of the secondary radical pair, $P_{700}^{\bullet+}A_1^{\bullet-}$, in plant Photosystem I, exhibiting photochemically induced quantum coherences. The power spectrum refers to the deuterated and $^{15}$N-substituted cyanobacteria *S. lividus* and $T = 70$ K. The quantum coherences were recorded in a pulsed X-band EPR experiment by monitoring the FID as a function of the delay, $t$, between the laser flash and a $\pi/2$ microwave pulse. Sequence: laser – $t$ – $(\pi/2)_x$ – $\tau$. Microwave frequency: $\omega/2\pi = 9.4693$ GHz. Coherence transfer was optimized by using a matched $\pi/2$ microwave pulse, applied at $B_0 = 338.0$ mT. Except for a $^2$H matrix line (*), all other sharp peaks indicate $^{15}$N nuclear coherences in $P_{700}^{\bullet+}$. (b) Calculated power spectrum using the extracted $^{15}$N hyperfine coupling parameters summarized in Table 1.

ably compares with the predictions of the analytical model given in Eq. (6). From the separation of the lines, the effective hyperfine coupling constant, $A_{eff}$, can be extracted.

Figure 12b depicts the power spectrum of the EPR time profiles, calculated for a random distribution of the model radical pair in the laboratory frame. To ensure convergence, 250000 different orientations were considered in the powder average. Surprisingly, one observes again two narrow lines as in the case of a single orientation (Fig. 12a). Analysis reveals that the line maxima can approximately be assigned to $^{15}$N nuclear transitions,

$$\nu_{\mp} = \left| \nu_N \mp \frac{1}{2} A_{\perp}/h \right| \quad (7)$$

whose hyperfine tensor is oriented perpendicular to the magnetic field (arrows in Fig. 12b). This allows direct extraction of $A_{\perp}$ from the powder spectrum. Because of the high detection sensitivity of the method, reliable values for $A_{\perp}$ are obtained. Moreover from the relative intensity of the lines, information on $A_{\parallel}$ can be extracted.

The outlined strategy has been used to determine the $^{15}$N hyperfine interactions in $P_{700}^{\bullet+}A_1^{\bullet-}$ (J. Lalevée, G. Link, J.-U. Weidner, M. Bechtold, T. Yago, S.L. Schlesselman, O.G. Poluektov, M.C. Thurnauer and G. Kothe, unpublished). In Fig. 13 we compare the experimental power spectrum of $P_{700}^{\bullet+}A_1^{\bullet-}$ (a) with the best simulation (b). The calculation is based on five different $^{15}$N nuclei for $P_{700}^{\bullet+}$ and the g-tensor orientation shown in Fig. 10. Generally, the agreement achieved is very good. In particular, the calculated peak positions deviate less than 0.1 MHz from the experimental counterparts. Values for the extracted $^{15}$N hyperfine coupling parameters are summarized in Table 1. The new information is based on the analysis of nuclear quantum beat oscillations observed for a deuterated and $^{15}$N-substituted sample.

## 2. Electronic Structure of $P_{700}^{\bullet+}$

Recently, a 2.5 Å resolution X-ray diffraction analysis confirmed the presence of a structural Chl *a* dimer in Photosystem I reaction center proteins (Jordan et al., 2001). The crystal structure revealed that $P_{700}$ is a Chl *a*/Chl *a′* 'heterodimer,' where Chl *a′* is the C13$^2$ epimer of Chl *a*. X-ray analysis, however, cannot provide information on the electronic coupling or electronic structure of the chlorophyll dimers in their active states. As for $P_{700}^{\bullet+}$, ENDOR data confirming either monomeric or dimeric electronic structures are very controversial (Käss et al., 1995; Mac et al., 1998). The observation of five different *anisotropic* $^{15}$N hyperfine couplings in the present experiment,

*Table 1.* $^{15}$N hyperfine coupling parameters of the primary donor $P^{\bullet+}_{700}$ in deuterated and $^{15}$N-substituted reaction centers of Photosystem I. The hyperfine parameters were extracted from a pulsed X-band EPR experiment using quantum beat oscillations of the spin-correlated radical pair $P^{\bullet+}_{700}A^{\bullet-}_{1}$ [a].

| $^{15}$N nucleus | $A_{\perp}$/MHz | $A_{\parallel}$[b]/MHz |
|---|---|---|
| I | 2.55 | 8.67 |
| II | 2.10 | 7.14 |
| III | 1.50 | 5.10 |
| IV | 0.85 | 2.89 |
| V | 0.25 | 0.85 |

[a]J. Lalevée, G. Link, J.-U.Weidner, M. Bechtold, T. Yago, S.L. Schlesselman, O.G. Poluektov, M.C. Thurnauer and G. Kothe, unpublished. [b]The values correspond to an average anisotropy ratio of $A_{\parallel} / A_{\perp}$ = 3.4, obtained by simulation of the quantum beat experiment.

however, is a strong argument in favor of a functional dimer. This conclusion holds independently of an *isotropic* histidine coupling to $P^{\bullet+}_{700}$, detected by ENDOR (Mac et al., 1996). *Isotropic* hyperfine couplings cannot be observed in nuclear quantum beat experiments (Kothe et al., 1998).

Thus, it appears that $P^{\bullet+}_{700}$ is a special pair (Norris et al., 1971) in which the unpaired spin is distributed over both chlorophyll molecules that constitute the primary donor. This spin delocalization can also account for the observed decrease in the anisotropy of the g-tensor in $P^{\bullet+}_{700}$ compared to Chl $a^{\bullet+}$ (Poluektov et al., 2002). Time-resolved D-band EPR spectra of $P^{\bullet+}_{700}$ and Chl $a^{\bullet+}$, recorded by the field-swept echo technique, indicate the largest shift for the $g_Z$-component (Poluektov et al., 2002). This finding cannot be explained in the framework of a monomer model. Rather, delocalization of the spin density of the unpaired electron between the two chlorophylls has to be invoked. Possibly, the large tilt angle of 29° between the Z-axis of the g-tensor and the chlorophyll normal (Fig. 10) can also be rationalized in terms of a delocalized electronic structure (Petrenko et al., 2005).

We have described a pulsed X-band EPR experiment, designed to probe nuclear quantum beat oscillations in the spin-correlated radical pairs of photosynthesis. Generally, formation of these coherences is associated with high nuclear spin polarization. This greatly enhances the detection sensitivity. Performing this experiment on $P^{\bullet+}_{700}A^{\bullet-}_{1}$ of Photosystem I sheds light on the nature of the electronic structure of the primary donor $P^{\bullet+}_{700}$. The results suggest that $P^{\bullet+}_{700}$ is a functional dimer. We expect that the new technique is of general interest, because the electronic structures of the oxidized donor and reduced acceptor can be studied in the active radical pair state.

## V. Concluding Remarks

Time-resolved EPR techniques comprise a suite of tools that can yield simultaneous structural and kinetic details of a dynamic chemical system. The power of these techniques is best illustrated by application to study photosynthetic energy conversion in reaction center proteins. In the foregoing we have described details of the quantum beat phenomena arising from spin-correlated radical pairs when monitored by both transient and pulsed time-resolved EPR (Kothe et al., 1991, 1998). The analysis of the quantum beat oscillations has allowed us to extract the three- dimensional structure of the charge-separated state, $P^{\bullet+}_{700}A^{\bullet-}_{1}$, of Photosystem I of oxygenic photosynthesis (Link et al., 2001) and to provide new insight into the electronic structure of $P^{\bullet+}_{700}$. Because these experiments can be performed on whole cells of bacteria, we could also obtain the arrangement of $P^{\bullet+}_{700}A^{\bullet-}_{1}$ in the membrane. This latter information required additional high-field time-resolved EPR experiments (Berthold et al., 1999).

Although our discussion was limited to reaction center proteins of Photosystem I, the spin-correlated radical pair nature of the charge-separated state is a common feature of systems that undergo efficient photo-induced charge separation. Apparently, these shared characteristics of natural and artificial systems, as revealed through time-resolved EPR, reflect underlying fundamental structural and energetic requirements for efficient charge separation. Therefore time-resolved EPR techniques continue to yield new insights into photochemical energy conversion in natural and artificial photosynthetic systems.

## Acknowledgments

We would like to thank our coworkers whose contributions are essential for our time-resolved EPR research of the photosynthetic systems: M. Bechtold, T. Berthold, U. Heinen, G. Link, J. Lalevée, E. Ohmes, S. Schlesselman, J. Tang, L. Utschig and S. Weber.

Research in the Freiburg laboratory was supported by a grant from the Deutsche Forschungsgemeinschaft (DFG) and by the DFG priority program "High-Field EPR in Biology, Chemistry and Physics" (SPP 1051). Work at the University of Chicago was supported by the US Department of Energy, Office of Basic Energy Sciences, Division of Chemical Sciences, Geosciences and Biosciences under grant DE-FG02-96ER14675. J.R.N. gratefully acknowledges support from the Alexander von Humboldt Foundation. Work at Argonne was supported by the US Department of Energy, Office of Basic Energy Sciences, Division of Chemical Sciences, Geosciences and Biosciences under contract W-31-109-Eng-38.

## References

Angerhofer A and Bittl R (1996) Radicals and radical pairs in photosynthesis. Photochem Photobiol 63: 11–38

Atkins PW and McLauchlan KA (1973) Electron spin polarization. In: Lepley AR and Closs GL (eds) Chemically Induced Magnetic Polarization, pp 41–93. John Wiley & Sons, New York

Berthold T, Bechtold M, Heinen U, Link G, Poluektov O, Utschig L, Tang J, Thurnauer MC and Kothe G (1999) Magnetic field induced orientation of photosynthetic reaction centers as revealed by time-resolved W-Band EPR of spin-correlated radical pairs. J Phys Chem B 103: 10733–10736

Bittl R and Kothe G (1991) Transient EPR of radical pairs in photosynthetic reaction centers: Prediction of quantum beats. Chem Phys Lett 177: 547–553

Bittl R, van der Est A, Kamlowski A, Lubitz W and Stehlik D (1994) Time-resolved EPR of the radical pair $P^{\bullet+}_{865}Q_A^{\bullet-}$ in bacterial reaction centers. Observation of transient nutations, quantum beats and envelope modulation effects. Chem Phys Lett 226: 349–358

Borovykh IV, Kulik LV, Dzuba SA and Hoff AJ (2001) Selective excitation in pulsed EPR of spin-correlated radical pairs: Electron-electron interactions, zero-, single-, and double-quantum relaxation and spectral diffusion. Chem Phys Lett 338: 173–179

Brettel K (1997) Electron transfer and arrangement of the redox cofactors in Photosystem I. Biochim Biophys Acta 1318: 322–373

Buckley CD, Hunter DA, Hore PJ and McLauchlan KA (1987) ESR of spin-correlated radical pairs. Chem Phys Lett 135: 307–312

Closs GL, Forbes MDE and Norris JR (1987) Spin-polarized electron paramagnetic resonance spectra of radical pairs in micelles. Observation of electron spin-spin interactions. J Phys Chem 91: 3592–3599

Dzuba SA, Gast P and Hoff AJ (1995) ESEEM study of spin-spin interactions in spin-polarised $P^{\bullet+}Q_A^{\bullet-}$ pairs in the photosynthetic purple bacterium *Rhodobacter sphaeroides* R26. Chem Phys Lett 236: 595–602

Dzuba SA, Bosch MK and Hoff AJ (1996) Electron spin echo detection of quantum beats and double-quantum coherence in spin-correlated radical pairs of protonated photosynthetic reaction centers. Chem Phys Lett 248: 427–433

Furrer R and Thurnauer MC (1981) Nanosecond time resolution in EPR transient nutation spectroscopy of triplet states. Chem Phys Lett 79: 28–33

Furrer R and Thurnauer MC (1983) Resolution of signals attributed to Photosystem I primary reactants by time-resolved EPR at K band. FEBS Lett 153: 399–403

Furrer R, Fujara F, Lange C, Stehlik D, Vieth H-M and Vollmann W (1980) Transient EPR nutation signals in excited triplet states. Chem Phys Lett 75: 332–339

Geacintov NE, van Nostrand F, Becker JF and Tinkel JB (1972) Magnetic field induced orientation of photosynthetic systems. Biochim Biophys Acta 267: 65–79

Heinen U, Berthold T, Kothe G, Stavitski E, Galili T, Levanon H, Wiederrecht G and Wasielewski MR (2002) High time resolution Q-band EPR study of sequential electron transfer in a triad oriented in a liquid crystal. J Phys Chem A 106: 1933–1937

Heinen U, Poluektov O, Stavitski E, Berthold T, Ohmes E, Schlesselman SL, Golecki JR, Moro GJ, Levanon H, Thurnauer MC and Kothe G (2004a) Magnetic field induced orientation of photosynthetic reaction centers as revealed by time-resolved D-band EPR of spin-correlated radical pairs. II. Field dependence of the alignment. J Phys Chem B 108: 4498–9504

Heinen U, Golecki JR, Poluektov O, Berthold T, Schlesselman SL, Frezzato D, Ohmes E, Moro GJ, Thurnauer MC and Kothe G (2004b) Magnetic field induced orientation of photosynthetic reaction centers as revealed by time-resolved W-band EPR of spin-correlated radical pairs. Development of a molecular model. Appl Magn Reson 26: 99–115

Hoff AJ (1984) Electron spin polarization of photosynthetic reactants. Q Rev Biophys 17: 153–282

Hoff AJ and Deisenhofer J (1997) Photophysics of photosynthesis. Structure and spectroscopy of reaction centers of purple bacteria. Phys Rep 287: 1–248

Isaacson RA, Lendzian F, Abresch EC, Lubitz W and Feher G (1995) Electronic structure of $Q_A^-$ in reaction centers from *Rhodobacter sphaeroides*. I. Electron Paramagnetic Resonance in single crystals. Biophys J 69: 311–322

Jeschke G (1997) Electron-electron-nuclear three-spin mixing in spin-correlated radical pairs. J Chem Phys 106: 10072–10086

Jeschke G and Matysik J (2003) A reassessment of the origin of photochemically induced dynamic nuclear polarization effects in solids. Chem Phys 294: 239–255

Jeschke G and Schweiger A (1996) Matched two-pulse electron spin echo envelope modulation spectroscopy. J Chem Phys 105: 2199–2211

Jordan P, Fromme P, Witt HT, Klukas O, Saenger W and Krauss N (2001) Three-dimensional structure of cyanobacterial Photosystem I at 2.5 Å resolution. Nature 411: 909–917

Kandrashkin YE, Salikhov KM, van der Est A and Stehlik D (1998) Electron spin polarization in consecutive spin-correlated radical pairs: Application to short-lived and long-lived precursors in type 1 photosynthetic reaction centers. Appl Magn Reson 15: 417–447

Käss H, Bittersmann-Weidlich E, Andréasson L-E, Bönigk B and Lubitz W (1995) ENDOR and ESEEM of the $^{15}N$ labelled radical cations of chlorophyll a and the primary donor $P_{700}$ in Photosystem I. Chem Phys 194: 419–432

Kiefer AM, Kast SM, Wasielewski MR, Laukenmann K and

Kothe G (1999) Exploring the structure of a photosynthetic model by quantum-chemical calculations and time-resolved Q-band electron paramagnetic resonance. J Am Chem Soc 121: 188–198

Kothe G, Weber S, Bittl R, Ohmes E, Thurnauer MC and Norris JR (1991) Transient EPR of light-induced radical pairs in plant Photosystem I: Observation of quantum beats. Chem Phys Lett 186: 474–480

Kothe G, Weber S, Ohmes E, Thurnauer MC and Norris JR (1994a) Transient EPR of light-induced spin-correlated radical pairs: Manifestation of zero quantum coherence. J Phys Chem 98: 2706–2712

Kothe G, Weber S, Ohmes E, Thurnauer MC and Norris JR (1994b) High time resolution electron paramagnetic resonance of light-induced radical pairs in photosynthetic bacterial reaction centers: Observation of quantum beats. J Am Chem Soc 116: 7729–7734

Kothe G, Bechtold M, Link G, Ohmes E and Weidner J-U (1998) Pulsed EPR detection of light-induced nuclear coherences in photosynthetic reaction centers. Chem Phys Lett 283: 51–60

Krabben L, Schlodder E, Jordan R, Carbonera D, Giacometti G, Lee H, Webber AN and Lubitz W (2000) Influence of the axial ligands on the spectral properties of $P_{700}$ of Photosystem I: A study of site-directed mutants. Biochemistry 39: 13012–13025

Link G, Berthold T, Bechtold M, Weidner J-U, Ohmes E, Tang J, Poluektov O, Utschig L, Schlesselman SL, Thurnauer MC and Kothe G (2001) Structure of the $P_{700}^{\bullet+}A_1^{\bullet-}$ radical pair intermediate in Photosystem I by high time resolution multifrequency electron paramagnetic resonance: Analysis of quantum beat oscillations. J Am Chem Soc 123: 4211–4222

Link G, Poluektov OG, Utschig LM, Lalevée J, Yago T, Weidner J-U, Thurnauer MC and Kothe G (2005) Structural organization in photosynthetic proteins as studied by high-field EPR of spin-correlated radical pair states. Magn Reson Chem 43: 103–109

Mac M, Tang XS, Diner BA, McCracken J and Babcock GT (1996) Identification of histidine as an axial ligand to $P_{700}^{\bullet+}$. Biochemistry 35: 13288–13293

Mac M, Bowlby NR, Babcock GT and McCracken J (1998) Monomeric spin density distribution in the primary donor of Photosystem I as determined by electron magnetic resonance: Functional and thermodynamic implications. J Am Chem Soc 120: 13215–13223

McLauchlan K (1989) Time-Resolved EPR. In: Hoff AJ (ed) Advanced EPR, Application in Biology and Biochemistry, pp 345–369. Elsevier, Amsterdam

Möbius K (2000) Primary processes in photosynthesis: What do we learn from high-field EPR spectroscopy? Chem Soc Rev 29: 129–139

Morris AL, Snyder SW, Zhang Y, Tang J, Thurnauer MC, Dutton PL, Robertson DE and Gunner MR (1995) An electron spin polarization model applied to sequential electron transfer in iron-containing photosynthetic bacterial reaction centers with different quinones as $Q_A$. J Phys Chem 99: 3854–3866

Norris JR, Uphaus RA, Crespi HL and Katz, JJ (1971) Electron spin resonance of chlorophyll and the origin of signal I in photosynthesis. Proc Natl Acad Sci USA 68: 625–628

Norris JR, Thurnauer MC and Bowman MK (1980) Electron Spin Echo Spectroscopy and the Study of Biological Structure and Function. In: Lawrence JH, Golfman JW and Hayes TL (eds) Advances in Biological and Medical Physics, pp 365–416. Academic Press, New York

Norris JR, Morris AL, Thurnauer MC and Tang J (1990) A general model of electron spin polarization arising from the interactions within radical pairs. J Chem Phys 92: 4239–4249

Pauling L (1979) Diamagnetic anisotropy of the peptide group. Proc Natl Acad Sci USA 76: 2293–2294

Petrenko A, Redding K, Kispert LD (2005) The influence of the structure of the radical cation dimer pair of aromatic molecules on the principal values of a g-tensor: DFT predictions. Chem Phys Lett 406: 327–331

Poluektov OG, Utschig LM, Schlesselman SL, Lakshmi KV, Brudvig GW, Kothe G and Thurnauer MC (2002) Electronic structure of the $P_{700}$ special pair from electron paramagnetic resonance spectroscopy. J Phys Chem B 106: 8911–8916

Poluektov OG, Utschig LM, Dubinskij AA and Thurnauer MC (2004) ENDOR of spin-correlated radical pairs in photosynthesis at high magnetic field: A tool for mapping electron transfer pathways, J Am Chem Soc 126: 1644–1645

Poluektov OC, Utschig LM, Dubinskij AA and Thurnauer MC (2005) Electron transfer pathways and protein response to charge separation in photosynthetic reaction centers: Time-resolved high-field ENDOR of the spin-correlated radical pair $P_{865}^{+}Q_A^{-}$. J Am Chem Soc 127: 4049–4059

Salikhov KM, Bock CH and Stehlik D (1990) Time development of electron spin polarization in magnetically coupled, spin correlated radical pairs. Appl Magn Reson 1: 195–211

Salikhov KM, Kandrashkin YuE and Salikhov AK (1992) Peculiarities of free induction and primary spin echo signals for spin-correlated radical pairs. Appl Magn Reson 3: 199–216

Schlodder E, Falkenberg K, Gergeleit M and Brettel K (1998) Temperature dependence of forward and reverse electron transfer from $A_1^{\bullet-}$, the reduced secondary electron acceptor in Photosystem I. Biochemistry 37: 9466–9476

Sétif P, Mathis P and Vänngard T (1984) Photosystem I photochemistry at low temperature. Heterogeneity in pathways for electron transfer to the secondary acceptors and for recombination processes. Biochim Biophys Acta 767: 404–414

Snyder SW and Thurnauer MC (1993) Electron spin polarization in photosynthetic reaction centers. In: Deisenhofer H and Norris JR (eds) The Photosynthetic Reaction Center, Vol 2, pp 285–329. Academic Press, New York

Stehlik D and Möbius K (1997) New EPR methods for investigating photoprocesses with paramagnetic intermediates. Annu Rev Phys Chem 48: 745–784

Stehlik D, Bock CH and Thurnauer MC (1989a) Transient EPR spectroscopy of photoinduced electronic spin states in rigid matrices. In Hoff AJ (ed) Advanced EPR, Application in Biology and Biochemistry, pp 371–403. Elsevier, Amsterdam

Stehlik D, Bock CH and Petersen J (1989b) Anisotropic electron spin polarization in photosynthetic reaction centers. J Phys Chem 93: 1612–1619

Stone AJ (1963) G-tensors of aromatic hydrocarbons. Mol Phys 7: 311–316

Tang J and Norris JR (1995) Multiple-quantum EPR coherence in a spin-correlated radical pair system. Chem Phys Lett 233: 192–200

Tang J, Thurnauer MC and Norris JR (1994) Electron spin echo modulation due to exchange and dipolar interactions in a spin-correlated radical pair. Chem Phys Lett 219: 283–290

Thurnauer MC and Norris JR (1980) An electron spin echo phase shift observed in photosynthetic algae. Possible evidence for dynamic radical pair interactions. Chem Phys Lett 76: 557–561

Thurnauer MC, Poluektov OG and Kothe G (2004) Time-resolved high-frequency and multifrequency EPR studies of spin-correlated radical pairs in photosynthetic reaction center proteins. In: Grinberg OY and Berliner LJ (eds) Biological Magnetic Resonance. Very High Frequency ESR/EPR, pp 165–206. Kluwer Academic Publisheres, New York

Torrey HC (1949) Transient nutations in nuclear magnetic resonance. Phys Rev 76: 1059–1072

Trifunac AD and Norris JR (1978) Nanosecond time resolved EPR spectroscopy. EPR time profile by electron spin echo CIDEP. Chem Phys Lett 59: 140–142

Trifunac AD and Thurnauer MC (1979) Time-resolved electron spin resonance of transient radicals. In: Kevan L and Schwartz RN (eds) Time Domain Electron Spin Resonance, pp 107–152. John Wiley & Sons, New York

Turro NJ and Khudyakov IV (1999) Applications of chemically induced dynamic electron polarization to mechanistic photochemistry. Res Chem Intermed 25: 505–529

Weber S, Kothe G and Norris JR (1997) Transient nutation electron spin resonance spectroscopy on spin-correlated racial pairs: A theoretical analysis of hyperfine-induced nuclear modulations. J Chem Phys 106: 6248–6261

Worcester DL (1978) Structural origins of diamagnetic anisotropy in proteins. Proc Natl Acad Sci USA 75: 5475–5477

Zech SG, Hofbauer W, Kamlowski A, Fromme P, Stehlik D, Lubitz W and Bittl R (2000) A structural model for the charge separated state $P_{700}^{\bullet+}A_1^{\bullet-}$ in Photosystem I from the orientation of the magnetic interaction tensors. J Phys Chem B 104: 9728–9739

Zwanenburg G and Hore PJ (1993) EPR of spin-correlated radical pairs. Analytical treatment of selective excitation including zero quantum coherence. Chem Phys Lett 203: 65–74

# Chapter 16

## Distance Measurements in Photosynthetic Reaction Centers by Pulsed EPR

Andrei V. Astashkin*
*Department of Chemistry, University of Arizona, Tucson, AZ 85721, U.S.A.*

Asako Kawamori*
*School of Science and Technology, Kwansei Gakuin University, Gakuen 2-1, Sanda 669-1337, Japan*



---

*Authors for correspondence, email: andrei@u.arizona.edu; kawamori-a@tulip.sannet.ne.jp

*Thijs J. Aartsma and Jörg Matysik (eds): Biophysical Techniques in Photosynthesis II, pp. 325–343.*

## Summary

The last decade has seen an enormous increase in interest to structural studies in biological systems using pulsed EPR techniques, in particular, to the distance measurements between pairs of paramagnetic centers using pulsed ELDOR and a whole arsenal of other techniques. Most notable applications are to the distance measurements between spin labels in double-labeled proteins and between electron transfer intermediates in photosynthetic reaction centers. In this chapter we will describe the theoretical and practical aspects of pulsed ELDOR and a related technique, 2+1, and demonstrate their application to measurements of distances between electron transfer cofactors in plant Photosystem II (PS II) and other photosynthetic reaction centers (RC). Since much of the distance information in photosynthetic RCs was obtained by pulsed ELDOR and 2+1, these are the main techniques we discuss. However, the results obtained by other pulsed EPR methods, the selective hole burning and out-of-phase electron-spin echo envelope modulation, will also be mentioned.

## I. Introduction

Measuring distances between paramagnetic centers is one of common applications of electron paramagnetic resonance (EPR). The parameter sensitive to the distance is the dipole interaction between the electronic magnetic moments. It is inversely proportional to the distance cubic, which means that rather accurate distance estimates can be obtained even from not so accurate measurements. Depending on the scale of the interaction, different EPR techniques are applicable. If the distance is short, the strong dipolar coupling may produce resolved splittings in continuous wave (CW) EPR spectra. If the coupling is weak, however, there is a wide variety of more subtle EPR techniques that can be applied, including the measurements of relaxation enhancement, instantaneous diffusion, spectral hole broadening, etc. For recent reviews of application of such EPR techniques to biological systems, see Berliner et al. (2000) and Chapter 14, Möbius and Goldfarb.

In this chapter, we will be mostly concerned with only one of these techniques, namely, pulsed electron-electron double resonance, or ELDOR (Milov et al., 1981), which is also often dubbed double electron-electron resonance (DEER). Some attention will also be given to the so-called '2+1' technique (Kurshev et al., 1989) that involves the same physical principles. Unlike relaxation enhancement techniques, pulsed ELDOR and 2+1 are direct methods to determine the dipole interactions, and they were successfully used to study model radical pairs (Milov et al., 1984, 1998; Kurshev et al., 1989; Larsen and Singel, 1993; Martin et al., 1998; Pannier et al., 2000; Jeschke, 2002; Weber et al., 2002), spatial distributions of radicals produced by irradiation and photolysis (Milov et al., 1981, 1984, 1998; Raitsimring and Samoilova, 1986; Kurshev et al., 1989, 1991; Bowman et al., 2005), protein structures by using nitroxide spin-labeling (Raitsimring et al., 1992; Sale et al., 2002, 2005; Milov et al., 2003; Schiemann et al., 2003, 2004; Jeschke et al., 2004; Chiang et al., 2005; Fajer, 2005; Hilger et al., 2005; Bennati et al., 2005), structure of the electron-transfer chain in photosynthetic systems (see below), and other biological systems (Codd et al., 2002; Elsässer et al., 2002, 2005; Bennati et al., 2003).

After describing, on a semi-qualitative level, the theoretical and practical aspects of pulsed ELDOR and 2+1, we will demonstrate their application to the measurement of distances between electron transfer intermediates in photosynthetic reaction centers (RC), primarily, in plant Photosystem II (PS II). Finally, for the sake of completeness, we will also mention some results obtained by other pulsed EPR methods for the photosynthetic RCs: the selective hole burning (Wacker et al., 1992; Dzuba et al., 1993) and out-of-phase electron-spin echo envelope modulation (ESEEM) (Tang et al., 1994).

*Abbreviations:* CW – continuous wave; DEER – double electron-electron resonance; ELDOR – electron-electron double resonance; EPR – electron paramagnetic resonance; ESE – electron spin echo; ESEEM – ESE envelope modulation; FT – Fourier transform; mw – microwave; OEC – oxygen-evolving complex; OOP ESE – out-of-phase ESE; PS II – Photosystem II; RC – reaction center; RIDME – relaxation-induced dipolar modulation enhancement; SCRP – spin-correlated radical pair

## II. ELDOR and '2+1' Basics

The basic theory of pulsed ELDOR is very simple, but finer points may rapidly increase the complexity of the discussion. In this section we will give a rudimentary description of the method, which is actually

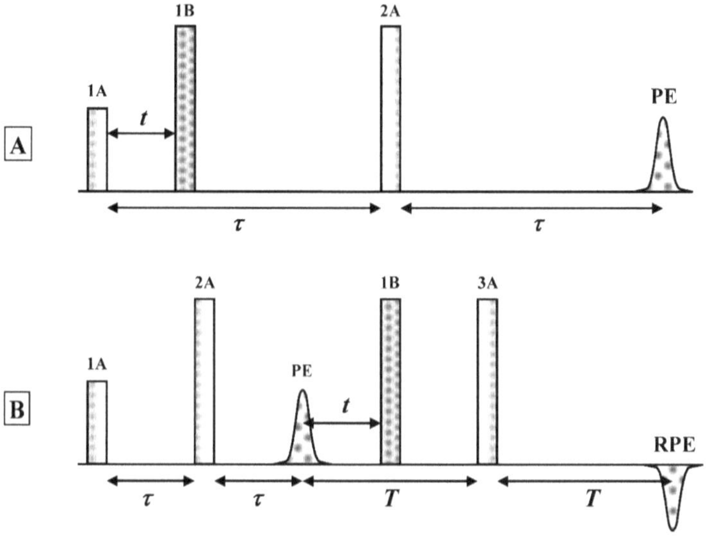


*Fig. 1.* The three-pulse (A) and four-pulse (B) sequences used for ELDOR. Pulses 1A, 2A and 3A are those of the observation sequence: they are in resonance with the observed spin A. Pulse 1B is the pumping pulse. It is in resonance with spin B. The 2+1 technique uses the same pulse sequence as three-pulse ELDOR, but the mw frequency of pulse 1B is the same as that of the observation sequence.

sufficient to understand simple experiments. In later sections we will mention some of the complications that may be appropriate to take into account in less trivial situations.

Let us consider a pair of paramagnetic centers, A and B, coupled by the magnetic dipole interaction with a coupling constant

$$D_o = g_A g_B \beta^2 / h R_{AB}^3 \qquad (1)$$

where $g_A$ and $g_B$ are the g-factors, β is the Bohr magneton, *h* is the Planck constant and $R_{AB}$ is the distance between the centers. Equation (1) implies that $R_{AB}$ is much larger than the characteristic size of the spin density distribution in either of the centers, and that the g-factors are fairly isotropic.

The simplest pulsed ELDOR technique (Milov et al., 1981) is based on monitoring a primary ESE signal from spins A ('PE' in Fig. 1A). The mw pulses of the observation two-pulse sequence, 1A and 2A, have the carrier frequency $\nu_A$ in resonance with spins A, and are separated by a fixed time interval τ. At some point before the formation of the ESE signal from spins A (usually, between pulses 1A and 2A) a pumping mw pulse is applied in resonance with spins B. The carrier frequency of this pulse is $\nu_B$, and $|\nu_A - \nu_B|$ is considered to be greater than the spectral coverage of the mw pulses. A theoretical analysis (Milov et al., 1981) shows that, as pulse 1B changes its position between pulses 1A and 1B, the amplitude of the primary ESE signal from spins A oscillates:

$$V(t) \propto \cos(2\pi D t) \qquad (2)$$

In this expression *D* is the magnitude of the dipole interaction for a given orientation of the pair:

$$D = D_0(1 - 3\cos^2\theta_{RB}) \qquad (3)$$

where $\theta_{RB}$ is the angle between the vector $\mathbf{R}_{AB}$ and the external magnetic field $\mathbf{B}_0$.

The physics behind Eq. (2) is as follows. The ESE signal in general represents a result of two processes. The first process is a magnetization defocusing, which is a precession of individual magnetic moments, with their individual precession frequencies, away from the state where they all were collinear and formed a non-zero macroscopic magnetization vector $\mathbf{M}_0$. In Fig. 1A this process is initiated by the first mw pulse, 1A, that orients an equilibrium magnetization vector $\mathbf{M}_0$ perpendicular to $\mathbf{B}_0$. The opposite process of refocusing, or convergence, is initiated by the mw pulse 2A. The refocusing occurs after the time interval τ after the second pulse. If the pumping mw pulse changes the orientation of spins B by 180°, the local field for spins A changes by *D*, and the precession phase difference of *Dt* appears at the convergence time.

Equation (2) assumes that each spin A has a neighboring spin B, and this spin B has been flipped by the pumping pulse. However, in many cases only a fraction $p_p$ of spins A will have spin B as a partner ($p_p \leq 1$). In addition, even if spin A is paired with B, the probability $p_f$ of flipping this spin B by the pumping pulse is usually less than unity. The most common reason for $p_f < 1$ is the width of the EPR spectrum being greater then the spectral coverage of the mw pulse 1B, $\Delta\nu_B$. One can thus write:

$$V(t) \propto 1 - p + p\cos(2\pi Dt) \tag{4}$$

where $p = p_p p_f$.

The three-pulse ELDOR sequence of Fig. 1A was historically the first. Recently also the four-pulse ELDOR sequence of Fig. 1B introduced by Martin et al. (1998) became very popular. In this pulse sequence the pumping pulse 1B crosses the primary ESE signal (formed by pulses 1A and 2A) and continues moving towards the refocusing observation pulse 3A. The ESE signal observed is the refocused primary ESE signal ('RPE' in Fig. 1B). The phenomenological description of the ELDOR effect for this signal is the same as in the case of 3-pulse ELDOR.

Another technique for distance measurements that employs the pulse sequence of Fig. 1A is called '2+1' (Kurshev et al., 1989). Unlike ELDOR, 2+1 uses the same mw frequency both for pumping and observation (that is, $\nu_A = \nu_B$), which is only meaningful if the paired spins have overlapping and reasonably narrow EPR spectra. The qualitative difference in the spin evolution between the three-pulse ELDOR and 2+1 is as follows. While in ELDOR spin B is flipped only once, by pulse 1B, in 2+1 the same spin can also be flipped back to the original state by pulse 2A. As a result, the 2+1 analog of Eq. 4 is:

$$V(t) = V_o + \left[ p_p p_{f2} - \frac{p_p p_{f23}}{2} \right] \cos(2\pi Dt) + \frac{p_p p_{f23}}{2} \cos(2\pi D(\tau - t)) \tag{5}$$

where

$$V_o = 1 - p_p p_{f2} - \left[ p_p p_{f3} - \frac{p_p p_{f23}}{2} \right] (1 - \cos(2\pi D\tau)) \tag{6}$$

and it is assumed that the EPR spectra of spins A and B are identical. In these expressions $p_{f2}$ and $p_{f3}$ are the flip probabilities for the second and third mw pulses (1B and 2A in Fig. 1A), $p_{f23}$ is the probability for a spin to be flipped by both, the second and third pulses, and $p_p = 2[AB]/([A]+[B])$. If $p_{f23} \ll p_{f2}$, the last term in Eq. (5) can be neglected, and the dipolar oscillation becomes similar to that in ELDOR. Physically, the situation also becomes similar to that in ELDOR, that is, one observes the echo from one sub-ensemble of spins while flipping another sub-ensemble. Thus, under the condition of $p_{f23} \ll p_{f2}$ the 2+1 technique works in an ELDOR limit.

Equations (4) and (5) correspond to a single crystal situation. In practice, however, one usually deals with paramagnetic centers stabilized in an orientationally disordered matrix. Equations (4) and (5) then need to be averaged over the orientations of $\mathbf{R}_{AB}$ relative to $\mathbf{B}_o$:

$$V_{\text{disordered}}(t) = \langle V_{\text{oriented}}(t) \rangle_\Omega \tag{7}$$

The subscripts in Eq. (7) are shown for clarity and will be dropped in the future discussion.

Sometimes this averaging may be tricky because, generally speaking, $p_f$ may depend on the orientation, but usually a simple averaging over $\theta_{RB}$ with the statistical weight of $\sin\theta_{RB}$ is adequate. In the case of ELDOR or 2+1 in the ELDOR limit this full orientational averaging results in the time dependence $V(t)$ being described by a damping oscillation shown in Fig. 2A. A Fourier transform (FT) of this oscillation will ideally produce a spectrum in the shape of the Pake pattern (see Fig. 2B).

## III. Elaborations and Possible Complications

In this section we will first elaborate in more detail on the pairing and flipping probability parameters, $p_p$ and $p_f$. Then we will mention some of the complications one may encounter when working with exchange-coupled complexes or the centers with large g-factor anisotropies, which is rather typical for biological systems. Unless specifically mentioned, we will be concerned with ELDOR, although many of the considerations below are applicable to 2+1 without any changes.

### *A. Pairing Probability, $p_p$*

A typical situation of a nearly perfect pairing, $p_p \approx 1$,

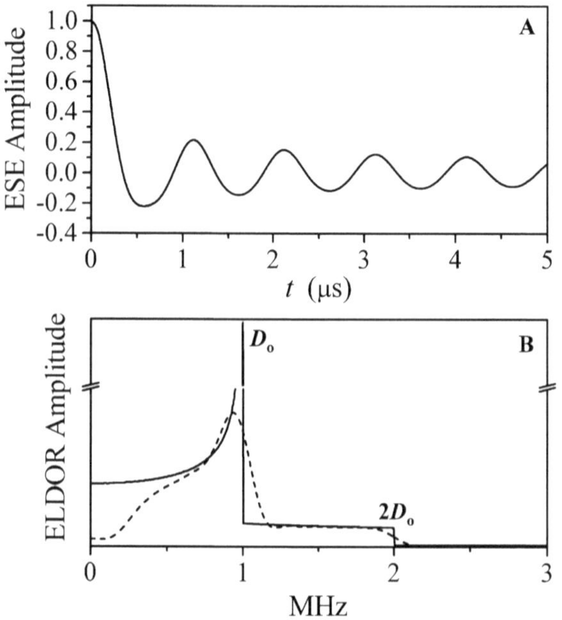


*Fig. 2.* Pulsed ELDOR trace in time domain (A) and after the FT to the frequency domain (B), numerically calculated for an orientationally disordered pair with $D_o = 1$ MHz and $p_f = p_p = 1$. Solid line in panel (B) is for the infinite time interval, while dashed line is for the finite time interval of 5 μs, after a cosine window apodization and subtraction of a background approximated by a smoothing cubic spline.

is that of model biradicals, while in biological applications one will often have $p_p < 1$. For example, in a photosynthetic RC the pairs of paramagnetic centers are generated by electron transfer reactions. Because of the short room temperature lifetimes, the trapping of these centers (by rapid freezing in liquid $N_2$) will usually result in probabilities for an average RC unit to have spins A and B of $p_A < 1$ and $p_B < 1$, respectively. If the only process that leads to losing spins A and B is the reverse electron transfer, the pairing probability is obviously $p_p = 1$. In the opposite limit of uncorrelated decays of spins A and B (e.g., if there are many electron transfer pathways available), $p_p = p_A p_B$.

### B. Inversion (Flipping) Probability, $p_f$

In an orientationally non-selective situation calculation of $p_f$ consists in averaging the flip probability over the EPR spectrum shape of spin B (Raitsimring et al., 1974):

$$p_f = \left\langle \frac{\nu_{1B}^2}{\nu_{NB}^2} \sin^2 \pi \nu_{NB} t_{p1B} \right\rangle_{g(\delta\nu_B)} \tag{8}$$

where $\nu_{1B} = g_B \beta B_{1B}$ is the pumping mw amplitude, $\nu_{NB} = (\nu_{1B}^2 + \delta\nu_B^2)^{1/2}$ is the spin B nutation frequency at an offset $\delta\nu_B$ from the exact resonance, and $t_{p1B}$ is the duration of the pumping pulse. In elaborate calculations one has to take into account the distribution of $\nu_{1B}$ in the sample. However, if one is only concerned with the distance between paired spins A and B, one usually does not bother to estimate $p_f$ at all.

In the context of the 2+1 technique $p_f$ corresponds to $p_{2f}$ (see Eqs. 5 and 6). The expression for $p_{3f}$ will be similar to Eq. 8, with $\nu_{1B}$, $\nu_{NB}$ and $t_{p1B}$ substituted, respectively, by $\nu_{1A}$, $\nu_{NA}$ and $t_{p2A}$. The expression for the double flip probability is

$$p_{f23} = \left\langle \left[ \frac{\nu_{1B}^2}{\nu_{NB}^2} \sin^2 \pi \nu_{NB} t_{p1B} \right] \cdot \left[ \frac{\nu_{1A}^2}{\nu_{NA}^2} \sin^2 \pi \nu_{NA} t_{p2A} \right] \right\rangle_{g(\delta\nu)} \tag{9}$$

where $\delta\nu = \delta\nu_A = \delta\nu_B$ (assuming identical spectra of A and B).

In some cases it is convenient to use the upper-limit estimates

$$p_f \le \sin^2 \frac{\theta_f}{2} \tag{10}$$

and

$$p_{f23} \le \left\langle \left[ \frac{\nu_{1B}^2}{\nu_{NB}^2} \sin^2 \pi \nu_{NB} t_{p1B} \right] \right\rangle_{g(\delta\nu)} \left\langle \left[ \frac{\nu_{1A}^2}{\nu_{NA}^2} \sin^2 \pi \nu_{NA} t_{p2A} \right] \right\rangle_{g(\delta\nu)} \le \sin^2 \frac{\theta_{f2}}{2} \sin^2 \frac{\theta_{f3}}{2} \tag{11}$$

where the various $\theta_f$ values are the nominal flip angles $\theta_f = \nu_1 t_p$.

### C. Distributed Paramagnetic Centers

From the viewpoint of spin A, there are two kinds of spins B. There is one spin B that makes a pair with this particular spin A, and there are numerous other spins B uniformly distributed throughout the sample. In ELDOR, these uniformly distributed (matrix) spins B will result in an exponential decay of the ESE signal of spins A (Raitsimring et al., 1974):

$$V(t) \propto \exp(-t/t_{dec}) \quad (12)$$

where $t_{dec}$ is given by:

$$t_{dec} = \sqrt{243}h / (16\pi^3 g_A g_B \beta^2 p_f \rho_B) \quad (13)$$

In this expression $\rho_B$ is the concentration of spins B, $\rho_B \sim \langle R_{BB}\rangle^{-3}$, where $\langle R_{BB}\rangle$ is the effective average distance between these spins B.

If $R_{AB} \geq \langle R_{BB}\rangle$, the spin B paired with spin A is indistinguishable from the matrix spins B and simply contributes to the exponential decay. In order to be able to measure $R_{AB}$, one should have $R_{AB} < \langle R_{BB}\rangle/2$ (Raitsimring and Salikhov, 1985), in which case $V(t)$ is described by the product of Eqs. (4) and (12). However, even with proteins, reaching this limit it is not always possible because of the aggregation of the proteins in frozen non-glassy samples (Codd et al., 2002; Milov et al. 2003).

### *D. Orientational Selectivity*

If $\mathbf{R_{AB}}$ is orientationally disordered with respect to the molecular frames of centers A and B, the ELDOR traces do not depend on the specific regions in the EPR spectra of A and B excited by the mw pulses (at least, when the spins are quantized along $\mathbf{B_0}$). The orientational averaging in Eq. (7) is a straightforward averaging over $\theta_{RB}$ with the statistical weight of $\sin\theta_{RB}$, and the flip probability $p_f$ can be estimated using Eq. (8).

If $\mathbf{R_{AB}}$ has a specific orientation with respect to the molecular frames of A or B, or both of them, the mw pulses will generally select some restricted set of angles $\theta_{RB}$ that contribute to the ELDOR spectra. As a result, several ELDOR experiments with different settings of $\nu_A$ and $\nu_B$ may be necessary to estimate $R_{AB}$. The interpretation of the ELDOR data will require numerical simulations with the orientational averaging in Eq. (7) taking into account the appropriate orientational selectivity. Even in this case the situation is often simplified by the fact that the EPR spectrum of B is usually narrow, and the range of $\mathbf{R_{AB}}$ orientations relative to the molecular frame of B selected by the pumping pulse is close to that of complete disorder.

The considerations presented above are of very general nature. Additional discussions and some numerical simulations can be found in the works dealing with specific systems: pairs of nitroxide radicals (Larsen and Singel, 1993) and [NiFe] center and the $[3Fe\text{-}4S]^+$ cluster in [NiFe] hydrogenase (Elsässer et al., 2002, 2005).

### E. Exchange Interaction

If the distance between A and B is not very large and the radial wavefunctions associated with these spins overlap, the exchange interaction can become comparable with, or exceed, the dipole interaction. This situation is facilitated by the presence of conjugated bonds between the centers, which helps propagating the spin density to rather large distances.

If the exchange and dipole interactions are much weaker than the Zeeman interaction ($J, D << g_A\beta B_0$, $g_B\beta B_0$, where $J$ is the isotropic exchange coupling constant), then the spin Hamiltonian is:

$$\begin{aligned}\hat{H} &= g_A\beta B_o S_{AZ} + g_B\beta B_o S_{BZ} + J\mathbf{S_A}\mathbf{S_B} + \mathbf{S_A}\mathbf{D}\mathbf{S_B} \\ &\approx g_A\beta B_o S_{AZ} + g_B\beta B_o S_{BZ} + (J+D)S_{ZA}S_{ZB} \quad (14)\end{aligned}$$

where again fairly isotropic $g_A$ and $g_B$ are assumed. One can see that in this approximation $J$ and $D$ act in the same way, and the frequency of the ELDOR oscillations will be simply $J$+$D$.

### *F. Spin Quantization Axes Different from* $\mathbf{B_0}$

Equation (3) implies that both spins in the AB pair quantize along the direction of $\mathbf{B_0}$, that is, the directions of the magnetic moments associated with these spins faithfully follow the direction of $\mathbf{B_0}$. This is a good approximation for organic radicals and other centers with $S = 1/2$ and small g-anisotropy, while for transition metal ions the situation is often much more complex. If, for example, the g-anisotropy of spin A is large and that of B is small (a conceivable situation for ELDOR), then instead of Eq. 3 we may write (Bencini and Gatteschi, 1990):

$$D = D_o\left[1 - 3\frac{r_X b_X g_{AX}^2 + r_Y b_Y g_{AY}^2 + r_Z b_Z g_{AZ}^2}{g_A^2}\cdot\cos\theta_{RB}\right] \quad (15)$$

where $r_X$, $r_Y$ and $r_Z$ are the direction cosines of $\mathbf{R_{AB}}$ in the principal axes frame {X,Y,Z} of the g-tensor of spin A; $b_X$, $b_Y$ and $b_Z$ are the direction cosines of $\mathbf{B_0}$, and $g_{AX}$, $g_{AY}$ and $g_{AZ}$ are the principal g-values.

The $g_A$ quantity in the denominator, as well as in the expression for $D_o$ (Eq. 1) now stands for the orientation-dependent g-value:

$$g_A = (g_{AX}^2 b_X^2 + g_{AY}^2 b_Y^2 + g_{AZ}^2 b_Z^2)^{1/2} \tag{16}$$

For an isotropic g-factor $g_{AX} = g_{AY} = g_{AZ} = g_A$, and the ratio in Eq. (15) reduces to $r_X b_X + r_Y b_Y + r_Z b_Z \equiv \cos\theta_{RB}$.

### G. Spin Delocalization and Structural Distribution

The point dipole approximation implied by Eq. (1) is good for a situation when the size of spin delocalization, $\Delta R_\rho$, is much smaller than $R_{AB}$. A simple analysis shows that neglecting the spin delocalization will result in a relative error of estimating the distance between the centers of gravity of spin density distributions from the experimental dipolar coupling constant on the order of:

$$\frac{\Delta R_{AB}}{R_{AB}} \le \frac{\Delta R_\rho^2}{2R_{AB}^2} \tag{17}$$

Assuming realistic $R_{AB} \sim 20$ Å and $\Delta R_\rho \sim 6$ Å, we can estimate $\Delta R_\rho / R_{AB} < 5\%$, which shows that the spin density distribution in most of practical ELDOR situations can be completely disregarded.

While the *quantum mechanical* spin density distribution results in a single well defined *D*-tensor, any *structural* inhomogeneity will result in a distribution of the dipolar coupling constants. Since $D_o \propto R_{AB}^{-3}$, even minor structural differences may lead to quite noticeable variations in $D_o$ values. In experiment such a distribution is recognized by an abnormally fast damping of the dipolar oscillations in an ELDOR trace. The interpretation of ELDOR data for such a situation usually involves a determination of the distance distribution function. This problem was mostly analyzed for photo-generated radical pairs (Milov et al., 1981, 1984, 1998; Raitsimring and Samoilova, 1986) and in the context of the measurements of distances between nitroxide spin labels in proteins (Steinhoff et al., 1997; Jeschke, 2002; Borbat et al., 2002; Sale et al., 2002, 2005; Jeschke et al., 2004; Schiemann et al., 2004; Chiang et al., 2005; Hilger et al., 2005), but it has also been encountered for the oriented PS II membranes (Astashkin et al., 1998; Kuroiwa et al., 2000; Kawamori et al., 2002) and other systems (Codd et al., 2002).

### H. Effect of Magnetic Nuclei

In addition to the electronic dipolar oscillations, the ESE envelope modulation (ESEEM) caused by the electron-nuclear magnetic interactions is usually observed when $|\nu_A - \nu_B| < \Delta\nu_A + \Delta\nu_B$. Although the techniques that use $\nu_A = \nu_B$ (2+1) (Kurshev et al., 1988, 1989; Raitsimring, 2000), SIFTER (Jeschke et al., 2000b) and DQC (Borbat and Freed, 1999, 2000)) are the worst in this respect, the $^1$H ESEEM is often observed in X-band ELDOR experiments with $|\nu_A - \nu_B|$ up to ~ 100 MHz. The partial remedy is to choose the constant time intervals, $\tau$ and $T$, as to minimize the ESEEM.

If a very strong ESEEM comes from a small number of nuclei, it can also lead to an orientational selectivity because it changes the contributions of different orientations to the ESE amplitude of spin A. Finally, the stronger the ESEEM, the weaker is the ESE signal at large separations between the mw pulses. Therefore, although the low-frequency mw bands may seem attractive for ELDOR applications because of reduction in the EPR linewidth determined by the g-anisotropy, the dramatic increase in ESEEM amplitude makes such ELDOR experiments extremely challenging. Therefore, the examples of pulsed ELDOR work at the mw frequencies below the X-band ones are very limited (Codd et al., 2002; Weber et al., 2002).

### I. Exchange-Coupled Clusters

In some systems one or both of the centers in a pair are exchange-coupled clusters. The total spin of such a cluster, *S*, is composed of spins of its constituent paramagnetic ions, $S_k$, and depends on $S_k$ and on the way how these $S_k$ are coupled with each other (which ions are coupled with which, and what kinds of couplings, ferromagnetic or antiferromagnetic, are they). As an example, let us consider a coupling of cluster A to an individual spin B. The individual spins forming cluster A are generally located at different distances from B, and therefore their dipolar couplings, $D_k$, to spin B are different. These $D_k$ values are not directly measurable in an experiment. Instead, one measures the total dipolar coupling *D* of the total spin *S*:

$$D = \Sigma K_k D_k \tag{18}$$

where $K_k$ are the spin projection factors that reflect,

how the spin functions of the individual ions contribute to the total spin function of the cluster. For example, for a dinuclear cluster one has (Bencini and Gatteschi, 1990):

$$K_{1(2)} = [S(S+1) \pm S_1(S_1+1) \mp S_2(S_2+1)]/[2S(S+1)] \quad (19)$$

where the top signs are taken for $K_1$, while the bottom signs are taken for $K_2$.

If $R_{AB}$ is significantly larger than the distances $R_{jk}$ between the ions in the cluster or, at least, the angles $\theta_{RB}$ for the individual ions are not very different, then a one-center point-dipole model with a single $\theta_{RB}$ can be used, and Eq. (18) is simplified to:

$$D = (\Sigma K_k D_{ok}) \cdot (1 - 3\cos^2\theta_{RB}) = D_o(1 - 3\cos^2\theta_{RB}) \quad (20)$$

which is similar to Eq. (3), with the only difference being that now $D_o$ stands for the dipole coupling constant for the whole cluster. In the limit of very large distances ($R_{AB} \gg R_{kj}$) all $D_{ok}$ values become similar and, since $\Sigma K_k = 1$, the composite nature of the cluster spin $S$ can be neglected. In the opposite case of $R_{AB} \sim R_{kj}$ an explicit treatment is necessary (Bertrand et al., 1994).

The situation of an exchange-coupled cluster was encountered in a pulsed ELDOR study of [NiFe] hydrogenase (Elsässer et al., 2002, 2005) where the distances between Ni and the individual Fe ions of the [3Fe-4S] cluster were known from X-ray crystallography, and the dipole interaction was used to assign the spin projection factors to the Fe ions.

## IV. Instrumentation Requirements

There are two obvious requirements to the 'ELDOR-enabled' pulsed EPR instrumentation as compared with the standard one used, say, in ESEEM or ENDOR measurements. First, one has to have an additional mw source that one could use for pumping. A usual makeshift solution here is to employ another pulsed (Milov et al., 1984) or CW (Hara et al., 1996) EPR spectrometer, or a mw synthesizer (Astashkin et al., 1998; Jeschke et al., 2000a). In addition, the resonator for these experiments should be able to accommodate the two mw frequencies. For this it whether should have a low quality factor $Q$ (and the price to pay is, of course, lower sensitivity), or a considerable multimode structure (we have to say that most of modern dielectric resonators designed for pulsed EPR are actually multimode, so one should not worry about this too much). Much work has also been done using bimodal resonators (Milov et al., 1998).

## V. Choosing the Right ELDOR Experiment

In order to decide, which of the ENDOR experiments to choose for a given problem, one has to consider the expected scale of the dipole interaction, the frequency/field separation between the A and B spins, the relaxation times of the spins, etc. The common rules are: (1) it is always better to use the three-pulse sequence of Fig. 1A in order to detect relatively weak dipolar couplings ($D_o \ll 1/t_p$), while (2) for stronger couplings it is better to use the four-pulse sequence of Fig. 1B.

Let us first explain (1). In order to accurately measure weak dipole interactions, one needs to increase the range of variation of the time interval $t$, $\Delta t$, to the limit. The limit is partly determined by the transverse relaxation time $T_2$, but since in the four-pulse sequence one can make the time interval $\tau$ almost arbitrarily short, the difference in the $T_2$ effects between the three- and four-pulse techniques will be insignificant.

A more important consideration is based on the dependence of the primary and refocused primary ESE signals on the flip angles. Assuming nominal flip angles of pulses 1A, 2A and 3A to be, respectively, $\theta_{1A}$, $2\theta_{1A}$ and $2\theta_{1A}$, one can write (Mims, 1965; Pannier et al., 2000):

$$V_{o(3\text{-pulse})} = A\sin^3\theta_{1A}\ ;\ V_{o(4\text{-pulse})} = A\sin^5\theta_{1A} \quad (21)$$

In a typical ELDOR experiment one tries to move $\nu_A$ very far from the resonator optimum. As a result, the mw field amplitude for spins A becomes very low, and $\theta_{1A}$ can become significantly smaller than 90°, leading to $V_{o(3\text{-pulse})} \gg V_{o(4\text{-pulse})}$ and a dramatic decrease of $\Delta t$ and dipolar resolution for the four-pulse sequence relative to the three-pulse one. This may not always be an issue for concentrated model nitroxides, but it is certainly something to take into account when working with broad EPR signals in dilute biological samples, e.g., with $Mn_4$ cluster and cytochrome *b*559 in PS II.

Let us now consider the second point. The three-pulse experiment of Fig. 1A has a dead time approximately equal to: $t_d \sim t_{p1A} + t_{p1B}$. If the first quarter-oscillation in the ELDOR trace falls within the dead time, the rest of the ELDOR trace may be almost useless because it has a considerably smaller amplitude, especially in systems where $D_0$ is statistically distributed. This results in a limitation of the maximal $D_0$ values that can be measured by this technique: $D_0 < 1/(4t_d)$. With $t_{p1A} = t_{p1B} = 10$ ns one estimates $D_0 < 12.5$ MHz, or $R_{AB} > 16$ Å.

In the four-pulse experiment there is no dead time, and the maximal measurable $D_0$ is mostly determined by the mw coverage of the observation mw pulses, $D_0 < 1/4t_{p1A}$ (for 90°–180°–180° observation sequence with the pulse durations $t_{p1A}$, $2t_{p1A}$, $2t_{p1A}$), simply because the changes in the local field that exceed this range will result in a loss of the ESE signal. For $t_{p1A} =$ 10 ns one thus estimates $D_0 < 25$ MHz, or $R_{AB} > 13$ Å. For $t_{p1A} = 16$ ns as used in measurements of model biradicals (Martin et al., 1998) the maximal $D_0$ value obtained was about 14.2 MHz ($R_{AB} \sim 15.4$ Å), which agrees with the limiting estimate $D_0 < 1/0.064$ MHz $= 15.6$ MHz.

## VI. Choosing the Pumping and Observation Frequencies

Usually, when performing a pulsed ELDOR experiment, one has to work in vicinity of a single optimal resonator frequency, $\nu_{opt}$. Since $\nu_A \neq \nu_B$, the question arises, which of the pulses, the pumping or the observation, should we give a preference? That is, should we set $\nu_A = \nu_{opt}$ and move $\nu_B$ away, or should we do it the other way around? There is also another dilemma: which EPR spectrum of those belonging to the spins in the pair is it better to use for observation, and which for pumping?

For nitroxide radical pairs these problems have been discussed by Larsen and Singel (1993) and by Jeschke (2000). The general rule is: assign the spectra for observation and pumping and set the mw frequencies relative to $\nu_{opt}$ in such a way as to provide for the best possible pumping conditions. This means: (1) set $\nu_B = \nu_{opt}$ and (2) use a more narrow and intense spectrum (if the spectra of the centers are different) or a more narrow and intense feature of the spectrum (if the spectra are identical) for pumping. This relative arrangement of the frequencies and the spectra is shown in Fig. 3.

The wisdom behind this rule is that it is better to have a large effect on a small signal than a small effect on a large signal. This is because in any real measurement system, apart from noises and artifacts independent of signal amplitude, there are usually distortions present that are proportional to the signal amplitude. In a pulsed ELDOR experiment the latter distortions include stochastic signal variations caused by instabilities of the mw source, temperature variations, microphonics, etc. There are also reproducible signal variations with changing the position of the pumping pulse, some of which are of technical origin and depend on the spectrometer design, while the others originate in the spin system itself (e.g., the nuclear ESEEM).

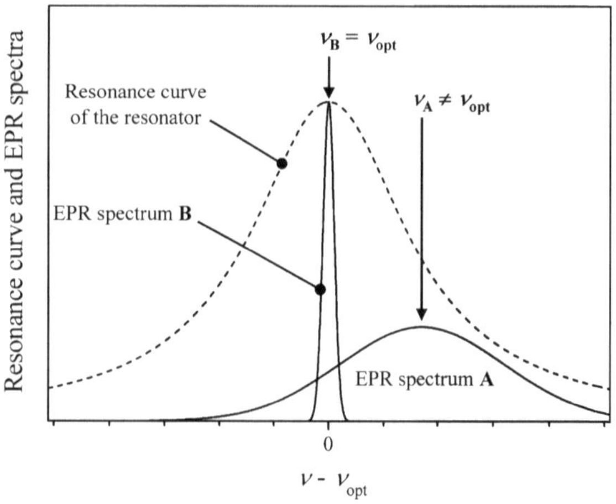


*Fig. 3.* Arrangement of the pumping and observation frequencies and the EPR spectra with respect to the resonance curve of the resonator typical for pulsed ELDOR. The narrow EPR spectrum is typically used for pumping, and the pumping frequency, $\nu_B$, is set equal to the optimum frequency of the resonator, $\nu_{opt}$. The other spectrum is used for observation, and the observation frequency, $\nu_A$, is shifted away from $\nu_{opt}$.

Of course, any rule has exceptions. For example, if the transverse relaxation time of the center with a broad EPR spectrum is too short, but the EPR spectrum is not too broad, it may be better to use the narrow EPR signal for observation and broad one for pumping.

The above discussion was related to the use of officially 'single mode' resonators in pulsed ELDOR experiments. We did not discuss the bimodal resonators for two reasons. First, the performances of such a resonator in both modes can be quite comparable, and therefore one is not confronted with the choices described above. On the other hand, the rectangular bimodal resonators used in the studies of model biradicals (Milov et al., 1998) often do not provide the sensitivity necessary for biological applications (mostly, because of much smaller filling factor).

## VII. Application Examples

In this section, we will illustrate the use of pulsed ELDOR and 2+1 techniques for evaluating distances between the electron transfer cofactors in photosynthetic RCs, mostly, of plant Photosystem II (PS II). These cofactors, while donating electrons down the electron transfer chain or accepting the electrons from up the chain, become transiently paramagnetic and can be detected by EPR. In PS II the process starts from the photoexcitation of a primary donor chlorophyll dimer, P680. The excited P680 donates an electron to the intermediate pheophytin acceptor, wherefrom it is transferred to the primary acceptor, plastoquinone $Q_A$. From $Q_A^-$ the electron is transferred to the secondary acceptor, plastoquinone $Q_B$. After receiving two electrons in two turnovers, $Q_BH_2$ leaves its binding site and moves away into the membrane, while its place is taken by a fresh neutral $Q_B$ from the quinone pool.

At the donor side the oxidized P680 is donated by an electron from the secondary donor, tyrosine Z ($Y_Z$), which then gets an electron from the oxygen-evolving complex (OEC) containing the $Mn_4$ cluster. The story with $Y_Z$ is actually more complicated and involves a deprotonation of $Y_Z^+$ with formation of a neutral radical $Y_Z^{\bullet}$ (Eckert and Renger, 1988). The reverse process consists in acquisition of an electron from the OEC concomitant with acquisition of the $H^+$, so that finally a neutral diamagnetic $Y_Z$ is restored.

The $Mn_4$ cluster represents a catalytic core that splits $H_2O$ and releases $O_2$. While donating electrons, the cluster cycles through five oxidation states denoted $S_0$ through $S_4$ (Kok et al., 1970). As the cluster reaches the $S_4$ state, it oxidizes two substrate water molecules (acquiring two electrons from each $H_2O$) and converts back to the $S_0$ state.

There are also some additional electron transfer cofactors of secondary importance that include cytochrome *b*559 (Cyt *b*559) and the so-called chlorophyll Z ($Chl_Z$). In addition, there is one more tyrosine residue, tyrosine D ($Y_D$), which will be at the center of our discussion of many ELDOR and 2+1 experiments below, although it is not a major figure in the electron transfer events in the PS II. Acting as a slow and inefficient electron donor, $Y_D$ converts to a neutral radical $Y_D^{\bullet}$ in essentially the same way as $Y_Z$ converts to $Y_Z^{\bullet}$ (Evelo et al., 1989). Unlike $Y_Z^{\bullet}$, however, $Y_D^{\bullet}$ is rather stable, and under illumination is easily accumulated in quantities approaching one radical per RC. Annealing the oxygen-evolving PS II sample in the dark, at the temperature of 0 °C for about 30 minutes results in decay of most of paramagnetic centers except $Y_D^{\bullet}$ and, notably, oxidized state of Cyt *b*559. This annealing procedure is called 'dark adaptation'.

Because the content of $Y_D^{\bullet}$ is about 100%, this center represents a natural reference point in ELDOR measurements. If the distance from $Y_D^{\bullet}$ is to be measured, the experimental efforts are reduced to trapping only one other cofactor, which is often much easier than simultaneously trapping two transient cofactors.

At the time when our work with PS II has started, the crystal structure of the PS II was not yet available, and we were trying to establish a spatial arrangement of the electron transfer cofactors. Of course, in an orientationally disordered system, the only thing one can realistically get from an experiment is the distance between the centers. However, there are two additional considerations. First, measuring the distances between various cofactors one obtains spatial constrains that might narrow down the range of possible structures. Second, it is possible to orient the PS II membranes on Mylar sheets (Rutherford, 1985). A study of such oriented system allows one to determine the angle between the radius vectors $\mathbf{R}_{AB}$ and the membrane normal. This helps to further decrease the structural uncertainties.

The third thing one has to take into account is that there was general knowledge of the PS II structure derived from protein sequencing and folding calculations. It was known, for example, that the PS II RC represents a heterodimer of so-called D1 and D2 proteins (isolated by Namba and Satoh (1987)) that correspond to L and M subunits in a bacterial RC. Most of the active electron transfer components are located at D1 side, except for $Q_A$ which belongs to D2 subunit together with $Y_D$. The two chlorophyll molecules that compose the primary electron donor P680 each belong to D1 and D2 subunits.

In addition to PS II we will mention some results obtained for PS I and for the bacterial photosynthetic RC. While the specific electron transfer cofactors for all of these systems may be different (Hillier and Babcock, 2001), the major principles remain the same, and even some of the names of the cofactors are also similar. We will not therefore describe these other systems in detail here.

In the following we will consider representative examples of ELDOR and 2+1 studies in PS II. Other studies that fall in similar categories will simply be mentioned. In addition, we have to note that in some experiments on trapped pairs also a selective hole burning technique (Wacker et al., 1992; Dzuba et al., 1993) was used, while transient spin-correlated radical pairs (SCRP) were studied using the out-of-phase ESEEM (Tang et al., 1994). Therefore, we will present the data obtained by those techniques as well. All of the data discussed below are summarized in Table I.

### A. '2+1' in Disordered PS II Membranes

We will consider the application of 2+1 using the distance measurement between the tyrosine radicals $Y_Z^{\bullet}$ and $Y_D^{\bullet}$ (Astashkin et al., 1994) as an example. In an oxygen-evolving PS II $Y_Z^{\bullet}$ rapidly converts back into a diamagnetic state by the electron transfer from the $Mn_4$ cluster in OEC. A solution how to stabilize $Y_Z^{\bullet}$ then is to remove the $Mn_4$ cluster from the PS II. Indeed, in a Mn-depleted PS II $Y_Z^{\bullet}$ can be trapped with a concentration comparable to that of $Y_D^{\bullet}$ (Kodera et al., 1992), and this technique was used to prepare the sample for the distance measurements (Astashkin et al., 1994). The concentration of trapped $Y_Z^{\bullet}$ was estimated to be ~ 0.7 per RC.

The two tyrosines give virtually identical EPR spectra with a width of about 2 mT, centered at g ≈ 2.0046 (see Fig. 4A). Thus, 2+1 was the technique of choice for the distance measurements, while for ELDOR the EPR linewidth was too small. The observation mw pulses used in the 2+1 experiment had durations of 16 ns and nominal flip angles $\theta_{f1} = \theta_{f3} \approx 30°$. The pumping mw pulse had a duration of 24 ns and a nominal flip angle $\theta_{f2} \approx 60°$. The flip probabilities were (see Eqs. 10, 11) $p_{f2} \leq 0.25$, $p_{f3} \leq 0.067$ and $p_{f23} \leq 0.017$. Thus, one had $p_{f23} << p_{f2}$, and the 2+1 experiment has actually corresponded to an ELDOR limit.

The dependence of the primary ESE amplitude on $t$ recorded for the sample with trapped $Y_Z^{\bullet}$ radicals is shown in Fig. 4B by open squares. This dependence reveals about two periods of low frequency oscillations and differs remarkably from the dependence obtained in the same sample after dark adaptation (filled triangles) where only $Y_D^{\bullet}$ radicals survive. As the average concentration of $Y_D^{\bullet}$ is very low (one radical per RC, at least 50 Å apart), there is no detectable dipole interaction between them, and the experimental 2+1 trace shows no oscillations. The difference between these two traces allows one to ascribe the oscillations observed for the sample with trapped $Y_Z^{\bullet}$ radicals to the dipole interaction between $Y_Z^{\bullet}$ and $Y_D^{\bullet}$.

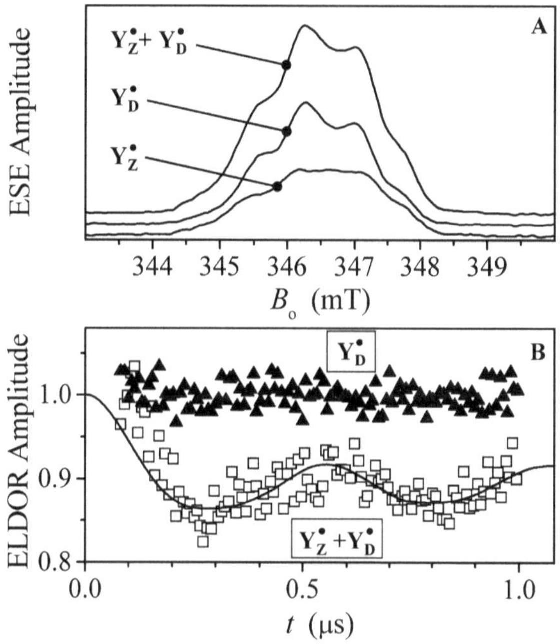


*Fig. 4.* (A) The field-sweep primary ESE spectra of oxidized tyrosine radicals in Mn-depleted PS II. The illuminated sample contains $Y_D^{\bullet}$ and $Y_Z^{\bullet}$ (top trace); the dark-adapted sample contains $Y_D^{\bullet}$ only (middle trace). The bottom trace is the difference between the top two ones, and corresponds to the spectrum of $Y_Z^{\bullet}$. Experimental conditions: τ = 200 ns; mw pulse durations, 16 ns (90°) and 24 ns (180°); boxcar integration gate, 180 ns; temperature, 80 K. (B) 2+1 traces at τ = 1080 ns, obtained in the illuminated (open squares) and dark-adapted (filled triangles) samples, using the mw pulses of 16 ns (30°) and 24 ns (60°) and 16 ns (30°). Solid line: calculated for $D_o$ = 2 MHz.

In order to determine $D_o$, numerical simulations were performed. The numerically simulated and experimental data were compared directly in time domain. A good agreement between them was obtained for $D_o$ = 2 MHz, which corresponds to the distance between $Y_Z^{\bullet}$ and $Y_D^{\bullet}$ of 29.6 Å.

The possibility of assignment of the observed oscillation to nuclear ESEEM was also carefully analyzed (Astashkin et al., 1994), and it was concluded that no such nuclear ESEEM was possible under the experimental conditions used.

In similar way, using 2+1, the distance between $Y_D^{\bullet}$ and $Q_A^-$ was found to be about 38.5 Å (Shigemori et al., 1998; Yoshii et al., 1999a), while the distance between $Y_D^{\bullet}$ and $Chl_Z^+$ was found to be 29.4 Å (Shigemori et al., 1998). The latter distance was interpreted as indication that this $Chl_Z$ molecule belongs to D2 protein (Kawamori et al., 2002).

### *B. ELDOR in Disordered PS II Membranes*

A representative example of this kind is given by the measurement of the distance between $Y_D^{\bullet}$ and the $Mn_4$ cluster in OEC. The EPR spectrum of the $Mn_4$ cluster is most easily observable in the $S_2$ state (Hara et al., 1996). In this state the cluster has $S = 1/2$, and several models are discussed as to how this spin is composed of those of the contributing Mn ions (Manchanda et al., 1995; Hasegawa et al., 1998). The EPR spectrum is over 150 mT wide and shows a partially resolved structure due to the *hfi* with the Mn nuclei (Fig. 5A). In spite of the large EPR width, there is very little anisotropy in *g*-factor or *hfi*.

The great width of the EPR spectrum of the $Mn_4$ cluster warranted the use of ELDOR. Naturally, the Mn ESE signal was used for observation ($B_0 = 345$ mT, $\nu_A = 9.79$ GHz), while the signal of $Y_D^{\bullet}$ was used for pumping ($\nu_B = 9.68$ GHz). The 110 MHz difference between $\nu_A$ and $\nu_B$ corresponds to the difference in the resonant magnetic fields of ~ 4 mT. The ELDOR trace recorded for the oxygen-evolving PS II in the $S_2$ state is shown in Fig. 5B by open squares. If $\nu_B$ was out of resonance with $Y_D^{\bullet}$, no oscillations were observed.

One has to note that the same field range is also contributed by the EPR signal of cytochrome *b*559 (see Fig. 5A where the $g_Z$ feature of the rhombic EPR signal of the Cyt *b*559 heme is clearly visible). Therefore, the oscillations observed are due to the dipole interaction between $Y_D^{\bullet}$ and some of the paramagnetic centers contributing to the ESE signal, i.e., the $Mn_4$ cluster or Cyt *b*559, or both of them. In order to decide, which of these centers is responsible for the ELDOR oscillations, the experiment was repeated at the temperature of 12 K, when the Mn multiline ESE signal is not observed because of very short transverse relaxation time, and the only center contributing to the ESE signal is Cyt *b*559. In this experiment, however, no oscillations were observed. The ELDOR trace of the PS II sample in the $S_1$ state (open circles in Fig. 5B) also did not show any oscillations. These results imply that the distance from $Y_D^{\bullet}$ to the heme iron of Cyt *b*559, $R_{DCyt}$, is fairly large, probably, greater than 47 Å as calculated from Eq. (1) (that is, assuming a complete orientational disorder, which is not really the case), as the oscillation frequency should be smaller than 0.5 MHz. In a later work the estimate $R_{DCyt}$ >50 Å was obtained (Kuroiwa et al., 2000). Thus, the oscillations in Fig. 5B are caused by the dipole interaction with the Mn cluster in OEC. The time domain ELDOR simulations result in $D_0 = 2.6$ MHz, which corresponds to the distance of $R_{DMn} = 27.1$ Å.

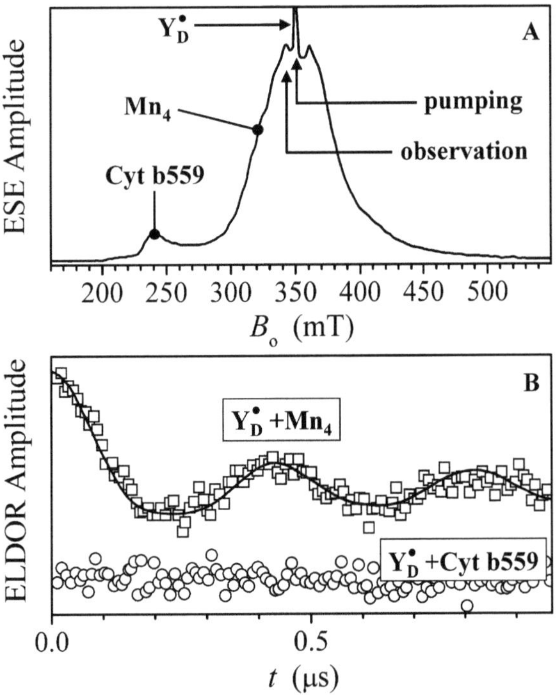


*Fig. 5.* (A) Primary ESE field sweep at $\tau = 200$ ns observed in the oxygen-evolving PS II in the $S_2$ state after illumination for 5 min at 230 K. Experimental conditions: mw pulses, 16 ns (90°) and 24 ns (180°); temperature, 6 K. (B) Open circles and squares, three-pulse ELDOR traces in the $S_1$ and $S_2$ states of the oxygen-evolving PS II, respectively. Experimental conditions: $B_0 = 345$ mT; $\tau = 1000$ ns; observation mw pulses, 16 ns (90°) and 24 ns (180°); observation mw frequency, $\nu_A = 9.79$ GHz (in resonance with the Mn multiline and Cyt b559 EPR signals); pumping mw pulse, 24 ns (180°); pumping mw frequency, $\nu_B = 9.68$ GHz (in resonance with $Y_D^{\bullet}$); temperature, 6 K. Solid line: calculated for $D_0 = 2.6$ MHz.

The EPR signal from the $Mn_4$ cluster in the $S_0$-state was observed in the $NH_2OH$ reduced PS II with added 3–5% methanol (Messinger et al., 1997). This signal was centered at g = 2, had an overall width of about 220 mT and revealed a complex hyperfine structure resembling that of the $S_2$ state multiline signal. The ELDOR experiment resulted in the distance between the $Mn_4(S_0)$ and $Y_D^{\bullet}$ of 34 Å (Kawamori et al., 2001). This distance estimate is significantly larger than 27

Å obtained for the $S_2$ state and 30 Å (between $Y_D$ and the center of the $Mn_4$ cluster) obtained by X-ray crystallography, which indicates that the simple point-dipole approach to the distance estimate should be corrected to account for the exchange-coupled nature of the $Mn_4$ cluster. An overall approach here can be similar to that taken by Elsässer et al. (2002) in their ELDOR study of [NiFe] hydrogenase, and should take into account the detailed analysis of the Mn–Mn exchange coupling schemes by Peloquin et al. (2000).

Pulsed ELDOR measurements were also used to estimate the distance between Cyt *b*559 and $Q_A^-$, $R_{QCyt} \approx 40$ Å (Kuroiwa et al., 2000). The distance from $Y_D^{\bullet}$ to the controversial species giving the so-called split signal in a modified $S_3$ state of the PS II was determined to be $R_{DS3} \approx 30$ Å (Hara et al., 1996). The distance of 34.5 Å between $Y_Z^{\bullet}$ and $Q_A^-$ was obtained in $Y_D$-less mutant using ELDOR with soft mw pulses (Kawamori et al., 2002) because the EPR spectra of the both species are narrow (≤ 2 mT wide). The same distance between $Y_Z^{\bullet}$ and $Q_A^-$ was determined in a three-spin system formed by $Y_D^{\bullet}$ and simultaneously trapped $Y_Z^{\bullet}$ and $Q_A^-$ in iron-depleted PS II (Kawamori et al., 2003). Similar distance of 34 Å was found for $Q_A^-$-$Chl_Z$ pair in a study of the three-spin system formed by $Y_D^{\bullet}$, $Q_A^-$ and $Chl_Z$ (Kawamori et al., 2003b).

We have to mention that the analysis of three-spin systems (ABC) is somewhat more complicated. For example, the ELDOR trace for spins A will include the oscillations with the frequencies of $D_{AB}$, $D_{AC}$, $D_{AB}+D_{AC}$ and $D_{AB}$-$D_{AC}$. Therefore, reliable results can only be obtained if all of the parameters, except one of the $D_{ij}$'s, are already known from other experiments.

### C. ELDOR and '2+1' in Oriented PS II Membranes

If a membrane suspension is painted and dried on a flat surface, e.g., Mylar sheets, the membranes tend to orient themselves parallel to that surface (Rutherford, 1985). However, the orientations of the membranes in the direction perpendicular to the surface normal are still arbitrary. The photosynthetic RCs are embedded in their membranes in a specific way, with the electron transfer vector pointing in a general direction across the membrane. Therefore, when the membranes are oriented, the RCs are oriented as well. For the photosynthetic membranes oriented in this way one can obtain, in addition to the distances between the electron transfer cofactors, the angles between the radius-vectors $\mathbf{R_{AB}}$ and the membrane normal, $\theta_{Rn}$.

Of course, one cannot distinguish between the angles $\theta_{Rn}$ and $180° - \theta_{Rn}$, and thus one cannot formally tell, which of the centers in the pair is closer to the acceptor side, and which is closer to the donor side of the RC. Therefore, in the interpretation of the angular information a degree of conjecture is involved, which is usually based on solid reasoning. In our discussion of the PS II we will assume that $\theta_{Rn} = 0°$ corresponds to the direction toward the stromal side of the membrane and a general direction toward the acceptor side of the PS II.

The 2+1 and pulsed ELDOR experiments with oriented PS II membranes (Astashkin et al., 1998) have confirmed the distances from $Y_D^{\bullet}$ to $Y_Z^{\bullet}$ ($R_{DZ} \approx 30$ Å) and to the $Mn_4$ cluster ($R_{DMn} \approx 27$ Å), but have also established that the angles of $\mathbf{R_{DZ}}$ and $\mathbf{R_{DMn}}$ with the membrane normal vector are about 100° and 110°, respectively. The angle between the radius-vector $Y_D^{\bullet}$ - $Q_A^-$ and the membrane normal was estimated as 13° ± 4° (Yoshii et al., 1999), which was corrected to 28° by Kawamori et al. (2003a). The angle for the radius vector $Q_A^-$-Cyt *b*559 was found to be whether 80° or 100° (Kuroiwa et al., 2000). The authors were undecided, which of the angles to prefer, but both of these angles place Cyt *b*559 on the stromal side of the membrane. The angle between the radius vector $Y_D^{\bullet}$ - $Chl_Z$ and the membrane normal was found to be 50° (Tonaka et al., 2000; Kawamori et al., 2002).

The angle from $Y_D^{\bullet}$ to the species giving the doublet split $S_3$ signal was determined to be 110° (Mino et al., 2000), which, together with the distance of 30 Å (Hara et al., 1996) and other evidence mentioned by the authors, indicates the involvement of the redox-active tyrosine $Y_Z^{\bullet}$ in formation of this signal.

The angle between the radius-vector from $P680^+$ to $Q_A^-$, $R_{PQ}$, was found to be about 21° (Yoshii et al., 1999b). Although this information was obtained using another pulsed EPR technique, out-of-phase ESEEM (see below), we mention it here along with all the other angular information for PS II.

### D. Selective Hole Burning

Let us consider a situation when one of the spins in the pair has a very broad EPR spectrum and the phase memory time, $T_2$, short compared with $1/D_0$. The broad EPR spectrum will prevent one from using this spin for pumping in an ELDOR experiment. On the other hand, the short $T_2$ does not allow one to use this spin for observation. The usual ELDOR experiment in such a situation will fail. However, there is still a trick one can play to measure the distance between this spin and its partner in the pair. In the experiment that follows this partner will be the spin to observe, spin A. The 'inconvenient' other spin will be spin B. The experiment proceeds as follows (Wacker et al., 1992; Dzuba et al., 1993).

A soft 180° mw pulse creates a hole in the EPR spectrum of spin A. After some time interval $T$ a two-pulse sequence generates the ESE signal. The ESE signal, roughly speaking, represents a Fourier image of the EPR spectrum shape, and thus it contains a broad weak negative component due to the hole burned in the spectrum by the first pulse and a strong narrow positive component due to all other spins B that were unaffected by the first pulse. If time interval $T$ is short, than obviously nothing happens to the original hole. However, if $T$ becomes comparable to $T_1$ of spin B, then for any spin A there is a probability that its neighboring spin B has flipped due to the longitudinal relaxation.

A flip of spin B (that is, the change of its orientation compared with the original one existing at the time when the hole was created) results in a change of the local magnetic field for spin A, which results in the broadening of the hole, which results in the narrowing of the component of the ESE signal produced by this hole. Analysis of this signal narrowing allows one to estimate $D_0$ and the distance $R_{AB}$.

Using this technique (Kodera et al., 1994) the distance from $Y_D^{\bullet}$ to the $Fe^{2+}$ ion at the acceptor side of the PS II was estimated to be less than 52 Å (see Fig. 6). The distance from $Y_D^{\bullet}$ to the $Mn_4$ cluster was estimated to be about 28 Å in the $S_2$ state and 30 Å in the $S_1$ state. We already described above the ELDOR measurement of the $Y_D^{\bullet}$ - $Mn_4$ distance for the $S_2$ state. In the $S_1$ state using ELDOR was impossible because the EPR signal of the $Mn_4$ cluster was observed at g ~ 5 (Dexheimer and Klein, 1992; Yamauchi et al., 1997). It is interesting to note that the distances from $Y_D^{\bullet}$ to the cluster in the $S_1$ and $S_2$ states were found to be different. One of possible explanations of this finding may be that the distances were estimated using Eq. (1), without taking into account the exchange-coupled nature of the $Mn_4$ cluster (Manchanda et al., 1995; Hasegawa et al., 1998). Another possible explanation includes the effect of the crystal field of the Mn ions on the spin quantization axes (Kodera et al., 1994).

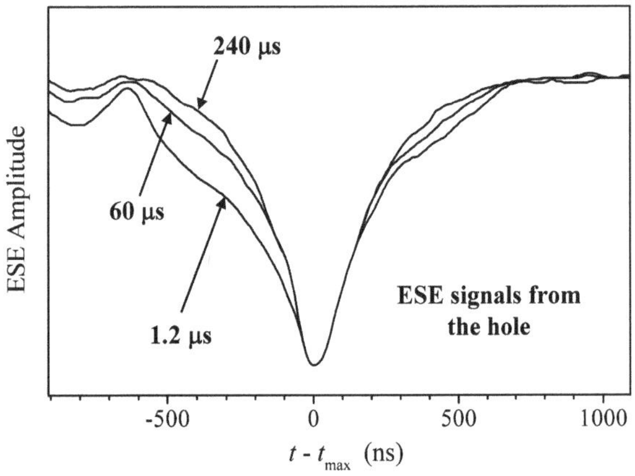


*Fig. 6.* The effect of the relaxation-induced spin flips of the acceptor $Fe^{2+}$ on the shape of the two-pulse ESE signal from the hole in the EPR spectrum of $Y_D^{\bullet}$ at different delays $T$ (indicated in the Figure) after the soft 180° preparation pulse. The horizontal axis shows time relative to the time of the ESE maximum, $t_{max}$. Experimental conditions: preparation pulse duration, 800 ns; observation pulses, 2 × 40 ns; time interval between the observation pulses, $\tau$ = 1200 ns; temperature, 4.2 K.

It is also appropriate to mention here another technique based on the same principle of relaxation-induced flips of spins B (Kulik et al., 2001). This technique (called by the authors 'relaxation-induced dipolar modulation enhancement,' RIDME) employs the three-pulse stimulated ESE sequence, all pulses being in resonance with spins A. The precession of spins A is started by the first 90° mw pulse and proceeds for the variable time interval $\tau$ before being stored by the second 90° pulse as a periodical magnetization pattern. During the time interval $T$ between the second and third mw pulses spins B have a chance to flip, which changes the local magnetic fields for spins A and corrupts (though in a regular way) the periodical magnetization pattern created by the first two pulses. The third 90° mw pulse again starts the precession of spins A, but since the magnetization pattern is corrupted, the amplitude of the stimulated ESE signal will generally be smaller than that obtained without the dipole interaction. The analysis shows that the dependence of the stimulated ESE amplitude on $\tau$ in this method is given by $\cos(2\pi D\tau)$ (Kulik et al., 2001). To our knowledge, this method has only been applied to nitroxide biradicals and nitroxide-labeled peptides (Kulik et al., 2001), but it obviously has a much broader range of potential applications.

### E. ESEEM from Spin-Correlated Radical Pairs (SCRP)

Sometimes trapping a specific radical pair to be studied by pulsed ELDOR may represent a challenge because of high efficiency of various pathways leading to disappearance of one or both of the radicals. Even such a doomed pair, however, may exist in good health for several microseconds after being generated by the laser pulse, a time scale sufficient for pulsed EPR. Although the pulsed ELDOR or 2+1 in this case are of no use, there is another ingenious technique that exploits the fact that the radical pair is born in a non-equilibrium (polarized) state.

This technique is called two-pulse ESEEM. There is, however, a very important difference from what one is used to see in usual ESEEM experiments with paramagnetic centers in thermal equilibrium. For radical pairs born in magnetic field from singlet precursors only the singlet ($|S\rangle$) and the triplet state with total $S_Z = 0$ ($|T_0\rangle$) is initially populated, while the populations of the triplet states with $S_Z = \pm1$, $|T_+\rangle$ and $|T_-\rangle$, are about zero. This non-equilibrium population distribution, along with the presence of the dipole and exchange interactions, results in formation of an ESE signal whose mw phase is shifted by 90° from that of a usual ESE signal one would obtain for a radical in thermal equilibrium. For this reason this echo signal is called out-of-phase ESE (OOP ESE) signal. The amplitude of the OOP ESE signal depends on the time interval $\tau$ between the mw pulses as $\sin(2\pi(J+D)\tau)$ (Tang et al., 1994). It is thus the same frequency as that obtained in an ELDOR experiment, and the magnitudes of the dipole and exchange couplings can be estimated whether from a time domain or frequency domain analysis (in this case a sine FT is used).

We have to note here that in the literature on SCRP it is customary to use a different definition for the exchange and the dipole interaction constants: $-2J$ and $(2/3)D$ instead of, respectively, $J$ and $D$ used in this work and in other works on stable doublet paramagnetic centers. So the OOP ESEEM frequency, using their notation is $-2J + (2/3)D$, instead of our $J + D$ (see above). One has to take these differences into account when comparing the data obtained by OOP ESEEM and by ELDOR.

The OOP ESEEM was actively applied to photosynthetic systems, mostly to obtain the distance between the primary donor and a primary acceptor. In this way the distance estimates of 27.2 Å (Hara et al., 1997) and 27.4 Å (Zech et al., 1997) were obtained for the pair of $P680^+$ and $Q_A^-$. In bacterial RC the distance between $P860^+$ and $Q_A^-$ was found to be 25.6 Å (Dzuba et al., 1995), 28.4 Å (Bittl and Zech, 1997) and 28.9 Å (Borovykh et al., 2003). In PS I, the distance between $P700^+$ and $A_1^-$ was evaluated as 25.4 Å (Dzuba et al., 1997; Zech et al., 1997). In addition to the primary donor-acceptor pairs, the distance between $Y_Z^{\bullet}$ and $Q_A^-$ in PS II was found to be 34 Å (Zech et al., 1999).

## VIII. Conclusion

The last decade has seen an enormous increase in the number of publication on distance measurements in biological systems by pulsed ELDOR and other pulsed and CW EPR techniques. However, taking into account the ever increasing number of biological systems being mapped by the X-ray crystallography, one may ask, why should we bother with the limited structural scope of the magnetic resonance techniques and not invest into the full structural picture provided by the X-ray? The answer is two-fold. First, achieving a good X-ray quality crystal for a particular protein may be an adventure in itself, and it may take many years. Meanwhile, the magnetic resonance techniques can be used in a disordered system, and their structural results, even incomplete, may be essential (in fact, sometimes they may be the only results one actually needs) for understanding the function of the particular system under investigation.

The second reason is related to the fact that while the X-ray is sensitive to the electron density, the magnetic resonance is sensitive to the spin density. Therefore, the comparison of the dipole coupling constant with that expected for the known X-ray distance allows one to make conclusions about the fine details of the electronic structure of the paramagnetic centers, possible mechanisms of structural disorder, etc. In other words, the combined use of magnetic resonance and X-ray can bring the investigation to a much more informative level. Although in this work we mentioned this aspect only briefly, this is an interesting emerging application of pulsed ELDOR and other magnetic resonance techniques.

Finally, specifically for the photosynthetic RCs, the ELDOR, SCRP ESEEM and other pulsed and CW EPR measurements actually help to identify the active electron transfer components involved in the primary photochemical processes.

*Table 1.* Distances $R_{AB}$ between the electron transfer intermediates in PS II and the angles $\theta_{Rn}$ between $\mathbf{R_{AB}}$ and the PS II membrane normal obtained by pulsed EPR and mentioned in this work. The angle $\theta_{Rn} = 0°$ corresponds to the direction towards the stromal side of the membrane and acceptor side of the PS II. The distances from X-ray analysis are obtained from the coordinates registered in Protein Data Bank by Zouni et al. (2001).

| Pair | EPR distance | X-ray distance | $\theta_{Rn}$ | Technique |
|---|---|---|---|---|
| $Y_D$–$Y_Z$ | 29.6[1] Å | 30 Å | 100°[2] | 2+1[1] |
| $Y_D$–$Q_A$ | 38.5[3] Å | 40 Å | 28°[4]* | 2+1[3] |
| $Y_D$–$Chl_Z$ | 29.4[3] Å | 26.8 Å | 50°[5],[6] | 2+1[4], ELDOR[5] |
| $Y_D$–$Mn_4(S_1)$ | 30[7] Å | 30.3Å | | Selective hole burning[6] |
| $Y_D$-$Mn_4(S_0)$ | 34[8] Å | 30.3Å | | ELDOR[7] |
| $Y_D$–$Mn_4(S_2)$ | 27.1[9] Å | 30.3Å | 110°[2] | ELDOR[8],[9] |
| $Y_D$ – Split signal$_3$ species | 30[9] Å | | 100°[10] | 2+1[10], ELDOR[11] |
| $Y_D$–non-heme Fe | < 52[7] Å | 37.6 Å | | Selective hole burning[12] |
| $Q_A$–Cyt $b_{559}$ | 40[11] Å | 47.8 Å | 80° or 100°[11] | ELDOR[13] |
| $Q_A$-$Chl_Z$ | 34[12] Å | | | ELDOR[14] |
| P680–$Q_A$ | 27.3[13],[14] Å | 26.9 Å | 21°[15] | SCRP ESEEM[15] |
| $Y_Z$–$Q_A$ | 34[16],[17] Å | 34.2 Å | | SCRP ESEEM[16], ELDOR[17] |

[1] Astashkin et al., 1994. [2] Astashkin et al., 1998. [3] Shigemori et al., 1998. [4] Yoshii et al., 1999a and Kawamori 2003. [5] Tonaka et al., 2000. [6] Kawamori et al., 2003. [7] Kodera et al., 1994. [8] Kawamori et al., 2001. [9] Hara et al., 1996. [10] Mino et al., 2002. [11] Kuroiwa et al., 2000. [12] Kawamori et al., 2002. [13] Hara et al., 1997. [14] Zech et al., 1997. [15] Yoshii et al., 1999b. [16] Zech et al., 1999. [17] Kawamori et al., 2002.

Table 1 summarizes the distances in PS II obtained by pulsed EPR. The distances from X-ray are listed in the same table for comparison. One can see that most of the distances obtained by these techniques are coincident within the X-ray resolution. However, the distances to the $Mn_4$ cluster and Cyt b559 are significantly different, which indicates that the details of the electronic structure of these centers (spin projection factors, g-tensors) should be taken into account when interpreting the ELDOR results.

## Acknowledgments

A.V.A. is grateful to Dr. A.M. Raitsimring for numerous stimulating discussions. He also gratefully acknowledges the funding from the Japan Society for Promotion of Science (JSPS) and the Kwansei Gakuin University during the time of his work in Japan, when some of the results cited in this work were obtained. A.K. is indebted to Professor Yu. D. Tsvetkov for his suggestion to apply ELDOR to photosystems.

## References

Astashkin AV, Kodera Y and Kawamori A (1994) Distance between tyrosines Z⁺ and D⁺ in plant Photosystem II as determined by pulsed EPR. Biochim Biophys Acta 1187: 89–93

Astashkin AV, Hara H and Kawamori A (1998) The pulsed electron–electron double resonance and '2+1' electron spin echo study of the oriented oxygen-evolving and Mn-depleted preparations of Photosystem II. J Chem Phys 108: 3805–3812

Bencini A and Gatteschi D (1990) Electron Paramagnetic Resonance of Exchange Coupled Systems. Springer, Berlin

Bennati M, Weber A, Antonic J, Perlstein DL, Robblee J and Stubbe J (2003) Pulsed ELDOR spectroscopy measures the distance between the two tyrosyl radicals in the R2 subunit of the *E. coli* ribonucleotide reductase. J Am Chem Soc 125: 14988–14989

Bennati M, Robblee JH, Mugnaini V, Stubbe J, Freed JH and Borbat PP (2005) EPR distance measurements support a model for long-range radical initiation in *E. coli* ribonucleotide reductase. J Am Chem Soc 127: 15014–15015

Berliner LJ, Eaton SS and Eaton GR (eds) (2000) Biological Magnetic Resonance. Vol 19. Distance Measurements in Biological Systems by EPR. Kluwer Academic Publishers, Dordrecht

Bertrand P, More C, Guigliarelli B, Fournel A, Bennett B and Howes B (1994) Biological polynuclear clusters coupled by magnetic interactions: From point dipole approximation to a local spin model. J Am Chem Soc 116: 3078–3086

Bittl R and Zech SG (1997) Pulsed EPR study of spin-coupled radical pairs in photosynthetic reaction centers: Measurement of the distance between $P700^+$ and $A_1^-$ in Photosystem I and between $P865^+$ and $Q_A^-$ in bacterial reaction centers. J Phys Chem B 101: 1429–1436

Borbat PP and Freed JH (1999) Multiple-quantum ESR and distance measurements. Chem Phys Lett 313: 145–154

Borbat PP and Freed JH (2000) Double-quantum ESR and distance measurements. In: Berliner LJ, Eaton SS and Eaton GR (eds) Biological Magnetic Resonance, Vol 19. Distance Measurements in Biological Systems by EPR, pp 383–459. Kluwer Academic Publishers, New York

Borbat PP, Mchaourab HS and Freed JH (2002) Protein structure determination using long-distance constraints from double-quantum coherence ESR: Study of T4 lysozyme. J Am Chem Soc 124: 5304–5314

Borovykh IV, Kulik LV, Gast P and Dzuba SA (2003) Conformation transition in the protein of a photosynthetic reaction center observed at the nanometer range of distances at cryogenic temperatures. Chem Phys 294: 433–438

Bowman MK, Becker D, Sevilla MD and Zimbrick JD (2005) Track structure in DNA irradiated with heavy ions. Radiat Res 163: 447–454

Chiang Y-W, Borbat PP and Freed JH (2005) The determination of pair distance distributions by pulsed ESR using Tikhonov regularization. J Magn Reson 172: 279–295

Codd R, Astashkin AV, Pacheco A, Raitsimring AM and Enemark JH (2002) Pulsed ELDOR spectroscopy of the Mo(V)/Fe(III) state of sulfite oxidase prepared by one-electron reduction with Ti(III) citrate. J Biol Inorg Chem 7: 338–350

Dexheimer SL and Klein MP (1992) Detection of a paramagnetic intermediate in the $S_1$ state of the photosynthetic oxygen-evolving complex. J Am Chem Soc 114: 2821–2826

Dzuba SA, Kodera Y, Hara H and Kawamori A (1993) The use of selective hole burning in EPR spectra to study spectral diffusion and dipolar broadening. J Magn Reson A 102: 257–260

Dzuba SA, Gast P and Hoff AJ (1995) ESEEM study of spin-spin interactions in spin-polarised $P^+Q_A^-$ pairs in the photosynthetic purple bacterium *Rhodobacter sphaeroides* R26. Chem Phys Lett 236: 595–602

Dzuba SA, Hara H, Kawamori A, Iwaki M, Itoh S and Tsvetkov YuD (1997) Electron spin echo of spin-polarised radical pairs in intact and quinone-reconstituted plant Photosystem I reaction centres. Chem Phys Lett 264: 238–244

Eckert H-J and Renger G (1988) Temperature dependence of $P680^+$ reduction in $O_2$-evolving PS II membrane fragments at different redox states $S_i$ of the water oxidizing system. FEBS Lett 236: 425–431

Elsässer C, Brecht M, and Bittl R (2002) Pulsed electron-electron double resonance on multinuclear metal clusters: Assignment of spin projection factors based on the dipolar interaction. J Am Chem Soc 124: 12606–12611

Elsässer C, Brecht M and Bittl R (2005) Treatment of spin-coupled metal-centers in pulsed electron-electron double resonance experiments. Biochem Soc Trans 33: 15–19

Evelo RG, Hoff AJ, Dikanov SA and Tyryshkin AM (1989) An ESEEM study of the oxidized electron donor of plant Photosystem II: Evidence that D• is a neutral tyrosine radical. Chem Phys Lett 161: 479–484

Fajer PG (2005) Site directed spin labelling and pulsed dipolar electron paramagnetic resonance (double electron-electron resonance) of force activation in muscle. J Phys Condens Matter 17: S1459–S1469

Ferreira KN, Iverson TM, Maghlaoui K, Barber J and Iwata S (2004) Architecture of the photosynthetic oxygen-evolving center. Science 303: 1831–1838

Hara H, Kawamori A, Astashkin AV and Ono T-A (1996) The distances from tyrosine D to redox-active components on the donor side of Photosystem II determined by pulsed electron-electron double resonance. Biochim Biophys Acta 1276: 140–146

Hara H, Dzuba SA, Kawamori A, Akabori K, Tomo T, Satoh K, Iwaki M and Itoh S (1997) The distance between P680 and $Q_A$ in Photosystem II determined by ESEEM spectroscopy. Biochim Biophys Acta 1322: 77–85

Hasegawa K, Kusunoki M, Inoue Y and Ono T-A (1998) Simulation of $S_2$-state multiline EPR signal in oriented Photosystem II membranes: structural implications for the manganese cluster in an oxygen-evolving complex. Biochemistry 37: 9457–9465

Hilger D, Jung H, Padan E, Wegener C, Vogel K-P, Steinhoff H-J, and Jeschke G (2005) Assessing oligomerization of membrane proteins by four-pulse DEER: pH-dependent dimerization of NhaA Na1/H1 antiporter of *E. coli*. Biophys J 89: 1328–1338

Hillier W and Babcock GT (2001) Photosynthetic reaction centers. Plant Physiol 125: 33–37

Jeschke G (2002) Distance measurements in the nanometer range by pulse EPR. Chem Phys Chem 3: 927–932

Jeschke G, Pannier M and Spiess HW (2000a) Double electron-electron resonance. In: Berliner LJ, Eaton SS and Eaton GR (eds) Biological Magnetic Resonance, Vol 19. Distance Measurements in Biological Systems by EPR. pp 493–512. Kluwer Academic Publishers, New York

Jeschke G, Pannier M, Godt A and Spies HW (2000b) Dipolar spectroscopy and spin alignment in electron paramagnetic resonance. Chem Phys Lett 331: 243–252

Jeschke G, Wegener C, Nietschke M, Jung H and Steinhoff H-J (2004) Interresidual distance determination by four-pulse double electron-electron resonance in an integral membrane protein: The Na+/proline transporter PutP of *Escherichia coli*. Biophys J 86: 2551–2557

Kamiya N and Shen J-R (2003) Crystal structure of oxygen-evolving Photosystem II from *Thermosynechococcus vulcanus* at 3.7 Å resolution. Proc Natl Acad Sci 100: 98–103

Kawamori A (2003) Electron transfer and structure of plant

Photosystem II, In Lund A and Shiotani M (eds) EPR of Free Radicals in Solid, pp 529–563, Kluwer Academic Publishers, Dordrecht

Kawamori A, Katsuta N, Arao S, Ishii A, Minagawa J, Mino H and Ono T (2001) Three-dimensional structure of Photosystem II studied by pulsed EPR. In: PS2001: Proceedings of 12th International Congress on Photosynthesis, S5-04. CSIRO, Melbourne (CD-ROM)

Kawamori A, Katsuta N, Mino H, Ishii A, Minagawa J and Ono T-A (2002) Positions of $Q_A$ and $Chl_Z$ relative to tyrosine $Y_Z$ and YD in Photosystem II studied by pulsed EPR. J Biol Phys 28: 413–426

Kawamori A, Katsuta N and Hara H (2003) Structural analysis of three-spin systems of Photosystem II by PELDOR. Appl Magn Reson 23: 557–569

Kodera Y, Takura K, Mino H and Kawamori A (1992) Pulsed EPR Study of Tyrosine-$Z^+$ in Photosystem II. In: Murata N (ed) Research in Photosynthesis, Vol II, pp 57–60. Kluwer Academic Publishers, Dordrecht

Kodera Y, Dzuba SA, Hara H and Kawamori A (1994) Distances from tyrosine $D^+$ to the manganese cluster and the acceptor iron in Photosystem II as determined by selective hole burning in EPR spectra. Biochim Biophys Acta 1186: 91–99

Kok B, Forbush B and McGloin M (1970) Cooperation of charges in photosynthetic $O_2$ evolution. 1. A linear 4 step mechanism. Photochem Photobiol 11: 457–475

Kulik LV, Dzuba SA, Grigoryev IA and Tsvetkov YuD (2001) Electron dipole-dipole interaction in ESEEM of nitroxide biradicals. Chem Phys Lett 343: 315–324

Kuroiwa S, Tonaka M, Kawamori A and Akabori K (2000) The position of cytochrome *b*559 relative to $Q_A$ in Photosystem II studied by electron-electron double resonance (ELDOR). Biochim Biophys Acta 1460: 330–337

Kurshev VV, Astashkin AV and Raitsimring AM (1988) Modulation effects in a 2+1 electron spin echo pulse sequence. J Struct Chem 29: 62–68

Kurshev VV, Raitsimring AM and Tsvetkov YuD (1989) Selection of dipolar interaction by the '2 + 1' pulse train ESE. J Magn Reson 81: 441–454

Kurshev VV, Raitsimring AM and Ichikawa T (1991) Spatial distribution of free radicals in γ-Irradiated alcohol matrices determined by the 2+1 electron spin echo method. J Phys Chem 95: 3564–3568

Larsen RG and Singel DJ (1993) Double electron–electron resonance spin-echo modulation: Spectroscopic measurement of electron spin pair separations in orientationally disordered solids. J Chem Phys 98: 5134–5146

Manchanda R, Brudvig GW and Crabtree RH (1995) High-valent oxomanganese clusters: Structural and mechanistic work relevant to the oxygen-evolving center in Photosystem II. Coord Chem Rev 144: 1–38

Martin RE, Pannier M, Diederich F, Gramlich V, Hubrich M and Spiess HW (1998) Determination of end-to-end distances in a series of TEMPO diradicals of up to 2.8 nm length with a new four-pulse double electron electron resonance experiment, Angew Chem Int Ed 37: 2834–2837

Messinger J, Robblee JH, Yu WO, Sauer K, Yachandra VK, and Klein MP (1997) The $S_0$ state of the oxygen-evolving complex in Photosystem II is paramagnetic: Detection of EPR multiline signal. J Am Chem. Soc 119: 11349–11350

Milov AD, Salikhov KM and Shirov MD (1981) Application of ELDOR in electron spin echo for paramagnetic center spin distributions in solids. Fiz Tverd Tela 23: 975–982

Milov AD, Ponomarev AB and Tsvetkov YuD (1984) Electron electron-double resonance in electron-spin echo model biradical systems and the sensitized photolysis of decalin. Chem Phys Lett 110: 67–72

Milov AD, Maryasov AG and Tsvetkov YuD (1998) Pulsed electron double resonance (PELDOR) and its applications in free-radicals research. Appl Magn Reson 15: 107–143

Milov AD, Tsvetkov YuD, Formaggio F, Oancea S, Toniolo C and Raap J (2003) Aggregation of spin labeled trichogin GA IV dimers: Distance distribution between spin labels in frozen solutions by PELDOR data J Phys Chem B 107: 13719–13727

Mims WB (1965) Electron echo methods in spin resonance spectrometry. Rev Sci Instr 36: 1472–1479

Mino H, Kawamori A and Ono T-A (2000) Pulsed EPR studies of doublet signal and singlet-like signal in oriented $Ca^{2+}$-depleted PS II membranes: Location of the doublet signal center in PS II. Biochemistry 39: 11034–11040

Nanba O and Satoh K (1987) Isolation of a Photosystem II reaction center consisting of D-1 and D-2 polypeptides and cytochrome *b*-559. Proc Natl Acad Sci USA 84: 109–112

Pannier M, Veit S, Godt A, Jeschke G and Spiess HW (2000) Deadtime free measurement of dipole-dipole interactions between electron spins. J Magn Reson 142: 331–340

Peloquin JM, Campbell KA, Randall DW, Evanchik MA, Pecoraro VL, Armstrong WH and Britt RD (2000) $^{55}$Mn ENDOR of the $S_2$-state multiline EPR signal of Photosystem II: Implications on the structure of the tetranuclear Mn cluster. J Am Chem Soc 122: 10926–10942.

Raitsimring AM, Slikhov KM, Umanskii BA and Tsvetkov YuD (1974) Instantaneous diffusion in the electron spin echo of paramagnetic centers stabilized in a solid host. Sov Phys Solid State 16: 492–497

Raitsimring AM (2000) '2+1' pulse sequence as applied for distance and spatial distribution measurements of paramagnetic centers. In: Berliner LJ, Eaton SS and Eaton GR (eds) Biological magnetic resonance, Vol 19. Distance Measurements in Biological Systems by EPR, pp 461–491. Kluwer Academic Publishers, New York

Raitsimring AM and Salikhov KM (1985) ESE method as used to analyze the spatial distribution of the paramagnetic centers. Bull Magn Reson 7: 184–195

Raitsimring AM and Samoilova RI (1986) Investigation of the pair radical distribution function by ESE method. The case of the long relaxation times. Khim Fizika 5: 1080–1084

Raitsimring AM, Peisach J, Caroline Lee H and Chen X (1992) Measurement of distance distribution between spin labels in spin-labeled hemoglobin. J Phys Chem 96: 3526–3531

Rutherford AW (1985) Orientation of EPR signals arising from components in Photosystem II membranes. Biochim Biophys Acta 807: 189–201

Sale K, Sár C, Sharp KA, Hideg K and Fajer PG (2002) Structural determination of spin label immobilization and orientation: A Monte Carlo minimization approach. J Magn Reson 156: 104–112

Sale K, Song L, Liu Y-S, Perozo E and Fajer PG (2005) Explicit treatment of spin labels in modeling of distance constraints from dipolar EPR and DEER. J Am Chem Soc 127: 9334–9335

Schiemann O, Weber A, Edwards TE, Prisner TF, and Sigurdsson ST (2003) Nanometer distance measurements on RNA using

PELDOR. J Am Chem Soc 125: 3434–3435

Schiemann O, Piton N, Mu Y, Stock G, Engels JW and Prisner TF (2004) A PELDOR-based nanometer distance ruler for oligonucleotides. J Am Chem Soc 126: 5722–5729

Shigemori K, Hara H, Kawamori A and Akabori K (1998) Determination of distances from tyrosine D to $Q_A$ and chlorophyll$_Z$ in Photosystem II studied by '2+1' pulsed EPR. Biochim Biophys Acta 1363: 187–198

Steinhoff HJ, Radzwill N, Thevis W, Lenz V, Brandenburg D, Antson A, Dodson G and Wollmer A (1997) Determination of interspin distances between spin labels attached to insulin: comparison of electron paramagnetic resonance data with the X-ray structure. Biophys J 73: 3287–3298

Tang J, Thurnauer MC and Norris JR (1994) Electron spin echo envelope modulation due to exchange dipolar interactions in a spin-correlated radical pair. Chem Phys Lett 219: 283–290

Tonaka M, Kawamori A, Hara H, and Astashkin AV (2000) Three-dimentional structure of electron transfer components in Photosystem II: '2+1' ESE of chlorophyll Z and tyrosine D, Appl Magn Reson 19: 141–150

Wacker T, Sierra GA and Schweiger A (1992) The concept of FID-detected hole-burning in pulsed EPR spectroscopy. Isr J Chem 32: 305–322

Weber A, Schiemann O, Bode B, Prisner TF (2002) PELDOR at S- and X-band frequencies and the separation of exchange coupling from dipolar coupling. J Magn Reson 157: 277–285

Yamauchi T, Mino H, Matsukawa T, Kawamori A and Ono T-A (1997) Parallel polarization electron paramagnetic resonance studies of the $S_1$-state manganese cluster in the photosynthetic oxygen-evolving system. Biochemistry 36: 7520–7526

Yoshii T, Kawamori A, Tonaka M and Akabori K (1999a) Relative positions of electron transfer components in Photosystem II studied by '2+1' pulsed electron paramagnetic resonance: $Y_D$ and $Q_A$. Biochim Biophys Acta 1413: 43–49

Yoshii T, Hara H, Kawamori A, Akabori K, Iwaki M and Itoh S (1999b) ESEEM study of the location of spin-polarized chlorophyll-quinone radical pair in membrane-oriented spinach Photosystem I and II complexes. Appl Magn Reson 16: 565–580

Zech SG, Kurreck J, Eckert H-J, Renger G, Lubitz W and Bittl R (1997) Pulsed EPR measurement of the distance between P680$^+$ and $Q_A^-$ in Photosystem II. FEBS Lett 414: 454–456

Zech SG, Kurreck J, Renger G, Lubitz W and Bittl R (1999) Determination of the distance between $Y_Z^{OX\bullet}$ and $Q_A^-$ in Photosystem II by pulsed EPR spectroscopy on light-induced radical pairs. FEBS Lett 442: 79–82

Zouni A, Witt H-T, Kern J, Fromme P, Krauß P, Saenger W and Orth P (2001) Crystal structure of Photosystem II from *Synechococcus elongatus* at 3.8 Å resolution. Nature 409: 739–743

# Chapter 17

## Spin Labeling of Photosynthetic Systems

Igor V. Borovykh
*Laboratory of Biophysics, Wageningen University, Dreijenlaan 3, 6703 HA Wageningen, The Netherlands*

Heinz-Jürgen Steinhoff*
*Department of Physics, University of Osnabrück, Barbarastraße 7, 49069 Osnabrück, Germany*



### Summary

Site-directed spin labeling EPR (SDSL-EPR) spectroscopy has become a powerful method for probing the structure and conformational dynamics of proteins. The strength of the method is obvious especially for those proteins which are not amenable to the atomic-resolution techniques as multidimensional NMR spectroscopy or X-ray crystallography. SDSL-EPR can provide information on the secondary, tertiary and quaternary structure of proteins at the level of the backbone fold with the extra advantage that the measurements can be performed under conditions relevant to function of proteins. Conformational changes that may accompany the function of the protein can be followed on a biologically relevant time scale. In addition, the application of SDSL-EPR is not limited by the size of the protein and large proteins and protein complexes can be studied.

In this chapter the application of SDSL-EPR to photosynthesis is described. Because of the fast development of this spectroscopic technique and due to the limited number of applications to photosynthetic systems the basics of SDSL-EPR spectroscopy will be described with references to and examples from the most recent publications where SDSL-EPR has been applied to different biological systems (Sections II–IV). Section V summarizes the application of spin labeling and SDSL-EPR to photosynthesis. The Section starts with a short historical overview and then summarizes the application of SDSL-EPR and molecular dynamics (MD) simula-

*Author for correspondence, email: heinz-juergen.steinhoff@uos.de

*Thijs J. Aartsma and Jörg Matysik (eds): Biophysical Techniques in Photosynthesis II, pp. 345–359.*

tions to probe the local structure and dynamics of the spin labeled photosynthetic reaction center protein from *Rhodobacter sphaeroides*. Information obtained by inter-spin distance measurements shine light on details of the global structure of the reaction center protein and of the N-terminal domain of a major light-harvesting chlorophyll *a/b* protein from higher plants.

## I. Introduction

Electron paramagnetic resonance (EPR) spectroscopy of site-directed spin labeled biomolecules (site-directed spin labeling, SDSL) has become a powerful method for probing the structure and conformational dynamics of water-soluble and membrane proteins of arbitrary molecular weight under conditions relevant to function (Hubbell et al., 1996, 1998, 2000; Feix and Klug, 1998; Steinhoff, 2002). The basic strategy of SDSL involves the substitution of native residues at specific sites within a protein with cysteines, followed by modification of the reactive SH groups with a selective nitroxide reagent, typically methanethiosulfonate spin labels (MTSSL) (Fig. 1). The continuous wave (cw) EPR spectrum yields information about physical characteristics at that site including the mobility of the spin label, its solvent accessibility, the polarity of its immediate environment and the distance between the nitroxide and another paramagnetic center in the protein. A scheme of the information contained in the EPR spectra is shown in Fig. 2. EPR data analysis of a series of spin labeled variants of a given protein thus provides the protein topology and the orientations of individual segments of the protein. A complete analysis allows modeling of protein structures with a spatial resolution at the level of the backbone fold (Hubbell et al., 1998, 2000; Perozo et al., 1998; Koteiche and Mchaourab, 1999; Mchaourab and Perozo, 2000; Wegener et al., 2001a). The method is applicable to any protein that retains its function after spin labeling. Furthermore, the method is sensitive to molecular dynamics: protein equilibrium fluctuations and conformational changes of functional relevance can be followed on a wide time scale ranging from picoseconds to seconds. Several detailed reviews of SDSL methods have been published (Berliner, 1976, 1979, 1998; Berliner and Reuben, 1989; Mchaourab et al., 1996; 2000 Feix and Klug, 1998; Steinhoff et al., 2000b; Steinhoff, 2002; ). The present report reviews applications of spin labeling to photosynthetic systems: the bacterial reaction center (RC) protein of *Rhodobacter* (*Rb.*) *sphaeroides* and the major light-harvesting chlorophyll *a/b* protein (LHCIIb) of higher plants. Due to a limited number of applications of spin labeling in photosynthesis and taking into account its fast development the basic ideas of SDSL will be described with reference to the most recent publications.

*Abbreviations:* BtuB – *Escherichia coli* outer membrane cobalamin transporter; CrOx – chromium oxalate; cw – continuous wave; DEER – double electron-electron resonance; EPR – electron paramagnetic resonance; LHCIIb – light-harvesting chlorophyll *a/b* protein; MD – molecular dynamics; MTSSL – methanethiosulfonate spin label; NiAA – nickel (II) acetylacetonate; *Rb.* – *Rhodobacter;* RC – reaction center; SDSL – site-directed spin labeling; SIFTER – single-frequency techniques for refocusing; Π – accessibility parameter

## II. Spin Label Dynamics and Solvent Accessibility

The relationship between the nitroxide side chain (R1) mobility and the protein secondary and tertiary structure has been extensively reviewed (Berliner, 1976, 1979, 1998; Berliner and Reuben, 1989; Mchaourab et al., 1996). In general the term 'mobility' is agreed to include effects due to the rate, amplitude and anisotropy of the nitroxide reorientational motion. Changes of the mobility are reflected in changes of the EPR spectral shape. In short, the energy of the electron spin transitions and therefore the EPR spectrum is determined both by the g tensor, which

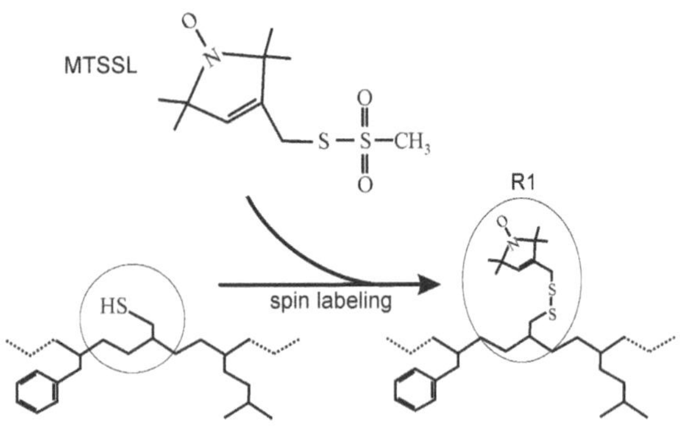


*Fig. 1.* The reaction of the methanethiosulfonate spin label (MTSSL) (Berliner et al., 1982) with a sulfhydryl group generates the spin label side chain R1.

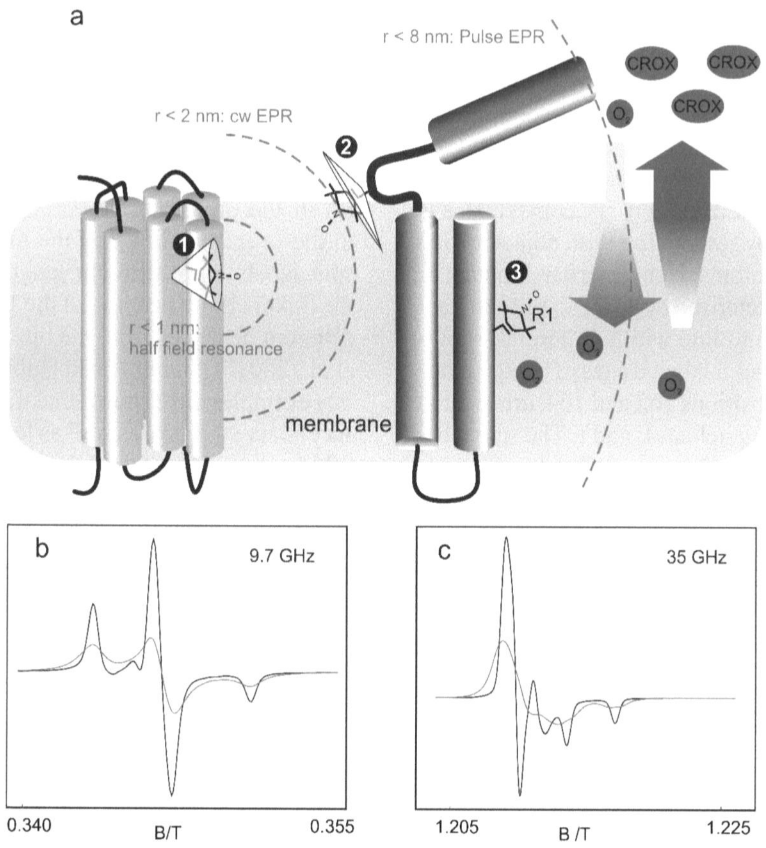


*Fig. 2.* (a) Structural information contained in EPR spectra of spin labeled membrane proteins. Secondary and tertiary structural elements with high (1) or small (2) restriction of the reorientational motion of the nitroxide side chain are reflected in the degree of averaging of the anisotropic components of the hyperfine interaction (see Section II). Dipolar interaction between spins of neighboring nitroxide side chains leads to considerable line broadening of continuous wave (cw) EPR spectra if the distance between the nitroxides is less than 2 nm. Normalized spectra (1st derivative representation) of doubly labeled samples with inter-spin distances of 1.2 nm (grey) and 2 nm (black) are shown for frequencies of 9.7 GHz (b) and 35 GHz (c). For spin-spin distances less than 1 nm half field resonance spectra yield precise distance values, for inter-spin distances between 2 and 8 nm pulse EPR methods have been applied successfully (see Section IV). The collision of paramagnetic ions or molecules of different solubility in the lipid or aqueous phases, as e.g. chromium oxalate or molecular oxygen, modulate the longitudinal relaxation time (3). This can be measured by means of cw EPR saturation methods and enables determination of the position of the nitroxide side chain with respect to the water-lipid interface (see Section II). High-field EPR spectroscopy (95 GHz, 3.4 T) reveals the polarity and proticity of the spin label micro-environment (see Section III).

defines the interaction between the electron and the applied static magnetic field, and by the hyperfine coupling tensor $A$, which defines the interaction between the electron spin and the spin of the nearby nitrogen nucleus. Both of these interactions are anisotropic, thus the transition frequencies depend on the orientation of the nitroxide ring with respect to the external magnetic field. Reorientational motion of the nitroxide leads to averaging of the anisotropic components of both tensors. Hence, the shape of the EPR spectrum varies with the mechanism and rate of spin label reorientation (Colombus et al., 2001; Columbus and Hubbell, 2004). In turn, this reorientational motion is restricted by interaction of the spin label side chain with nearby atoms and thus depends on the location of the spin label within the protein structure, i.e., on the secondary and tertiary structure in the vicinity of the binding site and on the dynamics of the whole protein. Methods for interpreting the R1 mobility in terms of protein structure are still under development. However, it is already well documented that a simple mobility parameter like the line width of the central line or the second moment of the spectra can provide a 'fingerprint' of the tertiary fold of a protein (Mchaourab et al., 1996; Fanucci et al., 2002; Isas et al., 2002; Koteiche et al., 2003; Bordignon et al., 2005; Kim et al., 2005). For example, R1 residues at exposed sites or in loop regions reveal high mobility due to weak interaction with neighboring side chains or backbone atoms. In this case the apparent hyperfine splitting and the line width are small. On the other hand, the motion of R1 residues at buried and tertiary contact sites is strongly restricted by interaction with adjacent side chains or backbone atoms. Hence, tertiary contact or buried sites are characterized by an increased apparent hyperfine splitting and line width. Figure 3b shows an example of three spin labeled sites of bacteriorhodopsin. Two sites, positions 163 and 164, are located in the loop connecting helices E and F. The spin label side chain at position 164 is oriented to the aqueous phase. Nearly unrestricted reorientational motion of the nitroxide is possible within a large motion cone as revealed by molecular dynamics (MD) simulations (Steinhoff et al., 2000b; Beier and Steinhoff, 2006). This motion leads to considerable averaging of the anisotropic contribution of the hyperfine interaction resulting in a small apparent hyperfine splitting. The side chain at position 163 is oriented away from the aqueous phase into the direction of helices B and C. The reorientational motion of the nitroxide is more restricted. Consequently, averaging of the anisotropic hyperfine interaction is less complete. The R1 side chain at position 170 is buried within the protein resulting in complete restriction of its motion. The corresponding EPR spectrum resembles that of a nitroxide powder spectrum. In addition, the dynamics of the protein backbone itself also contributes to the overall mobility of the nitroxide. The use of spin labels with different length and flexibility of the linker allows probing of the local environment at different distances from the binding site (Colombus et al., 2001; Columbus and Hubbell, 2004). Numerous examples have shown that the identity of regular secondary structure as well as features of the tertiary fold can be extracted from the periodic variation of the R1 mobility along a sequence in nitroxide scanning experiment (Hubbell et al., 1998; Fanucci et al., 2002; Isas et al., 2002; Koteiche et al., 2003; Bordignon et al., 2005; Kim et al., 2005).

Determination of the collision rates of nitroxide side chains with diffusible paramagnetic agents such as molecular oxygen $O_2$ and water soluble paramagnetic Ni(II) complexes or chromium oxalate (CrOx) is a complementary approach to map the protein topology (Fig. 2a). An accessibility parameter (Π),

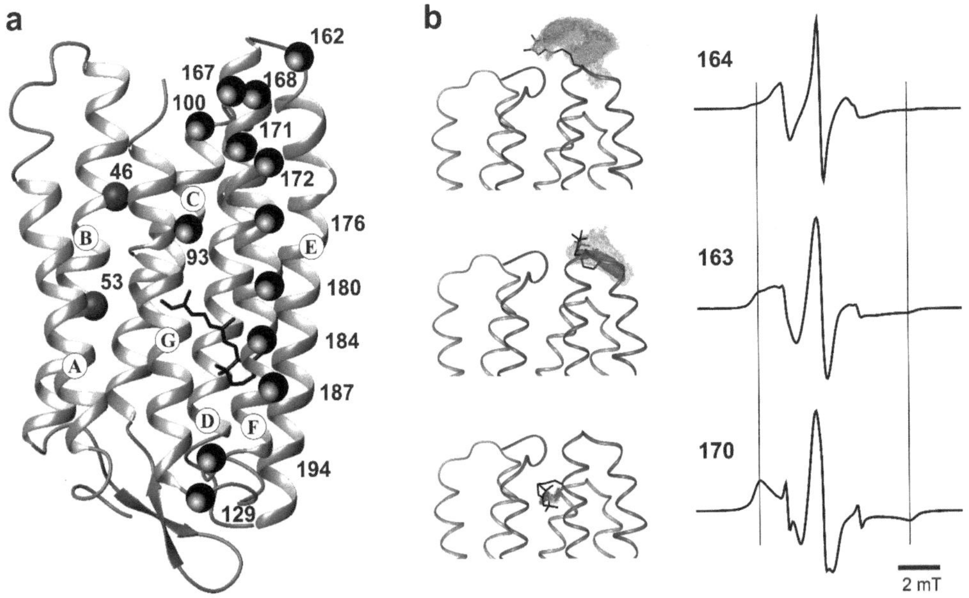


*Fig. 3.* (a) Structure of bacteriorhodopsin. A selection of spin labeled positions discussed in this text is highlighted. (b) Motional freedom of the spin label side chains at different positions in the E-F loop and in helix F of bacteriorhodopsin. (Only the upper section of (a) is depicted in each case.) The dotted area represents the accessible space of the nitroxide covered within a high temperature (600 K) molecular dynamics simulation of 6 ns length (Steinhoff et al., 2000b; Beier and Steinhoff, 2006). The different degree of mobility of the nitroxides is reflected in the variation of the apparent hyperfine splitting of the corresponding experimental spectra (T = 293K) (modified from Beier and Steinhoff, 2006).

proportional to the collision rate, is a direct measure of the nitroxide accessibility from either the water or lipid bilayer phase. Accessibility mapping allows identification of the side chain orientations with respect to the protein-water or protein-lipid interface. Polar metal complexes preferentially partition into the aqueous phase, whereas apolar oxygen exhibits a maximum value of the product of concentration and diffusion coefficient in the center of the membrane bilayer (Altenbach et al., 1994, 2005; Pyka et al., 2005). Continuous wave power saturation has been shown to provide an easy and reliable means for the quantification of the collision frequencies (Altenbach et al., 1994). The pattern of accessibility to $O_2$ or CrOx in a scanned sequence generally mirrors that of the side-chain mobility (see Fig. 4). The combination of mobility and accessibility studies allows determination of the secondary structure of the protein in different environment for example in soluble conformation or reconstituted into a membrane (Isas et al., 2003; Buchaklian and Klug, 2005; Nelson et al., 2005; Oh et al., 2005). As an example, a study of the structure and dynamics of the β-barrel of the transmembrane cobalamin transporter BtuB is shown in Fig. 4 (Fanucci et al., 2002). This protein is involved in an active transport of vitamin $B_{12}$ in the outer membrane of gram-negative bacteria. Figure 4 shows the behavior of the nitroxide side chain mobility (a) and accessibility for molecular oxygen and nickel (II) acetylacetonate (NiAA) (b) obtained by the analysis of EPR spectra corresponding to spin labeled single cysteine substitutions scanned throughout two putative β-strands of BtuB (Fanucci et al., 2002). Both parameters, reciprocal second moments (Fig. 4a) and accessibility Π, (Fig. 4b) show an $i$, $i+2$ periodicity. This periodicity is maintained throughout the scanned regions and indicates that these segments have a β-strand structure in and beyond the lipid bilayer. In addition, when plotted as function of residue number, $\Pi(O_2)$ and Π(NiAA) are typically in-phase when the corresponding parts of the protein are located in the aqueous phase. This is visible at the ends of both putative strand regions scanned (for example, positions 148–151). On the contrary, residues that are located within the membrane domain usually have Π values for $O_2$ and NiAA that are out of phase. The out of phase pattern is due to differences in the $O_2$ and NiAA concentrations in the hydrocarbon core compared to the aqueous phase: the concentration of oxygen increases but the concentration of NiAA decreases when approaching the center of the membrane. Such an out of phase behavior of accessibilities is observed for residues in the central region of the strand, for example for positions 156–160, Fig. 4b (Fanucci et al., 2002).

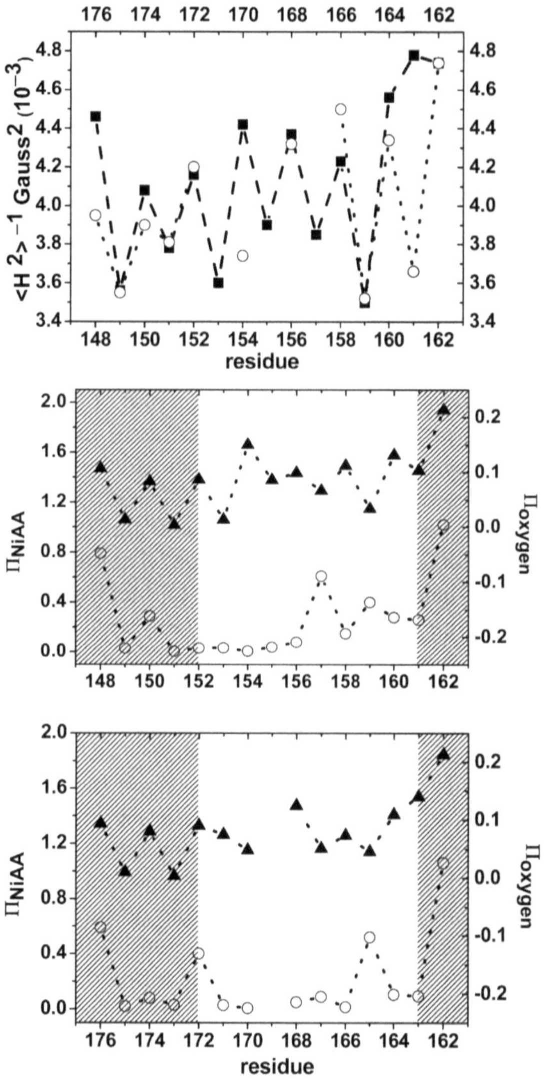


*Fig. 4.* (a) Plot of the inverse second moment of each of the EPR spectra measured (adapted from Fanucci et al., 2002, with permission). Residues 176–162 (○) and residues 148–162 (■) are plotted (left to right) from the extracellular to periplasmic side of the protein. (b) Collision accessibilities (Π) for oxygen (▲) and NiAA (○) for two putative β-strands in BtuB determined using air (20% oxygen) and 20 mM NiAA, respectively, with a nitrogen gas purge as background. The shaded regions correspond to those residues that are in the aqueous phase. A β-strand ($i$, $i$ + 2) periodicity is clearly revealed in these data. The values of Π(NiAA) and Π(oxygen) are in phase for portions of the protein exposed to the aqueous phase but are out of phase for protein within the membrane domain (Fanucci et al., 2002).

## III. Polarity of the Spin Label Micro-Environment

Two important parameters that can be deduced from the analysis of EPR spectra measured at temperatures below 200 K are the polarity and proticity of the local environment of spin labels (Wegener et al., 2001b; Savitsky et al., 2004). A polar environment shifts the tensor component $g_{xx}$ of a nitroxide to smaller values whereas the hyperfine tensor component $A_{zz}$ is increased (Stone, 1963; Plato et al., 2002) (Fig. 5). Hence, both tensor components can be regarded as polarity parameters. In addition, a plot of $g_{xx}$ vs. $A_{zz}$ reveals protic and aprotic properties of the micro-environment due to the different sensitivities of these tensor components towards the influence of hydrogen bonding to the NO group (Möbius et al., 2005; Plato et al., 2002; Steinhoff et al., 2000a). The application of high-field EPR techniques with Larmor frequencies exceeding 90 GHz has considerably enhanced the Zeeman resolution of rigid-limit spectra of disordered spin labeled samples and allows determination of the g-tensor and therefore of the polarity parameter with high accuracy (Möbius et al., 2005).

Details of the polarity profile along the bacteriorhodopsin proton channel were probed by g- and hyperfine tensor components from a series of site-directed spin-labeled bacteriorhodopsin mutants with spin labels located along the proton channel (Steinhoff et al., 2000a; Möbius et al., 2005). At 95 GHz the variation of $g_{xx}$ with the nitroxide binding site is revealed by the shift of the position of the low-field maximum of the EPR spectra. The plot of $g_{xx}$ vs. nitroxide position along the proton channel reveals distinct variations in the polarity of the nitroxide micro-environment. The high polarity in the environment of residues at the cytoplasmic and extracellular surfaces is clear evidence that these nitroxides are accessible to water. The environmental polarity reaches its minimum close to position 46 between the proton donor D96 and the retinal, in agreement with the protein structure. The $g_{xx}$ behavior provides quantitative information about the hydrophobic barrier the proton has to overcome on its way through the protein. In addition, the polarity variation along the lipid-exposed surface of transmembrane protein elements can be used to locate the spin labeled sites with respect to the lipid-water interface. These data provide information on the topology of membrane protein elements, their orientation with respect to the lipid phase (Steinhoff et al., 1999; Kveder et al., 2003) and possible water penetration along the protein-lipid interface. As an example, the hyperfine splitting components $A_{zz}$ of spin labels attached to helix F of bacteriorhodopsin are plotted as a function of their location with respect to the cytoplasmic water-lipid interface (Fig. 6) (Wegener, 2002). The corresponding values were determined by fitting of simulated to experimental spectra recorded at 170 K. All spin label side chains are either oriented towards the lipid or towards the aqueous phase. The distances were measured from the projections of the $C_\beta$ positions of the respective residues to a line connecting residues R164, the position of which is arbitrarily set to 0, and G73 on the extracellular surface of the protein. An error bar of 0.7 nm length indicates the uncertainty of the location of the N-O group in the polarity plot. The largest hyperfine splitting is revealed for positions 162, 168 and 129. The determined $A_{zz}$

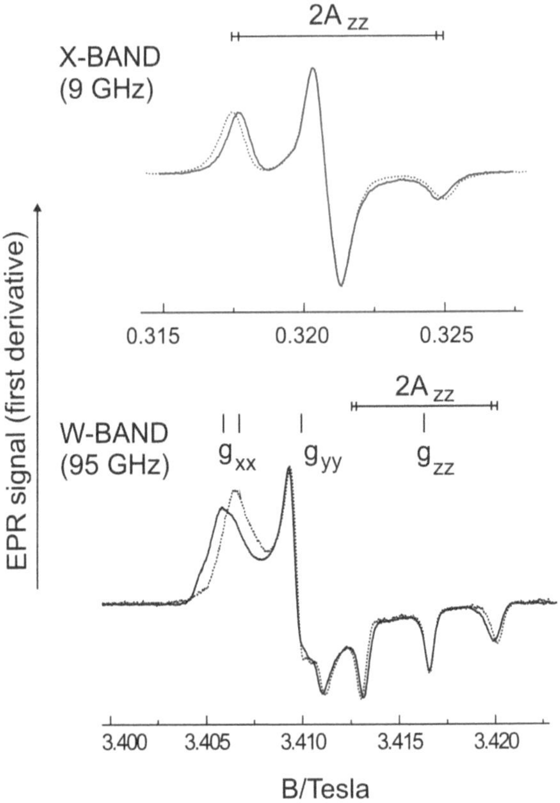


*Fig. 5.* EPR spectra of MTSSL in micro-environments of low (continuous line) and high (dotted line) polarity: *Top:* X-band spectra of bacteriorhodopsin variants L93R1 (protein interior, continuous line) and K129R1 (extracellular surface, dotted line) (Steinhoff et al., 1999). *Bottom:* W-band spectra of bacteriorhodopsin variants T46R1 (protein interior, continuous line) and M163R1 (cytoplasmic surface, dotted line) (Steinhoff et al., 2000a).

values are similar to that determined for unbound MTSSL in aqueous solution, which was found to be 36.5 G. According to the molecular model of bacteriorhodopsin residue 162 is located in the E-F loop at the cytoplasmic surface of bacteriorhodopsin. The experimental data prove that this nitroxide is exposed to the aqueous phase. The nitroxide side chain at position 168 reveals a hyperfine splitting which is even slightly higher than that of 162. The nitroxide moiety of this spin label side chain is located in or close to the polar headgroup region of the lipid bilayer. The environmental polarity of the nitroxides at position 172 is significantly less. The capability of forming hydrogen bonds between water molecules and the N-O group is restricted for the nitroxide in this location, which is still close to the water covered protein and membrane surfaces. The polarity decreases along the lipid-protein interface and exhibits a minimum close to positions 180–187. On the extracellular side represented by position 129 the hyperfine splitting component $A_{zz}$ reaches a value similar to that of position 162, which proves the accessibility of this site to bulk water. The polarity profile registered by the spin-label hyperfine splitting constants can, at least in part, be related to water penetration into the lipid membranes (Marsh, 2001). It thus provides means for characterization of the protein-lipid interface. For transmembrane proteins of unknown topology the inspection of the apparent hyperfine splitting of spin label side chains located in the protein-lipid interface allow location of these protein segments with respect to the membrane boundaries.

One of the attractive features of SDSL EPR is the possibility to follow changes in the parameters discussed above. In order to demonstrate this, polarity changes at spin labeled sites of colicin A, a bacterial toxin (Lakey et al., 1994; Stroud et al., 1998), were followed upon adsorption to lipid bilayers (Savitsky et al., 2004; Möbius et al., 2005). A large change of the polarity in the vicinity of position 176 in the hydrophobic core of the protein revealed a conformational change upon interaction with DMPG vesicles and confirmed the penknife model for the membrane associated structure of the colicin A channel forming domain (Savitsky et al., 2004; Möbius et al., 2005).

## IV. Inter-Spin Distance Measurements

The measurement of distances between two site-specific probes in biomolecules enables to determine or constrain structures, particularly for large or dynamic complex systems (Hustedt and Beth, 1999; Eaton et al., 2000). The spin-spin interaction between two spin labels attached to a protein is composed of static dipolar interaction, modulation of the dipolar interaction by the residual motion of the spin label side chains and exchange interaction. The static dipolar interaction leads to considerable broadening of the cw EPR spectrum and distances up to 2.0–2.5 nm can be measured by this technique (Fig. 2). Distances and distance distributions can be extracted from cw-EPR spectra of dipolar coupled nitroxides by a variety of approaches (Fig. 2). In the case two nitroxides adopt unique orientations relative to each other as found for spin labels introduced at buried sites in the rigid limit a rigorous solution of the spin Hamiltonian of the system can be obtained. In this case line shape simulations yield the distance between the nitroxides and the Euler angles describing their relative orientation and that of the inter-spin vector relative to the magnetic field (Hustedt et al., 1997; Hustedt and Beth, 1999). In other cases, distances may be estimated by the analysis of line height ratios (Kokorin et al., 1972), by determination of the homogeneous line broadening (Mchaourab et al., 1997), or by Fourier deconvolution/convolution methods (Rabenstein and Shin, 1995; Steinhoff et al., 1997; Altenbach et al., 2001). These methods have been applied and compared experimentally (Persson et al., 2001). Metal ion-nitroxide interactions in metallo-proteins or engineered copper-ion-binding sites allow estimation of intramolecular distances

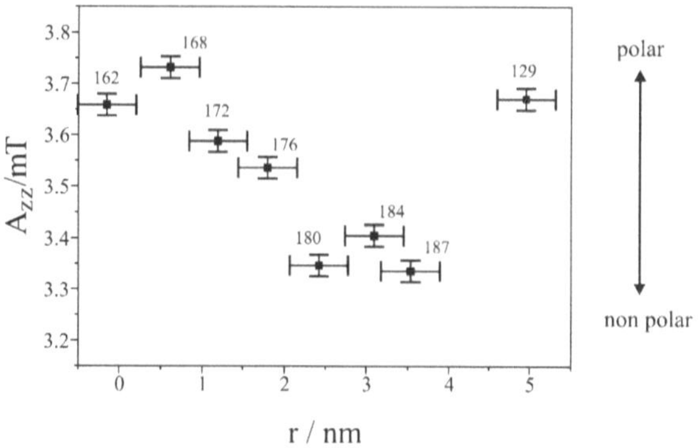


*Fig. 6.* The hyperfine interaction component $A_{zz}$ as a measure of the environmental polarity is depicted as a function of the nitroxide side chain location for a series of spin labeled bacteriorhodopsin mutants (Fig. 3). All nitroxide side chains are oriented either to the aqueous or to the lipid phase. The lipid-water interface is close to position 168 for the cytoplasmic side and close to position 129 for the extracellular side (data taken from Wegener, 2002; Savitsky et al., 2004).

also at room temperature (Leigh, 1970; Voss et al., 1995b). A single metal ion provides a reference site for the estimation of distances to multiple nitroxide sites (Voss et al., 1995a). The uncertainty in the precision of the distances depends upon the method used but frequently is stated to be of the order 0.1–0.2 nm. However, a major uncertainty in the interpretation is the appearance of distance distributions due to the flexibility of the spin label side chain (Hustedt et al., 2006).

For inter-spin distances exceeding 2 nm the line broadening due to dipolar interaction is much less than the influence of other homogeneous and inhomogeneous contributions. For measuring such distances the pulse EPR techniques are applied. These techniques include four-pulse double electron-electron resonance (DEER) (Pannier et al., 2000), the 2+1 pulse sequence (Kurshev et al., 1989), multiple-quantum EPR (Borbat and Freed, 1999), and single-frequency techniques for refocusing (SIFTER) electron-electron couplings (Jung et al., 1995; Jeschke et al., 2000) (for review, see Chapter 14, Möbius and Goldfarb).

The method of inter-spin distance determination has been successfully applied to a number of proteins. Most recent results include study of the structure of the sensory rhodopsin–transducer complex reconstituted in purple membrane lipids, the HAMP domain of the sensory rhodopsin II/transducer complex (Bordignon et al., 2005), the solution structure of the cytoplasmic domain of the anion exchange protein (cdb3) (Zhou et al., 2005) and the structure of the $Na^+$/proline transporter of *Escherichia coli* (Jeschke et al., 2004). Changes in the protein secondary structure, protein tertiary fold or domain movements could be followed with up to 0.1 ms resolution with conventional EPR instrumentation and detection schemes (field modulation) (Steinhoff et al., 1994; Hubbell et al., 1996; Rink et al., 2000; Xiao et al., 2000; Radzwill et al., 2001; Wegener et al., 2001a).

## V. Application of Spin Labeling to Photosynthetic Systems

Despite of the fact that the method of spin labeling has been applied to a broad variety of biological systems, we are aware of only a few publications where this method has been applied to study the RC protein (Solov`ev and Erokhin, 1985; Poluektov et al., 2003; Borovykh et al., 2006; Gajula et al., 2007) and the LHCIIb complex of plants (Jeschke et al., 2005). Due to this limited number of publications, in this chapter we will also present part of our unpublished result on spin labeled RC.

### A. Historical Overview

First applications of spin labeling to the study of photosynthetic systems goes back to 1974. Maruyama and Ohnishi (1974a,b) investigated the photosynthetic bacterium *Rhodospirillum rubrum* to obtain structural insight into membrane systems involved in light-energy conversions. Using hydrophobic spin labels, these authors made first attempts to spin label *Rhodospirillum rubrum* cells by biosynthetic incorporation and to study spin labeled chromatophores in vitro. In those studies, spin labeled stearic acids and lecithin bind to chromatophores. The EPR spectra consist of two components, one corresponding to spin labels located in the lipid bilayer portion of the membrane and the other due to spin labels directly or indirectly associated with membrane proteins. These labels were reduced by the nitroxide reductase system which exists in the cytoplasm. The reduced labels were rapidly photooxidized by chromatophores. The photooxidation and the dark-reduction were suggested to be coupled to the electron transport system of the chromatophores (Maruyama and Ohnishi, 1974b). In addition, it was suggested that the nitroxides influence the photophosphorylation and photoreduction activities of chromatophores (Maruyama and Ohnishi, 1974a). Subchinski et al. (1975a) investigated the photosynthetic redox conversions of various paramagnetic probes during their interaction with chloroplasts. These authors demonstrated that the kinetics of the photosynthetic redox conversions depends on the structure of the probe and on its location within the photosynthetic apparatus, as well as on the intensity and spectral composition of light. In addition, it was demonstrated that probes that bind to the membrane are indicators of conformational reorganization of the membrane under various physico-chemical influences, such as pH, temperature, presence of ATP and $MgCl_2$. It was demonstrated, e.g., that the rigidity of chloroplast membranes increases in the presence of $MgCl_2$ (Subchinski et al., 1975b). On the other hand, the presence of atebrin and $NH_4Cl$ increased the mobility of spin labels in the membranes (Subchinski et al., 1975b). Also, the presence of KCN and the availability of $K_3Fe(CN)_6$ facilitated the penetration of probes into the hydrophobic region of chloroplast membranes.

The effect of light-induced restriction of the spin label motion in the lumen of spinach thylakoids was studied by Nesbitt and Berg (1980). The observed effect was attributed to the presence of protons in the aqueous lumen of the thylakoids. Transfer of chlorophyll derivatives from one side of the bilayer to the other was studied by Berrell et al. (1980) using spin labels attached to porphyrin rings.

Giangrande and Kevan (1981) probed the transfer of electrons from the surface of the reaction center protein of *Rb. sphaeroides* R26 during the electron transfer cycle. Using different length *n*-doxylstearic acids these authors observed different electron transfer efficiencies from electron transfer components of RC to the spin label. Based on these results the position of the iron-quinone acceptor complex in the protein structure was predicted. This prediction placed the iron-quinone complex about 1.4 nm away from the protein surface. The result correlates well with that obtained later by X-ray analysis (Stowell et al., 1997).

To a large extent, the studies discussed above deal with processes taking place in photosynthetic membranes but are not directly related to the structure and function of proteins involved in photosynthetic energy conversion. First studies with the application of spin labels in this direction appeared in 1984. Solov`ev and Erokhin (1985) applied a mercurial spin label to probe the structure of the RC protein from *Rb. sphaeroides* R26. According to their results all 5 cysteine residues of the RC protein were accessible for this spin label. Using the same approach, Kochetkov et al. (1984) studied chromatophores and reaction centers from photosynthetic nonsulfur bacteria *Rhodospirillum rubrum*. First evidence for a correlation of the kinetics of direct electron transfer in the system of quinone acceptors of the RC with the mobility of spin label side chains was presented.

### *B. Probing Local Structure and Dynamics*

Recently, Poluektov et al. (2003) continued the study of photosynthetic RCs using spin labels. In this work, MTSSL was covalently attached to the RC protein and EPR spectra were measured as a function of temperature. The analysis of EPR spectra recorded at different frequencies suggested a restricted reorientational motion of the spin label side chain. By using the model of fast side chain librations in a cone with correlation times less than $10^{-9}$ s, the dependence of the cone angle $\theta$, on temperature was obtained. The angle $\theta$ was found to increase from only a few degrees up to about 25 degrees when the temperature was raised from 50 to 300 K. It was suggested that the spin label was attached to cysteine residue 156 of subunit H.

Gajula et al. (2006) combined EPR spectroscopy and molecular dynamics (MD) simulations. Comparison of spin labeled intact RC molecules with those that lack subunit H showed that the spin label was attached to one of the two cysteines of subunit H (Cys 156 or 234). MD simulations with spin labels attached to both cysteines were performed. Figure 7 shows the subset used for MD simulations. The structural model is based on chains L and H of 1AIJ.pdb. The space covered by the nitroxide of spin label C234R1 during a 6 ns time period of the MD simulation reveals considerable motional restriction. This is even more pronounced in the presentation of the Euler angle population distributions (Fig. 8a and b). The three Euler angles, $\alpha$, $\beta$ and $\gamma$, define the orientation of the nitroxide group with respect to the protein fixed reference frame. The population density in Euler space for C234R1 shows a single narrow peak (Fig. 8a). Consequently the EPR spectrum calculated for an isotropic distribution of protein orientations and with the potential determined from the shown orientation

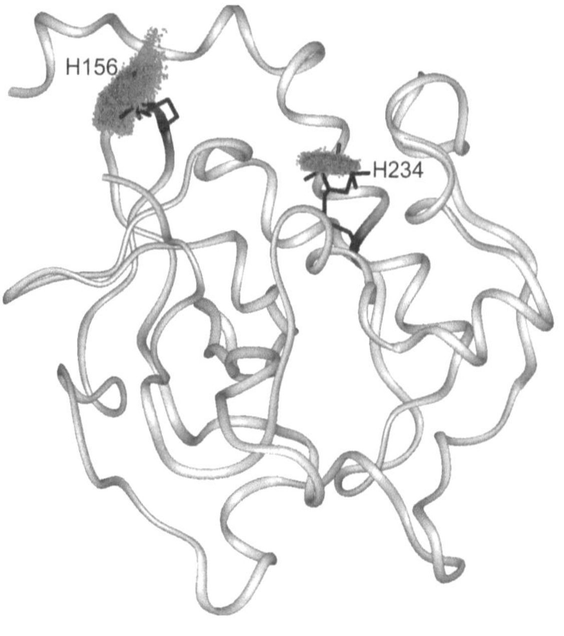


*Fig. 7.* Subset of the RC with spin labels attached to positions 156 and 234. Both spin labels, C156R1 and C234R1, are shown in bond presentation. The dotted areas define the space the nitroxides cover during a molecular dynamics run of 6 ns duration.

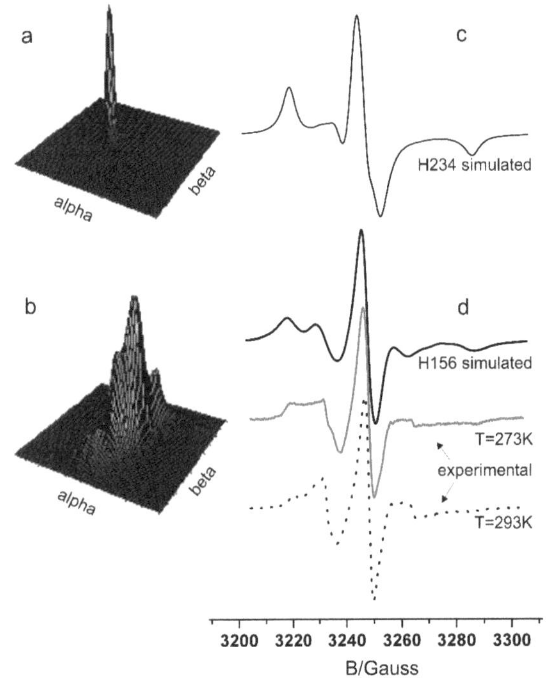


*Fig. 8.* Orientation distribution (a, b) and calculated EPR spectra (c, d) for spin labels attached to position 234, C234R1 (a, c), and 156, C156R1 (b, d) (Gajula et al., 2007). The population distributions projected onto the alpha-beta Euler angle plane reveal the different restrictions of the reorientational dynamics of the nitroxides in the two sites. This different nitroxide dynamics leads to distinguishable degrees of motional averaging of the g- and hyperfine tensor anisotropies and is reflected in the EPR spectral shapes. For comparison two experimental spectra of spin labeled RC protein from *Rb. sphaeroides* measured at 273 K and 293 K are also shown (d).

distribution of the nitroxide coincides with that of a powder spectrum (Fig. 8c). In contrast, the motion cone covered by the nitroxide of the spin label side chain C156R1 reveals less restricted dynamics. The population density shows a broad distribution of the angle β with distinct population maxima (Fig. 8b). The reorientation of the nitroxide within this limited cone leads to partial averaging of the anisotropic components of the g- and hyperfine tensors. The EPR spectrum calculated with the potential determined from the shown angle distribution consequently reveals inward shifts of the outer hyperfine maxima. As a result, the MD simulations and EPR spectra calculations show that the spin label side chain at position 234 is immobile (Fig. 8c) whereas that of position 156 is more mobile. The calculated spectrum of C156R1 (Fig. 8d) seems to consist of two components quite similar to the behavior of the experimental spectrum (Fig. 8d). The apparent hyperfine splitting of C156R1 is identical to that of the experimental spectrum, whereas the calculated spectrum of C234R1 resembles that of a powder spectrum. Based on this, Gajula et al. (2006) concluded that cysteine 156 of subunit H was labeled, whereas position 234 was not accessible for MTSSL. A deeper inspection of the simulated spectrum of C156R1 and comparison with the experimental data reveal differences of the spectral shapes which are due to a slightly larger restriction of the nitroxide motion in the simulations. One explanation could be that loops located near position 156 were restrained during the MD simulations but are flexible in the experiment.

### *C. Inter-Spin Distance Determination by Two-Frequency Pulsed EPR and Molecular Dynamics Simulations*

Recently, Jeschke et al. (2005) obtained information on unknown structural details of the N-terminal domain of a major light-harvesting chlorophyll *a/b* protein (LHCIIb) from higher plants using the combination of EPR distance measurements and triangulation. The N-terminal domain is involved in regulatory functions such as balancing the energy flow through photosystems I and II. The conformational distribution of the N-terminal domain of LHCIIb was characterized by DEER spectroscopy yielding distances between spin labels placed in various parts of the protein. Distance distributions involving residue 3 near the end of the N terminus, which is not resolved in the crystal structures, turned out to be bimodal. Hence, this domain exists in at least two conformational states. Only models with residue 3 located above the core of the protein and extending into the aqueous phase on the stromal side fit the data. On the other hand, in monomers the state with the N-terminal domain extending sideways from the protein core is populated. The authors suggested that these two conformational states may correspond to two functional states of LHCIIb, namely trimeric LHCIIb associated with PS II in stacked thylakoid membranes, and monomeric LHCIIb associated with photosystems I in nonstacked thylakoids. The switch between these two conformations is known to be triggered by phosphorylation of Thr-6. A similar phosphorylation-induced conformational change of the N-terminal domain has been observed by others in bovine annexin IV which, due to the conformational switch, also loses its membrane-aggregating property.

Borovykh et al. (2006) applied DEER in combina-

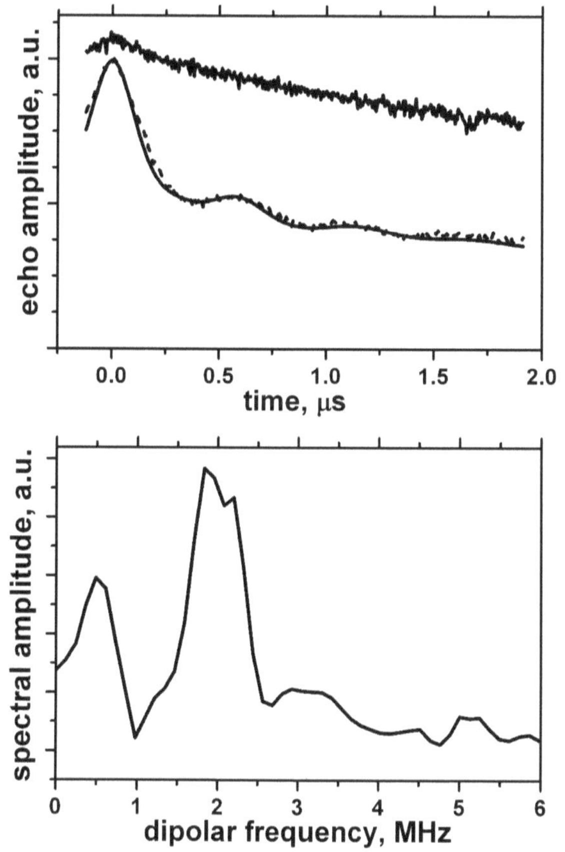


*Fig. 9.* Results of DEER experiments on spin labeled Zn-RC's: (a) DEER time traces of dark (top) and light (bottom) samples: dotted line-experimental trace and thin solid line-fit using Gaussian distance distribution. (b) Fourier transform of DEER time trace of light sample (data taken from Borovykh et al., 2006).

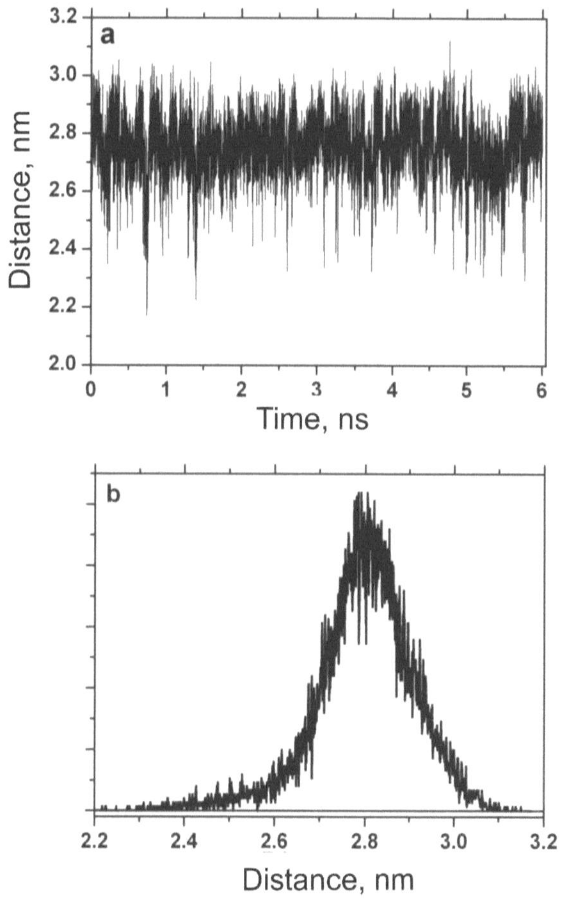


*Fig. 10.* Distance between O-5 of $Q_A$ and the center of the NO bond of the spin label over the length of the MD run (a) and (b) distance distribution of (a) (data taken from Borovykh et al., 2006).

tion with MD simulations to determine the distance between the native cofactor, $Q_A$, and a spin label covalently attached to cysteine residue 156 of subunit H in the Zn-reconstituted reaction center (Zn-RC) of *Rb. sphaeroides* R26. Fig. 9a shows the DEER time traces of samples frozen in the light in the presence of sodium ascorbate (with $Q_A^-$ radical) and in the dark (no $Q_A^-$). Whereas the time trace in the dark frozen sample reveals no distinct modulation, the trace of the light exposed sample shows a pronounced modulation. Since only spin-spin interactions give rise to modulation in the DEER experiment, the signal reveals the interaction between the two paramagnetic centers, the spin label and $Q_A^-$. By Fourier transformation of the time trace the frequency of the modulation (Fig. 9b), which is related to the distance, can be obtained. It shows a peak at 2 MHz, which corresponds to a distance of 3.0 nm. Analysis of the time traces using Gaussian distance distributions (Jeschke et al., 2002) yield an average distance of 3.05 nm and a distribution width (full-width at half-height) of 0.24 nm. The absence of modulation in the dark sample confirms that the distance of 3.0 nm is due to the interaction of the spin label with the light-induced $Q_A^-$ radical and not with spin labels of neighboring RC proteins or a second attached spin label side chain.

The results of MD simulations are shown in Fig. 10. The distances between the geometrical center of the N- and O-atoms of the nitroxide and the O-5 atom of the $Q_A$ are shown (Fig. 10a). The distance distribution for the distance from O-5 of the $Q_A$-ring to the center of the N-O-bond in the MD simulations reveals an approximately Gaussian shape with an average distance of 2.8 nm and distribution width of 0.17 nm (Fig. 10b). The small width of the distribution derives from the fact that changes of the linker conformation are restricted by interaction of the nitroxide with a protein loop. The average distance of 2.8 nm from the atom O-5 of $Q_A$ to the center of the NO-group of the spin label in the MD simulations is significantly smaller than the distance obtained from experiment. The disagreement must be even larger, since the other centers of spin density on $Q_A^-$ are closer to the nitroxide than O-5. For example, the smallest distance, 2.45 nm, is found for O-2, which certainly also carries a considerable spin density. A quantitative analysis of this factor will only be possible once the assignment of spin densities over $Q_A^-$ is known. Two main factors that could account for the discrepancy between the distance obtained by MD simulations and by DEER experiment are (i) a possible difference between the structure of the RC protein, in particular of subunit H, in the crystal and in frozen solution, in particular in the presence of glycerol, or (ii) a difference in the structure and position of the H-subunit due to the native Fe(II) being replaced by Zn(II) and the absence of $Q_B$ (Borovykh et al., 2006). With the approach described above it should be possible to study the effect of both factors on the structure of the RC protein and provide new information about the RC protein function. Such studies are important due to the fact that in the last decade evidence start to accumulate that the RC structure should not be considered as a fixed but rather as a flexible structure that depends on many external factors (i.e., temperature, type of detergent used for RC solubilization, illumination, pH and cryosolvent) (McMahon et al., 1998; Paschenko et al., 2001; Palazzo et al., 2002) and on the functional state of the RC (Kleinfeld et al., 1984).

The results summarized above show that EPR spectroscopy of site-directed spin labeled photosynthetic proteins in combination with MD simulation will provide a powerful method for the future study of photosynthesis — the biological process of the conversion of sunlight energy into chemical free energy, which is utilized by the organisms for synthesis, growth and replication. This method provides new insight into the structure and dynamics of photosynthetic proteins and is extremely sensitive to structural re-arrangements of the protein structure induced by various factors. Correlation of protein dynamics related to various photosynthetic events, in particular, to electron transfer in RC protein that forms the heart of photosynthetic machinery, will be a next step in the understanding of the function of this energy-transducing system.

## Acknowledgments

We gratefully acknowledge the support of the Deutsche Forschungsgemeinschaft (SFB431-P18) and of the Volkswagenstiftung (Az I/78 668).

## References

Altenbach C, Greenhalgh DA, Khorana HG and Hubbell WL (1994) A collision gradient method to determine the immersion depth of nitroxides in lipid bilayers: Application to spin-labeled mutants of bacteriorhodopsin. Proc Natl Acad Sci USA 91: 1667–1671

Altenbach C, Oh KJ, Trabanino RJ, Hideg K and Hubbell WL (2001) Estimation of inter-residue distances in spin labeled proteins at physiological temperatures: Experimental strategies and practical limitations. Biochemistry 40: 15471–15482

Altenbach C, Froncisz W, Hemker R, Mchaourab H and Hubbell WL (2005) Accessibility of nitroxide side chains: Absolute Heisenberg exchange rates from power saturation. Biophys J 89: 2103–2112

Beier C and Steinhoff, HJ (2006) A structure-based simulation approach for electron paramagnetic resonance spectra using molecular and stochastic dynamics simulations. Biophys J 91: 2647–2664

Berliner LJ (1976) Spin Labeling: Theory and Applications. Academic Press, New York

Berliner LJ (1979) Spin Labeling II: Theory and Applications. Academic Press, New York

Berliner LJ (1998) Spin Labeling: The Next Millenium. Academic Press, New York

Berliner LJ and Reuben J (1989) Biological Magnetic Resonance. Vol. VIII: Spin Labeling Theory and Applications. Plenum Press

Berliner LJ, Grunwald J, Hankovszky HO and Hideg K (1982) A novel reversible thiol-specific spin label: Papain active site labeling and inhibition. Anal Biochem 119: 450–455

Berrell GB, Boyd SA, Keana JFW and Hayes O (1980) Transverse motion of chlorophyll derivatives in phospholipid bilayers. Biochim Biophys Acta 603: 213–219

Borbat PP and Freed JH (1999) Multi-quantum ESR and distance measurements. Chem Phys Lett 313: 145–154

Bordignon E, Klare JP, Doebber M, Wegener AA, Martell S, Engelhard M and Steinhoff H-J (2005) Structural analysis of a HAMP domain: The linker region of the phototransducer in complex with sensory rhodopsin II. J Biol Chem 280: 38767–38775

Borovykh IV, Ceola S, Gajula P, Gast P, Steinhoff H-J and Huber M (2006) Distance between a native cofactor and a spin label in the reaction center of *Rb. sphaeroides* by a two-frequency pulsed electron paramagnetic resonance and molecular dynamics simulations. J Magn Reson 180: 178–185

Buchaklian AH and Klug CS (2005) Characterization of the walker A motif of MsbA using site-directed spin labeling electron paramagnetic resonance spectroscopy. Biochemistry 44: 5503–5509

Columbus L and Hubbell WL (2004) Mapping backbone dynamics in solution with side-directed spin labeling: GCN4–58 bZip free and bound to DNA. Biochemistry 43: 7273–7287

Colombus L, Kálai T, Jekö J, Hideg K and Hubbell WL (2001) Molecular motion of spin labeled side chains in α-helices: Analysis by variation of side chain structure. Biochemistry 40: 3828–3846

Eaton GR, Eaton SS, and Berliner LJ (2000) Distance Measurements in Biological Systems. Kluwer Academic Publishers, New York

Fanucci GE, Ladieux N, Piedmont CA, Kadner RJ and Cafiso OS (2002) Structure and dynamics of the β-barrel of the membrane transporter BtuB by site-directed spin labeling. Biochemistry 41: 11543–11551

Feix JB and Klug CS (1998) Site-directed spin labeling of membrane proteins and peptide-membrane interactions. In: Berliner LJ (ed) Spin Labeling: The Next Millennium, pp 251–281. Plenum Press, New York

Gajula P, Borovykh IV, Beier C, Shkuropatova T, Gast P and Steinhoff H-J (2007) Spin-labeled photosynthetic reaction centres from *Rhodobacter sphaeroides* studied by electron paramagnetic resonance spectroscopy and molecular dynamics simulations. Appl Magn Reson 31: 167–178

Giangrande M and Kevan L (1981) On the spatial correlation between the protein surface and the electron transfer cycle in bacterial photosynthetic reaction centers. Photochem Photobiol 33: 721–726

Hubbell WL, McHaourab HS, Altenbach C and Lietzow MA (1996) Watching proteins move using site-directed spin labeling. Structure 4: 779–783

Hubbell WL, Gross A, Langen R and Lietzow MA (1998) Recent advances in site-directed spin labeling of proteins. Curr Opin Struct Biol 8: 649–656

Hubbell WL, Cafiso DS and Altenbach C (2000) Identifying conformational changes with site-directed spin labeling. Nature Struct Biol 7: 735–739

Hustedt EJ and Beth AH (1999) Nitroxide spin-spin interactions: Applications to protein structure and dynamics. Annu Rev Biophys Biomol Struct 28: 129–153

Hustedt EJ, Smirnov AI, Laub CF, Cobb CE and Beth AH (1997) Molecular distances from dipolar coupled spin-labels: The global analysis of multifrequency continuous wave electron paramagnetic resonance data. Biophys J 72: 1861–1877

Hustedt EJ, Stein RA, Sethaphog L, Brandon S, Zhou Z and DeSensi SD (2006) Dipolar coupling between nitroxide spin labels: The development and application of a tether-in-a cone model. Biophys J 90: 340–356

Isas JM, Langen R, Haigler HT and Hubbell WL (2002) Structure and dynamics of a helical hairpin and loop region in annexin 12: A site-directed spin labeling study. Biochemistry 44: 1464–1473

Isas JM,. Patel DR, Jao C, Jayasinghe S, Cartailler J-P, Haigler HT and Langen R (2003) Global structural changes in annexin 12: The roles of phospholipids, $Ca^{2+}$ and pH. J Biol Chem 278: 30227–30234

Jeschke G, Pannier M, Godt A and Spiess HW (2000) Dipolar spectroscopy and spin alignment in electron paramagnetic resonance. Chem Phys Lett 331: 234–252

Jeschke G, Koch A, Jonas U and Godt A (2002) Direct conversion of EPR dipolar time evolution data to distance distributions. J Magn Reson 155: 72–82

Jeschke G, Bender A, Schweikardt T, Panek G, Decker H and Paulsen H (2005) Localization of the N-terminal domain in light-harvesting chlorophyll *a/b* protein (LHCIIB) by EPR measurements. J Biol Chem 280: 18623–18630

Jeschke G, Wegener C, Nietschke M, Jung H and Steinhoff H-J (2004) Inter-residual distance determination by four-pulse DEER in an integral membrane protein: The $Na^+$/proline transporter PutP of *Escherichia coli*. Biophys J 86: 2551–2557

Jung K, Voss J, He M, Hubbell WL and Kaback HR (1995) Engineering a metal binding site within a polytopic membrane protein, the lactose permease of *Escherichia coli*. Biochemistry 34: 6272–6277

Kim YE, Isas JM, Haigler HT and Langen R (2005) A helical hairpin region of soluble annexin B12 refolds and forms a continuous transmembrane helix at mildy acidic pH. J Biol Chem 280: 32398–32404

Kleinfeld D, Okamura MY and Feher G (1984) Electron-transfer kinetics in photosynthetic reaction centers cooled to cryogenic temperatures in the charge-separated state: Evidence for light-induced structural changes. Biochemistry 23: 5780–5786

Kochetkov VV, Likhtenshtein GI, Kol`tover VK, Knox PP, Kononenko AA, Grishanova NP and Rubin AB (1984) Relationship of intramolecular dynamics to electron transfer in photosynthetic reaction centers. Translated from Izvestiya akademii nauk SSSR 4: 572–578

Kokorin AI, Zamaraev KI, Rozantse EG, Grigorya GI and Ivanov VP (1972) Measuring of distances between paramagnetic groups in solid solutions of nitroxide radicals, biradicals and spin-labeled proteins. Biofizika [in Russian] 17: 34–41

Koteiche HA and Mchaourab HS (1999) Folding pattern of the a-crystallin domain in aA-crystallin determined by site-directed spin labeling. J Mol Biol 294: 561–577

Koteiche HA, Reeves MD and Mchaourab HS (2003) Structure of the substrate binding pocket of the multidrug transporter EmrE: Site-directed spin labeling of transmembrane segment 1. Biochemistry 42: 6099–6105

Kurshev VV, Raitsimring AM and Tsvetkov YD (1989) Selection of dipolar interaction by the 2+1 pulse train ESE. J Magn Reson 81: 441–454

Kveder M, Kriško A, Pifat G and Steinhoff H-J (2003) The study of structural accessibility of free thiol groups in human low-density lipoproteins. Biochim Biophys Acta 1631: 239–245

Lakey JH, van der Goot GF and Pattus F (1994) All in the family: The toxic activity of pore forming colicins. Toxicology 87:85–108

Leigh JS (1970) ESR rigid lattice line shape in a system of two interacting spins. J Chem Phys 52: 2608–2612

Marsh D (2001) Polarity and permeation profiles in lipid membranes. Proc Natl Acad Sci USA 98: 7777–7782

Maruyama K and Ohnishi S (1974a) Effect of stearic spin-labels on the photochemical activities of chromatophores from *Rhodospirillum rubrum*. J Biochem 75: 1165–1168

Maruyama K and Ohnishi S (1974b) A spin-label study of the photosynthetic bacterium, *Rhodospirillum rubrum*: Reduction and regeneration of nitroxide spin-labels. J Biochem 75: 1153–1164

Mchaourab HS, Lietzow MA, Hideg K and Hubbell WL (1996) Motion of spin-labeled side chains in T4 lysozyme. Correlation with protein structure and dynamics. Biochemistry 35: 7692–7704

Mchaourab HS, Oh KJ, Fang CJ and Hubbell WL (1997) Conformation of T4 lysozyme in solution — hinge-bending motion and the substrate-induced conformational transition studied by site-directed spin labeling. Biochemistry 36: 307–316

Mchaourab HS and Perozo E (2000) Determination of protein folds and conformational dynamics using spin-labeling EPR spectroscopy. In: Berliner LJ, Eaton SS, Eaton GR (eds) Distance Measurements in Biological Systems by EPR. Kluwer Academic Publishers, New York

McMahon BH, Müller JD, Wraight CA and Nienhaus GU (1998) Electron transfer and protein dynamics in the photosynthetic reaction center. Biophys J 74: 2567–2587

Möbius K, Savitsky A, Wegner C, Plato M, Fuchs M, Schnegg A, Dubinskii AA, Grishin YA, Grigor`ev IA, Kühn M, Duchè D, Zimmermann H and Steinhoff H-J (2005) Combining high-field EPR with site-directed spin labeling reveals unique information on proteins in action. Magn Reson Chem 43: S4–S19

Nelson WD, Blakely SE, Nesmelov YE and Thomas DD (2005) Site-directed spin labeling reveals a conformational switch in the phosphorylation domain of smooth muscle myosin. Proc Natl Acad Sci USA 102: 4000–4005

Nesbitt DM and Berg SP (1980) Proton involvement with the light-induced hindrance of spin label motion in the lumen of spinach thylakoids. Biochim Biophys Acta 593: 353–361

Oh KJ, Barbuto S, Meyer N, Kim R-S, Collier RJ and Korsmeyer SJ (2005) Conformational changes in BID, a pro-aprotic BCL–2 family member, upon membrane binding. J Biol Chem 280: 753–767

Palazzo G, Mallardi A, Hochkoeppler A, Cordone L and Venturoli G (2002) Electron transfer kinetics in photosynthetic reaction centers embedded in trehalose glasses: Trapping of conformational substates at room temperature. Biophys J 82: 558–568

Pannier M, Veit S, Godt A, Jeschke G and Spiess HW (2000) Dead-time free measurement of dipole-dipole interactions between electron spins. J Magn Reson 142: 331–340

Paschenko VZ, Knox PP, Chamorovsky SK, Krasilnikov PM, Mamedov MD, Semenov AY, Zakharova NI, Renger G and Rubin AB (2001) Effect of D2O and cryosolvents on the redox properties of bacteriochlorophyll dimer and electron transfer processes in *Rhodobacter sphaeroides* reaction centers. Bioelectrochemistry 53: 233–241

Perozo E, Cortes DM and Cuello LG (1998) Three-dimensional architecture of a $K^+$ channel: Implications for the mechanism of ion channel gating. Nature Struct Biol 5: 459–469

Persson M, Harbridge JR, Hammarström P, Mitri R, Mårtensson LG, Carlsson U, Eaton GR and Eaton SS (2001) Comparison of electron paramagnetic resonance methods to determine distances between spin labels on human carbonic anhydrase II. Biophys J 80: 2886–2897

Plato M, Steinhoff H-J, Wegener C, Törring JT, Savitsky A and Möbius K (2002) Molecular orbital study of polarity and hydrogen bonding effects on the g and hyperfine tensors of site directed NO spin labeled bacteriorhodopsin. Mol Phys 100: 3711–3721

Poluektov OG, Utschig LM, Dalosto S and Thurnauer MC (2003) Probing local dynamics of photosynthetic bacterial reaction center with a cysteine specific spin label. J Phys Chem B 107: 6239–6244

Pyka J, Ilnicki J, Altenbach C, Hubbell WL and Froncisz W (2005) Accessibility and dynamics of nitroxide side chains in T4 lysosyme measured by saturation recovery EPR. Biophys J 89: 2059–2068

Rabenstein MD and Shin YK (1995) Determination of the distance between two spin labels attached to a macromolecule. Proc Natl Acad Sci USA 92: 8239–8243

Radzwill N, Gerwert K and Steinhoff H-J (2001) Time-resolved detection of transient movement of helices F and G in doubly spin-labeled bacteriorhodopsin. Biophys J 80: 2856–2866

Rink T, Pfeiffer M, Oesterhelt D, Gerwert K and Steinhoff H-J (2000) Unraveling photoexcited conformational changes of bacteriorhodopsin by time resolved electron paramagnetic resonance spectroscopy. Biophys J 78: 1519–1530

Savitsky A, Kühn M, Duché D, Möbius K and Steinhoff H-J (2004) Spontaneous refolding of the pore-forming colicin A toxin upon membrane association as studied by X- band and W-band high-field EPR spectroscopy. J Phys Chem 108: 9541–9548

Solov`ev AA and Erokhin YE (1985) Issledovanie kolichestva, lokalizacii isravnitel'noj himicheskoj aktivnosti SH-grupp sub'edinic reakcionnih centrov *Rb. sphaeroides*. (Study of the quantity, localization and the comparative chemical activity of SH-groups in the reaction centers of *Rb. sphaeroides*) [in Russian] Mol. Biol. 19: 1100–1106

Steinhoff H-J (2002) Methods for study of protein dynamics and protein-protein interaction in protein-ubiquitination by electron paramagnetic resonance spectroscopy. Frontiers Biosci 7: c97–110

Steinhoff H-J (2004) Multi-frequency EPR spectroscopy studies of the structure and conformational changes of site-directed spin labeled membrane proteins. Pifat-Mrzljak G (ed) Kluwer Academic/Plenum Publisher, New York

Steinhoff H-J, Mollaaghababa R, Altenbach C, Hideg K, Krebs M, Khorana HG and Hubbell WL (1994) Time-resolved detection of structural changes during the photocycle of spin-labeled bacteriorhodopsin. Science 266: 105–107

Steinhoff H-J, Radzwill N, Thevis W, Lenz V, Brandenburg D, Antson A, Dodson G and Wollmer A (1997) Determination of interspin distances between spin labels attached to insulin: comparison of electron paramagnetic resonance data with the X- ray structure. Biophys J 73: 3287–3298

Steinhoff H-J, Pfeiffer M, Rink T, Burlon O, Kurz M, Riesle J, Heuberger E, Gerwert K and Oesterhelt D (1999) Azide reduces the hydrophobic barrier of the bacteriorhodopsin proton channel. Biophys J 76: 2702–2710

Steinhoff H-J, Savitsky A, Wegener C, Pfeiffer M, Plato M and Möbius K (2000a) High-field EPR studies of the structure and conformational changes of site-directed spin labeled bacteriorhodopsin. Biochim Biophys Acta 1457: 253–262

Steinhoff H-J, Müller M, Beier C and Pfeiffer M (2000b) Molecular dynamics simulation and EPR spectroscopy of nitroxide side chains in bacteriorhodopsin. J Mol Liq 84: 17–27

Stone AJ (1963) Gauge invariance of the g tensor. Proc Roy Soc London A271: 424–434

Stowell MH, McPhillips TM, Rees DC, Soltis SM, Abresch E and Feher G (1997) Light-induced structural changes in photosynthetic reaction center: Implications for mechanism of electron-proton transfer. Science 276: 812–814

Stroud RM, Reiling K, Wiener M and Freymann D (1998) Ion-channel-forming colicins. Curr Opin Struct Biol 8: 525–533

Subchinski VK, Ruuge EK and Tikhonov AN (1975a) Interaction of paramagnetic probes with membranes of the chloroplasts of higher plants. Translated from Fiziologiya Rastenii 22: 882–890

Subchinski, VK, Ruuge EK and Tikhonov AN (1975b) Structural reorganization of chloroplast membranes investigated by the method of paramagnetic probes. Translated from Fiziologiya

Rastenii 23: 660–665
Voss J, Hubbell WL and Kaback HR (1995a) Distance determination in proteins using designed metal ion binding sites and site-directed spin labeling — application to the lactose permease of *Escherichia coli*. Proc Natl Acad Sci USA 92: 12300–12303
Voss J, Salwinski L, Kaback HR and Hubbell WL (1995b) A method for distance determination in proteins using a designed metal ion binding site and site-directed spin labeling — evaluation with T4 lysozyme. Proc Natl Acad Sci USA 92: 12295–12299
Wegener AA, Klare JP, Engelhard M and Steinhoff H-J (2001a) Structural insights into the early steps of receptor-transducer signal transfer in archaeal phototaxis. EMBO J 20: 5312–5319
Wegener C (2002) Multi-frequency (9 GHz and 95 GHz) ESR spectroscopy studies of the dynamics and polarity of site-directed spin labeled bacteriorhodopsin [in German]. PhD thesis. Ruhr-Universität Bochum, Bochum
Wegener C, Savitsky A, Pfeiffer M, Möbius K and Steinhoff H-J (2001b) High-field EPR-detected shifts of magnetic tensor components of spin label side chains reveal protein conformational changes: The proton entrance channel of bacteriorhodopsin. Appl Magn Reson 21: 441–452
Xiao W, Brown LS, Needleman R, Lanyi JK and Shin Y-K (2000) Light-induced rotation of a transmembrane alpha-helix in bacteriorhodopsin. J Mol Biol 304:715–721
Zhou Z, DeSensi SC, Stein RA, Brandon S, Dixit M, McArdle EJ, Warren EM, Kroh HK, Song L, Cobb CE, Hustedt EJ and Beth AH (2005) Solution structure of the cytoplasmic domain of erythrocyte membrane band 3 determined by site-directed spin labeling. Biochemistry 44: 15115–15128

# Chapter 18

# Magic Angle Spinning (MAS) NMR for Structure Determination in Photosynthesis

Huub J. M. de Groot*
*Leiden University, Leiden Institute of Chemistry, Gorlaeus Laboratories, P.O. Box 9502, 2300 RA Leiden, The Netherlands*



*Email: ssnmr@chem.leidenuniv.nl

*Thijs J. Aartsma and Jörg Matysik (eds): Biophysical Techniques in Photosynthesis II, pp. 361–383.*

## Summary

Using solid-state nuclear magnetic resonance (NMR), structure and its underlying details, such as the rigidity, order, dynamics or electrostatic heterogeneity of the large membrane protein complexes involved in photosynthesis can be resolved at specific spots with a resolution well beyond the capabilities of X-ray and other diffraction methods. Following a brief explanation of the theoretical background of the magic angle spinning NMR methods, applications are presented involving specific and extensive labeling methods. The specific labeling methods provide access to the structure of the large reaction center protein complexes involved in photosynthesis and can be used to study and resolve details of the structure such as rigidity or dynamics of cofactors, H-bonding of the quinones, ground state charge effects on histidines and ligation characteristics of the $Ca^{2+}$ binding site in Photosystem II. In this way spots of physical frustration in the spatial, protonic and electronic structure of the ground state that are important for the biological mechanisms can be identified. In addition, the first *de novo* structure determination of systems comprised of smaller building blocks was made for the chlorosomes in the green photosynthetic bacterium *Chlorobium tepidum*. When the MAS data for the chlorosomes are compared with data collected from the isolated labeled BChl *c* aggregated in *n*-hexane, it can be shown that the major component of the MAS signals in the chlorosomes is from aggregated BChl *c* and a bilayer tube model for the structure was obtained from MAS NMR correlation spectroscopy and molecular modeling. In addition molecular mechanisms steering the suprastructure were identified by model studies. They provide a view on evolutionary selection and may be of interest for future design of artificial photosynthesis structures. Finally it is shown how pattern labeling of the LH2 protein leads to sequence specific assignments in a strategy that can be used to resolve structure of small membrane proteins and complexes.

## I. Introduction

### A. Magic Angle Spinning NMR in Photosynthesis Research

Magic angle spinning (MAS) NMR is a relatively new biophysical technique in photosynthesis research. It can be applied in the determination of structure for ordered systems without translation symmetry, such as proteins, macromolecular complexes, aggregates or membrane systems. MAS NMR structure determination critically depends on the range, resolution and stability that can be obtained with the NMR instrumentation. In both aspects, important improvements have been realized in recent years, initially driven by the fundamental research needs in the study of photosynthesis (Kiihne and de Groot, 2001). The resolution was improved by fast MAS, high field and modern pulse technology. The stability required to perform the new generation of experiments was achieved by improvements in the spectrometer electronics hardware, allowing fast, precise and coherent phase and frequency switching during the NMR experiment (van Rossum et al., 1997a). As a result the technology has matured and can be applied routinely in photosynthesis research. In section II, it will be shown how a variety of structural details can be investigated, illustrating the versatility of the MAS NMR as a biophysical technique in photosynthesis. For instance, in photosynthesis, primary charge separation is followed by secondary charge rearrangement events that are unique to the protein environment and play a crucial role in preventing wasteful back reactions to the ground state (Sham et al., 1998). These rearrangements are thought to involve protons, and involve changes between the ground state and the charge-separated state. The charge separated states have been investigated extensively with optical and EPR methods, while the MAS NMR can provide additional insight into the ground state structure. For a single label in a protein the assignment of its response is trivial, but also for multiply or uniformly labeled systems characterizing the structure by analy-

*Abbreviations:* 1-MeIm – 1-methylimidazole; BChl – bacteriochlorophyll; Chl – chlorophyll; CP – cross-polarization; $CP^3$ CHHC – $^{13}C$-$^{13}C$ polarization transfer via three-step cross-polarization; DFT – density functional theory; FSLG – frequency switched Lee-Goldburg; HETCOR – heteronuclear correlation; Im – imidazole; LG – Lee-Goldburg; LH2 – light harvesting complex 2; MAS – magic angle spinning; NICS – nucleus independent chemical shift; NOE – nuclear Overhauser effect; OEP – octa-ethyl-porphyrin; PDSD – proton driven spin diffusion; Pheo – pheophytin; Photo-CIDNP – Photochemically induced dynamic nuclear polarization; PMLG – phase modulated Lee Goldburg; ppm – parts per million; PS II – Photosystem II; Q – quinone; *Rb.* – *Rhodobacter;* RC – reaction center; RFDR – radio frequency driven recoupling; TPPM – two-pulse phase modulation; WISE – wideline separation

sis of the shift is becoming increasingly important in solids NMR, since shift assignments can be done both rapidly and rigorously with multidimensional spectroscopic methods.

Section III shows how to perform *de novo* and comprehensive structure determination of self-assembled solids that are neither accessible to diffraction, nor to solution NMR techniques (Boender, 1995a). These efforts were initiated to develop the methodology for solving the structure of a chlorosome light-harvesting antenna described in section IIIB. The chlorosome systems may be regarded as a truly functional nanoscopic supramolecular dye assembly developed by evolution in nature. The observation that a biological function can be realized by evolution without active participation of protein was initially considered anomalous, since the central dogma of molecular biology states that all function originates from the DNA code via protein: DNA → RNA → protein → function. In real life, biological constructs are the result of interplay between diversification at the basis and selection according to selection rules that originate from higher levels in the biological hierarchy. The chlorosome represents a beautiful opportunity to investigate the molecular basis of this process, since the self-assembly is directed by the physico-chemical properties of the constituting molecules in a balance between molecular interactions and emergent properties at the length scale of the suprastructure (de Boer and de Groot, 2005).

Since there exists a considerable demand for new methodology for structure determination of membrane proteins and complexes in general, the last subsection of this chapter is devoted to work into this direction, using the LH2 antenna system as a model protein. Since microcrystalline or paracrystalline structures are often encountered in biology, the most prominent examples being the fibrils involved in misfolding diseases, it is not surprising that the methods that were developed were quickly taken over by other investigators and are now rapidly improved and extended. Recent studies demonstrate that modern MAS NMR techniques allow the structure determination of a wide range of biosolids, including microcrystalline protein preparations (van Rossum et al., 2001; Castellani et al., 2002; Petkova et al., 2002; Rienstra et al., 2002; Goward et al., 2003).

## B. Theoretical Background

In magic angle spinning NMR, a sample is rotated rapidly around an axis at the magic angle $\theta_m = 54.74°$ with the static field (Andrew et al., 1958; Lowe, 1959). To describe the MAS NMR experiment, the Hamiltonian

$$H_S = H_{CS} + H_{II} + H_{IS} \tag{1}$$

is used. $H_{CS}$ is the chemical shift term, $H_{II}$ contains the homonuclear dipolar couplings and $H_{IS}$ represents the heteronuclear dipolar couplings. In the solid state, the terms in the Hamiltonian (1) are tensor interactions that are anisotropic and cause large NMR line broadening. The chemical shift, for example, depends on the orientation of the chemical environment of the nucleus in the magnetic field. This is what makes solids NMR special, since a high resolution can only be obtained if these interactions are averaged by rapid motion of the sample or the spins involved.

### 1. MAS Averaging of Anisotropic Interactions

To calculate the effects of sample spinning, the Hamiltonian can be expressed in irreducible spherical tensors (Rose, 1967) according to

$$H = \sum_k \sum_{q=-k}^{+k} (-1)^q A_{kq} T_{k(-q)} \tag{2}$$

The $A_{kq}$ represent the spatial part, while the $T_{k(-q)}$ contain the spin terms. For the chemical shift

$$H_{CS} = A_{00}T_{00} + A_{20}T_{20} \tag{3}$$

with

$$\begin{aligned} A_{00} &= -\frac{1}{\sqrt{3}}\sigma_i \\ T_{00} &= -\frac{1}{\sqrt{3}}\omega_0 I_z \\ A_{20} &= \sqrt{\frac{3}{2}}(\sigma_{zz} - \sigma_i) \\ T_{20} &= \sqrt{\frac{2}{3}}\omega_0 I_z \end{aligned} \tag{4}$$

where $\sigma_{zz}$ is a shift tensor element, $\sigma_i$ is the isotropic shift, $\omega_0$ is the nuclear frequency, $\theta$ is the angle between the rotor axis and the magnetic field $B_0$, and $I_z$ is a spin operator. If the sample is rotated with frequency $\omega_r$, $A_{20}$ becomes time dependent through $\sigma_{zz}$ (Duer, 2004). To illustrate this procedure for the simplest case of a single spin, $A_{20}$ can be transformed from the principal axis system of the chemical shift to the laboratory system via the rotor frame in two steps,

$$A_{20}^{\mathrm{PAS}} \xrightarrow{R(\alpha,\beta,\gamma)} A_{20}^{\mathrm{Rotor}} \xrightarrow{R(\varphi_0+\omega_r t,\theta_m,\psi)} A_{20}^{\mathrm{LAB}} \quad (5)$$

The first rotation is determined by the orientation of the chemical environment with respect to the rotor, while the second rotation introduces the spinning of the rotor. This yields

$$A_{20}^{\mathrm{LAB}}(t) = \sum_{q,q'=-2}^{2} A_{2q'}^{\mathrm{PAS}} \left[ e^{-i\alpha q'} d_{q',q}^{2}(\beta) e^{-i\gamma q} \right] \times \left[ e^{-iq(\varphi_0+\omega_r t)} d_{q,0}^{2}(\theta) \right] \quad (6)$$

The reduced Wigner rotation matrices $d$ are listed in many text books (e.g., Rose, 1967). The MAS signal that derives from Eq. (6) consists of a narrow center peak at the isotropic shift and spinning sidebands at multiples of $\omega_r$ from the center peak. The sideband pattern can be very informative about the chemical environment of the nuclear spin in a solid. For fast rotation Eq. (6) leads to a time average of

$$\langle H_{CS} \rangle = \sigma_{\mathrm{i}} \omega_0 I_z + \tfrac{1}{2}\left(3\cos^2\theta - 1\right)\left(\sigma_{zz} + \sigma_{\mathrm{i}}\right)\omega_0 I_z \quad (7)$$

The second term vanishes for $\theta = \theta_m$ and only the isotropic term survives.

For MAS to be efficient, the spinning rate $\omega_r$ needs to be much larger than the strength of the anisotropic interactions. In high field the dispersion of the signals increases relative to the broadening effects of other interactions and the resolution is enhanced (van Rossum et al., 1996). Since the chemical shift anisotropy scales with $B_0$, fast MAS is necessary in high field. MAS rates of ~15 kHz are now common practice and rates up to ~50 kHz or even higher are possible, although at the expense of smaller sample volumes (Samoson et al., 2005).

For acquiring spectra with MAS of proteins and other biological assemblies, cross-polarization (CP) is commonly used. The CP technique exploits the high abundance, high sensitivity and short relaxation times of the protons by transferring transverse $^1$H magnetization to another spin species (Hartmann and Hahn, 1962). The maximum enhancement for a $^{13}$C signal compared to direct $^{13}$C excitation is $g_{^1\mathrm{H}}/g_{^{13}\mathrm{C}} \approx 4$. In addition, a short recycle delay can be used between scans since the proton $T_1$ is determined by the short relaxation time, from rotating methyl groups, of the strongly dipolar coupled proton system. This leads to a significant gain in sensitivity. During the detection of the signal, heteronuclear decoupling is applied to achieve a high resolution. In particular the robust two-pulse phase modulation (TPPM) sequence is now widely used for this purpose (Bennett et al., 1995). It uses $\pi$ pulses with alternating phases for efficient decoupling.

## 2. Homonuclear Correlation Spectroscopy

The CP/MAS experiment with TPPM decoupling is the starting point for many advanced pulse sequences. To resolve signals and for *de novo* structure determination of solids, homonuclear correlation NMR spectroscopy of multi-spin labeled molecules is necessary. The polarization transfer between spins is governed by the high-field truncated Hamiltonian for the homonuclear dipolar coupling (Ernst et al, 1987)

$$H_{\mathrm{II}} = \omega_{\mathrm{D}}(3I_{1z}I_{2z} - \boldsymbol{I}_1 \cdot \boldsymbol{I}_2) \quad (8)$$

with

$$\omega_{\mathrm{D}} = -\frac{\mu_0 \gamma^2 \hbar}{8\pi r_{12}^3}\left(3\cos^2\theta - 1\right) \quad (9)$$

Here $\gamma$ is the gyromagnetic ratio, $r_{12}$ the distance between the spins and $\theta$ the angle between the internuclear distance vector and the external field. Dipolar couplings are averaged by MAS and can be reintroduced during a mixing interval to generate correlated spin states. The sequence of a $^{13}$C-$^{13}$C radio frequency driven recoupling (RFDR) MAS correlation experiment is shown in Fig. 1A (Bennett et al., 1992). Following CP, the $^{13}$C spins precess under heteronuclear decoupling during $t_1$ to give a high resolution. During $\tau_m$, however, the dipolar $^{13}$C-$^{13}$C couplings have to be reintroduced to promote transfer of magnetization. The magnetization is first stored along $z$ by a $\pi/2$ pulse. The actual recoupling is achieved by a series of $\pi$ pulses, which are synchronized with the rotor period. The evolution of the spin state $\rho$ is described by the commutator

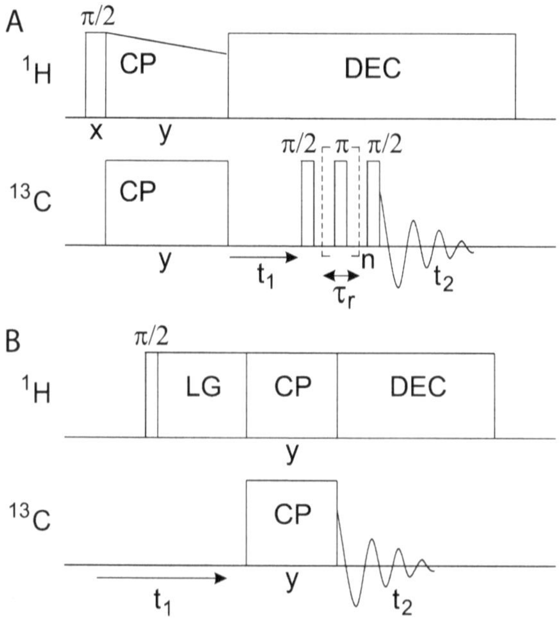


*Fig. 1.* (A) RFDR Pulse sequence for 2D homonuclear correlation spectroscopy. Following CP, the $^{13}C$ spins precess during $t_1$. During a mixing period, $^{13}C$-$^{13}C$ couplings are reintroduced by a rotor-synchronized train of π pulses. The NMR signal is collected during $t_2$. (B) 2D $^1H$-$^{13}C$ LG-CP HETCOR experiment. Following $^1H$ excitation, homonuclear decoupling (LG) is applied during the $^1H$ precession period $t_1$. After CP transfer the $^{13}C$ signal is detected in $t_2$ with heteronuclear decoupling (DEC) to decouple the $^1H$ spins (de Boer, 2004).

$$\frac{d\rho(t)}{dt} = -i[\tilde{H}_{II}, \rho(t)] \tag{10}$$

Here $\tilde{H}_{II}$ is a spin Hamiltonian describing the effective reintroduced homonuclear dipolar interactions. This yields

$$\rho(t) = I_{1z}\cos^2(\tau_m \tilde{d}/2) + I_{2z}\sin^2(\tau m \tilde{d}/2) + \frac{1}{2}(I_{1y}I_{2x} - I_{1x}I_{2y})\sin(\tau_m \tilde{d}) \tag{11}$$

where $\tilde{d}$ depends on the dipolar coupling strength (Bennett et al., 1992). The first two terms indicate transfer of longitudinal magnetization, while the third term represents double quantum states, which can be eliminated by phase cycling. If a short mixing time $\tau_m$ ~1 ms is used, only correlations between spins separated by one bond are promoted, which is the optimal condition for the assignment of the chemical shifts.

The RFDR sequence with a long mixing time has also been used to provide longer range $^{13}C$-$^{13}C$ correlations in chlorophylls (Boender et al., 1995b). Intermolecular transfer between $^{13}C$ spins, however, is generally difficult due to rapid relayed spin diffusion along the multispin $^{13}C$-labeled molecular network. An alternative is to generate $^{13}C$-$^{13}C$ correlations by $^1H$ spin diffusion (Mulder et al 1998; Wilhelm et al., 1998). In a $CP^3$ CHHC experiment the $^{13}C$ magnetization is transferred back to $^1H$ after the first precession interval. Next, $^1H$ spin diffusion is allowed to take place during a mixing period. Finally, the signal is transferred again to $^{13}C$ by a third CP step and detected. In this way, mixing by the strong $^1H$ dipolar interactions is combined with the high resolution of a $^{13}C$ MAS spectrum. An effective transfer range $d_{max}$ can be determined for short mixing times and intermolecular distance constraints can be resolved with this sequence (de Boer et al., 2002). In practice, a limited number of such constraints can already be very useful for elucidating the structure of solids.

## 3. Heteronuclear Correlation Spectroscopy

Another class of experiments that is widely used in biological solid state NMR is $^1H$-$^{13}C$ heteronuclear correlation spectroscopy. A straightforward $^1H$-$^{13}C$ correlation experiment consists of the CP scheme, where $t_1$ is inserted after the first $^1H$ π/2 pulse and the CP interval constitutes the mixing step. This is known as wide-line separation, since broad $^1H$ lines in the indirect dimension are separated by correlation with $^{13}C$ shifts in the direct dimension (Schmidt-Rohr and Spiess, 1994).

Modern frequency switched Lee-Goldburg (FSLG) MAS NMR methods also provide a direct correlation of proton signals of the protein with $^{13}C$ responses (van Rossum et al., 1997a). For heteronuclear transfer of magnetization, Lee-Goldburg cross polarization (LG-CP) methods are most convenient to improve the $^1H$ resolution (Fig. 1B). The LG homonuclear decoupling effectively produces isolated heteronuclear spin pairs for efficient transfer and is very stable since the performance is determined by the frequency or phase switching, which can be done very accurately with a modern NMR spectrometer (van Rossum et al., 2000; Ladizhansky et al., 2003). For instance, the FSLG techniques employ off-resonance radio frequency (rf) irradiation to generate an effective rf field inclined at the magic angle ( Lee and Goldburg, 1965; Bielecki et al., 1989). With the 2D LG/MAS experiment in Fig. 1B spectra can be obtained with a good resolution in both dimensions (van Rossum et al., 1997a). A recent version uses phase modulated Lee-Goldburg (PMLG) decoupling, which is easy to implement (Vinogradov et al., 1999). The effective

$$\tilde{H}_{IS} = \frac{\delta}{4}\left[ I_+S_-e^{i\varphi} + I_-S_+e^{-i\varphi} \right] \tag{12}$$

was introduced to describe a coupled $^1$H-$^{13}$C spin pair during LG-CP (van Rossum, 2000). Here $I_\pm$ and $S_\pm$ are spin operators in a tilted frame for the $^1$H and $^{13}$C spin, respectively. The dipolar coupling $\delta$ is given by

$$\delta = -G_1 \sin\theta_m \frac{\mu_0}{4\pi}\frac{\gamma_I\gamma_S\hbar^2}{r_{IS}^3} \tag{13}$$

with $G_1$ a geometrical factor and $r_{IS}$ the distance between the spins. The coherent build-up of the $^{13}$C signal $S(t)$ is then described by (van Rossum 2000)

$$S(t) = -\tfrac{1}{4}(Zk_BT)^{-1}\omega_{0I}\left(1-\cos\tfrac{1}{2}\delta t\right) \tag{14}$$

From the build-up of $S(t)$ the dipolar coupling can be determined. This technique yields accurate distances up to a few Å. Since the dipolar couplings scale with $r^{-3}$, the effects of long-distance interactions are obscured by strong short-range interactions. For longer CP times, the magnetization transfer is incoherent due to the many spin interactions and due to relaxation. Although accurate intermolecular distances are difficult to determine in chlorophylls, incoherent long-range transfer proceeds over an effective maximum transfer range $d_{max}$, which depends on the length of the mixing period (van Rossum et al., 2002).

## II. MAS NMR for Structure Determination in Large Complexes

### A. Cofactors

#### *1. Spheroidene Configuration in Rb. sphaeroides 2.4.1*

The photosynthetic reaction center of *Rb. sphaeroides* 2.4.1 contains one carotenoid that protects the protein complex against photodamage. The structure around the central (15,15′) double bond of the bound spheroidene carotenoid was investigated with low-temperature $^{13}$C MAS NMR to average the chemical shift anisotropy (de Groot et al., 1992). This allows an in situ characterization of the configuration of the central double bond in the carotenoid, a detail that could not be resolved unambiguously by diffraction methods. Carotenoidless reaction centers of R26 were reconstituted with spheroidene specifically labeled at the C-14′ or C-15′ position, and the signals from the labels were separated from the natural abundance background using $^{13}$C MAS NMR difference spectroscopy. The resonances shift 5.2 and 3.8 ppm upfield upon incorporation in the protein complex, similar to the 5.6 and 4.4 ppm upfield shift occurring in the model compound β-carotene upon trans to 15,15′-*cis* isomerization. Hence the MAS NMR favors a *cis* configuration, as opposed to the *trans* configuration deduced from X-ray data. It can be emphasized that the structural information is derived here from the isotropic chemical shift alone, without determining the sideband pattern or distances.

#### *2. Hydrogen Bonding and Dynamics of $Q_A$ and $Q_B$ in Rb. sphaeroides R26*

Photosynthetic enzymes convert light energy into chemically stored energy in a photochemical cascade of reactions with almost 100% quantum yield. The hydrogen bonding environment and dynamic structure of the quinones $Q_A$ and $Q_B$ that act as a redox switch between the electrons and the protons have been examined with MAS NMR and selective isotope labeling (Fig. 2). *Rb. sphaeroides* R26 reaction centers (RCs) were reconstituted at the $Q_A$ site with ubiquinone-10, selectively $^{13}$C-enriched at positions 1, 2, 3, 4, and 3-methyl (van Liemt et al., 1995; van Rossum et al., 1995). RCs dispersed in LDAO detergent were studied with $^{13}$C CP/MAS NMR spectroscopy at temperatures between 180 and 240 K, while RCs precipitated by removal of the detergent were investigated at ambient temperature and at temperatures down to 180 K. The upper left panel in Fig. 2 shows a contour plot of a 2D $^1$H-$^{13}$C WISE heteronuclear dipolar correlation spectrum of [1-$^{13}$C] $Q_A$ RCs, i.e., with a label at the 1 keto position (van Rossum, 2000). The [1-$^{13}$C] $Q_A$ resonance around 183.8 ppm has a correlation signal with a proton with $\sigma_i$ =9.5 ± 0.5 ppm, indicated with an arrow in the 2D dataset. The $T_2$ of the protons in the vicinity of the 1-$^{13}$C label is very short (≤10 μs), revealing a rigid structure around the isotope label. The linewidth of the 1-$^{13}$C signal at room temperature is 150 Hz, and at 230 K it is 250

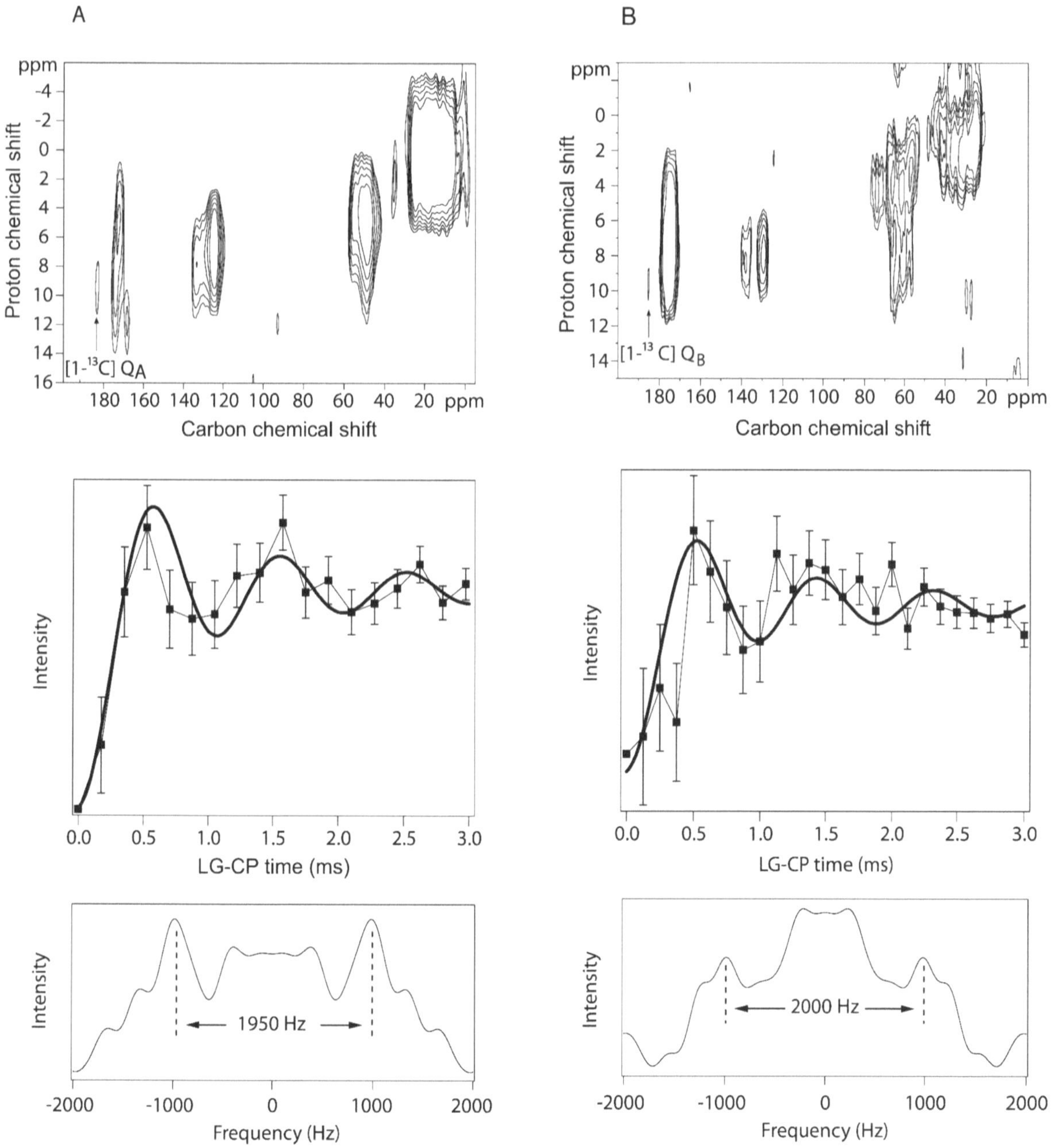


*Fig. 2.* Resolving H-bond lengths in large proteins with 2D HETCOR methods for the [l-$^{13}$C] in the $Q_A$ (left) or the $Q_B$ (right) reconstituted into R26 RCs. In the upper panels contour plots of 2D $^1$H-$^{13}$C heteronuclear dipolar correlation spectra are shown. The data for the $Q_A$ were collected with the CP/WISE technique in a magnetic field of 14.1 T. The dipolar interactions are truncated by the high field. Together with rapid spinning (15 kHz) this yields narrow signals. The HETCOR spectrum of [1-$^{13}$C] $Q_B$ was collected with LG homonuclear decoupling during $t_1$ and with a spinning rate of 12.0 kHz. The LG-CP contact time was 4.5 ms. The arrows indicate the correlation signals with the $^{13}$C labels. In the middle panels the LG-CP build-up curves are shown, with the Fourier transform in the lower panels. The thick solid lines are from simulations of the build-up for a spin-pairs with CH distances $r_{CH}$= 2.03Å (A) and $r_{CH}$= 1.96 Å (B) (van Rossum et al., 2007).

Hz. The signal from the other keto-position, 4-$^{13}$C, can only be observed below ~255 K; at 190 K the linewidth is ~220 Hz. The chemical shifts of other positions are 144 and 137 ppm for labels at positions 2 and 3, and 12 ppm for the 3-methyl $^{13}$C. The other line widths are between 150 and 300 Hz and inhomogeneous broadening is observed upon cooling. This contrasts with the MAS data for the labels in the vicinity of the special pair mentioned below. When [5-$^{13}$C]-and [6-$^{13}$C]-ubiquinone-10 isotopomers were incorporated into the $Q_A$-site, MAS NMR revealed an unperturbed 6-position, while the signal of the 5-position was absent (Boers et al., 2002).

Electrostatic charge differences in $Q_A$ induced by polarization from the protein are less than 0.02 electron equivalents for any of the labeled positions, including the 4-carbonyl. The principal components of the chemical shift tensors for the ring labels in $Q_A$ can be estimated using difference spectroscopy and indicate a decrease of the 4-C=O bond order upon binding to the protein, in contrast with the 1-C=O. Hence the data reveal an asymmetric structure of the $Q_A$ binding site with a dynamic perturbation involving the 4-carbonyl functionality, indicating local physical frustration in the structure.

The temperature-dependent structural asymmetry between the two carbonyls in $Q_A$ was subsequently characterized by CP-MAS NMR relaxation studies (van Rossum, 2000). The $^{13}$C $T_1$ is longer than 1 s for both carbonyls which confirms a rigid protein environment. The longitudinal relaxation time in the rotating frame $T_1^\rho$ of the 1-$^{13}$C signal is nearly temperature independent (8 ± 1 ms at RT and 7.2 ± 0.8 ms at 250 K), while the $T_1^\rho$ of the 4-$^{13}$C signal (3.8 ± 0.9 ms at 250 K and 8 ± 2 ms at 230 K) appears to decrease as the temperature approaches 255 K. This confirms that the H-bond interaction with the protein at position 4 involves dynamic character. It may help to lower the energy barrier for the $Q_A$ redox chemistry.

The correlation signal in Fig. 2A predominantly involves intermolecular polarization transfer, i.e., between the 1-C=O of the $Q_A$ and protons from surrounding amino acids. The middle panel in Fig. 2A shows a LG-CP build-up curve of the [1-$^{13}$C] $Q_A$ response in a series of 1-D spectra. The LG-CP build-up curve is slowly oscillating, and its Fourier transform provides the LG-CP carbon spectrum with two resolved maxima (Fig. 2A, lower panel). The build-up of the [1-$^{13}$C] $Q_A$ signal was simulated and the best correspondence between the experimental and simulated data is found for a spin pair with $r_{CH}$ = 2.03 Å. The splitting $\Delta\omega_r/2\pi$ = 1950 Hz between the maxima in the Fourier transform corresponds with a short distance $r_{CH}$ = 2.1 ± 0.1 Å (van Rossum et al., 2000).

In order to compare the NMR spectral characteristics of the [1-$^{13}$C] $Q_A$ site with the signals from the same label in the $Q_B$ site, the upper right panel in Fig. 2 shows a 2D $^1$H-$^{13}$C FSLG HETCOR spectrum of RCs reconstituted with [1-$^{13}$C] $Q_B$. Its $^{13}$C signal at 182.4 ppm correlates with a proton signal around 9.5 ± 0.5 ppm, indicated with an arrow in the spectrum. The LG-CP build-up curve recorded for the [1-$^{13}$C] $Q_B$ translates into a distance $r_{CH}$ = 1.96 Å. From the analysis of the Fourier transform of the [1-$^{13}$C] $Q_B$ signal build-up, a weak splitting $\Delta\omega_r/2\pi$ of ~ 2000 Hz is observed, which translates into a distance $r_{CH}$ of approximately 2.1 Å.

### *3. The Ligation of $Ca^{2+}$ in Photosystem II*

Instead of using isotope enrichment, substitution of an ion cofactor that is insensitive for NMR by an NMR sensitive ion offers an attractive route to probing protein structure. For instance, $Ca^{2+}$ is an essential cofactor for photosynthetic oxygen evolution in Photosystem II (PSII). Its ligand surrounding has been investigated with replacement of $Ca^{2+}$ by $Cd^{2+}$, followed by study with $^{113}$Cd MAS NMR spectroscopy (Matysik et al., 2000). A single $^{113}$Cd NMR signal 142 ppm downfield from $Cd(ClO_4)_2{\cdot}2H_2O$ was recorded from $Cd^{2+}$ present at the $Ca^{2+}$-binding site (Fig. 3). The line width of the proton signal in a WISE two-dimensional $^1$H–$^{113}$Cd NMR experiment demonstrates that the signal arises from $Cd^{2+}$ in a solid and magnetically undisturbed environment. The chemical shift, the small anisotropy, and the narrow line of the $^{113}$Cd NMR signal provide convincing evidence for a 6-fold coordination, which is achieved partially by oxygen and partially by nitrogen or chlorine atoms in otherwise a symmetric octahedral environment. The absence of a $^{113}$Cd signal below −70 °C has provided early structural evidence that the $Ca^{2+}$ binding site is close to the tetramanganese cluster, which was recently validated by X-ray data (Ferreira et al., 2004). Similar substitution experiments have been performed to determine a five-fold surrounding for the central $Mg^{2+}$ ion in self-assembled BChl *c* type aggregates (de Boer et al., 2004).

### *4. Multispin Labeling for Cofactor Analyses*

By using the improved range and resolution that is obtained in high field with rapid MAS, it is possible to use multiple labeled cofactors for the analysis of the structural embedding in the protein matrix. Bacterial photosynthetic reaction centers of *Rhodobacter sphaeroides* R26 were reconstituted with uniformly

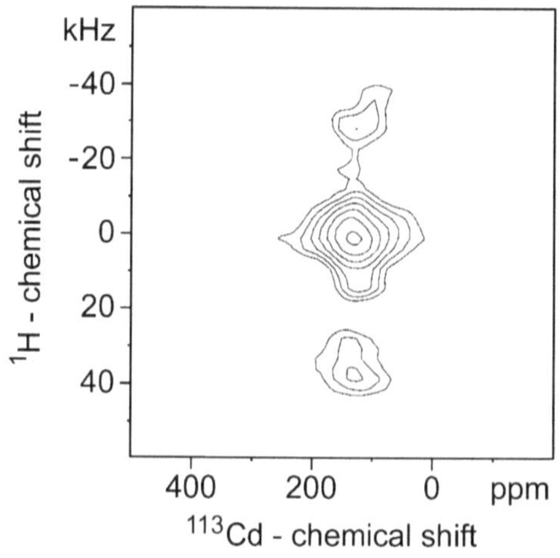


*Fig. 3.* 2D $^{1}$H-$^{113}$Cd WISE HETCOR of PSII membranes with $^{113}Cd^{2+}$ substituted into the $Ca^{2+}$ binding site. The $Ca^{2+}$ is close to the Mn cluster, which was recently confirmed by X-ray analysis of the structure (Matysik et al., 2000, with permission).

$^{13}$C biosynthetically labeled (plant) pheophytin *a* in the two pheophytin (Pheo) binding sites (Egorova-Zachernuyk et al., 1997). From the multispin labeled samples 2D solid-state $^{13}$C MAS NMR spectra were collected and many $^{13}$C resonances could be assigned in a single set of experiments. The NMR indicates a strong overall similarity between the two Pheo *a*, which is of interest in view of the asymmetry of the electron transfer and is in contrast with the differences in functional structure between quinones discussed in Section II.A.2. Multispin labeling and correlation spectroscopy of cofactors is also important for MAS photo-CIDNP studies that are treated extensively in Chapter 19 (Daviso et al.). In these studies light-induced charge separation is used to enhance the NMR response of a variety of reaction centers.

Finally, the structures of the bacteriochlorophyll (BChl) *a* type B800 and type B850 in the light-harvesting complex II (LH2) of *Rhodopseudomonas acidophila* strain 10050 have been characterized by MAS dipolar $^{13}$C-$^{13}$C correlation NMR spectroscopy, in relation to the functional properties (van Gammeren et al., 2005a). Uniformly [$^{13}$C,$^{15}$N] enriched LH2 complexes were prepared biosynthetically, while [$^{13}$C,$^{15}$N]-B800 LH2 complexes were obtained after reconstitution of apoprotein with uniformly [$^{13}$C,$^{15}$N]-enriched BChl cofactors. Extensive sets of isotropic $^{13}$C NMR chemical shifts were obtained for each bacteriochlorin ring species in the LH2 protein. $^{13}$C isotropic shifts in the protein have been compared to the corresponding shifts of monomeric BChl *a* dissolved in acetone-$d_6$. DFT calculations were performed to estimate ring current effects induced by adjacent cofactors. By correction for the ring current shifts, the $^{13}$C shift effects due to the interactions with the protein matrix can be resolved. The chemical shift changes provide a clear evidence for a global electronic effect on the B800 and B850 macrocycles, which is attributed to the dielectrics of the protein environment, in contrast with local effects due to interaction with specific amino acid residues. Considerable shifts of $-6.2 < \Delta\sigma_i < +5.8$ ppm are detected for $^{13}$C nuclei in both the B800 and the B850 bacteriochlorin rings. Since the shift effects for the B800 and B850 are similar, the polarization of the electronic ground states induced by the protein environment is comparable for both cofactors and corresponds with a red shift of ~30 nm relative to the monomeric BChl dissolved in acetone-$d_6$. It appears that the additional red shift in the B850 has a structural basis since it is almost exclusively due to macrocycle overlap.

## B. Protein Matrix

### 1. Structural Integrity

At an early stage $^{13}$C MAS NMR has been used to investigate the structural integrity of the ~140 kDa *Rb. sphaeroides* R26 RC complexes, selectively enriched in [4-$^{13}$C]-tyrosine (de Groot et al., 1999; Fischer et al., 1992). At a low temperature of ~ 215 K the isotope labels yield a MAS response that consists of at least seven narrow lines superimposed on a broad doublet (Fig. 4A). A narrow dispersion of the label signals, <5 ppm, corresponds to electron density variations of less than 0.03 electron equivalents and this provides evidence for an electrostatic uniform and neutral complex at the tyrosine OH moieties. The narrow signals can be separated from the broad background by taking the second derivative of the data (Fig. 4B). Two signals are 28 Hz wide, which is very narrow and shows that the protein interior has the characteristics of a linearly elastic organic solid. In addition, the line width as well as the chemical shift of the narrowest label signals is essentially independent of temperature. In contrast, when the temperature is raised, broad signals from other labels are quenched and the natural abundance resonances from the detergent become narrower, revealing a transition from disorder at low temperature to flexibility at higher temperatures for the exterior of the protein complex

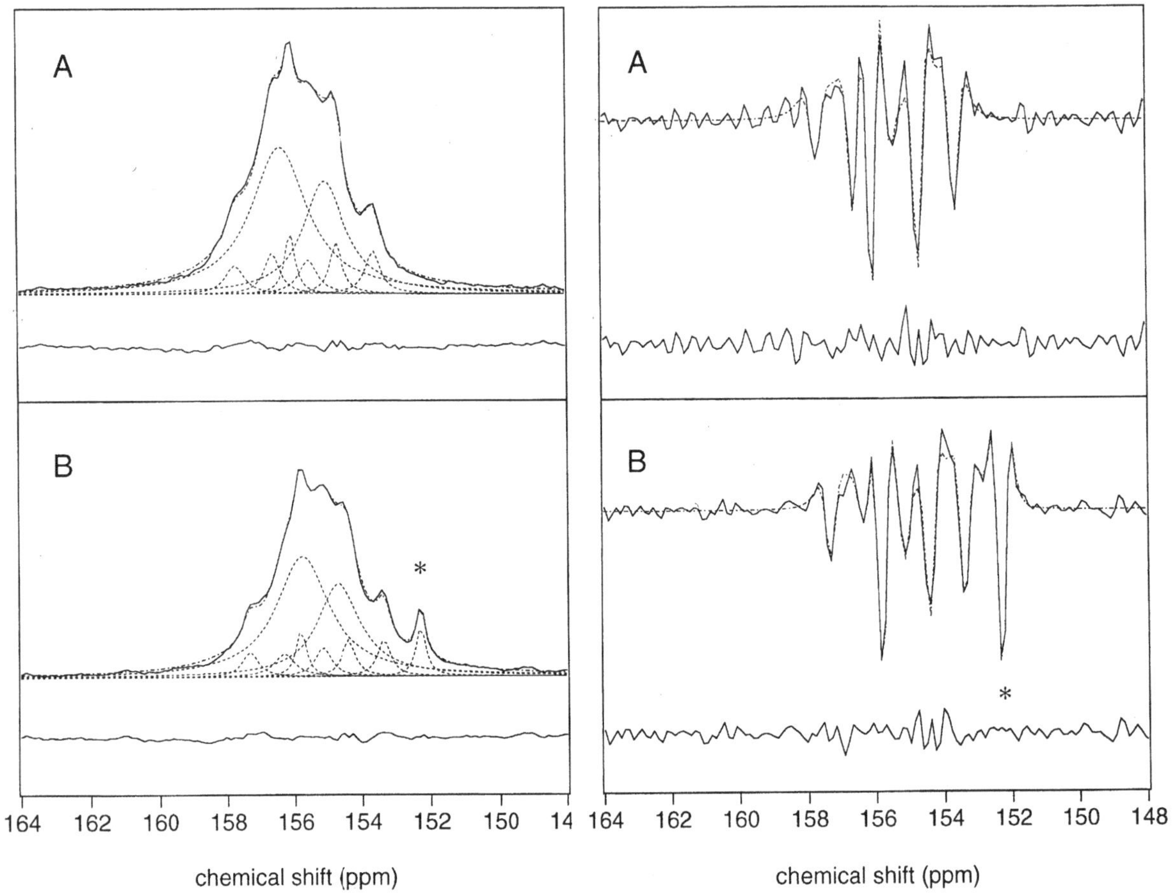


*Fig. 4.* The structural basis of the scaffolding provided by the protein complex probed with [4′-$^{13}$C]Tyr incorporated into R26. The left part shows difference spectra for the 2.4.1 (M)Y210W mutant (A) and for R26 (B), while the right panel shows the corresponding second derivative spectra. The solid lines are the data collected for the MAS centre band regions of the [4′-$^{13}$C]Tyr response while the dashed lines are the fits to the data by deconvolution. Below each band the residue, data minus fit, is shown. The asterisk indicates the signal from a single tyrosine site in the R26 complex, M210 (from Shochat et al., 1995, with permission of Elsevier).

and for the detergent belt. In this way a global assessment of the structure is obtained from a limited set of labels distributed over the protein complex.

The photosynthetic charge separation process is very sensitive to the replacement of Tyr M210, between the special pair and the pheophytin, by a tryptophan residue, and NMR data were collected from a *Rb. sphaeroides* 2.4.1 (M)Y210W mutant RC sample labeled with [4′-$^{13}$C]tyrosine (Shochat et al., 1994, 1995). The response of (M)Y210 could be assigned to the most upfield and narrowest narrow signal at $\sigma_i$ = 152.2 ppm (Fig. 4). This shows that M210 is in a structurally homogeneous region on the sensitivity scale of the MAS NMR technique and its side chain can be considered static with respect to rotational diffusion on time scales as long as $10^{-2}$ s (Fischer et al., 1992; Shochat et al., 1995). Hence it is unlikely that M210 is subject to conformational changes on the short time scale of the primary photoprocess.

The NMR data provide evidence for a very rigid environment of the special pair at the primary acceptor site and corroborate converging opinions that the main effect of amino acid side chains is to contribute to the tuning of the redox potentials of the prosthetic groups involved in electron transfer. The narrow signals from the remaining labels in the protein interior for the 2.4.1 (M)Y210W mutant are remarkably similar to those observed for R26, except for a narrow response at $\sigma_i$ = 156.4 ppm that sharpens slightly in the mutant (Fig. 4). Thus, the influence of the (M)Y210W mutation on the global structure of the protein is minimal. This effectively eliminates the possibility of explaining the slow and non-exponential electron transfer kinetics in the (M)Y210W RC in terms of loss of structural integrity upon mutation.

Another tyrosine, L162, is positioned between the special pair and its mutation decreases the re-reduction rate by the cytochrome (Farchaus et al., 1993). CP/MAS NMR data were collected from a tyrosine labeled (L)162YL mutant and the data reveal that L162 is in a structurally more heterogeneous section of the protein complex than M210 (van Rossum et al., 1997b). However, the overall structural changes in the mutant relative to R26 are again minimal. Most likely the decrease of the re-reduction rate of the oxidized primary donor P upon mutation originates from slow reorientation of the docked cytochrome as opposed to a structural change of the protein.

## 2. Electronic Structure and Protonation State

To analyze the imidazole side chains of histidine residues attached to the central $Mg^{2+}$ in the LH2 complex of *Rhodopseudomonas acidophila* histidines can be selectively labeled at both or one of the two nitrogen sites of the imidazole ring. Alternatively, the potential of the 2D NMR methods can be exploited for making comprehensive assays of complete amino acid side chains in a large protein matrix when used in combination with multispin labeling (Fig. 5). This eliminates the need for time-consuming selective labeling strategies involving total organic synthesis. For instance, two types of histidine residues can be resolved with MAS NMR in LH2: Type 1 (neutral) and Type 2 (positively charged). The imidazole τ-nitrogen is ligated to $Mg^{2+}$ of B850 BChl molecules (β-His30, α-His31). The π-nitrogens of these $Mg^{2+}$-bound histidines were found to be protonated and may be involved in hydrogen bond interactions (Alia et al., 2001).

For the interpretation of data on biological systems it is often important to have access to in vitro models to interpret the effects that are observed. Comparison of the 2D MAS NMR homonuclear ($^{13}C$-$^{13}C$) dipolar correlation spectrum of [$^{13}C_6$,$^{15}N_3$]-histidines in the LH2 complex with model systems in the solid state validates the two different classes of electronic structures for the histidines in the LH2 and confirms that the resonances of nitrogens that are interacting with B850 BChl *a* were assigned correctly (Alia et al., 2001, 2002).

Complementing the protein studies, imidazole coordination interactions have also been investigated by solution and solid-state MAS NMR spectroscopy (Van Gammeren et al., 2004a). $^1H$, $^{13}C$ and $^{15}N$ coordination shifts due to ring currents, electronic perturbations and structural effects have been assessed for imidazole (Im) and 1-methylimidazole (1-MeIm)

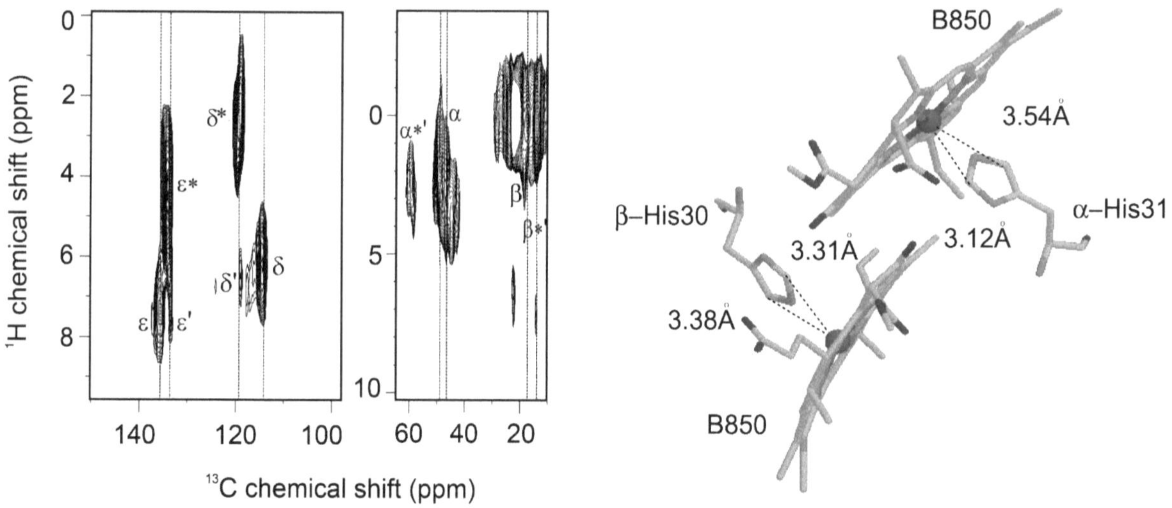


*Fig. 5.* $^1H$-$^{13}C$ HETCOR spectrum of uniformly $^{13}C$,$^{15}N$ labeled histidine in the LH2 complex. The data were collected with a spinning rate of 12 kHz in a magnetic field of 17.6 T. The left part shows the correlations for the aromatic carbons in the imidazole ring, while the right part shows the correlation signals for the aliphatic carbons. Cross peak from cationic histidines are indicated with primes. The signals from histidines that coordinate to the B850 as depicted on the right are indicated with an asterisk. The $\delta^*$ and $\varepsilon^*$ are from the carbons that are closest to the ring and the $^1H$ signals are shifted upfield due to the ring currents in the BChl. Such ring current shifts are very significant and can help both for assignment of signals and to provide information about the structure (Alia et al., 2004, with permission).

coordinated axially to Mg(II)-OEP and (B)Chl *a*. The coordination shifts depend on the distance of the nuclei to the porphyrin plane and the perturbation of the electronic structure. The signal intensities in the $^{1}H$ NMR spectrum reveal a five-coordinated complex, and the isotropic chemical shift analysis shows a close analogy with the electronic structure of the BChl *a*–histidine in natural LH2 complexes, while broad ligand signals provide evidence for a dynamic coordination bond in the models.

The efficient operation of the photosynthetic RC relies on a proper balance between forward and backward transfer rates to maintain a high quantum yield for optimal catalytic turnover. The key concept in enzyme mechanisms is that efficiencies are optimal since enzymes can lower the activation energy of their active elements (Pauling, 1946; Garcia-Viloca et al., 2004). It appears that histidine has the strongest effect in changing the midpoint potential in the ground state of chlorophylls involved in charge separation (Ivancich et al., 1998). The characterization of histidine signals from LH2 antenna systems and models provides the basis for a detailed structural analysis of the histidines interacting with chlorophyll donor molecules that are involved in charge separation.

## III. *De novo* Structure Determination

### A. Concept for Structure Determination with MAS NMR

#### 1. Assignment of Shifts with Dipolar Correlation Spectroscopy

Driven by the need to resolve the structure of the chlorosome antenna, a concept for MAS NMR structure determination was developed starting from uniformly $^{13}C$ enriched Chl *a*/$H_2O$ preparations (Boender et al., 1995b; van Rossum et al., 1998). First, dipolar correlation spectra are recorded to obtain an assignment of the $^{13}C$ and $^{1}H$ chemical shifts. With $^{15}N$ labeled material, nitrogen shifts can be obtained as well. Chemical shifts are generally quite sensitive to details of the electronic configuration, the molecular conformation, and intermolecular interactions and this can provide a model of the spatial structure that can be validated and refined with distance or torsional restraints, or by complementary low-resolution techniques like cryo-electron microscopy.

To work out a concept for structure determination, 2D MAS dipolar correlation spectra were collected from $^{13}C$ enriched Chl *a*/$H_2O$ aggregates. Application of TPPM decoupling and coherence pathway selection in a RFDR pulse scheme yields virtually pure 2D absorption line shapes and well-resolved spectra and all $^{13}C$ resonances of the Chl *a* molecule could be assigned (Boender et al., 1995b). The RFDR spectra with longer mixing times reveal through-space intermolecular polarization transfer. This work was the first demonstration that a concept for MAS NMR structure determination of uniformly $^{13}C$ enriched solids indeed exists and can be evaluated to perform *de novo* structure determination of ordered systems without translation symmetry.

Using Lee-Goldburg HETCOR spectra the $^{1}H$ shifts can be assigned. Dilution of the $^{1}H$ spins by $^{2}H$ has been explored as an alternative to reduce the $^{1}H$ line widths of chlorophylls (Schulten et al., 1998). Obviously, a reduction in sensitivity is a disadvantage of this approach. With a dilution level of 75%, the $^{1}H$ line widths in a 2D WISE experiment are narrowed significantly for several signals, although for the aliphatic protons there is little improvement. In the end, a 2D FSLG or PMLG experiment outperforms the 75% $^{2}H$ dilution, since a better resolution is obtained without losing sensitivity.

#### 2. Intermolecular Distance Restraints and Validation of Models

In the HETCOR spectra an effective maximum transfer range $d_{max}$ for $^{1}H$-$^{13}C$ transfer can be determined experimentally from the detection of a gradually decreasing series of intramolecular correlations with the $^{13}C$ along the molecular skeleton and this can be used for structure determination (van Rossum et al., 2002). To probe intermolecular contacts $d_{max}$ can be set to ~ 0.4 nm by choosing an LG-CP contact time of ~2 ms and long-range $^{1}H$-$^{13}C$ correlations were used in conjunction with carbon and proton aggregation shifts to probe the stacking in the Chl *a* aggregates. The MAS NMR analysis reveals a 2D stacking homologous to the molecular arrangement in crystalline solid ethyl-chlorophyllide *a*. A doubling of a small subset of the carbon resonances, in the 7-methyl region of the molecule, provides evidence for two marginally different well-defined molecular environments, while $^{1}H$ signals of neutral structural water molecules provide evidence for a hydrogen-bonded network to stabilize Chl *a* sheets. In line with the microcrystalline order observed for the rings, the long

$T_1$'s, and absence of conformational shifts for the $^{13}C$ in the phytyl tails, it was deduced that the Chl *a* form a rigid 3D space-filling structure by forming bilayers with interdigitation of elongated tails (van Rossum et al., 2002). Such a 3D space-filling organization of the aggregated Chl *a* from MAS NMR is well in line with models inferred from electron microscopy and low-resolution X-ray powder diffraction, while a micellar model based on neutron diffraction and antiparallel stacking observed in solution was rejected.

Calculations at the force field level have been validated in the past for large chlorophyll aggregates and models for the aggregate structure can be constructed and optimized using the MM+ force field (Holzwarth and Schaffner, 1994; Yagai et al., 2001). To validate the modeling, an improved 2D $^{13}C$-$^{13}C$ CP$^3$ CHHC MAS NMR correlation experiment with mixing by true $^{1}H$ spin diffusion for measuring polarization transfer on a length scale between 0.3 and 1.0 nm using short mixing times of 0.1 ms < $\tau_m$ < 1 ms was developed for this purpose (de Boer et al., 2002). With CP$^3$, correlations can be detected over a much longer range than with direct $^{1}H$-$^{13}C$ or $^{13}C$-$^{13}C$ dipolar recoupling by using a $^{1}H$ spin diffusion mixing period $\tau_m$ sandwiched between two cross-polarization periods (Mulder et al., 1998). The use of this experiment for genuine structure determination is demonstrated in Fig. 6 for the self-aggregated Chl $a/H_2O$. For a short $\tau_m$ = 0.1 ms, intermolecular correlations are detected between the ends of phytyl tails and ring carbons of neighboring Chl *a* molecules in the aggregate. In this way the model for the structure, with stacks of Chl *a* that are arranged back to back with interdigitating phytyl chains stretched between two bilayers, can be validated.

With the same approach a bacteriochlorophyllide model structure was validated by a CHHC experiment (de Boer et al., 2003). This experiment also shows close contacts between the 8-ethyl region and the 17-propionate region, revealing a head-to-tail arrangement of sheets of chlorophyllide. The microcrystalline structure was obtained from a 60-mer that was optimized with modeling (Fig. 7). The rings form parallel stacks with the 3$^1$-OH groups coordinated to the Cd of the neighboring molecule and sheets due to hydrogen bonding between the 3$^1$-OH functionalities of one stack and the 13-C=O groups of another, in line with the original modeling by Holzwarth and Schaffner (1994). The structure corresponds with a space group $P2_1$ with $a$ = 14.3 Å, $b$ = 27.3 Å, $c$ = 6.4 Å, $\beta$ = 147.2° and $Z$ = 2. This illustrates how a detailed structural model of a microscopically or-

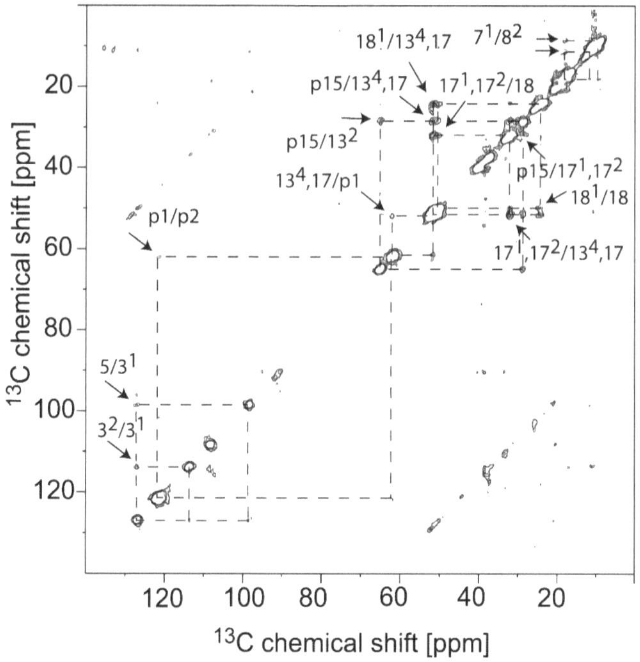


*Fig. 6.* CP$^3$ CHHC $^{13}C$-$^{13}C$ MAS spectrum of aggregated Chl $a/H_2O$, illustrating the detection of intermolecular cross peaks, next to intramolecular correlations, for validation of the structure (from de Boer et al., 2002, with permission of Elsevier).

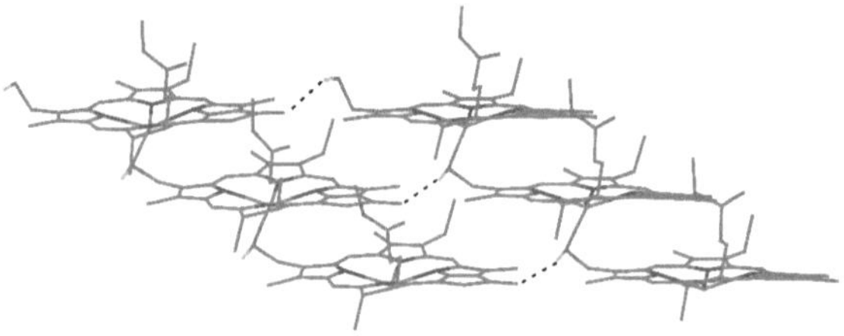

*Fig. 7.* Microcrystalline structure of a layer of parallel inclined chlorin rings forming stacks. The structure was determined by MAS NMR (de Boer et al., 2004).

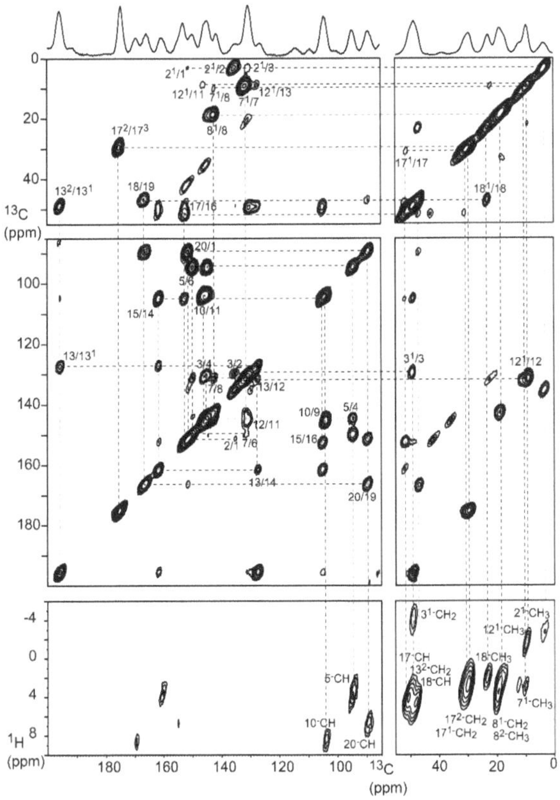


*Fig. 8.* The combined analysis of sections of a $^{13}$C-$^{13}$C PDSD dipolar correlation spectrum and a $^{1}$H-$^{13}$C MAS NMR PMLG dipolar correlation spectrum of aggregated chlorin recorded in a field of 17.6 T leads to a $^{13}$C-$^{13}$C and $^{1}$H-$^{13}$C connectivity network (dashed lines) and a full assignment of the $^{1}$H and $^{13}$C response. The arrows indicate a doubling of the 15/16-C correlation (de Boer et al., 2004).

dered structure can be obtained with MAS NMR in conjunction with molecular modeling.

## 3. Long Range Effects on the Chemical Shifts

Chemical shifts have been interpreted traditionally in terms of local paramagnetic and diamagnetic contributions. The importance, however, of long-range effects on the shifts in the solid state and their use for analyzing the packing of the molecular building blocks in a 3D solid can be illustrated by the analysis of ring current shifts for a uniformly $^{13}$C- and $^{15}$N-labeled Cd-bacteriochlorophyll *d* analogue, a derivative of the natural BChl *c* in the chlorosomes of green photosynthetic bacteria. An assignment of the $^{1}$H and $^{13}$C resonances can be obtained with 2D and 3D dipolar correlation spectroscopy (Fig. 8). The lines are narrow, reflecting local crystalline order. Similar to the Chl *a*, a splitting of resonances reveals a doubling of the microcrystalline unit cell induced by the long tails.

The NMR structures can be validated by a procedure that was implemented to calculate the long-range ring-current effects (de Boer et al., 2004). In this procedure every macrocycle is approximated by a circular loop with radius *a*. The isotropic ring-current shift $\sigma_{rc}$ is proportional to $B_z$, yielding

$$\sigma_{rc}(x,y,z) = c\int_0^{2\pi} d\varphi\left(1-\frac{x}{a}\cos\varphi-\frac{y}{a}\sin\varphi\right)\times\left(\left[\frac{x}{a}-\cos\varphi\right]^2+\left[\frac{y}{a}-\sin\varphi\right]^2+\left[\frac{z}{a}\right]^2\right)^{-3/2} \quad (15)$$

The ring current of a single loop with $a = 0.3$ nm can be estimated from a nucleus-independent chemical shift (NICS) DFT calculation for a ghost atom placed 0.4 nm above the center of a chlorin molecule, using a B3LYP/6-311G(d,p) level of theory. For these conditions, a ring-current shift of 3.90 ppm was obtained. Using Eq. (15), this provides $c = 2.87$ ppm as a calibration constant. The total intermolecular ring-current shift for each $^{1}$H atom was obtained by adding the contribution from all other rings in the lattice within a maximum range $r_{max} = 24$ Å

$$\sigma_{^1H}\left(x_{^1H},y_{^1H},z_{^1H}\right) = \sum_{i\neq 0}^{r<r_{max}} \sigma_{rc}\left(x_{^1H}-x_i, y_{^1H}-y_i, z_{^1H}-z_i\right) \quad (16)$$

where each $\sigma_{rc}$ given by the integral in Eq. (15) can be evaluated numerically.

The calculation can reproduce the experimental ring current shifts of $^{1}$H atoms with a standard deviation of the calculated versus the observed values of

1.2 ppm on a range of 11.6 ppm, and with a correlation coefficient of ~ 0.98 (de Boer et al., 2004). For instance, using shift data the unique densely packed structure of overlapping macrocycles arranged in sheets of parallel inclined stacks in Fig. 7 could be confirmed. In addition to ring current shifts, stabilizing interactions like electrostatic dipole moments can also give rise to long range effects on the chemical shifts and provides information that can help to determine the structure (van Gammeren et al., 2005b).

### B. Structure Determination of a Chlorosome Antenna

In photosynthesis, light is generally collected by light harvesting antenna complexes that contain 50–200 chlorophylls in well defined membrane bound protein structures. In contrast to the protein-based structures of all other light harvesting antennae in photosynthesis, the light harvesting complexes in green bacteria comprise a protein-free assembly of bacteriochlorophylls and are named chlorosomes (Blankenship et al., 1995).

Based on a variety of experimental data, a general model for the chlorosome structure was proposed by Holzwarth and Schaffner (1994). According to this model, BChl *c*, *d*, or *e* molecules are able to self-organize into tubular antennae superstructures yielding a strongly red-shifted *J*-type absorption band and extremely large exciton diffusion lengths covering up to 200 monomers at room temperature. These highly unusual optical and exciton properties were not yet found for any other organic material, which makes the study of chlorosomes relevant as a starting point for bio-inspired smart matrices for artificial photosynthesis or nanoelectronics.

#### 1. A Model for the Structure of the Chlorosome Antenna

$^{13}C$-Enriched chlorosomes were prepared from growth media containing $NaH^{13}CO_3$. The 2D response of the BChl *c* in intact chlorosomes is virtually indistinguishable from the data collected from in vitro aggregates prepared in hexane with respect to chemical shifts, line widths, and relative intensities of the cross-peaks (Fig. 9). This provides conclusive evidence that self-assembly provides the structural basis for the BChl *c* organization in vivo.

Although the resonances are inhomogeneously broadened and reveal considerable global structural heterogeneity, it was possible to determine a model for the structure by heteronuclear 2D and 3D dipolar correlation spectroscopy (van Rossum et al., 2001). 2D heteronuclear $^1H$-$^{13}C$ correlation data have been recorded using the frequency- and phase-switched Lee-Goldburg technique to assign the $^1H$ response. Aggregation shifts relative to the monomeric BChl *c* in solution are detected for protons attached to rings I, II, and III/V and to their side chains (Fig. 10). The 5-CH and the 7-methyl responses are doubled, which provides evidence for the existence of at least two relatively well-defined structurally different arrangements. *Ab-initio* quantum chemical modeling studies were performed to refine a trimer model for the local environment of self-assembled BChl *c* in two different types of BChl stacks (Fig. 11). The BChl in the

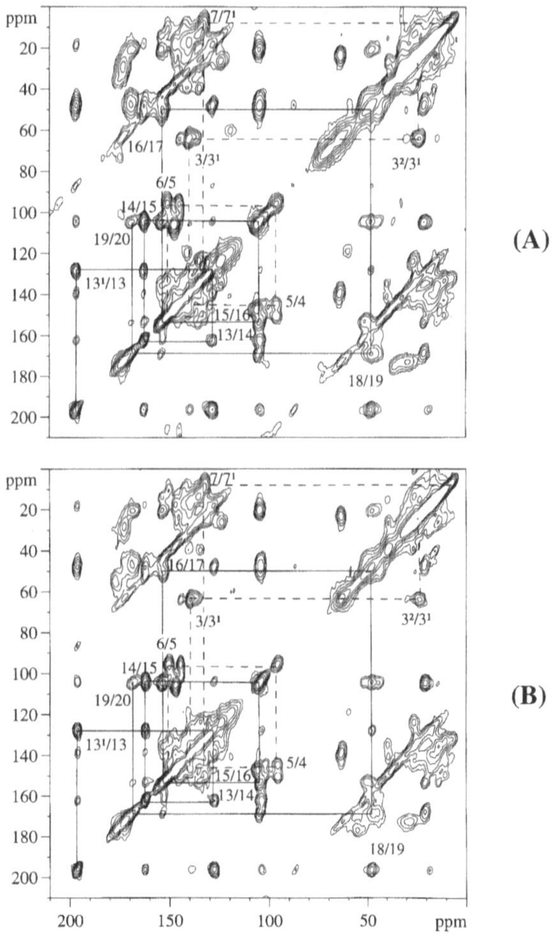


*Fig. 9.* Contour plots of 2D MAS dipolar correlation NMR spectra of uniformly $^{13}C$ labeled chlorosomes (A), and $^{13}C$ labeled BChl *c* aggregates (B). The lines indicate sequences of nearest neighbor correlations (de Boer and de Groot, 2005).

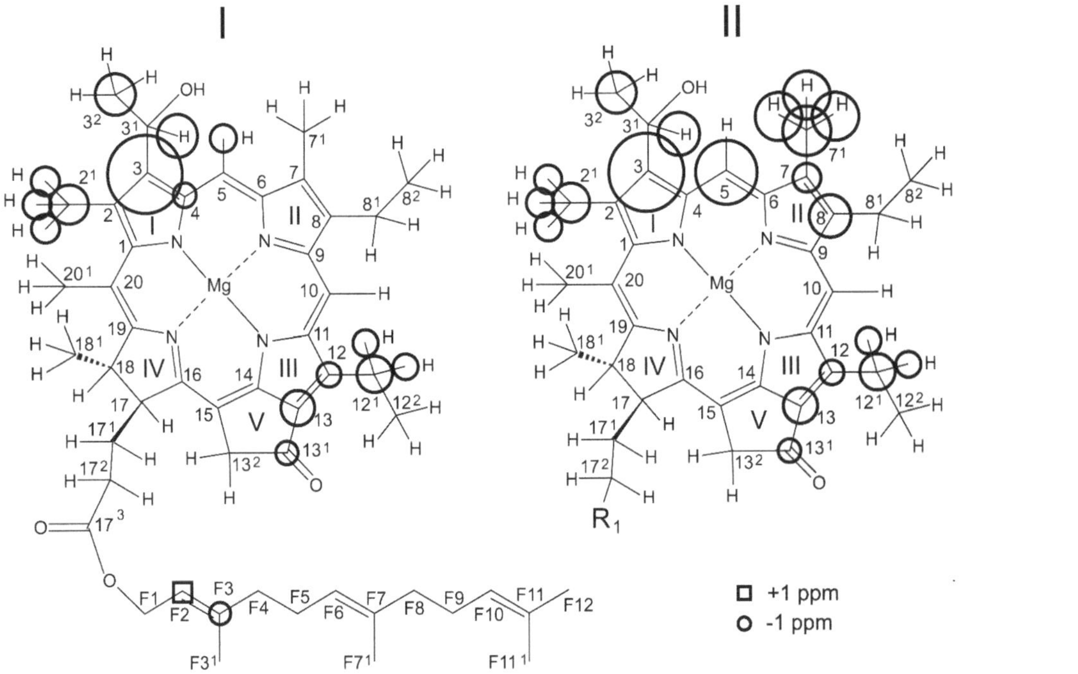


*Fig. 10.* The MAS NMR reveals two BChl *c* components in the chlorosomes (I and II). The circles around the carbon and hydrogen atoms represent upfield aggregation shifts, the squares downfield aggregation shifts. The size of a circle or square is proportional to the magnitude of the aggregation shift (van Rossum et al., 2001, with permission).

stacks can adopt either an *anti* or a *syn* configuration of the coordinative bond, where *anti* and *syn* designate the relative orientation of the Mg-OH bond relative to the direction of the 17-$17^1$ bond. This leads to a bilayer tube model based on antiparallel sheets of BChl *c* for the suprastructure, by elimination of other possibilities based on experiments and energetic considerations from modeling (Fig. 12). In our model structures we observe a cancellation of electric dipole moments between layers, and a bilayer tube would in principle allow for a similar cancellation at the level of the tube. The MAS NMR ring current shifts are very discriminative for validating the suprastructure up to ~ 2 nm. The variation of the radii probably forms a narrow Gaussian-like distribution around an average value. The bilayer tube model is currently under evaluation as a scaffold for the suprastructure of the chlorosome rods (Balaban et al., 1995; de Boer et al., 2003; Möltgen et al., 2002).

## 2. Molecular Control of Self-assembly in Chlorosome Antennae

From the early modeling, it was proposed that the self-assembly is driven by three noncovalent interactions without requiring proteins for structural control (Holzwarth and Schaffner 1994). These three interactions are coordination of the $3^1$-oxygen unit to the magnesium ion, hydrogen bonding between the $3^1$-alcohol to the 13-keto group and $\pi$–$\pi$ stacking between the extended chlorin chromophores. However, this does not explain the opposite curvatures of the inner and outer walls, and the fine-tuning of the macrocycle overlap to optimize for long exciton lifetimes.

Energetic stabilization of curved bilayer structures can occur when two components are present, due to strong interactions between the two components and due to stiffness of the spontaneous curvature in a single layer that reduces the free energy gain that can be obtained from entropy and interactions (Safran et al., 1991; van Rossum et al., 2001). MAS NMR spectroscopy has been used to investigate the role of symmetry breaking at the monomer level and how this leads to two components producing a robust suprastructure allowing for tuning of optical properties (de Boer et al., 2004). In the chlorosomes such processes should lead to phase separation on a nanoscale to establish two layers with opposite curvature and a bilayer tube (de Boer et al., 2005). The natural system was compared with the two model

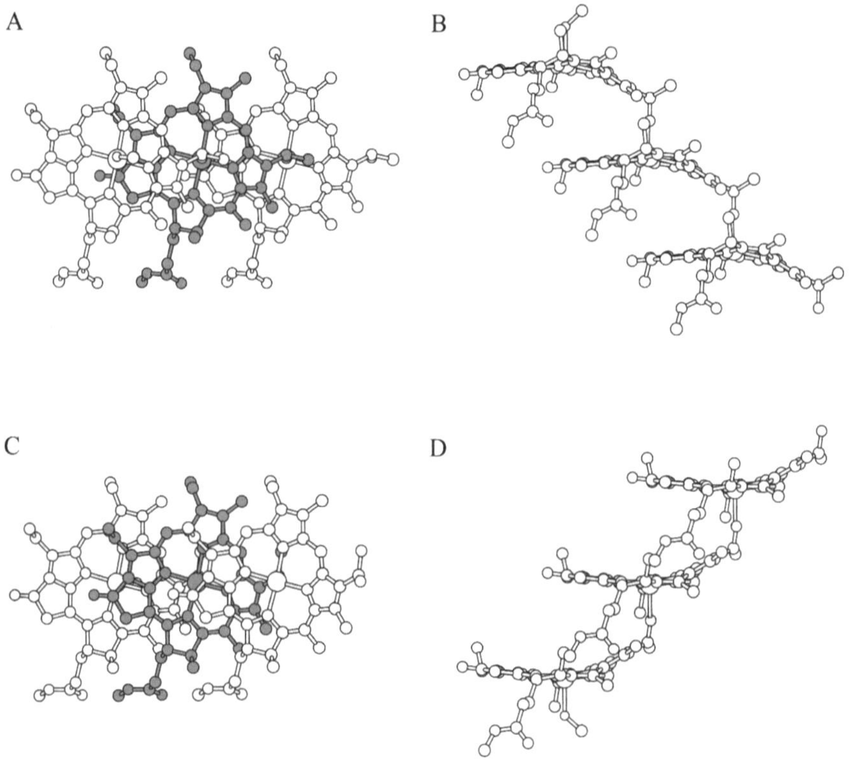


*Fig. 11.* The BChl *c* can form *anti* stacks (A and B) or *syn* stacks (C and D), which are shown here in two orthogonal views (van Rossum, 2000).

cadmium chlorins that were presented in the previous subsection (de Boer et al., 2004). The chlorin models differ from the natural BChl *c* in the central metal and the 3-, 12-, 17-, and 20-side chains. One model system has the farnesyl tail replaced by a methyl, while the other has a stearyl tail. The $^{113}$Cd MAS NMR signals indicate a HO···Cd coordination as a fifth ligand to the $Cd^{2+}$ metal ion, very similar to the HO···Mg coordination in the natural system. In contrast, the $^{1}$H ring-current shifts are up to 10 ppm, much larger than for the natural system, They reveal a dense orderly stacking in planar layers with a correlation length of at least 24 Å. The chlorin with the truncated tails forms microcrystalline aggregates with a monoclinic local symmetry (Fig. 7). In contrast, the stearyl tails give rise to a considerably disordered chlorin aggregate consisting of both *syn* and *anti* layers, similar to the chlorosomes.

With the help of these model structures, several key factors controlling the self-organization in the chlorosomes can be designated. The Mg···OH···O=C motif induces self-organization in a robust bottom-up process in essentially the same fashion as the Cd···OH···O=C moieties in the chlorin models. The packing of the molecules in sheets is less dense in the chlorosomes than for the models and this gives rise to moderate ring current shifts in the natural system. The lower density in the natural system can be attributed mainly to the $3^1$- and 20-substituents, which are bulkier than in the model systems. The model structures identify different 3-moieties in BChl *c* as a prerequisite for tubular suprastructure and validate the role of the [$3^1R$] and [$3^1S$] stereoisomers as the two essential components in the formation of the chlorosomal antennae, since tubular structures are not formed without this chirality. The *ab-initio* studies indicate that the [$3^1R$] and [$3^1S$] stereoisomers in the chlorosomes break the symmetry and produce sheets with opposite spontaneous curvature at longer length scales that are intrinsically robust over a wide composition range of the two components in a synergistic mechanism with the out-of-plane position of the central metal ion in its five-coordination, leading to the *syn* and *anti* conformations of the monomer (Fig. 11). The presence of *syn* and *anti* layers is associated with the presence of stearyl tails, similar to the chlorosomal antennae, and the long tails apparently help to stabilize a bilayer structure. The precise role

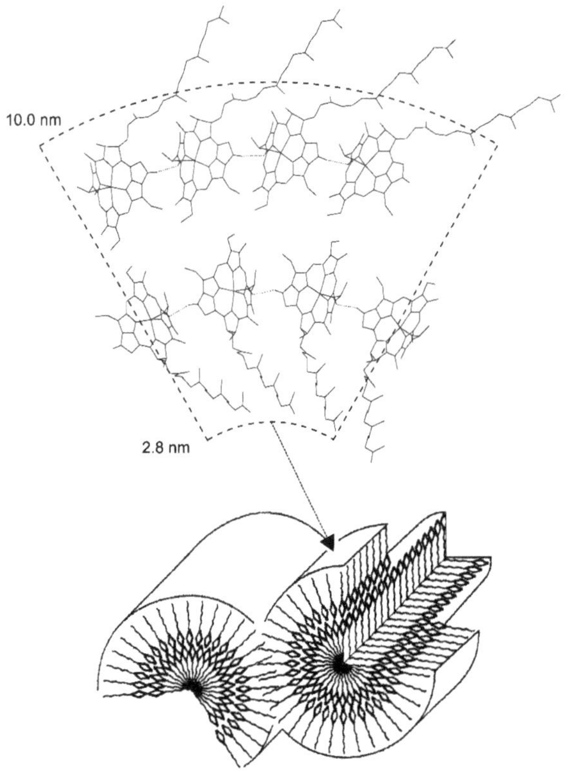


*Fig. 12.* In the chlorosome antenna the BChls form rods. According to the MAS NMR such rods can be formed from curved 2D sheets of *anti* (I) and *syn* (II) stacks (right panel) (de Boer and de Groot, 2005).

of the interactions between the two components and whether they stabilize or destabilize the structure is not yet fully known.

At an early stage the chlorosome of the green bacterium *Chlorobium tepidum* was studied by 1-D MAS NMR methods and antiparallel stacking of the BChl was proposed, in contrast with the parallel stacking emerging from the structural investigations that were performed afterwards (Nozawa et al., 1994). To investigate this point in more detail, ($3^1R$)-BChl *c* solid aggregates with an absorbance around 740 nm were formed from BChl *c* dimers, and 2D homonuclear $^{13}C$-$^{13}C$ radio frequency-driven dipolar recoupling as well as proton-driven spin diffusion dipolar correlation NMR spectra were collected in ultra high magnetic field, while $^{15}N$ chemical shifts were assigned from 2D heteronuclear $^{15}N$-$^{13}C$ correlation experiments using spectrally induced filtering in combination with cross polarization (Umetsu et al., 2004). The data clearly show that the stacking in $CH_2Cl_2$-treated aggregates is different from the stacking in the chlorosomes and hexane-treated aggregates. It was proposed that the $CH_2Cl_2$ precipitate represents a structural intermediate between the antiparallel dimer and the parallel stack as found in the chlorosome.

## *C. Structure Determination of Membrane Protein Complexes.*

### *1. Sequence Specific Assignments by Pattern Labeling*

Membrane proteins can form complex structures with primary, secondary, tertiary and quaternary structure. Such a high level of complexity requires specific strategies for NMR structure determination. An essential step for structure determination of membrane proteins by MAS NMR is to obtain a sequence-specific assignment of chemical shifts. This involves editing of spectra with pattern isotope labeling and implementation of N-C correlation pulse sequences for establishing correlations between adjacent amino acids in the sequence (Fig. 13). To achieve this, partly biosynthetic site-directed isotopically $^{13}C$ enriched LH2 complexes have been prepared from *Rhodopseudomonas acidophila* strain 10050 by using chemically labeled [1,2,3,4–$^{13}C$], [1,4–$^{13}C$] and [2,3–$^{13}C$] succinic acid as a precursor in the growth medium and 2D PDSD correlation spectroscopy was used to trace the $^{13}C$ isotopes from the labeled succinic acid precursors to their destination into the protein (van Gammeren et al., 2004). Carbonyl carbons in the protein backbone were labeled by [1,4–$^{13}C$]-succinic acid, while the Cα and Cβ carbons of the residues were labeled by [2,3 $^{13}C$]-succinic acid. Leucine and isoleucine residues were labeled using a uniformly labeled amino acid mixture in the medium.

The pattern labeling yields an increase of the resolution and less spectral crowding (Fig. 13). It provides an attractive route to resolve chemical shifts for α-helical transmembrane protein structures. Assignments were obtained from 2D PDSD $^{13}C$–$^{13}C$ correlation datasets with mixing times of 20 and 500 ms and band selective $^{13}C$–$^{15}N$ correlation spectroscopy (van Gammeren et al., 2005b). A decreased line width together with a reduced number of correlation signals of the selectively labeled samples compared to the uniformly labeled samples enable to resolve the narrowly distributed correlation signals of the backbone carbons and nitrogens involved in the long α-helical transmembrane segments. Correlations between nearby residues and between residues and the labeled BChl *a* cofactors, provided by the $^{13}C$–$^{13}C$ correlation experiments using a 500 ms spin diffusion period, are used to arrive at sequence specific chemical shift

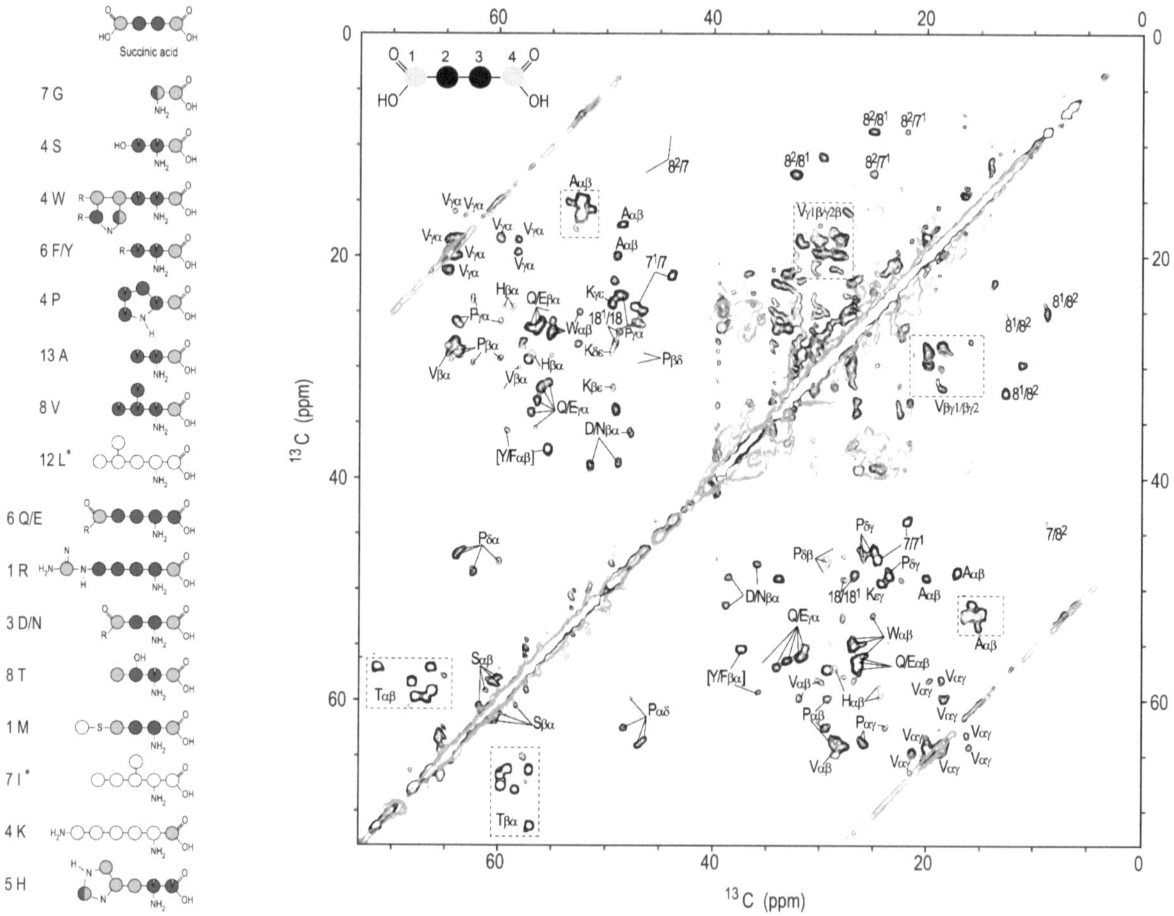


*Fig. 13.* Pattern labeling of LH2 starting from 1,4 (black) or 2,3 (grey) $^{13}$C labeled succinic acid in *R. acidophila*. On the left the enrichment of the residues is shown, while on the right a superposition of the aliphatic region of two $^{13}$C-$^{13}$C PDSD datasets is shown, illustrating the spectral simplifications that can be obtained with pattern labeling (van Gammeren et al., 2004b). See also Fig. 2, Color Plate 6.

assignments for 76 residues of the 94 residues of the monomeric unit of the LH2 complex. In this way it is demonstrated that MAS NMR methods combined with site-specific biosynthetic isotope labeling can be used for sequence specific assignment of the NMR response of transmembrane helical proteins.

## 2. Secondary Shifts for Structure Determination

In the practical applications of the solids NMR methods thus far, the distance information is generally scarce due to a competition between relayed transfer of magnetization along the molecular backbone with long range through space transfer. This contrasts with solution NMR, where abundant distance constraints from NOE's can be collected. To obtain a structure the resonance assignments obtained with MAS NMR on the LH2 can be compared with random coil values in the liquid state (Fig. 14). This is in line with the analysis of aggregation shifts in the previous sections for resolving structure of the aggregates and the chlorosome antenna system. The resulting chemical shift differences are called secondary chemical shifts and are evaluated in light of the backbone torsion angle $\psi$ previously reported using X-ray crystallography. Also for solids a correlation was reported suggesting that the concept of secondary chemical shifts, well established in the liquid state, can be of similar importance in the context of multiple-labeled polypeptides studied under MAS conditions (Luca et al, 2001). Since fast and reliable methods for structure determination are in great demand, several research teams are working on forging the tools for structure determination based on sequence specific assignments. For instance, probability maps for secondary structure can be obtained from chemical shift homology for strings of residues

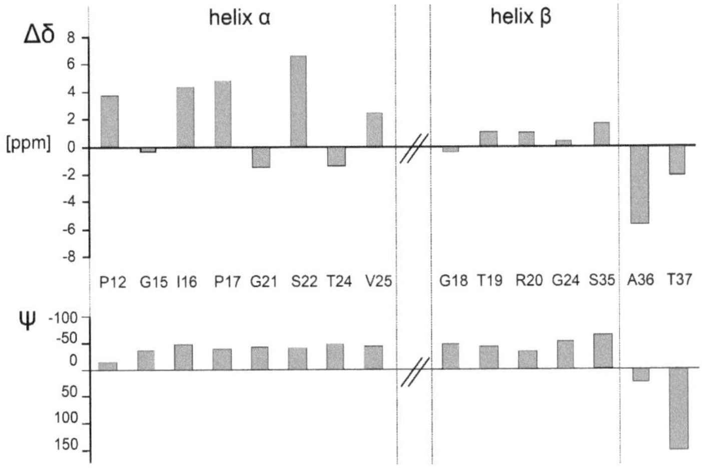


*Fig. 14.* Secondary chemical shifts for helix regions in LH2. Predominantly positive Δδ correspond with helical transmembrane regions, in line with the X-ray structure (Luca et al., 2001).

and such probability maps can be incorporated into structure calculation methods for membrane proteins (Cornilescu et al., 1999).

In the future the emphasis on analyses of NMR parameters like the chemical shift for structure determination may turn out to be the most convenient and robust method for solids NMR structure determination. While in solution the packing of molecules is of little interest, and "families" of molecular conformations are resolved, in a genuine solid state the space filling arrangement is an essential element of the structure. Thus, while the structure determination of molecules in solution can follow a purely bottom-up approach, in the microcrystalline or paracrystalline solid state the top-down restrictions following from the requirement to fill up space in an orderly fashion have to be taken into account in the construction of a structural model from MAS NMR data. Analysis of ring current shifts in the sections III and IV illustrate this point. Hence it can be envisaged that a database of shifts that include the longer range effects will become a versatile route for efficient structure determination of moderately ordered solid assemblies, including membrane proteins.

## Acknowledgments

This work could not have been possible without the invaluable scientific contributions of T. Adshiri, Alia, F. Buda, M. Amakawa, J. Ashurst, T.S. Balaban, M. Baldus, G.J. Boender, I. de Boer, R. Boers, J. van Boom, L. Bosman, R.J. Cogdell, T.A. Egorova-Zachernyuk, C. Erkelens, D. Fillipov, M.R. Fischer, H. Förster, H.A. Frank, E. Franken, N.J. Fraser, A. van Gammeren, P. Gast, R. Gebhard, H.J. van Gorkom, C. de Groot, K. van der Hoef, A.J. Hoff, J.G. Hollander, A.R. Holzwarth, F.B. Hulsbergen, S. Kiihne, V. Ladizhansky, S. van Leeuwen, W.B.S. van Liemt, S. Luca, J. Lugtenburg, J. Matysik, T. Miyatake, F.M. Mulder, G. Nachtegaal, T. Nozawa, D. Oesterhelt, H.Oschkinat, S. Prytulla, J. Raap, B.J. van Rossum, S. Sasaki, K. Schaffner, E.A.M. Schulten, S. Shochat, C. Soede-Huijbregts, D.B. Steensgaard, H.Tamiaki, U. Umetsu, S. Vega, E. Vijgeboom, C.A.Violette, E. Vinogradov, J. Wachtveitl, Z.Y. Wang, C. Winkel, S. Yagai and H. Yagai. In addition, N. Braakman is thanked for his help with editing the figures and Mrs. E. van der Velden for help with the preparation of the manuscript.

## References

Alia, Matysik J, Soede-Huijbregts C, Baldus M, Raap J, Lugtenburg J, Gast P, van Gorkom HJ, Hoff AJ and de Groot HJM (2001) Correlation spectroscopy of the histidine residues in Light-Harvesting Complex II from photosynthetic bacteria reveals partial internal charge transfer in the B850/His complex. J Am Chem Soc 123: 4803–4809

Alia, Matysik J, de Boer I, Gast P, van Gorkom HJ, de Groot HJM (2004) Heteronuclear 2D ($^{1}H$-$^{13}C$) MAS NMR resolves the electronic structure of coordinated histidines in light-harvesting complex II: Assessment of charge transfer and electronic delocalization effect. J Biomol NMR 28: 157–64

Andrew ER, Bradbury A and Eades RG (1958) Nuclear magnetic resonance spectra from a crystal rotated at high speed. Nature 182: 1659–1659

Balaban TS, Holzwarth AR, Schaffner K, Boender GJ and de Groot HJM (1995) CP/MAS $^{13}$C-NMR Dipolar Correlation Spectroscopy of $^{13}$C enriched chlorosomes and isolated bacteriochlorophyll *c* aggregates of *Chlorobium tepidum*: The self-organisation of pigments is the main structural feature of chlorosomes. Biochemistry 34: 15259–15266

Bennett AE, Ok JH, Griffin RG and Vega S (1992) Chemical-shift correlation spectroscopy in rotating solids — radio frequency-driven dipolar recoupling and longitudinal exchange. J Chem Phys 96: 8624–8627

Bennett AE, Rienstra CM, Auger M, Lakshmi KV and Griffin RG (1995) Heteronuclear decoupling in rotating solids. J Chem Phys 103: 6951–6958

Bielecki A, Kolbert AC and Levitt MH (1989) Frequency-switched pulse sequences — homonuclear decoupling and dilute spin NMR in solids. Chem Phys Lett 155: 341–346

Blankenship RE, Olson JM and Miller M (1995) Antenna complexes from green photosynthetic bacteria. In: Blankenship RE, Madigan MT and Bauer CE (eds) Anoxygenic Photosynthetic Bacteria (Advances in Photosynthesis and Respiration, Vol 2), pp 399–435. Kluwer Academic Publishers, Dordrecht

Boender GJ, Balaban TS, Holzwarth AR, Schaffner K, Raap J, Prytulla S, Oschkinat H and de Groot HJM (1995a) Comparison of the stacking of chlorophylls in chlorosomes versus aggregates of bacteriochlorophyll *c* and chlorophyll *a* using 2D MAS NMR spectroscopy. In: Mathis P (ed) Photosynthesis: From Light to Biosphere, Vol I, pp 347–350. Kluwer Academic Publishers, Dordrecht

Boender GJ, Raap J, Prytulla S, Oschkinat H and de Groot HJM (1995b) MAS NMR structure refinement of uniformly $^{13}$C enriched chlorophyll-*a*/water aggregates with 2D dipolar correlation spectroscopy. Chem Phys Lett 237: 502–508

Boers RB, Gast P, Hoff AJ, de Groot HJM and Lugtenburg J (2002) Synthesis and spectroscopic characterization of [5-$^{13}$C]- and [6-$^{13}$C]-ubiquinone-10 for studies of bacterial photosynthetic reaction centers. Eur J Org Chem 1: 189–202

Castellani F, van Rossum BJ, Diehl A, Schubert M, Rehbein K and Oschkinat H (2002) Structure of a protein determined by solid-state magic-angle-spinning NMR spectroscopy. Nature 420: 98–102

Cornilescu G, Delaglio F, Bax A (1999) Protein backbone angle restraints from searching a database for chemical shift and sequence homology. J Biomol NMR 13: 289–302

de Boer I (2004) Modeling of chlorosomal light-harvesting antennae: Molecular control of self-assembly of chlorins resolved by MAS NMR. PhD thesis. Leiden University

de Boer I and de Groot HJM (2005) Magic Angle Spinning NMR of the chlorosomes. In: Grimm B, Porra R, Rüdiger W, Scheer H (eds) Chlorophylls and Bacteriochlorophylls: Biochemistry, Biophysics, Functions and Applications (Advances in Photosynthesis and Respirations, Vol 25). Kluwer Academic Publishers, Dordrecht

de Boer I, Bosman L, Raap J, Oschkinat H and de Groot HJM (2002). 2D $^{13}$C-$^{13}$C MAS NMR correlation spectroscopy with mixing by true $^{1}$H spin diffusion reveals long-range intermolecular distance restraints in ultra high magnetic field. J Magn Reson 157: 286–291

de Boer I, Matysik J, Amakawa M, Jakai S, Tamiaki H, Holzwarth AR and de Groot HJM (2003) MAS NMR Structure of a microcrystalline Cd-bacteriochlorophyll *d* analog. J Am Chem Soc 125: 13374–13375

de Boer I, Matysik J, Erkelens K, Sasaki S, Miyatake T, Yagai H, Tamiaki H, Holzwarth AR and de Groot HJM (2004) MAS NMR structures of aggregated Cadmium chlorins reveal molecular control of self-assembly of chlorosomal bacteriochlorophylls. J Phys Chem B 108: 16556–16566

de Boer I, Matysik J, Sasaki S, Miyatake T, Yagai S, Tamiaki H, Holzwarth AR and de Groot HJM (2005) Self-organization of BChl *c* in chlorosomes studied by MAS NMR of aggregated Cd-chlorins. In: van der Est A and Bruce D (Eds) Photosynthesis: Fundamental Aspects to Global Perspectives, pp 121–122. Alliance Communications Group, Lawrence

de Groot HJM, Raap J, Winkel C, Hoff AJ and Lugtenburg J (1990) Magic Angle Spinning $^{13}$C NMR with atomic resolution of a photosynthetic reaction center enriched in [4′-$^{13}$C] tyrosine. Chem Phys Lett 169: 307–310

de Groot HJM, Gebhard R, van der Hoef K, Violette CA, Hoff AJ, Frank HA, Lugtenburg J (1992) $^{13}$C Magic Angle Spinning NMR evidence for a 15,15′-*cis* configuration of the spheroidene chromophore in the *Rhodobacter sphaeroides* photosynthetic reaction center. Biochemistry 31: 12446–12450

Duer MJ (2004) Introduction to Solid State NMR Spectroscopy. Blackwell, Oxford

Egorova-Zachernyuk TA, van Rossum BJ, Boender GJ, Franken E, Ashurst J, Raap J, Gast P, Hoff AJ, Oschkinat H and de Groot HJM (1997) Characterization of pheophytin ground states in *Rhodobacter sphaeroides* R26 photosynthetic reaction centers, from multispin pheophytin enrichment and 2D $^{13}$C MAS MR dipolar correlation spectroscopy. Biochemistry 36: 7513–7519

Ernst RR, Bodenhausen G and Wokaun A (1987) Principles of Nuclear Magnetic Resonance in One and Two Dimensions. Clarendon Press, Oxford

Farchaus JW, Wachtveitl J, Mathis P and Oesterhelt D (1993) Tyrosine 162 of the photosynthetic reaction center L-subunit plays a critical role in the cytochrome $c_2$ mediated rereduction of the photooxidized bacteriochlorophyll dimer in *Rhodobacter sphaeroides*. 1. Site-directed mutagenesis and initial characterization. Biochemistry 32: 10885–10893

Ferreira KN, Iverson TM, Maghlaoui K, Barber J and Iwata S (2004) Architecture of the photosynthetic oxygen-evolving center. Science 303: 1831–1838

Fischer MR, de Groot HJM, Raap J, Winkel C, Hoff AJ and Lugtenburg J (1992) $^{13}$C Magic Angle Spinning NMR study of the light-induced and temperature-dependent changes in *Rhodobacter sphaeroides* R26 reaction centers enriched in [4′-$^{13}$C] tyrosine. Biochemistry 31: 11038–11049

Garcia-Viloca M, Gao J, Karplus M and Truhlar DG (2004) How enzymes work: Analysis by modern rate theory and computer simulations. Science 303: 186–195

Goward GR, Sebastiani D, Schnell I, Spiess HW, Kim HD and Ishida H (2003) Benzoxazine oligomers: Evidence for a helical structure from solid-state NMR spectroscopy and DFT-based dynamics and chemical shift calculations. J Am Chem Soc 125: 5792–5800

Hartmann SR and Hahn EL (1962) Nuclear double resonance in the rotating frame. Phys Rev 128: 2042–2053

Holzwarth AR and Schaffner K (1994) On the structure of bacteriochlorophyll molecular aggregates in the chlorosomes of green bacteria — a molecular study. Photosynth Res 41: 225–233

Ivancich A, Artz K, Williams JC, Allen JP and Mattioli TA (1998) Effects of hydrogen bonds on the redox potential and electronic structure of the bacterial primary electron donor. Biochemistry 37: 11812–11820

Kiihne S and de Groot HJM (eds) (2001) Perspectives on Solid State NMR in Biology. Kluwer Academic Publishers, Dordrecht

Lee M and Goldburg WI (1965) Nuclear-magnetic-resonance line narrowing by a rotating rf field. Phys Rev A 140: 1261–1271

Luca S, Fillipov D, van Boom J, Oschkinat H, de Groot HJM and Baldus M (2001) Secondary chemical shifts in immobilized peptides and proteins: A qualitative basis for structure refinement under magic angle spinning. J Biomol NMR: 20:325–331

Matysik J, Alia, Nachtegaal G, van Gorkom HJ, Hoff AJ and de Groot HJM (2000) Exploring the calcium binding site in photosystem II membranes by solid state $^{113}$Cd-NMR. Biochemistry 39: 6751–6755

Möltgen H, Kleinermanns K, Jesorka A, Schaffner K and Holzwarth AR (2002) Self-assembly of [Et,Et]-bacteriochlorophyll c(F) on highly oriented pyrolytic graphite revealed by scanning tunneling microscopy. Photochem Photobiol 75: 619–626

Mulder FM, Heinen W, van Duin M, Lugtenburg J and de Groot HJM (1998) Spin diffusion with $^{13}$C selection and detection for the characterization of morphology in labeled polymer blends with MAS NMR. J Am Chem Soc 120: 12891–12894

Ladizhansky V, Vinogradov E, van Rossum BJ, de Groot HJM and Vega S (2003) Multiple-spin effects in fast magic angle spinning Lee-Goldburg cross-polarization experiments in uniformly labeled compounds. J Chem Phys 118: 5547–5557

Lowe IJ (1959) Free induction decays of rotating solids. Phys Rev Lett 2: 285–287

Luca S, Fillipov D, van Boom J, Oschkinat H, de Groot HJM and Baldus M (2001) Secondary chemical shifts in immobilized peptides and proteins: A qualitative basis for structure refinement under magic angle spinning. J Biomol NMR 20: 325–331

Nozawa T, Ohtomo K, Suzuki M, Nakagawa H, Shikama Y, Konami H and Wang ZY (1994) Structures of chlorosomes and aggregated BChl *c* in *Chlorobium tepidum* from solid state high resolution CP/MAS $^{13}$C NMR. Photosynth Res 41: 211–223

Pauling L (1946) Molecular architecture and biological reactions. Chem Eng News 24: 1375–1377

Petkova AT, Ishii Y, Balbach JJ, Antzutkin ON, Leapman RD, Delaglio F and Tycko R (2002) A structural model for Alzheimer's beta-amyloid fibrils based on experimental constraints from solid state NMR. Proc Natl Acad Sci USA 99: 16742–1674

Prokhorenko VI, Holzwarth AR, Nowak FR and Aartsma TJ (2002) Growing-in of optical coherence in the FMO antenna complexes. J Phys Chem B 106: 5761–5768

Rienstra CM, Tucker-Kellogg L, Jaroniec CP, Hohwy M, Reif B, McMahon MT, Tidor B, Lozano-Perez T and Griffin RG (2002) De novo determination of peptide structure with solid-state magic-angle spinning NMR spectroscopy. Proc Natl Acad Sci USA 99: 10260–10265

Rose ME (1967) Elementary Theory of Angular Momentum. Wiley, New York

Safran SA, Pincus PA, Andelman D and MacKintosh FC (1991) Stability and phase behavior of mixed surfactant vesicles. Phys Rev A43: 1071–1078

Samoson A, Tuherm T, Past J, Reinhold A, Anupõld T, Heinmaa I (2005) New horizons for magic angle spinning NMR. Topics in Curr Chem 246: 15–31

Schulten EAM, van Rossum BJ, Ashurst J, Oschkinat H, Raap J, Lugtenburg J and de Groot HJM (1998) Two-dimensional ($^1$H-$^{13}$C) solid state MAS NMR of isotope labeled chlorophyll *a*. In: Garab G (ed) Photosynthesis: Mechanisms and Effects, Vol I, pp 453–456. Kluwer Academic Publishers, Dordrecht

Sham YY, Muegge I, and Warshel A (1998) The effect of protein relaxation on charge-charge interactions and dielectric constants of proteins. Biophys J 74: 1744–1753

Shochat S, Arlt T, Francke C, Gast P, van Noort PI, Otte SCM, Schelvis HPM, Schmidt S, Vijgenboom E, Vrieze J, Zinth W and Hoff AJ (1994) Spectroscopic characterization of reaction centers of the (M)Y210W mutant of the photosynthetic bacterium *Rhodobacter sphaeroides*. Photosynth Res 40: 55–66

Shochat S, Gast P, Hoff AJ, Boender GJ, van Leeuwen S, van Liemt WBS, Vijgeboom E, Raap J, Lugtenburg J and de Groot HJM (1995) $^{13}$C MAS NMR evidence for a homogeneously ordered environment of tyrosine M210 in reaction centres of *Rhodobacter sphaeroides*. Spectrochim Acta A 51: 135–144

Schmidt-Rohr K and Spiess HW (1996) Multidimensional Solid State NMR and Polymers. Academic Press, London

Umetsu U, Matysik J, Hollander JG, Wang ZY, Adshiri Tada F, Nozawa T and de Groot HJM (2004) Magic-angle spinning nuclear magnetic resonance under ultrahigh field reveals two forms of intermolecular interaction within $CH_2Cl_2$-treated (31R)-type bacteriochlorophyll *c* solid aggregate. J Phys Chem B 108: 2726–2734

van Gammeren A, Hulsbergen FB, Erkelens K and de Groot HJM (2004a) Synthetic analogues of the histidine-chlorophyll complex: a NMR study to mimic structural features of the photosynthetic reaction center and the light-harvesting complex. J Biol Inorg Chem 9: 109–117

van Gammeren A, Hulsbergen FB, Hollander J, de Groot HJM (2004b) Biosynthetic site-specific $^{13}$C labeling of the light-harvesting 2 protein complex: a model for solid state NMR structure determination of transmembrane proteins. J Biomol NMR 30: 267–274

van Gammeren A, Buda F, Hulsbergen FB, Kiihne S, Hollander JG, Egorova-Zachernyuk TA, Fraser NJ, Cogdell RJ and de Groot HJM (2005a) Selective chemical shift assignment of both B800 and B850 bacteriochlorophylls of uniformly [$^{13}$C, $^{15}$N] labeled light-harvesting complexes from *Rhodopseudomonas acidophila* strain 10050 by solid state NMR at ultra-high magnetic field. J Am Chem Soc 127: 3213–3219

van Gammeren AJ, Hulsbergen FB, Hollander JG, de Groot HJM (2005b) Residual backbone and side-chain $^{13}$C and $^{15}$N resonance assignments of the intrinsic transmembrane light-harvesting 2 protein complex by solid state Magic Angle Spinning NMR spectroscopy. J Biomol NMR 31: 279–293

van Liemt W, Boender GJ, Gast P, Hoff AJ, Lugtenburg J and de Groot HJM (1995) $^{13}$C Magic Angle Spinning NMR characterization of the functionally asymmetric $Q_A$ binding in *Rhodobacter sphaeroides* R26 photosynthetic reaction centers using site-specific $^{13}$C-labeled ubiquinone-10. Biochemistry 34: 10229–10236

van Rossum BJ (2000) Structure refinement of photosynthetic components with multidimensional MAS NMR dipolar correlation spectroscopy. PhD thesis, Leiden University, Leiden

van Rossum BJ, van Liemt WBS, Boender GJ, Gast P, Hoff AJ, Lugtenburg J and de Groot HJM (1995) $^{13}$C MAS NMR relaxation study of the $Q_A$ binding in *Rhodobacter sphaeroides*-R26 reaction centers. In: Mathis P (ed) Photosynthesis: from Light to Biosphere, Vol 1, pp 899–902. Kluwer Academic Publishers, Dordrecht

van Rossum BJ, Boender GJ and de Groot, HJM (1996) High field for enhanced proton resolution in high speed CP/MAS heteronuclear $^{1}H$-$^{13}C$ dipolar correlation spectroscopy. J Magn Reson A120: 274–277

van Rossum BJ, Förster H and de Groot HJM (1997a) High-field and high-speed CP-MAS $^{13}C$ NMR heteronuclear dipolar-correlation spectroscopy of solids with frequency-switched Lee-Goldburg homonuclear decoupling. J Magn Reson 124: 516–519

van Rossum BJ, Wachtveitl J, Raap J, van der Hoef K, Gast P, Lugtenburg J, Oesterhelt D and de Groot HJM (1997b) $^{13}C$ MAS NMR evidence for structural similarity of L162YL mutant *Rhodobacter sphaeroides* R26 RC, despite widely different cytochrome $c_2$-mediated re-reduction kinetics of the oxidized primary donor. Spectrochim Acta A 53: 2201–2208

van Rossum BJ, Boender GJ, Mulder FM, Raap J, Balaban TS, Holzwarth A, Schaffner K, Prytulla S, Oschkinat H and de Groot HJM (1998) Multidimensional CP-MAS $^{13}C$ NMR of uniformly enriched chlorophyll. Spectrochimia Acta A 54: 1167–1176

van Rossum BJ, de Groot C, de Groot HJM, Ladizhansky V and Vega S (2000) A method for measuring hetronuclear ($^{1}H$-$^{13}C$) distances in high speed MAS NMR. J Am Chem Soc 122: 3465–3472

van Rossum BJ, Steensgaard DB, Mulder FM, Boender GJ, Schaffner K, Holzwarth AR and de Groot HJM (2001) A refined model of the chlorosomal antennae of the green bacterium *Chlorobium tepidum* from proton chemical shift constraints obtained with high-field 2D and 3-D MAS NMR dipolar correlation spectroscopy. Biochemistry 40: 1587–1595

van Rossum BJ, Schulten EAM, Raap J, Oschkinat H and de Groot HJM (2002) A 3-D structural model of solid self-assembled chlorophyll *a*/$H_2O$ from multispin labeling and MAS NMR 2D dipolar correlation spectroscopy in high magnetic field. J Magn Res 155: 1–14

van Rossum BJ, van Liemt WBS, Gast, P, Lugtenburg J and de Groot HJM (2007) $^{13}C$-$^{1}H$ heteronuclear dipolar correlation studies of the hydrogen bonding of the quinones in *Rhodobacter sphaeroides* R26 RCs. Appl Magn Reson, in press

Vinogradov E, Madhu PK and Vega S (1999) High-resolution proton solid-state NMR spectroscopy by phase-modulated Lee-Goldburg experiment. Chem Phys Lett 314: 443–450

Yagai S, Miyatake T, Shimono Y, Tamiaki H (2001) Supramolecular structure of self-assembled synthetic zinc-13′-oxo-chlorins possessing a primary, secondary or tertiary alcoholic 3(1)-hydroxyl group: Visible spectroscopic and molecular modeling studies. Photochem Photobiol 73: 153–163

# Chapter 19

# Photochemically Induced Dynamic Nuclear Polarization (Photo-CIDNP) Magic-Angle Spinning NMR

Eugenio Daviso[1], Gunnar Jeschke[2] and Jörg Matysik[1]*
[1]*Leiden Institute of Chemistry, Leiden University, P.O. box 9502, 2300 RA Leiden, The Netherlands;* [2]*Physikalische Chemie, Universität Konstanz, 78457 Konstanz, Germany*



## Summary

Photochemically induced dynamic nuclear polarization (photo-CIDNP) is non-Boltzmann nuclear magnetization which can be observed by NMR spectroscopy as enhanced absorptive (positive) or emissive (negative) signals. In solids, photo-CIDNP has been observed since its discovery in 1994 in various photosynthetic reaction centers (RCs) by magic angle spinning (MAS) solid-state NMR. The photo-CIDNP effect in solids can be explained by a combination of several mechanisms, e.g., three-spin mixing (TSM) and differential decay (DD), which transfer the electron spin polarization obtained in the initial singlet radical pair to nuclei via hyperfine interaction. Therefore, photo-CIDNP MAS NMR allows to study the electronic structure of the electronic

*Author for correspondence, email: j.matysik@chem.leidenuniv.nl

*Thijs J. Aartsma and Jörg Matysik (eds): Biophysical Techniques in Photosynthesis II, pp. 385–399.*

ground-state after the photocycle (chemical shifts) and of the radical pair (NMR intensities), providing insight into the photochemical machinery of RCs at the atomic scale. Recently, nuclear polarizations up to a factor of 10,000 above the Boltzmann equilibrium have been observed, opening new experimental possibilities for solid-state NMR.

## I. Theory of Photo-CIDNP in Solids

### A. Magnetic Field Effect

Photosynthetic reaction centers (RCs) are light-driven electron pumps that convert light energy into chemical energy. Upon photoexcitation, an electron is transferred from the primary electron donor, P, to a chain of acceptors and in a sequence of electron transfer steps a cascade of radical pairs is formed. In a bacterial RC the following sequence of radical pairs appears, involving $\Phi_A$, a bacteriopheophytin *a*, and the two quinones $Q_A$ and $Q_B$:

$$PQ_AQ_B \rightarrow P^* \Phi_AQ_AQ_B \rightarrow P^{\bullet+} \Phi_A^{\bullet-} Q_AQ_B \rightarrow$$
$$P^{\bullet+} \Phi_AQ_A^{\bullet-}Q_B \rightarrow P^{\bullet+} \Phi_AQ_AQ_B^{\bullet-}$$

The rate of the electron transfer decreases at each step. The first electron transfer step takes place within 3 ps, whereas the final step from $Q_A$ to $Q_B$ occurs in 60–100 ms (for review, see Hoff and Deisenhofer, 1997). When the electron transfer to $Q_A$ is blocked, the lifetime of $P^{\bullet+}\Phi_A^{\bullet-}$ increases ca. 1000-fold for the singlet (to ~20 ns) and 100-fold for the triplet state (to ~1 ns), allowing for evolution under hyperfine (hf) interaction between the electron pair and nuclei (Fig. 1). Hf interactions and the difference of the *g*-values of the two radicals, $\Delta g$, become evident by interconversion of the singlet-born spin-correlated radical pair (SCRP) to a triplet radical pair. When the magnetic field effect on the triplet yield was discovered in bacterial RCs (Blankenship et al., 1977; Hoff et al., 1977b), the connection to the photochemically induced dynamic electron polarization (CIDEP), which had been discovered shortly before in chloroplasts (Blankenship et al., 1975) and RCs of purple bacteria (Hoff et al., 1977a), has been made immediately. Both, triplet yield and CIDEP were traced back to magnetic-field dependent interaction of electrons with nuclei (Hoff et al., 1977b; Werner et al., 1978; Hoff, 1984). Based on this interpretation, 'new classes of experiments' were foreseen (Goldstein and Boxer, 1987), although the exact mode of electron-nuclear interaction was not known yet.

Photo-CIDNP is well-known from liquid NMR (for reviews, see Hore and Broadhurst, 1993; Goez, 1997) and was observed for the first time in 1967 (Bargon and Fischer, 1967; Bargon et al., 1967; Ward and Lawler, 1967). Soon, it was explained by the radical-pair mechanism (RPM) (Closs and Closs, 1969; Kaptein and Oosterhoff, 1969) which also inspired the interpretation of the electron-nuclear interaction in RCs (Haberkorn and Michel-Beyerle, 1979; Haberkorn et al., 1979; Boxer et al., 1983). Early experiments on RCs by Hoff and Kaptein using liquid $^1$H NMR did not succeed. In 1994, Zysmilich and McDermott succeeded for the first time to observe photo-CIDNP in RCs, applying magic angle spinning (MAS) NMR at low temperature to uniformly $^{15}$N-labeled quinone-blocked bacterial RCs of *Rhodobacter* (*Rb.*) *sphaeroides* R26 (Zysmilich and McDermott, 1994). This solid-state photo-CIDNP effect has indeed been shown to be field dependent (Fig. 2) (Prakash et al., 2005b). Using the signal from about 3300 methyl groups of the entire bacterial RC at 31 ppm as internal standard, enhancement factors of 60 (17.6 Tesla), 1000 (9.4 Tesla) and about 10,000 (4.7 Tesla) were computed. Such dramatic enhancement factor allows for an enormous increase of selectivity and sensitivity in solid-state NMR, enabling the new classes of experiments predicted by Goldstein and Boxer (1987).

*Abbreviations:* *A* – secular part of the hf interaction; *B* – pseudosecular part of the hf interaction; CIDEP – chemically induced dynamic electron polarization; *d* – electron coupling; *d′* – electron dipole-dipole coupling at current orientation; *D* – magnitude of the electron dipole-dipole coupling; DD – differential decay; DDG – digital delay generator; DR – differential relaxation; ENDOR – electron nuclear double resonance; EPR – electron paramagnetic resonance; hf – hyperfine; ISC – inter system crossing; MAS – magic angle spinning; P – primary electron donor; photo-CIDNP – photochemically induced dynamic nuclear polarization; PS I – Photosystem I; PS II – Photosystem II; PSU – photosynthetic unit; *Rb.* – *Rhodobacter*; rf – radiofrequency; RFDR – radio frequency dipolar recoupling; RPM – radical-pair mechanism; S – singlet state; SCRP – spin-correlated radical pair; $T_0$ – triplet state with magnetic quantum number 0; TPPM – two-pulse phase modulation; TSM – three spin mixing; WT – wild type; $\Delta\Omega$ – difference of the electron Zeeman frequencies; $\omega_I$ – nuclear Zeeman frequency; $\Phi$ – pheophytin

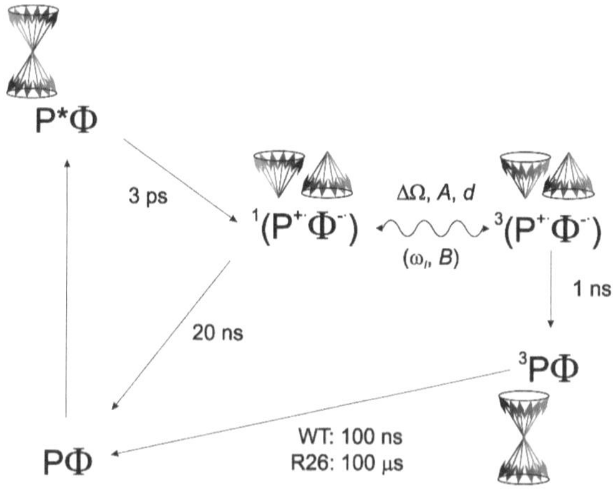


*Fig. 1.* Kinetics and spin dynamics of electron transport in quinone-blocked RCs of *Rhodobacter sphaeroides* wild-type (WT) and the carotenoid-less strain R26. After light-absorption, from the photochemically excited state of the primary donor P* an electron is transferred to the primary acceptor Φ, a bacteriopheophytin cofactor. This initial singlet radical pair $^1(P^{+\cdot}\Phi^{-\cdot})$ is fully *electron* polarized. An electron back-transfer leads to the initial electronic ground-state. Due to hyperfine interaction with nuclei, the singlet radical pair evolves into a triplet radical pair $^3(P^{+\cdot}\Phi^{-\cdot})$. Concomitantly to this process of spin intersystem crossing, electron polarization is transferred to nuclei by the three-spin mixing (TSM), which depends on $\Delta g$, $A$ and $d$. In the differential decay (DD) mechanism, a net photo-CIDNP effect is caused if spin-correlated radical pairs have different lifetimes in their singlet and triplet states. In RCs having a long lifetime of the donor triplet, as in R26 RCs, the differential relaxation (DR) mechanism occurs since nuclear spin relaxation is significant on the triplet branch causing incomplete cancellation of nuclear polarization of both branches.

## B. Primary Sources of Enhanced Nuclear Magnetization

The origin of photo-CIDNP in solution state is described by the RPM. In this process, either hf interaction or the $\Delta g$ mediates spin inter-system crossing (ISC) to the triplet state. The two different products, formed from singlets or triplets, have opposite polarization which does not cancel out if the chemical fate of singlet and triplet products is different, so that the nuclei with opposite polarization acquire different chemical shifts. The RPM, however, is not likely to be an explanation for solid-state photo-CIDNP in frozen RCs since no chemical branching occurs (Fig. 1). For the photo-CIDNP effect in solids, several new mechanisms have been proposed (for a recent discussion, see Jeschke and Matysik, 2003; Prakash et al., 2005b).

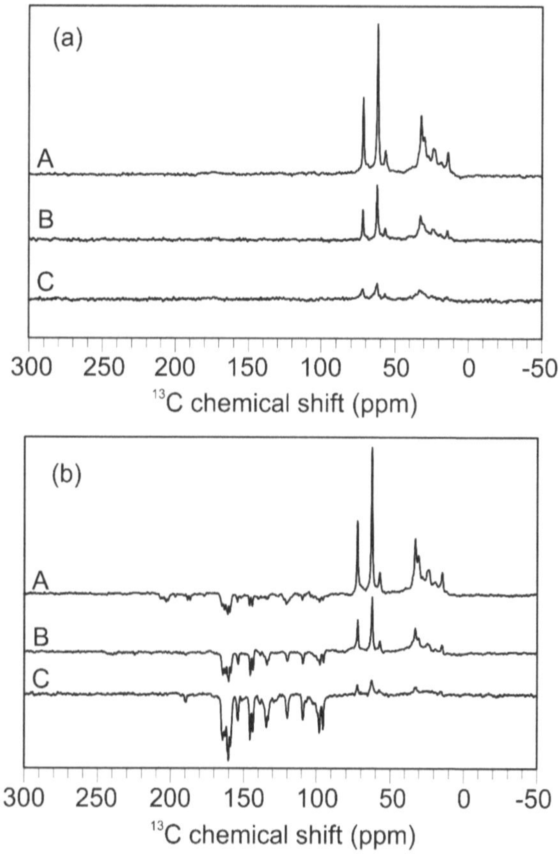


*Fig. 2.* $^{13}$C MAS NMR spectra of quinone-depleted RCs of *Rhodobacter sphaeroides* WT obtained in the dark (top) and under continuous illumination with white light (bottom) at different magnetic fields at 17.6 T (A), 9.4 T (B) and 4.7 T (C). Sample temperature was 223 K and MAS rotational frequency 8 kHz (Prakash et al., 2005b, with permission).

### 1. Three Spin Mixing

Photo-induced electron transfer in the RC leads to a spin-correlated radical pair (SCRP). The initial radical pair is born in a singlet state (S) and highly electron-polarized (Fig. 3). During the lifetime of the radical pair, the system oscillates between the singlet and the triplet state with magnetic quantum number $m = 0$ ($T_0$). The evolution to the $T_0$ state is accompanied by a transfer of polarization to the nuclei. The relevant parameters that drive the ISC are the secular part of the hf interaction ($A$), the difference of the $g$ value between the two electrons ($\Delta g$) and the coupling between the two electrons ($d$) (Fig. 4).

Net nuclear polarization is generated in SCRP due to the presence of both the pseudosecular part of hf interaction ($B$) and coupling between the two electron spins ($d$) (Fig. 5). The electron-electron-

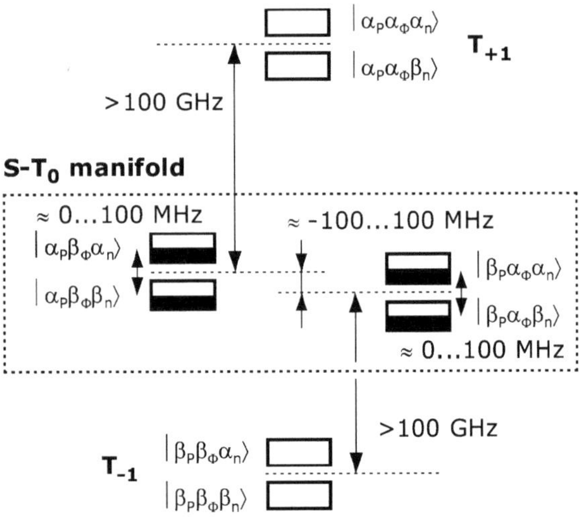


*Fig. 3.* Energy levels of the radical pair in bacterial RC. At thermal equilibrium, all four electron spin states in a correlated radical pair (S, $T_0$, $T_{+1}$ and $T_{-1}$) would be populated according to the Boltzmann distribution, however, immediately after generation only the S state is populated. Hence, the initial state is fully electron polarized. Intersystem crossing (ISC) by electron-electron-nuclear three-spin mixing (TSM) allows for equilibration between the S and $T_0$ states. The four quantum states of the S-$T_0$ manifold ($|\alpha_P\beta_\Phi\alpha_n\rangle$, $|\beta_P\alpha_\Phi\alpha_n\rangle$, $|\alpha_P\beta_\Phi\beta_n\rangle$ and $|\beta_P\alpha_\Phi\beta_n\rangle$) are split due to the difference of the Zeeman interaction of both electron spins, the hyperfine coupling to the nuclear spin, the nuclear Zeeman interaction, and the coupling between the two electron spins.

nuclear model system for three spin mixing (TSM) consists of two electron spins $S_1 = ½$, $S_2 = ½$, and one nuclear spin $I = ½$. A product operator basis for the S-$T_0$ subspace is constructed from operators of the nuclear spin $I = ½$ and of a fictitious electron spin $S' = ½$, with the fictitious spin corresponding to the zero-quantum transition of the two electron spins (S-$T_0$ manifold). Correspondence rules between spin operators of the two spins $S_1$ and $S_2$ and of the fictitious spin $S'$ have been given in another work (Jeschke, 1997). The density matrix at the instant of the radical pair formation is given by:

$$\sigma_{0(S_1T_0)} = \frac{1}{2}E_{S'} \mp S'_x$$

where $E$ is the identity operator. The Hamiltonian for the S-$T_0$ subsystem is:

$$H_{\text{S-T}_0} = \Delta\Omega S_z + \omega_I I_z + A_{zz} S_z I_z + B_{zx} S_z I_x + B_{zy} S_z I_y - d S_x$$

In this equation $\Delta\Omega$ is the difference of the electron Zeeman frequencies, $\omega_I$ the nuclear Zeeman interac-

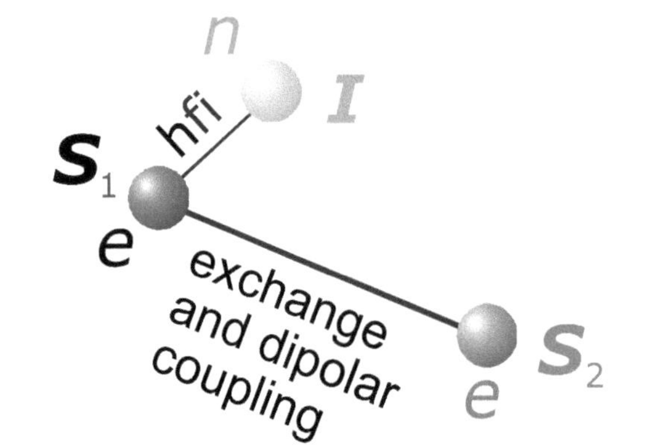


*Fig. 4.* The electron-electron-nuclear model system and the interaction forces occurring in the three-spin mixing (TSM) process. The interaction between electron $S_1$ and nucleus $I$, which may lead to a degeneracy of quantum states of the S-$T_0$ manifold, is described by the pseudosecular part $B$ of the hyperfine interaction (hfi) which introduces orientation dependence into the magnitude and sign of photo-CIDNP. The interactions between the two electrons are exchange coupling $J$ and dipolar coupling $\omega_{dd}$.

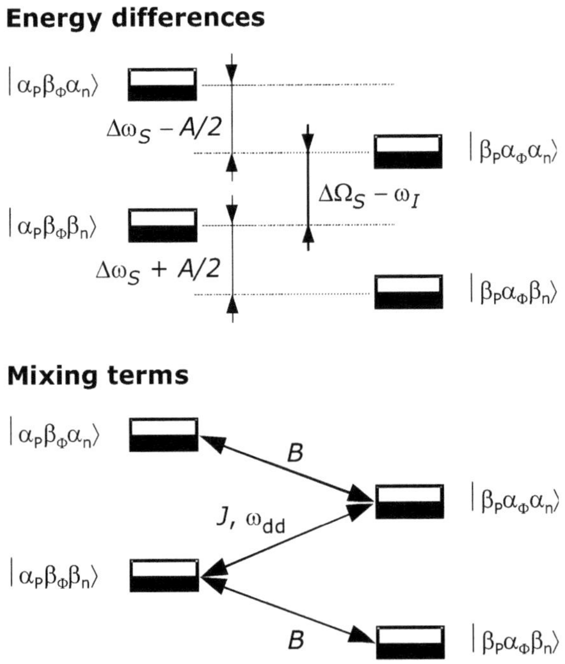


*Fig. 5.* Energy diagram (top) and mixing terms (bottom) allowing electron-electron-nuclear three-spin mixing. The four quantum states of the S-$T_0$ manifold ($|\alpha_P\beta_\Phi\alpha_n\rangle$, $|\beta_P\alpha_\Phi\alpha_n\rangle$, $|\alpha_P\beta_\Phi\beta_n\rangle$ and $|\beta_P\alpha_\Phi\beta_n\rangle$) are split due to the difference of the Zeeman interaction of both electron spins, the hyperfine coupling to the nuclear spin, the nuclear Zeeman interaction, and the coupling between the two electron spins. The mixing of the states is caused by pseudosecular part $B$ of the hyperfine interaction (hfi) as well as both exchange coupling J and dipolar coupling $\omega_{dd}$.

tion, $A$ the secular and $B$ the pseudosecular part of the anisotropic hf interaction.

A maximum of net nuclear polarization through the TSM is generated when the double matching condition is fulfilled:

$$2|\Delta\Omega| = 2|\omega_I| = |A|$$

In this situation, the three states $|S'_\alpha, I_\beta\rangle$, $|S'_\beta, I_\alpha\rangle$ and $|S'_\beta, I_\beta\rangle$ in the S-$T_0$ manifold have the same energy, but initially also have equal populations. The electron coupling $d$ mixes two of the states and thus creates a polarization of the fictitious electron spin $S'$ whose sign depends on the sign of $d$. This coupling $d$ between the two electron spins is the sum of the exchange coupling $J$ and the orientation-dependent dipole-dipole coupling $d'$, $d = 2J + d'/2$. The orientation dependence of $d'$ is given by $d' = D(3\cos^2\theta - 1)$, where D is the magnitude of the electron-electron dipolar coupling, and $\theta$ is the angle between the spin-spin-vector and the external magnetic field. The pseudosecular part $B$ is an off-diagonal element of the Hamiltonian that mixes one of the states affected by the $d$ term with the third state, so that the electron polarization is transferred to nuclei. The TSM contribution vanishes at the canonical orientation of the hf tensor where $B = 0$, while it persists for $\Delta\Omega = 0$ or $A = 0$. The polarization transfer is maximum when $\Delta\Omega = 0$. The TSM mechanism creates significant polarization only if the product of the electron-electron coupling $d$ and the radical pair lifetime is of the order of unity or larger. On the other hand, the contribution also vanishes if $d$ is so large that the S and $T_0$ states become eigenstates of the pair. TSM contributions are thus observed only for moderate electron-electron couplings.

## 2. Differential Decay

In the Differential Decay (DD) mechanism, photophysics and spin dynamics work together to build up polarization. Instead of the electron-electron coupling $d$, the different lifetimes $T_S$ of singlet and $T_T$ of triplet radical pairs establish a polarization of the fictitious spin $S'$. As in the TSM mechanism, the pseudosecular part $B$ of the hf interaction transfers this electron spin polarization to net nuclear polarization (Polenova and McDermott, 1999). The DD mechanism thus persists when $d = 0$ but vanishes when $\Delta\Omega = 0$ and $A = 0$ (Jeschke and Matysik, 2003).

The DD mechanism requires only a single matching of interactions,

$$2|\omega_I| = |A|$$

The time scale of radical pair decay, $(T_S T_T)^{1/2}$, is related to the decay time constant of recombination of the singlet radical pair to the ground-state $T_s$ and the formation of the donor triplet from the triplet radical pair $T_T$. The efficiency of the DD mechanism depends on a matching of the time scale of $\Delta g$- or hf-induced inter-system crossing with this time scale of radical pair decay (Fig. 6). It is quite astonishing that nature has chosen the time constants and magnetic parameters of the radical pair in bacterial RCs in such a way that the TSM and DD mechanisms make contributions of similar magnitude to the nuclear polarization. A predominant DD causes positive signals, while negative signals originate by a predominant TSM. This observation is in line with simulations (Jeschke and Matysik, 2003) as well as observations of $^{15}$N photo-CIDNP (Prakash et al., 2005c) and the temperature dependence of $^{13}$C photo-CIDNP (Daviso et al., 2005). Hence, the observed NMR envelope is a superposition of dark signals and photo-CIDNP signals produced by two competing mechanisms. The same conclusion has been made for $^{13}$C photo-CIDNP in Photosystem I (PS I) (Alia et al., 2004).

## 3. Differential Relaxation

In the Differential Relaxation (DR) mechanism the opposite polarization in singlet and triplet radical pairs does not cancel since part of the triplet-derived polarization relaxes during the lifetime of the donor

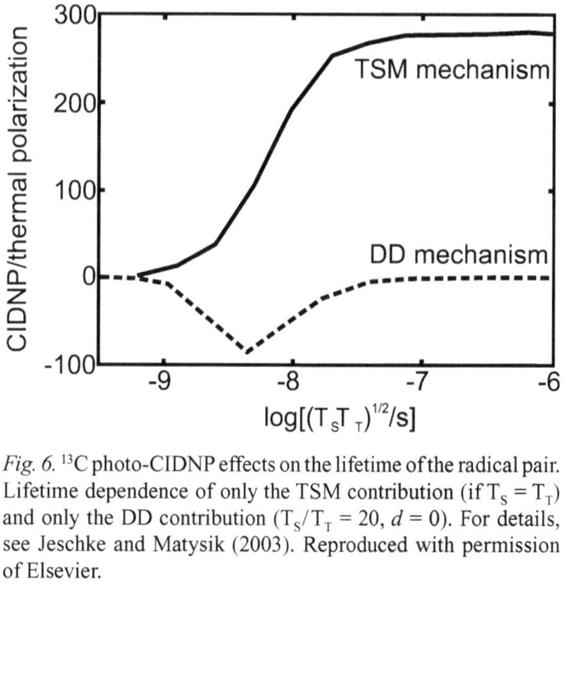


*Fig. 6.* $^{13}$C photo-CIDNP effects on the lifetime of the radical pair. Lifetime dependence of only the TSM contribution (if $T_S = T_T$) and only the DD contribution ($T_S/T_T = 20$, $d = 0$). For details, see Jeschke and Matysik (2003). Reproduced with permission of Elsevier.

triplet $^3$P (McDermott et al., 1998). This requires that the longitudinal relaxation time of the nuclei in the special pair triplet is shortened to such an extent that it is comparable to the triplet lifetime. The contribution of the DR mechanism to photo-CIDNP is difficult to predict since nuclear longitudinal relaxation times in the special pair triplet state are not known. However, as pointed out by Jeschke (1998) as well as Polenova and McDermott (1999), the observation of photo-CIDNP on the bacteriopheophytin acceptor (Zysmilich and McDermott, 1996; Schulten et al., 2002; Prakash et al., 2005b, 2006) can hardly be explained by the DR. Furthermore, a dramatic decrease of the DR contribution would be expected for WT (triplet lifetime of 100 ns) compared to R26 (triplet lifetime of 100 μs). In fact, longitudinal relaxation times as short as a few hundred nanoseconds can be excluded, since this would lead to broader ENDOR lines in the triplet state than have been observed on R26 centers (Lendzian et al., 1998). Assuming that nuclear relaxation times are the same in WT and R26 triplets, the DR mechanism is thus certainly negligible for WT bacterial RCs. However, the difference in polarization patterns between WT and R26 RCs can be explained by a DR contribution. The observed sign change for signals that were assigned to the special pair would be in line with such an explanation and the implied dependence of the nuclear spin relaxation times on the hyperfine couplings is in line with the expected efficiency of paramagnetic relaxation (Prakash et al., 2006).

### *4. Triplet-Born Radical Pairs*

Due to absence of a carotenoid cofactor close to the special pair in the R26 mutant, the lifetime of the molecular triplet state $^3$P increases to 100 μs compared to 100 ns in WT (Fig. 1). The $^{13}$C photo-CIDNP MAS NMR spectrum of R26 shows similar intensities as that of WT, however, the sign of the donor signals is inverted (Fig. 7) (Prakash et al., 2005b, 2006). This effect could be rationalized either by the DR mechanism or by significant back-transfer from the long-lived $^3$P state into the radical pair (Chidsey et al., 1985). Such triplet-born radical pairs undergo TSM and DD like singlet-born radical pairs, however the sign is inverted. Simulations indicate, however, that back-transfer would lead to a sign change for both acceptor and donor signals, while the DR contribution inverts the sign of only the donor signals, in better agreement with observations (Prakash et al., 2006).

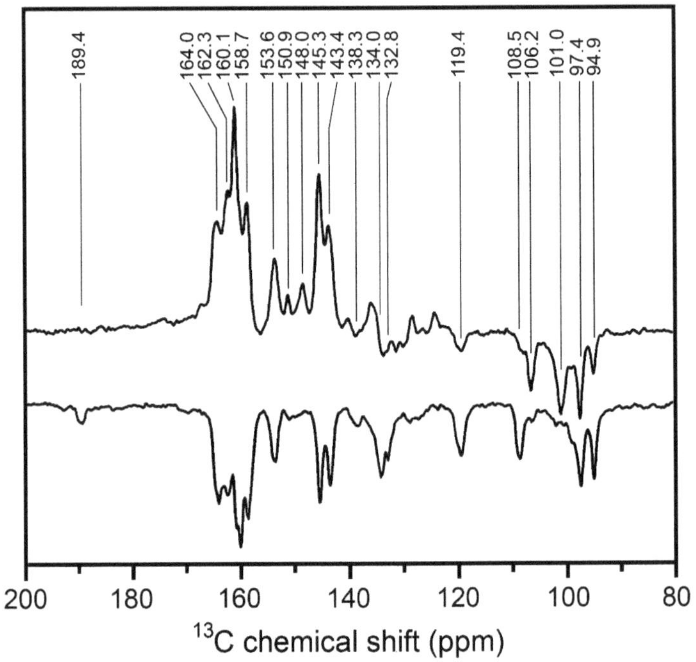


*Fig. 7.* Comparison of $^{13}$C photo-CIDNP MAS NMR spectra obtained from bacterial RCs: (A) Carotenoid-less mutant strain R26, (B) wild type (Prakash et al., 2005b, with permission).

### C. Secondary Effects

#### 1. Steady State Build-Up

Figure 8 shows the accumulation of light induced nuclear polarization to the steady state of *Rb. sphaeroides* R26 (Prakash et al., 2005a). Each point on the curve gives the ratio of intensities of the peak at 161 ppm ($I_{161}$) to the peak of the backbone signal at 31 ppm ($I_{31}$) at different light pulse durations. A fit by first-order build-up kinetics $I_{161}/I_{31} = 1 - \exp(-kt)$ results in $k = 0.266\ s^{-1}$. The initial point (no light) is at 0.067 due to a broad feature caused by aromatic carbons in the dark spectrum. 50% of the steady-state photo-CIDNP signal is reached after 4 s. After about 10 s, the steady state is reached to 90%. For comparable experimental conditions, an average value of 50 turnovers per second per RC has been estimated previously (Matysik et al., 2001a). Hence, the steady state is reached after about absorption of 500 photons per RC. Within the limits of current spectral quality, the build-up of emissive polarization can be fitted by the same time dependence as the build-up of absorptive polarization (data not shown). Similar results have been obtained by Polenova and McDermott (1999) by simulating the steady-state build-up of the DD mechanism. Thus, polarization from TSM and DD builds up with similar rate constants, as otherwise a biexponential build-up curve would be expected.

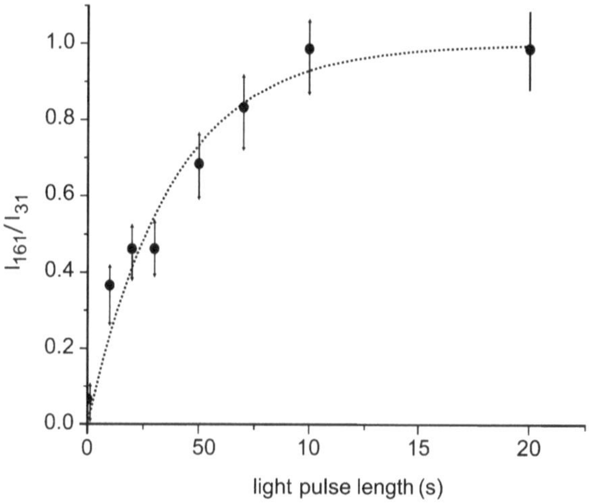


*Fig. 8.* The build-up curve of $^{13}C$ photo-CIDNP to the steady state as a function of different light pulse length. The ratio of the intensity of the light-induced signal at 161 ppm to the dark signal at 31 ppm approaches saturation at about 10 seconds of continuous illumination (Prakash, 2006).

#### 2. Polarization Transfer

In steady-state experiments on selectively $^{13}C$ isotope-labeled RCs, transfer of nuclear polarization to nuclei in close vicinity and concomitant equilibration of signal intensity have been observed (Fig. 9) (Matysik et al., 2001b). On one hand, such a secondary effect hides the true photo-CIDNP intensities; on the other hand, it allows to explore neighboring atoms selectively depending on the chosen label pattern. Polarization transfer into the protein matrix has also been observed in 4′-$^{13}C$-Tyr-labeled bacterial RCs (Matysik et al., 2001a). Hence, a molecule providing photo-CIDNP can be applied as 'spin-torch' in solid-state NMR to explore cavities and surfaces. By analogy to transfer of $^{13}C$ polarization between domains in polymer materials (Robyr et al., 1995), this transfer process is most likely due to proton-driven spin diffusion which would imply a through-space transfer mediated by dipole-dipole couplings between protons. For pairs of $^{13}C$ nuclei that are separated only by one to three bonds and whose shift difference is not much larger than the expected *J* coupling (corresponding to a few ppm), direct $^{13}C$-$^{13}C$ through-bond polarization transfer could be feasible. Finally, direct $^{13}C$-$^{13}C$ polarization transfer through space may be feasible when the chemical shift difference matches a multiple of the spinning frequency.

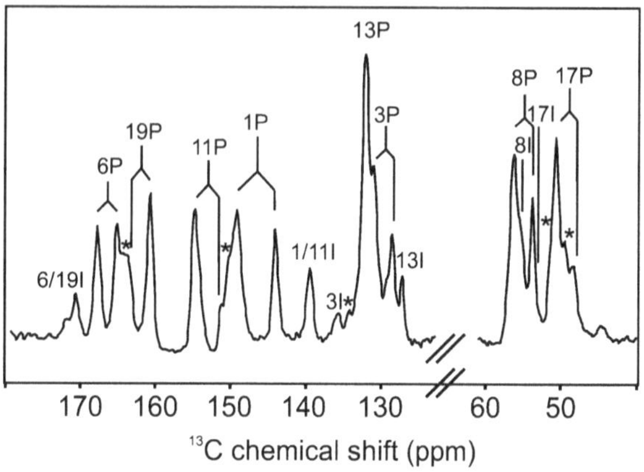


*Fig. 9.* $^{13}C$ photo-CIDNP MAS NMR spectrum of quinone-reduced [1,3,6,8,11,13,17,19-$^{13}C_8$]-BChl/BPhe labeled bacterial RC WT. All signals are emissive (negative), but shown inverted for sake of convenience (Matysik et al., 2001b; © Walter de Gruyter Verlag).

### D. Simulations

Numerical simulations of the photo-CIDNP effect are based on the theory described in (Jeschke and

Matysik, 2003), implemented in Matlab program (G. Jeschke, unpublished) for density matrix computation using the EasySpin library (Stoll, 2003). The program starts from a pure singlet state of the radical pair and computes the time evolution of the system using a Hamiltonian that includes electron Zeeman, nuclear Zeeman, and hf interaction, as well as dipole-dipole and exchange coupling between the two electron spins. The part of the density matrix that represents decay channels to the ground-state from either singlet or triplet radical pairs is projected out (diamagnetic part) and is further evaluated using a Hamiltonian including only the nuclear Zeeman interaction. Evolution is continued until the radical pairs have completely decayed (100 ns) and after that the nuclear polarization of the diamagnetic part of the density matrix is determined. As an extension to the approach described by Jeschke and Matysik (2003), this procedure is performed for a full powder average, describing all interactions by tensors, except for the nuclear Zeeman interaction whose anisotropy is negligible on a time scale of 100 ns. Nuclear polarization was normalized to the thermal polarization at the measurement temperature of 223 K (for details, see Prakash et al., 2005b).

Chemical shift values for simulating photo-CIDNP spectra have been compiled in Prakash et al. (2005b). The absolute value of nuclear polarization has been used in the input file of SIMPSON (Bak et al., 2000). For the SIMPSON simulation, a script for a Hahn echo sequence was written that implemented the same eight-step phase cycle (CYCLOPS and cycling of the (π) pulse) as used in the experiments. The simulated spectra of each resonance have been superimposed using SIMPLOT (Fig. 10). Considering that the protein matrix has not been included into the simulation, the good coincidence between experiment and calculation suggests that the continuous illumination $^{13}$C-photo-CIDNP MAS NMR spectrum can be understood in terms of a superposition of contributions from the TSM and DD mechanisms. Analogous simulations predict maximum enhancements for $^{1}$H as small as

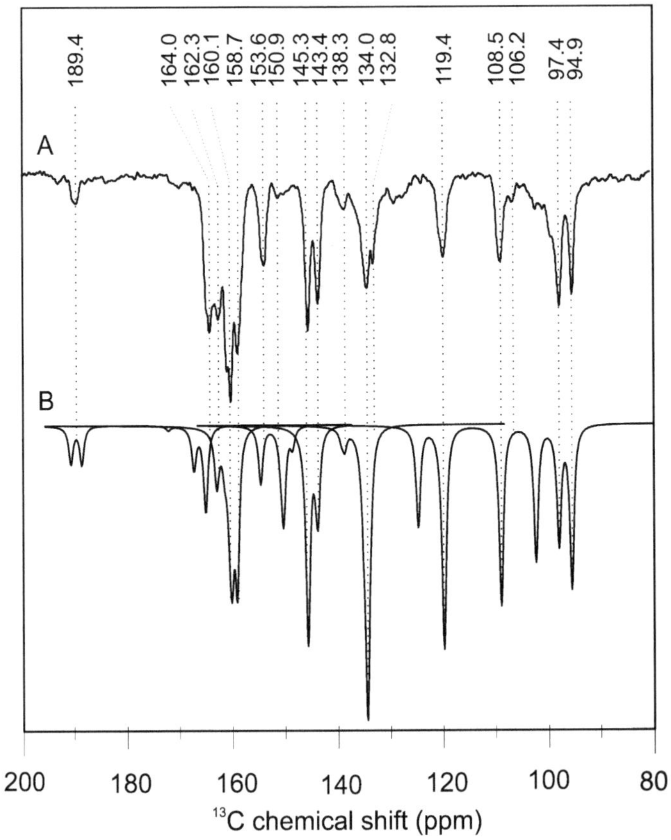


*Fig. 10.* $^{13}$C MAS NMR spectra RC of *Rb. sphaeroides* WT obtained at 200 MHz and 223 K under continuous illumination (A) and simulation of the polarization pattern (B).

2–3 times thermal polarization. This explains why attempts to detect solid-state photo-CIDNP on $^1$H failed.

## II. Experimental Setup

### A. Continuous Illumination Setup

The continuous light illumination setup for MAS NMR experiments comprise a 1000-Watt xenon arc lamp with collimation optics, a liquid filter and glass filters, a focusing element and a light fiber (Fig. 11). Since the emission spectrum of a Xe lamp is similar to sunlight, the full range of radiation from UV to IR is available for illumination. Disturbance of the spinning speed counter, working in the near-IR region, by the incident radiation is avoided by various liquid and glass filters. A fiber bundle is used to transfer the radiation from the collimation optics to the sample. A multi-mode light fiber bundle provides high optical transparency in a broad spectral range, as well as sufficient mechanical flexibility for being attached to the stator of the MAS probe. The MAS probes have been modified in order to illuminate the rotor from the side. This includes (a) a bore drilled into the most upper partition plate separating stator chamber and electronics, (b) drilling a small opening into the stator, and (c) winding a new coil from thin silver wire. The samples have been loaded into optically transparent sapphire rotors to ensure homogeneous illumination. Furthermore, a mechanical shutter can be easily incorporated into the setup, allowing for studies of build-up kinetics and spin-diffusion theoretically down to microsecond resolution. However, the photon rates achieved with a Xenon arc lamp are not sufficient to induce observable photo-CIDNP effects on time scales much shorter than a second.

### B. Laser Setup

Further improvement of time-resolution requires much higher photon rates and thus a laser setup. A nanosecond laser provides sufficient intensity for photo-CIDNP MAS NMR studies. Such a setup comprises a laser with 4–10 Hz repetition rate at 532 nm, a digital delay generator (DDG), a fiber aligner and a multimode optical light fiber (Fig. 12). The pulse width can be short as 6–8 ns. A DDG triggers both pump lamp, exciting the Nd:YAG crystal, and the laser Q-switch. The NMR spectrometer generates the triggering impulse for the DDG via NMR pulse programming. As determined by experiments with various Schott glass filters, green light produces the strongest photo-CIDNP effect since it penetrates deep into the optically very dense sample. The timing can be checked using a 500-MHz oscilloscope connected to the channel of the spectrometer that triggers the $^{13}$C pulse and to a home-made photodetector. The photodetector consists a fast FND-100 photodiode (EG&G) connected to an amplifier. The timing precision of the combined optical and NMR

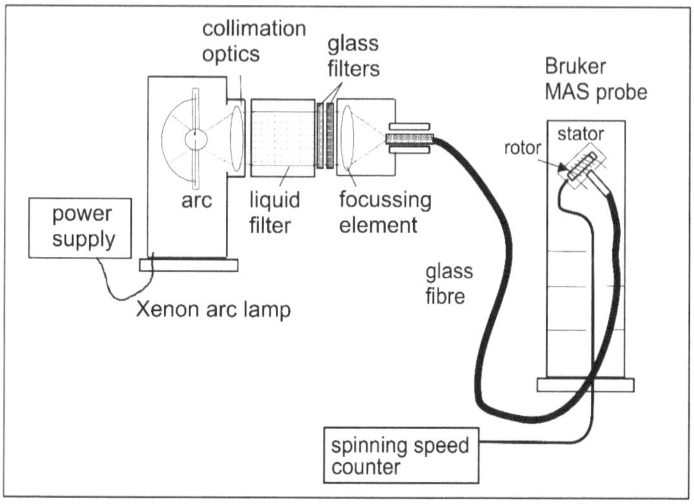


*Fig. 11.* Setup for the detection of continuous illumination photo-CIDNP. The setup is based on a 1000-W Xenon lamp, condenser optics and filters. The radiation is transferred by a multimode fiber bundle into the probe (Matysik et al., 2001a).

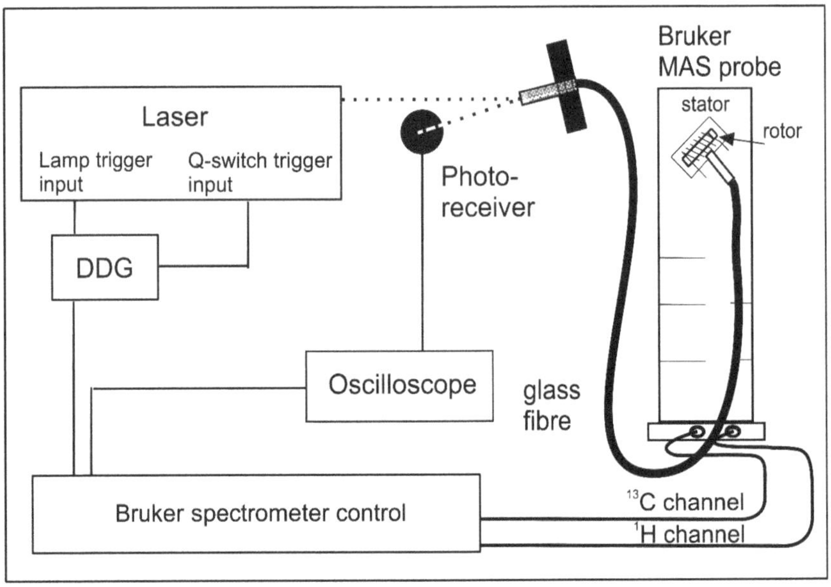


*Fig. 12.* Setup for the detection of laser pulsed photo-CIDNP. A digital delay generator (DDG) connects the NMR spectrometer console to the laser. Laser flashes are transferred by a fiber bundle into the probe. Radiation reflected at the surface of the fiber aperture is measured by a photoreceiver in order to check the performance of the laser during experiment.

experiment has been demonstrated to be about ±10 ns (Fig. 13). Note however that the time resolution of photo-CIDNP experiments is limited by the length of the ($\pi/2$) radiofrequency (rf) pulse used for NMR excitation. Detection of photo-chemically induced nuclear coherence, if feasible, would overcome this limitation of time resolution.

### *C. Pulse Sequences*

Standard steady-state photo-CIDNP MAS NMR spectra are collected under continuous illumination by recording simple Bloch decays with a Hahn-echo and under two-pulse phase modulation (TPPM) proton decoupling (Bennett et al., 1995). Depending on the static magnetic field, cycle delays from 4 to 15 s are used. In nanosecond laser flash experiments, primary polarization is weaker and sensitivity is too low for detection with such long cycle delays. Application of a $^{13}$C presaturation pulse destroys any polarization from previous laser flashes that is still present before the laser pulse and thus allows for a dramatic reduction of the cycle delay. The optimum length of the ($\pi/2$) carbon pulse, determined on uniformly $^{13}$C labeled tyrosine, varies from 4.8 to 5.1 μs under our experimental conditions using a rf power of ~250 W (Fig. 13A). Established multidimensional pulse sequences, for example radio-frequency dipolar

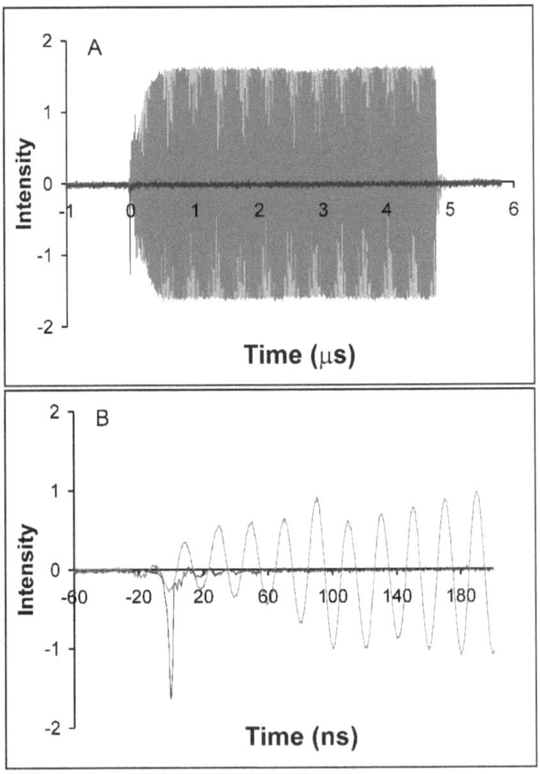


*Fig. 13.* (A) laser light pulse signal (black) and the $\pi/2$ $^{13}$C NMR pulse (grey). The length of the NMR pulse is exactly 4.85 μs. (B) Detailed view at the begin of the NMR pulse. The width at half height of the light pulse is 8 ns.

recoupling (RFDR) experiments (Gullion and Vega, 1992; Boender and Vega, 1998; Bak et al., 2000), can be applied to photo-CIDNP pulse schemes by exchange of the initial CP transfer by a direct ($\pi/2$) carbon pulse.

## III. Applications

### A. Electronic Structure of the Special Pair

The two-dimensional photo-CIDNP MAS NMR allows for an unequivocal chemical shift assignment of the nuclei of cofactors forming the radical pair. Since chemical shift information is related to the ground-state electronic structure, multidimensional NMR data allow mapping of electronic structures at the atomic resolution. Using RCs of *Rb. sphaeroides* with selectively $^{13}C$-labeled cofactors, a clear difference in the ground-state electronic structure of the two BChl *a* cofactors forming the special pair has been detected (Fig. 14) (Schulten et al., 2002; Prakash, 2006). Hence, the electronic symmetry of the special pair is already broken before charge separation. Comparison to 2-dimensional MAS NMR spectra, obtained in the dark by using cross-polarization at an ultra-high field, reveals identical chemical shifts at the isotope labeled positions of the cofactors, in other words, no light induced shifts have been observed. This is in line with a low reorganization energy facilitating fast electron transfer. Very recently, we have shown that the donor of the RC of the green sulfur bacterium *Chlorobium tepidum* is not formed by an asymmetric dimer (Roy et al., 2007)

### B. Entire Chromatophores and Cells

The enormous signal enhancement of photo-CIDNP allows the study of biological units larger than isolated RC proteins. Photo-CIDNP has been observed in entire photosystem units, so-called chromatophores (Prakash et al., 2003), as well as entire cells of *Rb. sphaeroides* R26 (Prakash et al., 2006). Such method enables to study the special pair BChl molecules directly in its native cellular environment at a concentration of ~100 nM without isotope enrichment. In bacterial cells, no significant changes of the electronic structure were detected in the ground-state or radical pair. Hence, isolation appears not to alter the donor side of bacterial RCs to a critical extent.

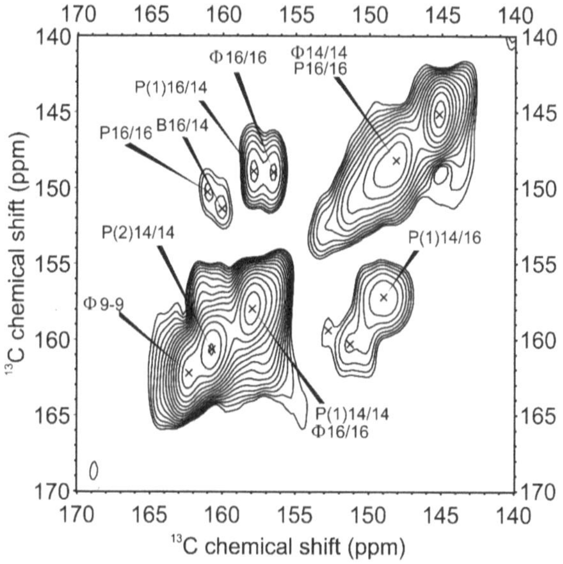


*Fig. 14.* 2-dim RFDR photo-CIDNP MAS NMR spectrum obtained at 200 MHz. The mixing time used is 4 ms with a spinning frequency of 8 kHz at 223 K (Prakash, 2006).

### C. Photosystems of Plants

In plant photosynthesis, the light-induced electron transfer is driven by Photosystem I (PS I) and Photosystem II (PS II). The oxidized primary electron donor of PS II, $P680^{\bullet+}$, is a very powerful oxidizing agent, whereas the electronically excited primary electron donor of PS I, P700*, is a strongly reducing agent. The origin of the different redox properties and the electronic structure of both RCs is not yet clear (for reviews, see: Webber and Lubitz, 2001; Barber, 2003; Witt, 2004). The $^{13}C$ photo-CIDNP MAS NMR spectra of the two plant photosystems look different in their sign patterns. As in bacterial RCs WT, in PS I preparations of PS I-110 particles all signals are emissive (Fig. 15) (Alia et al., 2004). On the other hand, the spectrum of a D1D2-preparation of PS II appears to have a sign pattern similar to R26: the signals between 120 and 175 ppm are enhanced absorptive while the methine signals between 90 and 110 ppm are emissive (Fig. 16) (Matysik et al., 2000; Diller et al., 2005), suggesting the absence of any triplet quencher close to P680. In both spectra, there is no evidence for signal doubling in the spectrum, which is in clear contrast to the spectra of bacterial RCs. In the PS I spectrum, all signals can be assigned to a single undisturbed Chl-*a* cofactor, presumable the donor cofactor. In the PS II spectrum, additional broad emissive signals appear at about 140 and 130

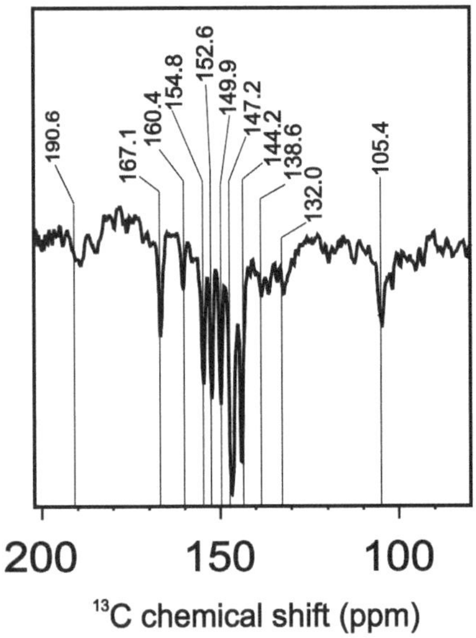


*Fig. 15.* $^{13}C$ MAS NMR spectra of PS I RCs obtained under continuous illumination with white light at 223 K, a magnetic field of 9.4 Tesla and a MAS frequency of 9.0 kHz. Assigned centerbands are visualized by the dashed lines (Alia et al., 2004).

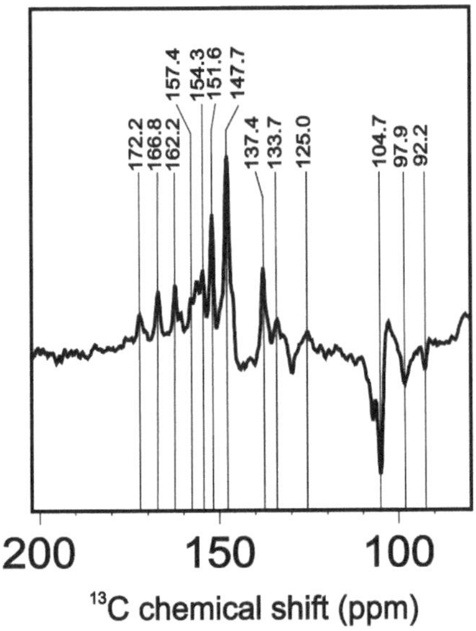


*Fig. 16.* $^{13}C$ MAS NMR spectra of PS II RCs obtained under continuous illumination with white light at 223 K, a magnetic field of 9.4 Tesla and a MAS frequency of 9.0 kHz. Assigned centerbands are visualized by the dashed lines (Diller et al., 2005).

ppm, which may arise either from amino acids or the Phe-*a* acceptor. In PS I, an undisturbed C-13[1] carbonyl carbon can be observed at about 190 ppm, while in PS II the only carbonyl carbon appears surprisingly at 172.2 pm. An assignment of this signal to the C-13[1] carbonyl carbon would imply a chemical modification of that function. On the other hand, this frequency is characteristic for the ester sidechain carbonyl carbons. However, from the sidechains no photo-CIDNP enhancement is expected, unless they would be located within the π-electron cloud of the macrocycle. The responses at around 100 ppm are assigned to methine carbons. It is remarkable that in the spectrum of PS II these signals have very different intensity. This asymmetry in electron spin density distribution has been interpreted in terms of a local electric field pulling the charge towards pyrrole rings III/IV (Matysik et al., 2000). The asymmetry has been confirmed by $^{15}N$ photo-CIDNP MAS NMR and as origin interaction with an axial histidine has been proposed (Diller et al., 2007).

### *D. Relaxation Measurement*

Photo-CIDNP build-up and steady-state polarization depend on the spin-lattice relaxation constant $T_1$, since the polarization decays by longitudinal relaxation. Relaxation times $T_1$ have been measured using the laser setup described in Fig. 13. A train of eighty pulses was applied in each scan in order to reach a quasi steady-state polarization. The delay between the last light flash and the detection pulse was varied from 0.1 s to 80 s in seven steps. Figure 17 shows the absolute intensities of the six most intense signals (Table 1) as a function of the delay. The relaxation time of approximately 17 s does not vary strongly between the different carbon nuclei. It is significantly longer than the time constant of polarization build-up observed under continuous illumination (~3.8 s, see Section III.C), which suggests that the photon rates used in the continuous illumination experiment are sufficient to saturate the system.

## IV. Concluding Remarks

Photo-CIDNP MAS NMR is expected to further improve our understanding of electron transfer and spin chemistry in photosynthetic RCs, especially on the exact mechanisms of electron-matrix interaction and on the precise electronic structure of the cofactors involved. Until now, the solid-state photo-CIDNP effect has been observed in five natural photosynthetic systems, while experiments on artificial systems failed. It appears to be possible that the occurrence of photo-CIDNP in solids is related to specific condi-

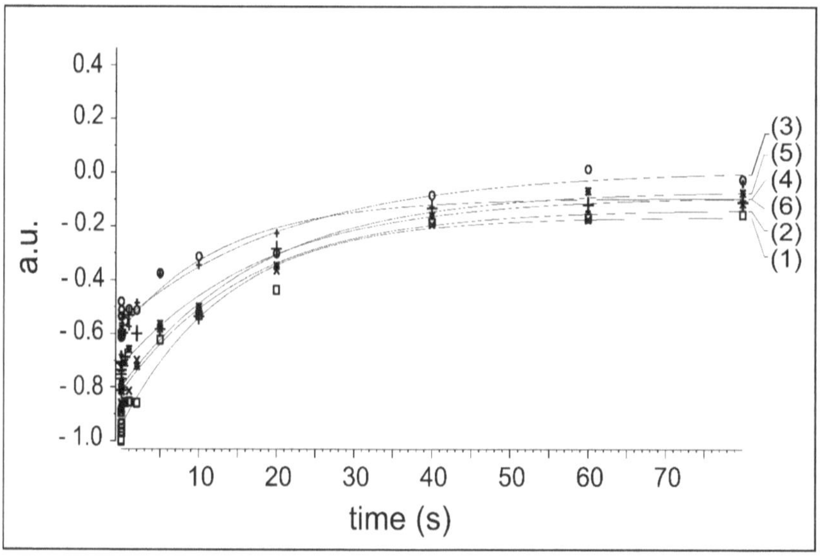


*Fig. 17.* Relaxation measurement of photo-CIDNP on a [4,5,9,10,14,15,16,20-$^{13}C_8$]-BChl/BPhe labeled bacterial RC WT. For symbols, see Table 1.

*Table 1.* Simulation of $T_1$ times for six $^{13}C$ photo-CIDNP MAS NMR signals (Fig. 17).

| Peak no. | Symbol | Chemical shift (ppm) | $T_1$ (s) |
|---|---|---|---|
| 1 | □ | 160.0 | 13.7 |
| 2 | × | 107.5 | 16.5 |
| 3 | ○ | 156.6 | 22.4 |
| 4 | ✚ | 101.5 | 13.5 |
| 5 | ✳ | 97.2 | 18.0 |
| 6 | + | 94.8 | 18.8 |

tions of the efficient electron transfer developed in nature. In fact, the occurrence of the DD mechanism is confined to a rather small kinetic window. Likewise, the TSM mechanism is operative only in a relatively small range of the electron-electron coupling. This and detection of the effect across a range of otherwise rather different photosynthetic RCs suggests that the occurrence of photo-CIDNP is related to an optimization of the first electron transfer step and may thus be a guide for the improvement of artificial systems. In any case, photo-CIDNP provides an enormous potential for solid-state NMR, a method which generally suffers of a lack of sensitivity and selectivity. The solid-state photo-CIDNP effect may also provide new schemes for NMR imaging and microscopy methods.

## Acknowledgments

We gratefully remember the late Prof. Arnold Hoff, who has been a very enthusiastic promoter of the emerging photo-CIDNP MAS NMR. Prof. H.J.M. de Groot is acknowledged for continuous support. Thanks to Prof. C. Glaubitz, Dr. Alia, Dr. P. Gast, Dr. H.J. van Gorkom, Dr. S. Prakash, A. Diller and E. Roy for stimulating discussions. This work has been financially supported by the Netherlands Organization for Scientific Research (NWO) through Jonge Chemici award (700.50.521), an open-competition grant (700.50.004) and a Vidi grant (700.53.423) as well as by the Volkswagen-Stiftung (I/78010) to JM. ED thanks EMBO for a short-term fellowship (ASTF-131-2005).

## References


Alia, Roy E, Gast P, van Gorkom HJ, de Groot HJM, Jeschke G and Matysik J (2004) Photochemically induced dynamic nuclear polarization in Photosystem I of plants observed by $^{13}C$ magic-angle spinning NMR. J Am Chem Soc 126: 12819–12826

Bak M, Rasmussen JT and Nielsen NC (2000) SIMPSON: A general simulation program for solid-state NMR spectroscopy. J Magn Reson 147: 296–330

Barber J (2003) Photosystem II: The engine of life. Q Rev Biophys 36: 71–89

Bargon J and Fischer H (1967) Kernresonanz-Emissionslinien während rascher Radikalreaktionen. 2. Chemisch induzierte dynamische Kernpolarization. Z Naturforsch A 22: 1556–1562

Bargon J, Fischer H and Johnson U (1967) Kernresonanz-

Emissionslinien während rascher Radikalreaktionen. 1. Aufnahmeverfahren und Beispiele. Z Naturforsch A 22A: 1551–1555
Bennett AE, Rienstra CM, Auger M, Lakshmi KV and Griffin RG (1995) Heteronuclear decoupling in rotating solids. J Chem Phys 103: 6951–6958
Blankenship RE, Schaafsma TJ and Parson WW (1977) Magnetic-field effects on radical pair intermediates in bacterial photosynthesis. Biochim Biophys Acta 461: 297–305
Blankenship RE, Babcock GT, Warden JT and Sauer K (1975) Observation of a new EPR transient in chloroplasts that may reflect electron-donor to Photosystem II at room-temperature. FEBS Lett 51: 287–293
Boender GJ and Vega S (1998) Phase sensitive detection of 2D homonuclear correlation spectra in MAS NMR. J Magn Reson 133: 281–285
Boxer SG, Chidsey CED and Roelofs MG (1983) Magnetic-field effects on reaction yields in the solid-state — an example from photosynthetic reaction centers. Ann Rev Phys Chem 34: 389–417
Chidsey CED, Takiff L, Goldstein RA and Boxer SG (1985) Effect of magnetic-fields on the triplet-state lifetime in photosynthetic reaction centers — evidence for thermal repopulation of the initial radical pair. Proc Natl Acad Sci USA 82: 6850–6854
Closs GL and Closs LE (1969) Induced dynamic nuclear spin polarization in photoreductions of benzophenone by toluene and ethylbenzene. J Am Chem Soc 91: 4549–4550
Daviso E, Prakash S, Alia, Gast P, de Groot HJM, Jeschke G and Matysik J (2005) The effect of temperature on photo-CIDNP in a bacterial RC. In: van der Est A and Bruce A (eds) Photosynthesis: Fundamental Aspects to Global Perspectives, Proceedings of the 13th International Congress on Photosynthesis, pp 134–135. Allen Press, Montréal
Diller A, Alia, Roy E, Gast P, van Gorkom HJ, Zaanen J, de Groot HJM, Glaubitz C and Matysik J (2005) Photo-CIDNP solid-state NMR on photosystems I and II: What makes P680 special? Photosynth Res 84: 303–308
Diller A, Roy E, Gast P, van Gorkom JH, de Groot HJM, Glaubitz C, Jeschke G, Matysik J and Alia A (2007) $^{15}$N-photo-CIDNP MAS NMR analysis of the electron donor of Photosystem II. Proc Natl Acad Sci USA 104: 12767–12771
Goez M (1997) Photochemically induced dynamic nuclear polarization. In: Neckers DC, Volman DH and von Bünau G (eds) Advances in Photochemistry, Vol 23, pp 63–163. Wiley, New York
Goldstein RA and Boxer SG (1987) Effects of nuclear-spin polarization on reaction dynamics in photosynthetic bacterial reaction centers. Biophys J 51: 937–946
Gullion T and Vega S (1992) A simple magic angle spinning NMR experiment for the dephasing of rotational echoes of dipolar coupled homonuclear spin pairs. Chem Phys Lett 194: 423–428
Haberkorn R and Michel-Beyerle ME (1979) Mechanism of magnetic-field effects in bacterial photosynthesis. Biophys J 26: 489–498
Haberkorn R, Michel-Beyerle ME and Marcus RA (1979) Spin-exchange and electron-transfer rates in bacterial photosynthesis. Proc Natl Acad Sci USA 76: 4185–4188
Hoff AJ (1984) Electron-spin polarization of photosynthetic reactants. Q Rev Biophys 17: 153–282
Hoff AJ and Deisenhofer J (1997) Photophysics of photosynthesis. Structure and spectroscopy of reaction centers of purple bacteria. Phys Rep 287: 2–247
Hoff AJ, Gast P and Romijn JC (1977a) Time-resolved ESR and chemically-induced dynamic electron polarization of primary reaction in a reaction center particle of *Rhodopseudomonas sphaeroides* wild-type at low-temperature. FEBS Lett 73: 185–190
Hoff AJ, Rademaker H, van Grondelle R and Duysens LNM (1977b) Magnetic-field dependence of yield of triplet-state in reaction centers of photosynthetic bacteria. Biochim Biophys Acta 460: 547–554
Hore PJ and Broadhurst RW (1993) Photo-CIDNP of biopolymers. Progr Nuc Magn Reson Spectrosc 25: 345–402
Jeschke G (1997) Electron-electron-nuclear three-spin mixing in spin-correlated radical pairs. J Chem Phys 106: 10072–10086
Jeschke G (1998) A new mechanism for chemically induced dynamic nuclear polarization in the solid state. J Am Chem Soc 120: 4425–4429
Jeschke G and Matysik J (2003) A reassessment of the origin of photochemically induced dynamic nuclear polarization effects in solids. Chem Phys 294: 239–255
Kaptein R and Oosterhoff JL (1969) Chemically induced dynamic nuclear polarization II: Relation with anomalous ESR spectra. Chem Phys Lett 4: 195–197
Lendzian F, Bittl R and Lubitz W (1998) Pulsed ENDOR of the photoexcited triplet states of bacteriochlorophyll *a* and of the primary donor P-865 in reaction centers of *Rhodobacter sphaeroides* R26. Photosynth Res 55: 189–197
Matysik J, Alia, Gast P, van Gorkom HJ, Hoff AJ and de Groot HJM (2000) Photochemically induced nuclear spin polarization in reaction centers of Photosystem II observed by $^{13}$C-solid-state NMR reveals a strongly asymmetric electronic structure of the P680$^{+}$ primary donor chlorophyll. Proc Nat Acad Sci USA 97: 9865–9870
Matysik J, Alia, Gast P, Lugtenburg J, Hoff AJ and De Groot HJM (2001a) Photochemically induced dynamic nuclear polarization in bacterial photosynthetic reaction centres observed by $^{13}$C solid state NMR. In: Kiihne SR and de Groot HJM (eds) Perspectives on Solid State NMR in Biology, pp 215–225. Kluwer Academic Press, Dordrecht
Matysik J, Schulten E, Alia, Gast P, Raap J, Lugtenburg J, Hoff AJ and de Groot HJM, (2001b) Photo-CIDNP $^{13}$C magic angle spinning NMR on bacterial reaction centres: Exploring the electronic structure of the special pair and its surroundings. Biol Chem 382: 1271–1276
McDermott A, Zysmilich MG and Polenova T (1998) Solid state NMR studies of photoinduced polarization in photosynthetic reaction centers: Mechanism and simulations. Sol State Nuc Magn Reson 11: 21–47
Polenova T and McDermott AE (1999) A coherent mixing mechanism explains the photoinduced nuclear polarization in photosynthetic reaction centers. J Phys Chem B 103: 535–548
Prakash S (2006) Photo-CIDNP studies on reaction centres of *Rhodobacter sphaeroides*. PhD thesis, Leiden University (https://openaccess.leidenuniv.nl/dspace/handle/1887/4555)
Prakash S, Alia, Gast P, Jeschke G, de Groot HJM and Matysik J (2003) Photochemically induced dynamic nuclear polarization in entire bacterial photosynthetic units observed by $^{13}$C magic-angle spinning NMR. J Molec Struct 661: 625–633
Prakash S, Alia, Gast P, de Groot HJM, Jeschke G and Matysik J

(2005a) Build-up kinetics of light induced nuclear polarization observed in RCs of Rhodobacter sphaeroides. In: A van der Est and A Bruce (eds), Photosynthesis: Fundamental Aspects to Global Perspectives, Proceedings of the 13th International Congress on Photosynthesis, pp 301–302. Allen Press, Montréal

Prakash S, Alia, Gast P, de Groot HJM, Jeschke G, and Matysik J (2005b) Magnetic field dependence of photo-CIDNP MAS NMR on photosynthetic reaction centers of *Rhodobacter sphaeroides* WT. J Am Chem Soc 127: 14290–14298

Prakash S, Tong SH, Alia, Gast P, de Groot HJM, Jeschke G and Matysik J (2005c) $^{15}N$ photo-CIDNP MAS NMR on reaction centers of *Rhodobacter sphaeroides*. In: A van der Est and A Bruce, Photosynthesis: Fundamental aspects to global perspectives, Proceedings of the 13th International Congress on Photosynthesis, pp 236–237. Allen Press, Montréal

Prakash S, Alia, Gast P, de Groot HJM, Matysik J, Jeschke G (2006) Photo-CIDNP MAS NMR in intact cells of *Rhodobacter sphaeroides* R26: Molecular and atomic resolution at nanomolar concentration. J Am Chem Soc 128: 12794–12799

Robyr P, Tomaselli M, Straka J, Grobpisano C, Suter UW, Meier BH and Ernst RR (1995) Rf-driven and proton-driven NMR polarization transfer for investigating local order — An application to solid polymers. Mol Phys 84: 995–1020

Roy E, Alia, Gast P, van Gorkom H, de Groot HJM, Jeschke G and Matysik J (2007) Photochemically induced dynamic nuclear polarisation in the reaction center of the green sulphur bacterium *Chlorobium tepidum* observed by $^{13}C$ MAS NMR. Biochim Biophys Acta 1767: 610–615

Schulten EAM, Matysik J, Alia, Kiihne S, Raap J, Lugtenburg J, Gast P, Hoff AJ and de Groot HJM (2002) C-13 MAS NMR and photo-CIDNP reveal a pronounced asymmetry in the electronic ground state of the special pair of *Rhodobacter sphaeroides* reaction centers. Biochemistry 41: 8708–8717

Stoll S (2003) Spectral simulation in solid-state EPR. Ph.D. thesis, ETH Zürich

Ward HR and Lawler RG (1967) Nuclear magnetic resonance emission and enhanced absorption in rapid organometallic reactions. J Am Chem Soc 89: 5518–5519

Webber AN and Lubitz W (2001) P700: The primary electron donor of Photosystem I. Biochim Biophys Acta 1507: 61–79

Werner HJ, Schulten K and Weller A (1978) Electron-transfer and spin exchange contributing to magnetic-field dependence of primary photo-chemical reaction of bacterial photosynthesis. Biochim Biophys Acta 502: 255–268.

Witt HT (2004) Steps on the way to building blocks, topologies, crystals and X-ray structural analysis of Photosystems I and II of water oxidising photosynthesis. Photosynth Res 80: 85–107

Zysmilich MG and McDermott A (1994) Photochemically induced dynamic nuclear-polarization in the solid-state $^{15}N$ spectra of reaction centers from photosynthetic bacteria *Rhodobacter sphaeroides* R26. J Am Chem Soc 116: 8362–8363

Zysmilich MG and McDermott A (1996) Photochemically induced nuclear spin polarization in bacterial photosynthetic reaction centers: Assignments of the $^{15}N$ SSNMR spectra. J Am Chem Soc 118: 5867–5873

# Chapter 20

# Calculations of Electrostatic Energies in Proteins Using Microscopic, Semimicroscopic and Macroscopic Models and Free-Energy Perturbation Approaches

William W. Parson*
*Department of Biochemistry, Box 357350, University of Washington, Seattle, WA 98195-7350, U.S.A.*

Arieh Warshel
Department of Chemistry, University of Southern California, Los Angeles, CA 90007, U.S.A.



## Summary

This chapter discusses computer models for evaluating electrostatic interactions in proteins, with emphasis on calculations of the free energies of electron-transfer states in photosynthetic bacterial reaction centers. We describe the microscopic Protein Dipoles Langevin Dipoles (PDLD) method, semimicroscopic approaches including the Poisson-Boltzmann, PDLD/S and Generalized Born models, a macroscopic model with a homogeneous dielectric medium, and microscopic free-energy-perturbation methods based on molecular dynamics simulations. We also describe the use of molecular dynamics simulations to obtain free energy surfaces of the reactant and product states as functions of the reaction coordinate for electron transfer.

*Author for correspondence, email: parsonb@u.washington.edu

*Thijs J. Aartsma and Jörg Matysik (eds): Biophysical Techniques in Photosynthesis II, pp. 401–420.*

## I. Introduction

Advances in spectroscopic and structural studies of photosynthetic reaction centers (RCs) have presented the challenge of understanding how these systems use light to drive electrons and protons against a steep thermodynamic gradient. Computer models provide a useful approach for exploring how the microscopic structure of the RC determines the $pK_a$'s of the groups that take up and release protons, and the energetics, dynamics and temperature dependence of the electron-transfer intermediates in these reactions. This chapter describes techniques for calculating the energies of electron-transfer states in proteins, with an emphasis on applications to photosynthetic RCs. We discuss the microscopic Protein Dipoles Langevin Dipoles (PDLD) method, several semi-microscopic and macroscopic approaches including the Poisson-Boltzmann (PB) and Generalized Born (GB) models, and microscopic free-energy-perturbation (FEP) methods based on molecular dynamics simulations.

That electrostatic interactions between charged groups in proteins are strongly screened has been shown by numerous experiments (Rees, 1980; Mehler and Eichele, 1984; Lockhart and Kim, 1993; Zhou and Swenson, 1995; Forsyth and Robertson, 2000; Johnson and Parson, 2002), and is well supported by conceptual considerations (Warshel and Russell, 1984; Sham et al., 1998). As an illustration, consider the mutations R(L135)L and R(L135)E in RCs of *Rhodobacter sphaeroides*. The first mutation removes a positive charge about 16 Å from the center of the bacteriochlorophyll dimer that serves as the primary electron donor (P); the second reverses the charge. In the absence of dielectric screening and other structural changes, the R(L135)L mutation would decrease the electrostatic energy difference between the oxidized and reduced forms of P by about 22 kcal/mol, and the R(L135)E mutation would decrease it by 44 kcal/mol. The measured changes in the midpoint reduction potential ($E_m$) of P/P$^+$, however, correspond to free energy changes of only −0.5 and −1.0 kcal/mol, respectively (Johnson and Parson, 2002). Charge-charge interactions between P and the ionizable residue thus are screened by a factor of about 40, even though P is buried in the hydrophobic, intramembrane region of the protein and the side chain of Arg (L135) also is largely inaccessible to the solvent.

As we discuss in the following sections, electrostatic screening in proteins can reflect changes in induced electric dipoles in the atoms surrounding the charged groups, movements of water and mobile ions, and structural reorganization of the protein. A computer model must take all these effects into account, but it does not necessarily have to evaluate each effect explicitly. Models that treat part of the system implicitly can still be instructive.

## II. The Microscopic Protein Dipoles Langevin Dipoles (PDLD) Model

Early models that treated proteins as spheres with a uniform low dielectric constant, such as the influential Tanford and Kirkwood (1957) model, were unable to account for the presence of charged groups and ion pairs in the interior of proteins, because they neglected both the partial charges of the protein atoms and the 'self-energies' of the charged groups (Warshel et al., 1984). As we discuss in more detail below, the self-energy of an individual charge is the solvation free energy of the charge in a medium with no other fixed charges. The first consistent treatment of electrostatic energies that took these factors into account in proteins was the Protein Dipoles Langevin Dipoles (PDLP) model of Warshel and Levitt, (1976). This model evaluated all the relevant electrostatic interactions microscopically but used a simplified model of dipoles on a grid to represent the solvent.

In the PDLD model (Warshel and Levitt, 1976; Warshel and Russell, 1984; Alden et al., 1995; Sham et al., 1997), the electrostatic partial molar free energy of a system is written

$$G_{elec} = V_{Q\mu} + V_{ind} + G_{wat} + G_{memb} + G_{ions} + G_{bulk} \quad (1)$$

The first term on the right-hand side of this expression, $V_{Q\mu}$ is the energy of unscreened (i.e., vacuum) electrostatic interactions among the partial charges of the atoms of the protein and any bound solvent molecules that are resolved in the crystal structure. This is given simply by Coulomb's law:

$$V_{Q\mu} = 332 \sum_{i=1}^{n-1} \sum_{j=i+1}^{n} Q_i Q_j / r_{ij} \quad (2)$$

*Abbreviations:* FEP – free energy perturbation; GB – generalized Born; LRA – linear-response approximation; MD – molecular dynamics; PB – Poisson-Boltzmann; PDLD – Protein Dipoles Langevin Dipoles; RCs – reaction centers; SCAAS – surface-constrained all-atoms solvent; US – umbrella sampling

where $Q_i$ is the partial charge of atom $i$, $n$ is the total number of atoms that are treated explicitly, $r_{ij}$ is the distance between atoms $i$ and $j$ (in Å), and $V_{Q\mu}$ has units of kcal/mol. Note that no dielectric constant appears in this expression. It often is convenient to divide the atoms into two groups, one for atoms that have two or more different charge states, and the other for atoms whose charges are constant. Interactions between atoms in the first group then are split out of $V_{Q\mu}$ and treated in a separate term ($V_{QQ}$). Electrostatic interactions of pairs of atoms separated by two or less covalent bonds are omitted in both terms and treated in a different way because they depend strongly on quantum mechanical effects.

The second term on the right-hand side of Eq. (1), $V_{ind}$, is the energy of interactions of the atomic partial charges with induced electronic dipoles in the protein. This energy is obtained by assigning each atom of the protein a polarizability (typically ~2 Å$^{-3}$ for non-hydrogen atoms), and calculating the electric field at each atom; the induced dipole is the product of the polarizability and the field. Induced dipoles are evaluated by an iterative procedure because the field at any atom depends on all the other induced dipoles and mobile electrolytes in addition to the partial charges of the protein and bound solvent.

The term $G_{wat}$ represents electrostatic interactions with free water surrounding the protein. Water molecules that are not resolved in the crystal structure are represented as 'Langevin dipoles' on a grid of points in which the protein is embedded. Points within the van der Waals radius of any atom are deleted to make the solvent grid complementary to the shape of the structure that is treated microscopically. The outer boundary of the solvent grid is trimmed to a sphere centered on the part of the system that is of interest, such as a bound electron carrier. The Langevin dipole at each grid point again depends on the total electric field at that point, but at high fields approaches a temperature-dependent limiting value in a way that mimics rotational orientation of a free solvent molecule. Because the fields usually are well below saturating values, a simple linear polarization often gives similar results (Alden et al., 1995). The Langevin dipoles can either be held fixed when charges at the site of interest are changed, or allowed to relax in response to the new charge distribution.

The next term, $G_{memb}$, represents interactions with induced dipoles in a nonpolar region such as the belt of detergents or phospholipids that surround a bacterial RC. The induced dipoles in this region also are treated as Langevin dipoles on a grid, but the polarizability is assumed to be lower than that of water. The polarizability ($\alpha$) corresponding to a given macroscopic dielectric constant ($\varepsilon$) can be obtained by the Clausius-Mossotti expression

$$\alpha = \frac{3}{4\pi}\frac{\varepsilon - 1}{\varepsilon + 2}v \tag{3}$$

where $v$ is the volume represented by a grid point.

The term $G_{ions}$ represents interactions with mobile ions in the water surrounding the protein. These ions can be treated as partial charges that distribute self-consistently on the solvent grid in response to the protein charges (Klein and Pack, 1983; Lee et al., 1993; Alden et al., 1995). Finally, $G_{bulk}$ represents interactions with the bulk solvent outside the spherical region of induced dipoles and ions. This term depends on the net charge, dipole and radius of the microscopic sphere and is obtained by expressions due to Born and Onsager (see below).

Now consider the change in the electrostatic free energy ($\Delta G_{elec}$) when one or more of the atomic partial charges is altered by a redox reaction or binding or release of a proton. Some of the protein and solvent atoms, induced dipoles, and electrolytes will move in response to the new charge distribution. The original PDLD model treated the reorganization of the Langevin solvent dipoles consistently, but used energy minimization to treat the effects of the protein reorganization in a more approximate way. More recent PDLD versions consider the effects of structural reorganization in a fully consistent way by using the expressions

$$\Delta G_{elec} \approx \frac{1}{2}\left(\left\langle \Delta G_{elec}\right\rangle_{\alpha} + \left\langle \Delta G_{elec}\right\rangle_{\beta}\right) \tag{4a}$$

$$\lambda \approx \frac{1}{2}\left|\left\langle \Delta G_{elec}\right\rangle_{\alpha} - \left\langle \Delta G_{elec}\right\rangle_{\beta}\right| \tag{4b}$$

Here $\lambda$ is the reorganization energy and $\langle \Delta G_{elec}\rangle_{\alpha}$ and $\langle \Delta G_{elec}\rangle_{\beta}$ denote averages of the calculated values of $\Delta G_{elec}$ over the equilibrium structural configurations of the reactant and product states, respectively (Lee et al., 1992). Equations (4a,b) are based on the Linear-Response Approximation (LRA), which assumes that the protein and solvent respond linearly to the electrostatic forces in the two states (Kubo et al.,

1985; Lee et al., 1992; Hummer and Szabo, 1996). (The LRA also is the basis for the well-known result that the free energy of charging an ion at position **r** in solution is $Q\phi(\mathbf{r})/2$, where $Q$ is the final charge of the ion and $\phi(\mathbf{r})$ is the potential created by fields from the solvent (Warshel and Russell, 1984).) Although it may seem surprising that it would provide a reliable way of describing the energetics of systems as complex as proteins, the LRA has proven to be a reasonably good approximation, particularly for electrostatic effects (Hwang and Warshel, 1987; Kuharski et al., 1988; Åqvist and Hansson, 1996). In some implementations of the PDLD model, a Monte Carlo procedure was used to generate families of low-energy structures with various orientations of the polar hydrogens in amino acid side chains and crystallographic water molecules (Alden et al., 1995; Johnson and Parson, 2002). However, the protein reorganization can be treated more effectively by molecular-dynamics simulations with free-energy perturbation as described in Section VI.

The PDLD model was used to calculate energies of the states in which the bacteriochlorophyll dimer (P) transfers an electron to a neighboring bacteriochlorophyll ($B_A$) or bacteriopheophytin ($H_A$) in *Blastochloris* (*Bl.*) *viridis* RCs (Creighton et al., 1988; Parson et al., 1990; Alden et al., 1995). The energy of the $P^+B_A^-$ state was of particular interest at the time because, although $B_A$ is located between P and $H_A$ in crystal structures of the RC, formation of $P^+B_A^-$ had not been detected experimentally. Initial considerations had suggested that $P^+B_A^-$ might be considerably higher in energy than the lowest excited state of P (P*), which would require electron transfer from P* to $H_A$ to occur by superexchange rather than by a two-step process with $P^+B_A^-$ as an intermediate.

To calculate the free energy of an ion-pair state such as $P^+B_A^-$ requires evaluating both the change in electrostatic free energy ($\Delta G_{elec}$) for forming this state from a reference state, and the change in the gas-phase molecular orbital energy of the system ($\Delta E_{gas}$). The vacuum energy difference between $P^+B_A^-$ and the ground state ($PB_A$) can be estimated either by quantum calculations (Scherer and Fischer, 1989; Thompson et al., 1991; Scherer et al., 1995; Blomberg et al., 1998; Hasegawa and Nakatsuji, 1998; Ivashin et al., 1998; Hughes et al., 2001), or by combining experimentally measured reduction potentials ($E_m$ values) with calculated solvation energies (Creighton et al., 1988; Parson et al., 1990; Alden et al., 1995, 1996). The latter approach seems preferable in this case because quantum calculations of energies for systems as large as a bacteriochlorophyll dimer are not highly accurate. The $E_m$'s are related to $\Delta E_{gas}$ by the expression

$$\Delta E_{gas} = -F\Delta E_{m}, -\Delta G^{ref}_{sol} \tag{5}$$

where $F$ is the Faraday constant, $\Delta E_m$ is the difference between the reduction potentials of the electron acceptor and donor, and $\Delta G^{ref}_{sol}$ is the sum of the changes in the solvation free energies for reduction of the acceptor and oxidation of the donor separately under the conditions of the redox measurements. $\Delta G^{ref}_{sol}$ can be calculated by the PDLD model in essentially the same manner as $\Delta G_{elec}$. In the calculations on *Bl. viridis* RCs, the $E_m$ values of P/P$^+$ and $H_A^-/H_A$ were used to obtain $\Delta E_{gas}$ for the formation of $P^+H_A^-$ from $PH_A$, and $E_m$ values of monomeric bacteriopheophytin-*a* and bacteriochlorophyll-*a* in polar solvents were used to find $\Delta E_{gas}$ for the reaction $B_A + H_A^- \rightarrow B_A^- + H_A$. Combining the two reactions gave $\Delta E_{gas}$ for $PB_A \rightarrow P^+B_A^-$.

The PDLD calculations put $P^+B_A^-$ slightly below P*, with an estimated uncertainty of about ±2 kcal/mol. Although they could not rule out the superexchange mechanism decisively, the results showed that the protein stabilizes $P^+B_A^-$ more effectively than had been anticipated, and favored the two-step pathway through $P^+B_A^-$ for charge separation. By contrast, similar PDLD calculations put the corresponding ion pair state involving the 'voyeur' or 'inactive' bacteriochlorophyll ($B_B$) about 5 kcal/mol above P*, providing a reasonable explanation for the observation that electron transfer strongly favors $B_A$ and $H_A$ (Parson et al., 1990).

One of the uncertainties in calculating the energies of $P^+B_A^-$ is how to treat the charges of ionizable amino acid side chains. There are two problems here: determining whether a given residue actually is ionized under the conditions of interest, and evaluating the interactions of an ionized residue with the electron carriers. As mentioned above and discussed in more detail in Section III, changing the charge of an ionizable residue by site-directed mutagenesis typically has only very small effects on the energies of other charged groups that are more than about 10 Å away. The best course, therefore, probably is to begin by modeling all the ionizable residues in their neutral charge states, and then to examine individually the potential effects of those that are likely to be charged. This was done in the PDLD calculations described above (Alden et al., 1995). Using calculated solvation energies in Eq. (5) can be advantageous here, because some of the uncertainties in $\Delta E_{gas}$ offset those in $\Delta G_{elec}$. For example, the interactions of a distant ionized side chain with $P^+H_A^-$ will be comparable to the sum of its separate interactions with $P^+$ and $H_A^-$ and will tend to cancel in the calculated total energy of $P^+H_A^-$ if $\Delta E_{gas}$ is obtained from the reduction potentials and calculated solvation energies of the electron carriers.

Calculations of the free energies of $P^+B_A^-$ and $P^+B_B^-$ using other models will be described in Sections III, IV, VI and VII. Those that take all the important factors into account are generally consistent with the PDLD calculations. Subsequent experimental work also has supported the two-step pathway for charge separation, and experimental estimates of the energy of $P^+B_A^-$ have converged on values within a few kcal/mol of the calculated value. However, calculations that neglected the self-energies of the charged species (Thompson et al., 1991; Marchi et al., 1993; Hasegawa and Nakatsuji, 1998) or assigned a low dielectric constant to the protein (Gunner et al., 1996), have put $P^+B_A^-$ at higher energies that appear to be at odds with experiment.

## III. Semimicroscopic Models

### A. Poisson-Boltzmann (PB) Models

The induced dipoles that contribute to the terms $\Delta V_{ind}$ and $\Delta G_{wat}$ in the PDLD model often depend on large numbers of pairwise interactions and must be evaluated by iterative procedures that converge relatively slowly. It is therefore useful to have semimicroscopic models that treat one or both of these terms implicitly. As mentioned above, early macroscopic models (Tanford and Kirkwood, 1957) treated the protein as a uniform dielectric medium without considering the partial charges of the protein atoms ($V_{Q\mu}$ in Eq. 1). The Tanford and Kirkwood (1957) model also omitted the contributions of the self-energies of the charges (see discussion in Warshel et al., 1984, 1997; Warshel, 1987). Some of the early semimicroscopic models of proteins (Warwicker and Watson, 1982; Gilson et al., 1985, 1987) had the same limitations. However, this work led to improved semimicroscopic models that resemble the PDLD model in including the protein's partial charges explicitly, while using a macroscopic-continuum approach to treat dielectric effects (Gilson and Honig, 1988; Sharp and Honig, 1990a,b; Nichols and Honig, 1991, 1995; Gunner et al., 1996, 1997; Lancaster et al., 1996; Alexov and Gunner, 1997, 1999; Rabenstein et al., 1998; Nielsen et al., 1999; Ullmann and Knapp, 1999; Gunner and Alexov, 2000; Georgescu et al., 2002; Rocchia et al., 2002; Voigt and Knapp, 2003; Kim et al., 2005).

Poisson-Boltzmann (PB) models, as implemented in the programs DelPhi (Gilson and Honig, 1988; Sharp and Honig, 1990a; Nicholls and Honig, 1991; Sitkoff et al., 1994; Rocchia et al., 2001), UHBDP (Madura et al., 1995) and APBS (Baker et al., 2001) map the protein-solvent system to a cubic grid with two or more dielectric regions. The region representing the solvent usually is assigned a macroscopic dielectric constant similar to that of water ($\varepsilon_s \approx 80$), while that representing the protein is given a lower dielectric constant ($\varepsilon_p$) between 2 and 40. The electric potential $\phi(\mathbf{r})$ at each grid point is evaluated by solving the Poisson-Boltzmann equation,

$$\nabla \cdot \varepsilon(\mathbf{r}) \nabla \phi(\mathbf{r}) = -4\pi\rho(\mathbf{r}) + \zeta \sinh[\phi(\mathbf{r})] \qquad (6)$$

subject to the boundary condition that the potential goes to the value given by the Debye-Hückel theory at long distances from the protein. In this expression, $\nabla$ is the gradient operator, $\rho(\mathbf{r})$ is the charge density at position $\mathbf{r}$, and $\zeta$ is a parameter that is proportional to the ionic strength. The change in electrostatic energy in a process such as oxidation of P then can be written

$$G_{elec} = V_{Q\mu} / \varepsilon_p + G_{RF} \qquad (7)$$

where $V_{Q\mu}$ again is given by Eq. (2) and $G_{RF}$ (the 'reaction field' free energy) represents interactions of the protein charges with the fields from the solvent and electrolytes. At low ionic strengths, the Poisson-Boltzmann equation can be linearized by making the approximation $\sinh[\phi(\mathbf{r})] \approx \phi(\mathbf{r})$. Some PB models also include sampling of multiple protonation states and rotational conformers of polar hydrogens (Alexov and Gunner, 1997, 1999; Ullmann and Knapp, 1999; Georgescu et al., 2002), which resembles the LRA treatment (Eq. 4a).

In general, semimicroscopic models give more stable (though not necessarily more accurate) results than the microscopic PDLD model. This is because microscopic treatments depend on differences between large numbers that are evaluated individually ($\Delta V_{Q\mu}$ on the one hand, and the sum of the other terms on the right-hand side of Eq. 1 on the other). By using a protein dielectric constant, $\varepsilon_p$, semimicroscopic models just assume implicitly that this compensation occurs. PB models thus often give excellent results for surface groups, whose overall solvation energies are similar to those in water. They become more problematic for groups that are sequestered from the solvent, where the results depend strongly on the assumed $\varepsilon_p$. It is important to recognize that $\varepsilon_p$ is not the actual protein dielectric constant, but rather just a factor that incorporates all the contributions that are not treated explicitly, including movements of water and counterions and structural relaxations of the protein around the charged groups (King et al., 1991; Warshel et al., 1997; Sham et al., 1998; Schutz and Warshel, 2001).

Because much of the available experimental information on electrostatic effects in proteins pertains to $pK_a$'s of ionizable amino acid residues, there is a natural tendency use these $pK_a$'s as benchmarks for electrostatics calculations. Using a PB model, Antosiewicz et al. (1994) found that an of $\varepsilon_p$ of 20 gave the best overall results for calculating $pK_a$'s of ionizable residues in a large group of proteins. This is quite different from the assumption that $\varepsilon_p = 4$, which these authors had held previously. Most ionizable residues, however, are located on protein surfaces, where they are largely surrounded by water. Such residues provide little basis for distinguishing between different electrostatic models because, as just mentioned, their $pK_a$'s usually are similar to the values in water. The focus on protein surface residues therefore may have impeded recognition of the nature of $\varepsilon_p$ (Schutz and Warshel, 2001). Studies that focus on interior groups, which should provide more discriminating benchmarks, indicate that the main difficulty with PB models is the proper selection of $\varepsilon_p$, and that the most appropriate value of $\varepsilon_p$ depends on the system and the experimental observable (Schutz and Warshel, 2001; Johnson and Parson, 2002). The value of $\varepsilon_p$ that is appropriate for charge-charge interactions, for example, is not necessarily the best value for calculating charge-dipole energies or self-energies of individual charged groups.

Gunner et al. (1996) have used Poisson-Boltzmann calculations with relatively small values of $\varepsilon_p$ (2 or 4) to estimate the free energy of $P^+B_A^-$ in *Bl. viridis* RCs. They took $\Delta E_{gas}$ from quantum calculations by Thompson et al. (1991), which gave a somewhat larger value than the method based on the measured reduction potentials (see Section II and a detailed comparison in Alden et al., 1996). All the ionizable residues except for Glu L104 were assumed to be ionized. With these assumptions, the calculations put $P^+B_A^-$ about 8 kcal/mol above P* in energy. Increasing $\varepsilon_p$ or decreasing $\Delta E_{gas}$ would move $P^+B_A^-$ downward, in better agreement with experiment and the PDLD results. Gunner et al. (1996) emphasize the importance of electric fields from ionized residues in favoring electron transfer to the acceptors on the 'A' side of the RC ($B_A$ and $H_A$) relative to those on the 'B' side ($B_B$ and $H_B$). As discussed in Section II, the estimated contributions of ionized residues are much smaller in the PDLD model, although electron transfer to $B_A$ and $H_A$ is strongly favored this model also. The free-energy perturbation calculations described in Section VI also reproduce this preference for the 'A' acceptors.

The PB model also has been used to calculate $E_m$ values of the four *c*-type hemes in the *Bl. viridis* RC (Gunner and Honig, 1991), to examine the effects of mutations that increase or decrease the number of hydrogen bonds to the bacteriochlorophylls of P (Muegge et al., 1996), and to explore effects of mutations near the quinones in *Rb. sphaeroides* RCs (Lancaster et al., 1996; Rabenstein et al., 1998; Alexov and Gunner, 1999; Gunner and Alexov, 2000; Alexov et al., 2000).

### *B. The PDLD/S and PDLD/S-LRA Models*

Semimicroscopic versions of the PDLD model, called PDLD/S (Warshel et al., 1989) and PDLD/S-LRA (Lee et al., 1993; Sham et al., 1997; Schutz and Warshel, 2001), have been developed in attempts to capitalize on the stability of semimacroscopic models while keeping the clear physical picture of the PDLD model. Dielectric constants $\varepsilon_p$ and $\varepsilon_w$ are assigned to the protein and water, as in PB models, and a thermodynamic cycle is used to relate the change in free energy for changing the charge of a protein group to the sum of the free-energy changes for three steps. To illustrate the cycle, suppose we change the charge of atom *i* ($Q_i$) from 0 to *Q*. The process begins with the protein surrounded by water and $Q_i = 0$. In the first step of the cycle, the dielectric constant of the solvent around the protein is changed from $\varepsilon_w$ to $\varepsilon_p$. The electrostatic free energy change for this step is

$$\Delta G_{sol(0)}^{w\to p} = G_{sol}^{0}\left(\frac{1}{\varepsilon_p} - \frac{1}{\varepsilon_w}\right) \tag{8}$$

where $G^0_{sol}$ is the total solvation free energy of the protein when atom $i$ is uncharged. This is calculated as the sum of $G_{wat}$ and $G_{bulk}$ in the Langevin-dipole model as described in Section II. Effects of induced dipoles in the protein and electrolytes in the solvent, however, are incorporated implicitly in the dielectric constants rather than being evaluated explicitly.

In the second step of the thermodynamic cycle, $Q_i$ is changed from 0 to $Q$. Since the protein and solvent now both have dielectric constant $\varepsilon_p$, the free energy change for this step is

$$\Delta G^{p}_{0\to Q} = \Delta G_{self(i)} + \Delta V_{Q\mu} / \varepsilon_p \quad (9)$$

where $\Delta V_{Q\mu}$ is the change in the unscreened electrostatic interactions of atom $i$ with the protein charges as in Eq. (2) and $\Delta G_{self(i)}$ is the change in the self-energy of atom $i$. The self energy of a single, spherical ion with radius $a_i$ (the 'Born radius' of atom $i$), charge $Q_i$, and internal dielectric constant 1 in an infinite medium with dielectric constant $\varepsilon_p$ can be obtained by the Born equation (Born, 1920), which reads

$$G_{self(i)} = -\frac{332Q_i^2}{2a_i}\left(1-\frac{1}{\varepsilon_p}\right) \quad (10)$$

The final step is to return the dielectric constant of the solvent around the protein to $\varepsilon_w$. This step changes the electrostatic free energy by

$$\Delta G^{p\to w}_{sol(Q)} = G^{Q}_{sol}\left(\frac{1}{\varepsilon_w}-\frac{1}{\varepsilon_p}\right) \quad (11)$$

where $G^Q_{sol}$ is the solvation free energy of the protein plus the solvent obtained by the Langevin-dipole model with $Q_i = Q$. Combining Eqs. (9–11), the overall free energy change associated with charging atom $i$ in the PDLD/S model is

$$\begin{aligned}\Delta G^{PDLD/S}_{elec} &= \Delta G^{w\to p}_{sol(0)} + \Delta V_{Q\mu}/\varepsilon_p \\ &+\Delta G_{self(i)} + \Delta G^{p\to w}_{sol(Q)} \\ &= \Delta G_{sol}\left(\frac{1}{\varepsilon_p}-\frac{1}{\varepsilon_w}\right) + \Delta V_{Q\mu}/\varepsilon_p \\ &-332\frac{Q^2}{2a_i}\left(1-\frac{1}{\varepsilon_p}\right)\end{aligned} \quad (12)$$

with $\Delta G_{sol} = G^Q_{sol} - G^0_{sol}$. The PDLD/S model thus still requires evaluating $\Delta V_{Q\mu}$, $\Delta G_{wat}$ and $\Delta G_{bulk}$ but avoids the need to evaluate induced dipoles in the protein ($\Delta V_{ind}$). The terms $\Delta G_{memb}$ and $\Delta G_{ions}$ can be treated explicitly as part of the solvation energies if necessary.

The PDLD/S-LRA model uses the same simplifications as PDLD/S, but reduces the unknown factors that contribute to $\varepsilon_p$ by averaging the PDLD/S energy over multiple configurations. The set of configurations is generated by MD simulations of both charge states of the system (see Eq. 4*a*). The Molecular Mechanics Poisson-Boltzmann Surface Area model (Kollman, 2000) also uses MD to generate a family of configurations, but includes only configurations generated with the charged solute (i.e., only the term $\langle \Delta G_{elec}\rangle_\beta$ in Eq. 4a).

Thermodynamic cycles similar to the cycle used by PDLD/S often allow one to replace challenging direct calculations of a physical process by calculations of several simpler processes, which can be either physical or non-physical. The difference between the binding constants with which the protonated and ionized forms of a small acid binds to a protein, for example, can be obtained simply by calculating the difference between the energies of ionizing the acid in the protein and in water. Since its introduction in the early 1980s (Warshel, 1981; Wong and McCammon, 1986), the use of thermodynamic cycles has become a major part of many microscopic calculations of free energies of biological processes (Kollman, 1993).

### C. Generalized-Born (GB) Models

A further simplification of semimicroscopic models is to treat $\Delta G_{wat}$ and $\Delta G_{bulk}$ implicitly along with $\Delta V_{ind}$ (Warshel, 1991). In an infinite homogeneous medium with dielectric constant $\varepsilon$, the electrostatic free energy of a set of charges can be written as the sum of the gas-phase energy ($V_{Q\mu}$, Eq. 2) and the total solvation free energy ($G_{sol}$):

$$G_{elec} = V_{Q\mu} + G_{sol} \quad (13)$$

Alternatively, we can regard the electrostatic free energy as the sum of the solvation free energies of the individual charges at infinite dilution ($G^\infty_{sol}$) plus the free energy of bringing the charges in from infinity to their actual positions ($V_{Q\mu}/\varepsilon$):

$$G_{elec} = G^\infty_{sol} + V_{Q\mu}/\varepsilon \quad (14)$$

Equating these two expressions and replacing $G_{sol}^{\infty}$ by the sum of the individual Born energies (Eq. 10) gives

$$G_{sol} = G_{sol}^{\infty} - V_{Q\mu}\left(1-\frac{1}{\varepsilon}\right) = -332\left\{\sum_{i=1}^{n}\frac{Q_i^2}{2a_i} + \sum_{i=1}^{n-1}\sum_{j=i+1}^{n}\frac{Q_iQ_j}{r_{ij}}\right\}\left(1-\frac{1}{\varepsilon}\right) \quad (15)$$

In 'generalized Born' (GB) models, the sums in Eq. (15) are manipulated further by writing

$$G_{sol}^{GB} = -166\left(1-\frac{1}{\varepsilon_{GB}}\right)\sum_{i=1}^{n}\sum_{j=1}^{n}\frac{Q_iQ_j}{f_{ij}}, \quad (16)$$

where $f_{ij}$ is an empirical function of distance. A commonly used function is

$$f_{ij} = \left[r_{ij}^2 + a_ia_j\exp\left(-r_{ij}^2/4a_ia_j\right)\right]^{1/2} \quad (17)$$

in which $a_i$ and $a_j$ are 'effective' Born radii for atoms $i$ and $j$ (Still et al., 1990). Models based on this approach (Constanciel and Contreras, 1984; Kozaki et al., 1988, 1989; Still et al., 1990; Hawkins et al., 1996; Schaefer and Karplus, 1996; Qiu et al., 1997; Ghosh et al., 1998; Bashford and Case, 2000; Onufriev et al., 2002; Gallicchio and Levy, 2004; Fan et al., 2005; Zhu et al., 2005) differ mainly in how they obtain the effective Born radii. Replacing the interatomic distance ($r_{ij}$) by $f_{ij}$ has the effect of decreasing the contribution of the term $(Q_iQ_j/r_{ij})(1 - 1/\varepsilon)$ to the solvation energy as $r_{ij}$ becomes small. The effective dielectric screening of the interactions of atoms $i$ and $j$ thus increases as the interatomic distance grows. If $r_{ij}^2 \gg 4a_ia_j$ for all $i$ and $j$, we recover the sum of the Born energies (Eq. 10), and the GB treatment becomes identical to the PDLD/S. The GB treatment focuses on charge-charge interactions, which usually will require a large $\varepsilon_{GB}$.

Generalized Born models have given encouraging results for solvation energies of simple ions in solution, and also in some MD simulations of proteins (Schaefer and Karplus, 1996; Bashford and Case, 2000; Fan et al., 2005). They have not yet been tested extensively for other problems in protein electrostatics. It is important to note that the GB model is not a mathematical solution to the problem of charges in a multicavity continuum, or of a system with multiple dielectric regions such as a protein surrounded by water (Warshel and Papazyan, 1998). Although an increase in dielectric screening with distance is consistent with experiment, there is no good theoretical basis for Eq. (16) or (17). Nor does the GB model prescribe the best value of $\varepsilon_{GB}$ to use with Eq. (16) in applications to proteins, although values of at least 20 probably will be needed in most cases.

### *D. Distance-Dependent Screening Factors*

Notwithstanding the remarks in the previous section concerning Eqs. (16) and (17), using a screening factor ($\varepsilon_{ij}$) that increases with the interatomic distance provides a simple but remarkably effective way to deal with dielectric effects in many problems involving proteins (Warshel et al., 1984). In this model, the electrostatic free energy is

$$G_{elec} = G_{sol}^{\infty} + 332\sum_{i=1}^{n-1}\sum_{j=i+1}^{n} Q_iQ_j/r_{ij}\varepsilon_{ij} \quad (18)$$

where $G_{sol}^{\infty}$ again is the sum of the solvation free energies (self-energies) of the individual charges at infinite separation. Several general forms other than Eq. (16) have been used for $\varepsilon_{ij}$. Warshel et al. (1984) used an exponential function, $\varepsilon_{ij} = 1 + 60[1 - \exp(-r_{ij}/r_0)]$ where $r_o$ is an empirical factor in the range of 5 to 10 Å; Mehler and Eichele (1984, Hingerty et al. (1985), Mehler and Guarnieri (1998) and Guenot and Kollman (1992, 1993) have used sigmoidal functions that give similarly large screening at distances greater than 10 Å. Equation (18) with either an exponential or sigmoidal scaling factor was found to reproduce the effects of mutations of ionizable residues on the $E_m$ of P/P$^+$ in bacterial RCs (Johnson and Parson, 2002) (see Fig. 1). The simple linear scaling $\varepsilon_{ij} = r_{ij}$, however, increases too slowly with distance.

Unlike the GB model, Eq. (18) does not try to capture the term $G_{sol}^{\infty}$ along with the charge-charge interaction term by a single sum and a single ε. $G_{sol}^{\infty}$ therefore must be obtained separately if one seeks $G_{elec}$. The methods used in PDLD, PDLD/S or PDLD/S-LRA are suitable for this. There are, however, many situations in which $G_{sol}^{\infty}$ either is constant or can be removed by a suitable thermodynamic cycle, leaving

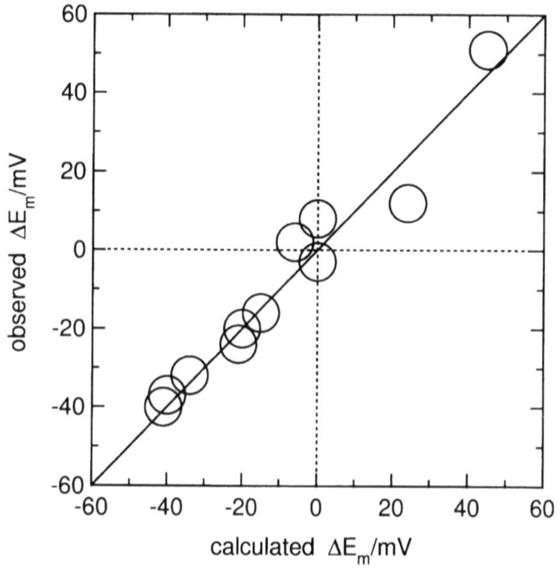


*Fig. 1.* Observed and calculated shifts in the $E_m$ of the special pair of bacteriochlorophylls (P/P$^+$) in RCs from 11 mutants of *Rb. sphaeroides* (Johnson and Parson, 2002, with permission). The calculated shifts in the $E_m$ were obtained by using a distance-dependent dielectric screening factor, $\varepsilon_{ij} = 1 + 60[1 - \exp(-r_{ij}/r_0)]$ with $r_o = 10$ Å (Eq. 18), and the LRA (Eq. 4a with averaging over five low-energy configurations of the polar hydrogens in each redox state). The changes in the atomic partial charges resulting from oxidation of bacteriochlorophyll-*a* were obtained with the program QCFF/PI (Warshel and Lappicirella, 1981), and divided equally between the two bacteriochlorophyll molecules of P (see Johnson and Parson, 2002; Johnson et al., 2002, for discussion of this division). The diameters of the symbols are approximately twice the standard error of the means of both the experimental and calculated values. From bottom to top, the mutations are C(L247)D, R(L135)E, R(M164)E, R(L135)Q, R(L135)L, R(M164)L, R(L135)K, Y(L164)F, R(L135E)/C(L247)K, D(L155)N, and C(L247)K.

the change in the screened charge-charge interactions as the factor of interest. The effects of mutations of an ionizable residue on the $pK_a$ of another residue or on the $E_m$ of a bound group such as P are such situations.

## IV. Macroscopic Models

We now consider a model with a homogenous macroscopic medium and a fixed dielectric constant, ε. Although this is obviously an oversimplified model for a protein in water, it helps to bring out the importance of the self-energy of a charge, and it provides an instructive illustration of how the free energy of a system depends on the assumed value of ε.

In the macroscopic model, the free energy change for a reaction such as $PB_A \rightarrow P^+B_A^-$ takes the form

$$\Delta G_{elec} = \Delta E_{gas} + \Delta G_{sol}^{\infty} + \Delta V_{Q\mu}/\varepsilon \quad (19)$$

where $\Delta E_{gas}$, as before, is the change in the vacuum molecular orbital energies of the electron donor and acceptor, $\Delta G_{sol}^{\infty}$ is the change in solvation free energies of the electron carriers at infinite separation in a homogeneous medium with dielectric constant ε, and $\Delta V_{Q\mu}$ is the change in unscreened electrostatic interactions, including direct interactions between the donor and acceptor ($\Delta V_{QQ}$). As discussed in Section II, $\Delta E_{gas}$ can be estimated either by quantum calculations or by combining measured $E_m$ values with calculated solvation energies (Eq. 5).

We next use the Born equation (Eq. 10) to relate $\Delta G_{sol}^{\infty}$ to the corresponding free energy change in water or another reference solvent with a high dielectric constant, $\Delta G_{sol}^{w,\infty}$:

$$\Delta G_{sol}^{\infty} = \Delta G_{sol}^{w,\infty}(1-1/\varepsilon)/(1-1/\varepsilon_w) \approx \Delta G_{sol}^{w,\infty}(1 - 1/\varepsilon) \quad (20)$$

This approach is particularly convenient if the same reference solution is used for calculating $\Delta E_{gas}$ by Eq. (5). By combining Eqs. (19) and (20) with Eq. (5), and equating $\Delta G_{sol}^{ref}$ with $\Delta G_{sol}^{w,\infty}$, the free energy change for the electron-transfer reaction becomes

$$\Delta G_{elec} \approx -F\Delta E_m + (\Delta V_{Q\mu} - \Delta G_{sol}^{ref})/\varepsilon \quad (21)$$

Calculations using Eq. (21) have been described for the initial ion-pair states in *Bl. viridis* RCs (Alden et al., 1996). Fig. 2 shows the calculated free energies of $P^+B_A^-$ and $P^+H_A^-$ relative to the ground state as functions of ε. The horizontal dashed line indicates the energy of the excited state (P*). Experimentally, $P^+H_A^-$ is found to lie about 6 kcal/mol below P*, or about 23 kcal/mol above the ground state. The macroscopic model returns this result when $\varepsilon \approx 2.9$. (The microscopic PDLD calculations described in Section II reproduce the measured energy of $P^+H_A^-$ without using an adjustable dielectric constant, but of course with much more effort.) More significantly, Fig. 2 shows that any value of ε that gives approximately the right free energy for $P^+H_A^-$ puts $P^+B_A^-$ between this state and P*. This is a useful result in spite of the limitations of the macroscopic model, because it provides a back-of-the-envelope check on the PDLD calculations, which also put $P^+B_A^-$ between

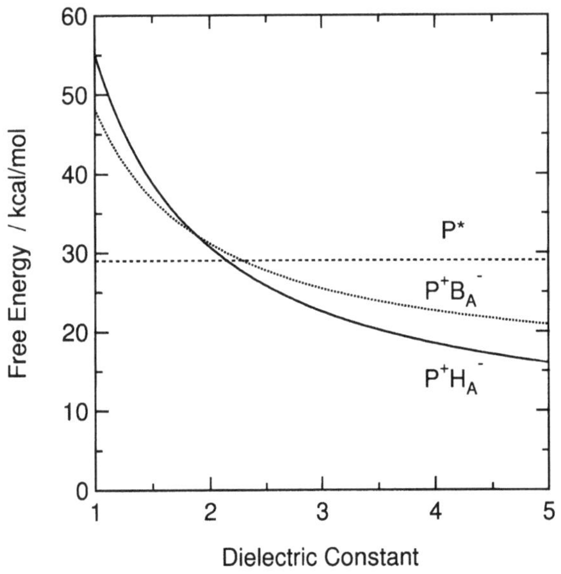


*Fig. 2.* Free energies of $P^+B_A^-$ and $P^+H_A^-$ in *Bl. viridis* RCs as functions of ε in a macroscopic model (Alden et al., 1996). All energies in this figure are expressed relative to that of the ground state. The horizontal *dashed line* shows the energy of the first excited singlet state of P (P*). $\Delta E_{gas}$ for the reaction $PH_A \rightarrow P^+H_A^-$ was obtained by Eq. (20) from the measured reduction potentials of P/P+ and $H_A^-/H_A$ in purified RCs and PDLD calculations of the change in solvation energy for each of the half-cell reactions in situ. $\Delta E_{gas}$ for $B_A + H_A^- \rightarrow B_A^- + H_A$ was obtained similarly from the reduction potentials of bacteriochlorophyll-*b* and bacteriopheophytin-*b* in polar solvents and calculations on the oxidized and reduced molecules in solution. $\Delta G_{elec}$, the calculated electrostatic free energy changes for $PB_AH_A \rightarrow P^+B_AH_A^-$ and $P^+B_AH_A^- \rightarrow P^+B_A^-H_A$ obtained by Eq. (21), were summed to obtain that for $PB_AH_A \rightarrow P^+B_A^-H_A$. Ionizable amino acid side chains were taken to be in their neutral ionization states. See Alden et al. (1996) for additional details.

P* and $P^+H_A^-$ (see Section II).

Perhaps the most important point that emerges from simple models of this type is that the self energies of charged groups cannot be dismissed as small corrections to the electrostatic free energy. Indeed, they often dominate the problem. In the calculations of the free energy of $P^+H_A^-$ shown in Fig. 2, $\Delta G_{sol}^{w,\infty}$ ($\Delta G_{sol}^{ref}$) is on the order of −77 kcal/mol, which is similar to $\Delta E_{gas}$ and about three times greater than $F\Delta E_m$ (Alden et al., 1996). Even for the charge-shift reaction $P^+B_A^-H_A \rightarrow P^+B_AH_A^-$, where there is no net change in electrical charge, $\Delta G_{sol}^{w,\infty}$ is about −12 kcal/mol, compared to 4 kcal/mol for $\Delta E_{gas}$ and −8 kcal/mol for $F\Delta E_m$ (Alden et al., 1996). Calculations that neglect the self-energies should not, therefore, be given much credence. The occasional ability of such calculations to reproduce an experimental result can be misleading in this regard, and in most cases probably can be traced to a fortuitous cancellation of errors or to the choice of a nondiscriminating benchmark. (See Alden et al., 1995, for a discussion of how calculations by Marchi et al. (1993), which omitted the self-energies and gave what is almost certainly much too high an energy for $P^+B_A^-$, fortuitously gave approximately the correct energy for $P^+H_A^-$.)

## V. Molecular-Dynamics Simulations

We now turn to calculations that use molecular dynamics (MD) simulations to treat motions of a protein and the surrounding solvent explicitly. In this approach, the total potential energy of a molecule is expressed in the form

$$V = V_{b,\theta}(\mathbf{b},\theta) + V_\phi(\phi) + V_{nb}(\mathbf{r}) \tag{22}$$

where **b**, θ and ϕ are vectors of bond lengths, bond angles and torsion angles, respectively, and **r** is the vector of Cartesian coordinates that are used to evaluate the distances between the nonbonded atoms. The terms on the right-hand side of Eq. (22) can be expressed as simple functions of the internal coordinates. The first term, for example, is written

$$V_{b,\theta}(\mathrm{b},\boldsymbol{\theta}) = \frac{1}{2}\sum_i K_{b,i}(b_i - b_{0,i})^2 + \frac{1}{2}\sum_i K_{\theta,i}(\theta_i - \theta_{0,i})^2 + \textit{cross terms} \tag{23}$$

where $b_i$ is the length of bond *i* at a given time, $b_{0,i}$ is the mean expected length of bond *i*, $\theta_i$ is the value of bond-angle *i*, $\theta_{0,i}$ is the mean expected value of this angle, and $K_{b,i}$ and $K_{\theta,i}$ are adjustable force constants (see, e.g., Warshel, 1991).

The last term on the right-hand side of Eq. (22) represents interactions of non-bonded atoms, including both van der Waals and electrostatic interactions. The electrostatic potential energies can be treated by any of the microscopic or semimicroscopic models described in the previous sections, but generally are the most difficult contribution to evaluate both rapidly and realistically.

The solvent can be included explicitly in the potential function by using either an all-atom solvent model

(Allen and Tildesley, 1987; King and Warshel, 1989; Kollman, 1993; Buono et al., 1994) or a simplified model such as the Langevin-dipole model described in Section II (Warshel and Levitt, 1976; Warshel and Russell, 1984). Implicit solvent models (Miertus et al., 1981) also can be used to represent the effects of the solvent.

The parameters of the potential functions (e.g. $K_{b,i}$, $K_{\theta,i}$, $b_{0,i}$ and $\theta_{0,i}$ in Eq. 23) can be adjusted by a least-squares procedure to minimize the difference between sets of calculated and observed molecular properties (Lifson and Warshel, 1968). Such a fitting to a set of measured energies, structures and vibrational frequencies generates a 'consistent force field.' The introduction of potential functions enabled conformational analysis via computers (Hendrickson, 1961; Burkert and Allinger, 1982), and formulation of the potential functions in terms of Cartesian coordinates (Lifson and Warshel, 1968) made it possible to extend conformational analysis and molecular dynamics simulations to macromolecules (Levitt and Lifson, 1969).

In MD simulations, one obtains the positions ('trajectory') and velocities of the atoms as a function of time. The trajectory is calculated by solving the classical Newtonian equations of motion for the molecule, which requires knowing all the forces that act on the atoms at any given time. These forces are easily obtained in numerical form from the Cartesian first derivatives of the analytical potential functions (Lifson and Warshel, 1968; Allen and Tildesley, 1987). The acceleration along a coordinate such as the length of a particular bond is simply the force divided by the reduced mass of the atoms. The positions, forces and velocities are recomputed at regular intervals on the order of 1 fs ($10^{-15}$ s) to generate a trajectory.

The atoms that are followed in a classical MD trajectory correspond to a single system with a fixed total energy determined by the initial conditions. Macroscopic molecular properties, however, usually represent statistical mechanical averages of very large numbers of molecules. To evaluate such properties, the system included in an MD simulation must be viewed as part of a large ensemble of systems whose atoms are not considered explicitly. The effects of the other parts of the ensemble can be treated by introducing a 'thermostat' that keeps the simulated system at a specified temperature. This can be accomplished by assuming equal partition of kinetic energy among all the degrees of freedom (Warshel, 1991) or by a variety of more sophisticated procedures (Allen and Tildesley, 1987). Some workers hold the total energy constant rather than the temperature (Day and Daggett, 2004).

Explicit, all-atom MD simulations of macromolecules embedded in a solvent are extremely demanding, even with current computers. The problem is partly the comparatively long-range nature of electrostatic interactions. The number of pair-wise interactions grows with the square of the number of atoms in the system, while the number of atoms in a spherical system increases as the cube of the radius. This leads to a vast number of interactions that must somehow be taken into account.

If electrostatic effects themselves are not of interest, one can simply assume that dielectric screening increases sufficiently rapidly with distance to justify disregarding interactions between atoms that are more than about 10 Å apart (Mezei and Beveridge, 1981; Chandrasekhar et al., 1984; Alagona et al., 1986; Belch et al., 1986; Steinbach and Brooks, 1994; Levitt et al., 1995). The contributions of electrostatic interactions to the potential function often is truncated smoothly over a distance of 0.5 to 1 Å, and a list of atom pairs that are within the cut-off distance is maintained and updated at regular intervals during the trajectory (e.g., every 10 steps) so that all the interatomic distances do not have to be evaluated at each step. However, cutting off electrostatic interactions at short distances can lead to major problems in treatments of the energy of charged groups in proteins (Lee and Warshel, 1992; Guenot and Kollman, 1993; York et al., 1993; Saito, 1994; Levitt et al., 1995; Åqvist, 1996). These problems depend partly on the procedure used for the truncation, and at least for peptides in water, can be minimized by using a 'force-shifted' spherical cutoff (Beck et al., 2005).

Several other ways of dealing with long-range interactions have been developed. One possibility is to use longer time steps for parts of the structure that are relatively far from the region of main interest (Marchi and Procacci, 1998). A more general approach for solvated systems is to represent the system by a spherical model that is treated microscopically, surrounded by a region of bulk solvent that is treated semimicroscopically or macroscopically (Warshel, 1979). The main conceptual advance based on this idea was the introduction of constraints that forced the solvent near the surface of the microscopic region to behave as though it were part of an infinite system. The Surface-Constrained All-Atom Solvent

(SCAAS) model (Warshel, 1979; King and Warshel, 1989) includes polarization constraints that force the surface region to have the angular polarization of a corresponding infinite system. Such constraints were found to be crucial for proper treatment of long-range electrostatic effects. The importance of proper electrostatic polarization of the surface region has not been widely appreciated, possibly because the commonly used 'stochastic boundary conditions' (Brooks III and Karplus, 1983) focus mainly on the thermostat for maintaining the temperature of this region.

Although the SCAAS model provides a fundamentally correct way of dealing with long-range solvent effects it still requires evaluating a large number of interactions if the microscopic model is of significant size and the use of a cut-off distance is avoided. Attempts to reduce the number of interactions that must be evaluated led to the Local Reaction Field (LRF) model (Lee and Warshel, 1992), in which the individual charges of 3 to 10 closely linked atoms are treated as a group. Interactions of a given group with the surroundings within a cutoff distance are evaluated explicitly on each time step. The effects of the surroundings outside the cutoff are treated by expanding the corresponding electrostatic potential to third order (up to and including the gradient of the field) and multiplying by a multipole expansion of the charges of the group. The expansion of the external potential is assumed to vary slowly, and is kept fixed for many time steps. The stability and reliability of this model have been validated for a variety of systems (Lee and Warshel, 1992; Åqvist, 1996; Sham and Warshel, 1998). Simulations using only the net charge and the dipole term of a multipole expansion have been described (Saito, 1992).

Another common way of handling long-range interactions is to use periodic boundary conditions, in which a cubic simulation system is made to behave as if it were replicated indefinitely in all directions. This allows electrostatic interactions to be treated by an approach that was first described by Ewald (1921) for electric fields in crystals. In the Ewald method, the charge of each atom is given a three-dimensional Gaussian distribution centered on the atom. The contribution of the atom to the electrostatic potential at a given point then can be represented by the sum of two functions, one of which falls off rapidly with distance and so can be truncated safely. The second function can be treated by Fourier transform techniques in a periodic system (Impey et al., 1983; Heinzinger, 1985; de Leeuw et al., 1986; Darden et al., 1993, 1997, 1999; Essmann et al., 1995; Hummer et al., 1996; Figueirido et al., 1997; Marchi and Procacci, 1998; Bogusz et al., 1998). Periodic Ewald treatments sometimes improve simulated structural properties but also can artificially stabilize certain structures at the expense of others (Figueirido et al., 1997; Hummer et al., 1997; Hunenberger and McCammon, 1999; Weber et al., 2000; Kastenholz and Hunenberger, 2004; Beck et al., 2005). Because periodic Ewald treatments have not been validated extensively for electrostatic free energies in proteins, it is not yet clear whether they can obtain these energies reliably (see discussion in Sham and Warshel, 1998). A spherical model devised by Kawajima and Warshel (1998) offers a possible way to incorporate the most useful part of the Ewald approach.

## VI. Free-Energy Perturbation and Umbrella-Sampling Methods

Evaluation of free energies by statistical mechanical approaches is extremely time-consuming because of the need to sample large regions of conformational space. Fortunately, in some cases it is possible to obtain meaningful results by perturbation approaches. Such calculations are usually done by the so-called free-energy perturbation (FEP) method (Zwanzig, 1954; Valleau and Torrie, 1977) and the related umbrella-sampling (US) procedure (Valleau and Torrie, 1977). In the FEP method, one evaluates the free energy change associated with moving a system from one state to another by breaking the process into small steps. This is achieved by running MD trajectories on a series of intermediate 'mapping' potential surfaces that force the coordinates (**s**) of the system to adapt progressively to the transition. The mapping potentials are given by

$$V_m(\mathbf{s}) = (1-\eta_m)V_\alpha(\mathbf{s}) + \eta_m V_\beta(\mathbf{s}) \tag{24}$$

where $V_\alpha(\mathbf{s})$ and $V_\beta(\mathbf{s})$ are the potential energies of the two states and $\eta_m$ is a parameter that is increased from 0 to 1 in $n$ equal steps ($m = 0, 1, \ldots, n$). The partial molecular free-energy change associated with incrementing $m$ is

$$\Delta G_{m\to m+1} = -k_B T \ln\left\{\left\langle \exp\left\{-\left(V_{m+1} - V_m\right)/k_B T\right\}\right\rangle_m\right\} \tag{25}$$

where $k_B$ is the Boltzmann constant and $\langle \ldots \rangle_m$ denotes the average value during MD trajectories on mapping surface $V_m$ (Valleau and Torrie, 1977). The overall free-energy change is the sum of the free-energy increments:

$$\Delta G^o_{\alpha\to\beta} = \sum_{m=0}^{n-1} \Delta G_{m\to m+1} \tag{26}$$

In an electron-transfer reaction, the mapping parameter $\eta_m$ is simply a fraction of an electronic charge that is removed from the donor and added to the acceptor. If the electronic interactions between the donor and acceptor are relatively weak (i.e., if the reaction is nonadiabatic), we can take $V_\alpha(\mathbf{s})$ and $V_\beta(\mathbf{s})$ to be the potential energies of the diabatic reactant and product states obtained by neglecting quantum mechanical mixing of these states. Implementing Eqs. (24–26) then is relatively straightforward if the distributions of charge over the atoms of the electron carriers are known. The situation is somewhat more complicated in an adiabatic process such as a ground-state proton-transfer reaction in which a bond is formed or broken, because the mapping potentials then must be obtained by diagonalizing the Hamiltonian matrix of the diabatic states and the off-diagonal interaction elements (for further discussion, see Warshel and Parson, 2001).

The FEP approach has been used extensively in studies of free energies of biological systems (for reviews, see Beveridge and DiCapua, 1989; Warshel and Parson, 1991; Kollman, 1993; Warshel and Parson, 2001). Fig. 3 illustrates its application to the reaction in which an electron moves from P* to $B_A$ in *Rb. sphaeroides* RCs. The potentials of the diabatic states (P* and $P^+B_A^-$) were calculated during the MD trajectories as

$$V_{p^*} = V_o + V_{elec(P^*)} \tag{27a}$$

$$V_{P^+B^-} = V_o + V_{elec(P^+B^-)} + \Delta E_{gas} \tag{27b}$$

where $V_o$ is a constant, $V_{elec(P^*)}$ and $V_{elec(P^+B^-)}$ are the electrostatic energies of the two states, and $\Delta E_{gas}$, as before, is the energy change for the hypothetical gas-phase reaction when P* and $B_A$ are infinitely far apart in a vacuum. The electrostatic energies were calculated essentially as in the PDLD model, and took into account induced dipoles in both the protein and solvent ($V_{ind}$, $G_{wat}$, $G_{memb}$ and $G_{bulk}$) in addition to the vacuum charge-charge interactions of P and B with each other ($V_{QQ}$) and with the protein atoms ($V_{Q\mu}$). $\Delta E_{gas}$ was calculated, as described briefly in Sections II and IV, by using measured $E_m$ values of P, $H_A$, and bacteriochlorophyll-*a* and bacteriopheophytin-*a* in solution, in combination with PDLD calculations of $\Delta G^{ref}_{sol}$ for oxidizing or reducing the individual components in the RC or in solution (Parson et al., 1990; Alden et al., 1995). For purposes of illustration here, we have adjusted $\Delta E_{gas}$ arbitrarily so that $\langle V_{P^+B^-} - V_{P^*}\rangle_{P^*} = 0$,

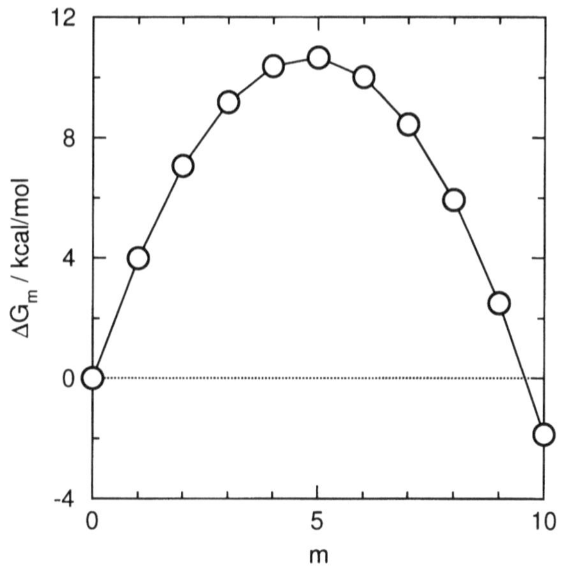


*Fig. 3.* Free energy perturbation calculations of the free energy change for the reaction $P^*B_A \to P^+B_A^-$ in *Rb. sphaeroides* reaction centers. Following a 1-ns MD trajectory on the potential surface of P (a reasonable model of P* for this purpose), FEP was used to convert the system to $P^+B_A^-$ in 10 steps. After a 1-ns trajectory on the potential surface of $P^+B_A^-$, the same procedure was used to return the system to P* in another 10 steps. Each FEP step consisted of a 100-ps MD trajectory on a mapping potential with a specified fraction ($m$/10) of a net positive charge on P and the same negative charge on $B_A$. $\Delta G_m$ is the cumulative change in free energy at mapping point $m$ relative to the free energy of P* (Eq. 26). The change of free energy for moving from mapping potential $m$ to $m$+1 was calculated as the average of $\Delta G_{m\to m+1}$ and $-\Delta G_{m+1\to m}$, with $\Delta G_{m\to m+1}$ as defined in Eq. (25). $\Delta G_m$ includes both the change in electrostatic free energy ($\Delta G_{elec}$), which becomes more negative with increasing $m$, and the change in $\Delta E_{gas}$, which increases linearly with $m$. For this system, the differences between $\Delta G_{m\to m+1}$ and $-\Delta G_{m+1\to m}$ were less than the height of the symbols, and calculating the overall free energy change by the LRA by (Eq. 4a) gave the same result as the FEP procedure within ±0.02 kcal/mol. Long-range electrostatic interactions were treated by the LRF method. Ionizable amino acid side chains were taken to be in their neutral ionization states. For other details, see Warshel and Parson (2001). With permission of Cambridge University Press.

which agrees with the calculated result within the estimated uncertainty of ±2 kcal/mol. The FEP calculations give an overall free energy change of −1.9 kcal/mol for formation of $P^+B_A^-$ from $P^*B_A$. This result is in accord with experiment, the PDLD calculations described in Section II, and our earlier FEP calculations (Creighton et al., 1988; Parson et al., 1990; Warshel et al., 1994; Alden et al., 1995, 1996). It also agrees with calculations by Blomberg et al. (1998) and some of the results presented by Ceccarelli and Marchi (2003a,b). Calculations that neglect the self-energies of the charged species (Thompson et al., 1991; Marchi et al., 1993; Hasegawa and Nakatsuji, 1998), and PB calculations using a low dielectric constant for the protein (Gunner et al., 1996), put $P^+B_A^-$ at higher energies that appear to be inconsistent with experiment.

Many biological processes can be formulated in terms of the probability of finding the system at a given value ($x'$) of the reaction coordinate ($x$) for a transition between two states, because this probability can be related straightforwardly to the relative free energy of the system. For this purpose it is convenient to define the reaction coordinate as the potential energy difference between the diabatic product and reactant states, $V_\beta - V_\alpha$. The free energy of the system at $x = x'$, $\Delta g(x')$, can be obtained by combining the FEP and US approaches (Warshel, 1982; Hwang and Warshel, 1987; Hwang et al., 1988; King and Warshel, 1990; Warshel and Parson, 2001; Blumberger and Sprik, 2006). For a nonadiabatic electron-transfer reaction, $\Delta g(x')$ is given by

$$\Delta g\left(x'\right) = \Delta G_k - k_B T \ln\left[\frac{\left\langle \delta\left(x - x'\right)\right\rangle_k}{\left\langle \delta\left(x - \langle x\rangle_k\right)\right\rangle_k}\right] \qquad (28)$$

The averages in this expression are taken on a mapping potential ($V_k$) that puts $x$ in the region of $x'$; $\delta(x-x')$ is the Dirac delta function (1 if $x$ is within a small region around $x'$, and zero otherwise); $\langle x\rangle_k$ is the mean value of $x$ during trajectories on $V_k$ as before, and $\Delta G_k$ is the cumulative free energy change for moving from the reactant potential ($V_\alpha$) to $V_k$:

$$\Delta G_k = \sum_{m=0}^{k-1} \Delta G_{m\to m+1} \qquad (29)$$

The delta functions simply count the number of time intervals during which $x$ is within a specified interval around $x'$ or $\langle x\rangle_k$, so the fraction in

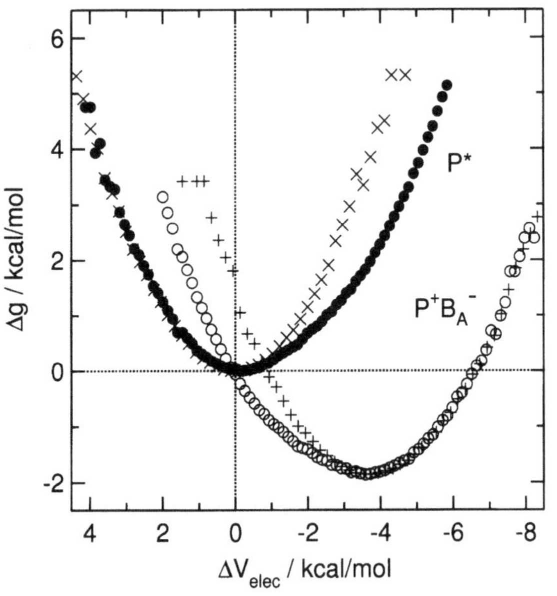


*Fig. 4.* Calculated free-energy functions for the diabatic reactant ($\Delta g_{P^*}$, × and Δ) and product ($\Delta g_{P^+B^-}$, + and o) states in the reaction $P^*B_A \rightarrow P^+B_A^-$ in *Rb. sphaeroides* reaction centers. The abscissa is the reaction coordinate, which is defined as the electrostatic energy difference between the two states. The free-energy functions plotted with × and + were obtained by Eq. (28) from separate MD trajectories on the two potential surfaces; those plotted with ● and ○ were obtained by combining the results from these trajectories (Eq. 30). For details see Warshel and Parson (2001). With permission of Cambridge University Press.

the logarithm on the right-hand side of Eq. 28 (the umbrella-sampling feature of the method) gives the relative probability of finding the system at $x'$ during trajectories on $V_k$.

Figure 4 shows the results of such calculations for electron transfer from P* to $B_A$ in *Rb. sphaeroides* RCs. The points plotted as '×' and '+' give the results obtained from separate trajectories on the potential surfaces of the reactant (P*) and product ($P^+B_A^-$) states, respectively. These free-energy curves are most reliable in the regions of their minima, where the system spends most of its time during the trajectory. Sampling of the region between the two minima can be improved by combining the results from the two trajectories. The free energy functions for P* and $P^+B_A^-$ then become

$$\Delta g_{P^*}(x') = -k_B T \ln\left\{\frac{1}{2}\left[\left\langle p\left(x'\right)\right\rangle_{P^*} + \left\langle p\left(x'\right)\right\rangle_{P^+B^-} \exp\left(\beta x' - \beta\Delta G^o_{P^*\to P^+B^-}\right)\right]\right\}$$
(30a)

$$\Delta g_{\mathrm{P^+B^-}}(x') = -k_BT\ln\left\{\frac{1}{2}\left[\left\langle p(x')\right\rangle_{\mathrm{P^+B^-}} + \left\langle p(x')\right\rangle_{\mathrm{P^*}}\exp\left(\beta\Delta G^o_{\mathrm{P^*\to P^+B^-}} - \beta x'\right)\right]\right\} + \Delta G^o_{\mathrm{P^*\to P^+B^-}} \quad (30b)$$

where $\langle p(x')\rangle_k = \langle\delta(x - x')\rangle_k / \langle\delta(x - \langle x\rangle_k)\rangle_k$, $\beta = 1/k_BT$, $\Delta G^o_{\mathrm{P^*\to P^+B^-}}$ is the overall free-energy change given by Eq. 26 (–1.9 kcal/mol). These functions are shown as the filled and empty circles in Fig. 4. If $|\Delta G^o_{\mathrm{P^*\to P^+B^-}}|$ were much larger, trajectories on additional mapping potentials would be required in order to sample the configurational space between the two minima adequately.

One way to check whether the conformational space has been sampled sufficiently well in a given region of the reaction coordinate is to see whether the free-energy functions for the reactant and product states satisfy the expression

$$\Delta g_\beta(x) = \Delta g_\alpha(x) + x \quad (31)$$

which follows from the relationship

$$\Delta g_\beta(x') = -k_BT\ln\left[\frac{\left\langle\delta(x - x')\exp(-x/k_BT)\right\rangle_\alpha}{\left\langle\delta\left(x - \left\langle x'\right\rangle_\alpha\right)\right\rangle_\alpha}\right] \quad (32)$$

(Warshel, 1982; Tachiya, 1989). The averaged free-energy curves plotted with filled and open circles conform to Eq. (31) in the region between the two minima, while the individual curves plotted with crosses and squares do not.

Equations (31) and (32) indicate that the entire free-energy curve for the product state could be obtained from a sufficiently long MD trajectory in the reactant state and vice versa, which may seem surprising. However, these expressions assume that the trajectory samples the entire configurational space adequately, which is unlikely to be the case for a complex system (Zhou and Szabo, 1995; Parson et al., 1998). Unless it is extremely long, a single trajectory on the reactant surface generally will not sample all the configurations that contribute significantly to the product surface, even if the overall free-energy change in the reaction is relatively small. Attempting to generate both free-energy surfaces from a single trajectory therefore is not to be recommended in most cases.

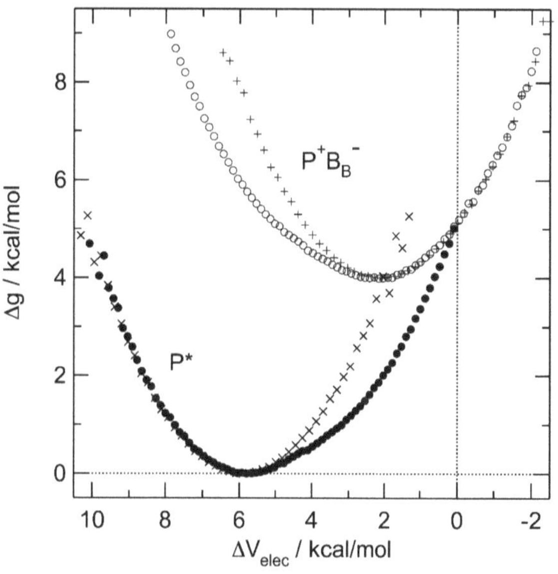


*Fig. 5.* Same as Fig. 4, but for P* (× and ●) and $\mathrm{P^+B_A^-}$ (+ and ○).

Figure 5 shows the results of similar FEP calculations for electron transfer from P* to the bacteriochlorophyll on the 'inactive' branch of electron carriers ($\mathrm{B_B}$). The product state ($\mathrm{P^+B_B^-}$) is found to lie about 4 kcal/mol above P*. This result agrees with the PDLD calculations on *Bl. viridis* RCs (see Section II and Parson et al., 1990), and is in good accord with the observation that electron transfer occurs predominantly to $\mathrm{B_A}$ and $\mathrm{H_A}$ in preference to $\mathrm{B_B}$ and $\mathrm{H_B}$. Mutations that should lower the energy of $\mathrm{P^+B_B^-}$ relative to $\mathrm{P^+B_A^-}$ have been shown to promote electron-transfer to $\mathrm{B_B}$ and $\mathrm{H_B}$ (see, e.g., Kirmaier et al., 2001, 2002, 2004; Haffa et al., 2004; Katilius et al., 2004).

The results shown in Fig. 5 disagree with MD simulations by Ceccarelli and Marchi (2003b), who put $\mathrm{P^+B_B^-}$ about 7 kcal/mol *below* P*. The discrepancy seems likely to stem in part from the use of periodic boundary conditions and Ewald sums in the latter study. As discussed in Section V, this approach has not been validated critically for calculating electrostatic free energies in proteins, but is known to lead to artifacts in proteins with ionized residues. The spherical SCAAS and LRF treatments that we used in Fig. 5 have been tested more thoroughly and probably provide a more consistent way of treating long-range effects in nonperiodic systems. Ceccarelli and Marchi (2003b) did use a longer simulation time (3.4 ns compared to 2.0 ns for the sum of our trajectories in the reactant and product states) and a larger model of the RC that included explicit detergent molecules. They propagated the trajectory only in the reactant state, however, and thus could not exploit the power of the FEP or LRA approaches for sampling the free energy of the charge-separation process. In our experience and as illustrated in Figs. 4 and 5, sampling only the initial state is problematic when applied to electrostatic free energies in proteins, even with very long simulation times. The authors' statement that experimental information on $\Delta G$ for charge separation was used to tune approximations of the medium dielectric response in our earlier calculations (Creighton et al., 1988; Parson et al., 1990; Warshel et al., 1994; Alden et al., 1995) is incorrect, as are their claims that previous studies did not model the RC and its environment explicitly, include both electronic and nuclear polarization, or calculate $\Delta G$ and $\lambda$ directly.

## VII. Concluding Comments

Unfortunately, the availability of a variety of simulation approaches, convenient software packages, and powerful computers does not guarantee reliable calculations of electrostatic energies in biological systems. The problems involve slow convergence, the need for proper treatments of long-range effects and the boundary conditions, and the difficulty of selecting an appropriate dielectric constant in semimacroscopic calculations. Important terms such as $\Delta E_{gas}$ cannot always be obtained reliably by quantum calculations, and if obtained from measured $E_m$ values and calculated solvation energies, are subject to experimental uncertainties as well as errors in the calculations. The underlying mathematical problem of a complex dielectric medium with multiple dielectric constants remains unsolved, and possibly may never be solved.

In this situation, it seems essential to explore several models and to examine the results critically before reaching firm conclusions. There is much to be gained from experience in related cases where the answer is known accurately from experiment. It also is important not to be misled by successes in reproducing nondiscriminating benchmarks such as $pK_a$s of surface groups in proteins, or by the seemingly rigorous nature of some treatments (see discussion in Kato and Warshel, 2005).

One also should not assume that a single MD simulation will justify anything beyond the most tentative conclusions, no matter how long the trajectory continues or how many atoms are included in the system. We would not, for example, subscribe to the claim that a 1-μs simulation of an ion traversing a transmembrane pore 'might give an enormous payoff in our understanding of what forces control the ion at various stages of permeation' (McClesky, 2000). Simulations of biological systems require treating many factors that might not be properly represented in any particular simulation program. A single trajectory obviously is not capable of ensuring that the results are robust to variations of the model or the approximations inherent in the treatment. Even if all the important factors are treated correctly, obtaining results that can be compared directly with experiment may require averaging over multiple trajectories.

But if the necessary attention is applied to the many factors that can affect calculations of electrostatic energies, calculations of this type are capable of providing useful answers to a broad range of challenging biophysical and biochemical questions that are not accessible experimentally. When combined with site-directed mutagenesis, they offer a powerful way to explore structure-function relationships in proteins, and in photosynthetic electron-transfer proteins in particular.

## Acknowledgments

Preparation of this chapter was supported in part by National Science Foundation grant MCB-9904618 to W.W.P. and National Institutes of Health grants GM24492 and 40283 to A.W.

## References

Alagona G, Ghio C and Kollman P (1986) Monte Carlo simulation studies of the solvation of ions. 1. Acetate ion and methylammonium cation. J Am Chem Soc 108: 185–191

Alden RG, Parson WW, Chu ZT and Warshel A (1995) Calculations of electrostatic energies in photosynthetic reaction centers. J Am Chem Soc 117: 12284–12298

Alden RG, Parson WW, Chu ZT and Warshel A (1996) Macroscopic and microscopic estimates of the energetics of charge separation in bacterial reaction centers. In: Michel-Beyerle ME (ed) The Reaction Center of Photosynthetic Bacteria: Structure and Dynamics, pp 105–116. Springer Verlag, Berlin

Alexov EG and Gunner M (1997) Incorporating protein conformational flexibility into the calculation of the pH-dependent protein properties. Biophys J 72: 2075–2093

Alexov EG and Gunner M (1999) Calculated protein and proton motions coupled to electron transfer: Electron transfer from $Q_A^-$ to $Q_B$ in bacterial photosynthetic reaction centers. Biochemistry 38: 8253-8270

Alexov E, Miksovska J, Baciou L, Schiffer M, Hanson DK, Sebban P and Gunner MR (2000) Modeling the effects of mutations on the free energy of the first electron transfer from $Q_A^-$ to $Q_B$ in photosynthetic reaction centers. Biochemistry 39: 5940–5952

Allen MP and Tildesley DJ (1987) Computer Simulations of Lipids. Oxford University Press Oxford

Antosiewicz J, McCammon JA and Gilson MK (1994) Prediction of pH-dependent properties of proteins. J Mol Biol 238: 415–436

Åqvist J (1996) Calculation of absolute binding free energies for charged ligands and effects of long-range electrostatic interactions. J Comput Chem 17: 1587–1597

Åqvist J and Hansson T (1996) On the validity of electrostatic linear response in polar solvents. J Phys Chem 100: 9512–9521

Baker NA, Sept D, Simpson J, Holst MJ and McCammon JA (2001) Electrostatics of nanosystems: Application to microtubules and the ribosome. Proc Natl Acad Sci USA 98: 10037–10041

Bashford D and Case DA (2000) Generalized Born models of macromolecular solvation effects. Annu Rev Phys Chem 51: 129–152

Beck DA and Daggett V (2004) Methods for molecular dynamics simulations of protein folding/unfolding in solution. Methods 34: 112–120

Beck DAC, Armen RS and Daggett V (2005) Cutoff size need not strongly influence molecular dynamics results for solvated polypeptides. Biochemistry 44: 609–616

Belch AC, Berkowitz M and McCammon JA (1986) Solvation structure of a sodium chloride ion-pair in water. J Am Chem Soc 108: 1755–1761

Beveridge DL and DiCapua FM (1989) Free energy via molecular simulation: Applications to chemical and biomolecular systems. Annu Rev Biophys Biophys Chem 18: 431–492

Blomberg MRA, Siegbahn PEM and Babcock GT (1998) Modeling electron transfer in biochemistry: A quantum chemical study of charge separation in *Rhodobacter sphaeroides* and Photosystem II. J Am Chem Soc 120: 8812–8824

Blumberger J and Sprik M (2006) Quantum versus classical electron transfer energy as reaction coordinate for the aqueous $Ru^{2+}/Ru^{3+}$ redox reaction. Theor Chem Acc 115: 113–126

Bogusz S, Cheatham III TE and Brooks BR (1998) Removal of pressure and free energy artifacts in charged periodic systems via net charge corrections to the Ewald potential. J Chem Phys 108: 7070–7084

Born M (1920) Volumen und Hydratationswärme der Ionen. Z Phys 1: 45–47

Brooks III CL and Karplus M (1983) Deformable stochastic boundaries in molecular dynamics. J Chem Phys 79: 6312–6325

Buono GS, Figueirido F and Levy RM (1994) Intrinsic $pK_a$'s of ionizable residues in proteins: An explicit solvent calculation for lysozyme. Proteins Struct Funct Gen 20: 85–97

Burkert U and Allinger NL (1982) Molecular Mechanics. American Chemical Society, Washington DC

Ceccarelli M and Marchi M (2003a) Simulation and modeling of the *Rhodobacter sphaeroides* bacterial reaction center: Structure and interactions. J Phys Chem B 107: 1423–1431

Ceccarelli M and Marchi M (2003b) Simulation and modeling of the *Rhodobacter sphaeroides* bacterial reaction center II: Primary charge separation. J Phys Chem B 107: 5630–5641

Chandrasekhar J, Spellmeyer DC and Jorgensen WL (1984) Energy component analysis for dilute aqueous solutions of $Li^+$, $Na^+$, $F^-$ and $Cl^-$ ions. J Am Chem Soc 106: 903–910

Constanciel R and Contreras R (1984) Self-consistent field theory of solvent effects representation by continuum models: introduction of desolvation contribution. Theor Chim Acta 65: 1–11

Creighton S, Hwang JK, Warshel A, Parson WW and Norris J (1988) Simulating the dynamics of the primary charge separation process in bacterial photosynthesis. Biochemistry 27: 774–781

Darden T, York DM and Pedersen L (1993) Particle mesh Ewald: An *N*.log(*N*) method for Ewald sums in large systems. J Chem Phys 98: 10089–10092

Darden T, Toukmaji A and Pedersen LG (1997) Long-range electrostatic effects in biomolecular simulations. J Chim Phys Physico-Chimie Biol 94: 1346–1364

Darden T, Perera L, Li LP and Pedersen L (1999) New tricks for modelers from the crystallography toolkit: The particle mesh Ewald algorithm. Struct Fold & Design 7: R55–R60

de Leeuw SW, Perram JW and Smith ER (1986) Computer simulation of the static dielectric constant of systems with permanent electric dipoles. Annu Rev Phys Chem 37: 245–270

Essmann U, Perera L, Berkowitz ML, Darden T, Lee H and Pedersen L (1995) A smooth particle mesh Ewald method. J Chem Phys 103: 8577–8593

Ewald PP (1921) Die Berechnung optischer und elektrostatischer Gitterpotentiale. Ann Phys 64: 253–287

Fan H, Mark AE, Zhu J and Honig B (2005) Comparative study of generalized Born models: Protein dynamics. Proc Natl Acad Sci USA 102: 6760–6764

Figueirido F, DelBuono GS and Levy RM (1997) On the finite size corrections to the free energy of ionic hydration. J Phys Chem B 101: 5622–5623

Forsyth WR and Robertson AD (2000) Insensitivity of perturbed carboxyl $pK_a$ values in the ovomucoid third domain to charge replacement at a neighboring residue. Biochemistry 39: 8067–8072

Gallicchio E and Levy RM (2004) AGBNP: An analytic implicit solvent model suitable for molecular dynamics simulations and high-resolution modeling. J Comput Chem 25: 479–499

Georgescu RE, Alexov EG and Gunner M (2002) Combining conformational flexibility and continuum electrostatics for calculating $pK_a$s in proteins. Biophys J 1731–1748

Ghosh A, Rapp CS and Friesner RA (1998) Generalized Born model based on a surface integral formulation. J Phys Chem B 102: 10983–10990

Gilson M and Honig B (1988) Calculation of the total electrostatic energy of a macromolecular system. Solvation energies, binding energies and conformational analysis. Proteins Struct Funct Gen 4: 7–18

Gilson M, Rashin A, Fine R and Honig B (1985) On the calculation of electrostatic interactions in proteins. J Mol Biol 503–516

Gilson M, Sharp KA and Honig B (1987) Calculating the electrostatic potential of molecules in solution: method and error assessment. J Comput Chem 9: 327–335

Guenot J and Kollman P (1992) Molecular dynamics studies of a DNA-binding protein. 2. An evaluation of implicit and explicit solvent models for the molecular-dynamics simulation of the *Escherichia coli* Trp repressor. Protein Sci 1: 1185–11205

Guenot J and Kollman P (1993) Conformational and energetic effects of truncating nonbonded interactions in an aqueous protein dynamics simulation. J Comput Chem 14: 295–311

Gunner M and Alexov EG (2000) A pragmatic approach to structure based calculation of coupled proton and electron transfer in proteins. Biochim Biophys Acta 1485: 63–87

Gunner MR and Honig B (1991) Electrostatic control of midpoint potentials in the cytochrome subunit of the Rhodopseudomonas viridis reaction center. Proc Natl Acad Sci USA 88: 9151–9155

Gunner M, Nicholls A and Honig B (1996) Electrostatic potentials in *Rhodopseudomonas viridis* reaction centers: Implications for the driving force and directionality of electron transfer. J Phys Chem 100: 4277–4291

Gunner M, Alexov EG, Torres E and Lipovaca S (1997) The importance of the protein in controling the electrochemistry of heme metalloproteins: Methods of calculation and analysis. J Biol Inorg Chem 2: 126–134

Haffa ALM, Lin S, Williams JC, Bowen BP, Taguchi AKW, Allen JP and Woodbury NW (2004) Controlling the pathway of photosynthetic charge separation in bacterial reaction centers. J Phys Chem B 108: 4–7

Hasegawa J and Nakatsuji H (1998) Mechanism and unidirectionality of the electron transfer in the photosynthetic reaction center of *Rhodopseudomonas viridis*: SAC-CI theoretical study. J Phys Chem B 102: 10420–10430

Hawkins GD, Cramer CJ and Truhlar DG (1996) Parametrized models of aqueous free energies of solvation based on pairwise descreening of solute atomic charges from a dielectric medium. J Phys Chem 100: 19824–19839

Heinzinger K (1985) Computer simulations of aqueous electrolyte solutions. Physica B 131: 196–216

Hendrickson JB (1961) Molecular geometry. I. Machine computation of the common rings. J Am Chem Soc 83: 4537–4547

Hingerty BE, Richie RH, Ferrell TL and Turner JE (1985) Dielectric effects in biopolymers. The theory of ionic saturation revisited. Biopolymers 24: 427–439

Honig B and Nicholls A (1995) Classical electrostatics in biology and chemistry. Science 268: 1144–1149

Hughes JM, Hutter MC and Hush NS (2001) Modeling the bacterial photosynthetic reaction center. 4. The structural, electrochemical, and hydrogen-bonding properties of 22 mutants of *Rhodobacter sphaeroides*. J Am Chem Soc 123: 8550–8563

Hummer G and Szabo A (1996) Calculation of free-energy differences from computer simulations of initial and final states. J Chem Phys 105: 2004–2010

Hummer G, Pratt LR and Garcia AE (1996) Free energy of ionic hydration. J Phys Chem 100: 1206–1215

Hummer G, Pratt LR, Garcia AE, Berne BJ and Rick SW (1997) Electrostatic potentials and free energies of solvation of polar and charged molecules. J Phys Chem B 101: 3017–3020

Hunenberger PH and McCammon JA (1999) Ewald artifacts in computer simulations of ionic solvation and ion-ion interaction: A continuum electrostatics study. J Chem Phys 110: 1856–1872

Hwang J-K and Warshel A (1987) Microscopic examination of free energy relationships for electron transfer in polar solvents. J Am Chem Soc 109: 715–720

Hwang J-K, King G, Creighton S and Warshel A (1988) Simulation of free energy relationships and dynamics of SN2 reactions in aqueous solution. J Am Chem Soc 110: 5297–5311

Impey RW, Madden PA and McDonald IR (1983) Hydration and mobility of ions in solution. J Phys Chem 87: 5071–5083

Ivashin N, Källenbring B, Larsson S and Hansson Ö (1998) Charge separation in photosynthetic reaction centers. J Phys Chem B 102: 5017–5022

Johnson ET and Parson WW (2002) Electrostatic interactions in an integral membrane protein. Biochemistry 41: 6483–6494

Johnson ET, Müh F, Nabedryk E, Williams JC, Allen JP, Lubitz W, Breton J and Parson WW (2002) Electronic and vibronic coupling of the special pair of bacteriochlorophylls in photosynthetic reaction centers from wild-type and mutant strains of *Rhodobacter sphaeroides*. J Phys Chem B 106: 11859–11869

Kastenholz MA and Hunenberger PH (2004) Influence of artificial periodicity and ionic strength in molecular dynamics simulations of charged biomolecules employing lattice-sum methods. J Phys Chem B 108: 774–788

Katilius E, Babendure JL, Lin S and Woodbury NW (2004) Electron transfer dynamics in *Rhodobacter sphaeroides* reaction center mutants with a modified ligand for the monomer bacteriochlorophyll on the active side. Photosynth Res 81: 165–180

Kato M and Warshel A (2005) Through the channel and around the channel: Validating and comparing microscopic approaches for the evaluation of free energy profiles for ion penetration through ion channels. J Phys Chem B 109: 19516–19522

Kim J, Mao J and Gunner M (2005) Are acidic and basic groups in buried proteins predicted to be ionized? J Mol Biol 348: 1283–1298

King G and Warshel A (1989) A surface constrained all-atom solvent model for effective simulations of polar solutions. J Chem Phys 91: 3647–3661

King G and Warshel A (1990) Investigation of the free energy functions for electron transfer reactions. J Chem Phys 93: 8682–8692

King G, Lee FS and Warshel A (1991) Microscopic simulations of macroscopic dielectric constants of solvated proteins. J Chem Phys 95: 4366–4377

Kirmaier C, He C and Holten D (2001) Manipulating the direction of electron transfer in the bacterial reaction center by swapping Phe for Tyr near BChlM (L181) and Tyr for Phe near BChlL (M208). Biochem 40: 12132–12139

Kirmaier C, Laible PD, Czarnecki K, Hata AN, Hanson DK, Bocian DF and Holten D (2002) Comparison of M-side electron transfer in *Rb. sphaeroides* and *Rb. capsulatus* reaction centers. J Phys Chem B 106: 1799–1808

Kirmaier C, Laible PD, Hanson DK and Holten D (2004) B-side electron transfer to form $P^+H_B^-$ in reaction centers from the F(L181)Y/Y(M208)F mutant of *Rhodobacter capsulatus*. J Phys Chem B 108: 11827–11832

Klein BJ and Pack GR (1983) Calculations of the spatial distribution of charge density in the environment of DNA. Biopolymers 22: 2331–2352

Kollman P (1993) Free energy calculations: Applications to chemical and biochemical phenomena. Chem Rev 93: 2395–2417

Kollman P (2000) Calculating structures and free energies of complex molecules: Combining molecular mechanics and continuum models. Acc Chem Res 33: 889–897

Kozaki T, Morihashi K and Kikuchi O (1988) An MNDO effective charge model study of the solvent effect. The internal rotation about partial double bonds and the nitrogen inversion in amine. J Mol Struct 168: 265–277

Kozaki T, Morihashi K and Kikuchi O (1989) MNDO effective charge model study of solvent effect on the potential energy surface of the SN2 reaction. J Am Chem Soc 111: 1547–1552

Kubo R, Toda M and Hashitsume N (1985) Statistical Physics II: Nonequilibrium Statistical Mechanics. Springer-Verlag, Berlin

Kuharski RA, Bader JS, Chandler D, Sprik M, Klein ML and Impey RW (1988) Molecular model for aqueous ferrous-ferric electron transfer. J Chem Phys 89: 3248–3257

Kuwajima S and Warshel A (1988) The extended Ewald method: A general treatment of long-range electrostatic interactions in microscopic simulations. J Chem Phys 89: 3751–3759

Lancaster CRD, Michel H, Honig B and Gunner M (1996) Calculated coupling of electron and protein transfer in the photosynthetic reaction center of *Rhodopseudomonas viridis*. Biophys J 70: 2469–2492

Lee FS and Warshel A (1992) A local reaction field method for fast evaluation of long-range electrostatic interactions in molecular simulations. J Chem Phys 97: 31003–3107

Lee FS, Chu ZT, Bolger MB and Warshel A (1992) Calculations of antibody-antigen interactions: microscopic and semi-microscopic evaluation of the free energies of binding of phosphorylcholine analogs to McPC603. Protein Eng 5: 215–228

Lee FS, Chu ZT and Warshel A (1993) Microscopic and semi-microscopic calculations of electrostatic energies in proteins by the POLARIS and ENZYMIX programs. J Comput Chem 14: 161–185

Levitt M and Lifson S (1969) Refinement of protein conformations using a macromolecular energy minimization procedure. J Mol Biol 46: 269–279

Levitt M, Hirshberg M, Sharon R and Daggett V (1995) Potential-energy function and parameters for simulations of the molecular dynamics of proteins and nucleic acids in solution. Comput Phys Commun 91: 215–231

Lifson S and Warshel A (1968) A consistent force field for calculation of conformations, vibrational spectra and enthalpies of cycloalkanes and n-alkane molecules. J Chem Phys 49: 5116–5129

Lockhart DJ and Kim PS (1993) Electrostatic screening of charge and dipole interactions with the helix backbone. Science 260: 198–202

Madura JD, Briggs JM, Wade RC, Davis ME, Luty BA and McCammon JA (1995) Electrostatics and diffusion of molecules in solution: Simulations with the University of Houston Brownian Dynamics Program. Comp Phys Commun 91: 57–95

Marchi M and Procacci P (1998) Coordinates scaling and multiple time step algorithms for simulation of solvated proteins in the NPT ensemble. J Chem Phys 109: 5194–5202

Marchi M, Gehlen JN, Chandler D and Newton M (1993) Diabatic surfaces and the pathway for primary electron transfer in a photosynthetic reaction center. J Am Chem Soc 115: 4178–4190

McClesky EW (2000) Ion channel selectivity using an electric stew. Biophys J 79: 1691–1692

Mehler EL and Eichele G (1984) Electrostatic effects in water-accessible regions of proteins. Biochemistry 23: 3887–3891

Mehler EL and Guarnieri F (1998) A self-consistent, microenvironment modulated screened Coulomb potential approximation to calculate pH-dependent electrostatic fields in proteins. Biophys J 77: 3–22

Mezei M and Beveridge DL (1981) Monte Carlo studies of the structure of dilute aqueous solutions of $Li^+$, $Na^+$, $K^+$, $F^-$ and $Cl^-$. J Chem Phys 74: 6902–6910

Miertus S, Scrocco E and Tomasi J (1981) Electrostatic interaction of a solute with a continuum. A direct utilization of ab initio molecular potentials for the provision of solvent effects. J Chem Phys 55: 117–129

Muegge I, Apostolakis J, Ermler U, Fritzsch G, Lubitz W and Knapp EW (1996) Shift of the special pair redox properties: Electrostatic energy computations of mutants of the reaction center from *Rhodobacter sphaeroides*. Biochemistry 35: 8359–8370

Nichols A and Honig B (1991) A rapid finite-difference algorithm utilizing successive over-relaxation to solve the Poisson-Boltzmann equation. J Comput Chem 12: 435–445

Nielsen JE, Andersen KV, Honig B, Hooft RV, Klebe G, Vriend G and Wade RC (1999) Improving macromolecular electrostatics calculations. Protein Eng 12: 657–662

Onufriev A, Case DA and Bashford D (2002) The effective Born radii in the generalized Born approximation: The importance of being perfect. J Comput Chem 23: 1297–1304

Parson WW, Chu ZT and Warshel A (1990) Electrostatic control of charge separation in bacterial photosynthesis. Biochim Biophys Acta 1017: 251–272

Parson WW, Chu ZT and Warshel A (1998) Reorganization energy of the initial electron-transfer step in photosynthetic bacterial reaction centers. Biophys J 74: 182–191

Qiu D, Shenkin PS, Hollinger FP and Still WC (1997) The GB/SA continuum model for solvation. A fast analytical method for the calculation of approximate Born radii. J Phys Chem 101: 3005–3014

Rabenstein B, Ullmann GM and Knapp E (1998) Calculation of protonation patterns in proteins with structural relaxation and molecular ensembles: Application to the photosynthetic reaction center. Eur Biophys J Biophys Lett 27: 626–637

Rees DC (1980) Experimental evaluation of the effective dielectric constant of proteins. J Mol Biol 141: 323–326

Rocchia W, Alexov EG and Honig B (2001) Extending the applicability of the nonlinear Poisson-Boltzmann equation: Multiple dielectric constants and multivalent ions. J Phys Chem B 105: 6507–6514

Rocchia W, Sridharan S, Nicholls A, Alexov EG, Chiabrera A and Honig B (2002) Rapid grid-based construction of the molecular surface and the use of induced surface charge to calculate reaction field energies: Applications to the molecular systems and geometric objects. J Comput Chem 23: 128–137

Saito M (1992) Molecular dynamics simulations of proteins in water without the truncation of long-range Coulomb interactions. Mol Simul 8: 321–333

Saito M (1994) Molecular dynamics simulations of proteins in solution: Artifacts caused by the cutoff approximation. J Chem Phys 101: 4055–4061

Schaefer M and Karplus M (1996) A comprehensive analytical treatment of continuum electrostatics. J Phys Chem 100: 1578–1599

Scherer POJ and Fischer SF (1989) Long-range electron transfer within the hexamer of the photosynthetic reaction center *Rhodopseudomonas viridis*. J Phys Chem 93: 1633–1637

Scherer POJ, Scharnagl C and Fischer SF (1995) Symmetry breakage in electronic structure of the photosynthetic reaction center

of *Rhodopseudomonas viridis*. Chem Phys 197: 333–341
Schutz CN and Warshel A (2001) What are the dielectric 'constants' of proteins and how to validate electrostatic models. Proteins Struct Funct Gen 44: 400–417
Sham YY and Warshel A (1998) The surface constrained all atom model provides size independent results in calculations of hydration free energies. J Chem Phys 109: 7940–7944
Sham YY, Chu ZT and Warshel A (1997) Consistent calculations of $pK_a$'s of ionizable residues in proteins: Semi-microscopic and macroscopic approaches. J Phys Chem B 101: 4458–4472
Sham YY, Muegge I and Warshel A (1998) The effect of protein relaxation on charge-charge interactions and dielectric constants in proteins. Biophys J 74: 1744–1753
Sharp KA and Honig B (1990a) Electrostatic interactions in macromolecules: Theory and applications. Ann Rev Biophys Biophys Chem 19: 301–332
Sharp KA and Honig B (1990b) Calculating total electrostatic energies with the nonlinear Poisson-Boltzmann equation. J Phys Chem 94: 7684–7692
Sitkoff D, Sharp KA and Honig B (1994) Accurate calculation of hydration free energies using macroscopic solvent models. J Phys Chem 98: 1978–1988
Steinbach PJ and Brooks BR (1994) New spherical-cutoff methods for long-range forces in macromolecular simulation. J Comput Chem 15: 667–683
Still WC, Tempczyk A, Hawley RC and Hendrickson T (1990) Semianalytical treatment of solvation for molecular mechanics and dynamics. J Am Chem Soc 112: 6127–6129
Tachiya M (1989) Relation between the electron transfer rate and the free energy change of reaction. J Phys Chem 93: 7050–7052
Tanford C and Kirkwood JG (1957) Theory of protein titration curves. I. General equations for impenetrable spheres. J Am Chem Soc 79: 5333–5339
Thompson MA, Zerner MC and Fajer J (1991) A theoretical examination of the electronic structure and spectroscopy of the photosynthetic reaction center from *Rhodopseudomonas viridis*. J Am Chem Soc 113: 8210–8215
Ullmann GM and Knapp E (1999) Electrostatic models for computing protonation and redox equilibria in proteins. Eur Biophys J Biophys Lett 28: 533–551
Valleau JP and Torrie GM (1977) A guide to Monte Carlo for statistical mechanics. 2. Byways. In: Bern BJ (ed) Modern Theoretical Chemistry, Vol 5, pp 169–194. Plenum Press, New York
Voigt P and Knapp E (2003) Tuning heme redox potentials in the cytochrome *c* subunit of photosynthetic reaction centers. J Biol Chem 278: 51993–52001
Warshel A (1979) Calculations of chemical processes in solutions. J Phys Chem 83: 1640–1650
Warshel A (1981) Calculations of enzymic reactions: calculations of $pK_a$, proton transfer reactions, and general acid catalysis reactions in enzymes. Biochemistry 20: 3167–3177
Warshel A (1982) Dynamics of reactions in polar solvents. Semiclassical trajectory studies of electron-transfer and proton-transfer reactions. J Phys Chem 86: 2218–2224
Warshel A (1987) What about protein polarity? Nature 333: 15–18
Warshel A (1991) Computer Modeling of Chemical Reactions in Enzymes and Solutions. John Wiley & Sons, New York
Warshel A and Lappicirella VA (1981) Calculations of ground and excited-state potential surfaces for conjugated heteroatomic molecules. J Am Chem Soc 103: 4664–4673
Warshel A and Levitt M (1976) Theoretical studies of enzymic reactions: dielectric, electrostatic and steric stabilization of the carbonium ion in the reaction of lysozyme. J Mol Biol 103: 227–249
Warshel A and Papazyan A (1998) Electrostatic effects in macromolecules: Fundamental concepts and practical modeling. Curr Opin Struct Biol 8: 211–217
Warshel A and Parson WW (1991) Computer simulations of electron transfer reactions in solution and photosynthetic reaction centers. Annu Rev Phys Chem 42: 279–309
Warshel A and Parson WW (2001) Dynamics of biochemical and biophysical reactions: Insight from computer simulations. Q Rev Biophys 34: 563–670
Warshel A and Russell ST (1984) Calculations of electrostatic interactions in biological systems and in solutions. Q Rev Biophys 17: 283–421
Warshel A, Russell ST and Churg AK (1984) Macroscopic models for studies of electrostatic interactions in proteins: limitations and applicability. Proc Natl Acad Sci USA 81: 4785–4789
Warshel A, Naray-Szabo G, Sussman F and Hwang J-K (1989) How do serine proteases really work? Biochemistry 28: 3629–3673
Warshel A, Chu ZT and Parson WW (1994) On the energetics of the primary electron-transfer process in bacterial reaction centers. J Photochem Photobiol A: Chem 82: 123–128
Warshel A, Papazyan A and Muegge I (1997) Microscopic and semimacroscopic redox calculations: What can and cannot be learned from continuum models. J Biol Inorg Chem 2: 143–152
Warwicker J and Watson HC (1982) Calculation of the electric potential in the active site cleft due to alpha-helix dipoles. J Mol Biol 157: 671–679
Weber W, Hünenberger PH and McCammon JA (2000) Molecular dynamics simulations of a polyalanine octapeptide under Ewald boundary conditions: Influence of artificial periodicity on peptide conformation. J Phys Chem B 104: 3668–3675
Wong CF and McCammon JA (1986) Dynamics and design of enzymes and inhibitors. J Am Chem Soc 108: 3830–3832
York DM, Darden T and Pedersen LG (1993) The effect of long-range electrostatic interactions in simulations of macromolecular crystals. A comparison of the Ewald and truncated list methods. J Chem Phys 99: 8345–8348
Zhou HX and Szabo A (1995) Microscopic formulation of the Marcus theory of electron transfer. J Chem Phys 103: 3481–3494
Zhou Z and Swenson RP (1995) Electrostatic effects of surface acidic amino acid residues on oxidation-reduction potentials of the flavodoxin from *Desulfovibrio vulgaris* (Hildenborough). Biochemistry 34: 3183–3192
Zhu J, Alexov EG and Honig B (2005) Comparative study of generalized Born models: Born radii and peptide folding. J Phys Chem B 109: 3008–3022
Zwanzig RW (1954) High-temperature equation of state by a perturbation method. I. Nonpolar gases. J Chem Phys 22: 1420–1426

# Chapter 21

# Theory of Excitation Energy Transfer and Optical Spectra of Photosynthetic Systems

Thomas Renger*
*Institut für Chemie und Biochemie (Kristallographie), Freie Universität Berlin, Takustraße 6, D-14195 Berlin, Germany*

Alfred R. Holzwarth*
*Max-Planck-Institut für Bioanorganische Chemie, Stiftstraße 34–36, D-45470 Mülheim an der Ruhr, Germany*



## Summary

In this chapter, we discuss the theories applicable for calculating energy transfer kinetics and optical spectra of photosynthetic pigment-protein complexes. The various theoretical approaches for obtaining expressions for rate constants and optical spectra are reviewed. At the extremes we distinguish weak and strong coupling between electronic excitations of the pigments. If the coupling is strong compared to the dynamic and static disorder introduced by the protein environment, then delocalized electronic states are formed after light excitation. The excitation energy relaxes between those delocalized states. In the weak coupling limit localized

*Authors for correspondence, email: rth@chemie.fu-berlin.de; holzwarth@mpi-muelheim.mpg.de

*Thijs J. Aartsma and Jörg Matysik (eds): Biophysical Techniques in Photosynthesis II, pp. 421–443.*

states are created by excitation and the excitation energy is transferred via a hopping mechanism. In general in photosynthetic antenna and reaction centers neither limit applies in a strict sense. Thus more sophisticated theories describing the intermediate cases have to be applied, and these should also account for the coupling of the excited states to the vibrational states of the environment. We discuss the recent attempts of solving the challenging problem to apply a non-perturbative description of both the pigment-pigment as well as the pigment-protein couplings. Applications of these theories to the spectroscopy of photosynthetic systems over the last decade are also reviewed.

## I. Introduction

Photosynthetic antennae absorb light and transfer the excitation energy efficiently to the photosynthetic reaction centers (RCs) where the primary electron transfer reactions convert the solar energy into an electrochemical gradient. The typical antenna systems are protein complexes that hold pigments (chlorophylls (Chls), bacteriochlorophylls (BChls), phycobilins, carotenoids etc.) in suitable positions for excitation energy transfer to occur efficiently and to reduce losses by fluorescence or non-radiative transitions. The excitation energy is transferred via an exciton mechanism, first proposed by Theodor Förster half a century ago (Förster, 1948, 1965). For many photosynthetic antennae and RCs the joint efforts of biochemists and crystallographers have made it possible to isolate and crystallize the pigment-protein complexes of the photosynthetic apparatus and to determine their three dimensional structures with atomic details. The first crystal structure of a photosynthetic antenna system, the water soluble Fenna-Mathews-Olson complex of green sulfur bacteria, appeared more than 30 years ago (Fenna and Matthews, 1975). Since that time, the number and size of structurally resolved light-harvesting antennae has steadily increased. Whereas the monomeric subunit of the FMO-complex contains only seven bacteriochlorophyll *a* (BChl *a*) molecules, the recently determined monomeric subunit of Photosystem (PS) I is made up of 96 chlorophyll *a* (Chl *a*) molecules (Jordan et al., 2001). One of the most challenging problems for understanding the spectroscopic properties of pigment-protein complexes is the determination of the transition energies (site energies) of the uncoupled chromophores. While the optical spectra and the energy transfer dynamics in the FMO complex are now quite well understood, the optical transition energies of the seven BChls are only determined as free parameters from a fit of optical spectra. It is still an open question how those transition energies are affected by specific pigment-protein interactions, despite the long known, highly resolved structure of this pigment-protein complex (Fenna and Matthews, 1975; Matthews and Fenna, 1980; Olson, 2004). Some progress was made recently by electrochromic shift calculations of BChl transition energies using the charged amino acid residues (Adolphs and Renger, 2006). For larger complexes, e.g., of the size of the PS I complex with 96 Chls, it is not possible to find the transition energies simply by fitting procedures due to many ambiguities in such a large system. Despite some recent attempts to determine the transition energies from the structure by quantum chemical methods (Damjanovic et al., 2002b) this remains a challenging problem, since the presently achieved accuracy for the site energies is insufficient for an exact quantitative modeling.

Except in the very weak coupling cases, the excited states of an array of pigments are delocalized over several chromophores and form so-called excitons. The spatial distribution of the exciton depends on the relative strength of the excitonic coupling and on the static and dynamic disorder introduced by the protein environment. In general, both types of coupling, i.e., the electronic interaction between pigments and the coupling of pigments with protein vibrations, are of the same order of magnitude in photosynthetic systems. Thus a partially delocalized excited state is formed, whose delocalization and spatial distribution may change in time. Whereas a treatment of the static disorder in the transition energies is numerically possible, it is one of the challenges for theory to describe the localization of excitons due to the coupling to the protein dynamics. How well the protein can exchange vibrational energy with

*Abbreviations:* BChl – bacteriochlorophyll; Chl(s) – chlorophyll(s); COP – chronological ordering prescription; CT – charge transfer; FMO complex – Fenna-Matthews-Olson complex; *J* – spectral density; LH(C) – light harvesting (complex); PES – potential energy surface; Pheo – pheophytin; POP – partial ordering prescription; PS – Photosystem; RC – reaction center; RP – radical pair; TDC – transition density cube; TrEsp – transition changes from electrostatic potentials

the excitons depends on one crucial property, the coupling weighted density of vibrational states of the protein (spectral density). This quantity can be extracted by various experimental techniques, like for example hole burning and fluorescence site selection (Pullerits et al., 1995; Peterman et al., 1997; Pieper et al., 1999; Renger and Marcus, 2002a), or photon echo experiments (de Boeij et al., 1996; Agarwal et al., 2002; Brixner et al., 2005). Thus the three essential quantities required to model excitation energy transfer and optical spectra of pigment-protein complexes are (i) the local transition energies of the pigments, the so-called site energies, (ii) the spectral density of the pigment-protein coupling, and (iii) the couplings between the electronic excitations of the pigments, the so-called excitonic coupling.

This chapter is organized as follows: The Hamiltonian used throughout this chapter to describe optical spectra and excitation energy transfer dynamics is introduced in Section II. The weak excitonic coupling limit is treated in Section III, where the standard theories of optical spectra by Lax (1952) and Kubo and Toyozawa (1955), and of excitation energy transfer by Förster (1948, 1965), are reviewed. The strong coupling limit of optical spectra and exciton relaxation between delocalized states is treated in the framework of multi-level Redfield theory in Section IV, using a second order perturbation theory with respect to the exciton-vibrational coupling. To study situations where an aggregate of strongly coupled pigments couples weakly to another chromophore or aggregate Förster's theory has been generalized in recent years, as explained in Section V. In Section VI, the case of comparable excitonic and exciton-vibrational coupling is described in terms of non-Markovian density matrix theories of optical spectra and the 'modified Redfield theory' of excitation energy transfer is introduced. Recent applications of these theories are presented in Section VII and an outlook discussing challenging open questions in the modeling of optical spectra and excitation energy transfer dynamics in pigment-protein complexes follows in Section VIII.

## II. The Hamiltonian of a Pigment Protein Complex

The Hamiltonian that is used to derive expressions for linear optical spectra and the rate constants of excitation energy transfer consists of three parts

$$H = H_{\text{ex}} + H_{\text{ex-vib}} + H_{\text{vib}} \tag{1}$$

i.e., the excitonic part $H_{\text{ex}}$, the exciton-vibrational coupling Hamiltonian $H_{\text{ex-vib}}$, and the vibrational Hamiltonian $H_{\text{vib}}$.

The excitonic part $H_{\text{ex}}$ contains the local transition energies of the pigments $E_{\text{m}}$ and the excitonic couplings $V_{\text{mn}}$

$$H_{\text{ex}} = \sum_m E_m |m\rangle\langle m| + \sum_{mn} V_{mn} |m\rangle\langle n| \tag{2}$$

The state $|m\rangle$ denotes a localized excited state of the pigment-protein complex for which the *m*th pigment is in the excited state and all others are in the ground state.

The transition energies $E_{\text{m}}$ are called 'site energies' since they depend on the local protein environment of the pigments. Various protein-chromophore interactions influence the site energies: (a) electrostatic interactions with charged amino acid side chains (Eccles and Honig, 1983), (b) specific protein-chromophore interactions like e.g., differences in $Mg^{2+}$-ligation, H-bonding and deviations from planarity of the chlorin ring induced by the protein binding pockets (Gudowska-Nowak et al., 1990), (c) electrostatic interactions with neutral amino acid side chains and the protein backbone (F. Müh, M. E. Madjet, J. Adolphs, A. Abdurahman, B. Rabenstein, H. Ishikita, E.-W. Knapp and T. Renger, unpublished), and (d) differences in the dielectric constant of the local environment. Because of the complexity of these interactions a microscopic calculation of site energies is difficult (Gudowska-Nowak et al., 1990; Damjanovic et al., 2002a; Adolphs and Renger, 2006) since, both the conformation of the pigment as well as the electrostatic coupling of the pigment with its protein environment have to be taken into account. In addition, some of the protein residues, the titratable groups, might occur in a protonated or unprotonated form and thus an average over the possible titration states of the titratable groups is necessary. In most cases therefore the site energies are used as free parameters that are determined from the fitting of the theoretical to the experimental optical spectra (Vulto et al., 1998a,b; Wendling et al., 2002; Raszewski et al., 2005; Adolphs and Renger, 2006).

The chromophore-chromophore coupling mainly involves the Coulomb coupling between the electronic excitations. This so-called excitonic coupling between the pigments leads to a non-radiative transfer of excitation energy between the pigments (Förster, 1948). The matrix element $V_{mn}$ responsible for this transfer reads

$$V_{mn} = <\phi_m^{(e)}\,\phi_n^{(g)} \mid V_{\text{Coulomb}} \mid \phi_m^{(g)}\,\phi_n^{(e)}> \qquad (3)$$

where $\phi_m^{(e)}$ denotes the excited state wave function of pigment $m$, $\phi_n^{(g)}$ is the ground state wave function of pigment $n$, and $V_{\text{Coulomb}}$ is the Coulomb-coupling between the electrons and nuclei of the two pigments.

If the extension of the electronic wave functions of the pigments is small as compared to the center to center distance between the pigments, $V_{\text{Coulomb}}$ can be expanded in a Taylor series and the first non-vanishing element of that series is the well-known point-dipole approximation (for a more detailed discussion, see van Amerongen et al., 2000; Madjet et al. 2006).

$$V_{mn} = \frac{1}{4\pi\varepsilon_0\varepsilon_r}\left\{\frac{\vec{\mu}_m\vec{\mu}_n}{R_{mn}^3} - 3\frac{(\vec{\mu}_m\vec{R}_{mn})(\vec{\mu}_n\vec{R}_{mn})}{R_{mn}^5}\right\} \qquad (4)$$

where $\vec{\mu}_m$ and $\vec{\mu}_n$ are the optical transition dipoles of pigments $m$ and $n$ and $\vec{R}_{mn}$ is the center to center spatial vector between the two pigments. The above formula expresses the excitonic coupling $V_{mn}$ as the dipole-dipole coupling between the optical transition dipoles. The orientations of transition dipoles $\vec{\mu}_m$ of the pigments and the distance vectors $\vec{R}_{mn}$ can be obtained from structural data, although a more accurate determination of the orientation of transition dipole moments may require quantum chemical calculations (Madjet et al., 2006) and suitable experiments (Bauman and Wrobel, 1980; Fragata et al., 1988; Kleima et al. 2000; Georgakopoulou et al., 2003). Recent quantum chemical calculations using time dependent density functional theory using a B3LYP exchange correlation energy functional confirm the orientation of the $Q_y$ transition dipole moment along the $N_B$-$N_D$ ($N_{21}$-$N_{23}$ in IUPAC nomenclature) axis for chlorophyll a and BChl *a* (Madjet et al., 2006). An estimation of the dipole strengths $|\mu_m|^2$ can be obtained from the optical spectra of the chromophores measured in suitable solvents (Knox and Spring, 2003). Therefore, Eq. (4) provides a very practical tool to estimate excitonic couplings. The screening effects of the environment on the Coulomb coupling are taken into account by the dielectric constant $\varepsilon_r$, whose precise value in proteins is however difficult to determine. In addition to screening effects there are also local field corrections (for a discussion of the latter see, van Amerongen et al., 2000; Hsu et al., 2001; Adolphs and Renger, 2006) that are neglected in the above equation. In actual applications the dipole strengths are thus often treated as effective dipole strengths that include the screening and local field corrections in an effective manner and in this case one would set $\varepsilon_r = 1$ in Eq. (4). A recent quantitative evaluation of those effects for the excitonic couplings in the FMO-protein gave a factor of 0.8 between the vacuum dipole strength and the effective dipole strength (Adolphs and Renger, 2006).

If the closest distance between pigments becomes comparable or smaller than the extension of their electronic wave functions the point dipole approximation becomes invalid and the excitonic coupling has to be calculated by quantum chemical methods. For this purpose the coupling $V_{mn}$ is expressed as Coulomb coupling between transition densities (e.g., Madjet et al., 2006)

$$V_{mn} = \frac{1}{4\pi\varepsilon_0\varepsilon_r}\int d\vec{r}_1 d\vec{r}_2 \frac{\rho_m(\vec{r}_1)\rho_n(\vec{r}_1)}{|\vec{r}_1 - \vec{r}_2|} \qquad (5)$$

where the transition density of pigment $m$ is defined as

$$\rho_m(r_1) = N\int d\vec{r}_2 \ldots d\vec{r}_N \left(\phi_m^{(e)}(\vec{r}_1,\vec{r}_2,\ldots\vec{r}_N)\right)^* \times \phi_m^{(g)}(\vec{r}_1,\vec{r}_2,\ldots\vec{r}_N) \qquad (6)$$

Here, the integration runs over $N-1$ electronic coordinates (and $N$ spin variables, not explicitly shown here for simplicity).

Once, those transition densities have been calculated by a quantum chemical method, different ways exist to evaluate the excitonic coupling. One way is to approximate the couplings by the Coulomb interaction between transition charges, the so-called transition monopoles, as derived for semi-empirical Pariser Parr Pople calculations (Weiss, 1972; Chang, 1977; Warshel and Parson, 1987; Sauer et al., 1996; Damjanovic et al., 1999). In the *ab-initio* transition density cube (TDC) method (Krueger et al., 1998) the coupling is obtained directly from the coupling of transition densities that is evaluated by performing a 3-D integration with finite volume elements the so-called transition density cubes.

The TDC method uses the full 3-D information contained in the transition density, whereas the simpler transition monopole method can be readily applied once the transition charges are determined. Recently, the advantages of both methods were combined in the so-called TrEsp (Transition charges from electrostatic potentials) method (Madjet et al., 2006). It was demonstrated that the complete 3-D information contained in the transition density of a Chl or BChl can be condensed in atomic partial charges if the latter are determined such as to fit the electrostatic potential of the transition density. Instead of 500000 × 500000 TDC couplings, necessary to reach convergence of the integral in Eq. (6), only 50 × 50 transition charge couplings need to be summed over in TrEsp, obtaining the same result for the excitonic coupling.

An approach that is almost as simple as the point dipole approximation is the extended dipole approximation (Pearlstein, 1991a) that approximates the transition density by an extended dipole. In calculations on special pairs dimers of the bacterial reaction center and the reaction centers of PS I and II the extension of the dipole was estimated on the basis of a comparison of the extended dipole coupling with the exact result obtained by TrEsp. An optimal extent of about 9 Å was inferred (Madjet et al., 2006), about 30% larger than assumed in the past (Pearlstein, 1991a,b).

We note that for small inter-pigment distances, besides the excitonic coupling, additional interactions between pigments occur. Intermolecular charge transfer states can be formed and have an important influence on the position and widths of optical lines (Warshel and Parson, 1987; Renger, 2004). The treatment of those effects is beyond the scope of this chapter, i.e. charge transfer interactions are not included in the excitonic Hamiltonian in Eq. (2).

Besides the static shift in site energies there is a dynamical modulation of pigment transition energies induced by the protein (and pigment) vibrations that is described by the exciton-vibrational Hamiltonian

$$H_{\text{ex-vib}} = \sum_m \sum_\xi \hbar\omega_\xi g_\xi^{(m)} Q_\xi |m\rangle\langle m| \qquad (7)$$

which assumes a linear modulation of the transition energy with the (dimensionless) vibrational coordinate $Q_\xi$ and contains the (dimensionless) coupling constant $g_\xi^{(m)}$ that describes the coupling between a vibrational mode with frequency $\omega_\xi$ and the optical transition of the $m$th pigment. The coordinate $Q_\xi$ can be expressed in terms of creation and annihilation operators of vibrational quanta, $Q_\xi = C_\xi + C_\xi^+$ (May and Kühn, 2000). In the expressions derived below for optical spectra and excitation energy transfer reactions the coupling constants $g_\xi^{(m)}$ appear in the spectral density $J(\omega)$, that can be extracted from high resolution optical spectra (Pullerits et al., 1995; Pieper et al., 1999; Renger and Marcus, 2002a). Due to lack of additional information usually the same spectral density is assumed for all similar chromophores in a system, although at present there is no way to validate this approximation.

The third part of the Hamiltonian describes the vibrational dynamics by harmonic oscillators

$$H_{\text{vib}} = \sum_\xi \frac{\hbar\omega_\xi}{4} Q_\xi^2 + T_{\text{nucl}} \qquad (8)$$

where $T_{\text{nucl}}$ denotes the kinetic energy of the nuclei. We note that the expressions derived here for harmonic oscillators may be also applied in the case of strongly anharmonic vibrational dynamics. In this case a second order cumulant expansion, which is a milder approximation, leads to an identical expression for the rate constant of the transition as the harmonic oscillator theory used here. From this equality it was concluded that a strongly anharmonic system may be described by a spectral density of effective harmonic oscillators (Georgievskii et al., 1999).

## III. Weakly Coupled Pigments

If the excitonic coupling between pigments is weak compared to the dynamic and static disorder due to pigment-protein coupling, the excited states of the pigment-protein complex are localized on single pigments. The excitation energy transfer then occurs between such localized states in a hopping like manner. In reality hardly any photosynthetic antenna or RC exists where the application of this theory for weak coupling leads to satisfactory results. Nevertheless the basic concept of how to describe the consequences of electron-vibrational coupling occurring in an optical transition and in excitation energy transfer dynamics is important and provides the foundation of the theory that takes into account also delocalized excited states, as discussed later. A notable exception of a system with weak excitonic coupling between chlorophyll molecules is the peridinin chlorophyll *a* complex (Hofmann et al., 1996).

To understand the consequences of weak excitonic coupling we introduce potential energy surfaces (PES) of localized states by rewriting the Hamiltonian in Eqs. (1), (2), (7), and (8) of the pigment-protein complex, using the completeness relation $|0\rangle\langle 0| + \sum_m |m\rangle\langle m| = 1$, as

$$H = U_0(Q)|0\rangle\langle 0| + \sum_m U_m(Q)|m\rangle\langle m| + \sum_{mn} V_{mn}|m\rangle\langle n| + T_{nucl} \quad (9)$$

with the PES of the ground state $|0\rangle$

$$U_0(Q) = \sum_\xi \frac{\hbar\omega_\xi}{4} Q_\xi^2 \quad (10)$$

and the PES of the local excited state $|m\rangle$

$$U_m(Q) = \tilde{E}_m + \sum_\xi \frac{\hbar\omega_\xi}{4}(Q_\xi + 2g_\xi^{(m)})^2 \quad (11)$$

The energy $\tilde{E}_m$ is the difference between the minimum of the ground state potential energy surface that occurs at $Q_\xi = 0$ and the minimum of the PES of the excited state $|m\rangle$ at $Q_\xi = -2g_\xi^{(m)}$. It holds $\tilde{E}_m = E_m - \sum_\xi \hbar\omega_\xi (g_\xi^{(m)})^2$, where the second term is called reorganization energy. In the weak excitonic coupling limit — after optical excitation of a localized excited state — the nuclei relax in the PES of the excited state and the change in free energy of the system during this relaxation equals the reorganization energy. An illustration of the potential energy surfaces of the ground and excited states of a pigment with site energy $E_m = \hbar\omega_e$ is shown in Fig. 1. The coupling of an optical transition to many vibrational coordinates is described by the spectral density $J(\omega)$

$$J(\omega) = \sum_\xi g_\xi^2 \delta(\omega - \omega_\xi) \quad (12)$$

It contains the square of the coupling constants $g_\xi = g_\xi^{(m)}$ that are usually assumed to be equal for all sites $m$. In a pigment protein complex, due to the large number of vibrational modes which couple to the optical transitions, $J(\omega)$ can be approximated by a continuous function. The spectral density is the key quantity that will appear in all expressions for optical spectra and energy transfer rate constants discussed in the following.

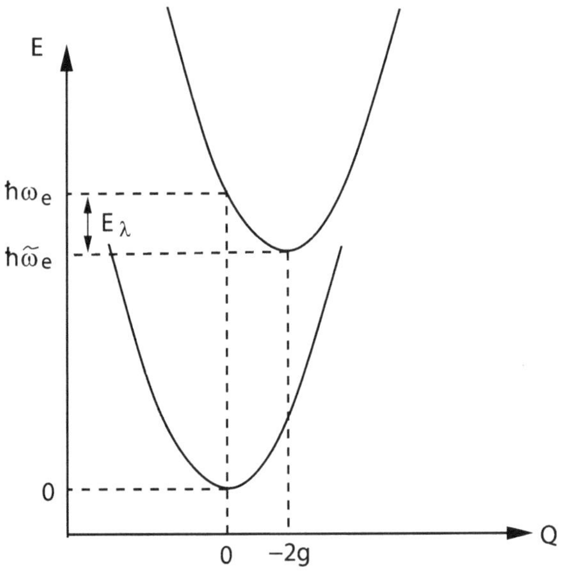


*Fig. 1.* Potential energy surfaces of the ground and excited state of a pigment in protein environment in dependence on an effective vibrational coordinate $Q$.

### *A. Optical Spectra—Theory of Kubo/Toyozawa and Lax*

An expression for the absorption line shape $D_\alpha(\omega)$ and the fluorescence line shape $D_I(\omega)$of a two-state electronic system that couples to a large number of harmonic oscillators was derived many decades ago (Lax, 1952; Kubo and Toyozawa, 1955). For a comprehensive treatment see (May and Kühn, 2000):

$$D_\alpha(\omega) = \frac{1}{2\pi\pm} \int_{-\infty}^{\infty} dt\, e^{-i(\omega-\tilde{\omega}_e)t} e^{G(t)-G(0)} \quad (13)$$

$$D_I(\omega) = \frac{1}{2\pi} \int_{-\infty}^{\infty} dt\, e^{-i(\omega-\tilde{\omega}_e)t} e^{G(t)-G(0)} \quad (14)$$

where the function $G(t)$ reads

$$G(t) = \int_0^\infty d\omega\, J(\omega)\{(1+n(\omega))e^{-i\omega t} + n(\omega)e^{i\omega t}\} \quad (15)$$

It contains the spectral density $J(\omega)$ and the mean number $n(\omega)$ of vibrational quanta that are excited in a vibrational mode of frequency $\omega$ at a given temperature $T$,

$$n(\omega) = \frac{1}{e^{\hbar\omega/kT} - 1} \tag{16}$$

Here $\hbar\tilde{\omega}_e$ denotes the energy difference between the minima of the PES of the excited and the ground state (see Fig. 1) which is obtained from the vertical optical transition energy $\hbar\omega_e$ (the so-called site energy) and the reorganization energy $E_\lambda$ by $\hbar\tilde{\omega}_e = \hbar\omega_e - E_\lambda$, where $E_\lambda$ is given now in terms of the spectral density as

$$E_\lambda = \hbar \int_0^\infty d\omega\, J(\omega)\omega \tag{17}$$

In a strict sense these equations only apply to a single chromophore coupled to a bath (solvent or protein environment.

## B. Förster Theory of Excitation Energy Transfer

After optical excitation, the excitation of a pigment may be transferred to a neighboring pigment by an exciton mechanism, as suggested more than half a century ago by Förster (1948, 1965). In the weak coupling limit considered here, the transfer starts from a vibrationally relaxed state where one pigment is in the excited electronic state and the other is in the electronic ground state, as illustrated in Fig. 2. A second order perturbation theory in the excitonic coupling is used to obtain a Förster type rate constant $k_{m\to n}$ for excitation energy transfer between pigments $m$ and $n$. The rate constant, which contains a non-perturbative description of the strong exciton-vibrational coupling

$$k_{m\to n} = 2\pi \frac{|V_{mn}|^2}{\hbar^2} \int_{-\infty}^{\infty} d\omega\, D_I^{(m)}(\omega) D_\alpha^{(n)}(\omega) \tag{18}$$

is given as the product of the square of the point dipole excitonic coupling $V_{mn}$ (eq. 4) and an overlap integral of the fluorescence lineshape function $D_I^{(m)}(\omega)$ of the donor and the absorption lineshape function $D_\alpha^{(n)}(\omega)$ of the acceptor, as described above (Eqs. 13 and 14). Noting that the experimental fluorescence signal of the donor $I_m(\omega)$ is proportional to $\omega^3 D_I^{(m)}(\omega)$ and the absorption signal of the acceptor $\alpha_n(\omega)$ is proportional to $\omega D_\alpha^{(n)}(\omega)$ (Lax, 1952; May and Kühn, 2000), the rate constant $k_{m\to n}$ can be expressed in terms of these experimentally accessible quantities as

$$k_{m\to n} = 2\pi \frac{|V_{mn}|^2}{\hbar^2} \frac{\int_0^\infty d\omega\, I_m(\omega)\alpha_n(\omega)\omega^{-4}}{\int_0^\infty d\omega\, I_m(\omega)\omega^{-3} \int_0^\infty d\omega\, \alpha_n(\omega)\omega^{-1}} \tag{19}$$

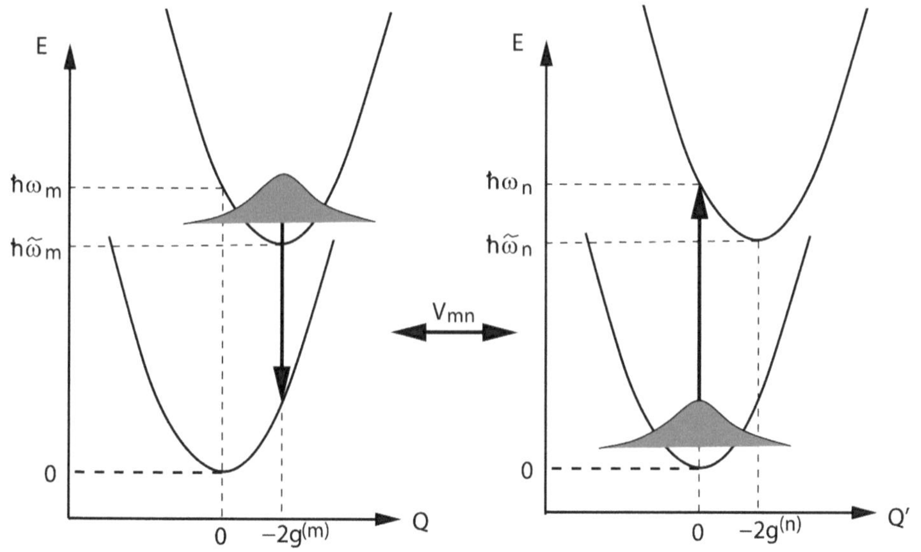


*Fig. 2.* Ground and excited state potential energy surfaces of two Coulomb-coupled pigments. The excitonic coupling matrix element $V_{mn}$ describes the interaction between the electronic transitions of the pigments $m$ and $n$.

In the original expression for the rate constant (Förster, 1948, 1965), the explicit calculation of the excitonic coupling $V_{mn}$ was avoided by expressing its square as

$$|V_{mn}|^2 = \frac{1}{4\pi\varepsilon_0\varepsilon_r} |\mu_m|^2 |\mu_n|^2 \kappa_{mn}^2 / R_{mn}^6$$

and including the dipole strengths $|\mu_m|^2$ of the donor and $|\mu_n|^2$ of the acceptor in the radiative lifetime $\tau_m$ of the donor and in the molar extinction coefficient $\varepsilon_n(\nu)$ of the acceptor, respectively. The orientational factor $\kappa_{mn} = \vec{e}_m\vec{e}_n - 3(\vec{e}_m\vec{e}_{mn})(\vec{e}_n\vec{e}_{mn})$ contains the unit vectors $\vec{e}_m$ along the transition dipole $\vec{\mu}_m$, $\vec{e}_n$ along $\vec{\mu}_n$, and $\vec{e}_{mn}$ along $\vec{R}_{mn}$, the vector connecting the centers of pigments $m$ and $n$. The rate constant of energy transfer is then obtained in dependence on the overlap between $\varepsilon_n(\nu)$ and the normalized fluorescence spectrum $I_m(\nu)$ as (Förster, 1948, 1965)

$$k_{m\to n} = \frac{\kappa_{mn}^2}{R_{mn}^6} \frac{9c^4 \ln(10)}{128\pi^5 n^4 N_L \tau_n} \int_0^\infty d\upsilon\, I_m(\upsilon)\varepsilon_n(\upsilon)\upsilon^{-4} \tag{20}$$

where the experimental spectra are given as a function of frequency $\nu=\omega/2\pi$, $c$, $n$, and $N_L$ denoting the velocity of light, the index of refraction, and the Loschmidt number, respectively.

## IV. Strongly Coupled Pigments

If the excitonic couplings $V_{mn}$ are large compared to the static and dynamic disorder, the excited states $|M\rangle$ are delocalized. Each of these exciton states is given as a coherent superposition of localized excited states $|m\rangle$, i.e., $|M\rangle = \sum_m c_m^{(M)}|m\rangle$where the coefficients $c_m^{(M)}$ and exciton-energies $\varepsilon_M$ are obtained by solving the eigenvalue problem

$$H_{\text{ex}}|M\rangle = \varepsilon_M |M\rangle \tag{21}$$

with the excitonic Hamiltonian $H_{\text{ex}}$ from Eq. 2. An illustration of the localized and delocalized states is given in Fig. 3. If the dynamic pigment protein coupling is neglected, the absorption spectrum is obtained as

$$\alpha(\omega) \propto \omega \sum_M |\mu_M|^2 \,\delta(\omega - \omega_M) \tag{22}$$

where $\omega_M = \omega_M/\hbar$ is the frequency and $|\mu_M|^2$ the absolute square of the transition dipole moment $\vec{\mu}_M$ of the excitation of $|M\rangle$. The transition dipole moment

$$\vec{\mu}_M = \langle M|\hat{\mu}|0\rangle = \sum_m c_m^{(M)} \langle m|\hat{\mu}|0\rangle = \sum_m c_m^{(M)}\vec{\mu}_m \tag{23}$$

is given in terms of the exciton coefficients $c_m^{(M)}$ and local transition dipole moments $\vec{\mu}_m$. The spectrum is sometimes called a 'stick spectrum', because it can

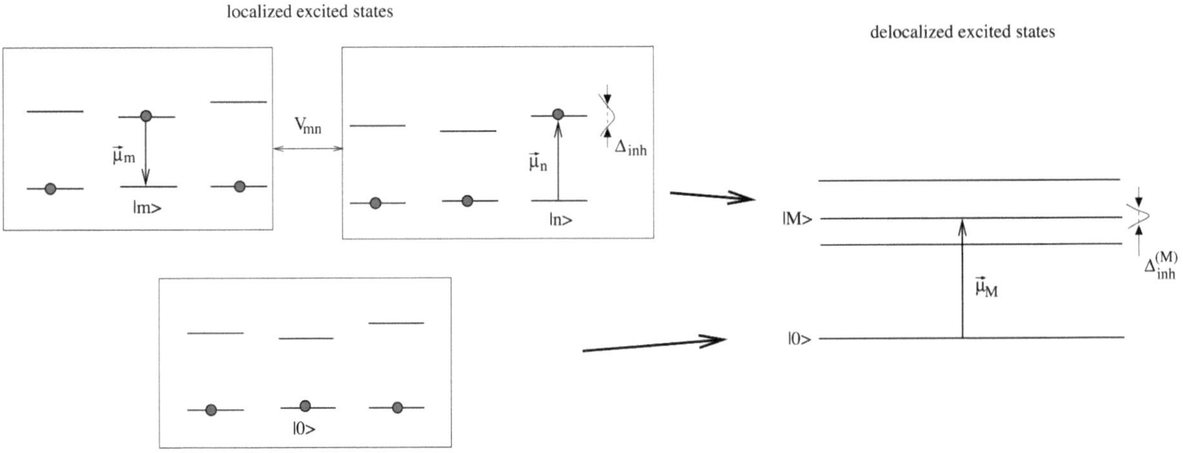


*Fig. 3.* Ground state and different excited states in the basis of localized states (left half) and in the basis of delocalized excited states (right half). Transition dipole moments between the ground state and the localized and delocalized excited states and the inhomogeneous width of the respective transition energies are also shown.

be visualized by vertical lines of height $|\mu_M|^2$ at positions $\omega = \omega_M$. No homogeneous or inhomogeneous broadening is included at that level. This treatment has been introduced by Davydov (1971).

### A. Gaussian Dressed Stick Spectra

Due to conformational motion of the protein, the site energies of pigments vary in time. If the conformational motion is slow compared to the excited state life time (100 fs to ns) the resulting disorder is termed static. In simple models of static disorder a Gaussian distribution function $P_{\text{inh}}(\omega_M - \bar{\omega}_M)$ of exciton energies, centered around a mean exciton energy $\hbar\bar{\omega}_M$ ,is assumed and the absorption $\alpha(\omega)$ in this case is obtained from an average of the homogenous absorption spectrum given above with respect to the distribution function $P_{\text{inh}}(\omega_M - \bar{\omega}_M)$,

$$\begin{aligned}\alpha(\omega) &\propto \omega \sum_M |\mu_M|^2 \int d\omega_M P_{inh}(\omega_M - \bar{\omega}_M)\delta(\omega - \omega_M) \\ &\propto \omega \sum_M |\mu_M|^2 P_{inh}(\omega - \bar{\omega}_M)\end{aligned} \tag{24}$$

For simplicity the same distribution function was assumed for all exciton states. Such a model was used to calculate optical spectra of the FMO-complex of the green sulfur bacterium *Chlorobium tepidum* in order to estimate the (mean) site energies of the 7 bacteriochlorophyll *a* molecules that are bound in the monomeric subunit of the trimeric FMO-complex (Vulto et al., 1998a). This approach does however neglect an important effect of disordered excitonic systems, the so-called 'resonance energy transfer narrowing' (Knapp, 1984) and is for a number of reasons generally unsuitable to provide proper spectral shapes. However for the special case of determining the site energies in the FMO complex it worked well, since a much more sophisticated theory yielded similar results (Adolphs and Renger, 2006).

In more realistic, although numerically much more expensive approaches a distribution of local site energies is assumed and a Monte Carlo scheme is used to calculate the spectra. Within this method different combinations of site energies of the pigments are generated randomly from a given distribution function. The optical spectrum then is calculated for every configuration of site energies and an average is performed. Usually, a Gaussian distribution function is used for every site and the fluctuations at different sites are assumed to be uncorrelated. In such calculations the mean site energies and the width of the distribution function for the site energies are fit parameters. The Monte Carlo approach includes the resonance energy transfer narrowing, i.e., a narrowing of the distribution of exciton energies with respect to the distribution of pigment energies due to the excitonic couplings, as illustrated in Fig. 3.

### B. Multilevel Redfield Theory of Optical Spectra and Exciton Relaxation

If the excitonic coupling is stronger than both the modulation of the site energies by vibrations and the static disorder, exciton relaxation between partially delocalized exciton states occurs, as illustrated in Fig. 4. Second order perturbation theory in the exciton-vibrational coupling, within the Markov approximation, yields the rate constant for exciton relaxation between states $M$ and $N$ (Renger et al., 2001; Yang and Fleming, 2002).

$$\begin{aligned}k_{M\to N} &= 2\pi\gamma_{MN}\omega_{MN}^2 \\ &\times\{J(\omega_{MN})(1+n(\omega_{MN})) + J(\omega_{NM})n(\omega_{NM})\}\end{aligned} \tag{25}$$

We note that $J(\omega)$ for negative $\omega$ is zero and

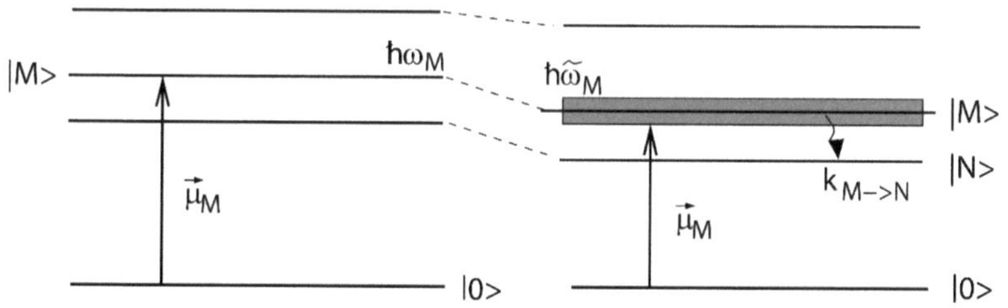


*Fig. 4.* Exciton states on the left half and shift of exciton energies and life time broadening due to relaxation on the right half.

$$\omega_{MN} = (\varepsilon_M - \varepsilon_N)/\hbar \tag{26}$$

is the transition frequency between the two exciton states. The function

$$\gamma_{MN} = \sum_{m,n} c_m^{(M)} c_m^{(N)} c_n^{(M)} c_n^{(N)} e^{-R_{mn}/R_c} \tag{27}$$

contains the coefficients of the exciton states $|M\rangle$ and $|N\rangle$ and a correlation radius $R_c$ of protein vibrations that describes in an effective way how the fluctuations of site energies at different sites are correlated (Renger and May, 1998; Renger et al., 2001; Renger and Marcus, 2002a).

In the limit $R_c \to \infty$, i.e., for strongly correlated fluctuations, we have

$$\gamma_{MN}|_{R_c\to\infty} = \sum_m c_m^{(M)} c_m^{(N)} \sum_n c_n^{(M)} c_n^{(N)} = \langle M | N \rangle \langle N | M \rangle = \delta_{M,N}$$

i.e., there is no exciton relaxation. The upper limit for the exciton relaxation rate is obtained if the site energies of the pigments fluctuate completely independently. In this case the function $\gamma_{MN}$ becomes

$$\gamma_{MN}|_{R_c\to 0} = \sum_m (c_m^{(M)})^2 (c_m^{(N)})^2$$

i.e., exciton relaxation occurs between such levels $|M\rangle$ and $|N\rangle$ that share contributions from the same pigments. From transient non-linear absorption data, a correlation radius of $R_c = 5$ was determined for the RC of Photosystem II (PS II) (Renger and Marcus, 2002b). Besides this electronic factor there is a vibrational factor $\omega^2_{MN} J(\omega_{MN})^2$ that describes how well the vibrational environment (bath) can dissipate (i.e., absorb) the excess energy $\hbar\omega_{MN}$ of the exciton during relaxation. The above two factors are illustrated in Fig. 5.

We finally show that the rate constant $k_{M\to N}$ in Eq. (25) fulfils the detailed balance condition. If we assume that $\hbar\omega_{MN} > 0$ we have

$$\frac{k_{M\to N}}{k_{N\to M}} = \frac{J(\omega_{MN})(1+n(\omega_{MN}))}{J(\omega_{MN})n(\omega_{MN})} = n(\omega_{MN})^{-1} + 1 = e^{\hbar\omega_{MN}/kT} \tag{28}$$

The approximations of the Redfield theory are (i) a perturbative treatment of the exciton-vibrational coupling and (ii) application of the Markov approximation. The latter assumes that vibrational relaxation in the states $M$ and $N$ is fast compared to the transfer. The first approximation neglects the reorganization effects of the vibrations during the transfer, i.e., the nuclei do not relax to a new equilibrium position during exciton relaxation. Thus the potential energy surfaces of states $M$ and $N$ are treated as unshifted along the vibrational coordinates. This approximation can be tested by using a theory that describes the diagonal part of the exciton vibrational coupling

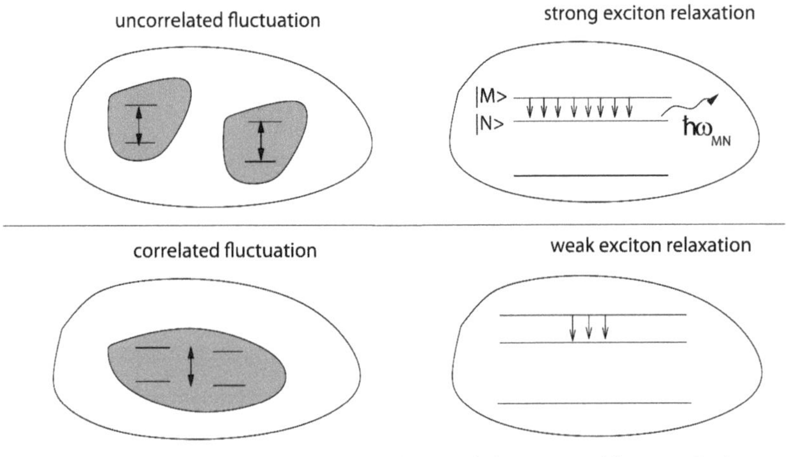


*Fig. 5.* Fluctuation of site energies by dynamic disorder due to modulation of pigment transition energies by protein vibrations. Uncorrelated fluctuation of different site energies (left upper part) leads to strong exciton relaxation between different delocalized states (right upper part), whereas a correlated fluctuation of site energies (left lower part) yields only weak exciton relaxation (right lower part). The transition energy $\hbar\omega_{MN}$ is dissipated into the protein environment.

in a non-perturbative manner as does the modified Redfield theory that is described further below.

Using standard Redfield theory, a Lorentzian lineshape function $D_M(\omega)$ is obtained for the optical transitions between the ground state and the exciton state $M$, (Renger et al., 2001)

$$D_M(\omega) = \frac{1}{\pi} \frac{\tau_M^{-1}}{(\omega - \tilde{\omega}_M)^2 + \tau_M^{-2}} \tag{29}$$

The homogeneous broadening is determined by the lifetime of exciton state $M$ that follows from the rate constants of exciton relaxation between this state and all other states

$$\tau_M^{-1} = \frac{1}{2} \sum_N k_{M \to N} \tag{30}$$

with the $k_{M \to N}$ given by Eq. (25).

The peak position of $D_M(\omega)$ occurs at $\omega = \tilde{\omega}_M$ where a shift of the excitation frequency $\omega_M$ by the diagonal (the $\gamma_{MM}$ term) and the off-diagonal part (the $\gamma_{MK}$ term) of the exciton-vibrational coupling is taken into account (Renger and Marcus, 2002a):

$$\tilde{\omega}_M = \omega_M - \frac{E_\lambda}{\hbar} \gamma_{MM} - \sum_{K \neq M} \gamma_{MK} \wp \int_{-\infty}^{\infty} d\omega \frac{\omega^2 \left\{ (1 + n(\omega)) J(\omega) + n(-\omega) J(-\omega) \right\}}{\omega_{MK} - \omega} \tag{31}$$

In the above equation $\wp$ denotes the principal part of the integral, and $\gamma_{MK}$ and $E_\lambda$ are given in Eqs. 27 and 17. An illustration of the energy shift and the lifetime broadening is given in Fig. 4.

## V. Weak Inter- and Strong Intra-Aggregate Coupling — Generalized Förster Theory

If the transfer occurs between two pigment aggregates $a$ and $b$ with strong excitonic couplings between the pigments in the same aggregate, and weak coupling between pigments in different aggregates, as illustrated in Fig. 6, the Förster theory can be generalized to describe this situation (Fetisova et al., 1996; Sumi 1999; Mukai et al., 1999; Scholes and Fleming, 2000; Jang et al., 2004; Raszewski et al., 2005). In this case the transfer starts from an electronically and vibrationally relaxed aggregate state and the rate constant contains a thermal distribution of exciton states of the donor aggregate, a generalized matrix element of the dipole-dipole interaction and a generalized overlap integral of fluorescence and absorption. Those quantities are generalized in the sense that they depend not only on the local pigment properties but also on the coefficients of excitonic wave functions of the aggregates. In particular, excitation energy transfer can involve optically dark states which is not the case in basic Förster theory, as explained below.

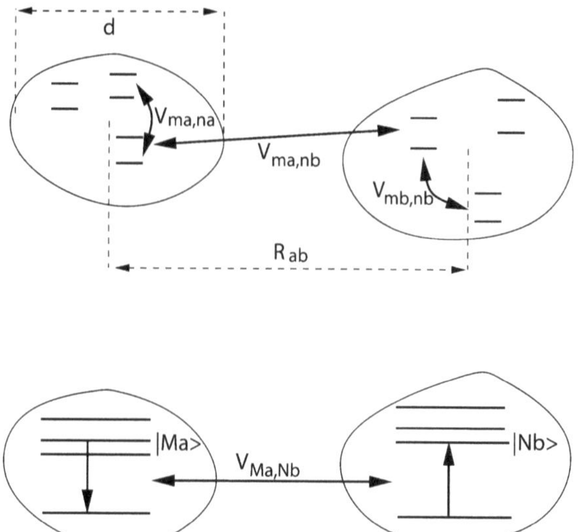


*Fig. 6.* In the upper part two aggregates formed by strongly coupled pigments are depicted. The two aggregates are coupled due to weak inter aggregate couplings between the pigments. $d$ is a measure of the spatial extent of the aggregates and $R_{ab}$ is the center to center distance between the aggregates. The coupling between the delocalized states of the two aggregates, in the lower part, is used to derive a rate constant of excitation energy transfer between the aggregates in generalized Förster theory.

The rate constant $k_{Ma \to Nb}$ for excitation energy transfer between exciton state $|Ma\rangle$ of aggregate $a$ and $|Nb\rangle$ of aggregate $b$ is obtained as (Fetisova et al., 1996; Mukai et al., 1999)

$$k_{Ma \to Nb} = 2\pi \frac{|V_{Ma,Nb}|^2}{\hbar^2} \int_{-\infty}^{\infty} d\omega \, D_I^{(Ma)}(\omega) D_\alpha^{(Nb)}(\omega) \tag{32}$$

where the excitonic coupling $V_{Ma,Nb}$ between exciton states $|Ma\rangle$ and $|Nb\rangle$ is given as a sum of the point-dipole couplings $V_{ma,nb}$ between the individual pigments (counted by $ma$ and $nb$) of the two aggregates (Eq. 4), weighted by the exciton coefficient $c_{ma}^{(Ma)}$ of pigment $m$ in aggregate $a$ and $c_{nb}^{(Nb)}$ of pigment $n$ in aggregate $b$

$$V_{Ma,Nb} = \sum_{m_a, m_b} c_{m_a}^{(Ma)} c_{n_b}^{(Nb)} V_{m_a n_b} \quad (33)$$

The intra-aggregate exciton relaxation can be assumed fast as compared to the transfer between the aggregates, i.e., the latter starts from an equilibrated manifold of exciton states of the initially excited aggregate. Hence, the rate constant of exciton transfer between the aggregates $k_{a\to b}$ is given as a Boltzmann weighted sum of the individual rate constants $k_{Ma\to Nb}$ in Eq. (32):

$$k_{a\to b} = \sum_{Ma,Nb} \frac{e^{-\varepsilon_{Ma}/kT}}{\sum_{Ka} e^{-\varepsilon_{Ka}/kT}} k_{Ma\to Nb} \quad (34)$$

If the center-to-center distance $R_{ab}$ between the aggregates is large as compared to their spatial extension $d$, the distances between individual pigments $R_{ma,nb}$ can be approximated by $R_{ab}$. In this case the sum occurring in Eq. (33) yields the transition dipole moments $\vec{\mu}_{Ma}$ of the exciton state $|Ma\rangle$ and similarly $\vec{\mu}_{Nb}$ of $|Nb\rangle$ (Eq. 23), and the coupling $V_{Ma,Nb}$ is simply given as the dipole-dipole coupling between the exciton transition dipole moments

$$V_{Ma,Nb}\,|_{d<<R_{ab}} = \frac{1}{4\pi\varepsilon_0\varepsilon_r}\left\{\frac{\vec{\mu}_{Ma}\vec{\mu}_{Nb}}{R_{ab}^3} - 3\frac{(\vec{\mu}_{Ma}\vec{R}_{ab})(\vec{\mu}_{Nb}\vec{R}_{ab})}{R_{ab}^5}\right\} \quad (35)$$

i.e., only optically allowed exciton transitions (those with non-zero transition dipole moments) contribute to the transfer. This case resembles the situation considered originally by Förster in the sense that the aggregate is treated formally like a supermolecule. However, when the extension $d$ of the aggregate is similar to the inter-aggregate distance $R_{ab}$, the above approximation is not valid and it is found that also dark exciton states (those with vanishing transition dipole moments) can contribute to the transfer. This idea was used to study excitation energy transfer in chlorosomes of green sulfur bacteria (Fetisova et al., 1996), to explain the rapid excitation energy transfer between the B800 and B850 states in the LH2 antenna of purple bacteria (Mukai et al., 1999; Scholes and Fleming, 2000; Jang et al. 2004), and to study excitation energy transfer between the peripheral Chl $z$ and the six strongly coupled core pigments in PS II reaction centers (Raszewski et al., 2005).

Whereas the calculation of the couplings $V_{Ma,Nb}$ is well understood, the calculation of the lineshape functions $D_I^{(Ma)}(\omega)$ and $D_\alpha^{(Nb)}(\omega)$ of the aggregates that enter the rate constant in Eq. (32) is a non-trivial task, because both the exciton vibrational coupling and the excitonic coupling need to be taken into account. Possible approaches are given by a Greens function technique (Davydov, 1971; Fetisova et al., 1996; Sumi 1999) or by non-Markovian density matrix theories (Renger and May, 2000; Renger and Marcus, 2002a; Jang et al. 2004). The line shape functions in general contain vibrational side bands of exciton transitions and a lifetime broadening that is due to exciton relaxation, as described below.

## VI. Strong Pigment-Pigment and Strong Pigment-Protein Coupling

In order to treat both strong excitonic and strong exciton-vibrational coupling we first introduce the PES of the exciton states. For this purpose we express the Hamiltonian in Eqs. (1), (2), (7) and (8) in terms of the exciton states $|M\rangle$ introduced above, using $|m\rangle = \sum_M c_m^{(M)}|M\rangle$ and the completeness relation $|0\rangle\langle 0| + \sum_M |M\rangle\langle M| = 1$, and obtain

$$H = \sum_\xi \left\{U_0(Q)|0\rangle\langle 0| + \sum_M U_M(Q)|M\rangle\langle M|\right\} + \sum_{M\neq N}\sum_\xi \hbar\omega_\xi g_\xi(M,N)Q_\xi|M\rangle\langle N| + T_{nucl} \quad (36)$$

with the PES $U_0(Q)$ of the ground state $|0\rangle$ in Eq. (10) and the PES

$$U_M(Q) = \varepsilon_M^{(0)} + \sum_\xi \frac{\hbar\omega_\xi}{4}(Q_\xi + 2g_\xi(M,M))^2 \quad (37)$$

of exciton state $|M\rangle$, where $\varepsilon_M^{(0)}$ is the energy difference between the minima of the PES of the $M$th exciton state and the ground state (Eq. 39). The exciton vibrational coupling constant $g_\xi(M,N)$ of the delocalized states is given in terms of the local coupling constants $g_\xi^{(m)}$,

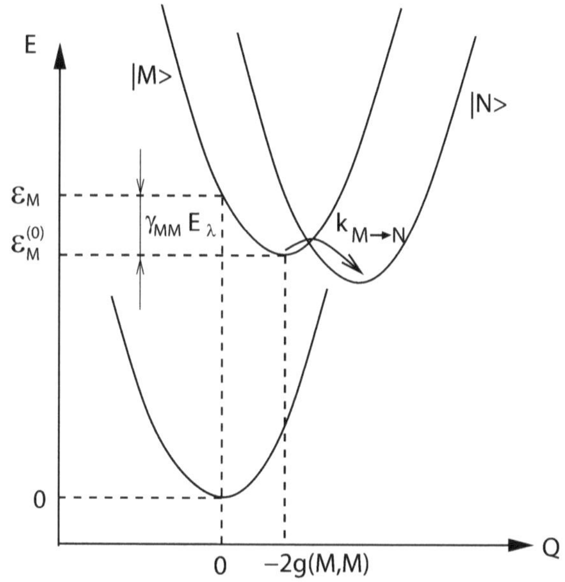


*Fig. 7.* Potential energy surfaces of the ground state and of exciton states $|M\rangle$ and $|N\rangle$. The mutual displacements of the different PES along the vibrational coordinate axis give rise to vibrational side bands of exciton transitions in optical spectra and reorganization effects of nuclei in exciton relaxation.

$$g_{\xi}(M,N)=\sum_{m} c_m^{(M)} c_m^{(N)} g_{\xi}^{(m)} \tag{38}$$

The local fluctuation of the site energies, in the basis of delocalized states, leads to diagonal and off-diagonal parts in the exciton vibrational coupling. The diagonal part $g_{\xi}(M,M)$ determines the shift of the excitonic PES of state $|M\rangle$ with respect to the ground state $|0\rangle$ along the coordinate axis, as illustrated in Fig. 7. The off-diagonal part $g_{\xi}(M,N)$ couples exciton states $|M\rangle$ and $|N\rangle$ (Eq. 4.5). As will be shown below, the off-diagonal parts are responsible for exciton relaxation, and the diagonal parts for the appearance of vibrational sidebands in the optical spectrum of exciton transitions. The diagonal parts describe the nuclear reorganization that occurs in response to optical excitation. An excitonic reorganization energy $E_{\lambda}^{(M)}$ may be introduced as the difference between the vertical transition energy $\varepsilon_M=\hbar\omega_M$ and the transition energy $\varepsilon_M^{(0)}$ between the minima of $U_0(Q)$ and $U_M(Q)$ as illustrated in Fig. 7

$$E_{\lambda}^{(M)}=\varepsilon_M-\varepsilon_M^{(0)}=\gamma_{MM}E_{\lambda} \tag{39}$$

Where $E_{\lambda}$ is the reorganization energy of the local pigment excitation given in Eq. (17) and the factor $\gamma_{MM}$ in Eq. (27) contains information about the delocalization of exciton states and the correlation radius of protein vibrations. The more delocalized the excited states are, the smaller is $\gamma_{MM}$, i.e., excitonic delocalization leads to smaller nuclear reorganization energies and thereby to smaller vibrational sidebands. Using the excitonic PES, introduced above and illustrated in Fig. 7, in the following expressions for optical spectra and excitation energy transfer are described that take into account the diagonal part of the exciton vibrational coupling non-perturbatively and use perturbation theory for the off-diagonal parts.

### *A. Non-Markovian Density Matrix Theory*

For the inclusion of vibrational sidebands a non-Markovian density matrix theory (Renger and Marcus, 2002a) is used for the diagonal parts of the exciton-vibrational coupling and the off-diagonal parts are described within the Markov approximation as before. In this case the lineshape function is obtained as (Renger and Marcus, 2002a)

$$D_M(\omega)=\frac{1}{2\pi}\int_{-\infty}^{\infty} dt\, e^{i(\omega-\tilde{\omega}_M)t} e^{G_M(t)-G_M(0)} e^{-|t|/\tau_M} \tag{40}$$

whose time dependent function $G_M(t)$ is

$$G_M(t)=\gamma_{MM}G(t) \tag{41}$$

with the $G(t)$ of the local line shape function (Eq. 15), the $\gamma_{MM}$ from Eq. (27), and the $\tau_M^{-1}$ from Eq. (30). This lineshape function contains both vibrational sidebands, described by $G_M(t)$, and life time broadening due to exciton relaxation, as described by $\tau_M$. The transition frequency is defined by Eq. (31). Comparison with the excitonic PES shows that $\tilde{\omega}_M$ equals the transition energy between the minima of the ground and excited PES plus an additional shift due to the off-diagonal parts of the exciton-vibrational coupling.

In the presence of static disorder the excited states are partially localized and the condition $\gamma_{MM}>\gamma_{MN}$, $(M\neq N)$ holds, i.e., the diagonal part of the exciton vibrational coupling dominates. If the off-diagonal parts are neglected (Zhang et al., 1998) the line shape function $D_M(\omega)$ just differs from the two level system result $D_{\alpha}(\omega)$ in Eq. (13) by the factor $\gamma_{MM}$ that describes the delocalization of exciton states. For localized states, $\gamma_{MM}=1$ as expected, for delocalized states $\gamma_{MM} < 1$ indicating that a delocalized electronic state couples more weakly to vibrations than a localized state. The inclusion of the off-diagonal part $\gamma_{MN}$ leads to lifetime broadening due to exciton relaxation between exciton states $|M\rangle$ and $|N\rangle$ (Ohta et al., 2001) and to an additional shift of the transition energy (Renger and Marcus, 2002a). The line shape function in Eq. (40) was derived by using a non-Markovian time local description (so-called partial ordering prescription (POP) theory) for the off-diagonal parts of the exciton vibrational coupling, and a Markov approximation for the diagonal parts. The alternative chronological ordering prescription (COP) theory (Mukai et al., 1999; Renger et al., 2001; Jang et al., 2004) was found to give less satisfactory results in calculations of fluorescence line narrowing spectra of B820 complexes (Renger and Marcus, 2002a) and comparison with experimental data (Creemers et al., 1999). We also note that the POP lineshape function $D_M(\omega)$ in Eq. (40) contains the limit of localized excited states, i.e., it reduces to the exact two-level system result in Eq. (13), whereas the alternative COP lineshape function does not. The two theories use a different partial summation of the exciton-vibrational coupling. At higher temperatures the perturbation theory of COP breaks down, whereas POP still describes the experimental data (Renger et al., 2007).

## B. Modified Redfield Theory

Whereas in a Markovian treatment of the diagonal part of the exciton vibrational coupling the mutual shift of the excitonic PES is neglected, modified Redfield theory allows to treat this shift and in this way to include the reorganization of nuclei that occurs upon exciton relaxation between different excitonic PES as illustrated in Fig. 7. If it is assumed that initially the system is relaxed in the PES of exciton state $|M\rangle$, second order perturbation theory in the coupling between this PES and the PES of state $|N\rangle$ yields a rate constant $k_{M\rightarrow N}$ for exciton relaxation between the two excitonic PES. The following rate constant is obtained when a harmonic oscillator description is used for the vibrations (Renger and Marcus, 2003)

$$k_{M\rightarrow N} = \int_{-\infty}^{\infty} dt\, e^{i\tilde{\omega}_{MN}t} e^{\phi_{MN}(t)-\phi_{MN}(0)} \times \left([\lambda_{MN}/\hbar + G_{MN}(t)]^2 + F_{MN}(t)\right) \quad (42)$$

where $\tilde{\omega}_{MN}$ is the transition frequency between the minima of the two excitonic potential energy surfaces, $\tilde{\omega}_{MN} = \omega_{MN} - (\gamma_{MN} - \gamma_{NN})\, E_\lambda$, with the $\gamma_{MN}$ and the $E_\lambda$ given by Eqs. (27) and (17) The time-dependent functions

$$\phi_{MN}(t) = \sum_m \left[\left(c_m^{(M)}\right)^2 - \left(c_m^{(N)}\right)^2\right]^2 \phi_0(t) \quad (43)$$

$$G_{MN}(t) = \sum_m \left[\left(c_m^{(M)}\right)^3 c_m^{(N)} - \left(c_m^{(N)}\right)^3 c_m^{(M)}\right] \phi_1(t) \quad (44)$$

$$F_{MN}(t) = \sum_m \left(c_m^{(M)}\right)^2 \left(c_m^{(N)}\right)^2 \phi_2(t) \quad (45)$$

contain the function $\phi_n(t)$ with $n = 0,1,2$

$$\phi_n(t) = \int_{-\infty}^{\infty} d\omega\, \omega^n e^{-i\omega t} (1 + n(\omega))[J(\omega) - J(-\omega)] \quad (46)$$

Independent fluctuations in pigment transition energies have been assumed to derive Eqs. (43)–(45). A generalization to the case $R_c > 0$ is straightforward (Raszewski et al., 2006). The above rate constant fulfils detailed balance and reduces to the standard Redfield result when the mutual shift of excitonic PES is neglected (Adolphs and Renger, 2006). This theory was applied recently to study exciton relaxation between the six strongly coupled core pigments in PS II reaction centers (Raszewski et al., 2005) and in the FMO-complex (Adolphs and Renger, 2006).

A similar, but somewhat more complicated, expression was derived by Mukamel and coworkers (Zhang et al., 1998) using a Brownian oscillator description. This theory called 'Modified Redfield Theory' (Yang and Fleming, 2002), was applied recently to study exciton relaxation among strongly coupled pigments in PS I (Yang and Fleming, 2003; Yang et al., 2003), exciton relaxation in the light-harvesting complex LHC-II (Novoderezhkin et al., 2004a) and in the FMO-protein (Cho et al., 2005).

The rate constant $k_{M\rightarrow N}$ in Eq. 42 and the Brownian oscillator result are equivalent (Renger and Marcus, 2003). We note that the limit of weak excitonic coupling, i.e., the Förster type rate constant of Eqs. (18) and (19) is not contained, since the modified Redfield theory uses delocalized states and does not describe a dynamical localization of excited states, i.e., a change of the exciton coefficients in time.

## VII. Applications to Photosynthetic Systems

We will in the following shortly review applications of advanced exciton theory for the description of optical spectra and energy transfer dynamics in photosynthetic systems. The spectral and exciton relaxation properties of light-harvesting complexes and reaction centers have been reviewed recently in several papers (Fleming and Jimenez, 1997; Krueger et al., 1999; Kühn et al., 1999; Sundström 2000; van Grondelle and Novoderezhkin, 2001, 2006; Holzwarth, 2004, 2007; Fleming and Scholes, 2004; Pullerits and Sundström, 2005). In recent years advanced exciton theories have been applied to several light-harvesting antenna, but the purple bacterial antenna have been the ones studied most extensively, in particular the B800-850 LH2 antenna and to a lesser extent the LH1 antenna, followed by the light-harvesting complex LHC II of higher plants.

### *A. Bacterial Antenna Systems*

The B800-850 antenna (LH2) complex consists of two rings of BChl a molecules containing 8 and 16 BChls in the outer and inner rings, respectively. The strongly excitonically coupled BChls in the inner ring absorb in the region from 820-880 nm, with a maximum at 850 nm, while the weakly coupled outer ring chromophores absorb around 800 nm. The dynamics in this system has been studied by many groups using ultrafast spectroscopic methods applying for example transient absorption, 1-D and 2-D photon echo and single-molecule spectroscopy. We will not review here the experimental data (for appropriate reviews see, Fleming and Jimenez ,1997; Sundström, 2000; van Grondelle and Novoderezhkin, 2001, 2006; Fleming and Scholes, 2004; Pullerits and Sundström, 2005), but only mention theoretical studies of the exciton spectra and dynamics. Standard Redfield theory has been found to provide a very good description of the femtosecond transient absorption kinetics, describing both the B800 to B850 as well as the intra-B850 exciton relaxation (Novoderezhkin et al., 2003). Dynamic single molecule data for the same complex have also been modeled in terms of standard Redfield theory and were found to provide good descriptions (Rutkauskas et al., 2004, 2005). Quite similar results were also found for the LH1 complex, which consists of a structure similar to the B850 ring in LH2, however with a larger number of coupled BChls (30–32 per ring instead of the 18 BChls in the B850 ring). Quantum mechanical calculations of the transition densities and analysis in terms of the generalized Förster theory were performed in detail for the B800-850 energy transfer (Krueger et al., 1999; Scholes and Fleming, 2000). It was found that the TDC method (Scholes et al., 2001), in combination with the generalized Förster theory, allows reliable descriptions of the energy transfer dynamics between molecular assemblies or aggregates, as is the case also for the B800-850 energy transfer. The static disorder was found to have a marked influence on the calculated rate of energy transfer and is responsible for a faster energy transfer time than would occur in its absence and furthermore accounts for the previously unexplained weak temperature dependence observed in the experiment. Large and long-lasting (1–2 ps) vibrational coherences were observed in femtosecond experiments on the B850 systems in LH2 and LH1 complexes. These coherences were also well described by theory (Chachisvilis et al., 1994, 1995a,b; Bradforth et al., 1995; Joo et al. 1996; Jimenez et al., 1997).

### *B. Light-Harvesting Complex II of Higher Plants (LHC II)*

The LHC II complex forms the peripheral antenna system of PS II in higher plants. LHC II occurs in its native state as a trimer that contains 8 Chl *a* and 6 Chl *b* molecules per monomer, arranged in an unregular manner in two layers, quite unlike the highly symmetric chromophore arrangements found in the light-harvesting complexes of purple bacteria (Kühlbrandt, 1994; Liu et al., 2004; Standfuss et al., 2005). The exciton dynamics in this complex has been studied extensively using femtosecond methods (see van Grondelle and Novoderezhkin, 2006, for a review). The Chl *b* to Chl *a* transfers range from 150–600 fs, while inter Chl *a* energy transfers occur in the picosecond range, up to 20 ps (Connelly et al., 1997; Salverda et al., 2003; Gradinaru et al., 1998a,b). The energy transfer dynamics and steady state spectra of LHC II have been modeled using modified Redfield theory (Novoderezhkin et al., 2004a,b) and higher excited states have been included in an extensive modeling of the exciton dynamics

(Renger et al., 1996; Renger and May, 1997). Based on the modified Redfield approach Novoderezihin et al. (2005b) proposed an exciton model of the LHCII trimer which allowed a simultaneous quantitative fit of the absorption, linear-dichroism, steady-state fluorescence spectra, and transient absorption kinetics upon excitation at different wavelengths. The fast *b* to *a* transfer is determined by Chls *b* coupled strongly to Chl *a* clusters arranged as dimers or trimers. Long-lived components of the energy-transfer kinetics are determined by a population of red-shifted Chl *b* and blue-shifted Chl *a* followed by a very slow (3 ps to 12 ps) flow of energy from monomeric bottleneck sites to the Chl *a* clusters. The dynamics within the Chl *a* region is determined by fast (with time constants down to sub-100 fs) exciton relaxation within a Chl *a* trimer, slower 200–300 fs relaxation within two Chl *a* dimers, even slower 300–800 fs migration between these clusters, and very slow transfer of up to 20 ps among a few distant Chl *a* sites. While the description of the steady state spectra and the energy transfer dynamics was generally quite satisfactory, more pronounced deviations between experiment and theory were found than for bacterial antenna systems. While the reasons for these deviations are unclear at present, further improvements in theory and modeling seem to be required to get a full understanding of the dynamic processes in this complex. A few other higher plant and algal antenna complexes have also been modeled. Thus, Förster theory has been applied recently to study excitation energy transfer between chlorophyll *a* molecules in the peridinin-Chl-*a*-protein (Kleima et al., 2000). The Peridinin-Chl-*a* protein is one of the few cases where only weak excitonic coupling among Chls seems to be realized.

## C. Reaction Centers of Higher Plants

The reaction center of PS II, the so-called D1-D2-Cyt-b559 complex has been biochemically isolated and has thus been studied spectroscopically by many groups (for recent reviews, see Dekker and van Grondelle, 2000; Renger and Holzwarth, 2005) The structure contains 4 Chl *a* and 2 Pheo *a* in the RC core which show intermediate to strong exciton coupling (Durrant et al., 1995). Two additional quasi-symmetrically arranged Chl *a* molecules are found at a distance of about 25 Å to the RC core pigments (Ferreira et al., 2004; Loll et al., 2005a,b). Many ultrafast studies of the energy and electron transfer processes in this complex have been performed (Durrant et al., 1992a,b, 1995; Hastings et al., 1992; Visser et al., 1995; Giorgi et al., 1996; Müller et al., 1996a,b; Greenfield et al., 1997, 1998, 1999; Konermann et al., 1997; Prokhorenko and Holzwarth, 2000; Groot et al., 2005; Holzwarth et al. 2006b). According to these data the exciton equilibration among the core RC pigments occurs on a time scale below 1 ps at room temperature, with a main component around 600 fs. Charge separation has been proposed to occur with lifetimes of about 3–6 ps (Gatzen et al., 1996; Holzwarth et al., 2006b; Miloslavina et al., 2006). In a different model much faster charge separation in the sub-ps range has been proposed (Groot et al., 1997, 2005; van Mourik et al., 2004). A multimer model for the RC core pigments has been developed using basic exciton theory (Durrant et al., 1995). A first attempt using higher level theory to explain spectra and kinetics has been made by Renger and Marcus (2002b). Subsequently, modified Redfield theory has been applied to theoretically treat the localization of the triplet state and the primary electron donor based on the X-ray structure data (Raszewski et al., 2005). Novoderezkhin et al. (2004b), using similar theory but also including the coupling to the charge transfer (radical pair) states, modeled the dynamics of both energy transfer and charge separation. In a further work these authors fitted simultaneously the theoretical dynamics to the transient absorption and the fluorescence kinetics data (Novoderezhkin et al., 2005a). The best fit has been obtained with a model implying that the charge separation occurs via an intermediate state with charge separation within the special pair (radical pair 1). This state is weakly dipole-allowed in the theory, due to mixing with the exciton states, and can be populated directly or via 100-fs energy transfer from the RC core pigments. This radical pair (RP) 1 state as well as the next two radical pairs with the electron transfer chain to the accessory Chl (RP2) and further to the pheophytin (RP3) are dominated in their properties by increased electron-phonon coupling and high energetic disorder. In the RP3 state, the hole was found in this modeling to be delocalized within the special pair, with a predominant localization at the inactive-branch Chl $P_{D2}$. These conclusions, based on the theoretical exciton calculations, are however in disagreement with the recent experimental findings which suggest that the charge separation starts with an initial electron transfer step from the accessory Chl to the pheophytin (Groot et al., 2005; Holzwarth et al., 2006b) and does not involve the 'special pair' Chls. Also the rate of charge separation continues to be a matter of debate (see above). In the theoretical calculations this rate was essentially determined by assumptions about the electronic couplings of the excited states and the CT states, which are essentially fitting parameters. Thus for a more thorough understanding further improved theoretical modeling of the exciton dynamics, and in particular the electron transfer dynamics in the isolated PS II reactions centers is required.

For the PS I reaction center, which can not be isolated without the surrounding antenna complex consisting of 90 Chl *a* molecules (Krauβ et al., 1996; Jordan et al., 2001), several theoretical models of the spectra and dynamics based on simple Förster theory have been published (Byrdin et al. 2002; Sener et al., 2002a,b, 2004). While providing some general structure-based insights into the overall dynamics, these approaches could not properly treat the finer details of the spectra and the dynamics of the PS I antenna/RC complex. A slightly more sophisticated approach to understand the spectra and dynamics has been made by Brüggemann et al. (2004). It provided a better agreement with the experimental data. One of the basic unresolved problems in this large antenna complex consists in the determination of the site energies of the pigments. A first attempt to calculate structure-based site energies for all the 96 Chls in the PS I core complex has been made recently (Damjanovic et al., 2002b). These data were used in a subsequent kinetic modeling to calculate spectra and the associated kinetics (Yang et al., 2003). Experimentally determined spectral densities were employed to calculate the energy transfer rates. The absorption spectrum and fluorescence decay time components of the complex at room temperature were reasonably reproduced. The roles of the special chlorophylls (red, linker, and reaction center, respectively) were also highlighted. One of the conclusions was an antenna-to-RC energy transfer bottleneck which contributes a large part to the total trapping time in the theoretical model. Thus essentially a diffusion-limited kinetics is predicted. This conclusion seems to be at variance however with recent experimental data of PS I core complexes which suggest a very fast antenna to RC energy transfer of around 1–2 ps and an overall trap-limited charge separation kinetics (Müller et al., 2003; Holzwarth et al., 2005, 2006a). At present it is unclear why the experimental and theoretical conclusions deviate. Again, more work is required to unravel this puzzle. A recent one- and two-color photon-echo peak-shift study confirmed the theoretically and experimentally found ultrafast (time scale up to 1 ps) antenna equilibration but did not provide any further clarification of the question of the antenna to RC transfer times (Vaswani et al., 2006).

### *D. Green Bacterial Antenna Systems*

Green bacteria contain two very special antenna complexes. The first one is a large antenna complex consisting of tens of thousands of tightly associated and strongly excitonically coupled BChl *c* molecules that are arranged in a supramolecular self-assembled manner without a protein scaffold (Balaban et al., 1995; van Rossum et al., 2001; for a recent review see Balaban et al., 2005), the so-called chlorosomes. The second one is a small water soluble pigment protein complex containing 7 BChl *a* molecules per monomeric unit, the so-called Fenna-Matthews-Olson (FMO) complex. The latter has been the first photosynthetic antenna complex to be resolved by X-ray structure to atomic resolution (Matthews and Fenna, 1980; Fenna and Matthews, 1975). For recent reviews, see Blankenship et al. (1995) and Olson (2004).

The FMO complex has long been, and continues to be, a fertile playing ground for the application of exciton theories on the one hand and novel experimental methods for the characterization of exciton coupling on the other hand in photosynthetic pigment-protein complexes. Apart from the highly resolved structure and the small number of chromophores, this interest is to a large part due to the fact that at low temperature the spectra are well resolved, showing more structure than other pigment-protein complexes. Early exciton studies have been performed by Pearlstein and co-workers (Pearlstein and Hemenger, 1978; Pearlstein 1991a,b) but did not lead to a satisfactory understanding of the spectra. An important step forward after long years of frustration for many theoretical approaches was taken by Aartsma and coworkers who found that a smaller effective dipole strength used in the calculation of excitonic couplings allows for a quantitative description of several optical spectra (Franken et al., 1998; Wendling et al., 2000, 2002; Vulto et al., 1998a, 1999). Recently, the value of the effective dipole strength was quantitatively explained by electrostatic calculations and first attempts were made for an explanation of the fitted site energies by calculations of electrochromic shifts of pigment transition energies by charged amino acids (Adolphs and Renger, 2006).

Likely more than any other method of optical spectroscopy, the recent developments of various methods in coherent two-dimensional spectroscopy, such as, e.g., two-color photon echo peak shift, two-color three-pulse photon echo, also in combination with heterodyne detection, are promising to provide deep insight into the dynamic properties of excitonically coupled systems (Fleming et al., 2003; Brixner et al., 2004a,b; Mancal and Fleming, 2004; Prall et al., 2004; Cho and Fleming, 2005; Cho et al. 2005, 2006; Engels et al., 2007). Consequently some of the first applications of these techniques were focused on studying the dynamic properties of the FMO complex (Brixner et al., 2005). A combination of generalized Förster and modified Redfield theory has been employed for calculations of the exciton transfer rates, the linear absorption spectrum, and the time dependent 2D spectra. This allowed to self-consistently describe the excited states as either localized or delocalized (excitonic) depending on the magnitude of the coupling constants. A good agreement between experiment and theory was obtained. Most of the exciton states are delocalized over two or three molecules, but the lowest exciton is essentially localized on BChl 3. The latter finding confirmed earlier assignments based on slightly less sophisticated theoretical models (Vulto et al., 1998b, 1999). Two-dimensional optical spectroscopic techniques and their applications to excitonically coupled photosynthetic systems are reviewed by Cho et al. (2006).

Exciton theory has also been applied to understand the spectroscopic properties and exciton relaxation in the supramolecularly organized BChls in the chlorosomal antenna system (Prokhorenko et al., 2000, 2002, 2003). Of particular interest in this context was the size-dependence of the previously not well understood CD spectra as well as other linear properties of these aggregates which could be clarified by using various levels of exciton theory (Prokhorenko et al., 2002; Didraga and Knoester, 2004a,b).

## VIII. Outlook and Challenging Problems

We are just at the beginning of a detailed understanding of the light harvesting function and the underlying building principles of photosynthetic antenna proteins that evolved through billions of years of evolution. From a theoretical viewpoint there are at least two main obstacles that limit our present understanding. The first is our present inability to derive all parameters of the theory directly from the structural data. A particularly challenging problem is to understand how the local protein environment tunes the excitation energies of the pigments and in this way controls excitation energy flow. There are at least two different ways to tackle this problem: One is a quantum chemical calculation of the pigment transition energies, including a small part of the local protein environment (Gudowska-Nowak et al., 1990; Damjanovic et al., 2002b), while the other one involves a simple electrostatic treatment that considers the Coulomb coupling between the charge density of the excited and ground states of the pigments with the charged amino acid side groups in the whole protein (Adolphs and Renger, 2006). Both methods contain only part of the answer. The former misses important long range electrostatic interactions, and the latter cannot describe details of the short range interactions as, e.g., hydrogen bonds and ligation effects. An approach that combines the advantages of both methods and includes also the neutral charge density of the whole protein is desirable.

Another important parameter of the theory that determines the optical line shape and the rate constants for excitation energy transfer is the spectral density $J(\omega)$ that characterizes the modulation of pigment transition energies by the protein dynamics. It can be expected that after the static site energy problem will be solved, our understanding of the pigment-protein coupling will also lead — in combination with molecular dynamics simulations — to realistic spectral densities. So far the only attempt to calculate $J(\omega)$ combined a quantum chemical with a molecular dynamics calculation (Damjanovic et al. 2002a,b) and resulted in a $J(\omega)$ that is very different from the $J(\omega)$ suggested by hole burning and fluorescence line narrowing studies. As in the case of static site energies it seems that a quantum chemical calculation alone is insufficient to treat the influence of the protein on the transition energies of the pigments.

A second obstacle that limits our understanding concerns the dynamical theory used for the calculation of optical spectra and excitation energy transfer. As noted earlier, the excitonic pigment-pigment coupling is of the same magnitude as the exciton-vibrational pigment-protein coupling. Hence there is no obvious small parameter that can be used for a perturbation theory and a non-perturbative description of both types of couplings is desired. A first step was taken in modified Redfield theory that considers the excitonic coupling and the diagonal part of the exciton-vibrational coupling beyond perturbation theory (Zhang et al., 1998; Yang and Fleming, 2002; Renger and Marcus, 2003). An important aspect is still missing however, namely a dynamical localization of exciton states. This dynamical localization of exciton leads to a time dependent contribution of pigments in the different excitons states, in contrast to the static delocalization assumed so far in modified Redfield theory.

A theory of optical spectra that can describe such a localization was presented recently (Renger, 2004), however, it does not include the lifetime broadening due to exciton relaxation. It remains a challenge for the theory to present a unified description of all three phenomena, i.e., vibrational sidebands, lifetime broadening, and dynamical localization of excited states.

## References

Adolphs J and Renger T (2006) How proteins trigger excitation energy transfer in the FMO complex of green sulfur bacteria. Biophys J 91: 2778–2797

Agarwal R, Prall BS, Rizvi AH, Yang M and Fleming GR (2002) Two-color three pulse photon echo peak shift spectroscopy. J Chem Phys 116: 6243–6252

Balaban TS, Holzwarth AR, Schaffner K, Boender GJ and de Groot HJM (1995) CP-MAS $^{13}$C-NMR Dipolar correlation spectroscopy of $^{13}$C enriched chlorosomes and isolated bacteriochlorophyll *c* aggregates of *Chlorobium tepidum*: The self-organization of pigments is the main structural feature of chlorosomes. Biochemistry 34: 15259–15266

Balaban TS, Tamiaki H and Holzwarth AR (2005) Chlorins programmed for self-assembly. In: Würthner F (ed) Supramolecular Dye Chemistry (Topics in Current Chemistry, Vol 258), pp 1–38. Springer Verlag, Heidelberg

Bauman D and Wrobel D (1980) Dichroism and polarized fluorescence of chlorophyll *a*, chlorophyll *c* and bacteriochlorophyll *a* dissolved in liquid crystals. Biophys Chem 12: 83–91

Blankenship RE, Olson JM and Miller M (1995) Antenna complexes from green photosynthetic bacteria. In: Blankenship RE, Madigan MT, Bauer CE (eds) Anoxygenic Photosynthetic Bacteria (Advances in Photosynthesis and Respiration, Vol 2), pp 399–435. Kluwer Academic Publishers, Dordrecht

Bradforth SE, Jimenez R, van Mourik F, van Grondelle R and Fleming GR (1995) Excitation transfer in the core light-harvesting complex (LH-1) of *Rhodobacter sphaeroides*: An ultrafast fluorescence depolarization and annihilation study. J Phys Chem 99: 16179–16191

Brixner T, Mancal T, Stiopkin IV and Fleming GR (2004a) Phase-stabilized two-dimensional electronic spectroscopy. J Chem Phys 121: 4221–4236

Brixner T, Stiopkin IV and Fleming GR (2004b) Tunable two-dimensional femtosecond spectroscopy. Opt Lett 29: 884–886

Brixner T, Stenger J, Vaswani HM, Cho M, Blankenship RE and Fleming GR (2005) Two-dimensional spectroscopy of electronic couplings in photosynthesis. Nature 434: 625–628

Brüggemann B, Sznee K, Novoderezhkin V, van Grondelle R and May V (2004) From structure in dynamics: Modeling exciton dynamics in the photosynthetic antenna PS1. J Phys Chem B 108: 13536–13546

Byrdin M, Jordan P, Krauss N, Fromme P, Stehlik D, Schlodder E (2002) Light harvesting in Photosystem I: Modeling based on the 2.5Å-structure of Photosystem I from *Synechococcus elongatus*. Biophys J 83: 433–457

Chachisvilis M, Pullerits T, Jones MR, Hunter CN and Sundström V (1994) Vibrational dynamics in the light-harvesting complexes of the photosynthetic bacterium *Rhodobacter sphaeroides*. Chem Phys Lett 224: 345–351

Chachisvilis M, Fidder H, Pullerits T and Sundström V (1995a) Coherent nuclear motions in photosynthetic pigments and dye molecules, probed by ultrafast spectroscopy. J Raman Spectrosc 26: 513–522

Chachisvilis M, Fidder H and Sundström V (1995b) Electronic coherence in pseudo two-colour pump-probe spectroscopy. Chem Phys Lett 234: 141–150

Chang JC (1977) Monopole effects on electronic excitation interactions between large molecules. I. Application to energy transfer in chlorophylls. J Chem Res 67: 3901–3909

Cho MH and Fleming GR (2005) The integrated photon echo and solvation dynamics. II. Peak shifts and two-dimensional photon echo of a coupled chromophore system. J Chem Phys 123: 114506-1–18

Cho MH, Vaswani HM, Brixner T, Stenger J and Fleming GR (2005) Exciton analysis in 2D electronic spectroscopy. J Phys Chem B 109: 10542–10556

Cho MH, Brixner T, Stiopkin I, Vaswani H and Fleming GR (2006) Two dimensional electronic spectroscopy of molecular complexes. J Chin Chem Soc 53: 15–24

Connelly JP, Müller MG, Hucke M, Gatzen G, Mullineaux CW, Ruban AV, Horton P and Holzwarth AR (1997) Ultrafast spectroscopy of trimeric light harvesting complex II from higher plants. J Phys Chem B 101: 1902–1909

Creemers TMH, De Caro CA, Visschers RW, van Grondelle R and Völker S (1999) Spectral hole burning and fluorescence line narrowing in subunits of the light-harvesting complex LH1 of purple bacteria. J Phys Chem B 103: 9770–9776

Damjanovic A, Ritz T and Schulten K (1999) Energy transfer between carotenoids and bacteriochlorophylls in light-harvesting complex II of purple bacteria. Phys Rev E 59:3293–3311

Damjanovic A, Kosztin I, Kleinekathöfer U and Schulten K (2002a) Excitons in a photosynthetic light-harvesting system: A combined molecular dynamics, quantum chemistry, and polaron model study. Phys Rev E 65: 031919-1–24

Damjanovic A, Vaswani H, Fromme P and Fleming GR (2002b) Chlorophyll excitations in Photosystem I of *Synechococcus elongatus*. J Phys Chem B 106: 10251–10262

Davydov AS (1971) Theory of Molecular Excitons. Plenum Press, New York

de Boeij WP, Pshenichnikov MS and Wiersma DA (1996) On the relation between the echo-peak shift and Brownian-oscillator correlation function. Chem Phys Lett 253: 53–60

Dekker JP and van Grondelle R (2000) Primary charge separation in Photosystem II. Photosynth Res 63: 195–208

Didraga C and Knoester J (2004a) Excitons in tubular molecular aggregates. J Luminesc 110: 239–245

Didraga C and Knoester J (2004b) Optical spectra and localization of excitons in inhomogeneous helical cylindrical aggregates. J

Chem Phys 121: 10687–10698
Durrant JR, Hastings G, Hong Q, Barber J, Porter G and Klug DR (1992a) Determination of P680 singlet state lifetimes in Photosystem 2 reaction centres. Chem Phys Lett 188: 54–60
Durrant JR, Hastings G, Joseph DM, Barber J, Porter G and Klug DR (1992b) Subpicosecond equilibration of excitation energy in isolated Photosystem II reaction centers. Proc Natl Acad Sci USA 89: 11632–11636
Durrant JR, Klug DR, Kwa SL, van Grondelle R, Porter G and Dekker JP (1995) A multimer model for P680, the primary electron donor of Photosystem II. Proc Natl Acad Sci USA 92: 4798–4802
Eccles J and Honig B (1983) Charged amino acids as spectroscopic determinants for chlorophyll in vivo. Proc Natl Acad Sci USA 80: 4959–4962
Engels GS, Calhoun TR, Read EL, Ahn TK, Mancal T, Chen Y-C, Blankenship RE and Fleming GR (2007) Evidence for wavelike energy transfer through quantum coherence in photosynthetic systems. Nature 446: 782–786
Fenna RE, Matthews BW (1975) Chlorophyll arrangement in a bacteriochlorophyll protein from *Chlorobium limicola*. Nature 258: 573–577
Ferreira KN, Iverson TM, Maghlaoui K, Barber J and Iwata S (2004) Architecture of the photosynthetic oxygen-evolving center. Science 303: 1831–1838
Fetisova Z, Freiberg A, Mauring K, Novoderezhkin V, Taisova A and Timpmann K (1996) Excitation energy transfer in chlorosomes of green bacteria: Theoretical and experimental studies. Biophys J 71: 995–1010
Fleming GR and Jimenez R (1997) Ultrafast spectroscopy of excitation transfer dynamics in biological systems. In: Sundström V (ed) Femtochemistry and Femtobiology: Ultrafast Reaction Dynamics at Atomic-Scale Resolution (Proceedings Nobel Symposium 101), pp 701–723. Imperial College Press, London
Fleming GR and Scholes GD (2004) Physical chemistry — Quantum mechanics for plants. Nature 431: 256–257
Fleming GR, Yang M, Agarwal R, Prall BS, Kaufman LJ and Neuwahl F (2003) Two dimensional electronic spectroscopy. Bull Korean Chem Soc 24: 1081–1090
Förster T (1948) Zwischenmolekulare Energiewanderung und Fluoreszenz. Ann Phys Leipzig 2: 55–75
Förster T (1965) Delocalized excitation and excitation transfer. In: Sinanoglu O (ed) Modern Quantum Chemistry. Part III. Action of Light and Organic Crystals, pp 93–137. Academic Press, New York
Fragata M, Norden B and Kurucsev T (1988) Linear dichroism (250–700 nm) of chlorophyll *a* and pheophytin *a* oriented in a lamellar phase of glycerylmonooctanoate/$H_2O$. Characterization of electronic transitions. Photochem Photobiol 47: 133–143
Franken EM, Neerken S, Louwe RJW, Amesz J and Aartsma TJ (1998) A permanent hole burning study of the FMO antenna complex of the green sulfur bacterium *Prosthecochloris aestuarii*. Biochemistry 37: 5046–5051
Gatzen G, Müller MG, Griebenow K and Holzwarth AR (1996) Primary processes and structure of the Photosystem II reaction center: 3. Kinetic analysis of picosecond energy transfer and charge separation processes in the D1-D2-Cyt-$b_{559}$ complex measured by time-resolved fluorescence. J Phys Chem 100: 7269–7278
Georgakopoulou S, Cogdell RJ, van Grondelle R and van Amerongen H (2003) Linear-dichroism measurements on the LH2 antenna complex of *Rhodopseudomonas acidophila* strain 10050 show that the transition dipole moment of the carotenoid rhodopin glucoside is not collinear with the long molecular axis. J Phys Chem B 107: 655–658
Georgievskii Y, Hsu C-P, Marcus RA (1999) Linear response in theory of electron transfer reactions as an alternative to the molecular harmonic oscillator model. J Chem Phys 110: 5307–5317
Giorgi LB, Nixon PJ, Merry SAP, Joseph DM, Klug DR, Durrant JR, De Las Rivas J, Barber J and Porter G (1996) Comparison of primary charge separation in the Photosystem II reaction center complex isolated from wild-type and D1-130 mutants of the cyanobacterium *Synechocystis* PCC 6803. J Biol Chem 271: 2093–2101
Gradinaru CC, Özdemir S, Gülen D, van Stokkum IHM, van Grondelle R and van Amerongen H (1998a) The flow of excitation energy in LHCII monomers: Implications for the structural model of the major plant antenna. Biophys J 75: 3064–3077.
Gradinaru CC, Pascal AA, van Mourik F, Robert B, Horton P, van Grondelle R, van Amerongen H (1998b). Ultrafast evolution of the excited states in the chlorophyll *a/b* complex CP29 from green plants studied by energy-selective pump-probe spectroscopy. Biochemistry 37:1143–1149
Greenfield SR, Seibert M, Govindjee and Wasielewski MR (1997) Direct measurement of the effective rate constant for primary charge separation in isolated Photosystem II reaction centers. J Phys Chem B 101: 2251–2255
Greenfield SR, Wasielewski MR and Seibert M (1998) Femtosecond PS II reaction center studies at 7K. In: Garab G (ed) Photosynthesis: Mechanism and Effects, pp 1029–1032. Kluwer Academic Publishers, Dordrecht
Greenfield SR, Seibert M and Wasielewski MR (1999) Time-resolved absorption changes of the pheophytin $Q_X$ band in isolated Photosystem II reaction centers at 7 K: Energy transfer and charge separation. J Phys Chem B 103: 8364–8374
Groot M-L, van Mourik F, Eijckelhoff C, van Stokkum IHM, Dekker JP and van Grondelle R (1997) Charge separation in the reaction center of Photosystem II studied as a function of temperature. Proc Natl Acad Sci USA 94: 4389–4394
Groot M-L, Pawlowicz NP, van der Wilderen LJGW, Breton J, van Stokkum IHM and van Grondelle R (2005) Initial electron donor and acceptor in isolated Photosystem II reaction centers identified with femtosecond mid-IR spectroscopy. Proc Natl Acad Sci USA 102: 13087–13092
Gudowska-Nowak E, Newton MD and Fajer J (1990). Conformational and environmental effects on bacteriochlorophyll optical spectra: Correlations of calculated spectra with structural results. J Phys Chem 94: 5795–5801
Hastings G, Durrant JR, Barber J, Porter G and Klug DR (1992) Observation of pheophytin reduction in photosystem two reaction centers using femtosecond transient absorption spectroscopy. Biochemistry 31: 7638–7647
Hofmann E, Wrench PM, Sharples FP, Hiller RG, Welte W and Diederich K (1996) Structural basis of light harvesting by carotenoids: Peridinin-chlorophyll-protein from *Amphidinium carterae*. Science 272: 1788–1791
Holzwarth AR (2004) Light Absorption and Harvesting. In: Archer MD and Barber J (eds) Molecular to Global Photosynthesis, pp 43–115. Imperial College Press, London
Holzwarth AR (2007) Ultrafast primary reactions in the pho-

tosystems of oxygen-evolving organisms. In: Zinth W (ed) Ultrafast Processes in Chemistry and Biology. Springer-Verlag, Heidelberg, Berlin, in press

Holzwarth AR, Müller MG, Niklas J and Lubitz W (2005) Charge recombination fluorescence in Photosystem I reaction centers from *Chlamydomonas reinhardtii*. J Phys Chem B 109: 5903–5911

Holzwarth AR, Müller MG, Niklas J and Lubitz W (2006a) Ultrafast transient absorption studies on Photosystem I reaction centers from *Chlamydomonas reinhardtii*. 2. Mutations around the P700 reaction center chlorophylls provide new insight into the nature of the primary electron donor. Biophys J 90: 552–565

Holzwarth AR, Müller MG, Reus M, Nowaczyk M, Sander J and Rögner M (2006b) Kinetics and mechanism of electron transfer in intact Photosystem II and in the isolated reaction center: Pheophytin is the primary electron acceptor. Proc Natl Acad Sci USA 103: 6895–6900

Hsu C-P, Fleming GR, Head-Gordon M and Head-Gordon T (2001) Excitation energy transfer in condensed media. J Chem Phys 114: 3065–3072

Jang SJ, Newton MD and Silbey RJ (2004) Multichromophoric Förster resonance energy transfer. Phys Rev Lett 92: 218301–1–4.

Jimenez R, van Mourik F, Yu JY and Fleming GR (1997) Three-pulse photon echo measurements on LH1 and LH2 complexes of *Rhodobacter sphaeroides*: A nonlinear spectroscopic probe of energy transfer. J Phys Chem B 101:7350–7359.

Joo T, Jia Y, Yu J-Y, Jonas DM and Fleming GR (1996) Dynamics in isolated bacterial light harvesting antenna (LH2) of *Rhodobacter sphaeroides* at room temperature. J Phys Chem 100: 2399–2409

Jordan P, Fromme P, Witt HT, Klukas O, Saenger W and Krau± N (2001) Three-dimensional structure of cyanobacterial Photosystem I at 2.5 Å resolution. Nature 411: 909–917

Kleima FJ, Wendling M, Hofmann E, Peterman EJG, van Grondelle R and van Amerongen H (2000) Peridinin chlorophyll *a* protein: Relating structure and steady-state spectroscopy. Biochemistry 39: 5184–5195

Knapp E-W (1984) Lineshapes of molecular aggregates. Exchange narrowing and intersite correlation. Chem Phys 85: 73–82

Knox RS and Spring BQ (2003) Dipole strengths in the chlorophylls. Photochem Photobiol 77: 497–501

Konermann L, Gatzen G and Holzwarth AR (1997) Primary processes and structure of the Photosystem II reaction center. 5. Modeling of the fluorescence kinetics of the D1-D2-Cyt-$b_{559}$ complex at 77K. J Phys Chem B 101: 2933–2944

Krauß N, Schubert W-D, Klukas O, Fromme P, Witt HT and Saenger W (1996) Photosystem I at 4Å resolution represents the first structural model of a joint photosynthetic reaction centre and core antenna system. Nature Struct Biol 3: 965–973

Krueger BP, Scholes GD and Fleming GR (1998) Calculation of couplings and energy-transfer pathways between the pigments of LH2 by the ab inito transition density cube method. J Phys Chem B 102: 5378–5386

Krueger BP, Scholes GD, Yu J-Y, Fleming GR (1999) The light harvesting process in purple bacteria. Acta Phys Pol A 95: 63–83

Kubo R and Toyozawa Y (1955) Application of the method of generating function to radiative and non-radiative transitions of a trapped electron in a crystal. Prog Theor Phys 13: 160–182

Kühlbrandt W (1994) Structure and function of the plant light-harvesting complex, LHC-II. Curr Opinion Struct Biol 4: 519–528

Kühn O, Renger T, May V, Voigt J, Pullerits T and Sundström V (1999) Exciton-vibrational coupling in photosynthetic antenna complexes: Theory meets experiment. Trends Photochem Photobiol 4: 213–256

Lax M (1952) The Franck-Condon Principle and its application to crystals. J Chem Phys 20: 1752–1760

Liu Z, Yan H, Wang K, Kuang T, Zhang J, Gui L, An X and Chang W (2004) Crystal structure of spinach major light-harvesting complex at 2.72 Å resolution. Nature 428: 287–292

Loll B, Kern J, Saenger W, Zouni A and Biesiadka J (2005a) Towards complete cofactor arrangement in the 3.0 Å resolution structure of Photosystem II. Nature 438: 1040–1044

Loll B, Kern J, Zouni A, Saenger W, Biesiadka J and Irrgang K-D (2005b) The antenna system of Photosystem II from *Thermosynechococcus elongatus* at 3.2 Å resolution. Photosynth Res 86: 175–184

Madjet ME, Abdurahman A and Renger T (2006) Intermolecular coulomb couplings from *ab initio* electrostatic potentials: Application to optical transitions of strongly coupled pigments in photosynthetic antennae and reaction centers. J Phys Chem B 110: 17268–17281

Mancal T and Fleming GR (2004) Probing electronic coupling in excitonically coupled heterodimer complexes by two-color three-pulse photon echoes. J Chem Phys 121: 10556–10565

Matthews BW and Fenna RE (1980) Structure of a green bacteriochlorophyll protein. Acc Chem Res 13: 309–317

May V and Kühn O (2000) Charge and Energy Transfer Dynamics in Molecular Systems: A theoretical introduction. John Wiley & Sons, Berlin

Miloslavina Y, Szczepaniak M, Müller MG, Sander J, Nowaczyk M, Rögner M and Holzwarth AR (2006). Charge separation kinetics in intact Photosystem II core particles is trap-limited. A picosecond fluorescence study. Biochemistry 45: 2436–2442.

Mukai K, Abe S and Sumi H (1999) Theory of rapid excitation-energy transfer from B800 to optically-forbidden exciton states of B850 in the antenna system LH2 of photosynthetic purple bacteria. J Phys Chem B 103: 6096–6102

Müller MG, Hucke M, Reus M and Holzwarth AR (1996a) Annihilation processes in the isolated D1-D2-Cyt-$b_{559}$ reaction center complex of Photosystem II. An intensity dependence study of femtosecond transient absorption. J Phys Chem 100: 9537–9544

Müller MG, Hucke M, Reus M and Holzwarth AR (1996b) Primary processes and structure of the Photosystem II reaction center: IV. Low intensity femtosecond transient absorption spectra of D1-D2 reaction centers. J Phys Chem 100: 9527–9536

Müller MG, Niklas J, Lubitz W and Holzwarth AR (2003) Ultrafast transient absorption studies on Photosystem I reaction centers from *Chlamydomonas reinhardtii*. 1. A new interpretation of the energy trapping and early electron transfer steps in Photosystem I. Biophys J 85: 3899–3922

Novoderezhkin V, Wendling M and van Grondelle R (2003) Intra- and interband transfers in the B800-B850 antenna of *Rhodospirillum molischianum*: Redfield theory modeling of polarized pump-probe kinetics. J Phys Chem B 107: 11534–11548

Novoderezhkin VI, Palacios MA, van Amerongen H and van Grondelle R (2004a) Energy-transfer dynamics in the LHCII complex of higher plants: Modified Redfield approach. J Phys

Chem B 108: 10363–10375
Novoderezhkin VI, Yakovlev AG, van Grondelle R and Shuvalov VA (2004b) Coherent nuclear and electronic dynamics in primary charge separation in photosynthetic reaction centers: A Redfield theory approach. J Phys Chem B 108: 7445–7457
Novoderezhkin VI, Andrizhiyevskaya EG, Dekker JP and van Grondelle R (2005a) Pathways and timescales of primary charge separation in the Photosystem II reaction center as revealed by a simultaneous fit of time-resolved fluorescence and transient absorption. Biophys J 89: 1464–1481
Novoderezhkin VI, Palacios MA, van Amerongen H and van Grondelle R (2005b) Excitation dynamics in the LHCII complex of higher plants: Modeling based on the 2.72 Å crystal structure. J Phys Chem B 109: 10493–10504
Ohta K, Yang M and Fleming GR (2001) Ultrafast exciton dynamics of J-aggregates in room temperature solution studied by third-order nonlinear optical spectroscopy and numerical simulation based on exciton theory. J Chem Phys 115: 7609–7621
Olson JM (2004) The FMO protein. Photosynth Res 80: 181–187
Pearlstein RM (1991a) Theoretical interpretation of antenna spectra. In: Scheer H (ed) Chlorophylls, pp 1047–1078. CRC Press, New York
Pearlstein RM (1991b) Theory of the optical-spectra of the bacteriochlorophyll *a* antenna protein trimer from *Prosthecochloris aestuarii*. Photosynth Res 31: 213–226
Pearlstein RM and Hemenger RP (1978) Bacteriochlorophyll electronic transition moment directions in bacteriochlorophyll *a*-protein. Proc Natl Acad Sci USA 75: 4920–4924
Peterman EJG, Pullerits T, van Grondelle R and van Amerongen H (1997) Electron-phonon coupling and vibronic fine structure of light-harvesting complex II of green plants: Temperature dependent absorption and high-resolution fluorescence spectroscopy. J Phys Chem B 101: 4448–4457
Pieper J, Voigt J, Renger G and Small GJ (1999) Analysis of phonon structure in line-narrowed optical spectra. Chem Phys Lett 310: 296–302
Prall BS, Parkinson DY, Fleming GR, Yang M and Ishikawa N (2004) Two-dimensional optical spectroscopy: Two-color photon echoes of electronically coupled phthalocyanine dimers. J Chem Phys 120: 2537–2540
Prokhorenko VI and Holzwarth AR (2000) Primary processes and structure of the Photosystem II reaction center: A photon echo study. J Phys Chem B 104: 11563–11578
Prokhorenko VI, Steensgaard DB and Holzwarth AR (2000) Exciton dynamics in the chlorosomal antennae of the green bacteria *Chloroflexus aurantiacus* and *Chlorobium tepidum*. Biophys J 79: 2105–2120
Prokhorenko VI, Holzwarth AR, Nowak FR and Aartsma TJ (2002) Growing-in optical coherence in the FMO antenna complexes. J Phys Chem B 106: 9923–9933
Prokhorenko VI, Steensgaard DB and Holzwarth AR (2003) Exciton theory for supramolecular chlorosomal aggregates. 1. Aggregate size dependence of the linear spectra. Biophys J 85: 3173–3186
Pullerits T and Sundström V (2005) From biological to synthetic light-harvesting materials — the elementary steps. In: Andrews DL (ed) Energy Harvesting Materials. World Scientific Publishing Company, Singapore
Pullerits T, Monshouwer R, van Mourik F and van Grondelle R (1995) Temperature dependence of electron-vibronic spectra of photosynthetic systems. Computer simulations and comparison with experiment. Chem Phys 194: 395–407
Raszewski G, Saenger W and Renger T (2005) Theory of optical spectra of Photosystem II reaction centers: Location of the triplet state and the identity of the primary electron donor. Biophys J 88: 986–998
Renger T (2004) Theory of optical spectra involving charge transfer states: Dynamic localization predicts a temperature dependent optical band shift. Phys Rev Lett 93: 188101
Renger G and Holzwarth AR (2005) Primary electron transfer. In: Wydrzynski TJ and Satoh K (eds) Photosystem II: The Light-Driven Water: Plastoquinone Oxidoreductase (Advances in Photosynthesis and Respiration, Vol 22), pp 139–175. Springer, Dordrecht
Renger T and Marcus RA (2002a) On the relation of protein dynamics and exciton relaxation in pigment-protein complexes: An estimation of the spectral density and a theory for the calculation of optical spectra. J Chem Phys 116: 9997–10019
Renger T and Marcus RA (2002b) Photophysical properties of PS-2 reaction centers and a discrepancy in exciton relaxation times. J Phys Chem B 106: 1809–1819
Renger T and Marcus RA (2003) Variable-range hopping electron transfer through disordered bridge states: Application to DNA. J Phys Chem A 107: 8404–8419
Renger T and May V (1997) Theory of multiple exciton effects in the photosynthetic antenna complex LHC-II. J Phys Chem B 101: 7232–7240
Renger T and May V (1998) Ultrafast exciton motion in photosynthetic antenna systems: The FMO-complex. J Phys Chem A 102: 4381–4391
Renger T and May V (2000) Simulations of frequency-domain spectra: Structure-function relationships in photosynthetic pigment-protein complexes. Phys Rev Lett 84: 5228–5231
Renger T, Voigt J, May V and Kühn O (1996) Dissipative exciton motion in a chlorophyll *a/b* dimer of the light harvesting complex of Photosystem II: Simulation of pump- probe spectra. J Phys Chem 100:15654–15662
Renger T, May V and Kühn O (2001) Ultrafast excitation energy transfer dynamics in photosynthetic pigment-protein complexes. Phys Reports Rev Sect Phys Lett 343: 137–254
Renger T, Trostmann I, Theiss C, Madjet ME, Richter M, Paulsen H, Eichler HJ, Knorr A and Renger G (2007) Refinement of a structural model of a pigment-protein complex by accurate optical lineshape theory and experiments. J Phys Chem B, in press
Rutkauskas D, Novoderezhkin V, Cogdell RJ and van Grondelle R (2004) Fluorescence spectral fluctuations of single LH2 complexes from *Rhodopseudomonas acidophila* strain 10050. Biochemistry 43: 4431–4438
Rutkauskas D, Novoderezhkin V, Cogdell RJ and van Grondelle R (2005) Fluorescence spectroscopy of conformational changes of single LH2 complexes. Biophys J 88: 422–435
Salverda JM, Vengris M, Krueger BP, Scholes GD, Czamoleski AR, Novoderezhkin V, van Amerongen H and van Grondelle R (2003) Energy transfer in light-harvesting complexes LHCII and CP29 of spinach studied with three pulse echo peak shift and transient grating. Biophys J 84: 450–465
Sauer K, Cogdell RJ, Prince SM, Freer A, Isaacs NW and Scheer H (1996) Structure-based calculations of the optical spectra of the LH2 bacteriochlorophyll-protein complex from *Rhodopseu-*

*domonas acidophila*. Photochem Photobiol 64: 564–576

Scholes GD and Fleming GR (2000) On the mechanism of light harvesting in photosynthetic purple bacteria: B800 to B850 energy transfer. J Phys Chem B 104: 1854–1868

Scholes GD, Jordanides XJ and Fleming GR (2001) Adapting the Förster theory of energy transfer for modeling dynamics in aggregated molecular assemblies. J Phys Chem B 105: 1640–1651

Sener MK, Lu DY, Park SH, Schulten K and Fromme P (2002a) Spectral disorder and excitation transfer dynamics in cyanobacterial Photosystem I. Biophys J 82: 292A

Sener MK, Lu DY, Ritz T, Park S, Fromme P and Schulten K (2002b) Robustness and optimality of light harvesting in cyanobacterial Photosystem I. J Phys Chem B 106: 7948–7960

Sener MK, Park S, Lu DY, Damjanovic A, Ritz T, Fromme P and Schulten K (2004) Excitation migration in trimeric cyanobacterial Photosystem I. J Chem Phys 120: 11183–11195

Standfuss J, van Scheltinga ACT, Lamborghini M and Kühlbrandt W (2005) Mechanisms of photoprotection and nonphotochemical quenching in pea light-harvesting complex at 2.5Å resolution. EMBO J 24: 919–928

Sumi H (1999) Theory on rates of excitation-energy transfer between molecular aggregates through distributed transition dipoles with application to the antenna system in bacterial photosynthesis. J Phys Chem B 103: 252–260

Sundström V (2000) Light in elementary biological reactions. Progr Quant Electr 24: 187–238

van Amerongen H, Valkunas L and van Grondelle R (2000) Photosynthetic Excitons. World Scientific, Singapore

van Grondelle R and Novoderezhkin V (2001) Dynamics of excitation energy transfer in the LH1 and LH2 light-harvesting complexes of photosynthetic bacteria. Biochemistry 40: 15057–15068

van Grondelle R and Novoderezhkin VI (2006) Energy transfer in photosynthesis: experimental insights and quantitative models. Phys Chem Chem Phys 8: 793–807

van Mourik F, Groot M-L, van Grondelle R, Dekker JP and van Stokkum IHM (2004) Global and target analysis of fluorescence measurements on Photosystem 2 reaction centers upon red excitation. Phys Chem Chem Phys 6: 4820–4824

van Rossum B-J, Steensgaard DB, Mulder FM, Boender G-J, Schaffner K, Holzwarth AR and de Groot HJM (2001) A refined model of chlorosomal antennae of *Chlorobium tepidum* from proton chemical shift constraints obtained with High-Field 2-D and 3-D MAS NMR dipolar correlation spectroscopy. Biochemistry 40: 1587–1595

Vaswani HM, Stenger J, Fromme P, Fleming GR (2006) One- and two-color photon echo peak shift studies of Photosystem I. J Phys Chem B 110: 26303–26312

Visser HM, Groot M-L, van Mourik F, van Stokkum IHM, Dekker JP and van Grondelle R (1995) Subpicosecond transient absorption difference spectroscopy on the reaction center of Photosystem II: Radical pair formation at 77 K. J Phys Chem 99: 15304–15309

Vulto SIE, de Baat MA, Louwe RJW, Permentier HP, Neef T, Miller M, van Amerongen H and Aartsma TJ (1998a) Exciton simulations of optical spectra of the FMO complex from the green sulfur bacterium *Chlorobium tepidum* at 6 K. J Phys Chem B 102: 9577–9582

Vulto SIE, Neerken S, Louwe RJW, de Baat MA, Amesz J and Aartsma TJ (1998b) Excited-state structure and dynamics in FMO antenna complexes from photosynthetic green sulfur bacteria. J Phys Chem B 102: 10630–10635

Vulto SIE, de Baat MA, Neerken S, Nowak FR, van Amerongen H, Amesz J, Aartsma TJ (1999) Excited state dynamics in FMO antenna complexes from photosynthetic green sulfur bacteria: A kinetic model. J Phys Chem B 103: 8153–8161

Warshel A and Parson WW (1987) Spectroscopic properties of photosynthetic reaction centers. 1. Theory. J Am Chem Soc 109: 6143–6152

Weiss C (1972) The π electron structure and absorption spectra of chlorophylls in solution. J Mol Spectrosc 44: 37–80

Wendling M, Pullerits T, Przyjalgowski MA, Vulto SIE, Aartsma TJ, van Grondelle R and van Amerongen H (2000) Electron-vibrational coupling in the Fenna-Matthews-Olson complex of *Prosthecochloris aestuarii* determined by temperature-dependent absorption and fluorescence line-narrowing measurements. J Phys Chem B 104: 5825–5831

Wendling M, Przyjalgowski MA, Gulen D, Vulto SIE, Aartsma TJ, van Grondelle R and van Amerongen H (2002) The quantitative relationship between structure and polarized spectroscopy in the FMO complex of *Prosthecochloris aestuarii*: Refining experiments and simulations. Photosynth Res 71: 99–123

Yang M and Fleming GR (2002) Influence of phonons on exciton transfer dynamics: Comparison of the Redfield, Förster, and modified Redfield equations. Chem Phys 275: 355–372

Yang MN and Fleming GR (2003) Construction of kinetic domains in energy trapping processes and application to a photosynthetic light harvesting complex. J Chem Phys 119: 5614–5622

Yang M, Damjanovic A, Vaswani HM and Fleming GR (2003) Energy transfer in Photosystem I of cyanobacteria *Synechococcus elongatus*: Model study with structure-based semi-empirical Hamiltonian and experimental spectral density. Biophys J 85: 140–158

Zhang WM, Meier T, Chernyak V, Mukamel S (1998) Exciton-migration and three-pulse femtosecond optical spectroscopies of photosynthetic antenna complexes. J Chem Phys 108: 7763–7774

# Chapter 22

## Molecular Dynamics Methods for Bioelectronic Systems in Photosynthesis

Ioan Kosztin*
*Department of Physics and Astronomy, University of Missouri-Columbia, Columbia, MO 65211, U.S.A.*

Klaus Schulten
*Beckman Institute and Department of Physics, University of Illinois, Urbana, IL 61801, U.S.A.*



### Summary

With the widespread availability of high performance computer clusters and efficient parallel molecular modeling software, molecular dynamics (MD) simulations became an indispensable tool for the study of the structure-function relationship in proteins with known crystal structures. However, understanding at atomic level the functioning of membrane bound pigment-protein complexes (PPCs), which in photosynthetic organisms convert the energy of the absorbed light into electronic excitations and electrochemical potential gradients, continues to remain a challenging problem. Indeed, the theoretical description of PPCs at physiological temperature in their native environment is a complicated stochastic quantum mechanics problem that requires determining and characterizing the quantum states of the interacting pigment molecules in the presence of thermal fluctuations.

*Author for correspondence, email: KosztinI@missouri.edu

*Thijs J. Aartsma and Jörg Matysik (eds): Biophysical Techniques in Photosynthesis II, pp. 445–464.*

Until recently most theoretical approaches for calculating the optical spectra and the electronic transfer rates of PPCs were based on empirical stochastic models in which several fitting parameters are adjusted to simulate the corresponding experimental results. In this chapter a general approach, which combines MD simulations, quantum chemistry (QC) calculations and quantum many-body theory, for predicting and characterizing charge transfer, spectral and optical properties (e.g., linear absorption and circular dichroism spectra) of PPCs is presented. The method requires only atomic-level crystal structure information and consists of three major steps: (i) the conformational dynamics of the protein matrix embedded into a fully solvated lipid bilayer is followed by means of classical MD simulations; (ii) the lowest energy quantum states of each pigment molecule are determined along the MD trajectory by means of QC calculations; and (iii) the transfer rate and/or optical spectra are determined in terms of a lineshape function which, within the cumulant approximation, can be calculated from the results of the QC calculations. To demonstrate its features, the combined MD/QC method is applied to calculate the linear optical spectra of the light harvesting complex LH2 from *Rhodospirillum molischianum* and the electron transfer rates in photosynthetic reaction center from *Rhodobacter sphaeroides*.

## I. Introduction

Photosynthetic systems contain a variety of large photoactive pigment-protein complexes (PPCs) that carry out important functions necessary for maintaining the life cycle of photosynthetic organisms (e.g., light-harvesting, quinone redox reactions and ATP synthesis). Being involved in electron and electronic excitation transfer processes PPCs have been the subject of numerous experimental and theoretical studies (Renger et al., 2001). In PPCs the photoactive pigment molecules are held in well defined spatial configuration and orientation by a scaffold of proteins. At physiological temperature the dynamics of the protein matrix are governed mainly by classical physics and, therefore, can be studied by means of molecular dynamics (MD) simulations provided that a sufficiently high resolution molecular structure of the system is available. However, key to photosynthesis is the interaction of light with the electronic degrees of freedom of the pigment molecules which is quantum mechanical in nature. In general, the properties of PPCs are determined by the chemical nature of the pigment, the electronic interactions between the pigment molecules, and the interactions between pigment molecules and their environment (e.g., protein, lipid and solvent molecules). In photosynthetic organisms PPCs function at physiological temperature and, therefore, their electronic and optical properties are strongly affected by thermal fluctuations which represent the main source of dynamic disorder in these systems. Thus, the theoretical description of thermally disordered PPCs is a complicated stochastic quantum mechanics problem that requires determining and characterizing the quantum states of the interacting pigment molecules in the presence of a fluctuating environment. Even in the simplest theoretical models of PPCs the simultaneous treatment of the electronic coupling between the pigments and the effect of thermal disorder can be done only approximately (Mukamel, 1995a; Chernyak et al., 1998; May and Kühn, 2000; van Amerongen et al., 2000). Most of the currently used theoretical approaches for calculating the electronic transfer rates and optical spectra of PPCs are based on empirical stochastic models in which several fitting parameters are adjusted to simulate the corresponding experimental results (Koolhaas et al., 2000).

In this chapter a general approach for predicting and characterizing charge transfer, spectral and optical properties, e.g., linear absorption (OD) and circular dichroism (CD) spectra, of PPCs is presented (Damjanovic et al. 2002a,b; Janosi et al., 2006). The approach that combines MD simulations, quantum chemistry (QC) calculations and quantum many-body theory is based solely on atomic-level crystal structure information. The conformational dynamics of the protein matrix embedded into its natural environment (a fully solvated lipid bilayer) are followed by means of classical MD simulations. Next, for each pigment molecule, modeled as a quantum two level system, the energy gap (and, in case of optical spectra, the transition dipole moment) time series are determined along a properly chosen segment of the MD trajectory by means of QC calculations. Finally, the transfer rate and/or optical spectra are

*Abbreviations*: CD – circular dichroism; DOS – density of states; LH2 – light-harvesting complex II; MD – molecular dynamics; OD – optical density (linear absorption); PPC – pigment protein complex; QC – quantum chemistry; *Rb.* – *Rhodobacter*; *Rps.* – *Rhodopseudomonas*; *Rs.* – *Rhodospirillum*; TDM – transition dipole moment

determined in terms of a lineshape function which, within the cumulant approximation, can be calculated from the sole knowledge of the energy gap time series. The authors made every effort to present the material in a manner that is accessible to the general biophysics readership, including both theorists and experimentalists.

The chapter is organized as follows. In the first part the theoretical background of the combined MD/QC method for calculating optical spectra and electron transfer rates is presented by following the same general strategy. The relationship of the presented method to the polaron model and spin-boson model calculations is also clarified. To make the presentation self contained, a brief description of the considered model systems, i.e., the light harvesting complex LH2 from *Rhodospirillum* (*Rs.*) *molischianum* (for calculating OD and CD spectra), and the photosynthetic reaction center from *Rhodobacter* (*Rb.*) *sphaeroides* (for calculating electron transfer rates) is provided. In the second part of the chapter the presented theory is demonstrated by calculating the optical spectra of the LH2 and the electron transfer rate in the reaction center.

## II. Optical Transitions at Finite Temperature

Following their crystal structure determination, LH2 complexes from *Rs. molischianum* (Koepke et al., 1996) and *Rps. acidophila* (McDermott et al., 1995) have been extensively studied both experimentally (Somsen et al., 1996; Beekman et al., 1997; Wu et al., 1997; Sundström et al., 1999; Scholes and Fleming, 2000; Yang et al., 2001; Georgakopoulou et al., 2002; Hu et al., 2002) and theoretically (Meier et al., 1997; Hu et al., 1998b; Linnanto et al., 1999; Ray and Makri, 1999; Sundström et al., 1999; Ihalainen et al., 2001; Yang et al., 2001; Damjanovic et al., 2002a; He et al., 2002; Hu et al., 2002; Jang and Silbey, 2003). LH2 from *Rs. molischianum* (Fig. 1) is an octamer of αβ-heterodimers arranged in a ring-like structure (Hu and Schulten, 1997; Hu et al., 1998a). Each protomer consists of an α- and a β-apoprotein which binds non-covalently one BChl-*a* molecule that absorbs at 800 nm (referred to as B800), two BChl-a molecules that absorb at 850 nm (referred to as B850) and at least one carotenoid that absorbs around 500 nm. The total of 16 B850 and 8 B800 BChls form two circular aggregates, both oriented parallel to the surface of the membrane. The excitonic coupling between the B800s is negligible because of their large spatial separation (~22 Å). Therefore, the optically active $Q_y$ excited electronic states of the B800s are almost degenerate. On the other hand, the tightly packed B850s (with and average Mg–Mg distance of ~9.2 Å within the αβ-heterodimer and ~8.9 Å between the neighboring protomers) are strongly coupled and the corresponding $Q_y$ excited states form an excitonic band in which the states that carry most of the oscillator strength are clustered about ~850 nm (1.46 eV).

Another important difference between the two BChl rings is that while the B800s are surrounded by mostly hydrophilic protein residues the binding pocket of the B850s is predominantly hydrophobic (Koepke et al., 1996). Thus, although both B800s and B850s are chemically identical BChl molecules, their specific spatial arrangement and the nature of their protein environment shape differently their spectral and optical properties. For example, it is quite surprising that the two peaks, due to the B800 and B850 BChls, in the experimental OD spectrum of LH2 from *Rs. molischianum* at room temperature (Zhang et al., 2000; Ihalainen et al., 2001) have comparable widths although, as mentioned above, the B800 levels are almost degenerate while the B850 levels form a ~0.2 eV wide excitonic band.

In order to calculate the linear optical absorption of a PPC one assumes that the electronic properties of individual pigment molecules can be described in terms of a two-level system, formed by the ground state and the lowest excited singlet state (e.g., the $Q_y$ state in the case of BChl) involved in the optical absorption process. Neglecting for the moment the direct interaction between the pigments (e.g., by assuming a sufficiently large spatial separation between them as in the case of the B800s in LH2), one denotes these two states for pigment $n = 1 \ldots N$, as $|0\rangle \equiv |0_n\rangle$ and $|n\rangle \equiv |1_n\rangle$, respectively. Once the interaction between the pigment and its environment composed of the protein matrix, lipid membrane and solvent molecules is taken into account these two levels transform into still well separated energy bands $|0;\lambda_0\rangle \equiv |0\rangle|\lambda_0\rangle$ and $|0;\lambda_n\rangle \equiv |n\rangle|\lambda_n\rangle$. Here the quantum numbers $\lambda_0$ and $\lambda_n$ specify the state of the $n^{th}$ pigment on the ground- and excited-state potential energy surface, respectively. Because the exact quantum mechanical treatment of the eigenstates $|0;\lambda_0\rangle$, $|n;\lambda_n\rangle$ and of the corresponding energy eigenvalues $E_{0,\lambda_0}$, $E_{n,\lambda_n}$ is out of question, usually the quantum numbers $\lambda_0$ and $\lambda_n$ are associated with the vibronic

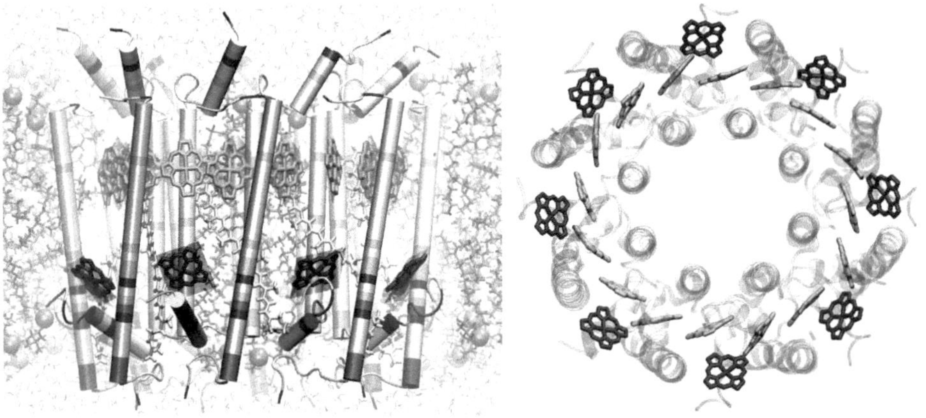

*Fig. 1.* Left: Side view of the LH2 complex from *Rs. molischianum* (entry code 1LGH in Protein Data Bank) embedded in a fully solvated POPC lipid bilayer. The transmembrane helices of the apoprotein subunits are shown as cylinders (cartoon representation) and are colored by residue type; dark (light) colors represent hydrophilic (hydrophobic) residues. For clarity only the BChl macrocycles and the back half of the lipids are shown. The clearly visible B800 (B850) ring is surrounded mostly by polar and charged (nonpolar) protein residues. Right: Tilted side view of the quantum system formed by the optically active B800 and B850 rings. Graphics rendered with the program VMD (Humphrey et al., 1996). See also Fig. 1, Color Plate 7.

states of the PPC that can be modeled within the harmonic approximation as a phonon heat bath. An alternative approach is to follow the dynamics of the nuclear degrees of freedom of the PPC by all-atom MD simulations, and determine the energy gap time series $\Delta E_n(t) = E_n(t) - E_0(t)$ at each MD time step by means of QC calculations. The main assumption of this approach is that the obtained energy gap time series $\Delta E_n(t)$ can be used to calculate approximately equilibrium quantities (such as energy gap density of states and time autocorrelation functions) of the original system without the knowledge of the exact energy-gap spectrum $\Delta E_{n,\lambda_n,\lambda_0} = \Delta E_{n,\lambda_n} - \Delta E_{n,\lambda_0}$.

In the absence of the excitonic coupling between the pigment molecules, the Hamiltonian of the system can be written as $H = H_0 + H'$, where

$$H_0 = \sum_{\lambda_0} | 0;\lambda_0\rangle E_{0,\lambda_0} \langle 0;\lambda_0 | \tag{1}$$

and

$$H' = \sum_n H_n = \sum_{\lambda_n} | n;\lambda_n\rangle E_{n,\lambda_n} \langle n;\lambda_n | \tag{2}$$

The electric dipole moment operator through which the incident light field couples to the $n^{\text{th}}$ pigment molecule is given by

$$\hat{\mu}_n = \sum_{\lambda_n,\lambda_0} \mathbf{d}_{n,\lambda_n,\lambda_0} | n;\lambda_n\rangle\langle 0;\lambda_0 | \tag{3}$$

where the transition dipole moment (TDM) matrix element $\mathbf{d}_{n,\lambda_n,\lambda_0}$ in the Condon approximation (May and Kühn, 2000) can be written

$$\mathbf{d}_{n,\lambda_n,\lambda_0} \approx \mathbf{d}_n \langle \lambda_n | \lambda_0 \rangle \tag{4}$$

Here $\mathbf{d}_n = \langle 1|\hat{\mu}_n|0\rangle$ is the real TDM vector whose time series can be determined from the same combined MD/QC calculations as $\Delta E_n(t)$. Note that $\langle 1|0\rangle = 0$, while in general the Franck-Condon factors $\langle \lambda_n|\lambda_0\rangle$ are finite (May and Kühn, 2000).

### A. Linear Absorption and Line Shape Function

Because, in general, the wavelength of the incident light is much larger than the size of PPCs, in leading approximation, the electric component of the light field can be regarded as being uniform throughout the system. Thus, according to standard linear response theory, the corresponding OD spectrum $I(\omega)$ is proportional to the dipole-dipole correlation function (Mukamel, 1995a; May and Kühn, 2000)

$$I(\omega) \propto \omega \sum_{n,m} \mathrm{Re}\left[\int_0^\infty dt e^{i\omega t}\left\langle \hat{\mu}^{\dagger}_{m,i}(0)\hat{\mu}_{n,i}(t)\right\rangle\right] \quad (5)$$

where $\hat{\mu}_{n,i}(t) = e^{-iH't}\hat{\mu}_{n,i}(0)e^{iH_0 t}$ is the $i \in \{x,y,z\}$ component of the time dependent electric dipole operator, and $\langle \ldots \rangle = \mathrm{Tr}\{Z_0^{-1}\exp(-\beta H_0)\ldots\}$ with $\beta = 1/k_B T$ the usual temperature factor and $Z_0$ the corresponding partition function. To simplify notation, throughout this paper we use units in which $\hbar = 1$, and apply the convention of implicit summation over repeated indices. By employing Eqs. (1), (2) and (3), the quantum dipole correlation function in Eq. (5) can be expressed as

$$\left\langle \hat{\mu}^{\dagger}_{m,j}(0)\hat{\mu}_{n,i}(t)\right\rangle = d_{n,i} d_{m,j} \delta_{nm} \left\langle e^{iH_0 t} e^{-iH_n t}\right\rangle \quad (6)$$

where $\delta_{nm}$ is the usual Kronecker delta symbol. With Eq. (6) the OD spectrum of an aggregate of non-interacting pigments in their native environment can be written in terms of the lineshape function

$$A_n(\omega) \equiv \mathrm{Re}\int_0^\infty dt e^{i\omega t}\left\langle e^{iH_0 t} e^{-iH_n t}\right\rangle \quad (7)$$

as

$$I(\omega) \propto \omega \sum_n d_n^2 A_n(\omega) \quad (8)$$

The main difficulty in calculating the quantum time correlation function in Eq. (7) is due to the fact that the Hamiltonians $H_0$ and $H_n$ do not commute. If they would, then the lineshape function could be expressed in terms of the energy gap density of states (DOS). Indeed, in this case $\langle e^{iH_0 t}e^{-iH_n t}\rangle \approx \langle \exp(-i\Delta H_n t)\rangle$, with $\Delta H_n = H_n - H_0$, and by calculating the time integral in Eq. (7) would follow

$$A_n(\omega) \approx \pi N(\omega) \quad (9)$$

$$N(\omega) \equiv \langle \delta(\omega - \Delta H_n)\rangle \approx \langle \delta(\omega - \Delta E_n(t)) \rangle \quad (10)$$

where the density of states $N(\omega)$ is approximated by the binned histogram of the energy gap fluctuations $\Delta E_n(t)$ obtained from combined MD/QC calculations (Mercer et al., 1999; Damjanovic et al., 2002a; Janosi et al., 2006). In general, Eqs. (9) and (10) overestimate the broadening of the lineshape function. Indeed, the Fourier transform of the exact spectral representation of the correlation function

$$\left\langle e^{-iH_0 t} e^{iH' t}\right\rangle = \sum_{\lambda_0,\lambda_n} \rho_{\lambda_0} \,|\langle \lambda_0 | \lambda_n \rangle|^2\, e^{-i(E_{n,\lambda_n} - E_{0,\lambda_0})t} \quad (11)$$

Where $\rho_{\lambda_0} = Z_0^{-1}\exp(-\beta E_{0,\lambda_0})$ is the statistical matrix of the electronic ground state, yields

$$A(\omega) = 2\pi \sum_{\lambda_0,\lambda_n} \rho_{\lambda_0} \,|\langle \lambda_0 | \lambda_n \rangle|^2\, \delta(\omega - \Delta E_{n,\lambda_n,\lambda_0}) \quad (12)$$

which can be regarded as a Franck-Condon weighted and thermally averaged density of state (May and Kühn, 2000). By setting the Franck-Condon factors $\langle \lambda_0 | \lambda_n \rangle$ equal to unity in (12) one obtains Eqs. (9) and (10). Since it is not possible to determine all these factors, it is often convenient to use Eqs. (9) and (10) as a rough estimate of $A_n(\omega)$ for calculating the OD spectrum.

A systematic way of calculating the correlation function in (7) is the cumulant expansion method (Mahan, 1990; Mukamel, 1995b; May and Kühn, 2000). Within the second order cumulant approximation that is often used in optical spectra calculations (Mukamel, 1995a) one has

$$\left\langle e^{iH_0 t} e^{-iH_n t}\right\rangle = \left\langle \mathrm{T}\exp\left[-i\int_0^t d\tau \Delta H_n(\tau)\right]\right\rangle$$
$$\approx \exp\left[-i\langle \Delta H_n \rangle t - \int_0^t d\tau (t-\tau) \mathcal{C}_n(\tau)\right] \quad (13)$$

where T is the time ordering operator, $\Delta H_n(t) = e^{iH_0 t}\Delta H_n e^{-iH_0 t}$, $\mathcal{C}_n(t) = \langle \delta H_n(t)\delta H_n(0)\rangle$, and $\delta H_n(t) = \Delta H_n(t) - \langle \Delta H_n \rangle$. The quantum statistical averages in Eq. (13) can be approximated by the corresponding classical ones, involving the energy gap time series $\Delta E_n(t)$, as follows

$$\langle \Delta H_n \rangle \approx \langle \Delta E_n(t) \rangle \equiv \omega_n \quad (14)$$

$$\mathrm{Re}[\mathcal{C}_n(t)] \approx C_n(t) \equiv \langle \delta E_n(t)\delta E_n(0)\rangle \quad (15)$$

where $\delta E_n(t) = \Delta E_n(t) - \langle \Delta E_n \rangle$. It should be noted that although the approximation of the real part of a quantum time correlation function by the corresponding classical correlation function as in Eq. (15) is generally accepted (Schulten and Tesch, 1991; Makri, 1999; Mercer et al., 1999;), other approximation schemes have also be used and tested (Egorov et al., 1999).

Next, by invoking the fluctuation dissipation theorem (Mukamel, 1995a), $\tilde{C}_n(-\omega) = \exp(-\beta\omega)\tilde{C}_n(\omega)$, where $\tilde{C}_n(\omega) = \int_{-\infty}^{\infty} dt C_n(t)\exp(i\omega t)$ is the Fourier transform of $C_n(t)$, the quantum correlation function in terms of the real spectral density

$$J_n(\omega) = \frac{1}{2}\left[\tilde{C}_n(\omega) - \tilde{C}_n(-\omega)\right] = \frac{1}{2}\left(1 - e^{-\beta\omega}\right)\tilde{C}_n(\omega) \tag{16}$$

can be written as

$$\begin{aligned} C_n(t) &= C_n'(t) - iC_n''(t) \\ &= \int_0^{\infty} \frac{d\omega}{\pi} J_n(\omega)\left[\coth(\beta\omega/2)\cos\omega t - i\sin\omega t\right] \end{aligned} \tag{17}$$

By identifying the real part of Eq. (17) with Eq. (15) one can determine both the spectral density and the imaginary part of the quantum correlation function, i.e.,

$$J_n(\omega) = 2\tanh(\beta\omega/2)\int_0^{\infty} dt\, C_n(t)\cos\omega t \tag{18}$$

and

$$C_n''(t) = \int_0^{\infty} \frac{d\omega}{\pi} J_n(\omega)\sin\omega t \tag{19}$$

Thus, the lineshape function within the second cumulant approximation is

$$\begin{aligned} A_n(\omega) \equiv \bar{A}_n(\omega - \omega_n) = \int_0^{\infty} dt\, e^{-\phi_n(t)} \\ \cos[(\omega - \omega_n)t + \varphi_n(t)] \end{aligned} \tag{20}$$

where the broadening and frequency shift functions are given by

$$\phi_n(t) = \int_0^t d\tau (t - \tau) C_n(\tau) \tag{21}$$

and

$$\varphi_n(t) = \int_0^{\infty} \frac{d\omega}{\pi} J_n(\omega) \frac{\omega t - \sin\omega t}{\omega^2} \tag{22}$$

Formally Eqs. (20) and (8) for the lineshape function and the OD spectrum remain valid in the case of $N$ excitonically coupled pigment molecules as well provided that the site index $n$ is replaced with the excitonic index $J$. In principle the energies $E_{J\lambda_J}$ and TDMs $\mathbf{d}_J$ of the excitonic states $|J;\lambda_J\rangle$, $J = 1,\ldots,N$, ought to be determined from QC calculations by considering all $N$ pigments as a single quantum system. Since such computational approach is still prohibitively expensive a practical alternative is to use an effective Hamiltonian for determining the time series $\Delta E_J(t) = E_J(t) - E_0(t)$ and $\mathbf{d}_J(t)$ from $\Delta E_n(t)$ and $\mathbf{d}_n(t)$ of the individual pigments. Assuming that the latter interact via a point dipole-dipole interaction

$$V_{nm} = \frac{1}{4\pi\varepsilon_0\varepsilon_r}\left[\frac{\mathbf{d}_n\mathbf{d}_m}{r_{nm}^3} - 3\frac{(\mathbf{d}_n\cdot\mathbf{r}_{nm})(\mathbf{d}_m\cdot\mathbf{r}_{nm})}{r_{nm}^5}\right] \tag{23}$$

where $\varepsilon_r$ is the relative dielectric permitivity of the medium, $\mathbf{r}_n$ is the position vector of pigment $n$, and $\mathbf{r}_{nm} = \mathbf{r}_m - \mathbf{r}_n$, the eigenvalue equation one needs to solve at every MD timestep is

$$\sum_m [(\Delta E_n \delta_{nm} + V_{nm}) - \Delta E_J \delta_{nm}] c_m^{(J)} = 0 \tag{24}$$

In term of the coefficients $c_n^{(J)} = \langle J|n\rangle$ the excitonic TDMs are

$$\mathbf{d}_J = \sum_n \langle J \mid n\rangle \mathbf{d}_n \tag{25}$$

Next, by rewriting the Hamiltonian (2) in diagonal form (i.e., in terms of noninteracting excitons) $H' = \sum_J H_J = \sum_{J\lambda_J} |J;\lambda_J\rangle E_{J\lambda_J} \langle J;\lambda_J|$, it follows

$$\left\langle \hat{\mu}_{m,j}^{\dagger}(0)\hat{\mu}_{n,i}(t)\right\rangle = \sum_J \langle J \mid n\rangle d_{n,i} d_{m,j} \langle m \mid J\rangle \left\langle e^{iH_0 t} e^{-iH_J t}\right\rangle \tag{26}$$

and

$$\sum_{n,m}\left\langle \hat{\mu}_{m,j}^{\dagger}(0)\hat{\mu}_{n,i}(t)\right\rangle = \sum_J d_{J,i} d_{J,j} \left\langle e^{iH_0 t} e^{-iH_J t}\right\rangle \tag{27}$$

Inserting Eq. (27) into Eq. (5) one obtains the desired OD spectrum of the excitonic system

$$I(\omega) \propto \omega \sum_J d_J^2 A_J(\omega) \tag{28}$$

where

$$A_J(\omega) = \text{Re}\int_0^\infty dt e^{i\omega t}\left\langle e^{iH_0 t} e^{-iH_J t}\right\rangle \tag{29}$$

## B. Polaron Model

An alternative approach for calculating the OD spectrum of a PPC at finite temperature is based on the so-called polaron model, according to which the excitonically coupled pigment molecules (excitons) interact with the vibronic modes (phonons) of the system (Meier et al., 1997; Damjanovic et al. 2002a). The corresponding model Hamiltonian (Holstein, 1959), in the second quantized, site representation reads

$$H = H_{ex} + H_{ph} + H_{int} \tag{30}$$

$$H_{ex} = \sum_n \varepsilon_n B_n^\dagger B_n + \sum_{n\neq m} V_{nm} B_n^\dagger B_m = \sum_J E_J B_J^\dagger B_J \tag{31}$$

$$H_{ph} = \sum_{n,\alpha} \omega_\alpha b_{n\alpha}^\dagger b_{n\alpha} \tag{32}$$

$$H_{int} = \sum_{n\alpha} g_\alpha \omega_\alpha B_n^\dagger B_n (b_{n\alpha}^\dagger + b_{n\alpha}) \tag{33}$$

In the exciton Hamiltonian $H_{ex}$ the operator $B_n^\dagger$ ($B_n$) creates (destroys) an electronic excitation $\varepsilon_n$ on the $n^{th}$ pigment (modeled as a two level system), while $V_{nm}$ describes the excitonic coupling between the pigments $n$ and $m$. The eigenstates of $H_{ex}$ are excitons (delocalized electronic excitations) characterized by the creation (annihilation) operator $B_J^\dagger$ ($B_J$) and energy $E_J$. The index $J = 1,\ldots,N$ labels the excitonic states upon increasing energy. Similarly, the operator $b_{n\alpha}^\dagger$ ($b_{n\alpha}$) in the phonon Hamiltonian $H_{ph}$ creates (annihilates) a vibronic mode $\omega_\alpha$ at site $n$. Finally, $H_{int}$ describes the interaction between excitons and phonons. For simplicity one assumes that both the phonon spectrum $\omega_\alpha$ and the exciton-phonon coupling constant $g_\alpha$ are site independent. The stationary states corresponding to (31) are excitons 'dressed' with a phonon cloud and are referred to as polarons. The parameters in the polaron model, i.e., $\varepsilon_n$, $V_{nm}$, $g_\alpha$ and $\omega_\alpha$, ought to be determined either empirically or from combined MD/QC simulations.

To illustrate the calculation of the OD spectrum within the framework of the polaron model, for simplicity, one assumes that all pigments are identical (i.e., $\varepsilon_n$ and the TDMs $d_n$ are the same for each pigment). Formally, by replacing the site index $n$ with $J$, the OD spectrum is given by Eqs. (7), (8) and (13), with $\Delta H_J = H_{ex} + H_{int}$. Using Eq. (31) one finds

$$\langle \Delta H_J \rangle = E_J \equiv \omega_J \tag{34}$$

and

$$\begin{aligned} C_J(t) &= \left\langle \delta H_J(t)\delta H_J(0)\right\rangle = \left\langle H_{int}(t) H_{int}(0)\right\rangle \\ &= \sum_{n,\alpha} g_\alpha^2 \omega_\alpha^2 \left\langle \rho_n(t)\rho_n(0)\right\rangle \left\langle A_{n\alpha}(t) A_{n\alpha}(0)\right\rangle \end{aligned} \tag{35}$$

where, by definition, the electronic excitation number operator $\rho_n(t) = e^{iH_{ex}t} B_n^\dagger B_n e^{-iH_{ex}t}$, and the phonon field operator $A_{n\alpha}(t) = b_{n\alpha}^\dagger e^{i\omega_\alpha t} + b_{n\alpha} e^{-i\omega_\alpha t}$. In terms of the Bose-Einstein distribution function, $N(\omega) = 1/(e^{\beta\omega} - 1)$, the (site independent) phonon field correlation function in Eq. (35) can be expressed as (Mukamel, 1995a; May and Kühn, 2000; Damjanovic et al., 2002a)

$$\begin{aligned} D_{\omega_\alpha}(t) &= \left\langle A_{n\alpha}(t) A_{n\alpha}(0)\right\rangle \\ &= \left[N(\omega_\alpha)+1\right] e^{-i\omega_\alpha t} + N(\omega_\alpha) e^{i\omega_\alpha t} \\ &= \coth(\beta\omega_\alpha/2)\cos(\omega_\alpha t) - i\sin(\omega_\alpha t) \end{aligned} \tag{36}$$

The excitonic factor in the correlation function (35) can also be readily calculated with the result

$$\begin{aligned} F_J(t) &= \sum_n \left\langle \rho_n(t)\rho_n(0)\right\rangle \\ &= \sum_{J'} e^{i(E_J - E_{J'})t} \sum_n |\langle n|J\rangle|^2 |\langle n|J'\rangle|^2 \end{aligned} \tag{37}$$

In general, $|F_J(t)| < 1$ and this factor is responsible for the so-called *exchange narrowing* of the lineshape function. In the absence of excitonic coupling between pigments, i.e., $V_{nm} \approx 0$, the index $J$ in (37) identifies with a particular site index $m = 1,\ldots,N$, resulting in $F_{J=M}(t) = 1$, i.e., as expected, there is no exchange narrowing. Next, by introducing the *phonon spectral density*

$$J(\omega) = \pi \sum_{\alpha} g_{\alpha}^2 \omega_{\alpha}^2 \delta(\omega - \omega_{\alpha}) = \pi \omega^2 \sum_{\alpha} g_{\alpha}^2 \delta(\omega - \omega_{\alpha}) \tag{38}$$

the sought quantum correlation function (35) assumes the general form

$$C_J(t) = F_J(t) C_n(t) \tag{39}$$

where

$$C_n(t) = \int_0^{\infty} \frac{d\omega}{\pi} J(\omega) \left[ \coth(\beta\omega/2) \cos(\omega t) - i \sin(\omega t) \right] \tag{40}$$

As mentioned before, in the absence of excitonic coupling $F_J(t) = 1$, and Eqs. (39)-(40) formally coincides with Eq. (17), thus suggesting that in this case the polaron model approach and the correlation function method described in Sec. A for calculating OD spectra are equivalent. However, in principle, the latter approach is more general than the former because it treats the environment beyond the harmonic heat bath approximation, albeit within the cumulant approximation. The determination of $J(\omega)$ from Eq. (38) requires the seemingly unattainable knowledge of the energies $\omega_\alpha$ of *all* phonons, together with their corresponding coupling constants $g_\alpha$. This problem is similar to the spin-boson model description of the coupling between protein motion and electron transfer processes (Xu and Schulten, 1994) that can be solved by evaluating the spectral function from the energy gap fluctuations $\delta E_n(t)$ as described in the previous sections. On the other hand, Eq. (38) provides a simple physical interpretation of the spectral function. Indeed, if one regards the environment in a PPC as an equivalent harmonic-phonon heat bath then one can interpret the magnitude of the spectral functions as a measure of the coupling strength to phonons of that particular frequency. In general, the complex structure of the spectral function (determined from the combined MD/QC calculations according to Eq. (18) (see also Fig. 5b) indicates that all inter and intra molecular vibronic modes within a wide range of frequencies will contribute to the lineshape function. Hence, attempts to use simplified model spectral functions appear to be unrealistic even if these may lead to absorption spectra that match the experimental results.

### C. Circular Dichroism

By definition, the CD spectrum $I_{CD}(\omega)$ is the difference between $I_L(\omega)$ and $I_R(\omega)$, the OD spectra for left and right circularly polarized light, respectively. Unlike in the case of the OD spectrum, the calculation of $I_{CD}(\omega)$ even within the leading order approximation requires taking into account the spatial variation of the light field across the PPC as well as the excitonic coupling between the pigment molecules regardless how small this may be (Somsen et al., 1996). The sensitivity of the CD spectrum to geometrical and local details of the PPC makes it a quantity difficult to predict by theoretical modeling. The CD spectrum is given by (van Amerongen et al., 2000)

$$I_{CD}(\omega) = \frac{1}{4}[I_L(\omega) - I_R(\omega)] \propto \omega \,\mathrm{Re} \int_0^{\infty} dt\, e^{i\omega t} \times \sum_{n,m} \frac{\pi}{\lambda} \varepsilon_{ijk} (\mathbf{r}_n)_k \left\langle \hat{\mu}_{m,i}^{\dagger}(0) \hat{\mu}_{n,i}(t) \right\rangle \tag{41}$$

where $\lambda$ is the wavelength of the incident light and $\varepsilon_{ijk}$ is the unit antisymmetric tensor of rank 3. Inserting Eq. (26) into (41) and making use of Eq. (29), one obtains

$$I_{CD}(\omega) \propto \omega \sum_J R_J A_J(\omega) \tag{42}$$

where

$$R_J = \frac{\pi}{\lambda} \sum_{n,m} \langle J | n \rangle [\mathbf{r}_n \cdot (\mathbf{d}_n \times \mathbf{d}_m)] \langle m | J \rangle \tag{43}$$

is the so-called *rotational strength* of the excitonic state $J$. It should be noted that in the absence of the excitonic coupling all $R_J = 0$ (because for a given $J$ only one coefficient $\langle J|n \rangle$ is nonzero) and the CD spectrum vanishes. The rotational strength plays the same role for the CD spectrum as the TDM strength for the OD spectrum. Specifically, $R_J$ gives the coupling between the TDM of the excitonic state $J$ and the orbital magnetic moment of the other excitons. The coupling to the local magnetic moment is assumed to be small (Cotton effect) and usually is discarded (Somsen et al., 1996; Amerongen et al., 2000).

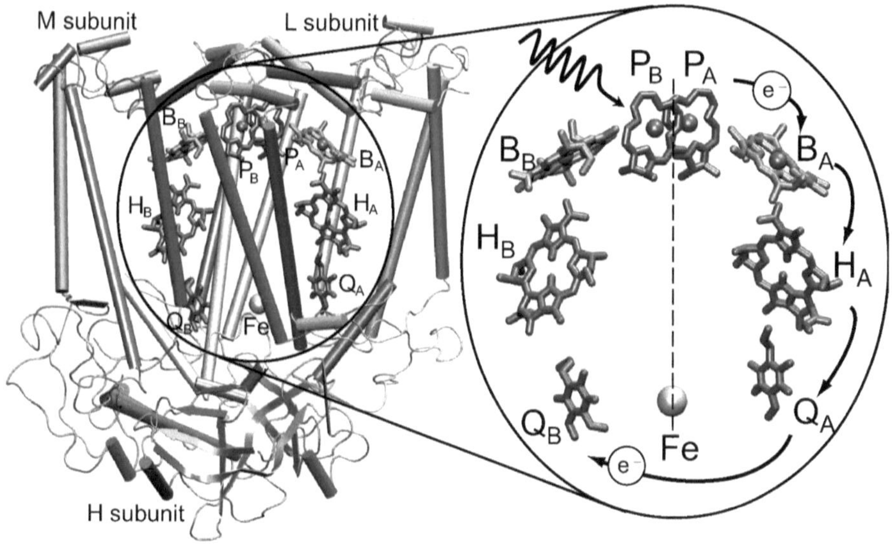


*Fig. 2.* Left: Structure of the RC of *Rb. sphaeroides* (entry code 1PCR in Protein Data Bank). For clarity only the backbone of the protein subunits L, M and H are shown. The protein helices are represented by cylinders. The cofactors are labeled, and for clarity their phytyl tails are not shown. Right: Spatial distribution of the cofactors in the RC. The path of the electrons through the RC is indicated by the arrows. Graphics rendered with the program VMD (Humphrey et al., 1996). See also Fig. 2, Color Plate 7.

## III. Electron Transfer at Finite Temperature

Another important class of quantum processes in PPCs involves electrons switching between two states. Two examples are electron transfer reactions in PPCs when an electron moves from an orbital on the donor moiety *D* to an orbital on the acceptor moiety *A* and bond formation or bond breaking in an enzyme when electrons shift from a non-bonding state to a bonding state or vice versa. Here only the electron transfer processes will be considered. An ideal PPC system for studying the electron transfer process is the photosynthetic reaction center (RC) in photosynthetic bacteria. For quite some time high resolution crystal structures are available for RCs from several species of photosynthetic bacteria, such as *Rhodobacter (Rb.) sphaeroides* (Deisenhofer et al., 1985) shown in Fig. 2.

In general, RCs are formed by three protein subunits, denoted L, M and H. The H subunit is located on the cytoplasmic side of the cell membrane and is anchored to it by a single transmembrane helix. The homologous L and M subunits have five transmembrane helices and display a quasi-twofold rotational symmetry. Several photoactive pigment molecules (termed as cofactors or prosthetic groups) are bound by the L and M subunits in a symmetric fashion and being labeled A and B, respectively. These cofactors, which play a key role in the electron transfer processes within the RC, consist of (Fig. 2): two BChls that form a strongly interacting dimer called the special pair ($P_A$, $P_B$) two monomeric BChls ($B_A$, $B_B$) located near the special pair; two bacteriopheophytins ($H_A$, $H_B$); a pair of ubiquinone molecules ($Q_A$, $Q_B$); and one non-heme iron atom (Fe) (Deisenhofer et al., 1985). The main steps involved in photosynthetic charge separation in the RC are as follows: (i) the special pair (the primary electron donor) is excited through the absorption of an incident photon, and then relaxes by transferring an electron through $B_A$ to $H_A$ in ~3 ps; (ii) the electron is transferred from $H_A^-$ to $Q_A$ in ~200 ps; (iii) $Q_A^-$ transfers an electron to $Q_B$ in ~ 200 μs by converting the latter into a semiquinone radical; (iv) during this time the positively charged special pair is neutralized by extracting an electron from a monoheme *c*-type cytochrome (Cyt $c_2$) on the periplasmic side of the membrane (this soluble cytochrome shuttles the electron between the Cyt $bc_1$ complex and the RC); (v) when another incoming photon is absorbed by the special pair, a second electron flows to $Q_B^-$ via the same pathway; (vi) this time $Q_B^{2-}$ is converted into *quinol* (hydroquinone $Q_BH_2$) by the simultaneous uptake of two protons from the cytoplasmic side of the complex; (vii) the quinol diffuses out of the RC and into the cell membrane that is rich in mobile quinones; (viii) the cycle of electron transfer is completed by the oxidation of the quinol by the Cyt $bc_1$ complex which results in the transfer of protons across the membrane and the re-reduction of Cyt $c_2$. The transmembrane proton gradient drives the synthesis of ATP, the universal fuel molecule, by ATP synthase.

The electron transfer process can also be formally regarded as a chemical reaction $AD \rightarrow A^+D^-$, where $AD$ and $A^+D^-$ are the reactant and product states, respectively. The energy expectation values of the two states, $E_1(t)$ and $E_2(t)$, vary in time due to motions along a reaction coordinate, but also due to thermal fluctuations of the remaining degrees of freedom of the PPC. Often the interaction energies which couple the two electronic states involved in the reaction are small compared to the temporal variations of $E_1(t)$ and $E_2(t)$. In this rather typical case the actual reaction process is confined to moments when the two electronic states become energetically degenerate ($E_1(t) = E_2(t)$). Such curve crossing processes in PPCs are strongly dependent on the thermal motion of the entire system including the protein matrix, the solvent and the lipid membrane in case of membrane bound PPCs. In a quantum mechanical description, one defines the Hamiltonians $H_1$ and $H_2$ that describe the collective motion of the system in the reactant (initial) and product (final) electronic states of the PPC. The weak coupling between the two can be described by a tunneling matrix element $V$.

Just like in the case of optical absorption spectra, the effect of dynamic disorder on the electron transfer processes in PPCs can also be determined by employing the combined MD/QC method described in the previous section. Once the $A$ and $D$ moieties have been identified the time series of the electronic ground state energies $E_{1,2}(t_i)$, $i = 0,1,\ldots$, of the two redox states can be determined by QC calculation for each snapshot $t_i$ along the MD trajectory.

### *A. Cumulant Approximation of the Electron Transfer Rate*

Assuming that the tunneling matrix element $V$ does not change significantly due to the thermal motion of the protein matrix (Condon approximation), within the lowest order of perturbation theory in $V$ the electron transfer rate $k_{ET}$ in a PPC can be expressed as (May and Kühn, 2000)

$$k_{ET} = |V|^2 \int_{-\infty}^{\infty} dt \left\langle e^{iH_1 t} e^{-iH_2 t} \right\rangle \quad (44)$$

Similarly to Eq. (13), by employing the cumulant approximation

$$k_{ET} \approx |V|^2 \int_{-\infty}^{\infty} dt \exp\left[-i\langle \Delta H \rangle t - \int_0^t d\tau (t-\tau) \mathcal{C}(\tau)\right] \quad (45)$$

where $\Delta H = H_2 - H_1$, $\Delta H(t) = e^{iH_1 t} \Delta H e^{-iH_1 t}$, $\mathcal{C}(t) = \langle \delta H(t) \delta H(0) \rangle$, and $\delta H(t) = \Delta H(t) - \Delta H$. By following the same methodology as in the derivation of the OD spectrum of a PPC (cf. Eqs. (14), (15) and (20)) the electron transfer rate can be brought to the form

$$k_{ET} = 2|V|^2 \int_0^{\infty} dt\, e^{-\phi(t)} \cos[\varepsilon t - \varphi(t)] \quad (46)$$

where $\varepsilon = \langle \Delta H \rangle \approx \langle \Delta E(t) \rangle$ is the mean energy gap,

$$\phi(t) = \int_0^t dt' (t - t') C(t) \quad (47)$$

with $C(t) = \mathrm{Re}\, \mathcal{C}(t) \approx \langle \delta E(t) \delta E(0) \rangle$, and

$$\varphi(t) = -\mathrm{Im} \int_0^t dt' (t-t') \mathcal{C}(t') = \int_0^{\infty} \frac{d\omega}{\pi} J(\omega) \frac{\omega t - \sin \omega t}{\omega^2} \quad (48)$$

where the spectral function has the usual form

$$J(\omega) = 2 \tanh(\beta\omega/2) \int_0^{\infty} dt\, C(t) \cos \omega t \quad (49)$$

Thus, the calculation of $k_{ET}$ and $J(\omega)$ requires in fact only the knowledge of the energy gap time series $\Delta E(t)$ and not those of the individual energies $E_{1,2}(t)$. This simple observation is rather important because most QC methods do not permit the accurate determination of individual energy levels but they can provide energy differences with fairly high precision.

A good estimate of the electron transfer rate in PPCs can also be obtained if the energy gap time series $\Delta E(t)$ are determined purely classically instead of combined MD/QC calculations. Knowing the atomic partial charges corresponding to AD and to $A^-D^+$ one can evaluate $E_1[\mathbf{R}(t)]$ and $E_2[\mathbf{R}(t)]$ along the $\mathbf{R}(t)$ MD trajectory as the Coulomb energies of the acceptor and donor moieties with the protein matrix, to which one adds the redox energies of the states AD and $A^-D^+$.

In the high temperature limit, one can easily show

that the expression of the electron transfer rate derived in this section by applying the cumulant approximation yields in leading approximation the result corresponding to the classical Marcus theory. To show this, one assumes that the correlation function of the energy gap fluctuations has a simple exponential form $C_M(t) = \Delta^2\exp(-t/\tau)$, where $\Delta^2 = \langle \delta E^2 \rangle = \langle \Delta E^2 \rangle - \langle \Delta E \rangle^2$ is the variance of the energy gap fluctuations and $\tau$ is the corresponding relaxation time. Thus, the integrals in (47) and (49) can be performed exactly with the results

$$\phi_M(t) = \Delta^2\tau\left[t - \tau\left(1 - e^{-t/\tau}\right)\right] \approx \frac{\Delta^2}{2}t \quad \text{for} \quad t \ll \tau \tag{50}$$

and

$$J_M(\omega) = 2\tanh(\beta\omega/2)\frac{\beta\Delta^2}{1+(\omega\tau)^2} \approx \frac{\beta\Delta^2\omega\tau}{1+(\omega\tau)^2} \tag{51}$$

where the high temperature limit $\beta\omega \ll 1$ has been assumed. Note that the maximum of the spectral function $J_M(\omega)$ corresponds to the energy gap

$$\varepsilon_M = \frac{\beta\Delta^2}{2} \tag{52}$$

Within the same range of approximations the phase factor (48) becomes

$$\varphi_M(t) \approx t\int_0^\infty \frac{d\omega}{\pi}\frac{J_M(\omega)}{\omega} = \varepsilon_M t \tag{53}$$

Inserting Eqs (53), (50) into (46) and performing the Gaussian integral, one obtains the well known Marcus formula (Marcus, 1956a,b; May and Kühn, 2000)

$$k_M = 2\sqrt{2\pi}\frac{|V|^2}{\Delta}\exp\left[-\frac{\left(\varepsilon - \varepsilon_M\right)^2}{2\Delta^2}\right] \tag{54}$$

Since in general the spectral function (49) has a complex structure (Damjanovic et al., 2002a,b; Janosi et al., 2006) the simple exponential approximation of the correlation function $C(t)$ may not be justified so that differences between the electron transfer rates calculated with the simple Marcus formula (54) and with the cumulant approximation (46) may be expected even at high temperatures.

### B. Spin-Boson Model of Coupling to a Heat Bath

Electron transfer between donor and acceptor moieties in a PPC can conveniently be described in terms of the *spin-boson model.* Similarly to the polaron model in the case of optical absorption in PPCs, in the spin-boson model the electronic degrees of freedom are treated as a two state system (one for the reactants and one for the products) and the nuclear degrees of freedom of the protein matrix are approximated by a harmonic heat bath. The name of spin-boson model stems from the fact that the two state electron transfer system is equivalent to a spin-½ system while the atomic motion is described by a set of independent bosons. Fore example, this model has been successfully applied to investigate the primary electron separation process in the photosynthetic reaction center of *Rhodopseudomonas* (*Rps.*) *viridis* by focusing on how the thermal oscillations of the protein atoms couple to the various transfer steps of an electron moving along the prosthetic groups (Xu and Schulten, 1994). The key new aspects of the spin-boson model description of electron transfer in PPCs is two-fold: first, all model parameters are determined from molecular dynamics simulations, thus requiring only the knowledge of the atomic resolution crystal structure of the PPC; second, the spin-boson model accounts for all vibrational modes of the PPC by means of the phonon spectral function that, similarly to the polaron model, can be determined from the time autocorrelation function of the energy gap corresponding to the product and reactant states. The spin-boson model may not only yield qualitatively different predictions than models involving a small number of vibrational modes coupled to the electron transfer, but it certainly makes the role of the medium surrounding an electron transfer reaction appear in a new light: essentially all motions of the environment are coupled significantly to the reaction. The reason is surprisingly simple and applies clearly to the case of a PPC: the coupling between electron transfer and medium is due to the Coulomb interaction. This interaction, however, is long range and encompasses a very large volume. The coupling results then from small additive contributions of many motions rather than from a few dominant modes.

A detailed review of the theory of the spin-boson model can be found in (Leggett et al., 1985). In the case of a PPC the electron transfer reaction is described in terms of the two–state Hamiltonian, written both in first and second quantized forms

$$\hat{H}_{el} = V\sigma_x - \frac{1}{2}\varepsilon\sigma_z = \sum_{n=1}^{2}\varepsilon_n B_n^\dagger B_n + \sum_{n\neq m=1}^{2} V B_n^\dagger B_m \tag{55}$$

Where $\sigma_x$, $\sigma_z$ are the usual $2 \times 2$ Pauli matrices, $\varepsilon = \varepsilon_2 - \varepsilon_1$ is the difference of product state $\varepsilon_2$ and reactant state $\varepsilon_1$ energies (energy gap), $V$ accounts for the coupling between reactant and product states (the coupling originating from tunneling of the electron between electron donor and electron acceptor moieties), and $B_{1,2}$ ($B_{1,2}^\dagger$) are the annihilation (creation) fermionic operators of the two redox states. The medium thermal motion is described through an ensemble of independent harmonic oscillators (phonons) with the Hamiltonian

$$H_{ph} = \sum_\alpha \sum_{n=1}^{2} \omega_\alpha b_{n\alpha}^\dagger b_{n\alpha} \tag{56}$$

Here $b_{n\alpha}$ ($b_{n\alpha}^\dagger$) creates (destroys) a phonon (vibronic) mode with frequency $\omega_\alpha$ in the $n$-th redox state. The coupling between the vibrational degrees of freedom and the two–state system is linear

$$H_{int} = \sum_\alpha \sum_{n=1}^{2} g_\alpha \omega_\alpha B_n^\dagger B_n (b_{n\alpha}^\dagger + b_{n\alpha}) \tag{57}$$

Where $g_\alpha$ describes the strength of the coupling of the electron transfer to the $\alpha$-th mode. The spin-boson Hamiltonian is the sum of all three contributions, i.e.,

$$H_{sb} = H_{el} + H_{ph} + H_{int} \tag{58}$$

One may worry at this point that the many parameters which appear in the spin-boson model are impossible to specify uniquely and therefore, the model is either arbitrary or of limited use. However, just like in the case of the polaron model (Damjanovic et al., 2002a; Janosi et al., 2006) described above, the value of the spin-boson model (Leggett et al., 1985) lies in the fact that the electron transfer rate can be determined uniquely in terms of the spectral function

$$J_{sb}(\omega) = \pi\omega^2 \sum_\alpha g_\alpha^2 \delta(\omega - \omega_\alpha) \tag{59}$$

the energy gap $\varepsilon$ and the coupling $V$. Note that Eq. (59) is formally identical with (38), and it can be determined from the real time autocorrelation function $C(t) \approx \mathrm{Re}[\mathcal{C}(t)]$ of the energy gap fluctuations, $\delta\varepsilon(t) = \varepsilon(t) - \langle\varepsilon\rangle$, by means of Eq. (18).

The energy gap time series $\varepsilon(t)$ and the coupling $V$ can be computed either from classical MD simulations or from combined MD/QC calculations. Once $C(t)$ and $J_{sb}(\omega)$ have been determined, the electron transfer rate $k_{sb}$ can be readily calculated by means of Eqs. (46)–(48).

## VI. Simulation of Optical Excitations

According to the results presented in Sec. II in order to calculate the OD and CD spectra of the B800 and B850 BChls in a single LH2 ring from *Rs. molischianum* first one needs to determine the time series of the $Q_y$ energy gap $\Delta E_n(\ell\Delta t)$ and TDM $\mathbf{d}_n(\ell\Delta t)$, $\ell = 0,1,\ldots,N_t$, for all individual BChls. This requires two steps: (1) use all atom MD simulations to follow the dynamics of the nuclear degrees of freedom by recording snapshots of the atomic coordinates at times $t_\ell = \ell\Delta t$, and (2) use QC calculations to compute $\Delta E_n$ and $\mathbf{d}_n$ for each snapshot (Damjanovic et al., 2002a; Janosi et al., 2006).

### A. Molecular Dynamics Simulations

The first MD simulation of the LH2 antenna complex from *Rs. molischianum* embedded in a fully solvated lipid bilayer mimicking its native environment was reported in (Damjanovic et al., 2002a). A perfect 8-fold LH2 ring was constructed starting from the crystal structure (pdb code 1LGH) of *Rs. molischianum* (Koepke et al., 1996) (see Fig. 1). After adding the missing hydrogens, the protein system was embedded in a fully solvated POPC lipid bilayer of hexagonal shape. A total of 16 $Cl^-$ counterions were properly added to ensure electroneutrality of the entire system of 87,055 atoms. In order to reduce the finite-size effects, the hexagonal unit cell (with side length ~60 Å, lipid bilayer thickness ~42 Å and two water layers of combined thickness ~35 Å) was replicated in space by using periodic boundary conditions. The CHARMM27 force field parameters for proteins (MacKerell Jr. et al., 1998) and lipids (Schlenkrich et al., 1996) were used. Water molecules were modeled as TIP3P (Jorgensen et al., 1983). The ground state ESP partial charges for geometry optimized BChls without phytyl tail were determined with the program JAGUAR (Ringnalda et al., 1996). The force field parameters for BChls were taken from (Foloppe et al., 1992, 1995) and for lycopenes were determined using the program QUANTA. After energy minimization, the system was subjected to a 2 ns long equilibration in the NpT ensemble (Feller et al., 1995) at normal temperature (T = 300 K) and pressure ($p$ = 1 atm), using periodic boundary conditions and treating the full long-range electrostatic interactions by the PME method (Darden et al., 1993). All MD simulations were preformed with the program NAMD2 (Phillips et al., 2005), with a performance of ~8.5 days/ns on 24 CPUs of an AMD 1800+ Beowulf cluster. During equilibration an integration time step of 2 fs was employed by using the SHAKE constraint on all hydrogen atoms (Miyamoto and Kollman, 1992). After the 2 ns equilibration a 1 ps production run with 1 fs integration step was carried out with atomic coordinates saved every other time step, resulting in $N_t$= 500 MD snapshots with $\Delta t$ = 2 fs time separation. These configuration snapshots were used as input for the QC calculations.

### B. Quantum Chemistry Calculations

The time series of the $Q_y$ transition energies $\Delta E_n$ and dipole moments $\mathbf{d}_n$ of individual BChls can be determined only approximately from the configuration snapshots obtained from MD simulations. The level of approximation used is determined by: (i) the actual definition of the optically active *quantum system*, i.e., the part of the system that is responsible for light absorption and needs to be treated quantum mechanically; (ii) the actual choice of the QC method used in the calculations; and (iii) the particular way in which the effect of the (classical) environment on the quantum system is taken into account in the QC calculations. Because the optical properties of BChls are determined by the cyclic conjugated π-electron system of the macrocycle the quantum system was restricted to a truncated structure of the BChl restricted to the porphyrin plane (Cory et al., 1998; Mercer et al., 1999; Janosi et al., 2006). Although in general the different truncation schemes yield excitation energy time series with shifted mean values, the corresponding energy fluctuations, which play the main role in calculating the optical absorption properties of PPC at room temperature in their native environment, are less sensitive to the actual size of the truncated pigment. On the other hand, however, the required computational effort can be reduced dramatically through such truncation of the quantum system.

The preferred method for calculating the $Q_y$ excitations of the truncated BChls is Zerner's semiempirical intermediate neglect of differential overlap method parametrized for spectroscopy (ZINDO/S) within the single-point configuration interaction singles (CIS) approximation (Ridley and Zerner, 1973; Zerner et al., 1980). Because it is much faster and more accurate than most of the computationally affordable *ab initio* QC methods (e.g., the Hartree-Fock (HF) CIS method with the minimal STO-3G* basis set), ZINDO/S CIS has been extensively used in the literature to compute low lying optically allowed excited states of pigment molecules (Linnanto et al., 1999; Ihalainen et al., 2001; Damjanovic et al., 2002a; Linnanto and Korppi-Tommola, 2004; Janosi et al., 2006). The ZINDO/S method is integral part of standard QC program packages such as HyperChem 7.5 (HyperChem(TM), Hypercube, Inc.) and GAUSSIAN 98 (Frisch et al., 1998).

The effect of the environment on the quantum system can be taken into account through the electric field created by the partial point charges of the environment atoms, including those BChl atoms that were removed during the truncation process. Thus, the dynamics of the nuclear degrees of freedom (described by MD simulation) have a two-fold effect on the fluctuations of the $Q_y$ state, namely they lead to: (1) conformational fluctuation of the (truncated) BChls, and (2) a fluctuating electric field created by the thermal motion of the corresponding atomic partial charges. The relative importance of these two effects on the time series $\Delta E_n(t)$ was estimated by performing the QC calculations both in the presence and in the absence of the point charges (Janosi et al., 2006). For each case, a total of 12,000 (500 snapshots × 24 BChls) ZINDO/S calculations were performed with a performance of ~2.3 min/CPU (~0.7 min/CPU) for each calculation with (without) point charges on a workstation with dual 3GHz Xeon EM64T CPU.

### C. Energy Gap Density of States

The 1 ps long time series of the $Q_y$ excitation energies $\Delta E_n(t_\ell)$ and TDMs $\mathbf{d}_n(t_\ell)$, ($t_\ell = \ell\Delta t$; $\ell = 0,\ldots,N_t$; $N_t = 499$; $\Delta t = 2$ fs) computed with the described combined MD/QC method for both B850 ($n = 1,\ldots,16$) and B800 ($n = 17,\ldots,24$) BChls in a LH2 ring from *Rs. molischianum* are sufficiently long for calculating the DOS of the $Q_y$ excitation energies and the corresponding OD and CD spectra (Janosi et al., 2006).

Figure 3 shows the $Q_y$ energy gap DOS, $N(\omega)$ (Eq. (10)), of the individual B800 (solid-lines) and B850 (dashed-lines) BChls calculated as normalized binned histograms of the time series $\Delta E_{B800} \equiv \Delta E_n(t_\ell)$ with $n = 17,\ldots,24$, and $\Delta E_{B850} \equiv \Delta E_n(t_\ell)$ with $n = 1,\ldots,16$, respectively. In the absence of the point charge distribution of the environment $N(\omega)$ for B800 and B850 (thin-lines) are almost identical, having peak position at 1.51 eV (817 nm) and 1.515 eV (818 nm), and full width at half maximum (FWHM) 51 meV and 59meV, respectively.

It should be noted that essentially the same mean energy gap of 1.5 eV was obtained in similar MD/QC calculations (Mercer et al., 1999) for BChl solvated in methanol also at room temperature. These results indicate that the thermal motion of the nuclei in individual BChls lead to $Q_y$ energy gap fluctuations that are insensitive to the actual nature of the *nonpolar* environment. Since in LH2 from *Rs. molischianum* the B800s (B850s) are surrounded by polar (nonpolar) residues, it is not surprising that once the point charges of the environment are taken into account in the QC calculations $N(\omega)$ changes dramatically only in the case of B800. Indeed, as shown in Fig. 3, in the presence of the point charges (thick-lines) the peak of $N_{B850}(\omega)$ is only slightly red shifted to 1.502 eV (825 nm) and essentially without any change in shape with FWHM of $\approx$ 53 meV. By contrast, as a result of the point charges the B800 DOS is not only blue shifted but it becomes asymmetric and almost twice as broad with FWHM $\approx$ 100 meV. Thus, in spite of a small blueshift to 1.528 eV (811 nm) of the peak of $N_{B800}(\omega)$ the mean value of the energy gap $\langle\Delta E_{B800}\rangle = 1.556$ eV (797nm) is increased considerably, matching rather well the experimental value of 800 nm.

The excitonic energies time series $\Delta E_J(t_\ell)$, $J = 1,\ldots,16$, of the B850 BChls were determined by solving for each MD snapshot, within the point-dipole approximation, the eigenvalue equation (24). In calculating the matrix elements (23) $\mathbf{r}_n$ was identified with the position vector of the Mg atom in the $n$-th

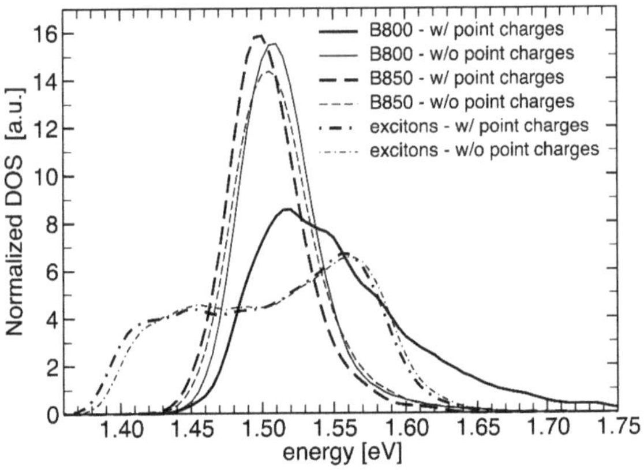


*Fig. 3.* Normalized DOS, $N(\omega)$, for individual B800 BChls (solid-line), B850 BChls (dashed-line), and B850 excitons (dashed-dotted-line) in LH2 from *Rs. molischianum* computed as binned histograms of the corresponding $Q_y$ excitation energy time series obtained from combined MD/QC simulations. Whether the charge fluctuations of the BChls' environment are included (thick-lines) or not (thin-lines) makes an important difference in $N(\omega)$ only for B800.

BChl. Consistent with the Condon approximation, the magnitude of the computed B850 TDM time series exhibited a standard deviation of less than 4% about the average value $\langle d_{B850}\rangle = 11.77$ D. The latter is by a factor of $k = 1.87$ larger than the experimentally accepted 6.3 D value of the $Q_y$ TDM of BChl (Visscher et al., 1989). By rescaling the TDMs from the ZINDO/S calculations to match their experimental value, and by setting $\varepsilon_r = 1.86$, one obtains for the mean value of the nearest neighbor dipolar coupling energies between B850s 27 meV $\approx$ 220 cm$^{-1}$ within a protomer and 24 meV $\approx$ 196 cm$^{-1}$ between adjacent heterodimers. As expected, the DOS of the excitonic energies (Fig. 3, dashed-dotted-line), computed as a binned histogram of $\Delta E_J(t_\ell)$, is not sensitive to whether the point charges of the environment are included or not in the B850 site energy calculations.

The mean excitonic TDMs, calculated from Eq. (25) and expressed in terms of $\langle d_{B850}\rangle$, are shown as an inset in Fig. 4. The error bars represent the standard deviation of the time series $d_J(t_\ell)$. As expected, most of the dipole strength is amassed into the lowest three excitonic states (Damjanovic et al., 2002a; Janosi et al., 2006).

According to Eqs. (8) and (9)–(10) a rough estimate of the OD spectrum of the B800 BChls and B850 excitons is given by the corresponding TDM strength weighted DOS

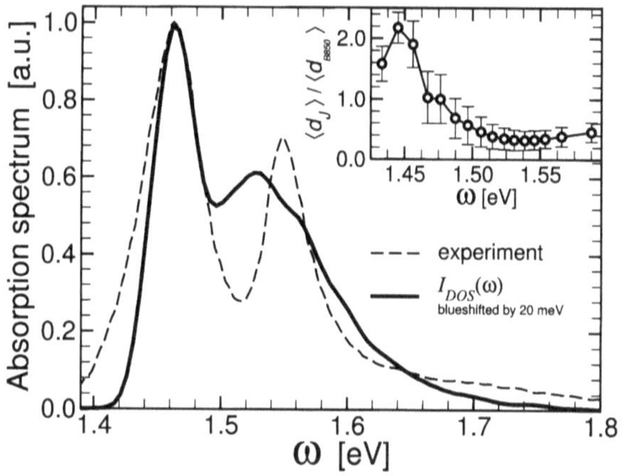


*Fig. 4.* Absorption spectrum $I_{DOS}(\omega)$ of LH2 for *Rs. molischianum* calculated as a combined DOS of B800 BChls and B850 excitons weighted by the corresponding dipole strengths (solid line). $I_{DOS}(\omega)$ was blueshifted by 20 meV in order to overlay its B850 peak with the corresponding one in the experimental OD spectrum (Zhang et al., 2000) (dashed line). Inset: Average transition dipole moments $<d_J>$ corresponding to the $J = 1,...,16$ B850 excitonic states. Both $<d_J>$ and the corresponding error bars are expressed relative to the mean dipole moment of individual B850s.

$$I_{DOS}(\omega) \propto \omega \left[ \sum_J d_J^2 \langle \delta(\omega - \Delta E_J) \rangle + \sum_{B800} d_{B800}^2 \langle \delta(\omega - \Delta E_{B800}) \rangle \right] \tag{60}$$

where the B800 index in the last term means summation over all B800 BChls. Figure 4 shows the calculated $I_{DOS}(\omega)$ blueshifted by 20 eV (solid-line) in order to match the B850 peak position with the one in the experimental OD spectrum (Zhang et al., 2000; Ihalainen et al., 2001) (dashed-line). While the B850 band and the relative heights of the two peaks in $I_{DOS}(\omega)$ match rather well the experimental data, the position and the broadening of the B800 peak do not. This result clearly shows that in general peak positions in optical spectra may be shifted from the corresponding peak positions in the excitation energy spectrum due to correlation effects between the ground and optically active excited states. The latter may also lead to different line broadening of the corresponding peaks. Therefore, methods for simulating optical spectra in which the position of the peaks are identified with the computed excitation energies (stick spectrum) are not entirely correct and using instead more sophisticated methods that include quantum correlation effects should be preferred.

### *D. Linear Absorption Spectrum*

According to Eq. (20) and (18) the lineshape functions of the individual B850 and B800 BChls is the (classical) autocorrelation function $C_n(t) = \langle \delta E_n(t) \delta E_n(0) \rangle$ of the energy gap fluctuation $\delta E_n(t) = \Delta E_n(t) - \langle \Delta E_n \rangle$ determined from combined MD/QC calculations. Since the LH2 ring from *Rs. molischianum* has an eight-fold symmetry, for best statistics one calculates a single time correlation function $C_{B800}(t)$ ($C_{B850}(t)$) by averaging over all B800 (B850) BChls according to the formula

$$C_\alpha(t_i) = \frac{1}{M} \sum_m \left[ \frac{1}{N_t - i} \sum_{k=1}^{N_t - i} \delta E_m(t_i + t_k) \delta E_m(t_k) \right] \tag{61}$$

where $M = 8$, $m = 17,\ldots,24$ for $\alpha = \text{B800}$, and $M = 16$, $m = 1,\ldots,16$ for $\alpha = \text{B850}$.

The normalized correlation functions $C_\alpha(t)/C_\alpha(0)$, $\alpha \in \{\text{B800,B850}\}$, are plotted in Fig. 5a. $C_\alpha(0) = \langle \delta E^2 \rangle$ represents the variance of the energy gap fluctuations with $C_{B800}(0) = 3.16 \times 10^{-3}$ eV$^2$ and $C_{B850}(0) = 8.68 \times 10^{-4}$ eV$^2$. Both correlation functions have a qualitatively similar behavior with the following features: (i) sharp decay to negative values in the first 9 fs, (ii) a ~18.5 fs period oscillatory component with uneven amplitudes, and (iii) vanishingly small magnitude after 400 fs. The spectral densities $J_\alpha(\omega)$ for B800 and B850, determined according to Eq. (18), are shown in Fig. 5b. The prominent peak about $\omega_P = 0.22$ eV is due to the fast initial decay of $C_\alpha(t)$ and it is most likely due to strong coupling of the pigment to an intramolecular C=O vibronic mode (Mercer et al., 1999; Damjanovic et al., 2002a). The complex structure of the spectral functions indicates that all inter and intra molecular vibronic modes with frequency below $\omega_P$ will contribute to the lineshape function. Hence, attempts to use simplified model spectral functions appear to be unrealistic even if these may lead to absorption spectra that match the experimental results.

The lineshape functions of individual B800 and B850, calculated from Eq. (20), are plotted in Fig. 6a. The origin of the frequency axis corresponds to the mean energy gaps $\omega_{B800}$ and $\omega_{B850}$, respectively. The highly polarized surrounding of the B800 BChls in *Rs. molischianum* renders $A_{B800}(\omega)$ twice as broad (FWHM $\approx$ 26 meV) as $A_{B850}(\omega)$ (FWHM $\approx$ 13 meV).

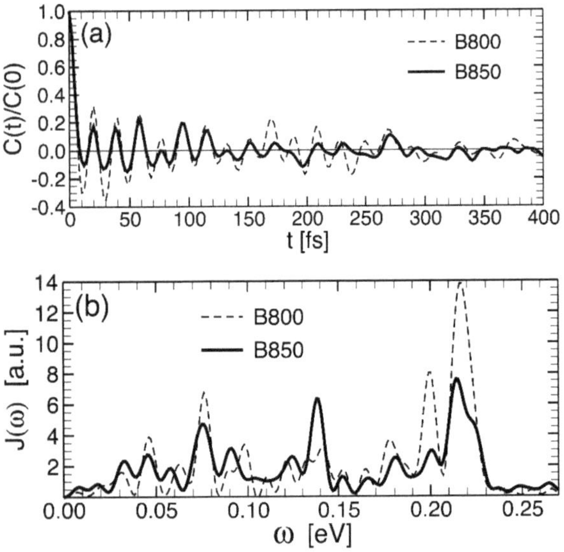


*Fig. 5.* (a) Normalized autocorrelation function $C(t)/C(0)$ of the energy gap fluctuations $\delta E(t) = E(t) - \langle E \rangle$ for individual B800 (dashed line) and B850 (solid line) BChls, calculated using Eq. (61). The mean square energy gap fluctuations are $C_{B800}(0) = 3.16 \times 10^{-3}\ eV^2$ and $C_{B850}(0) = 8.68 \times 10^{-4}\ eV^2$. (b) Spectral density function $J(w)$ for B800 (dashed-line) and B850 (solid-line) obtained according to Eq. (18). Reprinted with permission from the Journal of Chemical Physics. Copyright 2006, American Institute of Physics.

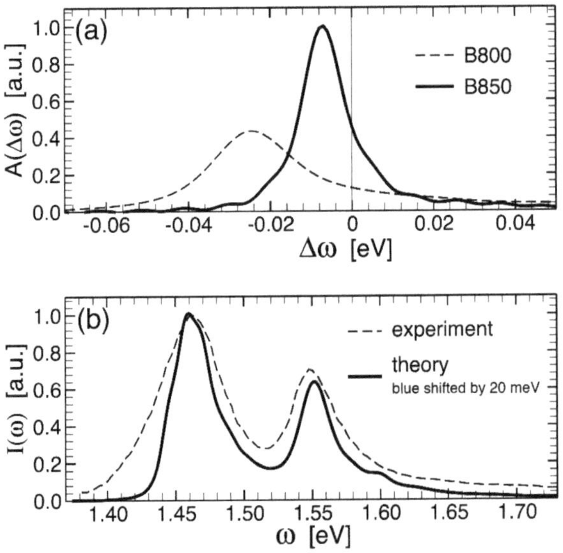


*Fig. 6.* (a) Lineshape functions $\bar{A}_{B800}(\Delta\omega)$ (dashed line) and $\bar{A}_{B850}(\Delta\omega)$ (solid line). (b) Computed (solid line) and experimental (dashed line) absorption spectra (in arbitrary units) of the BChl aggregate in *Rs. molischianum* LH2. The computed spectrum has been blue shifted by 20 meV for best match. Reprinted with permission from the Journal of Chemical Physics. Copyright 2006, American Institute of Physics.

Also, the redshift of the peak of the former ($\Delta\omega \approx 25$ meV) is more than three times larger than that of the latter ($\Delta\omega \approx 7$ meV).

Although the 1 ps long energy gap time series provide a proper estimate of the B800 and B850 lineshape functions, the same data is insufficient to determine with reasonable accuracy the individual excitonic lineshape functions $A_J(\omega)$. Thus, by neglecting the effect of exchange narrowing (Somsen et al., 1996; Amerongen et al., 2000), one can approximate $A_J(\omega) \approx A_{B850}(\omega)$, and the OD spectrum of the LH2 BChls becomes

$$I(\omega) \propto \omega \left[ \sum_J d_J^2 \bar{A}_{B850}(\omega - \omega_J) + 8 d_{B800}^2 \bar{A}_{B800}(\omega - \omega_{B800}) \right] \tag{62}$$

Where $\omega_J = \langle \Delta E_J \rangle$. As shown in Fig. 6b, $I(\omega)$ (subject to an overall blueshift of 20 meV) matches remarkably well the experimental OD spectrum, especially if we take into account that it was obtained from the sole knowledge of the high resolution crystal structure of LH2 from *Rs. molischianum* (Janosi et al., 2006). The reason why both B800 and B850 peaks of $I(\omega)$ are somewhat narrower than the experimental ones is most likely due to the fact that the effect of static disorder is ignored. Indeed, our calculations were based on a single LH2 ring, while the experimental data is averaged over a large number of such rings. While computationally expensive, in principle, the effect of static disorder could be taken into account by repeating the above calculations for different initial configurations of the LH2 ring and then averaging the corresponding OD spectra.

### *E. Circular Dichroism Spectrum*

Using the results from Sec. C, the calculation of the CD spectrum of the LH2 BChls proceeds along the following two steps (Janosi et al., 2006).

First, the rotational strength of both B850 excitons and B800 BChls is determined using Eq. (43). Here, just like in the case of the point-dipole interaction matrix elements (23), $\mathbf{r}_n$ represents the position vector of the Mg atom in the $n$-th BChl. The calculation of the rotational strength of the B800 BChls requires solving the corresponding excitonic Hamiltonian (24) regardless how small the dipole-dipole coupling is between these BChls. The calculation does not yield either noticeable corrections to the B800 excitation

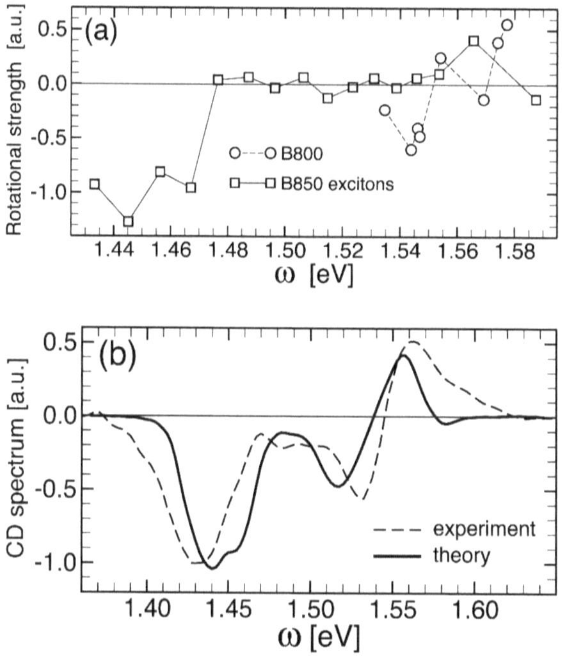


*Fig. 7.* (a) Mean rotational strength of the excitonically coupled B800 (circle) and B850 (box) BChls as a function of the corresponding excitonic energies. The purpose of the thin lines are to guide the eye. (b) Comparison between the computed (solid line) and experimental CD spectrum of the BChl aggregate in *Rs. molischianum* LH2. Reprinted with permission from the Journal of Chemical Physics. Copyright 2006, American Institute of Physics.

energies or admixture of the corresponding $Q_y$ states, however, it leads to sizable mean rotational strengths as shown in Fig. 7a (filled circles). Similarly to the TDM strengths, the largest (negative) mean rotational strengths are carried by the four lowest B850 excitonic states as shown in Fig. 7a (open squares). The second highest excitonic state also has a sizable rotational strength and is responsible for enhancing the positive peak of the B800 contribution to the CD spectrum.

Second, the CD spectrum is calculated from Eq. (42) where the summation index $J$ runs over all B850 and B800 excitonic states and $A_J(\omega) = \bar{A}_\alpha(\omega - \omega_J)$, with $\alpha \in \{B800, B850\}$. The obtained CD spectrum is shown in Fig. 7b (solid-line) and it appears to match fairly well the experimental spectrum (dashed line) (Ihalainen et al., 2001). It should be emphasized that apart from an overall scaling factor the CD spectrum was calculated from the same MD/QC data as the OD spectrum by following the procedure described above.

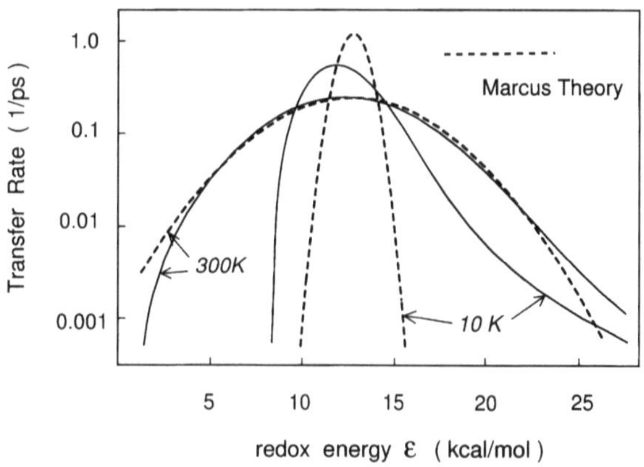


*Fig. 8.* Comparison of electron transfer rates $k(\varepsilon;T)$ shown as a function of ε evaluated in the framework of the spin-boson model (solid lines) and by Marcus theory (dashed lines) at temperatures 10 K and 300 K. The functions are centered approximately around $\varepsilon_M$.

## V. Calculation of Electron Transfer Rates

A detailed study of electron transfer rates $k_{ET}$ in the photosynthetic reaction center of *Rps. viridis* by employing the spin-boson model was reported in Refs. (Xu and Schulten, 1992, 1994). The model parameters Δ and τ were determined by means of all atom MD simulations. Due to large errors in calculating the mean redox energy gap ε the authors used this as a fitting parameter. The calculated $k_{ET}(\varepsilon;T)$ for temperatures T = 10 K and T = 300 K are shown in Fig. 8, and are compared with the corresponding results predicted by the Marcus theory (Marcus, 1956a,b). As expected, at high (physiological) temperature the rate evaluated from the Marcus theory in a wide range of ε values agrees well with the rate evaluated from the spin-boson model at T = 300 K. However the Marcus theory and the spin-boson model differ significantly at T = 10 K. At such low temperature the rate as a function of ε for the spin-boson model is asymmetrical. This result agrees with observations reported in (Gunn and Dawson, 1989) which show a distinct asymmetry with respect to $\varepsilon_M$ at low temperatures. Such asymmetry is not predicted by the models of Marcus and Hopfield (Hopfield, 1974; Marcus and Sutin, 1985; Sumi and Marcus, 1986).

If one makes the assumption that biological electron transfer systems evolved their ε values such that rates are optimized, one should expect that electron transfer rates in the photosynthetic reaction center are formed through a choice of $\varepsilon \rightarrow \varepsilon_{max}$, such that $k(\varepsilon_{max})$ is a maximum. In Fig. 9 the transfer rates $k(\varepsilon_{max})$ and

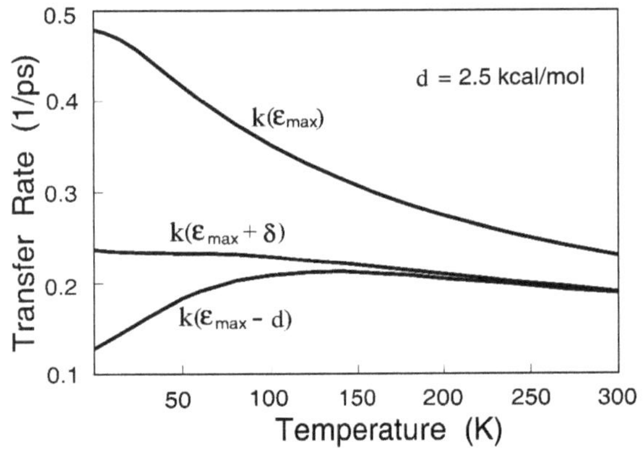


*Fig. 9.* Comparison of the temperature dependence of the maximum transfer rate of $k(\varepsilon_M)$ and off-maximum value $k(\varepsilon_{max} \pm \delta)$, where $\delta = 2.5$ kcal/mol. $k(\varepsilon_{max};T)$ represents the fastest transfer rate of the system, the rates $k(\varepsilon_{max} \pm \delta;T)$ are slower since their ε values deviate from the optimal value $\varepsilon_{max}$. (From Xu and Schulten (1994) with permission of Elsevier.)

$k(\varepsilon;T)$, corresponding for non-optimal values of $\varepsilon = \varepsilon_M \pm \delta$, with $\delta = 2.5$ kcal/mol, are shown.

Experimental data of electron transfer processes in the photosynthetic reaction center show increases similarly to those presented in Fig. 9 (Bixon and Jortner, 1986; Kirmaier and Holten, 1988; Martin et al., 1988; Nagarajan et al., 1990). However, Fig. 9 demonstrates also that electron transfer at ε values slightly off the maximum position can yield a different temperature dependence than that of $k(\varepsilon_M;T)$, namely temperature independence or a slight decrease of the rate with decreasing temperature. Such temperature dependence has also been observed for biological electron transfer (Nagarajan et al., 1990). The temperature dependence of the transfer rate resembles that of $k(\varepsilon_M;T)$ in photosynthetic reaction centers of native bacteria and in (M)Y210F-mutants with tyrosine at the 210 position of the M–unit replaced by phenylalanine. However, a replacement of this tyrosine by isoleucine ((M)Y210I-mutant) yields a transfer rate which decreases like $k(\varepsilon_M - \delta;T)$ shown in Fig. 9. This altered temperature dependence should be attributed to a shift of the redox potentials, i.e., $\varepsilon_M \rightarrow \varepsilon_M - \delta$.

It should be mentioned that there have been numerous similar investigations of biological electron transfer in the literature (Warshel and Hwang, 1986; Wolynes, 1987; Warshel et al., 1989; Zheng et al., 1989).

## Acknowledgments

This work was supported in part by grants from the University of Missouri Research Board, the Institute for Theoretical Sciences (a joint institute of Notre Dame University and Argonne National Laboratory), the Department of Energy (Office of Science contract No. W-31-109-ENG-38), the National Science Foundation (FIBR-0526854) for I.K., and the National Science Foundation (MCB02-34938) as well as the National Institutes of Health (NIH1-R01-GM067887-01, NIH-P41RR05969) for K.S. Supercomputer time provided by the National Center for Supercomputing Applications (MCB020036) to I.K. is also gratefully acknowledged.

## References

Beekman LMP, Frese RN, Fowler GJS, Picorel R, Cogdell RJ, van Stokkum, IHM, Hunter CN and van Grondelle R (1997) Characterization of the light-harvesting antennas of photosynthetic purple bacteria by Stark spectroscopy. 2. LH2 complexes: Influence of the protein environment. J Phys Chem B 101: 7293–7301

Bixon M and Jortner J (1986) Coupling of protein modes to electron transfer in bacterial photosynthesis. J Phys Chem 90: 3795–3800

Chernyak V, Zhang WM and Mukamel S (1998) Multidimensional femtosecond spectroscopies of molecular aggregates and semiconductor nanostructures: The nonlinear exciton equations. J Chem Phys 109: 9587–9601

Cory MG, Zerner MC, Hu X and Schulten K (1998) Electronic excitations in aggregates of bacteriochlorophylls. J Phys Chem B 102: 7640–7650

Damjanovic A, Kosztin I, Kleinekathofer U and Schulten K (2002a) Excitons in a photosynthetic light-harvesting system: a combined molecular dynamics, quantum chemistry, and polaron model study. Phys Rev E 65: 031919

Damjanovic A, Vaswani HM, Fromme P and Fleming GR (2002b) Chlorophyll excitations in photosystem I of *Synechococcus elongates*. J Phys Chem B 106: 10251–10262

Darden T, York D and Pedersen L (1993) Particle mesh Ewald. An N·log(N) method for Ewald sums in large systems. J Chem Phys 98: 10089–10092

Deisenhofer J, Epp O, Mikki K, Huber R and Michel H (1985) Structure of the protein subunits in the photosynthetic reaction centre of *Rhodopseudomonas viridis* at 3 Å resolution. Nature 318: 618–624

Egorov SA, Everitt KF and Skinner JL (1999) Quantum dynamics and vibrational relaxation. J Phys Chem A 103: 9494–9499

Feller SE, Zhang YH, Pastor RW and Brooks BR (1995) Constant pressure molecular dynamics simulation — the Langevin piston method. J Chem Phys 103: 4613–4621

Foloppe N, Breton J and Smith JC (1992) Potential energy function for photosynthetic reaction centre chromophores: energy minimisations of a crystalline bacteriopheophytin *a* analog. In: Breton J and Vermeglio A (eds) The Photosynthetic Bacterial Reaction Center II: Structure, Spectroscopy and Dynamics, pp 43–48. Plenum Press, New York/London

Foloppe N, Ferrand M, Breton J and Smith JC (1995) Structural model of the photosynthetic reaction center of *Rhodobacter capsulatus*. Proteins 22: 226–244

Frisch MJ, Trucks GW, Schlegel HB, Scuseria GE, Robb MA, Cheeseman JR, Zakrzewski JAM, Stratmann RE, Burant JC, Dapprich JMM, Daniels AD, Kudin KN, Strain MC, Farkas JT, Barone V, Cossi M, Cammi R, Mennucci B, Pomelli CA, Clifford S, Ochterski J, Petersson GA, Ayala QC, Morokuma K, Malick DK, Rabuck AD, Raghavachari JBF, Cioslowski J, Ortiz JV, Stefanov BB, Liu G, Liashenko A, Piskorz P, Komaromi I, Gomperts R, Martin DJF, Keith T, Al-Laham MA, Peng CY, Nanayakkara CG, Challacombe M, Gill PMW, Johnson BG, Chen MWW, Andres JL, Head-Gordon M, Replogle ES and Pople JA (1998) Gaussian 98. Gaussian Inc., Pittsburgh

Georgakopoulou S, Frese RN, Johnson E, Koolhaas C, Cogdell RJ, van Grondelle R and van der Zwan G (2002) Absorption and CD spectroscopy and modeling of various LH2 complexes from purple bacteria. Biophys J 82: 2184–2197

Gunn JR and Dawson KA (1989) Microscopic model of amphiphilic assembly. J Chem Phys 91: 6393–6403

He Z, Sundström V and Pullerits T (2002) Influence of the protein binding site on the excited states of bacteriochlorophyll: DFT calculations of B800 in LH2. J Phys Chem B 106: 11606-11612

Holstein T (1959) Polaron motion. I. Molecular-crystal model. Ann Phys 8: 325–342

Hopfield JJ (1974) Electron transfer between biological molecules by thermally activated tunnelling. Proc Natl Acad Sci USA 71: 3640–3644

Hu X and Schulten K (1997) How nature harvests sunlight. Physics Today 50: 28–34

Hu X, Damjanovic A, Ritz T and Schulten K (1998a) Architecture and function of the light harvesting apparatus of purple bacteria. Proc Natl Acad Sci USA 95: 5935–5941

Hu X, Damjanovic A, Ritz T and Schulten K (1998b) Architecture and mechanism of the light-harvesting apparatus of purple bacteria. Proc Natl Acad Sci USA 95: 5935–5941

Hu X, Ritz T, Damjanovic A, Autenrieth F and Schulten K (2002) Photosynthetic apparatus of purple bacteria. Q Rev Biophys 35: 1–62

Humphrey W, Dalke A and Schulten K (1996) VMD — Visual Molecular Dynamics. J Mol Graphics 14: 33-38

Ihalainen JA, Linnanto J, Myllyperkio P, van Stokkum IHM, Ucker B, Scheer H and Korppi-Tommola JEI (2001) Energy transfer in LH2 of *Rhodospirillum molischianum*, studied by subpicosecond spectroscopy and configuration interaction exciton calculations. J Phys Chem B 105: 9849–9856

Jang SJ and Silbey RJ (2003) Single complex line shapes of the B850 band of LH2. J Chem Phys 118: 9324–9336

Janosi L, Kosztin I and Damjanovic A (2006) Theoretical prediction of spectral and optical properties of bacteriochlorophylls in thermally disordered LH2 antenna complexes. J Chem Phys 124: 014903

Jorgensen WL, Chandrasekhar J, Madura JD, Impey RW and Klein ML (1983) Comparison of simple potential functions for simulating liquid water. J Chem Phys 79: 926–935

Kirmaier C and Holten D (1988) Temperature effects on the ground state absorption spectra and electron transfer kinetics of bacterial reaction centers. In: Breton J and Vermeglio A (eds) The Photosynthetic Bacterial Reaction Center: Structure and Dynamics, pp 219–228. Plenum Press, New York/London

Koepke J, Hu XC, Muenke C, Schulten K and Michel H (1996) The crystal structure of the light-harvesting complex II (B800-850) from *Rhodospirillum molischianum*. Structure 4: 581–597

Koolhaas MHC, van der Zwan G and van Grondelle R (2000) Local and nonlocal contributions to the linear spectroscopy of light-harvesting antenna systems. J Phys Chem B 104: 4489–4502

Leggett AJ, Chakravarty S, Dorsey AT, Fisher MPA, Garg A and Zwerger W (1985) Dynamics of the dissipative two-state system. Rev Mod Phys 59: 1–85

Linnanto J and Korppi-Tommola J (2004) Semi-empirical PM5 molecular orbital study on chlorophylls and bacteriochlorophylls: Comparison of semi-empirical, *ab initio*, and density functional results. J Comput Chem 25: 123–38

Linnanto J, Korppi-Tommopa JEI and Helenius VM (1999) Electronic states, absorption spectrum and circular dichroism spectrum of the photosynthetic bacterial LH2 antenna of *Rhodopseudomonas acidophila* as predicted by exciton theory and semi-empirical calculations. J Phys Chem B 103: 8739–8750

MacKerell Jr. AD, Bashford D, Bellott M, Dunbrack Jr RL, Evanseck J.D, Field MJ, Fischer S, Gao J, Guo H, Ha S, Joseph-McCarthy D, Kuchnir L, Kuczera K, Lau FTK, Mattos C, Michnick S, Ngo T, Nguyen DT, Prodhom B, Reiher III WE, Roux B, Schlenkrich M, Smith JC, Stote R, Straub J, Watanabe M, Wiorkiewicz-Kuczera J, Yin D, Karplus M (1998) All-atom empirical potential for molecular modeling and dynamics studies of proteins. J Phys Chem B 102: 3586–3616

Mahan GD (1990) Many-Particle Physics, 2nd Edition. Plenum Press, New York

Makri N (1999) The linear response approximation and its lowest order corrections: An influence functional approach. J Phys Chem B 103: 2823–2829

Marcus RA (1956a) Electrostatic free energy and other properties of states having nonequilibrium polarization. J Chem Phys 24: 979–989

Marcus RA (1956b) On the energy of oxidation-reduction reactions involving electron transfer. J Chem Phys 24: 966–978

Marcus RA and Sutin N (1985) Electron transfers in chemistry and biology. Biochim Biophys Acta 811: 265–322

Martin JL, Breton J, Lambry JC and Fleming G (1988) The primary electron transfer in photosynthetic purple bacteria: Long range electron transfer in the femtosecond domain at low temperature. In: Breton J and Vermeglio A (eds) The Photosynthetic Bacterial Reaction Center: Structure and Dynamics, pp 195-203. Plenum Press, New York/London

May V and Kühn O (2000) Charge and Energy Transfer Dynamics in Molecular Systems. Wiley-VCH, Berlin

McDermott G, Prince S, Freer A, Hawthornthwaite-Lawless A, Papiz M, Cogdell RJ and Isaacs N (1995) Crystal structure of an integral membrane light-harvesting complex from photosynthetic bacteria. Nature 374: 517-521

Meier T, Zhao Y, Chernyak V and Mukamel S (1997) Polarons, localization, and excitonic coherence in superradiance of biological antenna complexes. J Chem Phys 107: 3876-3893

Mercer I, Gould I and Klug D (1999) A quantum mechanical/molecular mechanical approach to relaxation dynamics: Calculation of the optical properties of solvated bacteriochlorophyll-*a*. J. Phys. Chem. B 103: 7720–7727

Miyamoto S and Kollman PA (1992) SETTLE: An analytical version of the SHAKE and RATTLE algorithm for rigid water molecules. J Comput Chem 13: 952–962

Mukamel S (1995) Principles of nonlinear optical spectroscopy. Oxford University Press, New York

Nagarajan V, Parson WW, Gaul D and Schenck C (1990) Effect of specific mutations of tyrosine-(M)210 on the primary photosynthetic electron-transfer process in *Rhodobacter sphaeroides*. Proc Natl Acad Sci USA 87: 7888–7892

Phillips JC, Braun R, Wang W, Gumbart J, Tajkhorshid E, Villa E, Chipot C, Skeel RD, Kale L and Schulten K (2005) Scalable molecular dynamics with NAMD. J Comput Chem 26: 1781–1802

Ray J and Makri N (1999) Short-range coherence in the energy transfer of photosynthetic light-harvesting systems. J Phys Chem A 103: 9417–9422

Renger T, May V and Kuhn O (2001) Ultrafast excitation energy transfer dynamics in photosynthetic pigment-protein complexes. Phys Rep 343: 138–254

Ridley J and Zerner M (1973) An intermediate neglect of differential overlap technique for spectroscopy: Pyrrole and the azines. Theor Chim Acta 32: 111–134

Ringnalda MN, Langlois JM, Murphy RB, Greeley BH, Cortis C, Russo TV, Marten B, Jr RED, Pollard WT, Cao Y, Muller RP, Mainz DT, Wright JR, Miller III WAG and Friesner RA (1996) PS-GVB v2.3. Schrödinger Inc, Portland, OR

Schlenkrich M, Brickmann J, MacKerell Jr AD and Karplus M (1996) Empirical potential energy function for phospholipids: Criteria for parameter optimization and applications In: Merz KM and Roux B (eds) Biological Membranes: A Molecular Perspective from Computation and Experiment, pp 31–81. Birkhäuser, Boston

Scholes GD and Fleming GR (2000) On the mechanism of light harvesting in photosynthetic purple bacteria: B800 to b850 energy transfer. J Phys Chem B 104: 1854–1868

Schulten K and Tesch M (1991) Coupling of protein motion to electron transfer: Molecular dynamics and stochastic quantum mechanics study of photosynthetic reaction centers. Chem Phys 158: 421–446

Somsen OJG, van Grondelle R and van Amerongen H (1996) Spectral broadening of interacting pigments: Polarized absorption by photosynthetic proteins. Biophys J 71: 1934–1951

Sumi H and Marcus RA (1986) Dynamical effects in electron transfer reactions. J Chem Phys 84: 4894–4914

Sundström V, Pullerits T and van Grondelle R (1999) Photosynthetic light-harvesting: Reconciling dynamics and structure of purple bacterial LH2 reveals function of photosynthetic unit. J Phys Chem B 103: 2327–2346

van Amerongen H, Valkunas L and van Grondelle R (2000) Photosynthetic Excitons. World Scientific, Singapore

Visscher K, Bergström H, Sundström V, Hunter C and van Grondelle R (1989) Temperature dependence of energy transfer from the long wavelength antenna BChl-896 to the reaction center in *Rhodospirillum Rubrum*, *Rhodobacter sphaeroides* (w.t. and M21 mutant) from 77 to 177 K, studied by picosecond absorption spectroscopy. Photosynth Res 22: 211–217

Warshel A and Hwang J-K (1986) Simulations of the dynamics of electron transfer reactions in polar solvent: Semiclassical trajectories and dispersed polaron approaches. J Chem Phys 84: 4938-4957

Warshel A, Chu ZT and Parson WW (1989) Dispersed polaron simulations of electron transfer in photosynthetic reaction centers. Science 246: 112–116

Wolynes PG (1987) Dissipation, tunneling and adiabaticity criteria for curve crossing problems in the condensed phase. J Chem Phys 86: 1957–1966

Wu HM, Ratsep M, Jankowiak R, Cogdell RJ and Small GJ (1997) Comparison of the LH2 antenna complexes of *Rhodopseudomonas acidophila* (strain 10050) and *Rhodobacter sphaeroides* by high-pressure absorption, high-pressure hole burning, and temperature-dependent absorption spectroscopies. J Phys Chem B 101: 7641–7653

Xu D and Schulten K (1992) Multi-mode coupling of protein motion to electron transfer in the photosynthetic reaction center: Spin-boson theory based on a classical molecular dynamics simulation. In: Breton J and Vermeglio A (eds) The Photosynthetic Bacterial Reaction Center: II. Structure, Spectroscopy and Dynamics, NATO ASI Series A: Life Sciences, pp 301–312. Plenum Press, New York

Xu D and Schulten K (1994) Coupling of protein motion to electron transfer in a photosynthetic reaction center: Investigating the low temperature behaviour in the framework of the spin-boson model. Chem Phys 182: 91–117

Yang M, Agarwal R and Fleming GR (2001) The mechanism of energy transfer in the antenna of photosynthetic purple bacteria. J Photochem Photobiol A 142: 107–119

Zerner M, Loew G, Kirchner R and Mueller-Westerhoff UJ (1980) An intermediate neglect of differential overlap technique for spectroscopy of transition-metal complexes. Ferrocene. J Am Chem Soc 102: 589–599

Zhang JP, Fujii R, Qian P, Inaba T, Mizoguchi T and Koyama Y (2000) Mechanism of the carotenoid-to-bacteriochlorophyll energy transfer via the $S_1$ state in the LH2 complexes from purple bacteria. J Phys Chem B 104: 3683–3691

Zheng C, McCammon JA and Wolynes PG (1989) Quantum simulation of nuclear rearrangement in electron transfer reactions. Proc Natl Acad Sci USA 86: 6441–6444

# Chapter 23

## Equilibrium and Dynamical Path Integral Methods in Bacterial Photosynthesis

Nancy Makri
*Departments of Chemistry and Physics, University of Illinois, 601 S. Goodwin Avenue, Urbana, Illinois 61801, U.S.A.*



## Summary

Classical molecular dynamics simulations have provided a wealth of information on processes in biological systems. In spite of the spectacular success and the insights gained from such simulations, processes involving electron or exciton transfer are inherently quantum mechanical and thus not amenable to a classical description. This chapter focuses on the use of path integral methods for simulating the equilibrium and dynamical properties of charged particles in dissipative environments.

Feynman's path integral theory is an exact formulation of time-dependent quantum mechanics and quantum statistical mechanics that circumvents the use of wave functions, whose storage requirements are prohibitive for systems of more than a few atoms. The calculation of thermal equilibrium properties in non-fermionic systems is possible with the well-established path integral Monte Carlo method. Dynamical processes involve multidimensional integrals of rapidly oscillating functions; Monte Carlo methods converge extremely slowly in such cases, and thus fully quantum mechanical simulations of real-time processes generally remain out of reach. In the case of electron (or exciton) transfer processes, the so-called linear response approximation makes it possible to replace the vast majority of nuclear coordinates by an effective dissipative environment of harmonic oscillators. This simplification allows a fully quantum mechanical treatment of the dynamics using an iterative decomposition of the path integral developed in the mid-1990s.

Applications of these methods to determine the exciton coherence length in the B850 ring of the light harvesting complex (LH2) and the mechanism of primary charge separation in the photosynthetic reaction center are reviewed. Path integral calculations, along with a visual inspection of statistically significant paths, led to the conclusion that the exciton is delocalized over two to three chlorophyll monomers at room temperature.

*Email: nancy@makri.scs.uiuc.edu

*Thijs J. Aartsma and Jörg Matysik (eds): Biophysical Techniques in Photosynthesis II, pp. 465–485.*

Iterative evaluation of the real-time path integral for a three-state model comprising the excited special pair, the reduced accessory bacteriochlorophyll, and the reduced bacteriopheophytin, offers evidence in support of the sequential mechanism, where the electron is first transferred to the accessory chlorophyll.

## I. Introduction

The molecular dynamics method has developed into an indispensable tool in the simulation of biological systems, providing valuable insights into their properties and function. The only severe flaw of molecular dynamics is its inability to account for quantum mechanical effects. While many biological processes can be understood on the basis of classical mechanics, the description of certain phenomena that involve electron transfer, and in some cases also proton transfer, require the use of quantum mechanics. Electron transfer is accompanied by a complex reorganization of atomic coordinates, whose motion must be treated explicitly. Fully quantum mechanical simulations of dynamical processes in biological systems with thousands of atoms remain out of reach, although significant progress can be made if some simplifying assumptions are introduced. Equilibrium properties, however, can be evaluated at the full quantum mechanical level in systems of arbitrary complexity using the path integral formulation of quantum mechanics.

This chapter reviews the use of path integral methods in the context of biological charge transfer, with special attention to those processes of importance in the early steps of photosynthesis. After a short overview of the path integral formulation of quantum mechanics, Section II discusses in general terms the path integral Monte Carlo method, i.e., application of the theory of path integration to calculate thermal equilibrium properties, and reviews an iterative methodology that allows simulation of dynamical properties for a subsystem coupled to a harmonic dissipative environment.

Photosynthesis commences with the absorption of visible light by light harvesting complexes, which can be thought of as ring-shaped biological antennas. The absorbed energy forms an exciton, which in an isolated, perfectly symmetric ring at zero temperature would be completely delocalized and retain its coherence indefinitely. Many external factors, primarily thermal fluctuations, static inhomogeneity, and dynamic disorder, affect the coherence of this exciton, generally favoring its localization. Section III reviews the calculation of the mean exciton coherence length using the path integral Monte Carlo method. By analyzing the contribution of various quantum paths representing the exciton one obtains a visual picture of exciton delocalization.

The energy captured by the light harvesting complexes is eventually transferred to the photosynthetic reaction center, where it excites a special chlorophyll pair. This excitation initiates a series of electron transfer reactions. The first few picoseconds in this series of events (the so-called primary charge separation) result in the transfer of an electron from the excited special pair to a pheophytin located about 17 Å away. The mechanism of this extremely fast and efficient electron transfer process has stimulated much research and controversy. Section IV of this chapter describes the use of real-time path integral methods to simulate this process. These simulations, in combination with experimental findings, have provided definitive evidence in support of a sequential process, where the electron is initially transferred to an accessory chlorophyll molecule, followed by a very rapid transfer to the pheophytin.

Length limitations in this chapter do not allow a comprehensive coverage of the very interesting literature surrounding the topics presented. Because the book is intended primarily as an aid to non-experts interested in learning and applying these algorithms to new problems, extensive references are given to articles describing the methodology and to those closely related to the simulations reviewed, while only key references are given to experimental work and to other theoretical methods that have been applied to these systems.

## II. Path Integral Methodology

### A. Overview

Path integral methods offer the only fully quantum mechanical simulation tool for studying equilibrium properties of many-particle systems and, in some

*Abbreviations:* $\ell$ – coherence length; LH1 – light harvesting complex I; LH2 – light harvesting complex II; MCL – Mean coherence length; *Rb.* – *Rhodobacter*; *Rps.* – *Rhodopseudomonas*; *Rs.* – *Rhodospirillum*; $\xi$ – order parameter

cases, dynamical properties as well. The basic idea of the path integral is very simple. Consider, for example, the coordinate matrix element of the time evolution operator for a system of one degree of freedom with mass $m$, $\langle x'|e^{-i\hat{H}t/\hbar}|x''\rangle$, which is usually referred to as the propagator. By splitting the time $t$ into $N$ shorter time increments of length $\Delta t = t/N$, the time evolution operator is written in the form of a product:

$$e^{-i\hat{H}t/\hbar} = \left(e^{-i\hat{H}\Delta t/\hbar}\right)^N = e^{-i\hat{H}\Delta t/\hbar} e^{-i\hat{H}\Delta t/\hbar} \ldots e^{-i\hat{H}\Delta t/\hbar} \quad (2.1)$$

This factorized form is now substituted in the propagator, and complete sets of coordinate states are inserted between each pair of operators to produce the following expression:

$$\langle x'|e^{-i\hat{H}t/\hbar}|x''\rangle = \int dx_1 \cdots \int dx_N \langle x'|e^{-i\hat{H}\Delta t/\hbar}|x_1\rangle$$
$$\langle x_1|e^{-i\hat{H}\Delta t/\hbar}|x_2\rangle \cdots \langle x_{N-1}|e^{-i\hat{H}\Delta t/\hbar}|x''\rangle \quad (2.2)$$

Equation (2.2) is a path integral representation of the propagator (Feynman, 1948; Feynman and Hibbs, 1965). The concept of 'paths' emerges by associating each auxiliary variable $x_k$ with the time $k\Delta t$. Thus, a particular realization of the coordinates $\{x_0 \equiv x'$, $x_1(\Delta t)$, $x_2(2\Delta t)$, ..., $x_{N-1}((N-1)\Delta t)$, $x_N \equiv x''\}$ forms a path defined on the discrete time increments $k\Delta t$ (see Fig. 1). According to Eq. (2.2), one must sum the integrand of Eq. (2.2) over all such paths with fixed endpoints $x'$ and $x''$. Note that no approximation has been introduced, so Eq. (2.2) holds exactly.

Equation (2.2) is useful in practice as long as one is able to identify a sufficiently accurate approximation to the short time propagator that enters the integrand. There are various approaches to this task, most of which are approximate forms valid only for small $\Delta t$ (Fujiwara et al., 1982; De Raedt and De Raedt, 1983; Ceperley and Pollock, 1986; Li, 1987; Kono et al., 1988; Makri and Miller, 1988a, 1989; Coalson et al., 1989; Bandrauk and Shen, 1991; Glasner et al., 1991; Lolle et al., 1991; Makri, 1992). The most commonly used approximations are obtained by partitioning the Hamiltonian into a sum of two operators $H_0$ and $H_1$ and replacing the exponential of the sum by a symmetrized product of exponentials (Trotter, 1959):

$$e^{-iH\Delta t/\hbar} \simeq e^{-i\hat{H}_1\Delta t/2\hbar} e^{-i\hat{H}_0\Delta t/\hbar} e^{-i\hat{H}_1\Delta t/2\hbar} \quad (2.3)$$

This last step requires neglecting commutators

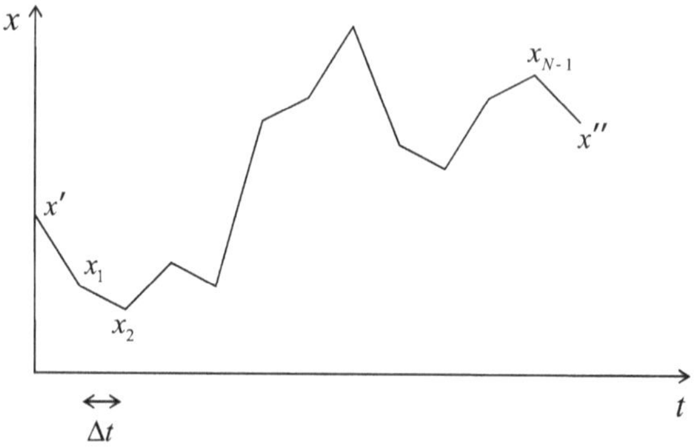


*Fig. 1.* A discrete path corresponding to a particular realization of the integration variables in Eq. (2.2).

that vanish in the limit $\Delta t \to 0$, thus the resulting expression is sufficiently accurate for short time increments. A particular splitting where the propagator is factorized based on the separation of kinetic and potential energy terms, i.e., $\hat{H}_0 = \hat{p}^2/2m$ and $\hat{H}_1 = \hat{V}$, leads to the approximate propagator

$$\langle x_{k_1}|e^{-i\hat{H}\Delta t/\hbar}|x_{k_2}\rangle \simeq \sqrt{\frac{m}{2\pi i\hbar\Delta t}}$$
$$\exp\left[\frac{i}{\hbar}\frac{m}{2\Delta t}\left(x_{k_2} - x_{k_1}\right)^2 - \frac{i\Delta t}{2\hbar}\left(V(x_{k_2}) + V(x_{k_1})\right)\right] \quad (2.4)$$

Use of this expression leads to the form

$$\langle x'|e^{-i\hat{H}t/\hbar}|x''\rangle \simeq \left(\frac{m}{2\pi i\hbar\Delta t}\right)^{N/2} \int dx_1 \cdots \int dx_N$$
$$\exp\left[\frac{i}{\hbar}\sum_{k=1}^{N}\left(\frac{m}{2\Delta t}\left(x_k - x_{k-1}\right)^2 - \frac{\Delta t}{2}\left(V(x_k) + V(x_{k-1})\right)\right)\right]$$
$$= \left(\frac{m}{2\pi i\hbar\Delta t}\right)^{N/2} \sum_{\text{paths } \{x_0 \equiv x', x_1, \ldots, x_N \equiv x''\}} \exp\left(\frac{i}{\hbar} S_N\left(x_1, \ldots, x_N; \Delta t\right)\right) \quad (2.5)$$

where

$$S_N\left(x_1, \ldots, x_N; \Delta t\right) =$$
$$\Delta t \sum_{k=1}^{N}\left(\frac{1}{2}\left(\frac{x_k - x_{k-1}}{\Delta t}\right)^2 - \frac{1}{2}\left(V(x_k) + V(x_{k-1})\right)\right) \quad (2.6)$$

is the trapezoid-rule discretization of the classical action functional

$$S\left[x(t')\right]=\int_0^t\left(\frac{1}{2}m\dot{x}(t')^2-V\left(x(t')\right)\right)dt' \quad (2.7)$$

In the limit $N\to\infty$ the short-time approximation to the propagator becomes exact and Eq. (2.5) reduces to Feynman's celebrated form of the path integral

$$\left\langle x'\left|e^{-i\hat{H}t/\hbar}\right|x''\right\rangle \propto \sum_{\text{paths }x(t')}\exp\left(\frac{i}{\hbar}S\left[x(t')\right]\right) \quad (2.8)$$

where the contribution of each path is proportional to a phase given by the classical action along the path in units of $\hbar$.

The path integral representation of the propagator can be extended to describe the statistical mechanical properties of systems at finite temperature (Feynman, 1972). The procedure is straightforward if one realizes that the Boltzmann operator $e^{-\beta\hat{H}}$ (where $\beta = 1/k_BT$) has the same mathematical form as the time evolution operator. An important difference of the two operators is that the exponent of the Boltzmann operator is real-valued, thus $\beta$ should be regarded as an imaginary time, and the integrand in the corresponding path integral expression is real and positive for all paths. For general many-particle systems the only available method for performing the sum over paths is the Monte Carlo method (Metropolis et al., 1953), which is very efficient when the integrand is a localized positive definite function, but extremely ill-converged otherwise. This feature makes path integral calculations the method of choice for studying the equilibrium properties of many-particle nonfermionic systems (Binder and Heermann, 1988). On the other hand, the real-time path integral converges exponentially slowly as the number of particles and/or the propagation time are increased (Doll et al., 1990; Mak and Egger, 1996; Makri, 1991), and thus dynamical properties of many-body systems continue to present a major computational challenge.

## B. Equilibrium Properties

Electron and energy transfer processes are often most conveniently described in terms of $n$ discrete states $|u_i\rangle$ that correspond to localized excitations of the particle of interest. Using Cartesian coordinates $\mathbf{R}$ (and momenta $\mathbf{P}$) for the nuclear degrees of freedom and assuming that coupling terms are independent of the nuclear coordinates, the Hamiltonian can be written in the form

$$\hat{H}=\hat{H}_0+\sum_{i=1}^{n}h_i(\hat{\mathbf{P}},\hat{\mathbf{R}})\left|u_i\right\rangle\left\langle u_i\right| \quad (2.9)$$

where

$$\hat{H}_0=\sum_{i=1}^{n}\varepsilon_i\left|u_i\right\rangle\left\langle u_i\right|+\sum_{i=1}^{n}\sum_{j=1}^{n}\Delta_{ij}\left|u_i\right\rangle\left\langle u_j\right| \quad (2.10)$$

and

$$h_i(\hat{\mathbf{P}},\hat{\mathbf{R}})=\frac{1}{2}\hat{\mathbf{P}}\cdot\mathbf{m}^{-1}\cdot\hat{\mathbf{P}}+V_i(\hat{\mathbf{R}}) \quad (2.11)$$

The canonical partition function is written in the path integral form

$$\begin{aligned}Z&=\mathrm{Tr}\,e^{-\beta\hat{H}}\\&=\sum_{k_1=1}^{n}\sum_{k_2=1}^{n}\cdots\sum_{k_N=1}^{n}\int_{-\infty}^{\infty}d\mathbf{R}_1\int_{-\infty}^{\infty}d\mathbf{R}_2\cdots\int_{-\infty}^{\infty}d\mathbf{R}_N\\&\left\langle u_N\mathbf{R}_N\left|e^{-\beta H/N}\right|\mathbf{R}_{N-1}u_{N-1}\right\rangle\cdots\left\langle u_2\mathbf{R}_2\left|e^{-\beta H/N}\right|\mathbf{R}_1u_1\right\rangle\\&\left\langle u_1\mathbf{R}_1\left|e^{-\beta H/N}\right|\mathbf{R}_Nu_N\right\rangle\end{aligned} \quad (2.12)$$

and various equilibrium averages are obtained by including the appropriate operator in the integrand of this expression and dividing by the partition function. For sufficiently large $N$, the high-temperature propagators can be approximated by the factorized form

$$\begin{aligned}&\left\langle u_i\mathbf{R}_i\left|e^{-\beta H}\right|\mathbf{R}_ju_j\right\rangle\\&\simeq\left\langle\mathbf{R}_i\left|e^{-\beta h_i(\hat{\mathbf{P}},\hat{\mathbf{R}})/2N}e^{-\beta h_j(\hat{\mathbf{P}},\hat{\mathbf{R}})/2N}\right|\mathbf{R}_j\right\rangle\left\langle u_i\left|e^{-\beta\hat{H}_0/N}\right|u_j\right\rangle\end{aligned} \quad (2.13)$$

Unless the Hamiltonians for the nuclear degrees of freedom have a special form, the first of these factors must be further approximated. Applying the Trotter splitting, one obtains

$$\begin{aligned}&\left\langle\mathbf{R}_i\left|e^{-\beta h_i(\hat{\mathbf{P}},\hat{\mathbf{R}})/2N}e^{-\beta h_j(\hat{\mathbf{P}},\hat{\mathbf{R}})/2N}\right|\mathbf{R}_j\right\rangle\\&\simeq\left\langle\mathbf{R}_i\left|e^{-\beta\mathbf{P}\cdot\mathbf{m}^{-1}\cdot\mathbf{P}/2N}\right|\mathbf{R}_j\right\rangle e^{-\beta V_i(\mathbf{R}_i)/2N}e^{-\beta V_j(\mathbf{R}_j)/2N}\end{aligned} \quad (2.14)$$

The first of these factors is the free particle propagator,

$$\langle \mathbf{R}_i | e^{-\beta \mathbf{P}\cdot\mathbf{m}^{-1}\cdot\mathbf{P}/2N} | \mathbf{R}_j \rangle =$$
$$\prod_l \left( \frac{m_l N}{2\pi\hbar^2\beta} \right)^{\frac{3}{2}} \exp\left( -\frac{N}{2\hbar^2\beta} (\mathbf{R}_i - \mathbf{R}_j)\cdot \mathbf{m} \cdot (\mathbf{R}_i - \mathbf{R}_j) \right) \tag{2.15}$$

Where $\mathbf{m}$ is the diagonal matrix of atomic masses $m_l$. Finally, the last factor in Eq. (2.13) can be calculated from the eigenstates of $\hat{H}_0$, which can be evaluated by straightforward matrix diagonalization.

Once the high-temperature propagators have been obtained, the sums or integrals in the path integral expression are evaluated by the Monte Carlo method. These algorithms are well-developed (Binder and Heermann, 1988) and will not be described here in detail. To summarize the procedure, one performs a random walk in the space of all integration variables guided by a sampling function that specifies the magnitude of the integrand. The sampling function corresponds to a probability distribution and must be normalized to unity. Thus, the sampling function typically includes the exponential factors of the integrand whose integral is known. In the case of statistical mechanical averages, where the entire partition function occurs in the denominator of the expression, one may use the entire integrand of Eq. (2.12) as the sampling function, since the integral of the latter automatically cancels upon division by $Z$ (Chandler, 1987).

These procedures may be used for computing the mean coherence length of the exciton, which gives a measure of exciton delocalization in the system. As will become clear in section III, additional insight is gained through visual inspection of paths that are most frequently sampled.

### C. Dynamical Properties

As discussed in the introductory part of this section, extracting dynamical information requires evaluation of the real-time path integral, which contains a delocalized, rapidly oscillatory phase that renders the convergence of Monte Carlo methods extremely difficult. For a given type of function and total number of sampled configurations, the extent of phase cancellation (and thus the statistical error of the calculation) increases exponentially with the dimension of the integral, i.e., the number of degrees of freedom and/or number of auxiliary path integral variables. While the computation of meaningful results may be possible at short times, the situation becomes disastrous when long-time information is desired.

Various algorithms have been reported for improving the convergence of the real-time path integral for short to intermediate times. These include the use of improved approximations to the short-time propagator (Fujiwara et al., 1982; De Raedt and De Raedt, 1983; Ceperley and Pollock, 1986; Li, 1987; Kono et al., 1988; Makri and Miller, 1988a; 1989; Coalson et al., 1989; Bandrauk and Shen, 1991; Glasner et al., 1991; Lolle et al., 1991; Makri, 1992), which allow the use of larger time steps and thus decrease the required number of path integral variables for a given total time, stationary phase filtering procedures (Filinov, 1986; Makri and Miller, 1987; Doll et al., 1988; Doll and Freeman, 1988; 1988b; Wang et al., 2001), the construction of smoother propagators (Makri, 1989; 1993; Hoffman and Kouri, 1992), the use of block summation algorithms (Mak, 1992), and information-guided noise reduction (Makri, 2004), a method for enhancing the convergence of the path integral. Long-time calculations over hundreds or thousands of time steps are possible only through the use of iterative decomposition schemes. These were developed initially for spin-boson or system-bath Hamiltonians (Makarov and Makri, 1994; Makri, 1995; Makri and Makarov, 1995a,b; Sim and Makri, 1996; 1997a), where a low-dimensional subsystem is linearly coupled to a harmonic dissipative environment that gives rise to a Gaussian influence functional (Feynman and Vernon, 1963), and later extended to some more general situations (Makri, 1999a). The basic theoretical framework is presented below in the case where degrees of freedom of the environment are assumed harmonic, and the reader is referred to the original articles and reviews for details.

To begin, the Hamiltonian describing collective effects of the nuclear degrees of freedom on the particle of interest (electron or exciton) is assumed to have the following quadratic form:

$$\sum_{k=1}^{n} \hat{h}_k |u_k\rangle\langle u_k|$$
$$= \sum_j \frac{\hat{p}_j^2}{2m_j} + \frac{1}{2} m_j \omega_j^2 \left( \hat{x}_j - \frac{c_j}{m_j \omega_j^2} \sum_{k=1}^{n} \sigma_k |u_k\rangle\langle u_k| \right)^2$$
$$= \sum_j \frac{\hat{p}_j^2}{2m_j} + \frac{1}{2} m_j \omega_j^2 \hat{x}_j^2 - c_j \hat{x}_j \sum_{k=1}^{n} \sigma_k |u_k\rangle\langle u_k|$$
$$+ \frac{c_j^2}{2m_j\omega_j^2} \sum_{k=1}^{n} \sigma_k^2 |u_k\rangle\langle u_k| \tag{2.16}$$

Thus, using $\sum_k |u_k\rangle\langle u_k| = 1$, the harmonic bath Hamiltonian is given by the expression

$$\hat{h}_k = \sum_j \frac{\hat{p}_j^2}{2m_j} + \frac{1}{2} m_j \omega_j^2 \left( \hat{x}_j - \frac{c_j}{m_j \omega_j^2} \sigma_k \right)^2 \tag{2.17}$$

Here $x_j$ are fictitious coordinates representing collectively the assumed Gaussian response of the environment to the dynamics of the charged particle, and $\sigma_k$ is a parameter that indicates the location of the particle of interest or excitation. The picture is very familiar from electron transfer theory (Marcus, 1956, 1993; Marcus and Sutin, 1985; Onuchic and Wolynes, 1988, 1993; Makri, 1999b), where the long-range of the Coulomb potential is often invoked to justify the Gaussian assumption (Makri, 1999b), and should also be applicable to exciton transport processes.

The evolution of observables pertaining to the system of interest is most effectively expressed in terms of the reduced density matrix,

$$\tilde{\rho}_{ij}(t) \equiv \mathrm{Tr}_{\mathrm{bath}} \left\langle u_i \left| e^{-i\hat{H}t/\hbar} \hat{\rho}(0) e^{i\hat{H}t/\hbar} \right| u_j \right\rangle \tag{2.18}$$

Here the trace is evaluated with respect to all the bath degrees of freedom $\{x_i\}$, and $\hat{\rho}(0)$ is the initial density operator. Its path integral representation takes the form

$$\begin{aligned}
\tilde{\rho}_{ij}(t) = &\sum_{k_0^+=1}^{n} \sum_{k_1^+=1}^{n} \cdots \sum_{k_{N-1}^+=1}^{n} \sum_{k_0^-=1}^{n} \sum_{k_1^-=1}^{n} \cdots \sum_{k_{N-1}^-=1}^{n} \\
&\int d\mathbf{x}_0^+ \int d\mathbf{x}_1^+ \cdots \int d\mathbf{x}_{N-1}^+ \int d\mathbf{x}_0^- \int d\mathbf{x}_1^- \cdots \int d\mathbf{x}_{N-1}^- \int d\mathbf{x}_N \\
&\times \left\langle u_i \mathbf{x}_N \left| e^{-i\hat{H}\Delta t/\hbar} \right| \mathbf{x}_{N-1}^+ u_{k_{N-1}^+} \right\rangle \left\langle u_{k_{N-1}^+} \mathbf{x}_{N-1}^+ \left| e^{-i\hat{H}\Delta t/\hbar} \right| \mathbf{x}_{N-2}^+ u_{k_{N-2}^+} \right\rangle \cdots \\
&\left\langle u_{k_1^+} \mathbf{x}_1^+ \left| e^{-i\hat{H}\Delta t/\hbar} \right| \mathbf{x}_0^+ u_{k_0^+} \right\rangle \left\langle u_{k_0^+} \mathbf{x}_0^+ \left| \hat{\rho}(0) \right| \mathbf{x}_0^- u_{k_0^-} \right\rangle \\
&\left\langle u_{k_0^-} \mathbf{x}_0^- \left| e^{i\hat{H}\Delta t/\hbar} \right| \mathbf{x}_1^- u_{k_1^-} \right\rangle \left\langle u_{k_1^-} \mathbf{x}_1^- \left| e^{i\hat{H}\Delta t/\hbar} \right| \mathbf{x}_2^- u_{k_2^-} \right\rangle \cdots \\
&\left\langle u_{k_{N-1}^-} \mathbf{x}_{N-1}^- \left| e^{i\hat{H}\Delta t/\hbar} \right| \mathbf{x}_N u_j \right\rangle
\end{aligned} \tag{2.19}$$

Following steps similar to those described in section IIb, the short time propagators are split as

$$\begin{aligned}
&\left\langle u_i \mathbf{x}_i \left| e^{-i\hat{H}t/N\hbar} \right| \mathbf{x}_j u_j \right\rangle = \\
&\left\langle \mathbf{x}_i \left| e^{-i\hat{h}_i t/N\hbar} e^{-i\hat{h}_j t/N\hbar} \right| \mathbf{x}_j \right\rangle \left\langle u_i \left| e^{-i\hat{H}_0 t/N\hbar} \right| u_j \right\rangle
\end{aligned} \tag{2.20}$$

It is often assumed that the system and bath are initially uncorrelated, such that the density operator factorizes:

$$\rho(0) = \tilde{\rho}(0)\, e^{-\beta \hat{H}_{\mathrm{bath}}} / \mathrm{Tr}\, e^{-\beta \hat{H}_{\mathrm{bath}}} \tag{2.21}$$

(The methodology can be extended to non-factorized initial conditions (Shao and Makri, 2001, 2002), e.g., $\hat{\rho}(0) = \exp(-\beta \hat{H})$, but this more general case will not be discussed in this chapter.) Further, one notices that the harmonic bath Hamiltonian in Eq. (2.17) describes displaced harmonic oscillators and thus the first factor of Eq. (2.20) is available analytically. Substituting the displaced oscillator propagator in the path integral expression and performing the resulting Gaussian integrals leads to the following path integral representation of the reduced density matrix:

$$\begin{aligned}
\tilde{\rho}_{ij}(t) = &\sum_{k_0^+=1}^{n} \sum_{k_1^+=1}^{n} \cdots \sum_{k_{N-1}^+=1}^{n} \sum_{k_0^-=1}^{n} \sum_{k_1^-=1}^{n} \cdots \sum_{k_{N-1}^-=1}^{n} \\
&\left\langle u_i \left| e^{-i\hat{H}_0\Delta t/\hbar} \right| u_{k_{N-1}^+} \right\rangle \left\langle u_{k_{N-1}^+} \left| e^{-i\hat{H}_0\Delta t/\hbar} \right| u_{k_{N-2}^+} \right\rangle \cdots \\
&\left\langle u_{k_1^+} \mathbf{x}_1^+ \left| e^{-i\hat{H}_0\Delta t/\hbar} \right| \mathbf{x}_0^+ u_{k_0^+} \right\rangle \\
&\left\langle u_{k_0^+} \left| \hat{\rho}(0) \right| u_{k_0^-} \right\rangle \left\langle u_{k_0^-} \left| e^{i\hat{H}_0\Delta t/\hbar} \right| u_{k_1^-} \right\rangle \\
&\left\langle u_{k_1^-} \left| e^{i\hat{H}_0\Delta t/\hbar} \right| u_{k_2^-} \right\rangle \cdots \\
&\left\langle u_{k_{N-1}^-} \left| e^{i\hat{H}_0\Delta t/\hbar} \right| u_j \right\rangle \\
&F\left( \sigma_{k_0^+}, \sigma_{k_1^+}, \ldots, \sigma_{k_{N-1}^+}, \sigma_i, \sigma_{k_0^-}, \sigma_{k_1^-}, \ldots, \sigma_{k_{N-1}^-}, \sigma_j \right)
\end{aligned} \tag{2.22}$$

In this $F$ is the Feynman-Vernon influence functional, given by the expression

$$F = \exp\left( -\frac{1}{\hbar} \sum_{k=0}^{N} \sum_{k'=0}^{k} \left( \sigma_k^+ - \sigma_k^- \right) \left( \eta_{kk'} \sigma_{k'}^+ - \eta_{kk'}^* \sigma_{k'}^- \right) \right) \tag{2.23}$$

Here $\eta_{kk'}$ are complex-valued coefficients which depend on the collective characteristics of the medium via integrals involving the bath spectral density,

$$J(\omega) = \frac{\pi}{2} \sum_j \frac{c_j^2}{m_j \omega_j} \delta(\omega - \omega_j) \tag{2.24}$$

and have been given explicitly by Makri and Makarov (1995b). The influence functional augments the integrand of the system's reduced density matrix with precisely those collective effects of the environment that affect its evolution (Caldeira and Leggett, 1983).

The spectral density function is the coupling-weighted density of states of the harmonic bath. If the environment consists of only a few degrees of freedom, as in the case of a small organic molecule, $J(\omega)$ is composed of $\delta$ function-like peaks. Broad spectral densities are characteristic of macroscopic media (solids, liquids or large biological molecules) and are responsible for loss. In order for system relaxation to occur via first-order processes, this function must overlap with the characteristic frequencies of the quantum system; otherwise, transitions can take place only via multiphonon effects and the ensuing decay is slow.

In the absence of system-bath coupling the influence functional is equal to unity. In that limit the path integral representation of the reduced density matrix is equivalent to the time-dependent Schrödinger equation for the reference Hamiltonian. The situation described by Eq. (2.22) is significantly more complicated. Although the evolution of the full-dimensional system-bath density matrix is still given by Schrödinger's first-order differential equation, the process of integrating out the bath to obtain the lower-dimensional reduced density matrix has destroyed the Markovian character of the dynamics, leading to an influence functional of the system paths with nonlocal interactions, i.e., couplings between the variables $\sigma_k^{\pm}$ and $\sigma_{k'}^{\pm}$. These couplings prevent decomposition of the reduced-dimension path integral into sequential integrations of the type familiar from wavefunction propagation.

The nonlocal terms in the path integral expression of the reduced density matrix arise from the correlation function of the medium (Feynman and Vernon, 1963),

$$\alpha(t-t') = \hbar^{-1}\sum_j c_j^2 \left\langle x_j(t')x_j(t)\right\rangle \tag{2.25}$$

which in the case of a harmonic bath takes the form

$$\begin{aligned}\alpha(t) &= \sum_j \frac{c_j^2}{2m_j\omega_j}\left[\coth\left(\tfrac{1}{2}\hbar\omega_j\beta\right)\cos\left(\omega_j t\right) - i\sin\left(\omega_j t\right)\right] \\ &= \frac{1}{\pi}\int_0^\infty d\omega\, J(\omega)\left[\coth\left(\tfrac{1}{2}\hbar\omega\beta\right)\cos\left(\omega t\right) - i\sin\left(\omega t\right)\right]\end{aligned} \tag{2.26}$$

The coefficients $\eta_{kk'}$ in Eq. (2.23) are closely related to the trapezoid-rule discretization of the function $\alpha(t\text{-}t')$ (Makri and Makarov, 1995a).

The correlation function for a dissipative bath also decays with the time separation index $\Delta k = k - k'$, implying that the nonlocality of the influence functional has in practice a finite span which is equivalent to $\Delta k_{\max}$ path integral time steps (Makarov and Makri, 1994). The last observation allows decomposition of the path integral into a series of lower-dimensional integrals (Makri and Makarov, 1995a,b; Sim and Makri, 1997a). This is best seen from the diagrammatic representation of the path integral expression, Eq. (2.22), shown in Fig. 2. Here each vertex represents a time point on the forward or backward time axis, while the loops indicate the factors (pairwise interactions) between these points in the path integral. The nearest-neighbor interactions correspond to the reference system propagators as well as the influence functional terms having $k' = k - 1$. All other loops indicate other influence functional terms. In this illustration it is assumed that the bath correlation function decays to zero within $\Delta k_{\max} = 3$ time steps, leading to omission of loops connecting points separated by a larger time interval. The reduced density matrix at a later time $t = N\Delta k_{\max}\Delta t$ is obtained by summing over the quadrature points corresponding to all previous points connected by the appropriate loops (multiplied by the proper initial condition). This operation can also be performed via the procedure outlined below.

First, each forward or backward path in Eq. (2.8) is decomposed into shorter segments that span the time intervals $0 \le t \le (\Delta k_{\max} - 1)\Delta t$, $\Delta k_{\max}\Delta t \le t \le (2\Delta k_{\max} - 1)\Delta t$, etc. The time points on which these path segments are defined are designated in Fig. 2 as vertices of different color. Next, one constructs an array of forward and backward path segment pairs, i.e.,

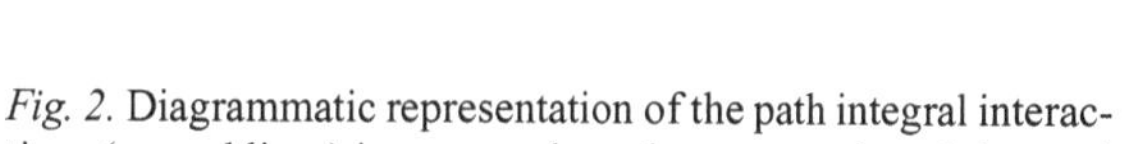


*Fig. 2.* Diagrammatic representation of the path integral interactions (curved lines) in a case where the memory length is equal to three time steps. The circles indicate time points separated by $\Delta t$. Black and white circles correspond to points defining the matrix **R** of path segments at times 0 and $k\Delta t$ ($k$ = 3), respectively. The solid lines indicate the interactions included in the propagator matrix.

sequences of $2\Delta k_{max}$ coordinate values $\sigma_1^{\pm},\ldots,\sigma_{\Delta k_{max}}^{\pm}$. The assumption that the length of influence functional interactions does not exceed $\Delta k_{max}$ time steps implies that each path segment is coupled only to its nearest neighbors. Therefore, a propagator matrix **T** can be constructed, whose matrix element $T_{ij}$ consists of all interactions that couple the *i*th path segment to its *j*th neighbor in the direction of increasing time. To take advantage of this structure, we introduce an array (vector) **R** of $L$ path segments (Makri, 1997) that is analogous to a multi-time reduced density matrix. The elements of this array are

$$R_i(m\Delta k_{max}\Delta t) \equiv \mathbf{R}((\sigma_1^{\pm},\ldots,\sigma_{\Delta k_{max}}^{\pm})_i;\, m\Delta k_{max}\Delta t),\quad i = 1,\ldots,L \tag{2.27}$$

defined on the time intervals $m\Delta k_{max}\Delta t \le t \le (m+1)\Delta k_{max}\Delta t$, where $m$ is an integer. This vector can be propagated forward in time via multiplication with the $L \times L$ propagator matrix **T**:

$$\mathrm{R}((m + 1)\Delta k_{max}\Delta t) = \mathbf{T}\cdot\mathbf{R}(m\Delta k_{max}\Delta t) \tag{2.28}$$

Notice that the matrix **T** connects pairs of 'earlier' and 'later' path segments. Finally, multiplication by an $n^2 \times L$ -dimensional endpoint propagator **S** yields (Sim and Makri, 1997a) the desired reduced density matrix at the discrete times $m\Delta k_{max}\Delta t$:

$$\tilde{\rho}(m\Delta k_{max}\Delta t) = \mathbf{S}\cdot\mathbf{R}((m - 1)\Delta k_{max}\Delta t) \tag{2.29}$$

The procedure has been described in detail (Sim and Makri, 1997a) and is summarized in the diagram of Fig. 3.

The $n$ states employed in the calculation give rise to a total of $n^{2\Delta k_{max}}$ forward and backward path segments. Although dramatically smaller than the total number $n^{2N}$ of paths that enter the path integral for $N >> \Delta k_{max}$, this size of the array **R** is in most cases prohibitive. The crucial observation (Sim and Makri, 1996; 1997a) that leads to a tractable scheme exploits the presence of a real part in the exponent of the influence functional. Due to this damping factor, the vast majority of paths enter the path integral with negligible weight and therefore could be neglected. Because of the product structure of the influence functional, a path that spans the long time $N\Delta t$ can have appreciable weight only if *all* of its constituent path segments have weights that are larger than some threshold. In general, for $\Delta k_{max} \gg 1$ the number of path segments with appreciable weight constitutes a very small fraction of the total number $n^{2\Delta k_{max}}$ of conceivable forward and backward path segments that span the memory length. Therefore, the iterative propagation is preceded by a Monte Carlo random walk in the space of paths with length equal to the memory time which identifies those with weight larger than the pre-selected threshold. The $L$ path segments selected from this procedure form the elements of the array **R** introduced above. In many situations typical of biological electron transfer the storage requirements of the scheme described above are modest.

During each step in the multiplication process, path segments are pieced together to form longer paths. After $N$ iterations of Eq. (3.11) the total number of forward/backward paths effectively included is equal to $L^{N-1}$. For example, a selection of just $10^4$ path segments leads after 100 iterations to a result equivalent to that obtainable through explicit summation over $10^{400}$ system paths. Clearly, inclusion of such an astronomical number of terms through global (non-iterative) summation is not possible.

## III. Exciton Coherence Length in Light Harvesting Complexes

The process of energy transduction in the membranes of photosynthetic bacteria begins with the absorption of visible light by light harvesting antennas which then funnel the energy into the reaction center (van Grondelle et al., 1994; Frank and Cogdell, 1996; Pullerits and Sundström, 1996; Kühn et al., 1997; Gnanakaran

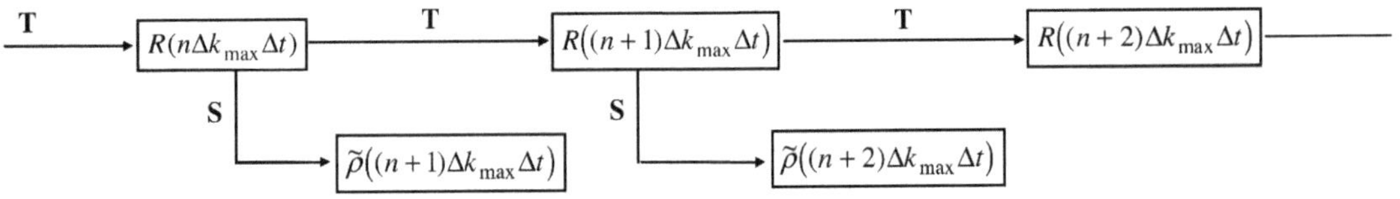


*Fig. 3.* Diagrammatic representation of the procedures involved in the iterative propagation of the reduced density matrix.

et al., 1999; Sundström et al., 1999). The light harvesting complexes are molecular aggregates composed of several units that contain peptides, chlorophyll molecules and carotenoids. These building blocks are organized in symmetric ring-shaped structures. The light harvesting complex I (LH1) immediately surrounds the reaction center, while a second type of light harvesting complex (LH2) channels energy to the reaction center through LH1.

The structure and electronic arrangement of *Rps. acidophila* and *Rs. molischianum* have been determined in experimental and theoretical studies (Hu et al., 1995; McDermott et al., 1995). The basic unit of LH2 is a heterodimer consisting of two small protein subunits referred to as the α- and β- apoproteins. The α heterodimers bind three bacteriochlorophyll (BChl) molecules, two of which are in close contact. These dimers form the ring, which absorbs around 850 nm (Fig. 4; see also Chapter 13, Fig. 3). The third BChl of the structural unit is located about 19 Å away on an outer ring whose absorption maximum lies at 800 nm. Carotenoid molecules in close proximity to the outer BChl ring harvest light in a different spectral range and also prevent photooxidation of the chlorophylls. The LH2 of the above species is composed of eight or nine such units.

This chapter is concerned with the energy transfer process within the inner (B850) ring of these LH2 systems. In the absence of static disorder, the eigenstates of the sixteen singly excited BChl molecules of *Rs. molischianum* form two exciton bands. Only the degenerate second and third excited states carry oscillator strength, and thus initial excitation at zero temperature should produce a delocalized linear combination of these states. Femtosecond pump-probe experiments that monitored the subsequent energy flow have found that intraband relaxation occurs within less than 50 fs at room temperature (Kumble et al., 1996; Nagarajan et al., 1996; Kennis et al., 1997; Arnett et al., 1998; Freiberg et al., 1998). Whether the energy deposited in the ring remains delocalized under biological conditions has been debated by several authors. Static disorder caused by structural inhomogeneity (Reddy et al., 1991, 1993; van Mourik et al., 1992; Visschers et al., 1993; Bradforth et al., 1995), but also exciton-vibration coupling and thermal disorder, are factors that favor localization. The energy transfer process within the LH2 system has been studied by fluorescence, transient absorption, and time-resolved fluorescence anisotropy techniques (van Mourik et al., 1992; Pullerits et al., 1994b; Bradforth et al., 1995; Hess et al., 1995; Monshouwer and van Grondelle, 1996). Because of the complexity of these systems, the interpretation of the measured superradiance and emission anisotropy has been elusive, and the reported estimates of the coherence lengths vary from 1–2 subunits to a substantial portion of the ring. Some groups have reported evidence of delocalized states at all temperatures (Xiao et al., 1994; Kennis et al., 1996; Wu et al., 1997a), while the findings of others indicate incoherent energy transfer between monomers or dimers even under cryogenic conditions (van Mourik et al., 1992; Monshouwer et al., 1997). These apparent discrepancies may arise (Meier et al., 1997a) from a strong temperature dependence of the superradiance enhancement factor not accounted for in the interpretation of these experiments. Theoretical investigations have assessed the effect of static disorder on the coherence of the exciton eigenstates in an isolated ring and found it to be small (Hu et al., 1997). The effects of static and dynamic disorder on the resulting superradiance have been examined using adiabatic and polaron models (Meier et al., 1997b). Other work has employed a multilevel Redfield description to follow the dynamics of the photoexcited state in the presence of coupling to vibrations, which were modeled in terms of a dissipative bath and estimated the exciton coherence domain (Chachisvilis et al., 1997; Kühn and Sundström, 1997a,b). A dis-

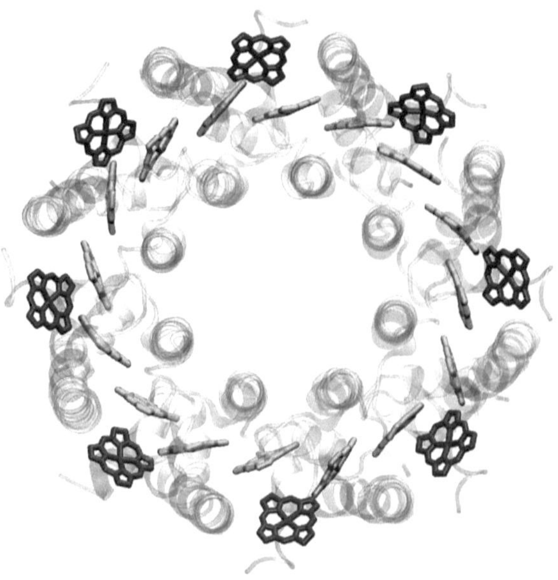

*Fig. 4.* Top view of the B850 ring of the LH2 in *Rs. molischianum*. (Courtesy of I. Kosztin.)

sipative single-exciton model of the LH2 complex has also been employed by Jang and Silbey (2003), who investigated the effects of dynamic and static disorder on spectral line shapes, in good agreement with recent single-complex experiments (Hofmann et al., 2003).

This section reviews the use of path integral techniques to investigate the equilibrium properties of the excited LH2 complex. The Monte Carlo path integral description of light harvesting systems is appealing for several reasons: it can take into account the effects of ensemble averaging at a given temperature, exciton-vibration coupling and static disorder automatically in a single calculation; in addition, it leads to simple pictures that allow easy visualization of the pertinent coherence length.

The photosynthetic antenna is modeled via the $n$-state Hamiltonian of Eq. (2.10), where $n = 16$ for the B850 ring of LH2 in *Rs. molischianum*. Here $|u_i\rangle$ denote single excitations of individual chlorophylls, $\varepsilon_i$ are the corresponding excitation energies, and $\Delta_{ij}$ are the couplings that are due primarily to dipolar interactions. Symmetry implies that $\Delta_{2k+i,2k+j} = \Delta_{ij}$. Further, in the absence of static or dynamic disorder the system is completely symmetric, thus $\varepsilon_i \equiv \varepsilon$. Hu et al. (1997) have parameterized this Hamiltonian by fitting the corresponding eigenvalues to results of ZINDO electronic structure calculations (Cory et al., 1998). The resulting parameters are given in Table 1.

The Hamiltonian describing the chlorophyll excitations is coupled diagonally to the vibrational and torsional modes of the pigments and their protein/water environment. A qualitative description of the effects of exciton-vibration coupling on the dynamics of energy transfer is obtained by modeling the nuclear motion of the complex in terms of a dissipative bath of effective harmonic modes, as described in the previous section. This effective harmonic bath is modeled by a spectral density that has the generic 'Ohmic' form,

$$J(\omega) = 2\pi\hbar\xi\omega e^{-\omega/\omega c} \tag{3.1}$$

This function has a maximum at $\omega_c = 100\ \text{cm}^{-1}$ (Kühn and Sundström, 1997b; Pullerits et al., 1994a). The reorganization energy between two adjacent chlorophyll monomers is related to the exciton-vibration coupling characterized by $\xi$ through the expression

*Table 1.* Parameterization of Eq. (3.1) as obtained by Hu et al. (1997). All parameters are in $\text{cm}^{-1}$.

| Parameters | Energy in $\text{cm}^{-1}$ |
|---|---|
| $\varepsilon$ | 13059 |
| $\Delta_{12}$ | 806 |
| $\Delta_{23}$ | 377 |
| $\Delta_{13}$ | −152 |
| $\Delta_{24}$ | −102 |

$$E_r = \frac{1}{\pi}\left(\sigma_k - \sigma_{k-1}\right)^2 \int_0^\infty \frac{J(\omega)}{\omega}\, d\omega = 2\hbar\xi\omega_c\left(\sigma_k - \sigma_{k-1}\right)^2 \tag{3.2}$$

A rough estimate of the reorganization energy is obtained from the dynamic Stokes shift, which has been found (Visser et al., 1995; Kumble et al., 1996) to be about $80\ \text{cm}^{-1}$. Ultrafast experiments (Kumble et al., 1996) also revealed that the electronic excitations are coupled to nuclear motions with frequencies ranging from 20 to $1000\ \text{cm}^{-1}$. We thus choose $\omega_c = 100\ \text{cm}^{-1}$. In the absence of more detailed information, $\sigma_k$ was set equal to the chlorophyll monomer index $k$ with periodic boundary conditions, i.e., $\sigma_{n+1} = \sigma_1$. The conclusions of this study were unaltered if a slightly nonuniform choice of $\sigma_k$ was made. Numerical calculations were performed for the values $\xi = 0.1$ and 0.25, which describe exciton-vibration coupling corresponding to reorganization energies equal to 20 and $50\ \text{cm}^{-1}$, respectively.

The calculation was performed in the site basis $u_k$. The partition function is given by the expression

$$Z = \sum_{k_1=1}^{n} \sum_{k_2=1}^{n} \cdots \sum_{k_N=1}^{n} \left\langle u_N \right| e^{-\beta H_0/N} \left| u_{N-1} \right\rangle \cdots \left\langle u_2 \right| e^{-\beta H_0/N} \left| u_1 \right\rangle \left\langle u_1 \right| e^{-\beta H_0/N} \left| u_N \right\rangle F\left(\sigma_1, \sigma_2, \ldots, \sigma_N\right) \tag{3.3}$$

Because paths entering the path integral expression of the partition function are closed, these paths can be thought of as a necklace whose beads correspond to the auxiliary variables of Eq. (3.3) (Chandler and Wolynes, 1981). In the present case each bead can occupy one of the $n$ lattice sites of the LH2 ring. This correspondence allows straightforward visual examination of the extent of coherence in the system.

Paths contributing to Eq. (3.3) are sampled via a Metropolis procedure using the entire integrand as the sampling function. The coherence length of a path is obtained from the expression

$$\ell = 2\sum_{k_1=1}^{n}\sum_{k_2=1}^{n}\cdots\sum_{k_N=1}^{n}\langle u_N|e^{-\beta H_0/N}|u_{N-1}\rangle\cdots$$
$$\langle u_2|e^{-\beta H_0/N}|u_1\rangle\langle u_1|e^{-\beta H_0/N}|u_N\rangle F(\sigma_1,\sigma_2,\ldots,\sigma_N)$$
$$\left[(\sigma_1-\sigma_c)^2+(\sigma_2-\sigma_c)^2+\cdots+(\sigma_N-\sigma_c)^2\right]^{\frac{1}{2}} \tag{3.4}$$

where $\sigma_c$ is the centroid of the closed path $\{\sigma_1, \sigma_2, \ldots, \sigma_N\}$.

The effects of static diagonal disorder can be evaluated by replacing the parameter $\varepsilon$ in Eq. (3.1) by the individual site energies $\varepsilon_i$ which are chosen from a Gaussian distribution. The mean coherence length (MCL) is then obtained from the ensemble average of $\ell$ with respect to the fluctuation of the site energy. Interestingly, the integration associated with the inhomogeneous broadening of the ensemble can be combined with the Monte Carlo search for cyclic paths, such that the effects of static disorder can be assessed within *a single random walk*. In this case the Boltzmann matrix elements of the reference Hamiltonian must be re-evaluated for each choice of site energies.

Figure 5 shows the coherence length in a 16-unit model of LH2 as a function of temperature for the isolated ring and also in the presence of small dissipation corresponding to $\xi = 0.1$ and $\xi = 0.25$. These calculations are performed without including static disorder, using the coupling parameters of Table 1 for $N = 32$ path integral slices and 50,000 Monte Carlo points per integration variable after an equilibration step of equal length. At absolute zero the eigenstates are entirely delocalized. Dominant paths reflecting this delocalized structure exhibit coherence. The centroid of a completely delocalized path ($\sigma_k = \text{int}(kn/N)$, $k = 1, \ldots, n$) is equal to

$$\sigma_c = \frac{n+1}{2} \tag{3.5}$$

and (taking $N = n$ for simplicity) the coherence length is

$$\ell = 2\sqrt{\frac{1}{n}\sum_{k=1}^{n}(k-\sigma_c)^2} = \sqrt{\frac{1}{3}(n^2-1)} \tag{3.6}$$

which for $n = 16$ equals 9.22. At low temperatures and in the absence of any disorder the system exhibits

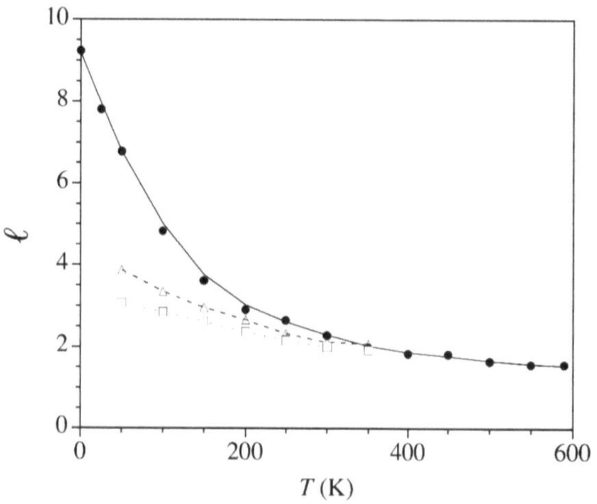

*Fig. 5.* Coherence length $\ell$ as a function of temperature for the LH2 model with couplings from Table 1. Solid circles: isolated ring ($\xi = 0$). Hollow triangles: very weak exciton-vibration coupling ($\xi = 0.1$). Hollow squares: weak exciton-vibration coupling ($\xi = 0.25$). The lines are guides to the eye.

moderate coherence, reflecting the delocalized nature of the eigenstates. This is illustrated in Fig. 6, which shows typical paths having coherence length within one unit of the calculated MCL at $T = 50$ K. Clearly, the delocalized exciton picture is relevant at or below this temperature. However, even in the case of an isolated, perfectly symmetric ring, only at very low temperatures does the path integral necklace begin to extend over the entire domain of the ring. As seen in Fig. 5, thermal averaging has a dramatic effect on energy localization. At room temperature the coherence length is only 2.3 in the absence of any disorder. Typical paths at $T = 300$ K are shown in Fig. 7. These paths now extend over only a few chlorophyll units. Note these paths appear more delocalized than they actually are, as significant portions of the path integral necklace have collapsed on a single chlorophyll unit, an effect that cannot be discerned in the graphs. Coupling the ring to a weakly dissipative environment induces further localization, leading to $\ell = 2.1$ with $\xi = 0.1$ and $\ell = 2.0$ if $\xi = 0.25$ at 300 K. Still, coherence spans one to two dimers at room temperature, so hopping between individual monomers does not provide a faithful representation of energy transfer. At low temperatures, where the perfect ring exhibits a high degree of coherence, the effects of dissipation are more dramatic, as can be seen from Fig. 5. Finally, inclusion of static disorder further enhances localization, although this effect is rather small since the standard deviation of the site energies is small

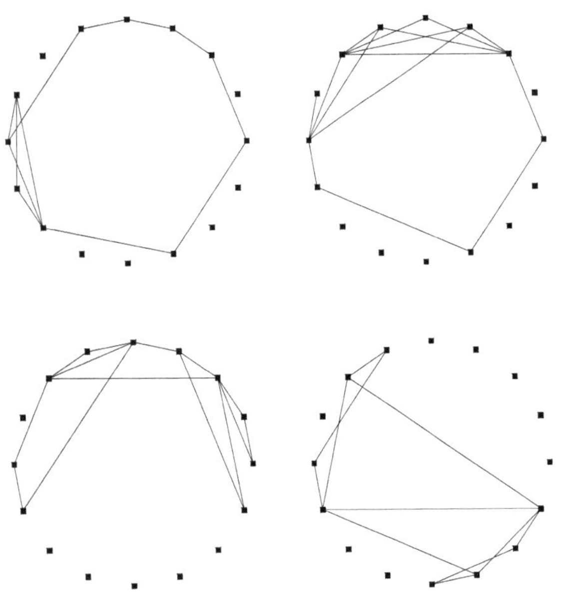

*Fig. 6.* Typical paths for the isolated LH2 system ($\xi = 0$, no static disorder) at $T = 50$ K with the parameters shown in Table 1. The markers represent the chlorophyll monomers.

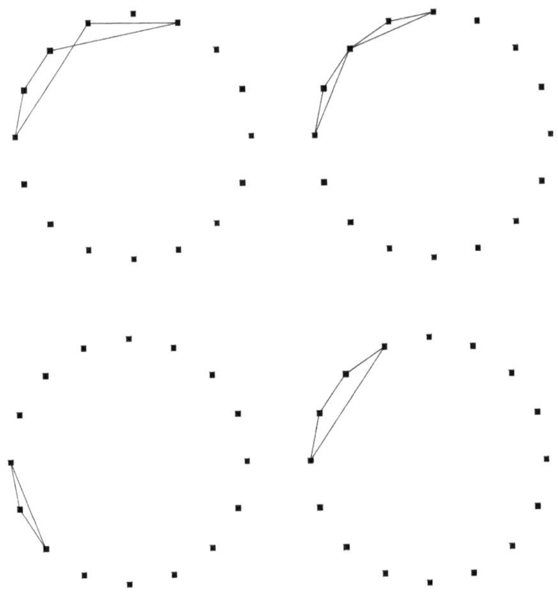

*Fig. 7.* Typical paths for the isolated LH2 system ($\xi = 0$, no static disorder) at $T = 300$ K with the parameters shown in Table 1. The markers represent chlorophyll monomers.

compared to the off-diagonal elements of the ring Hamiltonian (Hu et al., 1997; Wu et al., 1997b).

The off-diagonal coupling elements obtained by Hu et al. are likely to be too large (Krueger et al., 1998; Pullerits et al., 1997; Sauer et al., 1996). For this reason path integral calculations were also performed with couplings that were scaled down by twofold compared to those given in Table 1. Even though the MCL at zero temperature (where the ensemble is dominated by the ground state) is still given by Eq. (3.6), the MCL obtained with scaled down coupling parameters decreases faster with increasing temperature and becomes $\ell = 1.5$ at 300 K. The effects of dynamic disorder are again significant at low temperature, but given the already very small size of the paths at $T \geq 250$ K, this effect is negligible at room temperature.

From the results of these calculations it is evident that the coherence characterizing the exciton eigenstates of a perfect light harvesting ring prevails only at very low temperatures. The calculated low-temperature MCL is in agreement with the estimate of Fleming and coworkers (Jimenez et al., 1996). The major contributor to the localization observed at room temperature is thermal averaging, which destroys the coherent nature of individual eigenstates. Although static and dynamic disorder induce further localization, they are not crucial for disruption of coherence at biological temperatures. For this reason the conclusions reached on the basis of the simple model employing a harmonic dissipative bath appear robust and rather general.

In contrast to the localizing effects of the environment (temperature, static and dynamic disorder), the geometry of light harvesting rings favors coherent energy transfer at low temperatures. Compared to a linear aggregate, the circular arrangement of the chlorophyll molecules eliminates the penalty in the Boltzmann factor associated with closing the path integral necklace for a path extending over at least half of the ring units. Choosing for example $N = n$, a completely delocalized path with $\sigma_k = k$, $k = 1, \ldots, n$ enters the partition function with exactly the same weight as the path

$$\sigma_k = \begin{cases} k, & k = 1, \ldots, n/2 \\ n+1-k, & k = n/2+1, \ldots, n \end{cases} \tag{3.7}$$

This would not be the case if the pigments were arranged in an open shape. Since the coherence length of the first of these paths is larger, the overall MCL is increased. The geometric enhancement of coherence is modest under conditions which already favor delocalized paths, i.e., at low temperatures and weak disorder, but does not have an observable effect under biological conditions where coherence appears to span a small portion of the ring.

In the absence of static disorder, the energy absorbed by the B850 ring of the LH2 system is used to excite a pair of degenerate exciton states which are delocalized over the entire ring. However, coherence does not survive thermalization. At room temperature, the onset of thermal equilibrium is characterized by an incoherent ensemble of excitations that extend over two to three chlorophyll units. Thus, at room temperature the coherence length is reduced by a factor of four with regard to the zero temperature case. At very low temperatures coherence spans a significant portion of the ring and is enhanced by the circular arrangement of the pigments.

Although these conclusions were reached on the basis of a very simple model, they appear quite robust and insensitive to the details of the calculation. The path integral estimate of the room temperature MCL is in reasonable agreement with the conclusions reached by the groups of van Grondelle (van Mourik et al., 1992; Monshouwer et al., 1997) and Sundström (Pullerits and Sundström, 1996; Chachisvilis et al., 1997; Kühn and Sundström, 1997b; Meier et al., 1997b), which report a coherence domain of 3–4 monomers. Addition of exciton-vibration coupling enhances localization. In the present work this coupling was modeled in terms of a dissipative harmonic bath. Although anharmonicity of the vibrational degrees of freedom may weaken the localizing effects of dynamic disorder, such motion should nevertheless destabilize extended states. It is thus hard to imagine the more intricate motion of the pigments and protein atoms resulting in a different outcome. Finally, if static disorder is also taken into consideration, the resulting MCL becomes even shorter. Thus, energy delocalization over the entire ring appears extremely improbable under biological conditions, although within a dimer and probably across two dimers coherence appears to prevail.

## IV. Dynamics of Primary Charge Separation

The energy absorbed by the light harvesting complexes is funneled to the reaction center (see Fig. 8), where it excites a special chlorophyll pair. This excitation initiates a series of electron transfer events, of which the first and fastest involves charge separation, where an electron is transferred from the

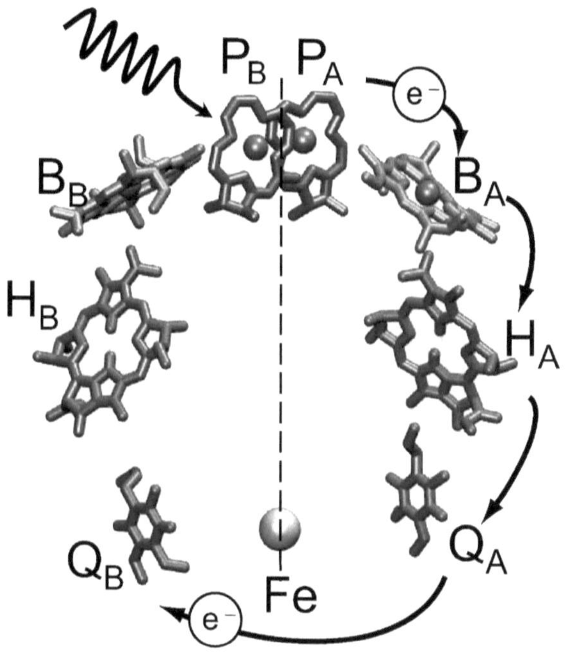


*Fig. 8.* Arrangement of chromophores in the photosynthetic reaction center of *Rb. sphaeroides*. The center-to-center distance between the bacteriochlorophyll special pair (donor) and the bacteriopheophytin (acceptor) is 17 Å. (Courtesy of I. Kosztin.)

excited special bacteriochlorophyll pair (the donor) to a bacteriopheophytin on the L branch (the acceptor). In spite of the large (17 Å center-to-center) distance between these two sites (Deisenhofer et al., 1984), this electron transfer process is very fast and efficient, characterized by a time constant of about 3 ps and a quantum yield that approaches unity. In addition, a weak inverse temperature dependence is observed. Understanding the specifics and mechanism of the primary charge separation has been the subject of numerous experimental and theoretical studies (Deisenhofer and Norris, 1993; Fleming and Grondelle, 1994).

A basic understanding of the steps in the electron transfer pathway emerged almost three decades ago. Early work by Rentzepis' group established the main sequence of events, while a series of experiments with subpicosecond resolution on the reaction centers of *Rps. viridis* and *Rb. sphaeroides* by Holten et al. (1980), Woodbury et al. (1985), Martin et al. (1986), Breton et al. (1986a,b, 1988), and Fleming et al. (Chan et al., 1991; Fleming et al., 1988) determined the kinetic features of the primary charge separation and its temperature dependence. Because of the large separation between the primary donor and acceptor, the electronic coupling between these units is estimated to be very small. It is now proven that a bacteriochlorophyll monomer located between the special pair and the bacteriopheophytin is involved by providing a bridge state. However, the mechanism of the process remained controversial for a long time (Bixon et al., 1987, 1991; Marcus, 1987,1988; Parson et al., 1987; Creighton et al., 1988; Warshel et al., 1988; Hu and Mukamel, 1989a,b; Sharp and Honig, 1990; Thompson and Zerner, 1991; Marchi et al., 1993; Egger and Mak, 1994; Alden et al., 1995; Mak and Egger, 1995). A bridge state energetically located below the donor is consistent with a simple two-step electron transfer process (in which the reduced bacteriochlorophyll participates as a true intermediate), while the opposite scenario implies superexchange (a process in which the bridge enhances the transfer rate without actually accommodating the electron). Determination of the energetics of the electron-occupied bridge state (i.e., the reduced bacteriochlorophyll) from first-principles calculations was not possible until the mid-1990s, and calculations including electron correlation on such large molecules remain out of reach. Further, the transient population of that state remained below experimental detection threshold. The lack of observable electron population on the accessory chlorophyll in *Rps. viridis* and *Rb. sphaeroides* implies that the transfer proceeds either via superexchange or via sequential hopping with a fast second kinetic step causing a rapid depletion of bridge population. The first concrete evidence in favor of the two-step interpretation came from experiments on modified reaction centers of *Rb. sphaeroides* by Zinth's group, which reported a substantial long-lived population of the bridge state (Huber et al., 1995; Schmidt et al., 1994). However, the validity of this interpretation of Zinth's experiments was challenged (Egger et al., 1994).

Real-time path integral calculations using the iterative decomposition of the path integral described in the previous section, which are feasible and accurate over long times, have helped settle the debate. These calculations (Makri et al., 1996; Sim and Makri, 1997b) employed a three-state model of charge transfer in wild-type and modified photosynthetic reaction centers using spectral densities obtained from molecular dynamics simulations (Marchi et al., 1993). Because this iterative procedure does not suffer from phase cancellation problems typical of real-time Monte Carlo path integral methods, it allowed calculation of the electronic state dynamics over 17 ps; this time interval was sufficiently long for transient populations to decay, offering a complete and unambiguous picture of the process.

The path integral calculations employed a model Hamiltonian that comprises three electronic states. The first of these states is the excited special bacteriochlorophyll pair $P^*$, while the charge distribution of the bacteriochlorophyll monomer $B_L$ and bacteriopheophytin $H_L$ remains unchanged. The transfer is mediated by the reduced state of the bacteriochlorophyll monomer, which couples to the electronic states of donor and acceptor. In short, the three relevant states are

| | | |
|---|---|---|
| State 1: | $P^* \; B_L \; H_L$ | (donor) |
| State 2: | $P^{*+} \; B_L^- \; H_L$ | (bridge) |
| State 3: | $P^{*+} \; B_L \; H_L^-$ | (acceptor) |

Ignoring direct donor-acceptor coupling, which is estimated to be very small due to the large separation of these units, the Hamiltonian takes the form of Eq. (2.10) with $\Delta_{13} = \Delta_{31} = 0$. In all cases the energy of the donor was set arbitrarily to zero. The value of $\varepsilon_3$ was determined from magnetic experiments and known redox potentials (Chidsey et al., 1985; Goldstein et al., 1988; Schmidt et al., 1994) and is shown in Table 1 for the various systems studied. The free energy of the accessory chlorophyll was varied in the range $-600$ to $+2000$ cm$^{-1}$, the first of these values being a lower bound implied by the experimental data reported by Schmidt et al. (1994). Similarly, the electronic couplings $\Delta_{12}$ and $\Delta_{23}$ were varied. For each set of these three unknown parameters the electron transfer dynamics were followed over a time interval sufficiently long to determine rates and population characteristics. Comparison with various experimental findings served as the acceptance criterion for realizable parameters.

The above Hamiltonian describes the energetics with all nuclear coordinates frozen at the minima of the corresponding diabatic potential curves. The collective effects of the protein and solvent coordinates were modeled as a Gaussian heat bath. Warshel and coworkers have pioneered the use of classical simulation methods for mapping anharmonic solvents onto effective harmonic oscillator baths by means of inverting energy gap correlation functions (Warshel and Hwang, 1986; Warshel et al., 1989). In the absence of more specific information, Makri et al. (1996)

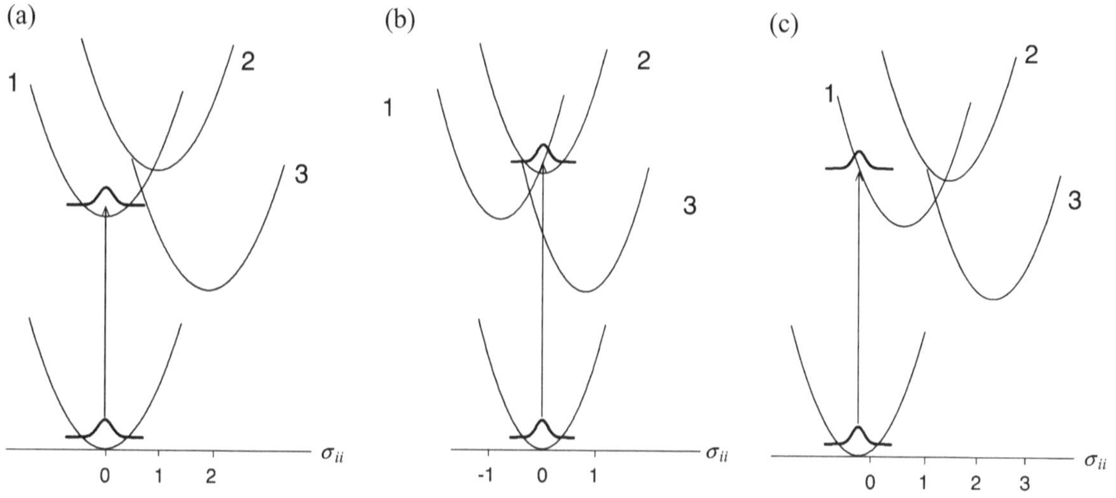


*Fig. 9.* Schematic representation of the diabatic free energy surfaces in the linear model discussed in section II. Three configurations of the electronic ground state of the bacteriochlorophyll special pair are shown, resulting in different initial states. (a) No displacement ($\delta = 0$), resulting in an excited state at thermal equilibrium. (b) The ground state is displaced toward the bridge ($\delta = 1$); the solvation energy of the initial donor state is $E_s = E_r/4$. (c) The ground state is displaced away from the bridge ($\delta = -1$); the solvation energy of the initial donor state is $E_s = E_r/4$.

assumed that the minimum energy configurations of the three electronic states lie on a straight line, being separated by equal distances, and that the frequencies and coupling characteristics of the effective medium modes have the same values for all three states. The parameter that specifies the relative positions of the nuclear coordinates in the three electronic states are given the values $\sigma_1 = -\delta$, $\sigma_2 = 1 - \delta$ and $\sigma_3 = 2 - \delta$, where $\delta$ is an auxiliary parameter that accounts for unrelaxed initial states of the bath. Adopting the convention that the electronic ground state, and thus the protein and solvent modes of the excited complex, are initially in the equilibrium state corresponding to $x_j = 0$, the choice $\delta = 0$ describes an initial state where all nuclear coordinates are equilibrated with respect to the excited special pair. Nonzero values of $\delta$ correspond to nonequilibrium initial states of the medium, where the nuclear coordinates are displaced, as in photoexcitation achieved by means of a vertical transition from the ground state P when the latter is displaced with respect to P* (Cho and Silbey, 1995). The solvation energy of the initial state is determined from the fluorescence Stokes shift. Typical arrangements of these states are depicted in Fig. 9.

The collective characteristics of the solvent or biological medium can be mapped onto those that specify the system-bath model by invoking the quantum-classical correspondence for harmonic oscillators. As the nuclear degrees of freedom (modeled as classical variables) fluctuate about their equilibrium configuration, the energy gap between the donor and acceptor states fluctuates about its average value as the function

$$\text{const.} + \mathfrak{d}\sum_j c_j x_j^{cl}(t) \tag{4.1}$$

Where $x_j^{\rm cl}(t)$ denote the classical coordinates and $\mathfrak{d}$ is a constant determined by the donor-acceptor distance in the Hamiltonian. The correlation function $C_{13}(t)$ of the donor-acceptor energy gap fluctuations about its equilibrium average is then

$$C_{13}(t) = \mathfrak{d}^2 \sum_j c_j^2 \left\langle x_j^{\rm cl}(0) x_j^{\rm cl}(t) \right\rangle_\beta \tag{4.2}$$

Noting that $C_{\rm DA}(t)$ is simply related to the classical limit of the response function,

$$C_{13}(t) = \mathfrak{d}^2 \lim_{\hbar \to 0} \hbar\alpha(t) \tag{4.3}$$

the gap fluctuation correlation function can be expressed in terms of the spectral density according to the relation

$$C_{13}(t) = \frac{2\mathfrak{d}^2}{\pi\beta} \int_0^\infty \frac{J(\omega)}{\omega} \cos\omega t \, d\omega \tag{4.4}$$

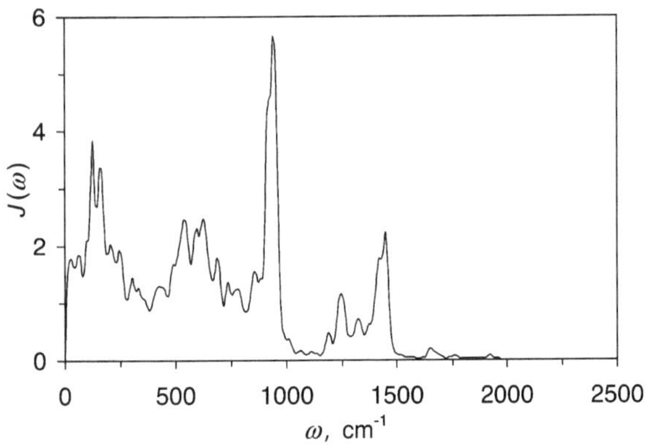


*Fig. 10.* The spectral density function of the donor-acceptor gap fluctuations.

This expression can be inverted to yield the spectral density as the Fourier integral

$$J(\omega) = \frac{\beta\omega}{\delta^2}\int_0^\infty C_{13}(t)\cos\omega t\,dt \tag{4.5}$$

Thus, the collective characteristics of the medium relevant to the electron transfer process can be found from the Fourier transform of the classical gap fluctuation correlation function. The latter was obtained by means of classical molecular dynamics simulations that included crystallization water molecules treated as simple point charges (Marchi et al., 1993).

This spectral density is depicted in Fig. 10. It should be noted that this function cannot be quantitatively accurate at very low frequencies, which correspond to the long time (tens of picoseconds) tail of the gap correlation function that is not accessible to the molecular dynamics simulation. Such frequencies have been estimated to account for a substantial fraction of the total reorganization energy (Gehlen, 1995) and have been shown to result in deviations from exponential kinetics (Gehlen et al., 1994). Although this deficiency of the spectral density employed in the path integral calculations may be of some concern, it is unlikely that it impacts the conclusions drawn from bulk features of the electronic population dynamics.

Finally, the energy $E_r$ for reorganization of the bath between the primary electron donor and acceptor states is given from the relation

$$E_r = \frac{4}{\pi}\int_0^\infty \frac{J(\omega)}{\omega}d\omega \tag{4.6}$$

while the reorganization energy between donor and bridge (and also that between bridge and acceptor) is equal to $E_r/4$. The solvation energy $E_s$ of the initial state was estimated from the measured Stokes shift. Most of the simulations used the value $E_s = 500$ cm$^{-1}$. Since 500 cm$^{-1}$ $= E_r/4$, the above solvation energy corresponds to $\delta = \pm 1$. Additional simulations performed with smaller values of $E_s$ confirmed that the main conclusions of this work are qualitatively insensitive to details of initial preparation.

From the weak temperature dependence of the experimental transfer rate in wild-type reaction centers (Fleming et al., 1988) it is concluded that the rate determining step is nearly activationless. With large positive values of the bridge free energy, for which the dominant mechanism is superexchange, activationless transfer implies that the reorganization energy between donor and acceptor must coincide with the donor-acceptor energy gap, i.e., $E_r \approx \varepsilon_1 - \varepsilon_3$. On the other hand, with values of $\varepsilon_2$ slightly below $\varepsilon_1$ the process is sequential with $P^*B_LH_L \rightarrow P^{*+}B_L^-H_L$ being the rate-determining step. In this case the reorganization energy between donor and bridge must be approximately equal to the energy gap of these states, i.e., $E_r/4 \approx \varepsilon_1 - \varepsilon_2$. The calculations used the value $E_r = 2000$ cm$^{-1}$. This satisfies the above requirement for the wild-type reaction centers if $\varepsilon_2 \gg 0$, in which case superexchange dominates, and also with $\varepsilon_2 \approx -500$ cm$^{-1}$ for which the transfer is sequential.

The computed electronic populations are shown in Fig. 11 for parameters that give rise to results compatible with the experimental data on the wild-type and modified reaction centers. The value of the bridge free energy for which best agreement was obtained is around $-400$ cm$^{-1}$, i.e., the reduced accessory chlorophyll lies substantially below the donor state. Even though a non-negligible transient population of this state can build up if the bridge lies within a few hundred wavenumbers above the excited special pair, that population is short-lived and decays within a few picoseconds. Thus, the iterative path integral simulations concluded that the detection of a significant long-lived population of the reduced chlorophyll monomer is not compatible with a superexchange mechanism. The coupling parameters for which the results of the path integral calculations reproduce the observed kinetics are $\Delta_{12} = 22$ cm$^{-1}$ and $\Delta_{23} = 135$ cm$^{-1}$ for the wild-type reaction center. Shortly after the path integral results were reported, Zhang and Friesner (1997) calculated these couplings by means of *ab-initio* electronic structure methods, obtaining values which are in very good agreement with those

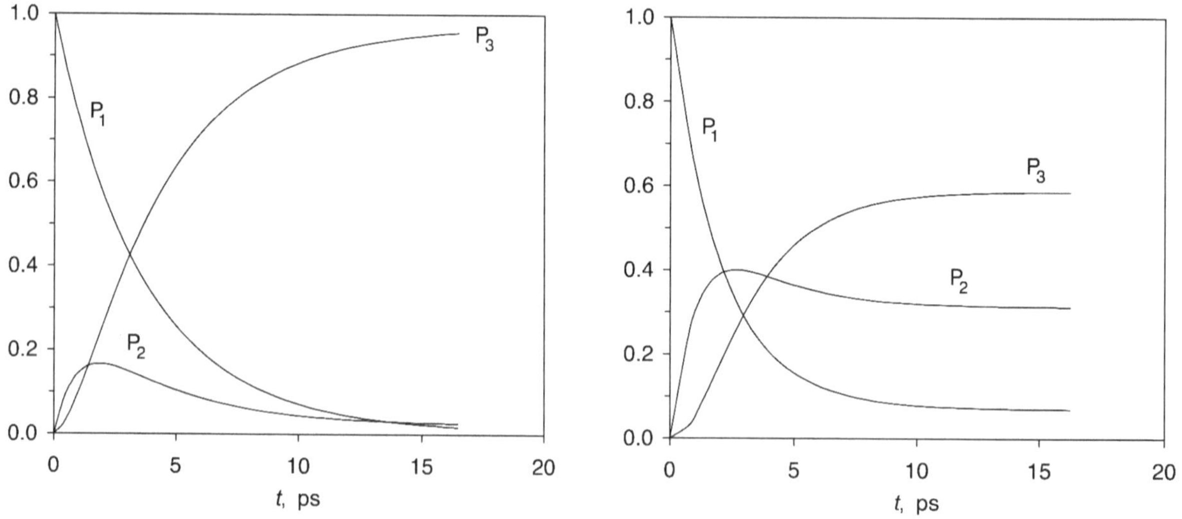


*Fig. 11.* Evolution of the electronic state populations for the wild-type reaction center with the parameters $E_1 = 0$, $E_2 = -400$ cm$^{-1}$, and $\Delta_{12} = 22$ cm$^{-1}$. The solvation energy of the initial state is $E_s = 500$ ($\delta = 1$). All results shown are converged to 5%. (left) Wild-type reaction center with $E_3 = -2000$ cm$^{-1}$, $\Delta_{23} = 135$ cm$^{-1}$. (right) Modified reaction center with $E_3 = -630$ cm$^{-1}$ and $\Delta_{23} = 22$ cm$^{-1}$.

determined by Sim and Makri (1997b). Additional path integral calculations explored the sensitivity of the electronic population dynamics with respect to various parameters and assumptions of the model, including the precise form of the initial unrelaxed state of the protein environment. These calculations concluded that the results are qualitatively insensitive to the above factors, thus confirming the robustness of the model and the validity of the derived conclusions.

## V. Concluding Remarks

It is well known that path integral methods provide the most powerful, robust and versatile simulation tool for investigating properties of large systems at thermal equilibrium. Although most applications of the path integral Monte Carlo method have focused on quantum fluids, biological systems present ideal candidates where path integral methods can lead to definitive answers and new insights. Less obvious is the utility of path integral methodology for time-dependent processes. The first viable real-time path integral algorithm emerged only in the mid-1990s, and while the method has already found wide application to quantum processes in harmonic dissipative media, its feasibility for simulating more complex systems in the absence of simplifying assumptions is yet to be demonstrated. Nevertheless, the path integral method has already enjoyed much success, and possibilities for improvement, extension and application are countless.

## Acknowledgments

The work reviewed in this chapter has been possible through the generous support of several agencies. I wish to thank the Arnold and Mabel Beckman Foundation for a Beckman Young Investigator Award, the David and Lucile Packard Foundation for a Packard Fellowship for Science and Engineering, the Alfred Sloan Foundation for a Sloan Research Fellowship, the Research Corporation for a Cottrell Scholar Award, the Camille and Henry Dreyfus Foundation for a Camille Dreyfus Award, and the National Science Foundation for a Young Investigator Award and for Grants No. 93-13603, 98-77050, 02-12640, 04-27082 and 05-18452.

## References

Alden RG, Parson WW, Chu ZT and Warshel A (1995) Calculations of electrostatic energies in photosynthetic reaction centers. J Am Chem Soc 117: 12284–12298

Arnett DC, Kumble R, Visschers RW, Dutton PL, Hochstrasser RM and Scherer NF (1998) Ultrafast studies of exciton dynamics in light-harvesting dimers. In: Scherer NF and Hicks JM (eds) Laser Techniques for Condensed Phase and Biological Systems, pp 244–255. SPIE Proc Vol 3273. International Society for Optical Engineering, Bellingham

Bandrauk AD and Shen H (1991) Improved exponential split operator method for solving the time-dependent Schrödinger equation. Chem Phys Lett 176: 428–432

Binder K and Heermann DW (1988) Monte Carlo Simulation in Statistical Physics. Springer-Verlag, Berlin

Bixon M, Jortner J, Michel-Beyerle ME, Ogrodnik A and Lersch W (1987) The role of the accessory bacteriochlorophyll in reaction centers of photosynthetic bacteria: Intermediate acceptor in the primary electron transfer. Chem Phys Lett 140: 626–630

Bixon M, Jortner J and Michel-Beyerle ME (1991) On the mechanism of the primary charge separation in bacterial photosynthesis. Biochim Biophys Acta 1056: 301–315

Bradforth SE, Jimenez R, van Mourik F, van Grondelle R and Fleming GR (1995) Excitation transfer in the core light-harvesting complex (LH-1) of *Rhodobacter sphaeroides*: An ultrafast fluorescence depolarization and annihilation study. J Phys Chem 99: 16179–16191

Breton J, Martin JL, Migus A, Antonetti A and Orszag A (1986a) Femtosecond spectroscopy of excitation energy transfer and initial charge separation in the reaction center of the photosynthetic bacterium *Rhodopseudomonas viridis*. Proc Natl Acad Sci USA 83: 5121–5125

Breton J, Martin JL, Petrich J, Migus A and Antonetti A (1986b) The absence of a spectroscopically resolved intermediate state $P^+B^-$ in bacterial photosynthesis. FEBS Lett 209: 37–43

Breton J, Martin J-L, Fleming GR and Lambry J-C (1988) Low-temperature femtosecond spectroscopy of the initial step of electron transfer in reaction centers from photosynthetic purple bacteria. Biochemistry 27: 9276–8284

Caldeira AO and Leggett AJ (1983) Path integral approach to quantum Brownian motion. Physica A 121: 587–616

Ceperley DM and Pollock EL (1986) Path integral computation of the low-temperature properties of liquid $^4$He. Phys Rev Lett 56: 351–354

Chachisvilis M, Kühn O, Pullerits T and Sundström V (1997) Excitons in photosynthetic purple bacteria: Wavelike motion or incoherent hopping? J Phys Chem 101: 7275–7283

Chan C-K, DiMagno TJ, Chen LX-Q, Norris JR and Fleming GR (1991) Mechanism of the initial charge separation in bacterial photosynthetic reaction centers. Proc Natl Acad Sci USA 88: 11202–11206

Chandler D (1987) Introduction to Modern Statistical Mechanics. Oxford University Press, New York

Chandler D and Wolynes PG (1981) Exploiting the isomorphism between quantum theory and the classical statistical mechanics of polyatomic fluids. J Chem Phys 74:4078–4095

Chidsey CED, Takiff L, Goldstein RA and Boxer SG (1985) Effect of magnetic fields on the triplet state lifetime in photosynthetic reaction centers: Evidence for thermal repopulation of the initial radical pair. Proc Natl Acad Sci USA 82: 6850–6854

Cho M and Silbey R (1995) Nonequilibrium photoinduced electron transfer. J Chem Phys 103: 595–606

Coalson RD, Freeman DL and Doll JD (1989) Cumulant methods and short-time propagators. J Chem Phys 91: 4242–4248

Cory MG, Zerner MC, Hu X and Schulten K (1998) Electronic excitations in aggregates of bacteriochlorophylls. J Phys Chem 102: 7640–7650

Creighton S, Hwang JK, Warshel E, Parson WW and Norris J (1988) Simulating the dynamics of the primary charge separation process in bacterial photosynthesis. Biochemistry 27: 774–781

De Raedt H and De Raedt B (1983) Applications of the generalized Trotter formula. Phys Rev A 28: 3575–3580

Deisenhofer J and Norris JR (eds) (1993) The photosynthetic reaction center. Academic Press, New York

Deisenhofer J, Epp O, Miki K, Huber R and Michel H (1984) X-ray structure analysis of a membrane-protein complex — Electron density map at 3Å resolution and a model of the chromophores of the photosynthetic reaction center from *Rhodopseudomonas viridis*. J Mol Biol 180: 385–398

Doll JD and Freeman DL (1988) Stationary phase Monte Carlo methods. Adv Chem Phys 73: 289–304

Doll JD, Freeman DL and Gillan MJ (1988): Stationary phase Monte Carlo methods: An exact formulation. Chem Phys Lett 143: 277–283

Doll JD, Freeman DL and Beck TL (1990) Equilibrium and dynamical Fourier path integral methods. Adv Chem Phys 78: 61–127

Egger R and Mak CH (1994) Dissipative three-state system and the primary electron transfer in the bacterial photosynthetic reaction center. J Phys Chem 98: 9903–9918

Egger R, Mak CH and Weiss U (1994) Rate concept and retarted master equations for dissipative tight-binding models. Phys Rev E 50: R655–R658

Feynman RP (1948) Space-time approach to non-relativistic quantum mechanics. Rev Mod Phys 20: 367–387

Feynman RP (1972) Statistical Mechanics. Addison-Wesley, Redwood City

Feynman RP and Hibbs AR (1965) Quantum Mechanics and Path Integrals. McGraw-Hill, New York

Feynman RP and Vernon J (1963) The theory of a general quantum system interacting with a linear dissipative system. Ann Phys 24: 118–173

Filinov VS (1986) Calculation of the Feynman integrals by means of the Monte Carlo method. Nucl Phys B 271: 717–725

Fleming GR and Grondelle RV (1994) The primary steps of photosynthesis. Phys Today Feb 1994: 49–55

Fleming GR, Martin JL and Breton J (1988) Rates of primary electron transfer in photosynthetic reaction centres and their mechanistic implications. Nature 333: 190–192

Frank HA and Cogdell RJ (1996) Carotenoids in photosynthesis. Photochem Photobiol 63: 257–264

Freiberg A, Jackson JA, Lin S and Woodbury NW (1998) Subpicosecond pump-supercontinuum probe spectroscopy of LH2 photosynthetic antenna proteins at low temperature. J Phys Chem 102: 4372–4380

Fujiwara Y, Osborn TA and Wilk SFJ (1982) Wigner-Kirkwood expansions. Phys Rev A 25: 14–34

Gehlen JN (1995) The Effect of High and Low Frequency Polarization Modes on the Kinetics of Electron Transfer. Ph. D. Thesis, University of California, Berkeley

Gehlen J, Marchi M and Chandler D (1994) Dynamics affecting the primary charge transfer in photosynthesis. Science 263: 499–502

Glasner M, Yevick D and Hermansson B (1991) Generalized propagation formulas of arbitrarily high order. J Chem Phys 95: 8266–8272

Gnanakaran S, Haran G, Kumble R and Hochstrasser RM (1999) Resonance energy transfer. In: Andrews DL, Demidov AA (eds) Resonance Energy Transfer, pp 308–365. John Wiley

and Sons, Chichester
Goldstein RA, Takiff L and Boxer SG (1988) Energetics of initial charge separation in bacterial photosynthesis: the triplet decay rate in very high magnetic fields. Biochim Biophys Acta 934: 253–263
Hess S, Chachisvilis M, Timpmann K, Jones MR, Fowler GJS, Hunter CN and Sundström V (1995) Temporally and spectrally resolved subpicosecond energy transfer within the peripheral antenna complex (LH2) and from LH2 to the core antenna complex in photosynthetic purple bacteria. Proc Natl Acad Sci USA 92: 12333–12337
Hoffman DK and Kouri DJ (1992) Distributed approximating function theory: A general, fully quantum approach to wave propagation. J Phys Chem 96: 1179–1184
Hofmann C, Ketelaars M, Matsushita M, Michel H, Aartsma TJ and Koehler J (2003) Single-molecule study of the electronic couplings in a circular array of molecules: Light harvesting-2 complex from *Rhodospirillum molischianum*. Phys Rev Lett 90: 013004
Holten D, Hoganson C, Windsor MW, Schenck CC, Parson WW, Migus A, Fork RL and Shank CV (1980) Subpicosecond and picosecond studies of electron transfer intermediates in *Rhodopseudomonas sphaeroides* reaction center. Biochim Biophys Acta 592: 461–477
Hu X, Ritz T, Damjanovic A and Schulten K (1997) Pigment organization and transfer of electronic excitation in the photosynthetic unit of purple bacteria. J Phys Chem 101: 3854–3871
Hu X, Xu D, Hamer K, Schulten K, Koepke J and Michel H (1995) Predicting the structure of the light harvesting complex II of *Rhodospirillum molischianum*. Protein Science 4: 1670–1682
Hu Y and Mukamel S (1989a) Superexchange and electron transfer in the photosynthetic reaction center. Chem Phys Lett 160: 410–416
Hu Y and Mukamel S (1989b) Tunneling versus sequential long range electron transfer: analogy with pump-probe spectroscopy. J Chem Phys 91: 6973–6988
Huber H, Meyer M, Nagele T, Hartl I, Scheer H, Zinth W and Wachtveitl J (1995) Primary photosynthesis in reaction centers containing four different types of electron acceptors at site $H_A$. Chem Phys 197: 297–305
Jang S and Silbey RJ (2003) Single complex line shapes of the B850 band of LH2. J Chem Phys 118: 9324–9336
Jimenez R, Dikshit SN, Bradforth SE and Fleming GR (1996) Electronic excitation transfer in the LH2 complex of *Rhodobacter sphaeroides*. J Phys Chem 100: 6825–6834
Kennis JTM, Streltsov AM, Aartsma TJ, Nozawa T and Amesz J (1996) Energy transfer and exciton coupling in isolated B800-850 complexes of the photosynthetic purple sulfur bacterium *Chromatium tepidum*. The effect of structural symmetry on bacteriochlorophyll excited states. J Phys Chem 100: 2438–2442
Kennis JTM, Streltsov AM, Permentier H, Aartsma TJ and Amesz J (1997) Exciton coherence and energy transfer in the LH2 antenna complex of *Rhodopseudomonas acidophila* at low temperature. J Phys Chem 101: 8369–8374
Kono H, Takasaka A and Lin SH (1988) Monte Carlo calculation of the quantum partition function via path integral formulations. J Chem Phys 88: 6390–6398
Krueger BP, Scholes GD and Fleming GR (1998) Calculation of couplings and energy transfer pathways between the pigments of LH2 by the *ab initio* transition density cube method. J Phys Chem 102: 5378–5386
Kühn O and Sundström V (1997a) Energy transfer and relaxation dynamics in light-harvesting antenna complexes of photosynthetic bacteria. J Phys Chem 101: 3432–3440
Kühn O and Sundström V (1997b) Pump-probe spectroscopy of dissipative energy transfer dynamics in photosynthetic antenna complexes: A density matrix approach. J Chem Phys 107: 4154–4164
Kühn O, Renger T, May V, Voigt J, Pullerits T and Sundström V (1997) Exciton-vibrational coupling in photosynthetic antenna complexes: Theory meets experiment. Trends Photochem Photobiol 4: 213–256
Kumble R, Palese S, Visschers RW, Dutton PL and Hochstrasser RM (1996) Ultrafast dynamics within the B820 subunit from the core (LH-1) antenna complex of *Rs. rubrum*. Chem Phys Lett 262: 396–404
Li X-P (1987) High-order correction to the Trotter expansion for use in computer simulation. J Chem Phys 86: 5094–5100
Lolle LI, Gray CG and Poll JD (1991) Improved short-time propagator for repulsive inverse-power-law potentials. Chem Phys Lett 177: 64–72
Mak CH (1992) Stochastic method for real-time path integrations. Phys Rev Lett 68: 899–902
Mak CH and Egger R (1995) On the mechanism of the primary charge separation in bacterial photosynthesis. Chem Phys Lett 238: 149–157
Mak CH and Egger R (1996) Monte Carlo methods for real-time path integration. Adv Chem Phys 93: 39–76
Makarov DE and Makri N (1994) Path integrals for dissipative systems by tensor multiplication: Condensed phase quantum dynamics for arbitrarily long time. Chem Phys Lett 221: 482–491
Makri N (1989) Effective non-oscillatory propagator for Feynman path integration in real time. Chem Phys Lett 159: 489–498
Makri N (1991) Feynman path integration in quantum dynamics. Comp Phys Comm 63: 389–414
Makri N (1992) Improved Feynman propagators on a grid and non-adiabatic corrections within the path integral framework. Chem Phys Lett 193: 435–444
Makri N (1993) On smooth Feynman propagators for real time path integrals. J Phys Chem 97: 2417–2424
Makri N (1995) Numerical path integral techniques for long-time quantum dynamics of dissipative systems. J Math Phys 36: 2430–2456
Makri N (1997) Path integral simulation of long-time dynamics in quantum dissipative systems. In: DeWitt-Morette C (ed) Path integrals: Basics and Applications. Plenum, New York
Makri N (1999a) Iterative evaluation of the path integral for a system coupled to an anharmonic bath. J Chem Phys 111: 6164–6167
Makri N (1999b) The linear response approximation and its lowest order corrections: An influence functional approach. J Phys Chem 103: 2823–2829
Makri N (2004) Information guided noise reduction for Monte Carlo integration of ascillatory functions. Chem Phys Lett 400: 446–452
Makri N and Makarov DE (1995a) Tensor multiplication for iterative quantum time evolution of reduced density matrices. I. Theory. J Chem Phys 102: 4600–4610
Makri N and Makarov DE (1995b) Tensor multiplication for

iterative quantum time evolution of reduced density matrices. II. Numerical methodology. J Chem Phys 102: 4611–4618
Makri N and Miller WH (1987) Monte Carlo integration with oscillatory integrands: Implications for Feynman path integration in real time. Chem Phys Lett 139: 10–14
Makri N and Miller WH (1988a) Correct short time propagator for Feynman path integration by power series expansion in $\Delta t$. Chem Phys Lett 151: 1–8
Makri N and Miller WH (1988b) Monte Carlo path integration for the real time propagator. J Chem Phys 89: 2170–2177
Makri N and Miller WH (1989) Exponential power series expansion for the quantum time evolution operator. J Chem Phys 90: 904–911
Makri N, Sim E, Makarov DE and Topaler M (1996) Long-time quantum simulation of the primary charge separation in bacterial photosynthesis. Proc Natl Acad Sci USA 93: 3926–3931
Marchi M, Gehlen JN, Chandler D and Newton M (1993) Diabatic surfaces and the pathway for primary electron transfer in a photosynthetic reaction center. J Am Chem Soc 115: 4178–4190
Marcus RA (1956) On the theory of oxidation-reduction reactions involving electron transfer. J Chem Phys 24: 966–978
Marcus RA (1987) Superexchange versus an intermediate BChl-mechanism in reaction centers of photosynthetic bacteria. Chem Phys Lett 133: 471–477
Marcus RA (1988) An internal consistency test and its implications for the initial steps in bacterial photosynthesis. Chem Phys Lett 146: 13–21
Marcus RA (1993) Electron transfer reactions in Chemistry: Theory and experiment (Nobel Lecture). Angew Chem Int Ed Engl 32: 1111–1121
Marcus RA and Sutin N (1985) Electron transfers in chemistry and biology. Biochim Biophys Acta 811: 265–322
Martin J-L, Breton J, Hoff AJ, Migus A and Antonetti A (1986) Femtosecond spectroscopy of electron transfer in the reaction center of the photosynthetic bacterium *Rhodopseudomonas sphaeroides* R-26: Direct electron transfer from the dimeric bacteriochlorophyll primary donor to the bacteriopheophytin acceptor with a time constant of 2.8 +/– 0.2 psec. Proc Natl Acad Sci USA 83: 957–961
McDermott G, Prince S, Freer A, Hawthornthwaite-Lawless A, Papiz M, Cogdell R and Isaacs N (1995) Crystal structure of an integral membrane light-harvesting complex from photosynthetic bacteria. Nature 374: 517–521
Meier T, Chernyak V and Mukamel S (1997a) Multiple exciton coherence sizes in photosynthetic antenna complexes viewed by pump-probe spectroscopy. J Phys Chem 101: 7332–7342
Meier T, Zhao Y, Chernyak V and Mukamel S (1997b) Polarons, localization, and excitonic coherence in superradiance of biological antenna complexes. J Chem Phys 107: 3876–3893
Metropolis N, Rosenbluth AW, Rosenbluth MN, Teller H and Teller E (1953): Equation of state calculations by fast computing machines. J Chem Phys 21: 1087–1092
Monshouwer R and van Grondelle R (1996) Excitations and excitons in bacterial light harvesting complexes. Biochim Biophys Acta 1275: 70–75
Monshouwer R, Abrahamsson M, van Mourik F and van Grondelle R (1997) Superradiance and exciton delocalization in bacterial photosynthetic light-harvesting systems. J Phys Chem B 101: 7241–7248
Nagarajan V, Alden RG, Williams JC and Parson WW (1996) Ultrafast exciton relaxation in the B850 antenna complex of *Rhodobacter sphaeroides*. Proc Natl Acad Sci USA 93: 13774–13779
Onuchic JN and Wolynes PG (1988) Classical and quantum pictures of reaction dynamics in condensed matter: Resonances, dephasing, and all that. J Phys Chem 92: 6495–6503
Onuchic JN and Wolynes PG (1993) Energy landscapes, glass transitions, and chemical reaction dynamics in biomolecular or solvent environment. J Chem Phys 98: 2218–2224
Parson WW, Creighton S and Warshel A (1987) In: Kobayashi T (ed) Primary Processes in Photobiology. Springer-Verlag, Berlin
Pullerits T and Sundström V (1996) Photosynthetic light-harvesting pigment-protein complexes: Toward understanding how and why. Acc Chem Res 29: 381–389
Pullerits T, van Mourik F, Monshouwer R, Visschers RW and van Grondelle R (1994a) Electron-phonon coupling in the B820 subunit form of LH1 studied by temperature-dependence of optical spectra. J Lumin 58: 168–171
Pullerits T, Visscher KJ, Hess S, Sundström V, Freibert A, Timpmann K and van Grondelle R (1994b) Energy transfer in the inhomogeneously broadened core antenna of purple bacteria — a simultaneous fit of low-intensity picosecond absorption and fluorescence kinetics. Biophys J 66: 236–248
Pullerits T, Hess S, Herek J and Sundström V (1997) Temperature dependence of excitation transfer in LH2 of *Rhodobacter sphaeroides*. J Phys Chem 101: 10560–10567.
Reddy NRS, Small GJ, Seibert M and Picorel R (1991) Energy transfer dynamics of the B800-B850 antenna complex of *Rhodobacter sphaeroides*—A hole burning study. Chem Phys Lett 181: 391–399
Reddy NRS, Cogdell RJ, Zhao L and Small GJ (1993) Nonphotochemical hole burning of the B800-B850 antenna complex of *Rhodopseudomonas acidophila*. Photochem Photobiol 57: 35–39
Sauer K, Cogdell RJ, Prince SM, Freer A, Isaacs NW and Scheer H (1996) Structure-based calculations of the optical spectra of the LH2 bacteriochlorophyll-protein complex from *Rhodopseudomonas acidophila*. Photochem Photobiol 64: 564–576
Schmidt S, Arlt T, Hamm P, Huber H, Nagele T, Wachtveitl J, Meyer M, Scheer H and Zinth W (1994) Energetics of the primary electron transfer reaction revealed by ultrafast spectroscopy on modified bacterial reaction centers. Chem Phys Lett 223: 116–120
Shao J and Makri N (2001) Iterative path integral calculation of quantum correlation functions for dissipative systems. Chem Phys 268: 1–10
Shao J and Makri N (2002) Iterative path integral formulation of equilibrium correlation functions for quantum dissipative systems. J Chem Phys 116: 507–514
Sharp K and Honig B (1990) Electrostatic interactions in macromolecules: Theory and applications. Annu Rev Biophys Biophys Chem 19: 301–332
Sim E and Makri N (1996) Tensor propagator with weight-selected paths for quantum dissipative dynamics with long-memory kernels. Chem Phys Lett 249: 224–230
Sim E and Makri N (1997a) Filtered propagator functional for iterative dynamics of quantum dissipative systems. Comp Phys Comm 99: 335–354
Sim E and Makri N (1997b) Path integral simulation of charge transfer dynamics in photosynthetic reaction centers. J Phys

Chem 101: 5446–5458
Sundström V, Pullerits T and van Grondelle R (1999) Photosynthetic light harvesting: Reconciling dynamics and structure of purple bacterial LH2 reveals function of photosynthetic unit. J Phys Chem B 103: 2327–2346
Thompson MA and Zerner MC (1991) A theoretical examination of the electronic structure and spectroscopy of the photosynthetic reaction center from *Rhodopseudomonas viridis*. J Am Chem Soc 113: 8210–8215
Trotter MF (1959) On the product of semi-groups of operators. Proc Am Math Soc 10: 545–551
van Grondelle R, Dekker JP, Gillbro T and Sundström V (1994) Biochim Biophys Acta 1187: 1–65
van Mourik F, Visschers RW and van Grondelle R (1992) Energy transfer and aggregate size effects in the inhomogeneously broadened core light-harvesting complex of *Rhodobacter sphaeroides*. Chem Phys Lett 193: 1–7
Visschers RW, van Mourik F, Monshouwer R and van Grondelle R (1993) Inhomogeneous spectral broadening of the B820 subunit form of LH1. Biochim Biophys Acta 1141: 238–244
Visser HM, Somsen OJG, van Mourik F, Lin S, van Stokkum IHM and van Grondelle R (1995) Direct observation of subpicosecond equilibration of excitation energy in the light-harvesting antenna of *Rhodospirillum rubrum*. Biophys J 69: 1083–1099
Wang H, Manolopoulos DE and Miller WH (2001) Generalized Filinov transformation of the semiclassical initial value representation. J Chem Phys 115: 6317–6326
Warshel A and Hwang J-K (1986) Simulation of the dynamics or electron transfer reactions in polar solvents: Semiclassical trajectories and dispersed polaron approaches. J Chem Phys 84: 4938–4957
Warshel A, Creighton S and Parson WW (1988) Electron-transfer pathways in the primary event of bacterial photosynthesis. J Phys Chem 92: 2696–2701
Warshel A, Chu ZT and Parson WW (1989) Dispersed polaron simulations of electron transfer in photosynthetic reaction centers. Science 246: 112–116
Woodbury NW, Becker M, Middendorf D and Parson WW (1985) Picosecond kinetics of the initial photochemical electron-transfer reaction in bacterial photosynthetic reaction centers. Biochemistry 24: 7516–7521
Wu H-M, Ratsep M, Lee I-J, Cogdell RJ and Small GJ (1997a) Exciton level structure and energy disorder of the B850 ring of the LH2 antenna complex. J Phys Chem B 101: 7654–7663
Wu H-M, Reddy NRS and Small GJ (1997b) Direct observation and hole burning of the lowest exciton level (B870) of the LH2 antenna complex of *Rps. acidophila*. J Phys Chem 101: 651–656
Xiao WH, Lin S, Taguchi KW and Woodbury NW (1994) Femtosecond pump-probe analysis of energy and electron transfer in photosynthetic membranes of *Rhodobacter capsulatus*. Biochemistry 33: 8313–8222
Zhang LY and Friesner RA (1997) *Ab initio* calculation of electronic coupling in the photosynthetic reaction center. Proc Nat Acad Sci USA 95: 13603–13605

# Chapter 24

## Density Functional Theory and Car-Parrinello Molecular Dynamics Methods

Francesco Buda*

*Leiden Institute of Chemistry, Leiden University, Einsteinweg 55, 2300 RA Leiden, The Netherlands*



### Summary

The recent progress in first-principles electronic structure calculations together with the steadily increase in computational power have considerably broadened the range and scope of application of these theoretical methods. Of particular interest is the combination of density functional theory calculations with classical molecular dynamics simulations, the so-called Car-Parrinello method. After twenty years since the publication of Car and Parrinello's seminal paper, this method has been proven to be a powerful tool for studying a large variety of problems, from condensed matter physics, to chemistry and more recently to highly complex systems of biophysics and biochemistry. This success is also linked to the considerable progress achieved in the development of more accurate approximations of the exchange-correlation functional in the density functional theory which are approaching the required chemical accuracy. The calculation of excitation energies and excited states has also become more manageable within the time-dependent density functional scheme developed in the last two decades. This development has paved the way for the study of photoactivated processes and photochemical reactions. Hybrid schemes combining quantum-mechanics with classical molecular mechanics (QM/MM) are

*Email: f.buda@chem.leidenuniv.nl

*Thijs J. Aartsma and Jörg Matysik (eds): Biophysical Techniques in Photosynthesis II, pp. 487–499.*

being also progressively improved and provide an excellent framework to study quite extended and complex biological systems with a high degree of accuracy. In this chapter, we introduce the most important concepts of these methods and describe the advancements that are relevant for the application of these methodologies to photo-biological processes and photosynthesis.

## I. Introduction

The rapid advances on the scientific frontiers of nanoscience and molecular biology provide a strong foundation for future breakthroughs in solar energy conversion. New materials to efficiently absorb sunlight, new techniques to harness the full spectrum of wavelengths in solar radiation, and new approaches based on nanostructured architectures can revolutionize the technology used to produce solar electricity. A promising strategy is the design of highly efficient, all artificial, molecular-level energy conversion machines exploiting the principles of natural photosynthesis. Progress in the field of solar energy depends critically on the development of new tools for the characterization of matter and on new theoretical tools.

On the experimental front, one major goal is to create probes that can reveal the structure and composition of nanoscale materials with atomic resolution. A second goal involves development of tools that can be used to follow the complete flow of energy through each primary step of the solar conversion processes, from absorption, to charge transfer, transport, harvesting, and chemical conversion and separation.

On the theoretical front, modeling, and computational tools are required to aid in the understanding of these elementary steps. The wide range of time and length scales spanned by these phenomena poses a significant theoretical challenge.

In this chapter, we will discuss the progress done in first-principles static and dynamic calculations both in the electronic ground state and in the excited state. The focus will be on density functional theory methods and Car-Parrinello molecular dynamics simulations, which represent a promising choice to tackle accurately the study of complex processes such as those mentioned above. It is outside the scope of this chapter to review all the numerous applications of these methods related to photosynthesis. We will only mention a few examples of applications to biomolecules, in order to give a flavor of the current potentiality of modeling in this field. We will also briefly underline the further advances needed in the computational tools in order to meet the challenges posed by the field of natural and artificial photosynthesis.

*Abbreviations:* BO – Born-Oppenheimer; CI – configuration interaction; CPMD – Car-Parrinello molecular dynamics; DFT – density functional theory; GGA – generalized gradient approximation; HK – Hohenberg-Kohn; KS – Kohn-Sham; LDA – local density approximation; MD – molecular dynamics; PES – potential energy surface; QM-MM – quantum mechanics–molecular mechanics; TDDFT – time dependent density functional theory; TDKS – time dependent Kohn-Sham; xc – exchange-correlation

## II. Density Functional Theory (DFT)

Density functional theory (DFT) is a theory of the electronic structure of atoms, molecules, and solids in their ground state, in which the electronic density $n(r)$ plays the central role. In traditional quantum chemistry approaches, this problem has been attacked by finding reasonable approximations for the solution of the Schrödinger equation for the system of $N$ interacting electrons in the field generated by the nuclei. The total wavefunction is however an object that depends on the $3N$ electronic coordinates, thus the complexity of the problem increases with the size of the system. The electronic density instead depends on three coordinates, independently of the number of electrons, making the scaling of DFT calculations much more favorable. In the traditional configuration interaction (CI) method, the trial wavefunction is written as a linear combination of Slater determinants with the expansion coefficients determined by using the variational principle that is by requiring that the energy should be a minimum. The CI method can in principle provide a very accurate ground-state energy of molecules, but the computing time grows so rapidly for large values of $N$, that this method can in practice handle only small chemical systems ($N \approx 10$).

The main reason for the large and increasing interest in DFT method is due to its computational simplicity, allowing to study larger systems ($N \approx 100$ or more), accompanied by a good accuracy. DFT has attracted initially mainly the interest of solid state physicists, but since the early nineties also theoretical chemists and others. Here I will sketch the basic elements of DFT and the current work done for extending its applicability and accuracy. For details the reader is referred to the excellent books by Parr and Yang (1989) and by Dreizler and Gross (1990).

### A. Theoretical Foundation

DFT is based on the mathematical proof (Hohenberg and Kohn, 1964) that for an interacting $N$-electron system in a general external potential $v(r)$, the knowledge of the ground-state density $n(r)$ uniquely determines $v(r)$:

$$n(r) \rightarrow v(r) \tag{1}$$

For a molecule or an extended system, the external potential is given by the Coulomb potential generated by the nuclei. The Hohenberg and Kohn (HK) theorem has important consequences: since $n(r)$ determines $v(r)$, it determines also the Hamiltonian $H$ and hence, implicitly, all properties derivable from $H$ such as the ground-state wavefunction and energy, excited state wavefunctions and energies, etc.

The total ground-state energy of the system can be written as a functional of the electron density $n(r)$

$$E \equiv E_{v(r)}[n(r)] \equiv \int v(r)n(r)dr + F[n(r)] \tag{2}$$

Using the Rayleigh Ritz principle for ground-state energy leads rather simply to the conclusion that, for a given external potential $v(r)$, the total energy is a minimum for the correct ground-state density $n(r)$ (Hohenberg and Kohn, 1964). Therefore, if the functional $F[n(r)]$ in Eq. (2) is known, the ground state energy and density for any electronic system, independently on the number of electrons, can be determined. The functional $F[n(r)]$ contains the kinetic energy term and the electron-electron interaction term. It is convenient to write this functional as follows:

$$F[n] \equiv T_s[n] + \frac{1}{2}\int \frac{n(r)n(r')}{|r-r'|}drdr' + E_{xc}[n] \tag{3}$$

Here $T_s[n]$ is the kinetic energy of non-interacting electrons of density $n(r)$, the integral is the classical electron-electron interaction energy (also known as the Hartree term), and $E_{xc}[n]$ is the exchange-correlation functional containing the remainder of the electron-electron interaction term and of the kinetic energy of the interacting system.

The problem of finding the minimum of the energy functional $E_{v(r)}[n(r)]$ can be translated in the problem of solving a set of one-electron self-consistent equations, the so called Kohn-Sham (KS) equations (Kohn and Sham, 1965):

$$\left[\frac{-\hbar^2}{2m}\nabla^2 + v(r) + \int \frac{n(r')}{|r-r'|}dr' + v_{xc}(r)\right]\psi_i(r) = \varepsilon_i\psi_i(r) \tag{4}$$

where the first term is the single-particle kinetic operator, the second term is the external potential, the integral corresponds to the classical (Hartree) electrostatic potential and the last term is the exchange-correlation potential defined as

$$v_{xc}(r) \equiv \frac{\partial E_{xc}\left[n(r)\right]}{\partial n(r)} \tag{5}$$

In the KS approach, the electron density is written in terms of the auxiliary single particle wavefunctions $\psi_i(r)$, representing the variational parameters of the problem:

$$n(r) = \sum_{1}^{N} |\psi_i(r)|^2 \tag{6}$$

In principle, if the exact $E_{xc}$ were used in the KS equations, the resulting self-consistent density and energy would be exact, including all many body effects.

### B. Exchange-Correlation Functionals

Unfortunately the exact form of the exchange-correlation functional is unknown and one has to rely on specific approximations for this functional. The simplest approximation, which has been proposed already in the original paper by Kohn and Sham (1965), is the so-called local density approximation (LDA):

$$E_{xc}^{LDA}[n(r)] \equiv \int \varepsilon_{xc}(n(r))n(r)dr \tag{7}$$

where $\varepsilon_{xc}(n(r))$ is the exchange-correlation energy per electron of a uniform electron gas of density $n$, which is known to a high accuracy from independent Quantum Monte Carlo studies of the uniform electron gas (Ceperley and Alder, 1980).

Although LDA appears to be a crude approximation, it has been successfully used for metallic and non-metallic solids, giving accurate cohesive energies and lattice constants. For molecules, however, the LDA gives large errors (~20%) in binding energies. Therefore since the late eighties, a considerable effort has been devoted in improving the quality of the exchange-correlation functional. Refinements of the LDA include the so-called generalized gradient approximations (GGA), where the functional depends not only on the local value of the density, but also on the gradient of the density:

$$E_{xc}^{GGA} = \int f(n(r), \nabla n(r))\, dr \qquad (8)$$

An appropriate choice of the function $f(n(r), \nabla n(r))$ that takes into account the correct $1/r$ asymptotic behavior of the exchange-energy density at long range, generally reduces the error of LDA considerably, up to an order of magnitude. Without going into the details of the GGA functionals, we mention here only some of the most widely used, such as the BP (from the initials of the authors, Becke and Perdew; Perdew, 1986; Becke, 1988) and the BLYP functional, combining the exchange functional due to Becke (1988) with the correlation functional of Lee, Yang, and Parr (1988).

A functional which has proven to perform remarkably well is the B3LYP (Becke, 1993; Stevens et al., 1994). This functional incorporates a certain amount of the exact Hartree-Fock exchange energy and for this reason is also called a hybrid functional.

The next step in climbing the density functional ladder is the generation of the so-called meta-GGA functionals. A meta-GGA functional uses the Laplacian (second derivative) of the density or the kinetic energy density in addition to the density and the magnitude of the gradient of the density (Baerends, 2001; Tao et al., 2003).

The DFT formalism can be easily generalized to spin polarized systems. In this case separate spin up and spin down densities are introduced and optimized independently in the self-consistent calculation.

### *C. Performance of DFT*

In practical calculations the single-particle KS wavefunctions need to be expressed as a linear combination of basis functions. The coefficients of the expansion are determined from the iterative solution of the self-consistent KS equations (4). The idea is to write the KS orbital, which are closely related to molecular orbitals (MO), as linear combination of atomic orbitals (LCAO). Mostly used basis set are atom-centered functions that resemble atomic orbitals, such as Slater-type or Gaussian-type orbitals. However, also plane waves are often used, especially for periodic or extended systems.

The main reason for the success of DFT calculations is the favorable scaling (no worse than $N^3$) with the number of basis functions used to represent the KS orbitals. This computational efficiency is combined with accuracy comparable to high level *ab initio* methods. In general it is found that, for a set of test reference cases, GGA methods give bonding energies, ionization potentials, geometries and vibrational frequencies of the same or better quality than MP2, a second order perturbation theory generally considered of a very good quality.

The drawback of DFT is that, at variance from CI methods, the accuracy depends on the specific functional used and cannot be improved systematically. Nevertheless, advances in exchange-correlation (xc) functionals coupled with parallel computation now enable accurate calculation of several properties of large molecular complexes, up to a few hundreds of atoms. Thus, static DFT calculations provide a valuable complementary tool to experiment in order to interpret and analyze geometries, harmonic vibrational frequencies, infrared and Raman spectra, electric dipole moment, electric polarizability, circular dichroism (CD) spectra, nuclear magnetic resonance (NMR) chemical shifts, electron spin resonance (ESR) g-tensor (Schreckenbach and Ziegler, 1998), among other properties. An overview on the accuracy achieved for several spectroscopic properties can be found, e.g., in the book of Cramer (2002).

A considerable amount of information is encoded in the multinuclear NMR spectra of a given compound: NMR data are used to extract structural parameters and moreover NMR properties reflect the particular electronic structure of a molecule and may help to rationalize other properties such as reactivity. In the last few years a number of NMR investigations of photosystems (Alia et al., 2004; Diller et al., 2005; Prakash et al., 2005) have been carried out providing a large amount of data that may help in understanding the electron transfer process. It is therefore of great interest to compute and analyze the observable quantities of an NMR spectrum from first principles. DFT is particularly robust for $^{13}C$ chemical shift that can be calculated for large molecules with good accuracy (Facelli, 1998; Bühl et al., 1999; Touw et al., 2004b). Another advantage in using DFT is that the spatial, electronic, and vibrational properties are accessible within a single theoretical framework and can then be used to elucidate sometime inconsistent data from different spectroscopic techniques. In this respect an interesting example is provided by the analysis of protonated retinal Schiff base models for visual photoreceptor (Touw et al., 2004a).

Recent examples of DFT applications relevant to photosynthesis (Lundberg and Siegbahn, 2005) and catalytic hydrogen production (Bruschi et al., 2004; Tard et al., 2005) clearly show the crucial role of DFT modeling.

## III. First-Principles Molecular Dynamics

The interatomic potential energy surface (PES) is the basic quantity that characterizes a specific material or complex molecule and is the essential ingredient in a molecular dynamics (MD) simulation (Allen and Tildesley, 1987; Frenkel and Smit, 1996). Indeed the PES determines the forces that drive the atomic motion, and hence the atomic trajectories. A statistical analysis of the numerical solution of the classical equations of motion can provide quantitative estimates of static or dynamical properties of the system. It is not surprising then that a major effort has been devoted to the development of accurate PES, either based on empirical data or on independent electronic structure calculations. In principle it is clear how to obtain a first-principles PES: It is indeed enough to compute the ground-state electronic energy by solving the Schrödinger equation for each nuclear configuration along the dynamical trajectory. The resulting Born-Oppenheimer molecular dynamics (BO-MD) can be defined by the coupled equations:

$$M_I \frac{d^2 R_I}{dt^2} = -\nabla_I \min_{\Psi_0} \langle \Psi_0 | H_e | \Psi_0 \rangle$$
$$H_e \Psi_0 = E_0 \Psi_0 \quad (9)$$

Here, $H_e$ is the total electronic Hamiltonian, which depends parametrically on the nuclear positions $R_I$. The first-principles molecular dynamics derived from these equations, implicitly assumes (i) the Born-Oppenheimer approximation that allows us to separate the electronic and the nuclear dynamics, (ii) the classical approximation for the nuclear motion. The solution of Eq. (9) for a large set of dynamical nuclear configurations has been however considered for a long time an extraordinary computational task. A major breakthrough came about twenty years ago with the development of the so-called Car-Parrinello method (Car and Parrinello, 1985). This approach is based on an efficient algorithm for solving the Schrödinger equation, and it takes advantage of the continuity of the dynamical trajectories in order to compute with a minimum computational effort the new electronic ground-state after each atomic step in the trajectory.

In the Car-Parrinello molecular dynamics (CPMD) method, DFT is used for computing the electronic ground-state energy. The gradient of the total energy with respect to the nuclear positions provides the forces that appear in the classical equation of motion for the nuclei. Though DFT is not necessarily the only possible choice (the same formalism can in fact be developed using, e.g., the Hartree-Fock approximation for the electrons), it is clearly the most convenient choice for the advantageous scaling property and accuracy discussed in the previous section. In the CPMD method therefore no empirical fitting is required for the PES and the only input required in the simulation is the atomic number of the atomic constituents.

The use of a first-principles PES has several advantages over the empirical potentials: (i) the PES is fully transferable, i.e., it can be used for a cluster as well as for an extended system in different condensed phases without the need for re-parameterization; (ii) chemical reactions can be simulated since bond breaking and forming are allowed by the rearrangement of the electronic density along the trajectory; (iii) increased predictive power of the simulation. Of course the price of using a first-principles PES is a much larger computational cost of the simulation. At present, CPMD simulations can handle systems consisting of a few hundred atoms, and can follow the trajectory for a time of the order of ten picoseconds.

### *A. The Car-Parrinello Lagrangian*

The basic idea of the Car-Parrinello approach is to exploit the quantum-mechanical adiabatic time-scale separation of fast electronic and slow nuclear motion by transforming that into a classical-mechanical adiabatic scale separation in the framework of dynamical systems theory. The two-component quantum–classical problem is mapped onto a two-component purely classical system with two separate components with different energy scales. In this way we loose the explicit time-dependence of the quantum subsystem dynamics. This two-component classical system is characterized by the following Lagrangian:

$$L_{CP} = \sum_I \frac{1}{2} M_I \dot{R}_I^2 + \sum_i \frac{1}{2} \mu_i \int dr \left| \dot{\psi}_i \right|^2 - E[\{\psi_i\},\{R_I\}] + constraints \tag{10}$$

where the first two terms are the classical kinetic energy of the nuclei and of the KS electronic orbitals $\psi_i$, which are interpreted here as classical fields. $E[\{\psi_i\},\{R_I\}$ is the DFT energy functional representing the potential energy of the two-component system. The last term in the Lagrangian ensures the orthonormality condition for the KS orbitals. The corresponding Newtonian equations of motion are obtained from the associated Euler-Lagrange equations like in classical mechanics, for both the nuclear positions and the orbitals. Note that the orthonormality constraints are holonomic. The Car-Parrinello equations of motion derived from the Lagrangian are of the form:

$$M_I \ddot{R}_I(t) = -\frac{\partial E[\{\psi_i\},\{R_I\}]}{\partial R_I}$$
$$\mu_i \ddot{\psi}_i(t) = -\frac{\delta E[\{\psi_i\},\{R_I\}]}{\delta \psi_i^*} + \frac{\delta}{\delta \psi_i^*}\{constraints\} \tag{11}$$

where $\mu_i$ are the fictitious masses or inertia parameter associated to the orbital degrees of freedom. The units of the mass parameter $\mu$ are energy times a squared time for reason of dimensionality. According to the Car-Parrinello equations of motion, the nuclei evolve in time at a certain (instantaneous) physical temperature proportional to the kinetic energy of the nuclei ($K_I$), whereas the electronic degrees of freedom evolve at a fictitious temperature proportional to their classical kinetic energy ($K_e$), which must be much lower than the physical nuclear temperature. Under the condition

$$K_e \ll K_I \tag{12}$$

the electronic subsystem remains close to its instantaneous energy minimum, i.e., very close to the Born-Oppenheimer surface. Therefore, provided that the electronic wavefunction has been optimized for the initial nuclear configuration, the electrons will stay close to their ground state along the trajectory.

The question is why and when does the Car-Parrinello method work? We want to achieve a decoupling of the two subsystems and a (quasi-)adiabatic evolution in which the fast electronic subsystem stay cold and still follows adiabatically the slow nuclear motion. This is possible if the power spectra of the two subsystems do not have substantial overlap in the frequency domain:

$$\omega_e \gg \omega_I \tag{13}$$

In this way we realize a metastable condition in which there is practically no energy transfer between electrons and nuclei on the time scale relevant for the simulation (Pastore et al, 1991). This condition on the frequencies is strictly related to the choice of the inertia parameter $\mu$ that should be much smaller than the physical mass of the nuclei. During a Car-Parrinello simulation satisfying this metastable condition we observe that:

(i) the fictitious electronic kinetic energy performs small oscillations around a constant, i.e. the electrons do not heat up systematically;

(ii) the physical total energy of the nuclear system, defined as the sum of the nuclear kinetic energy plus the potential energy ($E_{phys} = K_I + E[\{\psi_i\},\{R_I\}]$) is essentially constant on the relevant energy and time scale. This is particularly important since it implies that the energy of the nuclear system behaves approximately as the strictly conserved total energy in classical MD. This ensures that the resulting dynamics of the nuclei approximates very closely the microcanonical dynamics.

It has also been verified in specific simple examples, that the Car-Parrinello forces acting on the nuclei approach very closely the exact Born-Oppenheimer forces. Small and bound oscillations are observed around the exact forces that however do not disturb the correct time evolution because of an intrinsic averaging effect of these oscillations within a few molecular dynamics time steps (Pastore et al, 1991).

The next important question to address is under which circumstances the adiabatic separation can be achieved, and how it can be controlled. A simple harmonic analysis of the frequency spectrum of the orbital classical fields close to the minimum defining the BO ground state yields

$$\omega_{ij} = \left( \frac{2(\varepsilon_i - \varepsilon_j)}{\mu} \right)^{1/2} \quad (14)$$

where $\varepsilon_i$ and $\varepsilon_j$ are the eigenvalues of the occupied and unoccupied KS orbitals, respectively. Since this is in particular true for the lowest frequency, the analytical estimate for the lowest possible electronic frequency gives

$$\omega_{\min} \propto \left( \frac{E_{gap}}{\mu} \right)^{1/2} \quad (15)$$

Equation (15) shows that this frequency is not only related to the choice of the fictitious electronic mass, but also to the energy separation between the highest occupied (HOMO) and lowest unoccupied molecular orbitals (LUMO). In order to guarantee the adiabatic separation, the frequency difference between the maximum nuclear frequency and the minimum electronic frequency $\omega_{min}$ should be large. Decreasing the mass $\mu$ not only shifts the electronic frequency spectrum upwards, but also increases the maximum frequency. Since the time step is inversely proportional to the highest frequency in the system, decreasing the electronic mass would also force us to decrease the time step. It is therefore crucial to find a reasonable compromise in the choice of the inertia parameter: On one side it needs to be small enough to ensure the adiabatic behavior, but at the same time it should be as large as possible to provide the largest possible time step. Typical values for large gap systems are $\mu$ = 500–1000 a.u. and a time step of about 5–10 a.u. (0.12–0.24 fs).

More rigorous mathematical conditions for the adiabatic classical energy separation can be derived (see the review paper by Marx and Hutter (2000) and reference therein) but are outside the scope of the present work.

Although born in condensed matter physics, the Car-Parrinello method has been proven to be a powerful tool for the study of chemical reactions. In particular the combination of the CPMD method with constrained molecular dynamics and thermodynamic integration has allowed the calculation of activation free energies. The development of accurate GGA functionals has paved the way for the simulation of liquid water from first-principles (Laasonen et al., 1993) and of industrially relevant chemical reactions in solution (Ensing et al., 2001).

## IV. Time Dependent Density Functional Theory

In the previous sections we have seen that DFT can efficiently and accurately describe the electronic ground state of a complex molecular or extended system. The combination of DFT with MD simulations (CPMD method) has broadened considerably the potential applications of DFT including the calculation of thermodynamic properties, dynamical correlation functions, and the study of disordered systems and chemical reactions in solution. We are however also interested in describing excited electronic states, which are crucial if we want to study photochemical processes. A theoretical breakthrough came with the work of Runge and Gross (1984) in which a time-dependent analogue of the first part of the Hohenberg-Kohn theorem was demonstrated. They considered a system evolving from an initial state $\Psi(t_0) = \Psi_0$ in the presence of an external time-dependent potential $v(r,t)$ and proved that a unique correspondence exists between the time-dependent electron density $n(r,t)$ and the external potential. The external time-dependent potential $v(r,t)$ determines the total time-dependent wavefunction and therefore the expectation value of any time-dependent quantum-mechanical operator $\hat{O}(t)$ is a unique functional of the density:

$$O[n](t) = \langle \Psi[n](t) | \hat{O}(t) | \Psi[n](t) \rangle \quad (16)$$

In the ground state case, the Rayleigh-Ritz minimum principle for the total energy is used in the second part of the HK theorem to derive a formal method for obtaining the electronic density. In the time-dependent case, such an approach is not possible since no analogous minimum energy principle is available. Instead a stationary action principle is invoked. Solving the time-dependent Schrödinger equation is equivalent to finding the stationary point of the action integral $A$:

$$A = \int_{t_0}^{t_1} dt \left\langle \Psi(t) \middle| i\frac{\partial}{\partial t} - \hat{H}(t) \middle| \Psi(t) \right\rangle \qquad (17)$$

The action $A$, in view of the correspondence between the time-dependent density and wavefunction, can be regarded as a density functional $A[n]$ and must have a stationary point at the exact time-dependent density. This exact density can be found from the associated Euler-Lagrange equation:

$$\frac{\delta A[n]}{\delta n(r,t)} = 0 \qquad (18)$$

This equation can be used to derive a set of single-particle equations for time-dependent external potentials, the so-called time-dependent Kohn-Sham (TDKS) equations. We refer to some review papers (Gross and Kohn, 1990; Gross et al., 1996) on the subject for the details of the derivation and we give here only the resulting equations (in a.u.):

$$i\frac{\partial}{\partial t}\phi_j(r,t) = \left( \frac{-\nabla^2}{2} + v_s[n](r,t) \right)\phi_j(r,t) \qquad (19)$$

with the density written in terms of a set of noninteracting single-particle orbitals:

$$n(r,t) = \sum_1^N \left|\phi_j(r,t)\right|^2 \qquad (20)$$

The TDKS potential is subdivided in the same manner as in the static case:

$$v_s[n](r,t) = v(r,t) + \int \frac{n(r',t)}{|r-r'|} dr' + v_{xc}(r,t) \qquad (21)$$

where $v(r,t)$ is the external potential, the second term is the Hartree potential, and $v_{xc}(r,t)$ is the time-dependent exchange-correlation potential that is an unknown potential of the time-dependent electron density. Given a certain approximation for the time-dependent exchange-correlation potential, the TDKS equations can be solved iteratively, yielding the time-dependent density of the system. However, these calculations are very time consuming and have been so far performed only for atoms or small molecules (Gross et al., 1996). If we limit ourselves to properties that can be obtained within a perturbative method, then a more efficient approach is possible, allowing the treatment of large molecules.

### *A. Perturbative Solution of Time-dependent Kohn-Sham (TDKS) Equations*

Properties like polarizabilities and excitation energies can be obtained through a linear density response of the system. Therefore only a perturbative solution of the TDKS equations is required, providing a considerable simplification of the problem. Following Gross et al. (1996), we consider an external potential of the form:

$$v_{ext}(r,t) = \begin{cases} v_0(r) & ;\ t \le t_0 \\ v_0(r) + v_1(r,t) & ;\ t > t_0 \end{cases} \qquad (22)$$

Assuming a small perturbation, we can expand the density $n(r, t)$ as a functional of the external potential in a functional Taylor series:

$$n(r,t) = n_0(r) + n_1(r,t) + n_2(r,t) + \ldots \qquad (23)$$

where $n_0(r)$ is the unperturbed density of $t \le t_0$ which can be obtained from the ground-state KS equations in the potential $v_0(r)$. The first order time-dependent density for the KS system can be calculated from the linear response function $\chi_s$:

$$n_1(r,t) = \int dr' \int dt' \chi_s(r,t;r',t') v_{s,1}(r',t') \qquad (24)$$

where $\chi_s(r,t;r',t') = \delta n[v_{ext}](r,t)/\delta v_S(r',t')|_{v0}$ is a functional derivative evaluated at the initial potential $v_0$ and where $v_{s,1}(r,t)$ is the KS potential in first-order of the external potential, given by:

$$v_{s,1}(r,t) = v_1(r,t) + \int dr' \frac{n_1(r',t)}{|r-r'|} + \int dr' \int dt' f_{xc}[n_0](r,t;r',t') n_1(r',t') \qquad (25)$$

Here we have introduced the exchange-correlation (xc) kernel $f_{xc}$ which is the functional derivative of the time-dependent xc potential with respect to the time dependent density:

$$f_{xc}(r,t;r',t') = \frac{\delta v_{xc}(r,t)}{\delta n(r',t')} \tag{26}$$

It is convenient to Fourier transform in the frequency domain the equation for the first-order density change. The KS response function is known in terms of the unperturbed KS orbitals, their occupation numbers and their orbital energies:

$$\chi_s(r,r';\omega) = \sum_{j,k}(f_k - f_j)\frac{\phi_j(r)\phi_k^*(r)\phi_j^*(r')\phi_k(r')}{\omega-(\varepsilon_j-\varepsilon_k)+i\eta} \tag{27}$$

where $\eta$ is a positive infinitesimal.

Similar equations can be written for the nonlinear density response. This has been worked out in detail by Gross et al. (1996) for second- and third-order densities. From these nonlinear density changes, the nonlinear optical response of a system can be calculated. For details on the practical implementation of these perturbative solutions for molecular calculations, we refer to the work of Gisbergen et al. (1995), and Jamorski et al. (1996).

In the equation for the first order density response, we have introduced the xc kernel $f_{xc}$. This kernel determines the first-order change in the time dependent xc potential due to the applied electric perturbation. The exact form of this kernel is unknown and the most commonly used approximation is the so-called Adiabatic LDA (ALDA), in which this functional is assumed to be spatially local, frequency independent, a real function, evaluated at the local density $n_0(r)$:

$$f_{xc}^{ALDA}(r,r';\omega) = \delta(r-r')\left.\frac{dv_{xc}^{LDA}}{dn}\right|_{n=n_0} \tag{28}$$

In practice this approximation appears to be less severe than other approximations, such as the approximation for the ground-state xc potential, which determines the KS orbitals, and one-electron eigenvalues.

Time-dependent DFT (TDDFT) has proven to be very successful for low-energy excitations, and its accuracy is considerably better than e.g., Configuration Interaction Singles (CIS) method. The mean error on a set of organosulfur compounds was found around 0.2 eV by using the B3LYP functional (Fabian, 2001).

Recent applications of TDDFT for biomolecules also show that this approach offers great promise for future applications in photosynthesis (Sergi et al., 2001; Groenhof et al., 2002; Rosa et al., 2003; Thompson et al., 2003; Vaswani et al., 2003; van Wijk et al., 2005). TDDFT has been used, e.g., to elucidate the mechanism of the bathochromic shift in several chromophore-protein complexes and for studying photoisomerization reactions in photoactive proteins.

More recently a few implementations of force calculations within TDDFT have appeared (Hutter, 2003 and references therein). This is an important step forward since the availability of the forces allows for geometry optimization and even Molecular Dynamics simulation in the excited state (Sulpizi et al., 2003).

## V. Hybrid Quantum Mechanics (QM)/ Molecular Mechanics (MM) Methods

We have described above the first-principles methods that are available for simulating reactions and electronic processes with high accuracy in the ground-state and even in the electronic excited state. Still the systems of interest in computational biology and in particular in photosynthesis are quite complex with many thousands of atoms. In spite of the considerable progress done in DFT approaches, it is clear that, in order to treat complex biological systems, we still need to be able to combine different computational chemistry methodologies with different accuracy and cost. In this context, it is important to develop hybrid quantum mechanics-molecular mechanics (QM-MM) approaches embedding a quantum mechanics calculation in a classical molecular mechanics model of the environment. In the QM-MM scheme we can incorporate in the simulation the environmental effects at an atomistic level, such as mechanical constraints, electrostatic perturbations and dielectric screening. The idea of a QM-MM scheme is not new and the first published example appeared already thirty years ago (Warshel and Levitt, 1976). However, in the last few years this subject has developed very rapidly and QM-MM approaches have been implemented in the most commonly used computational packages. For recent reviews see e.g. Mordasini and Thiel (1998) and Sherwood (2000).

The first step in a QM-MM simulation is to divide the system in two subsystems: One ‘inner’ (usually small) region which is treated with quantum mechanics (QM) and an 'outer' region which is treated with molecular mechanics (MM). The basis for this separation is that the region of space where the QM approach is needed is usually limited to a relatively small region where the electronic structure changes significantly (bond-making and bond-breaking processes) during the simulation. For example we can consider an enzyme, whose active site is the region where the chemical reaction takes place and therefore must be treated with QM. The reminder of the enzyme is important for maintaining its proper structure, folding, etc., but can be treated with a classical force field approach (MM). Another typical example is the study of a chemical reaction in solution, where the solvent can be treated within an MM approach and only the reactants are described with QM.

We can in general write a total Hamiltonian of the QM-MM system as follows:

$$H = H_{QM} + H_{MM} + H_{QM/MM} \tag{29}$$

where $H_{QM}$ includes all the interactions between the particles treated with QM, $H_{MM}$ includes all the interactions between the classical particles, and $H_{QM/MM}$ accounts for all the interactions between one quantum particle and one classical particle.

### A. The Choice of the QM Method

The details of the first term, $H_{QM}$, depend upon the specific choice of the QM approach used. The choice of the QM method within a hybrid approach depends on the accuracy required and on the size of the QM region. Implementation of QM-MM methods have been reported with almost any QM approach, from semiempirical to sophisticated *ab initio* complete active space self-consistent field (CASSCF) methods (Hayashi et al., 2003; Andruniow et al., 2004). The first application of Warshel and Levitt (1976) employed a semiempirical method. More recently several implementation involving DFT (Stanton et al., 1993; Wei and Salahub, 1994; Lyne et al., 1999) and Car-Parrinello MD have been reported (Woo et al., 1997; Eichinger et al., 1999; Laio et al., 2002). This is a very interesting development since DFT can deal with relatively large QM regions and can be used also in combination with Molecular Dynamics (Sulpizi et al, 2003; Röhrig et al., 2005).

### B. The Choice of the MM Model

The second term, $H_{MM}$, is determined by the specific classical force field used to treat the MM part. The most popular force fields for hybrid QM-MM simulations are the same force fields mostly used for biomolecules, such as CHARMM (Brooks et al., 1983), AMBER (Cornell et al., 1995), and GROMOS96 (van Gunsteren et al., 1996).

### C. Handling the Hybrid Term

The third term of the Hamiltonian, $H_{QM/MM}$, is the most critical and the details of this interaction term may differ substantially in different implementations. In terms of classification, we can distinguish two possibilities: (i) the boundary separating the QM and MM region, does not cut across any chemical bond; (ii) the boundary cuts across at least one chemical bond.

In the first situation, the QM-MM coupling term in the Hamiltonian contains the non-bonded interactions, i.e., electrostatic and short-range (also called van der Waals) forces. The treatment of the electrostatic interactions varies for different implementations, but the most common is the electrostatic embedding, in which the classical part appears as an external charge distribution (e.g., a set of point charges) in the QM Hamiltonian.

In the second scenario where there are bonds between the QM and MM regions, it is necessary to introduce some termination of the QM part. For termination of sites where a covalent bond has been broken, addition of a so-called link atom is the most common approach. An extra nuclear centre is introduced together with the electrons required to form a covalent bond to the QM dangling valences that will mimic the bond to the MM region. The simplest and most used choice is to add a hydrogen atom as link atom. Of course there are chemical differences between hydrogen and the chemical group it replaces. One possible approach to adjust the link atom interaction is to place a pseudopotential at the MM site to mimic the electronic properties of the replaced bond.

## VI. Concluding Remarks

Currently, highly accurate quantum mechanical schemes, based on density functional theory, are well established to describe ground state structures of systems consisting of up to a few hundreds of atoms. In order to successfully describe the processes that are relevant to solar energy conversion, such as the electrochemical behavior of molecules, nanocrystal systems, catalytic hydrogen production, and biological light-harvesting systems, the capability of these approaches will need to be enhanced to deal with thousands of atoms: this will require the practical implementation of novel linear scaling methodologies. In addition, methods for excited-state potential energy surfaces will have to be developed and tested. Emerging techniques, such as time-dependent density functional theory are very promising in this respect. DFT-based methods combined with genetic algorithms can be also used as predictive tools, which starting from a wide range of simultaneous desired properties as inputs can yield materials arrangements as outputs.

Solar energy conversion processes, such as the processes that lead to photosynthesis, are characterized by activated catalytic processes, which cannot be simulated on the short time scale of first-principles molecular dynamics simulations. In this case, CPMD simulations need to be supplemented by approaches for finding chemical reaction pathways at finite temperature, such as constrained MD and transition path sampling. These approaches should allow us to characterize the reaction intermediates and transition states in chemical and photochemical reactions in processes like water-splitting, which is essential to solar hydrogen production by hydrolysis.

*Ab initio* quantum mechanical methods will need to be extended to deal with up to tens of thousands of atoms. This may be achieved by means of QM-MM approaches where a large part of the system not directly involved in the chemical reactions is described by parameterized empirical force fields. It is also desirable to develop new hybrid methods combining atomistic simulations with mesoscopic scale simulations.

Strong interaction between experiment and theory will be essential to develop the new theoretical tools: experiment will guide theory to identify basic physical processes and to validate the theoretical tools. Theory, in turn, will guide the interpretation of experiment and provide detailed models for energy conversion processes.

Moreover, when for a given goal we need specific material properties and system performance, the computational procedure can be used to design by computer a material composition and structure, a system architecture, or a process dynamic that meets a set of desired specifications. Of course this is a challenging program and requires large-scale computations, and the expected increase in computational power and improvement in parallel and/or grid computer architectures should also contribute to the success.

## Acknowledgment

Preparation of this chapter was supported in part through the NWO-CW Top grant on 'Ultrahigh field solid-state NMR of photosynthesis and artificial photosynthetic energy conversion systems.'

## References

Alia, Roy E, Gast P, van Gorkom HJ, de Groot HJM, Jeschke G and Matysik J (2004) Photochemically induced dynamic nuclear polarisation in Photosystem I of plants observed by $^{13}C$ magic-angle spinning NMR. J Am Chem Soc 126: 12819–12826

Allen MP and Tildesley DJ (1987) Computer Simulation of Liquids. Clarendon Press, Oxford

Andruniow T, Ferre N and Olivucci M (2004) Structure, initial excited-state relaxation, and energy storage of rhodopsin resolved at the multiconfigurational perturbation theory level. Proc Natl Acad Sci USA 101: 17908–17913

Baerends EJ (2001) Exact exchange-correlation treatment of dissociated $H_2$ in density functional theory. Phys Rev Lett 87: 133004-1–133004-4

Becke AD (1988) Density-functional exchange-energy approximation with correct asymptotic behavior. Phys Rev A 38: 3098–3100

Becke AD (1993) A new mixing of Hartree-Fock and local density-functional theories. J Chem Phys 98: 1372–1377

Brooks BR, Bruccoleri RE, Olafson BD, States DJ, Swaminathan S and Karplus M (1983) CHARMM—A program for macromolecular energy, minimization, and dynamics calculations. J Comput Chem 4: 187–217

Bruschi M, Fantucci P and De Gioia L (2004) Density functional theory investigation of the active site of Fe-hydrogenases. Systematic study of the effects of redox state and ligands hardness on structural and electronic properties of complexes related to the $[2Fe]_H$ subcluster. Inorg Chem 43: 3733–3741

Bühl M, Kaupp M, Malkina OL and Malkin VG (1999) The DFT route to NMR chemical shifts. J Comput Chem 20: 91–105

Car R and Parrinello M (1985) Unified approach for molecular dynamics and density-functional theory. Phys Rev Lett 55: 2471–2474

Ceperley DM and Alder BJ (1980) Ground state of the electron gas by a stochastic method. Phys Rev Lett 45: 566–569

Cornell WD, Cieplak P, Bayly CI, Gould IR, Merz KM, Ferguson DM, Spellmeyer DC, Fox T, Caldwell JW and Kollman PA (1995) A second generation force-field for the simulation of proteins, nucleic-acids, and organic molecules. J Am Chem

Soc 117: 5179–5197

Cramer CJ (2002) Essentials of Computational Chemistry — Theories and Models. John Wiley & Sons Ltd, Chichester

Diller A, Alia, Roy E, Gast P, van Gorkom HJ, Zaanen J, de Groot HJM, Glaubitz C and Matysik J (2005) Photo-CIDNP solid-state NMR on Photosystems I and II: What makes P680 special? Photosynth Res 84: 303–308

Dreizler RM and Gross EKV (1990) Density Functional Theory. Springer Verlag, Berlin

Eichinger M, Tavan P, Hutter J and Parrinello M (1999) A hybrid method for solutes in complex solvents: Density functional theory combined with empirical force fields. J Chem Phys 110: 10452–10467

Ensing B, Buda F, Blöchl P and Baerends EJ (2001) Chemical involvement of solvent water molecules in elementary steps of the fenton oxidation reaction. Angew Chem Int Ed 113: 2977–2979

Fabian J (2001) Electronic excitation of sulfur-organic compounds — performance of time-dependent density functional theory. Theor Chem Acc 106: 199–217

Facelli JC (1998) Density functional theory calculations of the structure and the N-15 and C-13 chemical shifts of methyl bacteriopheophorbide *a* and bacteriochlorophyll *a*. J Phys Chem B 102: 2111–2116

Frenkel D and Smit B (1996) Understanding Molecular Simulation — From Algorithms to Applications. Academic Press, San Diego

Gisbergen SJA van, Snijders JG and Baerends EJ (1995) A density functional theory study of frequency-dependent polarizabilities and Van der Waals dispersion coefficients for polyatomic molecules. J Chem Phys 103: 9347–9354

Groenhof G, Lensink MF, Berendsen HJC, Snijders JG and Mark AE (2002) Signal transduction in the photoactive yellow protein. I. Photon absorption and the isomerization of the chromophore. Proteins Struct Funct Gen 48: 20–211

Gross EKU and Kohn W (1990) Time-dependent density functional theory. Adv Quantum Chem 21: 255–291

Gross EKU, Dobson JF and Petersilka M (1996) Density functional theory. In: Nalewajski RF (ed) Density Functional Theory II: Relativistic and Time Dependent Extensions, Vol 181, pp 81–172 (Topics in Current Chemistry). Springer, Heidelberg

Hayashi S, Tajkhorshid E and Schulten K (2003) Molecular dynamics simulation of bacteriorhodopsin's photoisomerization using ab initio forces for the excited chromophore. Biophys J 85:1440–1449

Hohenberg P and Kohn W (1964) Inhomogeneous electron gas. Phys Rev B 136: 864–871

Hutter J (2003) Excited state nuclear forces from the Tamm-Dancoff approximation to time-dependent density functional theory within the plane wave basis set framework. J Chem Phys 118: 3928–3934

Jamorski CH, Casida ME and Salahub DR (1996) Dynamic polarizabilities and excitation spectra from a molecular implementation of time-dependent density-functional response theory: $N_2$ as a case study. J Chem Phys 104: 5134–5147

Kohn W and Sham LJ (1965) Self-consistent equations including exchange and correlation effects. Phys Rev A 140: 1133–1138

Laasonen K, Sprik M, Parrinello M and Car R (1993) Ab initio liquid water. J Chem Phys 99: 9080–9089

Laio A, VandeVondele J, Röthlisberger U (2002) A Hamiltonian electrostatic coupling scheme for hybrid Car-Parrinello molecular dynamics simulations. J Chem Phys 116: 6941–6947

Lee C, Yang W and Parr RG (1988) Development of the Colle-Salvetti correlation energy formula into a functional of the electron density. Phys Rev B 37: 785–789

Lundberg M and Siegbahn PEM (2005) Agreement between experiment and hybrid DFT calculations for O-H bond dissociation enthalpies in manganese complexes. J Comput Chem 26: 661–667

Lyne PD, Hodoscek M and Karplus M (1999) A hybrid QM-MM potential employing Hartree-Fock or density functional methods in the quantum region. J Phys Chem A 103: 3462–3471

Marx D and Hutter J (2000) Ab initio molecular dynamics: Theory and implementation. In: Grotendorst J (ed) Modern Methods and Algorithms of Quantum Chemistry, Vol 1, pp 301–449. NIC Series, Jülich

Mordasini TZ and Thiel W (1998) Combined quantum mechanical and molecular mechanical approaches. Chimia 52: 288–291

Parr RC and Yang W (1989) Density Functional Theory of Atoms and Molecules. Oxford University Press, Oxford

Pastore G, Smargiassi E and Buda F (1991) Theory of *ab initio* molecular-dynamics calculations. Phys Rev A 44: 6334–6347

Perdew JP (1986) Density-functional approximation for the correlation energy of the inhomogeneous electron gas. Phys Rev B 33: 8822–8824

Prakash S, Alia, Gast P , de Groot HJM, Jeschke G and Matysik J (2005) Magnetic field dependence of photo-CIDNP MAS NMR photosynthetic reaction centres of *Rhodobacter Sphaeroides* WT. J Am Chem Soc 127: 14290–14298

Röhrig UF, Guidoni L and Rothlisberger U (2005) Solvent and protein effects on the structure and dynamics of the rhodopsin chromophore. Chem Phys Chem 6: 1836–1847

Rosa A, Ricciardi G, Baerends EJ, Romeo A and Monsu Scolaro L (2003) Effects of porphyrin core saddling, *meso*-phenyl twisting, and counterions on the optical properties of *meso*-tetraphenylporphyrin diacids: The $[H_4TPP](X)2$ (X = F, Cl, Br, I) series as a case study. J Phys Chem A 107: 11468–11482

Runge E and Gross EKU (1984) Density-functional theory for time-dependent systems. Phys Rev Lett 52: 997–1000

Schreckenbach G and Ziegler T (1998) Density functional calculations of NMR chemical shifts and ESR g-tensors. Theor Chem Acc 2: 71–82

Sergi A, Grüning M, Ferrario M and Buda F (2001) Density functional study of the photoactive yellow protein's chromophore. J Phys Chem B 105: 4386–4391

Sherwood P (2000) Hybrid quantum mechanics/molecular mechanics approaches. In: Grotendorst J (ed) Modern Methods and Algorithms of Quantum Chemistry, Vol 1, pp 257–277. NIC Series, Jülich

Stanton RV, Hartsough DS and Merz KM (1993) Calculation of solvation free energies using a density functional/molecular dynamics coupled potential. J Phys Chem 97: 11868–11870

Stevens PJ, Devlin JF, Chabalowski CF and Frisch MJ (1994) *Ab initio* calculation of vibrational absorption and circular dichroism spectra using density functional force fields. J Phys Chem 98: 11623–11627

Sulpizi M, Carloni P, Hutter J and Rothlisberger U (2003) A hybrid TDDFT/MM investigation of the optical properties of aminocoumarins in water and acetonitrile solution. Phys Chem Chem Phys 5: 4798–4805

Tao J, Perdew JP, Staroverov VN and Scuseria GE (2003) Climbing the density functional ladder: Nonempirical meta-generalized gradient approximation designed for molecules and solids. Phys Rev Lett 91: 146401-1–146401-4

Tard C, Liu X, Ibrahim SK, Bruschi M, De Gioia L, Davies SC, Yang X, Wang LS, Sawersand G and Pickett CJ (2005) Synthesis of the H-cluster framework of iron-only hydrogenase. Nature 433: 610–613

Thompson MJ, Bashford D, Noodleman L and Getzoff ED (2003) Photoisomerization and proton transfer in photoactive yellow protein. J Am Chem Soc 125: 8186–8194

Touw SIE, de Groot HJM and Buda F (2004a) *Ab initio* modeling of the spatial, electronic, and vibrational structure of Schiff base models for visual photoreceptors. J Phys Chem B 108: 13560–13572

Touw SIE, de Groot HJM and Buda F (2004b) DFT calculations of the $^1$H chemical shifts and $^{13}$C chemical shift tensors of retinal isomers. J Mol Struct (Theochem) 711: 141–147

van Gunsteren WF, Billeter SR, Eising AA, Hünenberger PH, Krüger P, Mark AE, Scott WRP and Tironi IG (1996) Biomolecular simulation: GROMOS96 manual and user guide. BIOMOS b.v., Zürich

van Wijk AAC, Spaans A, Uzunbajakava N, Otto C, de Groot HJM, Lugtenburg J and Buda F (2005) Spectroscopy and quantum chemical modeling reveal a predominant contribution of excitonic interactions to the bathochromic shift in α-crustacyanin, the Blue carotenoprotein in the carapace of the lobster *Homarus gammarus.* J Am Chem Soc 127: 1438–1445

Vaswani HM, Hsu CP, Head-Gordon M and Fleming GR (2003) Quantum chemical evidence for an intramolecular charge-transfer state in the carotenoid peridinin of peridinin-chlorophyll-protein. J Phys Chem B 107: 7940–7946

Warshel A and Levitt M (1976) Theoretical studies of enzymic reactions: Dielectric, electrostatic and steric stabilisation of the carbonium ion in the reaction of lysozyme. J Mol Biol 103: 227–249

Wei D and Salahub DR (1994) A combined density functional and molecular dynamics simulation of a quantum water molecule in aqueous solution. Chem Phys Lett 224: 291–296

Woo TK, Margl PM, Blöchl PE and Zeigler T (1997) A combined Car-Parrinello QM/MM implementation for *ab initio* molecular dynamics simulations of extended systems: Application to transition metal catalysis. J Phys Chem 101: 7877–7880

# Index

# C

# D

## E

## H



## I

## J



## K



## L

# Q



# R

# S

# T

## Y



## Z

# Advances in Photosynthesis and Respiration

Series editor: Govindjee, University of Illinois, Urbana, Illinois, U.S.A.

1. D.A. Bryant (ed.): *The Molecular Biology of Cyanobacteria*. 1994 ISBN Hb: 0-7923-3222-9; Pb: 0-7923-3273-3
2. R.E. Blankenship, M.T. Madigan and C.E. Bauer (eds.): *Anoxygenic Photosynthetic Bacteria*. 1995 ISBN Hb: 0-7923-3681-X; Pb: 0-7923-3682-8
3. J. Amesz and A.J. Hoff (eds.): *Biophysical Techniques in Photosynthesis*. 1996 ISBN 0-7923-3642-9
4. D.R. Ort and C.F. Yocum (eds.): *Oxygenic Photosynthesis: The Light Reactions*. 1996 ISBN Hb: 0-7923-3683-6; Pb: 0-7923-3684-4
5. N.R. Baker (ed.): *Photosynthesis and the Environment*. 1996 ISBN 0-7923-4316-6
6. P.-A. Siegenthaler and N. Murata (eds.): *Lipids in Photosynthesis: Structure, Function and Genetics*. 1998 ISBN 0-7923-5173-8
7. J.-D. Rochaix, M. Goldschmidt-Clermont and S. Merchant (eds.): *The Molecular Biology of Chloroplasts and Mitochondria in Chlamydomonas*. 1998 ISBN 0-7923-5174-6
8. H.A. Frank, A.J. Young, G. Britton and R.J. Cogdell (eds.): *The Photochemistry of Carotenoids*. 1999 ISBN 0-7923-5942-9
9. R.C. Leegood, T.D. Sharkey and S. von Caemmerer (eds.): *Photosynthesis: Physiology and Metabolism*. 2000 ISBN 0-7923-6143-1
10. B. Ke: *Photosynthesis: Photobiochemistry and Photobiophysics*. 2001 ISBN 0-7923-6334-5
11. E.-M. Aro and B. Andersson (eds.): *Regulation of Photosynthesis*. 2001 ISBN 0-7923-6332-9
12. C.H. Foyer and G. Noctor (eds.): *Photosynthetic Nitrogen Assimilation and Associated Carbon and Respiratory Metabolism*. 2002 ISBN 0-7923-6336-1
13. B.R. Green and W.W. Parson (eds.): *Light-Harvesting Antennas in Photosynthesis*. 2003 ISBN 0-7923-6335-3
14. A.W.D. Larkum, S.E. Douglas and J.A. Raven (eds.): *Photosynthesis in Algae*. 2003 ISBN 0-7923-6333-7
15. D. Zannoni (ed.): *Respiration in Archaea and Bacteria*. Diversity of Prokaryotic Electron Transport Carriers. 2004 ISBN 1-4020-2001-5
16. D. Zannoni (ed.): *Respiration in Archaea and Bacteria*. Diversity of Prokaryotic Respiratory Systems. 2004 ISBN 1-4020-2002-3
17. D. Day, A.H. Millar and J. Whelan (eds.): P*lant Mitochondria*. From Genome to Function. 2004 ISBN 1-4020-2399-5
18. H. Lambers and M. Ribas-Carbo (eds.): *Plant Respiration*. From Cell to Ecosystem. 2005 ISBN 1-4020-3588-8
19. G. Papageorgiou and Govindjee (eds.): *Chlorophyll a Fluorescence*. A Signature of Photosynthesis. 2004 ISBN 1-4020-3217-X
20. Govindjee, J.T. Beatty, H. Gest and J.F. Allen (eds.): *Discoveries in Photosynthesis*. 2005 ISBN 1-4020-3323-0
21. B. Demmig-Adams, W.W. Adams III and A. Mattoo (eds.): *Photoprotection, Photoinhibition, Gene Regulation, and Environment*. 2005 ISBN 1-4020-3564-0

# Advances in Photosynthesis and Respiration

22. T.J. Wydrzynski and K. Satoh (eds.): *Photosystem II.* The Light-Driven Water: Plastoquinone Oxidoreductase. 2005 ISBN 1-4020-4249-3
23. R.R. Wise and J.K. Hoober (eds.): *The Structure and Function of Plastids.* 2006 ISBN 1-4020-4060-1
24. B. Grimm, R.J. Porra, W. Rüdiger and H. Scheer (eds.): *Chlorophylls and Bacteriochlorophylls.* Biochemistry, Biophysics, Functions and Applications. 2006 ISBN 1-4020-4515-8
25. R. Hell, C. Dahl, D. Knaff and T. Leustek (eds.): *Sulfur Metabolism in Phototrophic Organisms.* 2008 ISBN 978-1-4020-6862-1
26. T.J. Aartsma and J. Matysik (eds.): *Biophysical Techniques in Photosynthesis.* Volume II. 2008 ISBN 978-1-4020-8249-8

MIX
Papier aus verantwortungsvollen Quellen
Paper from responsible sources
FSC® C105338

If you have any concerns about our products,
you can contact us on
**ProductSafety@springernature.com**

In case Publisher is established outside the EU,
the EU authorized representative is:
**Springer Nature Customer Service Center GmbH**
**Europaplatz 3, 69115 Heidelberg, Germany**

Printed by Libri Plureos GmbH
in Hamburg, Germany